Shock Waves in Condensed Matter

Shock Waves in Condensed Matter

Edited by

Y. M. Gupta

Washington State University
Pullman, Washington

PLENUM PRESS • NEW YORK AND LONDON

Library of Congress Cataloging in Publication Data

American Physical Society Topical Conference on Shock Waves in Condensed Matter (4th: 1985: Spokane, Wash.)

Shock waves in condensed matter.

"Proceedings of the Fourth American Physical Society Topical Conference on Shock Waves in Condensed Matter, held July 22-25, 1985, in Spokane, Washington"—T.p. verso.

Includes bibliographies and index.

1. Condensed matter—Congresses. 2. Shock waves—Congresses. 3. High pressure (Science)—Congresses. 4. High temperatures—Congresses. I. Gupta, Y. M. (Yogendra M.) II. American Physical Society. III. Title.

QC173.4.C65A44 1985 530.4′1 86-8155

ISBN-13:978-1-4612-9296-8 e-ISBN-13:978-1-4613-2207-8

DOI: 10.1007/978-1-4613-2207-8

Proceedings of the Fourth American Physical Society Topical Conference on Shock Waves in Condensed Matter, held July 22-25, 1985, in Spokane, Washington

Softcover reprint of the hardcover 1st edition 1986

A Division of Plenum Publishing Corporation
233 Spring Street, New York, N.Y. 10013

PREFACE

The Fourth American Physical Society Topical Conference on Shock Waves in Condensed Matter was held in Spokane, Washington, July 22-25, 1985. Two hundred and fifty scientists and engineers representing thirteen countries registered at the conference. The countries represented included the United States of America, Australia, Canada, The People's Republic of China, France, India, Israel, Japan, Republic of China (Taiwan), United Kingdom, U.S.S.R, Switzerland and West Germany. One hundred and sixty-two technical papers, covering recent developments in shock wave and high pressure physics, were presented. All of the abstracts have been published in the September 1985 issue of the *Bulletin of the American Physical Society*. The topical conferences, held every two years since 1979, have become the principal forum for shock wave studies in condensed materials. Both formal and informal technical discussions regarding recent developments conveyed a sense of excitement.

Consistent with the past conferences, the purpose of this conference was to bring together scientists and engineers studying the response of condensed matter to dynamic high pressures and temperatures. Papers covering experimental, theoretical, and numerical studies of condensed matter properties were presented. A noteworthy feature of this conference was the participation by several leading scientists engaged in static high pressure research.

Donald Curran served as the Master of Ceremonies at the conference banquet, which was attended by two hundred and seventy-five conference participants and guests including Dr. Samuel Smith, the new President of Washington State University. Dr. Philip Abelson, internationally renowned scholar and recently retired Editor of *Science*, spoke at the banquet on the role of academic, industrial, and governmental participation in scientific research.

The organizing committee for the conference was comprised of the following people:

Chairman:	Yogendra M. Gupta Washington State University
Treasurer/Coordinator:	Debra R. Maggiora Washington State University
J.R. Asay	Sandia National Laboratories
M. Finger	Lawrence Livermore National Laboratory
J.E. Flinn	EG & G, Idaho, Inc.
J.N. Fritz	Los Alamos National Laboratory
J.W. Forbes	Naval Surface Weapons Center
R.A. Graham	Sandia National Laboratories
N.C. Holmes	Lawrence Livermore National Laboratory
T.R. Neal	Los Alamos National Laboratory
M. Ross	Lawrence Livermore National Laboratory
L. Seaman	SRI International
D.C. Wallace	Los Alamos National Laboratory

On behalf of the organizing committee, the chairman sincerely thanks the many people who contributed to the success of the conference: the session chairmen, the authors, the invited speakers, and the participants. The assistance of J.R. Asay, J.E. Flinn, J.W. Forbes, R.A. Graham, D.E. Grady, N.C. Holmes, W.J. Nellis and L. Seaman for organizing several of the sessions is acknowledged. The following individuals are acknowledged for their hard work at the conference: Ms. M. Daniels, Ms. M. Lafer, the many students of the Shock Dynamics Laboratory, and the staff of the Spokane-Sheraton Hotel.

The various individuals who reviewed the manuscripts deserve a special acknowledgement for their time and effort. They are: D.J. Andrews, J.R. Asay, L.M. Barker, N.S. Brar, W.J. Carter, D.P. Dandekar, G.E. Duvall, J.O. Erkman, M. Finger, J.E. Flinn, J.W. Forbes, J.N. Fritz, D.E. Grady, R.A. Graham, W. Herrmann, B.S. Holmes, N.C. Holmes, Y. Horie, G.I. Kerley, P.C. Lysne, R.G. McQueen, R.S. Miller, T.R. Neal, W.J. Nellis, M.F. Nicol, R. Rabie, J.T. Rosenberg, M. Ross, E.B. Royce, S.C. Schmidt, L. Seaman, J.W. Shaner, and D.G. Tasker.

I wish to express my gratitude to members of the organizing committee for their help, advice, and hard work to ensure a successful conference. Without their support, I could not have done my job. J.R. Asay and A. Roach are thanked for sharing many organizational details about the previous conference. Ms. Kjerstie Guenther provided superb assistance with the final preparation of the manuscripts for the proceedings. Finally, no single individual worked harder than Ms. Debra Maggiora towards making this conference a success. Her untiring and enthusiastic efforts are sincerely acknowledged and appreciated.

Y.M. Gupta

CONTENTS

CHAPTER I - PLENARY

CHAPTER II - EQUATION OF STATE

CHAPTER III - OPTICAL STUDIES

CHAPTER IV - MECHANICAL PROPERTIES (MICROMECHANICAL ASPECTS)

CHAPTER VII - EXPERIMENTAL DEVELOPMENTS

CHAPTER IX - HYDRODYNAMICS

CHAPTER X - MATERIAL SYNTHESIS

SHOCK WAVE RESEARCH: YESTERDAY, TODAY AND TOMORROW

G. E. Duvall

Shock Dynamics Laboratory
Department of Physics
Washington State University
Pullman, WA 99164-2814

I. Introduction

I was introduced to shock wave research in late 1953 by Dr. Thomas C. Poulter, Associate Director of Stanford Research Institute and head of a group which later became the Poulter Laboratories of SRI. For many years there was mounted on a wall in Poulter Laboratories a larger-than-life poster showing me standing in a firing pit at the Calaveras Test Site with shovel in hand while Poulter and Dr. Dan MacLachlan looked on. This picture, taken in late 1953 or early 1954, was symbolic of the state of shock wave research before World War II. It was changing in 1953, but it had not changed very much.

I remained at Poulter Laboratories until January 1964, and during those years we worked on many problems, most with a practical tilt, many of which are still of interest today. Some of them are listed in Table I.

TABLE I

Some Problems at Poulter Laboratories, 1953-1964

Fragmentation and Spall
Metal Cutting and Shaping with Explosives
Jetting and Penetration
Wave Shaping
Explosive Welding
Plastic Deformation
Metallurgy: Twinning and Transformation Bands
Phase Transformation in Steels
Powder Compaction
Synthesis of Diamond
Cracking of Tar Sands
Explosive Devices

A substantial amount of our work was concerned with development of products and devices. This involved fragmentation and spalling studies, metal cutting and shaping with explosives, jet formation from shaped charges and their penetration of solids. An interesting aside to this work was a series of studies on wave shaping of detonation and shock waves. We did some of the earliest work on explosive welding, a subject which I feel is still not well understood. We also did a fair amount of work on plastic deformation of metals and we did some metallographic studies of twinning in metals, transformation bands in iron, and phase transformations in some steels. We made a number of studies on powder compaction, and this led ultimately to synthesis of diamond by DeCarli and Jamieson. We were also involved for a brief time in the use of shock waves to crack the hydrocarbons in tar sands. Another activity which consumed a good deal of effort and produced some interesting observations was design and construction of explosive devices.

II. History 1953-1966

The true history of shock wave research in this period bears little direct relation to the problems listed in Table I. It was a time of simplification and of a rebuilding from fundamentals. It was a period marked by enthusiasm and innovation and the belief among its practitioners that here lay the seeds of important science, waiting only to be cultivated and the products harvested. Much of the production of the period represented the pentup outpourings of the years of World War II when some of the best minds in the world were turned to shock and related problems. It was fueled by the generally expansive attitudes of our society during that time. I've summarized in Table II some significant events of this period. The first written,

TABLE II
Some Major Developments in Shock Wave Research
(Real-Time Experiments)
1953-1965

PHENOMENA	TECHNIQUES
Equation of State	Pins
Phase Transitions	Streak Camera
Electrical Conduction	Quartz Gauge
Charge Release-Polarization	Piezoresistance Gauge
Shock Demagnetization	EMV Gauge
Shock-Induced Opacity	Laser Interferometer
Elastic Precursor and Decay	High Speed Scopes
Residual Stress	

public report of what can be properly called modern shock wave research was published by J.M. Walsh and R. H. Christian in March, 1955 [1]. It contained equations of state of aluminum, copper, and zinc derived from Hugoniot data at pressures between 100 and 500 kilobars. Shocks were generated by plane wave explosive lenses and explosive pads. Shock and free surface velocities were recorded using argon flash gaps and a rotating mirror streak camera. The Hugoniots were reduced to isotherms assuming constant Γ/V and constant C_v.

In 1955, pressure was still one of the most neglected state variables in condensed matter physics. There were measurements by Bridgman, and very little else. Here, in one stroke, the door was opened to the modern era: the accessible range of this neglected variable had been extended threefold; the measure was absolute, depending only on measurements of time and distance; a new technique was presented for gathering data in a hitherto inaccessible region of an important variable; and a procedure for converting the data to a generally useable form was described. Static high pressure research was expanding and there was a ready and eager audience clamoring for data. Moreover, given the concept of the experiment, other mechanical applications, such as elastic precursor and phase transition measurement, were logical extensions of the basic idea. (By mechanical parameter I mean one which produces a measurable change in the wave form of the shock. Since a transition to a denser phase causes a cusp to form in the Hugoniot curve, it is a mechanical variable.)

This work was quickly followed by other important papers. A complete account of phase transitions in shock was given in a paper on iron by Minshall, Bancroft, and Peterson of Los Alamos in 1956[2], and in a later paper by Russel Duff and Minshall on bismuth[3]. The fact

This drove him almost to distraction, but I was finally convinced, and my skepticism gave him excellent training for dealing with other doubters who came later.) The synthesis of inorganic material from the powdered forms of their elements was first reported separately in 1963 in two papers from Japan: one by Y. Horiguchi and Y. Nomura on titanium carbide [19] and another by Y. Kimura on zinc ferrite[20]. Later, in 1965, the synthesis of polymers from condensed monomers was reported by Adadurov, Barkalov, Gol'danskii, Dremin, Ignatovitch, Mikhailov, Tal'roze, and Yampol'skii of the USSR[21]. Thus, in this short time, the foundations were laid for what has since become a major activity in the chemistry of encapsulated substances compressed by shock waves.

The explosive compression of powders has been used commercially in various ways since the early 1900's, according to Rinehart and Pearson in their 1963 book on explosive metalworking[22]. It was given a new face in a watershed paper by Oswald Bergmann and J. Barrington published in 1966[23]. Current activities in powder compaction, shock sintering, and shock treatment of powders to be sintered are directly traceable to this paper.

We see then that in 1966 the stage had been set for the intensive activity of later years in two important areas of shock wave research: real-time experiments and recovery experiments.

III. The Structure of Shock Wave Research

Shock wave research is, in some ways, an odd field, as those of us who have been in it for a long time know. In particular it is odd because it encompasses so many different disciplines. In Fig 1 is shown a spider web At the center, like a giant spider, sit the fundamental bases of

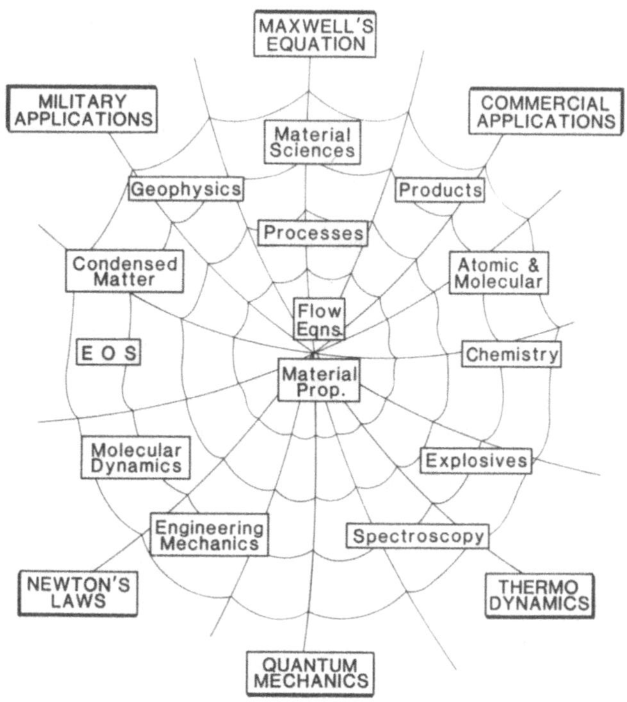

Figure 1.

all of shock wave research: the Nonlinear Flow Equations of Continuum Mechanics coupled with Material Properties. Caught in the web of these two sinister subjects are all of the disciplines which find the results of our experiments useful: atomic and molecular physics, chemistry, metallurgy, engineering mechanics, molecular dynamics, material science, solid state physics, spectroscopy, etc.

Shock wave research has a particular feature which is very much like that of the spider's web. When one insect caught in a real web moves, the motion is felt by all the others who are attached. Just so in shock wave research; when something momentous happens in one discipline or subject, the reverberations are felt by all the others. It is because of this transmission-of-influence effect that we of so many different backgrounds are able to come together in a meeting like this and talk to one another in meaningful ways, though our parent disciplines are poles apart.

Note also that the web is firmly anchored to the rest of the world by physical laws and the needs of the society in which we live. This is an important reminder that shock wave research requires funds, that its sponsors usually provide such funds because they want something from us, and that we must constantly tread a narrow path which produces something for the sponsor and also allows us to contribute to those aspects of human knowledge which we think are important.

If we dig more deeply, we find that the real tie which binds us is even more specific than the flow equations and material properties. It is THE EXPERIMENT, shown in Figure 2.

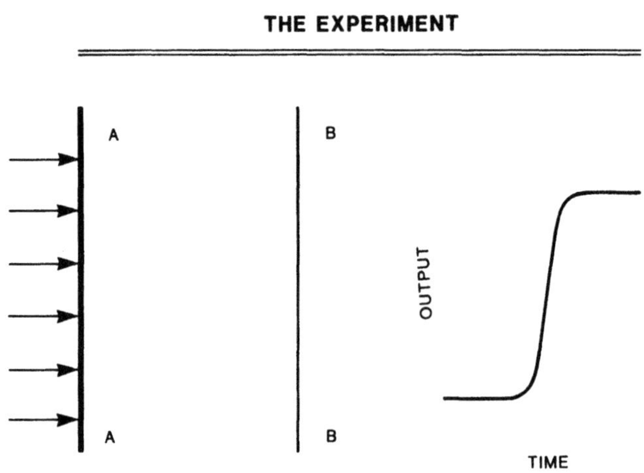

Figure 2.

The experiment tames the nonlinear flow equations and limits material response enough that we are able to make interpretable measurements. It is shown here in its most elemental form. It consists of a half space of a simple material: no elastic precursor, no phase transitions, no hysteresis, no time or rate dependent properties (except for some viscous-like force which gives form to the wave front), no internal variables, no lateral rarefactions. A uniform influence is applied in the form of a step to the surface AA at t=0, and held constant thereafter. It may be pressure, velocity, or any other suitable parameter. An ideal planar probe BB is inserted with plane parallel to the surface AA at some known distance from the surface. It takes up no space and produces no perturbations of the system. It just sits there and reports the value of one of the mechanical variables as the wave goes by: particle velocity, pressure, or density. There is no back surface, so there are no reflections. The probe records the arrival time of the shock, so we can obtain the shock velocity. From the amplitude we get one other parameter. We put these into the jump equations and determine the other three shock parameters, and our measurement is complete. In its interpretation we use the constraints supplied by the Rayleigh line and one dimensional strain with confidence, because the macroscopic flow is truly uniaxial.

that the iron transition was unknown and generated considerable controversy gave shock wave research an important boost outside the shock community.

The first report of shock-induced conductivity in insulators was given in 1956 by Alder and Christian of Livermore[4]. Walsh and Rice of Los Alamos reported observation of optical opacity in carbon tetrachloride in 1957[5]. In that same year Neilson of Sandia reported charge release measurements from ferroelectrics and, with George Anderson, shock demagnetization of iron[6,7].

The Walsh and Christian paper of 1955 treated metals as liquids. The question of residual shear became of considerable interest, and in 1961 G.R. Fowles of SRI was able to show that shear did exist behind the plastic wave [8]. He did this by comparing shock measurements to Bridgman's static measurements on aluminum. This interest in residual stress led to considerations of stress relaxation and decay of the elastic precursor, which was first reported by Rice and Taylor of Los Alamos in 1963[9].

At a conference in Paris in 1961, Hauver and Eichelberger of Ballistics Research Laboratory described polarization and charge release from unpoled dielectrics, apparently the result of dipole alignment or induction by the shock front[10]. At that same conference Neilson and others from Sandia described the thick quartz gage, which suddenly made shock research available to everyone and greatly accelerated progress in low pressure research[11]. The manganin piezoresistance gauge was reported in 1964 independently by Fuller and Price of AWRE in England and by Bernstein and Keough of SRI[12,13]. In the same year. Dremin and Adadurov of the USSR described the electromagnetic particle velocity gauge[14], and in 1965 Barker and Hollenbach of Sandia described the first incarnation of the laser interferometer[15].

While dramatic events were occurring in these real-time experiments, progress was also being made along a different line, namely recovery experiments. Some critical experiments in this area are listed in Table III.

TABLE III
Precedent Setting Recovery Experiments, 1953-1966

1956	**Decomposition of Inorganics**	**I.U. Riabinin, USSR**
1958	**Metallography of Fe**	**Cyril Stanley Smith**
1961	**Synthesis of Diamond**	**P.S. DeCarli and J.C. Jamieson**
1963	**Inorganic Synthesis**	**Y. Horiguchi and Y. Nomura Y. Kimura**
1965	**Polymerization**	**G.A. Adadurov, I.M. Barkalov, V.I. Gol'danskii, A.N. Dremin, T.N. Ignatovitch, A.N. Mikhailov, V.L. Tal'roze, and P.A. Yampol'skii**
1966	**Powder Compaction**	**O.R. Bergmann and J. Barrington**

In 1956 Riabinin of the USSR reported the results of a series of experiments in which he had placed various chemical substances in strong metal cylinders, encased the cylinders in high explosive, and detonated them. On recovering and opening the still-sealed cylinders, he found decomposition products - gaseous and solid - intermixed with remnants of the original substances[16]. This paper was the first in a long series from the USSR on various aspects of chemical reactions induced by shock waves in encapsulated samples.

In 1958 a paper by Cyril Stanley Smith reported on the metallography of the transformation and twin bands in iron which had been shocked explosively at Los Alamos[17]. This paper was a model for and a clearly recognizable ancestor of most metallographic shock experiments to this day.

Paul DeCarli of SRI and John Jamieson of Chicago reported in 1961 the first synthesis of diamond by shock waves[18]. This was an important achievement, and a very difficult one, as I can testify from personal experience. (I played a part in it. Every time Paul came to tell me about his progress, I would say "Paul, that can't be right. You must have made a mistake."

We never do this experiment, of course, but it represents an ideal which is always before us and provides a reference for our real experiments, which involve three modifications of the experiment, other than the obvious geometric ones and the properties of real probes.

In Modification 1, real mechanical properties are introduced. These cause the shape and amplitude of the probe output to change. Material properties information is contained in these changes, and we use whatever means and ingenuity we have to get it out. In case the sample has time-dependent properties on the scale of the experiment, multiple probes or multiple experiments are required.

In Modification 2, probes are introduced for measuring non-mechanical variables - temperature, electric and magnetic fields, epr, nmr, x-ray diffraction, Hall voltages, radiation, optical properties, etc. This represents a different kind of experiment from (1) inasmuch as the coupling between these variables and the wave form is weak and shock compression is being used to condition the material for measurement. The wave form does not, in general, reflect measurable changes in the properties being measured. But it must not be thought that shock compression is now only a substitute for static compression. Non-equilibrium situations exist on some time scale. If this time scale agrees with that of the experiment, we have an interesting and possibly valuable result which cannot be obtained statically.

A complication in this kind of experiment which is not present in modification 1 involves the necessity for sampling a finite volume of the material. Since the reason for making measurements of this kind may be to determine time-dependent effects, a careful balance between space and time resolution must be maintained.

In Modification 3. Electric, magnetic or other types of fields are applied to the sample in order to modify its intrinsic behavior. This kind of experiment has seldom been done, but it may have value which has not been realized.

When we were ignorant of the experiment, we did recovery experiments, which are very complicated, and tried to make sense of them, and we were usually not very successful. Now that we understand the experiment and its modifications very well, we are emboldened to try more complicated experiments again. We are aided in these efforts by wave propagation codes and sophisticated diagnostics, but such experiments are still very difficult to do and hard to interpret.

IV. History 1965-1980

With the experiment in mind we can review the progress of shock wave research in the 1965-1980 period in a reasonably coherent way.

The experiment with mechanical modifications has been wildly successful (modification 1). It has produced a wealth of data on equations of state, dynamic yield stress, and phase transitions. Variations have been developed which have opened new prospects and which promise to further expand our scope. Some examples are, the introduction of superposed shear waves by Abou-Sayed and Y.M.Gupta[24,25] and the introduction by Barker and Hollenbach of the fused quartz buffer to generate ramp waves[26]. Others include various efforts to generate and measure shocks at ultra high pressures[27].

Considerable effort has been devoted to the second modification of the experiment for measurement of non-mechanical variables, with substantial improvement over early efforts. These include new measurements of electric conductivity and polarization, a definitive study of shock demagnetization by Grady[28], and various efforts to measure temperature in the shock. New kinds of measurements have included index of refraction for visible light, Hall coefficients, stimulated Brillouin scattering, flash X-ray diffraction, transverse stress behind the shock, and, most recently, different kinds of spectroscopy. Until quite recently these techniques have represented solutions without problems. Spectroscopy may be an exception. None of this work has yet stimulated a rush of activity like that which followed the Walsh and Rice equation of state paper.

The second branch of our subject, recovery experiments, has experienced a similar swelling of activity since 1965. Shockwave chemistry and powder compaction related to sintering have been pursued with great vigor abroad, particularly in Japan and the USSR. The activity appeared to wane somewhat in the mid 70's, but has been renewed during the last five or so years, with increasing activity in the United States. Here too the principal thrust has been directed toward refining and extending the work of the 1955-65 period. There has been only limited progress toward quantifying the results of these experiments, partly because of the complex geometry involved, and sometimes because of the nature of the sample, eg powders. At least two products of these experiments, synthetic diamond and cubic boron nitride, are now commercial.

In brief, it appears fair to say that the 1955-65 period was one of great innovation and that it has largely set the pattern for activities of the intervening years. The period from 1965 to 1980 was one of consolidation, extension, and refinement. It was also a period when the boundaries of our subject were being explored and fundamental problems defined, particularly in the areas of mechanics and thermodynamics. The same cannot be said of non-mechanical subjects.

V. Fundamental Problems

There certainly are technical problems involving non-mechanical variables (i.e., modification 2), but the major barrier to expansion in these areas is lack of a set of problems of interest to other branches of science or technology for which shock experiments can provide the solution. I believe that such problems exist and that they will almost certainly be at the microscopic level. To discover them is going to take deep thought and collaboration of persons trained in condensed matter science with those knowledgeable in shock wave experiments. One area in which to look is that of time dependence and non-equilibrium effects. We must also remember that in elastic materials, the uniaxial strain condition is unique to shock experiments. Only if the shock environment offers an opportunity to make useful measurements that cannot be made in other ways, can it expect to be successful here.

I have listed Mechanical and Thermodynamic problems in Table IV. Of the Mechanical Problems, dynamic failure at very large strain rates is an important and difficult one. Our experiments at WSU with LiF certainly cast doubt on the ability of conventional dislocation

TABLE IV
Fundamental Problems in Mechanics and Thermodynamics
(Modification 1)

MECHANICAL
Dynamic Failure and Dislocations
Shock Front Structure
Phase Transitions: Speed, dP/dV, Shear
Deviations From Equilibrium
Mechanical Jitter and Time Resolution
Laser Generation of Shocks. Diagnostics
Constitutive Relations
THERMODYNAMICS AND EQUATIONS OF STATE
Temperature Measurement in the Shock
EOS Theory for Molecular Liquids and Solids
Specific Heat: Dependence on Density
Conversion of Internal to External Modes
Molecular Dynamics
Pair Potential Limits
Quantum Effects
Practical Equations of State

theory to account for elastic failure and measured values of precursor decay. Limited data on other materials suggest that the same statement can be more broadly made. It is possible that only a minor adjustment of current theory is required; or fundamental revisions may be necessary. In any case it is becoming ever more certain that shock induced processes are essentially heterogeneous, even in nominally homogeneous materials. It is incumbent on us to attempt to understand these heterogeneities.

As time resolution of shock experiments increases, the structure of the shock transition zone becomes of increasing interest. It is here also that we may expect to find microscopic effects which will be of interest for Modification 2. There are several reasons for this. One is that it affects experiments on a short time scale so that the permanent regime may not have been established. Another is that the shock transition itself is a very important event on a molecular scale, and if we can understand its details, new areas of investigation may appear.

There are three problems of some importance related to shock-induced phase transitions. One is the rate at which they appear to occur--much faster than one would expect from static measurements. A second is that the measured Hugoniot in the mixed phase region is usually much steeper than one calculates from the Clausius-Clapeyron equation; this may be related to the first problem. A third is the role that phase transition may play in the reduction of stress deviators in an elastic medium; can it preempt the role of plastic flow in uniaxial compression?

The shock transition is certainly a non-equilibrium process. There are also situations where deviations from equilibrium occur on a much longer scale, ie. behind the shock front. This situation offers a concern and an opportunity: The concern is that our measurements of equation of state, for example, may be distorted; the opportunity is that non-equilibrium measurements, if we understand them, provide new opportunities to acquire information that is otherwise inaccessible.

Mechanical jitter may seem an odd subject to include here. In fact it is quite important. It limits time resolution, and there is increasing evidence that improved time resolution is going to give us a window into new aspects of the shock compression process. The role of laser generated shocks is of more than passing interest for this reason. It offers the promise of avoiding the mechanical jitter problem because of the potential for synchronizing shock generation and diagnostics. Will this promise outweigh the problems associated with the method? You will hear later this week a progress report on an intensive effort to explore this problem.

Finally, constitutive relations are always with us. At present such relations are strongly ad hoc; the complexity of the problem is such that this may always be true. It is to be expected, however, that increasing order will be brought to the subject through combination of the logic of continuum mechanics with keen physical insight into the behavior of condensed matter.

Insofar as thermodynamics is concerned, equation of state is the primary problem. Theorists have made great contributions to both the interpretation of shock measurements and to the construction of equations of state from first principles. The principle additional contribution which the experimentalist can make to this problem is the measurement of temperature. Theory will still face several tasks. One is to extend the fundamental theory to molecular liquids and solids, and some progress is being made on this. A second is to look more closely at the specific heat problem. One question we constantly face is the effect on C_v of increasing density; a second, important in liquids, concerns the constraints applied to molecular degrees of freedom by the surrounding molecules and the correct treatment of the processes whereby molecular degrees of freedom are converted to those of the highly condensed state. It is well to include the problems and promises of molecular dynamics. It's contributions have been substantial and it promises to help us understand processes at the molecular and atomic levels. In order to do so it must seek out and face the constraints of the pair potential formula, and it must address quantum requirements in the intramolecular states.

Another problem is the construction of practical equations of state for limited ranges of pressure and temperature. Experiments are being performed today with many new substances. The experimenters have wave codes which they would like to run in order to help them interpret their experiments. We would like to do better than a Mie-Gruneisen equation of state in such situations.

There is more than one way to do science. I have been discussing the kind of approach to understanding which is familiar to most physicists: the number of variables is small, samples can be well characterized, and the experimental data are closely linked to a governing theory which provides a tight framework for interpretation. Persons engaged in shock chemistry and powder compaction have problems which are more nearly akin to those faced by biologists, geologists, or ceramists. The number of variables is very large; the details of the experiment cannot be controlled in a local sense, or are amenable to only modest control; the starting material may be impossible to characterize on a microscopic scale; the environment which determines the outcome of the experiment is not well known.

But when we think how much was learned about metallurgy and ceramics before those words were invented, we are encouraged to hope that the puzzles of shock-induced chemical reactions and sintering will be unravelled. The problems depend partly on whether the sample is homogeneous and non-porous, as, for example, the liquid monomer used as starting material for a polymerization experiment, or whether it is a partially compacted powder.

The common problem is this: The encapsulated sample is placed in an explosive (or target) assembly, where it is subjected to the impact of shock waves produced by the system. Reflections occur at all the boundaries of the capsule; diffraction occurs around its corners; the impedance of the sample changes as it is compacted, much more than does that of its container. The uniaxial strain condition no longer applies. If the sample is homogeneous and non-porous, one can, in principle describe its stress and temperature histories. In fact this is very hard to do because of uncertainties in equations of state, diffusion of heat within the sample and from the sample to the container, mass diffusion, changes of state resulting from chemical reaction, possible radiation from very hot regions, and the complexities of the very late stages of the entire process after pressures have started to decrease and before the recovered sample has been quenched.

There are additional problems with powder samples. It is very hard to characterize or even to reproduce the starting material, inhomogeneities in density imply very large inhomogeneities in temperature and stress distributions, microjetting and fracture occur, large electric fields are produced, and electrons, ions and photons are randomly emitted in quantity from the newly formed surfaces. If the object of the experiment is synthesis of compounds from their elements, mass transport is required to insure the necessary proximity of the components. This implies that the powder must be very fine and intimately mixed, thus compounding the problems of material characterization and reproducibility.

We may expect quasiempirical models to be generated which have predictive capabilities with both technical and scientific value. I am personally not optimistic about generating information which will change our microscopic physical models, but I recognize that such a possibility exists. Certainly such work may lead to valuable processes and products, and it has already influenced materials technology and science in a significant way.

VI. The 1980's

In the process of preparing for this talk, I made a rough analysis of the papers which are being given at this conference and of those which were given at the first topical conference in 1979. The results are shown in Table V.

There were 99 papers at the 1979 conference, there are 178 at this one. The top two rows of the table show a partition of the papers into experimental and theoretical or numerical. The fraction of experimental papers has increased somewhat. The section between the two horizontal dashed lines is a partition of papers by subject. There is an interesting reduction in the fraction of papers which address mechanical and thermodynamic problems; the remainder of the section shows the reason. There has been a large increase in optical papers, particularly spectroscopy, and in non-explosive chemistry. There has been a more modest increase in electric and magnetic papers, and a large drop in explosive papers. The last two lines of the table show the dramatic increase in recovery experiments since 1979. With one or two exceptions, these papers all involve shock chemistry or powder compaction, and, in fact, of the chemistry papers listed in the central section of the table, most were recovery experiments. Of the optics papers at this conference 13 are on spectroscopy, split evenly between vibrational and UV or visible. Five papers are devoted to temperature measurement and five to structure of the shock front.

These numbers confirm my belief that we are in a new period of exploration. It is different from the 1955-1965 period, because it is superposed on a reasonably mature and sophisticated background; but it is equally exciting. It is fueled by growing interest in the microscopic implications of shock wave experiments and the mysteries of the shock transition.

It comes at a time when there are new experimental techniques to be tapped, and it may be due in part to their existence. It seems to have had its origins in the wish to know more about the details of chemical reactions which are initiated and supported by the shock; but it has gone beyond that to a desire for more intimate knowledge of the behavior of electrons, atoms, and molecules in and behind the shock front. As for all explorations, there is more hope than progress at this point, but prospects are bright, and I shall be surprised if it doesn't initiate some continuing, productive programs, and if interest doesn't expand with progress.

TABLE V
An Analysis of the Contents of Papers at the Topical Conference on Shock Waves in Condensed Matter for 1979 and 1985.

	1979 %	1985 %
Experimental	61	64
Theoretical and Numerical	39	36
Mechanical and Thermodynamics	64	54
Electric and Magnetic	6	9
Optical	5	9
Spectroscopy	2	6
Chemistry (Nonexplosive)	1	10
Explosives	19	11
Shock Front Structure	2	2
Recovery Experiments	1	12
Powder Compaction	--	5

As a community, we are beginning to reach out to other disciplines for collaborators. Our relations with static high pressure have always been close and productive. It is becoming more so with introduction of the highly versatile diamond cell.

Much of the new activity is chemically oriented and spectroscopy plays an important role. I think this mix has the right ingredients for success, but it faces many problems. Both spectroscopy and shock are complicated subjects, and if they are to be successfully wedded, hard work, understanding, cooperation and forebearance will be required from each group. As our interest in the microscopic behavior of matter in non-equilibrium, time dependent regimes deepens, collaboration must be broadened. We will lose some of our sense of community, but we shall gain in greater understanding and more broadly useful research..

Shock recovery experiments will continue to be important because of their technological implications. Powder compaction is in a good position because of its promise for commercial applications in this materials-hungry world. Recovery shock chemistry is in a little less favorable position because no product is in sight. For all such experiments there is the hope and promise that they will reveal phenomena for exploration in real-time experiments.

Experimental papers and papers devoted to the mechanical and thermal aspects of shock wave research will continue to be prominent in future conferences of this kind, as will detonation and explosives research.The last may continue with new enthusiasm if explorations into the microscopic features of shock compression are successful.

I feel now, as I did in 1979, that electrical experiments have an untapped potential. I have no clear idea how this potential can be realized, but I am inclined to believe that an examination of fluctuations in the field may be more revealing of microscopic behavior than average field measurements.

In closing I shall remark on a phenomenon I've observed for the last 15 or so years.

Scientists in the shock wave field are becoming too closely programmed. There is a feeling among productive scientists that a minute in which progress is not made toward some programmed objective is a minute wasted. As a research manager and working scientist I'm aware that this feeling is often traceable to the attitudes of the manager. This philosophy produces a lot of data and a lot of papers, but it has some negative aspects. It deprives the scientist of the perspective required to evaluate what he's doing. It encourages the scientist to go on doing what he has done in the past for too long because study and thought take time. The net result may be a net loss to the science. Close programming also encourages the diffusion of responsibility for program decisions upward and may lead managers to ultimately believe that they know best what should be done. It is dangerous to proceed far without tapping the knowledge of the scientist in the laboratory. The greatest success will come from shared knowledge, shared responsibilities, and program decisions based on thorough study and careful thought by both manager and the working scientist.

It is difficult to restructure one's life so there is time for study and reflection; in the long run there is no alternative. It is necessary for our healthy survival in a promising but demanding future.

REFERENCES

1. J. M. Walsh and R. H. Christian, "Equation of State of Metals from Shock Wave Measurements," Phys. Rev. 97:1544 (1955).
2. D. Bancroft, E.L. Peterson, S. Minshall, "Polymorphism of Iron at High Pressure," J. Appl. Phys. 27:291 (1956).
3. R. E. Duff and F. S. Minshall, "Investigation of a Shock-Induced Transition in Bismuth," Phys. Rev. 108:1207 (1957).
4. B. J. Alder and R. H. Christian, "Metallic Transition in Ionic and Molecular Crystals," Phys. Rev. 104:550 (1956).
5. J. M. Walsh and M. H. Rice, "Dynamic Compression of Liquids from Measurements on Strong Shock Waves," J. Chem. Phys. 26:815 (1957).
6. F. W. Neilson, "Effects of Strong Shocks in Ferroelectric Materials," Bull. Am. Phys. Soc. 2:302 (1957).
7. G. W. Anderson and F. W. Neilson, "Effects of Strong Shocks in Ferromagnetic Materials," Bull. Am. Phys. Soc. 2:302 (1957).
8. G. R. Fowles, "Shock Wave Compression of Hardened and Annealed 2024 Aluminum," J. Appl. Phys. 32:1475 (1961).
9. J. W. Taylor and M. H. Rice, "Elastic-Plastic Properties of Iron," J. Appl. Phys. 34:364 (1963).
10. G. E. Hauver and R. J. Eichelberger, "Solid State Transducers for Recording of Intense Pressure Pulses," in Les Ondes de Detonation. Colloques Internationaux du Centre National de la Recherche Scientifique, Gif-sur-Yvette, 28 aout - 2 septembre, 1961, Paris, Editions du Centre National de la Recherche Scientfique, 1962.
11. F. W. Neilson, W. B. Benedick, W. P. Brooks, R. A. Graham, and G. W. Anderson, "Electrical and Optical Effects of Shock Waves in Crystalline Quartz," in Les Ondes de Detonation. Colloques Internationaux du Centre National de la Recherche Scientifique, Gif-sur-Yvette, 28 aout - 2 septembre, 1961, Paris, Editions du Centre National de la Recherche Scientfique, 1962.
12. P. J. A. Fuller and J. H. Price, "Dynamic Pressure Measurements to 300 Kilobars with a Resistance Transducer," Brit. J. Appl. Phys. 15:751 (1964).
13. D. D. Keough and D. Bernstein, "Piezoresistivity of Manganin," J. Appl. Phys. 35:1471 (1964).
14. A. N. Dremin and G. A. Adadurov, "The Behavior of Glass under Dynamic Loading," Sov. Phys.-- Solid State 6:1379 (1964).
15. L. Barker and R. E. Hollenbach, "Interferometer Tehnique for Measuring the Dynamic Mechanical Properties of Materials", Rev. Sci. Instrum. 36:1617 (1965).

16. Iu. N. Riabinin, "Certain Experiments on Dynamic Compression of Substances," Sov. Phys.-- Tech. Phys. 1:2575 (1956).
17. Cyril Stanley Smith, "Metallographic Studies of Metals After Explosive Shock," Trans. AIME 212:574 (1958).
18. P. S. DeCarli and J. C. Jamieson, "Formation of Diamond by Explosive Shock," Science 133:821 (1961).
19. Y. Horiguchi and Y. Nomura, "An Explosive Synthesis of Titanium Carbide," Bull. Chem. Soc. Jap. 36:486 (1963).
20. Y. Kimura, "Formation of Zinc Ferrite by Explosive Compression," Japan J. Appl. Phys. 2:312 (1963).
21. G. A. Adadurov, I. M. Barkalov, V. I. Gol'danskii, A. N. Dremin, T. N. Ignatovich, A. N. Mikhailor, V. L. Tal'roze, and P. A. Yampol'skii, "Polymerization of Condensed Monomers in a Shock Wave," Acad. Sci. USSR. Proceed. Phys. Chem. Sect. V. 165:835 (1965).
22. John Rinehart and John Pearson, Explosive Working of Metals. MacMillan, 1963.
23. O.R. Bergmann and J. Barrington, "Effect of Explosive Shock Waves on Ceramic Powders," J. Am. Ceram. Soc. 49:502 (1966).
24. A. S. Abou-Sayed and R. J. Clifton, "Pressure Shear Waves in Fused Silica," J. Appl. Phys. 47:1762 (1976).
25. Y. M. Gupta, "Shear Measurements in Shock-Loaded Solids," Appl. Phys. Lett. 29:694 (1976)
26. L.M. Barker & R.E. Hollenbach, "Shock Wave Studies of PMMA, Fused Silica, and Sapphire," J. Appl. Phys. 41: 4208 (1970).
27. Shock Waves in Condensed Matter - 1981 (Menlo Park), Chapter III, Am. Inst. Phys. (1982), W.J. Nellis, L. Seaman, R.A. Graham (Eds.).
28 D.E. Grady, G.E. Duvall, and E.B. Royce, "Shock-Induced Anisotropy in Ferromagnetic Materials. II. Polycrystalline Behavior and Experimental Results for YIG", J. Appl. Phys. 43 : 1948-55(1972).

RECENT DEVELOPMENTS IN STATIC HIGH PRESSURE RESEARCH

A. Jayaraman

AT&T Bell Laboratories

Murray Hill, New Jersey 07974

INTRODUCTION

The development of the diamond anvil cell for x-ray and spectroscopic studies has given a new impetus to high pressure static research.[1)] At the Geophysical Laboratory in Washington DC, pressures in excess of 2.5 megabars have been reached with the diamond cell recently, and high pressure x-ray studies in the megabar range are almost routine. In the same laboratory Raman measurements on solid hydrogen have been extended to 1.5 megabar pressure. This is the highest pressure at which any spectroscopic measurement has been performed. While these developments are at the boundaries of static high pressure range, it has now become possible to carry out x-ray and spectroscopic measurements to several hundred kilobars routinely. The interfacing of the diamond cell with synchrotron sources[2)] has revolutionized high pressure x-ray diffraction investigations. Compression measurements, phase transition and structural studies at high pressure are greatly facilitated by the rapidity with which high pressure x-ray diffraction data can be obtained with synchrotron sources.

A remarkable technical achievement with the diamond cell is the use of argon or xenon as pressure medium.[1)] This development has extended the hydrostatic regime to several hundred kilobars. Helium and hydrogen are found to be even better pressure transmitters, to pressures well over half a megabar. This has spurred high pressure studies at low temperatures using the diamond cell. Both low temperature and a high degree of hydrostaticity sharpen up the spectral features in spectroscopic studies, and in many cases the results of previous studies using a less satisfactory pressure media are being discounted.

In this paper we will review some recent studies under four headings (1) metallization under pressure (2) elemental systems at high pressure (3) molecular systems and (4) systems of interest to geophysics.

METALLIZATION UNDER PRESSURE

Metallization, induced by high pressure has been an active area in high pressure research, the prime candidate in recent years being hydrogen. However, more realistic candidates turn out to be Xe, CsI and Ba monochalcogenides. The interesting point about this set of compounds is that they are all isoelectronic with

Xe. Optical absorption studies on Xe,[3,4] optical absorption and x-ray diffraction studies on CsI,[5–9] and Ba chalcogenides[10] have been reported in very recent studies. Metallization of Xe and CsI has not yet been achieved, but BaTe has recently been shown to undergo the predicted metallization transition due to band overlap, at a pressure of about 20 GPa.[10] Pressure-volume studies on Xe, CsI and Ba chalcogenides have been carried out using high pressure x-ray diffraction, because volume is an important parameter in the theories predicting metallization. In this connection, the Herzfeld theory as well as band calculations have been used as a guide.[11,12] Table I summarizes the volume fraction for band overlap transitions predicted by these theories.

Ba-Chalcogenides

In Fig. 1 the pressure volume data for BaTe and BaSe are shown.[10] The drop in volume at the low pressure end is due to the NaCl to CsCl type (B1-B2) transition. In Fig. 2 are shown the shifts in the optical absorption edges of BaSe and BaTe as a function of relative volume. The extrapolated curves for BaTe shows that the band gap goes to 0 at $V/V_0 \sim 0.65$ and in the case of BaSe this would happen at a V/V_0 of ~ 0.58.[10] From the pressure-volume data these would correspond to 20 GPa and 43 GPa respectively for BaTe and BaSe band overlap. Resistance measurements[10] as a function of pressure show, that metallization is actually achieved in BaTe at V/V_0 of 0.65. In the case of BaSe the resistance data predict metallization around 40 GPa. It turns out that the experimentally obtained transition volumes are closer to the Herzfeld theory than to the rigorous band calculations.[12]

Cesium Iodide

Pressure-induced metallization in CsI has been the subject of intense experimental and theoretical investigations.[6–9,13–17] High pressure x-ray diffraction studies (up to 65 GPa)[8,14] and optical absorption studies (up to 60 GPa)[7] have been reported. The absorption data have been used to determine the pressure variation of the band-gap, from which the metallization pressure has

TABLE I

Metallic Transition Volumes and Pressures for Xe and Some Isoelectronic Compounds

Substance	Herzfeld Prediction V_m/V_0	Band Calculations V_m/V_0	Transition Pressure GPa
Xe	0.29	0.31 (a)	130
CsI	0.42	~ 0.40 (b)	~ 100
BaTe	0.62	0.74 (a)	30
BaSe	0.58	0.63 (a)	60
BaS	0.55	0.60 (a)	90
BaO	0.49	0.63 (c)	140

(a) Ref. 12 (b) Ref. 13 (c) Ref. 73

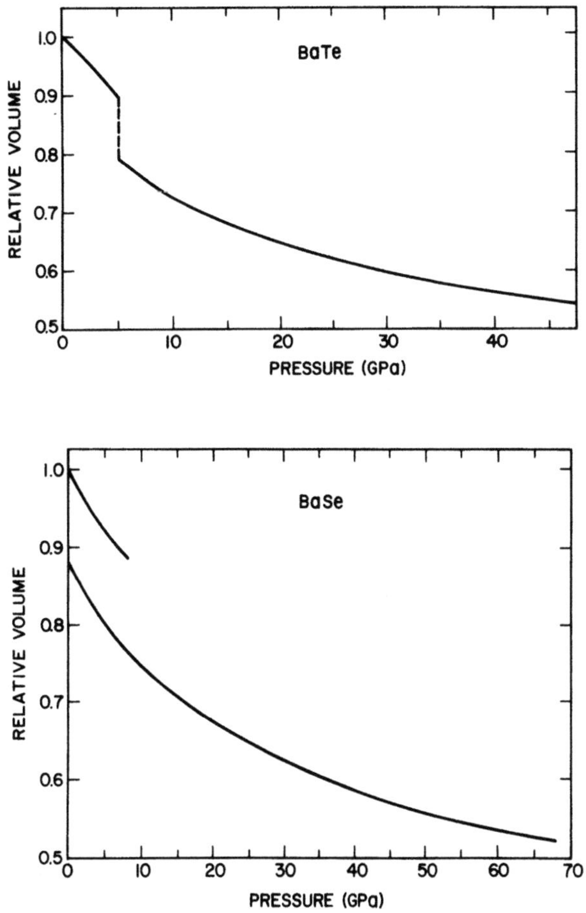

Fig. 1 Pressure-volume relationship for BaTe and BaSe. The abrupt decrease in volume at the low pressure-end is due to the B1 - B2 transition (ref. 10).

been estimated by extrapolation. Two sets of optical absorption data have been published,[5-8] namely, data obtained with polycrystalline CsI,[5,8] and data with single crystal[6,7] CsI[6] taken with xenon as pressure transmitting medium. These results differ in respect of the optical absorption feature as well as in the extrapolated transition pressures. There is also some difficulty in ascertaining the band gap correctly, because the gap cannot be obtained simply from the absorption edge, without a proper model for the transition (direct or indirect), or the absorption is due to an exciton. The optical density curves of CsI taken at different pressures are shown in Fig. 3. Absorption data obtained with single crystals by Makarenko et al[7] and the data obtained by Asaumi et al[5] on polycrystalline sample are shown. Makarenko et al have assigned the first absorption maximum to the exciton absorption peak corresponding to the $\Gamma_8 + \Gamma_8^+$ direct interband transition. This absorption is near 6 eV at ambient pressure and decreases to about ~ 1.5 eV near 60 GPa.[7] The energy gap obtained from this absorption study is plotted in Fig. 4 against pressure. The single crystal absorption data extrapolated to zero gap predicts a metallization pressure of ~ 110 GPa for

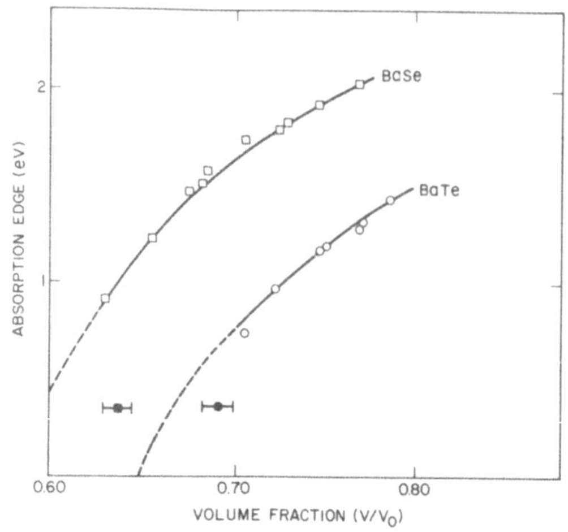

Fig. 2 Adsorption edge shift with pressure for BaSe and BaTe (ref. 10).

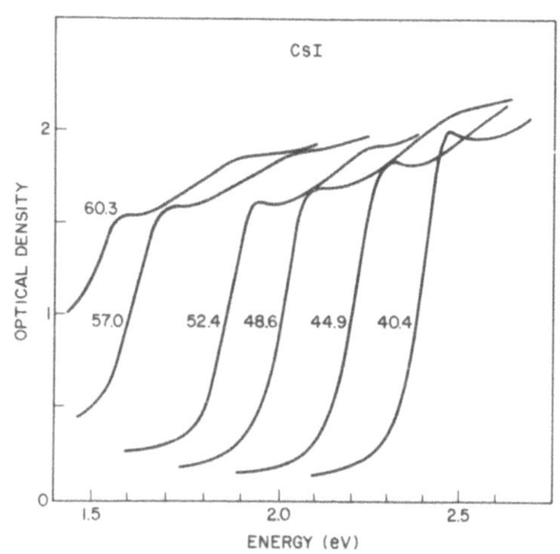

Fig. 3 Absorption edge shift with pressure in CsI (ref. 7).

CsI. The shaded area in Fig. 4 corresponds to the bandgap variation obtained by Asaumi et al, based on polycrystalline data. This extrapolates to zero gap in the range 70-80 GPa. The polycrystalline data of Knittle and Jeanloz extrapolate to 103 GPa.[9] The prediction based on single crystal study is closer to the theoretical estimate. Further, an independent study by Itie et al[6] on single crystal CsI with Xe as pressure medium is in reasonably good agreement with the results of Makarenko et al.[7] A linear extrapolation of this data gives 93 GPa for the band overlap,[6] as against the 110 GPa predicted by the nonlinear extrapolation of Makarenko et al.[7] The metallization pressure predicted by the single crystal absorption results should be preferred for obvious reasons. Cesium iodide is a promising candidate and its metallization may be expected to be achieved in the near future.

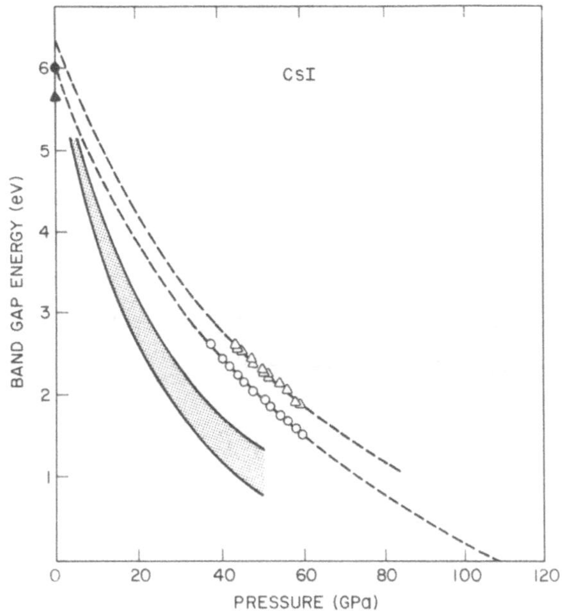

Fig. 4 Bandgap energy against pressure for CsI (ref. 7). The data points are from two different single crystal measurements (ref. 6 and 7). The shaded part is the polycrystalline data of ref. 5.

The high pressure x-ray diffraction studies have yielded not only the pressure-volume relationship, but also have revealed the existence of a phase transition in CsI from B2 structure to a tetragonal phase, near 40 GPa.[8,9,14–17)] (Fig. 5). The tetragonality increases continuously with pressure and the transition does not involve any volume change. There is good agreement between the x-ray data obtained in different studies. According to Asaumi[14] the tetragonal lattice changes to an orthorhombic one (see Fig. 5) around 56 GPa in CsI, but this is not confirmed in the study of Knittle and Jeanloz[8)] up to 61 GPa. Cesium bromide[17,18)] also undergoes the B2 to the tetragonal phase transition near 53 GPa. RbI and KI[15)] have been studied and they are found to be stable in the B2 structure up to 70 GPa, the limit of pressure investigated. The pressure-volume data for CsI, RbI and KI[15)] are shown in Fig. 6. The experimental equations of state of CsI and CsBr are in good agreement with a finite strain equation of state, based on the elastic constants measured at low pressures.[19)] The bulk modulus B_0 and B'_0 for these substances are collected in Table II. Asaumi et al[15)] predict metallization pressures of 85 and 115 GPa for RbI and KI, from their optical absorption data. However, these metallization pressures are likely to be lower estimates, in view of the fact that Asaumi et al[5)] have underestimated the transition pressure for CsI by 30-40 percent.

TABLE II

Bulk modulus B_0 and B'_0 for Ba chalcogenides and alkali metal iodides

	B_0 (GPa)	B'_0
CsI (a)	11.1 ± 0.8	6.9 ± 0.4
RbI (a)	16.0 ± 0.6	4.6 ± 0.1
KI (a)	24.2 ± 0.5	4.3 ± 0.1
BaSe (b)	43.0 ± 2.6	6.6 ± 1.4
BaTe (b)	29.4 ± 0.2	7.4 ± 0.2
Au (c)	166.6 ± 11	5.5 ± 0.8

(a) Ref. 15, (b) Ref. 11, (c) Ref. 69

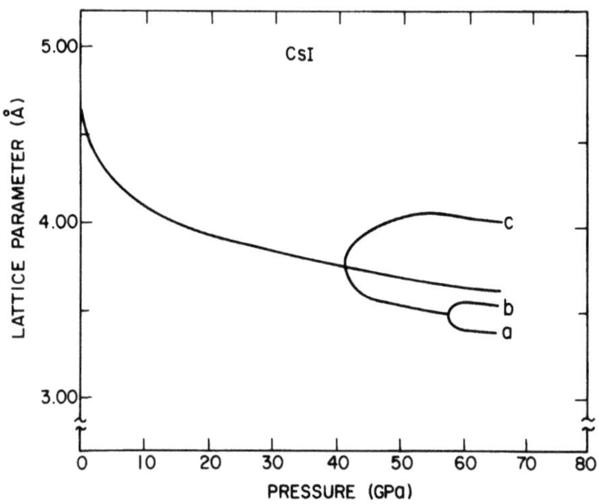

Fig. 5 Lattice parameter vs pressure for CsI from ref. 14. Near 40 GPa a B2 to tetragonal transition and near 56 GPa tetragonal to orthorhombic transformation takes place (ref. 14). a, b and c are the lattice parameters and the central line is $(abc)^{1/3}$.

Xenon

The metallization of Xe near 33 GPa reported by Nelson and Ruoff[20] some years ago has stimulated a great interest in this rare gas. Subsequent optical absorption studies[4,5] have invalidated metallization of Xe at such a low pressure, which led Chan et al[21] to revise their pressure of transition to pressures above 100 GPa, based on a more critical analysis of the original experiment. Optical absorption studies up to 63 GPa[4,5] has shown that Xe is a perfectly good insulator at this pressure. In two independent optical absorption measurements the bandgap energies are obtained as 3.9 eV at 44 GPa[22] and 3.7 eV at 55 GPa.[4] More

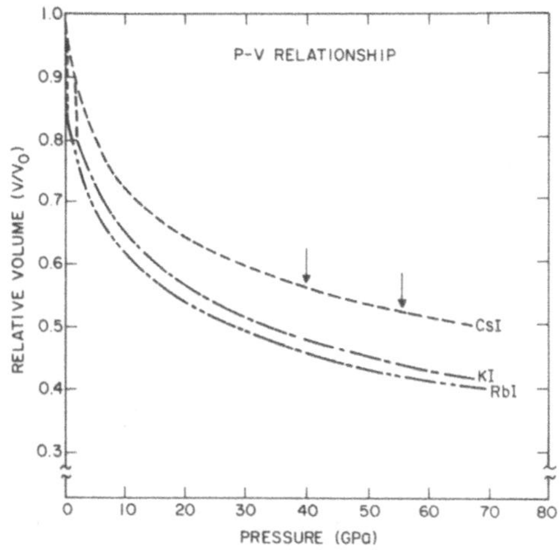

Fig. 6 Pressure-volume relationship for CsI, KI and RbI (ref. 15). Arrows designate the phase transition pressures for the B2-tetragonal and tetragonal orthorhombic transitions in CsI.

recently Asaumi[23] has carried out an x-ray diffraction study on solid Xe up to 32 GPa and Zisman et al[24] up to 55 GPa. Asaumi has reduced all available

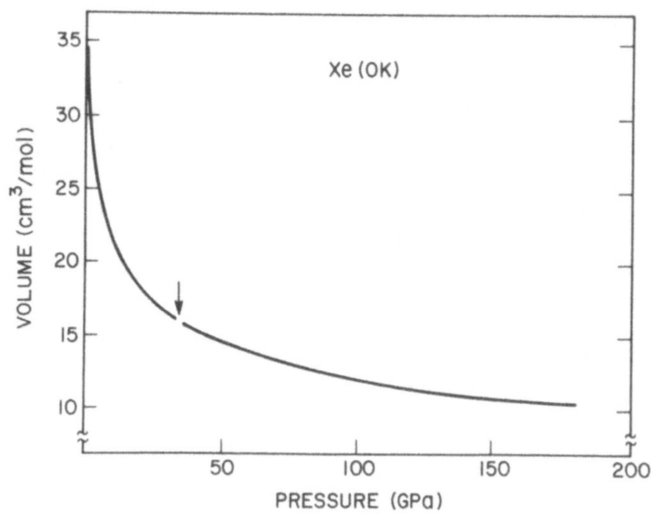

Fig. 7 Pressure-volume relationship for solid Xe, at O Kelvin, from ref. 23. Left of the arrow, experimental data; right of the arrow, extrapolated line using the Birch-Murnaghan equation of state.

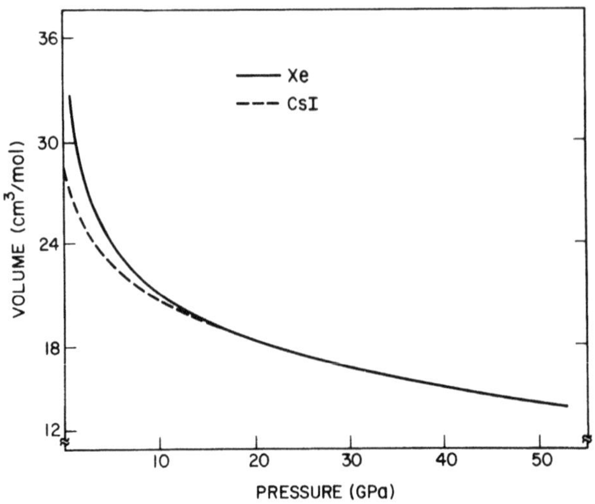

Fig. 8 P-V relationship for Xe and CsI determined in the same run by high pressure x-ray diff-raction(ref. 24). Above 15 GPa, the compression of CsI is identical with that of Xe.

pressure-volume data on Xe to O Kelvin using the Mie Gruneisen equation and Debye approximation and has fitted the reduced P-V data to the second-order Birch-Murnaghan equation. The fitted curve yields $B_0 = 5.18 \pm 0.32$ GPa and $B'_0 = 5.48 \pm 0.24$, whereas the bulk modulus B_0 of solid Xe at T = OK is found to be 3.65 GPa, from the piston-displacement data of Anderson and Swenson,[25] in the 20 kilobar range. It is believed that the higher value fits the presently obtained data better over a range of pressure an order of magnitude greater than that of Anderson and Swenson (23 GPa compared to 2 GPa). Asaumi has extended the fitted curve using the Birch-Murnaghan EOS to 200 GPa which is shown in Fig. 7. By combining the optical absorption data with the above EOS data the bandgap closing in Xe is estimated to be ~200 GPa, provided no phase transition intervenes. The corresponding metallization volume would be 10 cm^3/mol, which is in close agreement with the Herzfeld theory.[12] In Fig. 8, the P-V relationship for CsI and Xe obtained by Zisman et al are shown.[24] In this study the lattice parameter data were collected for CsI and Xe simultaneously in the same loading of the diamond cell, at several pressures determined by ruby fluorescence. The interesting point is that the compression curves of CsI and Xe become indistinguishable above 15 GPa. This is significant and suggests that the effective charge on Cs and I is decreasing drastically and CsI is becoming like Xe, with a hard core repulsion similar to that of the rare gas solid.

Halogens

Metallization of iodine[26] by continuous band overlap at pressures close to 18 GPa, and the phase transition to the monatomic metallic lattice near 21 GPa have evoked interest in Cl and Br, both of which have been recently studied[27–29] by high pressure x-ray diffraction[28] and high pressure Raman techniques.[29] No

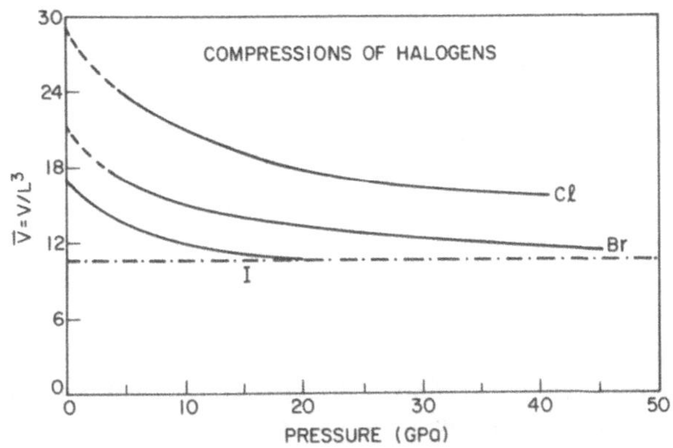

Fig. 9 Compression of Cl, Br and I (ref. 28). See text for the explanation of $\bar{V}$. The line corresponds to the $\bar{V}$ for the metallization transition in iodine (ref. 26).

phase transition was encountered up to 35 GPa in Br, and 45 GPa in Cl, the limits of pressure reached in the study. By comparing the high pressure behavior of Cl and Br with that of iodine, Düsing et al[28] have given a rough estimate for a monatomic metallic transition (similar to that of iodine) in Cl and Br as ~60 GPa and 150 GPa, respectively. This estimate is based on first putting the x-ray data for Cl, Br and I on the same basis, by scaling the lattice parameter with the bond length L of the free molecule. Then, the volume is also scaled to the bond length as $\bar{V} = V/L^3$. An asymmetry parameter Δ defined as $\Delta = [(\bar{b}/2)^2 + (\bar{c}/4)^2]^{1/2} - 1$,

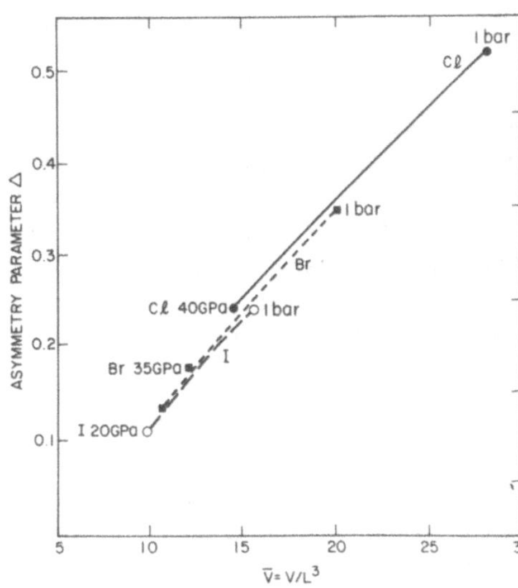

Fig. 10 Variation of the asymmetry parameter Δ for Cl, Br and I with respect to scaled volume $\bar{V}$ (see text) ref. 27.

where $\bar{b}$ and $\bar{c}$ are scaled lattice parameters with respect to the bond length, has been plotted against the scaled volume $\bar{V}$. This asymmetry parameter tends to zero when the inter- and intramolecular bonds become all equal in the b-c plane. The transition in iodine to the monatomic metallic phase takes place when $\Delta = 0.11$. If this value is taken as the critical value for the transition, then the variation of Δ with scaled volume for Cl and Br can be extrapolated to this Δ value to determine the volume at which transition will occur, and then the corresponding pressure can be obtained from the equation of state of solid Cl and Br. The transition presssures thus obtained (~60 GPa for Br and ~150 GPa for Cl) appear to be in reasonable agreement with high pressure Raman studies,[29] in which the authors have developed a bond charge model for describing the pressure dependence of the Raman active zone center phonons in Cl, Br and I. It is believed that with pressure covalency increases, and the conjecture is that at a critical value of the bond charge, the transition is triggered. Figure 9 shows the pressure-volume relationship for the halogens and Fig. 10 shows a plot of the asymmetry parameter against scaled volume.

PHASE TRANSITION IN ELEMENTS

Lanthanides and Actinides

The lanthanides and actinides have attracted attention because of the possibility of pressure induced f-electron delocalization and structural phase transitions occurring in them. In very recent years, many of the members of the series have been investigated by x-ray diffraction techniques to pressures up to 70 GPa.[31] It is well known that pressure favors the lower volume-higher valence state phase,[32] of which Ce has been a classic example for a long time. Cerium undergoes an isostructural transition with a large volume decrease at about 6 kbar pressure. Many monochalcogenide compounds involving Sm, Eu, Tm and Yb were studied at high pressures in the earlier years,[32] the classic case of this group being SmS, which exhibits a semiconductor to metallic gold-like phase transition at 7 kbar, due to 4f electron delocalization.[32] Elemental Eu and Yb are divalent and are, therefore, good candidates for pressure induced valence changes. A number of high pressure x-ray, optical, Mössbauer and LIII absorption edge studies have been reported in recent years.[1] The results indicate that both Yb and Eu undergo a continuous pressure-induced valence collapse with increasing pressure[33,34] (see Fig. 11 and 12). They exhibit anomalous compression and are much more compressible compared to their neighbors, trivalent Lu and Gd. For Eu, a mean valence greater than 2.5 is indicated at pressures near 20 GPa, if the P-V relationship of Eu and that of Gd are compared.

The trivalent rare earth elements undergo a sequence of pressure-induced structural transitions from hcp → Sm-type → dhcp → fcc → distorted fcc, in the pressure range 2-40 GPa.[32] The observed sequence for Yb is fcc-bcc-hcp at 4 GPa and 35 GPa and bcc-hcp-EuIII for Eu at 12.5 GPa and 32 GPa, respectively. Praseodymium undergoes a structural change (see Fig. 13) from distorted fcc to α-U structure, accompanied by a 9.3% volume change.[32] This is believed to involve a valence change towards the quadrivalent state. The behavior of Pr appears to be similar to that of Ce which assumes the α-U structure at pressures above 5 GPa.[35] Quite recently, new structural phase transitions in Nd and Sm[31] to low symmetry structures have been reported at 41 GPa and 33 GPa, respectively. However, in Gd and Tb there is no evidence for any further phase transition from the distorted fcc, in studies up to 50 GPa. High pressure experiments on lanthanides suggest

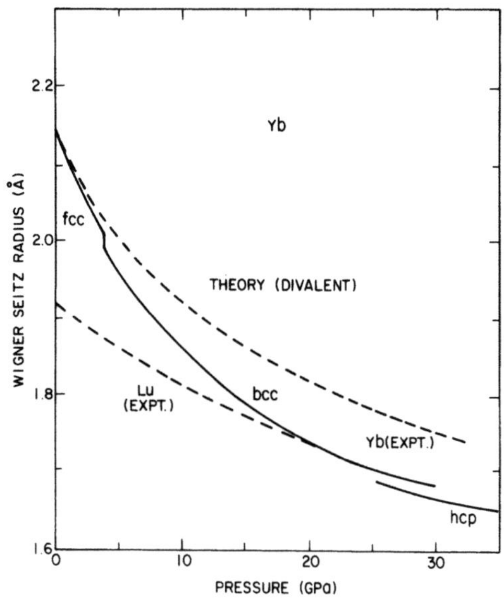

Fig. 11 Compression of Yb; divalent (theory) and experiment (solid line). fcc → bcc transition near 4 GPa, bcc → hcp near 35 GPa. The Lu experimental data shown for comparison. The anomalous compression of Yb (expt) is due to valence change (ref. 33).

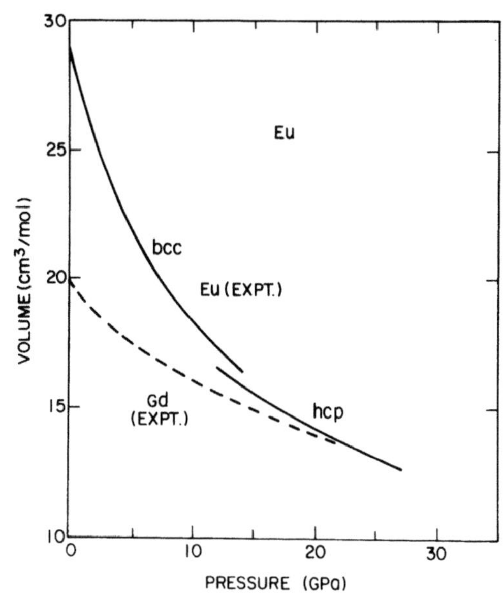

Fig. 12 Compression of Eu compared with Gd. Eu undergoes a continuous valence change and bcc → hcp transition near 12.5 GPa (ref. 34).

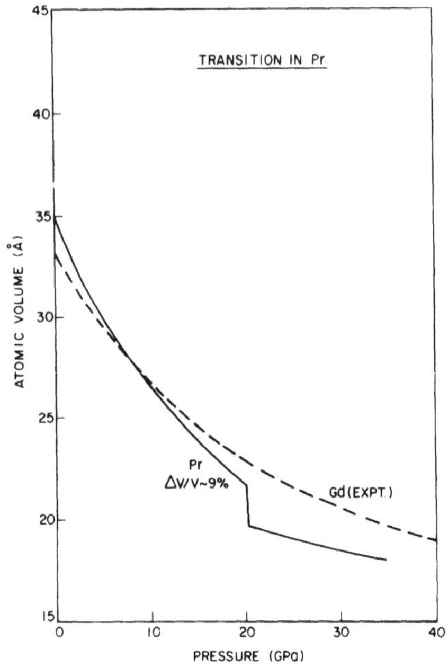

Fig. 13 Compression of Pr compared with Gd. Abrupt decrease in volume near 20 GPa is due to a transition from distorted fcc to the α - U structure (ref. 32).

that the lighter members assume low symmetry structures at high pressure and it is presumed that this behavior might be due to 4f electron participation in bonding. This is consistent with the structural behavior of lighter actinides, where the 5f electrons participate in bonding and complex structures are a rule.

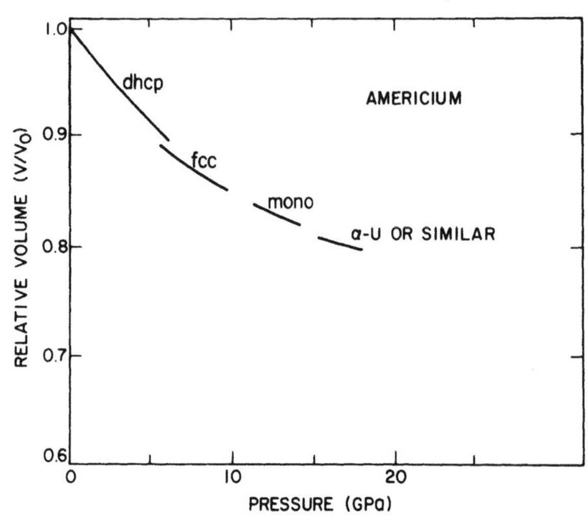

Fig. 14 Compression of Am and pressure-induced phase transitions (ref. 36, 37).

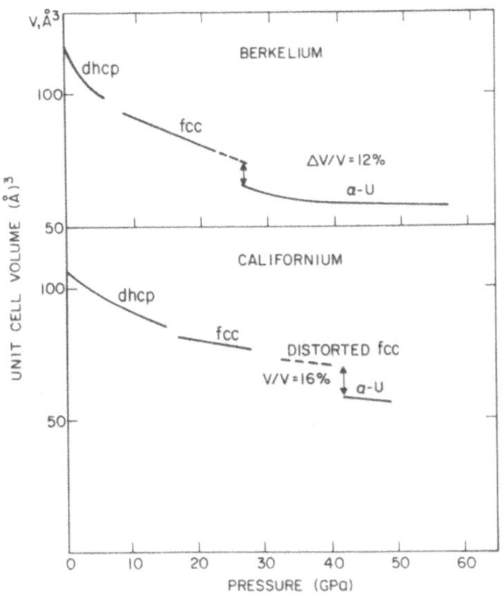

Fig. 15 Compression of Bk and Cf and pressure-induced phase transitions (ref. 36, 37).

The lighter actinides from Pa to Pu have a 5f band whereas in the heavier members from Am onwards, the 5f electrons are localized. This is evident from their magnetic behavior. The lighter actinides crystallize in low symmetry crystal structures, are less compressible and have smaller atomic volumes, whereas the heavier actinides exhibit structures similar to that of lanthanides, are more compressible and have larger atomic volumes.[36,37] Compression data up to 68 GPa on Th shows no phase transition. The metals Pa to Pu again don't show any phase changes,[36] although U may be undergoing a transition around 71 GPa. In these, less dense high temperature phases are replaced by denser low temperature ones, under the effect of pressure. The actinides from Am to Cf undergo pressure-induced transitions[36] from dhcp to fcc to α-U phase. This sequence is not unlike that of the lanthanides. Further, the high compression and large volume changes suggest 5f electron delocalizations.[37] Figures 14, 15 and 16 show the volume and structures of the actinides with increasing pressure.

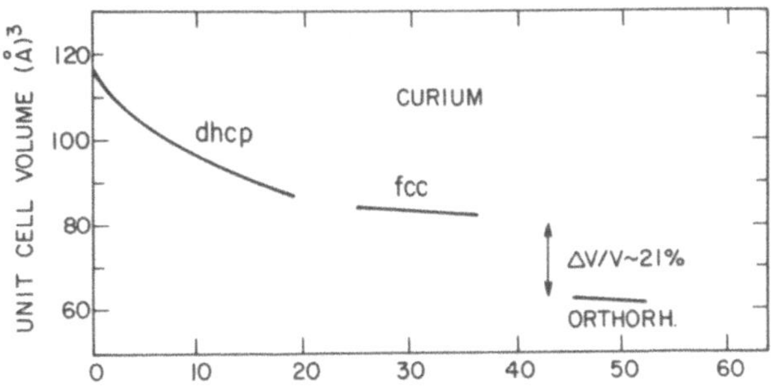

Fig. 16 Compression of Cm and pressure-induced phase transitions (36-37).

These two groups of elemental metals have been of great interest to high pressure research, both from the theoretical and experimental points of view.[38–40] The possibility of s-d transfer[38,39] and its effect on structural and electronic properties have been of particular interest. They have been studied to several hundred kilobars pressure in recent years.

The high pressure optical reflectivity studies on Cs, Rb and K are consistent with pressure-induced s-d transitions in them.[41,42] The structural sequence in Cs is bcc → fcc → tetragonal → Cs(V).[34] In the tetragonal phase Cs and Rb are eight fold coordinated, but at pressures above 10 GPa, Cs assumes a six-fold coordinated structure (Cs(V)). This decrease in coordination is believed to be due to a further progression of the s-d transfer, and the somewhat complex structure is due to the increasing importance of directional bonding in the d-band metal that Cs becomes at pressures above 10 GPa. McMahan[39] has predicted from theoretical calculations that Cs, Rb and K in the d-band metal would assume the hcp structure, followed by pressure-induced transitions to the Sm-type → dhcp → bcc structures, driven by the effects of hybridization induced at high pressures.

The crystal structures of Mg, Ca, Sr and Ba[40] have been investigated up to 50 GPa, by energy dispersive x-ray diffraction. Magnesium transforms around 50 GPa from hcp to the bcc structure. Both Ca and Sr are fcc at ambient pressure and transform to the bcc at 19.5 GPa and 3.5 GPa, respectively. A transformation to the simple cubic lattice has been reported for Ca at a pressure of 32 GPa. Strontium exhibits three pressure-induced transformations to undetermined structures, at 26 GPa, 36 GPa and 46 GPa. The 46 GPa transformation is to the same phase as "Ba-IV". These transformation sequences for the alkaline earth metals, and the phase transition in Mg and Be are shown in Fig. 17.[40] The transformations in the alkaline earths from simple structures to complex phases are again believed to be due to s-d transfer which turn the alkaline earths into d-band metals at high pressures. Theoretical calculations[43–45] indicate that the s-d

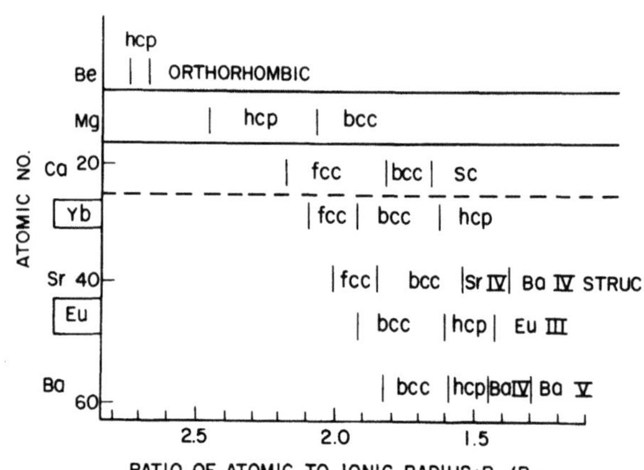

Fig. 17 Structural phase transitions in alkaline earth metal, Be and Mg shown against atomic to ionic radius ratio. Transitions in Yb and Eu are also included (ref. 40).

transfer in Ba is complete at about 50% compression, and this would indicate that Ba(IV) is a fully developed d-band metal. It is significant that Sr(V) has the same structure at the same radius ratio, and presumably is also a fully converted d-band metal.

Group IV, V and VI Elements

The theoretical advance in predicting the high pressure phases of group IV elements is phenomenal. Total energy calculations using the pseudopotential approach and density functional approximation[39,46,47] have been made for C, Si and Ge, and a number of related compound semiconductors. These calculations predict the high pressure phases remarkably well; in very good agreement with the experiment. Recently, new transitions in Si have been reported.[48] The β-tin phase of Si transforms to a primitive hexagonal structure near 15 GPa and then to an unknown intermediate phase designated Si(VI) around 35 GPa, and finally to the hcp structure at about 40 GPa. Thus, Si progressively increases its coordination with pressure and assumes the typically metallic hcp structure at very high pressures. Per contra Ge seems to be stable in the β-tin structure up to 54 GPa, the limit of pressure investigated.

A discontinuous transition in Sn to the bcc structure occurs near 35 GPa.[48] Before this phase change, the c/a ratio of the tetragonal tin-phase increases continuously with pressure. Table III summarizes the sequence of polymorphic transitions for group IV elements.

Theoretical calculations[49] indicate that diamond is stable up to 23 megabar with respect to a phase transition to any of the simple metallic structures.

TABLE III

Phase Transition Pressures and Volumes for Group IV Elements*

	CD → β-TIN	β-TIN → PH	PH → Si-VI	Si-VI → HCP	
Si	0.93 (0.92)	0.68 (0.69)		c 0.56 (0.54)	V/V_0
	9.9 (11.5)	14.3 (15)		41 (40)	$P[GP_a]$
Ge	0.895 (0.89)				V/V_0
	9.6 (10)				$P[GP_a]$
Sn		0.73 (0.88)	(0.70)	0.68	V/V_0
		(9.5)	(35)	40	$P[GP_a]$
		β-TIN → BCT	BCT → BCC	BCT → HCP	

* Ref. 48, PH = Primitive hexagonal. Experimental values are in parenthesis.

Group V

Phosphorus and antimony have been studied by high pressure x-ray diffraction, both at room temperature and elevated temperatures[50] (340°C), the higher temperature is to favor the kinetics of the phase transitions. Orthorhombic P transforms to the As-type rhombohedral (rh) form at about 4 GPa and 340°C, which is substantially lower than the P_{tr} of 5.5 GPa at room temperature. The (rh) form transforms to simple cubic (sc) near 10 GPa, without much hysteresis even at room temperature. This transformation occurs around 9 GPa at 270°C, and the (sc) structure is found to be stable up to 29 GPa, the limit of pressure in that study. In another high pressure x-ray study, the (sc) structure has been found to be stable up to 70 GPa (Fig. 18). This is surprising because the (sc) structure is not a close packed arrangement. The (rh) and (sc) black P are superconductors.[51,52] Wittig et al[52] have studied the pressure dependence of T_c up to 25 GPa and found that the T_c rises to 10K at 23 GPa and exhibits a double maxima in the pressure range 5-25 GPa (see Fig. 18). According to Wittig et al, metallic P represents the "best" superconductor within the class of sp metals.

In the high pressure x-ray studies[50] on Sb, the high pressure (sc) form reported to occur at 7 GPa was not confirmed. In experiments at 320°C, a transformation from rhombohedral to another form (not sc) occurs at about 6 GPa and the transition is almost complete at 8 GPa. Evidently, a lower symmetry structure becomes energetically favorable in the case of Sb which competes with the (sc) phase. The same may also be true for As and Bi.

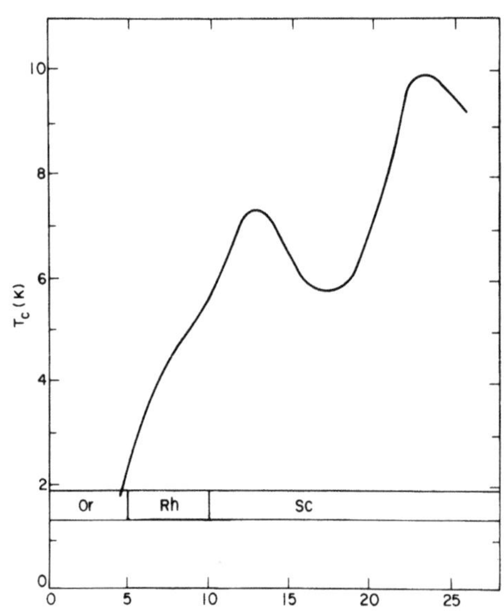

Fig. 18 Pressure dependence of the T_c in superconducting phosphorus (ref. 52).

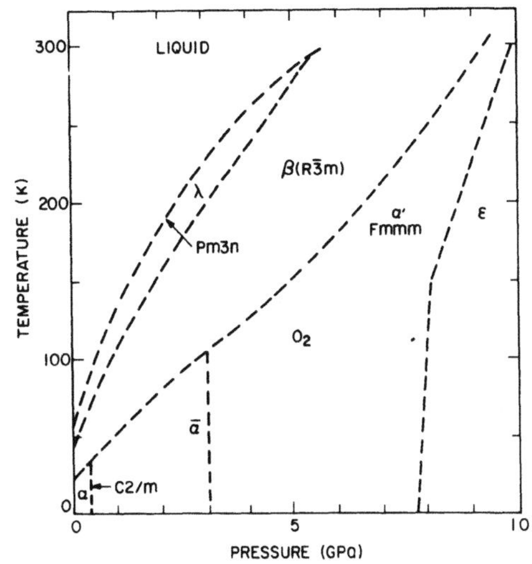

Fig. 19 P-T diagram of oxygen (ref. 53).

Group VI

Solid oxygen has been studied earlier[1)] but a recent pressure-temperature Raman investigation[53)] has delineated the phase boundaries more precisely than before. This P-T diagram is reproduced in Fig. 19. The structures are indicated in the diagram, wherever they are known. Recent high pressure x-ray studies to 13 GPa reveal the existence of two other high pressure phases, and also indicate that $\epsilon-O_2$ may be an orthorhombic distortion of $\alpha-O_2$ (C2/m). The phase diagram of O_2 seems to be very complex, with many low symmetry structures. Solid O_2 at high pressures is known to become opaque to visible radiation. The question of molecular breakdown and metallization of oxygen should be of great interest from both theoretical and experimental points of view.

Phase transitions to the metallic state in Se and Te are well established and the high pressure metallic forms of these two elements are superconducting.[54,55)] Sulphur at very high pressures is metallic, but whether it is superconducting or not is not clear.

MOLECULAR SYSTEMS

H_2O and D_2O

Pressure-induced phase transformations in ice have long been a subject of great interest. At least nine polymorphs of ice are known, and in all these the H_2O molecules are hydrogen-bonded. It has been predicted that at least one phase of ice would exist at high pressure (50-100 GPa range) where hydrogen would be located at the midpoint between two oxygens,[56,57)] and this is the so-called symmetric ice. Optical absorption measurements to 25 GPa on H_2O by Nicol and Holzapfel led them to estimate the transition to the symmetric ice from ice VII, as ~50 GPa. A subsequent high pressure Raman scattering study of ice VII up to 50 GPa led the authors to suggest that this transition might occur in the range 38-50 GPa. A recent Brillouin scattering[58)] study of H_2O up to 74 GPa has given some indications for a phase transition in ice, near 46 GPa. It has been suggested that this transition may be to the predicted symmetric-form of ice. The Brillouin frequency shift when plotted against pressure (see Fig. 20) shows a marked

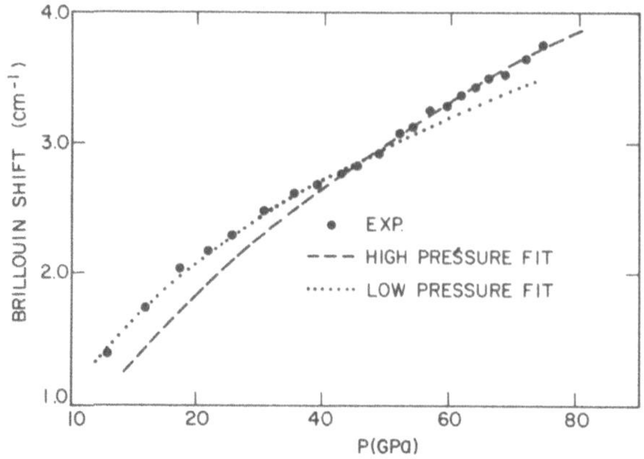

Fig. 20 Brillouin shift against pressure for H_2O (ice VII). Anomaly near 46 GPa is attributed to ice VII → ice X transition (ref. 58).

anomaly near 46 GPa. Likewise, the log of "effective" elastic constant obtained from the Brillouin shift when plotted against the log of relative density (ρ/ρ_0) shows an anomaly near the same pressure. These anomalies are regarded as evidence for the phase transition. It is hypothesised that ice X may have the cubic Cu_2O structure.[59)] However, the experimental evidence for ice X is rather tenuous and a high pressure x-ray diffraction, or a Raman study, is needed to settle this question. This is well within the possibility of the diamond anvil cell range. An interesting suggestion is that at still higher pressures another phase of ice, an ionic crystal[57)] in the antifluorite structure may exist. In Fig. 21 is shown a speculative phase diagram for H_2O. In the diagram even a metallic form of H_2O, in the range of several megabars, is given a place. This possibility should be examined, at least theoretically, for a metallic phase of H_2O at high pressures would indeed be as spectacular as metallic hydrogen.

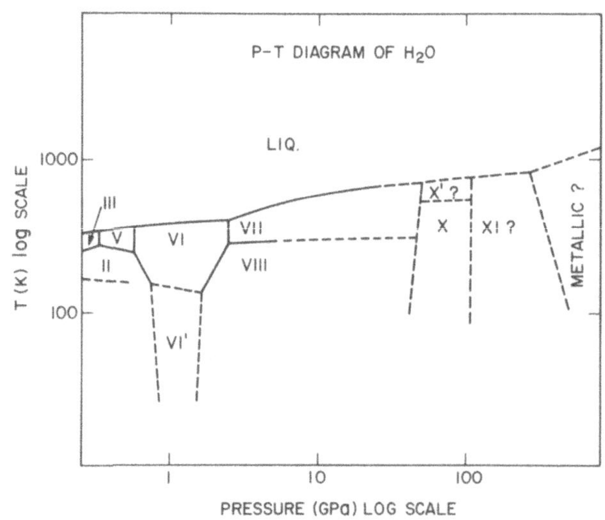

Fig. 21 Hypothetical P-T diagram of H_2O (ref. 59).

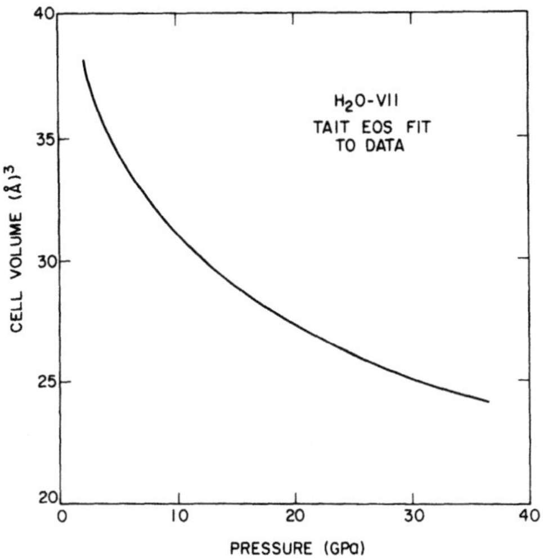

Fig. 22 Tait equation of state fit for the compression of H_2O (ref. 60).

The experimental equation of state for H_2O-VII and D_2O-VII has been extended to 36 GPa, in a recent high pressure x-ray diffraction study.[60,61] The pressure-volume data are shown in Fig. 21. All the available experimental P-V data seem to fit reasonably well with the different isothermal EOS. In Fig. 22, the data is a fit to the Tait equation of state

$$P = \frac{B_0}{r}[\exp\{r(1-V/V_0)\} - 1]$$

where $r = 1 + B'_0$. To choose any particular EOS for ice VII seems difficult according to Munro et al[60] and hence, their averaged values using seven different EOS' are shown in Fig. 23, summarizing the pressure dependence of the bulk modulus B_0 of H_2O-VII and D_2O-VII.

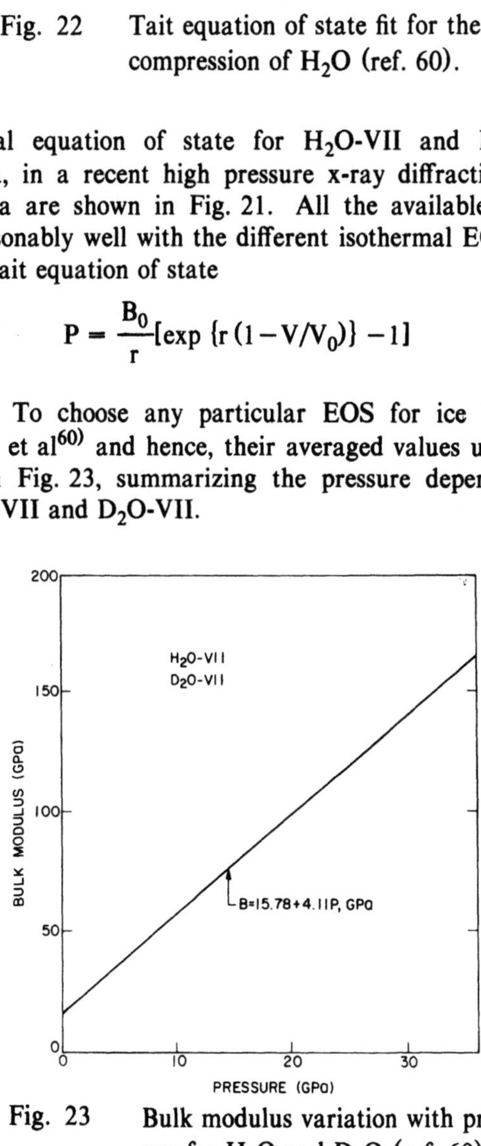

Fig. 23 Bulk modulus variation with pressure for H_2O and D_2O (ref. 60).

HIGH PRESSURE GEOPHYSICS

The structures and equations of state of mantle minerals and Fe at high pressure and temperature are of great importance to geophysics. In this connection, the experimental EOS for a number of oxides have recently been published and this list includes MgO, FeO, CaO, SrO[62–64] and BaO. These oxides crystallize in the B1 structure and undergo a pressure-induced transition to the B2 structure. This transition has been reported for CaO, SrO and BaO at the pressures of 60 ± 2 GPa, 36 ± 2 GPa and 14 GPa, respectively. In the case of MgO, the transition has not been observed yet, but on the basis of systematics for the B1-B2 transition, it is expected at pressures above 100 GPa. The B1-B2 transition[62] in CaO is of direct interest to the constitution of the mantle, for the observed density increase at this transition would allow a considerable proportion of CaO component within the lower mantle, at least below a depth of 1550 km corresponding to 65 GPa in pressure. However, the main component that is believed to dominate the lower mantle is the pervoskite structure oxides, rather than simple oxides. Both olivines and pyroxines end up with pervoskite structure, with some proportion of simple oxides, at the depths of the lower mantle right up to the core-mantle boundary ~2900 km.

The experimental EOS for Fe and MgO have been measured up to a megabar static pressure in earlier studies. More recently these have been extended to pressures in excess of 1.5 megabar.[65] The data tie in nicely with the earlier measurements, and these studies represent the highest-static pressure compression measurements.

The importance of high pressure – high temperature properties of minerals to geophysics has primarily stimulated the development of laser and external heating techniques for the diamond cell[1] and much progress has been made in recent years. Temperatures of ~5000K through laser heating has been achieved at pressures up to 100 GPa[65] and measured by optical radiometry (Using the black body radiation characteristics of the hot sample, recorded with fast spectroscopic techniques). In the method of external heating with resistance elements, the main problems are the graphitization of the diamond and the loss of the strength of the diamond anvil supports at high temperature. The graphitization has been reduced by operating in high vacuum.[67] The strength of material problem has been tackled by employing high temperature alloys. Temperatures close to 1200°C have been achieved by external heating.[67] However, the higher the temperature the smaller becomes the pressure range.

Another difficultly with high temperature experiments using the diamond cell is the question of pressure calibration. The ruby fluorescence method becomes rapidly useless at higher temperatures, because of line broadening. The only recourse is to perform the x-ray measurement of the lattice parameter of a standard substance, and get the pressure from high temperature EOS.[68,69] For this purpose, NaCl and Au have been developed as standards in recent years. The compression of Au has been extended to 70 GPa by high pressure x-ray diffraction studies and isothermal B_0 and B'_0 have been obtained as $B_0 = 166.6 \pm 11$ GPa and $B'_0 = 5.5$ (± 0.8) at room temperature. A thermal EOS for Au has been derived and using this high temperature EOS of Au, data have been generated to cover the 0 - 200 GPa and 300 to 3000K ranges (see Table IV).

More recently the pressure dependence of the first-order Raman mode of diamond has been suggested as a pressure calibrant, especially for the higher range of pressure.[70,72] This mode frequency is found to increase linearly with the ruby

TABLE IV*

THERMAL EQUATION OF STATE OF GOLD

Pressure (in GPa) as a Function of Compression (V/V_0) and Temperature (K)

$1-V/V_0$	300K	1000K	2000K	3000K
1.00	0.00	5.02	12.22	19.42
0.98	3.56	8.50	15.60	22.70
0.96	7.61	12.48	19.47	26.47
0.94	12.22	17.01	23.90	30.80
0.92	17.46	22.18	28.97	35.78
0.90	23.44	28.08	34.77	41.45
0.88	30.24	34.81	41.39	47.97
0.86	38.00	42.49	48.97	55.44
0.84	46.85	51.27	57.63	84.01
0.82	56.97	61.31	67.56	73.83
0.80	68.54	72.80	78.95	85.11
0.78	81.79	85.97	92.02	98.07
0.76	97.00	10.11	107.04	112.98
0.74	114.49	118.52	124.34	130.17
0.72	134.64	138.59	144.30	150.02
0.70	157.90	161.77	167.37	172.98
0.68	184.84	188.63	194.11	199.81
0.66	216.10	219.81	225.18	230.56

* Ref. 69

R_1 luminescence peak. Theoretical calculations of this mode frequency have been carried out to 600 GPa and the frequency varies sublinearly with pressure. However for pressures above 100 GPa diamond calibration may be very useful. Boppart et al[71)] have found means of cutting down the signal from the anvil diamond by focussing and spatial filtering.

SUMMARY

In recent years static high pressure research has made tremendous progress, both in terms of the limit of pressure as well as measurement techniques. Pressure-induced metallization, structural phase transitions and equation of state studies are the current topics of interest and in this connection a number of systems have been investigated at pressures close to a megabar. Such studies have stimulated theoretical calculations, which in turn have predicted new phase transitions, electronic structure and property changes. Experiments with the diamond cell continue to be challenging at megabar pressures. Pressure calibration, resistivity measurements, generation and high temperature calibration in the diamond cell are some of the experimental aspects that are currently receiving attention.

ACKNOWLEDGEMENTS

I wish to thank R. G. Maines Sr., for the painstaking effort in the preparation of the many figures reproduced in this paper.

REFERENCES

1) A. Jayaraman, Rev. Mod. Phys., *55*, 65 (1983).

2) E. F. Skelton, Physics Today, *37*, 44 (1984).

3) K. Asaumi, T. Mori and Y. Kondo, Phys. Rev. Lett., *49*, 837 (1982).

4) I. N. Makarenko, G. Weill, J. P. Itie and J. M. Besson, Phys. Rev., *B26*, 7113 (1982).

5) K. Asaumi and Y. Kondo, Sol. State Commun., *40*, 715 (1981).

6) J. P. Itie, A. Polian and J. M. Besson, Phys. Rev., *B30*, 2309 (1984).

7) I. N. Makarenko, A. F. Goncharov and S. M. Stishov, Phys. Rev., *B29*, 6018 (1984).

8) E. Knittle and R. Jeanloz, (unpublished).

9) E. Knittle and R. Jeanloz, Science, *223*, 52 (1984).

10) T. A. Grzybowski and A. L. Ruoff, Phys. Rev. Lett., *53*, 489 (1984) and A. L. Ruoff and T. A. Grzybowski, In Solid State Physics Under Pressure, Edited by S. Minomura, Terra Scientific Publishing Company, pp 69-74, (1985).

11) M. Ross and A. K. McMahan, In Physics of Solids Under High Pressure, Edited by J. S. Schilling and R. N. Shelton, North Holland Publishing Company, p 161, (1981).

12) A. E. Carlsson and J. W. Wilkins, Phys. Rev., *B29*, 5836 (1984).

13) J. Aidun, M. S. T. Bukowinski and M. Ross, Phys. Rev., *B29*, 2611 (1984).

14) K. Asaumi, Phys. Rev., *B29, 1118 (1984).*

15) K. Asaumi, Y. Kondo, T. Mori and T. Suzuki, In Solid State Physics Under Pressure, Edited by S. Minomura, Terra Scientific Publishing Company, pp 63-68, (1985).

16) T. L. Huang and A. L. Ruoff, Phys. Rev., *B29*, 1112 (1984).

17) T. L. Huang, K. E. Brister and A. L. Ruoff, Phys. Rev., *B30*, 2968 (1984).

18) E. Knittle, A. Rudy and R. Jeanloz, Phys. Rev., *B31*, 588 (1985).

19) G. Barsch and Z. P. Chang, in the National Bureau of Standards Special Publication #326, 173-189, (1971).

20) D. A. Nelson, Jr., and A. L. Ruoff, Phys. Rev. Lett., *42*, 383 (1979).

21) K. S. Chan, T. L. Huang, T. A. Grzybowski, T. J. Whetten and A. L. Ruoff, Phys. Rev., *B26*, 7116 (1982).

22) K. Syassen, Phys. Rev., *B25*, 6548 (1982).

23) K. Asaumi, Phys. Rev., *B29*, 7026 (1984).

24) A. N. Zisman, I. V. Aleksandrov and S. M. Stishov, JETP, Lett., *40*, 1029 (1984).

25) M. S. Anderson and C. A. Swenson, J. Phys. Chem. Solids, *36*, 145 (1975).

26) K. Takemura, S. Minomura, O. Shimomura, and Y. Fujii, Phys. Rev. Lett., 45, 1881 (1980).

27) P. G. Johannsen, E. F. Düsing and W. B. Holzapfel, In Solid State Physics Under Pressure, Edited by S. Minomura, Terra Scientific Publishing Company, pp 105-108 (1985).

28) E. Fr. Düsing, W. A. Grosshans and W. B. Holzapfel, J. de Physique, *45*, C8-203 (1984).

29) P. G. Johannsen and W. B. Holzapfel, J. de Physique, *45*, C8-191 (1984).

30) A. Jayaraman, In Handbook on the Physics and Chemistry of Rare earths, Edited by K. A. Gschneidner, Jr., and L. Eyring, North Holland Publishing Company, Vol. I, pp 707-741, 1978.

31) W. A. Grosshans and W. B. Holzapfel J. de Physique, *45*, C8-141 (1984) and references cited therein.

32) A. Jayaraman, In Handbook on The Physics and Chemistry of Rare-earths, Edited by K. A. Gschneidner, Jr., and L. Eyring, North Holland Publishing Company, Vol. 2, pp 575-611, (1979).

33) K. Syassen, J. de Physique, *45*, C8-123 (1984).

34) K. Takemura and K. Syassen, In Solid State Physics Under Pressure, Edited by S. Minomura Terra Scientific Publishing Company, pp 127-133 (1985).

35) W. H. Zachariasen and F. H. Ellinger Acta Cryst., *A33*, 155 (1977).

36) U. Benedict, J. Less. Comm. Met., *100*, 153 (1984).

37) U. Benedict, J. de Physique, *45*, C8-145 (1984).

38) H. L. Skriver, Phys. Rev. Lett., *49*, 1768 (1982).

39) A. K. McMahan, Phys. Rev., *B29*, 5982 (1984), A. K. McMahan and J. A. Moriarty, Phys. Rev., *B27*, 3235 (1983).

40) H. Olijnyk and W. B. Holzapfel, Phys. Lett., *100A*, 191 (1984), J. de Physique, *45*, C8-157 (1984).

41) H. Tups, K. Takemura and K. Syassen, Phys. Rev. Lett., *49*, 1776 (1982).

42) K. Syassen, K. Takemura and H. Tups, In Solids State Physics Under Pressure, Edited by S. Minomura, Terra Scientific Publishing Company, pp 55-62 (1985).

43) J. C. Duthie and D. G. Pettifor, Sol. State Commun., *27*, 613 (1978).

44) J. C. Duthie and D. G. Pettifor, Phys. Rev. Lett., *38*, 564 (1977).

45) J. F. Herbst, Phys. Rev., *B22*, 4600 (1980).

46) M. T. Yin and M. L. Cohen, Phys. Rev., *B26*, 5668 (1982).

47) M. L. Cohen, J. de Physique, *45*, C8-7 (1984).

48) H. Olijnyk and W. B. Holzapfel, J. de Physique, *45*, C8-153 (1984).

49) M. T. Yin and M. L. Cohen, Phys. Rev. Lett., *50*, 2006 (1983).

50) T. Kikegawa and H. Iwasaki, In Solid State Physics Under Pressure, Edited by S. Minomura, Terra Scientific Publishing Company, pp 201-206 (1985).

51) H. Kawamura, I. Shirotani and K. Tachikawa, ibid, pp 213-316 (1985).

52) J. Wittig, B. Bireckoven and T. Weidlich ibid. pp 217-222 (1985).

53) H. D. Hochheimer, H. J. Jodl and F. Bolduan, J. de Physique, *45*, C8-195 (1984).

54) F. P. Bundy and K. J. Dunn, In High Pressure Science and Technology, Edited by B. Vodar and Ph. Marteau, Pergamon Press, Vol. I, 461 (1980).

55) K. J. Dunn, In Mat. Res. Soc. Symp. Proc. Elsevier Science Publishing Company, Inc., High Pressure Science and Technology, Part I, p 321 (1984).

56) W. B. Holzapfel, J. Chem. Phys., *56*, 712 (1972).

57) F. H. Stillinger and K. S. Schweizer, J. Phys. Chem., *87*, 4281 (1983).

58) A. Polian and M. Grimsditch, Phys. Rev. Lett., *52*, 1312 (1984).

59) A. Polian, J. M. Besson and M. Grimsditch, In Solid state Physics Under Pressure, Edited by S. Minomura, Terra Scientific Publishing Company, pp 93-98 (1985) and references cited therein.

60) R. G. Munro, S. Block, F. A. Mauer and G. Piermarini, J. Appl. Phys., *53*, 6176 (1982).

61) Lin-gun-Liu, Earth and Planet. Sci. Lett., *61*, 359 (1982).

62) R. Jeanloz, T. J. Ahrens, H. K. Mao and P. M. Bell, Science, *206*, 829 (1979).

63) H. K. Mao and P. M. Bell, J. Geophys. Res., *84*, 4533 (1979).

64) Y. Sato and R. Jeanloz, ibid *86*, 11773 (1981).

65) P. M. Bell and H. K. Mao (Personal Communications).

66) R. Jeanloz and D. L. Heinz, J. de Physique, *45*, C8-83 (1984).

67) D. Schiefferl (Personal Communication).

68) L. C. Ming, M. H. Manghnani, J. Balogh, S. B. Qadri, E. F. Skelton and J. C. Jamieson, J. Appl. Phys., *54*, 4390 (1983).

69) D. L. Heinz and R. Jeanloz, ibid, *55*, 885 (1984).

70) M. Hanfland, K. Syassen, S. Fahy, S. G. Louie and M. L. Cohen, Phys Rev., *B31*, 6896 (1985).

71) H. Boppart, J. Van Straaten and I. F. Silvera, Bull. Am. Phys. Soc., *30*, 526, (JMIO) (1985).

72) P. M. Bell (Personal Communication).

73) M. S. T. Bukowinski and J. Hanser, Geophys. Res. Lett., *7*, 689 (1980).

COMPUTER SIMULATION OF NONEQUILIBRIUM PROCESSES

Duane C. Wallace

Los Alamos National Laboratory
Los Alamos, New Mexico 87545

ABSTRACT

The underlying concepts of nonequilibrium statistical mechanics, and of irreversible thermodynamics, will be described. The question at hand is then, How are these concepts to be realized in computer simulations of many-particle systems? The answer will be given for dissipative deformation processes in solids, on three hierarchical levels: heterogeneous plastic flow, dislocation dynamics, and molecular dynamics. Application to the shock process will be discussed.

I. STATISTICAL FOUNDATIONS

We begin by reviewing the underlying concepts of statistical mechanics, both nonequilibrium and equilibrium, and of thermodynamics, both irreversible and reversible. These concepts are rather old, and are hardly taught in school nowadays; undoubtedly many of the younger students will be surprised (startled?) to learn that there exists a well-established and logical framework for treating general nonequilibrium processes. While the essential points will be covered in the present paper, additional details can be found in two recent references of a review nature.[1,2]

Nonequilibrium Statistical Mechanics

In order to define the various statistical and thermodynamic theories, and to see their regions of applicability and their interconnections, it is easiest to start with the most general theory, namely nonequilibrium statistical mechanics. For this discussion, the system we consider is a nearly-ideal gas; it is composed of atoms whose only important interactions are binary elastic collisions. The gas is divided into a large number of cells, each with a constant total mass. (Alternatively, one can choose the cells to have constant volume, it is merely a convention.) Each cell contains a large number of atoms, and is treated as a statistical subsystem. Let $\vec{p}$ represent the atomic momenta, and let M be the mass of one atom. The momentum distribution for the cell located at position $\vec{r}$, at time t, is $f(\vec{r},\vec{p},t)$. The important mechanical properties of a cell are its density $\rho(\vec{r},t)$, its mean velocity $\vec{v}(\vec{r},t)$, and its total energy $\varepsilon(\vec{r},t)$:

$$\rho(\vec{r},t) = \int f(\vec{r},\vec{p},t)d\vec{p} \quad , \tag{1}$$

$$M\vec{v}(\vec{r},t) = \frac{1}{\rho(\vec{r},t)} \int \vec{p}f(\vec{r},\vec{p},t)d\vec{p} \quad , \tag{2}$$

$$\varepsilon(r,t) = \frac{1}{\rho(\vec{r},t)} \int \frac{\vec{p}^2}{2M} f(\vec{r},\vec{p},t)d\vec{p} \quad . \tag{3}$$

Since we are dealing with a nearly ideal gas, the energy is entirely kinetic energy.

We now inquire about a nonequilibrium process for a system, subject to initial conditions and boundary conditions on the momentum distribution $f(\vec{r},\vec{p},t)$. For a nearly ideal gas, the process will be completely described by the Boltzmann equation, which gives the evolution of $f(\vec{r},\vec{p},t)$. This evolution is consistent with conservation of mass, momentum, and energy, and also contains a time-irreversible term, namely the collision integral. For a nongaseous system, there does not at present exist a transport equation capable of describing the evolution of the momentum distribution. For such a system, a partial description of a nonequilibrium process is obtained from the equations of continuum mechanics, which are the equations for conservation of mass, momentum, and energy, locally applied to the densities $\rho(\vec{r},t)$, $\vec{v}(\vec{r},t)$, and $\varepsilon(\vec{r},t)$. The continuum mechanic equations are always correct, but they alone are not sufficient to determine a process.

Equilibrium

For uniform boundary conditions, e.g. an isolated system with a fixed volume, the system will approach an equilibrium state. In discussing equilibrium, we take the mean velocity of the system to be zero, or else transform to center-of-mass coordinates, thus eliminating one parameter. Then for a nearly ideal gas, the Boltzmann equation tells us that the equilibrium momentum distribution is constant in space and time, and is merely a function of the momentum magnitude $p = |\vec{p}|$, and of two scalar parameters β and ρ:

$$f_o(p;\beta,\rho) = c(\beta,\rho)\, e^{-\beta p^2/2M} \quad . \tag{4}$$

$c(\beta,\rho)$ is a normalizing function, such that the system density is ρ,

$$\rho = \int f_o(p;\beta,\rho)d\vec{p} \quad . \tag{5}$$

Then β turns out to be related to the kinetic energy density:

$$\varepsilon = \frac{3}{2\beta} \quad . \tag{6}$$

At this point, it is worthwhile to recall the relations among the whole family of equilibrium theories. First, the Boltzmann equilibrium limit is equivalent to the ensemble theory of equilibrium statistical mechanics. The Boltzmann distribution $f_o(p;\beta,\rho)$ corresponds directly to the canonical ensemble, in which phase-space averages are performed for a fixed density with the canonical weight function $e^{-\beta H}$. Equilibrium thermodynamics (which is often called "thermodynamics" for short) is made equivalent to equilibrium statistical mechanics through the identification of β with the temperature T:

$$\beta = \frac{1}{kT} \quad , \tag{7}$$

where k is Boltzmann's constant. Equilibrium thermodynamics then describes reversible processes, since by definition these are processes which are constrained to pass continuously through equilibrium states. Incidentally, the customary notion that a process will be reversible if it is done slowly enough is not always correct, as is shown by the example of elastic-plastic deformation in a solid.[2]

Irreversible Thermodynamics

Let us suppress the position and time variables, and write the momentum distribution simply $f(\vec{p})$. In the function space for $f(\vec{p})$, the equilibrium functions $f_o(p;\beta,\rho)$ occupy an extremely small part, only a two-dimensional surface (β and ρ) in an infinite-dimensional space. This is an important observation, because it is only on the equilibrium surface that the equilibrium quantities are defined. In particular, the temperature T and the entropy S are defined only on the equilibrium surface.

We now consider processes which are close to equilibrium. Specifically, this means that $f(\vec{p})$ is locally given by

$$f(\vec{p}) = f_o(p;\beta,\rho) + \delta f(\vec{p}) \quad , \tag{8}$$

where $\delta f(\vec{p})$ can be treated as a perturbation. Hence $f(\vec{p})$ is still restricted to a very small part of the total function space: it is close to the equilibrium surface. Within this region, however, the equilibrium quantities are defined approximately, and the thermodynamic relations among these quantities still hold approximately. This is the region of irreversible thermodynamics. In other words, an irreversible-thermodynamic process is one which passes through states which are sufficiently close to equilibrium so that the thermodynamic quantities can be meaningfully defined. Irreversible thermodynamics thus represents a very small part of the theory of nonequilibrium statistical mechanics.

Irreversible thermodynamics is characterized by dissipative force-flow processes. The force-flow relation is called the constitutive equation. For example, in heat conduction, the force is the temperature gradient $\vec{\nabla}T$, the flow is the heat current $\vec{J}$, and the constitutive equation is

$$\vec{J} = -\kappa\vec{\nabla}T \quad , \tag{9}$$

where κ is the thermal conductivity of the material in question. Note that constitutive equations are not contained in any equilibrium theory. The dissipation produces entropy; in the case of heat conduction this is given by

$$TdS = dQ \quad , \tag{10}$$

where Q is the quantity of heat transported. To calculate an irreversible process, one needs a complete set of equations. For a process such as a shock, where material is accelerated, the complete set of equations consists of the equations of continuum mechanics, the constitutive equations for whatever irreversible processes are driven by the shock (e.g. plastic flow, heat transport), the entropy-production equation, and some thermodynamic relations. Note that an entropy-production inequality, $TdS >$ something, is not sufficient for irreversible thermodynamics.

The question arises, how can one tell if irreversible thermodynamics is in fact a valid theory for a given process in a given material? For a nearly-ideal gas, this question can be answered by checking irreversible thermodynamics against the Boltzmann equation. It is known in general that the near-equilibrium solution of the Boltzmann equation is just the irreversible-thermodynamic solution for a viscous heat-conducting gas.[1] For other systems, computer simulations may eventually take the place of the Boltzmann equation. Or else one can first calculate the process from irreversible thermodynamics, then compare the spatial and temporal rates of change with appropriate relaxation lengths and times, to see if the system could in fact remain near equilibrium throughout the process.

II. MOLECULAR DYNAMICS

Ordinary MD

The procedure in ordinary MD is to place N particles in a computational cell, specify the potentials of interaction among the particles, and calculate the motion of the particles by numerically integrating the classical equations of motion (Newton's law). The calculation is usually done with periodic boundary conditions, in which the computational cell is surrounded by identical image cells, and particle interactions are allowed to cross cell walls. Periodic boundary conditions eliminate the large surface effects, of relative order $N^{-1/3}$, and introduce smaller effects of relative order N^{-1}. An extensive discussion of ordinary-MD techniques may be found in a recent review.[3]

There are two types of differences between an MD calculation and real nature. The first type of difference is in the interatomic potentials, which we never know exactly. The second type is due to computer artifacts, i.e. specific properties of the computer system which are not present in nature. This includes small-N effects: in natural systems N is enormous, while on

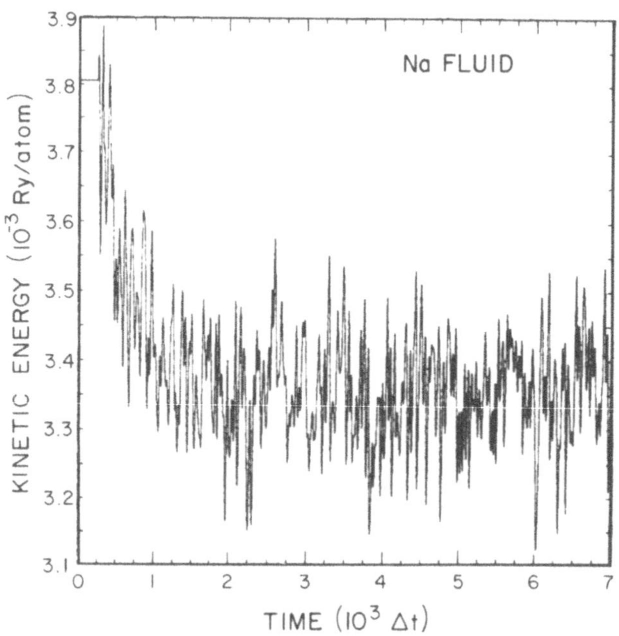

Fig. 1. Time development of the kinetic energy per atom for an MD system of sodium atoms. Initially the atoms are on bcc lattice sites, with a distribution of kinetic energies; the system evolves to an equilibrium fluid.

the computer we have N ~ 10^3. Another computer artifact is numerical integration with a finite timestep Δt; I do not know how nature integrates the classical motion of a many-particle system, but it is probably not with a finite Δt. A group of us at Los Alamos has carried out extensive calculations of equilibrium thermodynamic properties of solid and fluid metallic sodium,[4-12] a material for which we have a good representation of the interatomic potentials.[13] For equilibrium calculations, computer artifacts appear to be under control. The situation in nonequilibrium calculations is not yet known.

For an ordinary MD system of 686 sodium atoms, the evolution of the kinetic energy is shown in Figure 1. The calculation was started with the atoms at the equilibrium sites of a bcc lattice, and with a random Maxwellian velocity distribution. When the system comes to equilibrium, i.e. when the mean and bandwidth of the signal become essentially constant, the sodium is in the fluid phase. While this is an ordinary everyday graph of an ordinary MD run, it is to me a fascinating result. From a nonequilibrium initial state, the computer <u>finds</u> and <u>establishes</u> the correlations appropriate for the fluid equilibrium state. Furthermore, it is at least possible that the timescale of the approach to equilibrium, as seen in Figure 1, is a physically meaningful relaxation time for real metallic sodium.

<u>Exotic MD</u>

While it is a difficult problem to effectively simulate nature on a computer, it is no trouble at all for the computer to do things which are quite beyond nature's wildest imagination. It is this circumstance which has led to the statement,[3]

> "There used to be two realities in the world of physics: Experiment and Theory. Now there are three, and the third one is The Computer."

Of course, the computer's role is a legitimate one, and the computer will be of enormous help to us in many-particle problems. The point which will always require care is in properly interpreting what we do with the computer.

A major variation of molecular dynamics, which has achieved wide popularity in recent years, is to direct the computer to make specific changes in particle positions and/or velocities, at specific timesteps. Andersen[14] proposed "MD at constant temperature," achieved by altering the momentum of random particles at random instants of time, and "MD at constant pressure," achieved by rescaling particle positions at each timestep. By means of a more general rescaling of particle positions, involving changes in shape as well as volume of the computational cell, Parrinello and Rahman[15] extended Andersen's method to "MD at constant stress." Still another way of rescaling particle positions and velocities has been used by Abraham.[16,17] This general variation of molecular dynamics is fine as a technique for sampling phase space, in the same way that the systems of an ensemble represent a sampling of phase space. The specific rule according to which particle positions and/or velocities are changed will then define the type of ensemble which is represented. However, this variation of molecular dynamics does not give the real physical evolution of a nonequilibrium system.

An entire field of nonequilibrium molecular dynamics has been developed, largely from the pioneering work of Bill Hoover.[18-20] Hoover's technique is to apply macroscopic forces to the MD system, from an external source, or through the boundary conditions, and also to modify the equations of motion by the addition of frictional forces applied to the individual particles. These two effects are then balanced so that the MD system remains in a nonequilibrium steady state. In this way, the system behavior can be simulated in states far from equilibrium. Of course, the question arises as to the true source of irreversibility. In contrast to the usual point of view, where the equations of motion are reversible, and irreversibility presumably results from nonintegrability, Hoover uses equations of motion which are intrinsically irreversible. Since one does not have a physical justification for frictional forces in the atomic equations of motion, the physical meaning of Hoover's technique is at present unclear.

III. SOLID DEFORMATION ON THREE LEVELS

The three levels refer to different scales of size and time. The finest scale is the atomic scale, where the motions of individual atoms are studied; the intermediate scale is dislocation dynamics, where moving and interacting dislocations are studied; and the macroscopic scale is the continuum field theory of heterogeneous plastic flow. These scales form a hierarchy, because physical properties are related between neighboring levels.

Heterogeneous Plastic Flow

Let us begin by describing the "solid-dominant model" for plastic flow.[2] Consider a single crystal of material, containing one edge dislocation, as shown by the left drawings in Figure 2. This single crystal may be the entire sample under study, or it may be a small element inside a much larger sample. When a shear stress τ is applied to the surface of this crystal, it deforms elastically, as shown in Figure 2. Elastic forces are present throughout the crystal; these forces support elastic strain everywhere except at the dislocation core, and there the elastic forces drive the dislocation. When the dislocation moves, the top half of the crystal slides over the bottom half; the amount of this sliding is the plastic strain ϕ. The relation between the plastic strainrate $\dot{\phi}$ and the applied shear stress τ is the plastic constitutive relation for the crystal.

Now replace the entire slip plane of the dislocation by a shear band, as illustrated in the drawings on the right of Figure 2. When the same shear stress τ is applied, the crystal deforms elastically, with exactly the same elastic strain as before. Elastic forces are again present throughout the crystal, supporting elastic strain everywhere except in the shear band, where the elastic forces drive dislocations. Again the top half of the crystal slides over the bottom half, but now the sliding is presumably much faster than with only one dislocation present: $\dot{\phi}$ is much greater than before. Hence the essential difference between the presence of one dislocation, and a shear band, is a difference in the plastic constitutive relation.

We can summarize the main points of the solid-dominant model, as follows. Elastic strain and plastic strain are both present, and are independent variables. The shear stress satisfies simultaneously the elastic stress-strain relation, and the plastic constitutive relation. Work is done by stress driving strains; elastic work is reversible, and plastic work is mostly irreversible.

It is useful to contrast the behavior of a fluid with that of a solid. For a viscous fluid there is only one strain variable, namely the total strain, and the constitutive relation sets the shear stress proportional to the shear strainrate. The fact that fluid behavior does not contain an independent plastic strain variable is sufficient to show that an elastic-plastic solid cannot be represented by a fluid. Consider again the crystal illustrated in Figure 2. The drawings on the right are still correct if the shear band is replaced by a thin layer of viscous fluid; the crystal still has independent elastic and plastic strains, and the only thing that changes is the plastic constitutive relation. In order to eliminate the plastic strain as an independent variable, one has to go all the way to the "fluid-dominant model," in which the material is supposed to be a fluid with some bits of solid distributed in it. Only in this limit will the material behave more like a fluid than a solid.

With a computer, we can calculate (simulate) a macroscopic process involving heterogeneous plastic flow, by numerically integrating the complete set of coupled differential equations. This set is composed of the equations of continuum mechanics, the plastic constitutive equation, the entropy production equation, and some equilibrium thermoelastic relations. Such a computation is generally called hydrodynamics, although solid dynamics would be a more accurate term. Note that it is possible in principle to resolve the heterogeneities, such as shear bands or growing voids, which may appear in a solid-dynamic process. This will certainly be done in the future. In the meantime, the practice is to replace the heterogeneities by a homogeneous model. Such a model contains parameters which represent the properties of the inhomogeneities, for example their density and their constitutive behavior. This approach is represented by the work of Lee Davison and coworkers,[21,22] and of James Johnson.[23]

As a summary of the macroscopic theory of heterogeneous plastic flow, we list the essential input and output of this theory. The input is the complete set of coupled differential equations, together with the appropriate material properties, namely the thermodynamic and constitutive data. The output is a description of a macroscopic flow process, as e.g. a shock process.

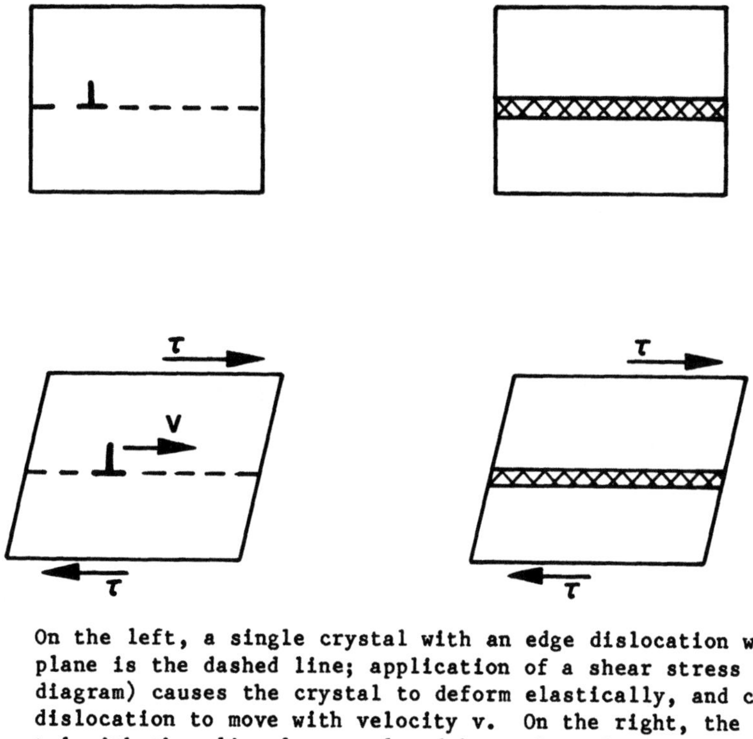

Fig. 2. On the left, a single crystal with an edge dislocation whose slip plane is the dashed line; application of a shear stress τ (lower diagram) causes the crystal to deform elastically, and causes the dislocation to move with velocity v. On the right, the same crystal with the slip plane replaced by a shear band; the same shear stress τ causes the same elastic deformation.

Dislocation Dynamics

We now want to simulate solid deformation on a different level, namely by studying the motion of individual dislocations. To do this, we construct a computer model of a system of dislocations and obstacles, as illustrated in Figure 3. The obstacles may be any physically appropriate barriers to dislocation motion, including other dislocations. To calculate the dislocation dynamics, we have to construct an equation of motion for the dislocations. The equation of motion contains a driving force, which results from a stress applied to the material, and a dislocation line tension, and terms representing interactions with the barriers, including barrier penetration probabilities. The equation of motion can also contain a dislocation drag term, and mechanisms to generate and annihilate dislocations. The procedure then is to numerically integrate the equation of motion, to simulate the dislocation dynamics. Realistic simulations for steady-state flows can be accomplished with current computer technology; such simulations are represented by the work of Schwarz and Labusch.[24-26]

In summary, the essential input and output for the computer simulation of dislocation dynamics is as follows. The input consists of a dislocation equation of motion, and the basic properties of a dislocation or dislocation system, namely its energy, its interaction with all sorts of barriers, and its generation and annihilation mechanisms. The output consists of the plastic constitutive properties of the material, in terms of dislocation density, the nature of the barriers, the temperature and so on. This output information becomes the input for a heterogeneous plastic flow calculation.

Atomic Dynamics

The ultimate level on which one can simulate solid deformation is the atomic level. Figure 4 illustrates the arrangement of atoms on the cross-sectional plane of an edge dislocation in a simple-cubic lattice. What we would like to do is construct a three-dimensional system of atoms on the computer, in the arrangement of a crystal lattice containing a single dislocation, then specify the potentials of interactions among the atoms, and calculate the motion of the atoms in the system. This is in fact an ordinary MD simulation, as we discussed in Section II. What can be learned from such a

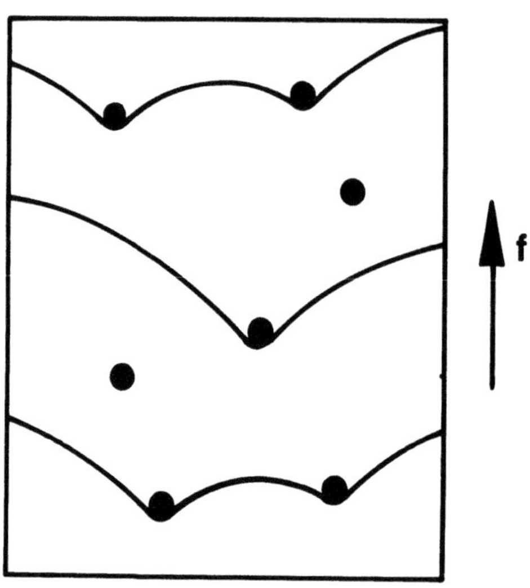

Fig. 3. Dislocations (lines) moving through an array of obstacles (dots). $\vec{f}$ is the force per unit length on a dislocation.

simulation? First of all, for the static lattice, which represents a crystal at zero temperature but without zero-point motion (since the theory is classical), one can find how the atoms relax their positions around a dislocation; hence the dislocation core structure and energy can be calculated. In the same way, the structure and energy can be calculated for related defect configurations, such as dislocation jogs and dislocation partials. Further, the effect of temperature on these defect configurations can be studied by doing ordinary MD simulations at finite temperatures, right up to the melting temperature of the material under consideration. One can apply a shear stress to the MD system, and watch the dislocation move. The interaction of a moving dislocation with the lattice can thus be studied. Finally, in principle, the interaction of a dislocation with an impurity, with a point defect, or with another dislocation can be determined by means of ordinary MD simulations.

Such calculations are enormously difficult. They are difficult for two reasons. In the first place, the elastic deformation of the lattice around a dislocation is long range, hence the atomic simulation of a dislocation requires an extremely large computational cell. Present-day computer technology is just barely able to do a realistic dislocation simulation. Secondly, we aren't quite sure of how to construct the interatomic potentials which operate in the region of a defect. To explain this problem, consider simple metals, for which pseudopotential perturbation theory applies.[27] The effective potential between ions operates through the conduction electrons, and the conduction electron density is nearly constant in space. This theory gives us good potentials for the vibrating ions in a nondefect crystal,[13] up to the melting temperature,[4,5,8] and even for the moving ions in the fluid.[7] However, if a defect is present in a crystal, the conduction electron density might have a large variation in the vicinity of the defect; if this happens, we do not know how to calculate the interionic potential in that region. The interionic potential then has to be found self-consistently with the ionic structure. This is a problem which I recommend to the theorists.

In spite of the difficulty of doing computer simulations of dislocations on the atomic level, this technique offers great promise for learning about

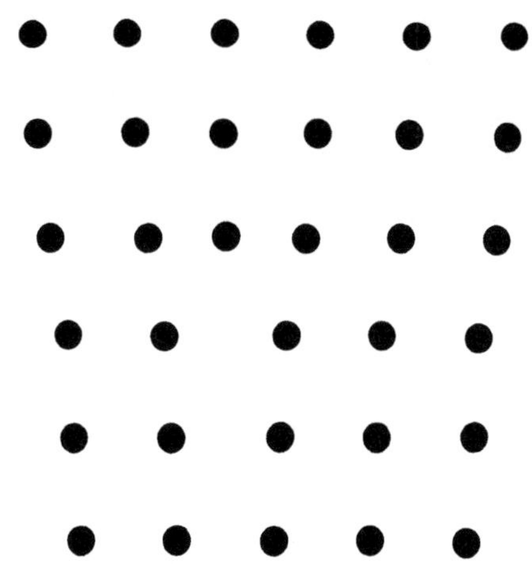

Fig. 4. View of the atoms in a planar cross section of an edge dislocation in a simple-cubic lattice.

atomic properties of dislocation in the future. Current state of the art is represented by the work of Vitek and collaborators, who have simulated the core structure and the motion of dislocations in bcc metals and in ordered alloys.[28-31] In these calculations, the long range deformation of the lattice in directions transverse to the dislocation line is represented by elastic-continuum boundary conditions for atoms outside the computational cell.

In summary, the essential input for the atomic simulation of dislocations is the atomic equation of motion and the interatomic potentials; the output is the basic properties of a dislocation in the crystal, its energy and its interactions with the lattice and with other defects. This output information becomes the input for dislocation dynamics simulations.

IV. APPLICATION TO SHOCKS

Nature of a Shock in a Solid

Let us first consider sound waves, which in first approximation are adiabatic elastic waves (where adiabatic means isentropic). When a small-amplitude elastic wave is generated in a solid, a large number of dissipative processes are driven. For example, because the local density and temperature vary in the presence of an elastic wave, the conduction electrons and the phonons are continually changing their distributions in an effort to stay in equilibrium, and this process is dissipative. Also because of the local temperature gradients, there is dissipative heat conduction. The sound wave scatters from defects, impurities, and grain boundaries; the free segments of pinned dislocations are caused to vibrate; domain walls are caused to move; these and many other dissipative processes are driven by an elastic wave in a solid. But the important point is, for a small-amplitude elastic wave in a solid, all of the dissipation effects are small. We know this from the following simple observation. A real sound wave, in the sonic or ultrasonic frequency range, is an adiabatic elastic wave clothed with all the accompanying dissipative processes. And real sound waves in solids normally travel for many wavelengths without significant attenuation.

The nature of a shock in a solid can be seen by contrasting a shock with a sound wave. For a sound wave, we can neglect dissipation entirely, and still have the essential physics, in terms of reversible thermoelastic theory. For a shock, the dissipative transfer of mechanical work into heat is an essential part of the process. Since in a solid this transfer takes place by plastic flow, we conclude that stress relaxation by plastic flow has to occur in the shock front. Here, plastic flow is used in the general sense, to include stress-relaxing structural changes, such as twinning. Other characteristics of a shock, such as the presence of an elastic (or partly elastic) precursor, and the presence of dissipation mechanisms in addition to plastic flow, should not obscure the fact that the existence of plastic flow is crucial to the shock process in a solid.

Heterogeneous Plastic Flow

A few years ago, we applied the continuum equations of thermoelastic-plastic flow, in the solid-dominant model, to the process of weak planar shocks in solids.[32,33] The assumption made in this application is that the space- and time-scales of the heterogeneities in the plastic flow are small compared to the space- and time-scales of the shock process. If this were not the case, the shock wave would break up into different structures at different locations, on a plane of constant distance from the impact surface. In the visar experiments on 6061 aluminum,[34] there was no indication of such macroscopically inhomogeneous wave structure, within the spatial and temporal resolution of the instrument. It was therefore appropriate to analyze these experiments with the above mentioned theory. In particular, by combining the experimental data with known thermoelastic properties, the plastic constitutive behavior of 6061 aluminum, averaged over the shock-induced heterogeneities, was obtained.[33] This analysis is difficult, and is subject to errors in the numerical integration, and errors in the input experimental data, but not significant errors in the underlying theory.

Grady has stressed the existence of shear-band structures in samples recovered from shock experiments.[35,36] In the case of aluminum, and 6061 aluminum, the spatial scale of these structures appears to be of order or less than the width of the plastic wave.[35] The existence of shear bands has led Grady to propose for the shock process a set of equations which are more representative of fluid behavior than they are of solid behavior. Let ε denote the total compression induced by the shock, the compression in the Hugoniot state. For a solid, ε contains both elastic and plastic contributions, hence ε is not the variable for specifying plastic strain, nor plastic dissipation, as Grady does (Ref. 35, eqs (1) and (3)). For the energy dissipated in a weak shock, Grady uses fluid theory (Ref. 35, eq. (3)), which gives a dissipation of order ε^3, while for a solid the dissipation is formally of order ε^2, as we have shown.[37]

In the weak shock regime, the decrease in shock risetime with increasing shock strength is a plastic strainrate effect: higher shear stresses drive plastic flow at higher rates. By extending the equations of thermoelastic plastic flow to overdriven shocks in metals,[38,39] we expect the risetime to be of order 10^{-12}s for shocks up to a few Mbar. We are also able to show that heat transport is necessary for the existence of an overdriven steady-wave shock.[38] Finally, for shock strengths above a few Mbar in metals, the electrons and phonons will not be able to remain near equilibrium, and hence irreversible thermodynamics will break down.[39,40]

MD Simulations

There have been a number of computer simulations of supported waves in one-dimensional chains of interacting atoms.[41-43] These waves are generated by simulating planar impact experiments; such experiments generate shocks in real materials. But the computer simulations generate nonsteady waves, characterized by a continually growing length of nonequilibrium material immediately following the wave front. This result has led to the suggestion that the steady-wave jump conditions cannot be used to interpret experiments on shocks in solids.[41] But such a conclusion is not warranted, because waves in a one-dimensional atomic chain have no plastic flow. The computer-simulated waves are not shocks, but are damped nonlinear elastic waves.

The same situation can appear in computer-generated waves in three-dimensional lattices. Thus, the computed nonsteady profiles of MacDonald and Tsai[44] are not shocks, and conclusions about real shocks in real materials cannot be drawn from these computer simulations. But a realistic shock <u>can</u> be generated in a three-dimensional lattice, as shown by the simulations of Holian and Straub.[45] For a sufficiently strong compressional wave, a highly specific mode of pastic flow is driven, namely a stress-relaxing dissipative structural transition. The resulting wave structure is a decaying elastic precursor, followed by a plastic compression to a steady state.[46]

Finally, we note that MD simulation of dissipative stress-relaxing flow in a viscous fluid is in principle much simpler than for an elastic-plastic crystalline solid. This is because viscous fluid flow occurs on the atomic scale, while plastic flow usually occurs on the scale of dislocations. Hence, computer simulations of shocks in fluids are already in reasonably good agreement with continuum irreversible thermodynamics.[47,48]

REFERENCES

1. H. J. Kreuzer, Nonequilibrium Thermodynamics and its Statistical Foundations (Clarendon Press, Oxford, 1981).
2. D. C. Wallace, Thermoelastic-Plastic Flow in Solids (LA-10119, Los Alamos National Laboratory, Los Alamos, 1985).
3. D. C. Wallace, in Electronic Structure, Dynamics, and Quantum Structural Properties of Condensed Matter, edited by J. T. Devreese and P. Van Camp (Plenum, New York, 1985), p. 521.
4. G. K. Straub, R. E. Swanson, B. L. Holian, and D. C. Wallace, in Ab Initio Calculation of Phonon Spectra, edited by J. T. Devreese (Plenum, New York, 1983), p. 137.
5. R. E. Swanson, G. K. Straub, B. L. Holian, and D. C. Wallace, Phys. Rev. B25, 7807 (1982).
6. B. L. Holian, G. K. Straub, R. E. Swanson, and D. C. Wallace, Phys. Rev. B27, 2873 (1983).
7. G. K. Straub, S. K. Schiferl, and D. C. Wallace, Phys. Rev. B28, 312 (1983).
8. G. K. Straub and D. C. Wallace, Phys. Rev. B30, 3929 (1984).
9. D. C. Wallace, S. K. Schiferl, and G. K. Straub, Phys. Rev. A30, 616 (1984).
10. D. C. Wallace and G. K. Straub, Phys. Rev. A27, 2201 (1983).
11. D. C. Wallace, B. L. Holian, J. D. Johnson, and G. K. Straub, Phys. Rev. A26, 2882 (1982).
12. S. K. Schiferl and D. C. Wallace, Phys. Rev. B31, 7662 (1985).
13. D. C. Wallace, Phys. Rev. 176, 832 (1968).
14. H. C. Andersen, J. Chem. Phys. 72, 2384 (1980).
15. M. Parrinello and A. Rahman, J. Appl. Phys. 52, 7182 (1981).
16. F. F. Abraham, in Proc. Intern. Conf. on Ordering in Two Dimensions, edited by S. K. Sinha (North-Holland, New York, 1980), p. 155.
17. F. F. Abraham, Phys. Reports 80, 339 (1981).
18. W. G. Hoover and W. T. Ashurst, in Theoretical Chemistry, edited by H. Eyring and D. Henderson (Academic, New York, 1975), Vol. 1, p. 1.
19. W. G. Hoover, Ann. Rev. Phys. Chem. 34, 103 (1983).
20. W. G. Hoover, Physics Today 37, 44 (1984).
21. L. Davison, in Shock Waves in Condensed Matter - 1983, edited by J. R. Asay, R. A. Graham, and G. K. Straub (North-Holland, Amsterdam, 1984), p. 181.
22. L. Davison, A. L. Stevens, and M. E. Kipp, J. Mech. Phys. Solids 25, 11 (1977).
23. J. N. Johnson, J. Appl. Mech. 105, 593 (1983).
24. R. B. Schwarz and R. Labusch, J. Appl. Phys. 49, 5174 (1978).
25. R. B. Schwarz, Phys. Rev. B21, 5617 (1980).
26. R. B. Schwarz, Acta Met. 29, 311 (1981).
27. D. C. Wallace, Thermodynamics of Crystals (Wiley, New York, 1972).
28. V. Vitek, Crystal Lattice Defects 5, 1 (1974).
29. V. Vitek, Proc. Roy. Soc. (London) A352, 109 (1976).
30. M. Yamaguchi, V. Paidar, D. P. Pope, and V. Vitek, Phil. Mag. A45, 867 (1982).
31. V. Paidar, M. Yamaguchi, D. P. Pope, and V. Vitek, Phil. Mag. A45, 883 (1982).
32. D. C. Wallace, Phys. Rev. B22, 1477 (1980).
33. D. C. Wallace, Phys. Rev. B22, 1487 (1980).
34. J. N. Johnson and L. M. Barker, J. Appl. Phys. 40, 4321 (1969).
35. D. E. Grady and J. R. Asay, J. Appl. Phys. 53, 7350 (1982).
36. D. E. Grady, in Shock Waves in Condensed Matter - 1983, edited by J. R. Asay, R. A. Graham, and G. K. Straub (North-Holland, Amsterdam, 1984), p. 363.
37. D. C. Wallace, Phys. Rev. B22, 1495 (1980).
38. D. C. Wallace, Phys. Rev. B24, 5597 (1981).
39. D. C. Wallace, Phys. Rev. B24, 5607 (1981).

40. D. C. Wallace, Phys. Rev. A25, 3290 (1982).
41. D. H. Tsai and C. W. Beckett, J. Geophys. Res. 71, 2601 (1966).
42. B. L. Holian and G. K. Straub, Phys. Rev. B18, 1593 (1978).
43. G. K. Straub, B. L. Holian, and R. G. Petschek, Phys. Rev. B19, 4049 (1979).
44. R. A. MacDonald and D. H. Tsai, Phys. Reports 46, 1 (1978).
45. B. L. Holian and G. K. Straub, Phys. Rev. Lett. 43, 1598 (1979).
46. G. K. Straub, private communication.
47. W. G. Hoover, Phys. Rev. Lett. 42, 1531 (1979).
48. B. L. Holian, W. G. Hoover, B. Moran, and G. K. Straub, Phys. Rev. A22, 2798 (1980).

THE ROLE OF SHOCK WAVE COMPRESSION IN MATERIALS SCIENCE

Akira B. Sawaoka
Center for Explosives Technology Research
New Mexico Institute of Mining and Technology
Socorro, NM 87801

Research Laboratory of Engineering Materials
Tokyo Institute of Technology
Yokohama 227, Japan

ABSTRACT

Shock wave compression technology has contributed much to the progress of materials science. The present author has tried to discuss the role of shock wave techniques with respect to very high pressure materials science studies. Shock wave techniques also unique very high temperature methods. The important role of shock wave technology seems to be the attainment of pressures above those which can be obtained by static techniques. A typical example is equation of state studies done up to pressures of a few megabars. Recently, a static technique utilizing a small pair of diamond anvils has been established which is capable of attaining a few megabar pressures. This technique is performing the majority of equation of state studies. Shock wave compression is accompanied by high shock temperature. Recently, new dynamic compression techniques, which are isentropic-like, are being developed and applied to equation of state studies. On the other hand, several shock studies suggested that due to the very high shock temperatures, accompanying shock compression, dynamic compression techniques could be utilized as a unique high temperature method. High inhomogeneous shock temperatures can be produced in porous and/or compressible materials during strong shock compression. These temperatures induce local melting and ionization. It is expected that a high dense plasma-like state is formed in extremely strong shock wave compression. After passage of the shock wave, the materials cool rapidly during pressure release to a residual temperature and pressure. Occasionally the materials recovered after having been converted to a dense liquid or gas state showed very unique microstructures. The present author believes that more precise, compact and easily operative shock wave compression apparatus should be developed and made available to general material scientists and engineers.

INTRODUCTION

Shock wave compression technology has recently played a major role in equation of state studies and this role seems to be unchanging also in the future. Especially the most reliable compression data in the extremely high pressure region can be obtained only by shock wave compression. However, detailed mechanism of shock compression of materials has not yet been clarified. Study of the compression mechanism of materials under shock loading is also a very important subject in relation to the mechanical properties of solids.

In this paper, the role of shock wave compression techniques as a method of material science study under extreme pressure and temperature conditions is discussed in comparison with static compression techniques. Extreme condition technologies related to high pressure, very high temperature, very low temperature and very high magnetic field are powerful tools for material research; however, the physical meaning of each method to the study of materials is different. There are two methods for very high pressure production; static and dynamic. Furthermore, dynamic compression techniques are divided into shock wave compression, which produces an entropy increase and isentropic compression methods. Shock wave compression techniques have been carrying a leading role in the study of materials science, especially the physics and chemistry of solids. A typical example is the ability to produce very highly compressed states in materials which can not be obtained by static techniques.

Generation of one megabar (100 GPa) pressure was a dream for static very high pressure men for a long time. For instance, two great high pressure pioneers, the late professor N. Kawai (Osaka University, Japan) and the late professor L.F. Vereshagin (Institute of High Pressure Physics, USSR) had separately challenged to produce 100 GPa in 1000 mm^3 pressure medium in 1970's. One difficulty in their study was uncertainty of pressure scale. Unfortunately, their reports were not yet confirmed.

Shock wave techniques were the only available methods to realize megabar pressures until the appearance of the improved diamond anvil apparatus, which can produce very high pressures, above 100 GPa in about 0.01 mm^3 (1-3). It is doubtfull that the compression state achieved in materials under shock compression is microscopically equivalent to the compressed state achieved under static compression. However, we have utilized shock technique as a unique method for megabar compression until the appearance of the 100 GPa diamond anvil apparatus. It is an interesting result that many pressure values obtained by shock and static methods at phase transitions are consistent with each other. Frankly speaking it is really a surprising result. Electronic type and diffusionless type phase transition pressures induced by shock compression are very close to these induced by static compression. Diffusion type phase transition such as graphite to diamond depends on kinetics. Thus pressure values of such transitions induced by shock wave compression are considerably higher than in the static case.

The development of the diamond anvil type apparatus is opening new possibilities to the study of materials science at very high pressures. The diamond anvil apparatus, consisting of a pair of small diamonds, can be operated by non-high pressure specialists who can produce pressures up to at least 50 GPa after training for a few weeks. The price of this apparatus, excluding instrumentation is a few thousand dollars. The above advantages of the diamond anvil apparatus opened very high pressure technology to the majority of materials scientists and engineers.

NEW ROLES OF SHOCK WAVE COMPRESSION TECHNOLOGY AS A UNIQUE METHOD OF VERY HIGH TEMPERATURE PRODUCTION

The potential of shock wave compression technique in the pressure range up to 100 GPa has decreased after the establishment and popularization of the diamond anvil high pressure apparatus. Shock wave compression technique still have much potential in the production of extreme high pressures. However, very high shock compression accompanied by very high shock and residual temperature generation. For instance, shock temperatures of iron at 70 GPa and 300 GPa are 1300 K and 10,000 K, respectively (4, 5). Iron is a rather incompressible material. Shock temperatures of compressive materials such as molecular crystals and porous substances, are very high. Shock temperatures of graphite are 3700 K at 70 GPa (5). Is such high temperature generation accompanying strong shock compression a disadvantage of shock compression techniques? The present author recognized that shock compression techniques are very useful methods for rapidly heating and quenching. Porous and low shock impedance materials which can be heated above a few thousand K by strong shock wave compression. Especially shock temperatures in inhomogeneous material are inhomogeneously distributed. It seems that local shock temperatures in porous materials easily reach 10,000 K during shock compression of 100 GPa. Such a very high temperature induces ionization of the material, and forms a plasma-like state. Material density at very high temperature is low compared with the starting density. On the other hand, shock compression can produce high density ionization or a plasma-like state. This state is very useful for causing condensation during pressure release and in the following process. Residual temperature is usually about 80% of the shock temperature. A residual temperature distribution also exists in inhomogeneous shock impedance materials. After passage of the shock wave occurs, rapid cooling of the low shock impedance-high residual temperature parts occurs by heat conduction to the high shock impedance-low residual temperature parts. There is not enough quantitative information concerning above rapid quenching, but some experiments showed very interesting results suggesting this phenomenon occurred.

Two examples, sphere diamond and whisker silicon nitride formation, are shown in the following section.

Shock wave compression technique of mixture of graphite and transition metal powder have been used for industrial production of shock-synthesized diamond. Relatively low shock temperature compressions resulted in the formation of a mixture of cubic and hexagonal phases of diamond whereas high shock temperatures produced only cubic diamond. Well crystallized graphite transformed easily to diamond. These results suggest that the transformation from graphite to hexagonal diamond is a martensitic/diffusionless type transformation (5). On the other hand, it was found by N. Setaka and Y. Seikawa (6) that diamond was formed by shock compression of glassy carbon. Spherical diamond particles with about 0.1 μm diameter were found. This fact suggested that the diamond was formed from molten carbon or a high dense gas phase of carbon. The density of glassy carbon is considerably lower than the density of graphite. Micropores exist in glassy carbon. Very high shock temperature is generated by shock compression of it. Formation of spherical diamond could have occurred during cooling from the shock temperature. This result gives us much new information about material synthesis.

Silicon nitride, with unique microstructures has been obtained by dynamic compaction of fine amorphous silicon nitride powder (7). Whiskers were grown in the powder compact during the post shock process. Total heating time above $1000^{\circ}C$ was less than 50 ms. It seems that the high density vapor of Si_3N_4 formed during shock compression condenses quickly during pressure release and following cooling process. It was a surprising fact that 10 μm whiskers were grown from the dense gas phase in such a very short time.

As mentioned above, shock wave compression techniques have excellent possibilities as a method for rapid heating to very high temperatures followed by rapid cooling of materials in a closed capsule. This new application to material science should open a new field. However there are some difficulties to be resolved; improvement of sample recovery techniques and precise information about local distribution and history of temperatures in the specimen.

DEVELOPMENT OF CONVENTIONAL IMPACT TECHNIQUE FOR

PRECISE MEASUREMENT OF MATERIALS UNDER SHOCK COMPRESSION

It is difficult for the majority of material scientists and engineers to conduct shock wave experiments in comparison with static very high pressure experiments, since such dangerous materials as powder and compressed gas are used and compression duration is extremely short. Especially explosive experiments require special facilities which are isolated from the general research areas. Sometimes, the explosive research facilities are located in a restricted security area. This situation obstructs the exchange of information between shock wave researchers and other material scientists. Opening shock wave technology to the bulk of material scientists and engineers is a key point in developing shock wave/material science. Recently relatively compact gas guns were designed for materials research by researchers themselves, as material scientists and geophysicists with the help of mechanical engineers. Some of them have been used in usual research buildings and producing interesting results. Typical sizes and power of these guns are 12 m long two-stage light gas gun with 20 mm bore diameter and a maximum velocity of 6 km/s, and 5 m powder gun with 20 mm bore diameter and a maximum velocity of 2 km/s. Precise instrument systems such as spectrometer and optical interferometer can be set up closely to the compact gun. A major advantage of these improved compact guns is easy operation. Usually it is possible to do two shots with a two-stage light gas gun per day or several shots with a powder gun by two researchers as a self service experiment. A 12 m conventional two-stage light gas gun is operating at the Tokyo Institute of Technology (8, 9). Operation and maintenance is performed by researchers themselves, including graduate students. Over-view of this gun is shown in the figure. Construction cost of this gun was \$100,000～200,000. The required specimen size for precise measurement during shock compression is being reduced with the development of better measurement techniques. Currently, minimum gunbore of 15 mm is required for electric and optical measurements of shock or post shock compression with acceptable accuracy. Effective shock duration time is decreasing with decreasing flyer diameter. Usually a shock duration of about 200 ns is required for observation of phase transitions and of equation of state measurements. Thus, the minimum size of the specimen is limited by physical measurement techniques. The size problems in recovery experiments are more serious. On the other hand, smaller guns are more convenient to operate and maintain. They are safe and cheaper to construct. As a general conclusion, the present author preferred the 10 mm smoothed bore gun for in situ shock measurement and the 15 mm smoothed bore gun for shock recovery experiments. This idea is suitable for both powder guns and two-stage light gas guns.

Figure 1. Photograph of two stage light gas gun in Tokyo Institute of Technology, HS-2B (6M long) and HS-3B (12M long).

Another advantage of gun usage in shock wave experiments is the acceleration of a flyer plate consisting of different shock impedance materials or whose shock impedance is gradually changing. Such a flyer plate will separate layer-by-layer during acceleration in explosive experiments. Impact by such a flyer having a gradual shock impedance increase in the flying direction can realize quasi isentropic dynamic compression of material above 100 GPa. Temperature increase accompanying this compression can be calculated by a simple adiabatic compression model, and its temperature produced is not so high. Isentropic dynamic compression techniques combined with compact two-stage light gas gun will be developed and used in the study of extremely compressed state of materials, above 100 GPa without very high temperature generation (10).

REFERENCES

1. H. Mao and P.M. Bell, Science 200, 145 (1978).
2. P.M. Bell, H. Mao and R.J. Hemley, presentation at Xth AIRAPT International High Pressure Conference at Amsterdam, July 9, 1985.
3. P.M. Bell, presentation at the Fourth APS Topical Conference on Shock Waves in Condensed Matter at Spokane, July 24, 1985.
4. R.G. McQueen, S.P. Marsh and J.N. Fritz, J. Geophysical Res., 72, 4999 (1967).
5. T. Sōma, A. Sawaoka, and S. Saito, proceedings of the 4th AIRAPT International High Pressure Conference, Kyoto-1974, 446 (1975).
6. N. Setaka and U. Sekikawa, J. Mat. Sci. Letters 16, 1728 (1981).
7. E.K. Beauchamp, M.J. Carr, presentation at the Fourth APS Topical Conference on Shock Waves in Condensed Matter at Spokane, July 24, 1985.
8. H. Sugiura, T. Mashimo, K. Kondo and A. Sawaoka, Rep. of Res. Lab. Eng. Materials, Tokyo Institute of Technology, No. 6, 93 (1981).
9. T. Mashimo and A. Sawaoka, Japan J. Appl. Phys, 20, 963 (1981).
10. L.M. Barker, J.L. Wise, T.G. Trucano and J.R. Asay, presentation at the Fourth APS Topical Conference on Shock Waves in Condensed Matter at Spokane, July 22, 1985.

THEORY OF THE HIGH TEMPERATURE SHOCKED METALLIC STATE

R.G. Dandrea and N.W. Ashcroft

Laboratory of Atomic and Solid State Physics
Cornell University, Ithaca, NY 14853

ABSTRACT

A unified theory is presented to describe the thermodynamics of a simple metal throughout a range of densities and temperatures appropriate to recent ultrahigh pressure shock experiments. Second order pseudopotential theory is used to express the free energy in terms of state-dependent ionic pair potentials. However, because temperatures attained along the Hugoniot are very high, proper evaluation of the pseudopotential Hamiltonian requires knowledge of the dielectric and thermodynamic functions of a semidegenerate electron gas. A major part of this work thus centers on new results for the high temperature properties of an interacting electron gas. Approximate inclusion of local field corrections to the random phase approximation yields an electron gas free energy that agrees well with Monte Carlo calculations both at zero temperature and in the classical regime. By combining the semidegenerate electron gas thermodynamics with a classical variational treatment of the ionic degrees of freedom, we are able to determine a Hugoniot curve for aluminum that compares favorably with experiment.

I. Introduction

In recent years, the maximum pressures attainable in dynamic shock compression experiments have been increasing notably. Hugoniot pressures as high as 400 TPa have been reported in convergent nuclear shock wave experiments.[1] Furthermore, improvements in the precision with which impedance matching measurements are done has resulted in the reporting of Hugoniot pressures for several materials up to 6 TPa with experimental uncertainties of the order of 4%.[2] The extreme conditions attained in these ultrahigh pressure experiments thus provide a stringent test for theoretical equation of state (EOS) models.

Because of irreversible heating that occurs during shock loading, these high pressure states also have commensurately high temperatures. As will be demonstrated for the case of simple metals, temperatures of the order of the Fermi temperature are currently reached in such shock wave experiments. At these extremes of metallic density and temperature, one common requirement of all the EOS models must be an accurate description of the semidegenerate electronic system. However, previous EOS models have generally considered only the ideal fermion contribution to the electronic pressure and free energy,[3] or have only included the Sommerfeld T^2 correction for finite temperature contributions.[4] The aim of this paper is therefore to present new results on the full thermodynamics and polarizability of an interacting electron gas at any degree of degeneracy, and to apply this information to the calculation of the Hugoniot curves of simple metals. As will be shown below, the exchange and correlation contributions to the electronic thermodynamics are indeed significant throughout the temperatures and densities probed in these shock wave experiments.

In section II we describe in detail the calculation of the dielectric and thermodynamic functions of a semidegenerate electron gas. Convenient fitting formulas are presented (Appendix) that interpolate between the results for a degenerate Fermi gas and those for a classical one-component plasma. In section III these electron gas results are applied in a straightforward manner to calculate the high temperature Hugoniot curves of simple metals. The electron-ion interaction is described by a local pseudopotential, which when treated perturbatively leads to a state-dependent ionic pair potential. As a first approximation to the complete problem we neglect both the thermal excitations of core electrons and the nonequilibrium aspects of irreversibility associated with the entropy production due to plastic flow. Within the approximation of thermodynamic equilibrium, the ionic degrees of freedom are traced over classically by application of the Gibbs-Bogoliubov variational principle. This model allows for determination of the EOS of simple metals up to temperatures where core excitations become significant. In section IV the results of calculations of the aluminum room temperature isotherm and high temperature Hugoniot are presented,[5] and compared to experiment. We conclude in section V.

II. The Semidegenerate Fermion Gas [6]

An accurate description of the dielectric and thermodynamic functions of a one component fermion plasma at any degree of degeneracy is a problem of fundamental significance. Besides its application to the high temperature metallic Hugoniot discussed below, the finite temperature EOS of the fermion gas is also needed for a proper treatment of high density inertially confined fusion plasmas (where quantum corrections become important), of the phase diagram of the electron-hole liquid in irradiated semiconductors, and of the finite temperature exchange-correlation potentials used in density functional calculations of atomic properties at high pressure and temperature.[7]

Accordingly, there have been several attempts to obtain this information, and these have been mainly of two types. First, Pokrant[8] has developed a theory that uses a finite temperature variational principle to evaluate the Slater sum, which he approximates as the product of an ideal fermion diagonal density matrix and a product of pair functions. The energy is then calculated by using a pair correlation function obtained from the hypernetted chain approximation. Second, several other groups[9–11] have concentrated on obtaining the exchange-correlation grand potential by using finite temperature perturbation theory within the random phase approximation (RPA). We also follow a calculational scheme along similar lines. It is known, however, that the RPA is a weak coupling theory that results in considerable error at coupling strengths appropriate to metallic densities. Thus, in order to obtain accurate results at such densities, we include an approximate treatment of the local field corrections to the RPA. Our treatment of local field corrections is similar to the zero temperature methods of Hubbard[12] and Geldart and Vosko,[13] and in the following we present a detailed account of this treatment, and compare the results of our calculations with those of Monte Carlo calculations both at zero temperature, and in the classical regime.

Consider what we have been calling the electron gas: a homogeneous system of N electrons at temperature T in a volume V, of density $n = N/V$ and chemical potential μ, together with a uniform positive background to ensure charge neutrality. As usual, the thermodynamic functions are expressed as a sum of their ideal fermion, exchange, and correlation parts, e.g. $F = F_o + F_x + F_c$. The basis for our calculation of the excess free energy is the fact that it is expressable as a coupling constant integration over the potential energy.[14] If this coupling constant integration is carried out in the grand canonical ensemble (at fixed T, V and μ), one obtains the excess grand potential. On the other hand, if it is carried out in the canonical ensemble (at fixed T, V and N), the excess Helmholtz free energy is obtained. Previous calculations within the RPA have all proceeded within the grand ensemble, where an approximate inversion of $n(\mu,T)$ to obtain $\mu(n,T)$ has to be performed in order to obtain EOS data. It can be seen however, from the dicussion of Perrot and Dharma-wardana,[11] that if this inversion is carried out to the same order in perturbation theory as the calculation of the grand potential itself, the results of the two ensembles are equivalent. In what follows, we shall be working within the canonical ensemble, (and using units such that $\hbar = 1$).

Before considering the exchange-correlation free energy, we review the ideal fermion contribution for the sake of completeness. The ideal fermion energy E_o and pressure p_o are

$$E_o(T,V,N) = \frac{3}{2} p_o V = 2V\,(2\pi)^{-3} \int d^3k\ \varepsilon_k^o f_k^o\ , \tag{1}$$

where $\varepsilon_k^o = k^2/2m$, $f_k^o = [1 + \exp(\beta\varepsilon_k^o - \eta_o)]^{-1}$, $\beta = 1/k_BT$, and $\eta_o(n,T) = \beta\mu_o$ is the ideal fermion activity, defined by

$$I_{1/2}(\eta_o) = \frac{2}{3t^{3/2}} \,. \tag{2}$$

Here we have introduced a reduced temperature $t = k_BT/\varepsilon_F$, where $\varepsilon_F = (3\pi^2 n)^{2/3}/2m$ is the zero temperature ideal Fermi energy, and the commonly used notation for the Fermi integral of order ν :

$$I_\nu(\eta) = \int_0^\infty \frac{x^\nu\, dx}{1 + e^{x-\eta}} \tag{3}$$

Using this notation, the ideal fermion energy, pressure, free energy and bulk modulus can be simply expressed (in Rydbergs per electron) as

$$\varepsilon_o(r_s,t) = \frac{3p_o}{2n} = \frac{3\, t^{5/2}}{2\alpha^2 r_s^2} I_{3/2}(\eta_o) \,, \tag{4}$$

$$f_o(r_s,t) = \frac{-2}{3}\varepsilon_o(r_s,t) + \frac{t}{\alpha^2 r_s^2}\eta_o(t) \tag{5}$$

and

$$\frac{B_o(r_s,t)}{n} = \frac{4}{3\alpha^2 r_s^2 \sqrt{t}\, I_{-1/2}(\eta_o)} \,, \tag{6}$$

where $\alpha = (4/9\pi)^{1/3}$, $4\pi(r_s a_o)^3/3 = 1/n$ and a_o is the Bohr radius. Accurate closed form fitting expressions for $I_{3/2}(\eta_o)$ and $I_{-1/2}(\eta_o)$ as functions of t are presented in the appendix.

We now turn to the exchange contributions to the thermodynamics. The coupling constant expression for the excess free energy is[14]

$$F_{xc}(T,V,N) = \int_0^1 d\lambda\lambda^{-1} \langle\lambda\hat{U}\rangle_\lambda \,, \tag{7}$$

where $\hat{U}$ is the potential energy operator and $\langle\cdots\rangle_\lambda$ denotes a canonical ensemble average in a system interacting via a Coulombic potential energy of $\lambda e^2/r$. In Eq. (7), the contribution first order in e^2 is the exchange free energy, $F_x = \langle\hat{U}\rangle_o$, which from a straightforward application of finite temperature perturbation theory can be written[15]

$$F_x(T,V,N) = -V(2\pi)^{-6} \int d^3k d^3q\, v(\vec{k}-\vec{q}) f_k^o f_q^o \,, \tag{8}$$

where $v(q) = 4\pi e^2/q^2$. In terms of the Fermi integrals defined in Eq. (3), this can be conveniently expressed (in Rydbergs per electron) as[16]

$$f_x(r_s,t) = \frac{-3\,t^2}{4\pi\alpha r_s} \int_{-\infty}^{\eta_o(t)} d\eta\, I_{-1/2}^2(\eta) \,. \tag{9}$$

Similarly, the exchange parts of the chemical potential and bulk modulus are

$$\mu_x(r_s,t) = -\sqrt{t}\, I_{-1/2}(\eta_o)/\pi\alpha r_s \tag{10}$$

and

$$\frac{1}{n}B_x(r_s,t) = \frac{-4}{3\pi\alpha r_s t} \frac{I'_{-1/2}(\eta_o)}{I_{-1/2}(\eta_o)} \,. \tag{11}$$

The exchange pressure is $p_x/n = \mu_x - f_x$, and because of the $1/r_s$ scaling in Eq. (9), the exchange energy can be expressed as $\varepsilon_x = 3\mu_x/2 - f_x$. It is worth noting that all of these quantities vanish as $1/t$ at high temperatures ($t \gg 1$). However, even at temperatures as high as $t = 1$, ($T = T_F$), the exchange contributions are still of a proportion similar to those at $T = 0$. A closed form fitting expression for $f_x(r_s,t)$ is given by Perrot and Dharma-wardana;[11] fitting expressions for Eq.'s (10) and (11) are presented in the appendix.

We now consider the correlation contribution to the free energy. If the exchange part is subtracted from the excess free energy defined by the coupling constant integration of Eq. (7), and the potential energy is expressed in terms of the static structure factor $S(q)$, the correlation free energy can be written

$$F_c(T,V,N) = \frac{1}{2}N(2\pi)^{-3} \int d\lambda\lambda^{-1} \int d^3q \; v_\lambda(q)[S_\lambda(q) - S_o(q)] \; . \tag{12}$$

Relating the static structure factor to the retarded polarizability $\Pi(q,\omega)$ by the fluctuation-dissipation theorem[17]

$$S(q) + N\delta_{q,0} = \frac{-1}{n} \int \frac{d\omega}{2\pi} \coth(\frac{\beta\omega}{2}) \; \mathrm{Im}\,\Pi(q,\omega) \; , \tag{13}$$

readily permits (12) to be rewritten as

$$F_c = \frac{-V}{2} \int_0^1 \frac{d\lambda}{\lambda} \int \frac{d^3q d\omega}{(2\pi)^4} \; v_\lambda(q) \coth(\frac{\beta\omega}{2}) \; \mathrm{Im}[\Pi_\lambda(q,\omega) - \Pi_o(q,\omega)] \; , \tag{14}$$

where $\Pi_o(q,\omega)$ is the ideal fermion polarizability.

As mentioned previously, the evaluation of (14) within the RPA, where $\Pi_{RPA} = \Pi_o/(1-v_q\Pi_o)$, leads to overly negative correlation energies at metallic densities. This is because the RPA pair correlation function [obtained by using Π_{RPA} in Eq. (13)] is actually quite negative at small distances, which can be attributed to the mean field type treatment of interactions in the RPA, and the concommitant inclusion of self-interaction effects. In order to include local field corrections in the pair correlation function, we follow an approximate procedure commonly used both at zero temperature[12,13,18] and in the classical regime,[18] where the polarizability is related to its ideal part by

$$\Pi(q,\omega) = \frac{\Pi_o(q,\omega)}{1 - [1-G(q)]v(q)\Pi_o(q,\omega)} \equiv \frac{\Pi_o(q,\omega)}{\xi(q,\omega)} \; . \tag{15}$$

Here G(q) represents the local field corrections to the RPA (G=0 in the RPA). In order to determine G(q), we make use of the two standard constraints upon it. First, the compressibility sum rule[19] implies that $G(q \to 0) = \gamma \, (q/2k_F)^2$, where the prefactor γ is related to the excess bulk modulus:

$$\gamma(r_s,t) = \frac{-3\pi\alpha r_s}{2} \frac{B_{xc}(r_s,t)}{n \; \mathrm{Ryd}} \equiv \gamma_x(t) + \gamma_c(r_s,t) \; . \tag{16}$$

Secondly, the cusp condition on short range correlations[20] implies that $G(q \to \infty) = 1 - g(0)$, where $g(0)$ is the pair correlation function at $r = 0$. These two constraints are conveniently included in G(q) by use of the Vashista-Singwi form[21] for the local field correction (LFC) function: $G(q) = A(1 - e^{-Bq^2})$, where A and B are determined by $\gamma(r_s,t)$ and $g(0)$. Here we have simply fixed $g(0)$ to a value appropriate[18] to the metallic densities considered here, i.e. $g(0) = 0.1$. The function $\gamma_x(t)$ is then calculated exactly from Eq. (11), while $\gamma_c(r_s,t)$ is obtained from an appropriate interpolation between known Monte Carlo results at zero temperature[22] and in the classical regime.[23] We believe the error at intermediate degeneracy caused by this *ad hoc* interpolation to be a second order effect, since we have found that the final F_c calculated is rather insensitive to the form of interpolation, largely because for metallic densities we expect $\gamma_x \gg \gamma_c$ for $t < 1$. Further evidence for this is the fact that Hubbard[12] obtained quite good results for the zero temperature correlation energy, even though his value of γ was in error by a factor of two.

In order to proceed with the evaluation of $f_c(r_s,t) = F_c/N$, one further approximation is made. Specifically, we observe that the coupling constant integration in Eq. (14) can be carried out in closed form if the λ dependence of G(q) is neglected. Since γ_x is independent of e^2, this is a very good approximation at $t < 1$. With this assumption, we obtain

$$F_c = \frac{V}{2} \int \frac{d^3q d\omega}{(2\pi)^4} \coth(\beta\omega/2) \; [1-G(q)]^{-1} \; [\arg(\xi) - \mathrm{Im}(\xi)], \tag{17}$$

where $\xi(q,\omega)$ is defined in (15) and arg(z) is the argument of the complex number z. Eq. (17) is evaluated numerically to obtain the correlation free energy. Note that it requires a three dimensional numerical integration, and that it is essentially a finite temperature generalization of the zero temperature form of Hubbard,[12] but with an improved form for G(q).[13,21]

In Fig. 1 we plot the results of our calculation of $f_c(r_s,T=0)$, and compare them to RPA and Monte Carlo results,[22] and also to the commonly used interpolation form of Hedin and Lundqvist.[24] The approximate inclusion of local field corrections has resulted in a large improvement of the RPA, and essentially has reproduced the Hedin-Lundqvist results. A similar conclusion applies to the classical regime as seen from Fig. 2, where the classical excess free energy is compared to Monte Carlo[23] and RPA (or Debye-Hückel) results. The classical regime is defined not only by requiring $t \gg 1$, but also by requiring the quantum correction parameter $\zeta \ll 1$, where ζ is the ratio of the thermal de Broglie wavelength to the classical Debye-Hückel screening length.[25] In order to compare our inherently quantum mechanical theory to classical results, we have thus fixed ζ at 0.05 and plotted the calculated values of f_c versus the classical coupling constant $\Gamma = e^2/(r_s a_o k_B T)$. Even at a coupling strength as large as $\Gamma = 4$, the LFC theory is still nearly exact, whereas the RPA results are a factor of two too negative. Although metallic densities are associated with only weakly coupled ($\Gamma \ll 1$) electronic plasmas in the classical regime (where the RPA results are accurate) the accuracy of the LFC results shown in Fig. 2 does clearly exhibit the strong coupling nature of our procedure. Also, quite seperate from the problem of high temperature liquid metals, these results should be useful for providing reliable quantum corrections to other nearly-classical plasmas.

Finally, in Fig. 3 we compare the results of our calculations to the RPA results of Perrot and Dharma-wardana[11] in the intermediate degeneracy region. As in the degenerate and classical regions, the excessive negativity of the RPA free energy, attributable to its overaccounting for the exchange-correlation hole, is corrected. Because they compare favorably to zero temperature and classical Monte Carlo calculations, it is believed that these LFC calculations will also give an accurate description of the semidegenerate fermion thermodynamics.

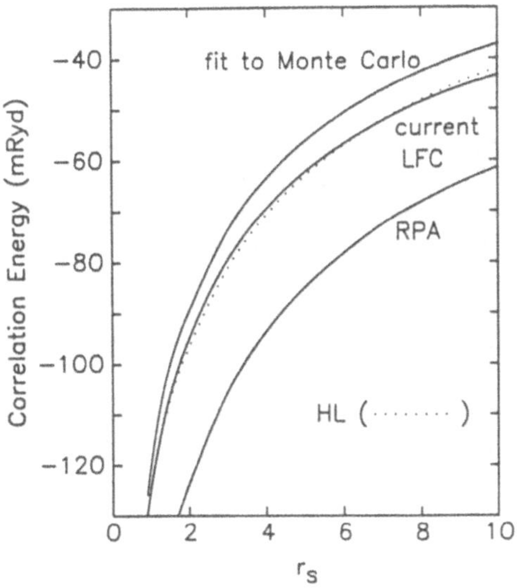

Figure 1. Correlation energy at T=0. Current LFC is from Eq. (17). Fit to Monte Carlo and RPA are from Ref. 22. Dotted curve marked HL is Hedin-Lundqvist interpolation form (Ref. 24).

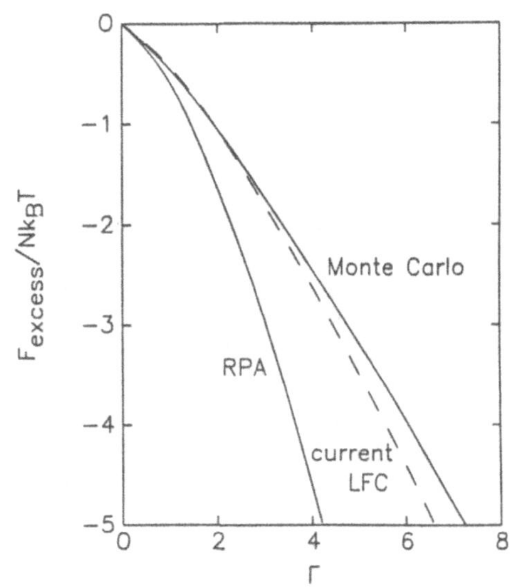

Figure 2. Excess free energy in classical regime, versus the classical coupling constant Γ. Current LFC is from Eq. (17). Monte Carlo results are from Ref. 23. RPA is classical Debye-Huckel formula.

In parallel with the ideal and exchange parts, a closed form fitting expression to $f_c(r_s,t)$, valid for all t and for r_s between 1 and 6, is presented in the appendix.

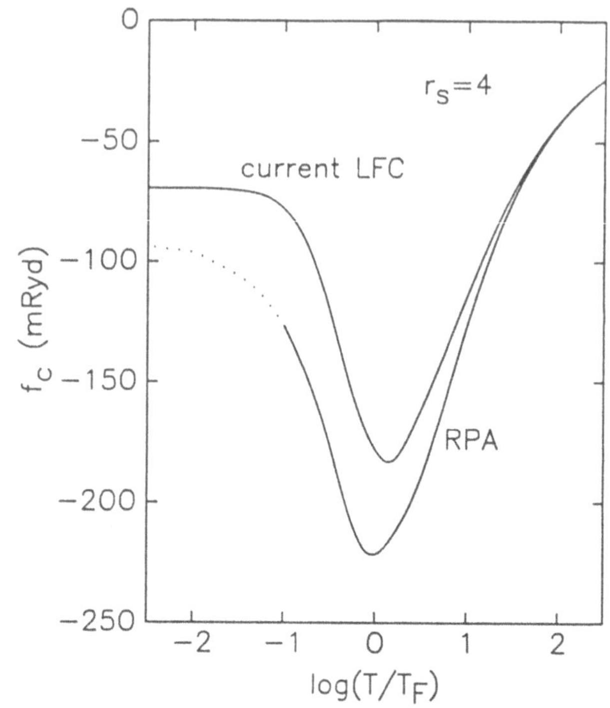

Figure 3. Correlation free energy at intermediate degeneracy for r_s=4. Current LFC is from Eq. (17). The RPA results for $T > 0.1\,T_F$ are from Ref. 11 (the dotted line for $T < 0.1\,T_F$ interpolates this to the $T=0$ RPA result).

III. The High Temperature EOS of Simple Metals

We now address the problem of calculating the metallic EOS at temperatures that range from those where the electronic system is degenerate ($T \ll T_F$), to those where the electrons are in the semidegenerate regime ($T \approx T_F$). Because we restrict out attention to simple metals, and as an initial approximation neglect the thermal ionization of core electrons, we treat the ions (nuclei plus core electrons) as unpolarizable particles, and consider the electron-ion interaction as describable by a local pseudopotential with fixed valence Z. The essence of this calculational scheme is to treat the pseudopotential perturbatively to obtain a state-dependent ionic pair potential, and then to evaluate the classical free energy of the ions variationally.

Following the method described by Ashcroft and Stroud,[26] the standard adiabatic approximation is invoked to reduce the metallic free energy to a form that describes a system of classical ions interacting via a state-dependent pair potential. Specifically, if second order pseudopotential theory is used to trace over the electronic degrees of freedom (i.e. if one assumes a *linear* response of the electrons to the "pseudo-ions" as a perturbation), the following pair potential expression is obtained for the Helmholtz free energy:

$$F(T,V,N) = F_e + F_v + \frac{3}{2}Nk_BT + \frac{1}{2}\sum_{i\neq j}\phi(R_{ij}) - TS_i\,, \tag{18}$$

where F_e is the free energy of a homogeneous electron gas (of density nZ and temperature T), F_v is a volume dependent term that contains the overall Hartree energy, and S_i is the entropy of a classical system of N ions interacting via a pair potential $\phi(r)$, with Fourier transform

$$\phi(q) = \frac{4\pi Z^2 e^2}{q^2} + |v_p(q)|^2\, \Pi(q)\,. \tag{19}$$

Here $\Pi(q) = \Pi(q,\omega{=}0)$ is the static polarizability of the electron gas, and $v_p(q)$ is the electron-ion pseudopotential. Note that the removal of the explicit electronic degrees of freedom from the problem has simply resulted in a screening of the bare Coulombic ion pair potential. The volume dependent term in Eq. (18) is

$$F_v = \frac{NZ^2 n}{2\,\hat{\Pi}(q{=}0)} + \frac{n}{2}\sum_{\mathbf{q}} |v_p(q)|^2\, \Pi(q) \equiv F_{v1} + F_{v2}\,. \tag{20}$$

Here $\hat{\Pi}$ is the irreducible part of the polarizability: $\Pi = \hat{\Pi}/(1-v_q\hat{\Pi})$.

From (18)-(20) it is seen that not only is an accurate description of $F_e(T,V,N)$ needed for proper evaluation of the metallic free energy at any degree of electronic degeneracy, but knowledge of $\Pi(q;n,T)$ is also needed. (The electronic contribution F_e was fully described in section II.) From the compressibility sum rule,[19] $\hat{\Pi}(q{=}0) = -n^2/B(r_s,t)$, where $B(r_s,t)$ is the electron gas bulk modulus. The first term of (20) is thus

$$F_{v1} = \frac{-NZ}{3\alpha^2 r_s^2}\,[1 + B_{xc}(r_s,t)/B_o(r_s,t)]\ \mathrm{Ryd}\,. \tag{21}$$

The functions $B_o(r_s,t)$ and $B_x(r_s,t)$ are given in Eq.'s (6) and (11), while $B_c(r_s,t)$ is obtained from the interpolation procedure described in section II.

In order to evaluate the ionic free energy in Eq. (18) and F_{v2} in Eq. (20), we have generalized to finite temperatures the zero temperature method described by Pettifor and Ward.[27] This method consists of using a constant core Heine-Abarenkov pseudopotential (where $v_p = -Ze^2/r$ for $r>r_c$, and $v_p = -\lambda Z e^2/r_c$ for $r<r_c$, λ and r_c being two constants) along with a rational fit to the static polarizability $\Pi(q)$. This fit is obtained by combining the Hedin-Lundqvist form[24] $G(q) = \gamma(r_s,t)Q^2$ (with $Q{=}q/2k_F$) for the LFC function in Eq. (15), with a Padé approximant for the ideal static polarizability:

$$v_q\Pi_o(q) \approx \frac{-\alpha r_s}{2\pi Q^2}\,\sqrt{t}\,I_{-1/2}(\eta_o)\,[1 + \sum_{j=1}^{4} a_{2j}(t)Q^{2j}]/[1 + \sum_{j=1}^{5} b_{2j}(t)Q^{2j}]\,. \tag{22}$$

The functions $b_i(t)$ are determined by considering the large and small q behavior of $\Pi_o(q)$, and the functions $a_i(t)$ are then obtained by fitting the form (22) to the exact $\Pi_o(q)$. This procedure is described in detail in the appendix. With these approximations for the pseudopotential and electronic polarization, F_{v2} and $\phi(r)$ can both be evaluated *in closed form,* and $\phi(r>2r_c)$ then turns out to be the sum of six complex Yukawa potentials, with exponential decay lengths and oscillation periods determined entirely by the poles of $\Pi(q)$.

In order to evaluate the ionic free energy throughout the range from $T = 0$ to $T{\approx}T_F$, we need to be able to treat both the solid and liquid metallic states. However, noting that the free energy of isochoric solid and liquid phases differ mainly in their respective entropic components, we evaluate the ionic pair potential energy by neglecting the long range positional order of the solid and using a liquid-like pair correlation function. This "supercooled-liquid" approximation has recently been applied to argon by Jones and Ashcroft.[28] Gibbs-Bogoliubov variational theory can then be used to evaluate the free energy in a fluid reference system, which for the choice of the hard sphere reference system reads

$$F = F_e + F_v + \frac{3}{2}Nk_BT + \min_{\eta}\left\{ \frac{N^2}{2V}\int d^3r\, \phi(r)\, g_{HS}(r;\eta) - TS_{HS}(\eta) \right\}, \tag{23}$$

where η is the packing fraction. The energy integral in (23) is then evaluated in closed form within a Perkus-Yevick approximation.[29] In the solid state the hard sphere entropy is obtained from molecular dynamics,[30] while in the liquid state we use the soft sphere model of Ross.[31]

In order to obtain EOS data, we first determine the two pseudopotential parameters r_c and λ by fitting the calculated room temperature isotherm to experimental data. The high temperature Hugoniot can then be straightforwardly calculated, as we show below for the case of aluminum.

IV. Application to Aluminum

Aluminum presents an ideal prototype system to test this EOS model both because of its well documented nature as a simple metal (whose triply charged ion cores remain small compared to the interparticle spacing even under substantial compression), and because of the wealth of experimental data available on it. It has, in fact, often been proposed as an EOS standard.[38] As mentioned above, in implementing this calculational scheme we first obtain the two pseudopotential parameters by fitting the calculated room temperature isotherm to experimental data.[32] For aluminum, this gives $r_c = 1.422a_o$ and $\lambda = 0.4246$. The resulting theoretical isotherm plotted in Fig. 4 is seen to agree quite well with experiment. Furthermore, the calculated value of the ground state binding energy of 1.374 Ryd/electron is very close to the experimental value of 1.38 Ryd/electron.[33]

In Fig. 5 we compare to experiment[2,34-36] our calculated aluminum Hugoniot curve, obtained by numerical solution of the standard Rankine-Hugoniot equation. In the insert is shown data up to 300 GPa obtained by impact loading, along with the experimental[37] and theoretical regions in which melting occurs. This region was calculated theoretically by plotting the Hugoniot pressures of both the solid and liquid phases against the melting curve as calculated by Moriarty, Ross and Young.[37] Up to 300 GPa, our calculated Hugoniot lies slightly above that calculated by Ross.[39] In the larger plot of Fig. 5, the Hugoniot curve is extended to 3 TPa, in order to compare to data obtained recently by nuclear explosive shock loading. The theoretical curve lies just under a recent data point[36] at 1.06 TPa, and is found to lie within the 5.4% error bars on compression about the Ragan data point[2] at 2.933 TPa. In general, the calculated Hugoniot agrees fairly well with experiment. The slight discrepancy in melting along the Hugoniot is to be expected, however, because of our neglect of the structural aspects of the free energy.

As mentioned earlier, accurate treatment of the semidegenerate nature of the electrons is critical in obtaining the proper Hugoniot pressures at the compressions considered in Fig. 5. To exhibit this, we have plotted in Fig. 6 our calculated values of the Hugoniot temperature versus compression. For example, consider the Hugoniot point at $V/V_o = 0.4$, where we calculate the pressure and temperature to be $p = 927$ GPa and $T = 48{,}300\ K = 0.194\ T_F$, where T_F is the electron Fermi temperature. The electronic contribution to the total pressure is calculated to be 374 GPa, with ideal,

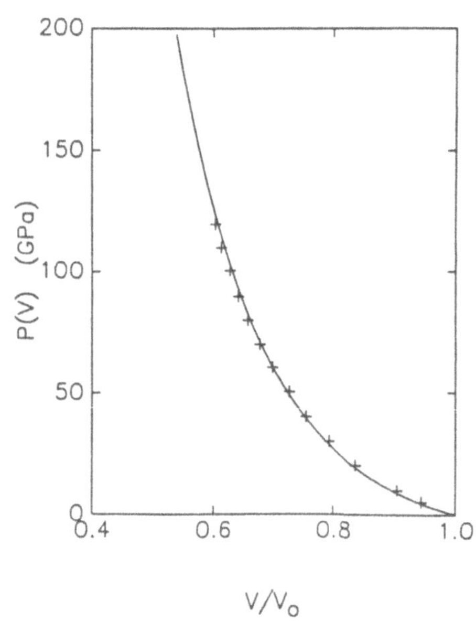

Figure 4. Fitted T=300K isotherm of aluminum. Theory (solid curve) vs. experiment (+'s) from Ref. 32.

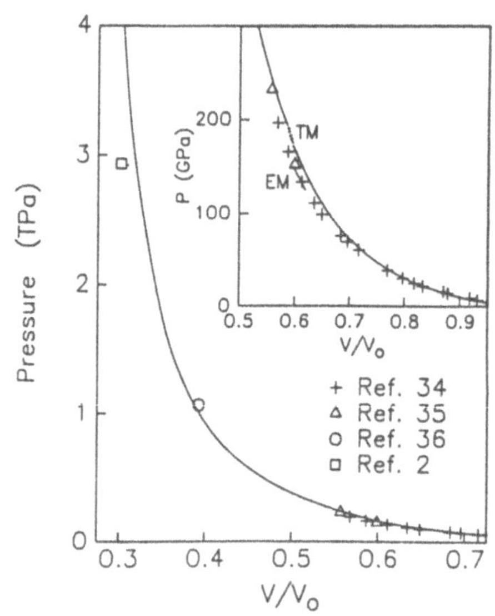

Figure 5. Aluminum Hugoniot. Theory (solid curve) vs. experiment. In the insert, TM and EM refer to the theoretical and experimental melting regions, respectively.

exchange and correlation parts $p_o = 710$ GPa, $p_x = -241$ GPa and $p_c = -95$ GPa. The exchange-correlation contribution is thus quite substantial. Furthermore, if finite temperature effects are taken into account only through the Sommerfeld T^2 correction to the ideal fermion contribution, an electronic pressure of 506 GPa is obtained, 132 GPa in excess of the value we calculate. As the temperature increases quite rapidly along the Hugoniot, these aspects of electron nondegeneracy become of increasing importance at higher pressures. At $V/V_o = 0.3$, we find $T = 0.77T_F$. At temperatures of this magnitude however, the core electrons cannot be assumed to remain in their ground states,[3] and the simplified theory presented here is inadequate.

At the extremes of compression, some error arises in our procedure because of the fact that our pair potential is expressable in complex Yukawa form (for which the Perkus-Yevick hard sphere results are applicable) only for $r>2r_c$. This is an artifact of the discontinuous form of the pseudopotential chosen. At high compression (for aluminum, when $V/V_o<0.38$) it turns out that the variational free energy described in section III is minimized at a packing fraction η that requires knowledge of $\phi(r<2r_c)$ to evaluate the energy integral in Eq. (23). Current work is thus concerned with extending this model to these higher compressions.

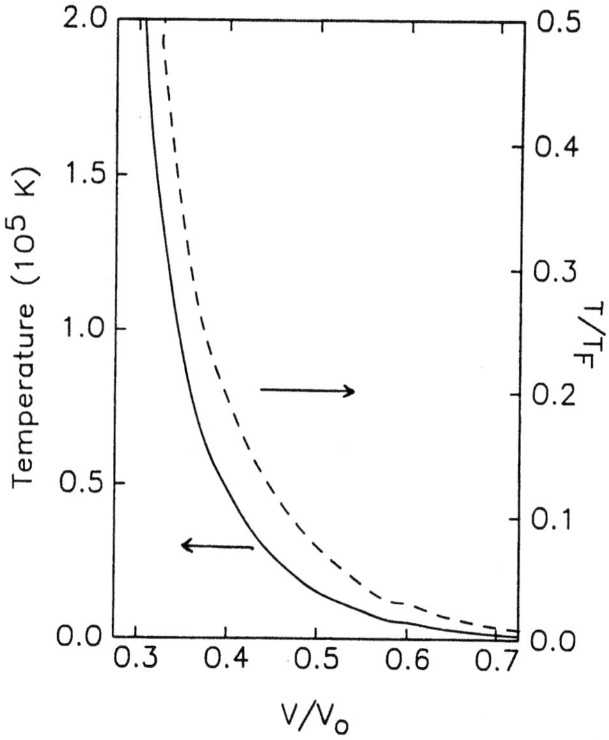

Figure 6. Calculated temperature along the aluminum Hugoniot. Note that melting is calculated to occur between $V/V_o = 0.585$ and 0.596.

V. Conclusion

The goal of this work has been to extend standard pseudopotential theory for the EOS of simple metals to temperatures where the electron system is not necessarily degenerate, and to apply this extended theory to the ultrahigh pressure Hugoniot of aluminum. Through an exact evaluation of the temperature dependent ideal fermion and exchange contributions, and an approximate treatment of correlation effects, we have presented a complete description of the thermodynamic and dielectric properties of an interacting electron system at any degree of degeneracy. We are thus able to obtain the metallic EOS at high temperatures by combining a perturbative pseudopotential treatment of the electron-ion interaction with a classical variational treatment of the ionic degrees of freedom. This calculational scheme should be accurate up to temperatures where thermal excitation of the core electrons becomes significant, and should thus bridge the gap between zero temperature EOS models and computer models[40] or Thomas-Fermi-Dirac type models valid at extremely high temperatures.

VI. Appendix

In this section we present the analytic fitting expressions that allow a closed form description of the electron gas thermodynamic and dielectric properties as functions of density and temperature, as described in section II. We begin by presenting the promised fits to the Fermi integrals $I_\nu(\eta_o)$ as funtions of $t = T/T_F$. By analyzing the low temperature behavior of $I_\nu(\eta_o)$ via a Sommerfeld expansion, and the high temperature behavior by a fugacity expansion, we are lead to the following Padé approximants, valid for all values of t:

$$I_{-1/2}(\eta_o) \approx \frac{2}{\sqrt{t}} \frac{1 + c_1t^2 + c_2t^4 + c_3t^6}{1 + (c_1+\pi^2/12)t^2 + c_4t^4 + (c_3/\sqrt{2\pi})t^{11/2} + (3c_3/2)t^7} , \tag{A1}$$

$$I_{-1/2}^{\prime}(\eta_o) \approx \sqrt{t} \frac{1 + c_1t^2 + c_2t^4 + c_3t^7}{1 + (c_1-\pi^2/6)t^2 + c_4t^4 + c_5t^6 + (3\sqrt{2}c_3/4\sqrt{\pi})t^{15/2} + (3c_3/4)t^9} , \tag{A2}$$

$$I_{3/2}(\eta_o) \approx \frac{2}{5}t^{-5/2} \frac{1 + c_1t^2 + c_2t^4 + c_3t^8}{1 + (c_1-5\pi^2/12)t^2 + c_4t^4 - (2c_3/15\sqrt{2\pi})t^{11/2} + (2c_3/5)t^7} \tag{A3}$$

and

$$I_{5/2}(\eta_o) \approx \frac{2}{7} t^{-7/2} \frac{1 + c_1t^2 + c_2t^4 + c_3t^{15/2}}{1 + (c_1-7\pi^2/6)t^2 - (\sqrt{2}c_3/35\sqrt{\pi})t^4 + (4c_3/35)t^{11/2}} . \tag{A4}$$

The constants c_i are given in Table 1. The forms (A1)-(A3) are found to be accurate to within 0.1%, while (A4) is accurate to within 1%.

A convenient fitting expression for our calculated values of the correlation contribution to the electron gas free energy is (in Rydbergs per electron)

$$f_c(r_s,t) \approx \frac{\varepsilon_c^o + d_1t^2 + d_2t^3 + d_3t^4}{1 + d_4t^2 + d_5t^3 - d_3\sqrt{3r_s/8\alpha^2}\, t^{9/2}} . \tag{A5}$$

Here $\varepsilon_c^o(r_s)$ is a Hedin-Lundqvist type[24] of interpolation to our calculated zero temperature correlation energy

$$\varepsilon_c^o(r_s) \approx -0.0448 \left[(1+y^3)\ln(1+1/y) + y/2 - y^2 - 1/3\right] , \tag{A6}$$

where $y = r_s/20.52$, and the functions $d_j(r_s)$ are given by

$$d_1(r_s) = c_1 + c_2r_s^{-c_3} \tag{A7}$$

$$d_2(r_s) = c_1 + c_2r_s^{-c_3} + c_4r_s^{-c_5} \tag{A8}$$

$$d_3(r_s) = c_1 + c_2r_s^{-c_3} + c_4r_s^{-c_5} \tag{A9}$$

$$d_4(r_s) = \frac{c_1 + c_2r_s + c_3r_s^2 + c_4r_s^3 + c_5r_s^4}{1 + c_6r_s^2 + 0.02206r_s^5} \tag{A10}$$

$$d_5(r_s) = (r_s+c_1)\,(c_2+c_3r_s^{-c_4}) \tag{A11}$$

The constants c_j are given in Table 1. This match to f_c reproduces our calculations for all t and for r_s between 1 and 6 with a maximun error of 3%. The largest errors in this parametrization occur for small t (t=0.05 to 0.30), because the form (A5) neglects the weak $t^2\ln(t)$ singularity in f_c, known to cancel a similar singularity in the exchange free energy.[11] Note that at large t it reduces to the Debye-Hückel form.

Finally, we present the closed form fitting expression used in section III to describe the electron gas static polarizability. This is obtained by relating the interacting polarizability to the noninteracting one via Eq. (15), and using the Padé approximant for $\Pi_o(q)$ given in Eq. (22). By analyzing the large q behavior of $\Pi_o(q)$, the functions $b_j(t)$ there are related to the functions $a_j(t)$ by

TABLE 1. Fitting parameters used to describe the finite temperature thermodynamic and dielectric properties of the electron gas. In the column marked c_j is the equation in which it appears.

c_j	j=1	j=2	j=3	j=4	j=5	j=6
(A1)	41.775	27.390	4287.2	50.605		
(A2)	2.2277	126.92	5248.0	97.720	861.30	
(A3)	5.3588	-2.5433	432.89	1.8800		
(A4)	-8.6164	-357.41	5711.1			
(A7)	0.14459	-5.4210	0.90023			
(A8)	-8.2583	11.196	0.13012	12.308	2.2634	
(A9)	0.01961	-13.837	2.2038	-4.6465	44.318	
(A10)	-328.63	316.63	99.064	8.0679	0.32713	36.371
(A11)	-0.61061	0.09812	-15.633	1.6810		
(A19)	3248.8	-691.47	-3202700	-4535.6	-462400	
(A20)	-4.8780	473.25	-2337.5	348.31	1517.3	
(A21)	-0.2280	0.4222	-0.6466	0.70572	5.8820	
(A22)	-3.0375	64.646	19.608	-96.978	423.66	-331.01
(A23)	-0.1900	0.36538	-2.2575	22.942	-43.492	106.40
(A24)	-7.1316	22.725	58.092	-436.02	-826.51	4912.9

$$b_{10}(t) = \frac{3}{2}\sqrt{t}\, I^o_{-1/2} a_8 \,, \tag{A12}$$

$$b_8(t) = \frac{3}{2}\sqrt{t}\, I^o_{-1/2} a_6 - \frac{1}{2} t^{5/2}\, I^o_{3/2} b_{10} \tag{A13}$$

and

$$b_6(t) = \frac{3}{2}\sqrt{t}\, I^o_{-1/2} a_4 - \frac{1}{2} t^{5/2}\, I^o_{3/2} b_8 - \frac{3}{10} t^{7/2}\, I^o_{5/2} b_{10} \,, \tag{A14}$$

where we have set $I^o_\nu \equiv I_\nu(\eta_o)$. Similarly, the small q behavior of $\Pi_o(q)$ implies that

$$b_2(t) = a_2 + 2J(t)/(3\sqrt{t}\, I^o_{-1/2}) \tag{A15}$$

and

$$b_4(t) = b_2^2 - a_2 b_2 + a_4 + 2K(t)/(15\sqrt{t}\, I^o_{-1/2}) \,, \tag{A16}$$

where

$$J(t) \equiv -\int_0^\infty dx\, x^{-2}\, [f(x) - f(0)] \,, \tag{A17}$$

$$K(t) \equiv -3\int_0^\infty dx\, x^{-4}\, [f(x) - f(0) - \frac{1}{2} x^2 f''(0)] \,, \tag{A18}$$

and $f(x) \equiv [1 + \exp(x^2/t - \eta_o)]^{-1}$. Analysis at high and low temperature leads to the following approximations for $J(t)$ and $K(t)$:

$$J(t) \approx \frac{1 + c_1 t^2 + c_2 t^4 + c_3 t^7}{1 + (c_1 - \pi^2/6)t^2 + c_4 t^4 + c_5 t^6 + (3\sqrt{2}c_3/4\sqrt{\pi})t^{15/2} + (3c_3/4)t^9} \tag{A19}$$

and

$$K(t) \approx \frac{1 + c_1 t^2 + c_2 t^4 + c_3 t^7}{1 + (c_1 - 3\pi^2/4)t^2 + c_4 t^4 + c_5 t^7 - (7\sqrt{2}c_3/8\sqrt{\pi})t^{17/2} - (3c_3/8)t^{10}} . \tag{A20}$$

Finally, the functions $a_i(t)$ are expressed as

$$a_2(t) \approx \frac{c_1 + t}{c_2 + c_3 t^{c_4} + c_5 t^2} , \tag{A21}$$

$$a_4(t) \approx \frac{1 + c_1 t + c_2 t^2}{c_3 + c_4 t + c_5 t^2 + c_6 t^3 + 20.833 c_2 t^4} , \tag{A22}$$

$$a_6(t) \approx \frac{c_1 + t}{c_2 + c_3 t + c_4 t^2 + c_5 t^3 + c_6 t^4} \tag{A23}$$

and

$$a_8(t) \approx \frac{0.91 - 6.4453t + 12.2324t^2}{1 + c_1 t + c_2 t^2 + c_3 t^3 + c_4 t^4 + c_5 t^5 + c_6 t^6} . \tag{A24}$$

The constants c_j in (A19)-(A24) appear in Table 1. By testing this approximation for $\Pi_o(q)$ over a wide range of t and q, the maximum error involved is found to be 2.5%. However, this occurs at q values sufficiently large that Π_o has fallen to about 10% of its maximum q=0 value; the fit is actually much better over the range of q where $\Pi_o(q)$ is substantial.

ACKNOWLEDGEMENTS

Part of this work (on the homogeneous electron gas) was carried out in conjunction with Anders Carlsson at Washington University (St. Louis). We gratefully acknowledge his help and the support of the Washington University physics department. This work was supported by the Army Research Office under grant DAAG-29-82-K-0170.

REFERENCES

[1] A.S.Vladimirov *et al*, JETP Lett. 39, 82 (1984).

[2] C.E.Ragan, Phys. Rev. A 29, 1391 (1984).

[3] B.K.Godwal, S.K.Sikka and R.Chidambaram, Phys. Rev. Lett. 47, 1144 (1981).

[4] D.A.Young and M.Ross, Phys. Rev. B 29, 682 (1984).

[5] A preliminary account of part of this work has appeared in the following: R.G.Dandrea and N.W.Ashcroft, J. de Phys. Colloq. (Supp.) 45, C8 (1984).

[6] A more detailed account of the work in this section, done in collaboration with A.E. Carlsson, will soon be published elsewhere.

[7] B.Rozsnyai, Phys. Rev. A 5, 1137 (1972).

[8] M.A.Pokrant, Phys. Rev. A 16, 413 (1977).

[9] U.Gupta and A.K.Rajagopal, Phys. Rev. A 22, 2792 (1980).

[10] W.Stolzmann and W.D.Kraeft, Ann. Phys. (Leipzig) 36, 388 (1979).

[11] F.Perrot and M.W.C.Dharma-wardana, Phys. Rev. A 30, 2619 (1984).

[12] J.Hubbard, Proc. Roy. Soc. (London) 243, 336 (1957).

[13] D.J.Geldart and S.H.Vosko, Can. J. Phys. 44, 2137 (1966).

[14] A.L.Fetter and J.D.Walecka, *Quantum Theory of Many Particle Systems* (McGraw Hill, New York, 1971), p.231.

[15] Reference 14, p.269.

[16] B.Horovitz and R.Thieberger, Physica 71, 99 (1974).

[17] D.Pines and P.Nozieres, *The Theory of Quantum Liquids* (W.A.Benjamin, New York, 1966), p.136.

[18] S.Ichimaru, Rev. Mod. Phys. 54, 1017 (1982).

[19] Reference 17, p.209.

[20] A.E.Carlsson and N.W.Ashcroft, Phys. Rev. B 25, 3474 (1982).

[21] P.Vashishta and K.S.Singwi, Phys. Rev. B 6, 875 (1972).

[22] S.H.Vosko, L.Wilk and M.Nusair, Can. J. Phys. 58, 1200 (1980).

[23] J.P.Hansen, Phys. Rev. A 8, 3096 (1973).

[24] L.Hedin and B.I.Lundqvist, J. Phys. C 4, 2064 (1971).

[25] H.E.DeWitt, J. Math. Phys. 3, 1216 (1962).

[26] N.W.Ashcroft and D.Stroud in *Solid State Physics,* ed. by H.Ehrenreich, F.Seitz and D.Turnbull (Academic Press, New York, 1978) vol. 33, p. 1.

[27] D.G.Pettifor and M.A.Ward, Solid State Comm. 49, 291 (1984).

[28] R.S.Jones and N.W.Ashcroft, J. Chem. Phys. 80, 3328 (1983).

[29] H.D.Jones, J. Chem. Phys. 55, 2640 (1971).

[30] B.J.Alder, W.G.Hoover and D.A.Young, J. Chem. Phys. 49, 3688 (1968).

[31] M.Ross, J. Chem. Phys. 71, 1567 (1979).

[32] R.G.McQueen *et al,* in *High Velocity Impact Phenomena,* ed. by R.Kinslow (Academic Press, New York, 1970) p.530.

[33] N.W.Ashcroft and D.C.Langreth, Phys. Rev. 155, 682 (1967).

[34] A.C.Mitchell and W.J.Nellis, J. Appl. Phys. 52, 3371 (1981).

[35] L.V.Al'tshuler *et al,* Sov. Phys. JETP 11, 766 (1960).

[36] A.P.Volkov *et al,* Sov. Phys. JETP Lett. 31, 588 (1980).

[37] J.A.Moriarty, D.A.Young and M.Ross, Phys. Rev. B 30, 578 (1984).

[38] A.C.Mitchell *et al,* in *Shock Waves in Condensed Matter - 1983,* ed. by J.R.Asay, R.A.Graham and G.K.Straub (Elsevier Science Publishers, Amsterdam, 1984), p.81.

[39] M.Ross, Phys. Rev. B 21, 3140 (1980).

[40] D.A.Young, J.K.Wolford, F.J.Rogers and K.S.Holian, Phys. Lett. 108A, 157 (1985).

SHOCK HUGONIOT EQUATION OF STATE - ELECTRON BAND THEORY APPROACH

S.K. Sikka

Neutron Physics Division
Bhabha Atomic Research Centre
Trombay, Bombay - 400 085, India

INTRODUCTION

Of the various high pressure experimental data on a given material, the equation of state (EOS) is most amenable to theoretical treatment. Its interpretation involves, in principle, the straightforward application of electron band and lattice dynamical techniques for the calculation of pressure and internal energy as a function of volume and temperature.[1] Though these formulations have been available since 1930's, it is only during the last decade that considerable progress has been made in EOS computations by using these. Two reasons can be attributed for this advance. One is the considerable development of experiments involving lasers, electrically driven foils and underground nuclear explosions to study EOS at pressures above 100 GPa[2-5]. Very recently, experimental data on iron and aluminium shocked to ~ 500 TPa have been reported[6]. These data have opened up a new area in the P-V-T surface for theoretical research, where the pressure and thermal ionisation effects are important. The second is the advent of large and fast computers and improvements in energy band methods themselves. These have made computations less time consuming and have permitted evaluation of total energies of crystalline solids to such a precision that not only can bulk properties like EOS be calculated accurately but energy differences between different phases can be determined to an accuracy of 0.1 ev per atom or better. It is now possible to obtain various properties assuming nothing more than the Born-Oppenheimer approximation and the accuracy of the local density approximation in the density functional formalism. Such EOS calculations have been performed for metals, semiconductors and insulators. Valuable insights have been obtained relating various anomalous features found in EOS to their microscopic physical origin. Here we intend to highlight some of these developments. First principle equation of state theories of condensed matter upto about 10TPa have been recently reviewed by us[7]. For more details, the reader should refer to this review. In the ensuing discussion, we shall restrict ourselves to the calculations of shock Hugoniot equation of state as most of the experimental data is this form.

GENERAL FORMULATION

Calculation of the shock Hugoniot requires solution of the equation

$$E-E_0 = 0.5(P+P_0)(V_0-V) \qquad \text{--(1)}$$

where P, V and E are the pressure, volume and internal energy respectively. The subscript refers to the initial state which for most cases is 1 atmosphere and 300 K. The variables P, V and E are related to each other through the Helmholtz free energy of the system

$$F = E - TS \quad \text{--(2)}$$

$$P = -(\partial F/\partial V)_T \quad \text{--(3)}$$

It is now well established that for most EOS work the total pressure and internal energy, under the neglect of electron-phonon interactions, can be separated into (i) a zero degree Kelvin contribution, (ii) a nuclear thermal vibration part and (iii) an electronic excitation component[8] (However, see caption of Fig. 7 below). Upto a pressure of ~5TPa along the shock Hugoniot, the upper limit of pressure considered in this paper, the various physical phenomena encountered are band reorderings, band gap closures, melting and pressure and thermal ionisation of core electrons. The theories described below appear to be adequate for these to about 10 percent within the experimental data. These may break down in other regions of pressure-volume-temperature surface.

DENSITY FUNCTIONAL THEORY OF 0 K ISOTHERM

In mid-sixties, Hohenberg and Kohn[9] and Kohn and Sham[10] showed that the ground state electronic energy of a many body system is a unique functional of the electron density and derived an effective one electron Schrodinger equation of the form

$$[-\nabla^2 + V_H(r) + V_{xc}(r)]\phi_a(r) = \varepsilon_a \phi_a(r) \quad \text{--(4)}$$

The electron density is related to the orbitals $\phi_a(r)$ by the formula

$$\rho(r) = \sum |\phi_a(r)|^2 \quad \text{--(5)}$$

The potential function

$$V_H(r) = 2\int \rho(r')/|r-r'|\, d^3r' - 2\sum Z_u/|r-R_u| \quad \text{--(6)}$$

where Z_u is the atomic number of the atom at position R_u. The exchange correlation potential

$$V_{xc}(r) = \partial E_{xc}/\partial\rho(r) \quad \text{--(7)}$$

and $$E_{xc} = \int \rho(r)\varepsilon_{xc}(r)\, d^3r \quad \text{--(8)}$$

where ε_{xc} is the exchange-correlation energy per electron, whose general form is not known. It is now conventional to make for this, the local density approximation by substituting the form for the energy of an interacting homogeneous electron gas. Thus, once Z_u's are specified, it is possible to solve the eigenvalue problem. Then the total energy at 0 K can be written as

$$E_c = \sum \varepsilon_a + \iint \rho(r)\rho'(r)/|r-r'|\, d^3r\, d^3r'$$
$$- \int \rho(r)[\varepsilon_{xc}(\rho(r)) - V_{xc}(\rho(r))]\, d^3r$$
$$+ \sum Z_u \sum Z_v / |R_u - R_v| \quad \text{--(9)}$$

An expression for cold pressure can then be obtained by

$$P_c = - dE_c/dV \qquad --(10)$$

or through the use of the virial theorem[11].

$$3P_cV = 2E_{ke} + U - 3\int \varrho(r)\,(\varepsilon_{xc} - V_{xc})\, d^3r \qquad --(11)$$

where E_{ke} and U are the kinetic and potential energy parts.

Equation (9) is more amenable to the pseudopotential method which substitutes a small effective electron-ion interaction (second term in eq.6) for the true one and converts the problem to that of a nearly free electron one. One can then use perturbation theory and write down the total cold energy, upto second order, as

$$E_C = Z\,(E_k + E_x + E_{CO}) - \alpha_M Z^2/R_A + E1 + E2 \qquad --(12)$$

Here E_k, E_x and E_{co} represent the kinetic, exchange and correlation energies of a homogeneous electron gas. The second term is the electrostatic energy, a function of the atomic radius R_a ($4\pi/3R_a$ = V/atom). α_M is the Madelung constant. E1 and E2 are the first and second order band structure energies. Detailed expressions for both solids and liquids may be found in our review[7]. This method is very fast to apply. However, this method is applicable to simple metals. Generalisations of the pseudopotential method for applications to both sp and d-band metals have been given by Ihm et al[12] and Moriarty[13].

The virial expression (11) has been used with band structure methods based on muffin-tin approximation for solids. In this, the electron charge density in each unit cell is replaced by its spherical average within an inscribed sphere about an ion and by a constant in the rest of the region. In APW and KKR methods, because it is difficult to find wave functions that are continuous across the boundary between two regions, the eigenvalue matrix becomes energy dependent. A trial and error method is resorted to find the desired eigenvalue in which the matrix is recalculated for each trial energy. This makes these methods rather time consuming. In 1975, Andersen[14] found that the wave function could be approximated with a special energy-independent function. This made it possible to construct the matrix once and then diagonalize it to obtain the eigenvalues with two orders of magnitude saving in time and with only slight loss of accuracy. Computer programms based on this formulation as applied to the KKR method (known as LMTO method) are now widely available[15]. One particular advantage with these muffin-tin methods (applicable to all kinds of electrons) is the decomposition of total pressure into individual s,p,d,f contributions. This is valuable for understanding the trends in cohesive properties.

LATTICE THERMAL CONTRIBUTION

For solids, it has been conventional to compute this contribution by quasiharmonic Mie-Gruneisen equation based on Debye's theory

$$E_{lt} = 3k_\beta T\, D_e(\theta_D/T) \qquad --(13)$$

$$\approx 3k_\beta T \text{ for } T>\theta_D \qquad --(14)$$

$$P_{lt} = \Gamma(V)/V\, E_{lt} \qquad --(15)$$

Here $D_e(\theta_D/T)$ is the Debye function and θ_D is the Debye temperature. Γ is the Grüneisen constant

$$\Gamma(V) = \langle - d \ln \nu_i / d \ln V \rangle \qquad \text{--(16)}$$

where $\nu_i(V)$'s are the lattice frequencies. In most EOS work, this has been estimated by Slater or Dugdale-MacDonald or free volume theory expressions, in terms of the zerodegree isotherms or a PV curve, appropriate to shocked state of the sample, derived from finite temperature band structure calculations.

However, more recently, Born-von Karman theory of lattice dynamics has been employed directly to evaluate the phonon free energy. Under harmonic approximation, the phonon frequencies are now given by

$$\left| D_{xy}(ll',q) - \nu^2 I \right| = 0 \qquad \text{--(17)}$$

where the dynamical matrix D is related to the potential energy function, which is again evaluated at 0 K. A comparison of such computed Γ values with the experimental data of Neal[16] on Be is shown in Fig. 1. It may be seen that though the precision of the experimental data is not very high (these are difficult experiments to do), the theoretical Γ's are in better quantitative agreement with the experimental data. One recent development for computing lattice frequencies for solids must be noted here. This is the so-called frozen-phonon approach[17A]. In this, one calculates the frequencies from the curvature of the total energy vs displacement curve. Such calculations have been done for semi-conductors, simple metals and transition metals. However, this method has not yet been applied for EOS work.

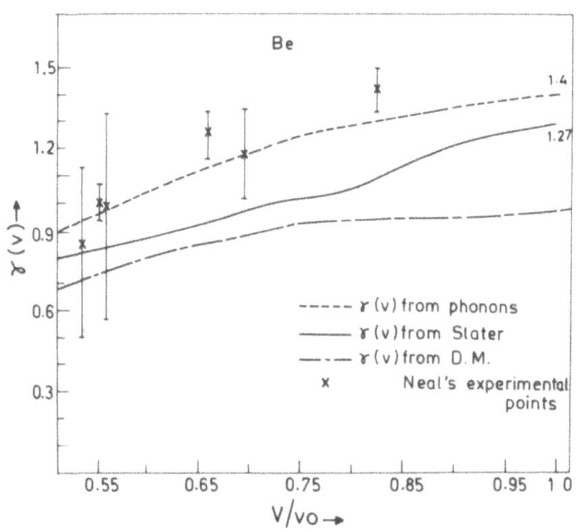

Fig.1 Gruneisen Γ versus V/V_0 for Be.

For liquids, variational theories based on Gibbs-Bogolyubov inequality have been developed.

$$F < F_0 + \langle \delta U \rangle_0 \qquad \text{--(18)}$$

Here F_0 is the free energy of a reference system and U is the potential energy of the real system. $\langle \delta U \rangle_0$ is the difference between the actual and reference potential energies, taken over all reference configurations. Various reference systems employed for liquid EOS calculations include hard sphere, soft sphere and one component plasma fluids. A comparison of all these reference systems has been, recently, presented for Al by Moriarty et al[18].

THERMAL ELECTRONIC CONTRIBUTION

This term begins to become appreciable at temperatures around 5000 K. In most band structure EOS calculations, the following model is used[19-20]. In this, the change in electronic internal energy is assumed to be caused by changes in 0 K one electron level occupancies

$$\Delta E_e = \sum \varepsilon_k(V)[n_k(T)-n_k(0)] \qquad \text{--(19)}$$

Here $n_k(0)$ and $n_k(T)$ are the occupancies at 0 K and temperature, T. The latter is now given by the Fermi - Dirac distribution function

$$n_k(T) = 1/[\exp((\varepsilon_k-\mu)/k_\beta T)+1] \qquad \text{--(20)}$$

Here μ is the chemical potential, which is fixed by conserving the total number of electrons. Using the entropy appropriate to the Fermi liquid, and differentiating the corresponding Helmholtz free energy, the electronic thermal pressure is

$$P_e = -\sum \delta\varepsilon_k/\delta V\ [n_k(T)-n_k(0)] \qquad \text{--(21)}$$

The validity of this model has been checked by MacMahan et al[20], who find that this gives excellent agreement with finite temperature SC-APW calculations for metallic iodine upto $T/T_F \approx 0.2$. Godwal[21] has used this model to check the earlier procedure employed in shock wave data interpretation, namely

$$\Delta E_e = 1/2\ \beta(V)T^2 \qquad \text{--(22)}$$

$$\text{and}\quad P_e = \gamma_e/V\ \Delta E_e \qquad \text{--(23)}$$

where β is the electronic specific heat and $\gamma_e = 0.5$, the electronic Gruneisen constant. Fig. 2 shows the variation of γ_e with T for Pb and Th, using equations 19-21. The calculation for Pb was based on the pseudopotential technique, while RAPW method was used for Th. It can be seen that for Th, γ_e lies between its limiting values of 2/3 in the Fermi-Dirac limit and 1/3 in an infinite atom limit. But for Pb the values are higher as well as the temperature variation is in opposite direction. The trends are similar for Al[22] and metallic iodine[20]. This suggests that the earlier procedure of setting $\gamma_e = 0.5$ is suspect, especially so for deriving the Hugoniot termperatures from experimental PV data.

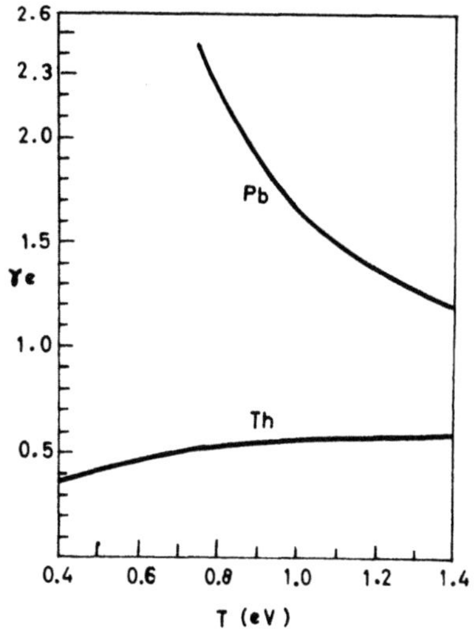

Fig.2 γ_e variation of Pb and Th versus T.

SOME RESULTS UPTO 200 GPa

Detailed shock Hugoniot calculations using, the above methods, have now been done for a number of materials (see Table 12 in reference 7). Some of these are compared with experimental data in figures 3 to 7. The agreement between theory and experiment is generally good. Some general points to be noted from these calculations are:

(1) The theory is able to discriminate between the various sets of experimental data, if they differ. The examples are in Al, iodine and Be[26]. In Al, a third set of data[27] lent further support to theoretical EOS because the two earlier sets were not in mutual agreement, one of them being much closer to the calculations.

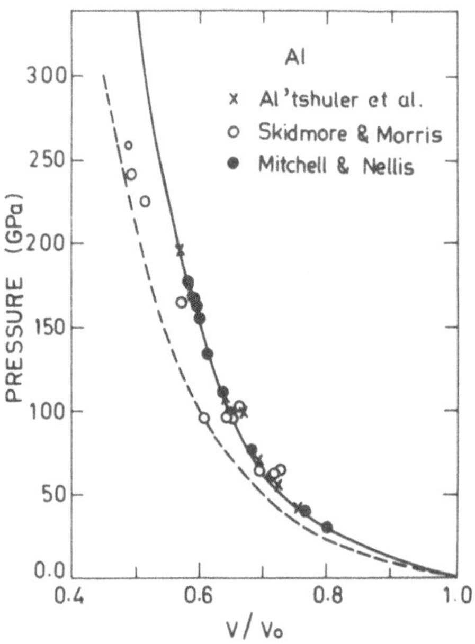

Fig.3 Shock Hugoniot of Al using solid state theory[22].

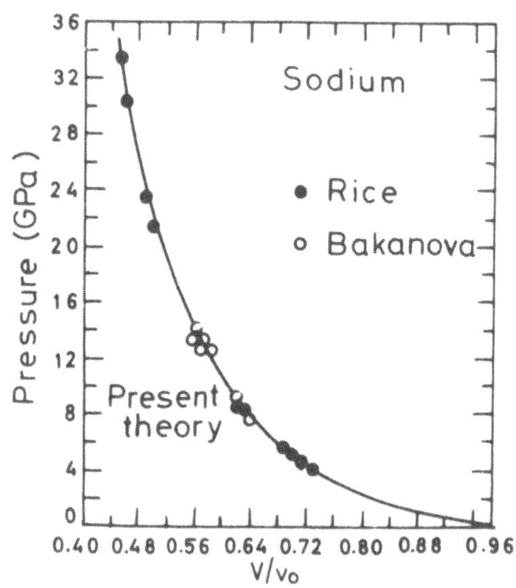

Fig.4 Liquid state shock Hugoniot calculation for Na[23].

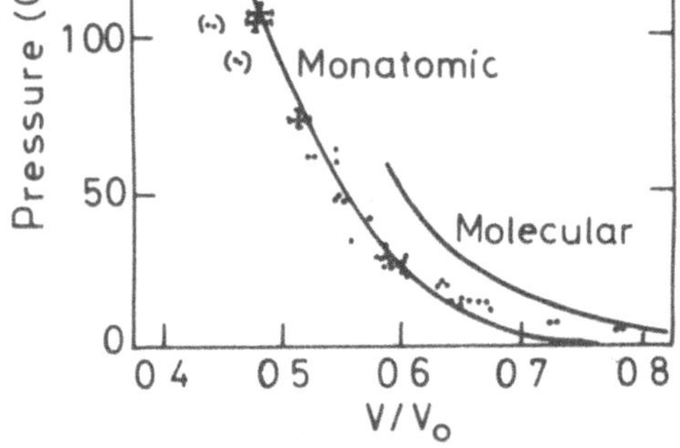

Fig.5 Iodine Hugoniot curve[20].

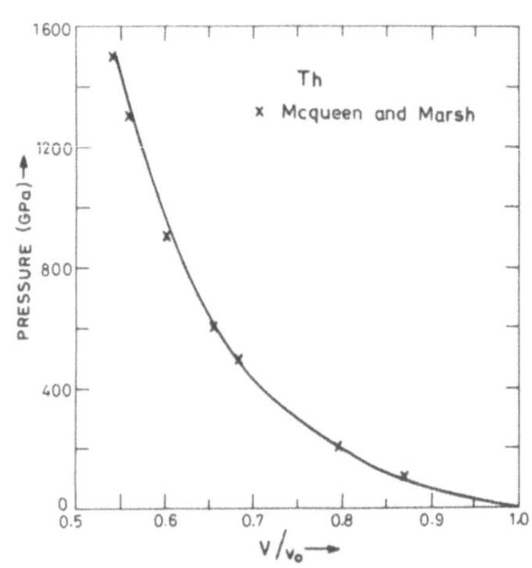

Fig.6 Shock Hugoniot of Th[24].

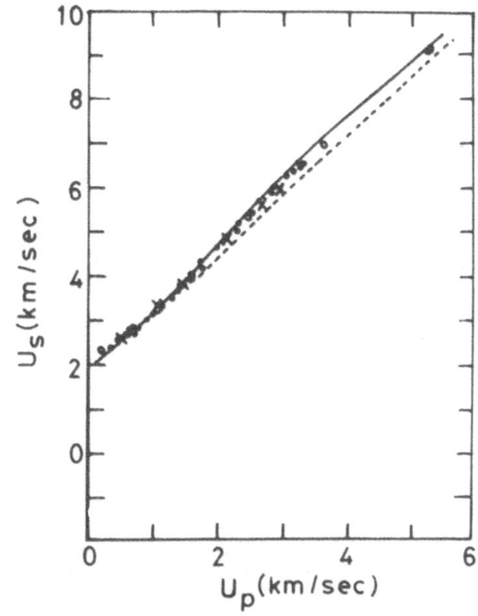

Fig.7 $U_s - U_p$ curve for La[25]. The dashed curve is for the temperature independent linear Γ. The solid curve is for temperature dependent Slater Γ. This implies a coupling between the electronic excitation and the latttice contributions.

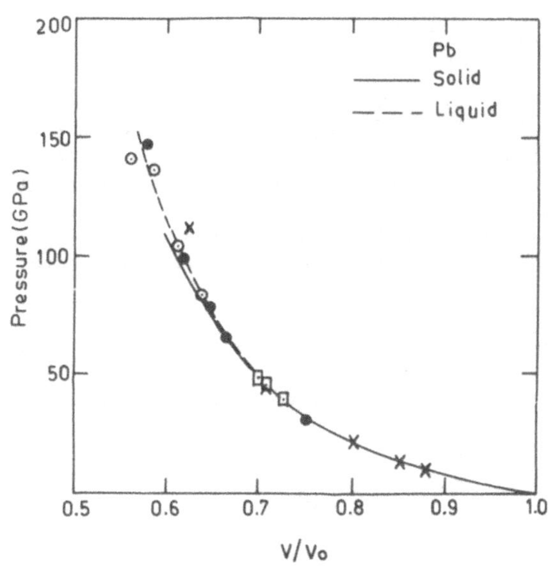

Fig.8 Shock Hugoniot for Pb in solid and liquid phases.

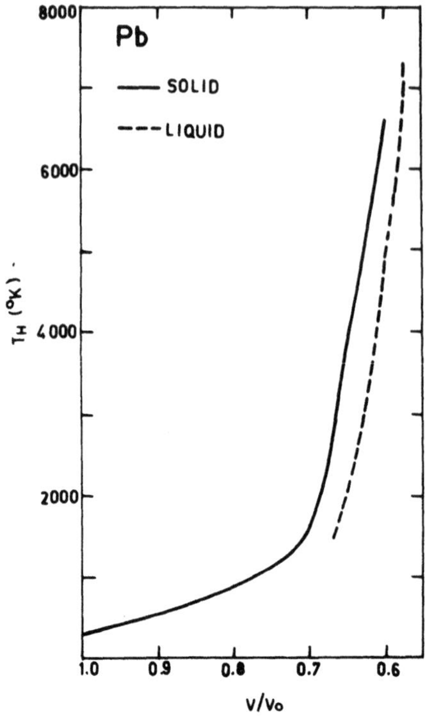

Fig.9 Shock temperature for Pb for solid and liquid calculations.

(2) The electronic potential for monoatomic solids can be split into

$$V(q) = s(q)\, V^{ion}(q) \qquad --(24)$$

where s(q) is the structure factor, which represent the geometrical structure and $V^{ion}(q)$, the ionic potential. The results on iodine, Be and La show that EOS is insensitive to s(q). It may be pointed out that in all these investigations fcc lattice has been assumed in place of true ones. This conclusion is re-enforced by our results on Pb, using both solid and liquid state theories (Fig.8). However, the calculated shock temperatures are quite different (Fig.9). The need for shock temperature measurements is very obvious.

(3) In some of the pseudopotential EOS calculations, the parameters used in electron-ion potentials have been fixed from experimental Fermi-surface data, phonon frequencies etc. at normal pressure. Apart from modifications due to pressure and thermal ionisation (see next section), these did not need any adjustments right upto the highest pressures reported. This justifies the earlier, widely prevalent, procedure of deriving EOS from zero pressure quantities like lattice constant, bulk modulus and its pressure derivative or Gruneisen constant. In fact the energy band methods do quite well (within 5 to 10 percent of experimental data) in predicting these (see Fig.10-11 and Table 1). Because of this, the equation of state data have some sometimes been used to derive the potential functions[29,30] and used to compute other thermodynamic quantities like Gruneisen parameter, melting temperature etc.

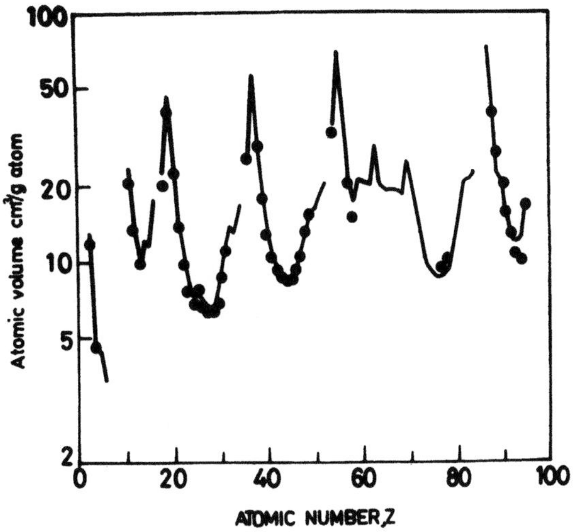

Fig.10 Zero pressure atomic volumes of the elements versus the atomic number. The experimental values lie along the quasi continuous lines. Those calculated by band theory are represented by points. The average discrepancy between theory and experimental data is 6 percent.

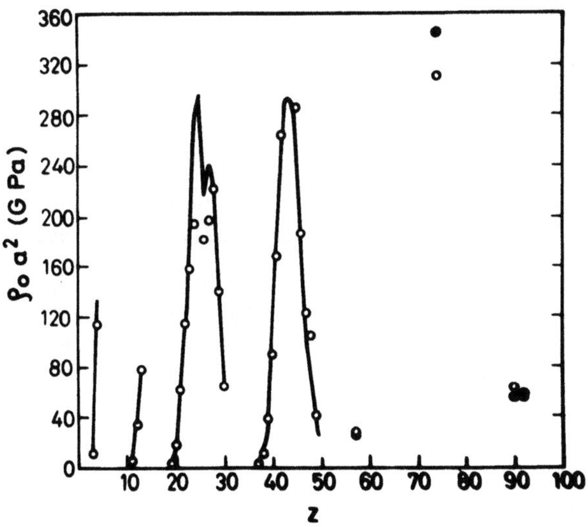

Fig.11 Bulk modulus of elements versus Z. The average discrepancy between theory and experimental data is 10 percent.

Table 1. Observed and calculated values of the b parameter

Element	b_{cal}*	b_{OB}+
Li	1.07	1.08
Na	1.14	1.25
K	1.15	1.175
Rb	1.18	1.21
Be	1.04	1.13
Mg	1.27	1.26
Ba	1.41	1.08
Al	1.29	1.34
In	1.40	1.53
Tl	1.41	1.38
Sn	1.46	1.58
Pb	1.49	1.528

* $b_{cal} \approx -[(V/4\ \delta^2 P_c/\delta V^2)/\delta P_c/\delta V]_{P_c \to 0}$

\+ see references for experimental b values in Sikka and Godwal[28].

SOME TRENDS IN SHOCK CONSTANTS

In absence of phase transitions, the Hugnoiot is described by a linear shock velocity-particle velocity relation

$$U_s = a + bU_p \qquad \text{--(25)}$$

The constant a is identified with the hydrodynamic sound speed and the quantity $\rho_0 a^2$ correlates with the bulk modulus. The bulk modulus, as the band calculations of Moruzzi et al[31] show, is related to the electron density in the intersitial region (the volume excluding the muffin-tin spheres) i.e on the density of outer nearly free electrons. In Fig.12, we have plotted $\rho_0 a^2$ against r_s, the mean electron spacing parameter derived from this intersitial density. The dependance of $\rho_0 a^2$ is now r_s**(-4.7) instead of the r_s**(-5) expected from the free electron theory.

The b parameter is also related to the electronic structure. Al'tshular[32] had, earlier, noted that its value is dependent upon the type of conduction electrons. This is evident from the values of b for materials having the same number but different type of conduction electrons (Fig.13). For simple sp metals, it has been shown by us[28] that it is related to the core radius R_c and the valence charge Z (Fig. 14). This is consistent with Alder's view[33] that b is tied to the repulsive potential of a solid.

For d electron materials, a simple analysis based on l-dependent pressure expression of Pettifor[34] has been used by us to show that for two hypothetical Z = 3 metals, in which n_s=3 in one case and n_s = 1.5 and n_d=1.5 in the other case, the b values are 1.07 and 0.87. This is in qualitative agreement with the trend in Fig. 13. In fact, from the decomposition of pressure in s and d components, the softness of b parameter in transition metals can be shown to be connected to the s+d electron transfer parameter[35].

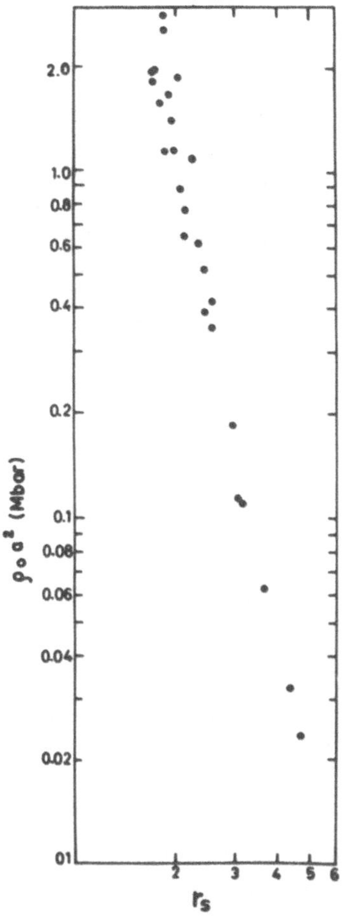

Fig.12 Shock bulk modulus ($\varrho_0 a^2$) versus the electron density parmeter r_s for elements.

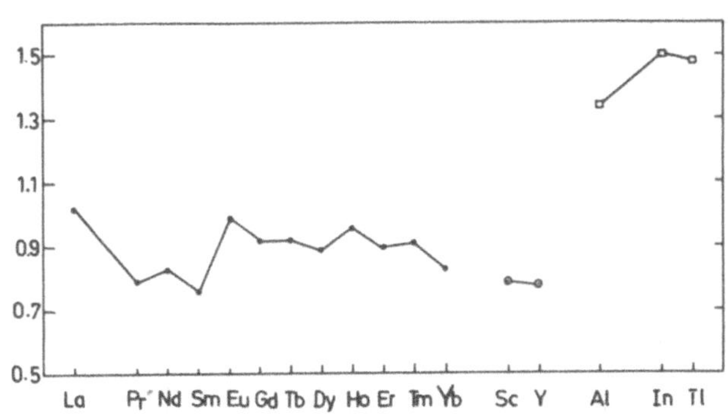

Fig.13 Experimental b values for trivalent elements

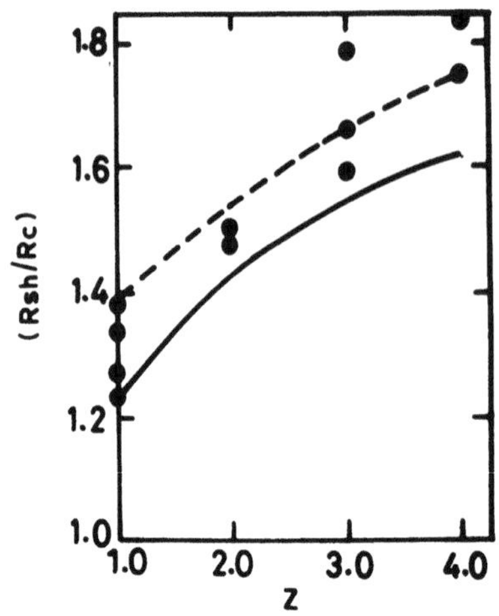

Fig.14 (R_{sh}/R_c) versus Z. The two theoretical curves are: - with Slater exchange and -- with Kohn-Sham exchange. $R_{sh} = R_a/(b/b-1)**(1./3.)$

PRESSURE AND THERMAL IONISATION EFFECTS

In normal state of matter, there are core levels and band electron states. Now under pressure, because of the reduction in atomic radius, a core state can change its character and become a band state (pressure ionisation). Under a rise of temperature, if the energy gap between core and conduction states becomes $\sim k_\beta T$, promotion of core electons can also take place (temperature ionisation). Under shock, both effects operate as the compression and heating occur simultaneously. This transfer of electons into the conduction band alters the ionic charge and the core radius, consequently altering the ion-electron potential as well as screening.

To estimate the valence charge on the atom due to pressure and thermal ionisation[36], we have used the screened-Coulomb modified form of Saha's equation (also called CSCP - IEEOS) which has been shown by Rouse[37] to be appropriate for both high and low densities. It may be interesting to point out that this theory correctly predicts the valence charges of a number of metals at normal density. Also, the valence 2 → 3 transition in Eu and Yb in static high pressure experiments is predicted fairly accurately by this theory[38] (see Fig.15).

Once the ionic charges are known, we employ the methods described in earlier sections to compute EOS. The modification in free energy due to these PT ionisations can be written as

$$F_{ion} = E_{ion} + E_{excit} - TS_{ion} \qquad \text{--(26)}$$

E_{ion} is the energy required for the ionisation of the $Z_{pt}-Z_0$ electrons, E_{excit} corresponds to ionic (bound state) excitation energies and S_{ion} is the entropy due to the ionic disorder because of the presence of different ionic species. However, we found that ionisation is sharply peaked around one ionic state at any V and T and the contribution of S_{ion} to the pressure is therfore very small (e.g. for Mo at $V/V_0 = 0.331$ and $P = 4.97$ TPa, this term contributes only about 0.02 TPa). The valence charges due to PT ionisations for Mo and Al at some compressions along the Hugoniot are

	V/V_o	Z_p	Z_{pt}	T_H	Pcal.	Pexp.
Mo	0.331	6.1	8.1	11.5ev	4.85TPa	4.97TPa
Al	0.32	3.0	3.01	7.9	1.81	1.8

Ross[23] has evolved an alternative procedure to estimate PT ionisation based on the APW method. Theories in which shell effects are incorporated into Thomas-Fermi-Dirac treatments are also in vogue[39-40].

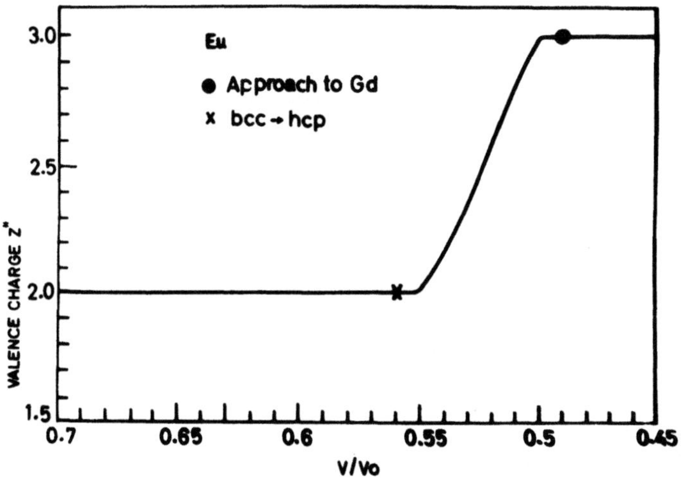

Fig. 15 Valence charge of Eu versus V/V_0. The experimental data is from reference 41.

ACKNOWLEDGEMENTS

The author is grateful to Dr. R. Chidambaram and Dr. B.K. Godwal for fruitful collaboration. Thanks are also due to Shri K.P. Rajasekharan for computer editing of the manuscript.

REFERENCES

1. E.P. Wigner and F. Seitz, Phys. Rev. 43 : 804 (1933).
2. R.J. Trainor, J.W. Shaner, J.M. Auerbach and N.C. Holmes, Phys. Rev. Lett. 42 : 1154 (1979).
3. R.C. Weingart, Lawrence Livermore Laboratory Report UCRL - 52752 (1979).
4. C.E. Ragan III, Phys. Rev. A 21 : 458 (1980); Phys. Rev. A 25 : 3360 (1982);Phys. Rev. A29 : 1391 (1984)
5. A.P. Volkov, N.P. Voloshin, A.S. Vladimirov, V.N. Nogin and U.A. Simonenkov, JETP Lett. 31 : 588 (1981).
6. A.S. Vladimirov, N.P. Voloshin, V.N. Nogin, A.V. Petrovtsev and U.A. Simonenkov, JETP Lett. 39 : 82 (1984).

7. B.K. Godwal, S.K. Sikka and R. Chidambaram, Phys. Rep. 102 : 123 (1983).
8. Ya. B. Zel'dovich and Yu.P. Raizer in :"Physics of Shock waves and High Temperature Hydrodynamics Phenomena," Vols. 1&2, Academic Press, New York (1967).
9. P. Hohenberg and W. Kohn, Phys. Rev. 136 : B 864 (1964).
10. W. Kohn and L.J. Sham,Phys. Rev. 140 : A 1133 (1965).
11. J.C. Slater, J. Chem Phys. 57 : 2389 (1972).
12. J. Ihm, A. Zunger and M.L. Cohen, J. Phys. C12 : 4409 (1972).
13. J.A. Moriarty, Phys. Rev. 16 : 2537 (1977); Phys. Rev. B 26 : 1754 (1982).
14. O.K. Adersen, Phys. Rev. B 12 : 3060 (1975).
15. H.L. Skriver, in "The LMTO Method," Springer - Verlag Berlin (1984).
16. T. Neal, in :"High Pressure Science and Technology", Vol 1 : 80 K.D. Timmerhaus and M.S. Barber ed., Plenium,New York (1978).
17. B.K. Godwal, Pramana 19 : 225 (1982).
17a. M.T. Yin and M.L. Cohen, Phys. Rev. B26 : 3259 (1982).
18. J.A. Moriarty, D.A. Young and M. Ross, Phys. Rev. B 30 : 578 (1984).
19. A.I. Voropinov, G.M. Gandelman and V.G. Podaval'ny, Sov. Phys. Usp. 13 : 56 (1970).
20. A.K. McMahan, B.L. Hord and M.Ross, Phys. Rev. B 15 : 726 (1977).
21. B.K. Godwal, J. Phys. F10 : 377 (1980).
22. B.K. Godwal, S.K. Sikka and R. Chidambaram, Phys Rev. B 20 : 2362 (1979).
23. M. Ross,Phys. Rev B 20 : 4891 (1979).
24. B.K. Godwal and S.K. Sikka, J. Phys. F 12 : 655 (1982).
25. A.K. McMahan,H.L. Skriver and B. Johansson, Phys. Rev. B23:5016 (1981).
26. F. Perrot, Phys. Rev. B21 : 3167 (1980).
27. A.C. Mitchell and W.J. Nellis, J. App. Phys. 52 : 3363 (1981).
28. S.K. Sikka and B.K. Godwal, Sol. State Comm. 38 : 949 (1981).
29. S.K. Sikka, Nuclear Physics and Solid State Phys.(India) 21 A : 83 (1978).
30. D.A. Young and M. Ross, Phys. Rev. B29 : 682 (1984).
31. L. Moruzzi, J.F. Jank and A.R. Williams, in :" Calculated Electric Properties of Metals", Pergamon, New York (1978).
32. L.V. Al'tshuler, Sov. Phys. Usp. 8 : 52 (1965).
33. B.J. Alder, in :" Solids under Pressure", W. Paul and D.M. Warschauer ed., McGraw Hill, New York (1963).
34. D.G. Pettifor, J. Phys. F7 : 613 (1977).
35. Y.K. Vohra, S.K. Sikka and W.B. Holzapfel, J. Phys. F 13 : L107 (1983).
36. B.K. Godwal, S.K. Sikka and R. Chidambaram, Phys. Rev. Lett. 47 : 1144 (1981).
37. C.A. Rouse, in :"Prog. in High Temp. Phys. and Chem.", Vol 4, C. A. Rouse, ed., Pergamon Press, New York (1971).
38. B.K. Godwal, Phys. Rev. A 28 : 1103 (1983).
39. R.M. More, Phys. Rev. A 19 : 1234 (1979).
40. D. Liberman, Phys. Rev. B 20 : 4891 (1979).
41. K. Takemura and K. Syassen, J. Phys. F 15 : 543 (1985).

THE OVERLAP OF ELECTRON CORE STATES FOR VERY HIGH COMPRESSIONS

Galen K. Straub

Equation of State and Mechanics of Materials Group
Los Alamos National Laboratory
Los Alamos, NM 87544

I. INTRODUCTION

At normal density and for modest compressions, the electronic structure of a metal can be accurately described by treating the conduction electrons and their interactions with the usual methods of band theory. The core electrons remain essentially the same as for an isolated free atom and do not participate in the bonding forces responsible for creating a condensed phase. As the density increases, the core electrons begin to "see" one another as the overlap of the tails of wave functions can no longer be neglected. The electronic structure of the core electrons is responsible for an effective repulsive interaction that eventually becomes free-electron-like at very high compressions.

The electronic structure of the interacting core electrons may be treated in a simple manner using the Atomic Surface Method (ASM).[1] The ASM is a first-principles treatment of the electronic structure involving a rigorous integration of the Schrödinger equation within the atomic-sphere approximation.[2] Solid phase wave functions are constructed from isolated atom wave functions and the band width W_ℓ and the center of gravity of the band C_ℓ are obtained from simple formulas. The ASM can also utilize analytic forms of the atomic wave functions and thus provide direct functional dependence of various aspects of the electronic structure. Of particular use in understanding the behavior of the core electrons, the ASM provides the analytic density dependence of the band widths and positions.

The Atomic Surface Method is discussed in Sec. II with a complete derivation and applications to transition metals and f-shell metals given in Ref. 1.

The process whereby core states interact with one another is best viewed as the formation of narrow electron bands formed from atomic states. As the core-core overlap increases, the bands increase in width and mean energy. In Sec. III this picture is further developed and from the ASM one obtains the analytic dependence on density of the relative motion of the different bands. Also in Sec. III is a discussion of the transition to free-electron bands. As an example, the results for the 3s, 3p, and 3d core states and their evolution into bands for nickel are given.

II. THE ATOMIC SURFACE METHOD FOR ELECTRONIC STRUCTURE

It is now possible to make a simple calculation of the electronic structure for overlapping core orbitals using the Atomic Surface Method (ASM).[1] The ASM is a first-principles treatment of the electronic structure that is computationally simple and can provide immediate analytic dependence of the principal band quantities. The theory has been derived in detail in Ref. 1, and only the main results of the theory will be presented here.

The ASM makes use of the properties of linear combinations of atomic orbitals (LCAO) wave functions. The theory can be roughly characterized in three main steps. The first step consists of using the atomic-sphere approximation[2] and a rigorous integration of the radial Schrödinger equation for the ℓ th orbital to obtain an exact expression for the bandwidth W_ℓ in terms of the condensed phase wave function evaluated at the atomic sphere radius r_0. Next, the condensed phase wave function is constructed in terms of the wave functions of an isolated atom. This then allows one to use analytic forms for atomic wave functions that are accurate in the asymptotic regions and which remain valid in the solid even for large compressions. The final step consists of writing the position of the band center C_ℓ in terms of the overlap integrals from adjacent sites using once again LCAO results and the atomic sphere approximation. A complete description of the electronic structure is obtained with the Friedel model[3] for the density of states $n_\ell(E)$.

The result for the band width is

$$W_\ell = \frac{-\hbar^2}{m} \frac{A_0}{N} \left.\frac{\partial \rho(r)}{\partial r}\right|_{r_0} \quad . \tag{II.1}$$

ρ_ℓ is the electron density and N is the normalization integral equal to

$$N = \int_0^{r_0} 4\pi r^2 \rho_\ell(r)dr \quad , \tag{II.2}$$

and is near unity at the zero pressure density. N departs significantly from one only at extremely high compressions. The position for the center of gravity is given by

$$C_\ell = \varepsilon_\ell + 1/2\ S_\ell\ W_\ell = \varepsilon_\ell + \Omega_0\ \rho_\ell\ W_\ell \quad , \tag{II.3}$$

where $S_\ell = 2\rho_\ell\ \Omega_0$ is the overlap integral. Ω_0 is the atomic sphere volume equal to $4\pi r_0^3/3$ and A_0 is the surface area of the atomic sphere equal to $4\pi r_0^2$. The Friedel model for the density of states is

$$n_\ell = \frac{2(2\ell+1)}{W_\ell} \quad \text{for} \quad C_\ell - \frac{1}{2} W_\ell < E < C_\ell + \frac{1}{2} W_\ell \quad , \tag{II.4}$$

and zero otherwise. ε_ℓ is the atomic term value.

As expressed above, the entire problem of describing the electronic structure has been reduced to finding the wavefunction and its first derivative at a single position, the atomic sphere radius r_0. In addition, at r_0 the wave function has been expressed in terms of the wave function of an isolated atom and may be evaluated from the Hartree-Fock tables of Mann.[5] This procedure has worked remarkably well for the transition-metal and f-shell-metal systems and the results are contained in Ref. 1. We will now proceed to the use of analytic expressions for the atomic wave functions.

At sufficiently large distances, the potential in the Hartree-Fock calculation of the electronic structure of a neutral atom approaches $-e^2/r$. The asymptotic form of the solution of the radial Schrodinger equation is given by

$$R_\ell\ (r) \sim r^\gamma\ e^{-\mu r} \quad , \tag{II.5}$$

where the exponential coefficient is related to the atomic term value ε by

$$-\varepsilon = \hbar^2\mu^2/2m \quad , \tag{II.6}$$

and

$$\gamma = e^2 m/\hbar^2 \mu - 1 \quad . \tag{II.7}$$

For $-\varepsilon$=1 Ry, is identically zero. We set γ=0 for all energy levels since the wave function is dominated by the exponential term and nonzero γ makes only a small quantitative difference. In the transition metal series, the atomic term values are near 1 Ry and the approximation is an excellent one. Normalizing the wave functions to 1 gives a radial wave function

$$R_\ell(r) = 2\ \mu^{3/2}\ e^{-\mu r} \quad . \tag{II.8}$$

This form of the wave function may be combined with Eqs. (II.1) and (II.3) to give the band width as

$$W_{\ell\infty} = \frac{16\ (-\varepsilon)(\mu r_0)\ e^{-2\mu r_0}}{1 - 2\mu^2 r_0^2\ e^{-2\mu r_0}} \quad , \tag{II.9}$$

and the center of gravity of the band as

$$C_{\ell\infty} = \varepsilon_\ell + \Omega_0\ W_{\ell\infty}\ \mu^3/\pi\ e^{-2\mu r_0} \quad . \tag{II.10}$$

The ∞ symbol is used to designate the large-r, asymptotic analytic form of the wave function. The band width and center of gravity are still dependent upon the angular momentum quantum number ℓ thru the value of the atomic term value ε_ℓ. The approximation works best with ℓ = 0,1,2 (s,p,d-states) but gives quite reasonable results for l=3 (f-states) as well.

III. ATOMIC CORE LEVELS UNDER COMPRESSION

A. Core States From A Band-Structure Picture

In a metal at zero pressure, we can divide the electrons into two groups, the conduction electrons and the core electrons. The conduction electrons are primarily responsible for the cohesive energy in a metal. The core electrons are considered to be those whose orbitals are initially unchanged from the orbitals in an isolated atom and are much more tightly bound to the nucleus. We now wish to consider the situation whereby the core electrons begin to "see" one another when the overlap of their wave functions becomes finite as the compression increases. One should picture the process as the evolution of zero width atomic states on individual sites changing into narrow width band states that extend over the entire system.

Strictly speaking, whenever core electrons interact with one another they also interact with all the conduction electrons plus all of the other core electrons. A quite reasonable approach is to first treat the interaction for a given angular momentum core state between ions on adjacent sites and then add the hybridization between different core angular momentum states. This is similar to the procedure of band theory. For instance, the s-s, p-p, d-d, etc. interactions are treated separately and only when there is actual overlap of the resulting bands will s-p, s-d, etc. hybridization be important since that interaction goes as the inverse of the energy difference between states. The evolution of core states from zero width atomic levels into broadened bands is a continuous process and one would expect that LCAO's would be an appropriate expansion set. Also, the ASM is well suited to this problem because the approximations involved are such that it limits exactly to Hartree-Fock atoms at infinite separation.

The results for nickel using the ASM and Hartree-Fock tables in the formulae in Section II are shown in Fig. 1 for the band widths and Fig. 2 for the relative positions of the bands. Both sets of curves are plotted as a function of r_0. The ASM results are in excellent agreement with full band calculations. For a value of r_0 just greater than 1 a_o (a_o= Bohr radius), the band widths for the 3s, 3p, and 3d orbitals equal one another and are also equal to the band width of free electrons. The 3s, 3p, and 3d bands also overlap one another near r_0=1.3 a_o. The zero pressure density corresponds to a value of r_0= 2.61 a_o.

Missing from Fig. 2 is the s-p conduction band that is strongly coupled to the 3d band thru hybridization. The ASM can be used to provide all the parameters for the entire band structure in the solid using the method of Harrison and Froyen.[7] The results for Ni at normal density are given in Ref. 1. To do the full band calculation, one must determine the average d-band energy and the hybridization matrix elements between the d-orbitals and the s-p conduction band. The average d-band energy may be determined entirely from neutral-atom atomic term values (and r_0). The procedure involves using the renormalized atom approach of Watson, Ehrenreich, and Hodges[8], the energy of the bottom of the conduction band as given by Froyen[6], and the coulomb energy required to shift an atomic electron from a

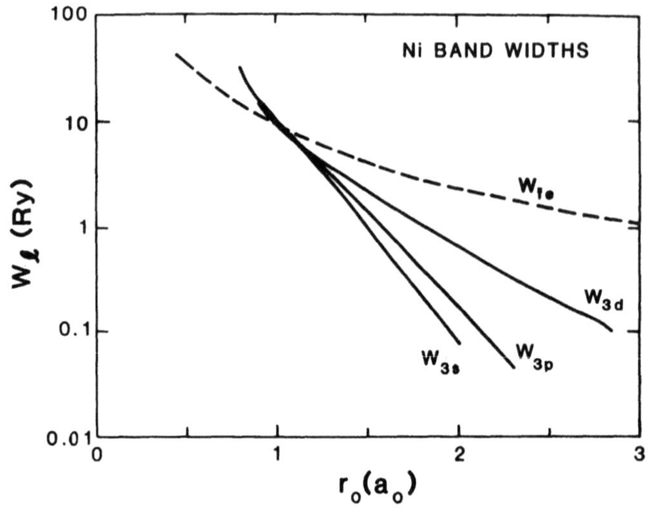

Fig. 1. The widths for the 3s, 3p, and 3d bands of nickel as a function of r_0. The dashed line is the band width for a system of free electrons. All of the band widths including the free electron width become equal at the same value of r_0.

s-orbital to a d-orbital. (See Eq.(27) of Ref. 1) The hybridization matrix element can be evaluated using the asymptotic form of the radial wave function with all of the resulting parameters again determined from atomic term values.

B. Analytic Treatment of the Relative Movement of Different Angular Momentum Bands

The formulas of Sec. II can be used to determine the analytic dependence of the relative movement under compression for bands with different orbital angular momentum quantum numbers. The overlap integral may also be obtained from Eq. (II.3), (II.5), and the electron density $\rho = R_\ell^2/4\pi$. The result is

$$S_\ell = 8/3\ \mu_\ell^3\ r_0^3\ e^{-2\mu_\ell r_0} \quad . \tag{III.1}$$

The exponential term dominates the r_0 dependence in all cases with W_ℓ and S_ℓ varying as $\exp(-2\mu_\ell r_0)$ and C_ℓ varying as $\exp(-4\mu_\ell r_0)$. Values for a sampling of elements are given in Table I. For all elements, $\mu_s > \mu_p > \mu_d$

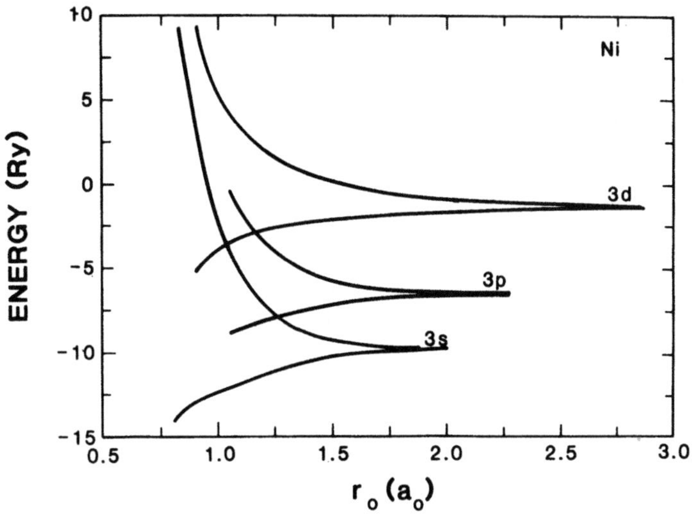

Fig. 2. The bottom, center, and top of the 3s, 3p, and 3d bands of nickel. The 3d band was calculated in the same manner as the 3s and 3p bands and does not include any hybridization with the conduction band.

for the same value of n, the principal quantum number. Thus, the s-band will rise in energy with respect to the p-band as r_0 decreases, the p-band will rise with respect to the d-band, and so forth. This result is easily seen in Fig. 1 for the band widths as well as the domination of the exponential term. In the f-shell elements the situation is more complicated. For instance, in tungsten μ_{4f} lies between μ_{5s} and μ_{5p}, but the predictions for the density dependence of the bands should still be reliable for small band widths. Improved numerical results are possible using atomic wave functions that include relativistic corrections.

TABLE I

Values of μ_ℓ for selected core states. The values of r_0, in units of Bohr radii a_o, are for zero pressure. The μ_ℓ are calculated from the tables of Mann[5] and are in units of $1/a_o$. The number preceding the angular momentum label is the principal quantum number.

	r_0				
		2s	2p		
Na	3.94	2.37	1.74		
Al	2.98	3.13	2.54		
		3s	3p	3d	
Mn	2.70	2.78	2.24	1.06	
Fe	2.67	2.89	2.35	1.10	
Ni	2.61	3.13	2.56	1.18	
		4f	5s	5p	5d
Eu	2.27	1.10	1.91	1.47	
W	2.95	2.10	2.42	1.93	0.90
		5f	6s	6p	6d
U	3.21	1.13	1.83	1.44	0.73

C. Transition to Free-Electron Bands

At extremely large compressions, the bands will completely overlap as shown in Fig. 2 and may be considered free-electron bands. For s and p bands the free-electron band width can be written (bcc structure)

$$W_{fe} = 2\pi^2 (3/8\pi)^{1/2} h^2/m\, r_0^2 \quad . \tag{III.4}$$

In Fig. 1, the s, p, d widths become equal and also equal to the free-electron width near the same value of r_0, and all may be treated as free-electron bands.[1] Another aspect of the identification with free-electron bands is their energy ordering, which should be s,p,d. In sodium and aluminum this same energy ordering also occurs as the 3s band moves above the 3d band (and above the Fermi energy) and the bands become ordered 2s, 2p, 3d. It should be noted that free-electron behavior occurs at extremely large compressions (a factor of 50 to 100) and corresponds to pressures of over 1000 Mbars. Perhaps more important is the result that an entire shell of electrons with the same principal quantum number n becomes free-electron-like at the same compression rather than one ℓ-orbital at a time. For instance, in sodium and aluminum all 8 electrons in the n=2 shell become free-electron-like instead of first the 6 p-electrons and later the 2 s-electrons. The n=2 electrons are then describable by simple metal theory, but with an ionic charge of Z=11 for aluminum and Z=9 for sodium. Similarly for nickel, the 16 n=3 electrons become free-electron-like together.

IV. SUMMARY

The electronic structure for core states of nickel has been calculated using the Atomic-Surface-Method. The evolution of the core electrons from isolated free-atom states to narrow bands and finally to free-electron behavior has been described in a simple theory that allows the determination of the analytic dependence of the electronic structure upon density. One finds that the band widths of an entire shell of electrons with the same principal quantum number become free-electron-like at the same density. In the case of nickel, this means that all 16 of the n=3 shell of electrons approach free electron behavior together.

REFERENCES

1. G. K. Straub and W. A. Harrison, Phys. Rev. B 31, 7668 (1985).
2. O. K. Andersen, Solid State Commun. 13, 133 (1973).
3. J. Friedel, in The Physics of Metals, edited by J. M. Ziman (Cambridge University Press, New York, 1969).
4. W. A. Harrison, Electronic Structure and the Properties of Solids (Freeman, San Francisco, 1980).
5. J. B. Mann, Los Alamos Scientific Laboratory Reports No. LA-3690 (1967) and No. LA-3691 (1968) (unpublished).
6. S. Froyen, Phys. Rev. B 22, 3119 (1980).
7. W. A. Harrison and S. Froyen, Phys. Rev. B 21, 3214 (1980).
8. R. E. Watson, H. Ehrenreich, and L. Hodges, Phys. Rev. Lett. 24, 829 (1970).

THEORY OF MOLECULAR DISSOCIATION IN SHOCKED NITROGEN AND OXYGEN*

G. I. Kerley and A. C. Switendick

Sandia National Laboratories
Albuquerque, NM

INTRODUCTION

Recent measurements of Nellis, et. al.[1,2] provide evidence for shock-induced dissociation in molecular nitrogen and oxygen. Nitrogen shows a dramatic increase in compressibility above about 30 GPa, with densities in the range expected for a monatomic fluid. This effect is much less obvious in the oxygen data, but it is reasonable to assume that dissociation occurs in that case as well. Static measurements on the solid show no evidence of a transition from a molecular to a monatomic form in nitrogen up to 52 GPa[3] or in oxygen up to 13 GPa.[4] Calculations by McMahan and LeSar[5] predict the transition to occur at about 100 GPa in solid nitrogen. Hence, high temperatures are believed to drive shock-induced dissociation.

In this paper, we use a chemical equilibrium model to explain the shock experiments. The fluid is taken to be a mixture of the molecular and monatomic species. The equation of state (EOS) for this mixture is computed from EOS for the pure molecular and monatomic fluids, assuming ideal mixing.[6] The equilibrium composition, or degree of dissociation, is determined by minimizing the free energy at constant temperature and pressure. For the molecular species, we use EOS described previously, [7] that have been shown to agree with experimental data in the regime where no dissociation occurs. EOS for the monatomic fluids are discussed below. Our predictions of the shock data are most sensitive to two aspects of the EOS for the monatomic species - the binding energy and the electronic statistical weight.

EOS FOR MONATOMIC NITROGEN AND OXYGEN

For the monatomic fluids, we write the Helmholtz free energy as the sum of three terms, [8]

$$A = A_t + A_\phi + A_e \ . \tag{1}$$

*This work performed at Sandia National Laboratories supported by the U. S. DOE under contract DE-AC04-76DP00789.

A_t, the contribution from translational degrees of freedom, is given by the ideal gas expression. A_ϕ, the contribution from atomic interactions, is calculated using the CRIS model.[8] A_e, the contribution from electronic excitations, is discussed below.

Zero-Kelvin Isotherms

For the CRIS model, it is necessary to know the zero-Kelvin isotherm for the solid, which is used to estimate the average energy of an atom in the cage formed by its neighbors.[8] Band-theoretical calculations for nitrogen show that the solid EOS varies with lattice structure,[5] and it is not obvious which solid structure is appropriate for use in modeling the fluid EOS. We assume that isotherms for different solid structures represent a range of parameters that can be used in the fluid model.

The pressure-density curves for monatomic nitrogen and oxygen were computed using both the APW method[9] and the INFERNO model.[10] The APW calculations were done for the fcc lattice structure and should give a rough lower bound on the density.[5] The INFERNO model replaces the structure about an atom by a uniform electron gas. For nitrogen, these results are similar to APW calculations for the sc lattice[5] and should give a rough upper bound on the density. The zero-pressure densities predicted by the two methods differ by about 15% in nitrogen and by about 30% in oxygen. However, the difference decreases rapidly with increasing pressure. As discussed below, Hugoniot calculations using the two isotherms differ by less than 5% in density.

Determination of the binding energy, needed to complete the description of the isotherm, is a more critical problem. We find that our results are fairly sensitive to this parameter, and that the onset of dissociation is best predicted using a value that is roughly half of the molecular binding energy in both cases. The binding energies for nitrogen, obtained using the APW method, vary by a factor of two for different solid structures.[5] The value that we have selected is midway between these two limits.

Electronic Contributions

To understand the general features of the electronic free energy in nitrogen and oxygen, calculations were made using the INFERNO code.[10] The results are shown in Fig. 1. At low densities, the electrons are localized about the atomic nuclei; the atomic levels are unperturbed, and the free energy is nearly independent of density. At high densities, the electrons are delocalized, and the free energy is metallic in character. INFERNO predicts a rapid change in free energy due to the insulator-metal transition, at densities from 2 to 4 g/cc in both nitrogen and oxygen. Hence the transition is in the very range studied in the shock experiments. The theoretical predictions of these data are sensitive to free energy differences of this magnitude.

Unfortunately, there is no theoretical model of electronic structure that gives a satisfactory description of the free energy in the region of the transition. At low densities, A_e can be calculated exactly, using known energy levels of the isolated atom. As shown in Fig. 1, INFERNO gives much too large a value in this limit. The principal reasons for the discrepancy, common to other theories as well, are that it uses average populations instead of sampling actual atomic configurations, and that it does not account for multiplet splitting due to spin interactions.

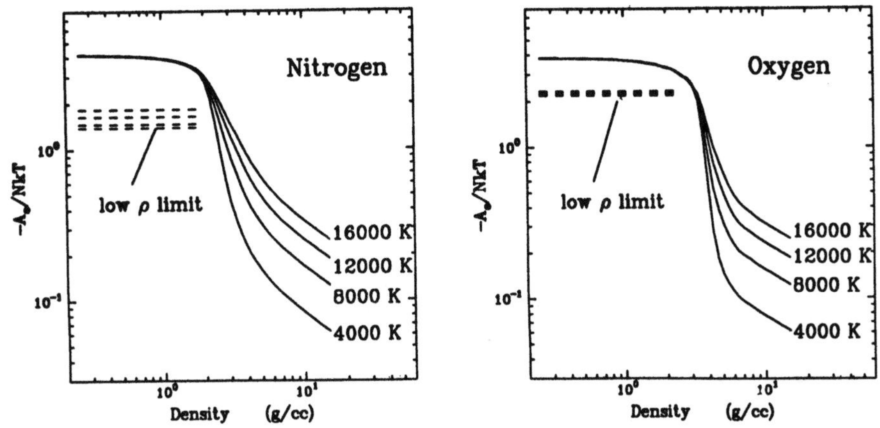

Figure 1: Electronic contribution to the free energy for monatomic nitrogen and oxygen. Solid curves were calculated using INFERNO. Dashed curves were computed from the atomic energy levels.

In this work, the electronic free energy was approximated by the simple expression

$$A_e = -NkT \ln g_e \ , \tag{2}$$

where g_e is a constant statistical weight. A reasonable upper limit to g_e is the total weight of all ground state configurations - 20 for nitrogen and 15 for oxygen. The INFERNO results suggest a lower limit of about 3 in the density range of interest. In this work, we find that good agreement with the shock data is obtained with values of 8 for nitrogen and 5 for oxygen.

RESULTS

Hugoniots for nitrogen and oxygen are shown in Fig. 2. Circles are experimental data. [1,2,11–13] Three sets of theoretical curves are shown. When dissociation is not allowed, the results are in good agreement with the data for pressures below about 30 GPa, as noted previously.[7] When dissociation is included, the model predicts a significant softening of the Hugoniot at high densities, in agreement with experiment. This effect is due in part to the higher density of the monatomic species and also to the fact that dissociation absorbs energy and lowers the temperature.

The figures also show the sensitivity of the theory to differences in the density on the isotherm. Curve A, using the APW isotherm, and curve B, using the INFERNO isotherm, differ by less than 5%. The reason is that the shock calculations are most sensitive to the high pressure portion of the solid isotherm, where the two models agree fairly well.

The calculations also predict that the increase in compressibility due to dissociation is greater for nitrogen than for oxygen, in agreement with the measurements. One reason for this fact is that molecular oxygen is more compressible than molecular nitrogen; hence the density change in going to the monatomic species is less for oxygen. A second reason is that the dissociation energy for oxygen is only half that for nitrogen, so less energy is absorbed by the reaction, and there is less lowering of the temperature.

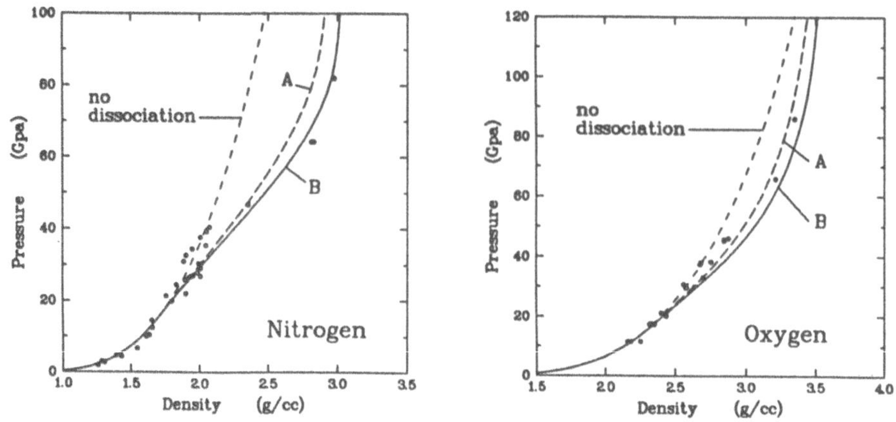

Figure 2: Hugoniots for liquid nitrogen and oxygen. See text for discussion of the curves.

The sensitivity of our theoretical predictions to the binding energy and statistical weight is illustrated for nitrogen in Figure 3. The cohesion of the condensed phase is important because it markedly decreases the energy gap between the molecular and monatomic forms, making dissociation more favorable at high densities than at low densities. Curve B is the same as in Fig. 2; the cohesive energy was decreased by 25% in curve C and increased by 25% in curve D. Changing the statistical weight does not affect the energy, but it changes the entropy, making dissociation more or less favorable. Parameter g_e was decreased from 8 to 4 in curve E and was increased to 12 in curve F. Variations in these two parameters have a greater effect than variations in the density of the monatomic form, but the general features of the calculations are preserved in all cases.

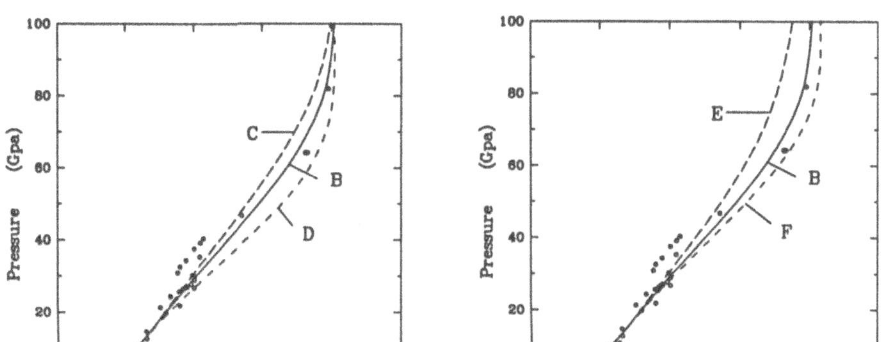

Figure 3: Sensitivity of nitrogen Hugoniot to variations in binding energy and statistical weight. See text for discussion of the curves.

Our theory is compared with reflected shock data for nitrogen[2] in Fig. 4. Agreement with the two high pressure points is good, but we do not predict the anomalous behavior, characteristic of a phase transition, of the two low pressure points.[2] A separation into two phases would indicate that the mixture of molecules and atoms is not as homogeneous as is assumed in ideal mixing. Another possibility is that the anomalous behavior is due to the transition between localized and delocalized electronic structure, as discussed above.

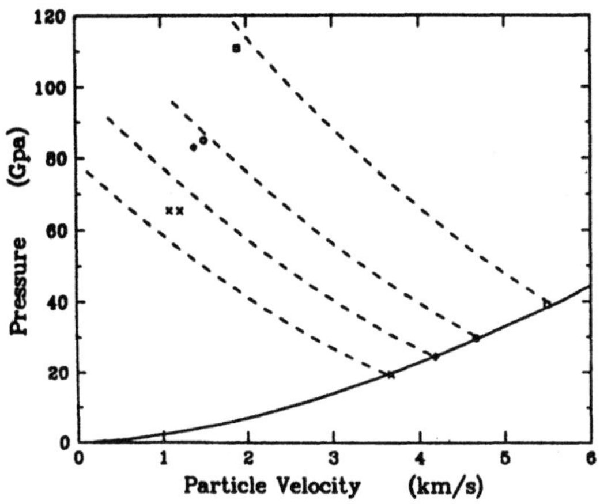

Figure 4: Reflected Hugoniots for nitrogen. Solid line is first shock curve, dashed lines are second shock curves. Data are from reference 2.

CONCLUSIONS

Shock-induced dissociation can be described by a chemical equilibrium model, using ideal mixing and the EOS of the molecular and monatomic species. The calculations agree very well with the general features of the data, although the details depend upon the values assumed for the EOS parameters of the monatomic species. A transition between localized and delocalized electronic structure occurs in the density range of the shock experiments. This transition could be the cause of the anomalous behavior seen in the reflected shock experiments. A better model for computing the electronic free energy is needed to describe this effect. Wave profile measurements in the anomalous region would also improve our understanding of this transition.

REFERENCES

1. W. J. Nellis and A. C. Mitchell, Shock Compression of Liquid Argon, Nitrogen, and Oxygen to 90 GPa (900 kbar), J. Chem. Phys., 73:6137 (1980).

2. W. J. Nellis, N. C. Holmes, A. C. Mitchell, and M. van Thiel, Phase Transition in Fluid Nitrogen at High Densities and Temperatures, Phys. Rev. Lett., 53:1661 (1984).

3. S. Buchsbaum, R. L. Mills, and D. Schiferl, Phase Diagram of N_2 Determined by Raman Spectroscopy from 15 to 300 K at Pressures to 52 GPa, J. Phys. Chem., 88:2522 (1984).

4. B. Olinger, R. L. Mills, and R. B. Roof, Jr., Structures and Transitions in Solid O_2 to 13 GPa at 298 K by X-Ray Diffraction, J. Chem. Phys., 81:5068 (1984).

5. A. K. McMahan and R. LeSar, Pressure Dissociation of Solid Nitrogen under 1 Mbar, Phys. Rev. Lett., 54:1929 (1985).
6. J. H. Hildebrand and R. L. Scott, "The Solubility of Nonelectrolytes," Dover, New York (1964).
7. G. I. Kerley and J. Abdallah, Jr., Theoretical Equations of State for Molecular Fluids: Nitrogen, Oxygen, and Carbon Monoxide, J. Chem. Phys., 73:5337 (1980).
8. G. I. Kerley, A Model for the Calculation of Thermodynamic Properties of a Fluid, in "Molecular-Based Study of Fluids," J. M. Haile and G. A. Mansoori, eds., Am. Chem. Soc., Wash., D.C. (1983) .
9. L. F. Mattheiss, J. H. Wood, and A. C. Switendick, A Procedure for Calculating Electronic Energy Bands Using Symmetrized Augmented Plane Waves, in "Methods in Computational Physics," 8:63 (1968).
10. D. A. Liberman, Self-Consistent Field Model for Condensed Matter, Phys. Rev. B 20:4891 (1979).
11. V. N. Zubarev and G. S. Telegin, The Impact Compressibility of Liquid Nitrogen and Carbon Dioxide, Sov. Phys. Dokl., 7:34 (1962).
12. R. D. Dick, Shock Wave Compression of Benzene, Carbon Disulfide, Carbon Tetrachloride, and Liquid Nitrogen, J. Chem. Phys., 52:6021 (1970).
13. J. Wackerle, W. L. Seitz, and J. C. Jamieson, Shock-Wave Equation of State for High-Density Oxygen, in "Behavior of Dense Media Under High Dynamic Pressures," Gordon and Breach, New York (1968).

HIGH-PRESSURE ION-THERMAL PROPERTIES OF METALS FROM *AB INITIO* INTERATOMIC POTENTIALS

John A. Moriarty

Lawrence Livermore National Laboratory
University of California
Livermore, California 94550

ABSTRACT

It has recently become possible for the first time to calculate *ab initio* two-ion and three-ion interatomic potentials in *d*-electron transition metals. The nature of these potentials for the 3*d* series metals is discussed. In the case of copper, where three-ion contributions may be neglected, application is made to the ion-thermal component of the equation of state and to the calculation of melting on the shock Hugoniot.

INTRODUCTION

Detailed knowledge about interatomic force laws in the condensed state represents an essential ingredient in a complete microscopic description of the materials properties of metallic solids and liquids. For example, interatomic potentials represent the primary physical input both in numerical simulations of materials properties by molecular dynamics and Monte Carlo techniques and in approximate calculations of the same by methods such as quasi-harmonic lattice dynamics in the solid and variational perturbation theory in the liquid. Heretofore, however, the first-principles theory of these potentials has only been developed for simple *sp*-bonded metals[1-5] (e.g., Al) and special limiting cases of *d*-electron metals near the beginning and end of the transition series (e.g., Ca and Zn).[3,6] Recently, we have succeeded in extending this capability to all *d*-electron transition metals,[7] opening up many new avenues of application including high-pressure structural and thermal properties relevant to shocked materials.

INTERATOMIC POTENTIALS

For any homogeneous metal with a well-defined atomic volume Ω, the total energy of the system may be developed as a rigorous multi-ion expansion in terms of the individual ion or nuclear coordinates $\mathbf{R}_i$:

$$E_{tot}(\mathbf{R}_1,\ldots,\mathbf{R}_N) = E_0(\Omega) + \frac{1}{2}\sum_{i,j}{}' v_2(R_{ij}) + \frac{1}{6}\sum_{i,j,k}{}' v_3(R_{ij},R_{jk},R_{ki}) + \ldots, \quad (1)$$

where E_0, v_2, v_3, etc. are volume, two-ion, three-ion, etc. functionals, with $R_{ij} = |\underset{\sim}{R}_i - \underset{\sim}{R}_j|$, etc., and the primes exclude all self-interaction terms from the sums. The interatomic potentials are implicitly volume dependent, but explicitly structure independent, that is they do not depend on the absolute ion positions $\underset{\sim}{R}_i$ and are thus completely transferable at fixed volume. For nontransition metals it has been possible to obtain both E_0 and the central-force, pair potential v_2 in terms of appropriate nonlocal pseudopotentials.[1-6] Furthermore, the neglect of v_3 and higher potentials has proven to be an excellent approximation in most applications. For pure transition metals, on the other hand, not only are the individual functionals necessarily more complicated, but one expects a priori that the angular forces associated with the multi-ion terms to be more important due to the stronger scattering nature of the d electrons.

To treat the general problem of d-electron metals, it has been possible to synthesize pseudopotential methods for s and p electrons with tight-binding methods for d electrons into a generalized pseudopotential theory (GPT) applicable to both simple and transition metals.[3,7] With regard to the total-energy expansion (1), however, the GPT has previously been developed only for the limiting cases of empty- and filled-d-band metals,[3] where simplifying approximations are possible. We have now obtained explicit expressions for the volume, two-ion, and three-ion components of Eq. (1) in the partially-filled-d-band case applicable to transition metals. We have applied these results to the calculation of normal-density potentials across the entire 3d series and to copper from 40% expansion to about 3-fold compression. In the present paper, we shall concentrate primarily on the latter application; more details on the theory and the former application will be published elsewhere.[7]

Although the GPT itself becomes rather complex for transition metals, the nature of the resulting potentials can be fairly readily understood. The pair potential v_2, for example, can be broken down as a sum of two major components. The first is a screened valence interaction analagous to what one encounters in simple metals. At sufficiently short distances, this reduces to a screened Coulomb repulsion between ions of charge Ze, where Z is the self-consistent valence of the metal, giving the average number of s and p electrons per atom. It turns out that Z is about 1.5 across the entire 3d series from Ca to Cu at normal density. In the case of copper, Z increases slowly with compression over the volume range we have considered, reaching a maximum value of about 2.2. The screened valence interaction at long range, on the other hand, has a slowly decaying oscillatory behavior, characteristic of metallic screening, where the wavelength of the oscillation decreases with increasing density. These short- and long-range features of v_2 are all clearly evident in Fig. 1 for copper.

The second component of v_2 includes all remaining direct interactions between overlapping d-electron states. Near the center of the series, this represents a strong bonding attraction, corresponding to the increased cohesion for a half-filled d band, and a deep negative minimum in v_2 occurs in the vicinity of near neighbors. The depth of this minimum recedes rapidly as one goes towards either end of the series, and at the end of the series in copper is actually pushed to positive energy due the increased kinetic energy contribution of nearly-full overlapping d shells. This near-neighbor well is also strongly volume dependent, in copper being moved to still higher energy under compression and eventually becoming only a shoulder on a steep repulsive curve, as shown in Fig. 1.

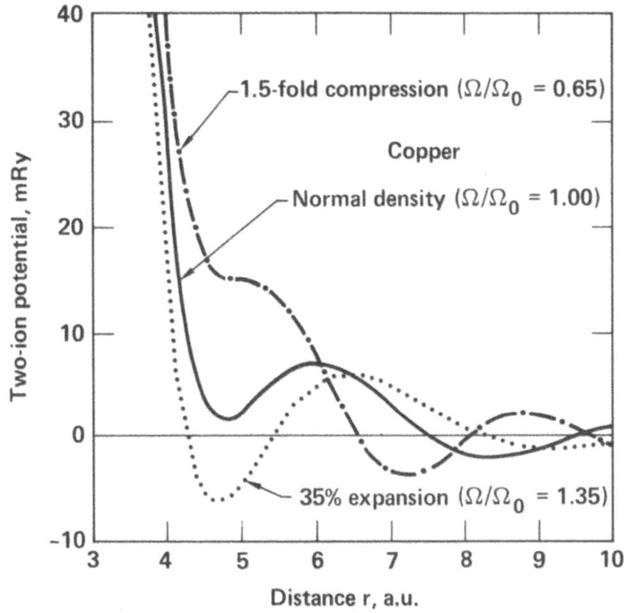

Fig. 1. GPT two-ion interatomic pair potential $v_2(r)$ in copper calculated at three different densities.

The three-ion potential v_3 in transition metals is dominated by d-electron terms which are the counterparts of those contributing to second component of v_2 mentioned above. At the ends of the series (i.e., Ca and Cu), v_3 turns out to be essentially negligible, although near the center of the series, v_3 is rather strongly repulsive and quite essential to properly balance the large attractive component of v_2. The fact that v_3 is negligible in copper makes this metal convenient for our present purposes, since calculating materials properties on the basis of Eq. (1) is then no more difficult than for simple metals.

APPLICATION TO SHOCK-COMPRESSED COPPER

We have investigated a number of materials properties of copper relevant to the shock-compressed state with our new interatomic potentials. The basic cohesive properties of this metal, which in the GPT depend on E_0 as well as v_2 in Eq. (1), are well described in comparison both to experiment and to more exact muffin-tin energy-band calculations,[8] which employ no expansion of the total energy. This is demonstrated in Table I. At high pressure, on the other hand, the GPT zero-temperature isotherm becomes somewhat too soft relative to the corresponding muffin-tin result for this quantity.[10] Important thermal properties such as phonons, the calculation of which is beyond current muffin-tin-band-theory capabilities, are obtained to about 10% accuracy in copper for individual modes at normal density with the present GPT pair potential. Phonon averages, however, such as the effective Debye temperature Θ_D and the Grüneisen constant γ, are, in this case at least, calculated more accurately, as also shown in Table I. (Here Θ_D is defined such that $9/8\ k_B\ \Theta_D$ equals the zero-point vibrational energy.)

Table I. Zero-temperature cohesive energy (E_{coh}), equilibrium atomic volume (Ω_0), bulk modulus (B), and normal-density Debye temperature (Θ_D) and Grüneisen constant γ in copper

	Muffin-tin band theory[a]	GPT	Experiment[b]
E_{coh} (Ry)	0.301	0.243	0.257
Ω_0 (a.u.)	77.5	77.0	78.9
B (Mbar)	1.55	1.50	1.42
Θ_D (K)	--	310	310
γ	--	2.1	2.0

[a] Ref. 8
[b] Ref. 9.

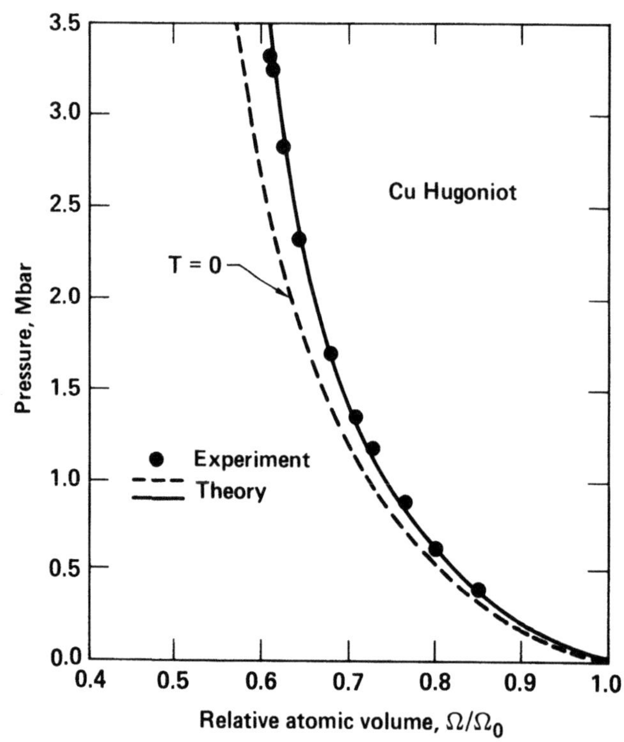

Fig. 2. First-principles theoretical Hugoniot (solid line) vs. experiment (solid dots, from Ref. 12) in copper. Zero-temperature isotherm (dashed line) is shown for comparison.

We have further calculated the principal Hugoniot, treating separately the three components to the energy and pressure, e.g., for the pressure we take

$$P(\Omega,T) = P_0(\Omega) + P_{ion}(\Omega,T) + P_{el}(\Omega,T) \quad , \tag{2}$$

where P_0, P_{ion}, and P_{el} are the zero-temperature, ion-thermal, and electron-thermal partial pressures, respectively. For P_0 we have used the first-principles T = 0 muffin-tin-band-theory result[10] mentioned above (which excludes the zero-point energy). The ion-thermal contribution has been calculated on the basis of our GPT pair potentials using quasi-harmonic lattice dynamics in the solid[5] and variational perturbation theory, with Ross's soft-sphere reference system, in the liquid.[5,11] The electron-thermal contribution has been obtained from the low-temperature formula

$$P_{el}(\Omega,T) = \frac{\pi^2}{6} (k_B T)^2 \frac{dN(\epsilon_F)}{d\Omega} \quad , \tag{3}$$

where $N(\epsilon_F)$ is the GPT-calculated density of states at the Fermi level ϵ_F. Our theoretical result is plotted in Fig. 2 and compared with the experimental two-stage light-gas-gun data of Mitchell and Nellis.[12] The agreement is clearly excellent and, indeed, is somewhat better than generally can be expected on the basis of uncertainties in the theory.

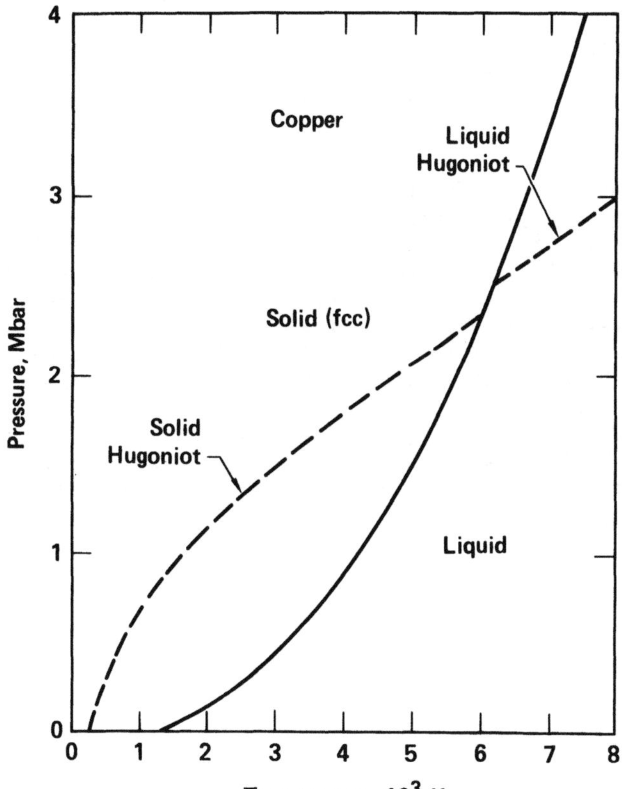

Fig. 3. Theoretical GPT melting curve (solid line) in copper and the predicted location of melting on the shock Hugoniot (dashed lines).

Finally, we have considered the question of melting on the Hugoniot. We have done a preliminary calculation of the melting curve in copper based on the location of the freezing line in the liquid. The latter is here approximated as the locus of points yielding a constant soft-sphere packing fraction η, as theoretically expected and as has been found to be true in the case of aluminum.[5] We have used the value $\eta = 0.477$, which is consistent with both the observed zero-pressure melting temperature (1356 K) and the measured liquid structure factor.[13] The calculated melting curve and its intersection with the Hugoniot are plotted in Fig. 3. The onset of melting is predicted to be at about 2.3 Mbar. Raising or lowering η by 0.025, respectively lowers or raises this prediction by about 0.3 Mbar.

ACKNOWLEDGEMENT

This work was performed under the auspices of the U. S. Department of Energy by Lawrence Livermore National Laboratory under Contract No. W-7405-ENG-48.

REFERENCES

1. V. Heine and D. Weaire, Solid State Phys. 24, 249 (1970). W. A. Harrison, Pseudopotentials in the Theory of Metals (Benjamin, New York, 1966).
2. E. g., L. Dagens, M. Rasolt, and R. Taylor, Phys. Rev. B 11, 2726 (1975); J. Hafner, J. Non-Cryst. Solids 61-62, 175 (1984).
3. J. A. Moriarty, Phys. Rev. B 26, 1754 (1982) and 16, 2537 (1977).
4. A. K. McMahan and J. A. Moriarty, Phys. Rev. B 27, 3235 (1983).
5. J. A. Moriarty, D. A. Young, and M. Ross, Phys. Rev. B 30, 578 (1984).
6. L. Dagens, J. Phys. F 6, 1801 (1976) and 7, 1167 (1977); Phys. Status Solidi B 84, 311 (1977).
7. J. A. Moriarty (to be published).
8. V. L. Moruzzi, J. F. Janak, and A. R. Williams, Calculated Electronic Properties of Metals (Pergamon, New York, 1978).
9. K. A. Gschneidner, Jr., Solid State Phys. 16, 275 (1964); R. M. Nicklow, G. Gilat, H. G. Smith, L. J. Raubenheimer, and M. K. Wilkinson, Phys. Rev. 164, 922 (1967). The room temperature data for Ω_0 and B have been adjusted to T = 0.
10. R. C. Albers, A. K. McMahan, and J. E. Müller, Phys. Rev. B 31, 3435 (1985). The nonrelativistic APW result given in Table I of this paper has been used here.
11. M. Ross, J. Chem. Phys. 71, 1567 (1979).
12. A. C. Mitchell and W. J. Nellis, J. Appl. Phys. 52, 3363 (1981).
13. O. J. Eder, E. Erdpresser, B. Kunsch, H. Stiller, and M. Suda, J. Phys. F 10, 183 (1980).

POSITRON ANNIHILATION AND PRESSURE-INDUCED ELECTRONIC s-d TRANSITION

A. K. McMahan

University of California
Lawrence Livermore National Laboratory
Livermore, California 94550

H. L. Skriver

Risø National Laboratory
DK-4000 Roskilde, Denmark

ABSTRACT

The polycrystalline, partial annihilation rates for positrons in compressed cesium have been calculated using the linear muffin-tin orbitals method. These results suggest that the pressure-induced electronic s-d transition in Cs should be directly observable by momentum sensitive positron annihilation experiments.

INTRODUCTION

Electronic s-d transition is a common phenomenon in high pressure physics, responsible for a variety of equation of state anomalies, insulator-metal transitions, and crystallographic phase transitions.[1] These are indirect manifestations of the pressure induced transfer of electrons from s-like (actually sp hybrid) valence states to those of predominately d character. We suggest here that positron annihilation measurements can permit direct observation of the electron transfer itself. This follows from the fact that s-band valence electrons tend to occupy states near the center of the Brillouin zone, while the d states to which they transfer are located near the zone boundaries. Thus, techniques which probe quantities related to the electron momentum density will detect this transfer.

Positron annihilation is suited to high pressure experiments since the positron source (e.g., Na^{22}) can be included in the pressure cell, while the annihilation γ rays pass through, and are detected outside of the high pressure apparatus. Annihilation rate[2] and angular correlation[3] measurements have been carried out for alkali metals under pressure, and the first positron annihilation (Doppler broadening) measurements in a diamond anvil cell were recently performed[4] on Fe at pressures up to 40.0 GPa.

In the present paper, we report calculations of the electron momentum density and partial annihilation rates (proportional to the two-photon momentum density) in polycrystalline compressed cesium. This material was chosen because its s-d transition is particularly dramatic. At 1 atm, Cs is a free-electron metal; yet by only about 15 GPa, its single valence electron has been transfered entirely from the 6s to the 5d band.[5]

CALCULATIONS

It is well known that electron-positron correlation enhances the electron density in the vicinity of the positron. Some success has been achieved, however, with a modified independent-particle model (IPM) in which electron states for the unperturbed crystal are used in the annihilation rate calculations along with enhancement factors to deal with the effects of the many-body electron-positron interaction.[6]

In the IPM, modified to include an energy-dependent[7] enhancement factor $\epsilon(E)$, the partial annihilation rate or probability per unit time and per unit volume of momentum space that an electron-positron annihilation will result in a pair of photons having a combined momentum of $\vec{p}$ is

$$R(\vec{p}) = r_0^2 c\pi \frac{2}{N} \sum_{\vec{k}j\vec{G}}^{occ} \epsilon(E_{\vec{k}j}) |A_{\vec{k}j}(\vec{G})|^2 \delta(\vec{p}-\vec{k}-\vec{G}) \tag{1}$$

where

$$A_{\vec{k}j} = \frac{1}{\sqrt{\Omega}} \int_\Omega d\vec{r}\, e^{-i(\vec{k}+\vec{G})\cdot\vec{r}}\, \psi_+(\vec{r})\, \psi_{\vec{k}j}(\vec{r}) \tag{2}$$

Here, r_0 is the classical electron radius, c is the speed of light, and N is the number of primitive cells in the crystal. The sums in Eq. (1) are over occupied electron states characterized by $\vec{k}$ in the first Brillouin zone and band index j, and over reciprocal lattice vectors $\vec{G}$. The integral in Eq. (2) is over the primitive cell of volume Ω on which both the positron and electron wavefunctions, $\psi_+(\vec{r})$ and $\psi_{\vec{k}j}(\vec{r})$, respectively, are normalized to unity.

The total annihilation rate λ is the reciprocal of the positron lifetime, and is given by the integral of the partial annihilation rate $R(\vec{p})$ over all momentum space. Setting $r_0^2 c\pi = \epsilon(E_{\vec{k}j}) = \psi_+(\vec{r}) = 1$ in Eqs. (1) and (2) converts $R(\vec{p})$ to the electron momentum density $\rho(\vec{p})$, which is customarily normalized so as to be unity for occupied, ideal, free-electron states, as will be done below.

Numerous calculations based on Eqs. (1) and (2) have been reported.[6] As in recent work by Singh and Jarlborg,[8] the present effort used the linear muffin-tin orbitals method[9] to obtain both the electron states for the unperturbed crystal, and the thermalized, $\vec{k}$=0, pure s, positron wavefunction. Our approach differs from that theirs, however, in calculation of the Fourier transform in Eq. (2). In the spirit of Andersen's combined correction,[9] we approximate the integral in Eq. (2) by

$$\langle ..\psi_+ \psi_{\vec{k}j}\rangle_\Omega \sim \langle ..\psi_+ \psi_{\vec{k}j}\rangle_S + \psi_+(S)[\langle ..\tilde{\psi}_{\vec{k}j}\rangle_\Omega - \langle ..\tilde{\psi}_{\vec{k}j}\rangle_S] \tag{3}$$

where $\langle\ \rangle_\Omega$ and $\langle\ \rangle_S$ indicate integration over the Wigner-Seitz polyhedra and sphere, respectively, $\psi_+(S)$ is the value of the positron wavefunction at the Wigner-Seitz radius S, and $\tilde{\psi}_{\vec{k}j}$ is the usual pseudowavefunction employed in the combined correction. To obtain the polycrystalline spherical average $R_{av}(p)$ we use the tetrahedral method[10] to integrate over surface segments of fixed $p = \hbar|\vec{k}+\vec{G}|$ for which $E_{\vec{k}j}$ lies below the Fermi energy.

The present calculations were scalar relativistic (spin-orbit omitted), used the von Barth-Hedin exchange-correlation potential,[11] and treated all electrons self-consistently in generating the one-electron potentials. The 5s, 5p, and higher lying levels were included as band states in evaluating Eqs. (1) and (2), with angular momentum components through d character retained. The Brillouin zone was sampled with 89 points in the irreducible wedge for the fcc structure. Each point was taken as the center of a small cube which was subdivided into 12 tetrahedra in obtaining the spherical average. The reciprocal lattice sum was truncated after 459 vectors. In one test calculation with 20 points per wedge and in another with 283 reciprocal lattice vectors, $R_{av}(p)$ changed slightly and negligibly, respectively, showing convergence with the above choices.

ELECTRON MOMENTUM DISTRIBUTION AND UNENHANCED PARTIAL ANNIHILATION RATES

Our results for the spherically averaged electron momentum density $\rho_{av}(p)$ are shown in Fig. 1 for reduced volumes V/V_0 = 1, 0.418, and 0.235, where V_0 = 795.5 bohr^3/atom. The corresponding pressures are 0, 4.2, and about 15 GPa, respectivley. At each volume, the solid curves give the total $\rho_{av}(p)$, while the dashed curves show just the 5s, 5p core contribution. The trivial volume dependence has been removed by using the dimensionless momentum pa/h where a is the fcc lattice constant corresponding to each volume (a^3=4V), and h is Planck's constant. Note that pa/h = 0.866, 1, and 1.732 at the L and X points and at the first reciprocal lattice vector, respectively.

For consistancy, we have reported all of our results for the fcc structure. While Cs exists in a bcc phase at 1 atm, we find in bcc calculations (140 points per irreducible wedge, 459 reciprocal lattice

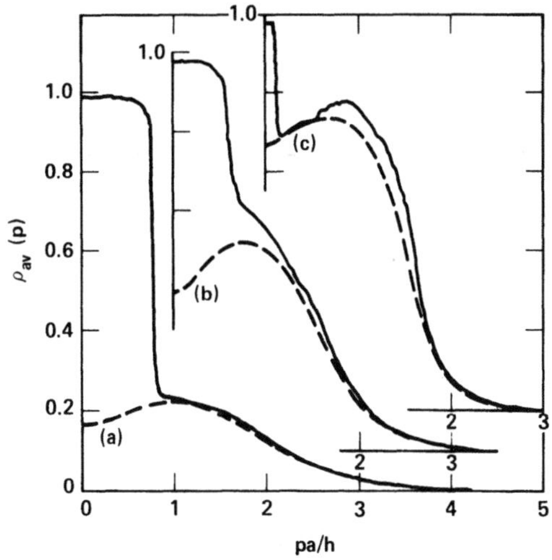

Fig. 1. Electron momentum density $\rho_{av}(p)$ vs. dimensionless momentum pa/h at a) V/V_0 = 1, b) V/V_0 = 0.418, and c) V/V_0 = 0.235. The solid and dashed lines show the total and core results, respectively.

vectors) that the valence $\rho_{av}(p)$ is nearly identical to that in Fig. 1(a), differing by less than about 1% of the solid curve, except in the small region of pa/h = 0.8 - 1.0, where the differences are still less than 0.02 on the scale given in the figure. Cesium is fcc at 4.2 GPa, and appears to be either tetragonal or orthorhombic at 15 GPa.[12]

The single valence electron in 1-atm Cs occupies sp-hybrid, 6s-band states near k = 0. This leads to the large valence density between the solid and dashed curves in Fig. 1(a) for pa/h < 0.78, as well as the small spherical average of umklapp echos from about pa/h = 1.4 to 2.2. Under compression, the rising 6s band causes the Fermi momentum p_F (where $\rho_{av}(p)$ is nearly vertical) to decrease. The total valence density beneath p_F drops from 0.73, to 0.22, and finally to 0.001 electrons in Figs. 1(a), 1(b), and 1(c), respectively. Near the conclusion of the s-d transition, Fig. 1(c) shows most of the valence $\rho_{av}(p)$ to be in the range pa/h ≈ 0.6 - 1.7, which arises from 5d states near the L and X Brillouin zone boundary points including a significant X-point umklapp contribution around pa/h = 1.414.

Even at 1 atm, there is significant overlap of the 5s, 5p cores in Cs. By 15 GPa, the 5p band is over 5 eV wide,[5] and so it should be no surprise that momentum density in these increasingly more itinerant states should shift to lower values of p as seen by the dashed curves in Fig. 3. Further reduction of volume be less than a factor of 2 results in a single, merged 5s, 5p, 5d valence band.[5]

The partial annihilation rates $R_{av}(p)$ are quite similar to Fig. 1; although, the relative importance of the 5s, 5p cores is reduced given the positron-nucleus repulsion. Given this similarity, we plot these results in an alternate form in Fig. 2, namely $4\pi p^2 R_{av}(p)$, so that the area is just the total annihilation rate λ. The counterpart of the 6s-band free-electron plateau in Fig. 1(a) is now the initial p^2 rise at small p in Fig. 2(a), followed, of course, by the rapid drop at p_F. This characteristic structure is greatly reduced in Fig. 2(b) at V/V_0 = 0.418, and is entirely gone by Fig. 3(c) at V/V_0 = 0.235 as the s-d transition is nearly completed. The growing partial annihilation rate from the newly occupied 5d-like states can be seen as

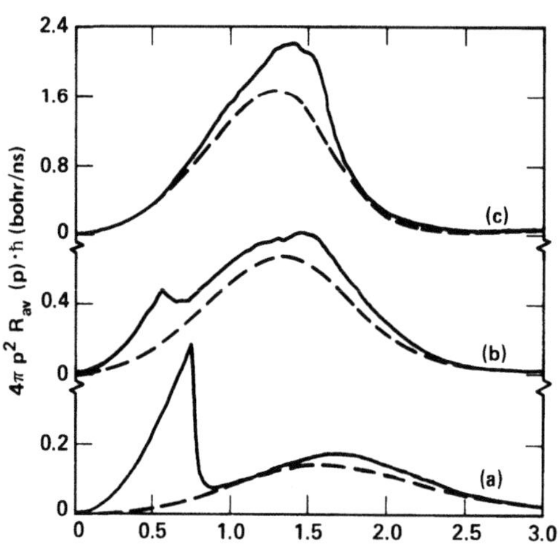

Fig. 2. Partial annihilation rates $4\pi p^2$ $R_{av}(p)$ vs. dimensionless momentum pa/h at a) V/V_0=1, b) V/V_0=0.418, and c) V/V_0=0.235. The solid and dashed lines show the total and core results, respectively. No enhancement was used.

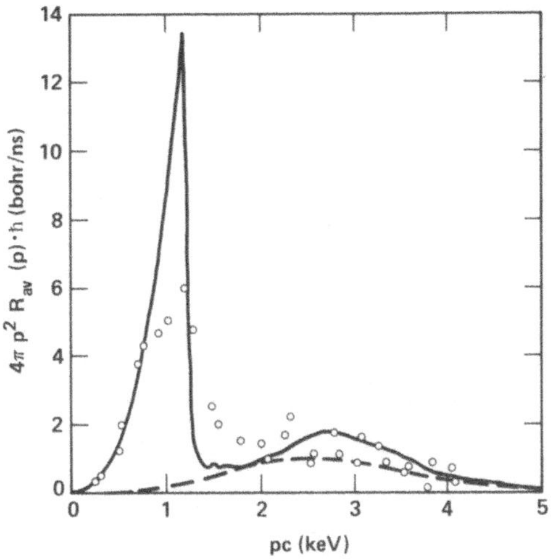

Fig. 3. Enhanced partial annihilation rate $4\pi p^2 R_{av}(p)$ vs. pc at $V/V_0=1$. The solid and dashed lines show the total and core results, respectively. The data points are from Ref. 14, and were normalized to agree with the solid curve at small momentum.

the increasing separation between the solid and dashed curves in the vicinity of $pa/h \approx 1.4$.

ENHANCED PARTIAL ANNIHILATION RATE

The question of what enhancement factor should be used for a given metal is still under study.[13] While it is beyond the scope of this preliminary work to deal with this question in any depth, we do provide in Fig. 3 an enhanced counterpart of Fig. 2(a) plotted against pc to give more familiar units. The factor $\epsilon(E)$ in Eq. (1) was taken from a recent paper by Sob who specified coeffients for Cs at 1 atm.[7]

Without enhancement, our results in Fig. 2(a) yield total annihilation rates of 0.093 and 0.068 ns^{-1} for core and valence electrons, respectively. With the enhancement, the curves in Fig. 3 yield values of 0.663 and 1.668 ns^{-1} for core and valence electrons, respectivley, in good agreement with Sob's analysis. The sum of core and valence contributions, 2.33 ns^{-1}, is within 3% of the experimental annihilation rate.[2,7] It should be noted that the extremely low electron density in Cs leads to both this unusually low annihilation rate as well as the very large enhancement in this material.

The data points in Fig. 3 are taken from a 1957 paper by Stewart.[14] They result from his differentiation of (instrumental-resolution corrected) long-slit angular correlation data, scaled by us to agree with the solid curve at small p. The disagreement between theory and experiment near the p_Fc = 1.25 keV is consistent with recent work by Chakraborty[15] which indicates that the Kahana-based enhancement used in Ref. 7 is likely to be too large just below p_F and too small just above p_F. Although there is considerable scatter in the range pc ≈ 2 - 4. keV, there is clear evidence for at least a shoulder to the main peak near pc ≈ 1 keV in agreement with the present calculations.

CONCLUSIONS

We suggest that the gross features of the s-d transition in compressed Cs should be clearly visible in high-pressure positron annihilation measurements even at resolutions obtainable with the Doppler-broadening technique. Specifically, the 6s-band dominated peak ($pa/h \approx 0.3 - 0.7$) seen in $4\pi p^2 R_{av}$ at 1 atm should decay under compression, as the shoulder region evolves into a new, broad, much larger peak ($pa/h \approx 0.8 - 1.7$) arising out of core and 5d zone-boundary contributions. Beyond these qualitative features, a quantitative comparison with experiment might prove highly informative in regard to the nature of enhancement effects in a range where core levels evolve into broad-band valence states.

ACKNOWLEDGMENTS

Part of this work (that of A.K.M.) was performed under the auspices of the U. S. Deparment of Energy by Lawrence Livermore National Laboratory under Contract No. W-7405-ENG-48, and part (that of H.L.S.) was supported by the Niels Bohr Foundation through the Royal Danish Academy of Sciences and Letters.

REFERENCES

1. See., e.g., M. Ross and A. K. McMahan, Phys. Rev. B 26, 4088, 1982; and H. L. Skriver, Phys. Rev. B 31, 1909 (1985), and references therein.
2. I. K. MacKenzie, R. LeBlanc, and B. T. A. McKee, Phys. Rev. Lett. 27, 580 (1971)
3. D. R. Gustafson and J. D. Willenberg, Phys. Rev. B 13, 5193 (1976).
4. H. B. Radousky, R. L. Reichlin and R. H. Howell, J. Phys. (Paris), C8, 369 (1984).
5. A. K. McMahan, Phys. Rev. B 29, 5982 (1984).
6. See, e.g., P. E. Mijnarends in Positrons in Solids, edited by P. Häutojarvi (Springer, Berlin, 1979), p. 25.
7. For 1-atm Cs, Sob [M. Sob, Solid State Commun. 54, 249 (1985)] suggests a=16.24, b/a=0.5, and c/a=0.4 in his Eq. (7) for the valence enhancement $\epsilon_v(E)$, and a constant core enhancement factor of 7.1.
8. A. K. Singh and T. Jarlborg, J. Phys. F 15, 727 (1985).
9. O. K. Andersen, Phys. Rev. B 12, 3060 (1975); H. L. Skriver, The LMTO Method (Springer, Berlin, 1984).
10. O. Jepsen and O. K. Andersen, Solid State Commun. 9, 1763 (1971). See last column in Table 1.
11. U. von Barth and L. Hedin, J. Phys. C 5, 1629 (1972).
12. K. Takemura, S. Minomura, and O. Shimomura Phys. Rev. Lett. 49, 1772 (1982); K. Takemura and K. Syassen in Proc. of the Int. Symp. on Solid State Physcis under Pressure, edited by B.I.N. Okai (Academic Press of Japan, Tokyo, 1984), and references therein.
13. See, e.g., Positron Annihilation, edited by P. G. Coleman, S. C. Sharma, and L. M. Diana (North-Holland, Amsterdam, 1982).
14. A. T. Stewart, Can. J. Phys. 35, 168 (1957).
15. B. Chakraborty, Phys. Rev. B 24, 7423 (1981).

SHORT RANGE INTERACTIONS IN DIATOMIC MOLECULES

P. J. Haskins and M. D. Cook

Royal Armament Research and Development Establishment

Fort Halstead, Sevenoaks, Kent TN14 7BP, England

Ab-initio molecular orbital calculations are being carried out to study many-body effects for diatomic molecules in the close-in repulsive region. Preliminary results are reported here for H_2, N_2 and CO. These results indicate that at least 4-body terms must be considered to obtain a good description of the potential for H_2 at separations below 1.8 A°. In N_2 and CO 3-body effects appear to become significant at separations below 2.8 A°.

1. INTRODUCTION

The short-range interaction between hydrogen molecules has been the subject of a number of theoretical studies and has been reviewed by McMahan et al. (1). High pressure hydrogen is both of interest in its own right and to theoreticians as the simplest molecule on which to test the accuracy of their methods. An accurate knowledge of the repulsive part of the potential is necessary if predictions regarding the molecular-metallic phase transition are to be made, and such potentials are also of considerable interest to astrophysicists since high pressure hydrogen is a major constituent of stars and certain planets.

Nitrogen and carbon monoxide, the other molecules considered in this paper, are both products of the detonation of many explosives. Consequently, intermolecular potentials appropriate to the high pressure and temperature regime behind the detonation front are of great importance in calculating equations of state for these product molecules.

In this paper we report some preliminary investigations into the influence of many-body effects on the interactions in H_2, N_2 and CO at high pressures. All calculations have been carried out at the Hartree-Fock SCF level using the code GAUSSIAN 76 (2). At short separations the SCF method is a good approximation and although the inclusion of electron correlation would improve the accuracy we did not consider it justified for preliminary studies. It is worth noting that in their study of 3 x H_2 Ree and Bender (3) found the SCF and CI estimates of the ratio of 3-body interaction energy to the pair energy almost identical. The calculations have all been carried out using fixed bond lengths. The bond lengths being (H_2 = 0.7326 A°, N_2 = 1.0920 A°, CO = 1.1340 A°) those obtained by minimising the energy of an isolated molecule using the same basis set as that employed in the many-body calculations.

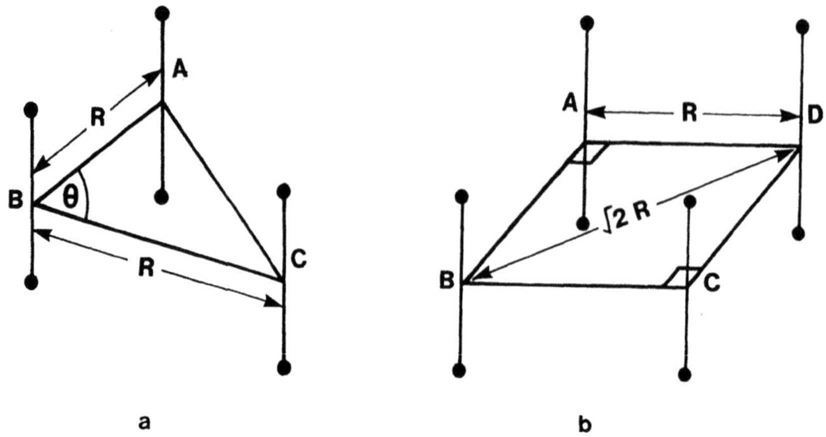

Fig 1. (a) 3 x H_2 and (b) 4 x H_2 geometries used in the calculations

2. MOLECULAR HYDROGEN

Ree and Bender (3) have carried out SCF and CI calculations for the 3 x H_2 system shown in Fig 1a. They showed that the pairwise additivity assumption fails at separations below 2.4 Å. By generating an 'effective' pair potential from their 3-body calculations they were able to obtain reasonable agreement with experimental pair potentials down to separations of $\approx$1.8 Å. However, below 1.8 Å, the inclusion of 3-body terms led to an excessive lowering of the potential, suggesting the existence of 4- and more-body terms with an overall repulsive nature.

In this work we have repeated Ree and Bender's SCF calculations for 3 x H_2 with $\theta = 60^\circ$ and 90°, and extended the calculations to look at one particular 4 x H_2 arrangement (Fig 1b) to assess the likely magnitude of higher order effects. The calculations employed a contracted gaussian basis set with two s and one set of 2p functions on each hydrogen atom and 6 gaussians per contraction. The calculations on 3 x H_2 are summarised in Fig 2, which shows the 3-body energy (E_{ABC}) as a percentage of the total pair interaction energy ($E_{AB} + E_{AC} + E_{BC}$) plotted against the separation (R). It can be seen that the 3-body term leads to a significant lowering of the intermolecular potential for these geometries.

The results for the 4 x H_2 arrangement (Fig 1b) are shown in Figs 3 and 4. In Fig 3 the 4-body term (E_{ABCD}) and the 3-body term ($E_{ABC} + E_{BCD} + E_{CDA} + E_{DAB}$) are both expressed as percentages of the total pair interaction energy. It is clear that for this geometry the 4-body term represents a significant repulsive correction which increases at short separations.

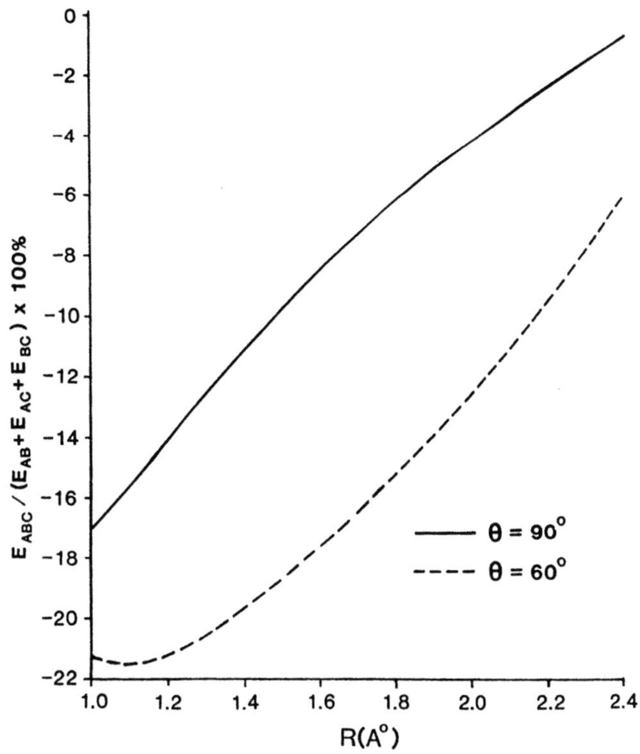

Fig 2. Calculations on 3 x H_2 showing the 3-body energy (E_{ABC}) as a percentage of the total pair interaction energy (E_{AB} + E_{AC} + E_{BC}) plotted against the separation (R).

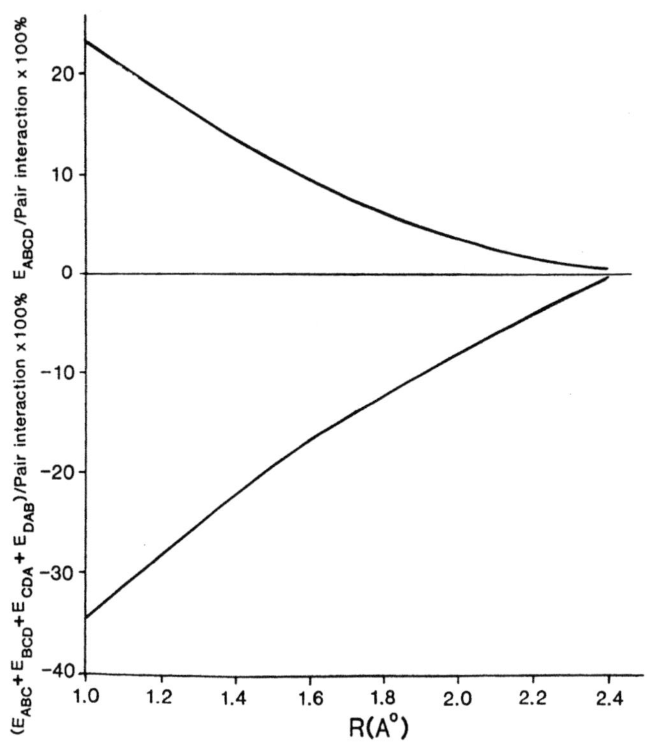

Fig 3. Calculations on 4 x H_2 showing the 4-body term (E_{ABCD}) and the 3-body term (E_{ABC} + E_{BCD} + E_{CDA} + E_{DAB}) expressed as percentages of the total pair interaction energy, plotted against separation (R).

Fig 4 shows the total interaction energy for the 4 x H_2 system as calculated by (i) a full 4-body calculation, (ii) a calculation incorporating 2 and 3-body terms only, and (iii) a calculation including pair interactions only. The 'exact' (ie full 4-body calculation) potential is clearly seen to lie between the pair-only and 2 and 3-body potentials at all the separations studied. Analysis of the results shows that at separations from 2.4 - 1.9 A° the inclusion of 3-body terms leads to a potential which, although low, is closer to the 'exact' potential than that based on pair interactions only. However, at separations of 1.8 A° and below the 'exact' potential lies closer to the pair-only potential than it does to that corrected for 3-body terms. This clearly supports the assertion by Ree and Bender (3) that 4- and more body-terms must significantly stiffen the potential in this close-in region.

In the near future we propose examining 4- and more-body terms in a range of geometries with a view to generating an 'effective' pair potential for the H_2 crystal. Such a potential may then be used in conjunction with an effective potential for metallic hydrogen (eg that due to Neece et al (4)) to predict the location of the 0° K insulator-metal transition. The metallic potential of Neece et al (4) has been used with various molecular potentials to yield transition pressures in the range 0.84 - 4.2 Mbar (1). The more recent experimental potentials and the 3-body calculations of Ree and Bender (3) suggesting the transition to be at the upper end of this range. However, since this corresponds to $R < 1.6$ A° it is clear that 4 and possibly more-body terms are important and an accurate theoretical determination of the transition pressure will not be possible without inclusion of these effects. The repulsive nature and magnitude of the 4-body term in the geometry studied here suggests that agreement between experimental and theoretical potentials should be improved by the inclusion of 4-body effects.

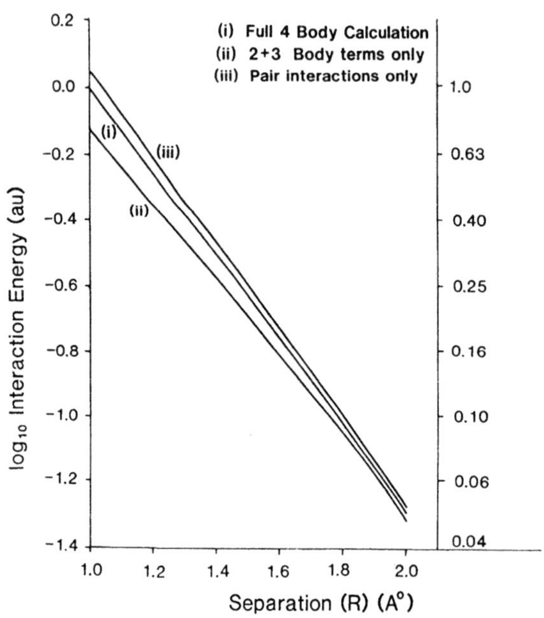

Fig 4. Total interaction energy for the 4 x H_2 system as calculated by (i) a full 4-body calculation, (ii) a calculation incorporating 2 and 3-body terms only and (iii) a calculation including pair interactions only.

3. NITROGEN AND CARBON MONOXIDE

The $N_2 - N_2$ interaction in the repulsive region has been the subject of recent study by SCF (5,6), CI (6) and many-body perturbation theory (6) methods. Hay et al (6) found that the SCF pair potential was more repulsive than that suggested by experiment. Whilst the inclusion of electron correlation softened the potential it was not adequate to obtain good agreement with the experimental data and they concluded that many-body forces were probably responsible for the remaining discrepancy.

Here we report the results of preliminary calculations at the SCF level to assess the likely magnitude of the 3-body interaction for N_2 and the iso-electronic molecule CO. To date calculations have been restricted to the equilateral triangle geometry (Fig 1a, with $\theta = 60^{\circ}$) as this was found to maximise the 3-body contribution for hydrogen. The calculations have all used a contracted gaussian basis with one 1s and two 2s and 2p sets on each atom, with 6 gaussians per contraction. The results for N_2 and CO are summarised in Fig 5 which shows the 3-body energy (E_{ABC}) as a percentage of the total pair interaction energy.

The results illustrated in Fig 5 feature two curves for carbon monoxide labelled CCC and COC, these refer to the two distinct arrangements which are possible with a heteronuclear diatomic, namely all molecules orientated the same way (CCC), or one molecule inverted with respect to the others (COC). It can be seen that in all three cases there is a significant reduction in the interaction energy as a result of the 3-body term. Measured as a percentage of the pair interaction it can be seen that N_2 and CO in the CCC orientation give rise to almost identical 3-body energies, while the energetically more favourable COC arrangement for CO gives rise to a slightly smaller 3-body correction.

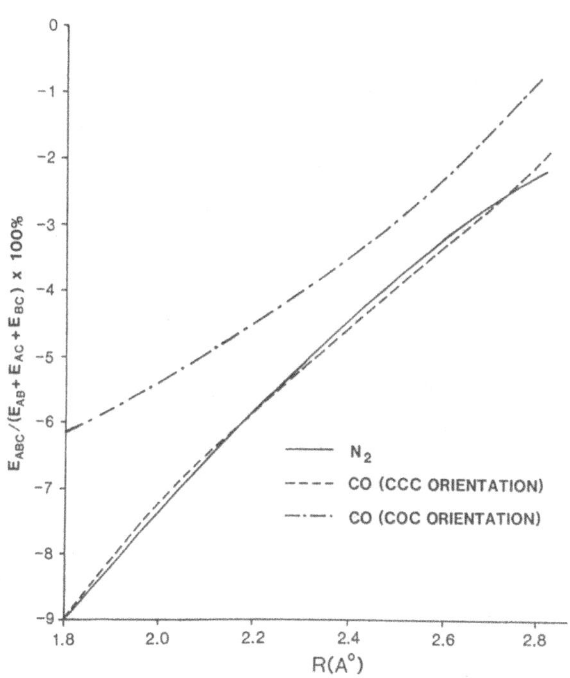

Fig 5. Plot of the 3-body energy (E_{ABC}) as a percentage of the total pair interaction energy against R.

These preliminary calculations indicate that, for both N_2 and CO, 3-body terms may be expected to be of importance at separations of 2.8 Å and below. It may be expected that the effect of the 3-body interaction will be to soften the intermolecular potential and improve agreement with experimental Hugoniot data for N_2 (7,8,9). A more complete study of many-body interactions in N_2 and CO is underway with the object of obtaining effective potentials in good agreement with experimental data.

4. CONCLUSIONS

It has been shown that for molecular hydrogen in the close-in repulsive region 4-body energies can be of comparable magnitude (but opposite sign) to the 3-body terms. A full analysis of such 4- and possibly more-body effects should lead to improved agreement between theoretical and experimental potentials in the region $R < 1.8$ Å.

Nitrogen and carbon monoxide have both been shown to exhibit significant 3-body interaction energies at separations below 2.8 Å. It therefore seems likely that many-body effects will need to be considered if potentials suitable for studies of the high pressure detonation product regime are to be calculated.

5. REFERENCES

1 McMahan, A K, Beck, H and Krumhansl, J A, Phys.Rev.A 9, 1852 (1974).

2 Pople, J A et al, QCPE 11, 368 (1978).

3 Ree, F H, and Bender, C F, Phys.Rev.Lett. 32, 85 (1974).

4 Neece, G A, Rogers, F J and Hoover, W G, J.Comput.Phys. 7, 621 (1971).

5 Ree, F H and Winter, N W, J.Chem.Phys. 73, 322 (1980).

6 Hay, P J, Pack, R T and Martin, R L, J.Chem.Phys. 81, 1360 (1984).

7 Marsh, S P, editor, LASL Shock Hugoniot Data, pp 113-114, University of California, Berkeley (1980).

8 Zubarev, V N and Telegin, G S, Sov.Phys.Dokl. 7, 34 (1962).

9 Nellis, W J and Mitchell, A C, J.Chem.Phys. 73, 6137 (1980).

A NEW FLUID INTEGRAL EQUATION

APPLICATION TO THE EQUATION OF STATE OF XENON

J.R. Jevais and G. Zerah

Commissariat à l'Energie Atomique
B.P. n°27
94190 Villeneuve-Saint-Georges (France)

INTRODUCTION

Fluid integral equations are a valuable tool for studying structural and thermodynamical properties of classical fluids at equilibrium. A renewed interest for this kind of theory stems from the appearance of new models able to treat accurately a wide spectrum of systems, in various conditions (neutral or charged pure fluids, mixtures).

Recently, such an integral equation has been devised[1], which combines an interpolation idea from Rogers and Young[2], and the cesure of the potentiel proposed by Weeks et al[3]. This integral equation, which reduces to the Rogers and Young closure in the case of a purely repulsive potentiel, yields an interpolation between the SMSA and HNC closure. Hence we called it HMSA equation. In reference 1, the capabilities of the model have been examined, and we use it here in the conditions generally encountered behind strong shock waves.

Two rather different systems will be examined : a rare gas (Xenon) and a model of a monovalent molten salt.

It is well known that the thermal electronic excitations play an important role and this quantity will be computed with the help of Liberman's self consistent field model A.

THEORETICAL DESCRIPTION

Given a pair potential $V_{ij}(r)$ between species i and j of respective densities ρi and ρj, one may compute the pair correlation function $g_{ij}(r)$ with the help of the following system of equations. It is solved with the help of the Newton's method proposed by one the authors (50 iterations are typically required in the case of a molten salt).

$$(1)\ g_{ij}(r) = \exp - \beta V_{ij}^{(1)}(\exp f(r)(-\beta V_{ij}^{(2)}(r)+h_{ij}-C_{ij})-1)/f(r)+1)$$

$$(2)\ h_{ij}(r) = C_{ij} + \sum_k \rho_k h_{ik} * C_{kj} \qquad (*\text{: convolution product})$$

$$(3)\ h_{ij}(r) = g_{ij}(r) - 1$$

In those equations, h_{ij}, C_{ij}, refer to the usual correlation and direct correlation function. $V^{(1)}_{ij}$, $V^{(2)}_{ij}$ are two complementary projections of the full potential V_{ij} defined according to WCA. Note that in the case of a coulombic system, only the "cation-anion" potential will be affected.

Once, these fucntions are computed, one may evaluate thermodynamical quantities with the help of the usal quadratures :

$$(4)\quad \frac{E}{NkT} = \frac{3}{2} + \frac{1}{2} \sum_{i,j} x_i\, x_j \int V_{ij}(r)\, g_{ij}(r)\, d^3r$$

$$(5)\quad \frac{PV}{NkT} = 1 - \frac{1}{6} \sum_{i,j} x_i\, x_j \int \frac{\partial V_{ij}(r)}{\partial r}\, r_{ij}\, g_{ij}(r)\, d^3r$$

$$(6)\quad \frac{1}{kT}\frac{\partial \rho}{\partial \rho} = 1 - \sum_{i,j} x_i\, x_j \int C_{ij}(r)\, d^3r$$

In these relations, x_i, x_j are the number concentrations of both species. ρ is the total number density.

Finally, the crossover function f (r), for which we take the form proposed by Rogers and Young, is settled to insure thermodynamical consistency between the ρ derivate of P/kT given by equation 5 and $\partial(p)/kT\partial\rho$ given by equation 6.

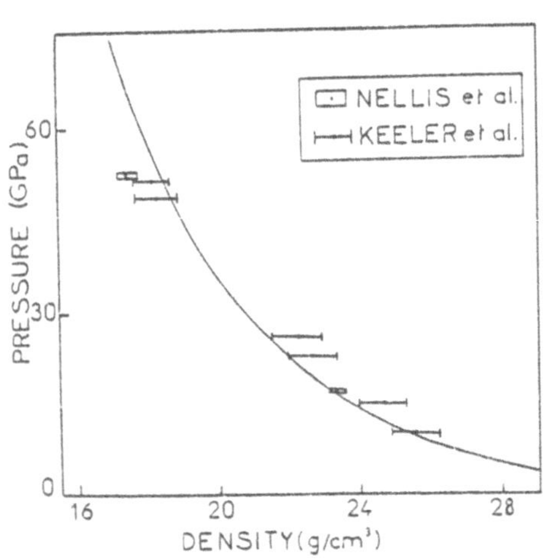

Fig.1 : Principal Hugoniot of Xenon computed with HMSA model.

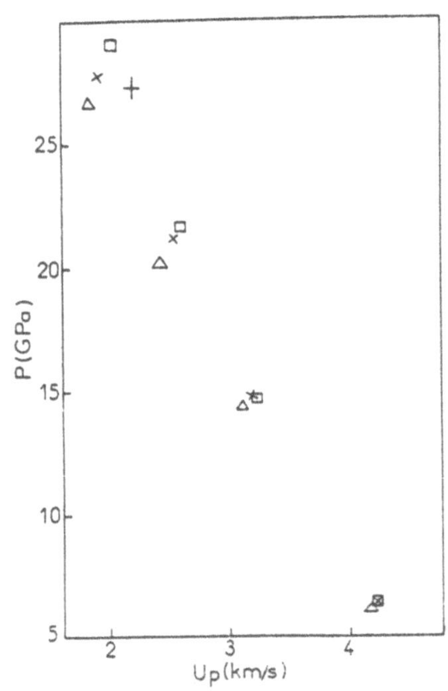

Fig.2 : Release isentrope of a high explosive reconstructed with our model (x), Ross model (Δ) and Kerley model (□). The cross (+) gives experimental uncertainty.

APPLICATION TO XENON

Principal Hugoniot

Xenon has been the object of a recent theoretical and experimental interest. one of the most commonly used experimental data, against which theories are tested is the principal Hugoniot of cooled Xenon. Here we have computed the Hugoniot with the help of the fluid model above defined, and liberman's SCF model A for the electronic excitations.

The pair potential used here is of the exp-6 form. The parameters , , have been obtained by a law of corresponding state frome those determined for Argon.

$$V = \varepsilon \left[\frac{6}{\alpha-6} \exp\left(\alpha(1 - \frac{r}{r^*}) \right) - \frac{\alpha}{\alpha-6} \left(\frac{r}{r^*}\right)^6 \right]$$

$\varepsilon/k = 235$ K , $\alpha = 13.0$, $r = 4.47$ Å

The results are shown in figure 1, together with the experimental data of Nellis et al[6] and Keeler et al[7]. A good agreement is observed below 60GPa

In fact, two other models have been proposed to study shocked Xenon. One model devised by Ross and Mac Mahan[8] combines a rather similar fluid model and a semi conductor model with a density dependent band gap for the electronic excitation. The other one, proposed by Kerley[9] combines CRIS fluid model with a combination of Liberman's SCF and Saha models. Both models are reported to yield good results along the principal Hugoniot. This fact calls for the use of a different approach to test Xenon EOS.

analyse of release isentropes of high explosives

Due to the high compressibility of gases, it is possible to get experimental informations along the intersections between the release isentrope of explosives and the principal Hugoniot starting with different densities of the gases. Such results have been recently reported in the case of Argon [10].

In order to compare the results yielded by the three models, we have generated a set of initial densities and shock velocities that would be encountered in a typical experiment of this type. This set was generated so that the Hugoniot computed with our EOS should match the release isentrope of a high explosive computed with the help the ETARC CODE[11]. It is listed in table 1. The release isentropes, reconstruded using the three different models for this set of four points, are shown in figure 2. At high pressure ($P > 20$ GPa) we observe differences which are larger than experimental or theoretical uncertainties. Moreover, due to the attained temperatures and densities, the electronic excitations play an important role and we see that both the fluid models and the electronic models disagree.

In fact, we are working in a rather difficult region of the gas, where many effects such as band gap narrowing or thermal ionization, come into play in a rather subtle way.

The magnification of rather small inaccuracies (W.R.T the Hugoniot) is such that experiments using Xenon as reference system and independent computations of release isentrope of explosives could be a useful test for rare gases EOS.

Table 1 : set of initial densities ρo and shock velocities Us in Xenon

To (K)	280	280	280	280
ρo (g/cm^3)	3.50	2.05	1.05	0.3
Po (GPa)	.42	$.13x10^{-2}$	$6x10^{-3}$	$4.2x10^{-3}$
Us (km/s)	4.091	4.085	4.327	4.989
Ti (K) [a]	7220	15100	20800	26200
ρi (g/cm^3) [a]	6.51	5.30	4.00	1.87

[a] T i, ρi are the temperature and the pressure of shocked Xenon on the Hugoniot of initial condition (ρo, To).

APPLICATION TO A MOLTEN SALT

Molten salts have been subject of a recent theoretical and experimental interest. This interest stems from the fact that they are isoelectronic to a rare gas (for example CsI is isolectronic to Xenon) and that they are an experimental realization of a strongly coupled two components plasma. Indeed it has recently been argued, that under pressure, their structure develops into a close packed like,controlled by repulsive forces. Hence along the freezing line, one will encounter a continuous shift from a NaCl like coulombic structure to a close packed structure.

This rather complex behaviour calls for the use of a fluid theory, able to cope with such rather dissimilar situations. We believe that the integral equation above presented is an interesting step in that direction.

We compared our calculations with M.D. simulation of Hansen and Mac Donald [12], for simple molten salts, in conditions very similar to those of triple point of NaCl.

Simple molten salt

The simple molten salt exhibits the features of more general salts, without, the well known pathologies of charged hard spheres. The interaction potential is of the form

$$V_{ij}(r) = \frac{e^2}{\lambda}\left[\frac{1}{9}\left(\frac{\lambda}{r}\right)^9 + Z_iZ_j\left(\frac{\lambda}{r}\right)\right]$$

Where Zi are the ionic charges in units of e, λ is a characteristic lenght at which the 'cation-anion' potential is a minimum.

HNC equation is known to overestimate the height of the first maximum of g_{+-} and underestimate the one of g_{++}[12]. As a consequence the pressure is too high. Similar remarks apply to the structure facter and inverse compressibility. The next set of values and figure 3 show how HMSA remedy those deficiencies.

P/NkT = 0.32 (MD), 0.96 (HNC), 0.30 (HMSA)

S (o) = 0.16 (MD), 0.50 (HNC), 0.143 (HMSA)

U/NkT = -33.52 (MD), -33.42 (HNC), -33.50 (HMSA)

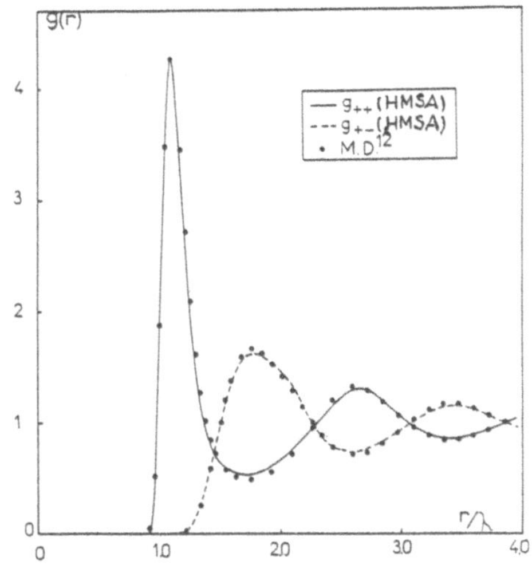

Fig.3 : Radial distribution function of a simple molten salt, from HMSA (solid curve and dashed curve) and from M.D. simulation (.)

CONCLUSION

Contrary to the couple HNC - PY which was previously studied, the couple HNC-SMSA always furnishes the way to a thermodynamicaly consistent interpolation. The result of this interpolation, which we call the HMSA equation leads to improved thermodynamical properties for a wide range of systems.

ACKNOWLEDGMENTS

We thank Dr. J.P Hansen for valuables discussions. We wish also to thank R. Chirat for having given us results from his ETARC code.

1. J.P. Hansen and G. Zerah, to be published
2. F.J. Rogers and D.A. Young, Phys. Rev. A,30:999 (1984)
3. J.D. Weeks, D. Chandler and H.C Andersen, J. chem. Phys., 54:5237 (1971)
4. W.G. Madden and S.A. Rice, J. chem. phys., 72:4208 (1980)
5. G. Zerah, to be published in J. comp. Phys.
6. W.J. Nellis, M. Van Thiel and A.C Mitchell, Phys. Rev. Let., 48:816 (1982)
7. R.N. Keeler, M. Van Thiel and B.J. Alder, Physica, 31:1437 (1965)
8. M. Ross and A.K. Mc Mahan, Phys. Rev. B, 21:1658 (1980)
9. SESAME 1983 : Report on the Los Alamos equations of state library, LALP - 83.4, February 1983
10. R. Chirat and J. Baute, 8th symposium on detonation, July 15 - 19, 1985, Albuquerque, USA
M. Pinegre , J. Aveille et al, conf. DEA 7304, Albuquerque (1984)
11. R. Chirat, G. Pittion - Rossillon, 7th symposium on detonation (1981)
12. J.P. Hansen and I.R. Mc Donald, Phys. Rev. A, 11:2111 (1975).

STATIC COMPRESSION OF GOLD AND COPPER AND CALIBRATION OF THE RUBY PRESSURE SCALE TO PRESSURES TO 1.8 MEGABARS (STATIC.RNO)

P. M. Bell, Ji-an Xu, and H. K. Mao

Geophysical Laboratory
Carnegie Institution of Washington
Washington, D.C. 20008

ABSTRACT

New experimental results on the static compression of gold and copper in the diamond-window, high-pressure cell have provided the basis for an accurate system of pressure calibration at ultrahigh pressures. The unit-cell volumes of gold and copper were measured at several points along the pressure curve by MoKα x-ray diffraction. Diffraction measurement was by film methods. At each point, the luminescent spectrum of ruby also contained in the high-pressure cell was measured. The shift with pressure of the luminescent *R*-line emission of ruby was calibrated to the shock-wave equation of state of gold and copper to 1.8 Mbar.

INTRODUCTION

The required range for calibration of pressures in the diamond-window high-pressure cell has been extended to maximum pressures exceeding 2.5 Mbar (Bell et al., 1984). Internal standards for diamond-cell experiments such as the ruby pressure scale have been related to shock-wave equations of state of metals for primary reference (Piermarini et al., 1975; Mao et al., 1978). Shock-wave data provide pressure, volume, and temperature values so that the equations of state of standard metals can be referenced in static experiments by means of *in situ* measurements of the metal unit cell volumes. These measurements are made by x-ray diffraction of the metal samples while they are held under pressure by the diamond-window cell, as described by Mao *et al.* (1978). Two of these metals, gold (Au) and copper (Cu), were employed in the present experiments to pressures up to 1.8 Mbar to calibrate the shifts with pressure of the ruby R_1 luminescent line, which are measured by optical spectroscopy at the same time on the x-ray diffraction measurements.

Gold and copper were selected as potential standards because these metals cover a broad range of compressibility and pressure without experiencing phase changes, because they are cubic, and because their unit-cell volumes can be measured in the diamond cell conveniently by x-ray diffraction techniques. These metals have been used extensively in the past as internal pressure-volume calibrants. Gold has been studied extensively in diamond-cell experiments (Jamieson et al., 1982), and its equation of state has recently been confirmed by experiment to 700 kbar for the purpose of calibration (Heinz and Jeanloz, 1984). Copper has also been studied in calibration experiments, but to higher static pressures (1 Mbar; Mao et al., 1978). The present experiments thus increase the working pressure calibration of the ruby scale by more than a factor of two.

EXPERIMENTS

Samples of the two experimental runs in this study consisted of a mixture of 20-mesh powdered gold (99.9999% purity; Alfa Products Inc.), and 10 μm grain diameter powdered copper (99.5% purity, Alfa Products Inc.), on top of which were placed crushed fragments of a ruby laser rod (0.5% Cr^{3+}, 1 μm grain size, Adolf Williams Co.). The experiments were performed in a recent design of the diamond cell (Bell et al., 1984) that had the following diamond-anvil parameters: experimental run 1, $\underline{A}$ = 0.61 mm, $\underline{B}$ = 0.11 mm, Θ = 7°; experimental run 2, $\underline{A}$ = 0.32 mm, $\underline{B}$ = 0.21 mm, Θ = 10°. (See explanation of symbols in Fig. 1.) The samples were contained in the sample chamber of a T301 (full hardened) stainless steel gasket. The procedure in each case was to pre-indent the gasket by pressurizing to 0.3 Mbar. Afterward, 50 and 75 μm sample chamber holes were drilled in the center of the indentations of the gaskets for experimental runs 1 and 2, respectively. Sample preparation included grinding a 5:1 mixture (by volume) of copper and gold and then filling the sample chamber with ground powder. Powdered ruby was then placed on the surface. The sample chambers expanded to diameters of 70 and 100 μm, respectively, in the two runs at pressures above 0.3 Mbar.

X-ray diffraction of the pressurized samples was performed with a MoKα radiation beam collimated to a diameter of 50 μm. A metal foil of pure Ta was used as an external standard to compute the sample-to-film distance by the method described by Mao et al. (1978). Exposure times of 2-4 days were required to obtain accurate $\underline{d}$ values of the (111) and (200) lines of gold and copper.

A discussion of the relationship between the x-ray beam diameter, the sample diameter, and the pressure gradient is appropriate. The diameter of the sample chamber (70 and 100 μm in the two experimental runs) is considerably larger than the x-ray beam diameter (50 μm), as can be seen in the diagrams shown in Fig. 1. Location of the beam at the center was adjusted to an accuracy of 2 μm. As discussed below, a pressure gradient existed within the 50-μm central zone, so the unit-cell volumes are averages of the sample over the gradient. Thus, it was necessary to measure the wavelength of the ruby emission at several points within the 50-μm diameter and obtain an average wavelength shift for the circular area sampled by the x-ray beam. This average shift is used to obtain the average pressure.

The wavelength of the ruby $\underline{R}_1$ luminescent lines was measured by two separate laser spectrometer systems. At pressures up to 1.8 Mbar the wavelengths were first measured by an epi-illuminated (vertical) optical laser-path system coupled to a 1/2 m, first-order Ebert spectrometer. The laser focal spot, which was 5 μm in diameter, followed a computer-controlled locator grid with 10-μm spacing between grid points (Mao et al., 1983). The system simply followed the locator grid automatically recording and storing a ruby emission spectrum at each point. The stored spectra were reduced in real time to values of pressure and plotted as contoured isobars corresponding to the surface of the diamond anvils.

The average pressure was determined by adding the pressures measured at grid points and by dividing by the number of points. This method of averaging was used because it was assumed that the x-ray diffraction technique resulted in an average measure of the unit cell volume over the same area. Because of the circular symmetry involved, the average pressure is thus close to the lowest pressure. The maximum pressure was measured at a point located approximately at the center of the cirular area. The profiles of the pressure gradients and maxima are similar to those described by Mao *et al.*, 1985. At pressures above 1.8 Mbar, fluorescence background from the diamonds was too strong for observation of the ruby luminescence with this system.

Table 1. Volume Compression of Gold (Au) and Copper (Cu)[a]

Gold			Copper		Ruby Fluorescence (Pressure)			
V/V_0	P,[b] Mbar	P,[c] Mbar	V/V_0	P,[b] Mbar	$\Delta\lambda_{ave}$, nm	P_{ave}, Mbar	$\Delta\lambda_{max}$, nm	P_{max}, Mbar
				Experimental Run 1				
0.867	0.38	0.35			13.0	0.37	14.9	0.43
0.819	0.61	0.58			21.3	0.62		
0.792	0.77	0.73	0.734	0.91	26.1	0.77	30.4	0.91
0.758	1.04	0.99	0.718	1.03	33.2	1.00	34.5	1.04
0.731	1.27	1.23	0.688	1.29	38.8	1.19	49.0	1.55
0.684	1.82	1.79	0.637	1.86	56.7	1.83	61.9	2.03
0.682	1.85	1.82	0.636	1.88	58.0	1.88	64.0	2.11
				Experimental Run 2				
			0.884	0.22	6.7	0.19	7.4	0.21
			0.788	0.59	20.2	0.59	22.4	0.66
			0.745	0.84	28.2	0.84	33.5	1.01
			0.737	0.89	29.5	0.88	33.6	1.12
0.757	1.05	1.00	0.709	1.10	34.4	1.04	40.9	1.26
			0.690	1.27	38.5	1.18	48.2	1.52
0.735	1.23	1.19	0.695	1.22	38.8	1.19	51.5	1.64
0.728	1.30	1.26	0.682	1.35	44.5	1.39	53.3	1.71
			0.669	1.48	47.1	1.48	58.5	1.90
			0.663	1.55	50.0	1.58	58.9	1.91

[a] V_0, volume at 1 atm; V, volume at pressure, P; $\Delta\lambda$, shift of ruby R_1 fluorescent line; subscripts ave, max, average or maximum $\Delta\lambda$ or P within the collimated spot of the x-ray beam. Experimental run 1, A = 0.61 mm, B = 0.11 mm, Θ = 7°; experimental run 2, A = 0.32 mm, B = 0.21 mm, Θ = 10°, are the dimensional parameters of the diamond anvils (see Fig. 2 for explanation).

[b] Gray, 1972.

[c] Heinz and Jeanloz, 1984.

The second ruby emission measurement system had an optical path whereby the incident laser beam was focused onto the pressurized sample at an angle of 45°. The wavelength of the ruby emission from the sample was measured with an imaging system employing an intensified diode array detector (Princeton Instruments Co., OSMA). A nonscanning, three-stage, polychromator spectrograph (Spex Industries, Triplemate) was coupled to receive the signal from the sample and aligned to the diode array detector. Ruby luminescence was measured for comparison with the scanning system to 1.8 Mbar with excellent agreement. The reported values are an average of the two techniques. The diode array system was considerably more sensitive, and the signal to noise ratio was enhanced because of greatly reduced background from diamond fluorescence at the 45° angle of laser incidence. This system was used by itself in the ruby measurements at pressures above 1.8 Mbar. The intensity of the ruby luminescence was weak, but not absent at pressures above 2 Mbar as previously believed (Bell et al., 1984). The strong fluorescence of diamond that appears at pressures of approximately 1.1 Mbar and increases in intensity as the pressure is raised to 2 Mbar diminishes at pressures above approximately 2.1 Mbar. The ruby emission lines, although weak, are thus more readily detectable at pressures in the range 2.1-3 Mbar than in the range 1.1-2.1 Mbar.

RESULTS

The data obtained in this study have been formulated in Table 1 as values of $\underline{V}/\underline{V}_0$ ($\underline{V}_0$ is the 1-bar value of specific volume calculated from measured unit-cell dimensions, $\underline{V}$ is the specific volume at pressure $\underline{P}$) of gold and copper. The pressure, $\underline{P}$, in Mbar for both metals for the equivalent $\underline{V}/\underline{V}_0$ is determined from two sources of shock-wave reduced data. One is from the American Institute of Physics Handbook (Gray, 1972); the other, the analysis of Heinz and Jeanloz (1984) for gold. Also listed for the two experimental runs are the measurements of wavelength shift of the ruby $\underline{R}_1$ line, both as averages over the x-ray focal spot diameter on the sample and as maximum values within the area of the x-ray beam focal spot. The pressures listed, corresponding to the average and maximum wavelength shifts, were determined from the ruby calibration scale of Mao et al. (1978) according to the following relationship:

$$\underline{P}(\text{Mbar}) = 3.808\{[1 + (\Delta\lambda/\lambda)]^5 - 1\} \qquad (1)$$

where λ is the wavelength of the ruby $\underline{R}_1$ line at 1 bar and $\Delta\lambda$ is the shift in wavelength measured at pressure, $\underline{P}$.

A plot of the data for the two experimental runs is provided in Fig. 2. The values for gold are based on the two equations of state referenced in Table 1. The curve plotted is calculated from equation 1. The close fit of all the data is a confirmation of the Mao et al. (1978) calibration. These results indicate that the calibration curve (equation 1) is valid to pressures of at least 2 Mbar.

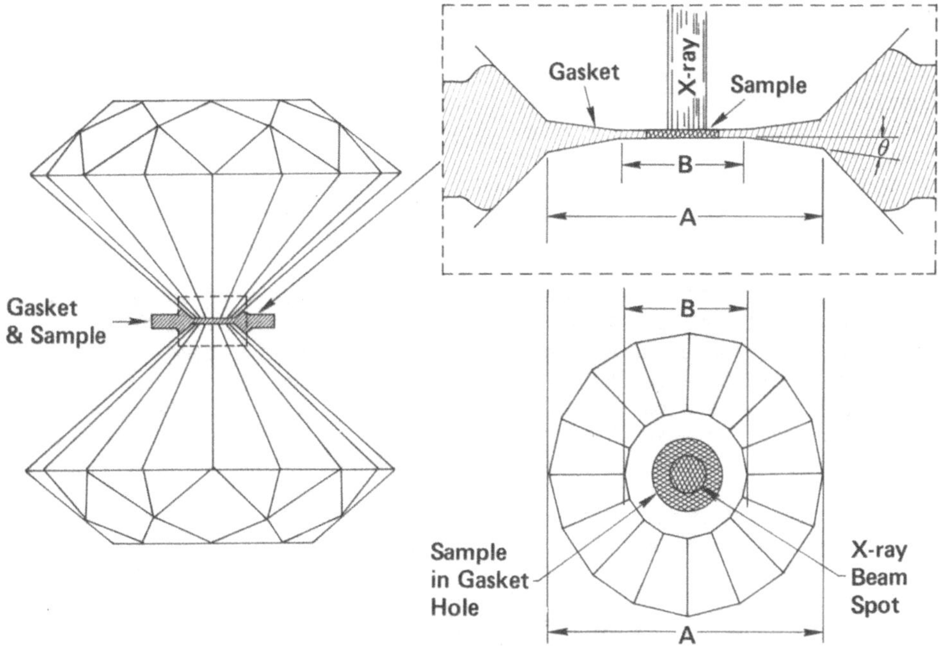

Fig. 1. Diagrammatic sketch of the diamond anvils, x-ray beam focal spot, sample diameter, and anvil surface bevel parameters, A, B, and θ. (See text for values and thus scale.)

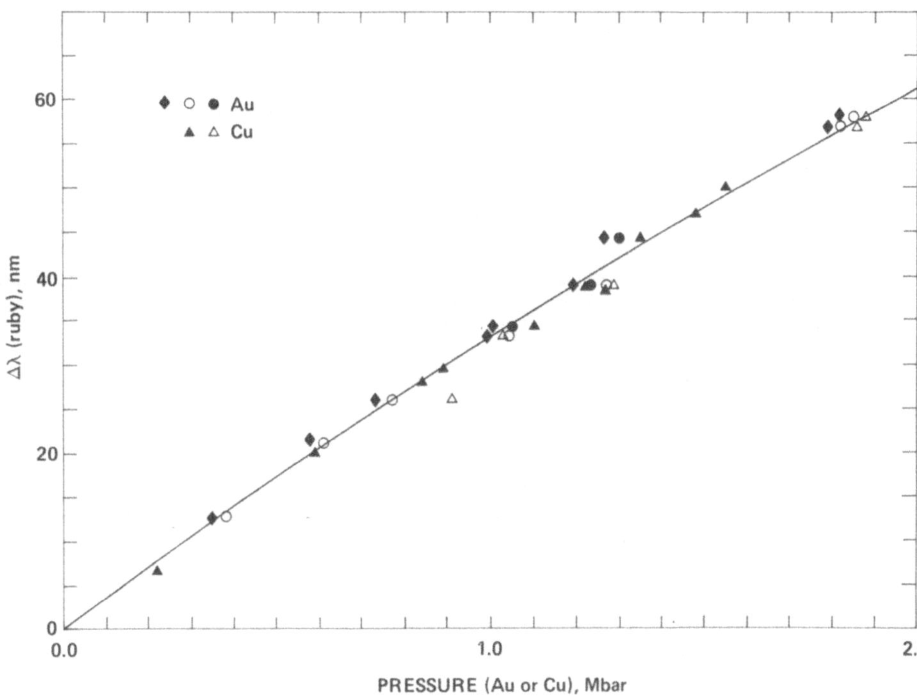

Fig. 2. Wavelenth shift (Δλ) of the ruby R_1 luminescent line with pressure. Solid diamond, gold points from experimental runs 1 and 2 based on the equation of state of Heinz and Jeanloz (1984); open and solid circles, gold points based on AIP Handbook equation of state (Gray, 1972); open and solid triangles, copper points based on AIP Handbook equation of state. The points are obtained from Table 1.

Conclusions

This calibration, which utilizes gold and copper as standards, is particularly valuable for static high-pressure experiments up to approximately 2 Mbar. The shock-wave Hugoniot data for these metals carry a relatively small error, the average maximum scatter being on the order of ±3 percent (Mao et al., 1978, Gray, 1972). An advantage of the present data is that the measurement of specific volume of both metals was made simultaneously in the same experiments, with the metals in mutual contact. The high pressure environment was virtually identical for both metals.

Static experiments are now being extended to pressures above 2.5 Mbar as stated above (Bell et al., 1984). The isotherm for copper has been determined to pressures of 4.5 Mbar, and thus copper will continue to be valuable for reference as static high-pressure technology is improved. The gold isotherm has been determined only to 2 Mbar. Ideally, high-pressure calibration requires several standards that can be employed simultaneously and separately. In some instances an interchange of internal standards is required to obtain sufficient resolution of x-ray diffraction over a range in pressure. Clearly, more equation of state isotherms are required for static high-pressure research than are now available.

This new calibration of the ruby pressure scale should provide for the first time an accurate standard for experiments to 1.8 Mbar. Extrapolation of equation (1) to the range 2-2.5 Mbar appears to be justified for the basis of a working calibration system, the absolute accuracy of which can be confirmed when static x-ray data become available.

REFERENCES

Bell, P. M., Mao, H. K., and Goettel, K., 1984, Ultrahigh pressure: beyond 2 megabars and the ruby fluorescence scale, Science, 226:542.

Gray, D. E., editor, 1972, "American Institute of Physics Handbook," McGraw-Hill Book Co., New York.

Heinz, D. L. and Jeanloz, R., 1984, The equation of state of the gold calibration standard, J. Appl. Phys., 55:885.

Jamieson, J. C., Fritz, J. N., and Manghnani, M. H., 1982, Pressure measurement at high temperature in x-ray diffraction studies: gold as a primary standard, in: "High-Pressure Research in Geophysics," p. 27, S. Akimoto and M. H. Manghnani, eds., D. Reidel Publishing Co., Boston.

Mao, H. K., Bell, P. M., Shaner, J. W., and Steinberg, D. J., 1978, Specific volume measurements of Cu, Mo, Pd, and Ag and calibration of the ruby R_1 fluorescence pressure gauge from 0.06 to 1 Mbar, J. Appl. Phys., 49:3276.

Mao, H. K., Goettel, K. A., and Bell, P. M., 1985, Ultrahigh-Pressure Experiments at Pressures Exceeding 2 Megabars, in "Solid State Physics Under Pressure: Recent Advance with Anvil Devices," S. Minomura, Ed., D. Reidel Publishing Co., Boston, 11.

Mao, H. K., Hadidiacos, C. G., Bell, P. M., and Goettel, K. A., 1983, Automated system for heating and for spectral measurements in the diamond-window, high-pressure cell, Carnegie Inst. Washington Year Book, 82:421.

Piermarini, G. J., Block, S., Barnett, J., and Forman, R. A., Calibration of the pressure dependence of the R_1 ruby fluorescence line to 195 kbar, J. Appl. Phys., 46:2774.

THE EQUATION OF STATE OF DENSE ARGON; A COMPARISON OF SHOCK AND STATIC STUDIES

M. Ross

Lawrence Livermore National Laboratory
University of California
Livermore, CA 94550

H. K. Mao, P. M. Bell, and J. A. Xu

Carnegie Institute of Washington
Geophysical Laboratory
Washington, DC 20008

ABSTRACT

In the present paper we report static high-pressure measurements for room temperature solid argon to 800 kbars. This isotherm is in excellent agreement with one predicted from a theoretical analysis of shockwave data.

INTRODUCTION

The static high-pressure and shockwave techniques for studying the properties of matter at very high pressure have so little in common that it may not be apparent to the nonspecialist that the two techniques can be used to generate equation of state data that are directly comparable.

In the diamond-window static high-pressure cell the faces of two opposing diamond anvils are squeezed together and isothermally compress a sample of a few microns in size to megabar pressures. The sample density is determined from a direct x-ray measurement of the unit cell volume. Pressure may be determined by measuring the density of a secondary standard such as NaCl or by recording the shift of the R1 ruby flourescence line. The pressure dependence of the ruby line has been calibrated against isotherms reduced from shock data. By contrast shockwave experiments create high pressures by introducing a rapid impulse into a sample. The pressure and compression are determined directly from a measurement of the two dynamic variables (the shock and particle velocity). In the present study we report new static pressure measurements for solid argon to 800 kbar at 293 K using the ruby scale calibrated for hydrostatic conditions[1] to determine the pressure. These are compared with the results from shockwave studies.

Shock compression experiments on liquid argon have been carried out by several workers.[2] The most recent and most accurate are those of Nellis and Mitchell.[3] Using the two stage light-gas gun at Livermore they achieved final pressures up to 900 kbars and temperatures calculated to be approximately 30,000 K.[4] In the range above 400 kbars the temperature is over 12,000 K and is sufficiently high to cause an appreciable degree of electronic excitation. Below this pressure the only contribution to the thermodynamic properties are from atom-atom interactions and atomic motion. It is these data, below 400 kbar and uncomplicated by thermal electron excitation, that are used herein to provide a determination of the interatomic potential and a direct comparison with the isotherm.

EXPERIMENTAL METHODS AND RESULTS

The static pressure measurements of argon were performed in a diamond-window high-pressure cell similar in design to the one described by Mao and Bell[5]. The experimental procedure involved filling the sample chamber of the apparatus with fluid argon pre-pressurized to 2 kbar, following the technique of Mills et al.[6]. Also included in the sample chamber was powdered ruby (0.5% Cr_2O_3). The sample chamber had a diameter of 200 μm and a thickness of approximately 15 μm. X-ray diffraction data were obtained by positioning a $MoK\bar{\alpha}$ x-ray beam, collimated to a diameter of 50 μ, in the center of the sample chamber. Data were recorded by film and by position-sensitive detector techniques[6].

Pressure calibration was based on a new ruby pressure scale for hydrostatic (and quasihydrostatic as in the case with solid argon) experiments when a gas, fluid, or loosley bound solid is the sample as the pressure-transmitting medium.[1] The pressure (P) was determined from the proportional shift of the ruby R_1 luminescent line ($\Delta\lambda$) from its one bar wavelength value (λo) according to the relationship:

$$P(\text{Mbar}) = A/B[(1 + \frac{\Delta\lambda}{\lambda o})^B - 1] \tag{1}$$

where A = 19.04 Mbar (A is the initial slope of the function, after Piermarini et al.,[7]) and B = 7.665.

The results are plotted in Fig. 1. Good agreement is found with the data of Zou et al.[9] to 70 Kbar and with data obtained in experiments where argon is employed as a pressure-transmitting medium (i.e. Jephcoat and Hemley,[10] and Xu et al.,[7]). The equation of state derived later in this paper may serve as a primary calibration standard to such experiments.

THEORETICAL CALCULATIONS AND RESULTS

Monte Carlo calculations[11] of the pressure and energy were made using the exponential-six (exp-6) potential:

$$\phi(r) = \epsilon\{(\frac{6}{\alpha-6})\exp(\alpha[1-r/r^*]) - (\frac{\alpha}{\alpha-6})(\frac{r^*}{r})^6\} \tag{2}$$

with parameters ϵ/k=122 K, r^* =3.85 Å and using two different values of α(13.2 and 13.0).

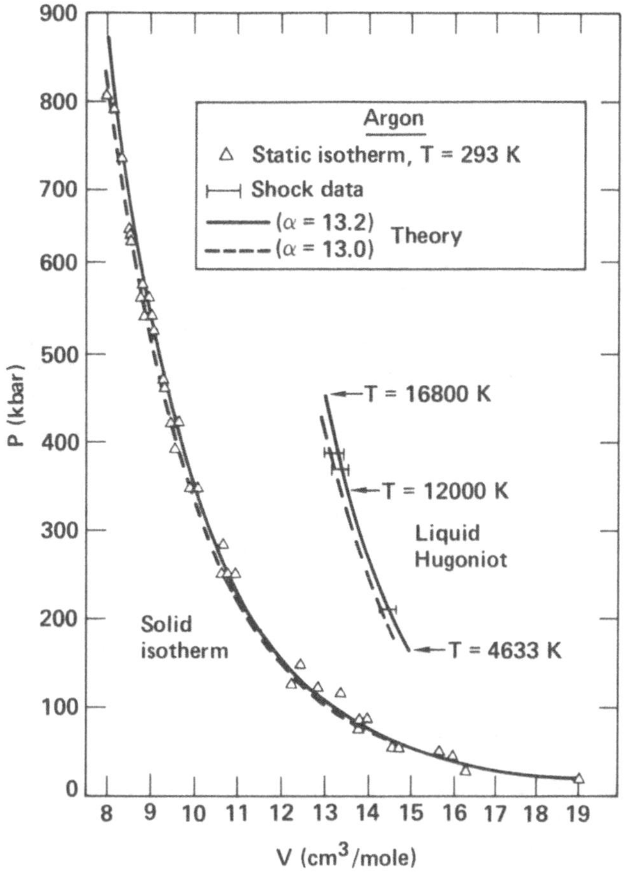

Fig. 1. A comparison of the static, shockwave and theoretical results. The shock temperatures were calculated at the pressures indicated.

The Hugoniots and isotherms calculated using these potentials are also shown in Fig. 1. The best fit to all the data is given by the exp-6 with an $\alpha = 13.2$. The curves calculated using $\alpha = 13.0$ are within the experimental precision but lie on the lower pressure side. Hence the curves calculated using these two α values provide a measure of the agreement between two very different sets of experimental data and of the uncertainty to which an effective potential can be determined. The agreement between shock and static data is excellent despite the appearance of some excessive scatter in the DAC results between 100 and 200 kbar. The goodness of the agreement lends added credence to the validity of the ruby scale at these pressures to an accuracy within 10%.

ACKNOWLEDGEMENTS

Work performed under the auspices of the U.S. Department of Energy by the Lawrence Livermore National Laboratory under contract number W-7405-ENG-48.

REFERENCES

1. H. K. Mao, J. Xu, and P. M. Bell, Calibration of the Ruby Pressure Gauge to 800 Kbar under Quasihydrostatic Conditions, in press, J. Geophys. Res. (Jameison volume) to be published in 1985.
2. M. van Thiel and B. J. Alder, Shock compression of Argon, J. Chem. Phys. 44, 1056 (1966); R. D. Dick, R. H. Warnes, and J. Skalyo, Jr., Shock compression of Solid Argon, J. Chem. Phys. 53, 1648 (1970).
3. W. J. Nellis and A. C. Mitchell, Shock Compression of Argon, Nitrogen, and Oxygen to 90 GPa, J. Chem. Phys. 73, 6137 (1980).
4. M. Ross, The Repulsive Forces in Dense Argon, J. Chem. Phys. 73, 4445 (1980).
5. H. K. Mao and P. M. Bell, Generation of Megabar Static Pressures in Solid Hydrogen, Carnegie Inst. Washington Year Book 81, p. 434-436.
6. R. L. Mills, D. H. Leibenberg, J. C. Bronson, and L. C. Schmidt, Procedure for Loading Diamond Cells with High-pressure Gas, Rev. Sci. Instr. 51, 891-895, 1980.
7. J. Xu, H. K. Mao, and P. M. Bell, Position Sensitive X-ray Diffraction: Hydrostatic Compressibility of Argon, Tantalum, and Copper to 769 kbar, High Temperature-high Pressures, 16, 495-499, 1984.
8. G. Piermarini, S. Block, J. D. Bennett, and R. A. Forman, Calibration of the Pressure Dependence of the R_1 ruby fluorescence line to 195 kbar, J. Appl. Phys. 46, 2774-2780, 1975.
9. G. Zou, L. W. Finger, R. M. Hagan, P. M. Bell, and H. K. Mao, Isothermal Equations of State of Neon and Argon, Carnegie Inst. of Washington Year Book 80, 295-299, 1981.
10. A. Jephcoat and R. J. Hemley, personal communication.
11. W. W. Wood and F. R. Parker, Monte Carlo Equation of State of Molecules Interacting with the Lennard-Jones Potential, J. Chem. Phys. 27, 720 (1957).

SHOCK WAVE OVERTAKE MEASUREMENTS ON CESIUM IODIDE[+]

C. A. Swenson

Department of Physics and Ames Laboratory, USDOE[*]
Iowa State University, Ames, IA 50011

INTRODUCTION

The present shock wave experiments on cesium iodide were carried out as a complement to the shock wave optical pyrometry measurements on cesium iodide by Radousky, et al.[1] Those experiments gave temperatures along the Hugoniot from 0.25 Mbar to 0.91 Mbar, and suggested that melting occurred at approximately 0.32 Mbar and 3700 K. One reason for the interest in melting in CsI is the suggestion by Ross and Rogers[2] that under these conditions the coordination number for liquid CsI will have changed from the normal 6 to 12, the value for a rare gas atom. The luminosity of the shock front for CsI makes it an ideal material for which to measure directly sound velocities along the Hugoniot using shock wave overtake methods.[3] In these measurements, the occurrence of melting along the Hugoniot is marked by a discontinuous decrease in the measured sound velocity.[4,5] In addition, CsI is isoelectronic with xenon, and is expected to begin to show metallic behavior along the Hugoniot near 0.9 Mbar.[6] The directly-determined sound velocities and corresponding elastic moduli would be expected to be more sensitive to this transition than either Hugoniot equation of state or optical pyrometry experiments. The following gives a brief description of the present experiments and results. A more complete discussion of the results will be published elsewhere.[7]

THE EXPERIMENT

Figure 1 shows in Lagrangian coordinates the principles of the shock wave overtake technique as described by McQueen, et al.[3] When a thin flyer plate (typically iron, aluminum or tantalum, 0.5 to 1 mm thick) travelling at high velocity (3 to 7 km/sec) strikes the target, shock waves propagate forward in the target and backward in the flyer. The leading edge of the rarefaction wave which is generated at reflection from the back surface of the flyer travels at the sound velocity in the compressed media, a velocity which is considerably greater than the shock velocity. When this leading

[+]Work done at Los Alamos National Laboratory and sponsored by the U. S. Department of Energy.
[*]Operated by Iowa State University for the U. S. Department of Energy under Contract No. W-7405-Eng-82. This work was supported in part by the Director for Energy Research, Office of Basic Energy Sciences.

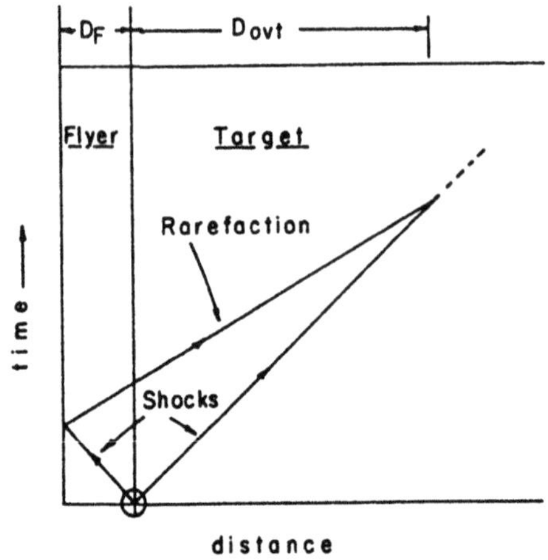

Fig. 1. A Lagrangian representation of the paths of the shock waves and the rarefaction wave in a shock wave overtake experiment.

edge overtakes the shock front, the shock pressure and the temperature, which are directly related (roughly proportional, see below) begin to decrease. Since the shock front luminosity for a transparent material such as CsI is roughly proportional to T^4, a small decrease in pressure will result in a much larger relative decrease in emitted light intensity.

These experiments are relatively straightforward for a transparent substance, but more complex for an opaque material.[3] The measured experimental quantity is the ratio of the distance the shockwave travels in the target before being overtaken (D_{OVT}) to the thickness of the flyer (D_F). This ratio is related to the properties of the flyer and the target by the relation[3]

$$R_{exp} = D_{OVT}/D_F = (u_{sT}/u_{sF})(R_T+1)[R_F/(R_F+1)]. \quad (1)$$

Here, the shock velocities, u_s, in the flyer and in the target, as well as the particle velocity, pressure and relative density for the experiment, ρ/ρ_o, are calculated by impedance-matching from the known flyer velocity and the Hugoniot parameters for the two materials. Since the overtake ratio for the flyer, R_F, has been determined in a separate experiment in which the flyer and the target were from the same material, R_{exp} gives the overtake ratio, R_T, for the target. The velocity of sound, c, and the corresponding elastic modulus now can be calculated from

$$c = u_s(V/V_o)(R+1)/(R-1), \quad (2)$$

and

$$MOD = \rho c^2. \quad (3)$$

The relevant elastic modulus for a solid is the longitudinal modulus, C_L, since the strength of the rarefaction wave is not sufficient to overcome the shear yield stress. A liquid, however, cannot support a shear stress, so the relevant elastic modulus is the adiabatic bulk modulus, B_S. These two moduli are related through Poisson's ratio, σ, by

$$(c_L/c_B)^2 = C_L/B_S = 3(1-\sigma)/(1+\sigma). \quad (4)$$

Since σ increases from approximately 0.3 to 0.5 on melting, the change in the

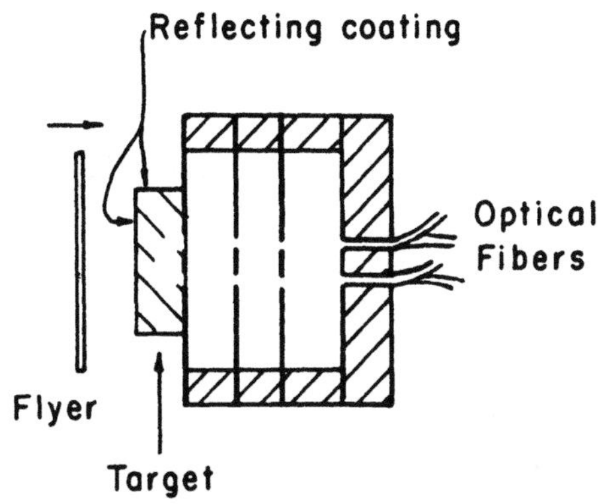

Fig. 2. The experimental set-up which was used to measure overtake times in CsI. Only 4 fibers out of 8 (2 per aperture) are indicated. The photomultiplier and oscilloscope which are associated with each fiber are not shown.

velocity of sound (and in R) is relatively large.

The experimental set-up is shown in Fig. 2. The flyer is accelerated to velocities of from 3 to 7 km/sec by the Los Alamos 2-stage light gas gun facility, for which flyer velocities can be determined to better than 1 percent. The rectangular CsI crystal is mounted on the flyer-side of a thin aluminum plate which contains four 1.0 mm diameter apertures which are arranged in a square pattern, roughly 8 mm on a side. A reflecting coating is placed on the 5 non-aperture sides of the sample to prevent pretriggering of the analysis system by the luminous "bow-wave" which is generated by the moving flyer. When the flyer strikes the target, light from the shock front which illuminates the apertures is monitored through the colliminating system by two independent quartz fibers for each aperture. This light is transmitted through the fibers to a photomultiplier, the output of which triggers a fast oscilloscope which gives a record of the intensity as a function of time using a fast camera.

Figure 3 shows oscilloscope traces taken in 3 different experiments. Figure 3(a) is typical of good data, with a relatively sharp rise in intensity and a clear indication of when overtake occurs. The results for a typical experiment will include 4 or more traces of the quality of Fig. 3(a), from which the overtake time can be calculated with a consistency (and accuracy) of from one to two percent. The overtake distance, D_{OVT}, then is calculated from the known shock velocity in the CsI. The other three traces represent unusual situations which will be discussed in the following sections. Figures 3(b) and 3(c) are traces from independent fiber channels which monitored the same aperture, but which had different oscilloscope intensity settings. Figure 3(d) is the record for an experiment in which both liquid and solid probably were present. The time marks in each case are at 50 ns intervals, with the total trace length corresponding to roughly one microsecond.

The use of so many (8) independent viewing channels allows a variety of oscilloscope sensitivities to be used to provide redundancy in the data, and to make allowance for unusual occurrences, such as, for instance, slower or faster than planned flyer velocities. The differing sensitivities of the photomultipliers and transmission efficiencies of the fibers (with unavoidable joins) require that each channel be "calibrated" against a stroboscopic light source during the set up of the experiment.

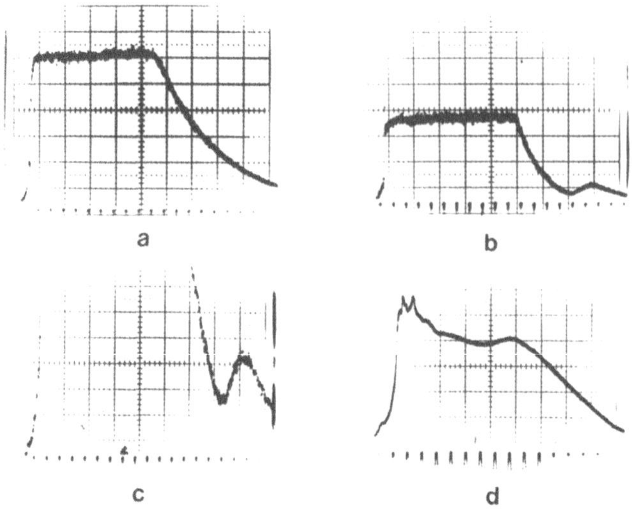

Fig. 3. Representative oscilloscope traces for three CsI experiments. See the text for details.

After an experiment, the magnitudes of the oscilloscope readings are used to obtain an average relative luminosity for the shock wave at the pressure (and, hence, temperature) of the experiment.

Figure 4 gives this information for the present experiment in the form

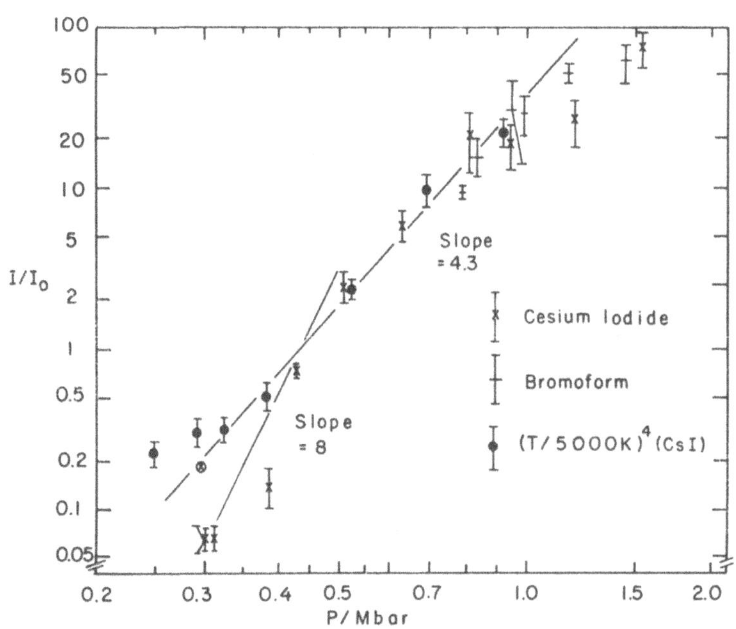

Fig. 4. Luminosity vs pressure results for two experiments, and as calculated from the pyrometry results for CsI.

of a double-logarithmic plot of relative intensity vs pressure (x). The fall-off in intensity at both high and low pressure is, we believe, due to the response of the photomultiplier tube. A single data point with an infrared-sensitive photomultiplier (the circled x) gives an intensity which agrees with an extrapolation of the mid-portion of this relation. The crosses (+) represent similar data for liquid bromoform which were obtained in an experiment on shocked graphite/diamond. The agreement is gratifying, if unexpected. Finally, the results for the pyrometry measurements on CsI are plotted as a function of pressure, with the assumption that the shock luminosity is proportional to T^4, with 5000 K chosen arbitrarily as a normalization temperature. These data have been adjusted slightly to correspond with our Hugoniot parameters.[7] The agreement again is good, and the "break" in these data at about 0.32 Mbar and 3700 K is associated with the onset of melting along the Hugoniot. The luminosity of the shock wave roughly is proportional to $P^{4.3}$, which supports our rationale in the Introduction for the sensitivity of these shock wave overtake measurements.

RESULTS

Figure 5 shows the experimental results for the pressure-dependence of the overtake ratio R for CsI. Three flyer materials were used (iron, aluminum and tantalum), with the iron data extending to sufficiently low pressures that the bcc-fcc phase transition unfortunately affects the results. The lowest pressure data correspond to those for the solid, with the solid line calculated from Eq. (4) and the liquid results using a reasonable value for Poisson's ratio, σ = 0.4. The two data points in parentheses in Fig. 5 were obtained in the same experiment, but from different apertures, suggesting an inhomogeneous sample, with discrete solid-like and liquid-like regions. The unusual trace in Fig. 3(d) came from this experiment.

The sound velocities calculated from the liquid results in Fig. 5 give the pressure-dependence of the adiabatic bulk modulus along the Hugoniot which is shown in Fig. 6. The Hugoniot bulk modulus, $B_H = -V(\partial P/\partial V)_H$, also is plotted in this figure for comparison, as is the pressure-dependence of C_L for the solid as calculated from Eq. (4).

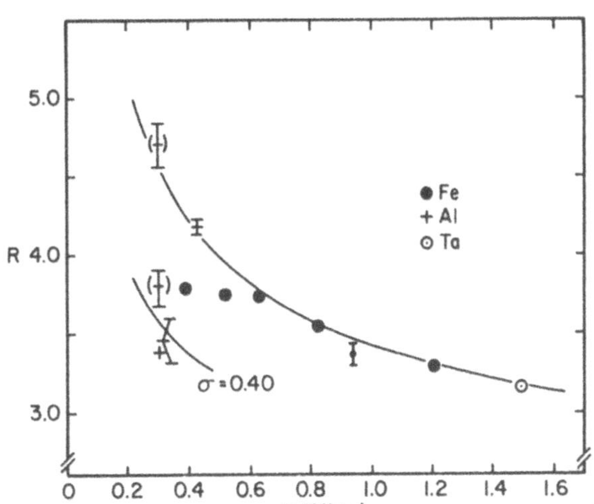

Fig. 5. Experimental results for the pressure-dependence of the overtake ratio, R, for CsI for three flyer materials.

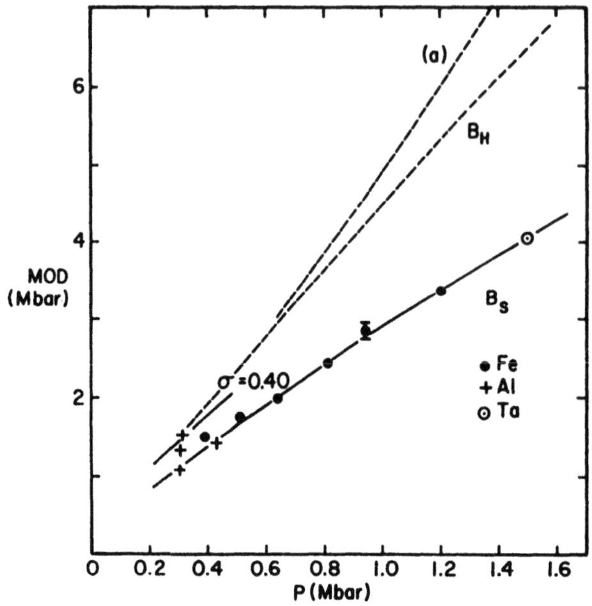

Fig. 6. Pressure-dependence of the adiabatic bulk modulus and the bulk modulus along the Hugoniot for CsI.

McQueen[8] gives a number of thermodynamic relationships with which the present results and those from the pyrometry experiments can be used to calculate various thermodynamic quantities. The sound velocities in the liquid and the Hugoniot relation (u_s vs u_p) can be used to obtain directly the pressure-dependence of the Grüneisen parameter, γ, as

$$\gamma = V(\partial P/\partial E)_V = \beta B_T V/C_V = \beta B_S V/C_P = (\partial P/\partial T)_V V/C_V \tag{5a}$$

$$= V\,\frac{(\partial P/\partial V)_S - (\partial P/\partial V)_H}{(\partial E/\partial V)_S - (\partial E/\partial V)_H} = 2\,\frac{B_S - B_H}{P - \eta(V_o/V)B_H}. \tag{5b}$$

Here, $\eta = (V_o-V)/V_o$. This calculation is sensitive to the shape of the Hugoniot relation through B_H, so we have fitted a quadratic expression in u_p to the existing u_s-u_p data to obtain an expression which we believe is most likely to be correct.[7]

The product of the density and γ, $\rho\gamma$, often varies more slowly with volume and temperature than γ, and is approximately a constant for many materials. Figure 7 gives the pressure-dependence of $\rho\gamma$ for CsI as determined in the present experiments. This quantity shows no evidence of unusual behavior, with its magnitude extrapolating to the P=0 value at 300 K. The lines marked (a) and (b) in Fig. 7 represent the effect of alternative choices for the Hugoniot relation,[7] and demonstrate the need for a reliable representation for the Hugoniot.

The existence of experimental temperatures for the Hugoniot[1] allows the calculation[8] of both C_V and B_T from our sound velocity results using purely thermodynamic relationships. The expression for C_V is relatively complex,

$$C_V = \frac{(\partial E/\partial V)_H + P)}{(\partial T/\partial V)_H + (\rho\gamma)T} = \frac{B_S - B_H}{\gamma[(\partial T/\partial V)_H + (\rho\gamma)T]}, \tag{6}$$

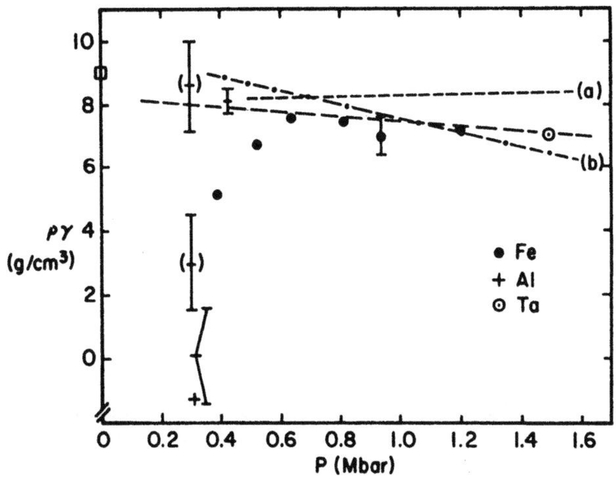

Fig. 7. The pressure-dependence of ργ for CsI.

but simplifies appreciably if, as Fig. 4 suggests, the Hugoniot temperatures for the liquid are expressed in terms of a power of the pressure as

$$T = T_0 (P/P_0)^{\alpha}. \qquad (7)$$

Here, T_0 = 12,410 K and α = 1.079 give a representation which reproduces the experimental results from 0.325 Mbar to 0.91 Mbar. Equation (6) now becomes

$$C_V = \frac{B_S - B_H}{T(\rho\gamma)[\gamma - \alpha(B_H/P)]}. \qquad (8)$$

Figure 8 gives the results of this calculation using our smooth values for the various variables, with C_V per atom expressed in dimensionless form using the gas constant, R. The "data points" in this figure have no meaning, since smoothed results have been used. The striking feature in this plot is the relatively large magnitudes for C_V/R, close to twice that (3) for a classical (Dulong and Petit) solid. These values depend primarily on T, B_H and B_S, with γ given by Eq. (5b).

Huang and Ruoff[9] give diamond anvil x-ray compression results for CsI at pressures to 0.650 Mbar at room temperature which can be combined with the accurate low pressure ultrasonic results of Barsch and Chang[10] to generate a room temperature isotherm. Huang and Ruoff report a phase transition at roughly 0.4 Mbar ($V/V_0 \approx 0.53$) which could have a large volume change, so the present discussion is restricted to lower pressures. Since the Hugoniot equation of state is known, including temperatures, the Mie-Grüneisen equation of state can be used with ργ (Fig. 7) to calculate an average heat capacity from room temperature to the Hugoniot using

$$P^* = P(T,V/V_0) - P(T_0,V/V_0) = \rho\gamma\,\Delta U = \rho\gamma(C_V)_{avg}(T_{Hug}-T_0). \qquad (9)$$

The results of this calculation also are plotted in Fig. 8 as a function of the Hugoniot temperature to demonstrate that this average C_V, which contains a spread-out latent heat of melting contribution for the liquid, is similar to the value which is obtained along the Hugoniot. Figure 8 also contains results calculated for the three lowest temperatures of Radousky, <u>et al</u>.,[1] for which $(C_V/R)_{avg}$ increases with increasing temperature from the classical value in the solid.

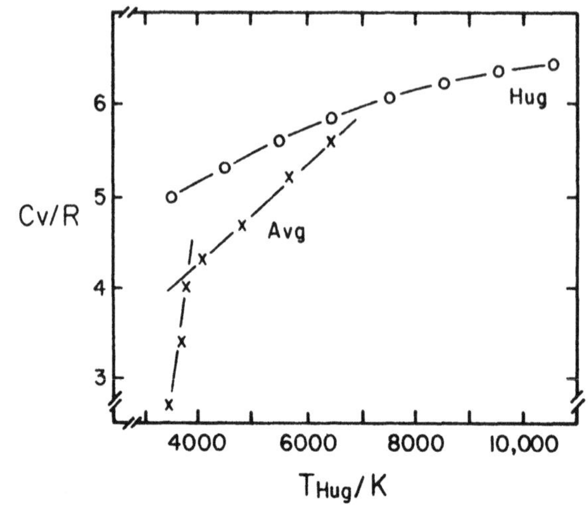

Fig. 8. The dimensionless constant volume heat capacity per atom as calculated for the Hugoniot compared with an average value, C_{avg}/R, for temperatures from room temperature to the Hugoniot.

Finally, the difference between the adiabatic and the isothermal bulk moduli can be calculated using

$$
\begin{aligned}
B_T &= B_S - \beta B_T \gamma T \\
&= B_S - (\rho\gamma)^2 (C_V/\rho) T \\
&= B_S - \gamma \frac{B_S - B_H}{[\gamma - \alpha(B_H/P)]}.
\end{aligned}
\qquad (10)
$$

The results of this calculation show that over the range of the liquid pyrometry data (that is, from 0.33 to 0.9 Mbar), the ratio B_S/B_T is roughly constant and is equal to 1.1. Thus, in Fig. 6, B_T for the liquid also would be a linear function of the pressure with roughly a 10% smaller slope. Over the range of overlap with the 300 K isotherm, B_T for a given compression is appreciably greater (25% or so) for the Hugoniot than for low temperatures.

DISCUSSION

The availability of conventional Hugoniot equation of state, temperature and sound velocity measurements allows a more detailed discussion of the thermodynamic properties of CsI along the Hugoniot than exists for any other material. Cesium iodide thus represents a unique situation. The results of this detailed description are, perhaps, most remarkable because they do not show obvious indications of the insulator-to-metal transition which is predicted by theoretical calculations.[1,6]

Melting

Our experiments have confirmed the observation from the pyrometry experiments that melting occurs in CsI along the Hugoniot at a pressure of approximately 0.32 (±0.01) Mbar and, using the pyrometry results, 3600

(±120) K. Our luminosity vs time traces show unusual behavior in this melting region [Figs. 3(b)-(d)]. The traces of Figs. 3(b) and 3(c) are independent records of the same event, and differ only the oscilloscope gain settings; they can be discussed qualitatively using the intensity vs pressure relation of Fig. 4. The calculated maximum pressure for this experiment was 0.428 Mbar, with a corresponding maximum temperature of 4970 K. The intensity minimum just before the rise corresponds to ("solid" at) 0.315 Mbar (3600 K), with the following intensity maximum corresponding to a ("liquid") pressure of 0.34 Mbar and a temperature of 3875 K. The energetics of this behavior are not at all clear, however. After the overtake, the trace represents a record of the shocking of the P=0 solid with an ever-weakening pulse, and, then, the appearance of a secondary shock wave which again is overtaken at longer times. Perhaps the liquid behind the shock front subcools as the pressure decreases during the overtake, and then rapidly freezes, generating a secondary, relatively weak, shock which raises the temperature of the shock front momentarily. This requires, however, that liquid and solid coexist in one region of the sample even when the liquid is subcooled. No easy explanation exists for the structure observed in Fig. 3(d).

Metallization

The present experimental results (Figs. 5-7) show no obvious indications of the insulator-to-metal transition which is believed to occur in CsI near 0.9 Mbar.[1,6] The behavior of $\rho\gamma$ in Fig. 6 is quite normal, and suggests continuity in behavior across the melting line to P=0. The magnitudes of the heat capacity for the liquid along the Hugoniot (Fig. 8) are unusually large, although similar data do not exist for a "normal" material. The average values of C_V/R from the 300 K isotherm to the Hugoniot also are large, but could be associated with uncertainties in the 300 K isotherm which was used. A less compressible 300 K isotherm would decrease P^* and, hence, $(C_V/R)_{avg}$. The diamond anvil x-ray results of Knittle and Jeanloz,[11] for instance, suggest a much harder solid, with the 300 K isotherm almost identical with the Hugoniot P-V relation. Additional room temperature data would assist in resolving these questions.

ACKNOWLEDGEMENTS

A number of persons made valuable contributions to this work, and their assistance is gratefully aknowledged. The experiment was suggested by John W. Shaner, with J. M. Brown instrumental in the initial planning. Bart Olinger, Denis Shampine and Robert Medina each at various times gave invaluable assistance in carrying out the experiments. J. N. Fritz contributed very much to the data analysis.

REFERENCES

1. H. B. Radousky, M. Ross, A. C. Mitchell and W. J. Nellis, Shock temperatures and melting in CsI, Phys. Rev. B 31:1457 (1985).
2. M. Ross and Forrest J. Rogers, Structure of dense shock-melted alkali halides: Evidence for a continuous pressure-induced structural transition in the melt, Phys. Rev. B 31:1463 (1985).
3. R. G. McQueen, J. W. Hopson and J. N. Fritz, Optical technique for determining rarefaction wave velocities at very high pressure, Rev. Sci. Instrum. 53:245 (1982).
4. R. G. McQueen, J. N. Fritz and C. E. Morris, The velocity of sound behind strong shock waves in 2024 Al, in "Shock Waves in Condensed Matter-1983," J. R. Assay, R. A. Graham and G. K. Straub, eds., Elsevier, New York (1984).

5. J. M. Brown and J. W. Shaner, Rarefaction velocities in shocked tantalum and the high pressure melting point, in "Shock Waves in Condensed Matter-1983," J. R. Assay, R. A. Graham and G. K. Straub, eds., Elsevier, New York (1984).
6. John Aidun, M. S. T. Bukowinski and M. Ross, Equation of state and metallization of CsI, Phys. Rev. B 29:2611 (1984).
7. C. A. Swenson, J. W. Shaner and J. M. Brown, Hugoniot overtake sound velocity measurements on CsI and the insulator-to-metal transition, to be published.
8. R. G. McQueen, Laboratory techniques for very high pressures and the behavior of metals under dynamic loading, in "Laboratory Techniques for Very High Pressures and the Behavior of Metals," K. A. Gschneidner, M. T. Hepworth and N. A. D. Parlee, eds., Metall. Soc. Conf. 22:44 (1964).
9. T. L. Huang and A. L. Ruoff, Equation of state and high-pressure phase transition of CsI, Phys. Rev. B 29:1112 (1984).
10. G. R. Barsch and Z. P. Chang, Ultrasonic and static equations of state for alkali halides, in "Accurate Characterization of the High Pressure Environment," E. C. Lloyd, ed., National Bureau of Standards Spec. Publ. 326, Washington, D. C. (1971).
11. E. Knittle and R. Jeanloz, Structural bonding changes in CsI at high pressures, Science 223:53 (1984).

HIGH PRESSURE STRENGTH OF SHOCKED ALUMINUM*

J. R. Asay, L. C. Chhabildas, G. I. Kerley, and T. G. Trucano

Sandia National Laboratories
Albuquerque, NM

ABSTRACT

Unloading wave experiments were used to determine the material strength and release adiabats in pure and 6061-T6 aluminum alloy over the stress range of 50-100 GPa. The experimental adiabats are in good agreement with theoretical calculations, assuming a constant $\rho\gamma$. The strength measurements show that a maximum occurs for shock states above 20 GPa and that a general softening of the strength is evident between 40 and 100 GPa.

INTRODUCTION

Early shock-wave equation-of-state (EOS) investigations neglected effects of material strength and assumed hydrodynamic material response[1]. Although strength effects were apparent from a variety of experiments, such as shock attenuation[2], they could not be measured accurately enough for implementation into theoretical models. The emergence of time-resolved wave profile techniques[3] has allowed routine measurements of strength properties and has shown that strength effects are important in shock-wave studies of the EOS.

Previous interferometer wave profile studies in aluminum have been limited to shock stresses of 40 GPa. This level has been increased to over 100 GPa in the present experiments, which show that a maximum in strength occurs for Hugoniot stresses between 20 and 40 GPa. This is followed by gradual softening for Hugoniot stresses to 100 GPa, which is due to either a decrease in the shear strength, resulting from the high average temperatures produced in the shocked state, or to a collapse of deviator stresses at the Hugoniot point. The latter could occur because of localized melting along shear bands.

The material studied in the present experiments was 99.99% pure aluminum with an initial density of 2.704 Mg/m^3[4]. A laser interferometer window of single-crystal LiF[5] in the <100> direction was bonded to the back surface of the specimen

*This work performed by Sandia National Laboratories supported by the U. S. Department of Energy under contract #DEAC04-76-DP00789.

to allow velocity interferometer measurements[6], and a two-stage light gas gun was used for the shock loading and unloading profile measurements[7]. For Hugoniot states to 50 GPa, a two-piece impactor was configured in the form of a "pillbox," which consisted of a 6061-T6 aluminum impacting plate backed by a central void to allow nearly full unloading. For the higher pressures, a tantalum impacting plate was inset directly in the lexan sabot, which provided partial unloading from the shocked state. The unloading response of the impactor was determined by impacting the projectile directly into the LiF window, which has been calibrated for shock stresses to about 180 GPa[5]. The initial Hugoniot states were determined from the measured projectile velocity, the particle velocity, and the known Hugoniot for pure aluminum[4].

DATA ANALYSIS

The wave profile data were converted to stress-volume data using the conservation equations in the incremental form[1]

$$d\sigma = \rho_o c du \ , \quad and \quad de = du/c \ , \tag{1}$$

where σ is stress, ρ_o is initial density, u is particle velocity, c is Lagrangian wave speed, and e is volumetric strain, defined as $1 - \rho_o/\rho$. Measurements of wave and particle velocities at a sequence of points on the wave profiles were used to integrate Equations 1.

Material strength in the shocked state was estimated using the techniques previously developed[8]. Basically, this involves using the conservation equations incrementally along the wave profile and the general relation between longitudinal stress, mean stress (assumed to be pressure), and shear stress. The shear stress change, $\tau_o + \tau_c$, estimated from the unloading wave structure can be expressed as,

$$\tau_o + \tau_c = \frac{3}{4}\rho_o \int_{e_1}^{e_2} (c^2 - c_B^2) de \ , \tag{2}$$

where τ_o is the shear stress immediately behind the shock front, τ_c is the shear strength for times well after shock loading, and c_B is the Lagrangian bulk velocity from the Hugoniot state, 1, to the plastic unloading state, 2. This technique was checked against wave code simulations and is estimated accurate to within 3%. However, real material strength effects (dispersion, Bauschinger effects, work hardening, etc.), limit the accuracy to about 15%. The experimental wave profiles were also substantiated with numerical simulations based on the Mie-Grüneisen EOS, using the finite-difference wave propagation code WONDY[9].

RESULTS AND DISCUSSION

The experimental parameters are summarized in Table 1 and the transmitted wave profiles are shown in Fig. 1. The stress-volume unloading adiabats determined from the wave profiles are shown in Fig. 2, along with the previously measured Hugoniot[5]; the errors in stress for a given strain are about ±2%. Although not readily apparent, the unloading curves are initially steeper than the Hugoniot, due to strength effects.

TABLE 1. Experimental Results

Exp.	Proj. Vel. km/s	Impactor Thick.(a) mm	Sample Thick.(b) mm	Hugoniot Stress(c) GPa	Hugoniot Volume m³/Mg	Particle Vel. km/s	$\tau_o + \tau_c$ GPa	Elastic Velocity(d) km/s	Bulk Vel.(d) km/s
A1	4.352	0.981(Ta)	3.932	85.9	0.2465	3.262	0.4-0.5	—	9.23 ± 0.14
A2	4.343	0.975(Ta)	—	83.1	0.0464	1.062	—	5.501 ± 0.20	—
A3	4.967	0.978(Ta)	3.935	103.8	0.2376	3.709	0.3-0.4	11.37 ± 0.36	9.59 ± 0.21
A5	4.411	2.074(Al)	3.987	49.7	0.2721	2.206	0.5-0.7	9.98 ± 0.16	8.25 ± 0.13
A6	4.320	2.027(Al)	—	47.2	0.2745	2.121	0.5-0.7	10.00 ± 0.33	7.96 ± 0.31

a. Aluminum impactors are 6061-T6 alloy.

b. Direct impact experiments on the LiF window are indicated by dashed lines.

c. The Hugoniot states were determined from the following equations of state:
Ta: ρ = 16.68 g/cm , c = 3.3062 km/s , s = 1.3033; Al: ρ = 2.704 g/cm , c = 5.386 km/s , s = 1.339

d. Eulerian velocities; bulk velocities were estimated by extrapolating the plastic velocities.

Sound speed measurements in 2024 aluminum[10] suggest that melting initiates for Hugoniot stresses of 120 GPa, so that none of the Hugoniot states in the present experiments are in the solid-liquid mixed phase region. Present theoretical calculations indicate that shock-induced melting could occur as low as 107 GPa, although the slight relaxations in particle velocity observed at shock stresses of 86 and 104 GPa may suggest the possibility of incipient melting, resulting from temperature inhomogenities. Our calculations also indicate that the unloading adiabat from 104 GPa could intersect the melt boundary for reasonable choices of EOS variables, although there is no clear evidence for this in the experimental profile.

The quantity $\tau_o + \tau_c$, which is related to the material strength $Y(2\tau_c = Y)$ in the shocked state, was estimated from Equation 2 and is presented in Fig. 3. There are two major points about the effects of shock compression on the strength of aluminum; (1) the strength of pure and 6061-T6 aluminum alloy are nearly identical

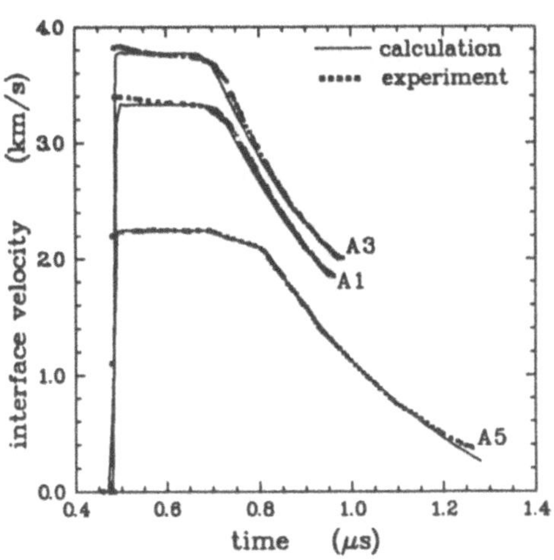

Figure 1: Experimental and calculated wave profiles. The profiles have been arbitrarily positioned to the arrival time for experiment A5.

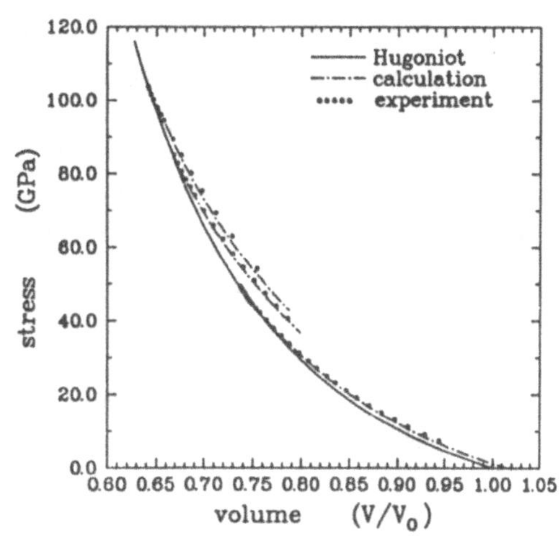

Figure 2: Unloading adiabats obtained in pure aluminum.

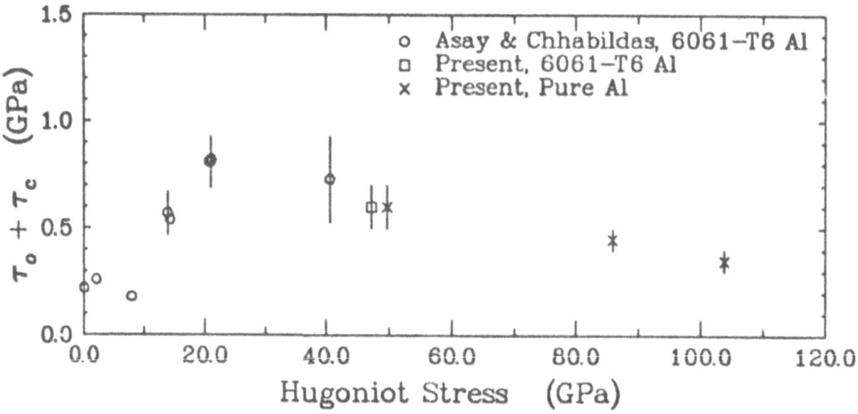

Figure 3: Change in Shear Stress vs. Hugoniot Stress. The data point at 40 GPa[8] has been reanalyzed using Equation 2.

at 50 GPa, which suggests that shock-induced strengthening mechanisms are similar in both materials, (2) the strength of 6061-T6 alloy appears to exhibit three regimes of hardening: the first extends to about 10 GPa, where no significant increase in strength occurs; the second from 10 to 20 GPa where rapid hardening is observed; the third above 20 GPa, where gradual softening is evident.

It is desirable to relate the observed changes in $\tau_o + \tau_c$ to changes in both the critical shear strength, τ_c, and the shear stress, τ_o, in the shocked state. Reloading experiments (which measure $\tau_c - \tau_o$) previously performed on 6061-T6 alloy to 20 GPa were combined with unloading measurements to show that τ_c increases over the region of 10-20 GPa and that τ_o remains fairly constant[8]. The observed decrease in $\tau_o + \tau_c$ at higher stresses indicates that either τ_c softens, due to the high temperatures achieved (the estimated temperatures range from 1100 K at 50 GPa to 3500 K at 104 GPa), or that τ_o collapses to the hydrostat, perhaps because of localized melting along shear bands during shock compression.[11] These questions could be resolved by reshock experiments at higher stresses.

The strength values reported here have implications to EOS measurements, since it is often desirable to estimate the hydrostatic pressure from shock-wave experiments. Although there is some uncertainty regarding the actual corrections, it is possible to put bounds on strength corrections in aluminum. For Hugoniot experiments, the strength correction to experimental data is $-4/3\tau_o$; the correction to unloading experiments is $+4/3\tau_c$; and the correction for reloading experiments is $-4/3\tau_c$. The approximate values for these corrections are summarized in Table 2.

The uncertainties indicated in Table 2 for the higher stresses reflect the range of values expected for τ_o (corresponding to $\tau_o = 0$ or $\tau_o = \tau_c$). For unloading and reloading experiments, the corrections represent approximate values for the stresses shown.

TABLE 2. Approximate Strength Corrections in Aluminum[a]

	10 GPa[b,c]	20 GPa[b,c]	40 GPa[b,d]	100 GPa[d]
Hugoniot	-0.1	-0.3	-(0.0-0.5)	-(0.0-0.3)
Unloading	+0.3	+0.8	+(0.5-0.9)	+(0.3-0.5)
Reloading	-0.3	-0.8	-(0.5-0.9)	-(0.3-0.5)

a. Corrections in GPa
b. 6061-T6 aluminum alloy
c. Using previous reloading data
d. Pure aluminum

CONCLUSIONS

Wave profile experiments performed on aluminum over the stress range of 50-100 GPa have been used to determine material strength in the shocked state and to estimate unloading adiabats from the Hugoniot states. The results show that the strength of shock-compressed aluminum decreases over the range of 50-100 GPa. The observed decrease could result from either a decrease in the shear strength of aluminum, due to high temperatures, or to a collapse of the Hugoniot shear stress to the hydrostat. Unloading adiabats measured over the range of 50-100 GPa agree with calculations of isentropic release based on Mie-Grüneisen theory.

ACKNOWLEDGMENTS

The authors would like to acknowledge Bill Nellis, Lawrence Livermore National Laboratory, for providing the specimens and shock Hugoniot data for pure aluminum. Acknowledgment is also due to Carl Konrad and the gun crew at the STAR Facility.

REFERENCES

1. Lee Davison and R. A. Graham, Phys. Reports, 55, 257 (1979).
2. D. R. Curran, J. Appl. Phys 36, 2591 (1965).
3. R. A. Graham and J. R. Asay, High Temp.-High Press., 10, 355 (1978).
4. Samples were provided by W. J. Nellis, LLNL, Livermore, CA.
5. J. L. Wise, Sandia National Laboratories, private communication.
6. L. M. Barker and R. E. Hollenbach, J. Appl. Phys., 43, 4669 (1972).
7. J. R. Asay, Sandia National Laboratories Report, SAND81-1901 (1981).
8. J. R. Asay and L. C. Chhabildas, in Shock Waves and High-Strain-Rate Phenomena in Metals, ed. by M. A. Meyers and L. E. Murr, Plenum Publ. (1981).
9. M. E. Kipp and R. J. Lawrence, Sandia National Laboratories Report, SAND81-0930 (1982).
10. R. G. McQueen, J. N. Fritz, and C. E. Morris, in Shock Waves in Condensed Matter - 1983, ed. by J. R. Asay, R. A. Graham, and G. K. Straub, Elsevier Science Publisher, p. 95 (1984).
11. J. W. Swegle and D. E. Grady, submitted to Explomet 1985, Portland, OR.

HUGONIOT MEASUREMENTS IN ALUMINUM TO 420 GPa USING THE LLNL TWO-STAGE LIGHT-GAS GUN

G. R. Gathers and A. C. Mitchell

Lawrence Livermore National Laboratory
University of California
Livermore, California 94550

ABSTRACT

Aluminum is a standard material for shock wave experiments. The increased velocity of our two-stage light-gas gun has been used to extend the range of previous aluminum measurements in order to reduce the gap between laboratory data and the data obtained with the nuclear driven shock impedance match methods. The experiments reported include direct collision Hugoniot measurements with tantalum impactors and reshock measurements using tantalum impactors and both platinum and copper anvils.

INTRODUCTION

Aluminum is an equation of state standard in shock compression experiments using two-stage guns and shock impedance match experiments performed close to underground nuclear explosions. The principal Hugoniot was measured in our laboratory to 160 GPa using tantalum impactors at velocities up to 6.9 km/s.[1] An extensive study of double-shock states in aluminum has also been performed using tantalum impactors at velocities up to 6.9 km/s.[2] The maximum velocity of our two-stage gun has been increased to 8 km/s.[3] The work reported here was undertaken to obtain single and double-shock gun data to the highest possible pressures achievable with 8 km/s impacts. In so doing, we can validate the equation of state of the aluminum standard at pressures closer to those achieved in the ultrahigh pressure experiments performed using nuclear explosives.[2]

Platinum is the highest density element readily obtainable that is suitable for use in fabricating high impedance anvils. The metal is both dense and relatively soft. Specimens thin enough to allow acceleration to 8 km/s become too distorted to allow acceptably planar shocks. Tungsten impactors are strong and more dense than tantalum, but considerably more difficult to machine and generally less uniform because of the metallurgical methods used in producing the material. The Hugoniot of tantalum has been accurately measured,[1] so it was selected for the impactor material. Since the Hugoniot of copper has also been accurately measured[1] it was chosen as a second anvil material. The direct collision measurements were made to verify the principal Hugoniot for copper,[4] aluminum,[4] and platinum.[5]

DESCRIPTION OF THE EXPERIMENT

The two-stage gun facility has been described elsewhere.[6] The projectile and targets used in this work are similar to those described by Mitchell and Nellis[1,6] but scaled down in dimension and mass to avoid distortions near the edge of the projectile. For the direct collision measurements, the pin detectors were in two planes, arranged in circles of radii 2.92 mm and 6.67 mm with six pins each. The separation between the two detector planes was 1.5 mm, 1.5 mm, and 1.3 mm for aluminum, copper, and platinum, respectively. The choice was small enough to assure that reverberations from the rear surface of the impactor did not overtake the shock wave before it reached the second plane of detectors, but otherwise was as large as possible to improve precision in the interval time measurement. A detector on axis was also used as previously.[1]

The targets for the reshock measurements were essentially the same but made of two pieces. The material to be studied is a simple plate called the baseplate with the outer circle of detectors resting on it. The anvil is a disk with a hole through the center to accomodate the central detector. The maximum thickness of the bonding material between the two pieces was 0.004 mm for platinum anvils and 0.002 mm for copper. The inner circle of detectors rests on the back of the anvil. The aluminum baseplates were 1.7 mm thick. The anvils were 1.7 mm and 1.3 mm thick for copper and platinum, respectively. For two of the reshock measurements with platinum anvils the pin circle radii were 3.70 mm and 6.50 mm. The tantalum impactors were 18.0 mm in diameter and 0.8 mm thick.

RESULTS

Direct collision measurements were made to check the platinum Hugoniot, because of its importance as an anvil in the double-shock experiments. The results are shown in Fig. 1, which also shows the results of McQueen, et al. at lower pressures and the fit to Morgan's data. The dashed curve corresponds to a Hugoniot with the slope increased from Morgan's value by about 4%. The discrepancy between our two points at a shock velocity of 9 km/s and Morgan's fit means that we will perform more platinum Hugoniot measurements to determine the platinum Hugoniot at these pressures.

The reshock measurements with copper anvils are shown in Fig. 2. Error bars have been omitted, since the results must be considered preliminary. The Hugoniot of the copper anvils was checked with direct collision measurements and found to be consistent with the results of Mitchell and Nellis[1] and Al'tshuler, et al.[4] The discrepancies observed are within the estimated error bars.

It should also be noted that the measurements of Mitchell and Nellis[1] were made using aluminum 1100 alloy, while the present measurements are for very pure aluminum. The difference cannot be distinguished.

Final results will be discussed in a future publication[8] after the platinum Hugoniot measurements at higher pressures are completed.

Work performed under the auspices of the U. S. Department of Energy by the Lawrence Livermore National Laboratory under contract number W-7405-ENG-48.

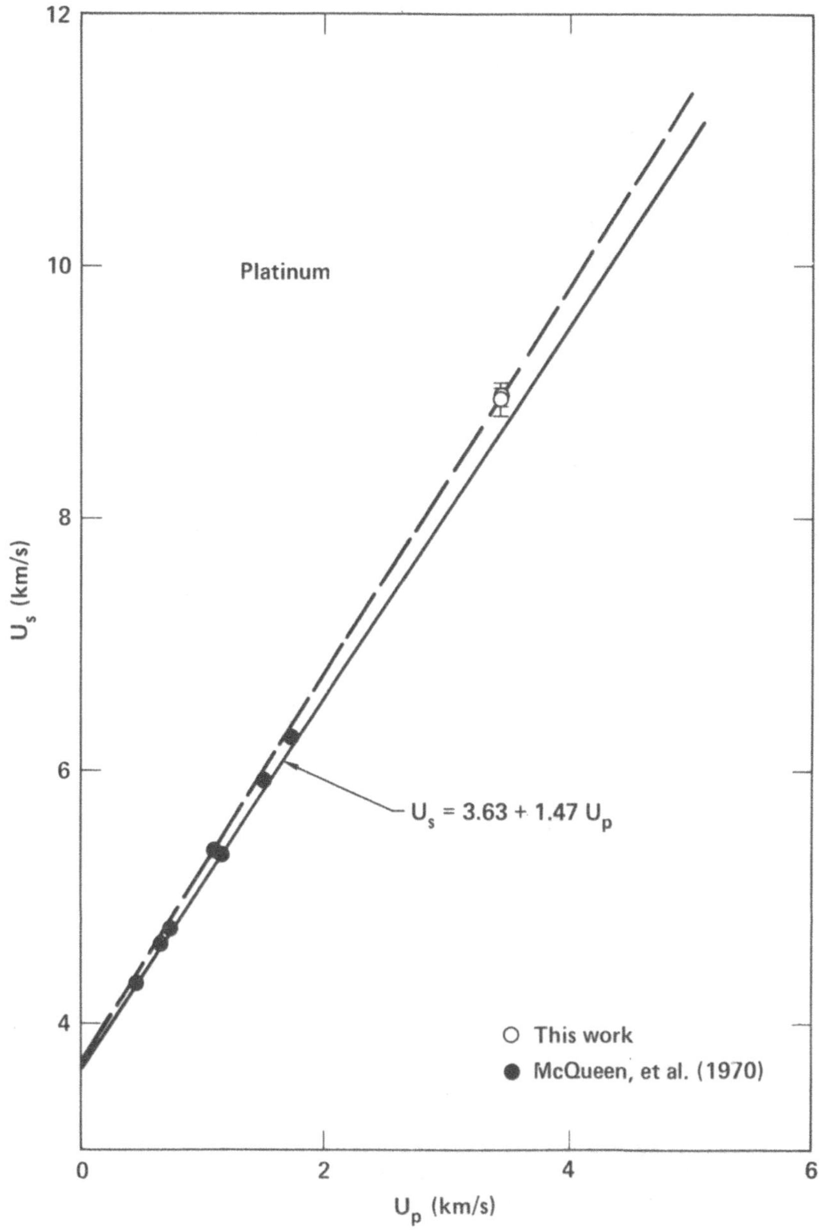

Fig. 1. Shock Hugoniot of platinum, plotted as shock velocity versus particle velocity. The solid curve is the fit reported by Morgan. The dashed curve corresponds to the stiffness coefficient s = 1.533.

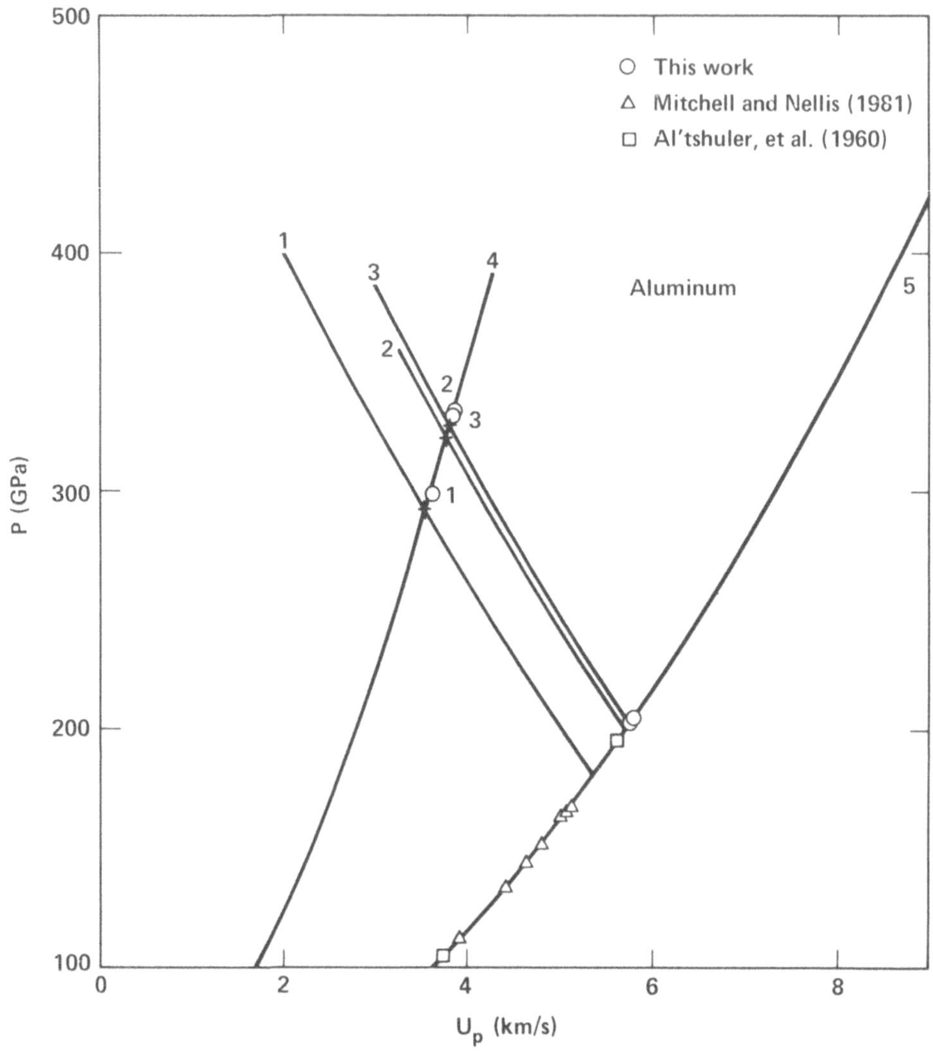

Fig. 2. Aluminum Hugoniot measurements with copper anvils, plotted as pressure versus particle velocity. The points and curves marked 1, 2, and 3 are reshock measurements. The reflection of the principal Hugoniot (curve 5) in the particle velocity of the initial shocked state for each reshock measurement is shown. The crosses indicate the impedance match solution using the anvil Hugoniot, shown as curve 4. The principal Hugoniot is that reported by Mitchell and Nellis[1].

REFERENCES

1. A. C. Mitchell and W. J. Nellis, J. Appl. Phys. 52, 3363 (1981).
2. A. C. Mitchell, W. J. Nellis, N. C. Holmes, M. Ross, G. W. Repp, R. A. Heinle, T. C. Valk, J. Rego, W. B. Graham, and R. J. Olness, in Shock Waves in Condensed Matter-1983, edited by J. R. Asay, R. A. Graham, and G. K. Straub (North Holland, Amsterdam, 1984), p. 81.
3. A. C. Mitchell, W. J. Nellis, and B. Monahan, in Shock Waves in Condensed Matter-1981, edited by W. J. Nellis, L. Seaman, and R. A. Graham (American Institute of Physics, New York, 1982), p. 184
4. L. V. Al'tshuler, S. B. Kormer, A. A. Bakanova, and R. F. Trunin, Sov. Phys. JETP 11, 573 (1960)
5. J. A. Morgan High Temp-High Press. 6, 195 (1974)
6. A. C. Mitchell and W. J. Nellis, Rev. Sci. Instrum. 52, 347 (1981)
7. R. G. McQueen, S. P. Marsh, J. W. Taylor, J. N. Fritz and W. J. Carter, in High Velocity Impact Phenomena, Ray Kinslow, ed., (Academic Press, New York and London, 1970)
8. W. J. Nellis, A. C. Mitchell, G. R. Gathers, and N. C. Holmes, to be published

USING THE HUGONIOT TO APPROXIMATE THE RELEASE ADIABAT

William C. Moss

Lawrence Livermore National Laboratory
Physics Department
Livermore, California 94550

INTRODUCTION

When analyzing data from uniaxial strain plate impact experiments, to obtain equation-of-state (EOS) information, material strength usually is assumed to be negligible, and consequently is ignored. However, anomalies in recent experimental data indicate that material strength may not be negligible. Quasi-isentropic (QI) compression (low strain-rate, long rise time uniaxial compression) data for aluminum[1] (0-9 GPa) exhibit a stress-volume path lying a few percent above the principal Hugoniot. Theories not accounting for strength predict the opposite behavior. The release adiabat of copper[2] shocked to $\approx$100 GPa shows equally peculiar behavior. The adiabat is below the Hugoniot for the initial $\approx$2/3 of the release, then crosses over for the remainder. Theories not accounting for strength predict that the release adiabat should lie above the Hugoniot. In a previous paper,[3] we showed how material strength can explain the QI loading data, and that material strength along QI compression paths can be large. Here we examine the effects of material strength on the release adiabat. In particular, we discuss the magnitude of the strength correction to data from laboratory impedance matching (IM) experiments. Our analysis differs from an earlier study by van Thiel and Cochran[4] in that we use experimental dynamic strength data to construct the release adiabat.

THEORY

A typical uniaxial strain plate impact configuration that is used for IM experiments is shown in Figure 1. The impactor (I), which has a velocity u_I, impacts the target (T). The target is backed by a specimen (S), whose EOS is to be determined. A Hugoniot point for the specimen is obtained using the EOS of the impactor and target, which are assumed to be known, and experimental measurements of the initial density and shock velocity in the specimen.

We will assume that the impactor and target are identical, so that the particle velocity of the shocked state (u_H) is $u_I/2$, as shown in the figure. The shock stress is σ_H. (σ is the longitudinal stress in a uniaxial strain configuration, and is related to the pressure (mean stress) P by $\sigma \equiv \sigma' + P$, where σ' is the longitudinal

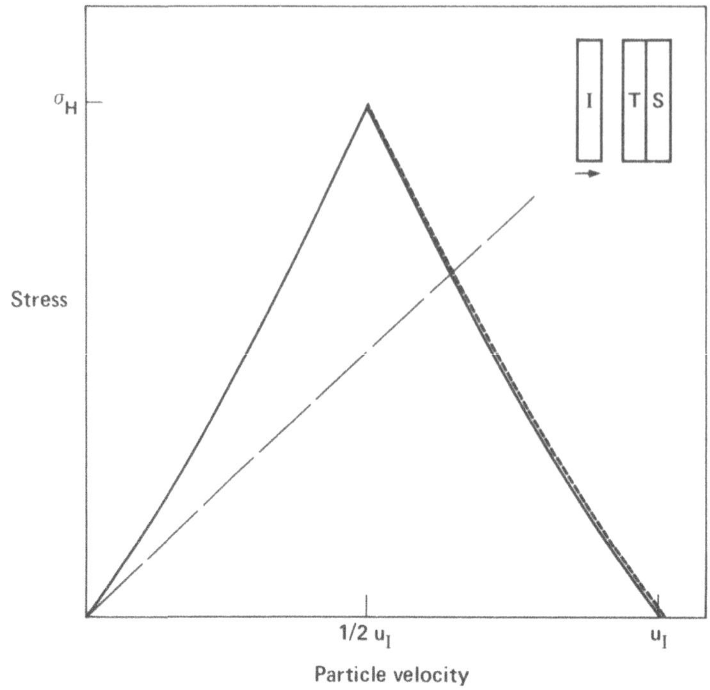

Fig. 1. Typical impedance matching experimental configuration and data.

deviatoric stress, and compressive stresses are positive. Thus, σ_H is a longitudinal stress.) The magnitude of the shock in the specimen is obtained by determining the intersection of the target release adiabat (target impedance > specimen impedance), and the line $\sigma = \rho_0 U_S u$ (large dashed line in the figure), which is the Rankine-Hugoniot equation for the shock stress as a function of initial specimen density, shock speed, and particle velocity. The release adiabat (small dashed line in the figure) is calculated using the known target EOS and neglecting the target strength. There is a small difference between the release adiabat calculated from the EOS and the reflected target Hugoniot (RTH), for low shock stresses, as shown in the figure. Consequently, the RTH is often used to represent the target release adiabat. This simplifies the data analysis, because the RTH is known analytically, whereas, the adiabat must be calculated from the EOS, using the constraint $dE = -P(V,E)dV$.

The preceding analysis has ignored the effects of material strength on determining the release adiabat. The results change appreciably, when strength is included. Specifically, we consider 6061-T6 aluminum releasing from 25 GPa. Three release curves are shown in Figure 2: RTH (thin solid line), σ_{ra} (thick solid line), which is the release adibat for the solid, and σ_{ra}^{fluid} (small dashed line), which the release adiabat assuming that the material releases from the shock state (σ_H, u_H) as a fluid (identical to the small dashed line in Fig. 1). We note that $\sigma_{ra}^{fluid} \equiv P_{ra}^{fluid}$. P_{ra} is the pressure along the release adiabat of the solid. P_{ra} is not shown in the main figure, but is shown for the regions "A" and "B" below the main figure (dotted line). P_{ra} differs from $\sigma_{ra}^{fluid} = P_{ra}^{fluid}$ in that P_{ra} is calculated subject to the constraint $dE = dQ - P(V,E)dV$, where dQ is the plastic work along the release adiabat. The release curves in the figure were constructed using experimental Hugoniot and dynamic strength data for 6061-T6 aluminum.[5,6,7]

We used a Mie-Grüneisen EOS with a linear U_s-u relation, volume dependent shear modulus,[8] and $\rho\gamma = \rho_0\gamma_0$, where γ is the Grüneisen gamma. Details of how the data are used to construct the curves in the figure are given by Moss.[3,9,10] At $u = u_H = 1.32$ (see "A"), $\sigma_H = \sigma_{ra} = \sigma_{ra}^{fluid}$ and $P_{ra} = P_H$. The difference between σ_{ra} and P_{ra} is σ'_{ra}. At the Hugoniot point (u = 1.32), $\sigma'_{ra} = \sigma'_H$. As the stress decreases, u increases. The unloading is elastic (σ_{ra} crosses the hydrostat P_{ra}, and σ'_{ra} becomes negative) until σ'_{ra} reaches the "other side" of the yield surface; subsequent unloading is plastic. The plastic work during unloading causes P_{ra} to cross over σ_{ra}^{fluid} at lower stresses. ($P_{ra} < \sigma_{ra}^{fluid}$ in "A", and $P_{ra} > \sigma_{ra}^{fluid}$ in "B".)

Copper shocked to ≈ 100 GPa has a release adiabat that is below the RTH initially, and then crosses over at lower stresses.[2] Figure 2 provides a qualitative explanation of this observation. Our calculations show that the initial unloading is elastic. The release adiabat will always be elastic initially, if the shocked material supports a nonzero deviatoric stress. The longitudinal elastic unloading modulus for the release adiabat is K + 4G/3, which is greater than the slope of the RTH (∿ K), thus, $\sigma_{ra} < \sigma_{RTH}$ during the initial loading. As unloading continues, there are two ways by which the adiabat may cross over the

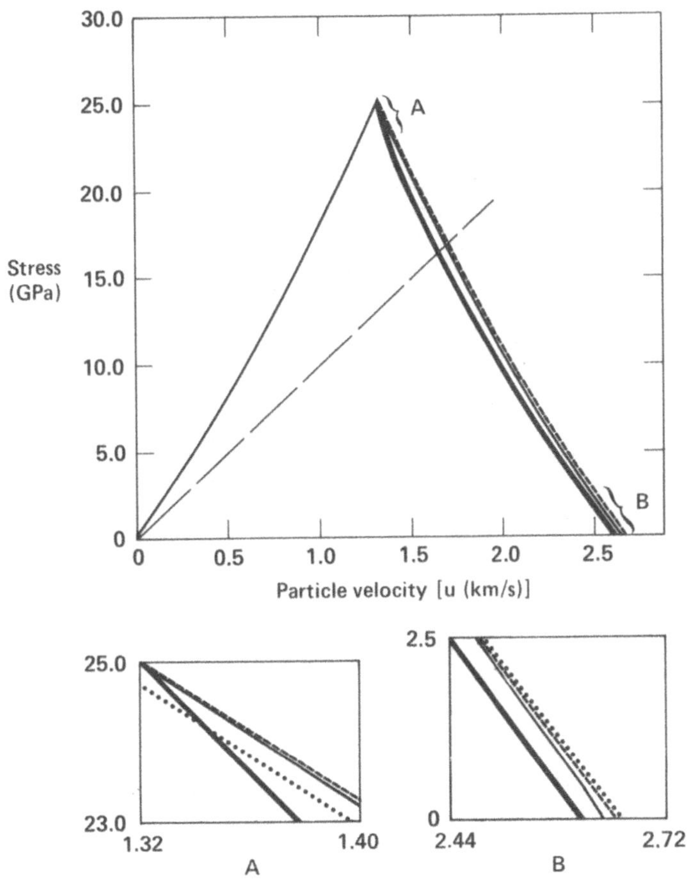

Fig. 2. Effect of different release curves on determining the specimen Hugoniot point. The large dashed line is the specimen Rayleigh line; the thick solid, thin solid, small dashed, and dotted lines are σ_{ra}, RTH, σ_{ra}^{fluid}, and P_{ra}, respectively.

RTH: (i) if σ'_{ra} becomes small, so that $\sigma_{ra} \approx P_{ra}$, or (ii) if P_{ra} is far enough above the Hugoniot, so that even with non-negligible values of σ'_{ra}, σ_{ra} still exceeds the Hugoniot. (Our calculations of a of a release adiabat from 50 GPa [not shown] exhibit the latter effect.)

RESULTS

The experimental accuracy in IM experiments is ~ 0.02.[11] This value includes all sources of error, except strength effects, which as we mentioned earlier, are ignored typically. Figure 2 shows that the error introduced by neglecting strength can be as large as the experimental accuracy. Consider the example shown in Figure 2, where a specimen impedance ($Z \equiv \rho_0 U_s$) of 10 GPa-s/km is assumed to be measured experimentally. (We ignore the Hugoniot Elastic Limit (HEL) of the specimen, to simplify the discussion. The analysis can be extended easily to include the HEL.) The relative errors are

$$(u_c - u^*)/u^* \equiv (\sigma_c - \sigma^*)/\sigma^* = -0.018, \tag{1}$$

where (σ^*, u^*) and (σ_c,u_c) refer to the intersection point of the line $\sigma = Zu$ with the RTH and release adiabat, respectively. The equivalence of the relative error in particle velocity and stress is due to the fact that the stress is related to the particle velocity by a constant (Z). The relative error is not dependent strongly on the specimen impedance, for the 25 GPa release shown in the figure. For example, impedances of 2.5, 8.0, or 15.0 GPa-s/km, produces a relative error of -0.014, -0.017, or -0.019, respectively.

Our analysis shows that the measured specimen Hugoniot is dependent on the choice of the target release adiabat. We may calculate the relationship between the specimen Hugoniots that result from using the RTH and the actual release adiabat. Consider the shock transmitted into the specimen. Using the Rankine-Hugoniot equations, we write

$$V_c = V_0(1 - \frac{u_c}{U_s}), \text{ and } \sigma_c = U_s u_c / V_0. \tag{2}$$

V_c and σ_c are the "correct" specific volume and longitudinal stress. Substituting $u_c \equiv u^* - \delta$ into Eq. (2), we obtain

$$V_c = V^* + \frac{\delta}{Z}, \text{ and } \sigma_c = \sigma^* - Z\delta . \tag{3}$$

Equation (3) shows that the correction to the specimen Hugoniot point is along the Rayleigh line:

$$(\sigma^* - \sigma_c)/(V^* - V_c) = -Z^2 , \tag{4}$$

which is the slope of the Rayleigh line. The sign of δ, that is, whether the adiabat is above or below the RTH determines the direction of the shift. Figure 3 shows both cases, schematically. The two dots represent specimen Hugoniot points obtained using the RTH to approximate the adiabat (for two hypothetical experiments). The dashed line through the points is the corresponding specimen Hugoniot. The solid line is the corrected Hugoniot. The magnitude of the shift is given by Eqs. (3) and (4). Thus, a reasonable correction to the specimen Hugoniot (obtained by using the RTH, or even σ_{ra}^{fluid}) can be made easily using

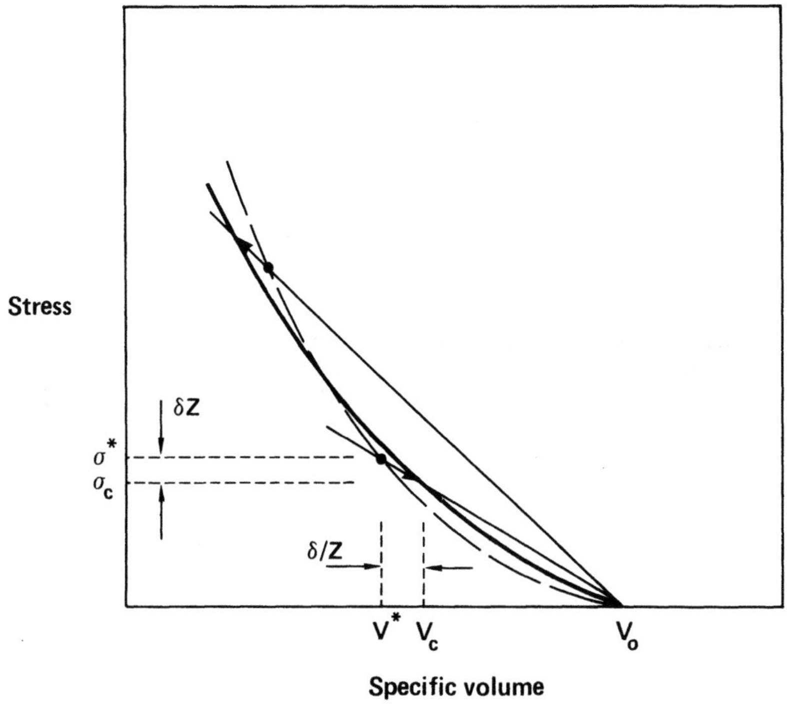

Fig. 3. Effect of target strength on measured specimen Hugoniot. Dashed and solid lines are specimen Hugoniots obtained by neglecting and including target strength. Shift is along the specimen Rayleigh line.

the graphical construction in Figure 3 and Eqs. (3)-(4), if δ can be estimated. If more accuracy is desired, then σ_{ra} must be obtained experimentally.

SUMMARY

We have analyzed the effects of material strength on determining the EOS of a material, using data from the impedance matching technique. The error introduced by neglecting the target strength is of the same order of magnitude as the total experimental accuracy, which is ~ 0.02, for the specific case we analyzed. We have also shown that the correct Hugoniot point is shifted along the Rayleigh line of the specimen, so that the measured specimen Hugoniot can be corrected easily if the error can be estimated. However, if more accurate EOS information is to be obtained from impedance matching experiments, then strength data must be included in the analyses.

ACKNOWLEDGEMENTS

The author would like to thank J. Sayer for computational assistance.

This work was performed under the auspices of the U.S. Department of Energy at Lawrence Livermore National Laboratory under Contract No. W-7405-ENG-48.

REFERENCES

1. L. M. Barker, High Pressure Quasi-Isentropic Impact Experiments, in "Shock Waves in Condensed Matter," J. R. Asay, R. A. Graham, and G. K. Strauss, eds., North-Holland, Amsterdam (1984).
2. H. Bernier, M. Valadon, and J. M. Lezaud, J. de Physique 45:C8-257 (1984).
3. W. C. Moss, J. Appl. Phys. 57:1665 (1985).
4. M. van Thiel and S. Cochran, Lawrence Livermore National Laboratory UCID-19745 (1983).
5. J. R. Asay and L. C. Chhabildas, Chapter 26, in "Shock Waves and High Strain Rate Phenomena in Metals," M. A. Meyers and L. E. Murr, eds., Plenum, New York (1981).
6. J. Lipkin and J. R. Asay, J. Appl. Phys. 48:182 (1977).
7. D. J. Steinberg, S. G. Cochran, and M. W. Guinan, J. Appl. Phys. 51:1498 (1980).
8. J. Moriarty, unpublished.
9. W. C. Moss, J. Appl. Phys. 55:2741 (1984).
10. W. C. Moss, unpublished.
11. A. C. Mitchell and W. J. Nellis, J. Appl. Phys. 52:3363 (1981).

SHOCK HUGONIOT MEASUREMENT ON TUNGSTEN

Jin Gang

Institute of Applied Physics, Chengdu University of

Science and Technology, Chengdu, Sichuan, China

Abstract

The Hugoniot of tungsten has been studied in the pressure range 30-450GPa. The linear fit between the shock wave velocity U_s and the particle velocity U_p is expressed as:

$$U_s = 4.049 + 1.21sU_p \ (\mathrm{MM \cdot \mu s^{-1}})$$

Introduction

The Hugoniot of tungsten material has been extensively studied and reported [1,2], but most of these results were obtained with impedance match solution. It is well known that tungsten is a high impedance material. In an attempt to make tungsten the standard for Hugoniot measuring of other materials, we systematically measured the Hugoniot of tungsten by the absolute method--free face (30-100GPa) and flyer impact (100-450GPa). A high explosive system was used to drive the shock wave into the target. The free face velocity, the flyer velocity, and the shock wave velocity were measured and recorded by electrical pin techniques and the oscilloscope. The average density (ρ_0) of tungsten material used in our experiments is 19.24 $g \cdot CM^{-3}$, and its purity is more than 99.95%.

Experiment

The experimental device is discussed briefly as follows:

High explosive system: The explosive assembly, of which the diameter is 200 MM, was used in our experimental device. The thickness of the explosive varies with the pressure, but does not exceed 400 MM. The flyer is curved when it flies a distance. The flatness of the flyer and the driven shock wave is about 50-100 ns in the central part, which is about 60 MM in diameter.

Target: The target, which simultaneously measures the shock wave velocity and the free surface velocity, or the shock wave velocity and the flyer velocity, is shown in Fig. 1 and 2, respectively. The "combined pin," which contains 14 single pins, is made of insulated copper wires that are first fixed in plexiglass and then are turned to the necessary shape and size. The flight time of the flyer and the

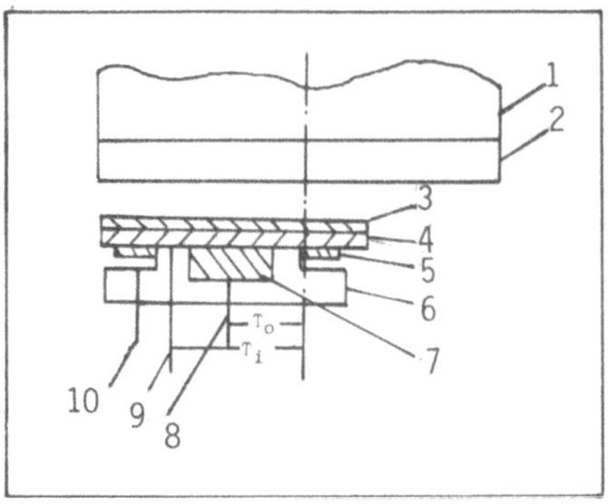

Fig. 1. Scheme of the target measuring the shock wave velocity and the free face velocity. 1) High explosive; 2) flyer; 3) plexiglass plate; 4) base plate; 5) free face; 6) plexiglass; 7) tungsten sample; 8) shock wave velocity signal pin; 9) base signal pin; 10) free surface velocity pin.

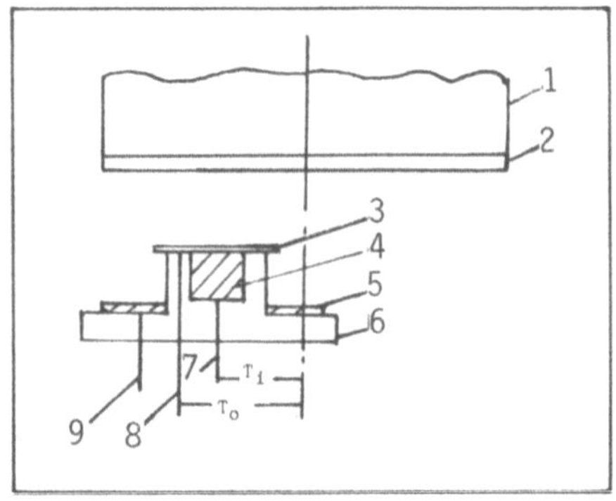

Fig. 2. Scheme of the target measuring the shock wave velocity and the flyer velocity. 1) High explosive; 2) tungsten flyer; 3) tungsten cover plate; 4) tungsten sample; 5) tungsten cover plate; 6) plexiglass; 7) shock wave velocity signal pin; 8) base signal pin; 9) flyer velocity signal pin.

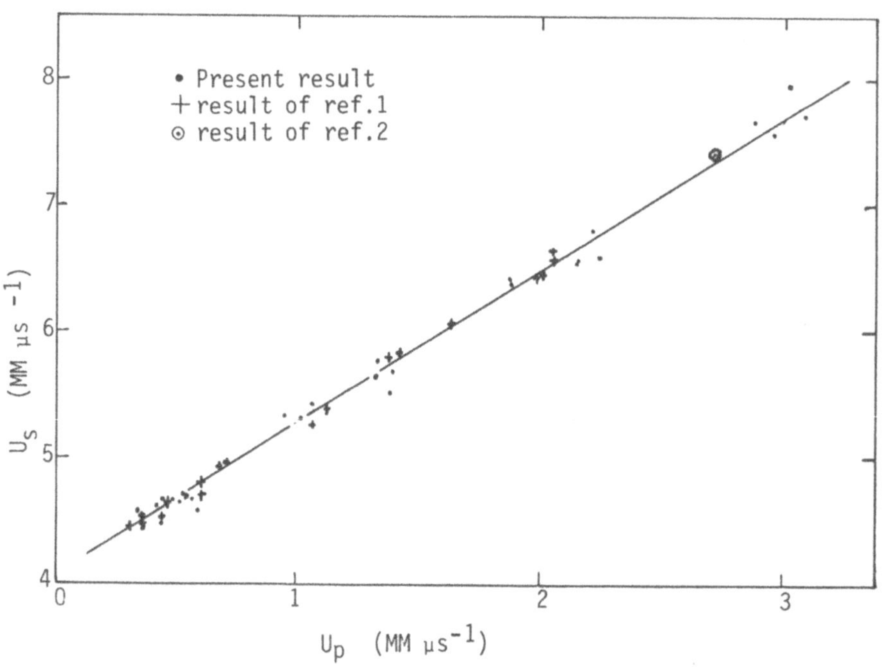

Fig. 3. U_s vs U_p curve for tungsten.

free face is about 500-1000 ns, and the traveling time of the shock wave is about 300-500 ns.

Catch-up ratio: The calculation shows that the catch-up ratio generally decreases as pressure increases. Therefore the questions arises: Does the catch-up ratio have a limit? We discuss this question as follows: When the Grüneisen equation of state is valid, and there is a linear relation between the shock wave velocity and the particle velocity ($U_s = a + sU_p$), then the sound velocity C behind the shock wave may be expressed as:

$$C = LP^{3/4} (1 + M/P^{1/2})^{1/2}, \qquad (1)$$
$$L = ((1 + S ((V_o - V)/V_o)) (1 - (\gamma/V) ((V_o - V)/2)))^{1/2} / (aV_o^{1/2}((V_o - V)/V_o)^{3/2})^{1/2},$$
$$M = V^2(\gamma/V)/2L^2,$$

Where a and S are constants in the equation: $U_s = a + SU_p$, γ is the Grüneisen parameter, V is the specific volume, P is the pressure and the subscript "o" represents the initial state. When the pressure P is very high, then

$$C = LP^{3/4} \qquad (2)$$

When the limit value of the catch-up ratio R_1 is derived with eq. (2) and the relations of the shock wave, we get:

$$R_1 = ((V_oS)t\ (V_oS)f^{-1})^{1/2}, \qquad (3)$$

where the subscripts t and f represent the target and the flyer, respectively. When the materials of the target and flyer are the same, $R_1 = 1$.

Experiment Results

Correction method: Experimental observations show that the curved shock wave may be approximated with a spherical surface; thus the correction formula is expressed as:

$$t_o = t_i + A(r_o^2 - r_i^2) + B\,(r_o^2 - r_i^2)^2, \qquad (4)$$

where t_o is the correction time; t_i is the record time; r_o and r_i are the distances of the base signal pin and signal pin to the axis of the experimental device, respectively; A and B are the fit constants. Moreover, if r_o and r_i satisfy the following relation:

$$\sum_{i=1}^{n} (r_o^2 - r_1^2) = o$$

then

$$t_o = \sum_{i=1}^{n} t_i/n \qquad (5)$$

Results and discussion: The measured results of the shock wave velocity U_s and the particle velocity U_p are shown in Fig. 3. The linear fit between them is expressed as follows:

$$U_s + 4.049 + 1.215U_p\ (MM\cdot\mu s^{-1}) \qquad (6)$$

Analysis of experimental results shows that the error of the shock wave velocity and the particle velocity is 1-2%. For comparison, the results in ref. [1] and [2] are also shown in Fig. 3. Because they are in accord with each other, we consider that the results of the present paper may be valid.

Acknowledgments

Thanks to my colleagues, especially Mr. Wang Quichao and Mr. Tang Falong, for their help.

References

1. R.G. Mcqeen, S.P. Marsh, J.W. Taylor, J.N. Frits and W.J. Carter, 1970, in High Velocity Impact Phenomena, edited by R. Kinslow.
2. A.H. Jones, W.M. Isbell and C.T. Maiden, 1966, J. Appl. Phys. 37(9):3492.

SIMPLIFIED EQUATION OF STATE $P = P(\rho,E)$ AND $P = P(\rho,T)$ FOR CONDENSED MATTER

Li Weixin
Institute of Applied Physics and Computational Matematics
P.O. Box 8009, Beijing Peopl's Republic of China

1. INTRODUCTION

In the present paper we consider the equation of state (EOS) in the following form

$$P = C_o^2(\rho - \rho_o) + (\gamma - 1)\rho E \tag{1}$$

where P, ρ, and E are the pressure, density and specific internal energy, respectively. C_o and ρ_o are the sound speed and density at normal state (room-temperature and zero-pressure), γ is a parameter to be determined. F. H. Harlow et al. call it a "stiff gas" equation of state. With this, they have calculated supersonic jets in metals and given $\gamma = 2.4$ for aluminum, and $\gamma = 2.9$ for iron.[1]

Here Eq. (1) represents a simplified EOS. We derive a formula for determining the parameter γ. When γ is known, the corresponding entropic equation of state, the expression for the sound speed, and the Riemann invariants etc. are easily obtained.

We establish also the simplified EOS $P = P(\rho,T)$ involving temperature T as a variable and the corresponding expression for the internal energy. They can be written in the following general forms:

$$P = P_x(\rho) + P_T(\rho,T), \qquad E = E_x(\rho) + E_T(\rho,T), \tag{2}$$

where P_x and E_x are the cold pressure and cold energy, respectively. P_T and E_T are the thermal pressure and thermal energy.

Using the obtained equation, we calculated thermodynamic quantities for a series of metals. As an example, the calculated results of Mach reflection are given here.

2. DETERMINATION OF THE COEFFICIENT

By adopting the EOS (1), the Hugoniot relation is written as

$$P_H = \frac{2\rho_o C_o^2(\rho_H - \rho_o)}{(\gamma+1)\rho_o - (\gamma-1)\rho_H} \tag{3}$$

where the subscript "H" refers to values on the Hugoniot. On the other hand, it is well known that the shock velocity D and particle velocity u satisfy the experimental relationship

$$D = C_o + su, \tag{4}$$

where C_o and s are constants, and C_o can be considered approximately as the sound speed at the normal state. By using Eq. (4), the Hugoniot equation may be written as

$$P_H = \frac{\rho_o C_o^2(1 - \rho_o/\rho_H)}{[1 - s(1 - \rho_o/\rho_H)]^2} \tag{5}$$

Equations (3) and (5) are to be equated. From this we derive the formula for γ along the Hugoniot:

$$\gamma = 4s - 2s^2(1 - \rho_o/\rho_H) - 1. \tag{6}$$

We see from this formula that γ depends on the density ρ of material as well as s, and varies inversely in ρ. In the limiting case of $\rho \to \rho_o$, Eq. (6) leads to $\gamma = \gamma_{max} = 4s - 1$. In the extreme case of strong shocks, Eq. (6) gives $\gamma = \gamma_{min} = 2s - 1$.

Equation (6) shows that γ is actually a function of entropy along the Hugoniot. Now, we assume that in any case γ is a function of entropy alone. Then γ is constant along the isentrope. Substituting Eq. (1) into $dE = -Pd(1/\rho)$ and taking the integral yields the entropic equation of state:

$$P = A\rho^\gamma - \rho_o C_o^2/\gamma, \tag{7}$$

where $A = A(S)$ is a function of entropy S. Then we obtain the expression for sound speed

$$C^2 = \gamma A\rho^{\gamma-1} = \gamma P/\rho + \rho_o C_o^2/\gamma. \tag{8}$$

According to the above assumption for the Hugoniot curve corresponding to weak shocks, γ is also approximately constant. It is known that through a forward-facing shock front the Riemann invariant β for the characteristic C_- is continuous. Taking into account Eq. (8) we obtain

$$\beta = u - \int (1/\rho C)dP = u - 2C/(\gamma - 1).$$

Thus, when the state ahead of the shock is known to be $P_o = u_o = E_o = 0$, we have $u - 2C/(\gamma-1) = -2C_o/(\gamma-1)$ in front of the shock wave. Substituting it and Eq. (8) into the shock relation $D - u = P/\rho u$ yields after some reduction

$$D = \frac{(\gamma+1)^2}{4\gamma} u + \frac{\gamma-1}{\gamma} C_o(1 + \eta) , \tag{9}$$

where

$$\eta = (1 - \rho_o/\rho)/2[(\rho/\rho_o)^{\frac{\gamma-1}{\gamma}} - 1].$$

This is a monotonic function of ρ. It is easily seen that $0 < \eta < 1/(\gamma-1)$. We express ρ by $\rho = \rho_o + \rho'$, where ρ' is small compared with ρ_o. Expanding η in powers of ρ' and neglecting terms of high order yields $\eta = 1/(\gamma-1)$. By inserting this in Eq. (9) we obtain

$$D = \frac{(\gamma+1)^2}{4\gamma} u + C_o. \tag{10}$$

This is a linear D—u relationship. By comparison with (4) we find $s = (\gamma+1)^2/4\gamma$. Then, this yields γ which depends on s only. Denoting it by γ_* we obtain

$$\gamma_* = 2s - 1 + \sqrt{(2s - 1)^2 - 1} \, . \tag{11}$$

It is obvious that $\gamma_{max} > \gamma_* > \gamma_{min}$. This result shows that γ can be taken approximately as a constant γ_* for shocks of moderate strength.

The calculated values of γ for some metals are listed in Table 1. We notice that the values of γ_*, in general, correspond to those given by Eq. (6) at $\rho_H \approx 1.4\ \rho_o$. This ρ_H is greater than what we presuppose. The basic parameters s, C_o and C_v listed in Table 1 are given by R. G. McQueen.[2]

Table 1. The Values of Parameter γ

	Al	Fe	Cu	W	U-Mo
γ_{max}	4.352	5.320	4.956	3.948	5.124
γ_{min}	1.676	2.160	1.978	1.474	2.062
γ_*	3.021	4.075	3.685	2.557	3.685
s	1.338	1.580	1.489	1.237	1.531
Co(km/s)	5.328	3.800	3.940	4.029	2.565
Cv(*)	8.380	4.387	3.718	1.310	1.149

$*10^6$ ergs/g^{o}k

3. EQUATION OF STATE $P = P(\rho,T)$

Substituting P and E from (2) into (1) and taking the expression for thermal pressure, we get

$$P_x = C_o^2(\rho - \rho_o) + (\gamma - 1)\rho E_x, \tag{12}$$

$$P_T = (\gamma - 1)\rho E_T. \tag{13}$$

It is known that when the temperature is less than, say, 2 ~ 3 (10^4 ^oK), the specific heat at constant volume is C_v = 3Nk, where N is the number of atoms per unit mass, and k is the Boltzmann's constant.[3] Thus, the thermal energy may be represented by

$$E_T = C_v(T - T_o) + E_{To}, \qquad C_v = 3Nk, \tag{14}$$

where E_{To} is the thermal energy at room temperature T_o. Taking account of (14), the thermal pressure can also be written as

$$P_T = (\gamma - 1)\rho C_v(T - T_o) + P_{To}, \tag{15}$$

where P_{To} is thermal pressure at $T = T_o$. Since $P = 0$ at normal conditions, we have $P_{To} = -P_{xo}$. Here P_{xo} is the cold pressure at $\rho = \rho_o$.

The cold pressure and energy are determined by the interaction between atoms. This pressure is independent of the temperature, so it is considered as the pressure at T = 0. Then from thermodynamics we know that

$$P_x = -\frac{dE_x}{dv}, \qquad \text{or} \qquad E_x = -\int_{v_{ok}}^{v} P_x dv, \tag{16}$$

where v is the specific volume and v_{ok} corresponds to T = 0.

On the other hand, we have relation (12) for P_x and E_x. By substituting E_x from (16) into (12), we obtain the integral equation for the cold pressure:

$$P_x = C_o^2(\rho - \rho_o) + (\gamma - 1)\rho \int_{v_{ok}}^{v} (-P_x)dv. \tag{17}$$

The curve of cold pressure is the isotherm T = 0, and is also the isentrope S = 0. Thus, γ in (17) is constant $\gamma = \gamma_o$ corresponding to S = 0. Now, we take the value of γ at the normal state $S = S_o$ as the approximation of γ_o. Then we have from (6), $\gamma_o = 4s - 1$. Substituting γ_o into Eq. (17) and taking its derivatives twice with respect to v, after some reduction, we finally obtain

$$\frac{d^2P_x}{dv^2} = -\frac{\gamma_o - 1}{v}\frac{dP_x}{dv}. \tag{18}$$

One of the initial conditions for the solution of this differential equation is $P_x = 0$ for $v = v_{ok}$. Furthermore, it is known that isothermal compressibility is $\kappa = -1/v(\partial v/\partial P)_T$, and the difference between the isothermal compressibility and the adiabatic one is very small for condensed matter. Supposing that they are equal at the normal state and neglecting the difference between v_{ok} and v_o, we have $dP_x(v_{ok})/dv = dP_s(v_o)/dv$. From the isentrope (7) we have $dP_s(v_o)/dv = -\rho_o^2C_o^2$, and, then, get the second initial condition: $dP_x/dv = -\rho_o C_o^2$ for $v = v_{ok}$.

Solving Eq. (18) under the above two initial conditions, we find the expression for cold pressure

$$P_x = \frac{\rho_o^2 C_o^2}{\gamma_o \rho_{ok}}\left[\left(\frac{\rho}{\rho_{ok}}\right)^{\gamma_o} - 1\right]. \tag{19}$$

By substituting (19) into (16) and taking the integral, we obtain the expression for cold energy

$$E_x = \frac{\rho_o^2 C_o^2}{\gamma_o \rho_{ok}^2}\left\{\frac{1}{\gamma_o - 1}\left[\left(\frac{\rho}{\rho_{ok}}\right)^{\gamma_o - 1} - 1\right] + \frac{\rho_{ok}}{\rho} - 1\right\} \tag{20}$$

As an example, the curves of P_x and E_x for aluminum are given in Fig. 1. For comparison, the corresponding results of R. G. McQueen[2] are also given. If we apply Eq. (19) and (20) to the case $v > v_{ok}$, we have $P_x < 0$ and $E_x > 0$. This is qualitatively rational until the material breaks. Some calculated results of P_x and E_x are given in Table 2.

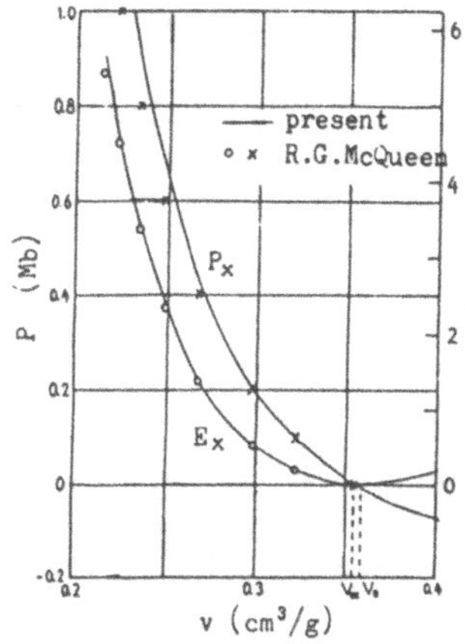

Fig. 1. The curves of P_x and E_x

4. ISENTROPES AND ISOTHERMS

Above we had obtained the isentropic equation (7). The value of γ is constant along a given isentrope, but different for different isentropes. If P and ρ on the isentrope are known, the temperature on it can be calculated by using (2) and (14) as follows:

$$T = T_o + \frac{1}{C_v}\left[\frac{P - C_o^2(\rho - \rho_o)}{(\gamma-1)\rho} - E_x + E_{xo}\right] . \tag{21}$$

The velocity u_f and density ρ_f on the free surface after shock compression are easily written as

$$u_f = u_H + \frac{2}{\gamma-1}(C_H - C_f), \quad C_f^2 = \frac{\rho_o}{\rho_f}C_o^2, \quad \rho_f = \rho_H\left[\frac{\rho_o C_o^2}{\gamma P_H - \rho_o C_o^2}\right]^{1/\gamma} . \tag{22}$$

Since $P = P_x + P_{To} = P_x - P_{xo}$ for $T = T_o$, the T_o isotherm is given by

$$P = \frac{\rho_o^2 C_o^2}{\gamma_o \rho_{ok}}\left[\left(\frac{\rho}{\rho_{ok}}\right)^{\gamma_o} - \left(\frac{\rho_o}{\rho_{ok}}\right)^{\gamma_o}\right] . \tag{23}$$

A part of the results calculated with the above equations are also given in Table 2.

Table 2. Calculated Results

		Hugoniot		Release Isentrope(P=0)			Zero Kelvin		T_o Isotherm	
	P Mb	T deg	C km/s	ρ g/cm³	T deg	U km/s	ρ g/cm³	E *	ρ g/cm³	C km/s
Al	0.0	293	5.328	2.785	293	0.000	2.817	0.000	2.785	5.328
	0.2	484	6.843	2.768	362	2.139	3.346	0.476	3.328	6.626
	0.4	927	7.754	2.726	568	3.723	3.687	1.283	3.674	7.616
	0.6	1581	8.440	2.676	848	5.076	3.947	2.165	3.937	8.346
	0.8	2403	9.007	2.623	1169	6.262	4.160	3.065	4.151	9.146
	1.0	3360	9.496	2.571	1516	7.341	4.341	3.966	4.334	9.777
Cu	0.0	293	3.940	8.930	293	0.000	9.017	0.000	8.930	3.940
	0.2	381	4.791	8.911	317	0.518	10.062	0.103	10.006	4.665
	0.4	573	5.365	8.851	403	1.731	10.799	0.301	10.757	5.253
	0.6	868	5.816	8.768	533	2.391	11.378	0.535	11.343	5.757
	1.0	1713	6.523	8.576	864	3.521	12.274	1.038	12.248	6.604
	1.4	2828	7.088	8.378	1245	4.495	12.968	1.557	12.947	7.313
W	0.0	293	4.029	19.224	293	0.000	19.276	0.000	19.224	4.029
	0.5	460	4.751	19.186	386	1.107	21.830	0.136	21.794	4.952
	1.0	996	5.235	19.062	730	1.993	23.721	0.404	23.692	5.061
	1.5	1904	5.610	18.885	1270	2.758	25.249	0.720	25.226	5.468
	2.0	3137	5.921	18.681	1948	3.444	26.545	1.056	26.525	5.830
	2.5	4649	6.190	18.463	2725	4.072	27.678	1.401	27.660	6.159

* 10^{10} ergs/g

5. SHOCK RELATIONS

By using the simplified EOS, the shock relations can be written in the following form, analogous to that for an ideal gas,

$$\frac{u}{C_o} = \frac{2}{\gamma+1}\left(\frac{D}{C_o} - \frac{C_o}{D}\right), \quad \frac{\rho}{\rho_o} = \frac{(\gamma+1)D^2}{(\gamma-1)D^2+2C_o^2}, \quad \frac{P}{\rho_o C_o^2} = \frac{2}{\gamma+1}\left(\frac{D^2}{C_o^2} - 1\right), \tag{24}$$

$$\gamma = 2s(1 + C_o/D) - 1 . \tag{25}$$

It is easy to see that by substituting (25) into the first form of (24), we obtain the linear relation $D = su + C_o$. So the results given by (24) will agree completely with experimental data. Once the above quantities are known, the temperature on the shock front is given as

$$T = T_o + [\tfrac{1}{2}P(v_o - v) - E_x + E_{xo}]/C_v . \tag{26}$$

The calculated temperature and sound speed on the Hugoniot are given in Table 2. In similar manner, the relations of the reflected shock wave are easily obtained.[4] Moreover, it is possible to treat analytically the Mach reflection for condensed matter. Figure 2 illustrates the "three-shock configuration" of Mach reflection. The velocity of the "incident" shock is denoted by D_1, then the branch point M moves with velocity $q_o = D_1/\sin\alpha_o$. The pressure of the Mach shock at point M is $P_M = P_2 = P_3$ and one at point Z on the rigid wall is given by P_Z, where $P_Z = 2\rho_o(q_o^2 - C_o^2)/(\gamma + 1)$. We used the obtained equations to evaluate the experiment of L.V. Al'tshuler.[5] The calculated results are satisfactory as shown in Fig. 3.

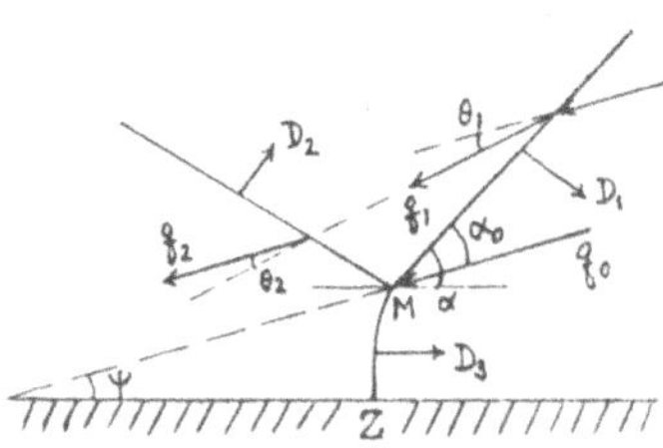

Fig. 2. The three-shock configuration of Mach reflection.

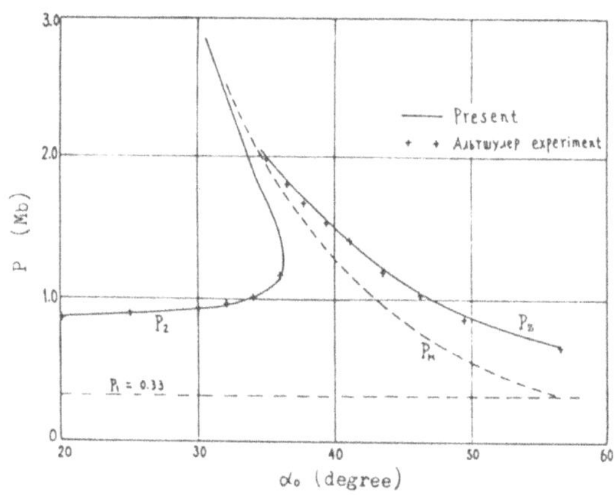

Fig. 3. The pressure of Mach reflection in aluminum as a function of the incident angle.

6. CONCLUSION

All of our results calculated by using the simplified EOS agree well with those given by R. G. McQueen, using Gruneisen equation. This EOS is very simple. In particular, by using it, a series of analytical relations is derived. It is necessary for us to study problems of condensed-matter dynamics. We have derived a useful formula for the parameter γ. The assumption that γ is constant in isentropic processes seems to be justified. A large number of calculations show that in the range where the linear relation $D = su + C_0$ is applicable, the simplified EOS gives satisfactory results.

REFERENCES

1. F. H. Harlow and W. E. Marsh, Phys. Fluid 9 (1966).
2. R. G. McQueen, S. P. Marsh, et al., The Equation of State of Solids from Shock Wave Studies in High-Velocity Impact Phenomena, edited by Ray Kinslow (Academic Press, New York and London, 1970).
3. Ya. B. Zeldovich and Yu. P. Rayzer, "Physics of Shock Waves and High-Temperature Hydrodynamic Phenomena" (Academic Press, New York, 1967).
4. Li Weixin, "A Simplified Equation of State in Condensed Medium," Explosion and Shock Waves 3, No. 2 (1983).
5. L. V. Al'tshuler et al., J. Exptl. Theoret. Phys. 34, 1382-1393 (1961).

A NEW METHOD FOR THE EVALUATION OF THE GRUNEISEN PARAMETER

FROM LAGRANGIAN GAGE RECORDS

Masatake Yoshida, Shyuzo Fujiwara, and Masao Kusakabe*

National Chemical Laboratory for Industry
Tsukuba, Ibaraki 305, Japan
* present address; Midori Anzen Kogyo Co. Ltd.
Inari, Soka, Saitama 340, Japan

INTRODUCTION

The Gruneisen's γ $(=v(\partial p/\partial e)_v)$ is an important thermodynamic parameter for extending an obtained Hugoniot equation of state -- a line in e-p-v space -- to the adjacent region. If the material, such as liquid, has a large linear thermal coefficient, it is possible to obtain a surface in e-p-v space by measuring several Hugoniots for different densities, which allow us to estimate not only γ but also other thermodynamic quantities (Lysne and Hardesty, 1973). Although this method can be applied to solids by changing the porosity, there are several problems such as non-equilibrium, error in the measurements, partial melting, and so on. If $(dp/dv)_S$ can be measured experimentally, the γ can be obtained from the measured Hugoniot e-p-v relationship and the following equation,

$$\frac{dp}{dv}_S - \frac{dp}{dv}_H = \frac{\partial p}{\partial e}_v \left(-p - \frac{de}{dv}_H \right), \tag{1}$$

where subscript S and H denote the isentrope and the Hugoniot respectively, (McQueen and Fritz, 1981; McQueen et al., 1982).

In this paper, a new method was proposed for the evaluation of Gruneisen γ using equation (1) and measuring the shock attenuation from multiple Lagrangian gage records in a strongly decaying shock wave, assuming that the Hugoniot is already known. An expression of γ by the Hugoniot parameter and the time derivatives of pressure or particle velocity along the shock and particle path was derived. Accuracy of obtained γ by this method was discussed under practical experimental conditions.

THEORETICAL

We focus our interest on particle velocity gages. It will be shown that time derivative of particle velocity u along the shock front is expressed by time derivative of u along the particle path and Gruneisen γ. This equation will be solved to give γ by measurable quantities. Only the results will be shown for pressure gages.

We consider the one-dimensional plane shock wave without body forces and heat transfer. Mass and momentum conservations across the shock front can be expressed as follows,

$$[u] = - M\,[v], \tag{2}$$

$$[p] = M\,[u], \tag{3}$$

$$[u_x] = \rho_0\,[\dot{v}], \tag{4}$$

$$[p_x] = - \rho_0\,[\dot{u}], \tag{5}$$

where x is Lagrangian (or material) coordinate and M is mass velocity ($=\rho_0 V$, V is the shock velocity in Lagrangian coordinate) of the shock wave. The notation [f] denotes the jump of f across the shock.

Time derivative of [f] along the shock path is

$$\frac{d[f]}{dt} = [\dot{f}] + V[f_x]. \tag{6}$$

From (5), (6) for p, and (3), we can obtain

$$[\dot{p}] = M\,\frac{d[u]}{dt} + [u]\,\frac{dM}{dt} + M[\dot{u}]. \tag{7}$$

From (4) and (6) for u,

$$[\dot{v}] = \frac{1}{M}\left(\frac{d[u]}{dt} - [\dot{u}]\right). \tag{8}$$

Combining equations (7) and (8), and solving it for d[u]/dt we obtain

$$\frac{d[u]}{dt} = \frac{\alpha + M^2}{\alpha - M^2}\,[\dot{u}] + \frac{M}{\alpha - M^2}\,\frac{dM}{dt}\,[u], \tag{9}$$

where

$$\alpha = \frac{[\dot{p}]}{[\dot{v}]} = \left.\frac{dp}{dv}\right|_S . \tag{10}$$

assuming the particle-isentropic flow behind the shock front.

We further assume that material does not move until the shock wave arrives, and introduce a Hugoniot parameter σ by $dU/du=\sigma$ ($U = V$ is the shock velocity). Then,

$$\frac{dM}{dt} = \rho_0\sigma\,\frac{d[u]}{dt}\,, \tag{11}$$

$$\left.\frac{dp}{dv}\right|_H = - \frac{U + \sigma u}{U - \sigma u}\,M^2. \tag{12}$$

From (1), (2), (3), (10), and (12),

$$\alpha = - \frac{U + (1 - \beta)\sigma u}{U - \sigma u}\,M^2, \tag{13}$$

where

$$\beta = (v_0 - v)\left.\frac{\partial p}{\partial e}\right|_v = \frac{u}{U - u}\,\gamma\,. \tag{14}$$

Substituting (11) and (13) into (9) we obtain the following relationship,

$$\frac{d[u]}{dt} = \frac{U(2 - \beta)\sigma u}{(2U - \sigma u)(U + \sigma u) - U\beta\sigma u}[\dot{u}]. \quad (15)$$

By solving the above equation for β and further for γ by equation (14), we can obtain expression of Gruneisen γ by the measurable quantities -- u, d[u]/dt and [u̇] -- from Lagrangian gage (particle velocity gage) records, which is

$$\gamma = \frac{U - u}{u}\left(\frac{(U - \sigma u)(2U + \sigma u)}{U\sigma u} \frac{d[u]/dt}{d[u]/dt - [\dot{u}]} + 2\right). \quad (16)$$

Similarly the identical equations to (15) and (16) can be obtained for pressure gage records, which are

$$\frac{d[p]}{dt} = \frac{U + \sigma u}{2U + \sigma u} \frac{(2 - \beta)\sigma u}{U + (1 - \beta)\sigma u}[\dot{p}], \quad (17)$$

$$\gamma = \frac{U - u}{u}\left(\frac{(2U + \sigma u)(U - \sigma u)}{u(U + \sigma u)} \frac{\frac{d[p]}{dt}}{\frac{2U + \sigma u}{U + \sigma u}\frac{d[p]}{dt} - [\dot{p}]} + 2\right). \quad (18)$$

DISCUSSION

Equation (15) (or (17)) has four solutions for d[u]/dt = 0 (d[p]/dt = 0) or steady propagation of one-dimensional shock wave. They are

(A) $[\dot{u}] = 0$ ($[\dot{p}] = 0$),

(B) $u = 0$,

(C) $\sigma = 0$, and

(D) $\beta = 2$.

The first solution is a square shock wave. As Lagrangian gage records do not give informations of the isentrope through the shocked state, it is impossible to estimate γ under the condition (A).

It is easy to show that (B), (C), and (D) are solutions of equation (1) under the condition,

$$\left.\frac{dp}{dv}\right|_S = - M^2. \quad (19)$$

Solutions (B) and (C) also satisfies the following condition,

$$\left.\frac{dp}{dv}\right|_H = - M^2. \quad (20)$$

The solution (B) corresponds to sound wave and (C) to well known Chapman-Jouget detonation wave. The estimations of γ under these two conditions are impossible.

It is not clear at present whether the existence of the solution (D) is possible or not. Even if it is possible, the pressure at which (D) is satisfied will be too high for experimental evaluation of γ.

As can be seen from the previous discussions, the evaluation of γ based on equation (16) or (18) is especially effective at high pressure region (far from (B)) and for strongly decaying shock waves (far from (A)). The condition (C) can be ignored for inert materials.

Differentiation procedures of experimentally obtained Lagrangian gage records will produce a large error in γ. Accuracy of γ obtained experimentally from particle velocity gage records can be estimated by the following,

$$\frac{\Delta\gamma}{\gamma} = E_1 \frac{\Delta\sigma}{\sigma} + E_2 \frac{\Delta R}{R} + E_3 \frac{\Delta Q}{Q} \qquad (21)$$

where $R = u/U$ and $Q = [\dot{u}]/(d[u]/dt)$, and

$$E_1 = \frac{1}{\gamma}\,\frac{1-R}{R}\,\frac{\sigma^2 R^2 - 2}{\sigma R}\,\frac{1}{1-Q}, \qquad (22)$$

$$E_2 = \frac{1}{1-R} + E_1, \qquad (23)$$

$$E_3 = \frac{1}{\gamma}\,\frac{1-R}{R}\,\frac{(1-\sigma R)(2+\sigma R)}{\sigma R}\,\frac{Q}{(1-Q)^2}. \qquad (24)$$

Assuming a proper value for γ, the value of Q can be estimated using equation (15) for a material of which Hugoniot is already known. For example, using shock wave parameters for NaCl at 30 GPa (Marsh, 1980) and assuming γ/v to be constant, we obtain 4.4, 2.9 and 4.0 for E_1, E_2, and E_3, respectively. These values become smaller as pressure increases.

REFERENCES

Lysne P.C. and Hardesty D.R., J. Chem Phys., 59, 6512 (1973)

Marsh S.P., ed., "LASL Shock Hugoniot Data," University of California Press, Berkeley (1980)

McQueen R.G. and Fritz J.N., in "Shock Waves in Condensed Matter -1981," p. 193, eds., Nellis W.J., Seaman L., and Graham R.A., American Institute of Physics, New York (1982)

McQueen R.G. Hopson J.W. and Fritz J.N., Rev. Sci. Instrum., 53, 245 (1982)

ANALYSIS OF TRANSITIONS BETWEEN FERROELECTRIC AND ANTIFERROELECTRIC STATES UNDER CONDITIONS OF UNIAXIAL STRAIN*

S. T. Montgomery
Computational Physics and Mechanics Division II
Sandia National Laboratories
Albuquerque, NM

The ferroelectric ceramic PZT 95/5, formed as a solid solution from the antiferroelectric, $PbZrO_3$, and the ferroelectric, $PbTiO_3$, is known to transform to an antiferroelectric state under pressure. This transition can be used to convert mechanical energy to electrical energy under shock conditions. A mathematical model for this transformation has been developed for conditions of uniaxial strain. It uses two order parameters as internal state variables to reflect the microscopic state of the material. One order parameter corresponds to the spontaneous lattice polarization. The other represents the microscopic alignment of the spontaneous polarization within the lattice with parallel alignment corresponding to a ferroelectric state and anti-parallel alignment corresponding to an antiferroelectric state. A numerical method for solution is used to illustrate dynamic material response resulting from the transition.

INTRODUCTION

The ferroelectric (FE) ceramic PZT 95/5 exhibits a stress-induced phase transition from a FE state to an antiferroelectric (AFE) state. The substance is formed as a solid solution from the AFE, lead zirconate, and the FE, lead titanate. The replacement of approximately 5% of the Zr^{+4} ions of the $PbZrO_3$ with Ti^{+4} ions of the $PbTiO_3$ results in a stable FE ceramic. As formed, the ferroelectric domains that make up the crystallites of the ceramic grains are randomly oriented, and the substance exhibits no net remnant polarization. However partial alignment of its domains can be achieved on poling the ceramic by the application of an electric field. This results in a net remnant polarization for the material along the direction of the poling field.

Neilson[1] first suggested that shock induced depolarization of a ferroelectric ceramic could be the basis for a pulsed power supply. The ease with which PZT 95/5 depoles due to the FE-AFE phase transition has made this material the subject of

* This work performed at Sandia National Laboratories supported by the U. S. DOE under contract DE-AC04-76DP00789.

numerous experimental studies. For example, Lysne and Percival[2] studied the charge release from bars of poled ceramic as a function of resistive load. Dick and Vorthman[3] measured transmitted wave profiles for both poled and unpoled ceramic bars. More recently, Chhabildas[4] measured wave profiles for uniaxial strain and pressure shear experiments with the unpoled ceramic.

Efforts to develop a mathematical model for the response of the PZT 95/5 in these experiments have been limited. Besancon, David and Vedel[5] presented a simple circuit model based on the assumption of instantaneous depolarization of the ceramic with different dielectric constants for the compressed and uncompressed material. The mechanical behavior of the material and the influence of the state of stress on the phase transition are important aspects that have not been considered. Here, a simple description for the phase transition is given and incorporated into the simplest constitutive theory appropriate for unpoled FE ceramics in the absence of electric fields.

PHASE DESCRIPTION

The FE phase in a crystallite is characterized by the existence of a spontaneous lattice polarization whose direction can be switched by the application of an electric field. Ferroelectricity usually disappears above a certain temperature called the transition temperature. Above the transition temperature, the crystallite has no spontaneous polarization and is said to be in a paraelectric (PE) state. The nature of the PE-FE phase transition can be investigated by considering an expansion of the thermodynamic potential in terms of the spontaneous polarization. Minimization of the potential with respect to the spontaneous polarization gives the equilibrium value of the spontaneous polarization and hence the equilibrium thermodynamic potential. Samara[6] has presented a discussion of the influence of hydrostatic pressure on the PE-FE phase transition and related measurements to the behavior of terms in the expansion of the thermodynamic potential.

The AFE state is similar to the FE state in that spontaneous lattice polarization exists for the crystallite lattice, yet different in that the distribution of the spontaneous polarization in the lattice is such that no net polarization exists for the crystallite. In order to allow for the possibility of a phase transformation between FE and AFE states it is necessary to use a description for the polarization state more general than just prescribing the value of the lattice polarization. One such description was obtained by Kittel[7], who used the concept of two identical interpenetrating sublattices, with each sublattice polarization being an order parameter in the expansion for the thermodynamic potential. An equivalent and somewhat simpler choice of order parameters is used here. One order parameter is, like in the FE case, the spontaneous lattice polarization, $\boldsymbol{P}_s$, while the other is chosen to be the degree of alignment of the spontaneous polarization in the lattice. This alignment variable, μ, then varies between ± 1, with the FE state characterized by $\mu = 1$ and the AFE state characterized by $\mu = -1$.

In general the thermodynamic potential for the material should depend on the state of stress, electric field, and temperature, as well as on the order parameters. The electric field dependence is neglected here since we restrict our attention to an unpoled FE ceramic with no applied electric fields. In addition it will also be presumed that the deviatoric stress components have negligible effect on the phase behavior. Hence, the thermodynamic potential, $\mathcal{G}$, is assumed to have the form

$$\mathcal{G}(\sigma,\theta,P_s,\mu) = \mathcal{G}_o(\sigma,\theta) + \frac{1}{2}[(b+g\mu)+(q+h\mu)\sigma+\lambda\mu^2]P_s^2 + \frac{1}{4}\xi P_s^4 + \frac{1}{6}\varsigma P_s^6 \quad (1)$$

where σ is the mean stress, θ is the temperature and, $\mathcal{G}_o$ is the part of the potential that does not depend on the order parameters. The parameters q, h, ξ, ς are constants, while the parameters b, g, λ depend linearly on temperature, i.e., $b = b_0 + b_1(\theta - \theta_0)$, where θ_0 is the Curie temperature for the FE ceramic.

Minimization of the thermodynamic potential with respect to the order parameters provides the values they take in equilibrium. It is easy to show that when $\lambda < 0$ the expression above has no local minima for $\mu^2 < 1$ except when $P_s = 0$, which corresponds to the PE phase. However, it is possible that the potential can also have minima when $\mu = 1$ or $\mu = -1$. In the pressure range of interest here, we note that, for fixed μ, it is possible to find a root of the equation $\frac{\partial \mathcal{G}}{\partial P_s} = 0$ such that $\frac{\partial^2 \mathcal{G}}{\partial P_s^2} > 0$. However, as mentioned above, this root never gives a local minimum for $\mu^2 < 1$, and it can only correspond to a local minimum at $\mu = 1$ if $\frac{\partial \mathcal{G}}{\partial \mu} < 0$. If $\mu = -1$ we must have $\frac{\partial \mathcal{G}}{\partial \mu} > 0$ in order for a local minimum to exist.

Figure 1 shows a phase diagram obtained from the thermodynamic potential above. The parameters used are indicated in Table 1 and are intended only to be representative of those required to describe PZT 95/5. Parameter determination from data is fairly complicated and will not be described here. As indicated in Fig. 1 by the solid lines, stress-temperature space is divided into three regions, where the designated phase has the lowest value for the thermodynamic potential. Here we are interested only in temperatures much lower than θ_0, where only the FE and AFE phases can exist. The dashed lines indicate the stability boundaries for the FE and AFE phases. As can be seen, there is a region for which one is metastable with respect to the other. Hence the phase transformation can exhibit hysteresis with respect to pressure.

PHASE TRANSITION UNDER UNIAXIAL STRAIN

We take the following simple constitutive equation for the stress, τ_{ij}, in the material

$$\tau_{ij} = 2GS_{ij} + [(K - \frac{2}{3}G)S_{kk} + \mathcal{V}(P_s^2 - P_0^2) + \mathcal{W}(\mu P_s^2 - P_0^2)]\delta_{ij} \quad (2)$$

where the strain, S_{ij}, is measured from the stress free reference state for which the ceramic is in the FE phase with spontaneous polarization P_o. The constants K and G are the bulk and shear modulus for the ceramic in the FE phase, and $\mathcal{V}$ and $\mathcal{W}$ are constants related to the pressure that would result in a confined specimen transformed to the PE and AFE phases. The order parameters that appear in the constitutive equation will be assumed to be governed by the following evolution equations

$$\dot{P}_s = -\Gamma_P \frac{\partial \mathcal{G}}{\partial P_s} \quad (3)$$

$$\dot{\mu} = -\Gamma_\mu \frac{\partial \mathcal{G}}{\partial \mu} \quad (4)$$

where the superposed dot denotes the time rate of change of the variable.

We now consider the situation of uniaxial strain loading generated by impacting a plate of the unpoled ceramic with a plate of some other material. This situation is conveniently treated numerically with the finite-difference wave propagation code WONDY[8]. Evaluation of the constitutive equation requires solution of the evolution equations. This task can be simplified if it is assumed that Γ_P is very large so that P_s may be regarded as relaxing to the appropriate root almost instantaneously. For the temperatures and stresses of interest, the spontaneous lattice polarization can be readily determined for any particular value of μ. Evaluation of the stress then simplifies to determination of μ through its evolution equation. When the slope $\frac{\partial g}{\partial \mu}$ changes continuously with stress, which is the case here, it is very simple to integrate this evolution equation with standard numerical methods and determine the stress.

A numerical calculation was done to simulate an experimental configuration used by Chhabildas to study the shock evolution in an unpoled sample of the FE ceramic PZT 95/5. The bulk and shear modulus were taken from the experimental work of Chhabildas, while the other parameters are estimates thought to be typical for PZT 95/5. The experimental configuration consisted of an unpoled sample of ceramic backed by a sapphire layer which was used as a window for interferometric measurement of the interface velocity between the ceramic and sapphire. Figure 2 shows the velocity for this interface as calculated for an impact of a fused quartz projectile onto the ceramic at approximately 111 m/s. Figure 2 shows that a two wave structure results from the phase tranformation between FE and AFE states. This is consistent with what was observed by Chhabildas for a similar impact stress.

SUMMARY

A simple model has been given for the FE-AFE phase transition like that observed in PZT 95/5. Once an appropriate thermodynamic potential govening the equilibrium of the possible phases was defined, the use of order parameters as internal state variables allowed the phase transformation to be treated as a dynamic process driven by the stress in the material. The model was used to simulate the dynamic response of the ceramic for conditions of uniaxial strain and produced results in reasonable accord with experimental measurements given the simplicity of the constitutive assumption. In conclusion, the model presented provides a simple representation for the FE-AFE phase transiton, and should be useful in the development of the more general constitutive equations required for more complicated electromechanical conditions.

Table 1. Parameters for the thermodynamic potential.
(MKS units are used with temperature in °C)

$\xi = -1.00 \times 10^{8}$	$b_0 = 7.34 \times 10^{6}$	$b_1 = 7.69 \times 10^{5}$
$\varsigma = 1.40 \times 10^{10}$	$g_0 = -7.20 \times 10^{6}$	$g_1 = -2.20 \times 10^{4}$
$\theta_0 = 220.0$	$\lambda_0 = -1.40 \times 10^{6}$	$\lambda_1 = 3.00 \times 10^{3}$
$q = -0.70 \times 10^{-2}$	$h = -1.80 \times 10^{-2}$	

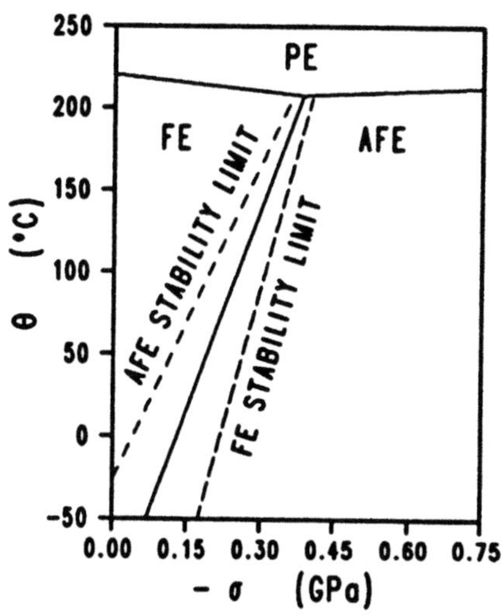

Figure 1: Phase diagram obtained from the thermodynamic potential.

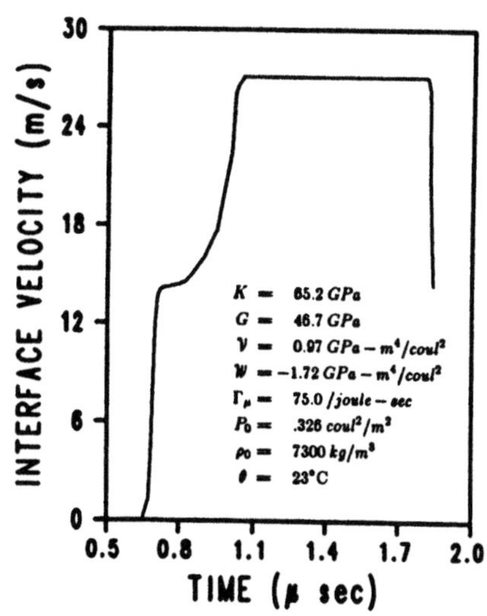

Figure 2: Ceramic-sapphire interface velocity calculated for a 111 m/s impact of fused quartz on the unpoled ceramic.

REFERENCES

1. F. W. Neilson, Effects of Strong Shocks in Ferroelectric Materials, Bull. Am. Phys. Soc., 2:302 (1957).
2. P. C. Lysne and C. M. Percival, Electric Energy Generation by Shock Compression of Ferroelectric Ceramics: Normal-Mode Response of PZT 95/5, J. Appl. Phys., 46:1024 (1977).
3. J. J. Dick and J. E. Vorthman, Effects of Electrical State on Mechanical and Electrical Response of a Ferroelectric Ceramic PZT 95/5 to Impact Loading, J. Appl. Phys., 49:2494 (1978).
4. L. C. Chhabildas, Dynamic Shock Studies of PZT 95/5 Ferroelectric Ceramic, Sandia National Laboratories Report SAND84-1729 (December 1984).
5. J. E. Besancon, J. David and J. Vedel, Ferroelectric Transducers, "Proc. of the Conf. on Megagauss Magnetic Field Generation by Explosives and Related Experiments," H. Knoepfel and F. Herlach, eds., Euratom 315 (1966).
6. G. A. Samara, The Effects of Hydrostatic Pressure on Ferroelectric Properties, in "Advances in High Pressure Research 3," R. S. Bradley, ed., Academic 155 (1969).
7. C. Kittel, Theory of Antiferroelectric Crystals, Phys. Rev., 82:729 (1951).
8. M. E. Kipp and R. J. Lawrence, WONDY V - A One-Dimesional Finite-Difference Wave Propagation Code, Sandia National Laboratories Report SAND81-0939 (June 1982).

SANDIA SHOCK COMPRESSION DATABASE

C. E. Anderson, J. S. Wilbeck, and J. C. Hokanson

Southwest Research Institute
San Antonio, Texas 78284

J. R. Asay, D. E. Grady, R. A. Graham, and M. E. Kipp

Sandia National Laboratories
Albuquerque, New Mexico 87185

ABSTRACT

An authoritative, updatable, and searchable bibliographic database whose topical area is shock compression is currently under development. The database is derived from published scientific articles and reports with current emphasis in the subareas of strength, phase transitions, spall, and stress-volume relations. The content of the cited work is described by keywords which are selected and reviewed by scientific advisors. Keywords are organized to provide major and minor descriptors which summarize the technical content in considerable detail. Details of theoretical and experimental approaches are described and major emphasis is identified. In addition to bibliographic information and abstracts, fields have been provided for figure captions, table titles and supplementary information.

INTRODUCTION

The desire to have a database in the discipline of shock physics arises from a number of considerations. It is apparent that in the scientific community today there exists a need for relatively rapid and organized procedures for locating and obtaining information applicable to a researcher's current experimental, analytical, computational, or modeling efforts. The problem is two-fold. First, there is the individual hunting for data/information who has problems because the collection or procurement of these data is periphery to his field of interest, or because he is a newcomer to the field. This requires a systematic procedure for scanning reports, many or most of which he may not be aware exist. Second, there is the veteran who has previewed, scanned and read countless documents of interest, but then no longer can remember where he actually saw the data. This individual also requires a systematic procedure for retrieving the authors, title, etc., of the source material. Both of these types of individuals will use a database if the descriptors/identifiers/keywords are pertinent and useful. Finally, there is concern that as researchers retire, information readily available from them would be difficult, and perhaps impossible, to obtain.

The utility and usefulness of a well organized database is seldom questioned. A brief discussion of one of the largest technical databases, the National Technical Information Service (NTIS), maintained by the United States Department of Commerce, is illustrative of problems which can be encountered with searchable databases. As useful as this database is in performing a search of technical reports by the U. S. Government and its contractors, it has two drawbacks for specialists in a particular discipline. By its very nature, documents listed in the database are quite varied, ranging from such diverse topics as fires on railroad tank cars, terminal ballistics, and combustion studies, to airplane safety, risk assessments and biological studies, etc. Second, authors assign their own keywords. Keyword assignment varies considerably with individuals. Some authors jot down the first few keywords that come to mind; other authors list many keywords, some only remotely associated with their document. General descriptors are often employed to allow the widest dissemination of their work. A user of the database, then, is faced with two problems in searching the database. General descriptors become almost useless in searching for relevant documents because of the extremely large number of entries that are returned. Guessing appropriate keywords is often an art within itself, as is the proper selection of appropriate keyword combinations. The list of reports retrieved may be unacceptably long; this list requires further discrimination, typically performed using a logical AND with two or more keywords (all keywords must appear or else the report is rejected). Thus, the reticent author's paper might not be found in a keyword search.

Other databases, either because they were limited in scope, not applicable, or too broad were found not suitable for the topical area of shock compression. Sandia National Laboratories decided to establish a database in shock physics, hereafter referred to as the SSC Database (Sandia Shock Compression Database). A major objective in organizing this database is the assignment of identifiers or keywords, which are **specific** descriptors to identify the technical content of a paper, thereby making it a very useful, searchable technical database.

The SSC Database has provisions for storing an extensive amount of information extracted from papers, reports, symposia proceedings, and books dealing with the discipline of shock physics. The information is presented in a logical format for the database user. It is organized into four categories: paper identification, bibliographic citation, technical keywords, and supplemental information. Each of these will be discussed briefly.

IDENTIFICATION

There are two parameters associated with identification of a document. The first parameter is the Chronological Paper Number. This is an unique number assigned to each document as it is processed for entry into the database. This number is the identifier for retrieval operations inside the database software. The second parameter, the Paper ID, is similar to reference citations of some journals, and is a function of the year the document was published, and the primary author's last name. The Paper ID has six characters: the first two are the last two digits of the year in which the document was published; the third is the first letter of the primary author's last name; and the last three represent a sequential assignment (as chronologically processed) of papers that year with a similar letter (e.g., 73G003 represents the third document processed into the database from the year 1973 with a primary author whose last name begins with a "G").

BIBLIOGRAPHIC CITATION

There are seven data fields with information which can be directly or indirectly associated with bibliographic information. These fields consist of the following: Authors; Title; Source; Alternate Publication; Submission Date; Affiliation; and Acknowledgements. The Authors and Title fields are self-explanatory. The Source field denotes the publication information (including such things as report number and publishing organization, journal name, volume number, publication date, etc.); that is, the Source field gives the information required to obtain a hard copy of the document. The intent of the Alternate Publication field is to denote the original journal if the article is from a translated journal (a common occurrence for Russian articles); that a paper was presented at a symposium or conference, though the published proceedings appear at some later date; or a preprint or reprint when assigned a Laboratory report number. The Submission Date is used primarily for journal articles which often have considerable time between submission and publication; and the dates for symposia or conferences which can precede a published proceedings by as much as a year. The Affiliation field gives the organizational affiliations of the authors of the paper, while Acknowledgements give the names of personnel acknowledged by the authors.

TECHNICAL KEYWORDS

Key to the database is the selection of technical keywords for categorizing the technical content of a paper. The technical keywords have been divided into ten distinct topical areas:

Special Emphasis	**Experimental Method**
Material Response	**Experimental Results**
Phenomena	**Theory/Model**
Properties	**Computational Method**
Materials	**Applications**

The first five technical keyword categories, in general, describe the technical thrust of the paper, while the second set of five words describes the methodology or manner in which the phenomena were studied. The primary emphasis of each paper, such as theoretical, experimental, analytical, numerical, review, or symposium proceedings, is given in the Special Emphasis field. This field also is used to state that experimental data are presented in the paper and whether these data are original or referenced. The next field distinguishes the primary material response under consideration, such as mechanical, thermodynamic, electrical, optical, chemical, etc. The third technical keyword field gives the physical phenomena the article is addressing. Phenomena of interest in shock compression physics currently in the database are:

Phenomena:	Strength	Porous
	Phase Transformation	Residual
	Spallation	Radiation
	Stress-Volume	Impact
	Equation of State	Stability

Obviously, some papers address more than one phenomenon, and more than one keyword can be used to describe an article (this is true of all the technical keyword fields). Measurements are made which characterize a phenomenon, and these physical attributes of a material are called properties:

Properties:	Strength	Residual
	Spallation	Porous
	Phase Transformation	Thermodynamic
	Stress-Volume	Elastic
	Equation of State	Rheologic

Note that while the keywords are similar, there is a distinction between studying a phenomenon and measured properties. Another method for subdividing articles in the database is by the material studied. After several iterations, the following material designators were defined:

Materials:	Metals	Organics
	Geophysical	Composites
	Polymers	Mixtures
	Ceramics	Liquids
	Inorganics	Explosives

Experimental work is divided into two distinct categories, methods and results. In the keyword field, Experimental Methods, the experimental methodology and instrumentation employed to collect the data are listed, e.g., the method for loading the material (flyer plate, etc.), the instrumentation used to monitor material response, and the geometry (plane strain, plane stress, expanding ring, etc.). Experimental Results are subdivided into active (or dynamic) measurements such as particle velocity measurements; and residual measurements, such as post-shock metallurgical examination. Some papers develop, describe and/or use specific theories or models. These are entered in the Theory/Model field to further describe the technical content of a paper. If computational methods are used in the analysis, then the code name, the type of code (Eulerian, Lagrangian, etc.), the numerical method (finite-difference, method of characteristics, etc.), and special numerical techniques (e.g., rezoning, artificial viscosity, etc.) which are employed and discussed, are listed in this field. Finally, one last field, Applications, specifies particular applications of the research material, such as ballistic impact, explosive initiation, rubblization, etc.

In each of the technical keyword categories, these major and sub-major descriptors are further qualified by minor descriptors. The intent is to describe, as best as possible with a finite set of keywords, the technical content of the paper. A dictionary of keywords has been prepared and lists the acceptable set of keywords. This dictionary, including the major and minor descriptors, is the result of several iterations with a test phase of approximately fifty reports covering a wide range of technical issues in the area of shock compression, including journal articles, articles from foreign journals, government reports, symposium presentations, and review papers. Besides a technical keyword dictionary, an author dictionary is being prepared which lists alphabetically all authors in the database. These dictionaries are not only of use to the user in the selection of descriptors, but since they are compiled from the database itself, misspellings are readily noted and corrected.

SUPPLEMENTAL INFORMATION

It has been found from experience that no matter how judicious the selection of keywords or identifiers, only so much information can be summarized by this method. While a judicious selection of keywords, and a combination of keywords can limit the number of articles to a manageable number, three supplemental information fields have been added to the database to supply further information on the technical content of a paper. An Abstract field contains an abstract of the paper content. Where possible, the original abstract of the paper is used. If no abstract was included by the original authors, a suitable abstract is written by the technical reviewer who assigns technical keywords. The Additional Information field is used by the technical reviewer to comment on the content of the paper with any information which might be relevant and informative to someone searching the database. Finally, a Data Summary field provides space for table headings and figure captions. At a later date, this field could be expanded to include experimental data.

SUMMARY

In principle, all fields in the database are searchable. Generally, though, searchable fields would be authors and acknowledgements, affiliations, and the technical keywords. The user will eventually have the opportunity to search the database via a "user friendly," menu-driven CRT display inquiry system, with an option to print the information on relevant reports retrieved by the database. An example of the information in a retrieved document is shown in Figure 1.

The authors would like to acknowledge the technical suggestions and the encouragement provided by Mr. Walt Herrmann and Dr. Lee Davison of Sandia National Laboratories. Their concern for retrieval of useful information and their perseverance in initiating this project were invaluable to this database endeavor.

Chrono Paper No.:	0017
Paper ID:	75B001
Authors:	Bertholf, L. D., Buxton, L. D., Thorne, B. J., Byers, R. K., Stevens, A. L., and Thompson, S. L.
Title:	Damage in Steel Plates from Hypervelocity Impact. II. Numerical Results and Spall Measurement
Source:	J. Appl. Phys. 46 (9), 3776-3783 (1975)
Original Publ.:	*
Submission Date:	Feb. 24, 1975
Affiliation:	Sandia Laboratories, Albuquerque, NM
Acknowledgements:	Bond, J. W., Shockey, D. A., Curran, D. R. and Herrmann, W.
Special Emphasis:	[numerical]
Mater. Response:	[mechanical]
Phenomena:	[phase transformation] mixed phase, polymorphic [impact] cratering, spall
Properties:	[spallation] spall strength
Materials:	[metals] steel [polymers] nylon

Fig. 1. Example of database information.

Exper. Method: [loading] flyer plate
[instrumentation] visar
[geometry] plane strain

Exper. Result: [active] free surface velocity
[residual] cross-sectional photograph

Theory/Model: [elastic-plastic] elastic-perfectly plastic

Comput. Method: [code] toody, csq, toorez
[type] lagrangian, eulerian
[techniques] rezone

Applications: penetration, cratering

Additional Info.: Limited spall strength measurement. Only one isolated hypervelocity impact experiment. Numerical prediction of spall depended upon mesh size

Abstract: Comparisons of calculated solutions with experimental solutions with experimental results for the impact of a 9.53-mm-diam nylon sphere on a 12.7-mm-thick steel plate at 5.182 km/sec are presented. Numerical solutions are difficult to obtain with existing Eulerian codes because of the need for accurate treatment of material strength and the definition of both the material interface and the free surfaces. The large distortion of the nylon sphere and the steel near the impact crater lip also make this a difficult problem for Lagrangian codes unless frequent rezoning is used. Newly developed Eulerian and Lagrangian rezoning codes are used. They give essentially the same solution. The computation results show that in order to obtain agreement with the experiment, it is necessary to account for the alpha (bcc)==epsilon (hcp) polymorphic phase change for iron. Calculations also indicate the need to obtain experimentally the correct spall strength for the steel and a description of this measurement is included. Treatment of elastic-plastic response, a good numerical model for fracture, and a high degree of numerical resolution are also shown to be necessary in order to predict accurately the back-surface spallation observed in the experiment.

Data Summary: Fig. 1 - Hugoniot states for iron (pressure vs. density).
Fig. 2 - Comparison of code calculations (Eulerian & Lagrangian) with experiment.
Fig. 3 - Effect of polymorphic phase change of iron.
Fig. 4 - Free surface velocity history of planar impact of two steel plates.
Figs. 5-6 - Comparison of code calculations with expert.
Table 1 - Material properties used for calculations.
Table 2 - Calculated spall strengths in steel for various calculational methods.

Fig. 1. Example of database information (continued).

RAMAN SPECTROSCOPY OF SHOCKED WATER

N. C. Holmes, W. J. Nellis, and W. B. Graham
Lawrence Livermore National Laboratory
Livermore, CA 94550

G. E. Walrafen
Howard University
Washington, D.C. 20059

ABSTRACT

We describe a new technique for recording spontaneous Raman spectra from molecules during the passage of strong shock waves. We have used this technique to study the OH-stretch band of liquid H_2O shocked to pressures up to 26 GPa and 1700 K. The shape of the band changes over the range 7.5-26 GPa, and is described well by a two-component mixture model, implying changes in the intermolecular coupling of shock compressed water molecules. We discuss the implications of the spectra on the mechanism responsible for the electrical conductivity of shocked H_2O.

INTRODUCTION

The passage of strong shock waves through molecular systems allows the study of these systems under extremes of pressure, density, and temperature. Most shock experiments measure bulk properties of molecules such as equation-of-state, conductivity, temperature, sound velocity, and so forth. These results for water are contained in Ref. 1 and references therein. However, our knowledge of the details on a molecular level in shock-compressed matter is most rudimentary. We generally do not know the molecular chemistry responsible for the bulk behavior of shocked substances. A detailed understanding requires a knowledge of the chemical species existing behind the shock front, what reactions occur, and the reaction kinetics, all of which may be functions of both time and position. During the past few years, modern spectroscopic techniques have just begun to be applied to this new and important regime of chemical physics.

For example, consider the case of water. Between 3 and 30 GPa (30-300 kbar) the electrical conductivity of shocked H_2O increases by over five orders of magnitude,[1-3] saturating at a value of about 20 $(\Omega\text{-cm})^{-1}$ above 30 GPa.[1] Hamann and Linton[4] proposed that water becomes fully ionized to H_3O^+ and OH^- at 15-20 GPa and that transport of these ions could account for the observations. This is the same process which occurs in water at ambient conditions, but greatly enhanced by the high temperature and density behind the shock front. The extensive studies of water ices at high pressure cast little light on this question, for the phase is different (shocked water is liquid above about 5 GPa), and the temperatures of 2000 K and higher far exceed what is currently possible in static compression experiments. What is clearly needed is a method to probe directly for the existence of new molecular species behind a strong shock wave front. It is precisely this problem which has motivated this work.

RAMAN SPECTROSCOPY

Raman spectroscopy is often applied to the study of molecular systems. The Raman effect is a type of inelastic light scattering in which a fraction of the energy of a photon incident on a molecule is exchanged with a molecular vibrational mode. When the process results in lower frequency (red-shifted) scattered light it is termed Stokes scattering; scattered light of higher frequency is called anti-Stokes scattering. The frequency shifts of the scattered light allow the determination of the molecules present, their structure, and often provide insight into the local environment of the scattering molecules. A great number of techniques have been developed which exploit this effect, and they have recently been summarized by Schmidt.[5] By far the most commonly used technique is spontaneous Raman scattering, a process for which the scattered intensity is linearly related to the incident light intensity. Spontaneous anti-Stokes Raman scattering has been used to determine the temperature of shocked liquids.[6] A variety of non-linear spectroscopic methods are now in use, which feature intense laser sources to produce stimulated scattering. Recently, stimulated Raman back-scattering has been used to measure the shift in the ν_1 vibrational mode of benzene shocked to 1.2 GPa.[7]

In the work reported here on shocked H_2O, we chose to use spontaneous Stokes Raman scattering to observe the OH-stretch vibrational mode for a variety of reasons. The bandwidth of this technique is large and limited only by the spectrograph, ideal for observing the OH-stretch for which the width of the frequency spectrum is typically 1000 cm^{-1}. Spontaneous Raman scattering occurs into 4π steradians, simplifying experiemental design. It does not require phase-matching at critical angles, an important advantage when dealing with large and moving refractive index gradients at the shock front. Most importantly, almost all previous Raman spectroscopic studies of water in liquid and ice phases have employed spontaneous scattering, so our results can be easily compared to the literature. Also, the molecular ions expected in shocked water should be distinguishable from the scattering from water molecules. The disadvantage of the spontaneous method is the weakness of the scattered light. Typical scattering cross-sections are of the order 10^{-29} cm^2 and lead to total scattering fractions of 10^{-7} or less. This makes detection difficult. Often spontaneous Raman spectra are obtained in minutes to hours while a line or band is carefully scanned. This is impractical in a shock wave experiment which lasts for about 10^{-6} seconds, precluding both signal averaging and scanning techniques. Recent advances in optical detector technology, described below, now make it possible to record braodband spectra in a much briefer interval.

EXPERIMENTAL DESCRIPTION

The experimental arrangement is shown schematically in Figure 1. Strong shock waves were generated by planar impactors accelerated to high velocities by a two-stage light-gas gun.[8] The water specimens were triply-distilled and deionized and were initially at 20 C and atmospheric pressure. The specimen holders were fabricated of 1100 aluminum and the baseplates of the holders were diamond-machined to minimize scatter of laser light into the detection optics. The dimensions of the target and impactor were chosen to provide a large volume of uniformly shocked water free of attenuation and edge effects.

An unpolarized KrF laser (248 nm, 0.2 J, 10 ns(FWHM) pulsewidth) was used to excite the Raman spectrum. The use of an ultraviolet laser exploits the fourth-power frequency dependence of the scattering cross-section to maximize the signal. The laser beam was aligned colinear with the centerline axes of the target and impactor. A fused silica lens was used to slightly focus the laser into the target, but the intensity was kept well below the threshold for stimulated scattering. The laser beam entered the water sample through an electronic-grade, z-cut, α-quartz window which displayed no detectable fluorescence on laser illumination. Raman scattered light was collected at 45 degrees to this axis by an f/4 fused silica lens and a matched lens formed an image on the input slit of the spectrograph of a small region 3.5 mm behind the shock front at roughly 1:1 magnification. Trigger pins installed in the water sample provided timing signals for the laser and detector electronics. A flat-field 0.3 m focal length spectrograph dispersed the

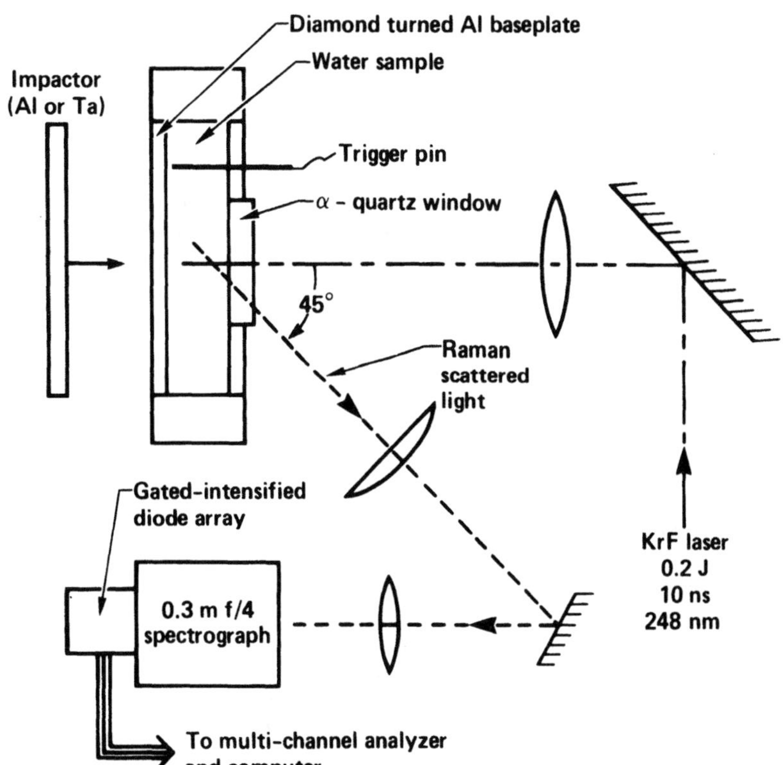

Fig. 1. The experimental arrangement for spontaneous Raman spectroscopy of shocked water. A two-stage light-gas gun is used to accelerate the impactor, generating strong shock waves in liquid water samples. The spectrograph and detector are located outside the target vacuum chamber.

scattered light for detection by a gated and intensified 1024-element linear diode array detector. The intensifier was required since the maximum amount of scattered light entering the spectrograph was about 10^{-10} J. The intensifier was gated on for 50 ns, allowing adequate discrimination against thermal emission from the hot, shocked water. The diode signals were digitized by a multichannel analyzer and stored in a small desk-top computer for analysis. Spectral resolution was measured to be about 15-20 cm^{-1}, limited mainly by the bandwidth of the KrF laser.

The experimental parameters for five experiments are listed in Table I. Pressures and densities were calculated using the impactor velocity,[9] the known equations-of-state of the impactor materials (Al, Ta),[10] the baseplate,[10] the water sample,[1,11] and the use of the impedance-matching method.[12] Shock temperatures were calculated using a constant C_v model.[13] In all cases the shocked water samples were in the liquid state based on the melting curve for ice VII.[14]

Table I. Experimental conditions for five spontaneous Raman experiments in shocked H_2O. The impactor material was aluminum for shots 8 and 9, tantalum for the others. Shock pressures and densities were calculated on the basis of known equations-of-state using the impedance-matching method. The shock temperatures were calculated using a constant C_v model.[13]

Shot #	Projectile Velocity (km/s)	Pressure (GPa)	Density (g/cm^3)	Calculated Temperature (K)
Raman 8	2.1	7.5	1.56	640
Raman 9	2.88	11.7	1.71	840
Raman 11	2.90	21.9	1.94	1480
Raman 12	2.86	21.4	1.93	1460
Raman 13	3.23	25.7	2.00	1710

RESULTS AND DISCUSSION

The OH-stretch spectrum of H_2O shocked to 7.5 GPa (75 kbar) and 640 K is compared with water at ambient conditions in Fig. 2. The band shape is strongly asymmetric in the shocked case, and the double-peak structure is no longer evident. More striking is the comparison with Lindner's data for statically compressed water at 0.4 GPa and 670 K[15] shown in Fig. 3. The effect of increased density alone is evident in the much more pronounced scattering in the low frequency range below 3400 cm^{-1}. As the shock pressure is increased to 22 GPa (Fig. 4) and 25.7 GPa (Fig. 5) the band gradually becomes more symmetric and the scattering intensity decreases substantially.

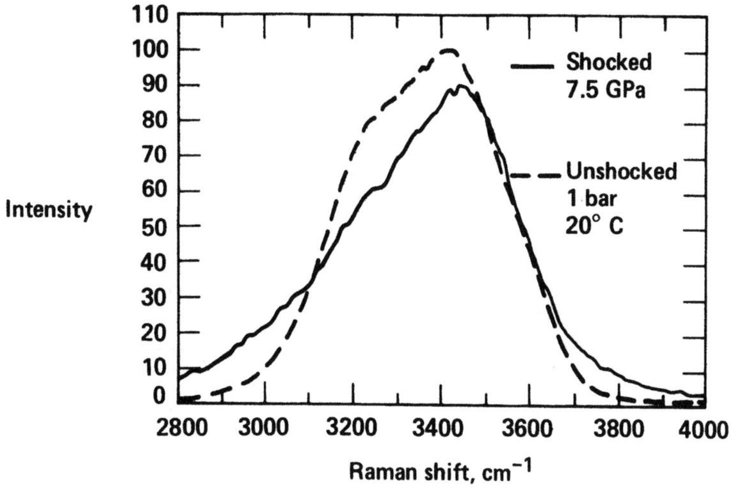

Fig. 2. Spontaneous Raman spectra of the OH-stretch band of liquid water at ambient conditions (dashed curve) and at 7.5 GPa and 640 K (solid curve).

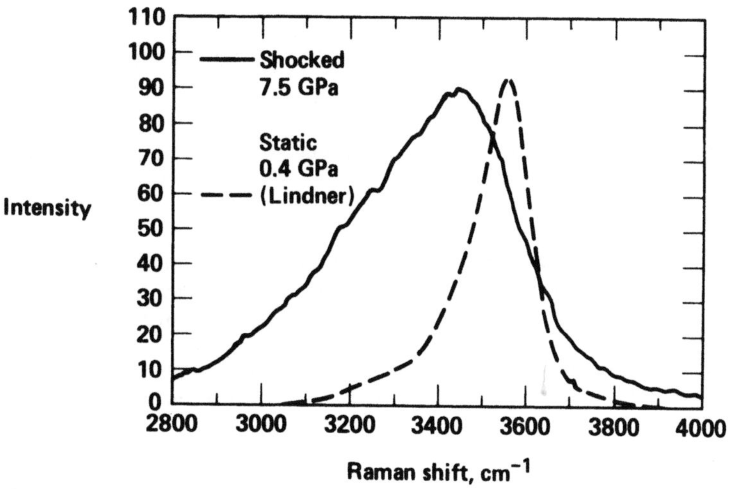

Fig. 3. A comparison of OH-stretch Raman spectra at nearly the same temperatures but at different pressures. The dashed curve is static data from Lindner[15] at 0.4 GPa and 670 K. The solid curve is the spectrum of shocked H_2O at 7.5 GPa and 640 K.

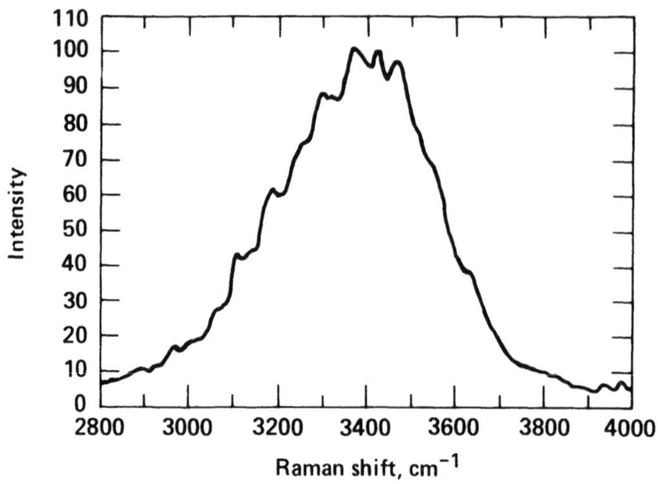

Fig. 4. Spontaneous Raman spectrum of shocked H_2O at 22 GPa and 1480 K.

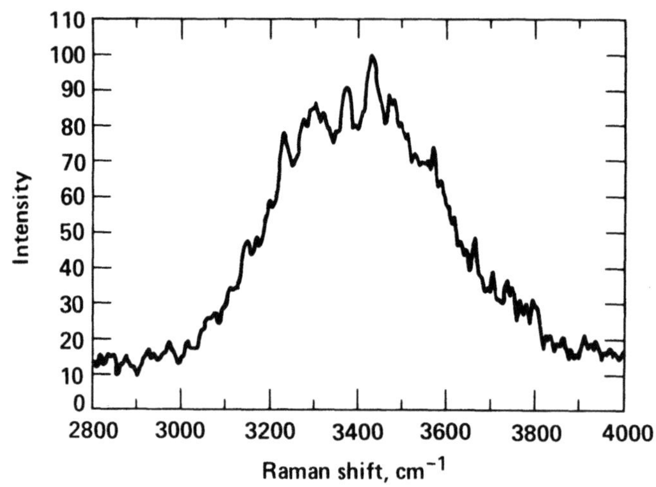

Fig. 5. Spontaneous Raman spectrum of shocked H_2O at 26 GPa and 1700 K.

The changes in the shape of the OH-stretch band can be described by a two-component mixture model for shocked water. That is, the bandshape is thought to arise from the superposition of scattering from two spectroscopically distinguishable classes of water molecules. One class of molecules is strongly hydrogen-bonded to its nearest neighbors in small tetrahedral clusters of the type shown in Figure 6. The strong intermolecular coupling results in a reduction of the OH-stretch frequency compared with isolated molecules in the liquid. The second class of molecules is thought to be either weakly bonded[16] or essentially monomeric water.[17] These molecules have a slightly higher OH-stretching frequency. This model is based on the observation of overlapping components in the Raman spectrum of liquid water at or near ambient conditions.[16,17] Studies of ice VII indicate an inverse correlation between the O-O distance and intermolecular O-H distance.[18] Under compression, as the O-O distance decreases, the OH-stretch frequency decreases reflecting the weakening of the longer O-H bonds.

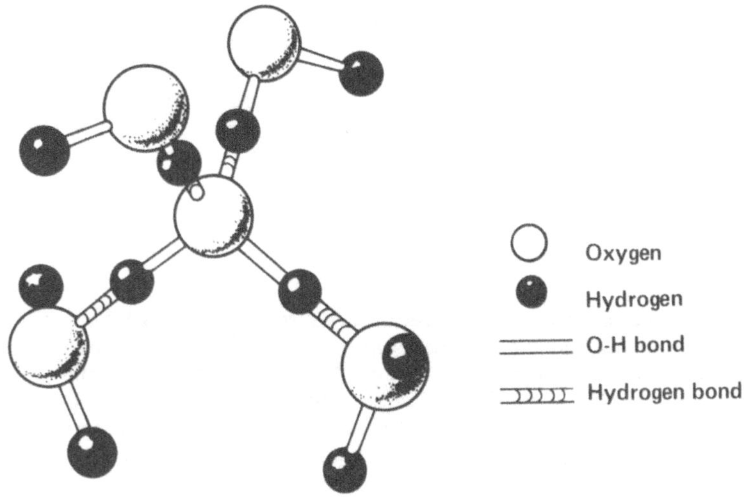

Fig. 6. The hydrogen-bonded structure of liquid water.

We used a least-squares technique[19] in an attempt to fit the data by a sum of two Gaussian components. The use of Gaussians seemed reasonable based on the observation that the wings of the band were of Gaussian shape for all cases, and because the broadening mechanisms expected behind the shock front lead to inhomogeneous broadening. A two-component fit is found to agree very well with all of the spectra of shocked H_2O, leading to rms errors of about 2%; the residual errors appear to be random in nature. The results of the fits are summarized below in Table II. We note that the positions of the components both show a gradual decrease in frequency with increasing pressure as expected.

Table II. Results of two-component Gaussian decompositions for OH-stretch spectra of shocked H_2O. Refer to Table I for physical conditions. I refer to the peak height of each component relative to a value of 1.0 at the maximum of the measured spectrum. The values of ν and $\Delta\nu$ are in cm^{-1} and refer to the center frequency and width (FWHM) of each component in the spectral decomposition.

Shot #	Component 1			Component 2			rms
	I	$\Delta\nu$	ν	I	$\Delta\nu$	ν	fit error
Raman 8	.62	490	3345	.43	208	3479	1.1 %
Raman 9	.68	466	3358	.33	208	3461	1.2 %
Raman 11	.49	353	3260	.66	300	3462	2.1 %
Raman 12	.36	240	3185	.84	325	3423	1.8 %
Raman 13	.14	166	3241	.72	415	3426	3.7 %

The total amount of scattered light in the two components changes markedly over the pressure range. The dependence of the lower-frequency, hydrogen-bonded component is plotted vs. pressure in Figure 7. We see from the figure that the hydrogen-bonded water structure dominates for pressures up to about 12 GPa, and then gradually decreases in intensity, diminishing to zero at about 26 GPa. On this basis, we claim that the spectra imply a destruction of hydrogen-bonded molecular complexes at very high shock pressures.

CONDUCTIVITY OF SHOCKED WATER

Contributions to the Raman spectrum from H_3O^+ should be distinguishable from the OH-stretch water band. The OH-stretch frequency of H_3O^+ has been measured to be between 2900-3000 cm^{-1} by both Raman[20] and infrared[21] techniques in solutions of strong acids. No such band is distinguishable in our data. This is perhaps not surprising. The conductivity in liquid water is believed to involve the transfer of electric charge through hydrogen-bonded molecular complexes of the form $H_9O_4^+$ and $H_7O_4^-$.[22] When the hydrogen-bonded complexes are broken at high shock pressures, this process is not likely to contribute significantly to the electrical conductivity. If large concentrations of the H_3O^+ ions were present, whether hydrated or not, they should be discernable in the spectra. An alternative hypothesis is that conduction occurs by the transport of essentially free H^+ and OH^- ions. The proton cannot be observed in our experiments, but the hydroxyl ion should be discernable unless its vibrational frequency is pressure shifted to overlap the OH-stretch band of water. It is also possible that the character of the OH bond becomes more ionic at high shock pressures in a way similar to that proposed for symmetric ice VII.[18] Since the OH^- ions cannot be distinguished in our spectra, further experiments are required to resolve this uncertainty.

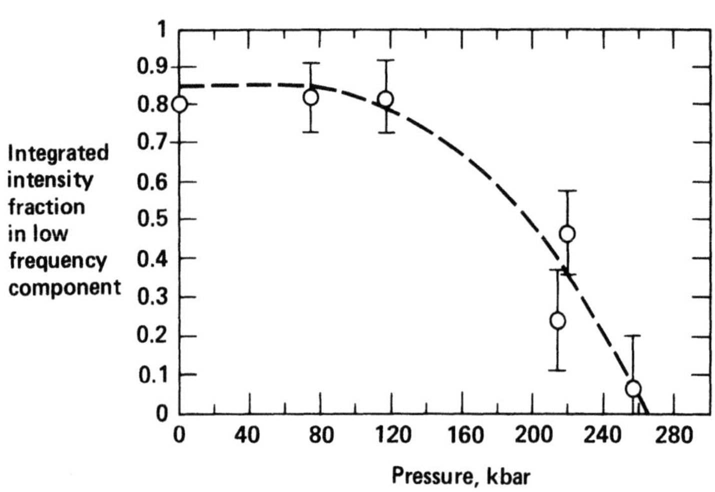

Fig. 7. Total scattering in the low-frequency component of the mixture model plotted vs. pressure. The decrease in the strength of this component indicates the gradual destruction of hydrogen bonds with increasing shock pressure and temperature.

CONCLUSION

We have observed the behavior of the OH-stretch band of liquid water to 26 GPa shock pressure and 1700 K by spontaneous Raman scattering behind the shock front. The technique we have described can record spontaneous Raman spectra in a single shot from shocked samples, and the method can be used up to the transparency limit of the material under study. The spectra we obtained show that the OH vibration persists up to the highest pressures studied. All of the shocked water spectra are consistent with a two-component mixture model, where the components are hydrogen-bonded and monomeric water molecules. No evidence for molecular ions was observed. Strong evidence is found that strong intermolecular hydrogen bonds persist to 12 GPa, decreasing in importance above that pressure. At 26 GPa, water is essentially monomeric.

We are pleased to thank A. C. Mitchell for his valuable contributions throughout the course of this work, and to B. J. Mysen who kindly supplied us with his Gaussian decomposition code. We are grateful to C. D. Wozynski, K. C. Pederson, and R. B. Schuldheisz for operation of the two-stage light-gas gun, and to J. I. Miller for projectile fabrication.

ACKNOWLEDGEMENTS

Work performed under the auspices of the U.S. Department of Energy by the Lawrence Livermore National Laboratory under contract number W-7405-ENG-48.

REFERENCES

1. A. C. Mitchell and W. J. Nellis, Equation of State and Electrical Conductivity of Water and Ammonia Shocked to the 100 GPa (1 Mbar) Pressure Range, J. Chem. Phys. 76:6273 (1982).

2. H. G. David and S. D. Hamann, The Chemical Effects of Pressure, Part 5, The Electrical Conductivity of Water at High Shock Pressures, Trans. Faraday Soc. 55:72 (1959).
3. S. D. Hamann and M. Linton, Electrical Conductivity of Water in Shock Compression, Trans. Faraday Soc. 62:2234 (1966).
4. S. D. Hamann and M. Linton, Electrical Conductivites of Aqueous Solutions of KCl, KOH, and HCl, and the Ionization of Water at High Shock Pressures, Trans. Faraday Soc. 65:2186 (1969).
5. S. C. Schmidt, D. S. Moore, and J. W. Shaner, Raman Spectroscopies in Shock-Compressed Materials, in: "Shock Waves in Condensed Matter-1983," J. R. Asay, R. A. Graham, G. K. Straub, ed., Elsevier, Amsterdam (1984).
6. A. Delpuech and A. Menil, Raman Scattering Temperature Measurement Behind a Shock Wave, ibid.
7. S. C. Schmidt, D. S. Moore, D. Schiferl, and J. W. Shaner, Backward Stimulated Raman Scattering in Shock-Compressed Benzene, Phys. Rev. Lett. 50:661 (1983).
8. A. H. Jones, W. M. Isbell, and C. J. Maiden, Measurement of the Very-High-Pressure Properties of Materials using a Light-Gas Gun, J. Appl. Phys. 37:3493 (1966).
9. A. C. Mitchell and W. J. Nellis, Diagnostic System of the Lawrence Livermore National Laboratory Two-Stage Light-Gas Gun, Rev. Sci. Instr. 52:347 (1981).
10. A. C. Mitchell and W. J. Nellis, Shock Compression of Aluminum, Copper, and Tantalum, J. Appl. Phys. 52:3363 (1981).

11. M. H. Rice and J. M. Walsh, Equation of State of Water to 250 Kilobars, J. Chem. Phys. 26:824 (1957).
12. Ya. B. Zel'dovich and Yu. P. Raizer, in: "Physics of Shock Waves and High-Temperature Hydrodynamic Phenomena", Plenum, New York (1967).
13. M. Cowperthwaite and R. Shaw, Cv(T) Equation of State for Liquids. Calculation of the Shock Temperature of Carbon Tetrachloride, Nitromethane, and Water in the 100-kbar Region, J. Chem. Phys. 53:555 (1970).
14. C. W. F. T. Pistorius, M. C. Pistorius, J. P. Blakey, and L. J. Admiraal, Melting Curve of Ice VII to 200 kbar, J. Chem. Phys. 38:600 (1963).
15. H. Lindner, doctoral dissertation, University of Karlsruhe, (1970).
16. J. R. Scherer, M. K. Go, and S. Kint, Raman Spectra and Structure of Water from -10° to 90°, J. Phys. Chem. 78:1304 (1974).
17. G. E. Walrafen, Raman Spectral Studies of the Effects of Temperature on Water Structure, J. Chem. Phys. 47:114 (1967).
18. G. E. Walrafen, M. Abebe, F. A. Mauer, S. Block, G. J. Piermarini, and R. Munro, Raman and X-ray Investigations for Ice VII to 36.0 GPa, J. Chem. Phys. 77:2166 (1982).
19. B. J. Mysen, private communication, (1985).
20. W. R. Busing and D. F. Hornig, The Effect of Dissolved KBr, KOH, or HCl on the Raman Spectrum of Water, J. Chem. Phys. 65:284 (1961).
21. P. A. Giguere and C. Madec, Spectroscopic Evidence for Ion-Pairs of Hydronium Halides in Liquid Sulphur Dioxide, Chem. Phys. Lett. 37:569 (1976).
22. M. Eigen and L. De Maeyer, Self-Dissociation and Protonic Charge Transport in Water and Ice, Proc. Royal Soc. London Ser. A 247:505 (1958).

MOLECULAR EMISSION SPECTRA FROM SHOCK-DECOMPOSED BENZENE

Mary L. Johnson,[a] Malcolm Nicol,[a] and Neil C. Holmes[b]

[a]Department of Chemistry and Biochemistry
University of California, Los Angeles, CA 90024
[b]Lawrence Livermore National Laboratory
University of California, Livermore, CA 94550

ABSTRACT

Benzene decomposes when shocked to pressures greater than 13 GPa. However, few products of the decomposition reactions have been identified in situ or by recovery methods. This paper reports the detection by time-resolved (50-to-120 nsec) chemiluminescence spectroscopy of excited molecular products of the decomposition of benzene at pressures from 24 to 63 GPa. Strong spectral bands characteristic of C_2 are readily identified among other features and a gray-body background. The dependence of the spectra on shock parameters are breifly discussed in terms of known decomposition mechanisms of highly excited benzene.

INTRODUCTION

The extent to which chemistry occurs in the shocked state and, if so, may differ qualitatively from chemistry at the same static temperatures and pressures has been actively considered in the shock wave community.[1] Benzene is typical of many organic molecules that react at shock and static pressures of the order of 10 GPa and moderate to high temperatures. From analyses of recovered products, some investigators [2] have concluded that the shock-induced chemistry differs qualitatively from the chemistry produced by the combination of high static pressures and temperatures. However, the specific chemical reactions that benzene undergoes at high shock pressures are not known.

The work described in the paper represents the first detection of molecules formed immediately behind the shock front during shock compression of benzene at pressures from 24 to 63 GPa. C_2, among other species, was detected by chemiluminescence, collected with 50-to-120-nsec resolution during passage of the shock front through the focus of an emission spectrographic system with gated diode-array detectors. In the

[a] Support and equipment provided by LLNL IGPP 84-19, NASA NAGW-104, and NSF DMR80-24620 are gratefully acknowledged.
[b] Work performed under the auspices of the U.S. Department of Energy under contract #W-7405-Eng-48, with partial support from LLNL IGPP 84-19.

following sections, the spectroscopic techniques used to detect these molecules are outlined, and typical spectra are described. The relevance of these observations to the known high temperature-low pressure [3-5 and references cited therein] and high static pressure chemistry of benzene [6-8] is discussed, and some directions for further work are suggested.

EXPERIMENTAL

The experimental techniques are briefly outlined in this section. Special features of the optical system are emphasized since space limitations preclude a complete description of the apparatus. A more complete description of the sample holder, sample, and optics will be be published elsewhere. [9] The shock techniques were similar to those used for optical pyrometry [10] and other optical studies of liquids. [11] Deoxygenated benzene samples (Mallinkrodt [12] analytical reagent grade) nominally at 295 K were confined in 32-mm diameter by 5-mm thick cylindrical cavities in Al targets, between Al base plates and quartz windows. Three self-shorting pins also were mounted in the benzene cavity and extended within 4.5 mm of the base plate. These pins detected the arrival of the shock and triggered (after appropriate electronic delays) the gated detectors.

Planar shock waves were generated by impacting the target with projectiles accelerated to velocities in the range 4-7 km/s. The velocities of the projectiles were measured by flash x-ray photography. Thermodynamic states in the samples were calculated by the shock impedence matching method from measured velocities (in the range 4-to-7 km/sec) of the metal impactors and published equations of state for the impactors, base plates, and benzene. [10, 13]

The spectrometers had four major components: a quartz collection collimator; a mirror that diverted the optical path by 90° from the gun axis; a broadband beamsplitter; and 1 or 2 monochromators with gated intensified diode-array detectors. The collimators were focused inside the benzene, 0.5-mm along the cavity axis from the surface of the quartz window. For most of the experiments to date, 25-mm diameter f/4 fused silica collimators were mounted inside the evacuated target chamber of the gas gun; but, for 3 experiments, the collimator was a 10-cm diameter f/10 CaF_2-quartz acromat, mounted outside the vacuum chamber. With the latter arrangement, the spectral response of the spectrometer was conveniently calibrated with a standard 3200 K lamp.

At various times, two Spex 0.25-meter grating monochromators were used to analyze the spectra. One monochromator had a 300 groove/mm grating with a 1 micron blaze and a Princeton Instruments IRY/700 array detector; the second monochromator was equipped with a similar 300 groove/mm grating (or, for some experiments, a 600 groove/mm grating blazed for 500 nm) and a Tracor Northern array detector. Each detector could cover a 120-to-250 nm range of wavelengths. The gate durations and delays of the detectors were set so that: (a) the detectors collected radiation emitted for a 50 or 120-nsec period during which the shock front passed through the focus of the optical system and (b) the gates were off before the shock front arrived at the quartz window.

For the initial experiments, the slit widths and detector gains were such that some intense features were cut off or detected with low resolution. Since the spectra were observed through unshocked benzene, wavelengths less than 270 nm could not be detected. No attempt was made to separate overlapping second-order features at first-order wavelengths longer than 550 nm.

RESULTS AND INTERPRETATION

Fig. 1 shows 9 spectra from 7 shock experiments at pressures from 24 to 63 GPa. These plots have not been corrected for the spectral responses of the optical systems; but comparison of this plot with the corrected plot of spectrum f in Fig. 2 implies that these corrections are relatively minor. In addition to broad gray-body backgrounds that become more intense with increasing pressure, several molecular features are apparent in these spectra. The violet-shaded bands whose red edges are near 470, 515, 550, and, in many spectra, 610 nm are the v"=v′+1, v"=v′, v"=v′-1, and v"=v′-2 bands of the Swan A $^3\Pi_g$ -> X $^3\Pi_u$ system of C_2. [14] Each band broadens toward higher wavelengths with increasing pressure as a result of emission from higher v" levels that, as one might expect, are more highly populated as the pressure, energy, and temperature behind the shock front increase.

The intensities of the bands are determined by competition among rates at which C_2 forms, radiates, decomposes, and is quenched and rates of other processes behind the shock front. Thus, intensities are not quantitatively equivalent to populations. However, the fact that, relative to the gray-body spectra collected with the same optics, the Swan bands are most intense at 38 GPa suggest that the yield of C_2 is highest near

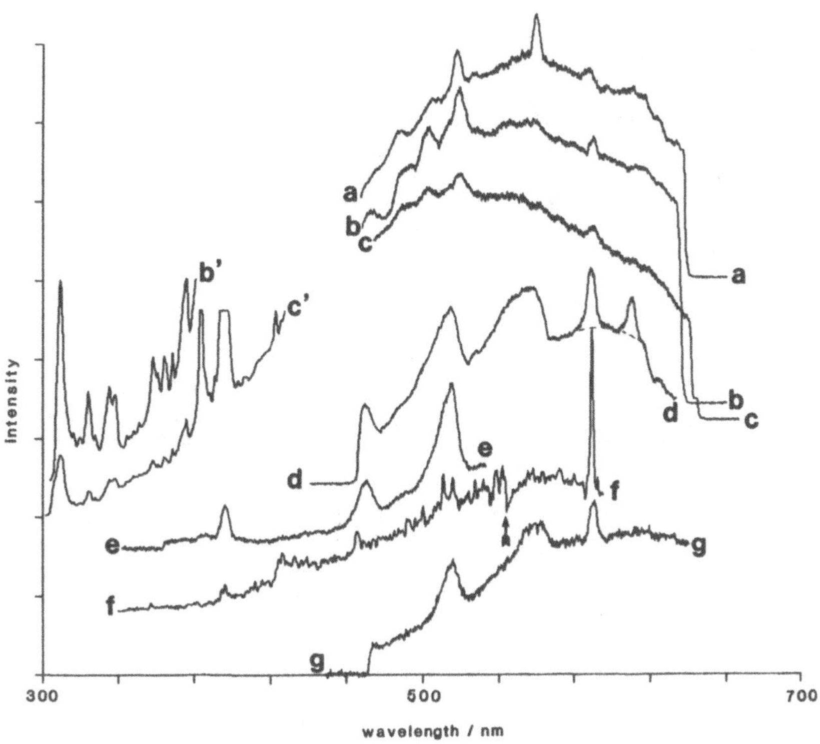

Fig. 1. Tribochemiluminescence of benzene shocked to: a, 63; b-b′, 58; c-c′, 59; d, 50; e, 38; f, 29; and g, 24 GPa. Except for f, f/4 lenses, the Princeton Instruments detector, and 5-nm instrumental resolution were used to collect the spectra. For clarity, the dark levels of these spectra (indicated by data from a few unintensified diodes); and the intensities have been multiplied by: a, 5; b, 1; c, 1; d, 5; e, 25; and g, 34. Spectra b′, c′, and f were obtained with the Tracor Northern detector at about 1-nm resolution. The dip above the arrow in spectrum f is an artefact.

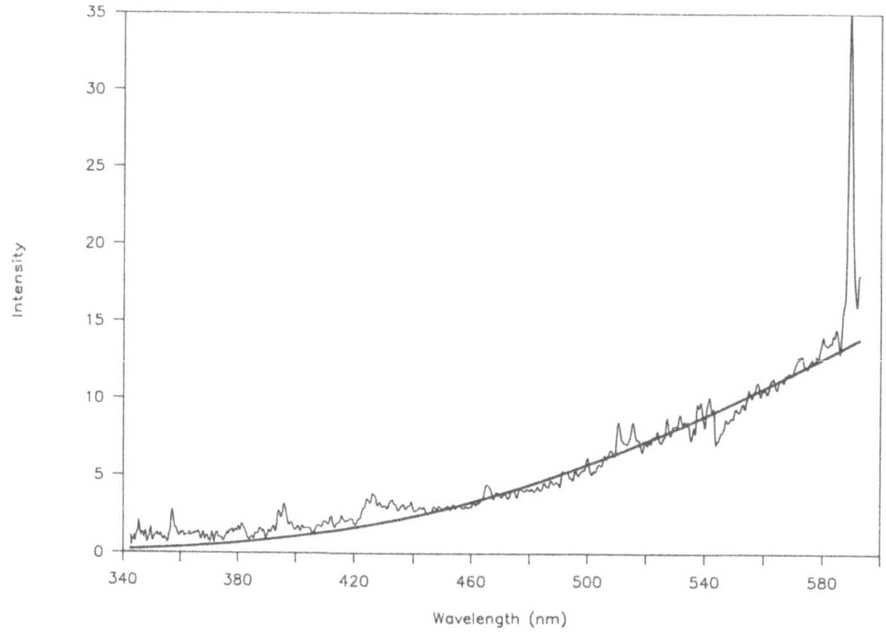

Figure 2. Spectrum f of Fig. 1, corrected for the spectral response of the optical system, is compared with a smooth 2600 K black-body spectrum.

this pressure. The molecular bands also appeared to be weaker (relative to the gray-body continuum) in the 3 spectra obtained with f/10 optics, including spectrum f. The f/10 optics have a greater depth of field, and the lower intensities of the molecular spectra may mean that the emitting molecules are in a relatively thin zone between the shock front and a zone of highly emissive material that produces the grey-body spectrum.

Another common feature of many spectra is the band near 395 nm of e in Fig. 1. In b′ and c′, this band saturates the detector, while it appears as two lines at 394.4 and 396.2 nm in f. Although possible changes of the wavelengths and shapes of spectra with pressure make it difficult to assign this band unambiguously, a possible assignment is as the 390-nm B -> X system of CH. [14] If so, the broad band near 430 nm in f, part of which also may be seen in spectrum c′, may be part of the A $^2\Delta$ -> X $^2\Pi$ system. [14]. In CH flame spectra, the 430-nm system usually is the stronger band as it appears to be in f. Although in some situations the relative intensities of these bands may be reversed, [15] the absence of a 430-nm feature in e raises doubts about this assignment. Both CH bands terminate in the ground state so this assignment might be tested by detecting laser-induced fluorescence from the shock front.

The persistent feature at 588 nm may be the yellow Na line, but a 294-nm band in second-order among other possibilities have not been excluded. Other features of these spectra have not been identified with known molecular emission bands. Promising coincidences with some uv features can be inferred from absorption spectra reported for species such as C_2H [16] and C_4H [17] and might be tested by laser-induced fluorescence.

The smooth curve in Fig. 2 represents the fit of a 2.6 x 10^2-K black-body spectrum to the corrected 29-GPa spectrum. Similarly good fits to gray-body spectra obtained at 26 and 42 GPa were obtained with black-body spectra for 2.25 and 3.4 x 10^2 K, respectively. These temperatures are in good agreement with the "reactive high C_p" fit to the 5-channel pyrometry of shocked benzene reported by Nellis et al. [10] Thus, it is likely that the observations described in this paper were obtained at temperatures and pressures between about 2000 K near 20 GPa and 4800 near 60 GPa. Extrapolation of this "reactive high C_p" fit to 13 GPa indicates that the temperature behind the shock front at this "transformation" pressure is 1350 $\pm$ 150 K.

The rapid formation of C_2 and other products of ring-opening chemistry behind the shock front and the 13 GPa transition pressure then can be understood in terms of what is known about benzene pyrolysis at lower pressures. This chemistry has been discussed in recent publications by Kiefer et al. [3] and by Smith and Johnson [4] among others. When benzene is pyrolyzed in shock tubes and flow reactors at pressures of the order of 10^5 Pa and temperatures below about 1300 K, polyphenyls are the major products. These products are formed by scission of C-H bonds on the benzenes to form phenyl radicals:

$$C_6H_6 \dashrightarrow C_6H_5\cdot + H$$

and subsequent free radical propagation and recombination reactions. This accounts for the often cited "expectation" that the products of benzene shock chemistry should be aromatic. However, when benzene is pyrolyzed occurs above 1300 K at low pressures, reactions that open the phenyl rings, for example:

$$C_6H_5\cdot \dashrightarrow C_4H_3\cdot + C_2H_2$$

predominate over the free radical reactions. C_2H_2 is the major product; but large amounts of H_2, C_4H_2, higher polyacetylenes, and carbons are also are produced by a complex series of reactions within micro- to milli-seconds. [2, 3, 5]

The rates of many of the reactions involved in benzene pyrolysis are reported to vary strongly with both pressure and temperature, with many of the reactions going more rapidly at higher pressures. Thus, it is not surprising that reactions, which occur in the milliseconds or microseconds near atmospheric pressure, proceed in nanoseconds at pressures 5 orders of magnitude higher nor that the "transformation" of product yields from aromatics to acyclics on the shorter time scale characteristic of high pressure shocks occurs near the same temperature of the low-pressure transformation.

SUMMARY

Emission spectra characteristic of excited C_2 and other small molecules generated by shock-induced pyrolysis of benzene at pressures from 24 to 63 GPa have been detected from a thin zone behind the shock. The species are similar to molecules found on the milli- to micro-second time scales in pyrolysis at comparable temperatures at low pressures. It, thus, appears that chemistry is the two pressure regiemes is similar; but, not surprisingly, rates differ. These rapid detection techniques, and extensions involving various laser-induced spectroscopies, may help to elucidate mechanisms of other fast reactions at high shock pressures.

ACKNOWLEDGEMENTS

We are grateful to K.D. Bayes, W. Graham, S.E. Kolden, A.C. Mitchell, P. McCandless, W.J. Nellis, K. Petersen, R.F. Schuldheisz, Y. Talmi, and C.D. Wozynski for valuable assistance in performing and interpreting these experiments.

REFERENCES

[1] See, for example, G.E. Duvall et al. "Shock Compression Chemistry in Materials Synthesis and Processing" (National Academy Press, Washington, 1985) and references cited therein.
[2] See, for example, L.V. Barabe, A.N. Dremin, S.V. Pershin, and V.V. Yakovlev, Shock-compression polymerization of hard-topolymerize organic compounds, Fizika Goreniya i Vzryva 5:528 (1969) and references cited therein.
[3] J.H. Kiefer, L.J. Mizerka, M.R. Patel and H.-C. Wei, A shock tube investigation of major pathways in the high-temperature pyrolysis of benzene, J. Phys. Chem. 89:2013 (1985).
[4] R.D. Smith, and A.L. Johnson, Mass spectrometric study of the high temperature chemistry of benzene, Combust. Flame 51:1 (1983).
[5] R.S. Slysh and C.R. Kinney, Some kinetics of the carbonization of benzene, acetylene, and diacetylene at 1200°, J. Phys. Chem. 65:1044 (1961).
[6] S. Block, C.E. Weir, and G.J. Piermarini, Polymorphism in benzene, naphthalene, and anthracene at high pressure, Science 169:586 (1970).
[7] J. Akella and G.C. Kennedy, Phase diagram of benzene to 35 kbar, J. Chem. Phys. 55:793 (1971).
[8] R.H. Wentorf, Jr., The behavior of some carbonaceous materials at very high pressures and high temperatures, J. Phys. Chem. 69:3063 (1965).
[9] M. Nicol, M.L. Johnson, and N.C. Holmes, (to be published).
[10] W.J. Nellis, F.H. Ree, R.J. Trainor, A.C. Mitchell, and M.B. Boslough, Equation of state and optical luminosity of benzene, polybutene, and polyethylene shocked to 210 GPa (2.1 Mbar), J. Chem. Phys. 80:2789 (1984).
[11] G.A. Lyzenga, T.J. Ahrens, W.J. Nellis, and A.C. Mitchell, The temperature of shock-compressed water, J. Chem. Phys. 76:6282 (1982).
[12] Reference to a company or product name does not imply approval or recommendation of the product by the University of California or the U.S. Department of Energy to the exclusion of others that may be suitable.
[13] A.C. Mitchell and W.J. Nellis, Diagnostic system of the Lawrence Livermore National Laboratory two-stage light-gas gun, Rev. Sci. Instr. 52:347 (1981).
[14] R.W.B. Pearse and A.G. Gaydon, "The Identification of Molecular Spectra" (Chapman & Hall, London, 1965) pp. 94-127.
[15] K.D. Bayes, private communication.
[16] W.R.M. Graham, K.I. Dismukes, and W. Weltner, Jr., C_2H radical: ^{13}C hyperfine interaction and optical spectrum, J. Chem. Phys. 60:3817 (1974).
[17] K.I. Dismukes, W.R.M. Graham, and W. Weltner, Optical and esr spectra of the C_4H radical in rare gas matrices at 4 K, J. Mol. Spectr. 57:127 (1975).

COHERENT ANTI-STOKES RAMAN SCATTERING IN BENZENE

AND NITROMETHANE SHOCK-COMPRESSED TO 11 GPA*

D.S. Moore, S.C. Schmidt, J.W. Shaner, D.L. Shampine, and W.T. Holt

The University of California
Los Alamos National Laboratory PO Box 1663
Los Alamos, New Mexico 87545

ABSTRACT

The shock-compression induced frequency shifts of the ring-stretching mode of liquid benzene and the CN-stretching mode of liquid nitromethane have been measured using coherent anti-Stokes Raman scattering (CARS). Shock pressures up to 11 GPa were achieved using a two-stage light gas gun. The frequency shifted CARS signals were generated using single-pulse Nd:YAG and broadband dye lasers. Where possible, the results of the dynamic experiments were compared to other dynamic high pressure experiments.

We have shown in previous work [1-5] that coherent optical scattering can be used to study the microscopic nature of shock-compressed molecular systems. In this paper, the results of using reflected broadband coherent anti-Stokes Raman spectroscopy (RBBCARS) to measure the vibrational frequencies of the ring-stretching mode of benzene and the CN-stretching mode of nitromethane at shock pressures up to 11 GPa are reported.

A schematic of the experimental apparatus is shown in Fig. 1. A two-stage light gas gun was used to accelerate a polycarbonate projectile with 11-mm-thick AZ31B magnesium or 2024 aluminum impactors to a desired velocity. The projectile struck an approximately 2.4-mm-thick 304 stainless steel target plate producing a shock wave which ran forward into the 2.7 to 3.3-mm-thick liquid sample. Lower pressures were achieved using previously described techniques.[1,3] Stainless steel was chosen as the target plate because of its ability to retain its reflectivity under shock compression. Reagent grade benzene (Mallinckrodt, Inc.) and commercial grade nitromethane were used. Standard data reduction techniques[6] using published shock-velocity, particle-velocity data[7] were used to determine the state of the shock-compressed samples. Design of the target assemblies was greatly facilitated using the MACRAME one- dimensional-wave-propagation computer code.[8]

---

*These experiments were performed under the auspices of the United States Department of Energy

The timing sequence for the CARS experiment was determined by the incoming projectile. Signals from three photodiodes that monitor three HeNe laser beams, which dissect the barrel ~ 2.2, 1.2 and 0.7 m from the target, in conjunction with appropriate time delays, served to trigger the Nd:YAG laser flashlamps approximately 300 μs prior to impact. A time-of-arrival pin activated just after the shock entered the liquid, along with another time delay, provided a trigger for the laser Q-switch such that its output light pulse arrived at the target approximately when the shock wave arrived at the quartz window.

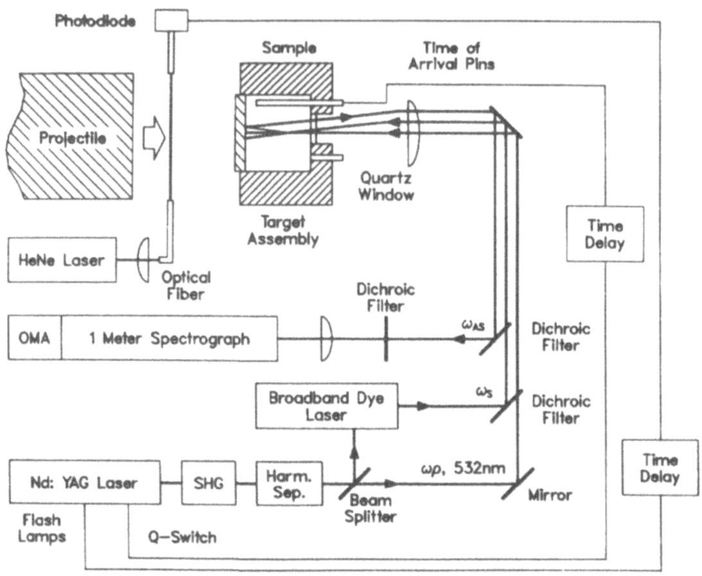

Fig. 1. Schematic representation of the reflected broadband coherent anti-Stokes Raman scattering experiment. SHG - second harmonic generator; Harm. Sep. - harmonic separator; OMA - optical multichannel analyzer; Sample - benzene or nitromethane.

The details of the CARS technique have been discussed previously both for this application[2-5] and others[9,10] and will not be repeated here. Suffice it to say that a broadband dye laser of width determined by the dye gain profile, which was pumped by a portion of the ~ 6 ns frequency doubled Nd:YAG laser pulse, and which provided a broadband of Stokes frequencies, was mixed in the sample with the remaining Nd:YAG pulse to produce the CARS signals. The CARS beam was reflected out of the shocked sample by the highly polished front surface of the target plate and along a path parallel to the two incoming beams. After being separated from the reflected pump and Stokes beams using a long-wave-pass dichroic mirror, the beam was focused into the 100 μm-wide entrance slit of an 8 Å/mm dispersion spectrometer. The signals were detected at the focal plane of the spectrometer using a gated, intensified, diode array (Tracor Northern Model TN-6133) coupled with a multichannel analyzer (Tracor Northern Model TN-1710).

Figure 2 shows the RBBCARS data for the ring-stretching mode of liquid benzene at ambient pressure and shock pressures up to 10.6 GPa. The position of each peak was measured relative to the 253.652 nm Hg emission line in second order. From these data, preliminary estimates of the frequency shifts

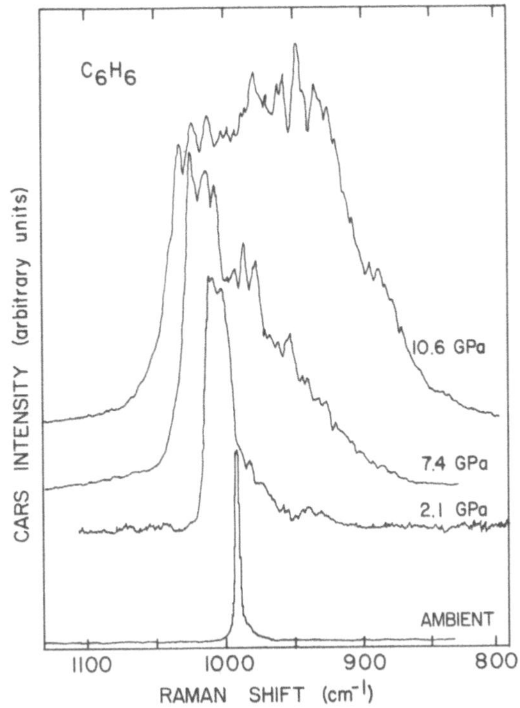

Fig. 2. RBBCARS spectra of ambient and shock-compressed benzene. The ambient benzene peak position is 992 cm^{-1}. Shock pressures are indicated. Wavelength calibration was done relative to the 253.652 nm Hg line in second order.

of the ring-stretching mode obtained using standard analysis procedures for CARS data containing significant non-resonant backgrounds[9] were made. These preliminary estimates ignore the presence of other Raman active lines in the region.[11] The estimates show a weakening of the initial linear dependence of frequency shift on pressure as the region near 13 GPa is approached. Previous work[12] suggests chemical reaction occurs at these pressures. The 10.6 GPa spectrum of Fig. 2 is significantly different from those at lower pressures. Final frequency shift values and interpretations of the spectra will appear in a future publication.

The RBBCARS spectra for the CN-stretching mode of liquid nitromethane at ambient pressure and shock pressures up to 7.6 GPa are shown in Fig. 3. Wavelength calibration was done as above. The existence of the CN mode for microsecond times after shocking to 5 GPa implies that decomposition of the nitromethane has not occurred in contrast to what was observed in static high temperature / high pressure studies.[13] Measurements are presently being extended to higher pressures where nitromethane is thought to become reactive after very short shock run distances.[12,14]

In these experiments we have used a fast, single-pulse, nonlinear optical method to measure the vibrational frequencies of the ring- stretching mode of liquid benzene and the CN-stretching mode of liquid nitromethane under conditions of high temperature and high pressure caused by shock compression. The technique has been developed as part of an effort directed toward understanding the dynamic behavior and energy transfer mechanisms in shock-compressed molecular systems. Of particular interest is the eventual use of similar optical techniques in processes, such as detonation waves,

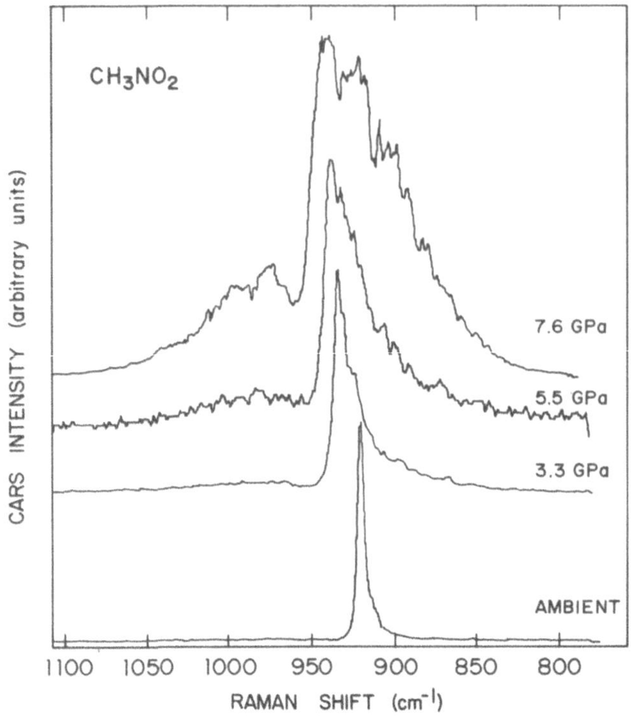

Fig. 3. RBBCARS spectra of ambient and shock-compressed nitromethane. The ambient peak position is 921 cm^{-1}. Shock pressures are indicated. Wavelength calibration was done with respect to the 253.652 nm Hg line in second order.

that are governed by transient and possibly nonequilibrium phenomena.

The authors wish to thank C. W. Caldwell, R. L. Eavenson, and R. S. Medina for their assistance in performing the shock-compression experiments and V. A. Gurule, C. N Gomez, and R. W. Livingston for machining and fabricating the target assemblies. Special thanks are due to J. N. Fritz for use of the MACRAME computer code.

REFERENCES

1. S. C. Schmidt, D. S. Moore, D. Schiferl, and J. W. Shaner, Phys. Rev. Lett. 50, 661 (1983).

2. D. S. Moore, S. C. Schmidt, and J. W. Shaner, Phys. Rev. Lett. 50, 1819, (1983).

3. S. C. Schmidt, D. S. Moore, and J. W. Shaner, in Shock Waves in Condensed Matter - 1983, J. R. Asay, R. A. Graham, and G. K. Straub, eds. (Elsevier Science Publishers B. V., 1984), p. 293.

4. D. S. Moore, S. C. Schmidt, D. Schiferl, and J. W. Shaner, in High Pressure in Science and Technology, Part II, C. Homan, R. K. MacCrone and E. Whalley, eds. (North Holland, New York, 1984), p. 87.

5. D. S. Moore, S. C. Schmidt, D. Schiferl, and J. W. Shaner, in Los Alamos Conference on Optics 83, R. S. McDowell and S. C. Stotlar, eds.

(Proceedings SPIE Volume 380, 1983), p. 208.

6. M. H. Rice, R. G. McQueen, and J. M. Walsh, in Solid State Physics: Advances in Research and Applications, F. Sietz and D. Turnbull, eds. (Academic Press, New York, 1958), Vol. 6, p. 1.

7. S. P. Marsh, "LASL Shock Hugoniot Data," (University of California Press, Berkeley, CA, 1980).

8. J. N. Fritz, in preparation for publication.

9. W. M. Tolles, J. W. Nibler, J. R. McDonald, and A. B. Harvey, Appl. Spectrosc. 31, 253 (1977).

10. W. B. Roh, P. W. Schreiber, and J.-P. E. Taran, Appl. Phys. Lett. 29, 174 (1976).

11. J. E. Griffiths, M. Clerc and P. M. Rentzepis, J. Chem. Phys. 60, 3824 (1974).

12. O. B. Yakusheva, V. V. Yakushev and A. N. Dremin, High Temp.-High Pres. 3, 261 (1971).

13. J. W. Brasch, J. Phys. Chem. 84, 2085 (1980).

14. A. Delpuech and A. Menil, in Shock Waves in Condensed Matter - 1983, J. R. Asay, R. A. Graham, and G. K. Straub, eds. (Elsevier Science Publishers B. V., 1984)

EFFECTS OF TEMPERATURE ON THE UV-VISIBLE SPECTRUM OF DYNAMICALLY COMPRESSED CS_2

G. E. Duvall, R. H. Granholm, P. M. Bellamy, and J. E. Hegland

Washington State University

The transmission spectrum of liquid CS_2 contained in a penny-shaped cavity approximately 150 μm thick has been measured in the range 320 to 700 nm. The samples were contained between sapphire plates and compressed in uniaxial strain induced by plane impact of a third sapphire plate on the front plate of the cell. The shock generated by the sapphire impact produced a series of reverberating shocks in the CS_2 which caused it to ring up to the impact pressure of the two sapphire plates.[1] Impact pressure ranged from 54 to 130 kilobars. The ringing-up process is shown in the x-t plane in Fig. 1 where the impact shock produces state 1 shown in the lower left.

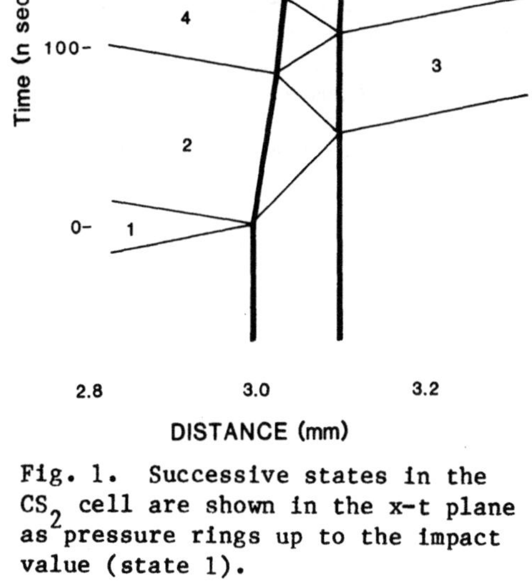

Fig. 1. Successive states in the CS_2 cell are shown in the x-t plane as pressure rings up to the impact value (state 1).

Several experiments were done with the sample initially at room temperature; one was done in a cell modified for cooling the sample to -72 °C; another was done in a cell heated to 125 °C. In each case it was found that the red edge of the absorption band centered at 320 nm was strongly shifted toward the red as the pressure rang up in the cell. By correlating shock pressures vs time, calculated from the known geometry and the Sheffield equation of state,[2] with $\Delta\lambda$ vs time measured from the spectrograms, it was possible to plot $\Delta\lambda$ vs pressure as shown in Fig. 2. A further application of the equation of state led to the graphs of temperature in the CS_2 vs pressure, shown in Fig. 3.

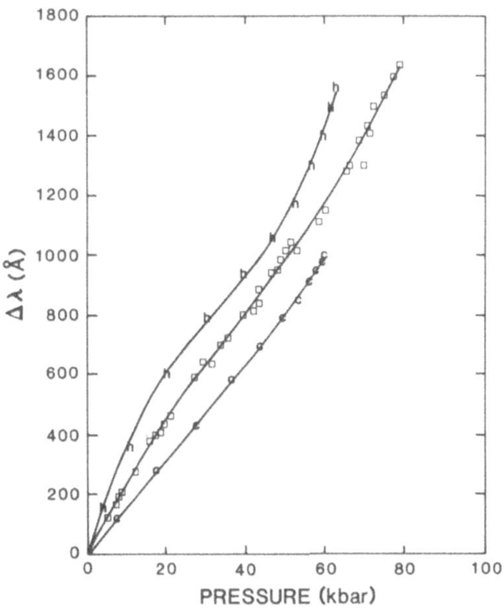

Fig. 2. Shift of the red band edge of the 3200 Å absorption band in CS_2 as a result of the pressure loading shown in Fig. 1. h - initial cell temperature = 398 °K; □ - initial cell temperature = 295 °K; c - initial cell temperature = 201 °K.

The wavelength of the red edge of the absorption band before shock arrival is known for each initial sample temperature. This, combined with the information contained in Figs. 2 and 3 enabled us to determine a set of points in the three dimensional space, λ, P, T, given in Table I. Additional measurements were made of band edge position at 25 °C and 125 °C at ambient pressure, and prepublication values of band edge position vs pressure from static measurements at room temperature were kindly supplied by S. F. Agnew of Los Alamos National Laboratories.[3]

With these data it was possible to produce a least squares fit of the λ, P, T data:

$$\lambda = \lambda_o + a_1(T - 295) + a_2(T - 295)^2 + b_1P$$

$$+ b_2P^2 + b_3P^3 + c_1 P(T - 295) + c_2P^2(T - 295)$$

$$+ c_3 P(T - 295)^2 + c_4 P^2(T - 295)^2 \qquad (1)$$

where the coefficients are

λ_o = 3550 Å $\qquad$ b_3 = 1.69×10^{-3}

$a_1 = 0.379$ $\qquad c_1 = .0492$

$a_2 = 7.47 \times 10^{-4}$ $\qquad c_2 = -5.11 \times 10^{-4}$

$b_1 = 12.15$ $\qquad c_3 = -4.38 \times 10^{-5}$

$b_2 = -0.132$ $\qquad c_4 = 5.58 \times 10^{-7}$

Pressure is in kilobars, T is temperature in ^{o}K, and λ is in Angstroms.

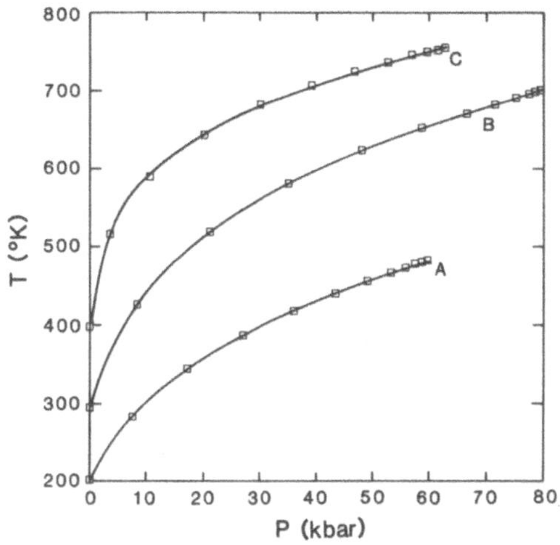

Fig. 3. Temperature vs pressure for the stepwise loading of Fig. 1 calculated from the Sheffield-Duvall equation of state: A - T_o (initial cell temperature) = 201°K; B - T_o = 295°K; C - T_o = 398°K.

Table I

Wavelengths, Temperatures, and Pressures for Hot, Cold, and Room Temperature Experiments in CS_2 (from Figs. 2 and 3).

λ,nm	P,kbars	T,°K
3683.	10.	302.
3845.	20.	358.
4001.	30.	398.
4165.	40.	432.
4325.	50.	459.
3790.	10.	441.
3988.	20.	512.
4177.	30.	561.
4357.	40.	597.
4526.	50.	628.
4692.	60.	656.
4923.	70.	679.
3943.	10.	586.
4201.	20.	644.
4388.	30.	681.
4549.	40.	709.
4719.	50.	731.

All terms but the third in Eq. (1) contribute substantially to λ at the highest pressures and temperatures. The equation reproduces the input data of Table I to better than one percent; but, like all such fits, there is no assurance that it can be extrapolated outside the fitting range.

Coefficients $(\partial\lambda/\partial T)_P$ and $(\partial\lambda/\partial P)_T$ are shown in Figs. 4 and 5 as functions of temperature. Pressure dependence can be inferred from these graphs or calculated from Eq. (1). It is interesting that both coefficients become relatively independent of both P and T at the higher pressures:

$$\partial\lambda/\partial T \simeq 1.5\ \text{Å}/^{\circ}\text{K}$$

$$\partial\lambda/\partial P \simeq 13\ \text{Å}/\text{Kb}$$

Considering the nature and quantity of the data, we would recommend that small variations in these coefficients not be taken too seriously at present, though the major trends should be reliable.

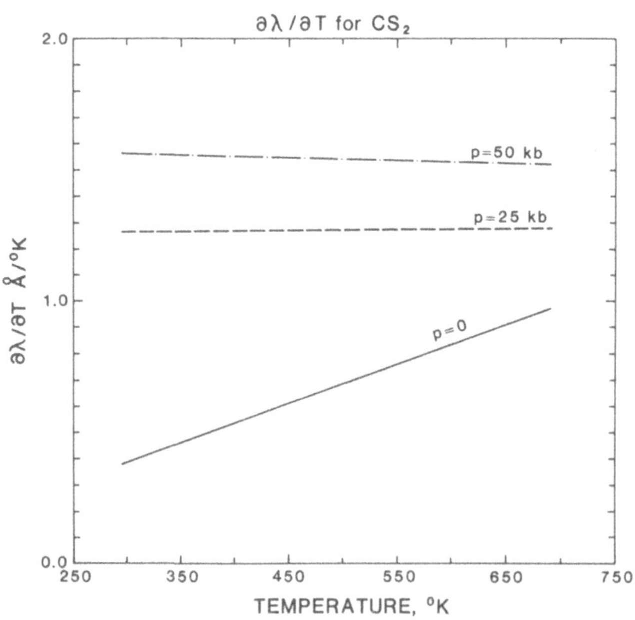

Fig. 4. $\partial\lambda/\partial T$ calculated from Eq. (1) vs temperature.

It is certainly reasonable to ask whether the observed band edge shifts are due to changes in pressure or temperature alone, or whether there may be non-equilibrium phenomena associated with the nearly discontinuous changes in pressure, temperature, and density which occur as the shock in CS_2 reverberates between the two sapphire plates. The data shown in Fig. 6 suggest very strongly that non-equilibrium effects, if they exist, are small compared with pressure and temperature effects.

The points labeled "SWL" in Fig. 6 represent the room temperature curve of Fig. 2 resulting from the stepwise loading of Fig. 1. The triangles in Fig. 6, labeled "Ramp 1" represent the shifts produced by isentropic loading with a rise time of 340 nsec. This was achieved by replacing the front sapphire of the target cell by a thick plate of fused quartz. The "Ramp 1" curve lies slightly below the "SWL" curve, but this is accounted for by a small difference in temperature between the two curves for a given pressure. Both of these curves may exhibit a dynamic compression effect which differs from that which would be seen in a

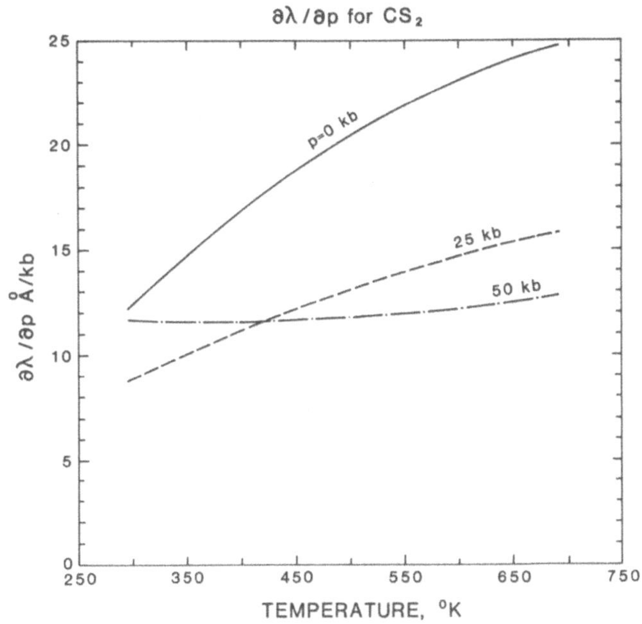

Fig. 5. ∂λ/∂P calculated from Eq. (1) vs temperature.

static experiment, but the difference will be due to the total rise time of 200-400 nsecs, not to the discontinuities in stepwise loading.

Finally, in Fig. 7, is shown the effect of diluting CS_2 with either ethanol or hexane. The band edge shift, $\Delta\lambda$, decreases linearly with solvent concentration to about 90% dilution. Further dilution causes the curve to turn down rapidly, apparently going to or very near to zero at infinite dilution. This result shows conclusively that the shifts we are recording are due to interactions among CS_2 molecules. The changes in electronic states represented by large values of $\Delta\lambda$ must therefore be presumed to result from the spreading of unimolecular electron states to encompass neighboring molecules. Since the effect is reversible below 90 kilobars in our experiments, it does not seem likely that this sharing of electron states can be due to the formation of polymer reported in static experiments.[4] It may well, however, be an intermediate reaction which is reversible in the times considered here and which plays a critical role in formation of the final product.

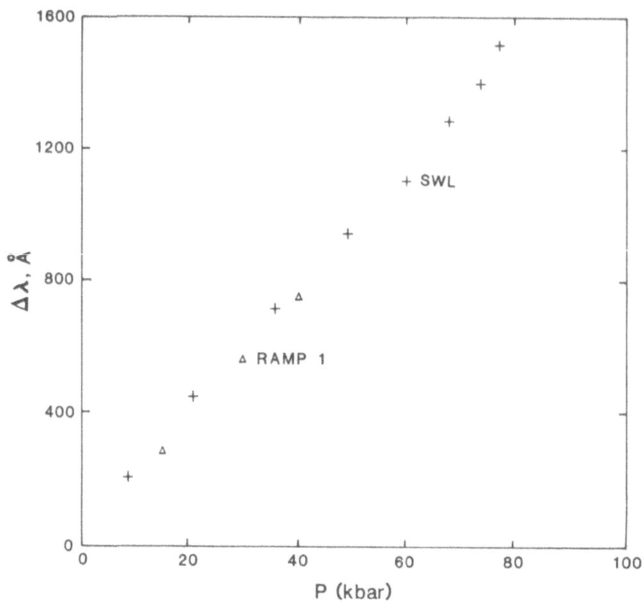

Fig. 6. Comparison of band edge shift for stepwise loading (SWL) with isentropic loading with 340 nsec rise time (Ramp 1). $T_o = 295^oK$.

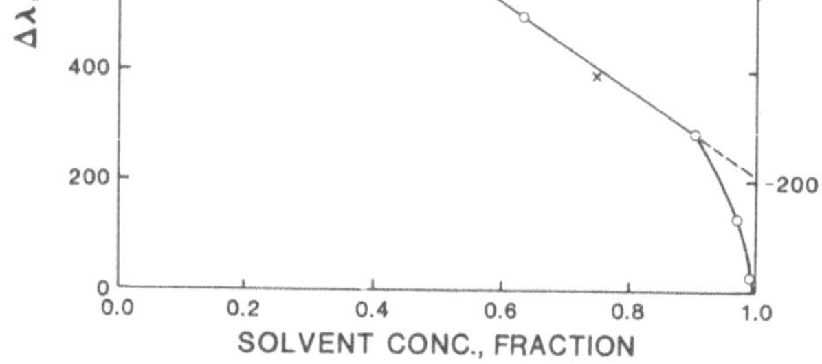

Fig. 7. Effect of dilution on band edge shift in CS_2. P = 50 kb, $T_o = 295^oK$, solvent is hexane-O or ethanol-X.

REFERENCES

1. K. M. Ogilvie and G. E. Duvall, J. Chem. Phys. 78, 1077 (1983).

2. S. A. Sheffield and G. E. Duvall, J. Chem. Phys. 79, 1981 (1983).

3. Agnew et al., these proceedings.

4. E. G. Butcher, M. Alsop, J. A. Weston, and H. A. Gebbie, Nature 199, 756 (1963).

SPECTROSCOPIC STUDIES OF CARBON DISULFIDE AT HIGH PRESSURE

Stephen F. Agnew*, Basil I. Swanson*, and Douglas G. Eckhart

Los Alamos National Laboratory,University of California
INC-4, mail stop C346
Los Alamos, New Mexico 87545

INTRODUCTION

Recent work on the transient absorption spectra induced by the dynamic shock-loading of carbon disulfide has shown[1] a shock-induced shift of the UV absorption edge up to a pressure of 8.0-9.0 GPa followed by irreversible chemistry in which the CS_2 becomes opaque. The authors suggested that this reaction was analagous to the Bridgeman's Black polymer which has been observed to occur when CS_2 was heated to 185°C at 5.5 GPa[2], 250°C at 3.3 GPa[3], or over a range of pressures and temperatures.[4] The latter work on this reaction at static high pressure has shown[4] infrared and visible/near-IR absorption spectra for the recovered CS_2 polymer and deduced that it was a mixture of carbon, sulfur, and an ill-defined polymer with the probable structure

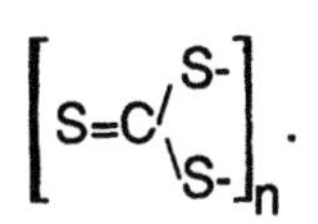

While it is certainly reasonable that the irreversible chemistry observed in the shock studies is analogous to the irreversible chemistry which has been observed in the recovered static high pressure and high temperature studies, there has as yet been no direct spectroscopic link between the shock-induced and the static high pressure experiments. We wish to report the first such spectroscopic link for CS_2. In addition, there has been no measurement of the absorption edge shift of CS_2 at static high pressure and temperature similar to those of the shock experiments. We have used diamond-anvil cells and a combination of Raman, infrared, and UV-vis-NIR absorption spectroscopies to analyze the behavior of CS_2 in the pressure range 0.2-10.0 GPa and have compared one of these spectroscopic observables, the UV- absorption edge shift prior to the onset of irreversible chemistry, with that measured in the shock experiment.

EXPERIMENTAL

Merrill-Bassett diamond-anvil cells[5] made of inconel 750 and equipped with type 2a diamonds were loaded with CS_2 purified by standard techniques, either neat or diluted with an appropriate amount of dichloromethane. Pressure was measured by means of the ruby fluorescence[6], assuming the R_2 line shift to be 0.1322 GPa/cm^{-1}.

Raman spectra were obtained on a SPEX 1403 double monchromator by means of a back-scattering technique. Resolution was three cm^{-1} and typically ten or more spectra were signal averaged using a Nicolet 1180E Raman data system. Spectra-Physics model 171 Kr^+ laser was used at 647.1 nm with 35 mW. Infrared spectra were obtained with a Digilab FTS-20CV evacuable FTIR equipped with specially designed condensing optics and a liquid N_2 cooled MCT detector. Typically, 100 scans were averaged at a resolution of 2 cm^{-1}. Absorption of the diamonds obscured the IR spectra over the rangle 1950-2250 cm^{-1}. Visible absorption spectra were recorded with a Perkin-Elmer 330 spectrometer equipped with a beam condensor and the model 2600 data station. Diamond absorption blocked wavelengths shorter than 230 nm.

A small heater was built which allowed UV-visible and pressure measurements up to 300°C. It consisted of an aluminum cylinder into which the cell was inserted and around which a nichrome heater was wound. Mica sheets provided electrical insulation for the heater wire from the aluminum and ceramic standoffs, which were turned from a machinable ceramic, provided thermal insulation from the mounts. The ruby fluorescence peak determined the pressure of the cell with the zero bar peak position adjusted for temperature.

RESULTS

The behavior of the phonons of neat CS_2 over the pressure range 1.8-8.3 GPa is shown in Fig. 1. The pressure dependence over the range 1.8-4.0 GPa agrees with that previously reported.[7] As can be seen, the phonon modes march up in frequency as a function of pressure and show no evidence for any phase transitions. This behavior is only evident for carefully grown single crystrals of a particular orientation. Figure 2 shows a series of spectra illustrating that the different crystal orientations evidently affect the relative intensities of the three phonon peaks quite severely. In addition, as the pressure is increased for multi-crystalline samples, severe pressure gradients develop in the cell as evidenced both by the variation of the ruby fluorescence peak as well as by differences in the frequencies of the phonon features, both as functions of position in the cell. This extreme anisotrophy is unusual for the relatively low pressure range 3.0-8.0 GPa and suggests that crystalline CS_2 is highly anisotropic at these pressures. This is equivalent to saying that the elastic constants vary along different crystallographic directions.

We also note that CS_2 becomes extremely photosensitive at pressures in excess of 4.0 GPa. However, it is possible to use red radiation of a low power level (647.1 nm, <30 mW) to obtain Raman spectra up until the point of pressure-induced chemistry, ca. 8.3 GPa. We found that neat CS_2 undergoes pressure-induced chemistry at or above this pressure which results in a solid that appears dark red when pressure is released to one bar. The absorption edge red-shifts markedly with increasing pressure, becoming opaque at pressures greater than 5.0 GPa. Partially reacted samples at 8.3 GPa show an inhomogeneous red color distribution throughout the now yellowish CS_2, the infrared spectrum of which is shown in Fig. 3b. The features at 1074 and 1150 cm^{-1} correspond to previously measured[3] peaks characteristic of the Bridgeman's Black polymer. Our material is not black, however, and is rather intensely red as is evident in either partially reacted samples or in thin layers of the recovered product as measured between diamond anvils without a gasket. The infrared spectrum in Fig. 3c is of such a thin layer of the CS_2 reaction product and also shows the two features mentioned above. Thus it seems rather clear that this pressure-induced reaction of CS_2 is producing a similar if not identical product (or products) to that produced at lower pressure but higher temperature (5.5 GPa and 185°C or 3.3 GPa and 250°C). The visible-NIR absorption of a thin layer of the recovered product isolated between two diamond anvils without a gasket is shown in Fig. 4a. This spectrum differs substantially from that measured[4] previously for the high temperature, high pressure reaction product, despite the fact that the infrared features measured on this product match ours very well. Whereas we have found a band edge that peaks at 22,500 cm^{-1} and is centered at 17,500 cm^{-1}, Chan and Jonscher measured[4] two features at 13,710 and 9678 cm^{-1}. It is not yet clear why their spectrum differs from

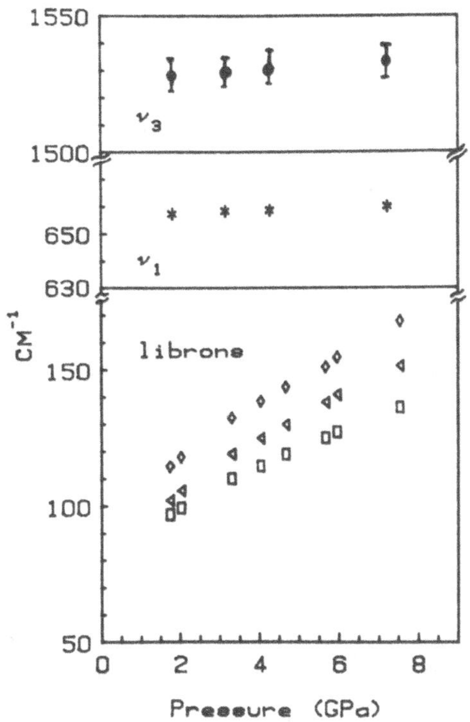

Fig. 1. Pressure dependence of librons and internal modes, ν_1 and ν_3, for CS_2.

ours. The spectrum in Fig. 4b is of a thin layer of the recovered product from our 2.4 GPa and 200°C reaction and shows a broad feature at 14,500 cm^{-1}. We believe that the dependence of this absorption feature on reaction conditions indicates a variation in the product distribution and/or polymer chain length. This question will be addressed further in a future publication.

The shift in the absorption edge of CS_2 as a function of pressure at ambient temperature is shown in Fig. 5 along with the results of Ogilvie and Duvall.[1] These latter results were obtained during shock-loading CS_2 with either step or ramp shocks. As can be seen, the results of the static measurements overlay those obtained from the ramp shock loading exactly, indicating that the ramp wave does indeed represent an isothermal compression in the thin cell (~1 μ), as was suggested by the authors. The results from step shock loading experiments shown in Fig. 5 differ substantially from those of the other two measurements. This difference can be explained, however, solely on the basis of the temperature induced shift of this same absorption edge. We have measured the absorption edge at 2.4 GPa for 28, 103, and 150°C as 366.1, 371.6, and 380.0 nm, respectively. On the basis of these temperature induced shifts at 2.4 GPa, we estimate the temperature of the step shock loaded CS_2 to be 218°C at 2.4 GPa, as compared to 353°C derived[7] from Sheffield and Duvall's equation-of-state. A peak appeared at 22,700 cm^{-1} for the 150°C measurement, evidently due to the accumlation of a product species induced in the still solid phase of CS_2. The CS_2 became very photosensitive at this point and even the 647.1 nm laser irradiation

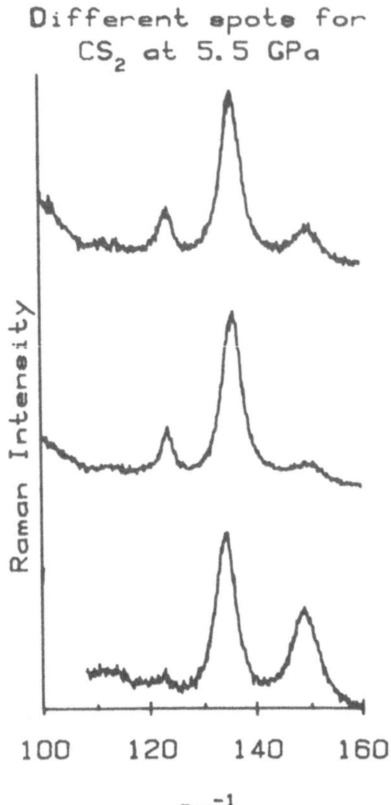

Fig. 2. Libron spectra of CS_2 for different crystal orientations.

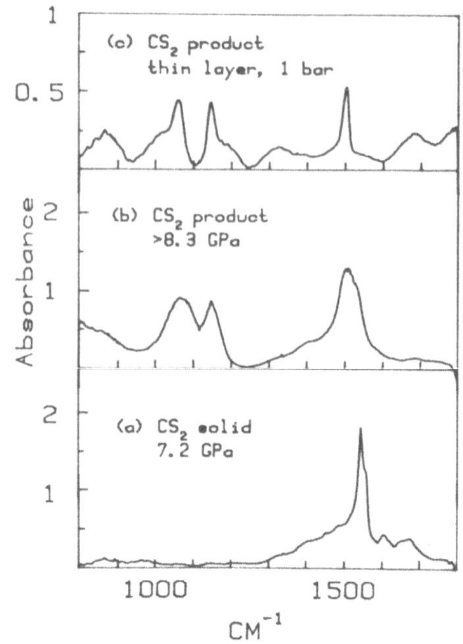

Fig. 3. Infrared spectra of: (a) single crystal CS_2 33 μ thick at 7.2 GPa; (b) Reacted sample at >8.3 GPa; (c) thin layer of preduct at one bar. Note that interference fringes underlie the product bands.

produced sample decomposition as noted by the appearance of a black spot in the sample. The melting point was reached at 181±6°C; and, although a ruby pressure measurement was not possible due to sample photosensitivity, we feel that the pressure was 2.4±0.2 GPa based upon the temperature-induced pressure shift up until that point. The temperature was raised to 200°C; and, during the course of a scan (ca. 10 minutes), the sample had reacted completely, this time forming a black solid with various infrared and visible-NIR features indicating that this product was the same as that measured[4] previously and known as Bridgeman's Black.

Figure 1. shows the shift of the symmetric and the antisymmetric stretches (ν_1 and ν_3) for neat CS_2. The pressure-induced shifts of both these bands are quite small. The Raman band shifts 2.6 cm^{-1}, 657.4 to 660.0 cm^{-1}, over the range 1.8 to 7.2 GPa (0.48 cm^{-1}/GPa) and, we estimate that ν_3 shifts at about 1.3 cm^{-1} /GPa, although the width of this feature (ca. 20 cm^{-1}) made the determination somewhat difficult. If the CS_2 is diluted with dichloromethane (CH_2Cl_2), however, the infrared band (Fig. 6) moves substantially, 30 cm^{-1} over the range 2.4 to 10.4 GPa (3.75 cm^{-1} /GPa).

Unfortunately, it was not possible to obtain the complimentary Raman band positions due to the low concentration (around 1%) needed to put the infrared feature on scale. At these concentrations, there was no reaction of the CS_2 either with itself or with the dichloromethane up to the highest pressure that was possible

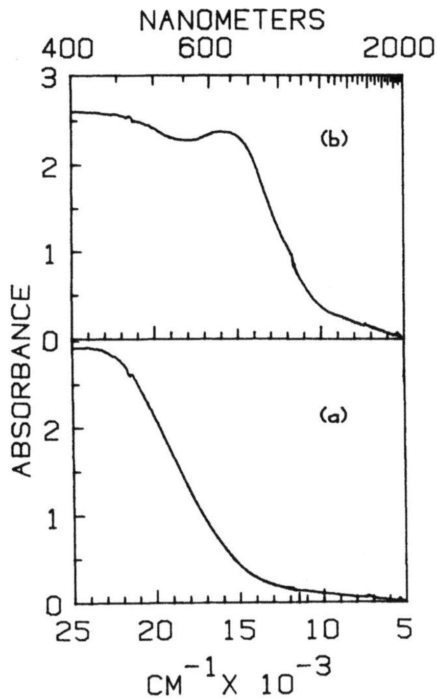

Fig. 4. Visible-NIR absorption of CS_2 product in thin layers at one bar. a) Reaction product from 8.3 GPa and ambient temperature. b) Reaction product from 2.4 GPa and 200°C.

with the cell used, 10.4 GPa. The UV absorption feature also red shifts significantly over the same pressure range, although the red-shifting tail of the dichloromethane obscured the exact band position of the CS_2 at pressures over 4.0 GPa.

We note that with concentrations higher than 10% CS_2, a brown color appeared upon solidification of the solution which reversed upon melting unless the frozen solution was subjected to either increasing pressure or excessive laser irradiation. The product of this irreversible chemistry showed infrared features identical to those mentioned above for the reaction of neat CS_2, and no evidence for a solute solvent reaction was found.

DISCUSSION

Our results for the shift of the phonons and internal modes of CS_2 over the range 1.8 to 8.3 GPa agree with the more limited results reported[7] by Shimizu and Ohnishi. However, they claimed that they could not obtain Raman data above 4.0 GPa due to sample decomposition. We have found, on the other hand, that with the 647.1 nm Kr^+ line at a low power, Raman spectra could be obtained up until the point of the pressure-induced chemistry at 8.3 GPa. It was not possible, though, to obtain spectra with either blue or green excitation above 4.0 GPa due to increased photochemical reactivity. Shimizu and Ohnishi further argued[7] that the Raman-active symmetric stretch (v_1) shifts very little with pressure because it is an internal mode, despite the fact that most internal modes shift far more than this over a similar pressure range. For instance, nitrogen shifts[8] 6 cm^{-1} (2336 to 2342 cm^{-1}) over the range 2.4 to 4.8 GPa (2.5 cm^{-1} /GPa), which is five times that of CS_2 over a

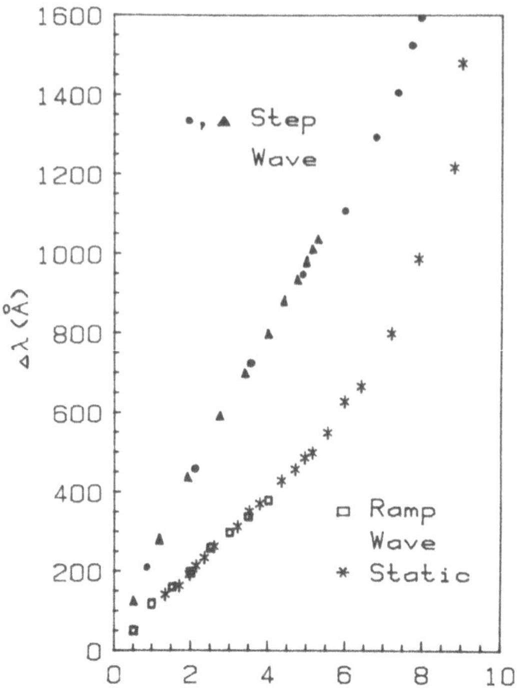

Fig. 5. Shift of absorption edge of CS_2 as a function of pressure for two step shocks (▲,•) and an isothermal ramp shock (□) from Duvall and coworkers[1] and for static high pressure at ambient temperature (*).

similar range. Nitrogen is stable to quite high pressure and the high frequency that results from its triple bond is representative of perhaps the prototypical isolated internal mode. It would, then, be quite surprizing if CS_2 turned out to have a more isolated internal mode than nitrogen.

The classical interpretation of the pressure induced shifts of vibrational transitions begins with the partition of the vibrational hamitonian as $H = H_o + H_b + \delta$ where $H_o = f_{har}q^2 + f_{anhar}q^3$, H_b is the bath, δ is the perturbation, and f_{har} and f_{anhar} are the harmonic and anharmonic force constants, respectively. A Taylor expansion of the perturbation along a normal coordinate followed by the application of first order perturbation theory results in the following expression for the frequency shift,

$$\Delta\nu = \nu_o - \nu_{gas} \propto \frac{1}{\nu_{gas}}\left[-\frac{3f_{anhar}}{f_{har}}\left[\frac{\partial\delta}{\partial q}\right]_o + \left[\frac{\partial^2\delta}{\partial q^2}\right]_o\right]$$

as per Benson and Drickamer.[10] The four types of interactions considered to be most important are dispersive (induced dipole-induced dipole), inductive (dipole-induced dipole), dipolar (dipole-dipole), all of which are attractive, and the repulsive $1/R^{12}$. Since CS_2 has no permanent dipole, there is no dipolar interaction in the potential. The pressure range of this study suggests that we are in a regime where the repulsive interactions should dominate the attractive ones. Indeed, the

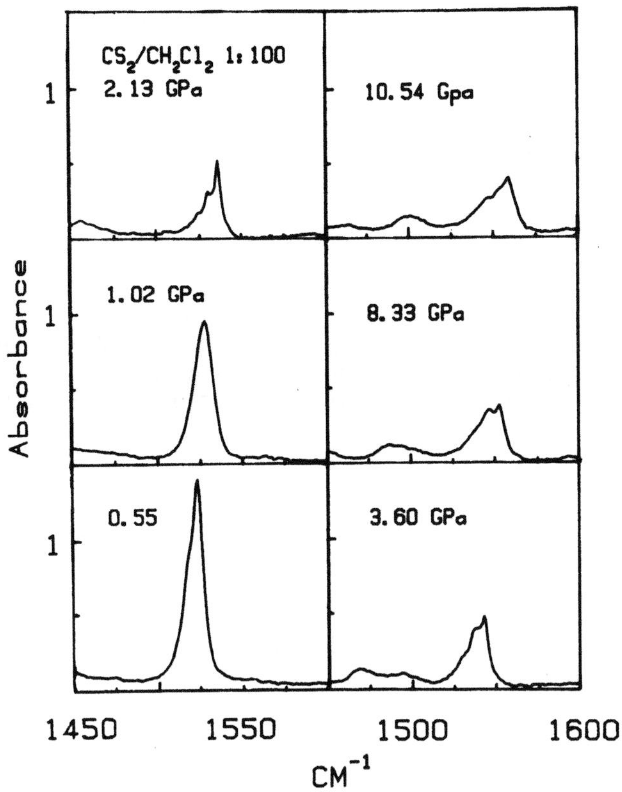

Fig. 6. Infrared spectra of CS_2 /CH_2Cl_2 solutions at the pressures noted and ambient temperature.

fact that the splitting between ν_1 and ν_3 (the symmetric and antisymmetric stretches) is not pressure dependent implies that the inductive interactions are not varying appreciably with pressure, since n_3 is dipole allowed in contrast to ν_1 and therefore affected more by the inductive interactions. Conversely, since the dispersive interactions depend on molecular polarizability, they should affect ν_1 more strongly that n_3. The repulsive interaction, on the other hand, should affect ν_1 and ν_3 in a similar way and therefore not influence the ν_1 /ν_3 splitting. Note that the above mentioned model for the vibrational shift as a function of pressure has assumed that the intramolecular potential function remains pressure invariant, i.e., that f_{har} and f_{anhar} are pressure invariant. If there is an electronic interaction which mixes the groundstate wavefunctions of adjacent molecules, the intramolecular potential could then become pressure or density dependent which is exactly what we suggest is occuring in CS_2.

Thus, we offer an alternative explanation for the small shift of ν_1 and ν_3 with pressure. It is that two competing effects, each driving the frequencies in opposite directions at nearly the same rate, cause the internal modes to shift up in frequency with increasing in pressure more slowly than is usual for internal modes that are subject to repulsive interactions. The first effect is due to decreasing intermolecular distances pushing the atoms of the molecules up against the repulsive wall of the stretching coordinates. This occurs, as we have stated above, when the repulsive intermolecular interactions predominate at high pressure and results in the frequency of the internal mode climbing with increasing pressure. The second, compensating effect is an electronic one involving an increasing overlap of the groundstate wavefunctions of adjacent molecules. The formation of this "extended bond" results in a decrease in the force constant of the internal modes which are sensitive to this change. Thus, while increasing intermolecular repulsion increases the frequency of the internal mode as per the equation shown above, the mixing of groundstate wavefunctions serves to decrease f_{har} and f_{anhar} thus decreasing the frequency. A similar situation has been noted[11] with H_2 where the internal mode first increases, then flattens out, subsequently decreasing as pressure increases. Oxygen at high pressure also shows[12] similar behavior in its e phase. Here the Raman-active stretch increases with increasing pressure while the infrared-active stretch decreases at twice the rate. Therefore, the authors argued that the oxygen's bond strength is actually decreasing with increasing pressure.

The ambient temperature pressure-induced chemistry at 8.3 GPa could very well result from the completion of the intermolecular overlap which we used to explain the insensitivity of the vibrational modes with pressure. The product is similar with that obtained[4] at lower pressure and higher temperature (3.3 GPa and 250°C). We have no evidence for either graphite or elemental sulfur as was found[4] in the harsher temperature, milder pressure CS_2 reaction, although we cannot entirely rule it out since our spectroscopic methods are not sensitive to either of these species (provided that the sulfur is amorphous). We believe, however, that the milder conditions of our ambient temperature study would not permit this more severe redox chemistry and that only the more simple addition chemistry takes place in this case. In support of this hypothesis is the fact that our visible/NIR absorption (Fig. 4a) for the ambient temperature pressure-induced reaction does differ from that reported earlier for the recovered CS_2 product. Dilution studies with dichloromethane also support this notion, since any enviornment that produces graphite and sulfur would most likely also produce a solvent reaction. However, we have observed no such reaction with dichloromethane as the solvent at ambient temperature. It is reasonable to assume, then, that this reaction of CS_2 is a concerted one and involves many CS_2 moeities leading to the proposed Bridgeman's Black structure as shown in the introduction.

The shift of the uv absorption edge as a function of static high pressure matches that obtained by means of an isothermal ramp shock exactly which means that the ramp shock is producing an isothermal compression and that the CS_2 is solid during the ramp shock as it is in our static high pressure measurement. A more suprising and unexpected similarity occurs between the step shock experiment and our static high pressure work. That is, while Duvall and Ogilvie noted[1] the onset of irreversible chemistry when their shock strength produced a pressure in the CS_2 in excess of 8.6 GPa and 600°C, we find that pressure-induced irreversible chemisty takes place in the solid at 8.3 GPa and ambient temperature. Although the similarity in the pressure dependence of the uv absorption edge between the isothermal ramp shock and the static high pressure experiment is not too suprising, the very similar pressures for the onset of irreversible chemisty in CS_2 between the static high pressure and the step shock experiments is very puzzling in that the temperatures are so very different. It is known, for instance, that increasing the temperature lowers[4] the pressure threshold for the CS_2 reaction markedly. For instance, as noted above, at 250°C, the reaction proceeds at 3.3 GPa. Indeed, we have found that the reaction occurs on a several minute time scale at 2.4 GPa and 200°C. The only possible way that we have to explain the similar pressures for the onset of irreversible chemistry between the step shock and the static high pressure work on CS_2 is for there to be a strong kinetic factor for the chemistry at high pressure. That is, since the step shock experiment takes the CS_2 up in both temperature and pressure over a very short time scale, a small rate constant for the subsequent reaction (i.e., a high activation barrier) could prevent the subsequent reaction from occuring until some limiting threshold was reached. The rate of the step-shock induced reaction has been observed[1] to accelerate dramatically with increasing pressure over the 8.6 GPa threshold, which is consistent with our kinetic argument. Our much lower predicted temperature (218 versus 353°C) for the step shock at 2.4 GPa could be used to calibrate the Sheffield EOS. This difference in temperature might indicate that thermal equilibrium had not yet been achieved in the sample were it not for the fact that Duvall, et al., performed an isentropic ramp shock measurement with a thick cell of CS_2. The shift of the absorption edge for this experiment followed the step shock curve, although the time scales for the two experiments are much different.

The dilution studies are also useful in isolating the two counteracting effects for neat CS_2. Thus, in dichloromethane the electronic interaction is effectively eliminated due to CH_2Cl_2's much higher electronic states as well as its very different electronic character. There is a substantial shift to higher frequency for n_3, the infrared active C-S stretch, with increasing pressure, which indicates that repulsive intermolecular interactions dominate over the attractive dispersive and inductive forces, the latter due to the solvent polarity. At higher concentrations than 1%, the CS_2 / CH_2Cl_2 turns brown reversibly when the solution solidifies at 2.5 GPa. The visible absorption spectrum does not show any distinct features, simply a tail underlying the UV band at 340 nm. There are no new infrared features accompanying this transformation either, although there was evidence for CS_2 in both the CH_2Cl_2 crystal as well as in aglomerates with itself in that different infrared features are discernable for these two enviornments, and further, laser illumination resulted in photodecomposition. We expect that the presence of the reversible color change that occurs in these more concentrated CS_2 / CH_2Cl_2 solutions upon solidification is due to the same electronic interactions we invoked above to explain both the onset of chemistry as well as the insensitivity of the internal modes to pressure.

In summary, we have identified the pressure for the onset of CS_2 chemistry at ambient temperature and have correlated it with the several other indicators of increasing electronic interaction. We have further explained the rather small shift of ν_1 and ν_3 with pressure as due to compensating effects of intermolecular repulsion and electronic ground-state mixing. Our measurement of the UV absorption edge with pressure matches that measured in an isothermal ramp shock and we have been able to estimate the temperature of a step shock at 2.4 GPa as well. There are further indications of various molecular and polymer species present as a result of our infrared work and the relative amounts of these species are affected by the temperature, pressure, and physical state of the CS_2.

REFERENCES

1. (a) Ogilvie, K.M.; Duvall, G.E. J. Chem. Phys 1983, 78, 1077-87. (b) Duvall, G.E.; Bellamy, P.M.; Hegland, J.; Granholm, R.; Eschbach, P. FY83 ONR annual Report.

2. Whalley, E. Can. J. Chem. 1960, 38, 2105-8.

3. Butcher, E.G.; Alsop, M.; Weston, J.A.; Gebbie, H.A. Nature 1963, 199, 756-8.

4. Chan, W.S.; Jonscher, A.K. Phys. Stat. Sol. 1969, 32, 749-61.

5. Merrill, L.; Bassett, W.A. Rev. Sci. Instrum. 1974, 45, 290-4.

6. Piermarini, G.J.; Block, S.; Barnett, J.D.; Forman, R.A. J. Appl. Phys. 1975, 46, 2774-80.

7. Schimizu, H.; Ohnishi, T. Chem. Phys. Lett. 1983, 99, 507-11.

8. LeSar, R.; Ekberg, S.K.; Jones, L.H.; Mills, R.L.; Schwabe, L.A.; Schiferl, D. Solid State Commun. 1979, 32, 131.

9. Adams, D.M.; Sharma, S.K.; J. Chem. Soc. Faraday Trans. 1974, 2, 1355.

10. Benson, A.M.; Drickamer, H.G. J. Chem. Phys. 1957, 27, 1164-74.

11. Sharma, S.K.; Mao, H.K.; Bell, P.M. Phys. Rev. Lett. 1980, 44, 886.

12. Agnew, S.F.; Swanson, B.I.; Jones, L.H. J. Chem. Phys. in preparation.

13. Weir, C.E.; Piermarini, G.J., Block, S.J. J. Chem. Phys. 1969, 50, 2089-93.

14. Wortman, J.; Gattow, G.Z. Z. Anorg. Allg. Chem. 1970, 376, 64-72.

ABSORPTION SPECTRA OF SHOCKED LIQUID CS_2

John C. Dallman

Reactive Hydrodynamics Group

Los Alamos National Laboratory

INTRODUCTION

The importance of shock initiation of high explosives (HE) was understood as early as 1863 when Alfred Nobel introduced the detonator as a means of detonating nitroglycerine. The critical pressure rise times required to achieve shock initiation and steady propagation of detonation are determined by the chemical and mechanical properties of an explosive. Although progress has been made in the understanding of the effects of mechanical properties, the detailed effects of high pressures on chemical reaction mechanisms are still only poorly understood. This paper reports the results of two experiments using CS_2, which is known to undergo electronic state transitions when shocked to high pressures.[1] The goal of these experiments was to examine the known shock-generated expansion of CS_2 absorption bands while generating the shocks with a flyer plate system driven by high explosives.

EXPERIMENT DESIGN

A. Shock Wave Generator Design

A common method for generating flat-topped shock waves in condensed media is the use of impacting flyers. The PAD[2] code was used as a design tool in determining the exact configuration of materials to be used in the flyer plate design. The design criteria used were: 1) the shock wave have a pressure in the CS_2 of at least 2.0 GPa, and 2) the shock be flat-topped with less than a 10% variation over 2 μs. Figure 1 shows the final system design. Note that it is a sandwich of four materials with two 6.3-mm vacuum gaps and a single air-space gap.

B. Spectroscopic Design

An explosively driven shock wave in Xenon was used as a light source. The Xe shock tube light source is known to have a blackbody equivalent temperature of about 35 000 K.[3] Although this light source is known to have abundant absorption structure,[4] the measurements of this study did not have sufficient resolution to observe this structure. A reflection optical path was used to pass the light through the CS_2. The reflecting surface was vapor deposited aluminum on a Plexiglas substrate, which formed the final plate in the flyer plate sandwich.

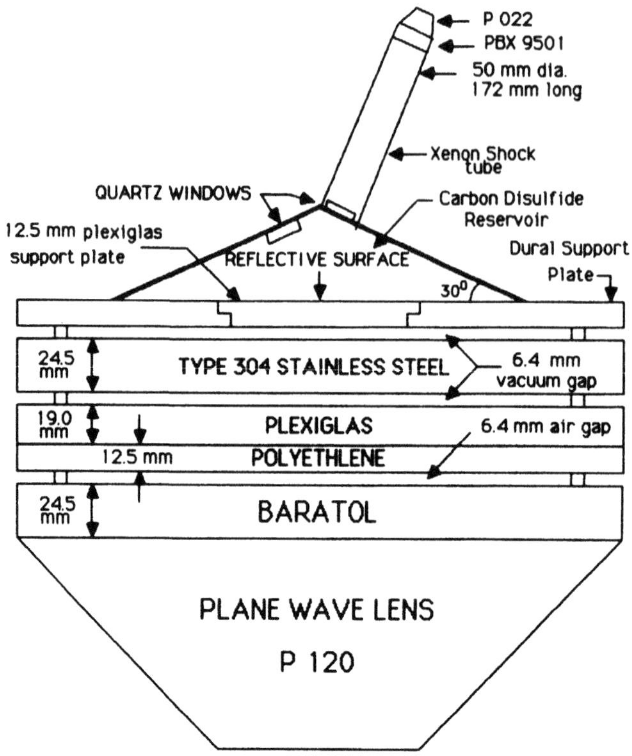

Fig. 1. Schematic of Shock Wave Generation and Absorption Spectroscopy System.

A three sided A-frame structure served as a support for the Xe light source and as the CS_2 holding cell. All windows were uv-grade fused silica.

C. Instrumentation

Four distinct types of instrumentation were employed in the experiments. These included capped coaxial pins mounted at 120° intervals flush with the surface nearest the explosive, and at a radial distance from the stainless steel plate center of 104.8 mm and 47.6 mm from the center of the CS_2 support plate, a spectrograph with gated intensified CCD array and optical multichannel analyzer, another spectrograph coupled with an electronic streak camera, two photomultiplier tubes with narrow band interference filters, and a manganin pressure gage.

The spectrograph with gated intensified CCD array and optical multi-channel analyzer (OMA) was a Tracor Northern 1710 system. This system covered a spectral range of 220-700 nm for experiment C5499 and 220-770 nm for C5519 with a resolution of about 1 nm. Overlap of spectral orders was reduced by using a No. 7910 Corning filter which removed wavelengths less than 220 nm. By gating the image intensifier this system takes a single snapshot spectrum.

The spectrograph with electronic streak camera consisted of a 200 line/mm glass transmission grating with a linear dispersion of 40 nm/mm.

The resulting spectra are imaged on the photocathode of an Imacon 790 electronic streak camera. Because a glass transmission grating was used, wavelengths shorter than about 360 nm were effectively filtered. The spectral range covered was from 360 to about 700 nm with a resolution of about 1.8 nm. The streak rate of the camera was 16.67 mm/μs with a total sweep time of 4.2 μs. This provided a time resolution of 60 ns.

Two Hamamatsu microchannel plate photomultiplier tubes (PMT) with interferance filters were used to observe the light transmitted through the CS_2. One filter had a 4 nm pass band at 391 nm and the other an 8 nm pass band at 780 nm. The output from these tubes was recorded by a Tektronix 7612AD (780 nm) with a sweep rate of 5 nm/point and by a 7912AD (391 nm) with 20 nm/point sweep rate. Pressure was measured using a manganin gage. This gage was glued to the aluminized surface of the Plexiglas support plate, i.e. the gage was immersed in the CS_2. The output from the gage was recorded by a Tektronix 7612AD at a 5 ns/point sweep rate.

RESULTS

The results of two experiments, runs C5499 and C5519, are presented. The gated spectrograph, manganin gage and pin closures were recorded in both runs; for run C5519 a spectrograph with streak camera and the output from the two PMT's were also recorded.

The pin measurements provided information about the planarity of flyer plate impact. Run C5499 showed a maximum variation of impact of 100 ns (equivalent to about 2.0 min. of arc) for the final flyer impact, while C5519 showed a 55 ns variation in the time of impact (equivalent to about 1.7 min. of arc).

The manganin gage results for the two experiments were slightly different. For C5499 a pressure of 2.1 GPa was measured (this run used all air gaps in the flyer plate system); while, for C5519, 2.67 GPa was the pressure at the time of the gating of the spectrograph. Using the temperature-pressure Hugoniot developed by Sheffield and Duvall,[4] the estimated temperature of the shocked CS_2 of C5499 is about 563 K and of C5519 is about 635 K.

Figure 2 shows the results of the gated spectrograph. These results are uncorrected for spectral sensitivity of the system. Post-test examination of C5499 indicated that the image intensifier tube showed strong gain distortion (i.e. spectrally) when it is gated for times less than 0.8 μs (0.1 μs was used for C5499). This was corrected in C5519 by using a gate width of 1.0 μs, as a result, these data are the more reliable. The blue cutoff in Fig. 2 does not occur abruptly, but approaches 10% of maximum intensity, as an arbitrary standard, at 435 nm for C5499 and 447 nm for C5519.

Figure 3 shows the recorded output for the two PMT's. Both records show an abrupt drop in intensity at 1.76 μs. This time corresponds exactly with the arrival of the shock at the Plexiglas-CS_2 interface as detected by the manganin gage. There appear to be elastic precursers starting at about 1.33 μs. The time resolution for the 391-nm record is about 20 ns/point. The relaxation time for the red expansion of the absorption bands at this wavelength appears to be about 60 ns. The 780-nm record shows an abrupt drop in intensity as the shock wave arrives at the reflective surface, but thereafter follows an almost linear decrease in intensity.

The spectrogram provided by the streak camera did not have sufficient intensity to be reproduced. However, densitometer scans of the

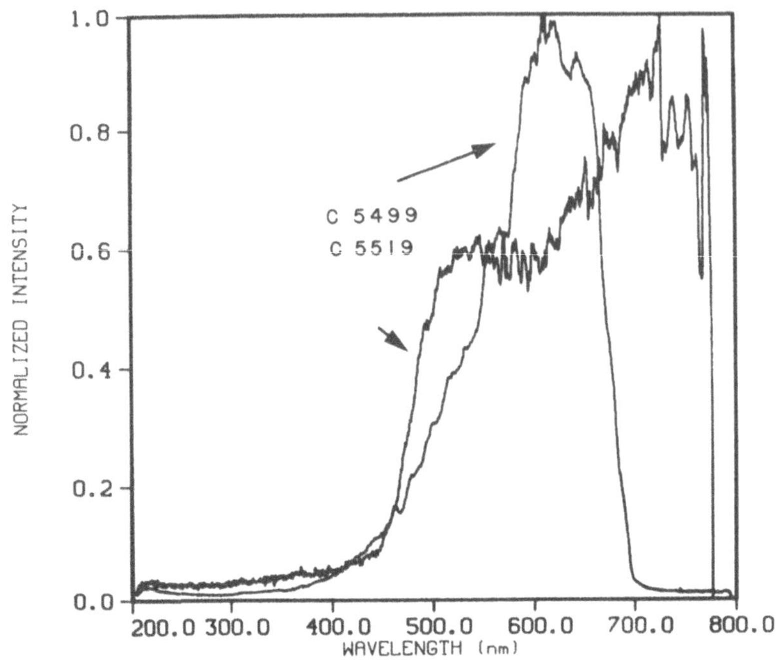

Fig. 2. Digital Spectrogram of Shocked CS_2.

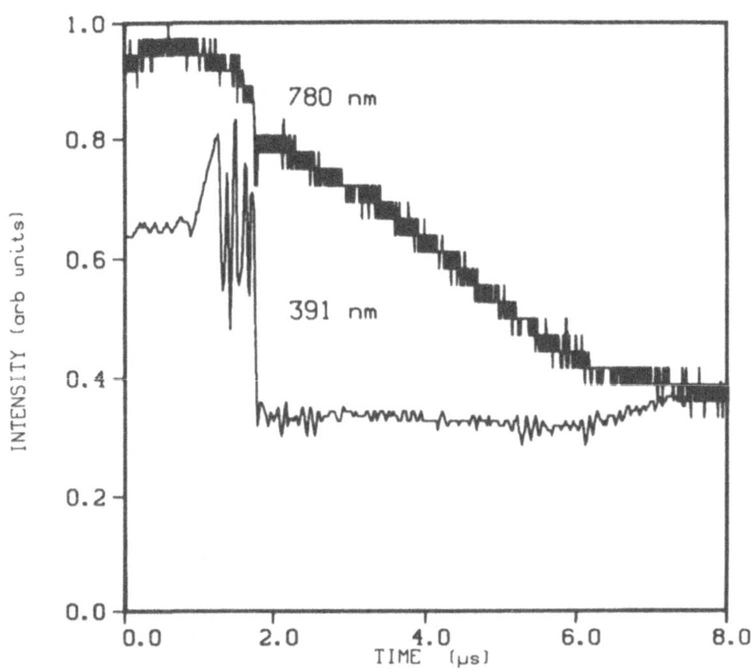

Fig. 3. Transmission of CS_2 before and after shock.

film indicated a red shift of the pass band to about 450 nm at the time of the gated spectrograph snapshot. The absorption band expansion occurs over a period of about 160 ns with a minimum resolution of 60 ns.

DISCUSSION

Measurements of the spectral transmission of liquid CS_2 at room temperature and pressure conditions in this laboratory and in others[1] indicate that there are two strong absorption features in the 200-800 nm range. Measurements were performed using a 1-mm-thick sample cell. The first of these bands extends from beyond the 220-nm limitation of these experiments to about 268 nm. The second feature extends from 279 to about 365 nm. Figure 2 indicates that the first of these absorption features expands under shocked conditions to entirely cut off transmission in the 268-279 nm region. The second absorption band expands to about 450 nm. These results indicate considerable excitation of upper vibrational states, but as was expected no chemical decomposition is observed.

The results of this study show several significant differences from the work of Ogilvie and Duvall.[1] The shocked CS_2 exhibited an expansion to the red of promiment unshocked absorption features to about 450 nm; while, for somewhat greater shock pressures (5.8 GPa) Ogilvie and Duvall have observed red shifts to 410 nm. One possible explanation for these differences is the greater initial temperature of the single shock in this experiment versus a reverberation experiment. Thus, there is the potential for the excitation of higher vibrational states, i.e., expansion of the CS_2 "hot bands". The linear decrease in transmittance at 780 nm indicates that some phenomena may be occurring other than just an increase in optical density. This may indicate the formation of C-S polymers. The unraveling of the interdependence of pressure and temperature effects in shocked CS_2 awaits the application of spectroscopic techniques to CS_2 in static pressure cells.

In conclusion, the time-resolved AS technique used in these experiments has been shown to have the potential to provide qualitative information about the time-dependent chemical processes occurring in strongly shocked transparent liquids. Application to homogeneous light transmitting explosives during the shock up to detonation continues to appear promising.

REFERENCES

1. K. M. Ogilvie and G. E. Duvall, "Shock-Induced Changes in the Electronic Spectra of Liquid CS_2", J. Chem. Phys. 78(3), (Feb. 1983).

2. W. Fickett, "PAD-A One-Dimensional Lagrangian Hydrocode", Los Alamos Scientific Laboratory Report LA-5910-MS (April 1975).

3. C. R. Jones and W. C. Davis, "Optical Properties of Explosive-Driven Shock Waves in Noble Gases", Los Alamos National Laboratory Report No. LA-9475-MS (Sept. 1982).

4. J. C. Dallman and D. H. Liebenberg, "Spectra of Strong Shock Waves in Argon, Argon with 2% Nitrogen and Xenon", Los Alamos National Laboratory Quarterly Progress Report, M-3-QR-83-3 (Oct. 1983).

5. S. A. Sheffield and G. E. Duvall, "Response of Liquid Carbon Disulfide to Shock Compression: Equation of State at Normal and High Densities", J. Chem. Phys. 79(4) (Aug. 1963).

TIME-RESOLVED INFRARED SPECTRAL PHOTOGRAPHY STUDIES OF SHOCK-INDUCED CHEMISTRY IN CS_2

A. M. Renlund, S. A. Sheffield and W. M. Trott

Sandia National Laboratories
Albuquerque, NM 87185

ABSTRACT

Experiments using time-resolved infrared spectral photography (TRISP) in the 6-8.5 μm range to study explosively driven shocks in liquid CS_2 are described. In this range, the temporal resolution of the TRISP system is <1 ns. We have shown that single-shot infrared absorption studies in shocked liquids are feasible with this technique; however, well-controlled planar shocks appear to be necessary for clear interpretation of dynamic spectral changes.

INTRODUCTION

Recently, there has been a growing emphasis on real-time spectroscopic measurements applied to shocked materials, and many such studies have been described at previous Topical Conferences[1] and at this meeting. One goal of these studies is to understand and measure chemical and physical changes induced by shock waves in condensed-phase materials. The work we describe in this paper is part of an ongoing program to develop and apply new spectroscopic techniques to the study of shock and detonation chemistry.[2,3] We have recently applied time-resolved infrared spectral photography (TRISP) to such studies. TRISP was developed by Sorokin and co-workers[4,5] as a means of obtaining single-pulse infrared absorption spectra with <5 ns temporal resolution, and a detailed description of the technique is found in their work. In brief, a fairly broad continuum in the infrared (IR) is generated via stimulated electronic Raman scattering (SERS) in alkali metal vapor. As the IR beam probes the sample, characteristic frequencies within the continuum are absorbed. Detection is facilitated by a four-wave mixing process, again in alkali metal vapor, which upconverts the IR probe beam to the visible, allowing sensitive photodetectors to be used. Sorokin and coworkers have utilized this technique to study the kinetics of several gas-phase photochemical reactions[5-7]. We report here the first use of TRISP to study a shocked, condensed-phase material.

There are several advantages for using TRISP in these studies. It is as sensitive as conventional IR absorption spectroscopy and, therefore, well-suited for detection of chemical species present in relatively small concentrations. This sensitivity and the temporal resolution of <5 ns make TRISP particularly suitable for detecting transient chemical species.

Because the IR probe is weak (<0.1 mJ) and thus unobtrusive, negligible heating or photochemical effects are induced. Finally, IR absorption is suitable for detecting most molecular species.

For our initial experiments we are investigating shock-induced chemistry in liquid CS_2. This molecule has been widely studied[8,9] and appears to react at modest shock pressures (~6 GPa). We began this study by characterizing the TRISP system in the 6-8.5 μm region where the CS_2 ν_3 asymmetric stretching mode and the CS IR absorption are active.[10,11] In preliminary experiments, we have used a small detonator to drive a shock impulse into the liquid sample, and have monitored the CS_2 absorption spectrum as a function of time after the detonator fire pulse.

EXPERIMENTAL

The experimental arrangement for this study is shown schematically in Fig. 1. The 308 nm output of a XeCl laser (Lambda Physik EMG 200E, 400 mJ, 8 ns FWHM) was used to pump two dye lasers. Approximately 75% of the output pumped a broadband dye laser[12] with DMQ/cyclohexane, giving 20-30 mJ centered at 360 nm. The dye laser output was mildly focused into a heatpipe filled with Rb vapor (12 Torr) to promote SERS on the 5s-7s levels and generating broadband IR in the 6-8.5 μm region. A ZnSe flat at Brewster's angle was used to separate the IR from the broadband ultraviolet. The IR beam was focused into the sample to be shocked, reflected back through the sample and then focused through a second ZnSe flat into a second Rb heatpipe. The remaining 25% of the output of the XeCl laser was used to pump a narrowband dye laser (Lambda Physik FL2002) operating on bis-MSB/ dioxane and tuned to 423 nm. Output of this laser, which was predominately vertically polarized, was passed through a Glan prism to reject any remnant of horizontal polarization. The beam was focused and reflected off the ZnSe flat to spatially overlap the IR beam in the second heatpipe. Temporal overlap of the two beams was achieved by a variable optical delay leg in the narrowband dye laser path. The four-wave mixing in the second heatpipe generated the upconverted IR, whose horizontal polarization allowed it to be discriminated from the narrowband laser by means of a second glan prism. The upconverted IR was appropriately filtered and focused to impinge on the

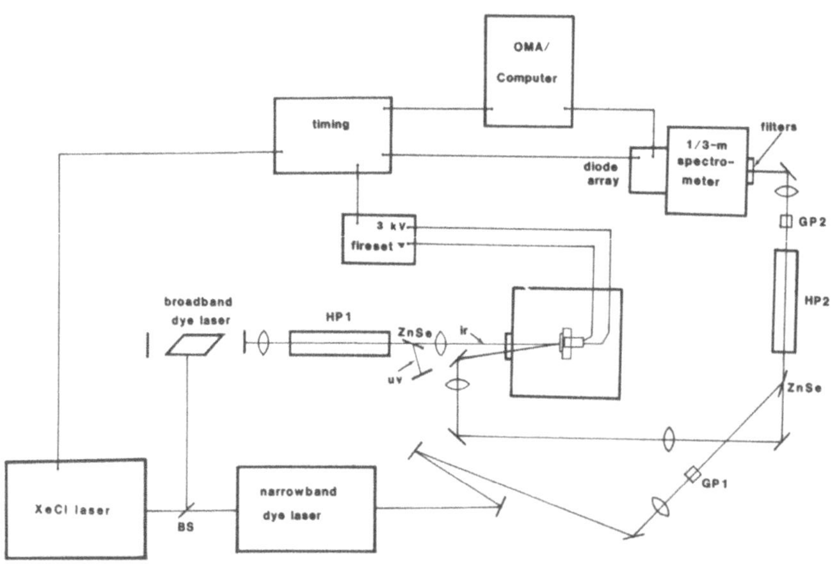

Fig. 1 Schematic of experimental arrangement.

slit of a 1/3-m spectrometer (Jarrell-Ash). The dispersed spectrum was viewed by an intensified diode array (Tracor Northern 6132) and processed by an optical multichannel analyzer (Tracor Northern 1710).

The sample cell which confined the liquid CS_2 consisted of a 0.2 mm-thick plastic ring backed on one side by a NaCl window and on the other side by a 0.2-mm thick aluminized Kapton foil. An RP-2 detonator (Reynolds Industries Inc.) was held in a PMMA fixture and the output face of the detonator was affixed to the Kapton foil with a thin layer of fast-setting epoxy. This entire fixture was inserted into an explosive chamber fitted with a NaCl window to admit the IR beam. A weak, visible laser beam at a frequency $\nu_L-2\nu_{IR}$, where ν_L is the dye laser frequency and ν_{IR} is the frequency of the generated IR beam, is also produced in the heatpipe (as a result of four-wave mixing of the dye laser output with two IR photons) and propagates with the IR beam.[4] This visible beam (at ~400 nm) was used for alignment in order to center the beam on the liquid cell at the approximate center of the detonator. The detonator was fired using a 3-kV fireset, and the laser and detector gate were delayed relative to the fire pulse using digital delay generators.

The temporal characteristics of this TRISP system at these wavelengths were examined by monitoring either the upconverted IR or, more commonly, by monitoring the weak visible beam at $\nu_L-2\nu_{IR}$. Signals were recorded from an appropriately filtered biplanar phototube (Hamamatsu R1193U-02) using a Tektronix 7104 oscilloscope (with 7A29 amplifier). The overall time constant of this phototube and oscilloscope combination was ~0.4 ns.

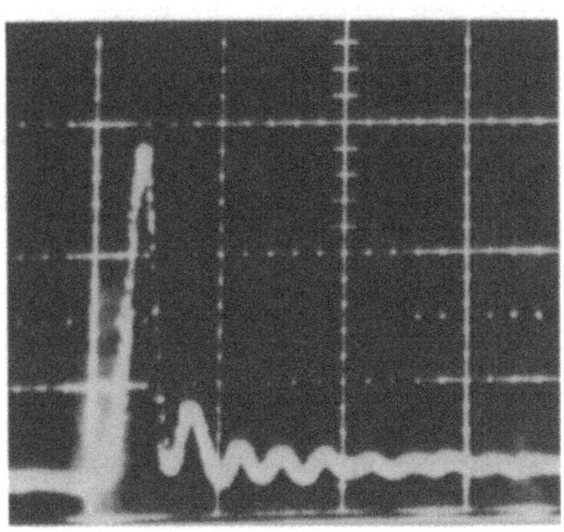

Fig. 2. Temporal profile of the $\nu_L-2\nu_{IR}$ beam generated via four-wave mixing at ~400 nm. The broadband IR generated by the SERS process is at 6-8.5 μm and should have a similar temporal profile.

RESULTS AND DISCUSSION

The temporal profile of the SERS-generated output from the interaction of the broadband dye laser beam with the Rb vapor in the first heatpipe is shown in Fig. 2. Since competing non-linear processes also occur due to the interaction of the dye laser with the alkali metal vapor[13], the SERS process which generates the broadband IR in the 6-8.5 μm region is active only in an early, high-intensity, portion of the dye laser pulse. As a result, the pulsewidth of the IR beam is <1 ns. Since reaction time scales are expected to be fast in shock-induced and detonation chemistry, the <1 ns temporal resolution defined by the IR pulsewidth is a distinct advantage.

Figure 3 shows the single-shot spectrum of the upconverted infrared beam with and without liquid CS_2 in a 0.2-mm thick IR cell backed by an optically flat Al mirror. Even over a pathlength of 0.4 mm (the beam propagates twice through the cell), the ν_3 absorption was saturated, but there was still good signal in the area of the CS absorption (~7.8 μm). In our experiments with the explosive fixture, however, the signal was reduced by an order of magnitude due to the poor flatness of the aluminized Kapton foil in contact with the detonator and this led to a lower signal-to-noise ratio for the shock experiments (see Fig. 4). Even though the signal level was considerably reduced, using the explosive fixtures gave us relatively a quick turn-around time on experiments.

Due to the small size of the detonator output face (3.86-mm diameter), we expect that the shock impulse delivered to the liquid sample was very irregular. No attempt was made during these studies to characterize the shock from the RP-2 detonator, which has RDX as the output explosive pellet. We estimate, however, that the shock delivered through the CS_2 was at least 5 GPa.

We examined the absorption spectrum of CS_2 in the 6-8.5 μm range as a function of time after the detonator fire pulse. The function time of the detonator at the operating voltage was 1.8 ± 0.03 μs and the transit time of the shock through the foil was expected to be <50 ns. We delayed the laser pulse from 50 to 400 ns after the breakout of the shock into the CS_2. Our results indicate that the liquid remains transparent to the IR during that interval, and the signal loss after 400 ns is due to destruction of the mirror. Figure 4 shows typical spectra of the absorption of unshocked CS_2 and of the sample 100 ns after the shock reaches the liquid. The subtle changes and the uncharacterized nature of the shock impulse do not permit an unambiguous interpretation of these results at the present time.

While these results are preliminary, they do offer encouragement for future work. We will soon be performing similar experiments using gas-gun driven shocks. In those experiments, the use of optically flat mirrors should improve the signal quality and the planarity of the shock will allow the results to be tractable to adequate interpretation.

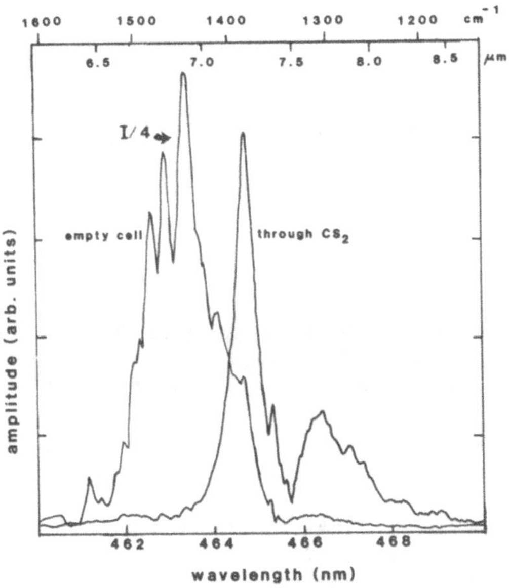

Fig. 3. Single-shot spectra of the upconverted IR through an empty cell and through 0.4-mm thickness of CS_2. Wavelength and energy scales at the top refer to the IR. The bottom wavelength scale is of the upconverted IR.

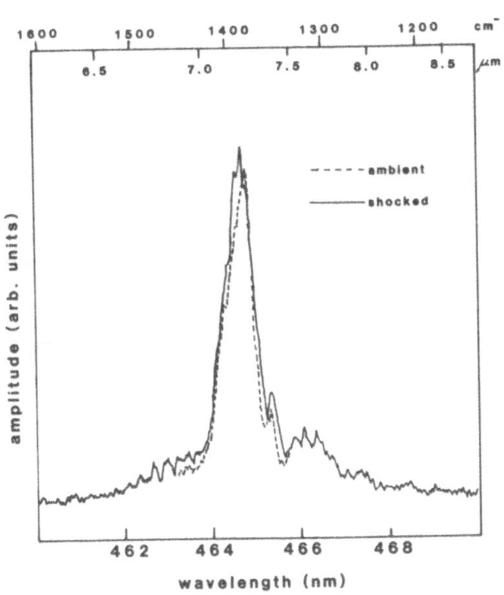

Fig. 4. Single-shot spectra of unshocked CS_2 and of shocked CS_2 in the explosive fixture. The laser was delayed 100 ns from the arrival time of the shock into the liquid sample. The two spectra overlap (within the noise) except where shown by the dashed line.

ACKNOWLEDGEMENTS

We express thanks to J. C. Pabst, D. W. Dugan and H. C. Richardson for their excellent technical assistance. We are also grateful to Drs. P. P Sorokin and J. H. Glownia for their encouragement and helpful discussions while we were setting up the TRISP system. This work was performed at Sandia National Laboratories, Albuquerque, NM, supported by the U. S. Department of Energy under contract number DE-AC04-76DP00789.

REFERENCES

1. See, e.g., "Proceedings of the American Physical Society Topical Conference on Shock Waves in Condensed Matter," J. R. Asay, R. A. Graham and G. K. Straub, eds., Santa Fe, NM (1983) Ch. 7.

2. W. M. Trott and A. M. Renlund, Single-pulse Raman scattering studies of heterogeneous explosive materials, Applied Optics 24:1520 (1985)

3. W. M. Trott and A. M. Renlund, Time-resolved spectroscopic studies of detonating heterogeneous explosives, in: "Proceedings of the 8th Symposium (International) on Detonation," Albuquerque, NM (1985) in press.

4. D. S. Bethune, J. R. Lankard, M. M. T. Loy and P. P. Sorokin, Time-resolved infrared spectral photography: a new technique, IBM J. Res. Dev. 23:556 (1979)

5. Ph. Avouris, D. S. Bethune, J. R. Lankard, J. A. Ors and P. P. Sorokin, Time-resolved infrared spectral photography: Study of laser-initiated explosions in HN_3, J. Chem. Phys. 74:2304 (1981)

6. D. S. Bethune, J. R. Lankard, P. P. Sorokin, A. J. Schell-Sorokin, R. M. Plecenik and Ph. Avouris, Time-resolved infrared study of bimolecular reactions between tert-butyl radicals, J. Chem. Phys. 75:2231 (1981)

7. A. J. Schell-Sorokin, D. S. Bethune, J. R. Lankard, M. M. T. Loy and P. P. Sorokin, Chlorine perchlorate. A major photolysis product of chlorine dioxide, J. Phys. Chem. 24:4653 (1982)

8. S. A. Sheffield and G. E. Duvall, Response of liquid carbon disulfide to shock compression: Equation of state at normal and high densities, J. Chem. Phys. 79:1981 (1983)

9. S. A. Sheffield, Response of liquid carbon disulfide to shock compression. II. Experimental design and measured Hugoniot information, J. Chem. Phys. 81:3048 (1984)

10. B. P. Stoicheff, High resolution Raman spectroscopy of gases. XI. Spectra of CS_2 and CO_2, Can. J. Phys. 36:218 (1958)

11. S. A. Suchard, ed., "Spectroscopic Data Vol. 1. Heteronuclear Diatomic Molecules" IFI/Plenum, New York, NY (1975) p. 354.

12. D. S. Bethune, Dye cell design for high-power low-divergence excimer-pumped dye lasers, Applied Optics 20:1897 (1981)

13. J. J. Wynne and P. P. Sorokin, Optical mixing in atomic vapors, in: "Nonlinear Infrared Generation," Y. R. Shen, ed. Springer-Verlag, Berlin (1977) p. 160.

SPECTRAL SHIFTS IN THE FLUORESCENCE OF ANTHRACENE AND LIFETIME CHANGES IN CRYSTAL VIOLET UNDER LASER DRIVEN SHOCK COMPRESSION: PROBES OF PRESSURE AND VISCOSITY

A.L. Huston#, B.L. Justus* and A.J. Campillo

Naval Research Laboratory
Code 6510
Washington, D.C. 20375

INTRODUCTION

Compressional shock waves produce rapid simultaneous jumps in temperature and pressure in materials that may lead to unusual chemical and physical behavior[1]. Optical techniques such as absorption spectroscopy[2] and stimulated raman spectroscopy[3] have been used to obtain information, at a molecular level, on changes in the material behind the shock front. Traditional techniques used to generate shock waves and optically probe the shocked regions involve the use of gas guns or explosive charges and mechanical synchronization of the interrogating light pulse. These techniques are well developed and have provided a vast amount of high-quality data on the physics and chemistry of shock loaded materials. These techniques do have certain drawbacks: they destroy the sample, the data rates are very low and the time resolution is limited.

Recent work in our laboratory[4] has demonstrated the feasibility of using a low-energy, picosecond laser source to generate multi-kilobar, near-planar compressional shock waves. Shock waves produced by this method are characterized by peak pressures that scale linearly with the energy of the laser pulse. Laser driven shock provides a versatile supplement to the conventional shock generation techniques for a variety of reasons: ease of synchronization of the arrival of the probe pulse with the passage of the shock front (subnanosecond resolution), much higher repetition rates, table-top setup and the ability to recover shocked material for further analysis.

We have recently developed a number of techniques using fluorescent probes to measure the pressure, temperature and viscosity of solutions under shock loading. In this paper, we report measurements of the spectral shifts of the fluorescence bands of anthracene in benzene as a function of shock pressure. We also report preliminary results on the determination of the viscosity of glycerol under shock loading by measuring the viscosity dependent fluorescence lifetime of the dye, crystal violet (CV). In addition, the nature of the viscosity dependence on pressure and temperature allows one to estimate the temperature change that accompanies passage of the shock wave through the solution. The viscosity of shock loaded solutions is potentially an important parameter for determining the rates of chemical reactions.

Sachs/Freeman Associates Inc., Landover, Md. 20785
* Joint NRL/NRC Postdoctoral Research Associate

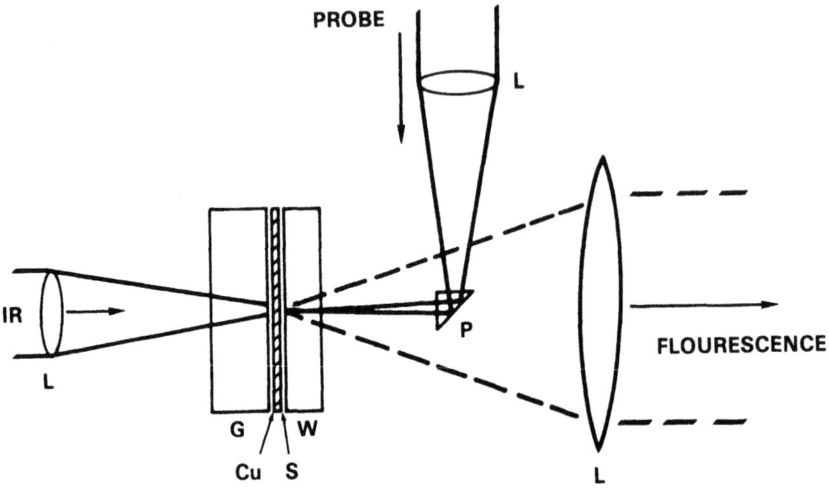

Fig. 1. Experimental apparatus for fluorescence probing of shocked samples.

LASER DRIVEN SHOCK EXPERIMENTS

Single 10 psec, 1054 nm pulses were selected from the output of a mode-locked Nd:glass laser and amplified to energies of up to 20 mJ. Near planar compressional shock waves were produced by focussing the gaussian IR pulse to a spot of 400 microns onto a thin copper foil (25-50 microns). The laser pulse is partially absorbed by the foil, creating a hot, confined plasma, which, upon expansion, launches a compressional shock wave into the copper. A portion of the IR laser pulse was split off prior to the shock cell and used to generate a probe pulse at the second and third harmonic frequencies (527nm and 351nm). The probe pulse used to excite the dye fluorescence was delayed 13 to 18 nsec (depending on the foil thickness) to allow for the transit time of the shock through the foil and pressure ring-up in the solution.[4] The sample cell and associated experimental apparatus is shown schematically in figure 1. The sample cell consisted of an ca. 1 micron thick solution sandwiched between a copper foil and a sapphire window. The copper foil was held in place with a glass window. The probe pulse was focussed to a spot (250 microns) which overlapped the shocked area so that only shocked material contributed to the observed signal. Fluorescence from the sample was focussed onto the slit of a 1/4 meter polychromator. Dispersed fluorescence spectra from single laser shots were detected with a SIT vidicon and then processed with an optical multichannel analyzer. Fluorescence lifetimes of CV were measured with a streak camera in place of the polychromator. The fluorescence data were fitted with single exponential decays using a nonlinear least squares fitting routine. Shock pressures were determined using an interferometric calibration technique described previously.[4] With this arrangement, data rates corresponding to 30 complete spectra per hour could be obtained at predetermined pressures.

SPECTRAL SHIFTS OF ANTHRACENE FLUORESCENCE

The fluorescence spectra of anthracene in benzene under ca. 10 kbar shock loading (dashed line) and at ambient pressure (solid line) are shown in figure 2. Each vibronic band of the anthracene fluorescence spectrum undergoes an ca. 13 nm red-shift at this pressure as well as some spectral broadening. The spectral shift of the 0-1 vibronic band is plotted in figure 3 vs. pressure for a series of laser shots (dots) along with

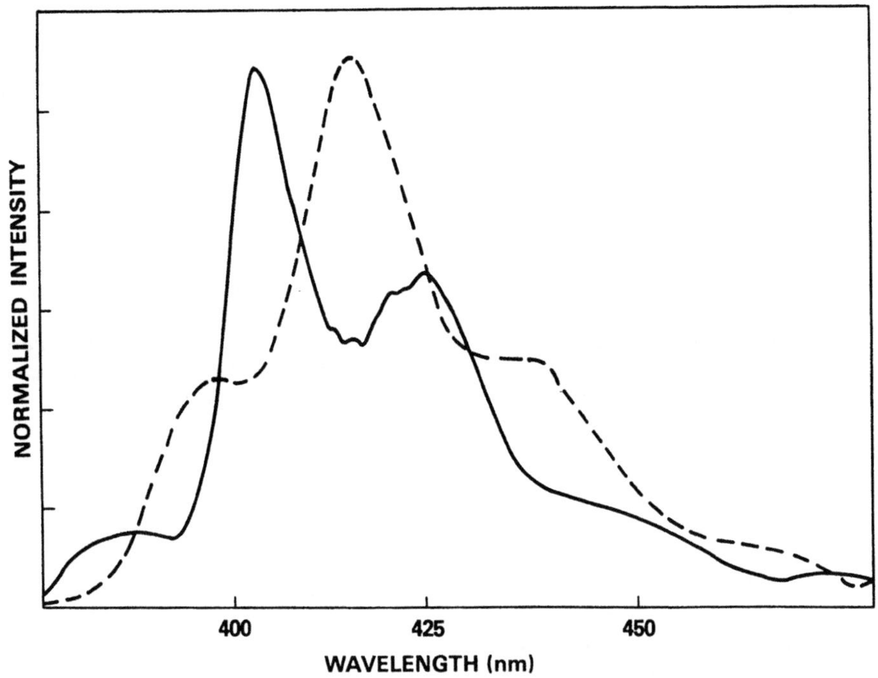

Fig. 2. Fluorescence spectra of anthracene in benzene solution at ambient pressure (solid line) and of shocked solution at ca. 10 kbar (dashed line).

the results of static pressure measurement[5] on anthracene in toluene (boxes). The band shits monotonically to decrease wavenumber at a rate of ca. 65 cm^{-1}/kbar over this pressure range. Spectra obtained from previously shocked regions displayed normal ambient pressure features.

The effects of static pressure and temperature on the photophysical properties of anthracene in the condensed phase have been previously studied.[5-7] The absorption and fluorescence spectra of anthracene shift to lower energies and broaden with the application of high static pressures. Changes in temperature have little effect on the spectral features.[7] Application of pressure alters such properties as the refractive index and specific volume of the solution. These properties determine the nature of the solvent-solute interactions between the anthracene and the solvent and are responsible for the changes observed in the spectroscopic parameters at high pressures. The principle effect of pressure is to cause a decrease in the average separation between solvent and solute molecules. The spectral shift is expected to vary as the inverse sixth power of the separation and the red shift has been shown[7] to be approximately linear in $V^{-2}(P)-V^{-2}(0)$. Using the specific volume data for benzene under shock conditions given in reference 3, the red shift vs. pressure is expected to be nearly linear over the pressure range 1 to 10 kbar. Under static conditions, the spectral shifts for anthracene in a variety of solvents have been observed to be fairly linear for pressures up to 30 kbar. The agreement between the magnitudes of the spectral shifts of the static data and our shock measurements is quite good. Indeed, this system and technique, having been characterized here, provides an excellent probe of pressure (see following paper, this volume).

VISCOSITY OF GLYCEROL

The viscosity of glycerol under shock loading has been determined by measuring the viscosity dependent fluorescence lifetime of a triphenylmethane dye, crystal violet (CV). To the best of our knowledge, this is the first direct measurement of the viscosity of a shock loaded sample. The photophysical properties of CV have been previously studied and are known to be very sensitive to the viscosity of the solvent[8]. The fluorescence lifetime of CV is determined primarily by the viscosity of the solvent through the effect of the viscous drag on the rate of internal conversion following singlet excitation. The fluorescence lifetime of CV in glycerol varies linearly from ca. 100 ps to 260 ps over the pressure range 0 to 18 kbar. In order to quantitatively correlate these lifetimes to the viscosity of the shocked glycerol, it was necessary to first obtain a calibration curve between CV lifetime and glycerol viscosity. The absolute viscosity of glycerol has been very accurately measured as a function of both temperature and pressure [9,10]. The fluorescence lifetime of CV was measured as a function of temperature at ambient pressure to provide the correlation between fluorescence lifetime and viscosity. This treatment assumes that the nonradiative decay rate of CV is independent of temperature over the temperature range of interest. The fluorescence lifetime of CV vs. viscosity raised to the 2/3 power, as predicted by theory[8], is plotted in figure 4. Our viscosity data obtained under shock loading is plotted as $\ln(\eta)$ vs. pressure in figure 5 in addition to static pressure data previously measured[10] for isotherms at 30 C and 70 C. Since the viscosity of glycerol is known to exponentially decrease with temperature and exponentially increase with pressure, the slope of the shock data curve is much less than that of the isotherms, reflecting the combined effect of the temperature and pressure on the viscosity of the shocked glycerol. The intersections of the shock data curve with the isotherms give a measure of the temperature of the shock loaded glycerol at those points.

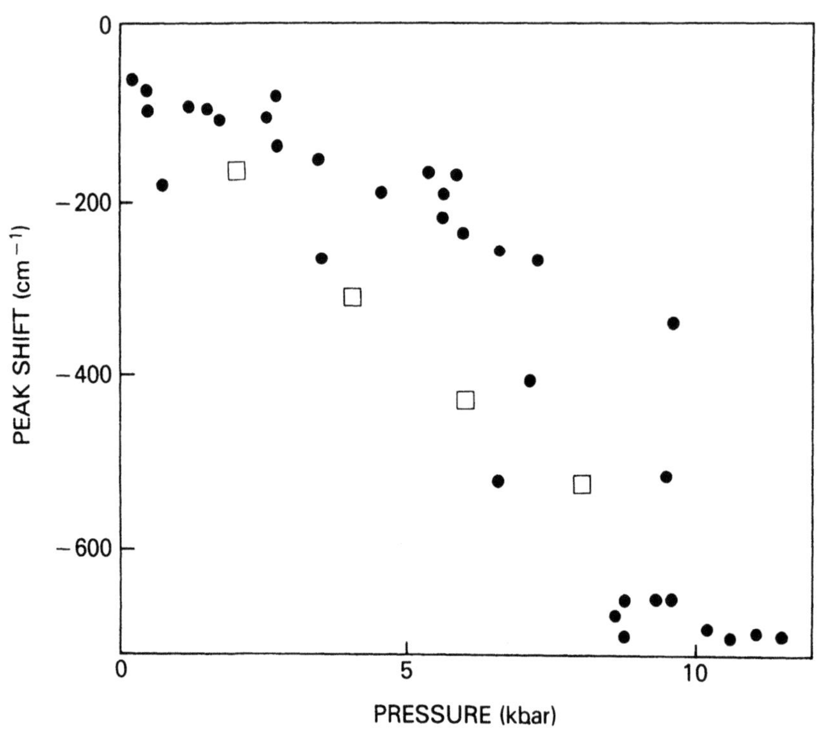

Fig. 3. Peak shift of the 0-1 vibronic band vs. pressure. Circles: shock data. Squares: static pressure data.

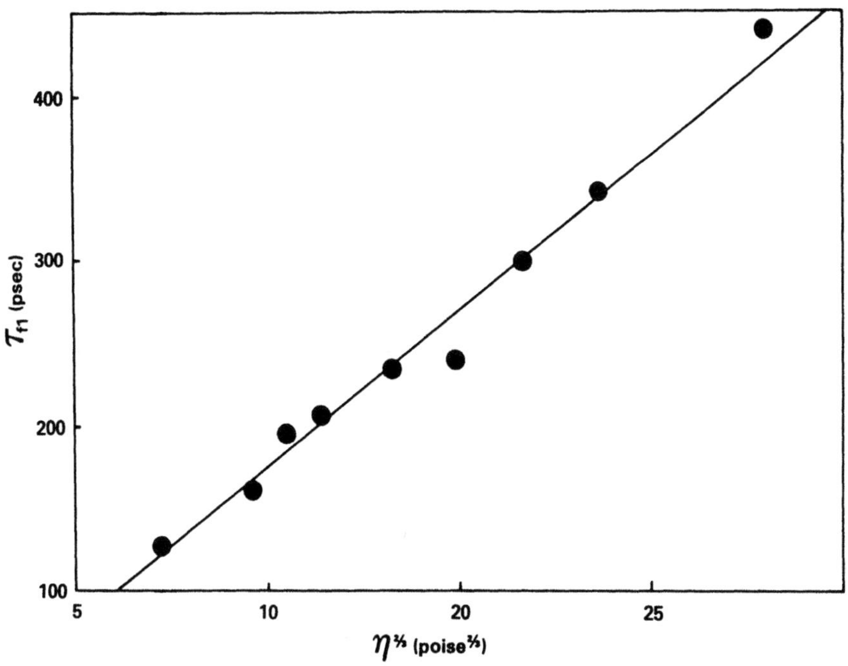

Fig. 4. Fluorescence lifetime of crystal violet in glycerol vs. viscosity$^{2/3}$.

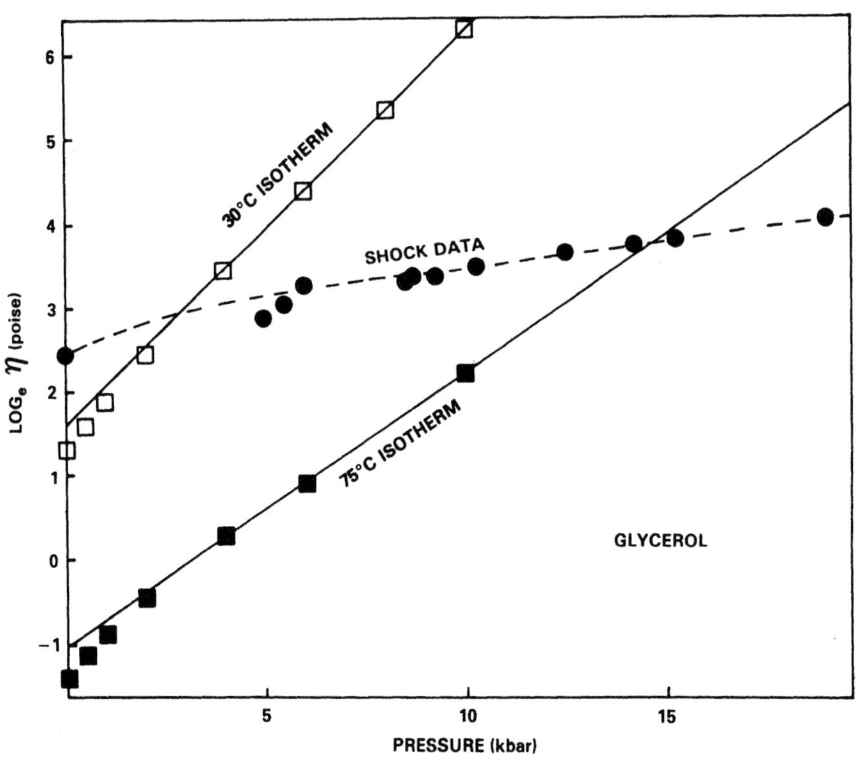

Fig. 5. Viscosity dependence of glycerol on pressure. Shock data as well as 30 C and 75 C static pressure isotherms are shown.

ACKNOWLEDGEMENT

This work was partially supported by the Office of Naval Research.

REFERENCES

1. D. W. Dodson and R. A. Graham, "Shock-Induced Organic Chemistry", in: "Shock Waves in Condensed Matter-1981", W. J. Nellis, L. Seaman and R. A. Graham, editors, Am. Inst. of Physics, New York (1982).
2. K. M. Olgilvie and G. E. Duvall, "Shock-Induced Changes in the Electronic Spectra of Liquid CS_2", J. Chem. Phys. 78(3):1077 (1983).
3. S. C. Schmidt, D. S. Moore, D. Schiferl and J. W. Shaner, "Backward Stimulated Raman Scattering in Shock-Compressed Benzene", Phys. Rev. Lett. 50:661 (1983).
4. P. E. Schoen and A. J. Campillo, "Characteristics of Compressional Shocks Resulting from Picosecond Heating of Confined Foils", Appl. Phys. Lett. 45(10):1049 (1984).
5. B. Y. Okamoto and H. G. Drickamer, "Evaluation of Configuration Coordinate Parameters from High Pressure Optical Data. 1. Phenanthrene, Anthracene, and Tetracene", J. Chem. Phys. 61(7):2870 (1974).
6. M. Nicol, "Fluorescence of Anthracene in Lucite at Very High Pressures", J. Am. Opt. Soc. 55:1176 (1965).
7. H. W. Offen, "Absorption and Luminescence of Aromatic Molecules at High Pressures", in: "Molecular Photophysics Vol. 1", J. B. Birks, editor, John Wiley and Sons, London, Chapt. 3 (1973).
8. T. H. Forster and G. Hoffmann, "Die Viskositatsabhangigkeit der Fluoreszenzquantenausbeuten einiger Farbstoffsysteme", Z. Phys. Chem. NF, 75:63 (1971).
9. M. L. Sheely, "Glycerol Viscosity Tables", Ind. Eng. Chem. 24(9):1060 (1932).
10. P. W. Bridgman, "The Effect of Pressure on the Viscosity of Forty-Three Pure Liquids", Proc. Am. Acad. Arts Sci. 61:57 (1926).

TEMPERATURE MEASUREMENTS OF SHOCKED WATER USING A FLUORESCENCE PROBE

B.L. Justus#, A.L. Huston* and A.J. Campillo

Naval Research Laboratory
Code 6510
Washington, D.C. 20375

INTRODUCTION

Shock experiments on condensed phase materials primarily yield shock and particle velocity data from which pressure, specific volume and internal energy are inferred using conservation relations[1]. Determination of a complete equation of state requires, in addition, thermal information. Unfortunately, there does not at this time exist a reliable and general method for measuring the temperature rise of a shock loaded liquid or solid. Limited success has been achieved experimentally using infrared detection for studies of transparent media in the temperature range 200 to 1500 C[2]. Visible pyrometry, utilized in many laboratories, is best suited for use at higher temperatures. Thermocouple[3] or thermistor[4] measurements, although more general, are beset with difficulties stemming from required miniaturization for fast response. Another interesting attempt has been the use of Raman spectroscopy[5] but this too is very restrictive in its application.

In this paper, we describe a fluorescence probe thermometer for measuring shock heating directly in suitable liquids. As a demonstration of the utility of this scheme, the temperature of shocked water was measured in a previously inaccessible pressure regime, 0 to 9 kbar, by observing the fluorescence enhancement from a fluorescein dye additive (10^{-2}M) following 527 nm excitation and shock loading. Good agreement between calculated temperatures[6] and our measured values were obtained. Temperature changes as small as 5 C were measured and accuracies of ±4 degrees demonstrated. It appears from our calibration data that much better accuracies (< 1 C) will be eventually achievable. We believe that dye additives can be incorporated in a straightforward manner into microparticles contacted to optical fibers to allow temperatures of opaque liquids or solids to be measured.

PRINCIPLE OF THE FLUORESCENCE THERMOMETER

In this form of thermometry, a fluorescing dye additive is mixed with the transparent liquid or solid whose temperature is to be measured. It is well known that the spectral features of the electronic absorption band of polyatomic molecules show a temperature dependence. There is often a shift

#Joint NRL/NRC Postdoctoral Research Associate
*Sachs/Freeman Associates Inc., Landover, Md. 20785

in the absorption peak as well as a thermal band broadening. These effects serve to significantly increase the absorption of light in the wings, especially the weak red band edge. Thus, if a probe beam is used to excite the dye in the red edge spectral region, the resulting fluorescence will show a marked increase with temperature. This effect is illustrated in figure 1 for fluorescein dye added to water. A C.W. spectrofluorimeter was used with the excitation wavelength set at 527nm to observe the integrated emission, F(T), in the range 533nm to 650 nm as a function of temperature, T. The data indicated by squares refers to a sample concentration of 10^{-4} M. F_o refers to the fluorescence at ambient temperature and the enhancement is plotted in normalized form, $(F(T)/F_o)-1$, versus T. As can be seen, the fluorescence increases with temperature, exhibiting a fourfold increase from 23 to 57 C. Indeed, based on observed spectral changes for temperatures up to 60 C, the enhancement is predicted to eventually increase to a maximum of 280 at ca. 600 C before decreasing. In practice, however, this form of thermometry will probably be limited by molecular decomposition above 300 C.

When applying fluorescence thermometry to shock measurements, the effects of pressure on the spectral properties of the probe dye must also be considered. However, as we show, pressure induced effects are small and can be easily incorporated into the calibration.

LASER DRIVEN SHOCK EXPERIMENTS

The shock experiments described in this work were performed using picosecond laser generation of the compressional shock wave[7] and simultaneous laser induced fluorescence probing of the shocked region. Both the IR driving pulse and UV/VIS probe pulses were derived from the output of a modelocked Nd:glass laser system. A single, 1054 nm, 10 psec laser pulse was selected from the output of the oscillator and amplified with two amplifier stages to an energy of 20 mJ. Following the second amplifier, the second and third harmonic frequencies (527 and 351 nm, respectively) were generated in two successive nonlinear crystals (labelled SHG and THG, respectively in figure 2). A UV pulse was used for pressure calibration purposes, a green VIS pulse was used for the fluorescein spectroscopy and an IR pulse was used to generate the shock in the sample (C).

The sample cell consisted of a thin layer (1 micron) of a 10^{-2} molar solution of disodium fluorescein (Exciton Chem. Co.) in water sandwiched between a 50 micron thick CuO-coated copper foil and a fused silica window. The copper foil was in turn placed between the sample/window combination and a second fused silica window. Near planar compressional waves were produced by focussing the gaussian IR laser pulse to a spot size of 430 microns on the copper foil. The IR laser pulse is partially absorbed at the CuO surface, creating a hot confined plasma which, upon expansion, launches a compressional shock wave into the copper where it is transmitted to the sample.

The probe beam was delayed so that it arrived at the sample 18 nsec after the arrival of the IR beam. This delay was necessary in order to allow the shock wave to travel through the copper foil and experience a pressure ring-up in the sample before interrogation. Considerable care was exerted to insure maximum overlap between the two beams. Several different spectroscopic techniques were used to probe the material behind the shock front in this work:

(a) Pressure Calibration

Previous work in our laboratory on the spectral shifts in the fluorescence spectrum of anthracene in benzene under shock loading[8] (see preceeding paper in this volume) found essentially linear shifts of approx.

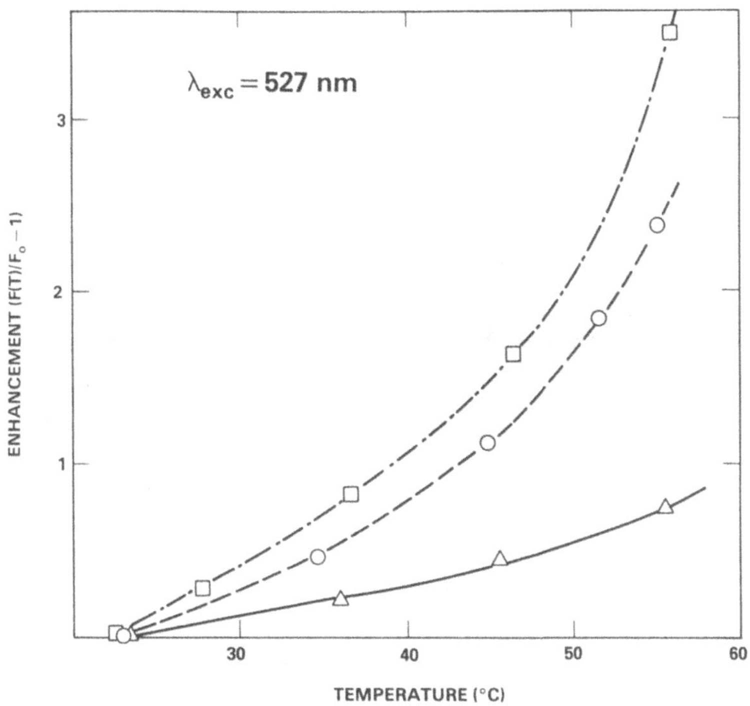

Fig. 1. Calibration plots showing fluorescence enhancement versus temperature at ambient pressure for a 1 cm sample of 10^{-4}M fluorescein in water (squares), a 1 cm sample of 10^{-2}M fluorescein in water (triangles) and a 1 micron sample of 10^{-2}M fluorescein in water (circles). The variation in enhancement among the three curves is due to self absorption effects.

65 cm^{-1}/kbar of the 0-1 vibronic peak. This experiment was repeated under the present experimental conditions in order to provide an accurate calibration of the pressure generated in the present study. The setup was similar to figure 2 except that a spectrograph/vidicon/optical multichannel analyzer was substituted for the PMT. The anthracene was probed with 351 nm radiation. By varying the intensity of the IR laser driver, shocks from 1 to 9 kbar, could be reliably generated under our present conditions. The pressures reported in this and the previous work are effective pressures, reduced somewhat from the peak pressure values due to the averaging effect resulting from the comparable sizes of the IR driver and probe beams, w and w', respectively. The effective pressure is related to the peak pressure by the factor $w^2/(w'^2 + w^2)$. Effective pressures calculated when probing the shocked region with the UV (spot size = 300 microns) or green (spot size = 175 microns) probe beams were reduced by a factor of 0.69 or 0.89, respectively, from the peak value.

(b) Measurement of Shock-Induced Fluorescence Enhancements

The experimental setup is as shown in figure 2 with the fluorescence emission detected using a cooled photomultiplier tube. The PMT output was displayed on a digital storage oscilloscope. The sample was rastered between shocked shots so that only clear, unblemished portions of the sample were interrogated. The fluorescence emission intensity was observed to increase by up to a factor of four over the intensity at ambient pressure for pressures up to 8 kbar.

(c) Static Measurement of Pressure Dependent Absorption Peak Shifts

A hydrostatic cell was used in conjunction with cw transmission spectroscopy to observe the effects of static pressure on the absorption spectrum of fluorescein in water. It was found that the peak of the absorption band shifted 24 cm^{-1}/kbar to the red. In the absence of dynamic shock data, we assumed that the pressure contribution to the absorption peak shift under shock loading was also 24 cm^{-1}/kbar. We found no band broadening under hydrostatic conditions. This was important, since our analysis shows that the major contribution to fluorescence enhancement appears to be due to band broadening rather than peak shifting. Consequently, the enhancement effect we observe in (b) is predominantly caused by temperature, while the shift in peak position due to pressure leads to a small easily accounted for correction.

(d) Shock Induced Changes in Fluorescein Fluorescence Lifetime

The effect of shock loading on the fluorescence lifetime of a 10^{-2}M solution of fluorescein in water was also investigated as this in principle could affect the observed enhancement. These experiments were performed with the setup shown in figure 2 with the photomultiplier replaced by a picosecond electronic streak camera coupled to a vidicon/OMA recording system. The fluorescence lifetimes of a dilute solution of fluorescein in water and that of our optically thin 10^{-2}M unshocked sample were found to be 3.8 and 2.6 nsec, respectively. The fluorescence lifetime of the 10^{-2}M sample under approximately 7 kbar was unchanged from that of its unshocked state indicating that a correction to our enhancement data due to lifetime change is unnecessary. This is perhaps not surprising since it is well known that the emission lifetimes of molecules such as fluorescein are affected by temperature and pressure primarily through the effect of the pressure and temperature on the solvent viscosity (see preceeding paper this volume). Under the conditions of our shock experiment, water is known to have a low viscosity which is not altered significantly by pressure or temperature. The viscosity is also in a range for which the lifetime of fluorescein is not expected to change.

DISCUSSION

The shock induced fluorescence enhancement data were corrected (a) for pressure induced shifts in peak absorption wavelength and (b) for the relative radii of the shocked region and the various probes and plotted in figure 3 in the form of temperature change above ambient, dT, versus effective pressure. Also shown for comparison are estimates of the temperature increase of shocked water calculated by Rice and Walsh[6] (dashed line). Their estimates were obtained by comparing the measured P-V Hugoniot and previously measured static pressure isotherms of Bridgman[9] and assuming the shock temperatures to be those of the isotherms at the P,V points of intersection with the Hugoniot.

Our data agrees with the calculated values of Rice and Walsh to within our present experimental accuracy (about ±4 degrees at 8 kbar). This preliminary accuracy, although superior to that available using other techniques in this pressure and temperature range, was greatly limited by our method of shock generation and is not indicative of the potential of this approach. The largest sources of error resulted from effects associated with the comparable sizes of the shock driver and probe beams. Consequently, misalignment errors and shot to shot variations in beam direction and diameter contributed to an approximately 20% error. Based on the data of figure 1, it appears that the ultimate accuracy of this scheme should be

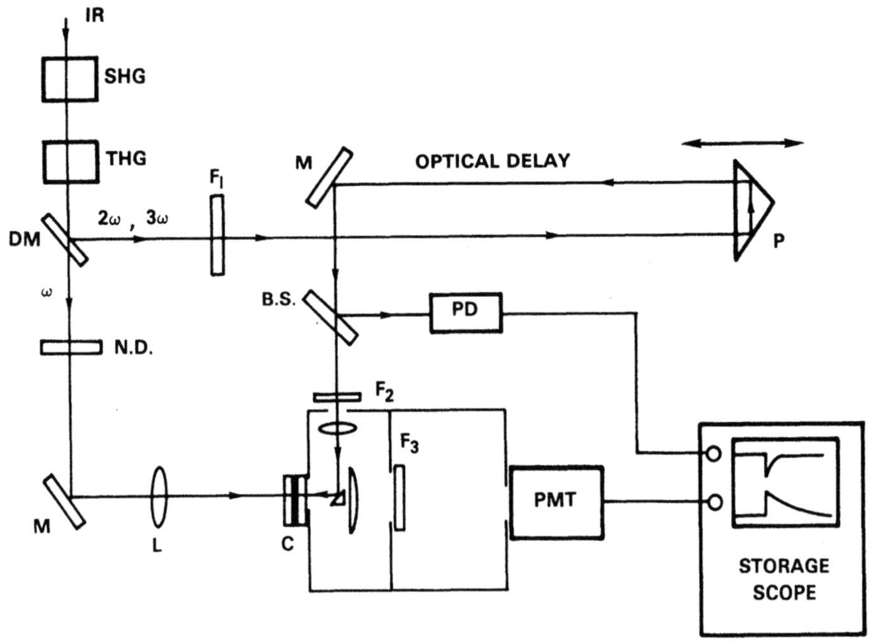

Fig. 2. Schematic of experiment to laser generate high pressure compressional shocks and to conduct fluorescence enhancement experiment.

better than 1 degree. Thus if one were to use this technique with gas gun shock generation, the planarity of the subsequent shock would eliminate the pointing and overlap errors.

Although our accuracy was limited by our method of shock generation, laser driven compressional wave studies have significant advantages over conventional generation schemes especially when used in conjunction with optical probing. These include safety, table top operation, low timing jitter and, most importantly, an experimental repetition rate orders of magnitude higher than either gas gun or explosive shock generation. Thus, laser shock generators, although less precise than gas gun techniques, allow rapid survey of phenomena under far more varied conditions.

In the present study our use of a solute dye is somewhat restrictive as the probe molecule will be insoluble in many systems of interest. However, preliminary work in our lab has indicated that it is possible to dissolve many dyes in suitable resistant plastics. The resulting fluorescence material, once characterized, can be interfaced to optical fibers for use as a generalized temperature probe of opaque as well as transparent solids and liquids.

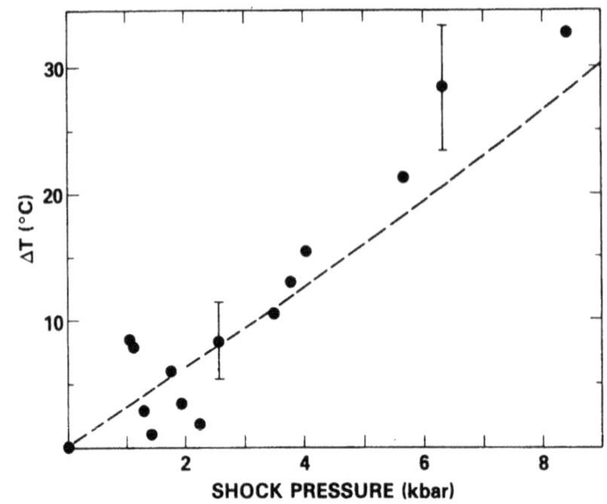

Fig. 3. Measured shock induced temperature rise in water versus shock pressure. Circles are single shot data. The dashed line follows calculated values from Rice and Walsh (Ref. 6).

ACKNOWLEDGEMENTS

This work was partially supported by the Office of Naval Research.

REFERENCES

1. L. Davison and R. A. Graham, "Shock Compression of Solids", Phys. Reports, 55:255 (1979).
2. W. G. von Holle, "Temperature Measurements of Shocked Translucent Materials by Time-Resolved Radiometry", in: "Shock Waves in Condensed Matter-1981", W. J. Nellis, L. Seaman and R. A. Graham, eds., Am. Inst. Physics, New York (1982).
3. D. D. Bloomquist and S. S. Sheffield, "Thermocouple Measurements in Shocked Compressed Solids", J. Appl. Phys., 51:5260 (1984).
4. Z. Rosenberg and Y. Partom, "Direct Measurement of Temperature in Shock-Loaded Polymethylmethacrylate with Very Thin Copper Thermistors", J. Appl. Phys., 56:1921 (1984).
5. A. Delpeuch and A. Menil, "Raman Scattering Temperature Measurement Behind a Shock Wave", in: "Shock Waves in Condensed Matter-1983", J. R. Asay, R. A. Graham and G. K. Straub, eds., North-Holland, Amsterdam (1984).
6. M. H. Rice and J. M. Walsh, "Equation of State of Water to 250 Kilobars", J. Chem. Phys., 26:824 (1957).
7. P. E. Schoen and A. J. Campillo, "Characteristics of Compressional Shocks Resulting from Picosecond Heating of Confined Foils", Appl. Phys. Lett., 45:1049 (1984).
8. A. L. Huston, B. L. Justus and A. J. Campillo, "Spectral Shifts in the Fluorescence Spectrum of Anthracene Under the Influence of Laser Driven Shock", Chem. Phys. Lett. (IN PRESS).
9. P. W. Bridgman, "The Effect of Pressure on the Viscosity of Forty Three Pure Liquids", J. Chem. Phys., 3:597 (1935).

REFLECTIVITY MEASUREMENTS ON SHOCK-UNLOADING SOLIDS

A. Ng, D. Parfeniuk, P. Celliers, and L. DaSilva

Physics Department
University of British Columbia
Vancouver, B.C.m V6T 2A6, Canada

We report on the first quantitative measurement of the reflectivity of a shock unloading solid. The reflectivity was observed to decrease much faster than that predicted by an one-dimensional model. The results suggest that the reflectivity measurement may provide a sensitive means for determining the electrical conductivity (or the electron-ion collision frequency) of a dense plasma.

The physical properties of high-density plasmas, such as the equation of state and transport coefficients, are of great importance to the study of many phenomenon in plasma physics, geophysics, planetary physics, nuclear explosions, and inertial confinement fusion. Equation of state data are generally derived from Hugoniot measurements using high pressure shock waves generated by "deceleration method"[1-6] or by neutrons from nuclear explosions[7]. In these experiments, accurate measurements of the shock speed and particle speed are obtained. Recently, we reported on the measurement of shock speed and temperature in aluminum[8,9] in an laser-generated shock wave experiment. The measured Hugoniot showed excellent agreement with theory and demonstrated its utility in equation of state studies. On the other hand, the electrical conductivity of a non-ideal and non-degenerate plasma has been measured by Ivanov et al.[10] using a probe method. However, measurements of transport coefficients of degenerate plasmas have not been reported.

In this paper, we wish to describe a novel diagnostic technique which may provide a new means for studying the physical properties of a dense, degenerate plasma. This is based on the measurement of the reflectivity of the shock-compressed, free surface of a solid when the shock emerges. By comparing the observed temporal evolution of the reflectivity of the unloading surface with results of numerical simulations, one can examine theoretical models of equations of state as well as transport coefficients. In particular, the electrical conductivity of a dense plasma is found to exhibit a strong effect on its reflectivity.

Figure 1 shows a schematics of the experimental set-up. Planar aluminum targets, typically 15mm x 15mm with thicknesses ranging from 2 5-50μm, were irradiated with a 0.53μm, 2ns (FWHM) laser pulse from a Nd-glass laser. The rear surface of the target foil was carefully polished to near optical quality with a residual surface roughness of ≤ 0.2μm. The laser beam was focussed onto the target front side with f/10 optics at an

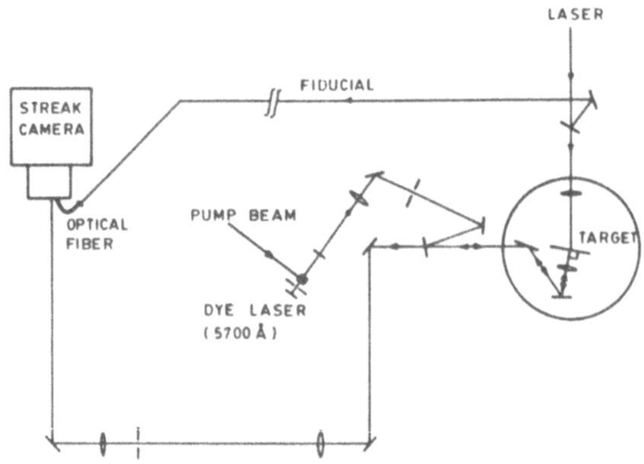

Figure 1. Schematics of the experimental set-up.

incident angle of 10° off target normal. The intensity distribution at focus is nearly Gaussian with 90% of the incident laser energy contained in a spot of 80μm diameter and 60% energy in a spot of 40μm diameter. The irradiance averaged over the central region containing 60% laser energy (Φ_{60}) was thus higher than that averaged over the 90% - energy area (Φ_{90}) by a factor of about 2.7. Earlier measurements of target rear surface luminescence[11] showed that the incident laser pulse produced a planar shock wave over a region of approximately 40μm diameter. The shock wave was also observed to propagate through aluminum targets with thicknesses up to 50μm without exhibiting any deleterious two-dimensional effects. Accordingly, the appropriate target irradiance characterizing the shock generation was Φ_{60}. For the measurements reported here, the maximum absorbed irradiance was $\Phi_{60} < 10^{14}$ W/cm^2. Modulations of the laser intensity at focus have also been characterized. Time-resolved (30ps resolution) measurements showed spatial intensity modulations (4μm resolutions) of <30% whereas time-integrated measurements revealed modulations of <10%.

To measure the reflectivity of shock-compressed aluminum, the target rear surface was illuminated at normal incidence by a dye laser probe beam. This was obtained by optical pumping of Rhodamine 6G using 0.53μm laser light from the Nd-glass laser. The resulting 2ns (FWHM), 0.57μm probe pulse was therefore automatically synchronized with the 0.53μm main beam. The target rear surface was imaged through the focussing lens of the probe beam (Figure 1) onto the entrance slit of a Hamamatsu C1370 streak camera. This yielded reflectivity measurements with a spatial resolution of 3μm and a temporal resolution of 10ps. Figure 2 shows a sample x-t streak record of the rear surface reflectivity near the time of shock breakout. A fiducial of the 0.53μm main laser pulse was also displayed on the streak camera for timing purpose. The shock wave was observed to emerge in a region of approximately 40μm. This is consistent with earlier luminescence measurements[11]. Variation in the shock speed was < 2% across the breakout region. This further verified that the shock front remained reasonably planar propagating through the target.

Also evident in Figure 2 is the rapid reduction in the reflectivity of the shocked target rear surface with the onset of the shock breakout. Quantitative analysis of the data required two calibrations. Absolute calibration of the measured reflectivity level was made in-stu using a first-surface mirror which was positioned at the plane of the target rear surface. To take into account the fine-scale temporal intensity modulations in the dye laser probe pulse, the streak signal in the central region of the shock front (region A in Figure 2) was normalized to that in an unshocked region of the target (region B in Figure 2). Furthermore, to avoid the complication of edge effects, we limited the reflectivity analysis to a central region of 10μm only. This also permitted the interpretation of the

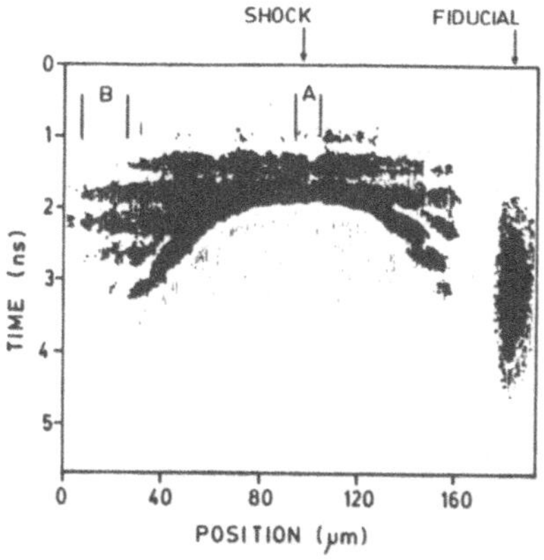

Figure 2. An x-t streak record of the reflectivity of the rear surface of a 50μm thick aluminum target for an absorbed irradiance of 7 x 10^{13} W/cm^2. The fiducial trace was time-delayed by 1.8ns relative to the reflectivity trace.

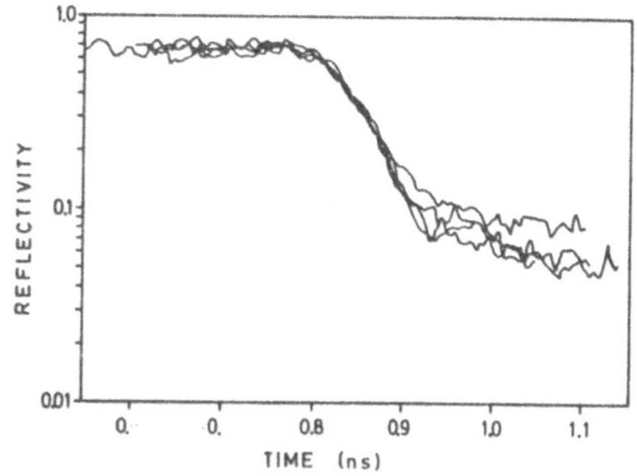

Figure 3. Temporal evolution of the reflectivity of the rear surface of a 50μm thick aluminum target for an absorbed irradiance of 5 x 10^{13} W/cm^2.

features described above. In the actual measurements, a faster streak speed (1.24ns streak time) was used to achieve better time resolution. The temporal evolutions of the rear surface reflectivity of aluminum targets corresponding to a shock speed of (2.2 ± 0.2) x 10^6 cm/s are presented in Figure 3. Each trace represents a single measurement. As the shock emerged, the rear surface reflectivity was observed to decrease nearly exponentially with time. Surface roughness of the target rear surface (typically < 0.2μm) would lead to a variation in the shock breakout time of < 10ps. This was much shorter than the observed time-constant of 150ps in the decrease in reflectivity. It should also be noted that similar results were obtained for the reflectivity measured over a 3μm region at the centre of the shock front. This again demonstrated the planarity of the shock wave observed.

To further understand the experimental results, one needs to examine the process of reflection of electromagnetic radiation by an inhomogeneous plasma. This problem has been treated in detail by Ginzburg[12]. First, let us consider the idealized situation of a shock front arriving at the target rear surface. The density, ρ, and the temperature, T, of the shocked material are determined by its equation of state along the principal Hugoniot. Figure 4 shows the SESAME data[13] for ρ and T as a function of shock speed in aluminum. (In this experiment, the shock speed rather than the shock pressure was determined directly from shock transit time measurements in targets of different thicknesses[8]). Also plotted in Figure 4 is the corresponding d.c. electrical conductivity, σ, from the SESAME data library. At the time of the shock arrival, the reflectivity of the target rear surface will be given by the Fresnel formula[12],

$$R(\omega) = \frac{4\pi\sigma/\omega + 1 - 2\sqrt{2\pi\sigma/\omega}}{4\pi\sigma/\omega + 1 + 2\sqrt{2\pi\sigma/\omega}} \qquad (1)$$

where ω is the frequency of the incident radiation. The Fresnel reflectivity calculated for a shocked aluminum surface is presented in Figure 5. However, direct measurements of such reflectivities are not feasible as the target rear surface will begin to unload as soon as the shock emerges. The incident radiation thus propagates up a density gradient in the unloading plasma.

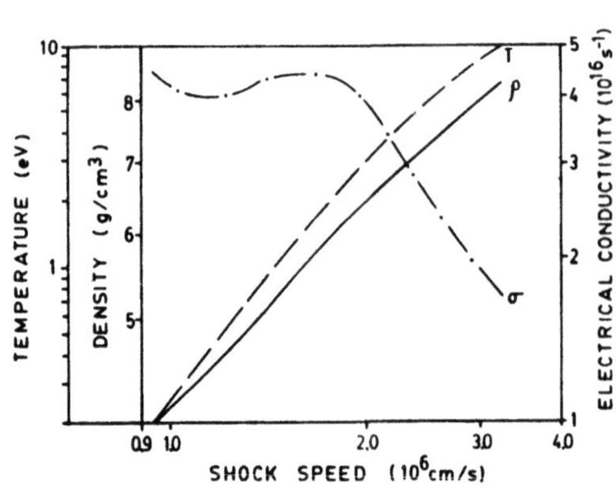

Figure 4. ρ, T and σ as a function of shock speed for aluminum given by the SESAME data.

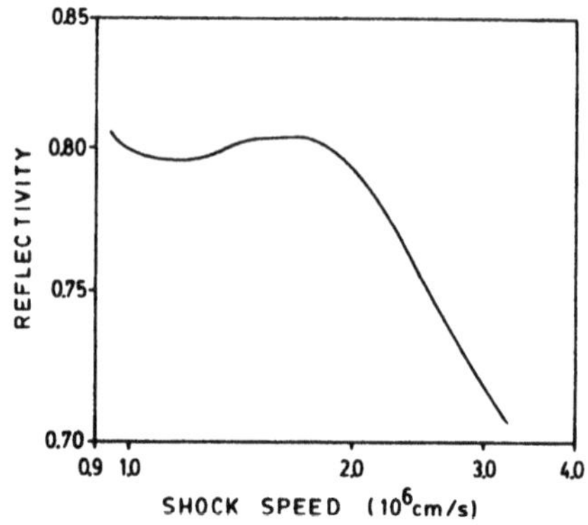

Figure 5. Fresnel reflectivity calculated for a shock aluminum surface as a function of the shock speed.

The hydrodynamics and thermodynamics of the unloading target rear surface were treated in a one-dimensional hydrodynamic code[14] incorpcorating flux-corrected transport[15] and SESAME equation of state[13]. The density and temperature profiles in the unloading wave were computed as a function of time. Since the characteristic unloading speed (isentropic sound speed) is much less than the speed of light, the reflectivity of the plasma at any instant can be determined by solving the wave equation[12] at a fixed time t,

$$\frac{d^2E}{dx^2} + \frac{\omega^2}{c^2}\,\varepsilon(\omega,x,t)\,E = 0 \qquad (2).$$

The dielectric function ε is given by the collisional, cold plasma dispersion relation

$$\varepsilon(\omega,x,t) = 1 - \frac{\omega_p^2(x,t)}{\omega^2}\left[\frac{1}{1 + i\nu_{ei}(x,t)/\omega}\right] \qquad (3)$$

where ω_p is the electron plasma frequency and ν_{ei} is the electron-ion collision frequency which is derived from the SESAME data for electrical conductivity[16] σ and average ionization[16] Z,

$$\nu_{ei}(x,t) = \frac{Z\,e^2\,n_i(x,t)}{m_e\,\sigma(x,t)} \qquad (4)$$

Exact solution of Equation (2) has not yet been obtained. However, when the density gradient scale length of the unloading plasma becomes larger than the wavelength of the incident light, propagation of the incident radiation in the plasma can be treated using the approximation of geometric optics (WKB) and the reflected radiation taken as a perturbation of this solution[12]. The reflectivity is then given by

$$R \cong \left\{ \tanh \left| \int_{-\infty}^{\infty} \frac{1}{2n}\frac{dn}{dx} \exp\left(-\frac{2i\omega}{c}\int_{-\infty}^{x} n(y)dy\right) dx \right| \right\}^2 \qquad (5)$$

where the refractive index $n = \sqrt{\varepsilon}$. This reflectivity calculation (Equation 5) is coupled to the hydrodynamic simulation as a post-processor. Details of the simulations will be published elsewhere. The predicted temporal evolution of the rear surface reflectivity for a shock speed of 2.2×10^6 cm/s in aluminum is indicated by the dot-dashed curve in Figure 6. Clearly, the calculated rate of decrease of the reflectivity is much slower than that observed in the experiment (Figure 3).

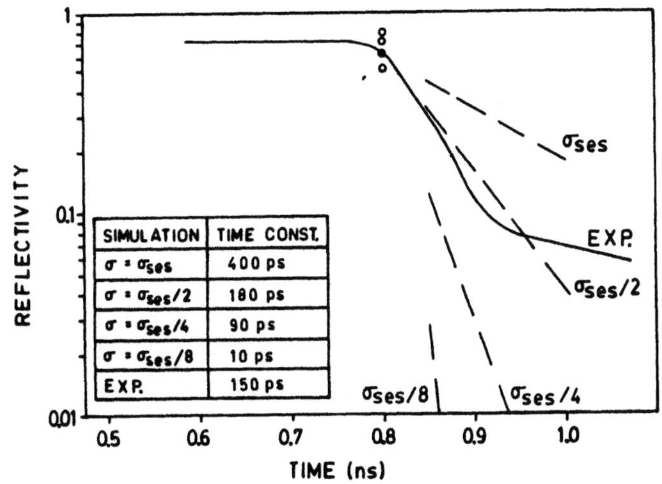

Figure 6. Calculated rear surface reflectivity for a shock speed of 2.2 x 10^6 cm/s in aluminum for different electrical conductivities. The open circles indicate the corresponding Fresenel reflectivities. Also shown are the time constants for the approximately exponential decrease of reflectivity with time.

From Equations (3), (4) and (5), it can be seen that the reflectivity depends on the density profile (and therefore $\omega_p(x)$, $n(x)$) of the unloading wave, as well as the electron-ion collision frequency ν_{ei}. The density profile is determined by the equation of state used in the hydrodynamic simulation. In this calculation, ν_{ei} is determined by the average ionization Z and the electrical conductivity σ. Earlier, it has been shown[8] that the measured shock speed and temperature Hugoniot in aluminum is in good agreement with calculations using the SESAME equation of state and ionization data. Therefore, one may conjecture that the apparent descrepancy between simulation and experimental results for the reflectivity arises from the values of σ (and hence, ν_{ei}) used. To examine the effect of σ on reflectivity, simulations were made using $\sigma = \sigma_{ses}/2$, $\sigma_{ses}/4$, and $\sigma_{ses}/8$, where σ_{ses} is the conductivity given by the SESAME data library. The results are shown in Figure 6. Evidently, the time rate of decrease in the reflectivity is strongly sensitive to the electrical conductivity of the unloading plasma. By comparing with the experimantal data (Figure 3), this also indicates that the SESAME conductivity might have been overestimated by a factor of 2 to 4. It is indeed most interesting to note that the σ given by the SESAME data is several times higher than that calculated recently by Lee and More[17].

In conclusion, we have obtained the first quantitative measurement of the reflectivity of a shock unloading plasma. The reflectivity was observed to decrease much faster than that predicted by an one-dimensional model using the SESAME equation of state and electrical conductivity data. The result suggested that the lower electrical conductivity calculated from the model of Lee and More might be more appropriate. More importantly, it is seen that the reflectivity measurement may provide a sensitive means for determining the electrical conductivity (or the electron-ion collision frequency) of a dense plasma.

Acknowledgement

This work is supported by the Natural Sciences and Engineering Research Council, and B.C. Hydro and Power Authority.

References

1. L.V. Al'tshuler, K.K. Krupnikov, B.N. Ledenev, V.I. Zhuchiklin, and M.I. Brazhnik, Sov. Phys. - JETP 7, 606 (1958).
2. L.V. Al'tshuler, A.A. Bakanova, and R.F. Trunin, Sov. Phys. - JETP 15, 65 (1962).
3. K.K. Krupnikov, A.A. Bakanova, M.I. Brazhnik, and R.F. Trunin, Sov. Phys. - Dokl. 8, 205 (1963).
4. A.H. Jones, W.M. Isbell, and C.J. Maiden, J. Appl. Phys. 37, 3493 (1966).
5. J.A. Morgan, High Temperature - High Pressures 6, 195 (1974).
6. A.C. Mitchell and W.J. Nellis, J. Appl. Phys. 52, 3363 (1981).
7. C.E. Ragan III, M.G. Silbert, and B.C. Diven, J. Appl. Phys. 48, 2860 (1977).
8. A. Ng, D. Parfeniuk, and L. DaSilva, Phys. Rev. Lett. 54, 2604 (1985).
9. A. Ng, D. Parfeniuk, and L. DaSilva, Opt. Commun. 53, 389 (1985).
10. Yu. Y. Ivanov, V.B. Mintsev, V.E. Fortov, and A.N. Dremin, Sov. Phys. - JETP 44, 112 (1976).
11. A. Ng, D. Parfeniuk, L. DaSilva, and D. Pasini, Phys. Fluids, September, 1985 (in print).
12. V.L. Ginzburg, The Propagation of Electromagnetic Waves in Plasmas, Chapter IV, (Pergamon Press, Oxford, 1970).
13. SESAME Library, LANL, Los Alamos, Material Number 3712.
14. L. DaSilva, A. Ng and D. Parfeniuk, Bull. Amer. Phys. Soc. 29, 1325 (1984).
15. D.L. Book, J.P. Boris, and K. Hain, J. Computational Phys. 18, 248 (1975).
16. SESAME Library, LANL, Los Alamos, Material Number 23713.
17. Y.T. Lee, and R.M. More, Phys. Fluids 27, 1273 (1984).

SHOCK INDUCED RADIATION FROM MINERALS

Douglas Schmitt, Bob Svendsen and Thomas J. Ahrens

Seismological Laboratory
California Institute of Technology
Pasadena, CA 91125

ABSTRACT

The shock-induced optical radiation from Al_2O_3, MgO, NaCl, KCl, x-cut and fused SiO_2, and LiF to 75 GPa may be characterized as thermal (greybody), characteristic (band), or some combination. During propagation through a mineral, emitted radiation (430 to 870 nm) is collected on a 500 channel, charge-couple detector spectrometer. In addition, framing camera photographs are taken and a photodiode records radiation intensity versus time. MgO, which does not undergo any resolvable shock-induced phase change below 200 GPa, displays a nonthermal band spectrum centered at 700 nm between 40 and 62 GPa. At 75 GPa, Al_2O_3, which also undergoes no known shock-induced phase change, displays nonthermal radiation centered at 500 nm, possibly combined with greybody radiation from sources with a temperature ~3000°K. X-cut SiO_2 displays band emissions centered between 500 and 550 nm combined with thermal radiation in the 6-16 GPa range. The processes responsible for nonthermal radiation are presently not well-understood, but the possibilities include triboluminescence due to microfracture and bond-breaking during brittle shock deformation. Fused SiO_2, NaCl, KCl and LiF exhibit low emissivity greybody radiation corresponding to temperatures much greater than calculated continuum (thermostatic) temperatures at high pressure. Framing camera photographs, taken simultaneously with the radiation spectrum, reveal a heterogeneous radiation field in x-cut SiO_2 at 7.6 GPa, fused SiO_2 at 9.8 GPa, and a nearly uniform radiation field in NaCl near 29 GPa, correlating with low (~0.02) and high (~0.5) effective emissivities, respectively. An important new result is that LiF, and perhaps Al_2O_3, are unique among all materials studied in that they emit heterogeneous greybody radiation but do not undergo any known solid-solid phase change in the range of pressure covered by our experiments. Greybody thermal radiation is probably due to localized inelastic deformation, producing locally high dissipation, temperatures, and in some cases melting. If the local rate of melting exceeds the local rate of dissipation, temperatures inferred from these radiation measurements should represent melting temperatures at high pressure.

INTRODUCTION

Previously, 500 channel charge-coupled detector spectrometry (Kondo and Ahrens[1]) of the light emitted upon propagation of shock waves through a number of initially transparent minerals implied most of these minerals emit radiation consistent with a greybody (thermal) source. Taking into account previous studies, Schmitt and Ahrens[2], following a suggestion by Grady[3], concluded that the type of optical emission produced during shock compression was dependent upon phase changes taking place during shock compression. In this paper, we present new observations of shock-induced optical radiation from Al_2O_3, MgO, NaCl, KCl, x-cut and fused SiO_2, and LiF at various pressures up to 75 GPa.

EXPERIMENTAL

Our experimental setup is similar to that of Schmitt and Ahrens[2]. A TRW Image Converter Camera with model 28B 3-frame plug-in unit has been added to take two or three 50 to 500 nsec-long exposures of the radiation field during shock wave propagation through the sample. These photographs reveal the structure of radiation-producing deformation and its degree of heterogeneity.

We used cut, polished and oriented samples of x-cut and fused SiO_2, MgO, Al_2O_3, NaCl, KCl, and LiF in these experiments. The impact side of all samples was silver-coated to minimize the intensity of recorded radiation from the sample-driver interface.

RESULTS AND DISCUSSION

Consistent with the results of Schmitt and Ahrens[2], single-crystal MgO displays an apparently nonthermal, broad-band spectrum centered at 700 nm which is insensitive to pressure in the 40-62 GPa range. We speculate that this radiation is due to Mn^{++} substituting for Mg^{++} in MgO (Rossman[4]) or previously unobserved triboluminescence (Walton[5]). A single experiment to 75 GPa on single-crystal Al_2O_3 showed that it emits nonthermal radiation centered at 525 nm possibly combined with greybody radiation from source(s) at ~3000°K; however, more experiments are required to substantiate or discredit this possibility. MgO and Al_2O_3 do not undergo any known, shock-induced phase change below 200 GPa[6] and 150 GPa[7], respectively.

Spectra from experiments #611 (10.6 GPa) and #615 (7.6 GPa) on x-cut SiO_2 presented in Fig. 1 appear to represent nonthermal radiation centered between 500 and 550 nm combined with greybody radiation. However, the shape of the spectra rule out an accurate determination of a greybody temperature. Previous work (Brannon et al.[8]) on x-cut SiO_2 in the pressure range 4.4 to 28.0 GPa showed the existence of two radiation peaks at 400 and 600 nm which apparently correlate with peaks at 430 and 630 nm in the spectra of silica fibers due to triboluminescence (Zink et al.[9]). The peak we observed between 500 and 550 may correlate with the 630 nm triboluminescence peak, possibly shifted by combination with greybody radiation in our observations. Fast framing photographs of x-cut SiO_2 (Fig. 2) exhibit a radiation field with numerous bright spots indicative of heterogeneous shock deformation in agreement with Brannon et al.[8]

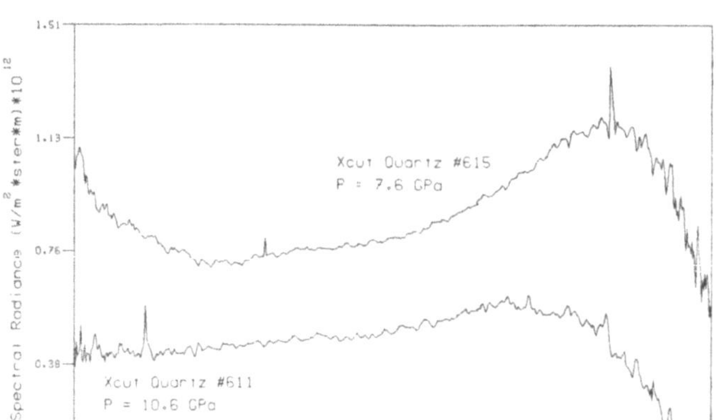

Figure 1: Shock Induced Spectra of X-cut SiO_2 at 10.6 GPa (#611) and 7.6 GPa (#615). Shape of spectra appears to be greybodies overlain by an intense characteristic emission which peaks between 500 and 550 nm making temperature calculation impossible.

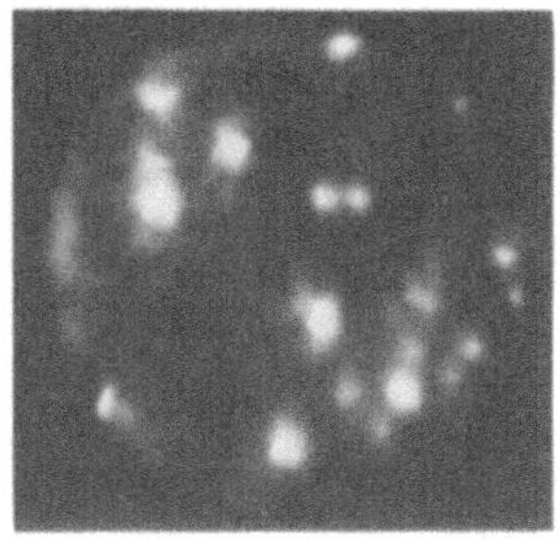

Figure 2: Framing camera photograph of x-cut SiO_2 at 7.6 GPa (#615) indicating heterogeneous shock deformation. Exposure was 100 ns starting 50 ns after sample shock entry.

Data for minerals emitting dominantly thermal radiation during shock compression are summarized in Table 1. In particular, two experiments on NaCl, #643 and #646 (Fig. 3) show a greybody response combined with line emissions centered at 770 nm. These line emissions may be due to impurities within the NaCl. The best fit greybody curves are plotted with the data. Note that temperatures inferred from these observations are ~1.5 to 10 times the corresponding calculated thermostatic (continuum) temperature (Table 1).

Table 1: Radiative characteristics of fused SiO_2, NaCl, KCl, and LiF.

Shot #	Sample	P (GPa)	$T_{experiment}$ (K)	ϵ	$T_{thermostatic}$ (K)
613	NaCl	25.0±0.5	3429	0.009	1180
646	NaCl	27.1±0.5	2216	0.319	1300
645	NaCl	28.0±0.7	3409	0.082	1370
612	NaCl	30.0±0.9	4313	0.029	1500
643	NaCl	31.4±0.8	2520	0.213	1590
652	KCl	14.6±0.3	3641	0.002	1250
640	KCl	19.9±0.5	3128	0.070	1890
651	KCl	30.0±0.6	3923	0.073	2950
644	LiF	42.0±1.0	2715	0.003	670
653	LiF	47.1±1.2	3152	0.005	750
634	Fused SiO_2	9.8±0.2	3938	0.023	400
635	Fused SiO_2	14.1±0.2	3462	0.006	480

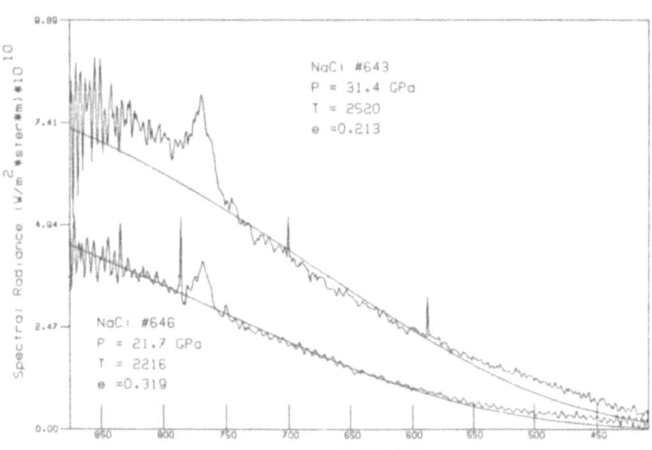

Figure 3: Greybody spectra of NaCl at 31.4 GPa #643) and 21.76 GPa (#646). Line emission at 770 nm is superimposed.

In most cases, the radiation intensity was too low to be photographed by the framing camera. Photographs of fused SiO_2 (shot #634, Table 1, not shown) display a spotty, heterogeneous radiation field similar to that shown for x-cut SiO_2 shot #615, thus implying that heterogeneous deformation also occurs in amorphous SiO_2. However, photographs of NaCl (shot #643 and #646, also not shown) exhibit a uniform radiation field, implying that localized deformation involves more of the sample as shear band density increases with pressure.

The results of the shock compression experiments conducted on single crystal NaCl are plotted (Fig. 3) in the temperature-pressure plane together with the results of previous studies. The inferred temperatures increase from 22 to 30 GPa, at which point a large drop followed by a general rise is seen. NaCl transforms from the B1 to B2 phase at 29 GPa and room temperature (Sato-Sorensen[10]) under static conditions, and presumably undergoes the same transition between 26 to 30 GPa (Ahrens et al.[11]) during shock compression. Grady et al.[12] presents a model of shear band deformation consisting of a region of high temperature melt surrounded by a zone of high pressure solid phase in turn surrounded by relatively undeformed low pressure phase. Assuming the rates of melting and formation of the new solid high pressure phase exceeds the rate of dissipation, this model implies that the inferred temperatures presented in Fig. 3 can be interpreted as the melting temperatures of the B1 phase of NaCl below 30 GPa and of the B2 phase of NaCl above 30 GPa.

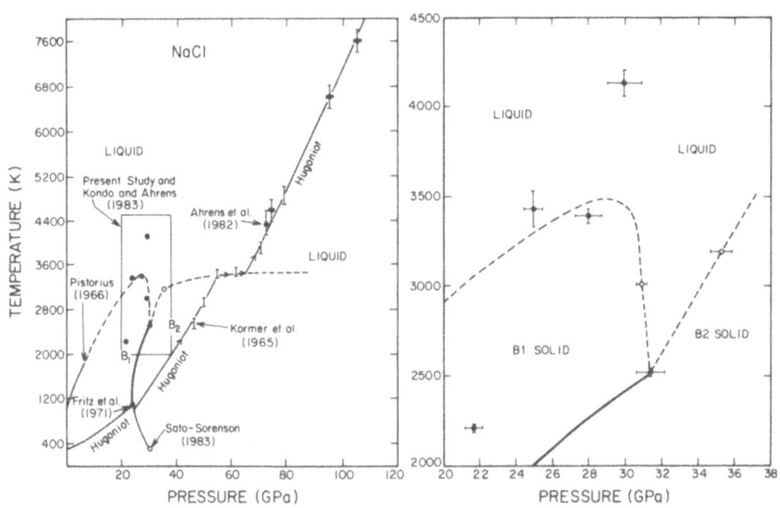

Figure 4: a) Proposed P-T diagram for NaCl based on present study and Sato-Sorenson[6], Kondo and Ahrens[2], Ahrens et al.[5], Fritz et al.[11], Pistorius[12], and Kormer[13]. Data of present study and Kondo and Ahrens[2] is in box. Dashed curve represents melting line. Heavy solid line represents B1-B2 boundary. Light dashed line represent estimated and observed continuum temperature along Hugoniot.
b) Expanded view of boxed area in a), (present study -- closed circles, Kondo and Ahrens[2] -- open circles).

CONCLUSIONS

1) Greybody emissions observed in LiF, which undergoes no known phase transition, imply that localized heating and perhaps melting occurs in this material during shock deformation. Consequently, a potential solid-solid phase change and resulting mechanical instability under dynamic conditions may not be prerequisite to the initiation of localized deformation as suggested previously (Schmitt and Ahrens[2]; Grady[3]). Since LiF emits the least-intense radiation of all thermal radiators studied so far, it probably develops fewer deformation features and/or less intense radiation sources at a given pressure during shock compression.

2) The framing camera photograph of fused SiO_2 #634 and simultaneous greybody spectrum indicate that inferred high temperatures correlate with heterogeneous shock deformation.

3) The new data for NaCl bounding the B1-B2 phase-transition region suggest that radiation measurements are a sensitive tool for the delineation of phase transformations and inference of the melting temperature with pressure.

Acknowledgements. The authors wish to thank E. Gelle, M. Long, C. Manning, L. Young, and L. Blayney for much needed technical assistance. J. Tyburczy, S. Rigden, and G. Rossman provided encouragement and ideas. This work supported under NSF Grant EAR-8313746 and NASA Grant NGL-05-002-105.

REFERENCES

1. K. E. Kondo and T. J. Ahrens,Heterogeneous shock-induced thermal radiation in minerals, Phys. Chem. Minerals 9:173 (1983).
2. D. R. Schmitt and T. J. Ahrens, Emission spectra of shock compressed solids, Geoph. Res. Lett. 10:1077 (1983).
3. D. E. Grady, Shock deformation of brittle solids, J. Geophys. Res. 85:913 (1980).
4. G. Rossman, personal communication.
5. A. J. Walton, Triboluminescence, Adv. Phys. 26:887 (1978).
6. M. S. Vassiliou and T. J. Ahrens, Hugoniot equation-of-state of periclase to 200 GPa, Geophs. Res. Lett. 8:729 (1981).
7. Al_2O_3, in: "LASL shock hugoniot data", S. P. Marsh, ed., University of California Press, Los Angeles (1980).
8. P. J. Brannon, C. H. Konrad, R. W. Morris, E. D. Jones and J. R. Asay, Spectral and spatial studies of shock-induced luminescence from quartz, Sandia Report SAND82-2469, (1983) 85 pp. .
9. J. I. Zink, W. Beese, and J. W. Schindler, Triboluminescence of silica core optical fibers, Appl. Phys. Lett. 40:110 (1982).
10. Y. Sato-Sorenson, Phase transitions and equations of state for the sodium halides: NaF, NaCl, NaBr, and NaI, J. Geoph. Res. 88:3543 (1983).
11. T. J. Ahrens, G. A. Lyzenga and A. C. Mitchell, Temperatures induced by shock waves in minerals: application to geophysics, in: "High Pressure Research in Geophysics," Advances in Earth and Planetary Sciences, V. 12, S. Akimoto and M. H. Manghnani, eds., Center for Academic Publications, Japan (1982).
12. D. E. Grady, W. J. Murri and P. S. DeCarli, Hugoniot sound velocities and phase transformations in two silicates, J. Geoph. Res. 80:4857 (1975).
13. J. N. Fritz, S. P. Marsh, W. J. Carter, and R. G. McQueen, The Hugoniot equation of state of sodium chloride in the sodium chloride structure, in: "Accurate Characterization of the High Pressure Environment", E. C. Lloyd, ed., Nat. Bur. Stand. Spec. Publ. 326, Washington D. C. (1971).
14. C. W. F. T. Pistorius, Effect of pressure on the melting points of the sodium halides, J. Chem. Phys. 45:3513 (1966).
15. S. B. Kormer, M. V. Sinitsyn, G. A. Kirillov and V. D. Urlin, Experimental determination of temperature in shock-compressed NaCl and KCl and of their melting curves at pressures up to 700 kbar, Sov. Phys. JETP 21:689 (1965).

MOLECULAR DYNAMICS STUDIES OF PLASTIC FLOW AT HIGH STRAIN RATES

Anthony J. C. Ladd
Physics Department
Lawrence Livermore National Laboratory
Livermore, California 94550

ABSTRACT

Non-equilibrium molecular dynamics techniques have been developed to investigate high strain rate plastic flow in simple two and three dimensional crystals. In these simulations the long-time, steady state, values of the stress tensor and internal energy are computed as a function of the independent variables, temperature and strain rate, which are constants of the motion. Our results show that the shear stress varies as a non-integer power of the applied strain rate. The power is approximately proportional to the temperature and reaches a value of about 0.5 just before melting.[1]

The molecular dynamics results can be at least semi-quantitatively reproduced by a model which involves the creation, motion and annihilation of edge dislocations. It is significant that the dislocation parameters in this model are all determined from separate calculations and represent closely the exact properties of dislocations in the material.

Our calculations are consistent with experimental data and suggest that the flow of close-packed metals is described by a single physical mechanism over a range of strain rates from 10 kHz to 1 THz.

INTRODUCTION

A fundamental understanding of the flow of crystalline solids can only emerge from models that incorporate the creation, interaction, motion and annihilation of defects. This work focuses on edge dislocations which can be created in large numbers during plastic deformation and which are the dominant contribution to plastic strain in shock loaded solids. The interaction of dislocations is described by linear elastic theory, except at small separations, but the creation, motion and annihilation of dislocations can only be understood by considering the underlying atomic structure of the crystal.

Atomistic computer simulations solve classical equations of motion for a given interparticle force law, generally pairwise additive. Although the interparticle forces are of necessity somewhat simplified from those existing in real materials, these simulations have considerably enhanced our understanding of the equation of state and, more recently, the transport properties of fluids.[2] Such simulations, applied to solids automatically include the essentials of defect motion, creation and annihilation, which are not obtainable in a fundamental way from a macroscopic continuum model. A better understanding of the link between microscopic defect physics and crystal plasticity should lead to improved constitutive modeling.

Computer simulations are limited to regions of linear dimension less than 100 atomic spacings and to times less than 10,000 vibration periods. These limitations make atomistic simulations unfeasible for plastic strain rates less than 10 GHz. Although much higher rates are undoubtedly reached, in strong shock waves[3] (> 1 Mbar), measurement of wave profiles by optical laser interferometry is limited to rise times greater than 1 nsec[4], i.e., strain rates less than 0.1 GHz, although x-ray lasers would make measurements at higher rates possible.

The scale gap, between microscopic computer experiments and macroscopic laboratory experiments, makes it worthwhile to develop "intermediate scale" simulations in which the defects themselves are the fundamental particles replacing the underlying atomic coordinates and momenta.[5] These "defect dynamics" simulations only have meaning if the assumed defect properties are firmly supported by atomistic calculations. Some encouraging preliminary results of a "dislocation dynamics" simulation are reported. Calculations in a similar spirit have been used to study vortex flow in fluids and combustion in turbulent flames.[6] In these calculations the "defects" are point vortexes which are used to represent the non-laminar flow field.

The Hooke's-Law Triangular Lattice--a Standard Plastic Material

In attempting to characterize and understand plastic flow, it is useful to have a standard material for study, both simple and reproducible. In atomic simulations of fluids, the hard-sphere and Lennard-Jones potential have generally been used in this role. Results of these calculations serve to highlight the universal features of liquids that are independent of the exact details of the interparticle forces, and provide a rigorous test of theoretical attempts to describe the same phenomena. In addition, the methodology developed to study simple systems is directly transferable to more complicated models. We have focused on a two-dimensional triangular lattice in which nearest-neighbors interact via truncated Hookes-Law forces. More complex models involving anharmonic forces and a three-dimensional crystal lattice are considered briefly.

The essential features of plastic flow, the creation, interaction, motion and annihilation of edge dislocations are contained in a two dimensional triangular lattice without the geometrical complexities involved in three dimensional crystals. The possible mechanisms of dislocation creation and annihilation are shown in Fig. 1.

The lattice is mechanically stable with just near-neighbor central forces and is elastically isotopic. Pairwise additive, piecewise linear interparticle forces are the simplest and are used in the bulk of this work. The zero-force separation is d and the force constant is k. The attractive force is greatest at $r = d + w$ and decreases linearly to zero at $r = d + 2w$. In the numerical results described here, $w = 0.15\ d$.

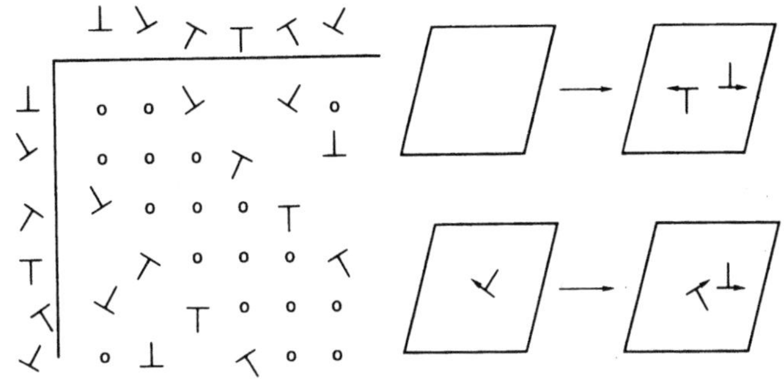

Fig. 1. Dislocation reactions in a triangular lattice There are three glide directions in a triangular lattice orientated at angles of 120°, 240°, and 360°, and thus six orientations for edge dislocations. The dislocation reaction matrix is shown on the left-hand side of the figure. The reaction can proceed in either direction. A zero indicates that it is not a geometrically possible reaction and a blank space indicates a perfect lattice. The right-hand side shows dislocation production in a sheared crystal. The direction of the shear is the one that is most favorable for the reactions shown. The arrows on the dislocations indicate the direction of glide of the dislocations if the external shear is larger than the attractive force between the pair.

Energy and Entropy of Interacting Dislocations

It is a reasonable assumption that the rate of formation of dislocations in a plastically flowing crystal is related to their free-energy of formation. This energy is balanced by the added plastic strain produced by the motion of the extra dislocations. Although dislocation creation is an atomic phenomenon, elastic theory provides an expression for the energy of an array of n edge dislocations. In two-dimensions for an elastically isotropic cryal, at constant pressure, it is

$$E_D = n\,E_c + D \sum_{i>j}^{n} -(\underline{b}_i \cdot \underline{b}_j)\ln(r_{ij}\phi/b) + (\underline{b}_i \cdot \underline{r}_{ij})(\underline{b}_j \cdot \underline{r}_j)/r_{ij}^2 \tag{1}$$

where b is the Burger's vector, E_c is the core energy and D is given in terms of the Lame constants (λ,η), by

$$D = (\eta/\pi)(\lambda+\eta)/(\lambda+2\eta) \tag{2}$$

The creation and annihilation of dislocations in the bulk of the crystal, away from free surfaces and grain boundaries, requires conservation of the total Burgers vector $B = \Sigma b_i$. The contribution to the energy of the term DBlnV, where V is the system size, is unchanged and therefore omitted from Eq. 1. The core energy E_c depends on the lattice structure and the interparticle forces, and cannot be calculated from elasticity theory. It can be determined by comparing the energies of atomic crystals containing dislocations with the predictions of continuum mechanics.[7] Consistency in the core energies for several different dislocation arrangmeents both demonstrates the validity of the elastic theory result (Eq. 1) and provides an estimate of E_c. The energy of crystals containing a single dislocation (or any other arrangement with $B \neq 0$) diverges logarithmically with system size and contains a constant term that depends on the exact nature of the imposed boundary conditions. Thus a calculation of the core energy requires arrangements of dislocations with a total Burgers vector B equal to zero. Periodic boundary conditions work best in these calculations. Although the dependence of the dislocation energy, E_D, on the number of atoms in the crystal is larger with periodic boundary conditions than with free boundaries, this number dependence can be accurately predicted by elastic theory.[7] Test calculations with both sets of boundary conditions gave similar dislocation energies in the large system limit.

Initial conditions are obtained by using linear elastic displacement fields. The crystals are relaxed to the equilibrium configuration (with forces less than 10^{-8}kd) with equations of motion that include viscous damping

$$\ddot{x} = F/m - \lambda\dot{x} \tag{3}$$

with $\lambda \sim (kd^2/mL^2)^{1/2}$ where F is the force, m is the atomic mass and L is the side length of the crystal. The two dislocation arrangements used are shown in Fig. 2.

We calculated dislocation energies for these arrangements for dislocation separations between 2b and 12b, with and without external shear strain, for crystals of about 10^2 to 10^3 atoms. In the cases examined, about 50 in

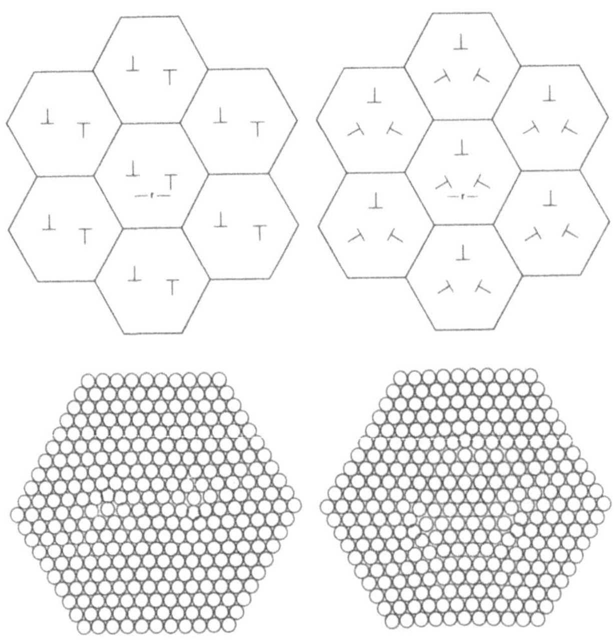

Fig. 2. Periodic arrays of dislocations. The upper half of the figure shows seven cells of the infinite periodic array. The dislocation separation is indicated. The lower half of the figure shows typical arrangements in fully relaxed unit cells.

all, elastic theory, suitably modified for finite size crystals with periodic boundary conditions, predicted the dislocation energy and the average stress in the crystal to within less than 1% for all except the smallest system sizes. The core energy, E_c, was found to be 0.26 ηb^2 for the triangular lattice with Hooke's-Law forces. For the case of the two-dimensional Coulomb crystal, the core energy is about 0.4 ηb^2. Results for three dimensional crystals have yet to be obtained.

The entropic contribution to the free energy of formation of dislocations has two parts. The configurational part is basically a combinatorial problem, but there is a vibrational part as well arising from the shift in normal mode frequencies caused by the dislocations. Our calculations of this shift indicate that this is a negligible contribution, about 1% of the free energy at the melting point. Thus the core energy, together with the elastic theory result of Eq. 1 is sufficient to calculate the free energy of formation of dislocations in this lattice.

Motion of Edge Dislocations

The motion of dislocations is still mysterious and no theory exists that can predict the velocity at which a dislocation will propagate in a sheared crystal lattice. Molecular dynamics simulations have been used to examine the propagation of edge dislocations in the triangular lattice in a quasi-experimental fashion.[8]

The calculations were carried out for long strips containing a single edge islocation in a fully relaxed crystal (Fig. 3). Dislocation motion can be initiated by applying either a constant force (Fig. 3) or a constant displacement boundary condition. The latter is easier to implement, especially with flat boundaries. It has been found that dislocations reach a steady velocity from rest over distances less than two lattice spacings.[8] The time spent in non uniform motion is short and an adequate equation of motion for dislocations expresses the propagation velocity as a function of the imposed shear strain. Results for both the Hookes-Law and Lennard-Jones potential are shown in Fig. 4.[8]

A finite strain is required to initiate dislocation motion. This strain, the Peirels strain, results from the atomic character of the lattice

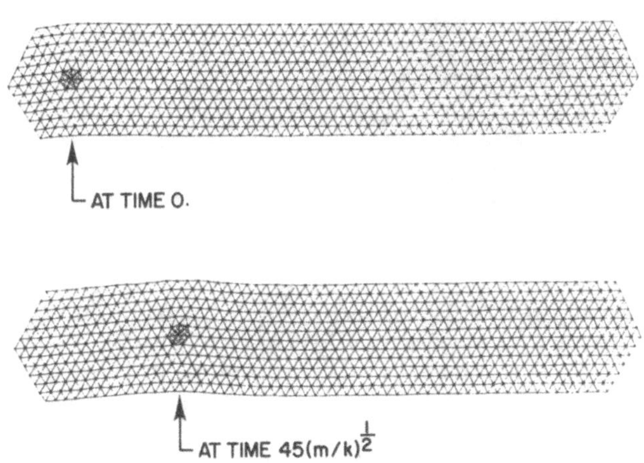

Fig. 3. Dislocation propagation for a Hookes-Law potential with w = 0.18 d, a constant stess 0.04 k. Rayleigh waves are generated by the moving dislocation.

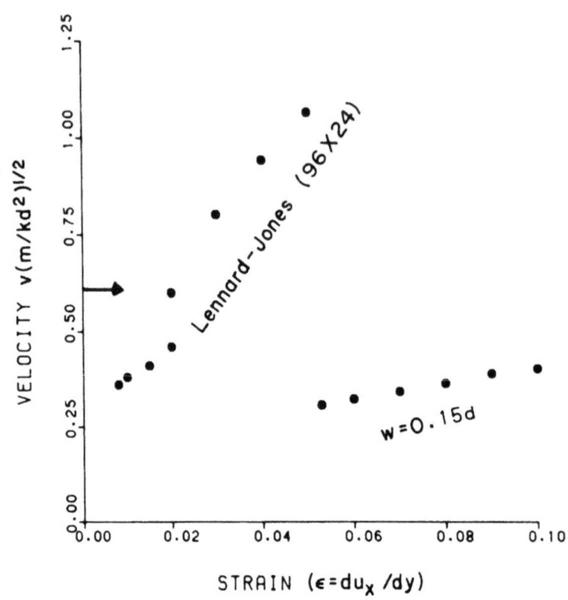

Fig. 4. Velocity versus strain for Lennard-Jones and Hookes-Law potentials, using fixed displacement, flat boundary conditions. The horizontal arrow indicates the transverse wave speed $c_t = 0.612\ (kd^2/m)^{1/2}$.

and is dependent on the interparticle forces. For the Hooke's-Law potential it is very large, about 0.05, but it is an order of magnitude smaller for the anharmonic Lennard-Jones potential. In real three-dimensional crystals it is often orders of magnitude smaller still. The dislocation propagation velocity in crystals with linear forces is only weakly dependent on the shear strain once the rather large Peirel's barrier is overcome. The dislocations move at about one-half the transverse sound speed. For the Lennard-Jones potential the velocities are higher and the behavior is more complex. The velocity undergoes a transition from subsonic to supersonic speeds at a strain of 0.02. None of these results are well represented by a linear or exponential dependence of the velocity on the shear strain.

Nonequilibrium Molecular Dynamics

Initial studies of transport processes in fluids by molecular dynamics simulation, focused on small deviations from equilbrium. The well known Green-Kubo fluctuation formulae were used to calculate the linear transport coefficients (shear viscosity, bulk viscosity and thermal conductivity) for hard-sphere and Lennard-Jones liquids. Subsequently methods were developed to set up non-equilibrium fluxes of momentum and energy in the system and measure the transport coefficients directly.[2] The early reservoir methods were replaced by homogeneous techniques, which use periodic boundary conditions in all directions, thus minimizing the number dependence of the results. The non-equilibrium methods were used to provide an independent check on Green-Kubo results and to probe the non-linear behaviour of fluids. The most recent methods set up nonequilibrium fluxes by applying perturbations to the Hamiltonian equation of motion.[2,9] In addition to being operationally convenient, these new "nonequilibrium equations of motion" have enabled some theoretical studies of systems far from equilibrium to be made. Although these results are somewhat restricted and less rigorous than equilibrium and linear transport theory, they provide evidence of the fundamental soundness of these nonequilibrium methods.[10]

If we define the "momentum" of a particle p, relative to the average local velocity, v, then the equation of motion for the coordinates of a system undergoing an homogeneous deformation is

$$\dot{\underline{q}} = \underline{p}/m + \dot{\underline{\underline{\varepsilon}}} \cdot \underline{q} \qquad \dot{\underline{\underline{\varepsilon}}} = \left[\underline{\nabla}\, \underline{v}\, (r)\right]^{t} \tag{4}$$

where the unsymmetrized strain rate tensor, $\dot{\varepsilon}$, is independent of location, though possibly time dependent. In molecular dynamics simulations of systems with periodic boundary conditions, images of particles in different unit cells are displaced relative to one another, but the "momenta" defined in Eq. 4 are the same for each image.

For a system of particles interacting with pairwise additive forces, the microscopic variables corresponding to the temperature, internal energy and stress tensor in continuum mechanics are

$$T = \frac{1}{3Nk_B} \sum_i \frac{p_i^2}{m}$$

$$E = \sum_i \frac{p_i^2}{2m} + \sum_{i>j} \phi_{ij} \tag{5}$$

$$\underline{\underline{\sigma}} = -\left\{\frac{1}{V} \sum_i \frac{\underline{p}_i \underline{p}_i}{m} + \sum_{i>j} \underline{q}_{ij} \underline{F}_{ij}\right\}$$

where k_B is Boltzmann's constant, $\phi_{ij} = \phi(q_{ij})$ is the pair potential and $\underline{F}_{ij} = -\nabla_{\underline{q}_{ij}} \phi_{ij}$ is the force. The velocities p/m appearing in these expressions, are the velocities of the particles relative to the local macroscopic stream velocity.

In an adiabatic volume change the temperature varies. It is natural, therefore, to expect any deformation of the coordinates to be accompanied by a change in "momentum". The equation of motion

$$\dot{\underline{p}} = \underline{F} - \dot{\underline{\underline{\varepsilon}}} \cdot \underline{p} \tag{6}$$

is consistent with microscopic analogue of the first law of thermodynamics,

$$\dot{E} = V\underline{\underline{\sigma}} : \dot{\underline{\underline{\varepsilon}}} \tag{7}$$

For flows containing a rotational component, i.e., plane couette flow, there is an alternative equation of motion also consistent with the first law,

$$\dot{\underline{p}} = \underline{F} - \underline{p} \cdot \dot{\underline{\underline{\varepsilon}}} \tag{8}$$

It turns out that either of these equations of motion lead to correct shear stresses, but the correct normal stress effects are only obtained from the first choice (Eq. 6).[10] The errors resulting from using the second set of equations show up in the kinetic contribution to the stress tensor and are thus small for solids and dense fluids.

These equations of motion convert external work into internal energy (heat) and it is convenient to use a thermostat to maintain a constant temperature. This can be done by adding a velocity dependent force

$$\dot{\underline{p}} = \underline{F} - \dot{\underline{\underline{\varepsilon}}} \cdot \underline{p} - \xi \underline{p} \tag{9}$$

where the friction coefficient ξ is a dynamical variable chosen to ensure that either $\dot{T}$ or $\dot{E}$ is equal to zero.[1,11] To maintain a constant internal energy for instance,

$$\xi = V\underline{\underline{\sigma}}:\dot{\underline{\underline{\varepsilon}}} \;/\; 3Nk_BT \tag{10}$$

where $\underline{\underline{\sigma}}$, E and T refer to the dynamical variables (Eq. 5). Other constraints can be used to maintain a steady state and make little numerical difference to the average properties of condensed phases under shear.

We have used non-equilibrium molecular dynamics to study two and three dimensional crystals of particles interacting via Hookes-Law forces with $w = 0.15\ d$.[1] A density of 1.1 times the stress-free density was used to minimize vacancy formation. The computer experiments set up a steady couette flow with a time independent strain rate

$$\dot{\varepsilon}_{xy} = \nabla_y v_x = \dot{\varepsilon}$$

the only non-zero element of the strain rate tensor. Crystals of 64-400 atoms were used in the two dimensional calculations and 216 and 512 atoms were used in the three dimensional ones. The calculations were run at constant temperature until a steady state was reached with fluctuations in the average shear stress and internal energy reduced to a few percent, typically requiring strains of at least 10. The number dependence of the results is small and in Fig. 5 the large system limit of the long-time steady state shear stress as a function of the strain rate, is plotted for three different temperatures. The temperatures correspond to $0.3T_m$, $0.6T_m$ and $0.9T_m$, where T_m, the melting temperature, is $0.0135\ kd^2/k_B$.

The straight lines correspond to a simple analytical fit to the data

$$\sigma_{xy}(\dot{\varepsilon},T) = \sigma_0 (\dot{\varepsilon}/\dot{\varepsilon}_0)^{\gamma(T)} \tag{11}$$

where

$$\gamma(T) = 3k_BT/Db^2 \tag{12}$$

and $\sigma_0 = 0.079\ k$ and $\dot{\varepsilon}_0 = 0.30\ (k/m)^{1/2}$ are fitted parameters. The constant D(T) involves the temperature dependent Lamé constants via the relation given in Eq. 2. The temperature dependence of the Lamé constants is small for this crystal[1]

$$D(0) = 0.084k,\ D(0.3\,T_m) = 0.083k,\ D(0.6T_m) = 0.081k,\ D(0.9T_m) = 0.072k \tag{13}$$

and so γ is approximately proportional to T. For a temperature $T = 0.9T_m$,

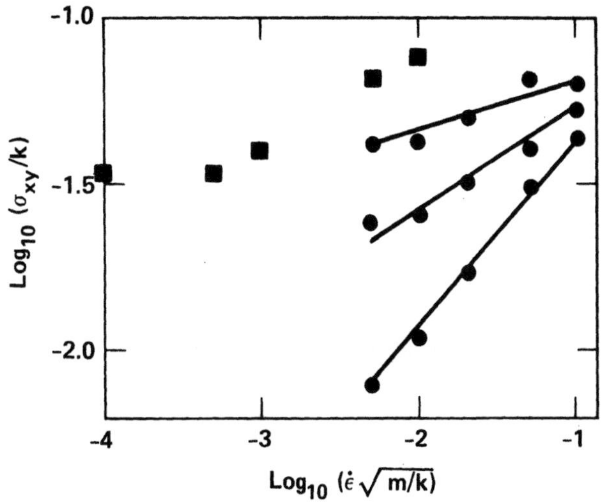

Fig. 5. Comparison of molecular dynamics and dislocation dynamics simulations of the shear stress as a function of strain rate. The molecular dynamics data (circles) are shown at temperatures of $0.3T_m$ (top line), $0.6\ T_m$ (middle line) and $0.9T_m$ (bottom line). The dislocation dynamics data (squares) were all calculated at zero temperature. The straight lines correspond to the expression given in Eq. (11) for the stress as a function of strain rate and temperature.

$\gamma = 0.55$ about half that in the liquid phase. Similar equations have been used to represent low strain rate data obtained from compression tests.[12] An explanation of this relatively simple dependence of shear stress on strain rate is an interesting theoretical challenge.

Dislocation Dynamics

The complexity of the short-range interaction between dislocations and the difficulty in predicting their response to an imposed shear stress has prevented a first-principles theory of plastic flow based on dislocation motion from being developed. Dislocation creation and motion are typically introduced in an emperical way by fitting theoretical predictions to experimental results.[13] Computer simulations which treat defects as point particles with well characterized properties can overcome many of the difficulties involved in an analytic theory describing the dynamics of interacting dislocations. The first goal of these simulations is to explain the results of the non-equilibrium molecular dynamics simulations in terms of the known properties of dislocation in the lattice. Since there are no arbitrary parameters in these calculations such a comparison is a rigorous test of the "dislocation dynamics" model. Some encouraging preliminary results of such a comparison are reported here. The introduction of dislocations instead of atoms, drastically reduces the number of degrees of freedom, and makes it practical to attempt simulations at laboratory shear rates, i.e., less than 10^8 Hz. The second goal, then, would be a comparison of dislocation dynamics simulations and shock wave experiments. Finally, a Boltzmann equation approach could be developed to calculate the evolution of dislocation densities, with the creation and annihilation of dislocations based on "collisional" processes observed in dislocation dynamics simulations.

In "dislocation dynamics", edge dislocations react to the local stress field. Molecular dynamics simulations indicate that they have negligible inertia[8], so that a first order equation of motion $\dot{r} = v(\sigma_L)$ can be used. The propagation velocity v is a function of the local shear stress, σ_L, resolved along the glide direction. In these calculations, which represent Hookes-Law forces, the data of Fig. 4 was approximated by a step function,

$$\begin{aligned} v(\sigma_L) &= 0 & \sigma_L &< 0.05\ \eta \\ v(\sigma_L) &= 0.5\ c_t & \sigma_L &> 0.05\ \eta \end{aligned} \tag{14}$$

where c_t is the transverse sound speed. The interaction between dislocations was approximated by superposing the static elastic stress fields from neighboring dislocations, using hexagonal periodic boundary conditions, and the "nearest-image" convention. A more accurate calculation would involve the stress fields from a periodic array of moving dislocations. Such a calculation is straightforward to implement, though more costly in computer time, and should be included in future work. The Burgers' vector is assumed to be infinitesimal, so that the shape of the crystal is unchanged by dislocation motion. External stress can be introduced in several quasi-experimental ways. For comparison with the atomistic nonequilibrium molecular-dynamics results we use a constant shear rate $\dot{\varepsilon}_{xy} = \dot{\varepsilon}$. The plastic strain, ε_p, resulting from dislocation motion is accumulated and the external shear stress is given by

$$\sigma_{xy} = \eta(\dot{\varepsilon}t - \varepsilon_p) \ . \tag{15}$$

We have assumed that the heat generated by plastic flow is removed by an homogeneous thermostat, as in the atomistic simulations, so that the temperature remains constant.

Whenever two dislocations approach each other closely, the interaction between them diverges. A dislocation in a triangular lattice can interact with another dislocation in one of six relative orientations. At short distances three of these interactions are repulsive and three are attractive (Eq. 1). The interaction between two dislocations that attract each other at short distances either leads to annihilation of the pair or to the formation of a new dislocation such that the total Burgers' vector is conserved. The matrix of possible reactions is shown in Fig. 1. In the dislocation dynamics simulations a reaction occurs whenever two dislocations approach within one Burgers'-vector length of each other.

The final part of the dislocation dynamics simulation, and the most difficult to implement, is dislocation creation. The matrix of reactions shown in Fig. 1 is reversible if the local stresses are large enough. To make this statement quantitative is more difficult. As an initial step we consider a solid at low temperature where thermal effects can be ignored, and we consider only those reactions involving the bifurcation of a dislocation into two others. The nucleation of a pair of dislocations from a perfect lattice is ignored. At each time step, the local stress tensor at each dislocation is computed and the resolved shear stress (including the external stress), along the direction of its possible bifurcation reactions is calculated. The elastic strain energy released by moving the two potential dislocations to a separation b along their glide direction is compared with the strain energy released in the same time by the original dislocation together with the energy of formation of the pair, calculated using Eq. 1. The core energy is that determined from the calculations described earlier, $E_c = 0.26\ \eta b^2$. If the energy balance is favorable, the reaction is assumed to have taken place, and the calculation is updated accordingly.

The multiplication of dislocations by such a mechanism for a very large and fixed external shear stress $\sigma_{xy} = 0.25\ \eta$ is shown in Fig. 6.

There is a tendency for repeated dislocation multiplication leading to the formation and separation of pairs of grain boundaries. At finite temperatures, fluctuations should result in more random structures. Dislocation multiplication at finite temperatures could be simulated by an Arrhenius rate law using the same calculation of the energy of formation. This has already been done for crystals at equilibrium, without any imposed shear,[14] to study the dislocation based melting theory of Halperin and Nelson.[15] A considerable amount of work remains to be done in checking this dislocation multiplication method with molecular dynamics, but we can still compare results for shear stress vs. strain rate obtained by dislocation dynamics and molecular dynamics simulations. The dislocation dynamics simulations were run for similar times as the atomistic calculations (of course this required far less computer time) for crystals containing about 50 dislocations, which corresponds to crystals of 3,000 to 30,000 atoms. The size dependence was small for crystals containing more than 10-20 dislocations, and the results are shown in Fig. 5.

It is encouraging that the average shear stress is in the expected range. For rates comparable to those used in the atomistic calculations the stress is rising too quickly with increasing shear rate, perhaps indicating that homogeneous nucleation is becoming important. At the lower rates through it, it is roughly constant, which is quite reasonable for a low temperature solid.

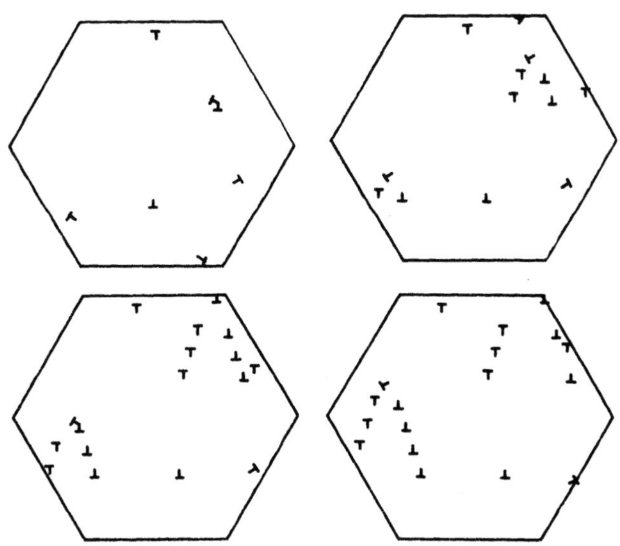

Fig. 6. Dislocation multiplication in a triangular lattice. The figures indicate the location of dislocations at 4 different times, $t(k/m)^{1/2} = 1$, 4, 7 and 10, for a crystal under shear $\sigma_{xy} = 0.25\ \eta$ and at zero temperature. The crystal initially contained six dislocations, one of each orientation, at random locations. The screening produced by a column of dislocations favors grain boundary formation.

Scaling Laws and Comparison with Experiment

For a plastically flowing solid undergoing large deformations, the most important constitutive law is the shear stress as a function of strain rate and temperature. By using an appropriate scaling of variables we can compare our simulation results in two and three dimensions with estimates deduced from shock wave experiments on beryllium[4] and aluminum.[16,17] The natural dimensionless variables are T/T_m, $\dot{\epsilon}d/c_t$ and σ_{xy}/η. For simplicity we have used the zero-temperature shear modulus in scaling the simulation data. The three dimensional simulations involve the shearing of basal planes. The appropriate shear modulus, $\eta = C_{55} - p = 0.408$ k is used to scale the shear stresses. The experimental materials are polycrystalline characterized by an average isotropic shear modulus. The results of the scaling are illustrated in Fig. 7.

The two dimensional results are shown by solid lines, corresponding to temperatures of $0.3T_m$, $0.6T_m$, and $0.9T_m$, and are extrapolated with dashed lines to lower strain rates where estimates are available from interpretation of shock-wave profiles. The "experimental results" include Wallace's calculations for aluminum.[16] These calculations are based on a continuum model, fitted to plastic-wave data and consistent with reasonable thermodynamic bounds on the temperature, stress, and entropy within non-equilibrium shock waves. The low-strain-rate data ($10^{-10} < \dot{\epsilon}d/c_t < 10^{-6}$) are taken from weak shock-wave experiments (21-89 kbar), where the change in thermodynamic state is small. The temperature, 300-400 K, corresponds to about one-third the melting temperature of aluminum. The agreement between

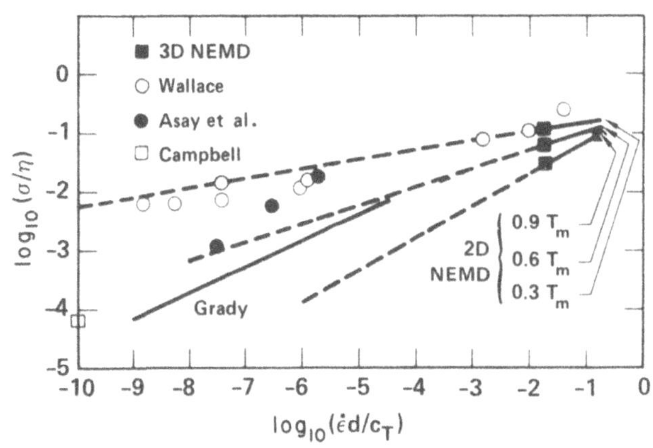

Fig. 7. Strain-rate dependence of the shear stress. η is the shear modulus, d the nearest-neighbor spacing in a closed-packed lattice, $\dot{\epsilon}$ the strain rate du_x/dy, and c_t the transverse sound velocity. The nonequilibrium molecular dynamics was generated for two dimensional solids at 1.1 times the stress-free density and temperatures of 0.3 (top solid line), 0.6 (middle line), and 0.9 (bottom line) times the melting temperature, using from 64 to 400 particles. The scaled three dimensional simulation results, at corresponding density and temperatures, are shown as solid squares. The melting temperature of the three-dimensional crystal was assumed to be $0.012\ kd^2/k_B$. The beryllium data from Asay et al., Wallace and Grady's estimates for aluminum, and Campbell's[18] older low-strain-rate estimate are shown. The uncertainty in the slope of the lines is a few percent.

these results and our $0.3T_m$ line is good. The high-strain-rate results are from strong shocks (∿ 1 Mbar) where the thermodynamic state is not well specified. The discrepancies between Wallace's[16] and Grady's[17] estimates, from the same experimental profiles, reflect the difficulty in estimating shear stress from measured longitudinal profiles. The upper bounds for shear stress, computed for aluminum, copper, and iron by Chhabildas and Asay[4], lie well above our low-temperature line.

The approximate methods used to estimate shear stress in Refs. 4, 16 and 17 are oversimplifications of complex phenomena, but they represent the best available experimental description of the dependence of shear stress on strain rate. The agreement between these results and the simulation data suggests that a fundamental understanding of the simulation results could represent a significant step towards explaining the mechanism of high strain rate ($\dot{\varepsilon} > 10^4$ Hz) plastic flow.

CONCLUSION

In this paper I have summarized a program of work aimed at providing an understanding of plastic flow in simple crystals by linking microscopic defect physics with macroscopic constitutive modeling. This work has been partially successful. We have demonstrated the use of atomistic simulations in modeling plastic flow, with results that correspond quite closely with extrapolations of experimental data. We have also shown that defect properties can be characterized by atomistic methods and that at least semiquantitative agreement between the "exact" atomistic results and the semi-macroscopic dislocation dynamics modeling of plastic flow can be obtained. A complete, temperature dependent characterization of the mechanics and kinetics of dislocation formation based on atomistic simulations is necessary for further progress.

ACKNOWLEDEMENTS

This work was done in collaboration with William Hoover at the Department of Applied Science, University of California, Davis, and at Lawrence Livermore National Laboratory. Bill Moran (LLNL) made many useful contributions. The work was supported by the Army Research Office, Research Triangle Park, North Carolina, under Contract No. W-7405-ENG-48, and by the Department of Energy.

REFERENCES

1. W. G. Hoover, A. J. C. Ladd and B. Moran, Phys. Rev. Lett., 48, 1818 (1982).
   A. J. C. Ladd and W. G. Hoover, Phys. Rev. B 28, 1756 (1983).
2. See for instance recent articles by D. J. Evans and W. G. Hoover in Physics Today 37, No. 1 (Jan. 1984) and in Nonlinear Fluid Behavior, edited by H. J. M. Hanley (Proceedings of the Conference on Nonlinear Fluid Behavior held at the University of Colorado, Boulder, 7-11 June 1982), Physica A 118 (1983).
3. D. C. Wallace, Phys. Rev. B 24, 5607 (1981).
4. L. C. Chhabildas and J. R. Asay, J. Appl. Phys. 50 2749 (1979).
   J. R. Asay, L. C. Chhabildas, and J. L. Wise, in Shock Waves in Condensed Matter - 1981 (Menlo Park), Proceedings of the Conference on Shock Waves in Condensed Matter, edited by W. J. Nellis, L. Seaman, and R. A. Graham (AIP, New York, 1982), pp. 417 and 427.
5. W. G. Hoover, A. J. C. Ladd, and N. E. Hoover, in Interatomic Potentials and Crystalline Defects, edited by J. K. Lee (Metallurgical Society, Warrendale, Pennsylvania, 1981).

6. A. J. Chorin, J. Fluid. Mech., 57, 785 (1973).
A. F. Ghoniem, A. J. Chorin and A. K. Oppenheim, Phys. Trans. R. Soc. Lond., A304, 303 (1982).
7. D. S. Fisher, B. I. Halperin, and R. Morf, Phys. Rev. B 20,
A. J. C. Ladd and W. G. Hoover, Phys. Rev. B 26, 5469 (1982).
8. W. G. Hoover, N. E. Hoover, and W. C. Moss, J. Appl. Phys. 50, 829 (1979).
9. W. G. Hoover, D. J. Evans, R. B. Hickman, A. J. C. Ladd, W. T. Ashurst, and B. Moran, Phys. Rev. A 22, 1690 (1980).
A. J. C. Ladd, Molec. Phys., 53, 549 (1984).
10. A. J. C. Ladd and W. G. Hoover, J. Statist. Phys., 38, 973 (1985).
D. J. Evans and G. P. Morriss, Phys. Rev., A 30, 1528 (1984).
11. D. J. Evans, W. G. Hoover, B. H. Failor, B. Moran and A. J. C. Ladd Phys. Rev. A 28, 1016 (1983).
12. L. D. Sokolov, Dokl. Akad. Nauk SSSR 67, 459 (1949) [Soviet Phys--Dokl. 67, 459 (1949)]; S. K. Samanta, Int. J. Mech. Sci. 11, 433 (1969).
13. J. J. Gilman, Micromechanics of Flow in Solids (McGraw-Hill, New York, (1969).
14. Y. Saito, Phys. Rev. Lett., 48, 1114 (1982).
15. D. R. Nelson and B. I. Halperin, Phys. Rev. B 19, 2457 (1979).
16. D. C. Wallace, Phys. Rev. B 22, 1477 (1980); 22 1487 (1980).
17. D. E. Grady, Appl. Phys. Lett. 38, 825 (1981).
18. J. C. Campbell, Mater. Sci. Eng. 12, 3 (1973).

WAVE PROPAGATION AND LONG-TIME BEHAVIOR

ON THE DRIVEN SINE-GORDON CHAIN

Michael D. Miller

Department of Physics
Washington State University
Pullman, WA 99164-2814

INTRODUCTION

Model one-dimensional lattices have been an important tool in the study of nonlinear wave motion in solids[1,2,3]. A particular model is specified by the choice of a potential function (usually restricted to nearest neighbors). The free end of a semi-infinite chain (the piston particle) is driven into the lattice at fixed speed to simulate some external impulse. The motion of the piston particle induces the creation of a propagating wave front. In these systems, the wavefront has the nature of a kink or soliton in that its speed depends on its amplitude and it can propagate (more or less) independently of radiation (phonons). A particularly careful study has been made of the Toda chain[3], the only known completely integrable lattice system. The dynamical behavior observed in the Toda chain is apparently generic to all of these hard-core nearest-neighbor lattices. At low piston particle speeds the asymptotic behavior of the chain is harmonic-oscillator-like in the sense that the particles eventually come to rest in the piston particle frame of reference. At high piston particle speeds (or equivalently large anharmonicity) The asymptotic dynamics is hard-rod-like in the sense that the oscillations in the piston particle frame of reference persist.

SINE-GORDON MODEL

We consider as a model for an adsorbed monolayer a harmonic chain of particles interacting with an external periodic (substrate) potential. The Lagrangian for the system can be written

$$L = \sum_{n=0}^{\infty} [\tfrac{1}{2} \dot{x}_n^2 - \lambda(1 - \cos(qx_n)] - \sum_{n=1}^{\infty} \tfrac{1}{2}(x_n - x_{n-1} - 1)^2, \quad (1)$$

where x_n is a dimensionless coordinate. The equations of motions obtained from (1) are

$$\ddot{x}_n - \Delta^2 x_n = -\lambda q \sin(qx_n) \quad n = 1,2... , \quad (2)$$

where the dots are time derivatives and $\Delta^2 x_n = x_{n+1} - 2x_n + x_{n-1}$ is the central second difference operator.

The equilibrium initial and boundary conditions are $\ddot{x}_n = \dot{x}_n = 0$ for $n \geq 1$, at $t = 0$ and $\dot{x}_o = u_p$ for all time. The asymptotic dynamical behavior of this system differs markedly from those mentioned in the Introduction.

The parameter q in Eqs. (1) and (2) is 2π times the ratio of the natural harmonic lattice spacing to the substrate spacing and is thus a measure of the mismatch in the two competing lengths. In this note we shall briefly discuss the role of q, the mismatch parameter, in determining the asymptotic dynamics.

The system with $q/2\pi = n$, n an integer, is known as a first order or 1:n commensurate system[5] (meaning simply that the equilibrium ground state has nearest neighbor adatoms spaced by n substrate lattice spacings and positioned in the potential well minimum). The system with n = 1 has been discussed in detail recently[6] by this author. For a system in the adiabatic regime ($u_p << 1$, $\lambda << 1$), the adatoms execute harmonic motion with a frequency $\nu_o = qu_p/2\pi$ at long times after the waterfront has past. As λ is increased at fixed u_p, higher harmonics of ν_o appear and at large enough λ disordered motion begins to occur. The mechanism for the transition from ordered to disordered motion is the pinning of the sine-Gordon kinks by the substrate lattice due to the Peierls barrier. For the details see Ref. 6.

The dynamic behavior of the rest of the 1:n family is simple to understand. Each time the piston particle moves unit distance, it winds n kinks onto the chain. The asymptotic frequency of vibration is $\nu_o = nu_p$. In Figs. 1 and 2 we illustrate this behavior for the n = 2 system with $u_p = 0.01$ and $\lambda = 0.001$. The structure shown in Fig. 1 is a propagating kink lattice. As a reflection of the Lorentz invariance of the continum limit, the kinks can only propagate with speeds < 1. The kink speed does show some mild q dependence: the system in Fig. 1 is propagating $\simeq$ 2% faster than its n = 1 cousin. In Fig. 3, we show the velocity distribution along the chain at t = 1000. The leading oscillations, the precursor phonons (cf. Fig. 1), have a maximum speed, v_{max}, determined by the value of ω_E the gap in the spectrum (where $\omega_E = \lambda^{\frac{1}{2}} q$). It is straightforward to show that

$$v_{max} \simeq 1 - \tfrac{1}{2}\,\omega_E \, , \tag{3}$$

for $\omega_E << 1$. Thus, $1 - v_{max} \propto n$. For n = 1, we found $1 - v_{max} \simeq 0.08$. From Fig. 3, we find $1 - v_{max} \simeq 0.16$ (as expected).

The second order commensurate systems (2:m with m odd) are the simplest examples of adatom lattices with a basis (the excitation spectrum thus shows additional optical branches). In Fig. 4, we show the phase diagram for the m = 1 system (with $u_p = 0.01$ and $\lambda = 0.01$). The piston particle winds a kink onto the chain each time it progresses through unit distance. Roughly speaking, the energy of a kink $E_s \propto 1/q$. Thus these kinks are far less "massive" than their 1:1 counterparts. Indeed in time = 1000 the kink lattice for the 1:1 system has only reached atom $\simeq$ 300 whereas for the 2:1 system shown in Fig. 4 the wavefront is almost to atom 800. In Fig. 5 we show the velocity profile for atom 10 (the counterpart to Fig. 2). The atom is executing periodic motion with frequency $\nu_o = u_p/n = 0.005$, as expected.

CONCLUSION

The asymptotic dynamical behavior of the higher commensurate states is qualitatively similar to the 1:1 state.

The considerations with regards incommensurate configurations are more involved and will be published elsewhere. The case of the first order

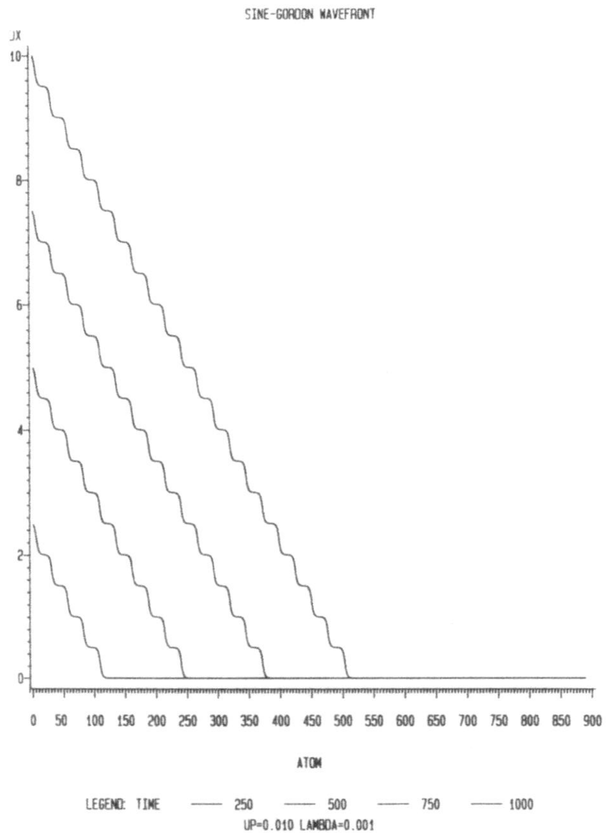

Fig. 1. Propagating kink lattice for $q = 4\pi$, 1:2 commensurate system.

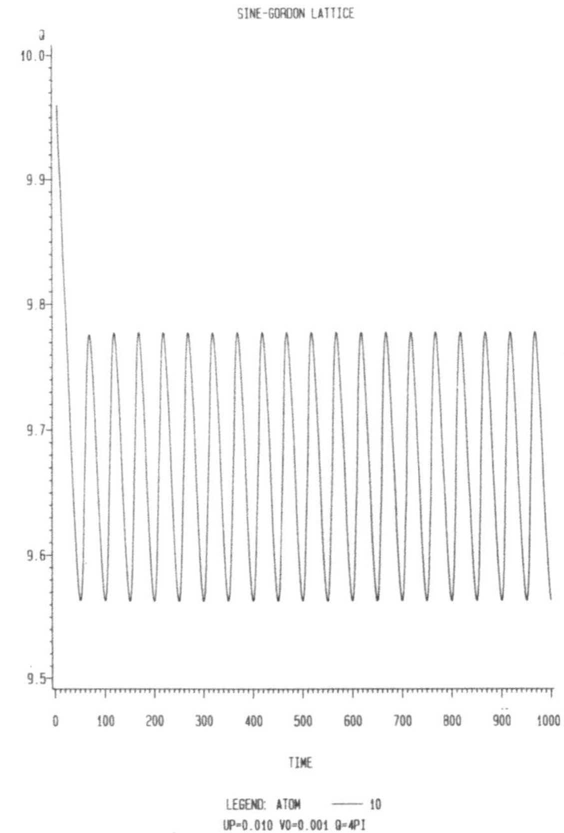

Fig. 2. Position as a function of time for atom 10 in the piston particle rest frame showing periodic motion with frequency $\nu_o = 0.02$.

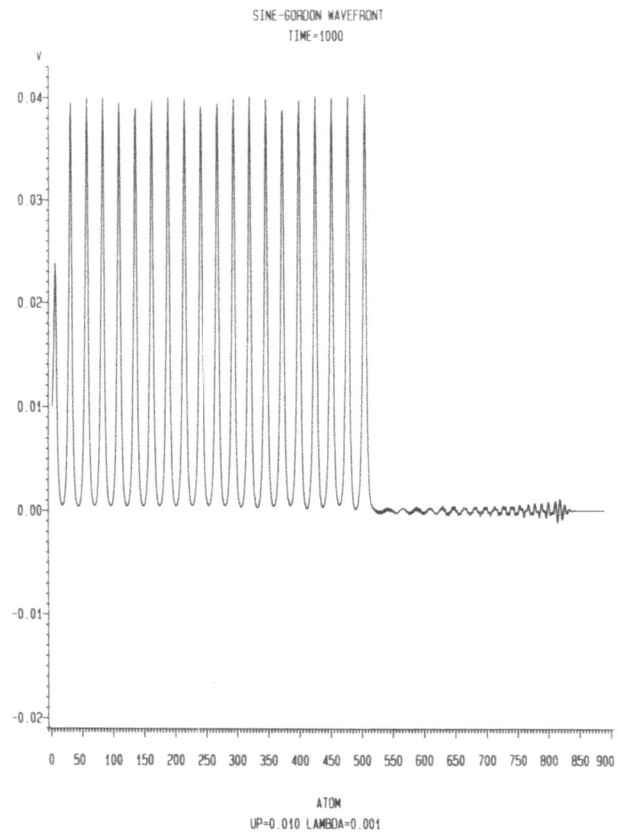

Fig. 3. Velocity profile of 1:2 lattice at time = 1000 showing the kinks and the precursor phonons.

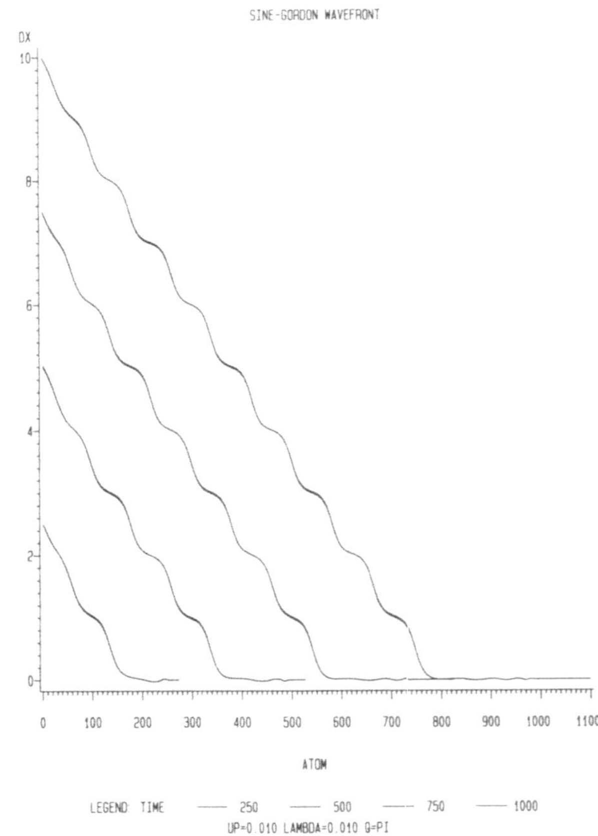

Fig. 4. Phase profile of 2:1 system at various times. There is one kink for each unit distance traveled by piston particle.

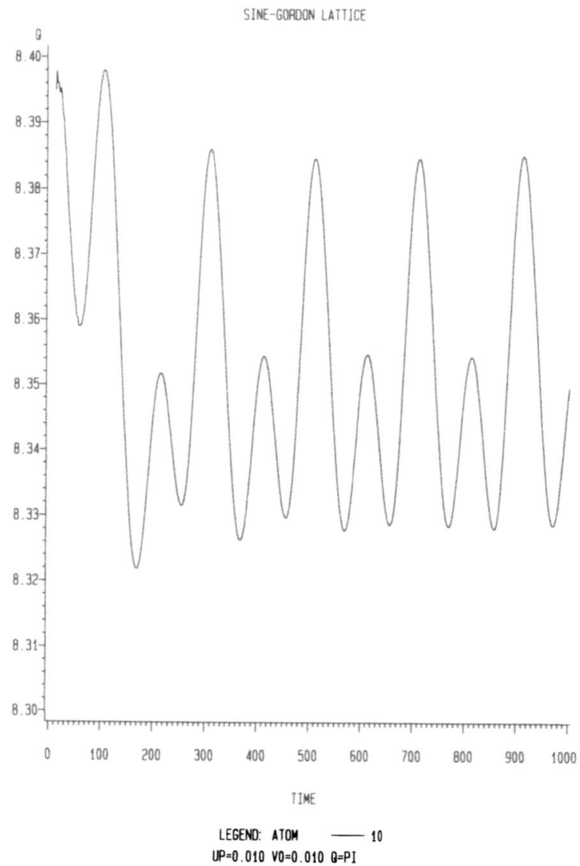

Fig. 5. Position as a function of time for atom 10 in the piston particle rest frame showing periodic motion with frequency $\nu_o = 0.005$.

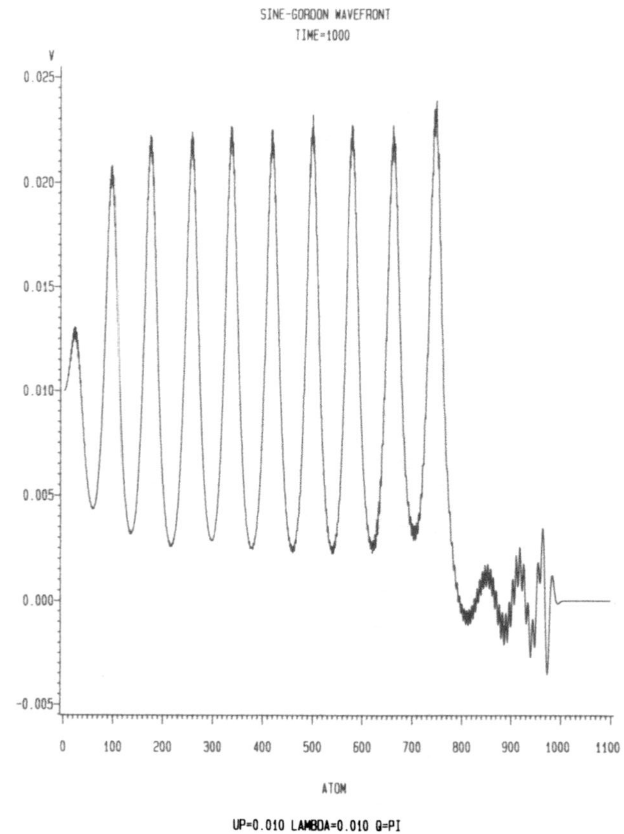

Fig. 6. Velocity profile of 2:1 system at t = 1000.

configurations was treated by Frank and van der Merwe[7] in a classic paper. They found that the equilibrium ground state was commensurate for $\lambda > \lambda_c$, where $\lambda_c = (\pi/4)^2$. However for $\lambda < \lambda_c$ the system was unstable to domain formation with kink boundaries. This latter state is incommensurate with the underlying substrate: it costs almost no energy to be rigidly translated. The incommensurate system has no gap in the lowest lying phonon mode[5,8]. The incommensurate ground states are not completely disordered and have been shown by Aubry[9] to correspond to smooth KAM trajectories.

REFERENCES

1. R. Manvi, G. E. Duvall and S. C. Lowell, Int. J. Mech. Sci. 11:1 (1969); G. E. Duvall, R. Manvi and S. C. Lowell, J. Appl. Phys. 40: 3771 (1969); R. Manvi and G. E. Duvall, J. Phys. D2: 1389 (1969).
2. J. Tasi, J. Appl. Phys. 43:4016(1972); ibid. 44:1414 (1973); ibid. 44:4569 (1973); ibid. 47: 5336 (1976).
3. B. L. Holian and G. K. Straub, Phys. Rev. B18: 1593 (1978); G. K. Straub, B. L. Holian and R. G. Petschek, Phys. Rev. B19: 4049 (1979); B. L. Holian, H. Flaschka and D. W. McLaughlin, Phys. Rev. A24: 2595 (1981).
4. M. Toda, "Theory of Nonlinear Lattices," Springer-Verlag, Berlin, (1981).
5. The terminology is that of S. C. Ying, Phys. Rev. B3: 4160 (1971).
6. M. D. Miller, submitted to Phys. Rev. B.
7. F. C. Frank and J. H. van der Merwe, Proc. Roy. Soc. London A198: 205 (1949).
8. G. Theodorou and T. M. Rice, Phys. Rev, B18: 2840 (1978).
9. S. Aubry in: "Solitons and Condensed Matter Physics," A. R. Bishop and T. Schneider eds., Springer-Verlag, Berlin (1978).

SIMULATION OF LATTICE DAMAGE DUE TO DYNAMIC LOADING†

Paul A. Taylor and Brian W. Dodson

Sandia National Laboratories
Albuquerque, New Mexico 87185

ABSTRACT

By means of a dynamic simulation technique developed by the authors and based upon the Monte Carlo method, we investigate the response of two-dimensional Lennard-Jones lattice systems to dynamic uniaxial strain paths that are either fast or slow with respect to a lattice equilibration time. The simulation of lattice response to various compression-release cycles indicates a dependence of the degree of sustained lattice damage (i.e., dislocations and defects) on the rate and maximum amplitude of applied strain. Specifically, greater lattice damage is predicted whenever the lattice is subjected to a fast-loading process and/or higher maximum strain amplitude as compared to that induced by a slow-loading process and/or lower maximum strain amplitude.

INTRODUCTION

The nucleation of microscopic material damage (i.e., dislocations and defects) by means of dynamic mechanical loading is of great interest to many researchers in the physics and engineering communities. Issues such as rate dependence and loading history play an important role in the generation of such damage. In addition, there exists evidence [1,2] to suggest that the highly dynamic process of shock loading sets up a unique environment, giving rise to the generation of microscopic damage at levels that are above those associated with finite strain-rate processes. The theoretical foundations describing such phenomena have yet to be layed. To this end, we have developed a dynamic simulation technique that allows us to investigate the mechanisms of dislocation and defect nucleation as they occur in materials subjected to finite strain-rate dynamic and shock loading processes.

In the next section, we describe the model that plays a central role in the simulation technique and the methodology associated with its application to various strain-rate phenomena. The following section reports on the simulation of atomistic lattice response to finite strain-

†This work performed at Sandia National Laboratories supported by the U.S. Dept. of Energy under contract DE-AC04-76DP00789.

rate and shock loading processes for various maximum strain amplitudes. Finally, the paper concludes with a discussion of relevancy of the results reported herein to those of experimental observation.

MODEL

The Monte Carlo method is typically used to model a system subjected to a set of static external conditions (T, V, P, etc.) [3]. We have extended this method to model lattice response to dynamic processes in which the external constraints are varying in time [4]. Due to the absence of an explicit time scale associated with the Monte Carlo method, the technique is limited to simulating dynamic processes whose rate of change is either fast or slow compared to a lattice equilibration time. If the constraints are varying slowly, then the overall process can be represented as a concatenation of differentially varying static conditions. The evolution between one set of differential constraints and the next is then calculated using the Monte Carlo method until the system has reached the new equilibrium state. The procedure is then repeated until the complete process of interest is modeled. If, instead, the constraints are rapidly varying relative to the lattice equilibration time, the system will show little response to the details of the process, which can then be modeled as a single jump from the initial to the final conditions. We call the first type of process a slow process, and the second type a fast process. By the modeling of processes as outlined above, we can investigate rate dependence using the Monte Carlo method. The slow process corresponds to the application of uniaxial strain at a rate sufficiently small such that configurational equilibrium is maintained throughout the process, whereas the fast process corresponds to the application of uniaxial strain at a rate greater than any reasonable equilibration time for the lattice. In view of this, the slow and fast processes may be associated with conventional dynamic (i.e., excluding shock loading) and shock wave loading, respectively.

The lattices used in this investigation are composed of 2-D arrays of atoms numbering 280 to 360 per lattice. It is assumed that the atoms interact in a pairwise fashion according to the Lennard-Jones (L-J) potential

$$\phi(r) = 4\varepsilon[(\frac{\sigma}{r})^{12} - (\frac{\sigma}{r})^{6}],$$

where the values of the parameters ε and σ are set for those of argon, i.e., $\varepsilon = 1.877\times10^{-14}$ ergs and $\sigma = 3.368$Å [5]. The zero temperature equilibrium structure is that of a hexagonal close-packed (HCP) lattice with a nearest-neighbor spacing of 3.747Å. Periodic boundary conditions were imposed to avoid surface effects, however, given the intent of the simulation, the boundaries are required to satisfy certain additional conditions (see [4]). In view of this, the present study involved lattices initially configured into 20x18, 20x16, and 20x14 atoms which allowed for maximum compressive strains of 19%, 36%, and 51% respectively.

We are interested in comparing the damage resulting from both the fast and slow processes for various maximum compressive strains. The objective of this work lies in determining the degree of microscopic damage (e.g. dislocations, point defects, etc.) resulting from shock wave loading as compared to the dynamic loading associated with much smaller strain rates. Many experimental results suggest that considerably more damage results from shock wave loading as compared to conventional dynamic loading [1,2]. The system is placed into uniaxial strain by

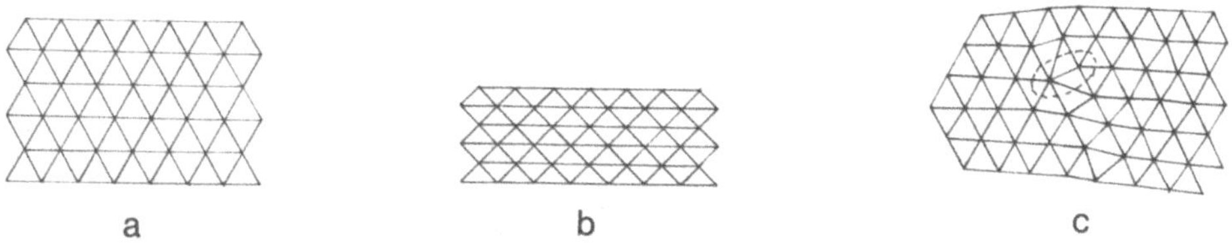

Figure 1. Schematic lattice configurations at various uniaxial strains. (a) The initial unstrained two-dimensional HCP lattice, in which every atom is located at the vertices of a nested collection of equilateral triangles. Here, the coordination number is 6. (b) Lattice distortion associated with uniaxial strain less than 19%. The lattice is now composed of isosceles triangles, but is otherwise unchanged. (c) A portion of a lattice which has been subjected to a uniaxial strain greater than 19%. Note the dislocation, lattice strain, (manifesting as irregular triangles) and the pair of atoms (inside the dashed contour) possessing coordination numbers of 7 and 5.

reducing the periodic distance along one axis, while maintaining a constant circumference in the perpendicular direction, resulting in a smaller area within which the particles reside and a change in the shape of that area. To simplify the examination of defective states resulting from dynamic loading, the simulation will be performed for zero temperature.

The methodology of the dynamic simulation technique used in the present study has been reported elsewhere [4] in great detail and thus will be omitted here for the sake of brevity.

RESULTS

The simulation of two-dimensional L-J lattice response to uniaxial compression shows the existence of three general classes of atomic configurations. The first is the initial perfect HCP lattice, pictured schematically in Fig. 1a. All bond lengths are equal, and all atoms are sixfold coordinated. The second class includes those systems which have been subjected to uniaxial compressive strain less than 19%. These systems exhibit a reduction in the lattice vector component along the axis of applied strain, as illustrated in Fig. 1b. Bond lengths are no longer equal, but result in a nested pattern of isosceles triangles. Here, the lattice has been homogeneously compressed in the vertical direction, with the atoms still displaying sixfold coordination. Finally, we consider systems subjected to larger uniaxial strains (>19%). In these systems, the benign uniaxial distortion of the lattice as displayed by the second class of atomic configurations becomes unstable, and evolves into a strained and defective HCP lattice (Fig. 1c). Here we show a piece of lattice containing a dislocation, which results in considerable lattice strain (manifesting as irregular bond lengths and angles) and in a pair of atoms with coordination numbers of 5 and 7. Simulation results show that the instability of the uniformly, uniaxially distorted lattice, as illustrated in Fig. 1b, results in massive lattice damage of the type displayed in Fig. 1c.

Simulation results on the evolution of the slow processes to 19%, 36%, and 51% were found to be qualitatively similar. In particular, the evolution of the 36% slow compression process is illustrated by the plot of lattice strain energy vs. uniaxial strain in Fig. 2. Accompanying

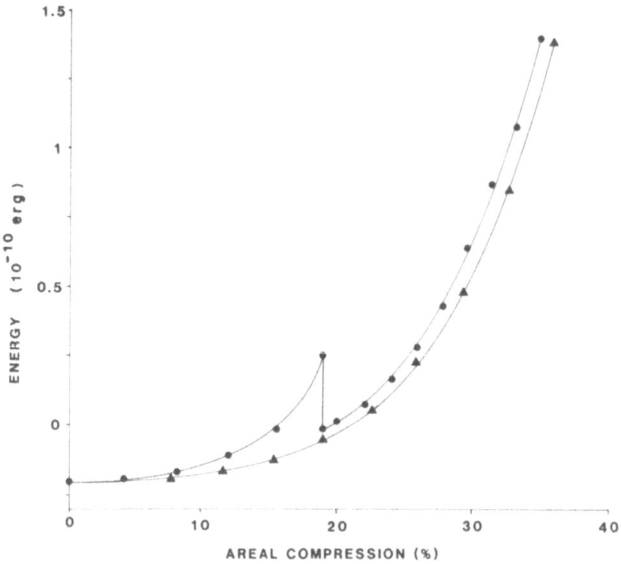

Figure 2. Lattice strain energy vs. areal compression for slow processes. Round points correspond to uniaxial compression, triangles to hydrostatic compression.

this plot is that of the corresponding hydrostatic compressional energy to the equivalent areal density. That the uniaxially strained curve always lies above the hydrostatic curve indicates, for compressions greater than 19%, the existence of an excess stored energy associated with lattice damage. A distinct feature of the energy plot is the abrupt reduction of energy occurring at 19% uniaxial strain. This abrupt transition to a lower energy state is preceded by a region of abnormally low compressibility. The transition is associated with the loss of stability of the uniaxial lattice deformation and subsequent formation of a locally relatively unstrained, but defective, HCP lattice. A more detailed investigation of the abrupt transition reveals its occurrence at 17.2% uniaxial strain. Since the initial lattice is perfect, a sufficient amount of strain energy must accumulate before these dislocations can be nucleated. The symmetry of the model initially precludes local focusing of the strain energy in the lattice, so that the strain field associated with the first nucleation event sets off other nucleations and a subsequent avalanche of dislocation generation and atomic transport. Initially, this avalanche results in a transient amorphous atomic state, which then equilibrates to a defective, but recognizable, HCP lattice. Although somewhat influenced by the choice of the lateral boundary conditions (simulations involving free lateral boundaries predict an abrupt transition at 14.3% uniaxial compressive strain [6]), the large elastic limit (17.2%) for the 2-D, L-J lattice primarily reflects the large energies necessary to nucleate dislocations in a perfect crystal. Continued slow compression of the lattice system to 36% uniaxial strain leads to a considerable degree of lattice damage and strain. Examination of the corresponding configuration data reveals the presence of two pairs of dislocations and their concomitant lattice strain. Here, the characteristic dislocation separation corresponds to a dislocation density of $6x10^{19}/cm^3$ for a three-dimensional system, indicating massive lattice damage. A simulation of lattice response to hydrostatic compression at the same areal densities gives a perfect HCP structure, in view of which, we conclude that the lattice damage is directly associated with the uniaxial nature of the applied strain. We now slowly expand and unload the lattice uniaxially from its maximum compressive state and search for a minumum in the strain energy which corresponds to a stress-free state. For the 36% strain case, the stress-free state occurs at the uniaxial tensile strain of 1.0%. The existence of this residual strain in the stress-free state indicates a departure from the perfect close-packed structure to a defective one. Indeed, configuration data on the 36% case shows the presence of two vacancies and a small area of disordered material in what is otherwise only a slightly strained lattice. The presence of residual strain in the stress-free state is observed in all model systems considered in this study.

The energy information also allows us to quantitatively examine the amount of irreversible damage experienced by the lattice when subjected to uniaxial strain. We define the excess energy E_x to be the energy per atom of the uniaxially strained lattice minus the energy per atom of the hydrostatically compressed lattice while subjected to the same areal compression. The residual defect energy E_D is defined as the residual excess energy at zero stress following completion of the compression-release cycle.

In Fig. 3, E_x divided by the hydrostatic strain energy $[E_{hydro} - E_{hydro}(0)]$ is plotted against uniaxial compression for both the slow and fast loading processes. In the slow case, the rapid accumulation of excess energy at strain values below 19% reflects stability of the perfect HCP lattice structure, and hence does not represent energy contributions due to a defective state, but rather is attributable to the uniaxial deformation of the local lattice geometry. At compressions above the onset of dislocation nucleation at 19%, the excess energy is 5-25% that of the hydrostatic strain energy. The fast loading process results in excess energy values roughly twice those calculated for the slow case.

Fig. 4 displays a plot of E_D divided by the initial lattice energy versus maximum uniaxial compressive strain for both the slow and fast loading processes. For both processes at maximum strain values less than 17.2%, the resulting E_D is zero, indicating that no irreversible damage has resulted from the overall compression-release cycle. At maximum strain values greater than the 17.2% transition, E_D becomes roughly 5% of the ground state energy for slow compression and release, and 8% for the fast cycle. Thus, rapid application of uniaxial strain results in a more defective lattice than that accompanying gradual application.

We have also simulated L-J lattice response to comparable hydrostatic areal compressions for both the slow and fast processes. In no instance was defect generation observed, where the perfect 2-D HCP lattice structure persisted for all values of hydrostatic compression considered in this investigation. This result suggests a fundamental microscopic difference in the natures of uniaxial and hydrostatic compression.

DISCUSSION

A considerable body of experimental evidence has been accumulated which suggests that the atomic environment in a shock wave, especially near the shock front, is a uniquely violent situation [1,2]. Many

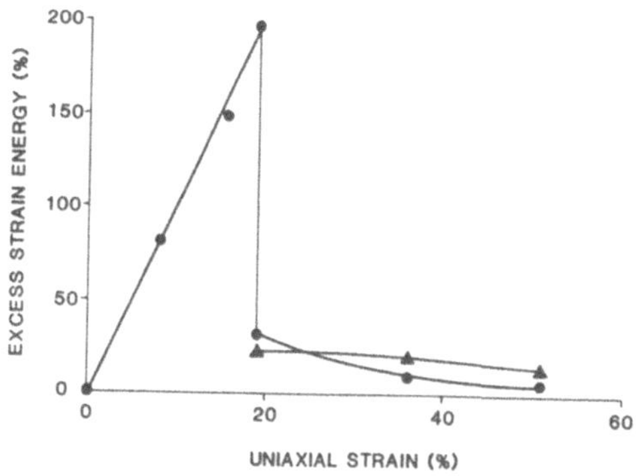

Figure 3. Normalized excess energy E_x vs. maximum uniaxial compression. Round points correspond to slow loading, triangles to fast loading.

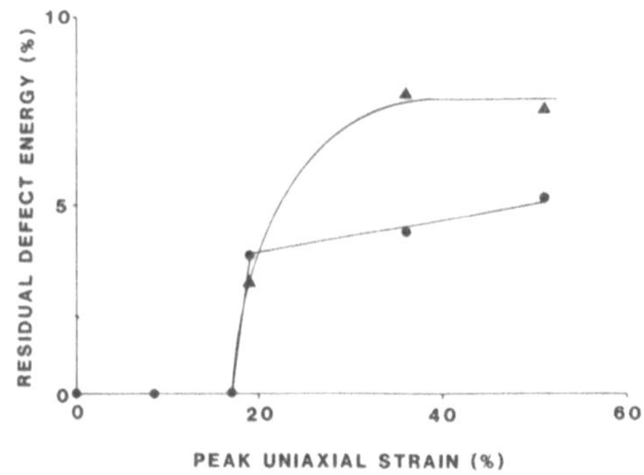

Figure 4. Normalized residual defect energy E_D vs. maximum uniaxial compression. Round points correspond to slow loading, triangles to fast loading.

unusual physicochemical events occur in the shock environment, which has prompted the "catastrophic shock" concept, wherein the shock wave is associated with generation of massive macroscopic and microscopic damage, which then act to enhance chemical reactivity [7].

We feel that our simulation of lattice response to dynamic uniaxial strain may provide some theoretical foundations for the concept of catastrophic shock. The physical and chemical consequences of the transition from a uniaxially distorted lattice to a relatively undistorted lattice are profound. Our simulations have predicted the formation of large densities of point defects and dislocations, which constitutes an abundance of active sites for chemical reaction if the lattice were composed of a less inert species than argon. In addition, the atomic rearrangement associated with the avalanche of dislocation generation can be viewed as a strongly driven enhanced diffusion process, which would greatly accelerate nonunimolecular chemical reactions. All of these effects are seen in practice, and form the experimental basis for the catastrophic shock concept [1,2]. There are certainly macroscopic instabilities associated with uniaxial shock loading which also contribute to the catastrophic effects, but the atomic-scale effects described herein provide a theoretical explanation of what may represent a major portion of the physicochemical consequences of shock loading.

REFERENCES

1. B. W. Dodson and R. A. Graham, in: "Shock Waves in Condensed Matter," W. J. Nellis, L. Seaman and R. A. Graham, Eds., AIP, New York (1982).
2. B. Morosin and R. A. Graham, in: "Shock Waves in Condensed Matter," Ref.1.
3. J. P. Valleau and S. G. Whittington, in: "Statistical Mechanics A," B. J. Berne, Ed., Plenum, New York (1977).
4. B. W. Dodson and P. A. Taylor, Phys. Rev. B, **30**, 1679 (1984)
5. W. A. Harrison, "Electronic Structure," McGraw-Hill, New York (1969).
6. P. A. Taylor and B. W. Dodson, to appear.
7. R. A. Graham, J. Phys. Chem. **83**, 3048 (1979).

DISLOCATION KINETICS BEHIND SHEAR SHOCKS*

R. B. Stout and G. D. Anderson

Lawrence Livermore National Laboratory
Livermore, CA 94550

ABSTRACT

High velocity oblique impact experiments result in both compression and shear shock waves. Behind the shear shock wave the particle velocity is transverse to the shock front. At large transverse particle velocities, dislocation kinetics can contribute a portion of the velocity. Based on a kinematic and thermodynamic model of dislocation kinetics, an analysis is made of the transverse strain and velocity behind a shear shock. Kinematics of dislocations in transverse motion behind the shock is formulated. A solution is given for an ideal case where the dislocation density function propagates as a pulse behind the shear shock.

INTRODUCTION

Oblique impact experiments have been performed to investigate metallic material response at high shear rates[1-4]. The inclined impact experiment results in both a compression shock wave and transverse or shear shock wave that propagate through the test specimen. The compression waves propagate faster than the transverse wave. The high shear rates occur across and behind the transverse wave. The particle velocity across a transverse shock wave has a discontinuity in the vector component perpendicular to the propagation direction of the shock front. For an elastic shock, the discontinuity in the transverse particle velocity results in a propagating discontinuity in the shear strain and in the shear stress; and the deformation after the shock is recoverable. For strong shocks which involve dislocations the deformation after the shock is not all recoverable. The following analysis considers the simplest case of dislocation kinetics across and behind a transverse shock. The field equations and the discontinuity conditions across a shock front for mass, momentum, and energy depend on a dislocation density function. In addition, there is a field equation and a discontinuity condition across a shock front for the dislocation density function. These equations and their associated discontinuity conditions are described elsewhere.[5-7]

*Work performed under the auspices of the U.S. Department of Energy by the Lawrence Livermore National Laboratory under contract number W-7405-ENG-48.

For a planar shear shock, coordinate axes can be selected such that the shock front propagates along the X_2 axis with velocity $\underline{V} = (0,V_2,0)$ and the particle velocity behind the shock front is along the X_3 axis. In the simplest idealized case, the nonrecoverable plastic deformations can occur from edge dislocations. For a statistical dislocation species distribution that depends only on the velocity of dislocations, the edge dislocation would have a fixed Burgers' vector $\underline{b} = (0,0,b_3)$, a velocity vector variable relative to the lattice structure $\underline{\nu} = (0,0,\nu_3)$, and a fixed line tangent vector $\underline{\xi} = (-\xi_1,0,0)$. The field equation and the discontinuity condition across a shock front[7] for the dislcocation density function $D(\underline{x},t,\underline{b},\underline{\nu},\underline{\xi})$ are given by

$$\dot{D} = \partial_t D + \underline{\nabla}\cdot((\underline{v} + \underline{\nu})D) = P + K\,\{\underline{\nu} \leftrightarrow \underline{\nu}^*) \tag{1}$$

and

$$[(\underline{V} - \underline{v} - \underline{\nu})D]\cdot\underline{n} = p + k\,\{\underline{\nu} \leftrightarrow \underline{\nu}^*\} \tag{2}$$

Equation (1) is a field equation of the Boltzmann type for the dislocation density function at spatial points without shock fronts, equation (2) is the discontinuity condition across a shock front propagating at velocity $\underline{V}$, and the other functions are for the mass velocity $\underline{v}$, the unit normal to the shock front $\underline{n}$, the volume production for dislocation species $(\underline{b},\underline{\nu},\underline{\xi})$ denoted by P, the shock surface production for dislocation species $(\underline{b},\underline{\nu},\underline{\xi})$ denoted by p, and K and k are the transition functionals for modelling dislocation interchanges between different velocities for species $(\underline{b},\underline{\nu},\underline{\xi})$ and $(\underline{b},\underline{\nu}^*,\underline{\xi})$ in the volume and at a shock surface, respectively. For a simple illustration, K and k are taken as zero and only a model for the production of dislocations is considered. A uniform density pulse of edge dislocation species (b_3,ν_3,ξ_1) that propagates at velocity V_2 can be represented with a step function as

$$D(\underline{x},t,b_3,\nu_3,\xi_1) = D_o(H(V_2t - x_2) - H(V_2t - x_2 - X_2)) \tag{3}$$

where the pulse width is X_2, D_o is a constant, and H is the step function which is unity when the argument is positive and zero otherwise. Equation (3) satisfies equation (1) at points away from the shock front and satisfies equation (2) at the "elastic-dislocation" shock front for $x_2 = V_2t$ and at the dislocation shock front for $x_2 = V_2t - X_2$. The production function creates dislocations at a rate of D_oV_2 at the elastic-dislocation shock front and annihilates dislocations at a rate of $-D_oV_2$ at the dislocation shock front.

From previous concepts developed for the deformation and kinematics due to dislocation kinetics,[5-7] the total relative particle velocity at points $\underline{\chi}$ behind a shock front currently located at $\underline{x}$ is given by

$$v_i\}(\underline{x}\to\underline{\chi},t) = v_i|(\underline{x}\to\underline{\chi},t) + \int_{\underline{x}}^{\underline{\chi}}\int_Q b_i e_{jk\ell}\xi_k\nu_\ell D + b_i e_{jk\ell}\xi_k b_\ell{}^*\dot{D}\,d\underline{q}\,dx_j(\underline{x}\to\underline{\chi},t) \tag{4}$$

For the transverse shock and edge dislocation species discussed above, index $i = 3$, index $j=2$, index $k = 1$, and index $\ell = 3$, this identifies the permutation tensor component $e_{213} = -1$. The total transverse particle velocity $v_3\}$ behind the shock has a lattice structure velocity term $v_3|$ and an edge dislocation velocity component represented by the integration over the dislocation species $\underline{q}(\underline{\nu})$ in domain Q and the integration over the spatial points in domain $(\underline{x} \rightarrow \underline{\chi})$. In previous work[5-7] the dislocation dependent contribution to the total relative velocity is denoted by $v_i]$ where "]" is used to identify the accumulated discontinuities from dislocation flux and dislocation density rate changes represented by the integrand. The parameter $\underline{b}^*$ in the integrand is the displacement that occurs when dislocation density changes occur, for example when new dislocations are created or when dislocations undergo transitions from $\underline{\nu}^* = 0$ to $\underline{\nu}^* = \underline{\nu}$.

For this simple illustration of deformation across and behind a transverse shock involving a propagating pulse of edge dislocations, equations (2), (3) and (4) imply that the relative transverse particle velocity can be integrated for $(x_2 - X_2) \leq \chi_2 \leq x_2$ to give

$$v_3\}(x_2 \rightarrow \chi_2, t) = v_3|(x_2 \rightarrow \chi_2, t) - b_3 e_{213} \xi_1 \nu_3 (\chi_2 - x_2) D_o - b_3 e_{213} \xi_1 b_3^* V_2 D_o \Big|_{x_2} \tag{5}$$

In equation (5) the discontinuities that contribute to the particle velocity from the dislocation flux term, $\nu_3 D_o$, behind the shock is represented as spatially linear by using a mathematical approximation that smooths the peaks in dislocation density that occur between atomic planes, and the contribution to the particle velocity from dislocation density changes at the shock front located at x_2 contains the term $b_3^* V_2 D_o$. The other contribution to particle velocity is the lattice structure term $v_3|$; which would have a discontinuity at the shock front in order for the shear strain, and consequently, shear stress discontinuities to occur.

As discussed in the previous thermodynamic analysis for dislocation kinetics the amount of work performed by the shear stress must be greater than the local thermodynamic chemical potential for dislocation kinetics [5-7] to occur at and behind the shock front. This thermodynamic chemical potential concept replaces the continuum plasticity concepts of a stress dependent yield condition that has numerous different forms in phenomenological approaches.

In order to illustrate some kinematic and thermodynamic aspects of dislocation kinetics for shear shocks, consider a shock impact that inputs a sustained and constant transverse particle velocity $v_3\}_o$ that propagates through a material which is initially at rest and stress free. Thus, the particle velocity $v_3\}_o$ must be attained from the lattice structure velocity $v_3|$ and the dislocation dependent velocity $v_3]$ of equation (5) evaluated at $\chi_2 = x_2 - X_2$, which is the end of the dislocation pulse. As discussed earlier, at the point, $x_2 - X_2$, a dislocation shock occurs to terminate the dislocation pulse; therefore, the relative velocity expression across the shock plus the dislocation pulse becomes

$$v_3\}_o = v_3\}(x_2 \rightarrow x_2 - X_2,t) = v_3|(x_2 \rightarrow x_2 - X_2, t) - b_3e_{213}\xi_1\nu_3X_2D_o \tag{6}$$

$$- b_3e_{213}\xi_1b_3{}^*V_2D_o|_{x_2} + b_3e_{213}\xi_1b_3{}^*V_2D_o|_{x_2-X_2}$$

For purposes of illustration, it is assumed that the lattice structure velocity $v_3|$ has a discontinuity at the shock front and is then constant behind the shock front so that the recoverable shear strain and the associate shear stress are also uniform in the domain ($x_2 \rightarrow x_2 - X_2$) of the dislocation pulse. Then neglecting the contributions to particle velocity from the dislocation density changes at x_2 and $x_2 - X_2$, equation (6) represents a linear increase in transverse particle velocity behind the shock due to the propagating pulse of edge dislocations behind the shock front. Thus, the shearing gradient of the nonrecoverable transverse velocity term from this dislocation pulse is constant and is given by

$$\Delta_2 v_3\} = - b_3e_{213}\xi_1\nu_3D_o \tag{7}$$

which is the "plastic strain rate" in dislocation theory.[8,9]

Finally, the nonrecoverable relative displacement from the dislocation flux term, ν_3D_o, over the pulse width can be obtained by a time integration; and for uniform shock propagation velocity V_2 it is a quadratic function of the pulse width, given by

$$u_3](x_2 \rightarrow x_2 - X_2,t) = - b_3e_{213}\xi_1\nu_3D_oX_2^2/2V_2 \tag{8}$$

The rate energy is dissipated by the dislocation flux term is the nonrecoverable work rate per unit volume performed by the shear stress[5,6] and is given by

$$\Delta\dot{W}|_{\text{dislocation flux}} = - \sigma_{32}b_3e_{213}\xi_1\nu_3D_o \tag{9}$$

where the shear stress for small strain elasticity is

$$\sigma_{32} = G\, v_3|/V_2 \tag{10}$$

Here, G is the elastic shear modulus and $v_3|/V_2$ is the recoverable or elastic strain across the shock front. Then the nonrecoverable work rate for the dislocation pulse volume behind the shock front is given by

$$\Delta\dot{W}\,\Big|_{\text{dislocation flux}} = - \sigma_{32}b_3e_{213}\xi_1\nu_3D_oX_2 \tag{11}$$

From the thermodynamics of dislocation kinetics,[5-7] this part of the work rate is converted to entropy production; and would result primarily in a localized temperature increase behind the shock front. For narrow dislocation pulse widths and high dislocation flux densities, a rapid temperature increase would occur and would dissipate the energy of a transverse shock front. Thus, for liquids containing dense sets of dislocations,[6] the propagation of shear shocks over long distances is not possible.

ACKNOWLEDGMENT

The authors wish to thank Lydia Grabowski for the excellent and timely typing of this paper.

REFERENCES

1. Gupta, Y.M., Shear measurements in shock-loaded solids, Appl. Phys. Lett., 29:694 (1976).

2. Kim, K.S., Clifton, R.J., and Kumar, P., A combined normal and transverse displacement interferometer with an application to impact of y-cut quartz, J. Appl. Phys. Res., 88:4304 (1983).

3. Gupta, Y.M., Shear and compression wave measurements in shocked polycrystalline Al_2O_3, J. Geophys. Res., 88:4304 (1983).

4. Clifton, R.J., Pressure-shear impact and the dynamic plastic response of metals, in "Shock Waves in Condensed Matter - 1983," J. R. Asay, R.A. Graham, and G. K. Straub, editors, Elsevier Science Pub. B.V. (1984).

5. Stout, R.B., Modelling the deformations and thermodynamics for materials involving a dislocation kinetics, Crys. Lattice Defects, 9:65 (1981).

6. Stout, R.B., A model for the deformations and thermodynamics of liquids involving a dislocation kinetics, Rheol. Acta, 21:659 (1982).

7. Stout, R.B., On kinematic and thermodynamic conditions for mass, dislocation, momentum, and energy densities across a propagating surface of discontinuity, Lawrence Livermore National Laboratory Report UCRL-92815, (1985).

8. Gilman, J.J., Dislocation dynamics and the response of materials to impact, Appl. Mech. Rev., 21:767 (1968).

9. Hirth, J.P. and Lothe, J., "Theory of Dislocations," McGraw-Hill, NY (1968).

TEMPERATURE DEPENDENCE OF THE PRECURSOR AMPLITUDE IN <111> LITHIUM FLUORIDE.[a]

S. E. Arione[b] and G. E. Duvall

Shock Dynamics Laboratory
Department of Physics
Washington State University
Pullman, WA 99164-2814

ABSTRACT

Undoped single crystals of lithium fluoride were shocked along the <111> direction with initial stresses above the dynamic yield point. Prior to impact, the samples were heated to a uniform temperature ranging from 151 to 723 K and the relation between the precursor amplitude and the sample temperature was found to be a rapidly decreasing function. This dependence on the temperature of dynamic strength of lithium fluoride under <111> loading is similar to the temperature dependence of the critical shear stress in lithium fluoride under quasi-static torsion. This similarity supports the view that in both dynamic and quasi-static regimes, the secondary slip systems were activated.

INTRODUCTION

In this paper we report the results of a set of experiments in which samples of single crystal, undoped LiF were impacted with a rapidly moving projectile. The experiments were designed in a way that made sample temperature the independent variable of interest. The principal goal was to find how the amplitude of the elastic precursor changed with temperature when the shock propagates along the <111> crystal direction. This geometry is of interest because there is no shear component of stress along any of the primary slip systems, the activation of which is believed to be responsible for the double wave profile in crystals impacted along the <100> direction.[1,2]

In the <111> impact case we attempted to find correlations between the precursor amplitude and the critical shear stress on the secondary slip systems [100]<100>. Specifically, we explored the relation between material strength and temperature at high strain rates and compared our results to those of Gilman[3] at low strain rates.

Gilman's experiments consisted of applying quasi-static torsion to a cleaved parallelepiped[4] of undoped material.[5] The sample temperature was increased up to 800 K and the critical shear stress at the yield point was recorded for each temperature. The results showed that brittle fracture, not yielding, occurred at temperatures less than roughly 450 K. Yielding occurred at higher temperatures and the critical shear stress decreased rapidly as temperatures increased from 450 K to 800 K. Because the applied torsion had no shear component of stress along the primary slip systems, Gilman concluded that the secondary slip systems were activated.

The experiments performed here were similar to Gilman's except that we used shock loading along the <111> crystal vector. This impact direction has nonzero shear components along the secondary slip system. If these slip systems are activated then we would expect a variation of the precursor amplitude with temperature which is similar, at least qualitatively, to the temperature dependence of the critical shear stress under quasi-static torsion tests.

DETAILS OF THE EXPERIMENTS

Small disks of undoped LiF were purchased from the Harshaw Chemical Company. The disks were cut by the manufacturer and their impact orientation was tested, by x-ray Laue diffraction, to be within 2 degrees of the <111> direction. Small pieces of a sample were used to test the static yield strength of the material (the method is described in Ref. 6) and dislocations were counted by the etch pit method.[7] The results were $Y \approx 10^{-2}\,kb$ and $N_o \approx 2 \times 10^5 cm^{-2}$ Samples that were impacted had thicknesses ranging from 1.57 to 4.33 mm. Sample thickness, however, was assumed not to influence the precursor amplitude because Rosenberg's experiments at room temperature showed the precursor amplitude to be independent of sample thickness between 0.4 and 3 mm.[8]

Each crystal was mounted on a target holder whose design differed depending on the temperature used for that experiment. Room temperature experiments had a simple target design with epoxy holding the sample to a target ring and mounted at the receiving end of a 4 inch gas gun. For heated experiments, a furnace was designed consisting of an aluminum body housing a heating nichrome coil. The target body was attached to the gas gun assembly by three extending legs which terminated in insulating lava pads. The sample was pressed by screws at the center of the target and was impacted by a projectile nose (usually made of copper). The low temperature target assembly was the same as that used and described by Vorthman.[9]

The data were collected using the VISAR[10] with the exception of the cold target experiment where a quartz gauge method was used. Table I gives a summary of the parameters used in each experiment and Table II gives the precursor amplitudes associated with each temperature. A graph of precursor amplitude versus temperature is given in Fig. 1.

TABLE I. - EXPERIMENTAL PARAMETERS

Shot No.	Sample Thickness (mm)	Projectile Velocity (mm/μs)	Projectile Type	Sample Temperature (K)	Window[a] Type
83-26	1.96	0.354	Cu	Room	Vacuum
83-39	1.567	0.638	Al	Room	Vacuum
83-38	2.39	0.637	Al	373	Vacuum
84-07	2.974	0.578	Cu	723	Fused Silica
84-16	3.086	0.594	Cu	573	Fused Silica
84-18	2.934	0.589	Cu	481	Fused Silica
84-23	4.000	0.54	Cu	382	Fused Silica
84-27	4.331	0.527	Cu	151	

a. Indicates type of window behind the sample for the VISAR set up.

ANALYSIS AND DISCUSSION

The amplitude of the precursor as given in Table II was obtained from the VISAR data reduction procedure but required two corrections: (a) correction for reflection at the VISAR window interface and (b) correction due to the temperature dependence of the elastic Hugoniot of both sample and VISAR window backing. The window reflection was corrected by assuming that the wave reflected into the LiF was purely elastic. The temperature dependence of the elastic Hugoniots for LiF was estimated using data available for the temperature dependence of the LiF elastic constants.[6,11] An analytical method was used to estimate the same correction for the fused

TABLE II - PRECURSOR AMPLITUDE DATA

Shot No.	Sample Temperature	Corrected Precurser Amplitude (kb)
85-26	Room	43.5
85-39	Room	42
83-38	373	37.5
84-07	723	10
84-16	573	21
84-18	481	29
84-23	382	36
84-27	151	39.5

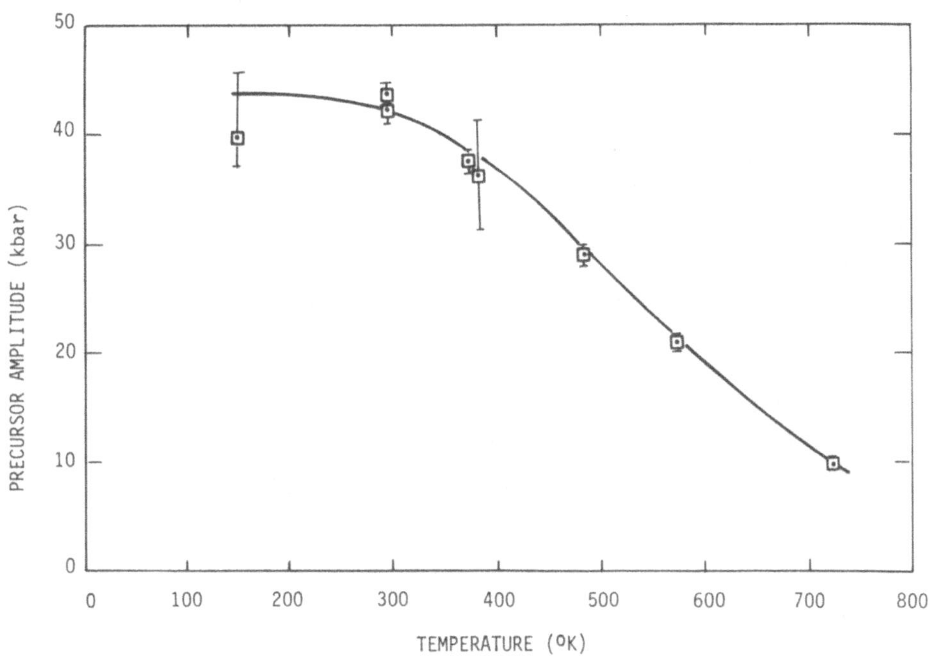

Fig. 1. Corrected Precurser Data. Error bars are maximum estimate (Ref. 6).

silica backing window but no significant deviation was noted from the room temperature Hugoniot curve.[13]

The results obtained in these experiments and summarized in the graph of Fig. 1 show a decrease in the precursor amplitude with temperature which has many qualitative similarities to the quasi-static torsion experiments.[3] Added to the already available evidence, especially the work of Kidron,[12] we conclude that activation of the secondary slip system is responsible for the observed precursors in <111> LiF.

The exact microscopic mechanism for slipping is not known. In the past it has been customary to use dislocation theory to explain the slip activation process, but the results have not been very successful with some serious difficulties still unresolved. We believe that under shock loading conditions dislocation theory plays a role, but the exact mechanisms are significantly different from those operating at low strain rates. This will be a subject of future work.

Finally, it should be pointed out that the VISAR records of our experiments were generally noisy. Test shots on samples of fused silica at room temperature were noticeably less noisy than those with lithium fluoride samples. We don't fully understand why this was so. The rapid jump of the LiF precursor amplitude and the changes in the sign of the particle acceleration in the stress relaxing region, may not be ideal features for the VISAR to record. We also recorded rise times for the precursor amplitude of typically 20 nsec, which is longer than one would expect when using the VISAR.

The experiments were difficult to perform and the amount of data collected was less than originally planned. It is recommended that these results be considered as preliminary until they are tested by either a different diagnostic method or by additional experiments.

ACKNOWLEDGMENTS

We are grateful to Mr. Paul Bellamy for his help with the experiments and Mr. Alan Ricker for contributing to the development of the WSU/VISAR data reduction program.

REFERENCES

a. Work supported in part by National Science Foundation grant number DMR 8212454 and the Sigma Xi Foundation.

b. Present Address: Boeing Aerospace Company, MS 87-68, Box 3999, Seattle, WA 98124.

1. J. R. Asay, G. R. Fowles, G. E. Duvall, M. H. Miles, and R. J. Tinder, J. Appl. Phys. 43:2132 (1972).
2. Y. M. Gupta, G. E. Duvall, G. R. Fowles, J. Appl. Phys. 46:532 (1975).
3. J. J. Gilman, Acta Met. 7:608 (1959).
4. In LiF this parallelepiped face is a [100] plane.
5. Gilman is not specific about the exact degree of purity of his crystals.
6. S. E. Arione, Ph.D. Thesis, Washington State University (1985).
7. J. J. Gilman and W. G. Johnston, Solid State Physics 13:147 (1962).
8. G. Rosenberg and G. E. Duvall, J. Appl. Phys. 51:319 (1980).
9. J. E. Vorthman and G. E. Duvall, J. Appl. Phys. 52:764 (1981).
10. L. M. Barker and R. E. Hollenbach, J. Appl. Phys. 43:4669 (1972).
11. J. E. Vorthman, Ph.D. Thesis, Washington State University (1979).
12. N. Kidron, Ph.D. Thesis, Washington State University (1983).
13. This was mainly due to the small value of the coefficient of thermal expansion for fused silica.

DISLOCATION MOBILITY IN HIGH PURITY LiF FROM 100°K to 300°K*

G. Meir

Department of Mechanical Engineering and Mechanics
Drexel University,
Philadelphia, PA 19104

R.J. Clifton

Division of Engineering
Brown University,
Providence, RI 02912

INTRODUCTION

Understanding of the dynamic plastic response of polycrystalline materials requires understanding of the stress and temperature dependence of dislocation mobility in single crystals. At low driving stresses dislocation mobility can be described by a thermal activation model, based on a mechanism of jerky glide of dislocations between obstacles. The average dislocation velocity increases with both stress and temperature. On the other hand, at sufficiently high stresses, such as those generated in plate impact experiments on high purity single crystals, the primary resistance to the motion of dislocations is that of the clear lattice. Under these conditions the dislocation mobility is usually regarded as described by a viscous-drag model of the form

$$\tau b = B\bar{v}_d \tag{1}$$

where τ is the resolved shear stress, b is the Burgers' vector, B is the so-called drag coefficient and $\bar{v}_d$ is the average dislocation velocity. According to currently accepted models for dislocation-phonon interaction, the drag coefficient B is expected to be proportional to the absolute temperature at temperatures from, say, 0.1 $_D$ to 1.0 $_D$ is the Deye temperature of the crystal.

*Based on results presented in a thesis submitted by the first author in partial fulfillment of the requirements for a Ph.D. degree at Brown University.

In order to investigate the validity of the viscous drag model (1) and the proportional dependence of B on temperature we use a so-called "plate impact recovery experiment" in which a plane stress pulse of a known shape is passed through a specimen. The specimen is recovered after the experiment and the resulting dislocation configurations are compared to the configuartions that existed in the original states. Recovery experiments have the attractive feature that microstructural phenomena can be observed directly and be related to the known stress history. Flinn et al.[1,2] introduced a technique in which a flat specimen is impacted by a flyer plate generating a sub-microsecond pulse. Kumar and Clifton[3] improved this method by using a gas-gun for driving the flyer plate, and by using a momentum trap and a star-shaped flyer plate to prevent repeated loading of the specimen by reverberating pulses. In the current work, the Kumar-Clifton technique was modified to allow for cooling of the specimen to liquid nitrogen (LN_2) temperature prior to impact. Room temperature experiments were performed for cross-reference with previous data on dislocation mobility.

EXPERIMENTAL TECHNIQUE

Experiments on pure single crystals of LiF were conducted in the 63.5mm gas gun facility at Brown University. In order to allow reliable interpretation of the stress history by means of linear elastic wave theory the impact velocities were kept as low as possible to avoid so-called "precursor-decay" phenomena and the associated stress relaxation which significantly affect the shape of the stress pulse. Impact velocities of 0.01-0.04mm/μsec were used, corresponding to normal stresses of 103-343 MPa. The resolved shear stresses on the 4 primary slip systems of LiF {110}<1$\bar{1}$0> that are active under [100] impact were 36.8-78.6 MPa. Since the quasistatic yield stress for these annealed LiF specimens is only 1 MPa (in shear) the given range of impact velocities causes stresses that are in the high stress or viscous drag regime.

Specimens of 22 x 22 mm and thickness of 2.5-5 mm were cleaved from larger crystals obtained from Harshaw Chemicals, as "vacuum ultraviolet" grade of 99.999% purity. A schematic of the experimental configuration is shown in Fig. 1. The specimen is sandwiched between a star-shaped plate at the front[3,4,5] and a momentum trap at the back; both are made of Aℓ6061-T6 which has nearly the same acoustic impedance as that of LiF. Cooling is provided by a front plate, through which LN_2 flows until the desired temperature is reached. Temperature is measured by thermocouples attached to the cooling plate and to the momentum trap. A sub-microsecond pulse is generated by a 1 mm thick aluminum flyer plate, mounted on a foam support. Simultaneous with the impact of the cooling plate by the flyer, the projectile is stopped by a steel anvil in which the target assembly is mounted. Consequently, the only momentum transferred to the target is that of the flyer plate. After each shot the LiF specimen is recovered, cleaved and etched, using the Gilman-Johnston technique. Fourteen experiments were conducted at temperatures ranging from 110°K to room temperature. Velocity-time profiles of the motion at the back of the momentum trap were recorded by using laser interferometry.

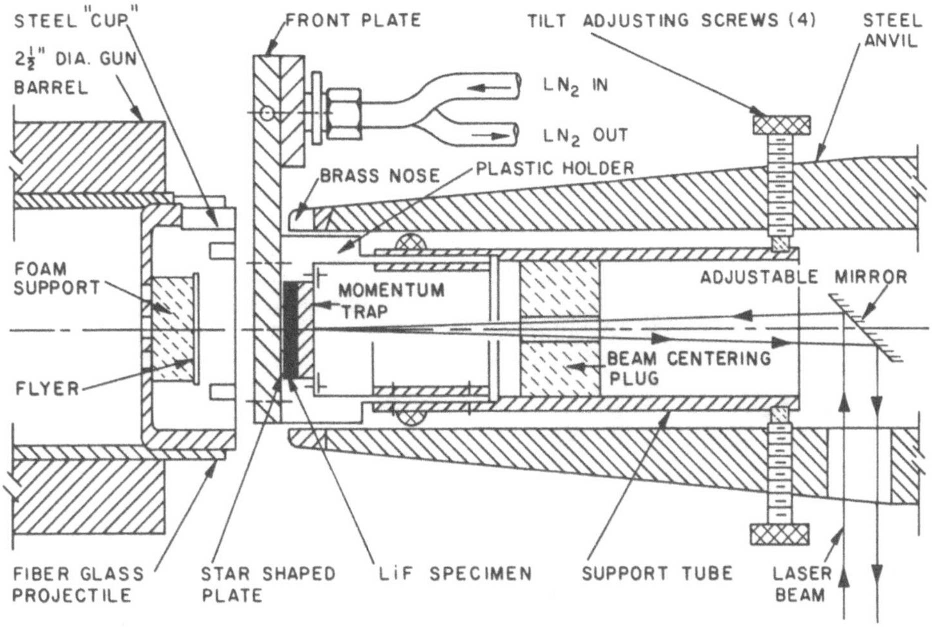

Fig. 1. Experimental Configuration for Plate Impact Recovery Experiments at Low Temperatures.

EXPERIMENTAL RESULTS AND DISCUSSION

A typical experimental record is shown in Fig. 2. Two oscilloscope traces are shown. They were obtained from two 180° out-of-phase normal displacement interferometers (NDI's) with overlapping windows of 190nsec. Dashed lines in the figure represent the velocity profile predicted for purely elastic behavior of the specimen and no vacuum gaps between the various plates. The "vacuum gap" shown was caused by a gap of less than 1μm between the specimen and the momentum trap. Because of this gap the leading part of the pulse was reflected and travelled a full round-trip in the specimen. The late pulse is a reverberation of the main pulse in the momentum trap, already separated from the specimen; hence, the late pulse is not part of the stress history of the specimen. The observed attentuation of the amplitude of the vacuum gap pulse compared to the amplitude of the original pulse, is an indication of the plastic flow in the specimen due to the main pulse.

Isolated glide bands, which appear as linear arrays of etch pits in recovered specimens, are assumed to have circular shapes and to nucleate at their centers. Dislocation velocities are obtained by dividing the band radius by the duration of the stress pulse. Average dislocation velocities in this investigation were 0.36-0.67 km/sec. The value obtained for the drag coefficient at room temperature is 2.9×10^{-5} Pa-sec, which is consistent with existing data on dislocation mobility. At 110°K the average value of B is $1.8\text{-}2.5 \times 10^{-5}$Pa-sec. The uncertainity in B at low temperatures is due largely to deviation of the stress pulses from the nominal square pulses. The values obtained for B at 110°K are twice as large as the value of 1.1×10^{-5} Pa-sec predicted by proportional scaling of the value for room temperature. This discrepancy suggests that

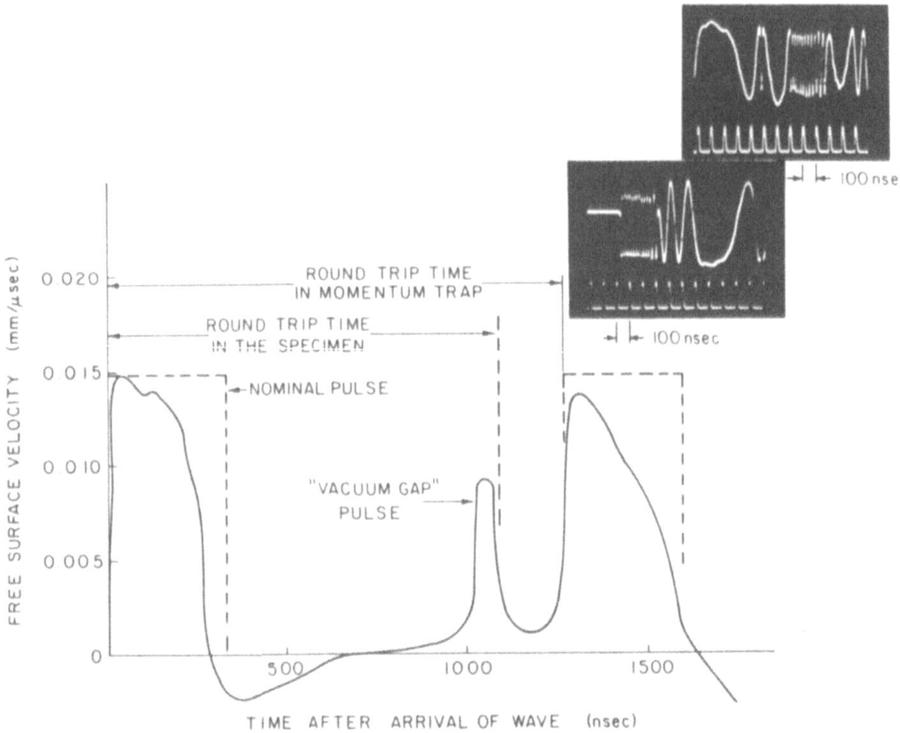

Fig. 2. Velocity-time Profile at the Rear Surface of the Momentum Trap.

mechanisms of phonon-dislocation interactions need to be reexamined.

Besides the increased length of glide bands there are other indications of the increased mobility of dislocations at low temperatures: (i) less cracking of recovered specimens in cold shots - 110°K specimens were recovered with no cracks at all; (ii) significantly larger precursor decay observed at low temperatures; (iii) larger attenuation of vacuum-gap pulses at low temperatures. The enhanced mobility of dislocations at low temperatures is consistent with the observations of Vorthman[6,7] who reported increased precursor decay at low temperatures in experiments in which specimens were not recovered. At room temperature the extensive generation of dislocations at surfaces and subgrain boundaries provides a reasonable explanation for the precursor decay anomaly in pure LiF (Meir and Clifton[8]).

The experiments suggest the need for further investigation of the mechanisms for the nucleation of dislocations at surfaces, and for the formation of glide bands. At room temperature, surface nucleated dislocations form bands that are longer than the average length of bands nucleated at subgrain boundaries and in the bulk; the dislocation velocities calculated for these bands are well below the corresponding shear wave speed. At higher stresses surface dislocations tend to form "heavier" and longer glide bands, but the general form of the dislocation configurations is essentially the same as at low stresses. On the other hand, at low temperature increasing the stress causes a marked change in dislocation configurations. Most surface-nucleated dislocations form heavy glide bands that appear to have propagated at supersonic velocities (~6km/sec), whereas very little nucleation of dislocations takes place in the interior and at subgrain boundaries. This suggests that low temperatures and higher stresses enhance the formation and propagation of glide bands, which are a form of unstable shearing deformation. Furthermore the experimental results suggest that glide bands can propagate at supersonic speeds, whereas supersonic velocities appear to be unlikely for the motion of a single dislocation.

ACKNOWLEDGEMENT

This research was supported by the NSF/Brown MRL at Brown University.

REFERENCES

1. J.E. Flinn and R.F. Tinder, Scripta Met. 8: 689 (1974).
2. J.E. Flinn, G.E. Duvall, G.R. Fowles and R.F. Tinder, J. Appl. Phys. 46: 3752 (1975).
3. P. Kumar and R.J. Clifton, J. Appl. Phys. 50: 4747 (1979).
4. P. Kumar and R.J. Clifton, J. Appl. Phys. 48: 4850 (1977).
5. K.S. Kim and R.J. Clifton, J. Mat. Sci. 19: 1428 (1984).
6. J.E. Vorthman and G.E. Duvall, J. Appl. Phys. 52: 764 (1981).
7. J.E. Vorthman and G.E. Duvall, J. Appl. Phys. 53: 3607 (1982).
8. G. Meir and R.J. Clifton, J. Appl. Phys. (accepted for publication, 1985).

THE DYNAMIC RESPONSE OF POROUS CALCIUM CARBONATE MINERALS*

D. S. Drumheller and D. E. Grady

Thermomechanical and Physical Division
Sandia National Laboratories
Albuquerque, NM

ABSTRACT

A theoretical study of the shock-loaded response of calcium carbonate materials is presented in which both dry and water-saturated samples with porosities up to 50 percent are considered. Data are presented for the unloading response from 15.0 and 18.5 GPa, and calculations from a mixture model using a Mie-Grüneisen equation of state with volume-dependent parameters are compared to both the Hugoniot and the isentropic unloading response.

INTRODUCTION

When porosity is present in mineral samples, shock loading results in greatly elevated temperatures and energies. It is recognized[1,2] that Hugoniot data from initially porous samples provide a method for the experimental probing of a more expanded region of the equation-of-state surface of the constituent mineral.

When a second substance, such as water, is introduced into the pores of the material, the high pressure constitutive response becomes more complex. The methods of mixture theory must be brought to bear, and attendant questions regarding mechanical and thermal interaction between components must be addressed.

The materials of interest in the present investigation are porous samples of the mineral calcium carbonate, in both a dry porous form and containing water to various degrees of saturation. This mineral is the primary constituent of limestone, marble and chalk, and as such, constitutes a significant portion of the earth's surface material.

*This work performed by Sandia National Laboratories supported by the U. S. Department of Energy under contract #DEAC04-76-DP00789.

The most extensive high-pressure study of this mineral has been reported by Kalashnikov, et al.[3]. Their work provides Hugoniot data to nearly 100 GPa for dry crystalline calcite, two porous calcites (chalk) tested in both the dry and water-saturated condition, and two other minerals of dolomitic composition. Their study also includes a preliminary analysis using a mixture theory.

The present study focuses on a more detailed application of mixture theory to this problem. The theory is compared with the data of Kalasnikov, et al.[3]. Recent high pressure shock and release measurements on porous calcite are also reported and preliminary comparisons with these data are made.

THEORY

Wet calcium carbonate is a mixture of three constituents, solid calcium carbonate, water, and void. It is modeled by using a theory of immiscible mixtures proposed by Drumheller and Bedford[4]. A constituent of this mixture is labelled C_ξ, where $\xi = s$, f, or v for solid, fluid, or void. The intrinsic mass density of C_ξ, $\bar{\rho}_\xi$, is the ratio of the mass of C_ξ divided by the portion of the mixture volume occupied by C_ξ. The volume fraction of C_ξ, ϕ_ξ, is the ratio of the volume occupied by C_ξ over the volume occupied by the mixture. The ϕ_ξ always satisfy the condition

$$\sum \phi_\xi = 1 \tag{1}$$

where the summation sign implies a sum over all constituents C_ξ. The partial mass density of C_ξ is denoted by ρ_ξ and is given by

$$\rho_\xi = \phi_\xi \bar{\rho}_\xi. \tag{2}$$

In situations where the C_ξ have independent motions, balance of mass and momentum equations are required for each constituent. In this analysis it is assumed that all the C_ξ have identical motions. As a result the summed balance of mass and momentum relations for the mixture are sufficient to specify the solution to the problem. These relations are identical to the traditional continuum relations in which the density ρ and the pressure P are identified with the mixture mass density $\rho = \sum \rho_\xi$ and the constituent pressures P_ξ, so that $P = P_\xi$. Because of the assumption of equal pressures between the C_ξ, $P_\xi = 0$ when void is present.

It is assumed that C_s has an equilibrium pressure given by the Mie-Grüneisen relation

$$P_s = \frac{\bar{\rho}_{s0} c_{s0}^2 \eta_s}{(1 - s_s \eta_s)^2} \left[1 - \frac{1}{2}\Gamma_{s0}\eta_s\right] + \bar{\rho}_s \Gamma_s e_s \tag{3}$$

and

$$\bar{\rho}_s \Gamma_s = \bar{\rho}_{s0} \Gamma_{s0} (1 - \eta_s)^n \tag{4}$$

where $\eta_s = 1 - \bar{\rho}_{s0}/\bar{\rho}_s$ and e_s is the specific internal energy. The material constants $\bar{\rho}_{s0}$, c_{s0}, Γ_{s0}, and n are intrinsic properties of the calcite mineral. A variable Grüneisen parameter Γ_s is assumed ($n = 0$ corresponds to the common assumption that $\bar{\rho}_s \Gamma_s = \bar{\rho}_{s0} \Gamma_{s0}$). Equation 4 is equivalent to that proposed by Al'tshuler[1].

The C_f has an equilibrium pressure given by a tabular equation of state[5] based on Hugoniot data for water to 100 GPa.

The traditional balance of energy relation is used to describe the conservation of the mixture energy, where this energy is given by $e = \sum \rho_\xi e_\xi / \rho$. Two possible thermal states are considered. In the first case the temperatures of the water and the solid are equal. In the second case the energy in the water is given by

$$\rho_f \dot{e}_f = \phi_f P_f \frac{\dot{\bar{\rho}}_f}{\bar{\rho}_f}. \tag{5}$$

This second condition implies that the dissipative heating of the mixture occurs within the solid and does not influence the water. In addition, local heat transfer between the solid and the water is neglected. Consequently all of the shock heating is produced and remains within the solid. These two cases represent reasonable extremes in the possible assumptions for the thermal behavior of the mixture. Comparison of the solutions for these two cases will illustrate the sensitivity of the model to these types of mixing assumptions.

These relations were studied using a steady-wave analysis; that is, the solution was assumed to depend only on the factor $\Omega = x - Ut$, where U is the velocity of the wave, x is the spatial position, and t is the time. For the equal-temperature case the resulting relations can be integrated to obtain jump conditions for the steady wave. When Eqn. 5 is used the equations must be integrated numerically.

RESULTS

Figure 1 contains comparisons of calculated Hugoniots to the data of Kalashnikov, et. al.[3] and Tyburczy and Ahrens[6] for various initial mass densities. The model parameters used in these calculations are: $\bar{\rho}_{s0} = 2710$ kg/m^3; $c_{s0} = 3.65$ km/s; $\Gamma_s = 0.8$; and $n = -2.25$. The porosities of these samples, which can be

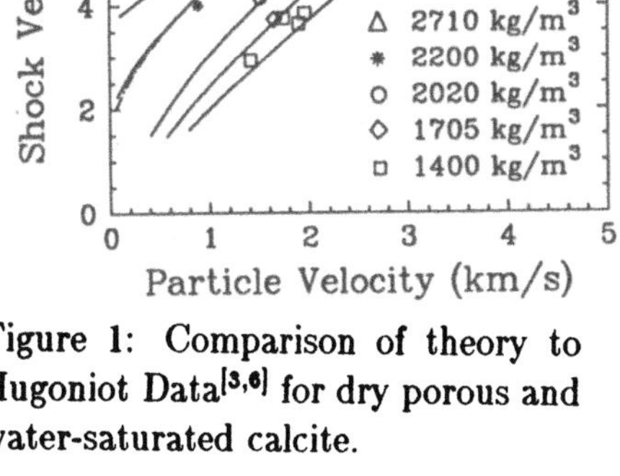

Figure 1: Comparison of theory to Hugoniot Data[3,6] for dry porous and water-saturated calcite.

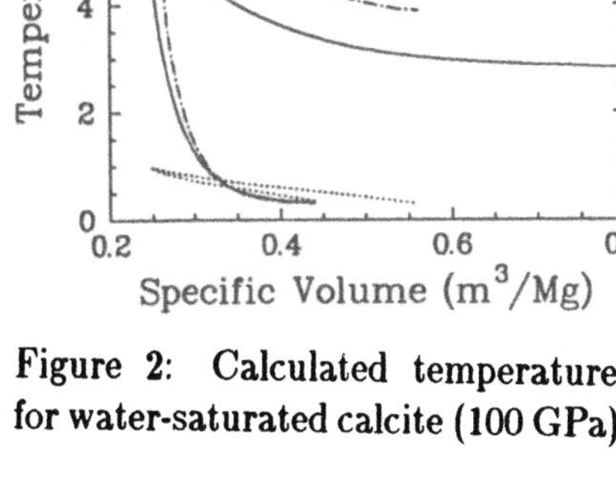

Figure 2: Calculated temperatures for water-saturated calcite (100 GPa).

computed from the listed initial densities range from zero to approximately fifty percent. With the exception of the water-saturated sample ($\rho_0 = 2200$ kg/m^3), all of these data are for dry chalk. Two calculations are shown for the water-saturated case. These curves correspond to the two thermal assumptions described in the previous section.

Good agreement is achieved for the dry samples; however, some discrepancy is observed in the comparison to the water-saturated sample. Kalashnikov, et al.[3] have also proposed a model for water-saturated chalk. They reported better agreement for the water-saturated case; however, they did not include a description of their equation of state for water.

The calculated Hugoniots in Figure 1 for the water-saturated case are relatively insensitive to the choice between the two thermal assumptions. The temperatures of the calcite and the water for these two cases is shown in Figure 2. Temperatures corresponding to the release paths from 100 GPa are also shown. These temperatures were computed by assuming a constant specific heat for calcite of 1200 J/kg K[7]. The porosity of the sample results in very high temperatures, and the values of the temperatures are strongly dependent upon the thermal assumption.

In Figure 3 the calculation is compared with release isentropes obtained from shock compression and unloading experiments on dry calcite samples with initial density of 1600 kg/m^3 and a porosity of approximately 40 percent[8]. Release paths from 15.0 and 18.5 GPa were determined from the measured wave profile data. The comparison in Figure 3 is based on the same volume dependent Grüneisen parameters used to model the Hugoniot data in Figure 1. Considering the current uncertainty in the release curves, a better fit was not attempted. However, for contrast a calculation based on constant $\rho\Gamma$ is shown in Figure 4. This assumption clearly leads to excessive volumetric expansion for the release isentropes, although virtually identical agreement with the Hugoniot data in Figure 1 can be achieved.

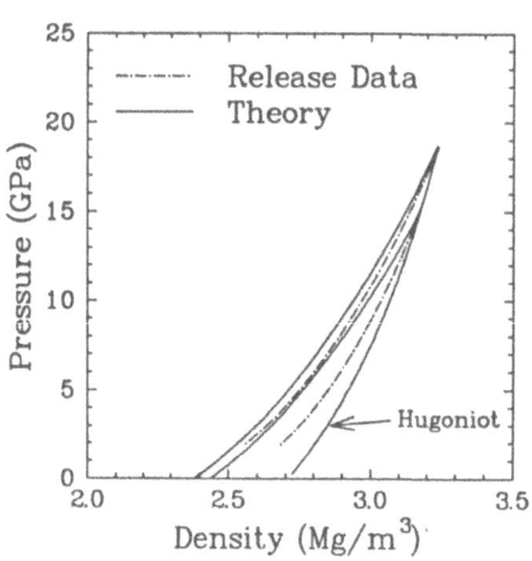

Figure 3: Comparison of the volume-dependent $\rho\Gamma$ theory to release data for 1600 kg/m^3 porous calcite.

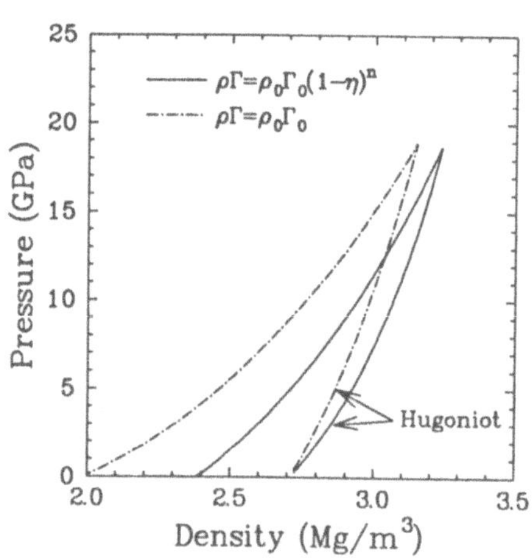

Figure 4: Comparison of Hugoniot and release paths calculated from two Mie-Grüneisen forms for 18.0 GPa data shown in Figure 3.

It is concluded that an overall better fit of a temperature-independent Grüniesen equation of state to both the high-pressure Hugoniot data and the release data is provided with a Grüneisen parameter which decreases with increasing specific volume. This would seem to be inconsistent with common models for solids from which the opposite behavior is predicted. Temperatures of 2500 K were calculated for the compressed state of the 1600 kg/m^3 sample. This strongly suggests that calcium carbonate may be a liquid at the Hugoniot state or on the release isentrope. Theoretical studies described by Royce[9] predict values in the liquid range which are significantly lower than those for the solid at the same specific volume. This is consistent with the trend observed in the present analysis.

ACKNOWLEDGEMENTS

The authors would like to thank Jeff Swegle for calculations needed to reduce the porous calcite data.

REFERENCES

1. L. V. Al'tshuler, K. K. Krupniskev, B. N. Ledenev, V. T. Zhnchikhin, and M. I. Brahgnik, Dynamic Compressibility and Equation of State of Iron Under High Pressure, Sov. Phys. JETP, 34, 606, 1958.
2. I. C. Skidmore and L. Morris, Experimental Equation of State for Uranium and its Interpretation in the Critical Region, Symposium on Thermodynamics of Nuclear Material, Vienna, May 21-25, 1962.
3. N. G. Kalashnikov, M. N. Pvlovskiy, G. V. Simakov, and R. F. Trunin, Dynamic Compressiblity of Calcite-Group Minerals, Izv., Earth Physics, 2, 23, 1973.
4. D. S. Drumheller, and A. Bedford, Arch. Rat. Mech. Anal., A Thermomechanical Theory for Reacting Immiscible Mixtures, 73, 257–284, 1980.
5. G. I. Kerley, Theoretical Equations of State for the Detonation Products of Explosives, Proc. Eighth Sym. on Detonation, July 15-19, 1985.
6. J. A. Tyburczy and T. J. Ahrens, Dynamic Compression and Volatile Release from Carbonates, J. Geophys. Res., to be published.
7. A. P. Dmitriyev, et al., NASA Technical Translation, NASA TT F-684, June 1972.
8. D. E. Grady and R. L. Moody, Shock and Release Equation of State of Calcite, Sandia National Laboratories Rept., SAND85–0947, May 1985.
9. E. B. Royce, High-Pressure Equations of State from Shock-Wave Data, in International School of Physics–Enrico Fermi edited by P. Caldirola and H. Knoepfel, Academic Press, New York, p. 80-91, 1971.

COMPUTATIONAL MODELS FOR NUCLEATION, GROWTH, AND COALESCENCE OF ADIABATIC SHEAR BANDS

D. R. Curran and L. Seaman

SRI International
333 Ravenswood Avenue
Menlo Park, California 94025

INTRODUCTION

Adiabatic shear bands are regions of intense plastic strain that often appear in solids exposed to high deformations at high rates. Nucleation and growth rate functions for adiabatic shear bands form the "damage evolution" relations that are required for any internal state variable description of damage.[1] In the following paragraphs we derive the nucleation and growth rate relations that are currently used in our multiplane plasticity computational model, SHEAR4. (SHEAR4 is described in detail in reference 1.)

NUCLEATION RATE

The conditions for localization of a planar flow perturbation of infinite extent have been studied, both numerically[2] and analytically[3] (also, T. W. Wright and R. C. Batra, personal communication). However, the adiabatic shear bands observed experimentally appear to nucleate at points and grow radially outward like macroscopic dislocations with edge and screw components.[1]

To better understand the nucleation process for such bands, we computationally simulated a situation in which a small planar crack-like imperfection was loaded remotely in pure shear across the crack face. Plane strain, zero friction conditions were assumed. The results are shown in Figure 1. The 4340, RC43 steel stress-strain curve used is shown in Figure 2. We see from Figure 1 that strain concentrations occur at the crack tips and extend about one crack half-length, R, from the tips. There is a low strain "dead zone" extending about R/2 from each crack face. For the 4340, RC43 steel properties used in the calculation, when the remotely applied engineering strain in the crack plane was 0.13, the strain magnification at the crack tips was about 2 for the cell sizes shown in Figure 1. The "jog," or differential slip at the center of the slit faces was about 0.1R, in agreement with observations.[1] Since there is no characteristic size in the constitutive relations, the size of the region of magnified strain at the crack tip, r, is a given fraction of R for a given remotely applied strain as long as material rate effects and inertia are negligible.

If the deformation occurs rapidly enough to maintain adiabatic conditions around the crack, the postulated threshold condition for nucleation is that the thermal softening at the crack tips is sufficient to cause the stress to drop with increasing strain.

Figure 2 shows static adiabatic yield curves for three hardnesses of 4340 steel. Most of the material strength has been lost for strains exceeding 1.0 to 1.4, and the stress begins to drop at strains of 0.2 to 0.3. The nucleation threshold condition is thus that the local strain at the crack tip reach 0.2 to 0.3.

Our proposed criterion for adiabaticity is that significant heat cannot flow a distance R during the loading time of interest. That is, the imperfection must remain thermally autonomous. In Figure 1 the remotely applied engineering strain in the crack plane is approximately related to the jog B by

$$\Delta\varepsilon_R \cong B/R \qquad (1)$$

Furthermore, the remote strain rate is given by

$$\dot{\varepsilon}_R = \Delta U/R \quad , \qquad (2)$$

where ΔU is the difference in particle velocity measured from R/2 above the crack to R/2 below it.

We now state our criterion for adiabaticity:

$$\Delta U > \nu_H \quad , \qquad (3)$$

where ν_H is a pseudovelocity of heat flow. That is, if particles on either side of the crack are moving past one another faster than heat can flow between them, adiabaticity will be approximately maintained.

The value of ν_H was derived from analysis of heat flow away from a hot slab, maintained at constant temperature, suddenly placed in contact with a cold material. The well known heat diffusion equation can be examined to determine the pseudovelocity of propagation of a given temperature. If the temperature chosen is one-fourth that of the hot slab temperature, one finds that

$$\nu_H \cong \frac{10k}{d} \quad , \qquad (4)$$

where k is the thermal diffusivity and d is the distance the heat has flowed, i.e., the distance into the originally cold material.

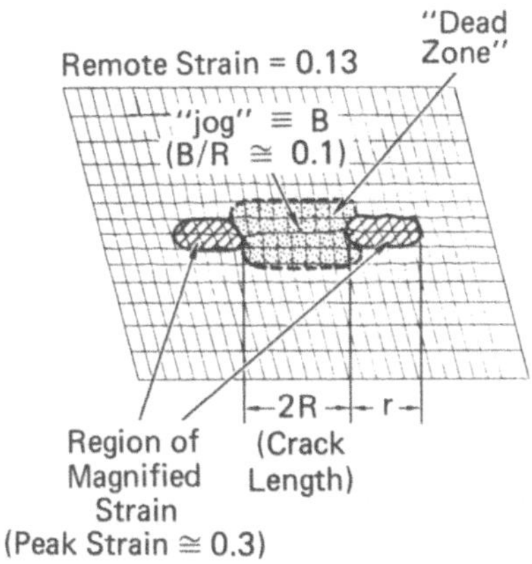

Fig. 1. Computational simulation of plane strain crack loaded in pure shear.

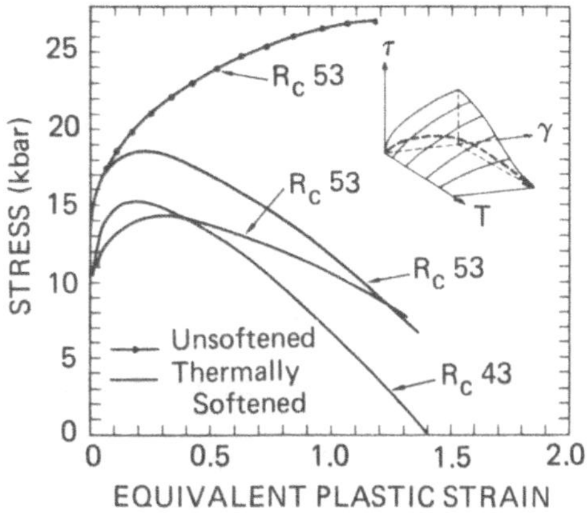

Fig. 2. Static adiabatic yield curves for 4340 steel.

If we now replace d with R in Eq. (4), and combine with Eqs. (2) and (3), we obtain for the adiabaticity criterion:

$$R > \sqrt{\frac{10k}{\dot{\varepsilon}_R}} \tag{5}$$

Inequality (5) provides information regarding the minimum size of crack-like imperfections that can be nucleated. That is, for the experiments discussed in Ref. 1 the strain rates during nucleation were about $10^4\ s^{-1}$. For steels, $k \cong 0.1\ cm^2/s$, so (5) tells us that cracks smaller than about 0.1 mm will not be nucleated (in agreement with observations).

We next discuss the size distributions of the nucleated imperfections. In our experimental observations of shear band size distributions, we found that they usually approximated the form

$$N_g = N_o \exp(-R/R_1) \quad , \tag{6}$$

where N_g is the number of bands per unit volume in a given material plane with sizes greater than R, N_o is the total number of bands per unit volume in the plane, and R_1 is the characteristic size of the distribution.

We let R_1 be the "adiabatic limit size" from (5):

$$R_1 = \sqrt{\frac{10k}{\dot{\varepsilon}_R}} \quad , \tag{7}$$

and we nucleate the bands with the distribution given by Eq. (6).

A problem with Eq. (6) is that it produces a few very large bands upon nucleation. Such large bands are not observed experimentally unless correspondingly large imperfections are initially present. Therefore, SHEAR4 contains an upper cutoff for the imperfection size, R_{max} . For values of $R > R_{max}$, N_g is set to zero. R_{max} must be determined experimentally.

We now derive the nucleation rate function, which we define to be:

$$\dot{N} = \frac{\text{Number of available nucleation sites}}{\text{Incubation time}}$$

$$= \frac{N_o f - N}{\tau} \quad , \tag{8}$$

where

$\dot{N}$ = number of bands nucleated per unit volume per unit time

N_o = density of nucleating imperfections

f = fraction of N_o that can become nucleation sites

τ = incubation time.

Thus,

$N_o f - N$ = current density of nucleation sites.

If the nucleating imperfections are widely spaced so no interaction between them is possible, then $f = 1$. However, if there is a high density of nucleating imperfections, band autonomy requires that the bands be spaced at least one heat flow spacing apart; i.e., the closest possible band spacing is given by the minimum from inequality (5):

$$\text{Minimum spacing} = \frac{10k}{\dot{\varepsilon}_R}$$

If the bands are spaced with the above minimum spacing, the volume per band is on the order of $(10k/\dot{\varepsilon}_R)^{3/2}$. Thus, the density of bands with minimum spacing is on the order of $(\dot{\varepsilon}_R/10k)^{3/2}$.

To cover both widely spaced and closely spaced nucleating imperfections, we take

$$f = \text{Min} \left| (\dot{\varepsilon}_R/10k)^{3/2}/N_o, \quad 1 \right| . \tag{9}$$

We next estimate the incubation time, τ. Referring again to Figure 1, we see that the local strain rate at the crack tip can be written

$$\dot{\varepsilon}_L = M \, \dot{\varepsilon}_R \quad , \tag{10}$$

where M was about equal to 2 in the example calculation. We estimate the incubation time to be the time required to bring the crack tip strain to the completely softened value, ε_m (1-1.4 for 4340 steel), from the nucleation value, ε_{cr}. Thus

$$\tau \cong \frac{(\varepsilon_m - \varepsilon_{cr})}{M \, \dot{\varepsilon}_R} \quad , \tag{11}$$

where ε_{cr} is the adiabatic strain at which the stress begins to drop.

Values of $\varepsilon_m - \varepsilon_{cr}$ can often be obtained from thermal softening data in handbooks. However, a rough estimate can be obtained from the formula

$$\sigma_y \, (\varepsilon_m - \varepsilon_{cr}) \cong \rho \, E_m \quad , \tag{12}$$

where σ_y is the yield strength at the onset of nucleation, ρ is the density, and E_m is the specific melt energy. Since ρ and E_m are easily available material properties and since Eq. (11) is only approximate in any case, we use Eq. (12) to replace (11) with

$$\tau = \frac{\rho E_m}{\sigma_y M \dot{\varepsilon}_R} \tag{13}$$

We now combine Eqs. (8), (9), and (13) to obtain the nucleation rate function:

$$\dot{N} = \frac{\sigma_y}{\rho E_m} M \dot{\varepsilon}_R \left[(\dot{\varepsilon}_R/10k)^{3/2} - N\right] H\left(\varepsilon_R - \frac{\varepsilon_{cr}}{M}\right) \tag{14}$$

for closely spaced nucleation sites, i.e., $(\dot{\varepsilon}_R/10k)^{3/2}/N_o < 1$, and

$$\dot{N} = \frac{\sigma_y}{\rho E_m} M \dot{\varepsilon}_R \left[N_o - N\right] H\left(\varepsilon_R - \frac{\varepsilon_{cr}}{M}\right) \tag{15}$$

for widely spaced nucleation sites, i.e., $(\dot{\varepsilon}_R/10k)^{3/2}/N_o >> 1$, where $H\left(\varepsilon_R - \frac{\varepsilon_{cr}}{M}\right)$ is a Heaviside function.

GROWTH RATE

The criteria for band growth (extension) are that adiabaticity be maintained and that the strain at the band tip attains ε_m. The time to reach ε_m is the incubation time from Eq. (11). The growth rate is thus

$$\dot{R} = \frac{r M}{(\varepsilon_m - \varepsilon_{cr})} \dot{\varepsilon}_R \quad . \tag{16}$$

As discussed earlier, in the absence of material rate effects or inertial effects, r is proportional to R:

$$r = \beta R \quad . \tag{17}$$

Thus, (16) becomes

$$\dot{R}/R = \frac{\beta M}{(\varepsilon_m - \varepsilon_{cr})} \dot{\varepsilon}_R \tag{18}$$

The values of β, M, ε_m, and ε_{cr} are material properties and must come from analyses or numerical simulations like that of Figure 1. This expression is expected to hold up to strain rates at which inertial effects become important. At that point, SHEAR4 limits the band velocity to the Rayleigh wave velocity. Recent work by Wu, Toulios, and Freund[4] also suggests a wave velocity limit.

SIZE OF JOG, B

From Figure 1 we obtained Eq. (1):

$$\Delta\varepsilon_R \cong B/R$$

We can also rewrite Eq (10) as

$$\Delta\varepsilon_L = M \, \Delta\varepsilon_R \quad ,$$

where $\Delta\varepsilon_L$ is the local strain at the band tip. The threshold condition for band nucleation and growth is

$$\Delta\varepsilon_L = \varepsilon_{cr} \quad . \tag{19}$$

Thus,

$$\Delta\varepsilon_R = \frac{B}{R} = \frac{\varepsilon_{cr}}{M} \quad ,$$

$$\text{or} \quad B \cong (\varepsilon_{cr}/M)R \quad . \tag{20}$$

This linear dependence of the jog, B, on band length, R, was observed in the experiments discussed in Reference 1, where B/R was observed to be about 0.1 for 4340 RC43 steel.

DISCUSSION

The derived nucleation and growth relations are approximate and can only serve as starting points for iterative computations with SHEAR4 to simulate experimental results. Nevertheless, comparisons with experimental data show reasonable agreement not only for the form of the equations, but for the numerical values as well.

ACKNOWLEDGEMENT

The calculation of Figure 1 was performed by Thomas Cooper. The research was supported by the U.S. Army Research Office.

REFERENCES

1. D. A. Shockey, D. R. Curran, and L. Seaman, "Development of Improved Dynamic Failure Models," Final Technical Report for U.S. Army Research Office, Research Triangle Park, NC 27709, Contract No. DAA-29-81-K-0123, (15 Feb. 1985).
2. A. M. Merzer, J. Mech. Phys. Solids 30, (1982) 323.
3. T. G. Shawki, R. J. Clifton, and G. Majda, Technical Report No. 3, U.S. Army Research Office Grant DAAG29-81-K-0121 (1983).
4. F. H. Wu, M. Toulios, and L. B. Freund, Brown University NSF Materials Research Laboratory Report No. MRL E-151 (1984).

MODELING OF INSTABILITY AT THE TIP OF A SHEAR BAND

Marc A. Meyers and Shinhou Kuriyama*
Center for Explosives Technology Research
New Mexico Institute of Mining and Technology
Socorro, New Mexico 87801

* Permanent Address:
Institute of Physical and Chemical Research
Wako, Saitama, Japan

INTRODUCTION AND ASSUMPTIONS

The present work focusses on the tip of the shear band, and assumes that the critical phenomena dictating the propagation or arrest of a shear band occur at the tip. This approach is analogous to fracture mechanics in which the crack tip is the region where the relevant processes are taking place, while the crack surfaces are merely the product. The driving energy for the extension of the tip comes from an increase of the imposed displacement, which generates shear stresses and strains. In the analysis presented in this paper the plastic deformation ahead of a shear band is calculated as a function of imposed displacement. A number of assumptions are required to render the problem tractable. The principal assumptions are given and justified below.

a) Negligible flow stress in shear band (behind the tip). Post-deformation measurements have shown that the hardness in the shear band can be very high and often exceeds that of the surrounding material. However, during the process of propagation, plastic deformation is highly localized in the shear-band region, leading to significant temperature increases. Temperatures can approach and possibly exceed the melting point.

b) An adiabatic stress-strain curve represents the plastic deformation process. An elasto-plastic model that would incorporate these variables and also include heat transfer would be exceedingly complex. Since plastic deformation is occurring at a high strain rate, the assumption of adiabaticity is a reasonable one. The use of an adiabatic stress-strain equation was introduced by Olson et al[1] and greatly simplifies computer calculations, allowing one single equation to represent the behavior of the material over a temperature range up to the melting point; the strain rate is assumed constant. For the model presented here the adiabatic stress-strain curve for quenched and tempered HY-TUF steel obtained by Olson et al.[1] is used; the computational predictions are compared with those of an isothermally deformed material.

c) The body is assumed to be in quasi-static equilibrium throughout the deformation process. As such, wave-propagation effects are absent. In order to express the dynamic movement of the body, the stresses are assumed to increase along the adiabatic stress-strain curve (thus, thermal conductivity is not considered) at the high imposed strain rate.

The adiabatic stress-strain curves are characterized by initial work hardening followed by work softening; a plastic instability strain defines the bounds between the two regimes. For the model herein developed, the adiabatic stress-strain curve for a high strength steel determined at a strain rate of $\dot{\gamma} = 10^2\ s^{-1}$ in a torsion test by Olson et al.[1] was used; it is shown in Fig. 1.

Converting Olson et al.'s [1] equation into effective stress versus effective strain:

$$\bar{\sigma} = \bar{\sigma}_o\ (1 + \alpha\bar{\varepsilon}_p)\ \exp(-\beta\bar{\varepsilon}_p) \qquad [1]$$

where $\bar{\sigma}_o$ = 1588.3 MPa, α = 13.6, β = 7.24, $\bar{\varepsilon}_p^i$ = 0.0646, and $\bar{\sigma}_{max}$ = 1868.9 MPa. The material constants for elastic deformation are E = 206.7 GPa (Young's modulus) and ν = 0.28 (Poisson's ratio). In order to compare the propagation of the shear band under an adiabatic condition with the progressive deformation produced under conditions where no instability occurs, a work hardening curve shown by the broken line (Fig. 1) was developed. It simulates the isothermal behavior of the material, if the assumption that instability is generated by thermal softening is a correct one. Up to the instability strain γ_p^i it is very close to the adiabatic curve. Beyond instability, the two curves diverge markedly, with the work-hardening curve being represented by:

$$\bar{\sigma} = 2240\ (\bar{\varepsilon}_p + 0.005)^{0.0649} \qquad [2]$$

These assumptions allow the problem to be modeled by the finite-element method for an elasto-plastic material. The mechanical behavior of the material is assumed to obey the von Mises flow criterion and the incremental theory of Prandtl-Reuss. Deformation of the rectangular body is treated as a plane-strain problem where displacements are given as the boundary condition. The form of elements is triangular, and the displacement function is expressed by a linear polynomial. The adiabatic stress-strain curve given by Equation 1 is used as a constitutive equation in the present code.[2-5]

Metallographic observations indicate that the band thickness, in steels, is in the range 1-3 μm. In the model developed herein the band width was assumed to have a thickness of 20 μm. The material within the band is assumed to have zero strength; the extremity of the band is taken to be approximately semi-circular. Figure 2(a) shows the mesh used to analyse the stresses and strains in the extremity of the shear band. The shape is a rectangle with a width of 400 μm and a length of 500 μm. The shear band is represented by a notch with a semi-circular extremity with a 10 μm radius (BAB') in Figure 2 (a); the notch has a depth of 60 μm and a width of 20 μm. The mesh contains 318 elements and 176 nodal

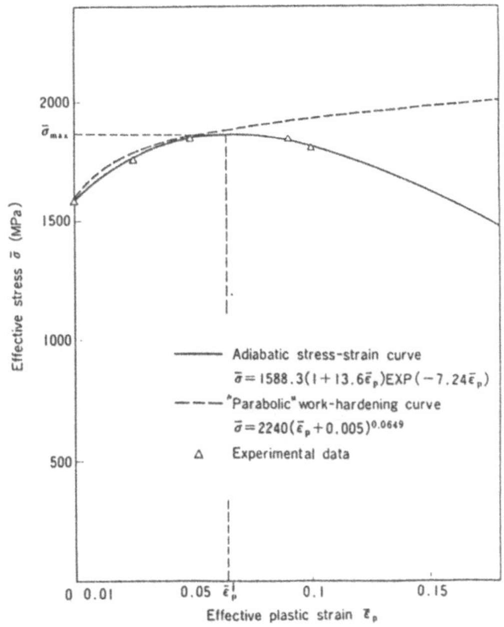

Fig. 1. Effective stress-strain curves for HY-TUF steel in quenched and tempered condition.

points. The shear band is considered to propagate into an infinite material in the x-direction, so that the displacements in x-direction given along a boundary DED' vary linearly as shown in the Figure 2(b) and its boundary condition in y-direction is fixed.

MODEL PREDICTIONS

Figure 3 shows the isostress and isostrain fields developed after a displacement d is given. The stresses and strains marked along the lines are effective values. It can be seen that, although the stress level is fairly high in the whole body (1588-1869 MPa), the plastic strain is concentrated on a narrow band ahead of the notch tip. A thicker solid curve in Figure 3 shows a contour line of $\bar{\sigma}=1869$ MPa and $\bar{\varepsilon}_p=0.0646$; these values correspond to the maximum stress $\bar{\sigma}_{max}$ and the instability strain $\bar{\varepsilon}_p^i$ at the transition point on the adiabatic stress-strain curve. The stress outside the contour line shown by the solid curve increases with increasing plastic strain due to strain-hardening; however, the stress inside the contour line decreases with increasing plastic strain due to strain-softening. This behavior is shown explicitly in Figure 4(a), where the distributions of stress and plastic strain are shown only in the vicinity of the notch tip. The region contained within the isostress line ($\bar{\sigma}=1869$ MPa) is considered to correspond to the shear band. In order to compare the propagation of shear band with the progressive deformation produced in a material which has no instability region, the deformation of the notched body is determined for the monotonic work-hardening curve, represented by Equation 2. Figure 4(b) shows the distribution of effective stress and plastic strain near the notch in the parabolic work-hardening material. The deformation of the body and the distributions of stress and strain are almost the same as those in the

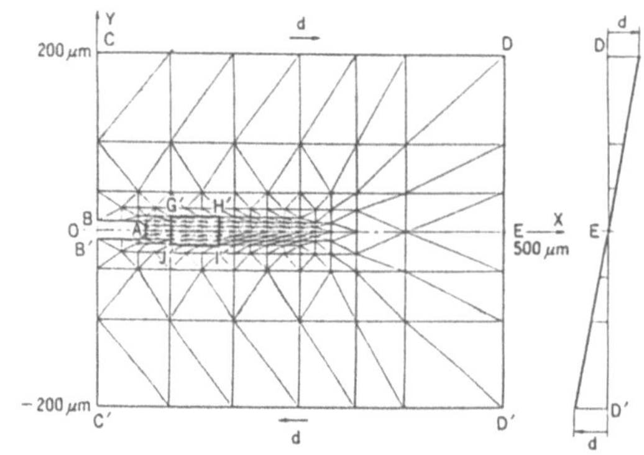

Fig. 2. Mesh division used for sample shear deformation; (a) notch in a rectangular body; (b) boundary displacements imposed along boundary DED.

material of the adiabatic curve except in the vicinity of the notch tip. The isostrain line for $\bar{\varepsilon}_p = 0.0646$ is shown by a solid curve in Figure 4(b); this value is equal to the instability strain $\bar{\varepsilon}_p^i$ of the adiabatic curve. The stress inside the contour line shown by the solid curve increases with increasing plastic strain. This behavior is the same as that outside the contour, but opposite to that of the adiabatic curve. This contour line reaches a distance of 50 μm from the notch tip when the tangential displacement is 6.87 μm. The concentration of strain within the $\bar{\varepsilon}_p^i = 0.0646$ isostrain line, with the attendant reduction of stress, is the critical feature responsible for the propagation of a shear band. By increasing the imposed displacement d, this behavior becomes more and more pronounced; the stress within the instability strain contour line will decrease as d is increased. This drastic difference between adiabatic and isothermal deformation within the $\bar{\varepsilon}_p^i = 0.0646$ envelope contrasts with the nearly identical isostress and isostrain contours outside the envelope. This shows that the overall stress distribution is very similar and explains the localization of the shear along a narrow band.

In order to more clearly assess the effect of the increasing tangential displacement d on the plastic strain distribution along the symmetry axis OE, the plot shown in Figure 5 was made. The symmetry axis OE is indicated by X^1X^3 in Figure 5. The dashed lines represent the isothermal (work-hardening) behavior, while the full lines indicate the adiabatic behavior. At imposed displacements below 3.60 μm, the two conditions deform identically. As d is increased, the plastic shear strain increases at a faster rate for the adiabatic than for the isothermal curve. The difference is highest at the notch tip. The length of the instability region is indicated as S in Figure 6. This region is defined as the length of the $\varepsilon_p^i = 0.0646$ envelope for purposes of comparison with the work-hardening curve. Again, the lengths S show marked differences for the two cases beyond a critical displacement d. The length of the instability region S increases very rapidly with increasing displacement d for the adiabatic behavior. Attempts were made at increasing the displacement d further, but serious problems arose regarding the divergence of the solution. Therefore, it was not possible to analyse the full range of the adiabatic stress-strain curve. However, the trends shown in Figures 3-6 can only accentuate themselves as displacement d is increased.

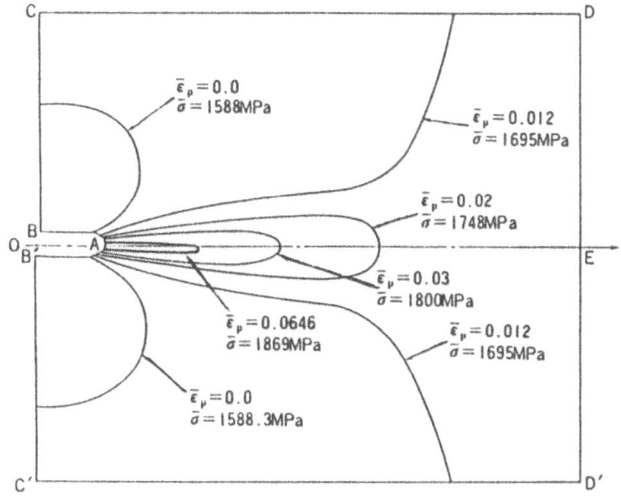

Fig. 3. Isostress (σ) and isostrain ($\bar{\varepsilon}_p$) contour lines in materials with adiabatic stress-strain curve.

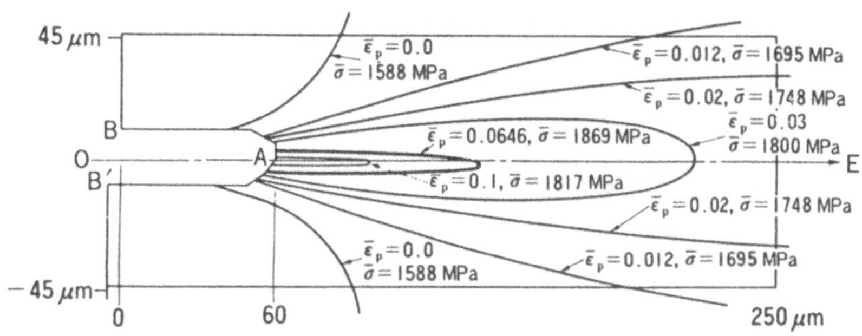

Fig. 4(a). Isostress ($\bar{\sigma}$) and isostrain ($\bar{\varepsilon}_p$) contour lines near the notch in material with adiabatic stress-strain curve. Shear band is produced inside an envelope of solid line $\varepsilon_p^i = 0.0646$.

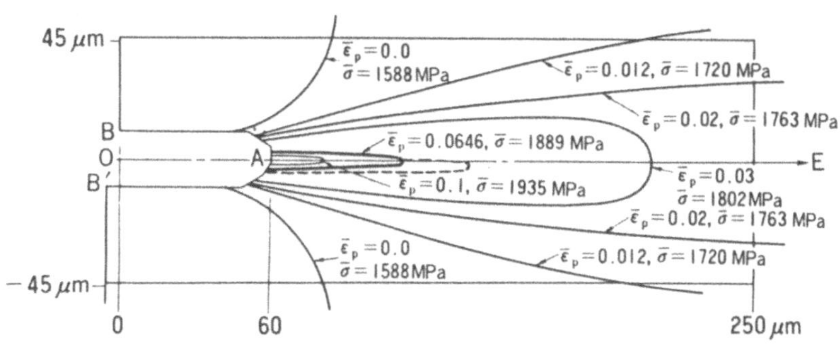

Fig. 4(b). Isostress ($\bar{\sigma}$) and isostrain ($\bar{\varepsilon}_p$) contour lines near the notch in material with workhardening curves.

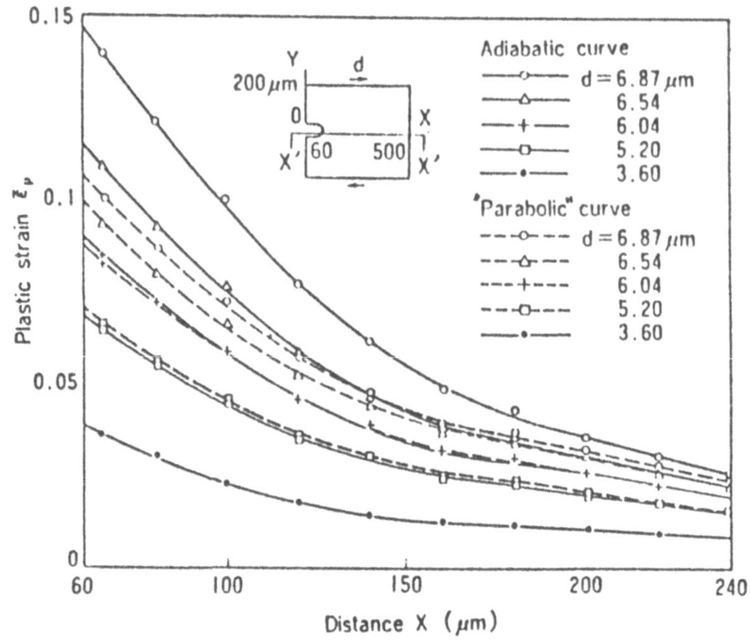

Fig. 5. Comparison of plastic strain distributions along x^1x^1 between material with adiabatic stress-strain curve curve and material with monotonic work-hardening curve.

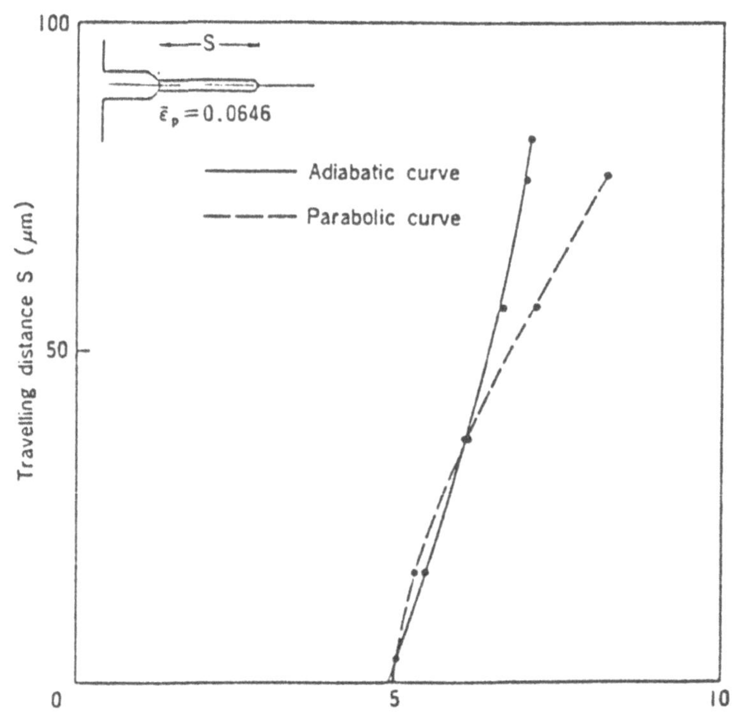

Fig. 6. Lenght of shear band for adiabatic stress-strain curve compared with corresponding distance S for monotonic work-hardening curve.

ACKNOWLEDGMENT

The authors wish to thank Mr. K. Miyauchi (The Institute of Physical and Chemical Research, Japan) for valuable discussions, and Dr. J.L. Swedlow (Carnegie-Mellon University) for permission to use the code. This research was supported by NSF Grant 8115127 and by the Center for Explosives Technology Research.

REFERENCES

1. G.B. Olson, J.F. Mescall, and M. Azrin, in: "Shock Waves and High-Strain-Rate Phenomena in Metals," M.A. Meyers and L.E. Murr, eds., Plenum, New York, (1981), p. 221.
2. J.L. Swedlow, Computers and Structures, 3:878 (1973).
3. S. Kuriyama, H. Hayashi, and S. Yoshida, in: "Proceedings of the 4th International Conference on Production Engineering," Tokyo, (1980), p. 38.
4. Y. Yamada, "Applications of the Finite Element Method," Tokyo, p. 149 (In Japanese).
5. O.E. Zienkiewicz, "The Finite Method in Engineering Science," McGraw-Hill, London, England, p. 50.

A MICROMECHANICAL MODEL OF POROUS-BRITTLE SOLIDS

L. G. Margolin[a], Barham W. Smith[b], and G. P. DeVault[b]

[a]Lawrence Livermore National Laboratory, Livermore, CA
[b]Los Alamos National Laboratory, Los Alamos, NM

INTRODUCTION

There are many advantages to micromechanical models of material response. These are models that predict the behavior under load of materials by considering the behavior of microscopic structures embedded in an equivalent continuum[1,2]. We have derived a theoretical model framework that allows quasistatic plastic pore crush and brittle fracture along three orthogonal sets of crack planes. The contribution of internal energy to the stress is included. We discuss the model, a numerical solution method, and a sample application problem, namely response of rock to an underground explosion.

DESCRIPTION OF THE MODEL

We describe a constitutive model for a solid containing spherical pores and penny-shaped cracks. We assume that pores and cracks are distributed uniformly in space, at least over the smallest lengths that we need to resolve in our calculations, i. e., within a computational cell. We further assume that the initial number density of cracks as a function of size and orientation can be characterized by a distribution, and that this distribution is a material property. Presently we assume that the pores are all initially the same radius, and have a number density consistent with the macroscopic porosity. We also assume that the constitutive law in the solid matrix is of a Gruneisen form

$$\sigma_{ij} = M_{ijkl}\, \varepsilon_{kl} - \rho_0\, \Gamma\, I\, \delta_{ij} \tag{1}$$

where σ is a stress, M is the modulus tensor, I is specific internal energy, ρ_0 is the solid material density, and Γ is a Gruneisen constant.

Under these assumptions, it is possible to calculate effective moduli M^e for the composite material. In this calculation, the boundary of the composite body is loaded with a spatially uniform stress σ^e. The total strain ε is calculated as the sum of the strain in the matrix plus the additional strain due to opening and shearing of cracks and closing of pores. These extra strain contributions are calculated from micromechanical models of cracks[3] and pores[4] in applied stress. The stress-strain analysis of those models ignores the energy term in eqn. (1) and thus applies only to the low-internal-energy regime, but we include the term in the composite model for generality.

The effective compliance C^e (the inverse of M^e) is then found from the relation between applied stress and resultant strain, $\varepsilon^e = C^e \sigma^e$. Details of the calculation for the cracks alone can be found in ref. 3. Due to limitations of space here, we can only summarize the results for the model described above.

All of the strains are linear functions of applied stress. Hence C^e is a constant tensor that dimensionally must be proportional to the compliance C of the solid matrix material. C^e is enhanced over C by functions of certain dimensionless numbers related to the crack and pore distributions. For pores the number is the number density of pores multiplied by the cube of the pore radius. This number is related to porosity and to the extension ratio α (> 1; called "porosity" in ref. 5). For the cracks the numbers are components of a tensor, calculated by integrating the number density of cracks multiplied by the cube of the crack radius over the distribution of crack size and orientation. These expressions are complicated[3].

We emphasize that effective moduli describe material response in terms of strain relative to the current unloaded state. This reference state depends on the microstructure and evolves with time. We model plastic pore collapse as a function of stress using the hollow sphere model of Carroll and Holt[5]. We model growth of cracks based on a generalized Griffith criterion[6]. These detailed models are used to follow changes in the pore and crack distributions, which in turn cause the effective compliances to vary. Macroscopically, these variations represent fracture and compaction and are directly related to the microscopic description of crack growth and pore collapse.

The effective moduli are defined with reference to elastic strain (unlike so-called tangent moduli sometimes used to model nonlinear response), while the constitutive relation most useful in a numerical computation such as we describe below is defined with reference to total strain. Therefore we must subtract the portion of the total strain rate that is due to irreversible pore crushup (expressed in terms of the change in α) in order to construct an elastic strain rate that can be equated with the rate of change of $C^e \sigma^e$.

The final form of the constitutive relation is

$$\left(1 - \frac{1}{\alpha D}\frac{d\alpha}{dt}\right) \varepsilon_{ij} = \frac{d}{dt}\left(C_{ijkl}\sigma_{kl}\right) + \frac{\rho_0 \Gamma}{3 \kappa_0}\left(\frac{\sigma_{kl}\dot{\varepsilon}_{kl}}{\rho} + \dot{I}_{ext}\right) \delta_{ij} \qquad (2)$$

where ε is the strain rate tensor, D is its trace, α is the extension ratio, and $\dot{I}_{ext}$ is the externally applied internal energy rate. This equation is integrated over a time step using straightforward time-centered differencing and solved directly using a matrix scheme, as follows. A modified effective compliance Ξ is defined by

$$\Xi_{ijkl} = C_{ijkl} + \frac{\langle\alpha\rangle \Gamma}{6 \kappa} \dot{\varepsilon}_{kl} \Delta t \, \delta_{ij} \qquad (3)$$

where $\langle\alpha\rangle$ is the average α over the time step Δt, and κ is the material bulk modulus. Eqn. 2 can be simplified using the inverse of Ξ to put the time-advanced stress alone on the left-hand side.

During the plastic phase of pore crush, the scalar equation

$$P_{crush} = \frac{2}{3} Y \ln \frac{\alpha}{\alpha - 1} + P_{ob} \qquad (4)$$

must be solved simultaneously with the stress-strain relation. Here, as in ref. 5, Y is a yield parameter, and we offset the crush curve by the local overburden pressure P_{ob} when doing a field-scale problem. A pressure is derived from eqn. 2 by computing its trace.

IMPLEMENTATION AND SAMPLE CALCULATION

The model, known as the Porous-Brittle Constitutive Model (PBCM), has been implemented in the SHALE computer code[7], a two-dimensional finite-difference program. We have used SHALE to numerically simulate ground motion due to a large buried explosion (see also ref. 8). Phenomena associated with such an explosion are complex, including a high stress phase during which rock is vaporized or melted (not discussed here), a phase in which large plastic deformation of rock takes place as pores are crushed out, and an elastic phase of wave propagation. The wave eventually reflects from the ground surface, creating tensile failure (spall) underground and projecting a spalled-off layer into the air. Closure of the underground spall gap thus formed occurs at slapdown, when the spalled layer falls back to earth.

We have calculated an idealized situation to study the phenomena produced by our new numerical model. The explosion is modeled as a simple Gaussian pressure pulse in time applied in a cell. The peak pressure was 3.5 kbar and the standard width was 1 ms. In the calculation presented here, we do not include macroscopic plastic shear deformation in the usual way. Parameters used include an initial α of 1.27, and average radius and number density of microcracks of 0.12 cm and 0.2 cm^{-3}. In Figure 1 we show the stresswave shape 50 ms after the explosion which was at 300 m depth, when the plastic phase is essentially over. Waves are traveling both upward and downward in this vertical slice through the problem. The ramp in vertical stress underlying the waveform is due to overburden stress, and the parallel, broad top of the waveform is at overburden plus the elastic limit for pore crushup. The pulse width if broadening by pore crush were omitted would be about 50 m. The trailing edge of the pulse in Fig. 1 is somewhat steepened, because material behind the pulse has less porosity, hence a higher density and sound speed. This is the classic condition for formation of a shock. However, since the wave is also dissipative, the final shape represents a balance between the competing effects of steepening and broadening.

Figure 2 shows the computed acceleration history at the ground surface directly above the explosion. The initial arrival of the compressive wave is modulated later by partial reflections from the spall gap (now a new underground free surface). Then those decay, and we see only the acceleration of gravity throughout the ballistic phase. High surface accelerations are again observed at slapdown.

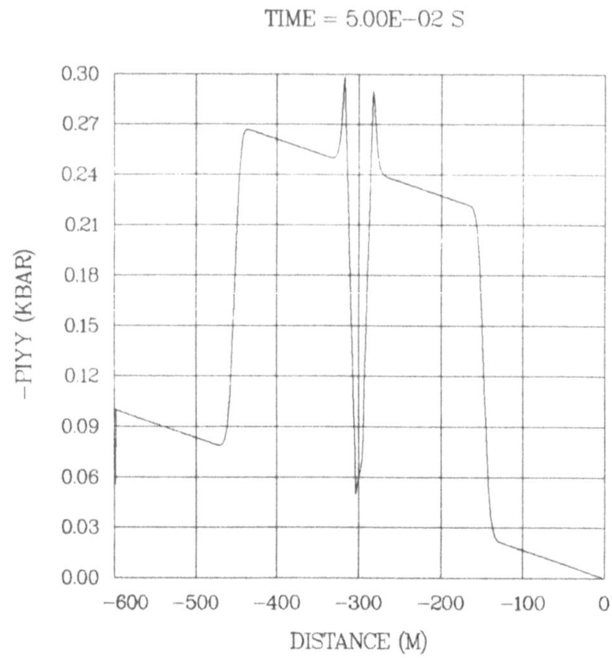

Fig. 1: Vertical stress as a function of depth.

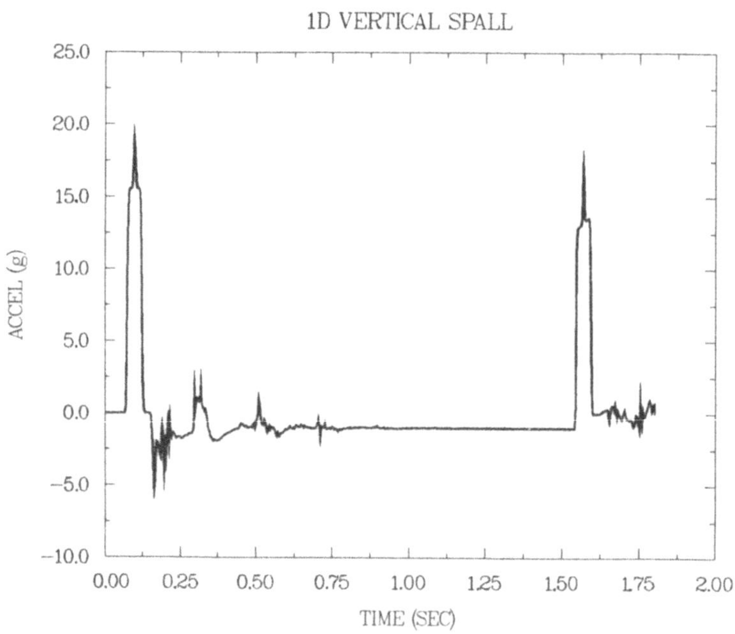

Fig. 2: Surface acceleration as a function of time.

CONCLUSIONS

We have constructed a new model for a porous-brittle solid such as rock. Because the model is micromechanical, its adjustable parameters are physically measurable material properties, rather than curve-fitting parameters from loading curves as is typical in phenomenological models. Our model has a numerical solution which is stable and quickly computable. We have calculated a hypothetical case to explore qualitative results of using PBCM, and we find it gives realistic ground motion simulations based on true physical behavior of the embedded microstructure. This computation should not be taken as an attempt to model any particular data set.

This work was supported by the United States Department of Energy.

REFERENCES

1. L. G. Margolin, "Microphysical Models for Inelastic Material Response", Int. J. Eng. Sci., 22:1171 (1984).
2. L. G. Margolin and B. W. Smith, "A Numerical Model for Simulating Dynamic Processes in Rock", Proceedings 5th Int. Conf. on Numerical Methods in Geomechanics, Nagoya, Japan (1985).
3. L. G. Margolin, "Elastic Moduli of a Cracked Body", Int. J. Fracture, 22:65 (1983).
4. J. K. Mackenzie, "Elastic Constants of a Material Containing Spherical Holes", Proc. Phys. Soc., B63:2 (1950).
5. M. M. Carroll and A. C. Holt, "Static and Dynamic Pore-Collapse Relations for Ductile Porous Materials", J. Appl. Phys., 43:1626 (1971).
6. L. G. Margolin, "A Generalized Griffith Criterion for Crack Propagation", Eng. Fracture Mech., 19:539 (1984).
7. R. B. Demuth, L. G. Margolin, B. D. Nichols, T. F. Adams, and B. W. Smith, "SHALE: A Computer Program for Solid Dynamics", Los Alamos National Laboratory report LA-10236 (1985).
8. B. W. Smith and L. G. Margolin, "A New Technique for Modeling Fracture and Spall", Proceedings 2nd Containment Symposium, Lawrence Livermore National Laboratory report CONF-830882, 165 (1983).

A MULTIPHASE MIXTURE MODEL FOR THE SHOCK INDUCED CONSOLIDATION OF METAL POWDERS

R. A. Berry and R. L. Williamson

Idaho National Engineering Laboratory
Idaho Falls, Idaho

INTRODUCTION

Recently, shock waves have been used to dynamically consolidate rapidly solidified metal powders. Because of the apparently low bulk temperatures and accompanying high surface temperatures, the plastically deformed particles consolidate to form fully dense monoliths which essentially retain the unique, metastable morphology of the original powder particles, e.g., Flinn et al., [1] and Korth et al., [2].

The modeling of consolidation processes is complex and can be approached in several ways. To model individual granules and their interactions with their neighbors in three-dimensions with hydrocodes is limited by insufficient computing resources. On the other hand, the gross scale, effective mixture approach, while adequate for scoping and design purposes, is inadequate for describing certain microscale phenomena. In this discussion we adopt a multiphase mixture model approach, intermediate to those mentioned. We include the gaseous phase in our model because there are indications that it may have non-negligible effects.

In powder compaction the solid and gaseous constituents retain their identities. To construct the multiphase model the notions of Truesdell [3] are followed; however, the equations must be modified somewhat to account for the immiscible nature of the mixture. To use a continuum description for such a mixture a "material particle" is identified as a group of several grains of the solid or their gaseous interstitial region. The averaged continuum equations describing the behavior of a "material particle" are modified to account, in an approximate way, for the discrete nature of the mixture.

BALANCE EQUATIONS

We confine our attention to plane wave propagation with its accompanying uniaxial strain. At a mixture continuum level our model is then one-dimensional. The bulk stress of each component is taken to be hydrodynamic with the shock loading condition. Extending the concepts of Goodman and Cowin [4] the local, phasic mass, momentum, energy and equilibrated force balance equations are written as

$$\rho_a' = -\rho_a \frac{\partial v_a}{\partial x} \tag{1}$$

$$\rho_a v_a' = -\frac{\partial \Pi_a}{\partial x} + m_a^+ \tag{2}$$

$$\rho_a e_a' = -\Pi_a \frac{\partial v_a}{\partial x} - \frac{\partial q_a}{\partial x} + \rho_a r_a - m_a^+ v_a + e_a^+$$

$$+ \frac{\rho_a k_a' \phi_a'^2}{2} + h_a \frac{\partial \phi_a'}{\partial x} - \rho_a f_a \phi_a' - g_a^+ \phi_a' \tag{3}$$

$$\rho_a (k_a \phi_a')' = \frac{\partial h_a}{\partial x} + \rho_a (\ell_a + f_a) + g_a^+ \tag{4}$$

$$\sum_a m_a^+ = 0 \quad , \quad \sum_a e_a^+ = 0 \quad , \quad \sum_a g_a^+ = 0 \tag{5}$$

where m_a^+, e_a^+, and g_a^+ are the growths of linear momentum, energy, and equilibrated force per unit volume and where the tick mark above a variable denotes the material time derivative of an "a" constituent quantity following the barycentric motion of the "a" constituent. Π_a is the force per unit area acting on the "a" constituent particle. Goodman and Cowin call k_a the equilibrated inertia, h_a the equilibrated stress, f_a the equilibrated force supply, and ℓ_a the equilibrated body force. Since the volume fractions and the motions are kinematically independent, the equilibrated force balance equations are necessary to describe the microstructural aspects of the mixture's response and corresponding, higher order stress and body force terms are included in the energy equations for the energy flux and energy supply associated with the time rate of change of volume fractions. Goodman and Cowin's "balance of equilibrated inertia" equation has been dropped in favor of a constitutive definition for k_a as described in Nunziato and Cowin [5]. Introducing the entropy, η_a, an absolute temperature, T_a, and the Helmholtz free energy, $\psi_a = e_a - T_a \eta_a$, the reduced dissipation inequality,

$$\sum_a \frac{1}{T_a} \Bigg[-\rho_a (\eta_a T_a' + \psi_a') - \Pi_a \frac{\partial v_a}{\partial x} - \frac{q_a}{T_a} \frac{\partial T_a}{\partial x}$$

$$- m_a^+ v_a + e_a^+ - \left(g_a^+ - \frac{\rho_a k_a' \phi_a'}{2}\right) \phi_a' + h_a \frac{\partial \phi_a'}{\partial x} \Bigg] \geq 0 \tag{6}$$

is used to impose restrictions on the mixtures response.

Imposing the saturation constraint, $\Sigma\, \phi_a = 1$, with reaction forces which do no work, we adopt the form of Passman, Nunziato, and Walsh [6] in which the power extended on constituent "a" due to the other constituents because of the constraint is

$$w_a^+ = P_i \frac{\partial \phi_a}{\partial t} = P_i (\phi_a' - v_a \frac{\partial \phi_a}{\partial x}) \tag{7}$$

where P_i is called the interface pressure. The entropy inequality (6) is then

$$\sum_a \frac{1}{T_a}\left[-\rho_a(n_a\dot{T}_a + \dot{\psi}_a) - \Pi_a \frac{\partial v_a}{\partial x} - \frac{q_a}{T_a}\frac{\partial T_a}{\partial x}\right.$$

$$\left. - m_a^{+e} v_a + e_a^{+e} - \left(g_a^{+e} - \frac{\rho_a \dot{k}_a \dot{\phi}_a}{2}\right)\dot{\phi}_a + h_a \frac{\partial \dot{\phi}_a}{\partial x}\right] \geq 0 \tag{8}$$

where

$$m_a^{+e} = m_a^{+} - p_i \frac{\partial \phi_a}{\partial x}, \qquad e_a^{+e} = e_a^{+} + w_a^{+}, \qquad g_a^{+e} = g_a^{+} + p_i + \rho_a f_a \tag{9}$$

CONSTITUTIVE AND REDUCED BALANCE EQUATIONS

Since we are imposing hydrodynamic bulk stress behavior the mixture model is basically that of a fluid suspension. Using the Principle of Phase Separation of Drew and Segel [7], we take the dependent variables ψ_a, Π_a, n_a, q_a, h_a, and k_a to depend upon the array $S_a = (\gamma_a, \phi_a, T_a, \partial\phi_a/\partial x)$ and the exchange variables m_a^{+e}, e_a^{+e} and g_a^{+e} depend on the S_a arrays for <u>both</u> phases. Applying the argument of Coleman and Noll [8] to the inequality (6) leads to the following:

$$n_a = -\frac{\partial \psi_a}{\partial T_a}, \qquad h_a = \phi_a \gamma_a \frac{\partial \psi_a}{\partial\left(\frac{\partial \phi_a}{\partial x}\right)}, \qquad \Pi_a = \phi_a p_a + h_a \frac{\partial \phi_a}{\partial x}, \tag{10}$$

$$\sum_a \frac{1}{T_a}\left[e_a^{+e} - m_a^{+e} v_a - \left(g_a^{+e} + \beta_a - p_a - \frac{\rho_a \dot{k}_a \dot{\phi}_a}{2}\right)\dot{\phi}_a - \frac{q_a}{T_a}\frac{\partial T_a}{\partial x}\right] \geq 0 \tag{11}$$

where

$$p_a = \gamma_a^2 \frac{\partial \psi_a}{\partial \gamma_a} \quad \text{and} \quad \beta_a = \phi_a \gamma_a \frac{\partial \psi_a}{\partial \phi_a} \tag{12}$$

are the phasic pressure and configuration pressure respectively. Expanding the summation in (11) and using (4), (5), (9) and regrouping terms gives

$$\frac{T_g - T_s}{T_s T_g}\left(e_s^{+e} - m_s^{+e} v_s\right) - \frac{1}{T_g}(v_s - v_g)m_s^{+e} - \frac{1}{T_g}\left(g_g^{+e} + \beta_g - p_g - \frac{\rho_g \dot{k}_g \dot{\phi}_g}{2}\right)\dot{\phi}_g$$

$$- \frac{1}{T_s}\left(g_s^{+e} + \beta_s - p_s - \frac{\rho_s \dot{k}_s \dot{\phi}_s}{2}\right)\dot{\phi}_s - \frac{q_s}{T_s^2}\frac{\partial T_s}{\partial x} - \frac{q_g}{T_g^2}\frac{\partial T_g}{\partial x} \geq 0 \tag{13}$$

A sufficient, but <u>not</u> necessary, condition for satisfying the inequality (13) is to require each term in the inequality to be non-negative whereby there exists non-negative functions $\bar{\lambda}_g$, $\bar{\lambda}_s$, h_T, μ_s, μ_g, δ which depend upon the thermodynamic state S_s, S_g such that

$$m_s^{+e} = - m_g^{+e} = \delta(v_g - v_s) \tag{14}$$

$$e_s^{+e} = - e_g^{+e} = h_T(T_g - T_s) + \delta(v_g - v_s)v_s \tag{15}$$

$$q_g = -\bar{\lambda}_g \frac{\partial T_g}{\partial x}, \quad q_s = -\bar{\lambda}_s \frac{\partial T_s}{\partial x} \tag{16}$$

$$\rho_s k_s \phi_s'' + \left(\mu_s + \frac{\rho_s k_s'}{2}\right)\phi_s' = p_s - \beta_s + \frac{\partial h_s}{\partial x} + \rho_s \ell_s - P_i \tag{17}$$

$$\rho_g k_g \phi_g'' + \left(\mu_g + \frac{\rho_g k_g'}{2}\right)\phi_g' = p_g - \beta_g + \frac{\partial h_g}{\partial x} + \rho_g \ell_g - P_i \tag{18}$$

For details see Berry and Williamson [9],

The empirical functions can be given familiar names: $\bar{\lambda}_a$, effective thermal conductivities, h_T, the heat transfer coefficient, μ_a, microviscosities, and, δ, the drag coefficient.

Since k_a is proportional to the amount of mass in our continuum "material particle" not moving at the phasic, barycentric velocity and is believed to be small (see Williamson and Berry [10], for example) we ignore inertia terms as a first approximation. We also neglect ℓ_s, ℓ_g, h_g, μ_g, and β_g terms. As a first approximation we also linearize h_s,

$$h_s = A_s \frac{\partial \phi_s}{\partial x} \quad \text{(with } A_s = \text{constant)} \ . \tag{19}$$

With all of these constitutive restrictions and approximations our final balance equations are

$$\frac{\partial \rho_s}{\partial t} + \frac{\partial(\rho_s v_s)}{\partial x} = 0 \tag{20}$$

$$\frac{\partial \rho_g}{\partial t} + \frac{\partial(\rho_g v_g)}{\partial x} = 0 \tag{21}$$

$$\rho_s\left[\frac{\partial v_s}{\partial t} + v_s \frac{\partial v_s}{\partial x}\right] = -\phi_s \frac{\partial p_s}{\partial x} + (p_g - \dot{p}_s)\frac{\partial \phi_s}{\partial x} + \delta(v_g - v_s) - 2A_s \frac{\partial \phi_s}{\partial x}\frac{\partial^2 \phi_s}{\partial x^2} \tag{22}$$

$$\rho_g\left[\frac{\partial v_g}{\partial t} + v_g \frac{\partial v_g}{\partial x}\right] = -\phi_g \frac{\partial p_g}{\partial x} - \delta(v_g - v_s) \tag{23}$$

$$\rho_s\left[\frac{\partial e_s}{\partial t} + v_s \frac{\partial e_s}{\partial x}\right] = -\phi_s p_s \frac{\partial v_s}{\partial x} + \frac{\partial}{\partial x}\left(\bar{\lambda}_s \frac{\partial T_s}{\partial x}\right) + h_T (T_g - T_s) + A_s \frac{\partial \phi_s}{\partial x}\frac{\partial \phi_s'}{\partial x} - (p_s - \beta_s)\phi_s' + \mu_s \phi_s'^2 \tag{24}$$

$$\rho_g\left[\frac{\partial e_g}{\partial t} + v_g \frac{\partial e_g}{\partial x}\right] = -\phi_g p_g \frac{\partial v_g}{\partial x} + \frac{\partial}{\partial x}\left(\bar{\lambda}_g \frac{\partial T_g}{\partial x}\right) - h_T (T_g - T_s) + \delta(v_g - v_s)^2 - p_g \phi_s' \tag{25}$$

$$\phi_s' = \frac{\partial \phi_s}{\partial t} + v_s \frac{\partial \phi_s}{\partial x} = \frac{1}{\mu_s}\left[p_s - \beta_s - p_g + A_s \frac{\partial^2 \phi_s}{\partial x^2}\right] . \tag{26}$$

The balance equations (21) through (26) must be supplemented with experimental determination of δ, h_T, $\overline{\lambda}_s$, $\overline{\lambda}_g$, μ_s, and A_s. As a first approximation for the drag coefficient we could utilize the correlation developed by Ergun [11], from an experimental study of high-speed gas flow in a packed granular bed of spherical particles, in which he found that δ could be approximated by

$$\delta = \frac{\mu_g \phi_s}{d_p^2 \phi_g^3}(1.75\,\mathrm{Re} + 150\,\phi_s) \quad , \quad \mathrm{Re} = \frac{\rho_g}{\mu_g}\left| v_g - v_s \right| d_p \tag{27}$$

where μ_g is the gas viscosity and Re is the Reynolds number based on the initial particle diameter d_p. For convective heat transfer between the gas and particles, Gelperin and Einstein [12] developed a correlation for h_T, which depends on particle size of the form

$$h_T = \frac{12\lambda_g}{d_p^2}\phi_g^{2/3}\left[1 + \frac{\mathrm{Re}^{2/3}\,\mathrm{Pr}^{1/3}}{5}\right] , \qquad \mathrm{Pr} = \frac{\mu_g c_p}{\lambda_g} \tag{28}$$

where the gas Prandtl number, Pr, is defined in terms of the specific heat of the gas, c_p. The solid viscosity μ_s may be estimated from Carroll and Holts' [13] work (neglecting inertial volume fraction effects) by

$$\mu_s = \frac{d_p K_T}{c} \tag{29}$$

where c is the sound speed in the solid and K_T is the bulk modulus of the solid phase. We are currently unaware of any method to evaluate the constant A_s from independent data so we recommend its value be chosen to give the best overall agreement with experimental data.

As state equation first approximations we take a Gruneisen form for the solid, $p_s(\gamma_s, e_s)$, and an ideal gas type for the gas, $p_g(\gamma_g, e_g)$.

We note, that in the absence of a gas pressure, viscous resistance, and an equilibrate stress effect, the volume fraction propagation equation (26) reduces to

$$p_s = \beta_s \; . \tag{30}$$

If we let $\beta_s = p_{porous}/\phi_s$ then (30) is just

$$p_{porous} = \phi_s p_s(\gamma_s, e_s)$$

which is Carroll and Holts' modification to the p-α compaction model. If we now include the viscous resistance, equation (26) becomes

$$p_s - \beta_s = \mu_s \phi_s' - p_g$$

which, if we take $\mu_s = \overline{\mu}_s/\phi_g$ and β_s to be the equilibrium porous pressure, is equivalent to the compaction model of Holt, Carrol, and Butcher [14]. We therefore adopt the idea of Baer et al. [15], and write

$$p_s - \left(\frac{p_{porous}}{\phi_s}\right)_{equil.} = \frac{\overline{\mu}_s}{\phi_g}\phi_s' - p_g \tag{31}$$

where the new configuration pressure, (p_{porous}) equilibrium, is defined using experimental data of pressure vs. volume fraction of statically-loaded powder columns assuming that the compaction is unidirectional. Rearranging (31) gives the volume fraction propagation equation

$$\phi_s' = \frac{\phi_g}{\bar{\mu}_s}\left(p_s - \frac{p_{porous}}{\phi_s} - p_g\right) . \qquad (32)$$

Equations (20) through (25) and (32) are in a form which can be solved numerically. We defer such discussion to a later date. In future work our intent is to couple this model with local temperature and loading history models to simulate bonding effects (or perhaps unbonding) in compaction and the consolidation of metallic powders.

REFERENCES

1. J. E. Flinn et al., "Dynamic Consolidation of Aluminum Powders," this proceedings.
2. G. E. Korth et al., "Dynamic Consolidation of Rapidly Solidified Type 304 SS Powders," Exploment '85: International Conference on Metallurgical Applications of Shock Wave and High-Strain-Rate Phenomena, Portland, Oregon, July 28-August 1, 1985.
3. C. Truesdell, "Rational Thermodynamics," McGraw-Hill, New York, 1969.
4. M. A. Goodman and S. C. Cowin, "A Continuum Theory for Granular Materials," Arch. Rat. Mech. Anal. 44, 249-266 (1972).
5. J. W. Nunziato and S. C. Cowin, "A Nonlinear Theory of Elastic Materials with Voids," Arch. Rat. Mech. Anal. 72, 175-201 (1979).
6. S. L. Passman, J. W. Nunziato, and E. K. Walsh, "A Theory of Multiphase Mixtures," Appendix 5C of "Rational Thermodynamics," 2nd Ed. by C. Truesdell, Springer-Verlag, New York, 1984.
7. D. A. Drew and L. A. Segel, "Averaged Equations of Two-Phase Flows," Studies in Appl. Math. L, 205-231 (1971).
8. B. D. Coleman and W. Noll,"The Thermodynamics of Elastic Materials with Heat Conduction and Viscosity," Arch. Rat. Mech. Anal. 13, 167-178 (1963).
9. R. A. Berry and R. L. Williamson, "A Multiphase Model for the Dynamic Compaction and Consolidation of Metal Powders," Idaho National Engineering Laboratory Report EGG-RST-6922, in progress.
10. R. L. Williamson and R. A. Berry, "Microlevel Numerical Modeling of the Shock Wave Induced Consolidation of Metal Powders," this proceedings.
11. S. Ergun, "Fluid Flow Through Packed Columns," Chem. Engrg. Progress 48, 89-94 (1952).
12. N. I. Gelperin and V. G. Einstein, "Heat Transfer in Fluidized Beds," Chapt. 10 in "Fluidization" Edit. by J. F. Davidson and D. Harrison, Academic Press, New York (1971).
13. M. M. Carroll and A. C. Holt, "Static and Dynamic Pore-Collapse Relations for Ductile Porous Materials," J. Appl. Phys. 43, 1626-1635 (1972).
14. A. C. Holt, M. M. Carroll, and B. M. Butcher, "Application of a New Theory for the Pressure Induced Collapse of Pores in Ductile Materials," Proceedings of the RILEM-IUPAC International Symposium, Pore Structure and Properties of Materials, Prague, September 18-21, 1973.
15. M. R. Baer, et al., " A Study of Deflagration-to-Detonation Transition (DDT) in the Granular Explosive, CP," Sandia National Laboratory SAND83-1929, February 1984.

MICROLEVEL NUMERICAL MODELING OF THE

SHOCK WAVE INDUCED CONSOLIDATION OF METAL POWDERS

R. L. Williamson and R. A. Berry

Idaho National Engineering Laboratory
Idaho Falls, Idaho

INTRODUCTION

Dynamic consolidation of powders using shock waves is a promising technique for obtaining high strength materials. It has been demonstrated experimentally that consolidated materials which retain the desirable nonequilibrium properties of the initial powder can be produced using this approach, e.g. Flinn et al. [1] and Korth et al. [2]. It has been suggested by Raybould [3], that the consolidation process is achieved as a result of very short duration high energy deposition at particle surfaces through various mechanisms such as rapid plastic deformation and friction between particles. Because of this selective heating at particle surfaces, it is possible to obtain densification and particle bonding without subjecting the bulk material to high temperatures which create undesirable microstructural changes. To achieve a better exploitation of the dynamic consolidation technique, it will be necessary to obtain a greater understanding of the important physical mechanisms involved and the dynamic behavior of the particles during compaction.

In this paper we investigate, from a numerical simulation viewpoint, the phenomena occurring during dynamic powder consolidation. A model is developed at the particle level based on a solution of the continuum mechanics equations using conventional constitutive relations. We examine shock wave propagation through a powder material, the resulting stress/strain distribution and plastic flow, and the resulting temperature distribution and surface melting and freezing.

This work was done with two goals in mind: (1) to gain a better phenomenological understanding of the processes involved in dynamic compaction, and (2) to provide a basis for microstructural mixture models currently under development e.g. Berry and Williamson [4].

MODEL DESCRIPTION

The geometry and materials considered in the numerical simulation are shown in Figure 1. For this investigation, the model is limited to two spatial dimensions and thus the individual particles are assumed to be cylindrical rather than spherical. The powder material is

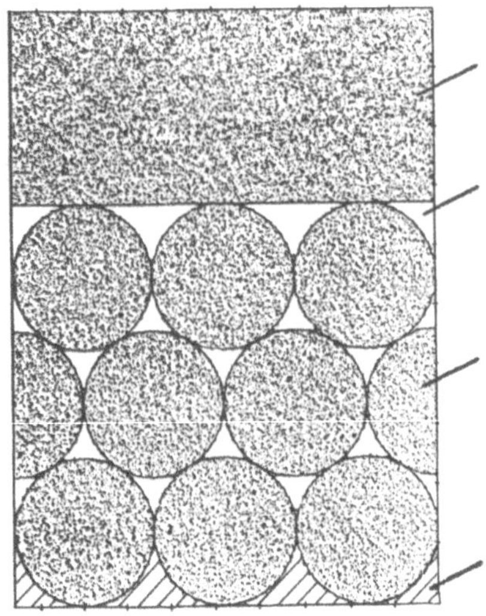

Figure 1. The Geometry and Materials Considered in the Numerical Simulation

304 stainless steel (rapidly solidified) with each particle having a diameter of 75 μm. It is assumed that the particles are contained within rigid boundaries on the bottom and sides, and thus the model can be thought of as a very thin layer of powder material held within a rigid cavity. To investigate the effects and response of trapped gases during the compaction process, the four interstitial regions at the bottom of the cavity were assumed to be air filled; all other interstitial regions were voided. Compaction is initiated by a high velocity flyer plate impact; for the results presented below, a stainless steel flyer having an initial velocity of 1 km/s is assumed.

The CSQII code, written by Thompson [5], is utilized to simulate the material response during compaction. CSQII is a two-dimensional finite difference code capable of handling multiple materials. CSQII is basically an Eulerian or fixed mesh code, but employs a unique computational approach. During each time cycle, the mesh is treated as material fixed and a complete Lagrangian calculation is performed. The entire Lagrangian mesh is then rezoned or mapped back onto the fixed Eulerian mesh. To properly handle edges of solids and interior cracks with this type of scheme, an interface tracking algorithm is built into the code. Elastic and plastic material response is handled using an elastic perfectly plastic model and thus it is assumed that material flows plastically under a constant stress without work hardening. For this investigation, the shear stress across material interfaces is assumed to be zero and thus friction between particles is neglected. Conduction heat transfer effects are included.

Conventional constitutive equations are used to describe the thermodynamic behavior of both the stainless steel and air, and are discussed in detail by Thompson [6]. Relations are included to describe the solid to liquid melt transition and temperature dependent material strength properties.

The computational mesh consists of 80 x 82 rectangular cells with the particle region (which is more finely nodalized than the flyer plate) constructed from cells approximately 2.8 μm on a side.

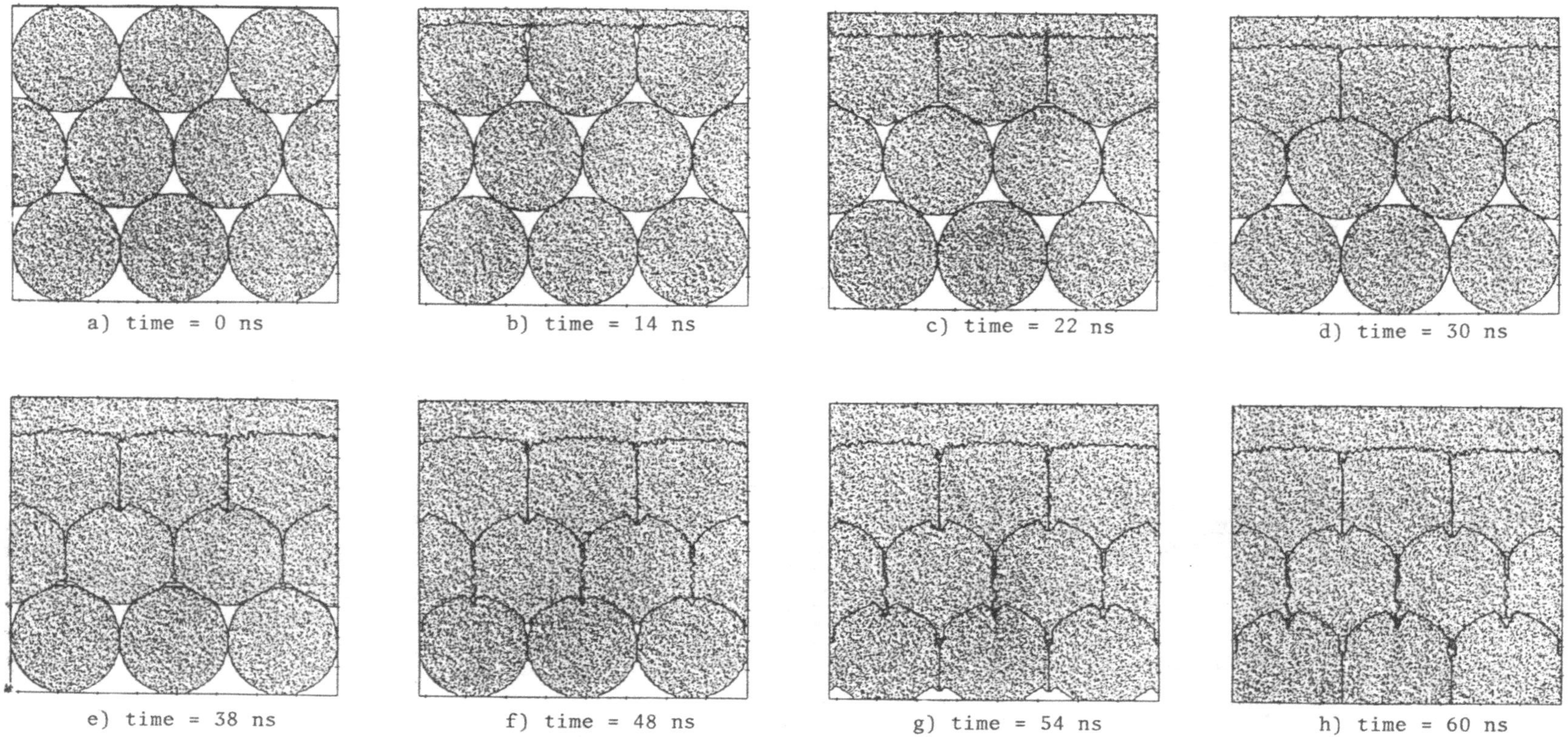

Figure 2. A Time Series Showing Particle Deformation During Compaction.

RESULTS

The transient simulation was run for approximately 100 ns following flyer plate impact. For the impact velocity used, an initial shock pressure of approximately 15 GPa is obtained in the powder; note that this is the pressure in an individual particle and does not represent the average pressure in the powder mixture.

Figure 2 presents a time series of the compaction process and shows how the particles plastically deform to fill the interstitial regions. The passage of the first shock wave results in full compaction of the powder which is complete within 60 ns. It is evident from the figure that individual particles do not undergo significant relative movement, however a great deal of material deformation at or near particle surfaces is required to achieve compaction. Photomicrographs of actual consolidation experiments such as those obtained by Korth et al. [2], show a final particle shape similar to those shown in Figure 2-h.

Figure 3 describes the temperature history during compaction at various locations within an individual particle. Initial shock arrival is evident between 25 and 30 ns. Noteworthy is the observation that particle interior regions, which are subjected to much less deformation than the surface regions, do not achieve the high temperatures experienced at or near the particle surface. For this simulation, the peak temperature in the particle interior regions is approximately 450 K which is sufficiently low such that undesireable microstructural changes cannot occur.

In Figure 4, we present a magnified view of an interstitial region during the final moments of collapse. Surface plots of temperature for this same region are also shown. The arrows in the figure indicate

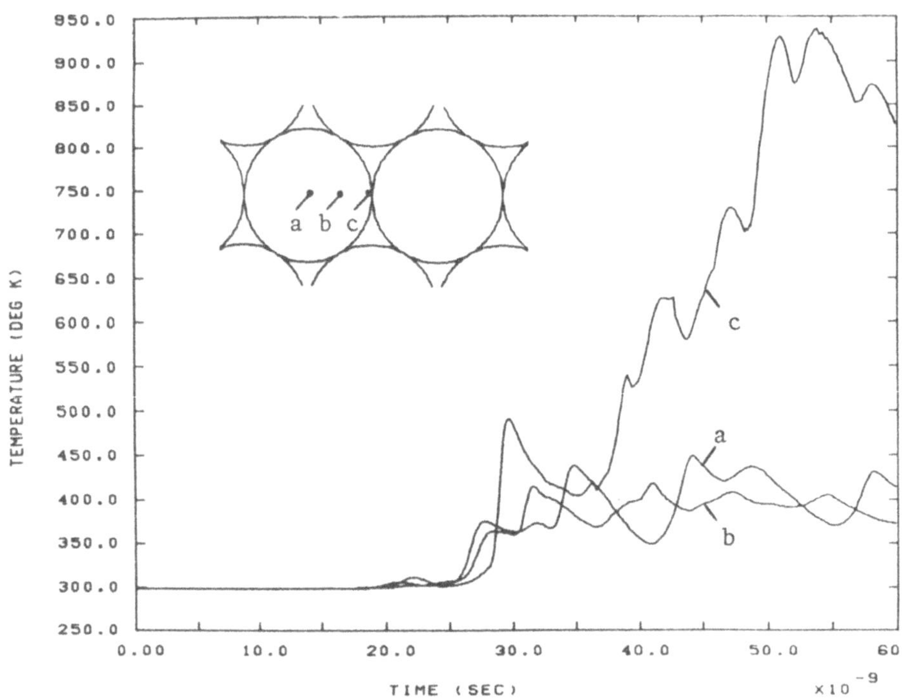

Figure 3. The Temperature History During Compaction at Various Locations Within a Particle.

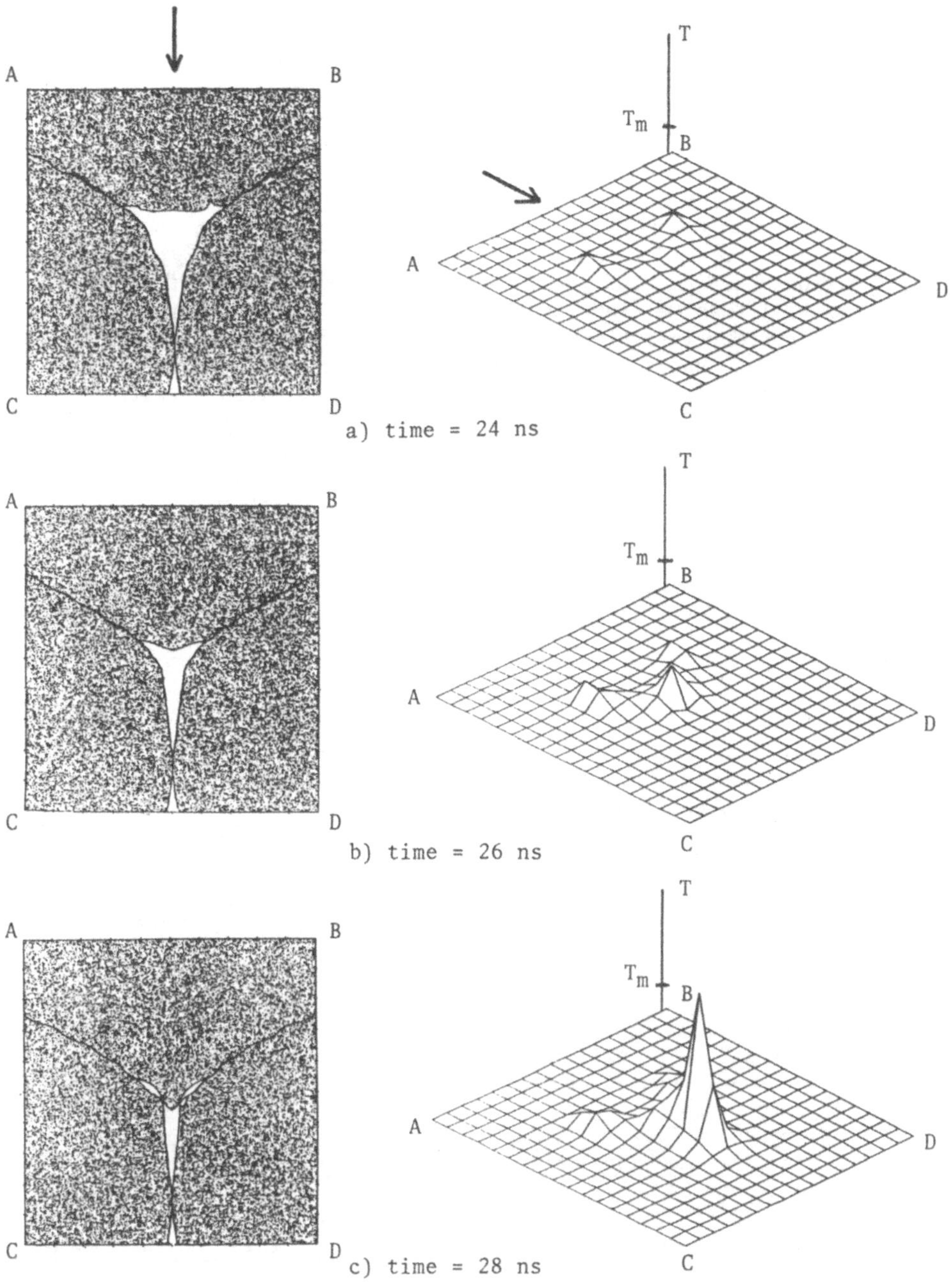

Figure 4. An Interstitial Region During the Final Moments of Collapse, Accompanied by Surface Plots of Temperature.

the planar wave propagation direction. Full scale on the temperature axis is approximately 7000 K; the static melt temperature of the stainless steel (T_m) is indicated. From this figure it is evident that during the interstitial collapse, temperatures well in excess of the melt point are obtained in regions of rapid material deformation, and liquid metal is formed. Extremely high particle surface heating rates (10^{10} - 10^{12} K/s) are predicted.

Space limitations permit presentation of only a limited selection of the results generated in this simulation; consult Williamson and Berry [7] for a more detailed coverage.

CONCLUDING REMARKS

As noted in the model description section, significant simplifications such as the constraint to two dimensions, the absence of an interparticle friction model, and the neglect of work hardening effects in the material behavior model, were required for this simulation. These simplifications must be kept in mind when interpreting results. Future work will involve removing these constraints and applying the model to further investigate such areas as particle surface melting and solidification, shock duration effects, and compaction rate dependencies. To the extent possible, a comparison of our results with experimental data and with existing models such as the energy deposition approach of Gourdin [8] and Schwarz et al. [9] (as well as other models for the dynamic response of porous materials), will also be performed.

It is evident that the shock compaction process at the particle scale is, at present, inadequately understood. We feel that numerical simulation at this level is a valuable tool for increasing this understanding.

REFERENCES

1. J. E. Flinn et al., "Dynamic Consolidation of Aluminum Powders," this proceedings.
2. G. E. Korth et al., 1984, "Dynamic Consolidation of Rapidly Solidified Type 304 SS Powders," Explomet '85: International Conference on Metallurgical Applications of Shock Wave and High-Strain-Rate Phenomena, Portland, Oregon, July 28 - August 1 (1985).
3. D. Raybould, "The Production of Strong Parts and Non-Equilibrium Alloys by Dynamic Compaction," Shock Waves and High-Strain-Rate Phenomena in Metals, edited by M. A. Meyers and L. E. Murr, Plenum, New York (1981).
4. R. A. Berry and R. L. Williamson, "A Multiphase Flow Model for the Shock Induced Consolidation of Metal Powders," this proceedings.
5. S. L. Thompson, CSQII - An Eulerian Finite Difference Program for Two-Dimensional Material Response Part 1. Material Sections, SAND 77-1339 (1979).
6. S. L. Thompson, 1972, Improvements in the CHARTD Radiation-Hydrodynamic Code III: Revised Analytic Equations of State, SC-RR-71-0714 (1972).
7. R. L. Williamson and R. A. Berry, Microlevel Numerical Modeling of the Shock Wave Induced Consolidation of Metal Powders, Idaho National Engineering Laboratory Report, EGG-RST-6923, in progress.
8. W. H. Gourdin, "Energy Deposition and Microstructural Modification in Dynamically Consolidated Metal Powder," Journal of Applied Physics, 55, 172-181, January 1984.
9. R. B. Schwarz et al., "A Theory for the Shock-Wave Consolidation of Powders," Acta Metall, 32, 1243-1252, 1984.

CONSTITUTIVE RELATION FOR 6061T6 ALUMINUM UNDER SHOCK LOADING CONDITIONS

Davis L. Tonks

Applied Theoretical Division
Los Alamos National Laboratory
Los Alamos, N.M.

ABSTRACT

We publish for the first time a constitutive relation of Wallace and Straub based on plastic flow data extracted from 6061T6 Al shocks by Wallace. The data involves strain rates up to 10^7 and shear stresses up to 5kbar. We interpret the relation in terms of a dislocation-kinetics model.

INTRODUCTION

Meaningful dislocation modeling of the plastic front of shock waves has become possible in recent years with the achievement of accurate particle velocity[1] and dislocation velocity measurements[2]. This work has consisted mostly of fitting the shock velocity profile using a paramaterized dislocation model and a computer hydrodynamics code.

Recently, D.C. Wallace[3,4,5] devised a rigorous thermodynamic scheme for weak shocks in solids. This scheme was superior to earlier efforts in that entropy contributions to the stresses were included, nonlinear elasticity was treated, and the anisotropy of the stress response was taken into account. He used this scheme in Ref. 4 to extract plastic constitutive information from the shock profiles of Johnson and Barker[1]. This constitutive information depended only on the thermodynamics and elastic properties and not at all on any dislocation model. Wallace and Straub[6] assembled the information into a constitive relation. It is the purpose of this paper to publish the relation for the first time and interpret it in terms of a dislocation kinetics model. We explain the constitutive relation in terms of the Orowan equation and a dislocation generation model. We also explain why, in this situation, the plastic strain is a state variable to a good approximation.

THE CONSTITUTIVE RELATION

The open symbols of Fig. 1 show the constitutive data plotted in terms of $10^6(t' - t_o)/\dot{\psi}$ and $\psi\,\dot{\psi}$, where ψ and $\dot{\psi}$ are the effective plastic strain and plastic strain rate, respectively. Also, t' is the effective shear stress scaled according to a linear pressure and temperature dependence of the shear modulus of Aluminum, G[7]. See Ref. 3 for the definitions of ψ and t' in terms of the tensor quantities. The

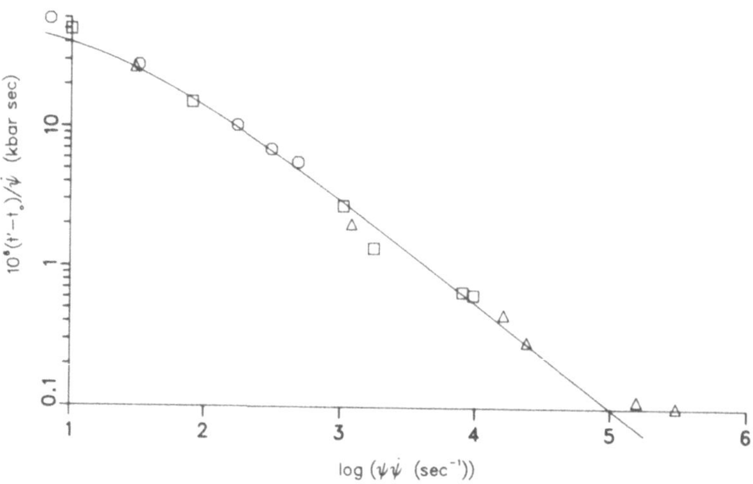

Fig. 1. The circles, squares, and triangles are data from 21, 37, and 89 kbar shocks, respectively. The line is an analytic fit.

line is the constitutive relation of Wallace and Straub[6], given by Eq.(1) when 8.33 is added to the denominator of the r.h.s. :

$$10^6(t'- t_o)/\ \dot\psi= c/(\psi\ \dot\psi)^{.75} \qquad (1)$$

where c equals 558. This 8.33 was used to interpolate to lower strain rates. Eq.(1) is the high strain rate asymptote which we use to study the high strain rate behavior. In Eq. (1), t' and t_o are in kbar and t_o is the scaled internal stress given by 1+22.5 ψ. The term 1 is due to the precipitation hardening of the alloy, which is taken as constant, and the other term is due to work hardening or dislocation accumulation. This expression was obtained by a straight line fit to the final shock stresses and strains. See Fig. 2. The fit is not prefect, which suggests that a better t_o could be found. The data from the three shocks fits the constitutive relation remarkably well, which suggests that the results are fairly accurate and that a single dislocation process is in operation for most of the data.

DISLOCATION MODEL

In Fig. 2, t' is usually more than .5 kbar above the internal stress. This suggests that the dislocations are running freely in viscous glide. The situation resmbles that found for diamond-lattice materials by Haasen and co-workers[8] where initial dislocation densities were small, the overstress large, and strong dislocation generation occurred. They found a behavior seen in Eq. (1),: at constant strain rate, the overstress, τ = t' - t_o, fell with increasing plastic strain since lower dislocation velocities were needed when more dislocations were present.

Based on the above ideas, we will model ψ, $\dot\psi$, τ, and the total dislocation density, N, by combining the Orowan Eq. (2), with an equation for dislocation generation, Eq. (3). Here, all dislocations are assumed mobile and contribute to $\dot\psi$:

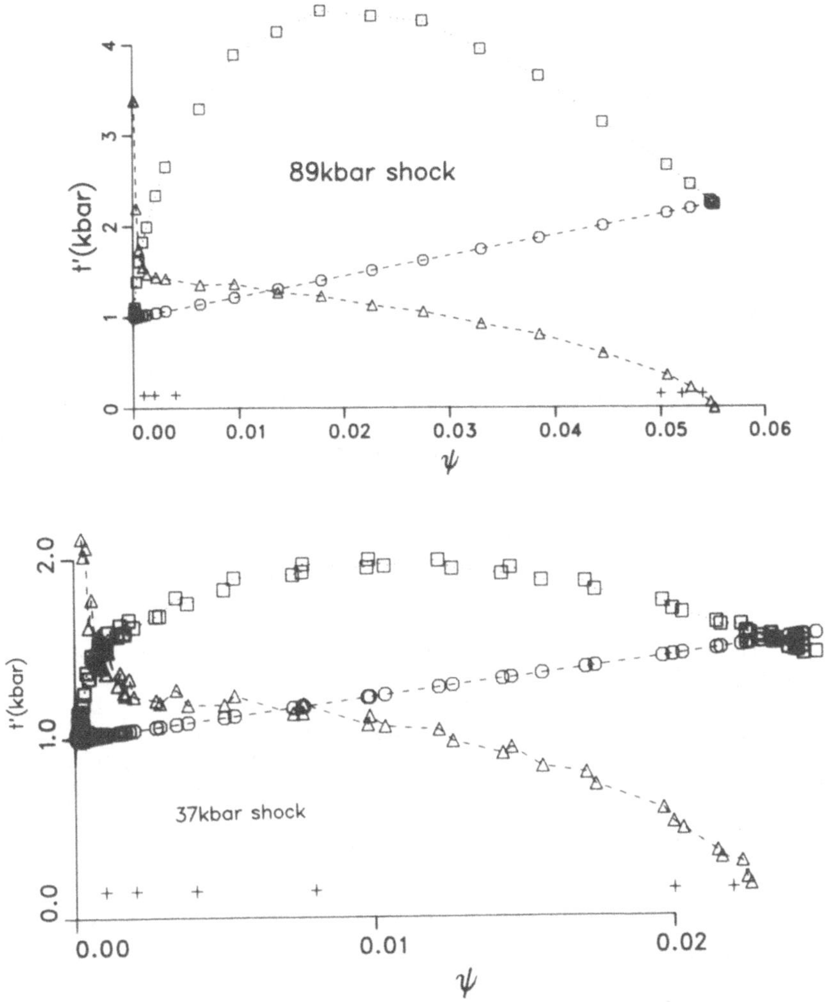

Fig. 2. The squares are t'-data. The circles and triangles are t_o and F_s values. The y-axis scale applies to F_s if taken to be unitless. The pluses show what ψ-values were used in the constitutive relation.

$$\dot{\psi} = b\, N\, V \tag{2}$$

$$\dot{N} = \alpha\, N^{\delta}\, \tau^{\beta} V \tag{3}$$

We used the value $2.86\text{x}10^{-8}$ cm [1] for b, the Burgers vector length, and α is a parameter to be fitted. V is the average dislocation velocity, which we take to be $D\, \tau^{\varepsilon}$, where ε and D are parameters. The dot superscripts above denote differentiation with respect to time. Later we take ε = 1 as the most reasonable choice for freely moving dislocations. In this case, D becomes equal to b/B, where B, the viscosity coefficient, was taken to be 2.66e-4 dyne-sec/cm^2, a pure Al value. See Fig. 6 of Gorman et. al.[9]

We solved for N from Eqs. 2 and 3 by eliminating the time differential between them and integrating on N and ψ. The result, assuming the initial N was zero, was:

$$(2 - \delta)(\alpha/b) \int \tau^{\beta} \, d\psi = N^{2-\delta}. \qquad (4)$$

This result can be substituted into the Orowan Eqn. and with further algebra the following equations obtained:

$$\tau = F_s [(2-\delta) \, D^{2-\delta} \alpha \, b^{1-\delta}]^{-\mu} \, \psi^{(2-\delta)\mu} \, \psi^{-\mu}, \qquad (5)$$

$$F_s = (\tau^{\beta} \psi \, / \int \tau^{\beta} \, d\psi)^{\mu}, \qquad (6)$$

where μ is $1/[\varepsilon(2-\delta) + \beta]$. Eq.(5) is of the same form as the constitutive Eq.(1) except for the factor F_s. Fitting the exponents of Eq.(5) with those of Eq.(1) results in the unique value 5/3 for δ and the relation $\varepsilon + 3\beta = 4$. Hence, no matter what the right velocity law is, i.e. the value of ε, we learn that dislocation multiplication depends on the 5/3 power of N. We note that the law found to work in Ge by Haasen and others was N τ V. Our N-term differs, possibly because our source density might depend on the node density of the dislocation network, $N^{1.5}$, instead of the dislocation line length, N.

In order to assess the variation of the factor F_s, we calculated this quantity along the shock process using the τ-ψ data of Fig. 2. ε was set to 1. The results were not sensitive to varying ε by .5 or so. The Taylor factors drop out of the product $\tau \, d\psi$. From the calculated values of F_s can be seen that F_s is usually around one for the points used to construct the constitutive relation, and otherwise varies from about 2 to .2 or so. This variation of about 10 over the shock process is much less that the variation of about 1000 of the ordinate of Fig. 1. Hence, F_s is constant to a good approximation by comparison and the constitutive relation (1) holds to a good approximation except at the beginning and end of the shock. This means that ψ is approximately a state variable for the loading paths encountered here. More generally, N would be used instead.

As a check on our dislocation model, we should compare the dislocation densities at the end of the shock, calculated by Eq.(4), with those measured in the recovery experiments of Rhode et. al.[10], who found a density, N, of about 1.5×10^{10} after the passage of a 90 kbar shock in 6061T6 Aluminum. Eq.(4) predicts densities of about 2.4×10^{9}, 6.1×10^{10}, and 3.8×10^{13} for the 21, 37, and 89kbar shocks, respectively. Considering that no parameters were adjusted, the present formulation gives tolerable densities, except for the 89kbar shock. We believe that the problem is due to using an unphysical model for the internal stress, i.e. the expression $1+22.5\,\psi$ for t_o. We have done preliminary calculations using the model $A\,N^{1/2}$ for t_o, where A is $G\,b/(2\pi)$; the t_o successfully used by Haasen and others in modeling the diamond-lattice materials. It is based on the average stress between individual dislocations. Without adjusting any parameters, N for the 89kbar shock was reduced at least an order of magnitude. Further work is underway.

In calculating N from Eq. (4), we used a value for α obtained by equating $c \times 10^{-6}/(f)^{1/2}$, where the c is that of Eq.(1), with the collection of constants in Eq.(5). The square root of f=2.06, the Taylor factor, is necessary to account for polycrystaline effects.

We calculated dislocation velocities from $V = (b/B)\tau$, rather than from the Orowan equation because the dislocation densities were suspect. The highest velocity found was .3 times the shear velocity, which is reasonable.

In our model, temperature effects appear in the shear modulus scaling and in the temperature dependence of B, which is known to be roughly linear near room temperature, the initial shock condition. In Eq.(5), D and thus its T-dependence, occur to the 1/4 power after $\delta=5/3$ is used. This constitutes weak T-dependence. Since the temperature increases calculated in Ref. 4 were less than 32K for all points used in the constitutive relation (except the three end points of the 89kbar shock, where the increases were about 85K),we can only say that the dislocation model predicts only small temperature effects and those actually experienced were not really large enough to test them.

CONCLUSIONS

We have modeled shock loading conditions in which the driving stress exceeds the internal barrier or frictional stresses sufficiently to cause the dislocations to run freely and multiply rapidly. The internal stress or work hardening model used here is very special to the large overstress-small initial work hardening situation. For large overstresses and strong initial work-hardening, the present model should work if the t_o -model is generalized to include stored dislocations.

We thank U.S. D.O.E. for support and D. C. Wallace for making his notebooks available.

REFERENCES

1. J. N. Johnson and L.M. Barker, J. Appl. Phys. 40,4321 (1969).
2. W. G. Johnston and J.J. Gilman, J. Appl. Phys. 30,129 (1959).
3. D. C. Wallace, Phys. Rev. B22, 1477 (1980).
4. D. C. Wallace, Phys. Rev. B22, 1487 (1980).
5. D. C. Wallace, Phys. Rev. B22, 1495 (1980).
6. D. C. Wallace and G.K. Straub, private communication, (1979).
7. J. F. Thomas Jr.,Phys. Rev. 175, 955 (1968).
8. H. Alexander and P. Haasen in Solid State Physics, vol.22, F. Seitz, D. Turnbull, and H. Ehrenreich (eds.), Academic Press, New York (1968) p.28.
9. J. A. Gorman, D. S. Wood, and T. Vreeland, Jr., J. Appl. Phys. 40, 833 (1969).
10. R. W. Rohde, J. L. Wise, J. G. Byrne, and S. Panchanadeeswaran in Shock Waves in Condensed Matter-1983 (AIP), J. R. Asay, R. A. Graham, G. K. Straub (eds.) North Holland, Amsterdam (1984) p.407.

SHOCK VISCOSITY AND THE CALCULATION OF STEADY SHOCK WAVE PROFILES*

J. W. Swegle and D. E. Grady
Computational Physics and Mechanics Division 1531
Thermomechanical and Physical Division 1534
Sandia National Laboratories
Albuquerque, NM

A model is developed which incorporates experimental data on viscous-like behavior of solids during large amplitude compressive stress-wave propagation. Maximum strain rate in the plastic wave has been determined for wave profiles in eight materials obtained with velocity interferometry methods. A plot of strain rate versus Hugoniot stress for each material is described by $\dot{\eta} = a\sigma^4$. The model is capable of reproducing the experimental data for steady-wave profiles and correctly predicts the evolution of the waves as they approach steady state.

INTRODUCTION

With the emergence of ultra-short time resolution instrumentation techniques, shock compression profiles, heretofore identified as shocks, are being resolved into finite risetime compressive wave profiles with wave structure features which depend sensitively on the shock amplitude and propagation distance. A feature unique to the realm of large amplitude waves is the steady-wave or permanent-regime plastic-shock profile.[1,2,3] The steady-wave profile, as its name implies, propagates without change in form and is a consequence of a stable balance between the competing processes of stress-strain nonlinearity and dissipative or viscous material behavior.

In the present study, a collection was made of steady-wave shock viscosity data on six metals and two insulating solids obtained by velocity interferometry techniques.The single significant observation to note in the trend of the data is the very rapid increase in steady-wave strain rate (or decrease in risetime) with increasing peak stress. This increase is substantially greater than would be predicted by a simple Newtonian proportionality between viscous stress and strain rate. The trend is emphasized in Fig. 1 by comparing the data for each material with a line segment of slope $\dot{\eta} \sim \sigma^4$. It should be noted that the stress dependence of strain rate does not differ significantly from material to material.

---

*This work performed by Sandia National Laboratories supported by the U. S. Department of Energy under contract #DEAC04-76-DP00789.

A related issue is the numerical description of stress wave propagation. The difficulty of treating both discontinuous shocks and the regions of continuous flow between them was solved through a method introduced by von Neumann and Richtmyer.[4] This numerical technique makes use of an artificial viscosity which smears discontinuous shocks into fast but smooth transitions over several mesh intervals. This method is, however, incapable of predicting the experimental risetime data, since the calculated wave thickness depends on the zone size and the artificial viscosity coefficients, and is independent of the shock stress. A model was thus developed which incorporates the observed trends of the shock viscosity data. The model includes the viscous behavior, through a rate-dependent formulation, in a rational three-dimensional framework. The model successfully reproduces the experimental wave profile data from which the shock viscosity data were extracted. The model has also been shown to correctly predict the evolution of the waves as they approach steady state.

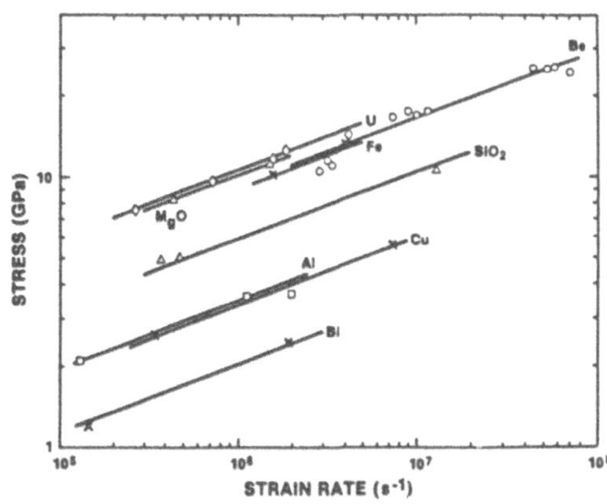

Figure 1: Shock stress versus strain rate for materials in this study.

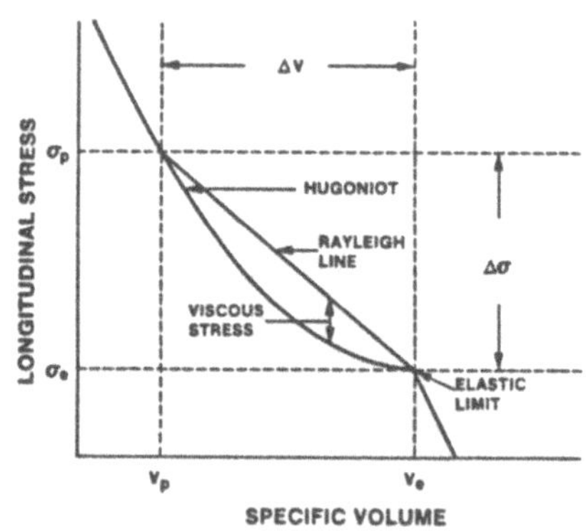

Figure 2: Hugoniot and Rayleigh line for an elastic-plastic material.

CONSTITUTIVE MODEL

Consider a uniaxial-strain shock wave propagating into an elastic-plastic material. For a restricted range of shock amplitudes, a two-wave structure develops with the first wave compressing the material to the elastic limit, with density ρ_e and particle velocity u_e. The second wave, or plastic wave, further compresses the material to the final stress amplitude. As shown in Fig. 2, the locus of possible end states attainable in the plastic wave is given by the Hugoniot. However, when a steady wave is achieved, the total stress in the plastic wave lies on the Rayleigh line. Thus, the viscous stress can be identified as the stress on the Rayleigh line minus the equilibrium stress. In general, the equilibrium stress is not known precisely, but it can be approximated by the Hugoniot stress.

By assuming a linear relation between shock velocity, D, and particle velocity, u, of the form $D - u_e = C_0 + S(u - u_e)$, an expression can be obtained for the stress on the Hugoniot. Combining this with the equation for the Rayleigh line yields the viscous stress as a function of specific volume through the plastic wave. The maximum viscous stress is then given, to lowest order in the stress jump $\Delta\sigma$, by[5]

$$\sigma_v = \frac{S(\Delta\sigma)^2}{2\rho_e C_0^2} + \cdots . \tag{1}$$

The assumption is made that there is no bulk viscosity, so that the viscous stress is a purely deviatoric stress. Thus,

$$S_v \equiv \sigma_v , \tag{2}$$

where S_v is the longitudinal component of the deviatoric viscous stress. The experimental data in Fig. 1 suggest an empirical dependence for the plastic strain rate of the form,

$$\dot{\eta}_p = A(\Delta\sigma)^4 , \tag{3}$$

where A is a constant of proportionality determined from the data. Combining this with Eqs. (1) and (2) yields,

$$\dot{\eta}_p = \frac{4\rho_e^2 C_0^4 A}{S^2} S_v^2 . \tag{4}$$

We have made the further constitutive assumption that Eq. (4) applies to the instantaneous value of viscous stress, in addition to the maximum viscous stress, in the plastic wave.

These expressions involve quantities related to the uniaxial strain shock wave situation. However, the theory should describe the general three-dimensional case and be frame indifferent, which is the requirement that the constitutive relations be independent of the reference frame in which the stress and strain tensors are represented. Thus the stress and strain measures will be functions only of invariants of the stress and strain tensors. Converting to the octahedral shear stress and strain, τ and γ, Eq. (4) becomes,[5]

$$\begin{aligned} \dot{\gamma}_p &= \frac{8\sqrt{2}\rho_e^2 C_0^4 A}{3S^2}\tau_v^2 \\ &= A'\tau_v^2 . \end{aligned} \tag{5}$$

The deviatoric constitutive relation for an elastic-plastic material assumes that the stresses are supported only by the elastic strains,

$$d\tau = 2\mu\dot{\gamma}_e dt , \tag{6}$$

where μ is the shear modulus. The total strain rate is decomposed into elastic and plastic components, so that the the elastic strain rate, $\dot{\gamma}_e$, is the total strain rate, $\dot{\gamma}$, minus the plastic strain rate, $\dot{\gamma}_p$. Therefore,

$$d\tau = 2\mu(\dot{\gamma} - \dot{\gamma}_p)dt . \tag{7}$$

Substituting Eq. (5) for the plastic strain rate in Eq. (7) yields

$$d\tau = 2\mu\dot{\gamma}dt - 2\mu A'\tau_v^2 dt . \tag{8}$$

Equation (8) is the basis of the model used in the calculations described in the next section.

RESULTS

A finite-difference formulation of the above model[5] has been implemented in the wavecode TOODY,[6] which allows comparison of calculations with the experimental wave profiles used to generate Fig. 1. Calculations have been made for aluminum, beryllium, bismuth, copper, iron, magnesium oxide, fuzed silica and uranium at various stress levels.[5] All calculations showed good agreement with the experimental wave profiles, as exemplified by Fig. 3, which shows the comparison between calculations (solid line) and experimental data (dotted line) for uranium. The figure clearly shows the dependence of plastic wave risetime on stress level.

Another intriguing aspect of the comparisons is the question of whether or not all wave profiles in the study represent steady waves. A rudimentary analysis[5] indicates that some of the waves may not have propagated far enough to reach steady state. However, calculations agree well with all profiles, including those whose attainment of steady state is questionable. Since the calculations were all done for propagation distances which matched experimental sample thicknesses, the comparisons indicate that the model predicts the correct wave profiles as the waves evolve toward steady state, and thus is capable of predicting nonsteady as well as steady-wave risetimes. A confirmation of this is provided by Fig. 4, which compares calculations with experimental data for .95 GPa wave profiles in aluminum at different sample thicknesses. Since the waves had clearly not reached steady state, these profiles were not included in the original set of data used to produce Fig. 1. However, the model still predicts the correct risetime for these waves. Thus, although the model was developed on the basis of an observed regularity in steady-wave profile data, it also applies to nonsteady waves as they evolve toward steady state.

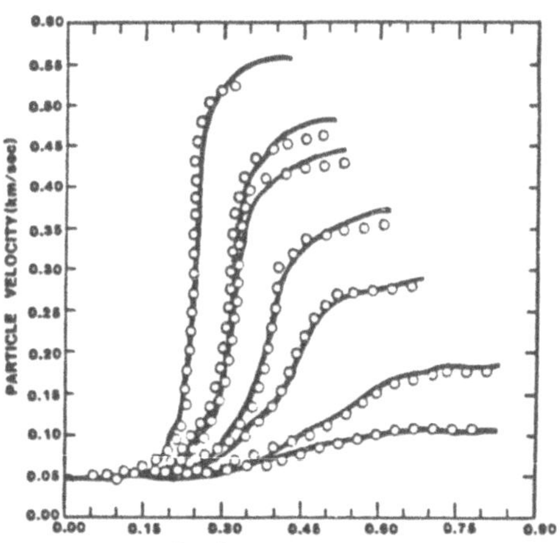

Figure 3: Comparison of experiment (dotted line) and calculation (solid line) for wave profiles in uranium. Calculated stress levels are 16.7, 14.4, 13.2, 11.2, 8.7, 5.7, and 3.3 GPa.

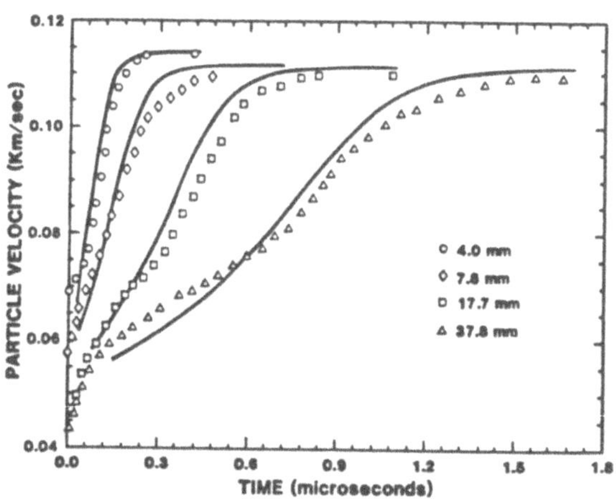

Figure 4: Comparison of experiment (dotted line) and calculation (solid line) for .95 GPa nonsteady wave profiles in aluminum at different sample thicknesses.

DISCUSSION AND CONCLUSIONS

The underlying mechanisms governing shock viscosity and their behavior in the shock process are not yet well understood. Viscous-like flow within the shock is thought to be associated with the microscopic processes of dislocation multiplication and motion, twinning, vacancy production, precipitate alteration, etc. Most probably it is a complex event involving the collective participation of several of these mechanisms.

The present study puts forth what may be a useful clue in terms of an ultimate understanding of the viscous behavior of solids. In comparing one aspect of steady-wave plastic shocks (Hugoniot stress-maximum strain rate dependence) for a number of different materials, we have observed an interesting and perhaps significant trend. Namely, the decrease in steady-wave risetime with increasing Hugoniot stress is approximately the same for all of the materials examined. This observation is clearly tentative considering the limited data, but it does suggest an intriguing material independent feature in the steady shock process.

We have used the observed viscous behavior as the motive for a phenomenological description of the shock deformation process. Such an approach is self-consistent and eminently practical in application. The numerical model readily predicts the wave profile data available, and these data are sufficiently diverse to lend confidence to a limited amount of extrapolation, both to higher pressure for the materials studied and to other materials not yet investigated.

The model also provides a heretofore unavailable capability: performing calculations with correct strain rates in both steady and nonsteady-wave fronts. This allows more accurate results to be obtained in wave interaction and stress attenuation analyses and provides a foundation for the development of micromechanical models in which the evolution of the parameters is controlled by the strain rate.

REFERENCES

1. W. Band, Studies in the Theory of Shock Propagation in Solids, J. Geophys. Res., 65:695 (1960).

2. W. Band and G. E. Duvall, Physical Nature of Shock Propagation, Amer. J. Phys., 29:780 (1961).

3. D. R. Bland, On Shock Structure in a Solid, J. Inst. Math. Applic.,1:56 (1964).

4. J. von Neumann and R. D. Richtmyer, A Method for the Numerical Calculation of Hydrodynamic Shock, J. Appl. Phys., 21:232 (1950).

5. J. W. Swegle and D. E. Grady, Shock Viscosity and the Prediction of Shock Wave Risetimes, J. Appl. Phys., in press.

6. J. W. Swegle, TOODY IV – A Computer Program for Two-Dimensional Wave Propagation, Sandia National Laboratories Report, SAND78-0552 (1978), unpublished.

STRAIN RATE DEPENDENCE IN STEADY, PLASTIC SHOCK WAVES*

J. E. Dunn and D. E. Grady
Thermomechanical and Physical Division 1534
Sandia National Laboratories
Albuquerque, NM

INTRODUCTION

One-dimensional plane shock waves have been widely used to study the mechanical response of solids to high velocity deformation. For many materials, under a certain range of impact pressures, there exists a two-wave structure of which the first wave, the so-called elastic precursor, travels at the velocity of sound waves, while the second wave, the so-called plastic shock wave, travels at a slower speed which increases with the impact pressure. While the full two-wave structure is thus not steady (does not propagate without change of form), each component wave may be treated as steady after sufficient propagation distance. With regard to the steady plastic wave, Grady[1] noted that the wave profile data of Barker[2] for aluminum suggested the empirical relationship

$$\dot{\epsilon}_M = k\, p_D^4 , \tag{1}$$

where k is a material dependent constant, and where $\dot{\epsilon}_M$ is the maximum of the absolute value of the strain rate in the plastic wave, while p_D is the driving or impact pressure required to create the wave. Within the class of materials meeting

$$\sigma = f(\epsilon) + g(\dot{\epsilon}) , \tag{2}$$

where σ is the axial stress and where $f(\epsilon)$ and $g(\dot{\epsilon})$ are certain functions of, respectively, the uniaxial strain ϵ and its material time rate ϵ, Grady further observed that (1), along with a standard assumed form for $f(\cdot)$, implied that, to a first approximation,

$$g(\dot{\epsilon}) = \mu_o sgn(\dot{\epsilon}) \mid \dot{\epsilon} \mid^{\frac{1}{2}} \tag{3}$$

where μ_o is a material dependent constant.

More recently, Swegle and Grady[3] noted that the empirical relation (1) in fact seemed to hold not only for the plastic waves in aluminum but also for the plastic

*This work performed by Sandia National Laboratories supported by the U. S. Department of Energy under contract #DE-AC04-76-DP00789.

waves in uranium, beryllium, magnesium oxide, iron, fuzed silica, copper, and bismuth. Additionally, they showed that the constitutive model (2) and (3) (with $f(\epsilon)$ as given below) produced wave profiles in good agreement with the reported data on these materials.

The present note is a brief summary of our further attempts to understand the constitutive implications of (1) for the rather broader class of materials given by

$$\sigma = f(\epsilon) + g(\epsilon, \dot{\epsilon}) \ , \tag{4}$$

where the elastic part of the stress, $f(\epsilon)$, is as in (2), while the viscous part, $g(\epsilon, \dot{\epsilon})$, is now permitted to be sensitive to strain as well as strain rate and need only meet $g(\epsilon, 0) = 0$. As we shall see, the constitutive structure (4) is significantly richer than (2); indeed so rich that relations like (1) impose only very mild restrictions on the form of $g(\cdot, \cdot)$. In fact, we show that any functional relation $\dot{\epsilon}_M = \dot{\epsilon}_M(p_D)$ is, in general, compatible with an infinity of viscous stress forms $g(\cdot, \cdot)$. We illustrate this by showing that even within the restricted class of power law models

$$g(\epsilon, \dot{\epsilon}) = sgn(\dot{\epsilon}) h(\epsilon) \mid \dot{\epsilon} \mid^m \ , \tag{5}$$

where the positive number m is called the *strain rate sensitivity*, there is, for every $m > \frac{2}{9}$, a strain dependent viscosity function $h_m(\epsilon)$ such that (1) is satisfied. Thus, once we broaden our choice of constitutive forms beyond the special structure of (2), empirical relations like (1) give, by themselves, only a crude guidance toward detailed constitutive structure. The full wave profiles from which (1) was extracted provide, of course, more detailed guidance and we close by comparing the experimental wave profiles with the theoretical wave profiles predicted by (4) and (5) with (1) for various values of m.

PRELIMINARY NOTIONS

Let $\mathcal{B}$ be a one-dimensional body whose particles in a fixed reference configuration are labelled X, X $\in (-\infty, \infty)$. If $x = \chi(X, t)$ is a *motion* of $\mathcal{B}$, then let

$$\begin{aligned} v &= v(X, t) \equiv \dot{\chi}(X, t) \equiv \frac{\partial \chi(X,t)}{\partial t} \ , \\ \epsilon &= \epsilon(X, t) = \frac{\partial \chi(X,t)}{\partial X} - 1 \ , \end{aligned} \tag{6}$$

denote, respectively, the *velocity* of X and the *strain* at X at time t. We recall that balance of linear momentum for $\mathcal{B}$ can be written in the form

$$\frac{\partial \sigma}{\partial X} = \rho_R \dot{v} \ , \tag{7}$$

where $\sigma = \sigma(X, t)$ is the *axial stress* at X at time t, and where $\rho_R = \rho_R(X)$ is the density at X in the given reference configuration (and is here taken to be independent of X). We note that in compression $\sigma \leq 0$ and ϵ is in the interval (-1,0].

A *structured steady shock wave* is any smooth triad (σ, v, ϵ) meeting (6) and (7) and such that

$$\sigma = \hat{\sigma}(X - \mathcal{V}t) \ , \qquad \hat{v} = \hat{v}(X - \mathcal{V}t) \ , \qquad \epsilon = \hat{\epsilon}(X - \mathcal{V}t) \ , \tag{8}$$

where the number $\mathcal{V}$ is called the *referential wave speed.* Upon inserting $(8)_{1,2}$ into (7), it is easy to see that the profiles $\hat{\sigma}(\cdot)$ and $\hat{v}(\cdot)$ must satisfy

$$\hat{\sigma}(\xi) = -\rho_R \mathcal{V} \hat{v}(\xi) + c_1 \tag{9}$$

for all $\xi \in (-\infty, \infty)$, where c_1 is a constant of integration. Further, $(8)_{2,3}$ are compatible with $(6)_{1,2}$ if and only if

$$\hat{v}(\xi) = -\mathcal{V}\hat{\epsilon}(\xi) + c_2 \tag{10}$$

for all $\xi \in (-\infty, \infty)$, where c_2 is a second constant of integration. Upon entering (10) into (9), we find then that in a structured shock wave

$$\hat{\sigma}(\xi) = \rho_R \mathcal{V}^2 \hat{\epsilon}(\xi) + c_3 \tag{11}$$

for some constant c_3. (Of course, (11) is just the assertion of the Rayleigh line construction: in a structured steady shock wave the stress-strain pairs experienced throughout $\mathcal{B}$ fall on a straight line of slope $\rho_R \mathcal{V}^2$.) Finally, when the constitutive equation (4) is entered into (11), we arrive at

$$-g(\epsilon, -\mathcal{V}\epsilon') = f(\epsilon) - \rho_R \mathcal{V}^2 \epsilon - c_3 \,, \tag{12}$$

since now $\dot{\epsilon} = -\mathcal{V}\epsilon'$, $\epsilon' \equiv \dfrac{d\hat{\epsilon}}{d\xi}$. We thus have a differential equation to be solved for the profile shape $\epsilon = \hat{\epsilon}(\xi)$ subject to the boundary conditions we now outline.

We are first of all interested in solutions to (12) such that $\hat{\epsilon}'(\pm\infty) = 0$. Since $g(\epsilon, 0) \equiv 0$, we may use (12) to solve for c_3 and $\rho_R \mathcal{V}^2$; indeed

$$c_3 = f(\epsilon^{\pm}) - \rho_R \mathcal{V}^2 \epsilon^{\pm} \quad and \quad \rho_R \mathcal{V}^2 = \frac{f(\epsilon^+) - f(\epsilon^-)}{\epsilon^+ - \epsilon^-} \,, \tag{13}$$

where, of course, $\epsilon^+ = \hat{\epsilon}(+\infty)$ and $\epsilon^- = \hat{\epsilon}(-\infty)$. Now the steady plastic wave we wish to model sits atop the elastic precursor. We thus view the plastic wave as propagating into material excited to a velocity $v^+ = \hat{v}(+\infty)$ and a strain ϵ^+ by the prior passage of the elastic wave. In an obvious notation then, $v^+ = v_{el}^-$, $\epsilon^+ = \epsilon_{el}^-$, while $v_{el}^+ = \epsilon_{el}^+ = 0$. But, $f(0) = 0$, so applying (10) and $(13)_2$ to the elastic precursor, we see that $v_{el}^- = -C_L \epsilon_{el}^-$ and $\rho_R C_L^2 = f(\epsilon_{el}^-)/\epsilon_{el}^-$ where C_L is the elastic longitudinal wave speed. Taking v^+ and C_L as known, we thus see that our plastic wave meets $\epsilon^+ = -v^+/C_L$ and $\sigma^+ \equiv f(\epsilon^+) = -\rho_R v^+ C_L$. We thus may evaluate the constant c_2 in (10) for the plastic wave or, equivalently, we may write (10) in the form

$$\hat{v}(\xi) - v^+ = -\mathcal{V}\{\hat{\epsilon}(\xi) - \epsilon^+\} \,. \tag{14}$$

Finally, *given the functional form of* $f(\cdot)$, we may, if $\sigma^- = f(\epsilon^-)$ is known, solve for ϵ^- and then compute $\mathcal{V}$ from $(13)_2$; or, if $v^- \equiv \hat{v}(-\infty)$ is known (and this is more often the case), we may use (14) to write that $v^+ - v^- = -\mathcal{V}(\epsilon^+ - \epsilon^-)$, which together with (13) gives two equations to be solved for the two unknowns ϵ^- and $\mathcal{V}$. In either case, we see that the back and front states of our plastic wave, along with the wave speed $\mathcal{V}$, are completely determined and that (12) is now a well-posed ordinary differential equation for the wave profile $\hat{\epsilon}(\cdot)$. Once $\hat{\epsilon}(\cdot)$ is obtained (14) may be used to obtain $\hat{v}(\cdot)$.

Lastly, although much of our analysis is independent of a specific functional form for $f(\cdot)$, the elastic portion of the stress, for numerical calculations we employ the well-known[4] empirical relation that, for many materials, the plastic shock speed $\mathcal{V}$ and the back state particle velocity v^- obey a linear relationship of the form

$$\mathcal{V} = C_o + s(v^- - v^+) , \tag{15}$$

where C_o is usually identified with the bulk elastic sound speed and where s is a material dependent constant, usually between 1 and 2. If we apply (14) to (15), we see that

$$\mathcal{V} = \frac{C_o}{1 + s(\epsilon^- - \epsilon^+)} \tag{16}$$

and, upon entering this last into $(13)_2$, we arrive at

$$f(\epsilon^-) = f(\epsilon^+) + \frac{\rho_R C_o^2(\epsilon^- - \epsilon^+)}{\{1 + s(\epsilon^- - \epsilon^+)\}^2} . \tag{17}$$

Thus, the linearity of the plastic $\mathcal{V}$ *vs* v^- relationship, requires that $f(\cdot)$ be of the form

$$f(\epsilon) = \sigma^+ + \frac{\rho_R C_o^2(\epsilon - \epsilon^+)}{\{1 + s(\epsilon - \epsilon^+)\}^2} . \tag{18}$$

for $\epsilon \leq \epsilon^+(\leq 0)$. The form of $f(\cdot)$ on the interval $(\epsilon^+, 0)$, usually taken to be linear, is not required in the present analysis, since we are not concerned with the structure of the elastic precursor.

PEAK STRAIN RATE ANALYSIS

Since $\dot{\epsilon} = -\mathcal{V}\epsilon'$, $\epsilon' > 0$, the absolute value of the strain rate in the plastic wave takes on its maximum value when $\hat{\epsilon}'(\xi)$ takes on its maximum value. Let $\xi^* \in (-\infty, \infty)$ denote the place where this occurs and set $\epsilon^* = \hat{\epsilon}(\xi^*)$, $\dot{\epsilon}_M = \mathcal{V}\hat{\epsilon}'(\xi^*)(\geq 0)$. Since $\hat{\epsilon}''(\xi^*)$ vanishes, we may differentiate (12) to find that ϵ^* and $\dot{\epsilon}_M$ together meet

$$\begin{aligned} -g(\epsilon^*, -\dot{\epsilon}_M) &= f(\epsilon^*) - f(\epsilon^+) - \rho_R \mathcal{V}^2(\epsilon^* - \epsilon^+) , \\ -g_\epsilon(\epsilon^*, -\dot{\epsilon}_M) &= f_\epsilon(\epsilon^*) - \rho_R \mathcal{V}^2 . \end{aligned} \tag{19}$$

Now, for *given* functions $f(\cdot)$ and $g(\cdot,\cdot)$, (19) constitutes two equations for the two unknowns $\epsilon^* = \epsilon^*(\epsilon^-)$ and $\dot{\epsilon}_M = \dot{\epsilon}_M(\epsilon^-)$, where the dependence on ϵ^- is due to the dependence of, by $(13)_2$, $\rho_R \mathcal{V}^2$ on ϵ^-. If on the other hand, we regard only the elastic part, $f(\cdot)$, of the stress as known then (19) tells us very little about the viscous portion, $g(\cdot,\cdot)$. Indeed, we may append to (19) *any* postulated or empirical relation $\dot{\epsilon}_M = \dot{\epsilon}_M(\epsilon^-)$ (and (1) is certainly of this form) as well as *any* postulated or empirical relation $\epsilon^* = \epsilon^*(\epsilon^-)$ (and we have very little knowledge of the form of such a relationship) and (19) then serves only to tell us the values of $g(\cdot,\cdot)$ and $g_\epsilon(\cdot,\cdot)$ on the single curve $(\epsilon^*, -\dot{\epsilon}_M)(\cdot)$; off this curve $g(\cdot,\cdot)$ is completely arbitrary. (Of course, within a sufficiently special family of functions, (19) and an assumed form for, say, $\dot{\epsilon}_M(\epsilon^-)$ may well serve to define g. Indeed, as the analysis of Grady[1] showed, this is exactly what happens if one restricts oneself to viscous stresses g that depend on strain rate alone: now $g_\epsilon \equiv 0$, $(19)_2$ can be solved alone for $\epsilon^* = \epsilon^*(\epsilon^-)$, and this, $(19)_1$, and the function $\dot{\epsilon}_M(\epsilon^-)$ completely determines the function $g(\dot{\epsilon})$.)

To illustrate the above remarks and to show how the full wave profile can be used to refine the implications of gross empirical measures like (1), we have studied numerically the class of power law models given by (4) and (5). For these models, if we differentiate $(19)_1$ with respect to ϵ^- and apply $(19)_2$, we arrive at

$$mh(\epsilon^*)\dot{\epsilon}_M^{m-1}\frac{d\dot{\epsilon}_M}{d\epsilon^-} = -(\epsilon^* - \epsilon^+)\frac{d}{d\epsilon^-}(\rho_R \mathcal{V}^2) \ .$$

If we now divide this last into $(19)_1$, we see that

$$\frac{\dot{\epsilon}_M}{m\left(\frac{d\dot{\epsilon}_M}{d\epsilon^-}\right)} = \frac{\left\{\frac{f(\epsilon^*)-f(\epsilon^+)}{\epsilon^*-\epsilon^+}\right\} - \rho_R \mathcal{V}^2}{-\frac{d}{d\epsilon^-}(\rho_R \mathcal{V}^2)} = \frac{(\epsilon^- - \epsilon^+)\left[\left\{\frac{f(\epsilon^*)-f(\epsilon^+)}{\epsilon^*-\epsilon^+}\right\} - \rho_R \mathcal{V}^2\right]}{\rho_R \mathcal{V}^2 - f_\epsilon(\epsilon^-)} \ , \tag{20}$$

where we have used $(13)_2$ to compute $\frac{d}{d\epsilon^-}(\rho_R \mathcal{V}^2)$. The result (20) is useful in that it allows us to calculate $\epsilon^*(\epsilon^-)$ (or its inverse $\epsilon^-(\epsilon^*)$) *without foreknowledge of the structure of* $h(\cdot)$ *if* the functions $f(\epsilon)$ and $\dot{\epsilon}_M(\epsilon^-)$ are known. Once $\epsilon^*(\epsilon^-)$ is thus found $(19)_1$ serves to determine $h(\cdot)$; indeed

$$h(\epsilon^*) = \frac{f(\epsilon^*) - f(\epsilon^+) - \rho_R \mathcal{V}^2(\epsilon^* - \epsilon^+)}{\dot{\epsilon}_M(\epsilon^-)^m}\Big|_{\epsilon^- = \epsilon^-(\epsilon^*)} \ . \tag{21}$$

Thus, like $g(\dot{\epsilon})$ in[1], these power law models are extremely special: knowledge of $g(\epsilon, \dot{\epsilon})$ on the curve $(\epsilon^*, -\dot{\epsilon}_M)(\cdot)$ determines $g(\epsilon, \dot{\epsilon})$ everywhere. Their interest for us here lies in the fact that they afford an infinite class of models meeting relations like (1).

Specifically then, if we adopt the form (18) and the empirical relation (1), so that now $\dot{\epsilon}_M(\epsilon^-) = kp_D^4$, where $p_D = \mid \sigma^- - \sigma^+ \mid = \mid f(\epsilon^-) - f(\epsilon^+) \mid$, then for each $m > 0$ (actually there are technical reasons that require $m > \frac{2}{9}$) (21) defines an associated $h_m(\cdot)$ such that the constitutive model $\sigma = f(\epsilon) + g(\epsilon, \dot{\epsilon}), g(\epsilon, \dot{\epsilon}) = sgn(\dot{\epsilon})h_m(\epsilon) \mid \dot{\epsilon} \mid^m$, automatically satisfies (1) in every plastic compression wave.

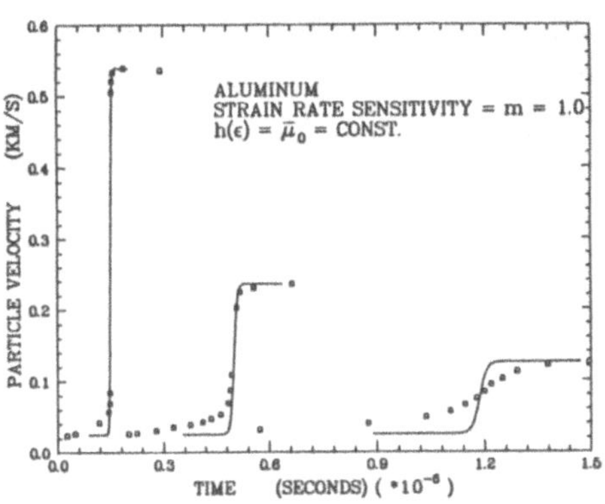

Figure 1: Wave profiles for aluminum. $g(\epsilon, \dot{\epsilon}) = \bar{\mu}_o\dot{\epsilon}$

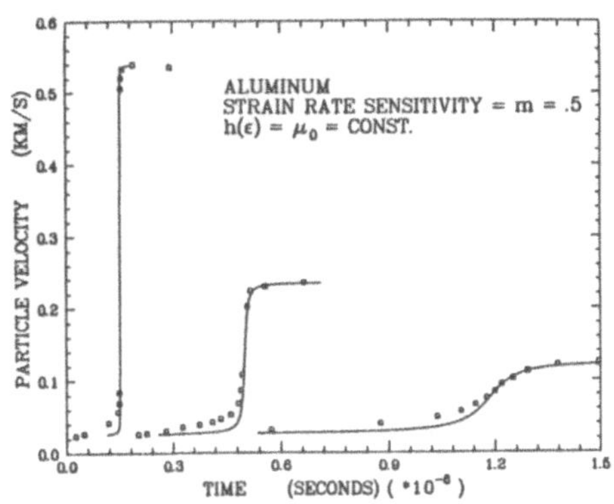

Figure 2: Wave profiles for aluminum. $g(\epsilon, \dot{\epsilon}) = \mu_o sgn(\dot{\epsilon}) \mid \dot{\epsilon} \mid^{\frac{1}{2}}$

Adopting the values $s = 1.337$, $\rho_R = 2{,}700$ kg/m^3, $C_o = 5.38$ km/sec, $C_L = 6.39$ km/sec, and $v^+ = .025$ km/sec, Figures 1 and 2 present a comparison of the theoretical (solid line) and experimental (squares) steady wave profiles for 6061-T6 aluminum, for respective driving pressures of $|\sigma^-| = 8.86$, 3.65, and 1.93 GPa, under the constitutive assumption (4). Here $g(\epsilon, \dot{\epsilon})$ is taken to be of the respective forms $\bar{\mu}_o \dot{\epsilon}$ (Newtonian viscous stress) and $\mu_o sgn(\dot{\epsilon})\, |\dot{\epsilon}|^{\frac{1}{2}}$ (Grady's form (3)), where $\bar{\mu}_o$ and μ_o are material dependent constants which were determined by requiring that, in accord with experiment, $\dot{\epsilon}_M = 2 \times 10^6\ s^{-1}$ for the $|\sigma^-| = 3.65$ GPa wave. As can be seen, the square root strain rate model fits the experimental data significantly better than the linear strain rate model, but even it still fails to capture the more gradual slope changes taking place in the lower portion of the profiles.

In Figure 3, we present a comparison of the above experimental profiles with the theoretical (solid line) steady wave profiles for the power law model (4) and (5), where $m = .69$ with $h(\epsilon) = h_m(\epsilon)$ determined according to (21), (1), and the normalization $\dot{\epsilon}_M = 2 \times 10^6\ s^{-1}$ when $|\sigma^-| = 3.65$ GPa. As is readily observed, this model captures much more thoroughly the asymmetry between the upper and lower portions of the wave. The microstructural mechanisms responsible for this asymmetry are, of course, only implicit in the form of $h_{.69}(\epsilon)$ which we present in Figure 4. They evidently involve a high degree of strain softening.

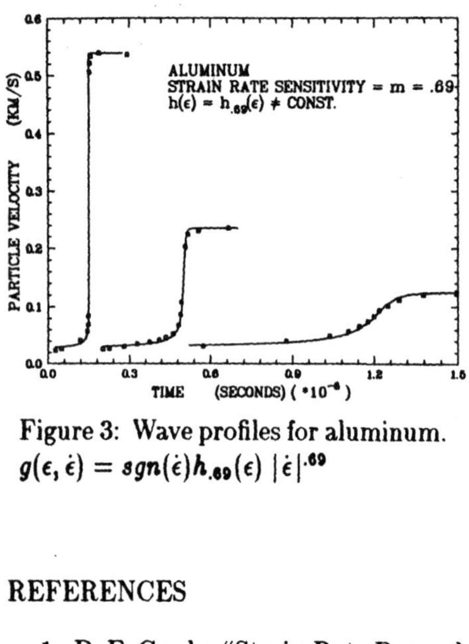

Figure 3: Wave profiles for aluminum. $g(\epsilon, \dot{\epsilon}) = sgn(\dot{\epsilon}) h_{.69}(\epsilon)\, |\dot{\epsilon}|^{.69}$

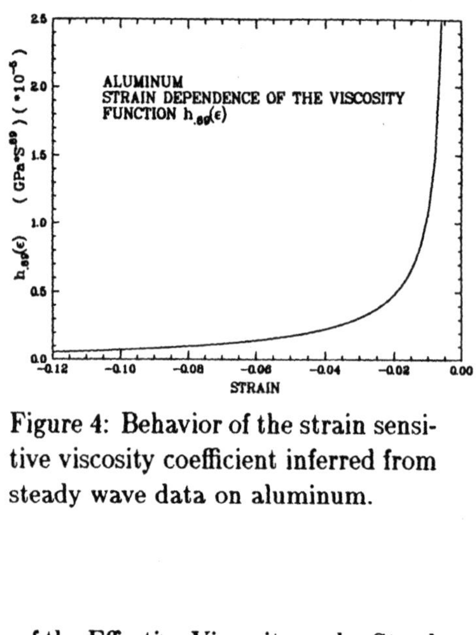

Figure 4: Behavior of the strain sensitive viscosity coefficient inferred from steady wave data on aluminum.

REFERENCES

1. D. E. Grady, "Strain-Rate Dependence of the Effective Viscosity under Steady-Wave Shock Compression," Appl. Phys. Lett. 38:825 (1981).

2. L. M. Barker, "Fine Structure of Compression and Release Wave Shapes in Aluminum Measured by the Velocity Interferometer Technique," in *Behavior of Dense Media under High Dynamic Pressure*, (Gordon and Breach, New York, 1968).

3. J. W. Swegle and D. E. Grady, "Shock Viscosity and The Prediction of Shock Wave Risetimes," J. Appl. Phys. (1985).

4. M. H. Rice, R. G. McQueen, and J. M. Wash, "Compression of Solids by Strong Shock Waves," in *Solid State Physics*, Vol. 6, (Academic Press, New York and London, 1958).

SHOCK RESPONSE OF SiC/2014-T4 ALUMINUM COMPOSITE

Dattatrya P. Dandekar and Craig M. Lopatin

U.S. Army Materials and Mechanics Research Center
Watertown, Massachusetts

INTRODUCTION

Composites form a class of materials which has the potential to be tailored to possess specific desirable physical and mechanical properties sometimes exceeding the desirable characteristics of its constituents. Interest in the composite material of this investigation, namely, silicon carbide whisker reinforced 2014 aluminum alloy arose from a desire to determine if this composite would have better mechanical properties than either of its two constituents. One of the properties of interest is the nature of the deformation of this composite, hereafter referred to as SiC/2014-T4 Al, under shock loading. This paper presents the preliminary results of a limited number of shock compression experiments performed on SiC/2014-T4 Al of two orientations to 5.2 GPa. The two orientations of SiC/2014-T4 Al result from the fabrication process of this composite by the method of hot-isostatic compression, usually called the HIP method. This aspect of the material orientation is briefly described in the following section. The results of this work, even though only preliminary in nature, do provide some pertinent information regarding both the nature of the deformation of SiC/2014-T4 Al under shock compression and its spall threshold.

MATERIAL

The silicon carbide aluminum composite used in the present work consists of randomly dispersed 25 volume percent silicon carbide whiskers in a 2014 aluminum alloy matrix. The whiskers were 0.2 to 0.5 micron in diameter and have a length to diameter ratio of approximately 80. The billets of the above mentioned composition were fabricated by the HIP method. The HIP material was subsequently given the conventional heat treatment bringing 2014 aluminum to the T4 condition. For 2014 aluminum alloy T4 condition involves : (i) heating the alloy to 769±6k, (ii) keeping it at this temperature for 1 to 2 hours, (iii) quenching the alloy in water, and (iv) finally aging the alloy at room temperature. This treatment basically homogenizes the material and tends to eradicate most of the detrimental microstructural variations in 2014 aluminum alloy. The above treatment was given to the present composite material (SiC/2014-T4 Al) under the hypothesis that it would reduce the variations in the material property obtainable between the billets. The billets of SiC/2014-T4 Al obtained in the above manner were in the form of thick walled cylinders. The discs of SiC/2014-T4 Al used in the shock experiments were obtained so as to have their faces either normal to the axis of the cylindrical billets (z axis) or normal to the radial direction (r) of the cylindrical billets. The specimens of SiC/2014-T4 Al with face normal in z and r directions are, respectively, identified as SiC/Al(z) and SiC/Al(r) in this paper.

The average density of SiC/2014-T4 Al was determined to be 2.91±0.01 Mg/m^3, the expected theoretical density of the composite. The average values of the longitudinal and shear elastic wave velocities determined by ultrasonic technique were 7.31±0.03 km/s and 3.89±0.01 km/s, respectively. These measurements were performed on 9 different specimens of SiC/Al(z) and SiC/Al(r).

Specimens of SiC/2014-T4 Al used in the shock wave experiments were circular discs with 31.75 mm diameter and thickness varying between 2 and 8 mm. The specimens were flat to 5 micron and opposing faces of the discs were mutually parallel within 0.4 parts per thousand over their lateral dimensions.

EXPERIMENTS

The shock experiments were performed with a 10.2 cm diameter light gas gun facility at the Army Materials and Mechanics Research Center. Direct impact and transmission shock experiments were performed to determine the Hugoniot of SiC/2014-T4 Al to 5.2 GPa. In addition, tension inducing shock wave experiments were conducted to determine the spall threshold in SiC/2014-T4 Al. The details of direct impact and transmission experiments and the procedure for the reduction of experimental data to deduce the shock compressed state of the material of interest can be found in Ref. 1 and 2. In every spall experiment a 2 mm thick disc of SiC/2014-T4 Al impacted a 4 mm disc of SiC/2014-T4 Al. Existence or absence of spall in the composite was inferred by examining the sliced discs of SiC/2014-T4 Al recovered in these experiments under microscope. All experiments were conducted in vacuum pressure of less than 7 Pa.

The parameters measured in the direct impact and transmission experiments were impact velocity and stress profile as recorded by X-cut quartz gauge. The X-cut quartz gauges were either used in shunted or shorted mode according to the presciption suggested by Graham[3]. The uncertainties in the measurements of impact velocity and stress by X-cut quartz gauges were 0.5% and 2.5%, respectively. During the spall experiments only the impact velocities were measured. The planarity of the impact in the experiments of the present investigation never deviated by more than 2 mrad.

RESULTS

The results of direct impact and transmission experiments are given in Tables 1 and 2. Fig. 1 provides a graphic representation of stress(σ) versus particle velocity (u) of SiC/Al(z) and SiC/Al(r). Table 2 shows that the Hugoniot Elastic Limit (HEL) values of SiC/Al(z) and SiC/Al(r) vary between 0.3 and 0.335 GPa, and 0.275 and 0.368 GPa, respectively, Calculation of the magnitude of HEL (5_{HEL}) estimated from the static proportional elastic limit (Y_o) data on SiC/2014-T4 Al from the relation

$$5_{HEL} = \left[(1 - \nu)/(1 - 2\nu)\right] Y_o \qquad (1)$$

yield a value of 0.398 ± 0.113 GPa. In the above relation, the value of Y_0 has been found to be 0.225 ± 0.064 GPa for SiC/2014-T4 Al[4], and ν is the Poisson's Ratio and has a value of 0.303, as obtained from the elastic

Table 1. Summary of direct impact experiments in SiC/2014-T4 Aluminum.

Experiment	Buffer	Impact velocity (km/s)	Final compression Stress (GPa)	Final compression Particle velocity (km/s)	Final shock velocity[a] (km/s)
SiC/Al(z)					
8410	WC	0.361	5.18	0.311	5.64
8418	Sapphire	0.348	4.05	0.255	5.34
SiC/Al(r)					
8504	None	0.206	1.7	0.094	6.00

[a] These velocities were determined assuming a HEL of 0.314 GPa.

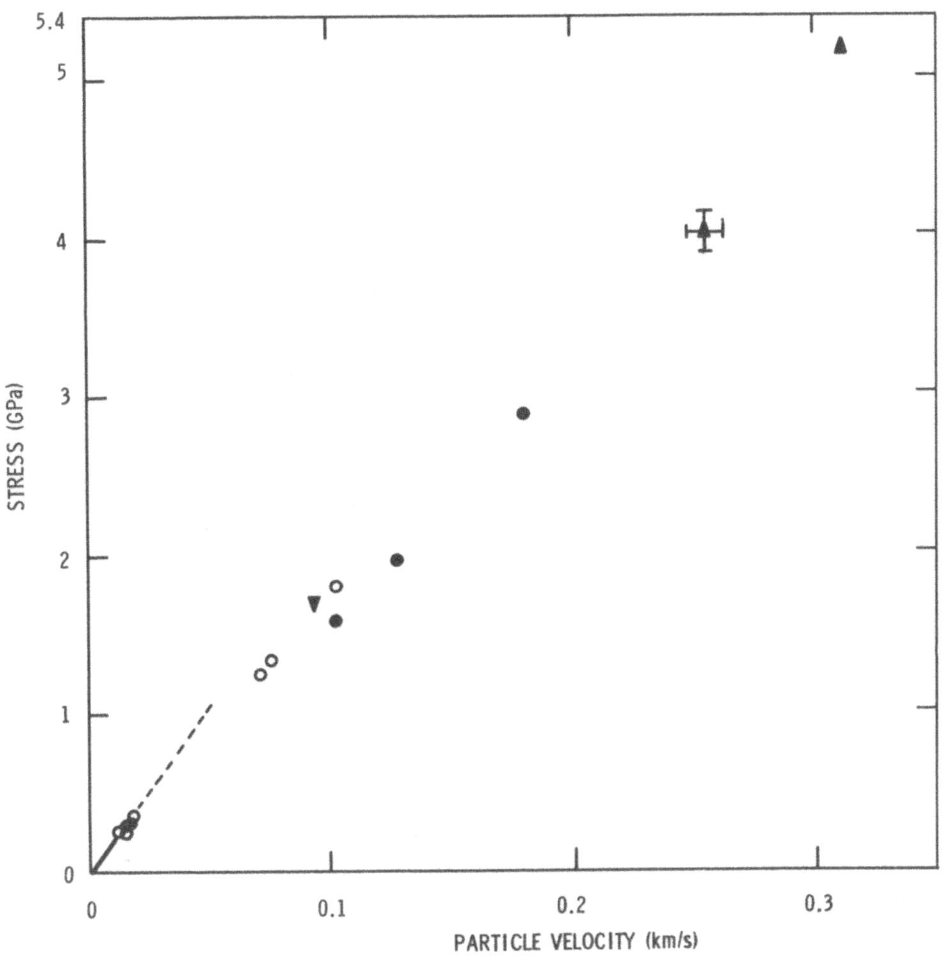

Figure 1. Stress-Particle Velocity Diagram of SiC/Al(z) and SiC/Al(r) Transmission: SiC/Al(z) ●, SiC/Al(r) O, Direct Impact: SiC/Al(z) ▲, SiC/Al(r) ▼.

Table 2. Summary of transmission experiments in SiC/2014-T4 Aluminum.

Experiment	Specimen thickness (mm)	Impactor	Impact velocity (km/s)	Elastic compression Stress (GPa)	Elastic compression Particle velocity (km/s)	Elastic compression Density (Mg/m^3)	Final compression Shock velocity (km/s)	Final compression Stress (GPa)	Final compression Particle velocity (km/s)	Final compression Density (Mg/m^3)
SiC/Al(z)										
8410	8.00	SiC/Al(z)	0.361	0.335	0.0158	2.912	5.33	2.89	0.181	3.02
8509	7.78	SiC/Al(z)	0.256	0.311	0.0150	2.912	5.05	1.96	0.128	2.97
8518	7.89	SiC/Al(z)	0.204	0.300	0.0141	2.912	5.02	1.58	0.102	2.96
SiC/Al(r)										
8504	7.452	SiC/Al(r)	0.206	0.275	0.013	2.911	5.92	1.82	0.103	2.96
8522	7.920	2014-T4 Al	0.167	0.296	0.014	2.912	5.9	1.34	0.0747	2.94
8522	6.300	2014-T4 Al	0.167	0.368	0.0173	2.913	5.8	1.27	0.071	2.94

Table 3. Summary of spall experiments on SiC/2014-T4 Aluminum.

Experiment	Impact velocity (km/s)	Spall	Experiment	Impact velocity (km/s)	Spall
SiC/Al(z)			SiC/Al(r)		
8425	0.153	Yes	8501	0.133	No
8501	0.133	No	8506	0.151	No
			8508	0.191	Yes
			8520	0.175	Yes

wave velocities for SiC/2014-T4 Al given earlier. The values of HEL obtained in these shock experiments appear to be lower than the estimates of HEL obtained from the static proportional elastic limit data. One possible reason for this discrepancy may lie in the fact that the stress-strain curves for SiC/Al(z) and SiC/Al(r) did not show a sharp yield stress like in 2014 Aluminum alloy. The wave profiles obtained in the transmission experiments listed in Table 2 do not show any elastic precursor decay in 6-8 mm thick specimens of SiC/Al(z) or SiC/Al(r). Thus the values of HEL may be taken as the range of HEL for SiC/2014-T4 Al, and the scatter in the HEL values are probably real and thus indicative of material variability.

The deformation of SiC/2014-T4 Al above the HEL presented in terms of (5,u) coordinates (Fig. 1) appear to be consistent with each other irrespective of whether these coordinates were obtained from direct impact or from transmission experiments for each z and r oriented from direct impact or the non-elastic deformation of SiC/Al(z) and SiC/Al(r) differ in regard to the estimates of shock velocities following the elastic precursors in these specimens. The estimates of final shock velocities in SiC/Al(z) are consistently lower than the estimates of these velocities in SiC/Al(r) and the bulk sound speed of 5.76$\pm$ 0.03 km/s in SiC/2014-T4 Al. The reason for this consistent difference in the estimates of final shock velocities in SiC/Al(z) and SiC/Al(r) is unknown. The deformation behavior of SiC/Al(z) and SiC/Al(r) above the HEL will be investigated in further detail in the future.

There appears to be some difference in the spall thresholds of SiC/Al(z) and SiC/Al(r) as indicated by Table 3. It shows that whereas SiC/Al(z) spalls when it is impacted by itself with a velocity of 0.153 km/s, SiC/Al(r) does not spall when impacted with similar velocity. However, a definite conclusion about the implied difference in the spall threshold of SiC/Al(z) and SiC/Al(r) must await performance of more spall experiments.

REFERENCES

1. P. C. Lysne, C. M. Percival, R. R. Boade, and O.E. Jones, Determination of Release Adiabats and Recentered Hugoniot Curves by Shock Reverberation Techniques, J.Appl.Phys.40:3786 (1969)
2. D. P. Dandekar, A. G. Martin, and J. V. Kelley, Deformation of Depleted Uranium 0.78 Ti under Shock Compression to 11.0 GPa at Room Temperature, J.Appl.Phys. 51:4784 (1980)
3. R. A. Graham, Piezoelectric Current from Shunted and Shorted Guard-Ring Quartz Gauges, J.Appl.Phys. 46:1901 (1975)
4. S. C. Chou, J. L. Green, and R. A. Swanson, To be presented at Testing Technology of Metal-Matrix Composites, First Symposium Nov. 1985

THRESHOLD STRESS MEASUREMENTS IN SHOCK-DEFORMED COPPER

Paul S. Follansbee and George T. Gray

Los Alamos National Laboratory
Los Alamos, N.M.

INTRODUCTION

Shock recovery experiments provide a useful technique to investigate the response of materials to extreme conditions of strain rate, imposed hydrostatic pressure, and temperature. Significant findings of some previous investigations have been reviewed by Leslie[1], Murr[2], and Mikkola and Wright[3]. It has been well documented in FCC metals that shock wave deformation leads to higher hardness values[2], dislocation densities[2,4], stored energies[5] and resistivities[6] and decreased cell sizes[2] than quasi-static deformation to the same "equivalent" strain. One difference noted between microstructures formed by shock-deformation and those formed by quasi-static deformation is the increased contribution of deformation twinning in the former[1,3]. Another microstructural difference is the observed increase in the levels of vacancy generation in shock-deformation; evidence for this has been reviewed by Graham[7].

Summarizing these previous studies of the shock-deformation behavior in FCC metals, one would conclude that at moderate shock pressures the deformation mechanisms involve dislocation multiplication and dislocation/dislocation interactions similar to those documented at quasi-static strain rates. As the shock pressure, and thus the imposed strain rate, is increased, deformation twinning becomes increasingly important. The influence of the observed increased levels of vacancy production on deformation mechanisms remains an open question.

In FCC metals the resistance that a dislocation experiences as it moves along a glide plane is illustrated in Fig. 1.[8] This figure shows schematically two obstacles separated by a distance λ_2. These obstacles could be forest dislocations, solute atoms or even vacancy clusters. The obstacle height and density determine the mechanical threshold $\hat{\tau}$, which represents the stress that must be applied in the absence of thermal activation to push a dislocation past the obstacles. At quasi-static strain rates thermal activation assists the applied stress and allows dislocation motion at stresses below $\hat{\tau}$. Under dynamic conditions the applied stress may actually exceed $\hat{\tau}$; in this case, the velocity of the dislocation as it moves the distance λ_2 between obstacles becomes rate controlling. During transit between obstacles the stress driving the dislocation velocity is essentially the applied stress. In other crystal structures, a lattice friction stress may oppose the applied stress.

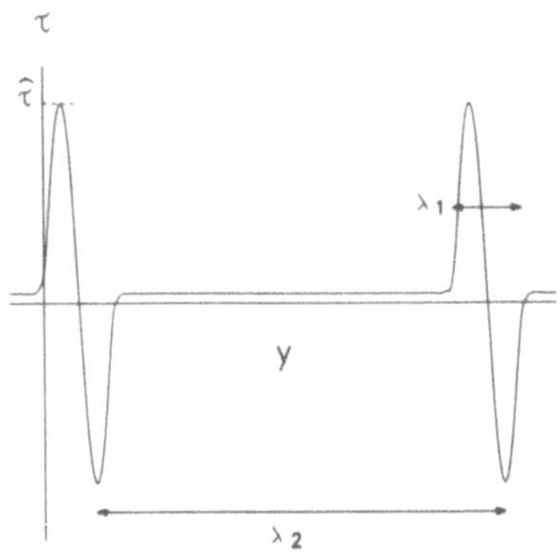

Figure 1. Glide resistance profile showing the interaction between a mobile dislocation and two short range ($\lambda_1 << \lambda_2$) obstacles separated by a distance λ_2.

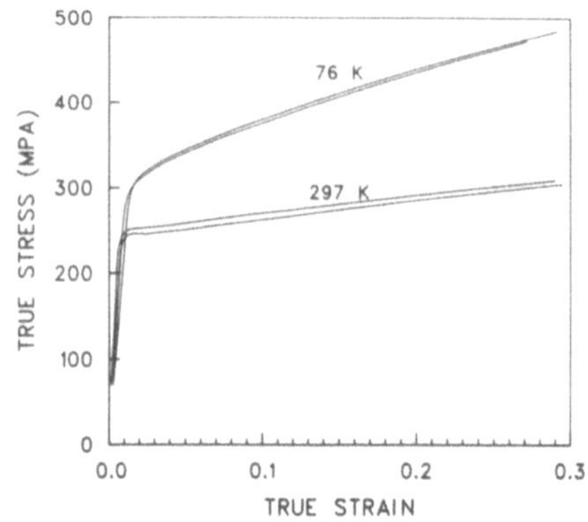

Figure 2. Stress strain curves for quasi-static reloading at 297 K and 76 K following 10 GPa shock deformation.

The height and shape of the obstacle profile for the dominant obstacle provide information on the nature of the deformed microstructure. These quantities can be measured on recovered shock-deformed material using a technique outlined previously[9,10]. Measurements of the mechanical threshold on shock-deformed material, their implications regarding deformation mechanisms and correlation with the microstructure characterized with TEM are described in this paper.

EXPERIMENTAL

Oxygen-free-electronic (OFE) copper was selected for this study since its shock-deformation behavior has been extensively studied and since these shock-deformation results complement extensive measurements[10] at strain rates up to 10^4 s^{-1}. Specimens for the shock recovery experiment were machined from the starting plate material and recrystallized at 600°C in vacuum for one hour to yield equiaxed grains of 40 μm average linear dimension.

The shock recovery experiments were performed utilizing a 40 mm single-stage gas gun. The specimen configuration consisted of a 4.72 mm thick, 12 mm diameter tapered (10°) sample tightly fitting into two concentric copper momentum trap rings with outside diameters of 25 mm and 42 mm. The sample surface was protected from impact and the entire sample from spallation by a close-fitting copper cover plate (2.5 mm thick) and spall plate (12 mm thick), respectively. The copper sample was shocked to 10 GPa for 1 μs pulse duration by impacting a 2.26 mm thick copper flyer plate, fixed to a low-impedance aluminum honeycomb-filled projectile, at 518 m/s with the specimen assembly in a vacuum (< 2 Pa). This shock pressure for copper translates into a transient strain of 0.0825 using the relation $\varepsilon_s = 4/3 \ln(V/V_0)$. Thickness measurements on the recovered sample revealed a final sample thickness varying over the sample between 4.63 mm and 4.70 mm compared to a pre-shot thickness of 4.72 mm.

Four compression specimens of average dimensions 4.5 mm in length by 4.3 mm in diameter were electro-discharge machined (EDM) from the recovered sample and reloaded at a strain rate of 0.0015 s^{-1} with a screw-driven mechanical testing machine equipped with a specially designed subpress which could be completely immersed in liquid nitrogen. Two specimens were reloaded at room temperature (297 K) while the remaining two were reloaded at liquid nitrogen temperature (76 K). It should be emphasized that for these experiments no precautions were taken to minimize recovery processes in the time period between the shock-deformation and the quasi-static reloading.

Samples for optical metallography and transmission electron microscopy (TEM) were sectioned from the pieces of the shock-deformed disc which remained after the compression specimens were removed. Discs 3 mm in diameter were punched and electropolished in a solution of 25% H_3PO_4 and 75% H_2O at 0^oC utilizing a current density of 80 mA/mm^2. Observation of the foils was made using a Phillips 400 at 120 KV utilizing a goniometer-tilt stage.

RESULTS

Mechanical Threshold Measurements

The four stress strain curves for the reloads at 297 K and 76 K are shown in Fig. 2. These results are similar to those reported by Appleton and Waddington using tensile reloads[11]. The yield points, determined by back-extrapolating the work hardening behavior, are 245 MPa and 249 MPa at 297 K and 307 MPa and 313 MPa at 76 K.

The mechanical threshold is the yield stress at 0 K, which we determine by the extrapolation procedure described previously[9,10]. For the quasi-static strain rate of the reload experiments, the deformation is thermally activated and can be represented by an equation of the form $\dot{\varepsilon} = \dot{\varepsilon}_o \exp - \Delta G/kT$, where $\dot{\varepsilon}_o$ is a constant of the order $10^8\ s^{-1}$ and ΔG is the activation energy. When short range obstacles are rate controlling, the activation energy can be approximated by[8]

$$\Delta G = \mu b^3 g_o \left\{1 - \left(\frac{\sigma}{\hat{\tau}}\right)^{1/2}\right\}^{3/2} , \qquad (1)$$

where g_o is the normalized activation energy. The exponents 1/2 and 3/2 in Eq. 1 are chosen to represent the obstacle profile for short range obstacles[8]. Extrapolation to 0 K requires that the temperature dependence of the activation energy ΔG be added to the intrinsic temperature dependence (kT) given by the Arhennius law. Assuming that the temperature dependence of the activation energy is equal to that of the shear modulus μ, $\Delta G(T) = \mu(T)b^3 g_o$, then the extrapolation to 0 K is given by

$$\left(\frac{\sigma}{\mu(T)}\right)^{1/2} = \left(\frac{\hat{\tau}}{\mu(T)}\right)^{1/2} \left[1 - \left(\ln\frac{\dot{\varepsilon}_o}{\dot{\varepsilon}} \frac{kT}{\mu(T)b^3 g_o}\right)^{2/3}\right] . \qquad (2)$$

If the normalization procedure with respect to temperature is correct, then a plot of $(\sigma/\mu(T))^{1/2}$ versus $(kT/\mu(T)b^3)^{2/3}$ for the reload experiments at constant strain rate but varying temperature should yield a straight line. The intercept at zero temperature in this plot gives the mechanical threshold normalized by the shear modulus while the slope is inversely related to the normalized activation energy. The data from the reload experiments are plotted on these coordinates in Fig. 3. The values of the mechanical threshold (referenced to 297 K) and the normalized free energy for the shock recovered material are listed in Table 1 along with previously obtained values for lower strain rates[10]. The latter data are grouped into two categories to show comparisons at constant strain as well as at constant mechanical threshold.

In the previous study, mechanical threshold values were obtained as a function of strain. By interpolating these results, we have plotted in Fig. 4 the mechanical threshold values at a single strain of 0.0825, which corresponds to the transient plastic strain during the shock-deformation. The actual strain rate for the 10 GPa shock is unknown and probably varies significantly between the loading and unloading portions of the wave. The strain rate during the shock is assumed in Fig. 4 to lie within the range 10^5 s^{-1} to 10^6 s^{-1}; the lower limit is set by the transient strain divided by the pulse duration.

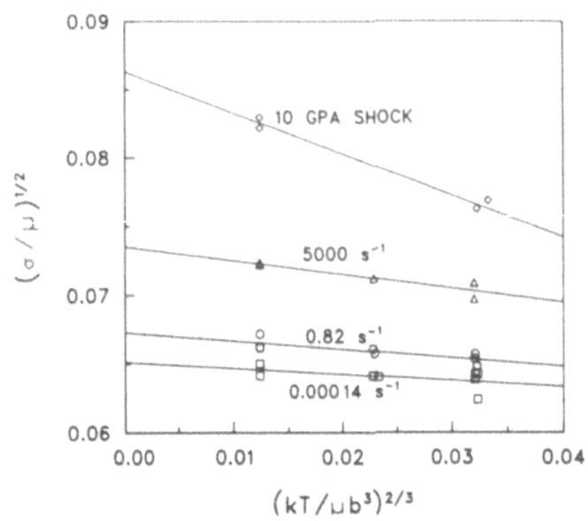

Figure 3. Normalized plot of reload yield stress versus test temperature for the 10 GPa shock-deformed sample and for three histories at lower strain rates.

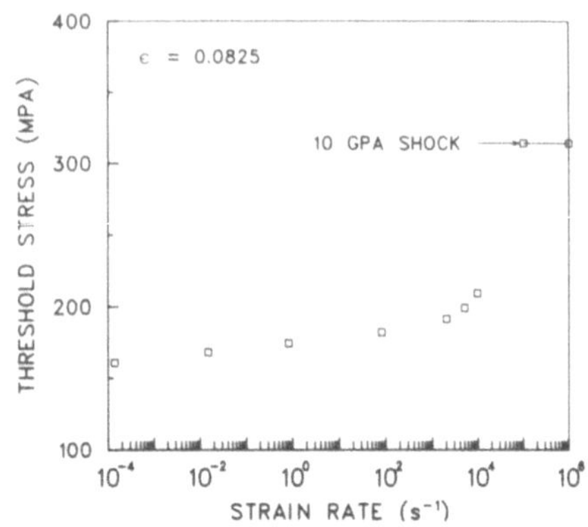

Figure 4. Variation of the mechanical threshold at a constant strain of 0.0825 with strain rate.

Table 1. Measured Values of $\hat{\tau}$ and g_0

	ε	$\dot{\varepsilon}$ [s^{-1}]	$\hat{\tau}$ [MPa]	g_0
10 GPa Shock	.0825	~ 10^6	314	3.26
Constant Strain	.10	.00014	179	38.1
	.10	.015	187	23.2
	.10	.82	191	24.4
	.10	81	204	7.7
	.107	1800	218	27.7
	.10	5000	228	13.3
	.087	9500	212	19.6
Constant Mechanical Threshold	.25	.82	309	8.4
	.25	81	329	7.7
	.209	1800	309	8.5
	.211	5000	325	5.8

Substructure Observations

TEM micrographs of the shock-deformed copper are shown in Figs. 5 and 6. The micrographs reveal evidence of a highly dislocated structure with a percentage of grains exhibiting fine parallel markings which TEM and selected-area diffraction (SAD) analysis verify as deformation twins. The predominant deformation substructural feature, that of dislocation cells, is consistent with numerous previous studies on shock-loading of copper[2,4,12,13] at moderate pressures. Dislocation loops were also observed within some cell interiors. Depending upon orientation, grains were found to exhibit entirely cellular, twinned, or a mixed morphology. A small fraction of grains exhibited bands of high dislocation density[4] lying on {111} planes; these may be features identified as microbands which have been observed in heavily cold rolled copper[14].

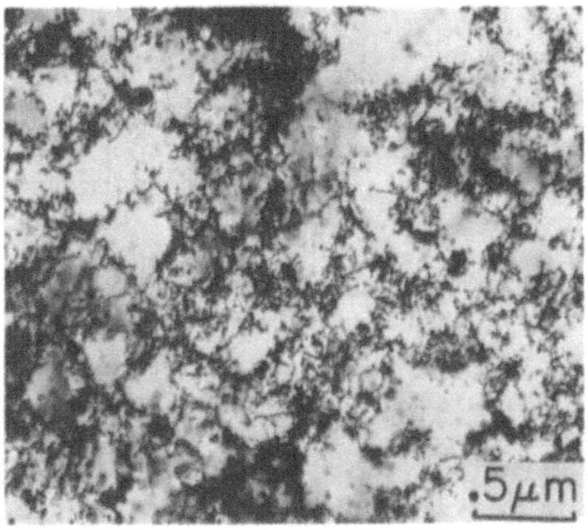

Figure 5. Bright field electron micrograph (<213> zone) of cellular dislocation substructure in shock-deformed copper.

Figure 6. Bright field electron micrograph and SAD diffraction pattern (<110> zone) of twin substructure.

DISCUSSION

The mechanical threshold measurements lead to several interesting conclusions. The comparisons shown in Table 1 and Fig. 4 indicate that at equivalent strain, the mechanical threshold increases dramatically in the shock-deformation regime. In fact, as shown in Fig. 4, the increase begins at strain rates closer to 10^3 s^{-1}. The mechanical threshold is a mechanical measure of structure, and for dislocation hardening alone the mechanical threshold is related to the total dislocation density ρ by[15]

$$\hat{\tau} = \mu b \rho^{1/2} \ . \tag{3}$$

If the sole difference between the quasi-static deformation and shock-deformation were in the rate of dislocation storage, then from Eq. 3 and from the data in Table 1 the dislocation density in a shock-deformed sample should be approximately three times that in a sample quasi-statically deformed to a strain of 0.10. While it is well documented using measurements of resistivity[6], energy storage[5], and actual dislocation counts in TEM foils[4] that the dislocation densities in shock-deformation exceed those in quasi-static deformation, the experimental techniques for measuring the dislocation density are probably not precise enough to verify this factor of three estimate.

The other comparison shown in Table 1 is that of the normalized activation energy at <u>constant threshold stress</u>. The value for the shock-deformed material, $g_0 = 3.26$, is roughly one-half the values for material deformed at strain rates less than 10^4 s^{-1}. This suggests that the obstacle in the former is slightly more rate sensitive than that in the latter. A factor of two difference in g_0 is small. However, we believe that this difference lies outside the uncertainty in the estimate of g_0 and, thus, that it represents a real variation in the obstacle character.

Microscopic characterization of the shock-deformed microstructure is consistent with the conclusions based on the mechanical threshold measurements. The observed structure was heavily dislocated and contained well developed cellular structures. At a shock pressure of 10 GPa, a few grains deform solely by deformation twinning while in other grains both twins and cellular dislocation structures are observed. It is unclear what influence the twins have on the mechanical threshold measurements.

Since the increase in the dependence of the mechanical threshold on strain rate begins at strain rates as low as 10^3 s^{-1}, where deformation twinning definitely has not been found, the observed twinning at 10 GPa probably has not influenced the mechanical threshold measurements. Although some observations of the shock-deformed microstructure are consistent with higher levels of vacancy production, this does not appear to be a dominant feature. The mechanical threshold measurements do show a slight change in the nature of the obstacle toward a more rate sensitive defect, which vacancy clusters might be expected to provide.

The conclusion based on the experiments described here is that the mechanical threshold measurements are consistent with measurements at lower strain rates. The mechanical threshold of shock-deformed copper exceeds that of quasi-statically and dynamically ($\dot{\varepsilon} = 10^4$ s^{-1}) deformed copper. Our interpretation is that the high strain rates achieved during the shock process lead primarily to an increased rate of dislocation storage, which is consistent with trends noted[10] at strain rates exceeding 10^3 s^{-1}.

ACKNOWLEDGEMENTS

The authors wish to acknowledge C. Frantz for his major contributions to the design of the gun facility and of these recovery experiments and B. Jacquez for his assistance in performing the latter. We also wish to thank U. F. Kocks for critically reviewing and commenting on the manuscript. This work was performed under the auspices of the U. S. Dept. of Energy with support from the Los Alamos National Laboratory Center for Materials Science.

REFERENCES

1. W. C. Leslie, in: "Metallurgical Effects at High Strain Rates," R. W. Rohde, B. M. Butcher, J. R. Holland, and C. H. Karnes, eds., Plenum Press, New York, 571 (1973).
2. L. E. Murr, in: "Shock Waves and High-Strain Rate Phenomena in Metals," M. A. Meyers and L. E. Murr, eds., Plenum Press, New York, 607 (1981).
3. D. E. Mikkola and R. N. Wright, in: "Shock Waves in Condensed Matter - 1981," W. J. Nellis, L. Seaman, and R. A. Graham, eds., American Institute of Physics, New York, 98 (1982).
4. J. George, Phil. Mag. 15:497 (1967).
5. D. C. Brillhart, R. J. De Angelis, A. G. Preban, J. B. Cohen and P. Gordon, AIME Trans. 239:836 (1967).
6. H. Kressel and N. Brown, J. Appl. Phys. 38:1618 (1967).
7. R. A. Graham, in: "Shock Waves and High-Strain-Rate Phenomena in Metals," M. A. Meyers and L. E. Murr, eds., Plenum Press, New York, 375 (1981).
8. U. F. Kocks, A. S. Argon and M. F. Ashby, "Thermodynamics and Kinetics of Slip," Prog. Mtl. Sci. 19, Pergamon Press, New York (1975).
9. P. S. Follansbee, G. Regazzoni and U. F. Kocks, in: "Mechanical Properties at High Rates of Strain," Inst. Phys. Conf. Ser. No. 70, J. Harding, ed., Institute of Physics, London (1984).
10. P. S. Follansbee, U. F. Kocks and G. Regazzoni, The mechanical threshold of dynamically deformed copper and Nitronic 40, to be published in: "Proc. of DYMAT - Int. Conf. on Mechanical and Physical Behavior of Material under Dynamic Loading," Paris (1985).
11. A. S. Appleton and J. S. Waddington, Phil. Mag. 12:273 (1965).
12. A. S. Appleton and J. S. Waddington, Acta Met. 12:956 (1964).
13. O. Johari and G. Thomas, Acta Met. 12:1153 (1964).
14. A. S. Malin and M. Hatherly, Metal Sci. 13:463 (1973).
15. H. Mecking and U. F. Kocks, Acta Met. 29:1865 (1981).

NON-LINEAR PROPAGATION OF ELASTO-PLASTIC WAVES IN RODS

G. F. Kuscher, V. Hohler and A. J. Stilp

Ernst-Mach-Institute, Eckerstr. 4
7800 Freiburg, Fed. Rep. of Germany

INTRODUCTION

The behaviour of rods subjected to loads beyond the elastic limit is theoretically described by von Karman[1] 1950 and other authors. This model assumes two waves propagating along the rod: an elastic wave, stressing the material up to the elastic limit, and travelling with longitudinal sound velocity, followed by a plastic wave which propagates more slowly and with non-linear velocity. Up to now, it has not yet been possible to experimentally verify these predictions, especially values of the dynamic yield strength according to this elasto-plastic theory could not be estimated. That is mainly due to the fact that the conventional diagnostic methods applied in this field of research (material testing machines, strain gauges [2, 3], high speed cameras[4]) do not have the necessary time resolution or cannot continuously record wave propagation processes, respectively.

Our experimental procedure for the study of the rear-side movement of a rod after impact loading is based on a VISAR-Laserinterferometer technique[8]. A continuous high-resolution time record of the rear side of the rod is obtained, which allows to derive the elasto-plastic wave propagation.

THEORY

The one-dimensional elasto-plastic theory, given by von Karman[1], predicts that an elastic wave which stresses the material up to the elastic limit Y and travelling with longitudinal sound velocity ($\beta = c_o$ in fig. 1), is followed by a plastic wave, exerting the constant stress level ε_1 of the material and propagating with a non-linear velocity v_{pl}. Between the elasto-plastic boundary and the elastic wave front, there is a regime of varying stress, strain, strain velocity and velocity of propagation β. According to this theoretical concept, the rod is accelerated by successive reflections of the elastic wave between the rear side of the rod and the elasto-plastic boundary. At each cycle the velocity of the rod material increases by the amount of the particle velocity associated with the elastic stress (fig. 2 - 4). This offers the possibility of estimating the dynamic yield stress Y by the measured particle velocity during a reflection of the elastic wave at the rear side of the rod, according to

$$Y = \frac{1}{2}\Delta v \rho \cdot c_o$$

Y : dynamic yield stress
c_o : longitudinal sound velocity
Δv : particle velocity
ρ : density of the rod material

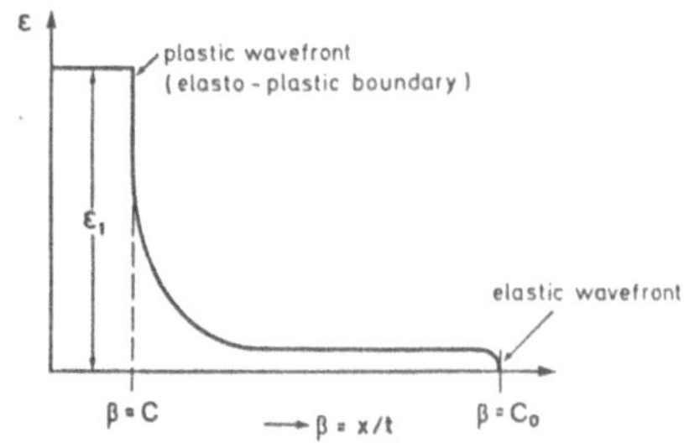

Fig. 1: Propagation velocity versus strain along the rod axis

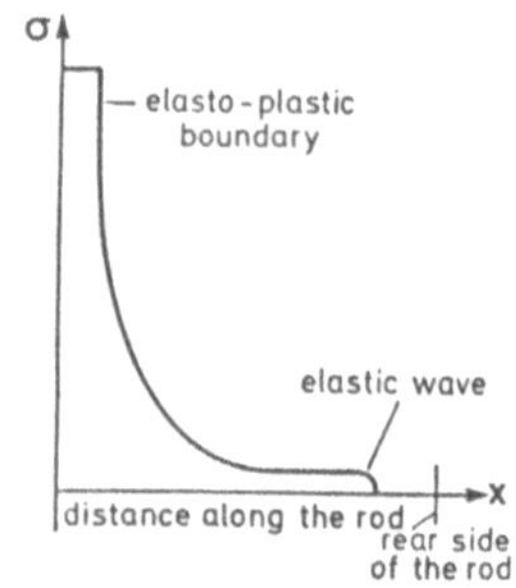

Fig. 2: Elasto-plastic two-wave-structure after impact loading

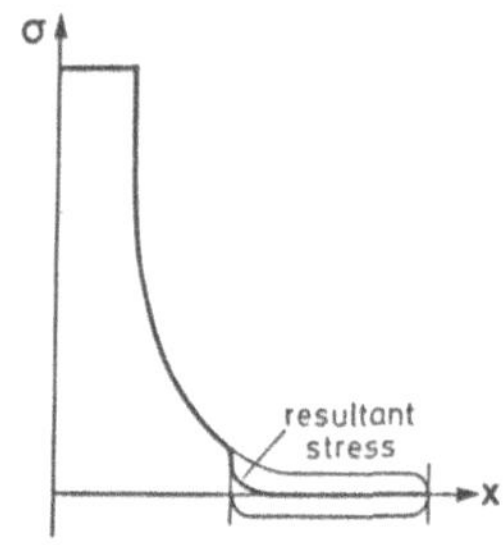

Fig. 3:

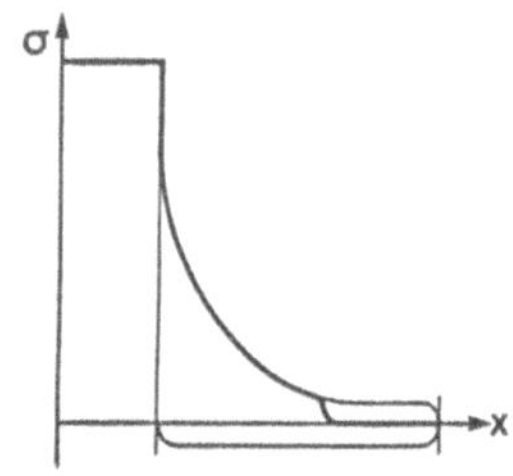

Fig. 4:

Stress distribution along the rod axis after the reflection of the elastic wave at the rear side of the rod (fig. 3) and at the elasto-plastic boundary (fig. 4), respectively.

It has not yet been possible to experimentally verify these predictions of the von Karman[1] elasto-plastic theory. However, to obtain rough estimates of material values at high impact velocities, approximation models have been developed by Taylor[5] and Tate[6, 7]. These models are based on the measured geometric dimensions of the plastically deformed rod after a deformation process. Taylor's[5] basic idea implies a constant velocity of propagation of the elasto-plastic boundary, and in the model developed by Tate [6, 7], any wave propagation is neglected. Up to now, it could not be examined to which extent these models yield realistic values of the dynamic yield strength Y.

EXPERIMENTAL

Inverse impacts have been carried out, in which the target was fired against the rod. The powder gun used (cal. 50 mm) allows the acceleration of masses up to 500 g to maximum velocities of about 1800 m/sec. Steel cylinders with a diameter of 40 mm and a thickness of 15 mm embedded in plastic sabots have been used as targets. In order to reduce possible interactions with the support system, the rod was suspended nearly force-free in a soft styrofoam cylinder. This cylinder itself is held within a steel cylinder through which the target is shot against the rod (fig. 5). The laser beam is focussed to the rear side of the rod, opposite to the impact site. Caused by the elasto-plastic wave motion, the laser light undergoes a Doppler shift which as analysed by a VISAR-Laserinterferometer[8]. The two VISAR-data traces were recorded by

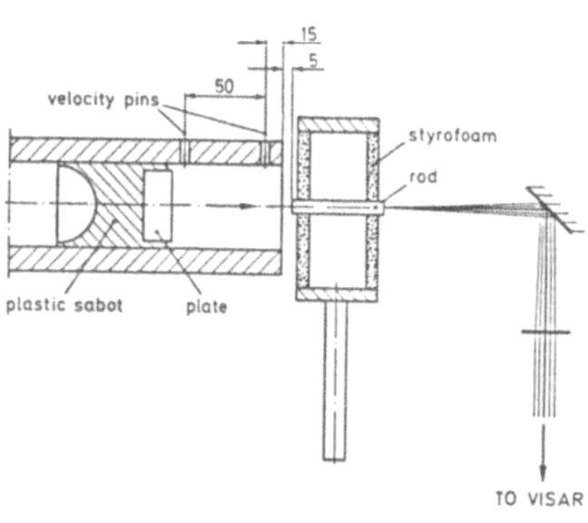

Fig. 5: Target/rod configuration

means of a 200 MHz transient digitizer (Tektronix 7612). Thus, a maximum time resolution of 5 nsec was achieved. The impact velocities of the cylindrical plates ranged between 150 m/sec and 1000 m/sec, covering the velocity regimes of pure bulging of the rod (< 400 m/sec) and bulging with beginning target penetration. The elasto-plastic deformation mechanism of different rod materials (steel 9S20k, tungsten sinter alloys, aluminium) were investigated. In this paper the results concerning steel 9S20k are reported as an example.

RESULTS

Fig. 6 shows typical velocity-time diagrams of the rear side of steel rods. All curves reveal a typical stepwise velocity increase as predicted

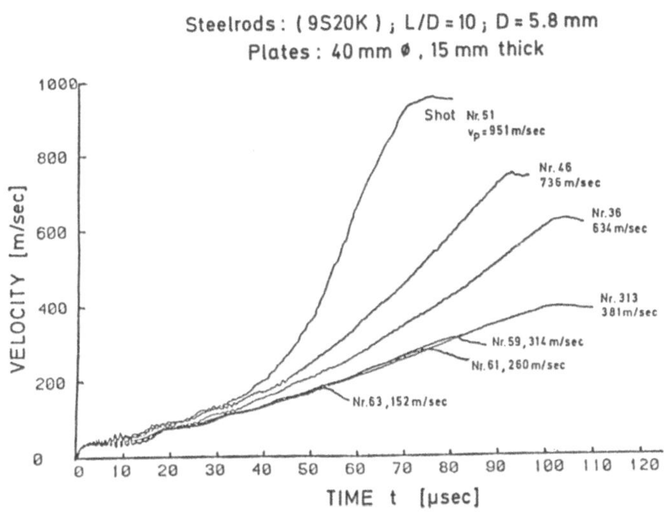

Fig. 6: Velocity-time diagrams of the rear side of equal geometry steel rods at different impact velocities.

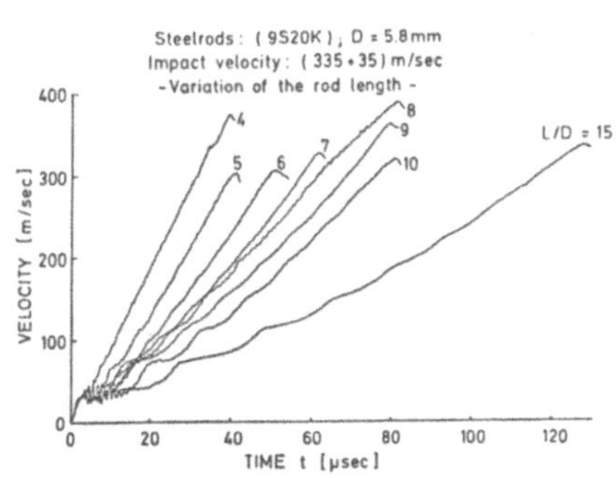

Fig. 7: Velocity-time diagrams of rods with different lengths but equaly impact velocity

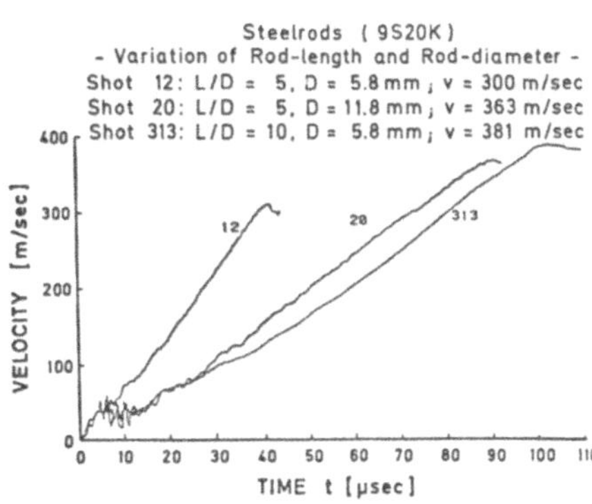

Fig. 8: Velocity-time diagrams of rods with different length and diameter, but equal impact velocity

by the von Karman[1] elasto-plastic theory, caused by respective reflections of the elastic wave. The rods used for the experiments shown in fig. 6 were all of the same geometry, and it is obvious that the measured particle velocity during each reflection is independent of the impact velocity. This behaviour was also found for rods with different lengths and diameters (fig. 7 and 8). The experiments shown in fig. 7 were carried out at approximately equal impact velocities with rods of the same diameter but with different lengths. In fig. 8 diameter and length of the steel rods were varied. The three diagrams (fig. 6 - 8) reveal that the measured particle velocity during each reflection of the elastic wave at the rear side of the rod appears to be material specific in the sense of the von Karman[1] theory. This allows the computation of the dynamic yield strength Y according to this theory. Together with the values c_o = 5200 m/sec, ρ = 7.85 g/cm^3 and Δv = 37.4 m/sec, the above given formula yields Y = 0.763 GPa. A comparison of this value, estimated according to the elasto-plastic theory, with values of Y for rods, given by Stilp and Tham[9], which were calculated by means of the Taylor[5] approximation model shows deviations of up to 50 %. This is presumably due to the assumption of a constant propagation velocity of the elasto-plastic boundary in the Taylor[5] model. This assumption does not satisfactorily describe the real bulging process of the rod, what will be shown below. Any neglecting of wave motion in the model developed by Tate[6, 7] yields values of Y showing an even greater deviation from the values estimated according to the elasto-plastic theory. This holds for the impact velocity regime of pure bulging ($v < 400$ m/sec).

Fig. 9 shows a velocity-time diagram of the rear side of steel rod. This figure indicates clearly the oscillations between the first reflections of the elastic wave which can already be seen in the curves of fig. 6 - 8. Fig. 10 depicts these oscillations in more detail for a rod with an L/D-ratio of 4. These phenomena obviously refer to radial oscillations of the rear side of the rod after the respective arrivals of the elastic wave. Conway[10] already predicted theoretically oscillations of this kind, and the computed oscillation period of 1.06 µsec for rods with a diameter of 5.8 mm agree very well with our measured data (fig. 9 and 10).

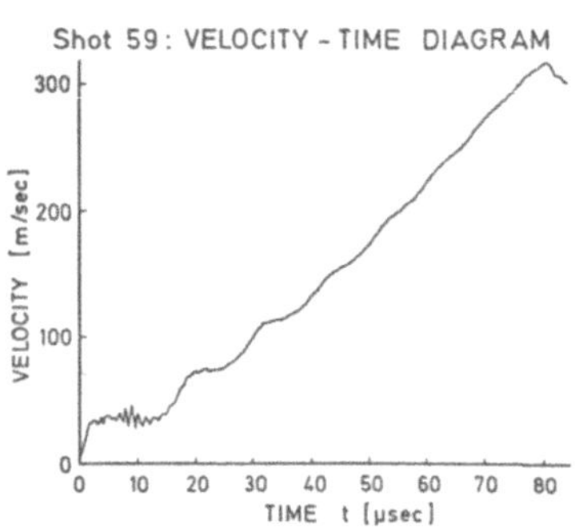

Fig. 9: Velocity-time diagram (exp. data in fig. 7)

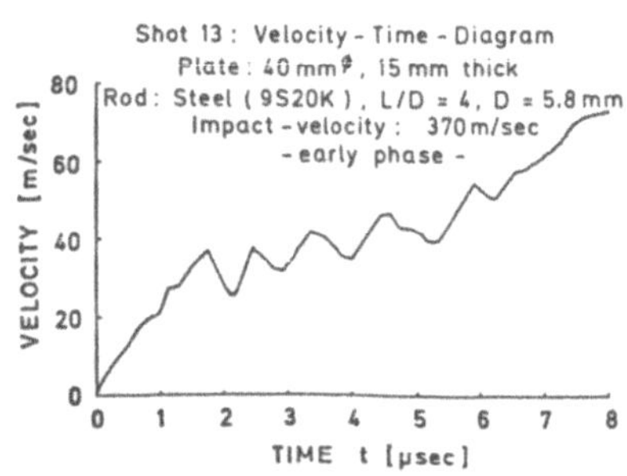

Fig. 10: Early phase of a rod acceleration showing radial oscillations

Figures 6 - 9 reveal a shortening of the duration between successive reflections of the elastic wave at the rear side of a rod, as a consequence of the further penetration of the elasto-plastic boundary into the rod. Based on this shortening of duration, velocity-time diagrams of the elasto-plastic boundary could be computed, using an exponential approximation, that is described in more detail by Kuscher[11]. A respective velocity-time of measured and computed data of a steel rod is given in fig. 11. The knowledge of the distance-time-diagram of the elasto-plastic boundary permits the illustration of the elasto-plastic wave propagation in form of Lagrange - diagrams. Fig. 12 shows such a diagram for a real bulging process of a steel rod (Shot 59).

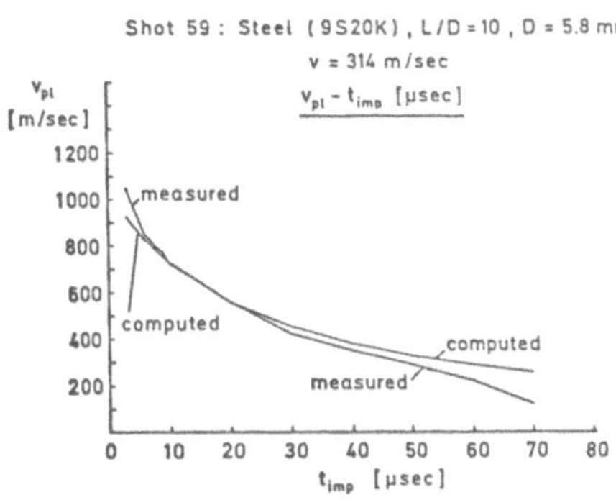

Fig. 11: Velocity-time diagrams of the elasto-plastic boundary

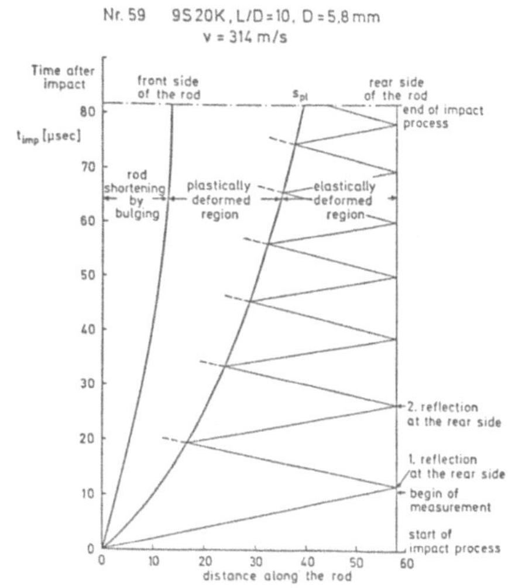

Fig. 12: Lagrange diagram of a real bulging process

ACKNOWLEDGEMENT

The authors gratefully acknowledge many helpful discussions and assistance by Prof. Dr. Adomeit and Prof. Dr. Wortberg from the Institut für Allgemeine Mechanik of the Rheinisch-Westfälische Technische Hochschule Aachen, 5100 Aachen, FRG.

REFERENCES

1. von Karman, T. and Duwez, P.; J. Appl. Phys. 21, 987 (1950)
2. Rosenberg, Z.; Yaziv, D.; Partom, Y.; J. Appl. Phys. 51, 4790 (1980)
3. Hauver, G. E.; Proc. 5th Int. Symp. Ball., Toulouse, France (1980)
4. Weihrauch, G.; ISL, St. Louis, France, Report 7/71
5. Taylor, G.; Proc. Roy. Soc. London A194, 300 (1948)
6. Tate, A.; J. Mech. Phys. Solids, 15, 387 (1967)
7. Tate, A.; J. Mech. Phys. Solids, 17, 141 (1969)
8. Barker, L. M.; Hollenbach, R. E.; J. Appl. Phys. 43, 4669 (1972)
9. Stilp, A. J.; Tham, R.; in "Shock waves in condensed matter - 1983", Elsevier Science Publisher B. V. (1984)
10. Conway, H. D.; Jakubowski, M.; J. Appl. Mech., 36, 809 (1969)
11. Kuscher, G. F.; Ernst-Mach-Institute, 7800 Freiburg, Report 2/85

USE OF THE BODNER-PARTOM VISCOPLASTIC CONSTITUTIVE MODEL TO DESCRIBE HY100 STEEL

A. M. Rajendran and S. J. Bless

University of Dayton
Research Institute
Dayton, Ohio 45469

1. INTRODUCTION

The objective of the present work was to develop a constitutive model for HY100 steel from a variety of high strain-rate experiments, encompassing a wide range of stress states and strain-rates. Data were obtained for compressive and tensile one-dimensional stress loading with a split Hopkinson bar, and for one-dimensional strain loading with plate-impact experiments. For the present work, the state variable based constitutive model developed by Bodner and Partom[1] (B-P model) was selected. The B-P model has several advantages. The model parameters can be determined from split Hopkinson bar and plate-impact experiments. Implementation of B-P model in finite difference codes are fairly straightforward. Since it is a three-dimensional constitutive theory, material response to a broad range of load histories can be calculated. In Section 2, the B-P model is briefly discussed. Experiments that are needed for the evaluation of model parameters, and the corresponding procedures, are discussed in Section 3. The B-P model simulations of the high strain-rate experiments are compared with the experimental data in Section 4.

2. BODNER-PARTOM MODEL

Bodner and Partom[1] proposed a unified, state variable based elastic-viscoplastic constitutive theory motivated by the equations of dislocation dynamics. The total strain rate is taken to be decomposable into elastic and inelastic components

$$\dot{\varepsilon}_{ij} = \dot{\varepsilon}^{e}_{ij} + \dot{\varepsilon}^{p}_{ij} \qquad (1)$$

The elastic strain rate, $\dot{\varepsilon}^{e}_{ij}$, is related to the stress rate by the time derivative of Hooke's Law. The inelastic strain rate $\dot{\varepsilon}^{p}_{ij}$ can be expressed as function of deviatoric stress, S_{ij}, a state variable, Z, and the second invariant of the stress deviator (J_2):

$$\dot{\varepsilon}^{p}_{ij} = D_{\emptyset} \exp\left[- \left(\frac{n+1}{2n}\right)\left(\frac{Z^2}{3J_2}\right)^{n} \right] \frac{S_{ij}}{\sqrt{J_2}} \tag{2}$$

where $D_{\emptyset}$ is the limiting value of plastic strain rate in shear, and n is a parameter that is related to strain-rate sensitivity. The state variable Z is a measure of the overall resistance of the material to plastic flow, and it depends on the loading history. The evolution equation for Z is given by,

$$\dot{Z} = m\ (Z_1 - Z)\ \dot{W}_p \quad \text{and} \quad Z = Z_1-(Z_1-Z_{\emptyset})\ e^{-mW_p} \tag{3}$$

where m is an empirical parameter and $\dot{W}_p$ is the rate of plastic work. $Z_{\emptyset}$ is the initial value of Z and Z_1 is the maximum value that Z can attain. The B-P model contains five principal material parameters, $D_{\emptyset}$, $Z_{\emptyset}$, Z_1, m, and n that have to be evaluated from high strain-rate experiments.

3. EVALUATION OF MATERIAL PARAMETERS

The Bodner-Partom model algorithm was developed and introduced through a series of subroutines into a finite difference code, called STEALTH[2]. Additionally, a program called BPSOLVE was developed to obtain uniaxial stress response by a direct solution of the B-P equations for a given strain-rate history.

Split Hopkinson bar (SHB) and plate-impact tests were the principal experimental techniques employed to evaluate the model parameters $Z_{\emptyset}$, Z_1, n and m. The SHB design and analysis procedure are described by Lindholm[3] and Nicholas[4]. Plate impact tests and the VISAR diagnostic techniques are described in References 5 and 6. Ten tensile and seven compression SHB tests were conducted. The stress-strain data at low ($150s^{-1}$), medium ($500s^{-1}$), and high ($>1000s^{-1}$) strain-rates were obtained. SHB data were extended to higher strains, strain rates, and mean stresses by high speed photography of necking specimens, using the techniques described in References 7 and 8. Three necking SHB tests were conducted on HY100 steel. Seven plate impact tests were conducted and the Hugoniot elastic limit (HEL), which is the principal stress component under one-dimensional strain at very high strain rates, was obtained.

D_0 in the model is the limiting strain-rate beyond which the strain rate has a minimal effect on the yield strength. In the present work, D_0 is arbitrarily set at $10^8 s^{-1}$ which is consistent with high strain rate regimes. The parameters n, Z_0, Z_1, and m were first estimated from tensile SHB data, using the procedure described in Reference 8. Plate impact tests were simulated through STEALTH using these estimated values. The HEL and the release were compared with the experimental data. The parameter values were modified until the calculated data matched both the SHB tensile and the plate impact test data. The B-P model parameters that were obtained for HY100 steel using this procedure were: $n=1.2$, $m = 1.0$ Kbar^{-1}, $Z_0 = 24.0$ Kbar, $Z_1 = 35.5$ Kbar, and $D_0 = 10^8 s^{-1}$.

4. B-P MODEL SIMULATIONS

In this section, we demonstrate the ability of the B-P model to describe all three types of tests. The SHB tests were simulated through the program "BPSOLVE". The strain-rate histories obtained from the tensile tests were used as input in the program. Using the B-P model parameters for HY100 steel, given in the preceding section, the stress response was computed and presented in Figure 1. Results are presented for low and high strain-rate tensile tests. It can be seen that the model simulation compared extremely well with the SHB test data. This material exhibited rate dependency. Several intermediate strain-rate tests were also simulated and the model results compared extremely well with the experimental data. The corresponding stress-strain data were within the low and high strain-rate curves shown. The compression tests were also simulated with BPSOLVE. The comparison between the model prediction and the test data are also shown in Figure 1. The model predicted the experimental results extremely well.

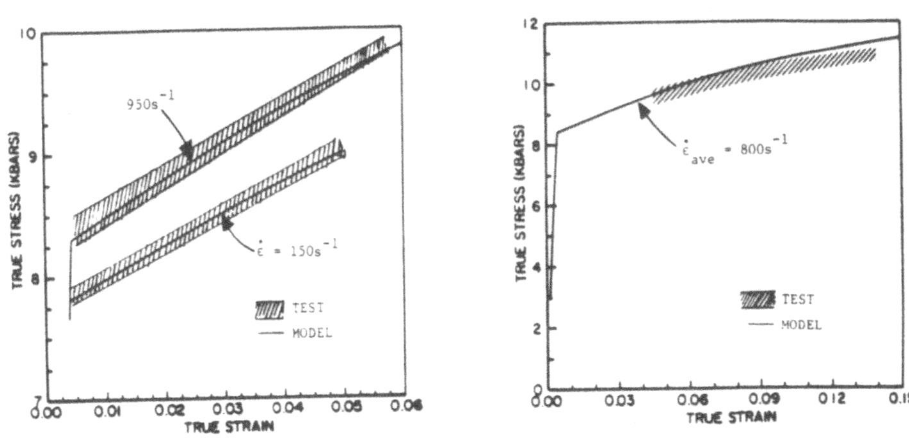

Fig. 1. Comparison of simulated and measured SHB data. (left - tension, right - compression)

The plate impact test was simulated with STEALTH using the same B-P model parameters that were used in the SHB test simulations. The simulated free surface velocity history of the target plate is compared with the VISAR data in Figure 2. The HEL and the release characteristics of these two plots matched extremely well. The peak velocity in the simulation is within 10% of observed value. However, the theoretical estimation of the maximum particle velocity using the shock relationship coincided with the impedance-match solution. The descrepancy in the VISAR data may have been caused by a drift in the VISAR signal (e.g., slow change in fringe contrast).

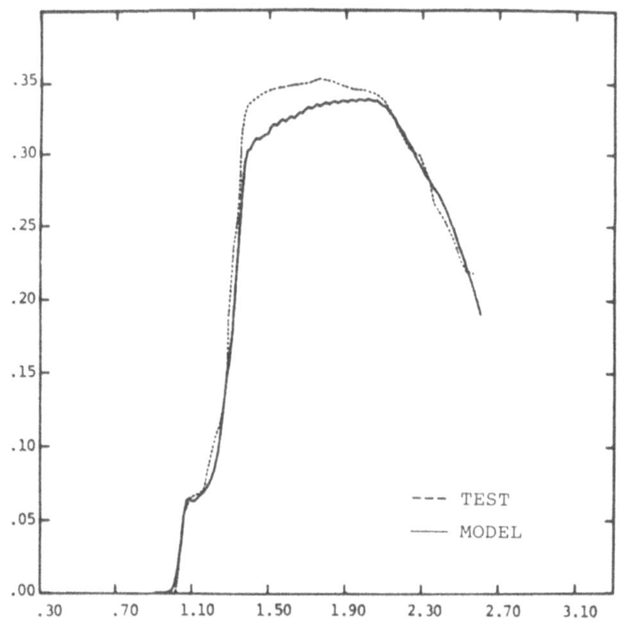

Fig. 2. Comparison of free surface velocity with model prediction.

5. EXTENSION TO VERY LARGE STRAIN

Finally, the results obtained from the necking SHB tests were used to construct the effective stress (σ_{eff}) vs. effective plastic strain (ε^p_{eff}) plot beyond the onset of necking instabilty. The radius of curvature (R) and the minimum radius of the neck (a) were obtained from the necking tests. The classical, approximate analysis of Bridgman provides a method to obtain the σ_{eff} vs. ε^p_{eff} plot through the following relationships:

$$\varepsilon^p_{eff} = 2 \ln \left(\frac{a}{a_o}\right) \tag{4}$$

$$\sigma_{eff} = \frac{\sigma_T}{(1 + \frac{2R}{a}) \ln (1 + \frac{a}{2R})} \quad . \tag{5}$$

Here a_o is the initial radius of the tensile specimen, and σ_T is equal to the load (as determined from the SHB strain gauge data) divided by the area of the minimum cross section of the necking specimen. Bridgman conducted quasi-static tests on various metals and measured 'a/R' at different effective plastic strains. His results and the results obtained from the three necking tests on HY100 steel are shown in Figure 3. The present dynamic data fall within the scatter band of Bridgman's data. The present authors have also validated the use of Bridgman's analysis for high rate of deformation in Reference 8. The σ_{eff} vs. ε^p_{eff} plot that was constructed using the above relationships is shown in Figure 4. The strain-rate beyond the onset of necking increased continually as the neck formation progressed. The increase in flow stress in Figure 4 beyond the onset of necking is significant in HY100 steel due to its rate dependent behavior. Since the strain rate increases rapidly, the flow stress also showed a steep increase. This further confirms the SHB tension test results that the HY100 steel is a rate dependent material. Attempts were not made to simulate the necking tests using STEALTH with Bodner-Partom routines. However, efforts are currently being focussed on accomplishing this aspect of the high strain-rate modeling.

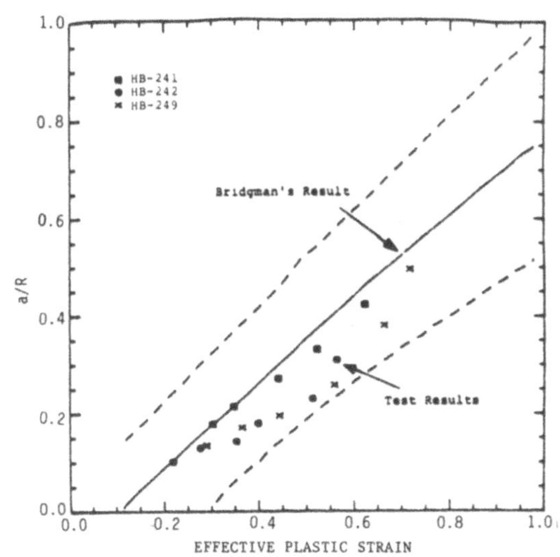

Fig. 3. Comparison of Bridgman's observed necking behavior (with scatter band) and the necking behavior of the HY100 steel.

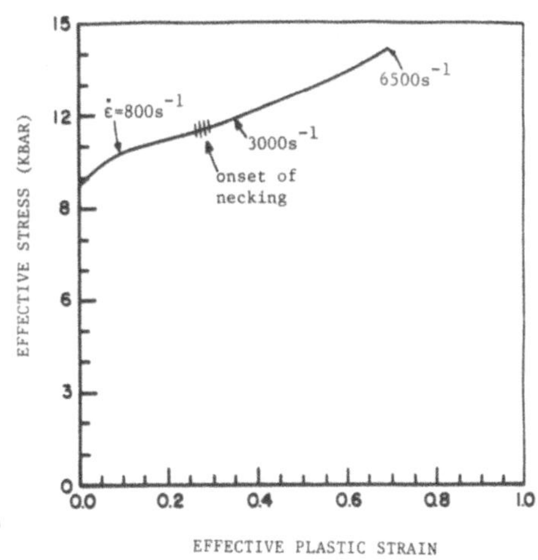

Fig. 4. The effective stress vs. effective plastic strain plot constructed using the Bridgman analysis and highspeed photograph.

6. SUMMARY

Data are presented for HY100 steel under dynamic compression and tension, including necking in SHB specimens. The ability of the Bodmer-Partom viscoplastic model to describe high strain-rate material behavior of HY100 steel is demonstrated. The important aspect of the successful modeling efforts was the use of variety of experimental data encompassing extremes of stress states and strain rates in the model parameters evaluation. Therefore, the developed material model parameters are general and can be applied to any geometrical configuration. The model is three-dimensional, viscoplastic and capable to include loading history effects.

REFERENCES

1. S. R. Bodner and Y. Partom, "Constitutive Equations for Elastic-Viscoplastic Strain Hardening Materials," J. Appl. Mech., 42, pp. 385-389, (1975).
2. R. Hoffman, "STEALTH, Lagrange Explicit Finite-Difference Code for Solids, Structural and Thermohydraulic Analysis," EPRI NP-2080, Nov. 1981.
3. U. S. Lindhom, "Some Experiments with the Split Hopkinson Pressure Bar," J. Mech. Phys. Solids, Vol. 12, pp. 317-335, (1964).
4. T. Nicholas, "Tensile Testing of Materials at High Rates of Strain," Exp. Mech., 21, pp. 177-185, (1981).
5. C. H. Karnes, The Plate Impact Configuration for Determining Mechanical Properties of Materials at High Strain Rates, in: "Mechanical Behavior of Materials Under Dynamic Loads," U. S. Lindholm, ed., Springer-Verlag, New York (1968).
6. J. R. Asay and J. Lipkin, "A Self Consistant Technique for Estimating the Dynamic Yield Strength of a Shock Loaded Material," J. Appl. Phy., Vol. 49, p. 4242, (1978).
7. L. A. Cross, S. J. Bless, A. M. Rajendran, E. A. Strader, and D. S. Dawicke, "New Technique to Investigate Necking in a Tensile Hopkinson Bar," Exp. Mech. 24, pp. 184-186, (1984).
8. A. M. Rajendran and S. J. Bless, "High Strain Rate Material Behavior," Air Force Report: AFL-TR-85-4009, April 1985.

DETERMINATION OF DYNAMIC MECHANICAL PROPERTIES FROM EXPLOSIVELY DRIVEN RINGS

I. Bransky, M. Hayek, D. Halevy, S. Miller and Y. Kivity

Rafael Ballistics Center, Israel Ministry of Defence
P.O. Box 2250, Haifa Israel

ABSTRACT

The expanding ring test is employed as a method of screening metals for shaped charge liners. The dynamic elongation of the liner metal at high strain rate ($\sim 10^4$ sec^{-1}) is found from the maximum elongation prior to rupture as determined from the voltage induced in a pickup coil by an expanding ring which is explosively driven in a magnetic field. The ring radii at various times as well as the mode of rapture are found from flash x-ray radiographs. The dynamics of the ring is analyzed by a two dimensional Hydro-Plasto-Dynamic code. The pickup coil signal is calculated from an analytical solution of an air core transformer, the primary of which is the ring with time dependent radius.

INTRODUCTION

It is well established that the ballistic performance of shaped charge penetrators either in the form of a jet or SFF (self forming fragment) are characterized by two important properties, i.e., the density and the maximum elongation prior to rapture. It is therefore of interest to measure the maximum elongation of metals pertinent to shape charge liners under conditions of shock acceleration by explosive. Strain rates of interest are above about $\dot{\varepsilon} = 10^4$ sec^{-1}.

In the present study we utilize the electric signal induced in a pickup coil due to the motion of an expanding ring in a magnetic field to determine the time of rupture(1). Flash X-ray radiography is used to obtain the ring radii at various stages. The results are compared with a two dimensional computer simulation which provides information concerning the dynamic behaviour of the expanding ring. Results of the computer simulation are also used as data for analytical calculation of the electric signal. Although copper is the most commonly used metal for shaped charge liners heavy metals are also attractive due to their higher density. In the present study rings made of two grades of copper, two tungsten alloys and tantalum were tested.

EXPERIMENTAL PROCEDURES

The maximum elongation prior to rupture is determined from the voltage induced in a pickup coil by an expanding metallic ring which is explosively

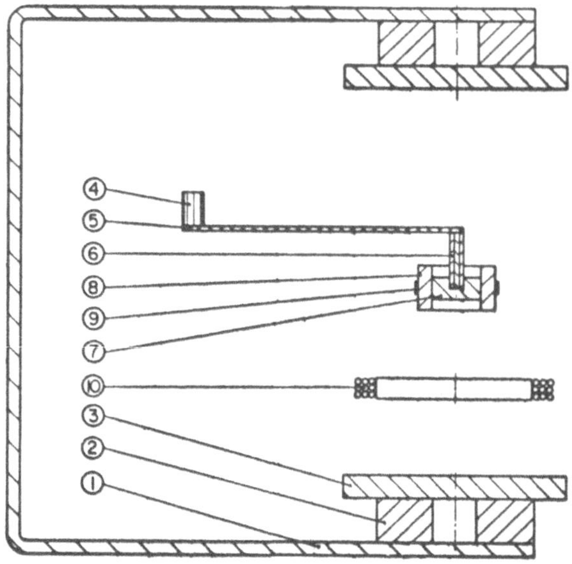

Fig. 1. Experimental setup for expanding ring in magnetic field.

driven in a magnetic field. The experimental setup is illustrated in Fig. 1. Two Ba ferrite ceramic magnets connected by a U shaped iron strip are used to produce a magnetic field of 100 Oe (± 2 pct) perpendicular to the ring. Rings of ID 35 mm height of 5 mm and equal mass per unit area of 1.43 gr/cm^2 are explosively driven as shown in Fig. 1. A mean radial velocity of 0.35 mm/μS is obtained with PBX pellet 20mm in diameter and 10mm height. The pellet is initiated at the center by one end of a thin plastic tube filled with PETN explosive. A long strip of Deta sheet is used to initiate the other end of the tube thus eliminating electrical noises due to the EBW detonator. The electrical signal is measured by a Tektronics 7612 Digitizer. Flash X-ray radiographs were taken at various stages of the ring motion.

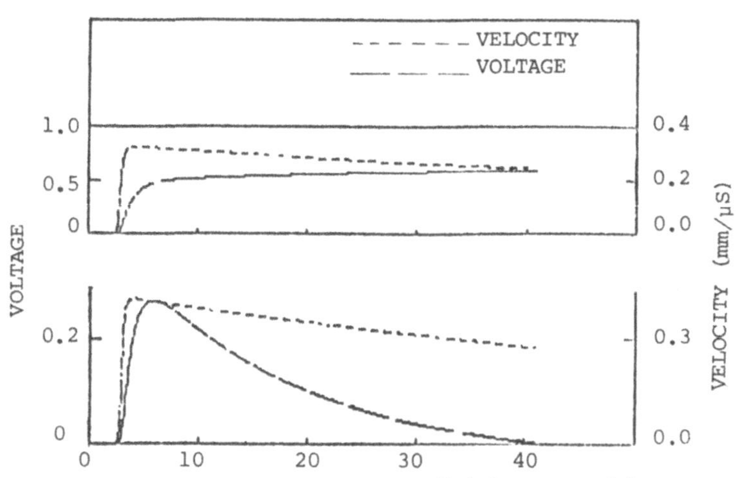

Fig. 2. Calculated time dependence of the radial velocity and pickup coil voltage for copper and tantalum rings.

COMPUTER SIMULATION AND ELECTRIC SIGNAL CALCULATION

The dynamics of the expanding ring is simulated by a two dimensional Hydro-Plasto-Dynamic code "Munroe"[2]. The code is based on a Lagrangian finite element formulation, and is capable of solving explosive metal interaction and impact problems.

For the constitutive relations of metals the Steinberg et al.[3] model was adopted. According to this model the dependence of the shear modulus and yield strength on temperature is taken into account. The equation of state is divided into a cold and thermal components. For the zero temperature isotherm we use a polynomial[4], and the thermal component assumes a constant Gruneisen coefficient and a constant heat capacity at a constant volume, c_V.

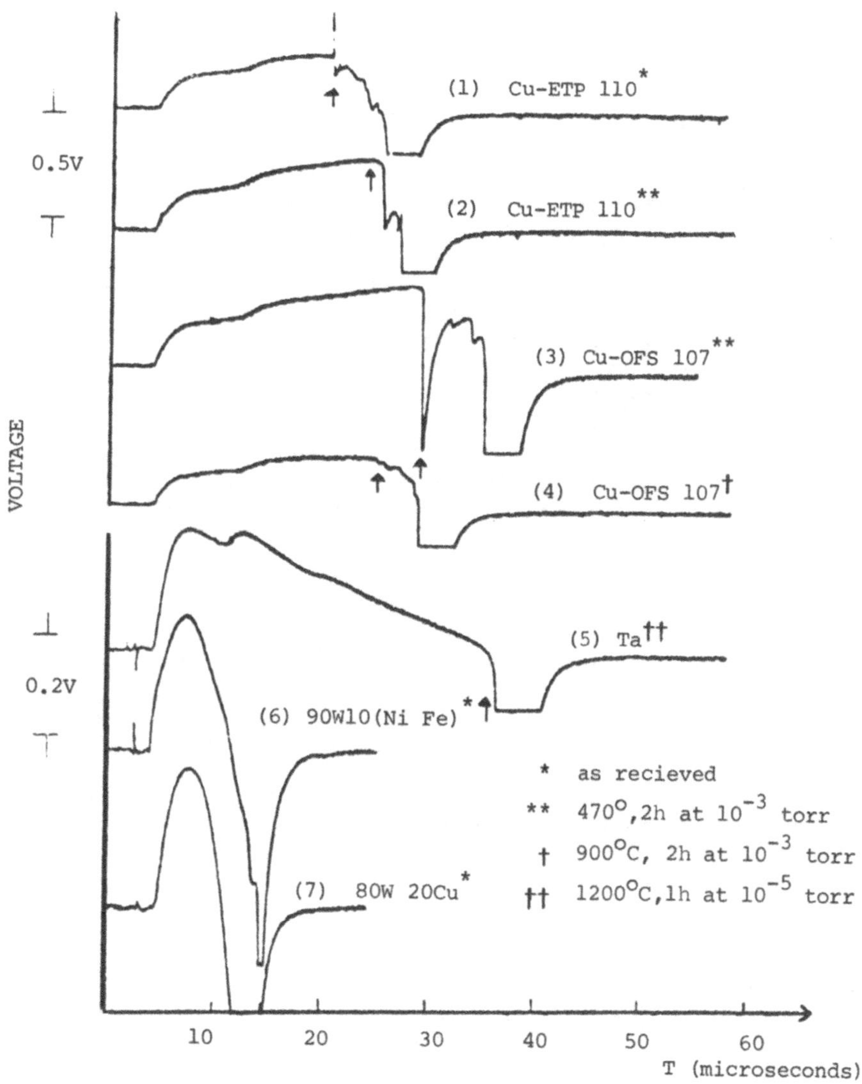

Fig. 3. Voltage vs. time history of several expanding ring experiments. Voltage is given in MKS units. Experiment numbers are lettered to Table 1. The arrows indicate the beginning of rupture.

It is observed that even with a thin ring of about 1mm thickness and height of about 5mm the strain rates arising from wave propagation along the radial direction ($\partial u/\partial r$) and along axial direction ($\partial u/\partial z$) predominate in the first 7 to 8 microseconds of the ring motion. Here u is the ring radial velocity. Only after about 8 microseconds all the waves are smoothed out and the ring attains a uniform radial velocity. The ring motion is then governed by the hoop uniaxial state of stress until rupture occurs.

The signal induced in the pickup coil is calculated from an analytical solution of an air core transformer. The primary of the transformer is the ring assumed to be a loop with time dependent radius, the secondary is the pickup coil. Skin effect is not considered. From the computer simulation of the ring motion the average radial velocity as well as the average radius of the ring are recorded at each time step on a file which serves as input data for the electric signal calculation. The time dependence of the average radial velocity and the calculated electric signal are shown in Fig. 2.

RESULTS AND DISCUSSION

The voltage vs. time for several expanding ring tests are shown in Fig. 3. In this figure the vertical scale for tantalum and tungsten alloys is expanded since their signal is lower than that of copper due to their higher electrical resistivity. The amplitude of the measured signal is close to that calculated analytically for an ideal expanding ring as is shown in Fig. 2. The shape of the signal in this case is influenced by the electrical resistivity of the metal. However close to rupture deviations from the ideal behaviour can arise due to the increase of metal resistance due to micro cracks and voids formation.

Measurements of the ring radii are found to be in good agreement with those found from the computer analysis. The maximum elongation prior to rupture, denoted here as dynamic elongation, was calculated from $(R_R-R_o)/R_o$ where R_o is the initial ring radius and R_R is the radius found from the computer simulation at the time of initial appearance of rupture as indicated by the voltage vs. time history detected by the pickup coil. A comparison with the static standard STM elongation measured for unidirectional tensile specimens at about 0.1mm/min is given in Table 1.

The significant result of this study is the ability to distinguish between static and dynamic elongations. We believe that the expanding ring test is a suitable method of screening metals for shaped charge liners. The dynamic elongation thus determined is closely related to one of the factors predominating the performance of a shaped charge projectile. In general the dynamic elongation is influenced both by the shock history during the first stage of the metal acceleration and by the mode of the dynamic rupture. For instance the large difference between the static and dynamic elongations found for annealed ETP 110 copper in experiment No. 2 is probably due to the strong influence of the high concentration of Oxide inclusions on the dynamic rupture. This rupture is similar to ductile spall fracture which is typically nucleated at non-metallic inclusions in low purity copper. In higher purity copper (OFS 107), on the other hand, the preferred nucleation sites for fracture are the grain boundaries(5). The influence of grain boundaries size is apparent from the difference of dynamic elongations of OFS 107 copper of different grain size (experiments No. 3 and 4). In shaped charge practice it is indeed found that fine grained OFS 107 copper is advantageous to ETP 110 copper although both have similar static elongations. The effect of shock history on the dynamic elongation is seen in experiments No. 6 and No. 7. For the high density metals tested it was found that the tungsten alloys cannot withstand the high strain rate plastic deformations

Table 1. Static elongation compared with dynamic elongation from expanding ring experiments

Exp. No.	Metal	Density (gr/cm^3)	Heat treatment	Static Elongation (%)	Dynamic Elongation (%)
1	Cu(99.9%) ETP 110	8.9	as received (cold rolled)	20-25	27±2
2	"	8.9	470°C, 2h, in 10^{-3} torr (15μm grains)	53-55	35±2
3	Cu(99.95%)OFS 107	8.9	470°C, 2h, in 10^{-3} (20μm grains)	54-55	45±2
4	"	8.9	900°C, 2h, in 10^{-3} torr (150μm grains)	48-50	37±2
5	Ta (99.7%), Murrex, England	16.6	1200°C, 1h, in 10^{-5} torr (15μm grains)	45	63±2
6	A-90, 90W10(Ni Fe) Ashot-Ashkelon, Israel	17.1	as received	30	<4
7	H30-W 80W20Cu, Hogen, Ind. U.S.A.	15.6	as received		<4

introduced during the first stages of acceleration. Flash x-ray radiographs show that they disintegrate into fine particles at this time. This effect is manifested by the large discrepancy between static and dynamic elongation found for 90W 10 (Ni Fe) in experiment No. 6. Jets of these metals are found to break up into fine particles at very early stages of formation. Tantalum on the other hand seems to be a better high density metal candidate due to its high dynamic elongation.

The flash x-ray radiographs taken shortly after rupture show that rupture in Cu and Ta takes place simultaneously and that fracture are uniformly distributed (mostly in 45°) around the ring. The length of the segments of copper rings several microseconds after rupture are in accordance with the dynamic elongation. ETP 110 copper (experiment No. 1) is also found to separate into rings in the axial direction. This happens due to axial strains arising from wave propagation along the z-axis. This effect is much smaller for the more ductile copper OFS 107. No such separation is found for Tantalum. The study is presently extended to higher strain rates than the range of 10^4 sec^{-1} of the present investigation.

REFERENCES

1. F. Olive, A. Nicaud and R. Loichot, Rupture Behavior of Metals in Explosive Expansion, Inst. Phys. Conf. Ser. No. 41 Chap. 2, 242, (1979).
2. S. Miller, "Munroe" Code, Rafael Ball. Center Internal Report (1985).
3. D.J. Steinberg, S.G. Cochran and M.W. Guinan, A Constitutive Model for Metals Applicable at High Strain Rates, J. Appl. Phys., 51(3), 1498, (1980).
4. Y.B. Zeldovich and Y.P. Raizer, "Physics of Shock Waves and High Temperature Hydrodynamic Phenomena", Chap. XI, Academic Press, 1967. Ed. W.D. Hayes and R.S. Probstein.
5. G.G. Brandon, M. Boss and Z. Rosenberg, 3rd Conf. Mech. Prop. High Rates of Strain, Oxford, 261, (1984).

COMPUTER SIMULATION AND ANALYSIS OF THE EXPANDING RING TEST

John E. Reaugh

Lawrence Livermore National Laboratory
Physics Department
Livermore, CA 94550

ABSTRACT

Explosively launched expanding ring tests have been used to extract material flow stress at strain-rates from 10^3 to 3×10^4 sec^{-1}. Computer simulations of such tests using a strain-rate independent, but work hardening constitutive model were analyzed by the same methods used experimentally. Results of this analysis also showed an apparent strain-rate effect. Further examination of the computer simulations has shown that much, if not all, of the measured strain-rate effect is an increased flow stress due to a plastic strain, ε_L, that is locked in during the initial stages of launch. At a larger launch velocity, hence a larger strain-rate, ε_L is also larger. The value of ε_L is sensitive to details of the launch dynamics.

INTRODUCTION

It has often been observed for many engineering materials that a structure loaded dynamically has a higher apparent strength than the same structure loaded more slowly. Most of the reported measurements of the strain-rate dependence of material flow stress rely on a quasi-static analysis of an experiment to connect the measured quantity with the material property. If a structure made of that material is subject to dynamic loads that are similar to the experiment, then the use of the inferred strength increase is appropriate for a similar quasi-static analysis of structural response.

Dynamic (wave-propagation) computer codes allow and require the separation of material behavior from stress-wave interactions. Thus a model for strength increase developed by quasi-static analysis cannot be used directly in such codes. Instead, the experiments that were the basis of the model are simulated to determine the portion of the apparent strength increase that is caused by stress wave interaction in the experimental geometry. The remainder, if any, may then be modeled as a material strain-rate dependence.

Expanding ring experiments have been used to extract material flow stress at strain rates in the range 10^3 to 3×10^4 sec^{-1}. Two experimenters[1,2] have used identical geometry to launch rings of various materials, including 6061-T6 aluminum. Since a model for the strain-rate-independent work

Table 1. Constitutive Model for 6061-T6 Aluminum

Property	Value
Bulk Modulus	0.728 Mbar
Shear Modulus	0.248 Mbar
Flow Stress	$0.00285\ (1 + 125\varepsilon)^{0.1}$ Mbar, where ε is the equivalent plastic strain
Density	2.7 g/cm^3

hardening of that alloy was determined previously[3], those experiments were selected for computer simulation with HEMP[4]. The two experimenters used different methods to analyze the experimental data. The HEMP simulations were analyzed both ways.

COMPUTER SIMULATIONS

The experimental geometry is shown in Fig. 1. The aluminum ring has 5.08 cm ID, 0.16 cm wall thickness and 0.64 cm width. The computer simulations were simplified by applying a constant pressure P_o to the inside surface of the ring for a time τ, where τ is approximately equal to the transit time of a shock wave through twice the ring thickness, h. The initial ring velocity, v_o, is given by

$$v_o = \frac{P_o \tau}{\rho h} ,$$

where ρ is the mass density of the ring material.

The constitutive model used for 6061-T6 aluminum is given in Table 1. Two simulations were performed with initial ring velocities of 100 m/sec (P_o = 8.1 kbar, τ = 0.53 μsec) and 200 m/sec (P_o = 16.2 kbar, t = 0.53 μsec). The velocity and displacement histories of the point located at the center of the outside surface are shown in Fig. 2 and 3 for the 100 m/sec simulation. The high-frequency oscillation seen in the early part of the velocity history (Fig. 2) is caused by the reflection of stress waves from the ring boundaries. The oscillation dies out as plastic flow continues.

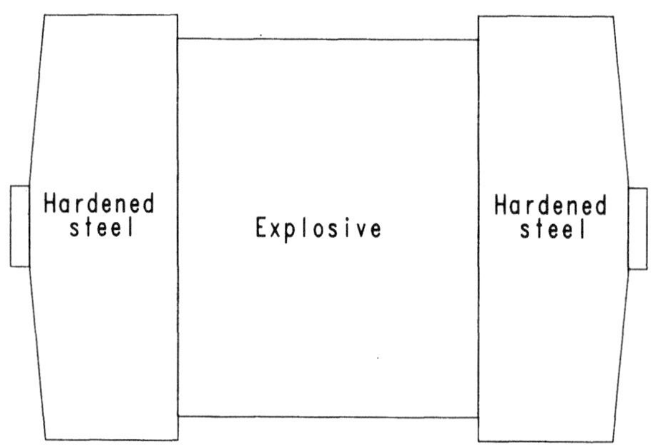

Fig. 1. Experimental geometry for expanding ring tests.

DATA ANALYSIS

In Ref. 1, the displacement history of the outside radius of the ring was recorded with a streak camera. The peak displacement, D_M, and the time between D_M and the first motion, t_M were measured. Data were analyzed by assuming constant deceleration, $a = 2D_M/t_M$. For that case, the flow stress Y was taken as $Y = \rho Raf$ where f is a correction factor for finite thickness. For the ring geometry used, f = 0.96. The outside radius R is given by $R = R_o + at_Mt - 1/2\ at^2$. At a time t, the strain, ε, and strain rate, $\dot{\varepsilon}$, are given by $\varepsilon = \ln(R/R_o)$, $\dot{\varepsilon} = \dot{R}/R$.

In Ref. 2, the velocity history was measured using an interferometer, and differentiated to obtain the deceleration, a, at time, t, when the strain was 2%. The flow stress was taken to be $Y = \rho Ra$. Data were reported as the flow stress Y at 2% strain, but at the initial strain rate v_o/R_o.

The two computer simulations were analyzed by both methods. Results are shown in Figs. 4 and 5 together with the flow stress used in the simulations. Figure 5 also includes other data. The scatter in the apparent strain-rate dependence of the 2% flow stress is similar to the difference between the results when the two methods are used to analyze the computer data.

Analysis of the Computer Simulations

Both methods of data analysis indicate the flow stress increases with strain rate, even though the constitutive model used in the simulations was rate independent. The cause of this discrepancy is that the ring has been work hardened by shock processing as it is launched. The two simulations show that the plastic strain, which determines the flow stress in the rate-independent model, is given by $\varepsilon = \varepsilon_L + \ln(R/R_o)$. In each simulation ε_L is approximately equal to the plastic strain produced in a single plane-strain cycle of loading to the launch pressure P_o followed by unloading to zero.

Other computer simulations that varied the pressure history on the inside surface of the ring show that ε_L is very sensitive to details of the launch process, which were not measured or reported in either Ref. 1 or 2. In the velocity range considered, 100-200 m/sec, the computed ε_L varied between 0.2 and 6%. At 2% strain, this results in a stress increase between 0.02 and 0.4 kbar, which encompasses essentially all of the data reported (Fig. 5) at strain rates less than 8000 sec^{-1}. The data in Fig. 5 include results from expanding ring tests and from other experimental techniques.

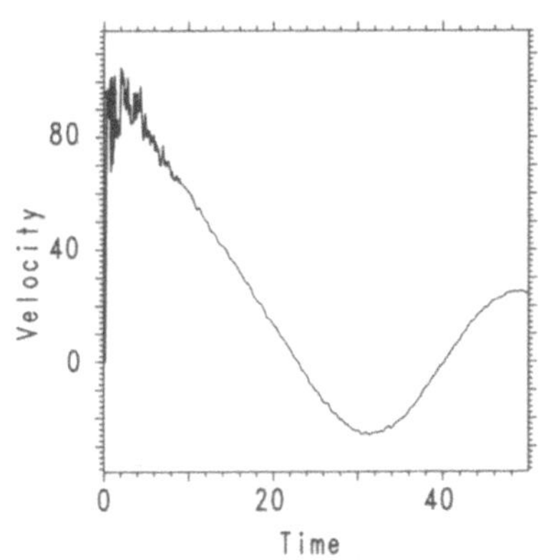

Fig. 2. Velocity (m/sec) vs. time (μsec) at the ring's outside surface.

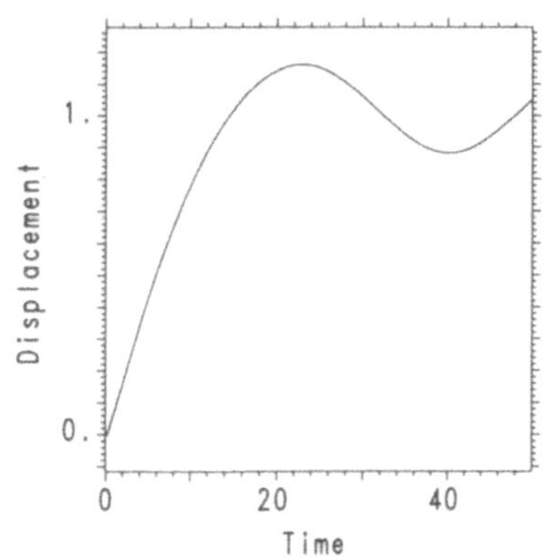

Fig. 3. Displacement (mm) vs. time (μsec) at the ring's outside surface.

Experimental Evidence of ε_L

It was reported in Ref. 1 that the 6061-T6 aluminum rings were observed to fracture at 12% strain. The computer simulations confirm that after the transients from the launch have decayed, the ring is in uniaxial tension. Comparison of simple tension tests of 6061-T6 aluminum (Ref. 3) with computer simulations show that the strain to fracture in the necked region is 52%. The strain in the 2-in. gauge length for the standard tensile test, which is an average measure including the necked region, is about 25%. Some of the discrepancy between the strain at ring fracture, 12%, and the nominal strain to fracture in simple tension, 25%, can be explained by ε_L.

Reported necking of ring specimens (Ref. 2) at 8% strain is also suggestive of ε_L, since the data presented in Ref. 3 show 10% uniform strain before necking begins, and a somewhat larger strain before it is apparent. The effect of necking on either the nominal strain to fracture or the kinematics of ring expansion is uncertain.

SUMMARY AND CONCLUSIONS

In explosively launched expanding ring tests, a portion of the reported strain rate dependence of flow stress is caused by shock processing as the ring is launched. (This was noted in Ref. 2 as a possible contribution to their observed results.) Computer simulations have shown that the plastic strain ε_L is sensitive to details of the shock wave interactions during launch. This strain, between 0.2 and 6%, depending on the launch conditions in the simulations, is sufficent to cause a 0.4 kbar spread in the apparent flow stress at 2% strain. This spread in flow stress encompasses most of the reported data on 6061-T6 aluminum. The observed fracture at 12% nominal strain in the ring tests is consistent with ε_L. The observed necking at 8% nominal strain is also consistent with ε_L.

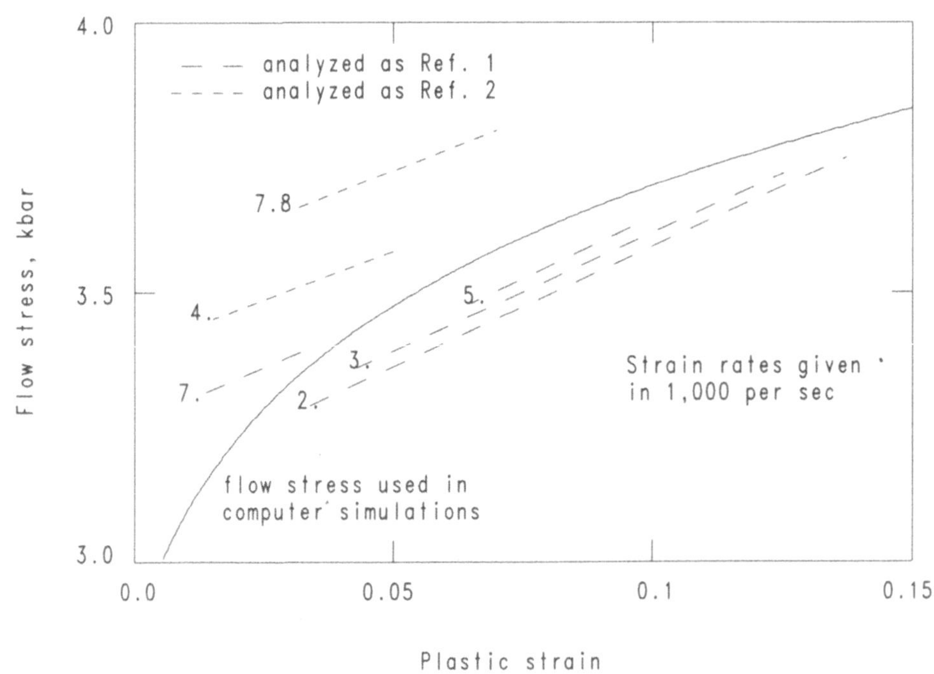

Fig. 4. Flow stress vs. plastic strain from the computer simulations.

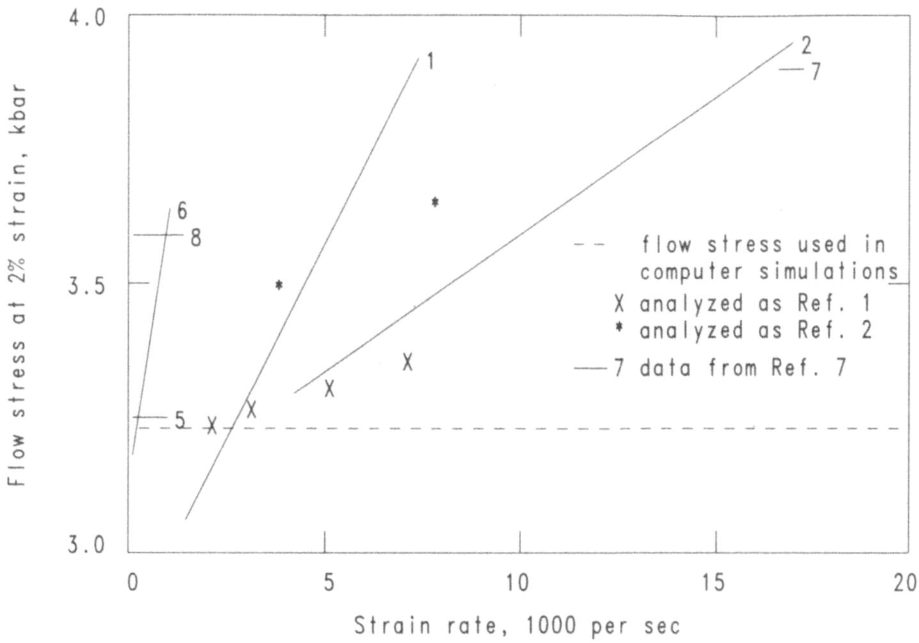

Fig. 5. Flow stress at 2% strain vs. strain rate from the computer simulations and from experimental data.

ACKNOWLEDGEMENTS

This work was performed under the auspices of the U.S. Department of Energy by Lawrence Livermore National Laboratory under Contract No. W-7405-ENG-48.

REFERENCES

1. C. R. Hoggatt and R. F. Recht, "Stress-Strain Data Obtained at High Rates Using an Expanding Ring", Experimental Mechanics, (9), 441-448, Oct. 1969.
2. A. E. Carden, P. E. Williams, R. R. Karp, Comparison of the Flow Curves of 6061 Aluminum Alloy at High and Low Strain Rates, in: Shock Waves and High-Strain-Rate Phenomena in Metals, M. A. Meyers and L. E. Murr, eds., Plenum, 1981, pp. 37-50.
3. M. L. Wilkins, R. D. Streit, J. E. Reaugh, "Cumulative-Strain-Damage Model of Ductile Fracture: Simulation and Prediction of Engineering Fracture Tests", Lawrence Livermore National Laboratory, Livermore, CA, UCRL-53058 (1980).
4. M. L. Wilkins, "Calculations of Elastic-Plastic Flow", in: Methods in Computational Physics, Vol. 3 Fundamental Methods in Hydrodynamics, B. Alder, ed., Academic Press, N. Y., 1964, pp. 211-263.
5. C. J. Maiden and S. J. Green, Compressive Strain Rate Tests on Six Selected Materials at Strain Rates from 10^{-3} to 10^{4} in./in./sec, J. Appl. Mech., p. 496, Sept. 1966.
6. C. H. Mok and J. Duffy, "The Dynamic Stress-Strain Relation of Metals as Determined from Impact Tests with a Hard Ball", Brown Univ. Rept. Nonr-562(20) 137, June 1964.
7. A. L. Austin and R. F. Steidel, Jr., A Method for Determining the Tensile Properties of Metals at High Rates of Strain, Proc. SESA. XVII (1), pp. 99-114, 1959.
8. U. S. Lindholm and L. M. Yeakley, High Strain-Rate Testing: Tension and Compression, Experimental Mechanics, 8(1), pp. 1-9, 1968.

RAPID CAVITATION INDUCED BY THE REFLECTION OF SHOCK WAVES

Philip L. Marston and Bruce T. Unger

Physics Department
Washington State University
Pullman, WA 99164-2814

ABSTRACT

Impact techniques were used to investigate rapid cavitation in moderately pure water and in ethylene glycol. A compressive pulse of duration $\approx$ 1.7 μs and amplitude $\geq$ 10 MPa was propagated into the liquid sample. The pulse was reflected at the rear surface to create a tensile pulse in the liquid. The displacement history of the rear surface indicates that cavitation typically occurred within 1 μs of the creation of tension. The cavitation was manifested as a rapid rebound of the surface velocity and by evidence of reverberation within a spalled layer. Though the inferred dynamic tensile strength typically exceeded 5 MPa, the experiments are suggestive of pre-existent, undissolved gas nuclei. Cavitation was not clearly indicated in two shots suggesting tensile strengths > 11 MPa for those samples.

INTRODUCTION

Liquids rupture or cavitate when subjected to tensions in excess of their tensile strength. This phenomenon has been of general interest because cavitation is a mechanism for damage in various acoustic and hydrodynamic technologies.[1] Cavitation in the water adjacent to a structure may also affect the damage inflicted by an underwater explosion.[2] There has been recent interest[1,3] in cavitation in response to tensions in the range 1 to 10 MPa (1 MPa = 9.87 atm.) One reason for this interest is that pulsed ultrasonic imaging devices (commonly used medical diagnostics) have been estimated to produce transient tensions in excess of 4 MPa. We investigated the response of liquids by applying impact techniques similar to those used to study the tensile loading of metals[4] and elastomers.[5] One noteworthy aspect of our technique is the unusually low value ($\approx$ 10 m/s) for the impact velocity.

Estimates[1,6] of the intrinsic tensile strength of pure water at room temperature give values in excess of 150 MPa. These estimates greatly exceed measured values and considerable evidence has accumulated[1,7] that rupture occurs at inhomogeneities generically described as "nuclei". These nuclei are thought to be localized regions of gas which are prevented from dissolving by various stablization mechanisms. The dynamic tensile strength or cavitation "threshold" depends on the size of the gas pockets, the surface tension of the liquid, and the pressure history. In water subjected to repeated acoustic bursts, cavitation from one burst may produce new nuclei which affect thresholds for subsequent bursts.[8] This acoustic generation of nuclei may be avoided by using short, well separated, acoustic bursts (See e.g., Ref. 3) or by our technique which uses a compressive-tensile pair of pulses. Additional background information was reported with a related previous experiment.[9]

MODEL AND ANALYSIS

Figure 1 is a diagram of the apparatus used. In this section, we model the waves induced by the impact of the projectile. The model assumes the strain is uniaxial and perpendicular to the surface of the buffer. Consequently, the effects of projectile tilt and edge waves are neglected as are the microscopic dynamics of cavity growth. Nevertheless, several features predicted by this idealized model were observed in displacement records of the rear surface of the liquid given by an interferometer. The impactor and buffer were lapped aluminum plated. The pressure p and particle velocity u for the states produced may be well approximated by taking the material impedance to be the appropriate acoustic impedance $z_n = \rho_n c_n$. Here ρ_n and c_n denote the unperturbed density and sound speed of the impactor ($n = 1$), buffer ($n = 2$), and liquid ($n = 3$).

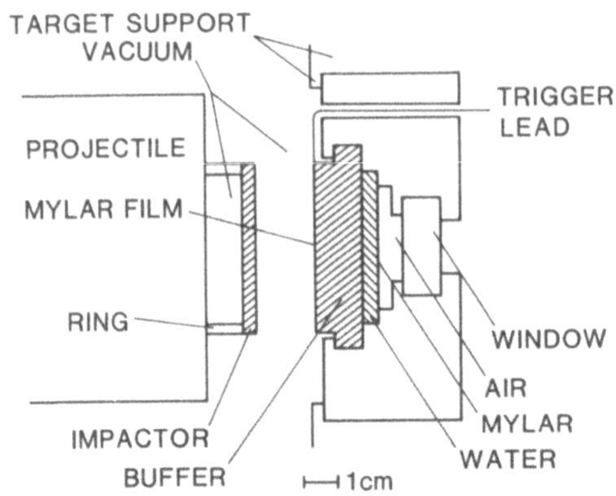

Fig. 1. Diagram of the experimental apparatus.

Figure 2 shows the states generated when the liquid is water and the projectile velocity $u(A_i)$ is 10 m/s. Figure 3(a) shows the states as a function of the longitudinal coordinate x and time t. Because $u(A_i)$ is $\ll c_n$, the choice between Eulerian and Lagrangian x would not be clearly manifested on the scale of Fig. 3(a). Figure 3(b) shows the predicted (u, t) history of the rear (free) surface of the water. The late time behavior in the absence of cavitation is indicated by the dot-dashed curve; idealized rupture of the water (discussed below) is illustrated by the dotted option. The material thickness Δx_n used in the construction of Fig. 3 were 5.4, 12.5, 3.8 mm for $n = 1,2$, and 3. These correspond to those of shot 3/16 and well approximate those of the other shots. The duration of the compressive pulse is $\Delta t = 2\Delta x_1/c_1 = 1.7\ \mu s$.

Behind the water (Fig. 1) was an aluminized Mylar film. The problem of separation of the film from the water, discussed in Ref. 9, was avoided here by using a thickness of 6 μm. The time for a wave to travel across the Mylar is about 3 ns; consequently, the ring-up and ring-down of the Mylar was not included in Fig. 2 and 3. The rarefaction which terminates G has been idealized and stretched in time for clarity of presentation. When the rarefaction which terminates D, reaches the buffer-impactor interface, a gap should open radiating a compressive wave back into the buffer. In the absence of cavitation, this should result in the $J \rightarrow L$ transition in Fig. 3(b). The associated pulse will be termed the "gap pulse."

Intersection of the $C \rightarrow G$ rarefaction with $C \rightarrow F$ rarefaction causes a transition to tensile state H if there is no cavitation. This intersection first occurs at a distance $c_3\Delta t/2 = 1.3\ mm$ from the rear surface and we will focus attention on the state in this plane. In the absence of cavitation, the duration of the tensile pulse here is also given by Δt. The transition to H is terminated at the intermediate state H' if the tensile strength is reached causing the liquid to cavitate. If we neglect the time for cavity growth, the pressure here is abruptly reduced to zero (state I'). A recompression wave is radiated which is assumed to be a plane wave. The transition for u_{fs} toward $u(J) = 0$ is terminated by this wave at $u(J')$. Inspection of Fig. 2 gives the well-known[4,9] expressions for the pulse pressure $p(C)$ and the tensile strength $-p(H')$:

$$p(C) = (1/2)z_3u(G); \qquad -p(H') = (1/2)z_3[u(G) - u(J')]. \tag{1a,b}$$

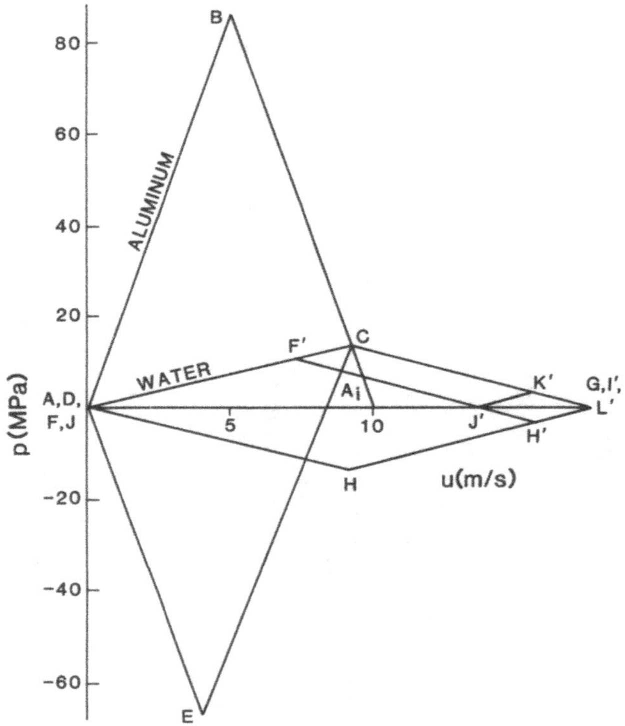

Fig. 2. Wave transistions in the (p, u) plane. The slopes are the acoustic impedances. The primed states would results from idealized rapid cavitation.

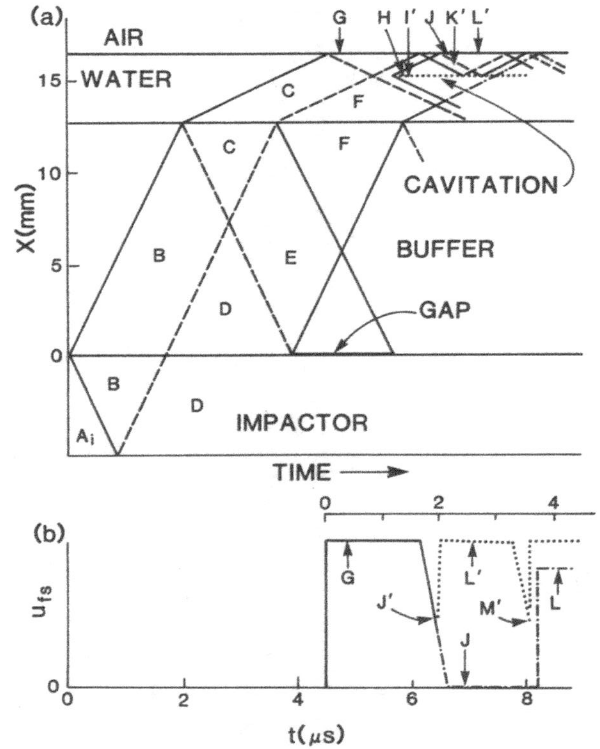

Fig. 3. (a) (x, t) diagram and (b) free-surface (u, t) diagram. The time axes in (a) and (b) are correlated. The model for the response to idealized cavitation is indicated by primes.

In this derivation we neglect the initial pressure of the liquid (0.1 MPa) which is also the pressure at the rear surface. These differ slightly from the assumed $p(I')=0$; however, they do not significantly affect the interpretation of our data.

EXPERIMENTAL PROCEDURE

The liquids studied were distilled water and reagent grade ethylene glycol. The liquids were degassed as described in Ref. 9. The target was filled through a 0.2 μm Millipore filter. The temperature was typically 22 C before impact.

The Mylar displacement was measured with a Fizeau interferometer diagrammed in Ref. 9. The reference beam of the interferometer was a reflection from the window (Fig. 1) which was stationary during the interval of interest. The output voltage V of a phototube cycled for Mylar displacements of $\lambda/2 \approx 316\ nm$ such that the signal frequency $F = 2u_{fs}/\lambda$ sometimes exceeded 40 MHz. Oscilloscopes were triggered when the impactor made electrical contract with a thin aluminized-Mylar film bonded uniformly to the front surface of the buffer. The spectral components of V having $F \geq 20\ MHz$ were used as a backup trigger. The impactor and target were optically aligned with a tilt of $\leq 1\ mrad$.

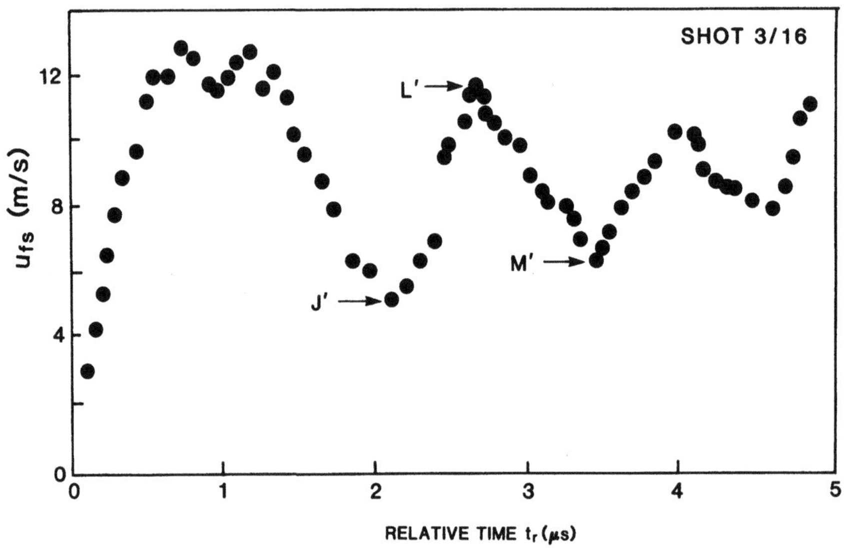

Fig. 4. Measured surface velocity for the response of water to a shock pulse.

RESULTS AND DISCUSSION

The record shown in Fig. 4 manifests effects of cavitation. The time scale t_r is relative to the arrival of the pulse at the Mylar as is the upper time scale in Fig. 3(b). Cavitation produces a recompression wave which arrives at the Mylar at $t_r = 2.0\ \mu s$. This is significantly ahead of the earliest edge wave which (from Fermat's principle and the radius of 26 mm for the impactor) could not arrive until $t_r = 2.5\ \mu s$. The minimum in u_{fs} at M' corresponds to the predicted arrival time of a reflection from the cavitation (or "spall") plane. A subsequent reverberation is also visible. The large amplitude of these reverberations suggests the rupture of water was nearly complete and that the void may have contained vapor which inhibited subsequent collapse. This would also explain why the gap pulse and the edge waves are not evident. If not affected by cavitation, the gap pulse is modeled to arrive at $t_r = 2\Delta x_2/c_2 = 3.9\ \mu s$

Shots 3/16, 4/20, 7/12, and 8/02 clearly manifest a recompression wave produced by cavitation. Equation (1b) was used to estimate the tensile strength, $-p(H')$, and the results are shown in Table I. The observed $u(G)$ was used with Eq. (1a) to predict the amplitude of the initial pulse $p(C)$. The table also gives the $u(G)$ predicted from impact velocity $u(A_i)$ and the z_n. The difference between the predicted and observed $u(G)$ may be attributed to the effects of impactor tilt which are enhanced[9] because of the smallness of $u(A_i)$ and because of the large change in the impedance at the buffer-liquid interface.

TABLE I. Summary of Shot Data

Liquid and Shot No.	$u(A_i)$ (m/s)	$u(G)_{pred.}$ (m/s)	$u(G)_{obs.}$ (m/s)	u(J') (m/s)	p(C) (MPa)	-p(H') (MPa)
Water:						
3/16	8.2	15.2	12.5	5.2	9.3	5.4
4/01	10.0	18.4	17.0	1.5	12.6	11.5
4/20	9.9	18.2	16.5	11.2	12.2	3.9
7/12	9.3	17.1	14.0	6.5	10.4	5.6
Ethylene Glycol:						
8/02	8.6	15.5	10.0	5.4	9.0	4.1
8/19	11.4	20.5	18.0	0.9	16.2	15.8

Shots 4/01 and 8/19 do not clearly manifest the recompression wave from cavitation; u_{f}, decreases to the tabulated values of $u(J')$ which were close to 0. An inflection anomaly was evident in u_{f}, for the $C \rightarrow J'$ rarefaction of shot 8/19. Shot 4/01 shows evidence of a gap pulse while in shot 8/19 the late arrival of a large pulse can be attributed to the gap pulse and/or reverberation within the whole liquid sample. Since cavitation appears to have been insignificant for shots 4/01 and 8/19 the local dynamic tensile strengths appear to have exceeded 11 and 15 MPa, respectively.

For the water samples exhibiting cavitation, we estimate a lower limit R on the radius of gas-filled nuclei. This limit follows from Blake's quasistatic stability calculation for nuclei (Sec. 7.2 of Ref. 1). In the present problem the "Blake threshold" reduces to $p(H') = -0.77\gamma/R$ where $\gamma = 0.071\ N/m$ is the surface tension. Taking $-p(H') = 5\ MPa$ gives $R = 1\ nm$. Dynamical effects necessitate that the nuclei be somewhat larger. The extent of the cavitation suggested by Fig. 4 indicates that numerous nuclei were present.

This work was supported by ONR. Acknowledgment is made to the Donors of the Petroleum Reseach Fund, administrered by the American Chemical Society, for the initial support of this research.

REFERENCES

1. R. E. Apfel, Acoustic Cavitation, in "Ultrasonics," P.D. Edmonds, ed., Academic, NY (1981).
2. M. R. Driels, Effect of a Non-Zero Cavitation Tension on the Damage, Sustained by a Target Plate Subject to an Underwater Explosion, J. Sound Vib. 73:533 (1980).
3. L. A. Crum and J. B. Fowlkes, Evidence for Cavitation Generated by Microsecond Pulses of Ultrasound (to be published).
4. S. Cochran and D. Banner, Spall Studies in Uranium, J. Appl. Phys. 48:2729 (1977).
5. Y. M. Gupta and W. J. Murri, SRI Project PYU-7802 Report (1981), unpublished.
6. R. J. Speedy, Stability-Limit Conjecture, An Interpretation of Properties of Water, J. Phys. Chem. 86:982 (1982).
7. L. A. Crum, Nucleation and Stabilization of Microbubbles in Liquids, Appl. Sci. Res. 38:101 (1982).
8. H. G. Flynn and C. C. Church, Mechanisms for the Generation of Cavitation Maximum by Pulsed Ultrasound, J. Acoust. Soc. Am. 76:505 (1984).
9. P. L. Marston and G. L. Pullen, Cavitation in Water Induced by the Reflection of Shock Waves, in "Shock Waves in Condensed Matter--1981," W. J. Nellis, et al. ed., AIP, NY (1982).

TENSION-RECOMPRESSION RESPONSE OF SHOCK LOADED POLYCARBONATE[a]

P. Majewski, and Y. M. Gupta,

Shock Dynamics Laboratory
Department of Physics
Washington State University
Pullman, WA 99164-2814

L. Seaman

SRI International
333 Ravenswood Avenue
Menlo Park, CA 94025

ABSTRACT

--An experimental method has been developed to study the tension-recompression response of samples in plate impact experiments. The normal configuration of a spall experiment was modified by placing a small gap followed by a recompression plate behind the rear surface of the target. Material velocity was measured using EMV gauges located at the impact surface and at the surface of the recompression plate. The time at tension, and thus the spall damage, was varied by varying the thickness of the gap; recompression amplitude can be varied by the choice of the recompression plate material. We have used this method to examine the behavior of Lexan[2] for two tensile durations. Using Lexan material properties obtained earlier[2], the full spall and low damage cases were numerically simulated. The agreement between the calculations and the experiments was good for the full spall case and fair for the low damage case. Potential applications for this method will be indicated.

Introduction

Studies of tensile fracture under uniaxial strain impact loading have been performed on several materials.[1-6] The work performed at SRI International has been particularly valuable in understanding the progression of tensile fracture from the onset of damage to full separation.[1-3] The SRI nucleation and growth (NAG) model has also been important for modeling fracture and fragmentation under multi-dimensional loading. Despite the many experiments on tensile fracture, there is little information on the compaction response of the fractured material. This information is valuable in many high strain-rate applications, particularly those involving energetic materials.

In this paper we describe an experimental method designed to study the recompaction response of the fractured material on the microsecond time scale. Time-resolved measurements under uniaxial strain loading and comparison with predictions from existing models are the central theme of the present work. We chose to examine the behavior of a polycarbonate developed by General Electric Company (trade name Lexan) because the tensile response of this material has been characterized in a previous study.[2]

The same overall experimental concept was independently arrived by D. R. Curran of SRI International.

Experiments

The experimental configuration for the tension-recompression experiment is shown in Fig. 1. The usual configuration of the tension experiment [2] is modified by incorporating a thin gap (order of 0.1 mm) followed by a recompression plate. In the usual tension experiment, the pulse amplitude and duration are controlled by the impactor velocity and the relative thicknesses of the impacting plates. Upon impact, tension is produced in the thicker sample by the intersection of the rarefactions from the free surfaces of the flyer and sample plates. In the experimental configuration shown in Fig. 1, the gap width provides additional control on the tensile pulse duration. However, the main feature of the present configuration is the recompression plate. The sample free surface impacts the recompression plate producing a backward facing compression wave in the sample and a forward facing compression wave in the recompression plate. By varying the recompression plate material, the compression pulse magnitude can be altered. This feature is expected to be important in understanding the compaction response of the damaged material containing voids or cracks introduced by the tensile pulse.

A key aspect of the experiment is the ability to make time-resolved measurements at the front surfaces of the sample and of the recompression plates as shown in Fig. 1. As discussed in the next section, different parts of the time-resolved profiles provide information about the kinetics of the tension and recompression processes.

Three experiments on Lexan with varying impact velocities are summarized in Table I. All three plates were machined from Lexan sheet stock. Electromagnetic velocity gauges were used to measured the particle velocity at the two surfaces indicated in Fig. 1. Only the particle velocity profiles from the second gauge are shown in this paper. The samples were recovered in all three of the experiments. Experimental results are described and compared with the numerical simulations in the Analysis and Discussion Section.

Numerical Calculations

The analysis of these experiments is sufficiently complicated that the use of a wave propagation program is required. In the present work, the Lexan response has been calculated using two different material models: BFRACT model of Seaman, *et al.*,[3] and the spall model of Cochran and Banner.[4] The model of Seaman *et al.* is based on microstructural observations and treats the nucleation, growth, and coalescence of cracks in the material. This overall approach has been widely used for a number of materials and can be generalized to multidimensional loading. The model of Cochran and Banner is essentially a cumulative damage model that contains no explicit kinetics for crack nucleation and growth. This model was developed to simulate time-resolved free surface measurement (pull back signals) in metals subjected to uniaxial strain loading; extension to multidimensional loading is not possible in the present form of the model. Further details about these models may be seen in Refs. 3 and 4.

The numerical calculations with the BFRACT model were performed at SRI using the PUFF code.[6] The material parameters in these calculations were taken from the paper of Curran, *et al.*[2] The numerical calculations with the Cochran and Banner model were performed at Washington State University using the COPS code.[7] This fracture model exists as the SPALL1 routine[8] in the COPS code. The compression model for the COPS calculation was taken to be linear elastic; the tensile parameters for the SPALL1 routine were obtained by iteratively matching the calculations to the experiment.

Analysis and Discussion

The measured particle profiles at the front surface of the recompression plate are shown for all three experiments in Fig. 2. In Ref. 9 it has been shown that the magnitude of the first drop in the particle velocity (Δu) is a fairly good measure of the dynamic tensile stength. If the first drop reaches zero particle velocity then it indicates a lack of tensile damage for that particular experiment. As can be seen from the three profiles in Fig. 2, the fractional drop (ratio of velocity decrease to initial jump) increases with decreasing tensile damage as expected. The magnitude of the second peak in Experiment 3 is most likely an experimental artifact.

Comparisons between the predictions of the BRACT model and the measured profiles for the three experiments are also shown in Fig. 2. The agreement between the model prediction and the measured profiles is reasonably good for Experiment 1 (full spall case) but is poor for Experiments 2 and 3. The model calculation based on the profile in Fig. 2c suggests no tensile damage in Experiment 3. The measured profile and the recovered sample indicate minimal damage. The material parameters used in these calculations were taken directly from Curran, *et al.*[2] and no attempt was made to improve the agreement by varying the parameters.

TABLE I - EXPERIMENTAL PARAMETERS

Experiment Number	Impactor Velocity (Km/sec)	Impact Tilt (mrad)	Impactor Thickness (mm)	Sample Thickness (mm)	Gap Width (mm)	Backing Thickness (mm)
1. (84-568)	0.302	1.2	3.06	6.22	0.54	12.02
2. (85-552)	0.199	1.1	2.37	4.73	0.28	11.74
3. (85-518)	0.156	0.5	2.94	6.14	0.17	11.66

TABLE II
LEXAN BEST FIT VALUES TO COCHRAN AND BANNER MODEL

Spall Strength	0.161	GPa
Crush Stength	0.150	GPa
Critical Damage	0.011	cm

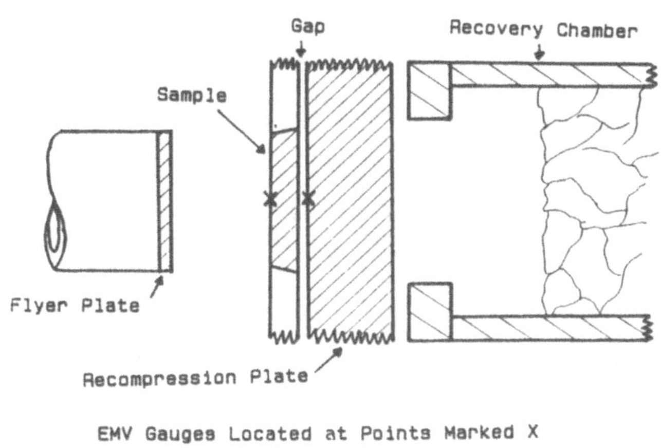

1. Experimental system for tension/recompression experiments.

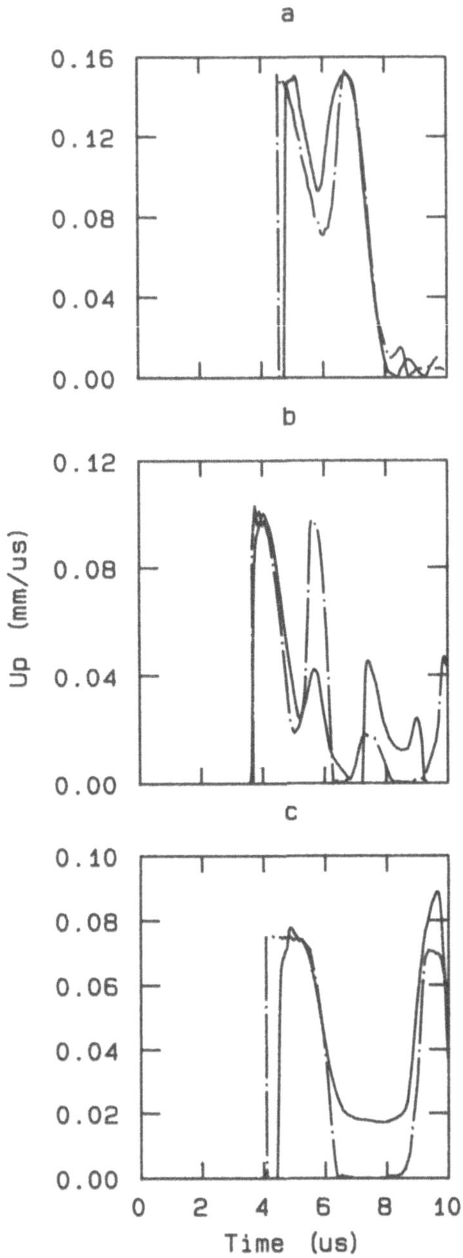

2. Comparison of observed (solid) and calculated wave profiles. Calculations based on PUFF code with BFRACT model.
a) Exp. 1, sample completely seperated on spall plane. b) Exp. 2, sample spalled across 90% of diameter, did not seperate. c) Exp. 3, a large number of individual cracks appear in the central region of the sample.

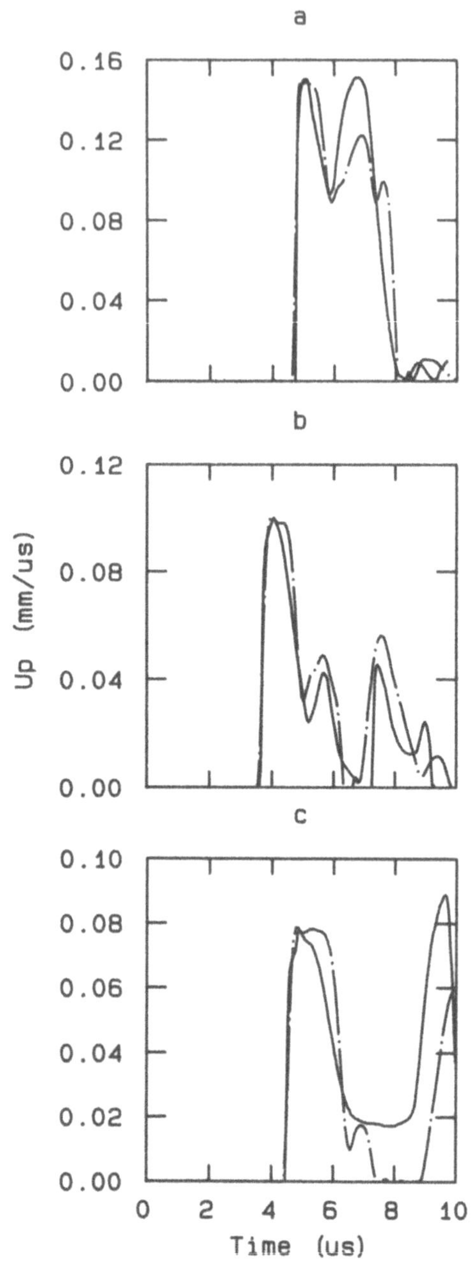

3. Comparison of observed (solid) and calculated wave profile. Calculation based on COPS code with SPALL1 model. a) Exp. 1, b) Exp. 2, c) Exp. 3.

Comparisons between the Cochran-Banner model predictions and the measured profiles are shown in Figs. 3a - 3c. The model parameters, presented in Table II, were obtained by performing a large number of calculations and selecting the parameters that gave the best overall fit to all three experiments. Again, the low velocity experiment has the poorest fit. The use of a linear elastic response during compression and release in the model leads to a flatter pulse than that observed in experiments.

From the comparison between theory and experiment, it appears that the experiments at low velocities provide a severe check on the tension-recompression description in the material models.

Conclusion

An experimental method has been developed to examine the tension and recompression response of a solid subjected to uniaxial strain impact loading. Numerical calculations have been performed with two different fracture models to interpret the experimental data. With the simpler tensile fracture model, a set of material parameters were iteratively obtained to provide an overall fit to several experiments. Experiments performed at velocities close to the threshold values required to produce tensile damage provide a severe check for the fracture models. Work is currently under progress to study energetic materials particularly with large recompression wave amplitudes.[10]

References

1. T.W. Barbee, L. Seaman, R.C. Crewdson, and D.R. Curran, J. of Materals 1:393 (1972).
2. D.R. Curran, D.A. Shockey and L. Seaman, J. Appl. Phys. 44:4025 (1973).
3. L. Seaman, D.R. Curran, and D.A. Shockey, J. Appl. Phys. 47:4814 (1976).
4. S. Cochran and D. Banner, J. Appl. Phys. 48:2729 (1977).
5. D.E. Grady and M.E. Kip, Int. J. Rock Mech., Min. Sci. and Geomech. 16:293 (1979).
6. L. Seaman, SRI PUFF 8 Computer Program for One-Dimensional Stress Wave Propagation (SRI International, Menlo Park, California, 1978), unpublished.
7. Y.M. Gupta, COPS Computer Program for One-Dimensional Compression and Shear Wave Propagation (SRI International, Menlo Park, California, 1978), unpublished.
8. One of us (Y.M. Gupta) acknowledges the contribution of W.C. Moss for incorporating the Cochran and Banner mode in the COPS program.
9. Y.M. Gupta and W.J. Murri, Response of a Plain and Filled Elastomer to High Strain-Rate Compression, Shear, and Tension Loading, Techical Report under ONR Contract N00014-78-C-0549 (SRI International, Menlo Park, California, 1981), unpublished.
10. Y.M. Gupta and M.C. Cowperthwaite, unpublished.

THE STRESS AND STRAIN-RATE DEPENDENCE OF SPALL STRENGTH IN TWO ALUMINUM ALLOYS*

D. R. Ek and J. R. Asay

Sandia National Laboratories
Albuquerque, NM

ABSTRACT

The spallation behavior of 6061-T6 and AMg-6 aluminum alloys was studied as a function of loading stress and initial pulse duration for stress amplitudes ranging from 5-20 GPa and for pulse durations spanning the range of 0.06-0.9 μs. These experiments show that the spallation strength varies from about 0.7-1.4 GPa for strain rates varying from 0.15-1 μs^{-1} and for impact stresses from 5-20 GPa.

INTRODUCTION

The dynamic fracture of materials is fundamental to a variety of applications, including ballistic penetration in conventional ordnance applications, energy recovery in mining operations, and dynamic fragmentation in hypervelocity particle target interactions. A special case of dynamic fracture concerns the spallation of materials exposed to high-rate tensile loading. This fracture mode occurs in the interior of a body when tensile stresses of sufficient magnitude are generated to cause internal rupture. These conditions result from the interaction of rarefaction waves and occur in a variety of experimental conditions, including plate impact, explosive loading, and rapid energy deposition.[1]

In the present experiments, plate impact techniques and laser interferometry were used to determine the spall strengths of 6061-T6 and AMg-6 aluminum alloys over the stress range of 5-20 GPa and for pulse durations ranging from 0.06-0.90 μs. The spall strengths were determined from the magnitude of the "pull-back" in free-surface particle-velocity experiments and compared to numerical computations based on a simple fracture criterion. These investigations indicate a dependence of spall strength on strain rate (which depends on pulse duration) but not on stress amplitude in the alloys. Large discrepancies are observed between the experimental

*This work performed by Sandia National Laboratories supported by the U. S. Department of Energy under contract #DEAC04-76-DP00789.

results and wavecode calculations with a simple maximum-stress fracture criterion, indicating the importance of cumulative damage effects over the time regime studied. Preliminary numerical simulations with a cumulative damage model produce qualitatively the features observed experimentally (reduced amplitude of the ringing signal).

EXPERIMENTAL TECHNIQUE

The aluminum alloys studied in the present experiments were 6061-T6, which is a precipitation-hardened material containing about 1% magnesium, and annealed AMg-6, which contains approximately 6% magnesium, with minor amounts of other metallic constituents. The initial densities of the materials studied were 2.69 Mg/m^3 for 6061-T6 and 2.64 Mg/m^3 for AMg-6.

Symmetric plate impact, using either a 100-mm gas gun or an 89-mm propellant gun[2], was employed in the study of spallation in these samples. The impactor thickness was about half of that for the specimen in all experiments to ensure that the interaction of release waves occurred within the specimens. The thicknesses of both impactor and sample were varied in order to study the effects of pulse duration (defined as the double shock transit time in the impactor) or strain rate on spall strength. Impact velocity was also varied to study the effect of shock stress on spall strength. The back-surface particle velocity was determined with VISAR laser interferometry techniques[3], and projectile velocity was measured with shorting pins to an accuracy of about 0.2%.

RESULTS AND DISCUSSION

The present data can be used to address several important topics concerning the spallation strength of these two alloys, including (1) the similarity of spall strength for the two materials, (2) the inapplicability of rate-independent spall theories over the time regime studied, (3) the dependence of spall strength on stress amplitude, and (4) the dependence of spall strength on pulse duration or, equivalently, strain rate. Experimental results are presented in the following figures to evaluate these points.

Figure 1 presents the particle velocity profiles for each of the experiments, along with the technique used to determine the spall strength. Romanchenko and Stepanov[4] show that the change in free surface velocity, u_{fs}, (shown for Exp. AMG6-2 as the peak at A to the first valley, B) can be used to estimate the spall strength, using the relation $\sigma = \frac{1}{2}\rho_o c_e u_{fs}$, where ρ_o is the initial density and c_e is an effective wave velocity[4], assumed to be constant. Although the elastic velocity is usually used in the above equation, Romanchenko shows that a better approximation is a combination of the elastic longitudinal and bulk velocities, $2(c_l c_o)/(c_l + c_o)$. The spall strengths determined from this relation are presented in Table I.

In Table I, the experiment AMG6-9 shows an excessive degree of impact tilt. It is presently uncertain how this might distort the spallation data. Therefore, the results of this experiment will be neglected in discussing the trend of the data with strain rate and impact stress.

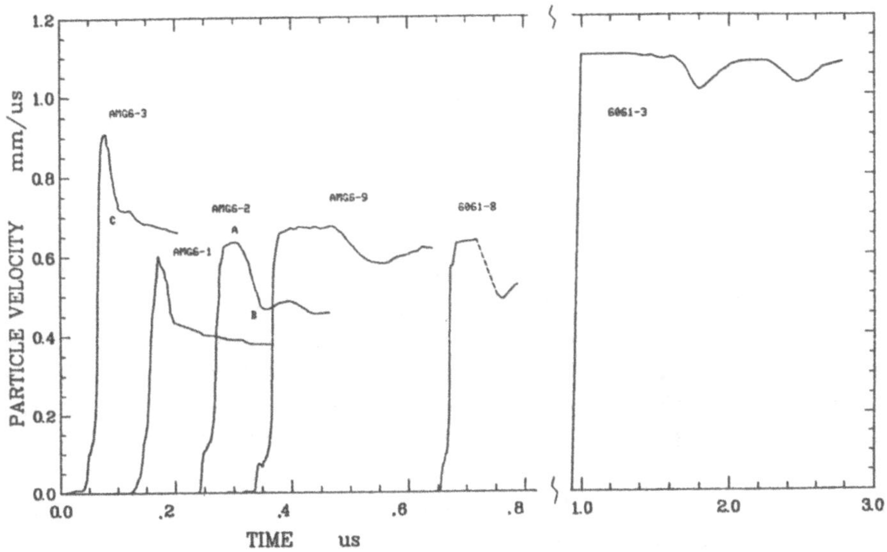

Figure 1: Free surface velocity records for AMg-6 and 6061-T6 aluminum alloys. The dashed line for Exp. 6061-8 is estimated. The free surface velocity for experiment 6061-3 has been reduced by a factor of 2 to allow plotting with the other profiles.

Table I. Experimental Results

Exp.[a] No.	Impact Vel. (km/s)	Impactor Thick. (mm)	Sample Thick. (mm)	Impact[b] Stress (GPa)	Impact Tilt (mrad)	Pulse[c] Duration (μsec)	Pullback[d] Vel. (km/s)	Spall Strength (GPa)	Strain[e] Rate (μs^{-1})	HEL (GPa)
AMG6-1	0.628	0.162	0.319	4.71	0.2	0.057	0.17	1.30	0.81	1.03
AMG6-2	0.621	0.309	0.613	4.66	1.4	0.109	0.17	1.30	0.42	0.93
AMG6-3	0.893	0.167	0.327	6.92	0.5	0.057	0.18	1.38	0.89	0.80
AMG6-9	0.716	0.544	1.260	5.56	4.5	0.189	0.09	0.70	0.15	0.62
6061-3	2.204	3.111	6.335	20.3	0.7	0.904	0.16	1.22	0.31	—
6061-8	0.658	0.301	0.600	5.07	0.3	0.105	0.15	1.15	(0.4)[f]	—

a. Symmetric impact.
b. The Hugoniot relationships used were:
6061-T6: U = 5.26 + 1.47 u, km/s
AMG-6: U = 5.25 + 1.40 u, km/s
c. Estimated from the shock transit time in the impactor.
d. Initial minimum after unloading.
e. Determined from unloading rate.
f. Estimated.

Figure 1 illustrates the major trends in the wave profiles observed for the spallation behavior of these two materials. For the short duration experiments (*e.g.*, Exp. AMG6-3), the free-surface velocity records show that ringing of stress waves in the spalled piece is suppressed and a well-defined minimum does not exist. As the pulse duration increases, the evidence for a minimum in the pullback signal becomes apparent. For Exp. 6061-3, the pulse duration is about 0.9 μs and ringing is clearly evident. When a definite minimum could not be defined, the pullback amplitude was determined as the initial pullback from the peak free surface velocity (point C).

Although not reported in this paper, numerical simulations of the experiments have been initiated to address questions of rate-dependent spallation at the high rates achieved in these two materials. These results will be reported in a forthcoming paper, although the main features will be mentioned here. A major discrepancy between the experimental results and numerical simulations using a maximum tension criterion is the absence of ringing in the spalled piece for the thinner samples. The rate-independent calculation results in several large-amplitude wave reverberations within the spall piece, which are not observed experimentally. The ringing is caused by residual stress gradients trapped in the spall piece after spallation. These experimental results indicate that the rate of damage growth in the spall plane is comparable to the tensile loading rate for thin samples. As a result, the residual stress gradients between the spall plane and the free surface are greatly reduced at the time of complete separation. Such effects have been observed previously[1], although enhanced behavior is observed in the present experiments because of the short pulse durations.

The spall strengths of both materials are plotted versus strain rate in Figure 2. The tensile strain rate at the time of fracture, which is plotted in the figure, was estimated from the free-surface velocity records, using the relation $\dot{\epsilon} = \frac{1}{2}\dot{u}_{fs}/c_o$.

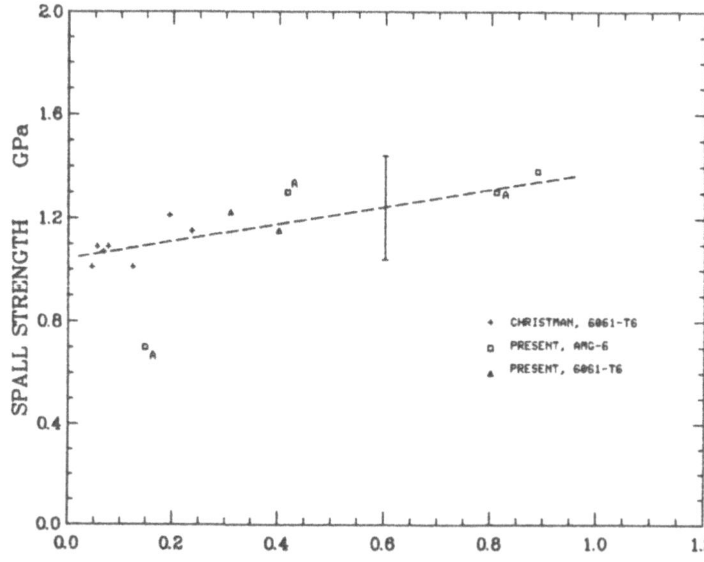

Figure 2: Spall Strength versus tensile strain rate in AMg-6 and 6061-T6 aluminum alloys.

This approximation does not account for the complex wave interactions occurring during the fracture process or for wave propagation effects, which will cause the strain rate appropriate to the spallation zone to vary with propagation distance.

As illustrated in Figure 2, the spall strength increases with increased strain rate. The increase appears linear, although the line shown does not represent a statistical fit to the data. No attempt has been made in the figure to plot the data versus strain rate at a constant stress, because of the limited data available. However, the data identified by the letter A were all obtained for impact stresses of about 5 GPa. The results also indicate that the spall strengths of 6061-T6 and AMg-6 agree to within experimental uncertainty and that the present data on 6061-T6 are in good agreement with the previous work of Christman, et al.[5] (which were obtained over the stress range of 1.4-2.6 GPa). Similar comparisons cannot be made for AMg-6, since time-resolved spallation experiments have not been performed on this material. However, the present results are in overall agreement with previous estimates of the spall strength of AMg-6, based on shock recovery experiments[6,7].

Figure 3 indicates virtually no dependence of spall strength versus impact stress for the two materials, although experimental uncertainties and varying strain rates obscure the conclusion. The two sets of data marked by the letters A and B represent strain rates of about 0.85 μs^{-1} and 0.35 μs^{-1}, respectively. To within experimental error, these data have the same spall strength, indicating no dependence on Hugoniot stress.

Additional experiments are planned for a wider range of both impact stress and strain rate in order to define these relationships more explicitly.

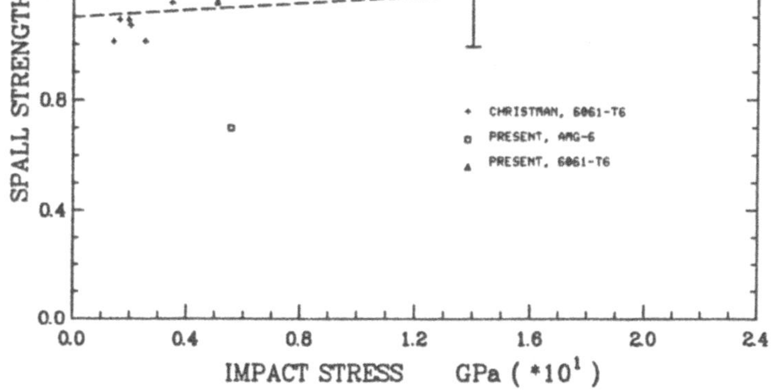

Figure 3: Spall strength versus Hugoniot stress in AMg-6 and 6061-T6 aluminum alloys.

CONCLUSION

Spallation experiments were conducted on 6061-T6 and AMg-6 aluminum alloys over the stress range of 5-20 GPa and for pulse durations of about 0.06-0.9 μs. These results indicate that: (1) rate-dependent models of spallation, based on maximum failure stress, are in significant disagreement with experimental data; (2) the spall strengths of these two alloys are similar; (3) the spall strength is strain-rate sensitive over the range studied; and (4) the spall strength appears to be independent of Hugoniot stress, although more data are required to define this relationship more explicitly. Finally, theoretical efforts are in progress[8] to apply damage-evolution models[9] to describe the present results.

REFERENCES

1. Lee Davison and R. A. Graham, Shock Compression of Solids, in Physics Reports 55, 257 (1979).
2. J. R. Asay, The Sandia National Laboratories Shock Thermodynamics Applied Research Facility," SAND81-1901 (1981).
3. L. M. Barker and R. E. Hollenbach, J. Appl. Phys., 43, 4669 (1972).
4. V. I. Romanchenko and G. V. Stepanov, J. Appl. Mech, and Tech. Phys., 21, 555 (1980).
5. D. R. Christman, W. M. Isbell, S. G. Babcock, A. R. McMillan, and S. J. Green, DNA Report DASA 2501-3, AD735966 (1971).
6. B. A. Tarasov, J. Mech. Tech. Phys., 14, 850 (1973).
7. V. K. Golubev, S. A. Novikov, V. A. Sinitsyn, and Uy S., Sovolev, J. Appl. Mech. and Tech. Phys., 21, 433 (1980).
8. J. E. Dunn, Sandia National Laboratories, private communication.
9. L. Davison, A. L. Stevens, and M. E. Kipp, J. Mech Phys. Solids, 25, 11 (1977).

SPALL ZONES IN POLYCRYSTALLINE CERAMICS

S.J. Bless, D. Yaziv and Z. Rosenberg

University of Dayton Research Institute
300 College Pk.
Dayton, OH 45469

ABSTRACT

The nature of spall in ceramics was studied by plate impact experiments using VISAR and manganin gauge measurements and a new recovery technique. Three polycrystalline ceramics were included: 97.7% alumina, 85% alumina, and 99.9% beryllia. Spall strengths were determined to be a few kbar below and negligible above the HEL. Analysis of spall signals and recovered targets shows that spall occurs over finite regions. A "spall zone" model is proposed.

MATERIALS

The materials investigated were WESGO alumina Al-300, Coors Porcelain alumina AD-85, and Ceradyne beryllia Ceralloy 418. Density, sound speed, elastic constants, and metallographic measurements were conducted on all materials. Measured properties are summarized in Table 1. (In this table, ρ is density, C_L is longitudinal sound speed, E is Young's modulus, and ν is Poisson's ratio.) In addition, the Al-300 was sent to Coors for a chemical analysis, which was within specifications. The porosity of the BeO targets was 0.6%. Targets were also checked ultrasonically for major flaws.

Table 1. Properties of Ceramics

Material	ρ g/cm^3	C_L mm/μs	E Kbar	ν	Average Grain size (μm)
Al-300	3.750	10.20	3172	0.21	20
AD-85	3.421	8.80	2207	0.20	5
BeO	2.99	11.44	NM	NM	NM

MEASUREMENTS OF SPALL STRESS

A 50-mm gun was used to launch 50-mm diameter metal flyer plates at ceramic targets. Impact velocity was measured to within 1% by contact pins.

Tilt was less that 0.05°. Targets were usually backed with PMMA windows.

Measurements were made at the target window interface with a VISAR or with manganin gauges[1-2]. The UDRI VISAR-data analysis program was modified using the usual window corrections[3]. The rear surfaces of the windows were cut at a wedge angle to avoid specular reflections into the VISAR.

A typical experiment is diagrammed in Figure 1; (a) shows the x,t diagram, (b) illustrates the appearance of the interface velocity history, and (c) shows the impedance match solution. The impact produces only a single shock wave (1) which reflects as release waves at (2) and (3). Reflection of the copper release wave from the alumina surface causes the stress to increase at (4). Intersection of the crossed release waves in the ceramic should lower the stress to (5"). However, the spall stress is (5'); the material rebounds from (5') to (5) when spall occurs, which shows up as a transition from (6') to (6) at the PMMA interface. As shown, for example, by Yaziv[4], the spall stress in the target is given by:

$$\sigma_s = (Z_{ceramic} + Z_{PMMA})\, u_{SI}/2 \qquad (1)$$

Here, $Z = \rho U$, is the shock impedance (U being shock velocity). The time interval between the arrivals at (6) and (7) can be used to determine the distance between the target rear surface and the spall, as $c_L(t_7 - t_6)/2$.

The shock impedances for loading and unloading in all three materials was determined for peak stresses below and just above the HEL. Detailed results are reported by Yaziv[4]; here we summarize: For Al-300, the HEL was 80 ± 1 kbar for 6.5-mm thick samples. Above the HEL, the Hugoniot was found to be U = 7.44 + 1.15u (where u is the particle velocity). Release was elastic below the HEL. The HEL of AD-85 was 59 ± 2 kbar for 6.4-mm samples. The HEL decayed in thicker samples. Above the HEL, the Hugoniot was described by U = 6.62 + 0.88u.

Determination of properties of BeO was complicated by the variable porosity. The Hugoniot of zero-porosity BeO was derived and is described elsewhere[4-5]. The HEL of the 0.6% porosity material was 95 kbar for 5-mm samples.

Table 2 lists the shots in which spall stress was determined. The spall stresses below the HELs were small but easily detectable. For shocks below 30 kbar, the spall strength of Al-300 was about 4 kbar, and of AD-85, about 3 kbar (in agreement with Rosenberg[6]). The spall strength of BeO shocked to 85 kbar was only 0.8 kbar. Spall strength decreased as the HEL was approached in these materials, and spall signals were not observed for shots above the HEL.

The spall mechanism also appeared different for shocks above the HEL. Figure 2 shows data from Shot 853 (AD-85 shocked to 103 kbar). The "step down" expected in the reverberation signal (as in Figure 1) did not occur. Instead, there was a relatively late "step up" caused by reflection from the copper flyer plate. This arrival indicates that there is no distinct spall boundary in the target, and that the shock velocity in the target is considerably less than its normal value. It is deduced that fracture above the HEL occurs during compression, as discussed below.

RECOVERY EXPERIMENTS

It was desired to recover spalled targets to aid interpretation of the spall signals. Star shaped flyer plates[6-7] were used in order to reduce

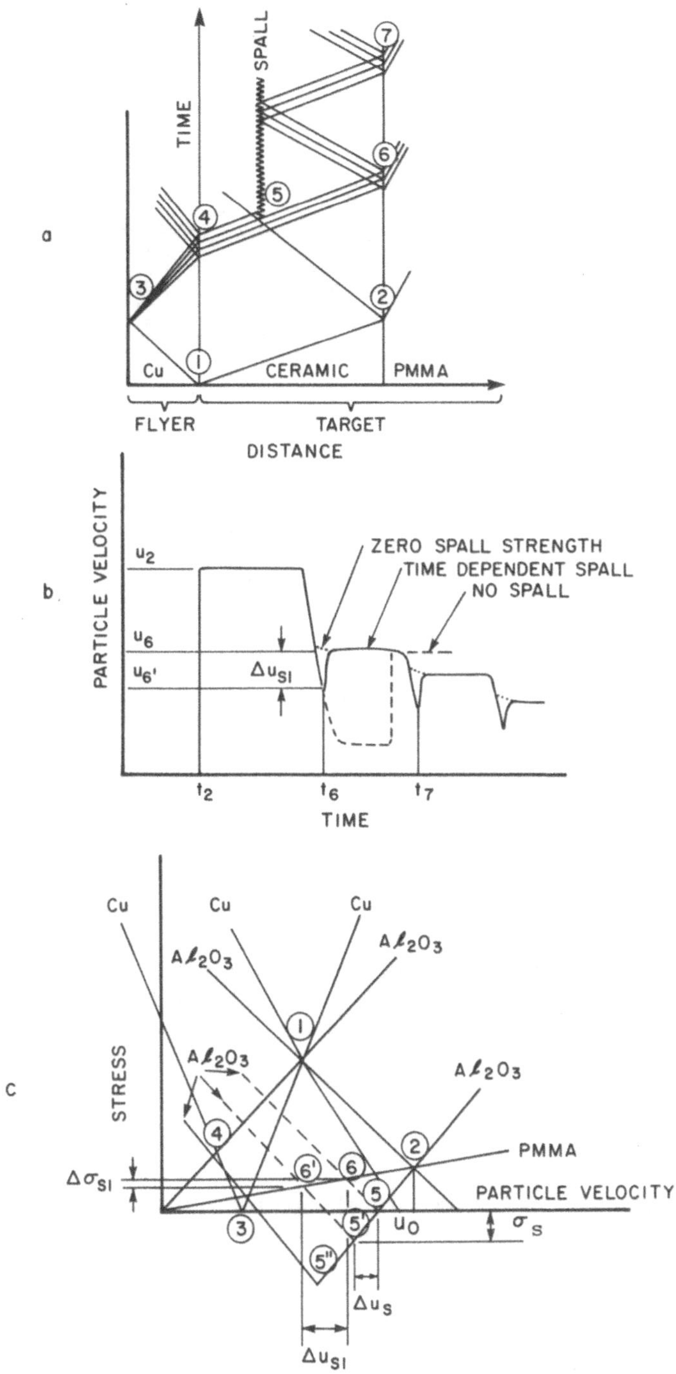

Fig. 1. Spall in PMMA backed target (see text).

Table 2. Spall Experiments

Shot No.	Flyer Material	Flyer Thickness (mm)	Sample Alumina	Sample Thickness (mm)	Impact Velocity (mm/μs)	Exp. Tech.	Shock Stress (Kbar)	Spall Strength (Kbar)	Spall Position Δx (mm)
817	Al	2	AD-85	6.4	0.119	M	12	2.75	----
782	Al	2.03	AD-85	6.4	0.156	V	15	3.3	----
861	Al	0.9	AD-85	6.4	0.198	V	20.5	2.75	0.25
796	Al	1.95	AD-85	6.4	0.203	V	20.5	3.2	0.65
798	Al	1.90	AD-85	6.45	0.253	V	26	----	0.8
784	Al	2.03	AD-85	6.4	0.282	V	29.5	2.8	0.85
831	Al	2	AD-85	6.4	0.293	V	31	3.0	0.85
823	Al	2	AD-85	6.4	0.335	M	31	----	1.05
802	Cu	1.95	AD-85	6.4	0.345	V	56	1.0	----
853	Cu	3	AD-85	6.4	0.805	M	103	0	0.25
792	Al	2.03	Al-300	6.1	0.105	V	11.5	4.2	----
820	Al	2	Al-300	6.9	0.201	M	22.5	3.9	0.8
790	Al	2.03	Al-300	6.1	0.251	V	28.5	----	1.15
705	Cu	2.03	Al-300	6.5	0.292	V	54	1.6	0.55
710	Cu	2.06	Al-300	6.5	0.310	V	57.5	0	----
806	Cu	1.95	Al-300	6.3	0.315	V	58	0	----
711	Cu	2.03	Al-300	6.5	0.396	V	71.5	0	----
730	Cu	1	BeO	4	0.477	V	85	0.8	1.1
731	Cu	1	BeO	8	0.512	V	89.5	0	----
732	Cu	1	BeO	4	0.635	V	106	0	----

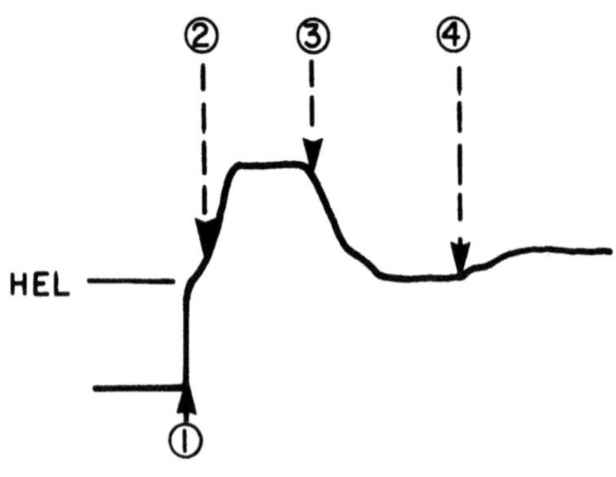

Fig. 2. Tracing of manganin gauge record, Shot 853 (1-elastic shock; 2-plastic shock; 3-first release; 4-compression from flyer.

breakup of target material from lateral release waves in the flyer plate. Lateral release waves in the target were suppressed with a tight circumferential clamp.

Ordinarily, recovery experiments are conducted with a momentum trap that carries away the flyer plate momentum, but this would have prevented spall. Therefore, targets were designed so that the momentum in the spall plate (defined as the distance from spall initiation to the target rear surface) was the same as the momentum in the flyer plate. We have called this a "self trap" technique[4].

Successful recovery experiments were conducted with 12.4-mm AD-85 targets (without windows) struck at almost 100 m/s. Figure 3 illustrates the appearance of these targets after the impact.

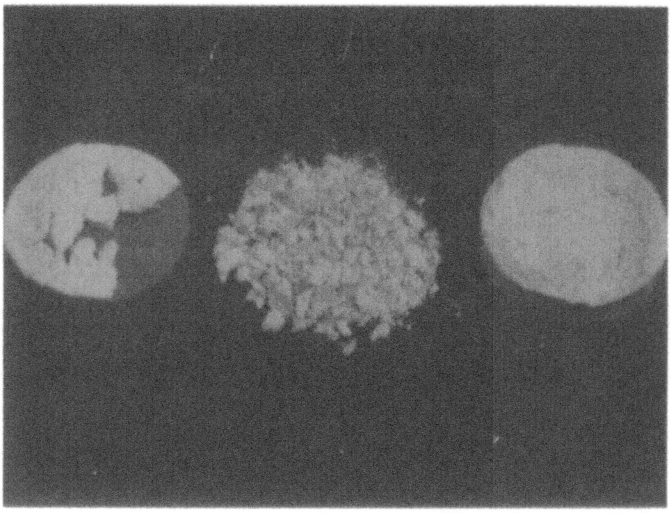

Fig. 3. Recovered AD-85 target.

SPALL ZONES

It can be deduced from the dynamic records and the recovered targets that spall below the HEL occurs over a zone, not in a plane. Spall initiates at x_5 (Figure 1), which is determined by time t_6. The ringing period of the spall signal determines the location of the spall arrest, x_a. The difference $\Delta x = x_a - x_5$ is the thickness of the back section of the spall zone. The ringing period is observed to remain constant, indicating that once the spall is arrested, it does not propagate further. Observed values of Δx are also listed in Table 2. The dynamic data do not define the location of the front half of the spall zone, but measurements on one recovered target show it is about 3-mm.

This observation of spall zones raises the theoretical questions: Why does the spall propagate, and why does it arrest? In metals, $|\sigma_s|>|\sigma_{HEL}|$. When spall develops, a compression wave is generated that propagates at the elastic wavespeed. The release wave that initiated spall travels at a plastic velocity. The compression wave quickly overtakes the release wave, and the spall is quenched. In ceramics, $|\sigma_s|<|\sigma_{HEL}|$ and all waves have approximately the same elastic velocity. Figure 4 illustrates this situation. Here the spall strength is exceeded at time t_s, and the spall occurs after a short incubation time, Δt_i, after which cracks nucleate and start to grow. However, the compression wave that originates after time Δt_i does not overtake the release waves that created the tension. The shaded areas in Figure 4 all experience the same tension; the spall propagates in both directions, and a spall zone is generated.

Unlike the case in ductile metals, fracture of the ceramic involves propagation of brittle cracks that initiate at the spall nucleation sites. The speed of crack propagation is relatively slow, and the propagating cracks are arrested by later incoming compression waves (Figure 4). Hence, the comminuted material is limited to a spall zone. The propagation velocity of the spall cracks was inferred from the spall initiation times and width measurements. In AD-85, it was found to be 25 to 60% of c_L, and in Al-300, 20 to 30% of c_L.

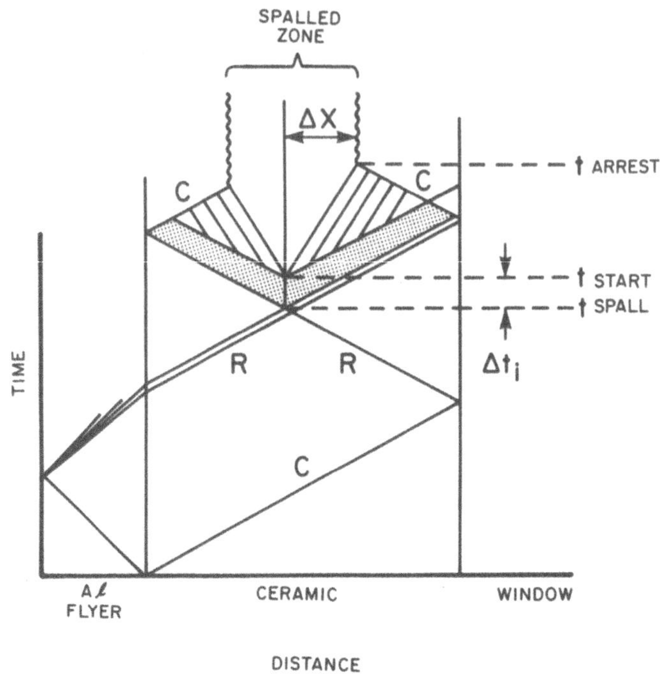

Fig. 4. Development and quenching of spall zones

ACKNOWLEDGEMENTS

These experiments were conducted with the assistance of Mark Laber, Range Technician in the UDRI Impact Physics Laboratory. This work was supported by the U.S. Air Force under contract F33615-81-C-5082. Dan Yaziv also acknowledges support from the Government of Israel while he completed his Ph.D studies.

REFERENCES

1. Z. Rosenberg, D. Yaziv, Y. Partom, 1980, J. Appl. Phys, 3702.
2. D. Yaziv, Z. Rosenberg, Y. Partom, 1980, J. Appl. Phys. 51. 6055.
3. L.M. Baker, R.E. Hollenback, 1972, J. Appl. Phys. 43, 4669.
4. D. Yaziv, Ph.D Thesis, University of Dayton, June 1985.
5. D. Yaziv, S.J. Bless, D. Dandekar, to be presented at EXPLOMET 85: INT'L CONF. METALLURGICAL APPLICATIONS OF SHOCK WAVE AND HIGH-STRAIN-RATE PHENOMENA, Portland, OR July 1985.
6. Z. Rosenberg, Y. Yeshurun, D.G. Brandon, to be presented at INT'L CONF. MECHANICAL AND PHYSICAL BEHAVIOR OF MATERIALS UNDER DYNAMIC LOADING.
7. P. Kumar, R.J. Clifton, 1972, J. Appl. Phys. 50, 4747.
8. J.E. Vorthman, G.E. Duvall, 1982, J. Appl. Phys. 53, 3607.

SHOCK FRACTURE AND RECOMPACTION OF CERAMICS

D. Yaziv, S.J. Bless, Z. Rosenberg and D. Jurick

University of Dayton Research Institute
300 College Pk.
Dayton, OH 45469

ABSTRACT

A new double impact technique for studying the dynamic properties of shock damaged materials is presented. Two flyer plates are separated by a small gap. The first plate causes spall in the target. The second plate recompacts the material and allows determination of the properties of the damaged zone. Two polycrystalline commercial aluminas were studied using VISAR and manganin gauges. A theoretical model that treats the spall zone homogeneously was used in a modified SWAP code.

INTRODUCTION

Spall fracture in metals caused by plate impact has been extensively studied. Analysis of spall signals and of recovered targets has led to several analytical models for accumulation of damage that culminates in complete spall rupture. However, spall fracture in highly brittle materials, such as ceramics, has been less thoroughly investigated[1-4], and a general model of dynamic fracture in highly brittle materials has not yet been developed. Understanding of dynamic fracture of brittle materials has lagged behind that of metals chiefly because of the relative difficulty in recovering brittle targets for post-impact inspection. The double flyer plate technique that we have developed circumvents this problem.

EXPERIMENTAL TECHNIQUE

The basic plate impact technique, VISAR and manganin gauge measurements are described in a companion paper[5]. The experimental set up is shown in Figure 1. The materials were Wesgo Al-300 and Coors AD-85 alumina. Physical and shock properties were measured and are also reported in our companion paper[5].

Conventional plate impact techniques provide little information about the properties of the damage once spall fracture occurs. In order to obtain more data, two flyer plates were used, separated by a small gap. The second flyer plate has a higher shock impedance than the first. Impact of the first flyer plate induces spall damage in the target. Impact of the second flyer plate causes a higher stress than the initial impact. The shock wave from the second impact recompacts the spall-damaged target material.

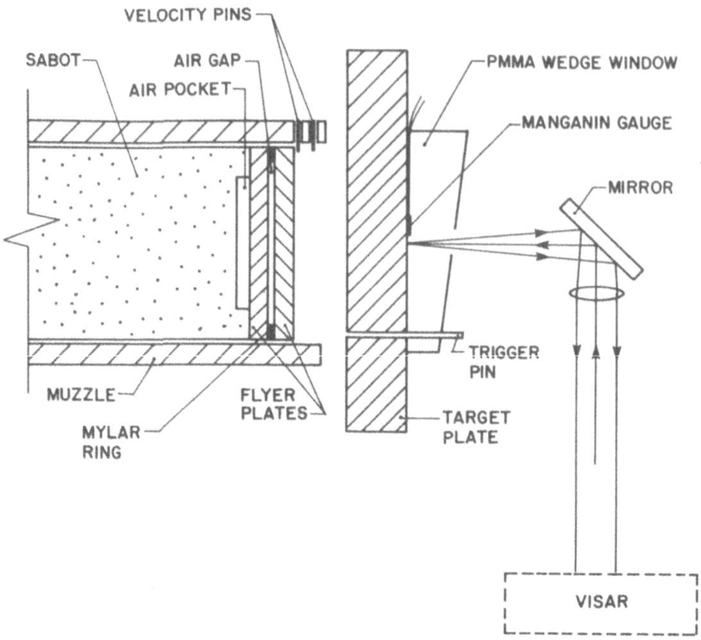

Fig. 1. Experimental set up: The double impact technique together with VISAR and Maganin gauge.

Development shots were conducted on metal targets, since it is relatively easy to determine the damage from a post-impact metallographic analysis (preliminary results from these tests were reported in Reference 6). The first and second flyer plates were aluminun and copper, respectively. The targets were OFHC copper. The gaps between the two flyer plates were controlled by a thin ring made of mylar (Fig.1). The air gap was 0.13 or 0.25 mm. The rear surface velocity of the target was measured with the VISAR. The targets were recovered, cut and examined after impact. The impacts were also simulated with the SWAP 9[7] method of characteristics code using a simple spall stress criterion.

In order to demonstrate the action of the second flyer plate, we performed experiments with dual impactors. In these shots, only half of the first aluminum plate was backed by a second copper flyer, the other half was air backed. Figure 2 shows post impact photographs of targets. The two halves of the target display different damage, as predicted by the SWAP simulation. The single flyer plate impact produced partial spall (a region containing porosity), while no damage is readily visible opposite the two-flyer plate impact. Following the first impact, when the shock wave from the second flyer impact is transmitted to the target, the spall is closed (at the lower section) and a new spall starts to develop, as predicted by the code.

In order to confirm the transient behavior, the VISAR was used to determine the free surface velocity of the portion of the target struck by the two flyer plates. The free surface velocity record from Shot 719 is shown in Figure 3. Superimposed on the conventional "spall signal" (starting at point 3), we see the compression wave from the second flyer impact (point 4) that has "leaked" through the damaged region.

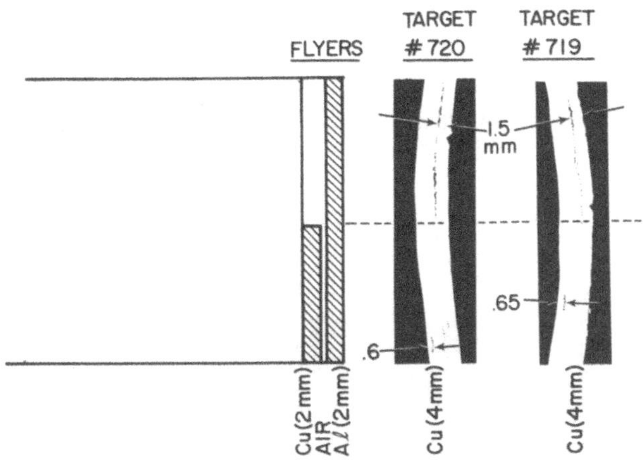

Fig. 2. Double Impact. Photographs of two post-shot copper targets.

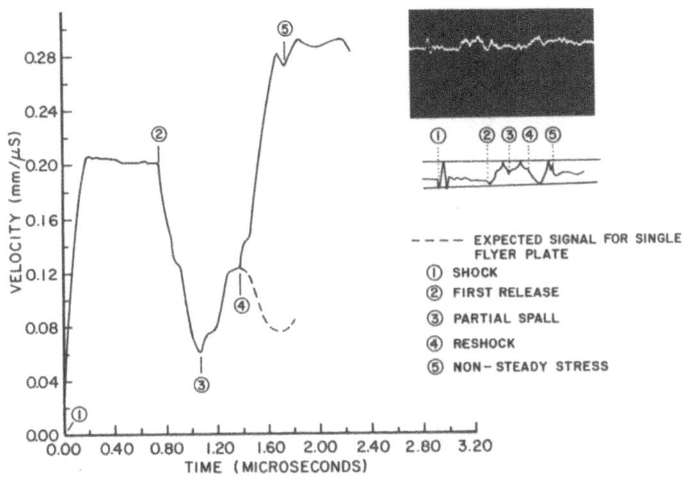

Fig. 3. VISAR data of double impact shot on a copper target (Shot 719).

EXPERIMENTS WITH CERAMICS

The double impact technique was applied to ceramic targets. The alumina Al-300 targets were 5.6 to 6.9 mm thick and the alumina AD-85 was 6.35 and 6.5 mm thick. Most of the flyers were aluminum plates followed by copper plates. Since aluminum has lower impedance than the target materials, the flyer plate rebounds from the target after impact. This causes a sharply falling release wave in the ceramic targets. The gap that develops between the flyer plate and the target causes an additional calculable delay in the arrival of the reshock wave from the second flyer. In some shots we used a copper first plate followed by a tungsten second plate or copper second plate. The thickness of the first flyer was around 2 mm, and the thickness of the second flyer was 2 to 4 mm. The impact velocities were 100 to 360 m/s. All the stresses involved were below the HEL of the target materials. Since the spall strength of alumina is 0.4 GPa or less[1,2,8], spall occurred in all the shots. A typical manganin gauge record of the stress history at the interface of Al-300 target is shown in Figure 4. A typical

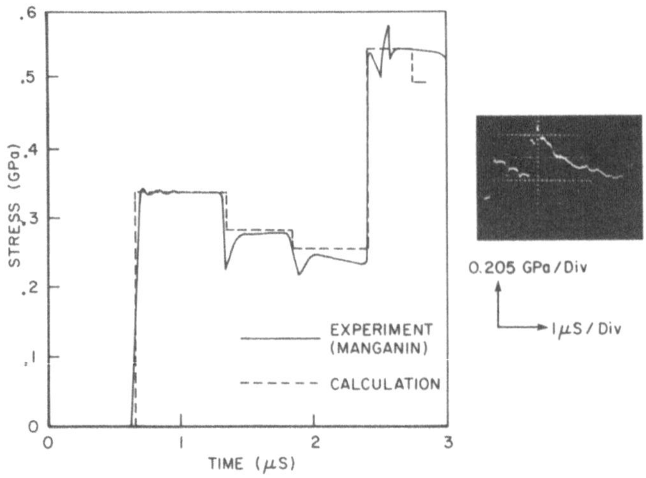

Fig. 4. Measured and calculated stress history at the interface of Al-300 Alumina (Shot 820).

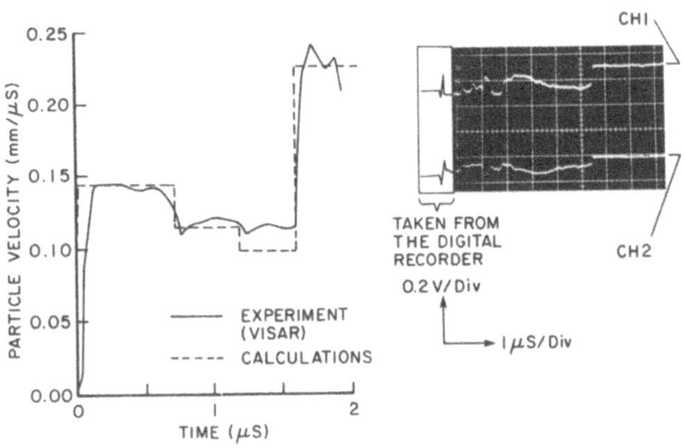

Fig. 5. Measured and calculated stress history at the interface of AD-85 alumina (Shot 790).

VISAR record of the velocity history at the interface of an AD-85 target is shown in Figure 5. In these figures, we observed shock and release caused by the first flyer, followed by the well-known ringing signal of the spalled target. The impact of the second flyer is manifested by a second compression wave.

THEORETICAL MODEL

In our companion paper[5], we observed that when tensile fracture occurs in a ceramic plate that has been shocked below its elastic limit, spall occurs not at a plate but in a finite zone. The double flyer technique enables us to study the dynamic response of the spalled zone.

The SWAP code was used to compute the stress and particle velocity histories. The goal was to simulate the timing and the level of the recompaction wave in order to characterize the average properties of the spall zone. It was necessary to modify the code in order to introduce time-dependent material properties. To do this, we stopped the calculation when the spall zone was established. The original material in this zone was replaced by a "new" material with different properties. Then we restart the calculations keeping most of the previous states and waves of all the other materials.

We assumed that the elastic Hugoniot for tension is symmetric to the elastic Hugoniot for compression, as shown in Figure 6. Spall occurs at $\sigma = -\sigma_s$ and $V = V_{oo}$. After spall, the stress jumps to $\sigma = 0$, but the specific volume remains at V_{oo}. This model satisfies conservation of mass. After spall, the damaged material contains a pore volume given by $V_{oo} - V_o$. The transformation is calculated from:

$$\rho_{oo} = \frac{\rho_o}{1-\varepsilon}$$

where ε is the strain, ρ_o and ρ_{oo} are the initial densities of the original and the spalled materials.

Once the spall zone was established, we restart the calculations and treat it homogeniously. The restart time is shown in Figure 7, the shaded zone is the spalled material (having the new properties). The "new" material is governed by the Hugoniot equations and by the SWAP formulation. The spalled zone restarts with zero stress, as do the neighbor zones to the left and to the right of the spall zone (signed N in Figure 7). The particle velocity of the spall zone and also the neighbor zones have the average velocity that the two neighbors had prior to the transformation. For the Gruneisen parameter of the spalled material, we assumed:

$$\gamma_{oo} = \frac{\gamma_o \rho_o}{\rho_{oo}}$$

where γ_o is the parameter of the original material. We also assumed zero compressed strength and zero spall strength. The two parameters remaining to be evaluated are C_o and S, the two spalled material constants in the shock velocity-particle velocity equation. The width of the spall zone was determined from the experimental record[5] for each shot and was included into the program.

The later structure of the wave train was not analysed or simulated. The instantaneous spall failure criterion used in SWAP for the initiation of the spall (prior to restart time) prevents simulation of the pull back signals observed at the window interface. The profile computed by SWAP has the appearance of rectangular steps.

The calculated recompaction wave profile was made to conform to the measured one by changing the constants C_o and S of the spall zone. Typical fittings of simulation to a manganin record (Shot 820) and to a VISAR (Shot 790) are shown in Figures 4 and 5. The initial densities of the spalled zones (ρ_{oo}) were just 1 to 2 percent lower than the initial densities of the target materials (ρ_o). However, the sound speeds, C_o, are distributed between 2.2 to 3.4 mm/μs. High density and low sound velocity mean that the spall zone is made up of dense material containing large cracks which delay the propagating waves. This conclusion is consistant with what we observed at the cross section of the recovered targets[5].

Calculations of the stress in the spall zone indicate that the recompactions achieve just 80 to 90 percent of the compression in the original material. This result is consistent with our copper-copper flyer (Shot 802) in which the reshock level measured at the window interface was lower than the first shock level. This means that full recompaction was not

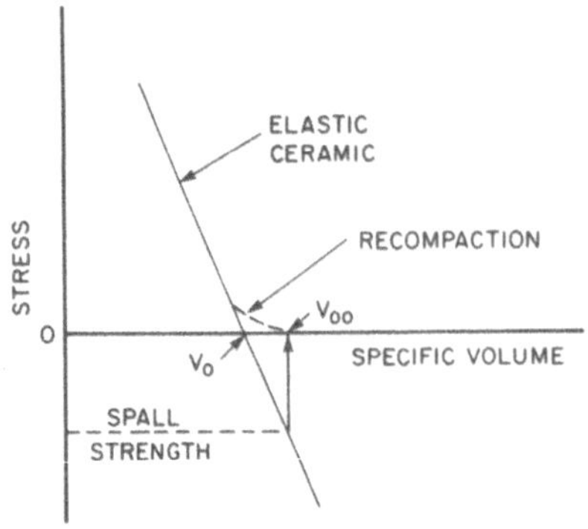

Fig. 6. The physical model for the SWAP calculations.

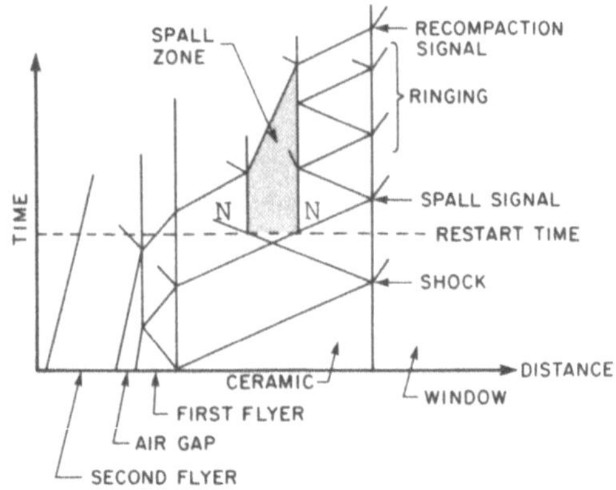

Fig. 7. An x-t diagram for spall and recompaction model.

reached. The non-steady level of the second shock signal (Figures 4,5) reflects the dynamics of the recompaction process of the spalled zone.

ACKNOWLEDGEMENTS

The authors wish to thank M. Laber of the UDRI Impact Physics Laboratory for his valuable assistance. This work was supported by the U.S. Air Force under contract number F33615-82-C-5126. D. Yaziv was also supported by the Government of Israel.

REFERENCES

1. D.E. Munson and R.J. Lawrence, J. Appl. Phys, 50, 6272 (1979).
2. Z. Rosenberg, Y. Yeshurun and D.G. Brandon, to be presented in THE INTERNATIONAL CONFERENCE ON MECHANICAL AND PHYSICAL BEHAVIOR OF MATERIALS UNDER DYNAMIC LOADING, France 1985.
3. D.E. Grady and R.E. Hollenbach, Geophys. Res. Letters, G. 73 (1979).
4. D.A. Shockey, C.F. Peterson, D.R. Curran and J.T. Rosenberg in NEW HORIZONS IN ROCK MECHANICS, ASCE, New York 709-738 (1972).
5. S.J. Bless, D. Yaziv and Z. Rosenberg in the FOURTH APS TOPICAL CONFERENCE ON SHOCK WAVES IN CONDENSED MATTER, Spokane, WA (1985).
6. D. Yaziv and S.J. Bless in PROCEEDINGS OF THE ARMY SYMPOSIUM ON SOLID MECHANICS - 1984, AMMRC AD-A150-042.
7. L.M. Barker and E.G. Young, Sandia Report, SLA-74-0009 (1976).
8. D. Yaziv, Ph.D Thesis, University of Dayton, 1985.

MEASUREMENTS OF COMPRESSIVE AND TENSILE WAVE IN A SHOCK LOADED PYREX GLASS

Y. Yeshurun, G. Rosenberg and Z. Rosenberg

Rafael, A.D.A., P.O.Box 2250, Haifa 31021, Israel

ABSTRACT

Shock loading experiments were conducted in order to determine failure modes of Boro silicate (Pyrex) glass under impact.

Projectiles accelerated by our 2½" gas and powder guns were used to generate 1-D strain in Pyrex discs. Manganin gauges were used to monitor the stress history in the material, from which values of the Hugoniot Elastic Limit (H.E.L) and spall strength were obtained. From our results, a ramping up to 2.7GPa was observed and considerably high H.E.L of 7.8±0.1GPa was found. However an almost zero spall strength was recorded for impact loading over the H.E.L. The above results strongly support our expectation that the Pyrex glass undergoes a complete loss of tensile strength when impacted at stress higher than its H.E.L.

1. INTRODUCTION

The failure mechanizm of Boro silicate glass (Pyrex) under shock loading was previously investigated by Cagnaux [1] who used manganin stress gauges to monitor the longitudinal and deviatoric stress components. His data shows the exsistance of deviatoric stress at compression states above the H.E.L., thus a viscoplastic deformation with Von-Mises yeild condition was suggested. In earlier work by Kanel et al [2] the stress anisottropy of K-8 glass and the cleavage strength of window glass under dynamic loading were measured. For impact stresses above the H.E.L their data shows stress anisotropy that vanishes after unloading and a considerable cleavage strength with some spall indications. These results were considered as fracture of brittle material so that the particles are free to slip, but they are still connected so that the bulk maintains a finite tension strength.

Another point related to the results obtained for the Boro silicate glass [1], is that the HEL increases with driving stress. Such effect previously discussed for <111> loaded LiF by Rosenberg and Duvall(3) who suggested a coulombic fracture mechanizm for it.

In our work recently submitted for publication [5] we shocked Al_2O_3 samples and measured the spall strength. This material has intergranullar glassy phase which we modified by several heat treatments [10]. The impact results [5] clearly show that the spall strength is reduced when the compression shock intensity was increased. We suggested a failure model of microcracks generated by the shock front.

Comparing all the above information together we followed the recomendation given earlier by Cagnaux [1], to conduct spall strength measurements in Pyrex glass.

We used manganin gauges to measure the longitudinal stress and applied the evaluation method described earlier by Grady and Hollenbach [4] to study the failure mode. A small number of shots was sufficient to find that the Pyrex does not generate a spall signal, thus a fracture due to loading over the H.E.L is suggested.

2. EXPERIMENTAL

Planar impact experiments were conducted in our 2.5 inches gas and poweder guns [6]. Impact velocities were between 300 and $1300mS^{-1}$ and measured to 0.5%. A scheme of the impact assembly is shown in Fig. 1.

The manganin gauge (type LM-SS-125CH-048, manufactured by Micro-Mesurements) were calibrated both under loading [7] and unloading [8]. Cauge emplacement and data reduction procedures were identical to those previously described [7,8]. The 5μm thick, grid-like, foil gauges mounted on 0.04mm epoxy sheet, were embedded between specimen and a backing disc of poly (metyl-methacrylate) (PMMA). This configuration used to measure the spall strength and HEL of the Pyrex glass by means of a technique previously described in [9].

Pyrex discs were cut and lapped from rectangular tiles 6.4mm thick. The static modulus of rupture quoted by the manufacturer is 42MPa and the density is 2.23g/cc.

From our ultrasonic measurements a values of C_L=5.6$Kmsec^{-1}$ for the longitudinal wave velocity was recorded. The Young's and shear module, as given by the manufacturer are E=65GPa, G=26GPa respectively, with a Poisson's ratio of ν=0.20.

3. RESULTS

The data recorded in our experiments was used to obtain both Hugoniot data for the Pyrex glass and the fracture strength as well.

The experimental configuration is illustrated in Fig. 1, already explained above. Eleven shots were conducted, nine of which below the HEL and two above it. The stress levels for the main compression waves, together with the impactor impedance and velocity, were used to plot the Hugoniot curve. This Hugoniot for Pyrex glass in given in Fig. 2. Shot numbers are listed by the data points were 2024Aℓ impactors were, used in shots: G-3, 589, 602 4/25, 4/26, 569 and Pyrex impactors were used in all other shots.

The Hugoniot Elastic limit was found as 7.8GPa, also indicated on the curve. This value is an average of precursor amplitudes recorded by the manganin gauges. The elastic impedance which obtained from the Hugoniot agrees with our expectation from previous density and ultrasonic measurements. The value of 7.8±0.1GPa measured for the HEL is also in good agreement with previous results given by Cangnaux [1].

All shots used to determine spall had a configuration with PMMA backing as illustrated in Fig. 1. Evaluation of fracture states was made primarily by determation of the shape of the stress profile at the release waves as earlier explained by Grady and Hollanbach [4]. These shots can be divided into two groups, blow HEL and above it.

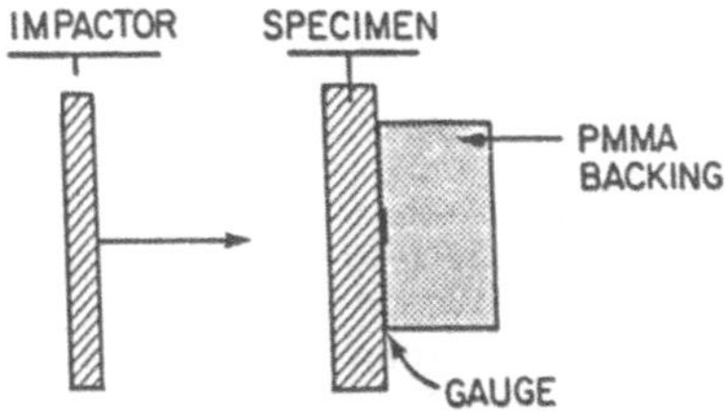

Fig. 1: Scheme for back surface planar impact experiment.

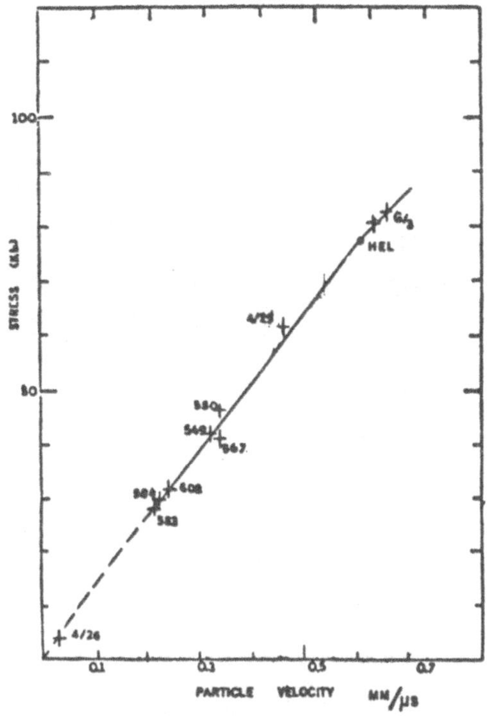

Fig. 2: The measured Hugoniot curve of pyrex glass.

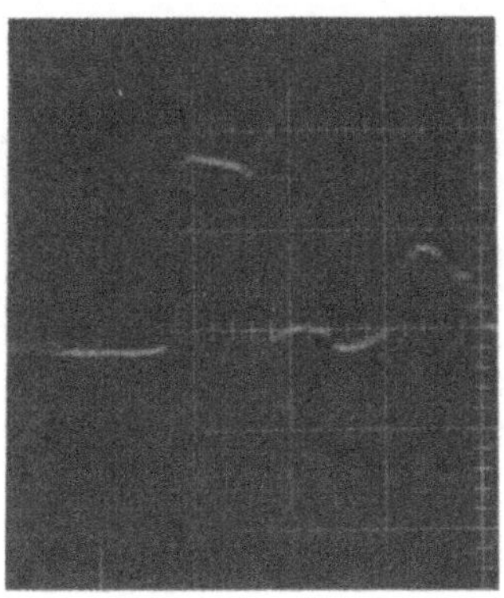

Fig. 3: Typical gauge signal at stress lower then the H.E.L (Shot No.602).

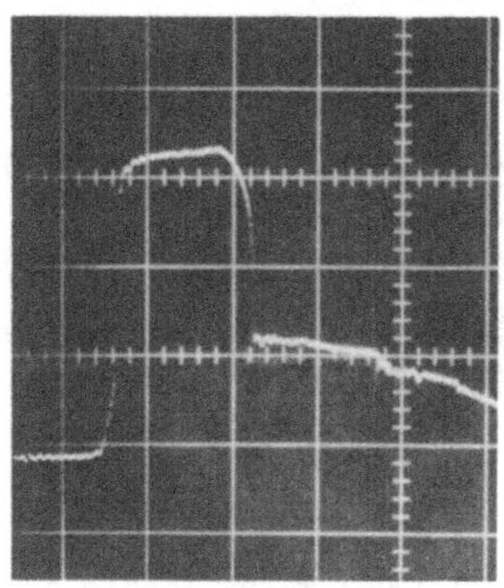

Fig. 4: Typical gauge record at stress level above the H.E.L (Shot No.G/3).

At the lower stress levels, blow the HEL, the signals recorded by the manganin gauges clearly show zero stress amplitude at the release wave. Such typical gauge record is shown in Fig. 3. Following Grady and Hollenbach, a signal like that is a typical example of material in which the tensile stress is insufficient to cause fracture.

None of our signals at the loading levels above HEL shows a second stress rise following the realse waves (Pull Back Signal). A typical gauge record is given in Fig. 4.

Using the same evaluation method it clearly indicates a state of instantaneous fracture. It means that this material, when compressed over it's H.E.L, goes through a failure which do not allow any tensile strength.

4. SUMMARY AND DISCUSSION

The dynamic response of Pyrex glass was determined using implanted manganin specimen - PMMA interface and enabled the HEL and spall strength to be determined.

We found that there is no spall in impact pressure below the dynamic yield strength (HEL), which means that this Pyrex glass can sustain a considerable tensile stress after such loading.

On the other hand a compression wave above the HEL reduces the spall strength almost to zero, which is a result of instantanous fracture.

Comparing these results with our previous work on Al_2O_3 which contained glassy phase, we intend to believe that the changes in the spall strength of brittle materials like these are due to microcracking which occurs during the dynamic yielding. To confirm this assumptions, we intend to perform recovery shots with Scaning Electron Microscopty (S.E.M) observation.

ACKNOWLEDGEMENTS

The authors wish to thank Mr. R. Krief and Mr. D. Mazar for their thechnical asistance in performing the shots.

REFERENCES

[1] J. Cagnoux, Shock waves in condensed matter, Ed. J. Nellis, L. Seaman, R. Graham, p. 392, (1981).

[2] G. Kanel, A.M. Molodest and A.N. Dermin, Comb. Explos. Shock Waves 13, 772 (1977).

[3] G. Rosenberg and G.E. Duvall, J. Appl. Phys 51 (1) (1980).

[4] D.E. Grady and R.E. Hollenbach, Geo. Res. Lett. Vol. 6, No. 2, p. 73 (1979).

[5] Z. Rosenberg, Y. Yeshurun and D.G. Brandon, Submitted to Dymat. Conf. Paris 2-5 Sept. (1985).

[6] G. Rosenberg, Rafael Terminal Ballistic Laboratory, AiP Proc. #78 on Shock Waves (Menio park) (1981).

[7] Z. Rosenberg, D. Yaziv and Y. Partom, J. Appl. Phys. 51, 3703 (1980).

[8] D. Yaziv, Z. Rosenberg and Y. Partom, J. Appl. Phys. 51, 6055 (1980).

[9] Z. Rosenberg, G. Lattwak, Y. Yeshurun and Y. Partom, J. Appl. Phys. 54, 2147 (1983).

[10] Y. Yeshurun, Z. Rosenberg, D.G. Brandon, N. Travitzky to be Published in Mater. Sci. Eng. 71 (1985).

AN APPLICATION OF GEOMETRIC STATISTICS TO DYNAMIC FRAGMENTATION*

M. E. Kipp and D. E. Grady

Solid Dynamics Department
Sandia National Laboratories
Albuquerque, NM

ABSTRACT

One-dimensional fragmentation, viewed as a random Poisson process, results in an exponential distribution of particle sizes which compares well with expanding ring fragmentation data. Extending geometric concepts of random Poisson processes to areas and volumes results in theoretical distributions that are construction dependent, although there are indications that an exponential distribution is applicable. If we represent homogeneous fragmentation processes with the exponential distribution, then this distribution can be used as a basis to describe statistical heterogeneity in more complex fragmentation processes.

INTRODUCTION

The present study is focused on achieving a better understanding of the distributions in particle size which result from catastrophic fragmentation of a solid body. A substantial literature exists in which the statistical and physical principles underlying the processes of dynamic fragmentation have been explored (See the review in Grady and Kipp[1]). Studies range from purely geometric statistical methods to concepts involving physically founded laws governing nucleation, growth and coalescence of fractures driven by stress wave propagation.

In the present work we consider further the problem of geometric statistical fragmentation —namely the random partitioning of a line, a surface, or a volume. The limitations of such constructions should be kept clearly in mind. In actual dynamic fragmentation events these partitions correspond to growing, propagating, bifurcating, stress relieving, energy consuming, and interacting cracks and fractures. The apparent simplicity of geometric statistical fragmentation, on the other hand, lies in the fact that no assumptions regarding material response are necessary other than some statement regarding a natural or random fragmentation process. Consequently, if an underlying theory of geometric statistical fragmentation could be found, then deviations of observed fragment distributions from those predicted from geometric statistics would be expected to result from the material or physical conditions noted above.

*This work performed at Sandia National Laboratories supported by the U.S. Department of Energy under contract number DE-AC04-76DP00789.

In this paper, a problem of random fragmentation in one dimension is discussed and compared with experimental data. A method to extend the one-dimensional statistical theory to two- and three-dimensional fragmentation is described and contrasted with earlier theoretical work. Geometric statistical fragmentation studies are used to explore alternative theoretical approaches in two dimensions. A concept of strain-rate inhomogeneity is then introduced, which incorporates the effect of different fracture load intensities on observed fragment distributions in complicated fragmentation events.

RANDOM FRAGMENTATION OF AN INFINITE LINE (ONE DIMENSION)

The present description of the application of geometric statistics to the fragmentation of a linear body parallels that of other authors (e.g., Lienau[2]). For points distributed at random on an infinite line, the Poisson distribution provides the probability of finding n points in a length ℓ,

$$P(n/\ell) = (N_0\ell)^n \, e^{-N_0\ell}/n! \,, \tag{1}$$

where N_0 is the average number of points per unit length. The cumulative probability distribution of fragment lengths larger than length ℓ is determined to be

$$N(\ell) = N_0 e^{-N_0\ell} \,. \tag{2}$$

Recently Grady and Benson[3] have performed dynamic fragmentation studies on thin ductile metal rings with relatively small cross sections. The rings were forced to undergo rapid radial expansion to fracture through the application of an impulsive magnetic load. At the highest expansion velocities, fracture into 10 to 12 randomly sized fragments was observed. In Figure 1, a cumulative number distribution, obtained by combining the results of four tests in which exactly ten fragments each were produced, is compared with an exponential fit to the data. The fit provides a reasonably good description of 35 or 36 of the 40 fragments.

EXTENSION TO AREA AND VOLUME FRAGMENTATION

Although there is general agreement on the application of geometric statistical fragmentation in one-dimension, there is no consensus on the extension to random partitioning of bodies with dimension greater than one. The initial step of random placement of points in space is Poisson. The subsequent partitioning of the space depends upon the construction method chosen. We propose a method for randomly partitioning an area which avoids the selection of a particular geometric description. If the fragment area is viewed as a scalar variable, then random fragmentation of an area is analogous to the one-dimensional problem. Fragmentation is determined by breaks distributed randomly on the scalar measure of area. These points define a Poisson variate and lead to a cumulative fragment number distribution similar to Equation 2,

$$N(a) = N_0 e^{-N_0 a} \,, \tag{3}$$

where N_0 is the number of fragments per unit area and a is the fragment area. This distribution will be called a *linear exponential distribution*. In contrast, in the commonly used Mott distribution[4], the argument of the exponential depends on the square root of fragment area:

$$N(a) = N_0 e^{-\sqrt{2N_0 a}} \,. \tag{4}$$

Through a logical extension of the same argument, random fragmentation of a three-dimensional region or volume would lead to a cumulative fragment number distribution,

$$N(v) = N_0 e^{-N_0 v} \,, \tag{5}$$

where v is the fragment volume. The arguments of Mott and Linfoot[4] would suggest a distribution of the form

$$N(v) = N_0 e^{-\sqrt[3]{3N_0 v}} \ . \tag{6}$$

COMPARISON OF THEORETICAL AND COMPUTER-GENERATED GEOMETRIC FRAGMENT DISTRIBUTIONS.

Certain geometric statistical fragmentation problems can be posed for which analytic solutions of the fragment distribution can be obtained. However, those problems amenable to analytic solutions are somewhat limited in the randomizing geometries allowed. Geometric fragmentation problems without an analytic solution can, in principle, be studied with numerical algorithms (e.g. Crain and Miles[5]). Initially, we numerically reproduced the geometric problem of random horizontal and vertical lines solved analytically by Mott and Linfoot, and determined that satisfactory distributions could be obtained (Grady and Kipp[1]). In the natural extension of this geometric construction to both randomly oriented and distributed infinite lines, initially suggested by Mott and Linfoot[4], the Mott distribution also provides a satisfactory fit.

Our next effort focused on formulating a random geometric fragmentation problem of the unit square which avoided the use of infinite lines and, consequently, would be more in keeping with a Poisson process: define a point with random coordinates; divide the square with a randomly oriented line through this point; for each subsequent random point chosen, the fragment in which the point occurred was further divided by a line of random orientation (Figure 2a). The fragment distributions created by this algorithm are shown in Figure 2b and compared with the Mott and linear exponential distributions. Good agreement with the linear exponential distribution is clearly observed.

There are, of course, other constructions for randomly fragmenting an area which begin with a Poisson point process. The most popular is the Voronoi construction, in which perpendicular bisectors of line segments joining the random points partition the area. Although it has been suggested as a model for mechanical fragmentation, most applications have been in other areas of science. In Figure 2b, the strongly centered cumulative fragment distribution for the Voronoi construction is compared with both the linear exponential and the Mott distribution.

STATISTICAL INHOMOGENEITY AND FRAGMENT AREA DISTRIBUTIONS

In the geometric fragmentation problems considered earlier, statistical homogeneity over the unit area was tacitly assumed: that is, the average fragment size at a point within the unit area does not differ from any other point. Normally, due to the complexity of the dynamic loading conditions leading to fragmentation, the intensity of fracture will vary from point to point within the body, and consequently the average fragment size will also be a function of position. The uniformly expanding ring discussed earlier or the uniform expansion of a spherical shell represent unique experimental geometries in which nearly homogeneous fragmentation is realized. Most experimental fragmentation geometries will lead to a situation of ***statistically inhomogeneous fragmentation.***

If we assume that the linear exponential distribution provides the basis for homogeneous random fragmentation, then heterogeneous random fragmentation could be described with a superposition of Poisson mixtures. The distribution would take the form

$$N(a) = \sum_{1}^{n} g_i N_{io} e^{-N_{io} a} \ , \tag{7}$$

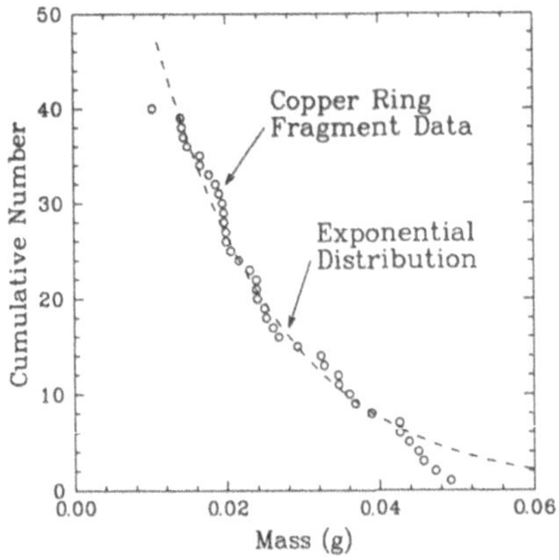

Figure 1: Cumulative number distribution of fragments from four expanding copper ring experiments.

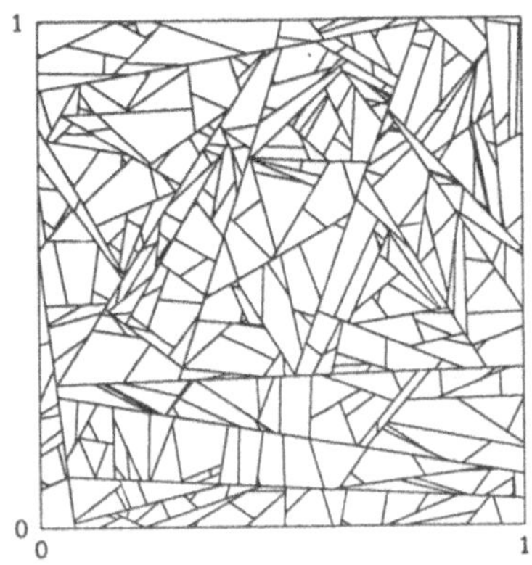

Figure 2a: Fragmentation of unit area with randomly oriented line segments.

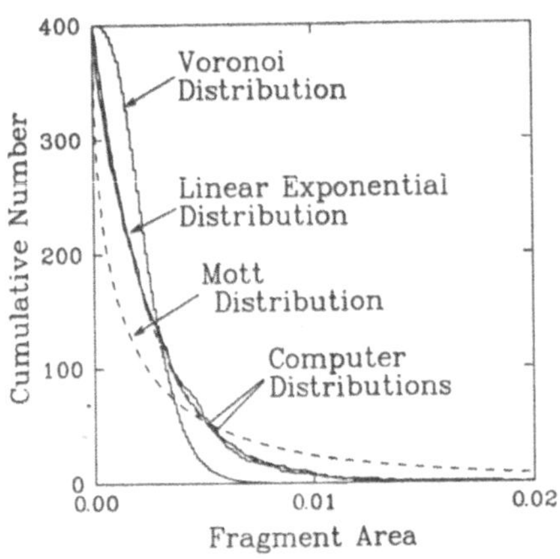

Figure 2b: Computer, Mott, Voronoi and linear exponential distributions.

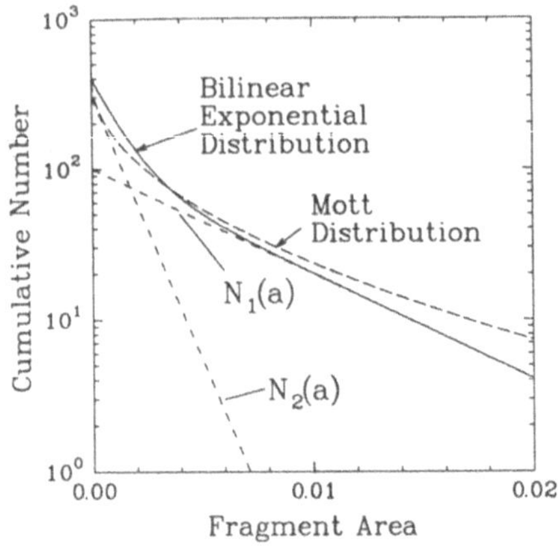

Figure 3: Cumulative number fragment distributions with the heterogeneous bilinear exponential ($N_{10} = 160, N_{20} = 800, g_1 = 5/8, g_2 = 3/8$).

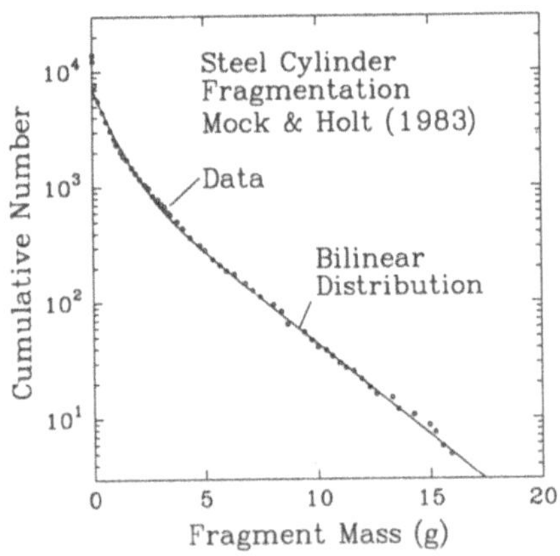

Figure 4: Exploding steel cylinder fragmentation data and representation with a bilinear exponential.

obtained from n statistically homogeneous distributions with corresponding fragment densities N_{i_0} and fragment area fractions, g_i. A plot of a heterogeneous bilinear exponential fragment distribution is illustrated in Figure 3, along with the two homogeneous distributions of which it is composed. Observe that the Mott distribution (Equation 4), in a linear semilogarithmic representation is concave upward, similar to the bilinear curve in Figure 3.

Recent data have been published by Mock and Holt[6] in which explosive-filled cylinders of armco iron and several heat-treated steels were detonated, and the fragments collected and analyzed. Of the more than 10,000 fragments collected in this test, fewer than 80 retained segments of both the inner and outer surface, so the fragment mass should be expected to scale approximately as the volume rather than the area. Hence, a linear exponential distribution for volumetric fragmentation (Equation 5) would provide the basic theoretical curve for homogeneous fragmentation. As noted by Mock and Holt, the data are quite linear with the exception of the small fragment end of the distribution. The data can be represented with a bilinear exponential curve, although there are indications that the small particle portion of the distribution is more complex than this.

SUMMARY AND DISCUSSION

In the context of dynamic fragmentation, the Mott and Linfoot theory of statistical fragmentation has played a dominant role. We have argued, however, that the introduction of area as a scalar random variable is a more natural assumption. The application of a Poisson point process on the area then leads to a linear exponential fragment area distribution for homogeneous fragmentation which differs significantly from the distribution of Mott and Linfoot. Statistical area distributions produced through methods of geometric fragmentation were found to be dependent on the method of construction.

The concept of inhomogeneity in the statistical fragmentation process has also been considered in the present study. A mixing of fragments from different intensity fragmentation events will, in general, broaden and complicate the observed distribution. The Mott distribution, which is slightly broader than the linear exponential distribution, may have provided a reasonable fit to data in many cases because observed fragment distributions in munition studies are probably of heterogeneous origin. We have noted excellent agreement with some munitions data with a bilinear exponential fragment distribution.

REFERENCES

1. D. E. Grady and M. E. Kipp, Particle Size Statistics in Dynamic Fragmentation, J. Appl. Phys., (Accepted) (1985).

2. C. C. Lienau, Random Fracture of a Brittle Solid, J. Franklin Inst. (1936).

3. D. E. Grady and D. A. Benson, Fragmentation of Metal Rings by Electromagnetic Loading, Exp. Mech., 23:393 (1983).

4. N. F. Mott and E. H. Linfoot, A Theory of Fragmentation, Extra-Mural Research No. F72/80, Ministry of Supply, A. C. 3348 (Jan 1943).

5. I. K. Crain and R. E. Miles, Monte Carlo Estimates of the Distributions of the Random Polygons Determined by Random Lines in the Plane, J. Statist. Comput. Simul., 4:293 (1976).

6. W. Mock, Jr. and W. H. Holt, Fragmentation Behavior of Armco Iron and HF-1 Steel Explosive-Filled Cylinders, J. Appl. Phys., 54:2344 (1983).

LASER INTERFEROMETER MEASUREMENTS OF REFRACTIVE INDEX IN SHOCK-COMPRESSED MATERIALS*

J. L. Wise and L. C. Chhabildas

Thermomechanical and Physical Division
Sandia National Laboratories
Albuquerque, NM

ABSTRACT

Laser interferometer systems provide a means for probing the refractive index of transparent specimens subjected to dynamic compression. Previous interferometer measurements of optical properties under shock loading are reviewed for polymethyl methacrylate, fused silica, sapphire, nitromethane, and an aqueous solution of zinc chloride; various degrees of departure from Gladstone-Dale behavior are noted for these materials. In addition, a detailed summary of recent optical studies of lithium fluoride (LiF) is provided. Interferometer data from plate-impact experiments verify sustained LiF transparency for Hugoniot stresses to at least 160 GPa, and establish the variation of LiF refractive index for shock amplitudes ranging from 1.58 to 115 GPa. The refractive-index data for LiF agree with earlier static and shock-wave data, and exhibit a pronounced deviation from predictions based on the Gladstone-Dale, Lorentz-Lorenz, and Drude relations. A modified form of the Gladstone-Dale relation is presented which correctly models the latest LiF measurements. Potential applications of LiF and other window materials to dynamic high-pressure experimentation are discussed.

INTRODUCTION

Using Doppler-shifted light backscattered from a moving reflector, the laser velocity interferometer (VISAR) developed by Barker and Hollenbach[1] permits time-resolved measurements of the reflector velocity. Application of this technique to impact-response tests on selected target materials often involves a transparent window which is bonded to the rear surface of the target so as to permit velocity measurement at the reflective target/window interface. Analysis of data obtained from such experiments relates the observed interference-fringe count to the reflective-interface velocity in a manner that

*This work performed by Sandia National Laboratories supported by the U. S. Department of Energy under contract #DEAC04-76-DP00789.

must account for a predetermined Doppler-shift correction arising from the strain-induced change in refractive index of the window[1–4]. Conversely, if the dynamic mechanical properties (*i.e.*, Hugoniots) of the impactor, target, and window have been independently determined, then the interferometer data yield a direct measurement of the Doppler-shift correction and refractive index for the shock-compressed window. The present paper begins with an outline of relevant principles governing laser interferometry, then reviews existing interferometer measurements of refractive index for several window materials. Recent results are summarized for lithium fluoride (LiF), and potential window developments and applications are considered.

ANALYSIS OF VISAR RECORDS

Interferometer systems of the VISAR (velocity) or Michelson (displacement) varieties combine two input beams of light to produce a single output beam exhibiting an interference fringe frequency equal to the frequency difference between the input beams. One of the input beams consists of Doppler-shifted light which is reflected from the moving test surface; the other input beam is reference light. For the Michelson interferometer, the reference beam typically has the same frequency as the source laser; however, in some cases[3], the frequency of the reference beam may also be Doppler shifted due to reflection from a surface moving at a known (usually steady) velocity. For the VISAR, the reference beam is a portion of the Doppler-shifted light which has been reflected from the test surface and then separated from the remaining light by means of a beam splitter; this reference beam is then subjected to a time delay τ before recombination with undelayed light. If incident laser light with a wavelength λ_0 is used, then the frequency shift (*i.e.*, Doppler shift) of the backscattered light is $\nu_0 = 2u/\lambda_0$ for reflection from a free surface (*i.e.*, no window) moving at a velocity u. When a reflective surface moving at the same velocity is covered by a window, the frequency shift of the backscattered light is $\nu = \nu_0 + \Delta\nu$, where the correction term, $\Delta\nu$, arises from strain-induced changes in the window refractive index which are a consequence of stress-wave propagation. The expression $(1 + \Delta\nu/\nu_0) = \nu/\nu_0$ represents the ratio of the Doppler shift observed with a window to the frequency shift which would have been observed without a window for a reflector moving at the same velocity. All optical measurements and related analyses discussed in ensuing paragraphs of this paper implicitly apply to experimental times prior to that at which first motion of the window free surface is initiated by stress-wave arrival.

As shown by Barker and Schuler[4], the most general equation expressing the velocity, u, of a reflective interface as a function of the VISAR fringe count, F, is

$$u(t - \frac{1}{2}\tau) = \frac{\lambda_0 F(t)}{2\tau(1+\delta)(1+\Delta\nu/\nu_0)}\,, \tag{1}$$

where t is time, τ is the user-selected interferometer delay time, λ_0 is the wavelength of the laser light, δ is a correction term for the dependence of etalon refractive index on wavelength of the Doppler-shifted laser light, and $\Delta\nu/\nu_0$ is the frequency correction related to the strain dependence of window refractive index. If fused-silica etalons are used, then $\delta = +0.0239$ for $\lambda_0 = 632.8$ nm (He-Ne laser), and $\delta = +0.0339$ for $\lambda_0 = 514.5$ nm (argon-ion laser).

Characterization of an interferometer window material involves a determination of the strain-dependent behavior of $\Delta\nu/\nu_0$. As discussed by Hardesty[5], the refractive index, n_H, in the uniaxially strained portion of a window behind a steady, planar shock which propagates into undisturbed material may be derived from $\Delta\nu/\nu_0$ measurements and existing Hugoniot data using the expression

$$n_H = n_0 \frac{U_S}{U_S - u_P} - (1 + \Delta\nu/\nu_0) \frac{u_P}{U_S - u_P} , \qquad (2a)$$

where U_S and u_P are, respectively, the shock velocity and Hugoniot particle velocity in the window, and n_0 is the refractive index of unshocked window material. Letting v_H and v_0 represent, respectively, the specific volume of shocked and unshocked window material, the Hugoniot compressive strain is given by $\epsilon_H = (v_0 - v_H)/v_0$, and mass conservation through the shock leads to the following reformulation of equation (2a):

$$n_H = \frac{1}{1 - \epsilon_H} [n_0 - \epsilon_H (1 + \Delta\nu/\nu_0)] . \qquad (2b)$$

If a material satisfies the Gladstone-Dale relation, $d\rho/\rho = (n-1)^{-1} dn$ where ρ is the material density, then $\Delta\nu/\nu_0$ must be zero on the basis of equation (2b). Hence, the magnitude of $\Delta\nu/\nu_0$ is a measure of the degree of departure of the refractive index from the Gladstone-Dale model.

PREVIOUS EXPERIMENTAL STUDIES

Solid-state windows consisting of polymethyl methacrylate (PMMA), fused silica, and sapphire have been characterized optically and mechanically in shock-wave studies by Barker and Hollenbach[3]. For the optical studies, a symmetric-impact configuration was employed such that the target specimen and gun-launched impactor were both fabricated from the same window material. In each experiment, a Michelson interferometer yielded a direct measurement of $\Delta\nu$ for a shock-induced particle velocity exactly equal to one-half the measured impact velocity.

For PMMA, the values of $\Delta\nu/\nu_0$ observed by Barker and Hollenbach[3] lie between −0.0022 and −0.0117 for particle velocities ranging from 0.0320 to 0.3173 km/s and corresponding Hugoniot stresses ranging from 0.108 to 1.19 GPa. Subsequent VISAR data obtained by Asay and Hayes[6] show that $\Delta\nu/\nu_0$ for PMMA is zero (to within experimental accuracy) for a shock stress of 5.3 GPa, and work by Lee[7] demonstrates that the $\Delta\nu/\nu_0$ correction is negligible in the 9.5–12.5 GPa range of shock stresses. Later VISAR measurements performed by Chhabildas and Asay[8] with PMMA windows indicate that $\Delta\nu/\nu_0$ is less than 0.02 at 22 GPa. As a consequence of the consistently small magnitude of the $\Delta\nu/\nu_0$ correction for PMMA, the refractive index is closely modeled by the Gladstone-Dale relation for window stresses up to 22 GPa. Above this level, the results of Chhabildas and Asay[8] reveal a loss of transparency (for Hugoniot stresses of 22.5–25 GPa) which has been attributed to an apparent phase transition or molecular dissociation. This loss of transparency precludes any VISAR measurements with PMMA windows subjected to shock stresses in excess of ~22 GPa.

The results reported by Barker and Hollenbach[3] for fused silica show values of $\Delta\nu/\nu_0$ which decrease monotonically from a maximum of 0.0627 at a particle velocity of 0.0713 km/s (stress = 0.905 GPa) to a minimum of 0.0335 at a velocity of 0.571 km/s (stress = 6.53 GPa). These $\Delta\nu/\nu_0$ measurements have been subjected to additional analysis by Setchell[9] to determine the explicit dependence of refractive index on the density of dynamically compressed specimens. This analysis was motivated by ramp-wave VISAR experiments in which fused silica windows were subjected to time-dependent strain gradients. These gradients created a spatial variation of refractive index within the window which altered the frequency correction needed for proper interpretation of the fringe records. As shown by Setchell, the VISAR equation (1) may be rewritten in terms of a velocity correction, Δu, as

$$u(t - \frac{1}{2}\tau) = \frac{\lambda_0 F(t)}{2\tau(1+\delta)} - \Delta u \ , \tag{3}$$

where the velocity correction is given by

$$\Delta u = (\Delta\nu/\nu_o)\, u(t - \frac{1}{2}\tau) \ . \tag{4}$$

Setchell noted that the Δu versus Hugoniot particle velocity, u_P, data of Barker and Hollenbach[3] can be accurately represented by a power law:

$$\Delta u = c_1 {u_P}^{c_2} \ , \tag{5}$$

where the coefficients for a least-squares fit to the fused silica data are $c_1 = 0.02600$ and $c_2 = 0.6680$.

Following Barker[2], Setchell suggested that an appropriate expression linking refractive index to material density may be generated by incorporating a density-dependent correction term $\xi(\rho)$ in the Gladstone-Dale relation; the result is

$$\frac{n-1}{n_0-1} = \frac{\rho}{\rho_0}[1 - \xi(\rho)] \ . \tag{6}$$

Setchell assumed a power-law dependence of $\xi(\rho)$ on density of the form

$$\xi(\rho) = \alpha[(\rho/\rho_0) - 1]^\beta = \alpha[\epsilon/(1-\epsilon)]^\beta \ , \tag{7}$$

where the parameters α and β are to be determined for a given window material. This determination involves an iterative approach[9] which couples the index/density relation (using trial values of α and β) with conservation and constitutive expressions governing the mechanical response of the material to obtain the density distribution and resultant refractive-index profile through the compressed region of the window. Knowledge of the refractive-index profile then permits calculation of a frequency correction, $\Delta\nu/\nu_0$, and a velocity correction, Δu, corresponding to the trial values of α and β. For a material which supports a single, planar shock, these corrections are obtained directly from equations (2) and (4); however, in the case of fused silica the corrections must be computed on the basis of the combined ramp/shock waveforms produced in the experiments. Different (α, β) trial pairs are selected and used to calculate sets of (u_P, Δu) pairs until a least-squares, power-law fit of the form shown in equation (5) yields coefficients c_1' and c_2' for a calculated (u_P, Δu) set that are equal to the coefficients c_1 and c_2 found earlier by a direct fit to the measured Δu versus u_P data. The values determined in this fashion by Setchell[9] for fused silica are $\alpha = 0.02996$ and $\beta =$

0.6571. Similar to PMMA, a change in response mechanisms places an upper limit on dynamic stresses for which fused silica is a useful window. In particular, Setchell[10] has observed a serious loss in VISAR fringe contrast whenever fused-silica windows are shocked to stresses near or above the elastic limit of approximately 9.8 GPa reported by Wackerle[11] or 8.7 GPa reported more recently by Chhabildas and Grady[12].

Barker and Hollenbach[3] discovered that the optical correction for sapphire samples with axis parallel to the crystallographic Z-axis shows a monotonic decrease from $\Delta\nu/\nu_0 = 0.7771$ for a shock-induced particle velocity of $u_P = 0.0616$ km/s (stress = 2.76 GPa), to $\Delta\nu/\nu_0 = 0.7500$ for $u_P = 0.2875$ km/s (13.15 GPa). Over this range, sapphire exhibits elastic behavior which is slightly nonlinear[3]. As with fused silica, Setchell[9] transformed the frequency corrections to velocity corrections for the sapphire data and applied the power-law fit shown in equation (5) to find the coefficients $c_1 = 0.7307$ and $c_2 = 0.9755$. Using these coefficients and the iterative approach outlined above, Setchell arrived at values of $\alpha = 0.8280$ and $\beta = 0.9606$ to be used in equations (6) and (7) to describe the index/density behavior of sapphire. Barker and Hollenbach[3] detected erratic optical behavior for two shots producing a stress of about 15 GPa in sapphire windows; they concluded that the upper limit of sapphire's usefulness as an elastic window is between 13 and 15 GPa—a stress range which coincides with that predicted for the onset of plastic yielding based on measurements by Graham and Brooks[13] of a Hugoniot elastic limit between 14 and 21 GPa for Z-cut sapphire. However, the possibility of using sapphire windows for stress levels substantially above the elastic regime is supported by VISAR data from a single test performed by Lee[7] which indicates a value of $\Delta\nu/\nu_0 = \sim 0.65$ for a Hugoniot stress of 42.4 GPa in a sapphire window.

The VISAR technique has also been applied to optical measurements on shocked liquids, including studies of nitromethane by Hardesty[5] and an aqueous solution of zinc chloride by Wise[14]. For nitromethane, the experiments by Hardesty[5] over the stress range from 7.5 to 9.5 GPa indicate negative values of $\Delta\nu/\nu_0$ which vary linearly with the density, ρ_H, of the shocked liquid. As determined by Hardesty, a linear, least-squares fit to the data yields $\Delta\nu/\nu_0 = -1.288 + 0.717\rho_H$, and the corresponding refractive-index data are within 2% of the values predicted by the Gladstone-Dale relationship. The results reported by Wise[14] for a 9.1-molar aqueous solution of zinc chloride show positive values of $\Delta\nu/\nu_0$ which vary linearly with the Hugoniot particle velocity, u_P, for shock stresses from 2.18 to 24.1 GPa. A linear, least-squares fit to the data yields $\Delta\nu/\nu_0 = 0.020 + 0.029u_P$. The resultant refractive-index data exhibit increasing departure from the Gladstone-Dale model as the shock amplitude in the zinc-chloride solution is increased; however, the measured refractive index is only 3.1% lower than the Gladstone-Dale value for the highest stress level achieved (24.1 GPa). Subsequent observations on this solution at higher stress levels reveal a loss of solution transparency for shock amplitudes in excess of 25–30 GPa.

STUDIES ON LITHIUM FLUORIDE

As part of an ongoing effort to develop additional window materials for high-pressure interferometry, the optical response of lithium fluoride to shock compression has been investigated in a series of planar (uniaxial strain) impact experiments. These tests have been conducted with several different guns[15] to achieve impact velocities ranging from 0.2 to 5.5 km/s, resulting in LiF window stresses ranging from 1.6 to 115 GPa.

The experimental configuration for these tests is illustrated in Figure 1. An impactor and backing plate are mounted on the forward face of a projectile that is accelerated to a preselected velocity before striking a target which consists of a buffer/window assembly. Prior to target construction, the forward surface of the LiF window is lightly lapped and then vapor plated with an aluminum film ~0.5 μm thick to provide a diffuse reflective surface for the incident laser beam. A thin epoxy bond is used to attach the window to the buffer rear surface, and this assembly is, in turn, potted concentrically within a support ring. Pins mounted around the periphery of the support ring provide measurements of impact velocity and tilt (typically a few milliradians[15]).

The shock impedance (product of shock velocity times initial density) of the backing plate behind the impactor is typically different than that of the impactor itself; consequently, this configuration produces initial Hugoniot states in both the buffer and window which persist for ~1 μs before release or recompression due to arrival of the wave reflected from the backing/impactor interface. If the VISAR fringe count, F, and actual particle velocity, u_P, for the initial Hugoniot state in the window are known, then equation (1) may be solved to determine the frequency correction, $\Delta\nu/\nu_0$. The fringe count is monitored continuously during each experiment and an average value is determined for readings obtained during the initial Hugoniot state; the particle velocity at the reflective interface must be determined independently. For those tests involving a symmetric impact (*i.e.*, LiF buffer and impactor), the particle velocity in the initial shock state is exactly one-half the impact velocity. Otherwise, the particle velocity of the buffer/window interface following arrival of the incident shock may be predicted on the basis of an impedance-matching technique using the measured impact velocity and existing Hugoniot equation-of-state (EOS) data for the impactor, buffer, and window.

All lithium fluoride specimens used in this work are optical-quality single crystals obtained from the Harshaw Chemical Company[16]. The axis of each LiF specimen is oriented parallel to a ⟨100⟩ crystallographic direction. In addition to LiF, other impactor and buffer materials include aluminum (6061), beryllium, copper, sapphire, and tantalum. Relevant Hugoniot EOS parameters used for analysis of the present results are summarized in Table 1 which also includes a description of the initial condition for each material. The parameters c_0 and S listed in Table 1 are coefficients in a linear fit, $U_S = c_0 + S u_P$, to the shock-velocity/particle-velocity Hugoniot for each material.

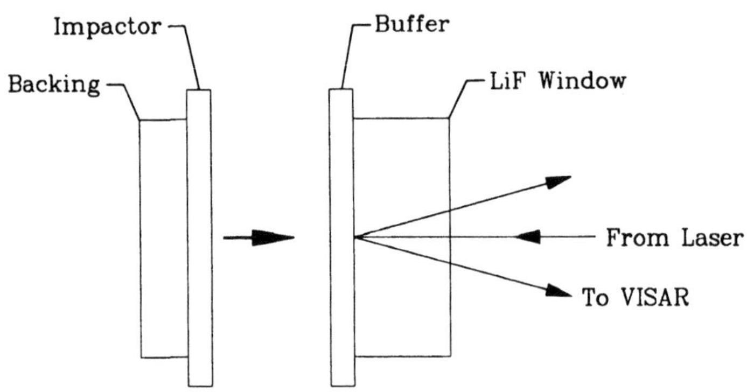

Figure 1: Experimental Configuration for VISAR Studies of Shock-Compressed LiF Windows.

Table 1. Material Specifications and Hugoniot Parameters

Material and Condition	Initial Density (Mg/m^3)	c_0 (km/s)	S	Reference
Al-aluminum (6061-T6)	2.70	5.25	1.35	Herrmann[17]
Al_2O_3-sapphire (single crystal, axis ‖ Z-axis)	3.985	11.19	1.00	Barker, *et al.*[3]
Be-beryllium (polycrystalline)	1.851	7.998	1.124	McQueen, *et al.*[18]
Cu-copper (oxygen-free, high-conductivity)	8.930	3.940	1.489	McQueen, *et al.*[18]
LiF-lithium fluoride (single crystal, axis ‖ ⟨100⟩ direction)	2.640	5.148	1.353	Carter[19]
Ta-tantalum (polycrystalline)	16.654	3.414	1.200	McQueen, *et al.*[18]

Details relating to impact parameters and resultant optical data for individual experiments in this study are displayed in Table 2. The shots are listed in order of increasing window stress and strain. In several cases, two sets of data are reported for a single shot in which two VISARs were used simultaneously; these individual sets are distinguished by appending a "P" (Primary VISAR) or "S" (Secondary VISAR) to the shot number. Except for the first two listings (shots LIF13 and LIF14), all values for window stress and strain have been calculated from the Rankine-Hugoniot relations for a single shock propagating into the window material at a velocity calculated from the linear U_S–u_P relation reported by Carter[19] for stresses ranging from 6.9 to 107.8 GPa in LiF. The highest estimates of window stress, strain, and particle velocity (reported for test LIF12) are based on an extrapolation of Carter's data. At the other extreme, shots LIF13 and LIF14 involve low shock stresses and the effects of the elastic precursor observed in LiF are no longer negligible; hence, the window stress and strain in the initial shock state are obtained by direct integration of the particle-velocity records obtained for these two shots using the respective values of $\Delta\nu/\nu_0$ shown in Table 2.

A review of the information contained in Table 2 reveals that the $\Delta\nu/\nu_0$ data for shock-loaded lithium fluoride show a mean value of 0.281, with a standard deviation of 0.009, for experiments producing window stresses from 10.3 to 115.3 GPa. Over this range, all of the individual $\Delta\nu/\nu_0$ measurements lie within two standard deviations of the mean. Only the two tests (LIF13 and LIF14) involving stresses below 10 GPa indicate $\Delta\nu/\nu_0$ values which are slightly, but measurably, outside the scatter band defined by the remaining data. In general, the main sources of error stem from uncertainties in: (1) projectile-velocity measurement; (2) VISAR fringe count; and, (3) Hugoniot equation-of-state (EOS) parameters. For the low and intermediate projectile velocities which were produced, respectively, by a single-stage gas gun and a propellant (powder) gun, the estimated uncertainty is ±0.2%. The highest projectile velocities (in excess of 2.3 km/s) were attained on a two-stage, light-gas gun and detected with an estimated uncertainty of ±1.0%. Following Barker and Hollenbach[1], estimated uncertainty in the VISAR fringe count is ±0.02. For those shots utilizing a symmetric-impact configuration, uncertainties in EOS data are not a factor. Since the observed scatter in $\Delta\nu/\nu_0$ data for symmetric-impact experiments is comparable to that for impedance-matching

Table 2. Summary of VISAR Experiments on Lithium Fluoride Windows.

Shot Number	Buffer / Impactor Material	Impact Velocity (km/s)	Window Stress (GPa)	Window Strain	Window Particle Velocity (km/s)	Optical Correction $\Delta\nu/\nu_0$	Refractive Index n_H
LIF13	LiF / LiF	0.2247	1.576[a]	0.02136[a]	0.1124	0.260	1.3968
LIF14	LiF / LiF	0.6021	4.295[a]	0.05586[a]	0.3010	0.247	1.4026
BE3P	Be / Al_2O_3	0.8713	10.29	0.1074	0.6471	0.269	1.4089
LIF5P	none[b] / LiF	1.322	10.54	0.1094	0.6610	0.282	1.4076
LIF5S	LiF / LiF	1.322	10.54	0.1094	0.6610	0.268	1.4094
LIF4	LiF / LiF	1.396	11.23	0.1146	0.6980	0.291	1.4072
BE19P	Be / Al_2O_3	0.9462	11.28	0.1150	0.7008	0.268	1.4103
BE19S	Be / Al_2O_3	0.9462	11.28	0.1150	0.7008	0.269	1.4101
LIF3	LiF / LiF	1.616	13.31	0.1295	0.8080	0.277	1.4113
CU17P	Cu / Cu	1.186	13.88	0.1333	0.8373	0.273	1.4125
CU17S	Cu / Cu	1.186	13.88	0.1333	0.8373	0.270	1.4130
AL10	Al / Al	1.755	14.96	0.1403	0.8916	0.281	1.4123
BE1P	Be / Be	1.968	16.70	0.1511	0.9777	0.283	1.4136
BE1S	Be / Be	1.968	16.70	0.1511	0.9777	0.275	1.4151
BE9S	Be / Be	1.984	16.86	0.1520	0.9853	0.274	1.4154
BE14P	Be / Al	1.969	17.15	0.1538	0.9995	0.297	1.4115
LIF2	LiF / LiF	2.021	17.37	0.1550	1.010	0.278	1.4152
BE4P	Be / Be	2.040	17.41	0.1553	1.012	0.294	1.4123
LIF1	LiF / LiF	2.247	19.79	0.1685	1.124	0.271	1.4188
AL20	Al / Al	2.216	19.81	0.1686	1.125	0.288	1.4154
AL5	Al / Al	2.224	19.89	0.1691	1.129	0.279	1.4173
AL4	Al / Al	2.253	20.20	0.1707	1.143	0.283	1.4167
BE13P	Be / Cu	1.962	25.12	0.1947	1.361	0.280	1.4214
BE13S	Be / Cu	1.962	25.12	0.1947	1.361	0.275	1.4226
BE10P	Be / Cu	1.970	25.24	0.1953	1.367	0.286	1.4201
BE10S	Be / Cu	1.970	25.24	0.1953	1.367	0.286	1.4201
BE15P	Be / Cu	1.987	25.51	0.1965	1.378	0.285	1.4205
BE6P	Be / Cu	2.000	25.71	0.1974	1.387	0.299	1.4172
BE6S	Be / Cu	2.000	25.71	0.1974	1.387	0.289	1.4197
CU19P	Cu /Cu	2.020	26.47	0.2007	1.412	0.286	1.4210
LIF9	none[b] / Ta	3.473	60.90	0.3030	2.644	0.286	1.4408
LIF11P	LiF / Ta	5.028	102.3	0.3682	3.778	0.277	1.4620
LIF11S	LiF / Ta	5.028	102.3	0.3682	3.778	0.272	1.4649
LIF12P	LiF / Ta	5.459	115.3	0.3829	4.090	0.288	1.4596
LIF12S	LiF / Ta	5.459	115.3	0.3829	4.090	0.292	1.4571

[a]Obtained by direct integration of particle-velocity record.
[b]VISAR measurement made at impact surface.

experiments, no attempt has been made to quantify EOS uncertainties affecting interpretation of results from the impedance-matching experiments.

Using equation (4), the $\Delta\nu/\nu_0$ data have been converted to velocity corrections, Δu, which are plotted as a function of the Hugoniot particle velocity, u_P, in Figure 2. Estimated error bars are displayed for the impedance-matching experiments which used tantalum impactors to achieve the highest shock stresses; error bars for symmetric-impact tests are contained by the plotted points. A least-squares, power-law fit of the form shown in equation (5) has been performed for all 35 Δu–u_P pairs, and the resulting coefficients are $c_1 = 0.2784$ and $c_2 = 1.031$ for lithium fluoride. This power-law fit is also plotted in Figure 2.

As discussed earlier, each $\Delta\nu/\nu_0$ datum provides a measurement of the refractive index. For the present study, the refractive index is obtained directly from equation (2) which, strictly speaking, applies only to the case of a single shock propagating into undisturbed material. In lithium fluoride, the refractive index determined in this manner is exact (to within experimental accuracy) for shock stresses in excess of ~18 GPa since the precursor is overdriven. At lower shock stresses, the observed frequency correction, $\Delta\nu/\nu_0$, is a consequence of changes in refractive index through both the precursor and the shock. To investigate the effect of elastic-plastic window response on the relationship between the refractive index and the Doppler-frequency correction, $\Delta\nu/\nu_0$, an analysis has been performed for a two-wave loading profile consisting of a steady elastic precursor and a trailing steady shock which propagate at velocities relative to undisturbed window material of c_l and U_S, respectively. Under these conditions, the refractive index, n_H, of the shock-processed material is given by

$$n_H = n_0 \frac{c_l}{U_S - u_P} - n_1 \frac{c_l - U_S}{U_S - u_P} - (1 + \Delta\nu/\nu_0) \frac{u_P}{U_S - u_P} , \tag{8}$$

where n_1 is the refractive index of material between the precursor and shock. As expected, equation (8) reduces to equation (2a) when $n_1 = n_0$ (no precursor), or when $c_l = U_S$ (precursor just overdriven). Thus, a determination of the refractive index in the final Hugoniot state requires an independent means (experimental or analytical) for specifying the refractive index, n_1, in material which has been processed by the precursor only. If an assumption is made that the change in refractive index through the precursor can be estimated using the Gladstone-Dale relation, existing data for the precursor velocity and amplitude in LiF (e.g., see Asay, et al.[20]) yield an

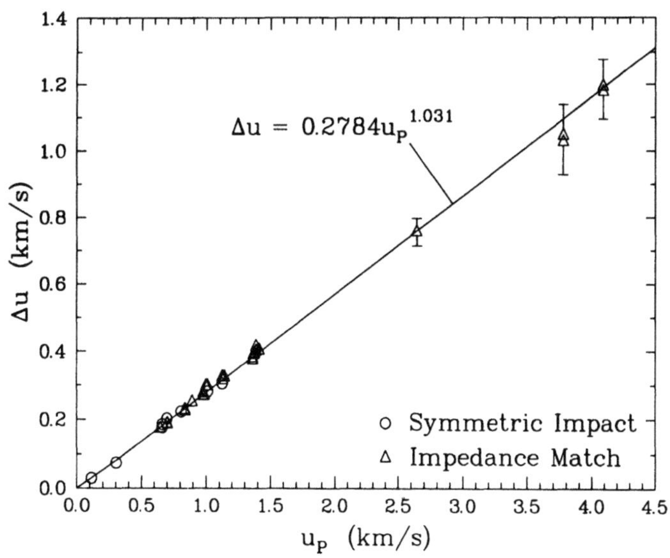

Figure 2: Velocity Correction, Δu, for VISAR Data Influenced by Refractive-Index Effects in Lithium Fluoride.

estimate for n_1 which is ~0.10% larger than n_0, and equation (8) may subsequently be solved to determine n_H for those experiments producing a shock stress below 18 GPa in the window. Values of n_H calculated in this fashion are within 0.00–0.05% of the values calculated using equation (2) where the effects of the precursor were ignored. This agreement is well within the estimated experimental uncertainty, and the use of equation (2) is consequently justified. The resultant refractive-index data are listed in Table 2 and are plotted as a function of compressive strain in Figure 3.

For comparison to the present data, predicted variations of refractive index with strain based on the Gladstone-Dale, Lorentz-Lorenz, and Drude models are also plotted in Figure 3. The integrated form of the Gladstone-Dale relation is

$$\frac{1}{\rho}(n-1) = constant\ . \tag{9}$$

In the case of the Lorentz-Lorenz model, the index/density relation is given by

$$\frac{1}{\rho}\frac{(n^2-1)}{(n^2+2)} = constant\ ; \tag{10}$$

whereas the Drude expression is

$$\frac{1}{\rho}(n^2-1) = constant\ . \tag{11}$$

An examination of Figure 3 quickly reveals that all of these models fail to provide a fit to the data for lithium fluoride; hence, a semi-empirical index/density relationship of the type shown in equations (6) and (7) has been attempted. However, the correction term $\xi(\rho)$ given by equation (7) results in an index/strain relation which, contrary to the

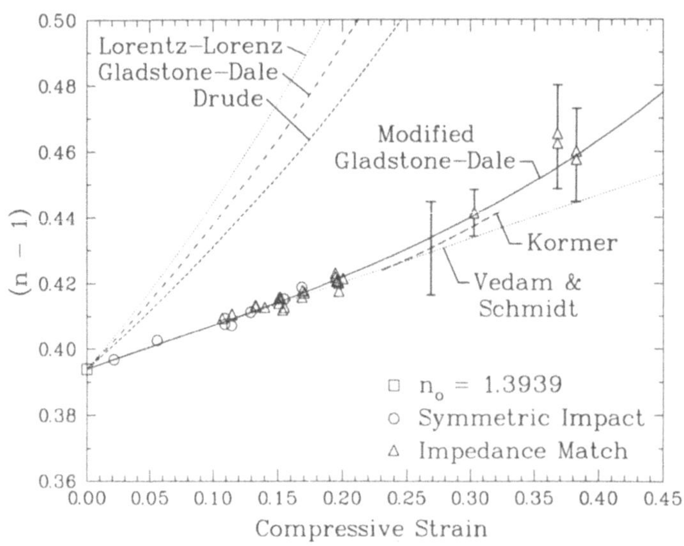

Figure 3: Strain Dependence of LiF Refractive Index.

present data, is convex upward and predicts decreasing values of refractive index for strains greater than ~0.38. An excellent fit to the present data has been achieved by adopting an alternate functional form for $\xi(\rho)$ given by

$$\xi(\rho) = \gamma[1 - (\rho_0/\rho)]^{\kappa} = \gamma\epsilon^{\kappa} , \tag{12}$$

where γ and κ are constants to be determined using the same iterative approach outlined earlier which assures that a power-law fit to the set of (u_P, Δu) values calculated on the basis of the index/density relations (6) and (12) yields the same coefficients obtained for a power-law fit to the actual (u_P, Δu) measurements. For the present data, the appropriate coefficients for LiF are $\gamma = 0.7630$ and $\kappa = 1.040$. Using the dispersion formula and constants reported by Herzberger and Salzberg[21], the initial refractive index, n_0, of LiF is 1.3939 for $\lambda_0 = 514.5$ nm. With this value of n_0 and the experimentally determined values of γ and κ, equations (6) and (12) have been used to calculate the locus of refractive index versus compressive strain (labelled "Modified Gladstone-Dale") which is plotted along with the individual data points in Figure 3.

Previous refractive-index data for shock-loaded LiF have been reported by Kormer[22] over a more limited range of strain ($\epsilon_H = 0.231$–0.324). For this work, an explosive driver system in direct contact with a flat surface of the test specimen generated a planar shock (parallel to the driver surface) which propagated into the transparent material. A collimated beam of light from an explosive source was transmitted into the unshocked portion of the specimen so as to be obliquely incident upon the moving shock front. This incident light was partially reflected by the shock itself; the remaining light entered the compressed portion of the specimen and was obliquely reflected by the rear surface which was in contact with the driver system. A high-speed streak camera recorded the relative locations of these two obliquely reflected beams after they emerged from the test assembly, and the refractive index under shock loading was subsequently inferred using geometric optics and the independently determined compressive strain in the transparent specimen.

For several alkali halides, including LiF, Kormer[22] found that the relationship between the refractive index and density may be written as

$$n_H(\sigma_H) = n_0 + \frac{dn_H}{d\sigma_H}(\sigma_H - 1) , \tag{13}$$

where $\sigma_H = \rho_H/\rho_0$ is the compression and $dn_H/d\sigma_H$ is a constant with a different characteristic value for each crystal. Over the range of his reported data ($\sigma_H = 1.30$–1.48), Kormer reports a value of $dn_H/d\sigma_H = 0.1$ for LiF. The initial index, n_0, listed by Kormer for LiF is 1.392 for the broadband light from his explosive source, whereas $n_0 = 1.3939$ for incident light from an argon-ion laser ($\lambda_0 = 514.5$ nm). Measurements made under broadband lighting with a Bausch & Lomb Abbe-3L refractometer confirm that $n_0 = 1.3925$ for representative LiF samples used in the present studies. However, for comparison to the VISAR results, equation (13) has been used with $n_0 = 1.3939$ to construct a locus for the Kormer data which is shown in Figure 3. The single error bar centered on this line corresponds to Kormer's estimated accuracy of 1–2% for n/n_0.

The refractive index of LiF under hydrostatic pressure has been investigated by Vedam and Schmidt[23]. Results from this study show that the index increases linearly with pressure to at least the maximum observed loading of 0.7 GPa, corresponding to a volume (compressive) strain, $\epsilon = (v_0 - v)/v_0$, of 0.01. A linear fit to the index/strain data yields $n - n_0 = 0.132\epsilon$ for $0 \leq \epsilon \leq 0.01$. Although the maximum strain achieved in the hydrostatic pressure tests is substantially below the range covered by the shock-compression studies ($\epsilon_H = 0.02$–0.38), an extrapolation of the linear fit to the Vedam and Schmidt data (using $n_0 = 1.3939$) shows very good agreement with shock-wave data for strains up to ~0.20–0.25 (see Figure 3). Above this level, the shock-wave measurements indicate an increasing departure from the linear index/strain relation.

The present measurements extend the available optical data for LiF to substantially higher and lower dynamic strain states than previously investigated. Sustained transparency of LiF has been confirmed to at least the highest shock amplitude (115.3 GPa) achieved in these studies, and a more recent VISAR experiment by Grady[24] verifies LiF transparency for an estimated Hugoniot stress of 160 ± 10 GPa. Comparison of the two existing sets of refractive-index data on shock-compressed LiF over the shared range of Hugoniot strain demonstrates an excellent agreement between the present results (based on interference techniques) and Kormer's results[22] (determined by geometric optics).

DISCUSSION

The laser interferometer measurements of refractive index described in preceding sections of this paper stem from efforts to develop appropriate window materials to support interferometric monitoring of dynamic response for a wide variety of test specimens. Solid-state materials sufficiently characterized for use as interferometer windows now include lithium fluoride, PMMA, sapphire, and fused silica; optically calibrated liquid-state materials include a 9.1-molar aqueous solution of zinc chloride and nitromethane (a liquid explosive). Transparency under anticipated levels of shock loading is the obvious factor governing selection of a window material for a particular application; from this standpoint, lithium fluoride is presently the most versatile candidate by virtue of its sustained transparency for shock stresses to at least 160 GPa.

A requirement typically placed on a window is that its shock impedance closely match the impedance of the target material which it contacts, thereby minimizing release or recompression of the target following wave interaction with the target/window interface. The intermediate shock impedance of lithium fluoride is a very close match to such technically important materials as aluminum and beryllium. High-impedance materials, such as copper, steel, and uranium, are best paired with sapphire windows for an impedance match, whereas low-impedance materials, such as polymeric compounds and condensed explosives, are best matched by fused silica, PMMA, and liquid windows. Liquid solutions offer the unique possibility of tailoring the shock impedance over a wide range by variation of the solute concentration. By directly wetting the surface of the target material, liquid windows may also prove advantageous for applications such as shock rise-time measurements where the epoxy bond typically found between the target and a solid-state window can create an impedance mismatch which substantially complicates data analysis.

The continuing development of new dynamic loading and measurement techniques will necessarily involve parallel efforts to optically characterize suitable window materials. New loading capabilities at ultra-high pressures are anticipated for rail-gun and laser facilities where optical measurements will undoubtedly benefit from windows for shock stresses well in excess of 100 GPa. The increasing use of experimental geometries which employ oblique impact[25] or anisotropic crystals[26] to generate coupled pressure-shear response in target materials will benefit from the development of high-shear-strength windows and bonding methods such as that reported by Chhabildas[26]. Windows have already been employed by Barker[27] in experiments involving the generation and observation of quasi-isentropic compressional waves. Barker[28] has suggested that the smaller temperature rise and strain rate associated with a quasi-isentropic wave as compared to a shock wave with the same peak pressure may lead to window transparency limits at stresses much higher than the limiting values under shock compression. Other window applications include, for example, the infrared radiometry and hot-spot photography reported by Von Holle[29] for studies of inert and reactive (explosive) materials experiencing shock compression. The rapidly expanding fiber-optic technology may also result in new diagnostic probes which require (or permit) optical characterization of additional window materials and geometries and/or the fiber itself under dynamic loading.

REFERENCES

1. L. M. Barker and R. E. Hollenbach, Laser Interferometer for Measuring High Velocities of Any Reflecting Surface, J. Appl. Phys. 43:4669 (1972).
2. L. M. Barker, Fine Structure of Compressive and Release Wave Shapes in Aluminum Measured by the Velocity Interferometer Technique, in: "Behaviour of Dense Media under High Dynamic Pressures," Gordon and Breach, New York (1968).
3. L. M. Barker and R. E. Hollenbach, Shock-Wave Studies of PMMA, Fused Silica, and Sapphire, J. Appl. Phys. 41:4208 (1970).
4. L. M. Barker and K. W. Schuler, Correction to the Velocity-per-Fringe Relationship for the VISAR Interferometer, J. Appl. Phys. 45:3692 (1974).
5. D. R. Hardesty, On the Index of Refraction of Shock-Compressed Liquid Nitromethane, J. Appl. Phys. 47:1994 (1976).
6. J. R. Asay and D. B. Hayes, Shock-Compression and Release Behavior Near Melt States in Aluminum, J. Appl. Phys. 46:4789 (1975).
7. L. M. Lee, Shock-Induced Index-of-Refraction Variations in PMMA, Sapphire, and Lithium Fluoride, Ktech Corp. Report TR-76-04 (1976).
8. Lalit C. Chhabildas and James R. Asay, Rise-time Measurements of Shock Transitions in Aluminum, Copper, and Steel, J. Appl. Phys. 50:2749 (1979).
9. Robert E. Setchell, Index of Refraction of Shock-Compressed Fused Silica and Sapphire, J. Appl. Phys. 50:8186 (1979).
10. Robert E. Setchell, Ramp-Wave Initiation of Granular Explosives, Combust. Flame 43:255 (1981).
11. Jerry Wackerle, Shock-Wave Compression of Quartz, J. Appl. Phys. 33:922 (1962).
12. L. C. Chhabildas and D. E. Grady, Shock Loading Behaviour of Fused Quartz, in: "Shock Waves in Condensed Matter–1983," J. R. Asay, R. A. Graham, and G. K. Straub, ed., North-Holland, Amsterdam (1984).
13. R. A. Graham and W. P. Brooks, Shock-Wave Compression of Sapphire from 15 to 420 kbar. The Effects of Large Anisotropic Compressions, J. Phys. Chem. Solids 32:2311 (1971).

14. J. L. Wise, Refractive Index and Equation of State of a Shock-Compressed Aqueous Solution of Zinc Chloride, in: "Shock Waves in Condensed Matter–1983," J. R. Asay, R. A. Graham, and G. K. Straub, ed., North-Holland, Amsterdam (1984).
15. Lalit C. Chhabildas, The Sandia Shock Thermodynamics Applied Research Facility, in: "Shock Waves in Condensed Matter–1981," W. J. Nellis, L. Seaman, and R. A. Graham, ed., American Institute of Physics, New York (1982).
16. The Harshaw Chemical Company, Crystal & Electronic Products, Solon, Ohio.
17. W. Herrmann, Development of a High Strain Rate Constitutive Equation for 6061-T6 Aluminum, Sandia Report SLA-73-0897 (1974).
18. R. G. McQueen, S. P. Marsh, J. W. Taylor, J. N. Fritz, and W. J. Carter, The Equation of State of Solids from Shock Wave Studies, in: "High-Velocity Impact Phenomena," Ray Kinslow, ed., Academic Press, New York (1970).
19. W. J. Carter, Hugoniot Equation of State of Some Alkali Halides, High Temp.-High Press. 5:313 (1973).
20. J. R. Asay, G. R. Fowles, G. E. Duvall, M. H. Miles, and R. F. Tinder, Effects of Point Defects on Elastic Precursor Decay in LiF, J. Appl. Phys. 43:2132 (1972).
21. Max Herzberger and Calvin D. Salzberg, Refractive Indices of Infrared Optical Materials and Color Correction of Infrared Lenses, J. Opt. Soc. Am. 52:420 (1962).
22. S. B. Kormer, Optical Study of the Characteristics of Shock-Compressed Condensed Dielectrics, Sov. Phys. Usp. 11:229 (1968).
23. K. Vedam and E. D. D. Schmidt, Effect of Hydrostatic Pressure on the Refractive Index of LiF, Solid State Commun. 3:373 (1965).
24. D. E. Grady, Sandia National Laboratories, private communication.
25. A. S. Abou-Sayed, R. J. Clifton, and L. Hermann, The Oblique-Plate Impact Experiment, Exp. Mech. 16:127 (1976).
26. Lalit C. Chhabildas, Dynamic Transverse Particle Velocity Measurements using Interferometric Techniques, in: "High Speed Photography, Videography, and Photonics," Dennis L. Paisley, ed., Proc. SPIE 427 (1983).
27. L. M. Barker, High-Pressure Quasi-Isentropic Impact Experiments, in: "Shock Waves in Condensed Matter–1983," J. R. Asay, R. A. Graham, and G. K. Straub, ed., North-Holland, Amsterdam (1984).
28. L. M. Barker, Sandia National Laboratories, private communication.
29. William G. Von Holle, Shock Wave Diagnostics by Time-Resolved Infrared Radiometry and Non-Linear Raman Spectroscopy, in: "Shock Waves in Condensed Matter–1983," J. R. Asay, R. A. Graham, and G. K. Straub, ed., North-Holland, Amsterdam (1984).

EXPERIMENTAL TECHNIQUE FOR MEASURING THE ISENTROPE OF HYDROGEN TO SEVERAL MEGABARS*

L. M. Barker, T. G. Trucano, J. L. Wise, and J. R. Asay

Thermomechanical and Physical Division
Sandia National Laboratories
Albuquerque, NM

INTRODUCTION

The experimental measurement of the EOS of hydrogen has been of interest for some time because of the theoretical expectation of a transition to the metallic state in the multi-megabar pressure regime. Previous experiments [1-4] have reported results which are consistent with a metallic transition, but experimental uncertainties have precluded positive identification of the metallic phase.

In this paper we describe a new experimental approach (Fig. 1) to the measurement of the high-pressure EOS of hydrogen. A cryogenic hydrogen specimen, either liquid or solid, is located in the muzzle of a gun barrel between a tungsten anvil and another tungsten disk called a shim. Helium gas in the gun barrel cushions the impact and allows nearly isentropic compression of the hydrogen. The time-resolved pressure in the specimen is calculated from a laser interferometer (VISAR) measurement [5] of the acceleration history of the anvil's free surface, and volume measurements at specific times are made by combining VISAR data, which define the position of the anvil, with flash x-ray photographs which define the shim position.

In order to successfully accomplish the experiment illustrated by Figure 1, major developments in gun technology are required in three areas. The first involves the placement of a cryogenic specimen in a precisely aligned position inside the gun muzzle at shot time. The second concerns the new use of helium in the gun barrel to cushion the impact and produce isentropic compression of the hydrogen. The third development area relates to the instrumentation, and particularly to the required state-of-the-art x-ray system. The following sections will briefly discuss each of these areas.

THE CRYOGENICS

We advance a new technique for preparation of cryogenic specimens by making a small portion of the end of the gun barrel of Figure 1 removable, so that it can be suspended inside a liquid-helium cryostat and filled with a condensed hydrogen specimen. The removable gun-barrel end is called the target fixture. Just prior to firing, an automatic mechanism

*This work performed by Sandia National Laboratories supported by the U. S. Department of Energy under contract #DEAC04-76-DP00789.

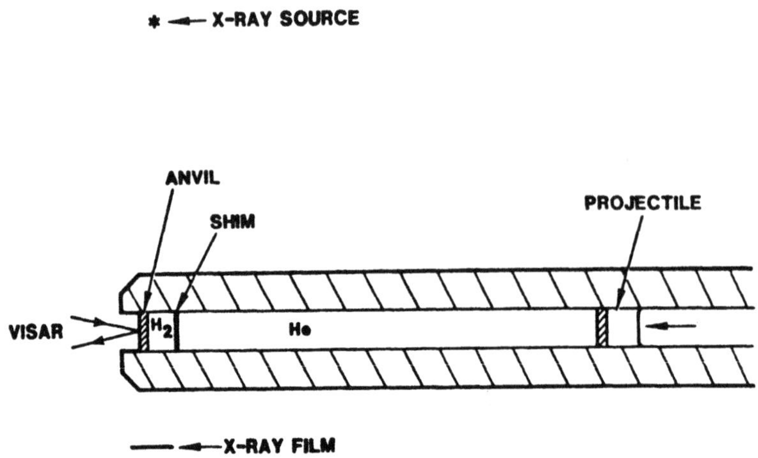

Figure 1: Schematic of the Experiment to isentropically compress hydrogen samples to multi-megabar pressures. X-rays measure sample volume, and the VISAR-measured anvil free surface velocity determines pressure history in the sample.

removes the target fixture from the cryostat and quickly inserts it into position in front of the gun barrel, so that the schematic configuration of Figure 1 is regained. The gun is then immediately fired before the temperature of the hydrogen rises appreciably.

To sufficiently cool the target fixture *in situ* at the end of the gun barrel would require heroic efforts in cryogenic apparatus and insulation, much of which would be destroyed with each shot [6]. The proposed system incorporating a separate dewar and a target insertion mechanism (TIM) allows the cryogenic system to be protected from the impact event. Also, the absence of cryogenic apparatus and insulation around the target fixture leaves it more accessible to VISAR, x-ray, and other types of instrumentation.

The target fixture shown in Figure 2 is made of a high-density alloy containing over 90% tungsten to enhance the lateral inertial confinement of the hydrogen during compression. Since tungsten is opaque to x-rays, thin slits are machined in the fixture and filled with sapphire windows to allow x-ray photographs of the shim position for specimen volume determination. Eight slots and windows allow four flash x-rays to be taken during an experiment.

The mechanical arm requires less than 3 sec to remove the target fixture from the dewar and insert it into the impact chamber in front of the gun muzzle. The mechanism is rather simple, requiring only a little more than one turn of a single shaft to perform the entire operation. However, it only places the target fixture close to its final position, where another automatic device drops hooks over the two "ears" on the target fixture (Fig. 2) and pulls it firmly into the final aligned position. An electronic sensor detects the arrival of the target fixture at its final position and immediately fires the gun.

The details of the cryogenic system will appear in a subsequent report [7].

DYNAMIC ISENTROPIC COMPRESSION

The experimental configuration of Fig. 1 employs a high shock impedance anvil behind the hydrogen to generate the desired multi-megabar pressures. The helium in the gun barrel serves to cushion the impact, and produces a shockless isentropic compression of the

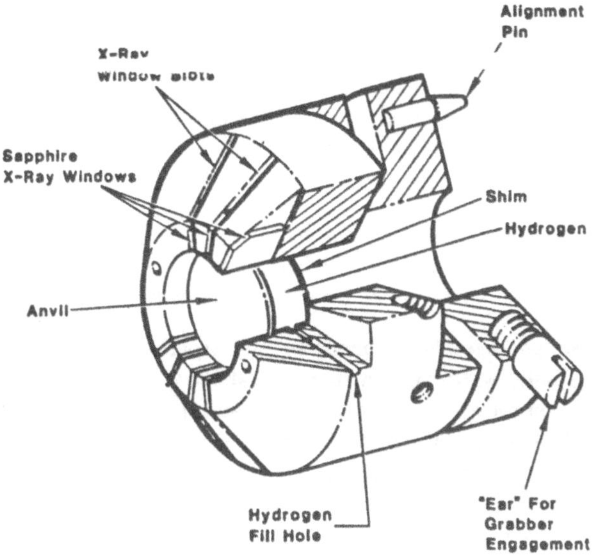

Figure 2: Cut-away view of the tungsten alloy target fixture.

hydrogen. In fact, finite difference computer calculations indicate that the pressure can be kept nearly uniform through the hydrogen specimen during the compression. The hydrogen pressure rises very slowly at first, requiring many tens of microseconds to reach a pressure of 1 GPa. Thereafter, however, the rate of increase picks up rapidly and 10 GPa is reached in only a few more microseconds. The rise from 10 GPa to peak pressure requires less than half a microsecond.

The mass of the shim plays an important role in the impact event. It shields the hydrogen from shocks in the helium which reverberate between the projectile nose and the specimen, and it introduces a dynamic phenomenon akin to a resonance which can be "tuned" to increase the peak pressure in the hydrogen by 30% or more.

There are a number of parameters which interact to determine the peak isentropic pressure attainable in the hydrogen experiment. Some of the more important ones are the shim mass, the projectile mass/velocity combination attainable with our gun facility, the initial helium pressure in the gun barrel, and the initial thickness of the hydrogen specimen. In order to determine the effects of parameter variations, and to choose near-optimum combinations for the initial experiments, it has been necessary to do a large number of finite difference computer calculations [8]. The ability to perform such calculations has greatly enhanced our understanding and optimization of the experiment.

INSTRUMENTATION

The primary goal of the instrumentation is to measure the pressure-volume states achieved in the experiment. The pressure history will be determined at the hydrogen-anvil interface by measuring the free surface velocity history of the anvil using VISAR instrumentation [5]. If the compression history is sufficiently gradual that no significant shocks arrive at the anvil's free surface, then a knowledge of the anvil material's EOS allows one to accurately translate the anvil free surface velocity history into the pressure history at the hydrogen-anvil interface. Since computer calculations indicate pressure uniformity along the axis of the specimen at any given time, the pressure history at the hydrogen-anvil interface applies to the whole hydrogen specimen. This assumes, of course, that no significant lateral edge effects arrive at the centerline of the specimen (where the pressure is measured) during the time of interest. Because of the very rapid compression of the specimen after the yield strength of the confining target fixture is reached, inertial confinement will be effective, and the pressure measurement will be nearly free of edge effects.

The volume measurement will be made in cooperation with Dr. J. Chang using his state-of-the-art developments in flash x-radiography [9,10]. The x-rays allow the shim-to-anvil distance to be determined, which establishes the hydrogen volume.

Four flash x-ray photographs are made by four separate Febatron 706 machines, each having about 3 ns pulse width and an energy of about 600 kev. Since the desired volume resolution at peak compression is ±5% or better, and since the distance between the shim and the anvil will be less than 1 mm at that time, the x-ray resolution must be better than 50 μm. Because of the extremely short 3 ns pulse time, blur due to motion of the disks will be only about 10 μm; while blur due to the finite x-ray source size can probably be made smaller than 50 μm by using ultra-small x-ray anodes developed by Chang (1982). These sources are only 100 μm in size. By making the source-to-specimen distance 400 mm and the specimen-to-film distance only 100 μm the blur should, in principle, be only 25 μm.

Figure 3 shows an artist's conception of the hydrogen experiment, in which the four x-ray machines are shown slung under rails for easy removal and replacement of the needle anodes in the x-ray heads after each shot.

SUMMARY

The project to measure the high-pressure EOS of hydrogen is challenging in its requirements for the best available techniques in several different disciplines, such as cryogenics, precision ultra-high velocity impact techniques, precision laser velocimetry, and flash

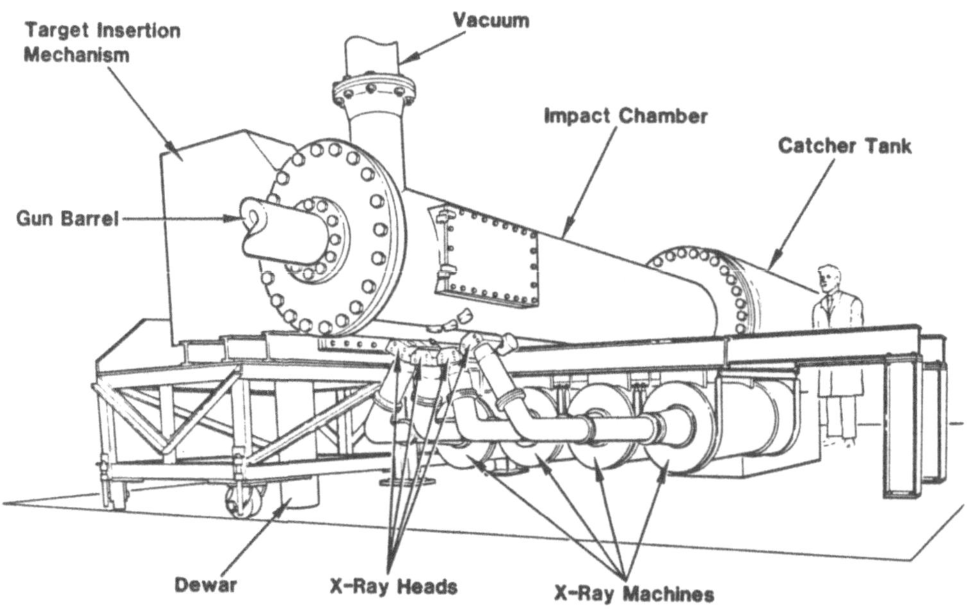

Figure 3: Hydrogen metallization experiment.

x-radiography. It has already led to innovations which will be useful in other studies as well, such as the isentropic loading of the specimen via helium in a dead-end gun barrel, and ease of specimen temperature control via the rapid target insertion and alignment mechanism. It is anticipated that the program will supply new and useful data on the EOS of hydrogen, and that the metallic hydrogen transition will be detected and characterized if it occurs within the 4 Mbar pressure range which we plan to achieve.

ACKNOWLEDGEMENTS

The authors are grateful to many who have contributed to this project. Special thanks go to Dave Cox, Bob Hardy, Doug Scott, Jim Chang, Chuck Martinez, Larry Azevedo, and Don Stewart.

REFERENCES

1. F. V. Grigor'ev, S. B. Kormer, O. L. Mikhailova, A. P. Tolochko, and V. D. Urlin, "Experimental Determination of the Compressibility of Hydrogen at Densities 0.5 - 2 g/cm^3. Metallization of Hydrogen," *JETP Lett.*, **16**, 201-203, 1972.
2. F. V. Grigor'ev, S. B. Kormer, O. L. Mikhailova, A. P. Tolochko, and V. D. Urlin, "Equation of State of the Molecular Phase of Hydrogen in the Solid and Liquid States at High Pressures," *Sov. Phys. JETP*, **42**, 378-381, 1975.
3. R. L. Hawke, D. E. Duerre, J. G. Hubbard, R. N. Keeler, and H. Klapper, "The Purpose, Technique, and Applications of Magnetic Compression," Lawrence Livermore Laboratory Report UCRL-74442, 1974.
4. R. L. Hawke, T. J. Burgess, D. E. Duerre, J. G. Huebel, R. N. Keeler, H. Klapper, and W. C. Wallace, "Observations of Electrical Conductivity of Isentropically Compressed Hydrogen at Megabar Pressures," *Phys. Rev.*, **41**, 994-997, 1978.
5. L. M. Barker and R. E. Hollenbach, "Laser Interferometer for Measuring High Velocities of Any Reflecting Surface," *J. Appl. Phys.*, **43**, 4669-4675, 1972.
6. W. J. Nellis and A. C. Mitchell, "Shock Compression of Liquid Argon, Nitrogen, and Oxygen to 90 GPa," *J. Chem. Phys.*, **73**, 6137-6145, 1980.
7. J. L. Wise, L. M. Barker, J. R. Asay, D. E. Cox, R. D. Hardy, D. D. Scott, L. J. Azevedo, D. T. Stuart, and C. E. Sisson, "A Cryogenic Sample-Preparation System for Dynamic Compression Studies on Condensed Hydrogen," to be published.
8. T. G. Trucano, L. M. Barker, J. R. Asay, and G. I. Kerley, "Numerical Studies of the Gasdynamic Precompression of Solid Molecular Hydrogen," to be published.
9. J. Chang, D. L. Fehl, L. Baker, and D. J. Johnson, "Three-Frame Flash X-Radiography System for Ion-Beam Pellet Implosion Studies," *SPIE*, **Vol. 348**, *High Speed Photography*, 696-699, 1982.
10. J. Chang, C. Martinez, T. L. Franklin, and L. M. Barker, "A Four-Frame Flash X-Ray System for Shock Physics Experiments," to be published.

NUMERICAL STUDIES OF THE DYNAMIC ISENTROPIC LOADING OF SOLID MOLECULAR HYDROGEN*

T. G. Trucano, L. M. Barker, J. R. Asay, and G. I. Kerley

Solid Dynamics Department
Sandia National Laboratories
Albuquerque, NM

INTRODUCTION

Barker, *et al* [1,2], have proposed a dynamic technique for isentropically compressing solid molecular hydrogen specimens to multi-megabar pressures. The reason for introducing a dynamic technique is to explore the equation of state (EOS) in the region where the theoretical metal-insulator transition might occur. Because of the many novel features of this experimental technique, computational simulations of the experiment have been used extensively in the experimental design effort. Our purpose in the present paper is to briefly summarize important conclusions resulting from this work.

The basic experiment, in which helium is used to cushion the impact of a high-velocity projectile against the hydrogen sample is described by Barker, *et al* [2], these Proceedings. The issues that numerical computations address for this work are many, including: Is quasi-isentropic (little associated temperature rise) dynamic loading of the hydrogen sample to multi-megabar pressures possible? What are crucial experimental parameters, and how can they be adjusted to achieve optimal pressures in the sample? What are the possible experimental diagnostics, and how can they be interpreted to yield equation of state information for the hydrogen? What influence does the equation of state of hydrogen exert on the general experimental design?

In Section 2, we will briefly describe the equation of state of solid molecular hydrogen currently in use in our calculations. We are, of course, interested in defining current uncertainties in the equation of state, and in understanding how these influence the calculated hydrodynamics. In Section 3, we discuss the computational approach and summarize some important conclusions of the work. Finally, in Section 4, we consider the relevance of these results to the proposed experiment.

EQUATION OF STATE CONSIDERATIONS

In Figure 1, we present the idealized uniaxial strain geometry used in our calculations. There are three general classes of materials present in the calculations: 1) solids which were

*This work performed by Sandia National Laboratories supported by the U. S. Department of Energy under contract #DEAC04-76-DP00789.

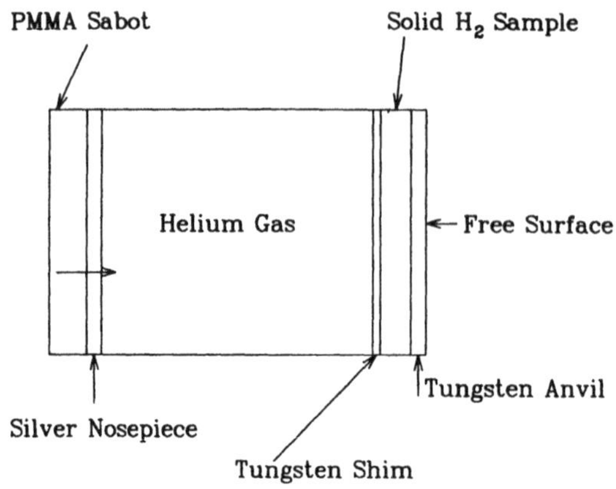

Figure 1: Geometry for the one-dimensional calculations.

modeled using a standard Mie-Grüneisen equation of state with the Hugoniot as a reference curve, 2) the helium gas, which was modeled using the ideal gas equation of state, and 3) the solid molecular hydrogen, which was modeled also using a Mie-Grüneisen equation of state, but with the zero Kelvin isotherm as a reference curve.

The standard Mie-Grüneisen materials are PMMA, silver, and tungsten. Standard EOS parameters are available for these materials [3]. All of them are assumed to be without strength in the computations we discuss. While this is for simplicity, it is also clear that strength effects have little influence on the details that we are most interested in.

Use of the ideal gas EOS for the helium in the launch tube has been rationalized based on the expected pressures in the tube. It is difficult to *a priori* judge that the pressures achieved in the experiment justify a more sophisticated EOS, especially since helium is the most "ideal" of all gases. Certainly, the temperatures achieved are not high enough to worry about non-ideal thermal components in the EOS. To gauge non-ideal volume effects, we have compared the performance of the ideal gas description of the helium with a van der Waals gas EOS, and we have found essentially no difference in computed results. We feel that this is strong evidence that the ideal gas EOS is appropriate to use in these circumstances.

The solid hydrogen EOS [3] uses a cold curve originally discussed by Kerley [4]. This cold curve is an analytic fit to experimental data at low pressures, and to theoretical calculations at high pressure. The analytic fit is improved by certain theoretical considerations. Lattice vibration are included via the Debye model, while the contributions from molecular vibrations and rotations are neglected. This is a reasonable assumption for the low temperatures achieved in the experiment. The Mie-Grüneisen EOS is well-justified as a low-temperature EOS for this substance [1]. The major uncertainties in this EOS of concern to us are in the pressure range of roughly 1 - 3 Mbars, and these influence the computed peak pressures [3]. This, of course, is the pressure range that the current experiment is being designed to probe. In Figure 2, we present a comparison of the cold curve used in our work with selected experimental and theoretical data [5–8].

NUMERICAL SIMULATIONS

To perform our simulations, we have used the 1-D, Lagrangian wavecode WONDY [9]. Specific details of the computational approach have been presented elsewhere [3]. We do note here that quite coarse zoning is being used for the helium layer of the calculations. The only relevant effect of this zoning is to lower the computed peak pressure in the hydrogen

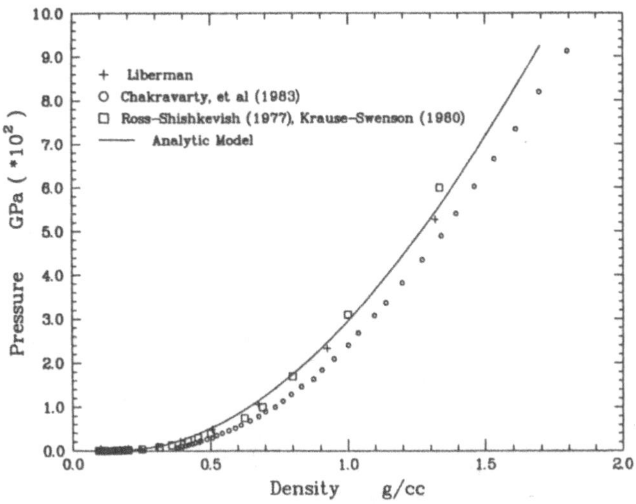

Figure 2: Cold curve for solid molecular hydrogen.

sample on the order of 5% . There is a major reduction in computational expense when coarse zoning is used in the gas. In all other layers of the problem, we have seen insensitivity of the results to the choice of zoning.

Typical pressure histories in the hydrogen are presented in Figure 3. Here, we have compared calculations in which the thickness of the tungsten shim was varied. The launch velocity of the projectile for these calculations was 6.0 km/s, the launch tube was 5.0 m long, and the thickness of the silver nosepiece was 3.66 mm. The total problem duration for these calculations was 840 μsec. The required CPU times on the CRAY 1-S computer at Sandia National Laboratories for these computations were on the order of 500 seconds for the coarse-zoned helium. For very fine zoning in the helium (zones 1.0-mm thick) as much as 6000 seconds of CPU time was required. Note that the time durations of pressures greater than 1 Mbar are typically 1 μsec or less, but that the duration of the experiment is more than two orders of magnitude larger.

DISCUSSION

It is impossible in the present work to discuss the hundreds of calculations that have been performed in an effort to understand and optimize the complicated experimental conditions. It is fruitful to mention issues of specific concern to us, and draw some broad conclusions.

Figure 3 shows a typical trend in optimizing the experimental design - namely, that the thickness of the tungsten shim strongly influences the maximum peak pressure achieved in the sample. The figure also shows the tradeoffs in the process of tuning the shim thickness. For example, thinner shims yield slower loading conditions which more closely approximate isentropic loading. They also yield more homogeneous pressure states in the sample. Increasing the shim thickness increases the sample peak pressures, and steepens the loading path, which ultimately precludes isentropic loading. By comparing the computed loading with the 4.2 K isentrope of our hydrogen EOS we conclude that isentropic loading is maintained, under the current conditions, for shims up to 1.0 mm thick. For the 1.2 mm thick shim, deviation of the loading path from the isentrope is observed. Our theoretical calculations indicate that such deviation from the isentrope is enough to cause temperature increases that would prevent the metal-insulator transition from occurring in the sample.

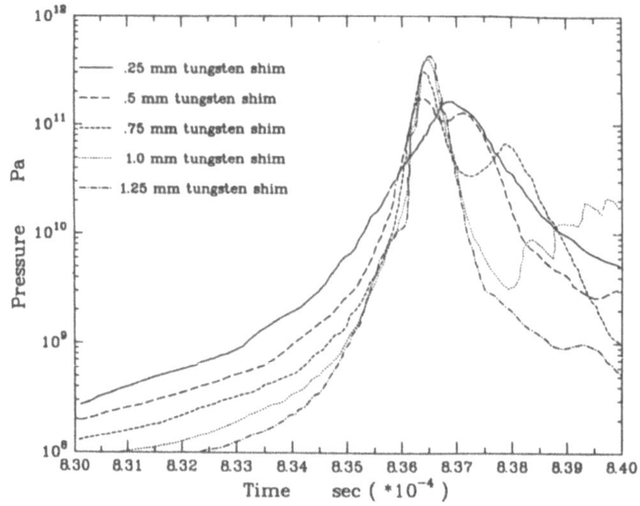

Figure 3: Pressure histories in the hydrogen as a function of shim thickness.

In the same fashion, all the variables in the experimental system must be understood and calibrated through computer simulation. These include the ambient pressure of the helium gas, the length of the launch tube, the thickness of the tungsten anvil, the projectile mass *vs* projectile velocity performance of the two-stage light gas gun, and the occurrence of two-dimensional effects, such as lateral expansion of the target fixture during the time of peak compression. (The latter is currently being examined using a two-dimensional Lagrangian wavecode, TOODY [10].) The studies to date, many of which are summarized in [3], indicate that pressures in excess of 3 Mbar can be achieved, via quasi-isentropic loading, with relative homogeneity of the pressures within the hydrogen sample of 5% to 10% . Pressures up to 4 Mbars seem possible, but require additional sacrifice in terms of pressure homogeneity in the sample. The significance of this is still under investigation.

It is a pleasure to acknowledge the people who have discussed these problems with us. They include L. W. Davison, J. L. Wise, M. E. Kipp, W. J. Nellis, M. Ross, A. C. Switendick, W. R. Davey, and D. S. Drumheller.

REFERENCES

1. L. M. Barker, T. G. Trucano, J. R. Asay, and J. L. Wise, "Preliminary Considerations for the Gasdynamic Precompression of Solid Molecular Hydrogen," SAND85-0111, Sandia National Laboratories, Albuquerque, New Mexico, March, 1985.
2. L. M. Barker, T. G. Trucano, J. L. Wise, and J. R. Asay, " Experimental Technique for Measuring the Isentrope of Hydrogen to Several Megabars," these Proceedings, 1985.
3. T. G. Trucano, L. M. Barker, J. R. Asay, and G. I. Kerley, "Numerical Studies of the Gasdynamic Precompression of Solid Molecular Hydrogen," to be published.
4. G. I. Kerley, "A Model for the Calculation of Thermodynamic Properties of a Fluid," in *Molecular-Based Study of Fluids*, edited by J. M. Haile and G. A. Mansoori, American Chemical Society, Washington, D.C., 107-138, 1981.
5. J. K. Krause and C. A. Swenson, "Direct Measurements of the Constant-Volume Heat Capacity of Solid Parahydrogen from 22.9 tp 16.19 cm ^{3}/mole and the Resulting Equation of State," *Phys. Rev. B*, **21**, 2533-2548, 1980.
6. M. Ross and C. Shishkevish, "Molecular and Metallic Hydrogen," DARPA Report, R-2056-ARPA, 1977.

7. S. Chakravarty, J. H. Rose, D. Wood, and N. W. Ashcroft, "Theory of Dense Hydrogen," *Phys. Rev. B*, **24**, 1624-1635, 1981.
8. D. A. Liberman, "Equation of State of Molecular Hydrogen at High Pressure," *Int. J. Quantum Chem.*; **10**, 297-303, 1976.
9. M. E. Kipp and R. J. Lawrence, "WONDY V – A One-Dimensional Finite-Difference Wave Propagation Code," SAND81-0930, Sandia National Laboratories, Albuquerque, NM, June, 1982.
10. J. W. Swegle, "TOODY IV - A Computer Program for Two-Dimensional Wave Propagation," SAND-78-0552, Sandia National Laboratories, Albuquerque, NM, September, 1978.

SHOCK TEMPERATURE MEASUREMENTS IN AMMONIA

H. B. Radousky, A. C. Mitchell, W. J. Nellis and M. Ross

Lawrence Livermore National Laboratory
Livermore, CA 94550

ABSTRACT

Our first shock temperature measurements on a cryogenic target are reported for NH_3. A new fast optical pyrometer and a cryogenic specimen holder for liquid NH_3 were developed to measure shock temperatures of 4400 and 3600 K at pressures of 61 and 48 GPa. These conditions correspond to those in the ice layers in Uranus and Neptune. The shock temperature data are in reasonable agreement with an an equation of state based on an intermolecular potential derived from NH_3 Hugoniot data.

INTRODUCTION

Temperature measurements on shock compressed materials can be extremely valuable for diagnosing the physics which occurs at these extreme conditions. They provide a much more stringent test of theoretical models than the Hugoniot alone, and give information on the partition of energy between internal energy and phenonena which might be occuring, such as dissociation-ionization. In this paper, we report shock temperatures of ammonia at 48 and 61 GPa, which represents our first such temperature measurements on liquid samples which require cryogenic specimen holders.

The work on ammonia was motivated by its relevance to planetary models. Such models of the outer planets generally consist of three layers.[1,2] The outer layer is assumed to be H_2 and He, probably in proportions similar to the solar mixture, a He mass fraction of 0.22. The middle layer might be either He and metallic H in Jupiter and Saturn, or H_2O, NH_3, CH_4, and possibly CO, CO_2 and N_2 "ices"[3] in Uranus and Neptune. The innermost layer is a rocky core. The condensed gases were compressed isentropically starting from low, atmospheric density and temperature. By virtue of the large masses of these planets the interior temperatures and pressures are quite high. In the outer layer of Jupiter, for example, they range up to about 400 GPa (4 Mbar) and 10,000 K. The conditions in the outer layer of Uranus and Neptune range up to about 20 GPa and 2000 K, while the middle layer ranges up to bout 600 GPa and 7000 K. Shock compression of liquids achieves virtually the same states in the outer planets as isentropic compression starting from the gas phase.

EXPERIMENTAL RESULTS

The shock temperatures in NH_3 were measured using a six channel optical pyrometer. The technique is based on observing the thermal radiation emitted from electrons behind the shock front, which are in thermodynamic equilibrium with the lattice. This radiation can escape in the forward direction through the unshocked and still transparent material. The experimental design is shown in Fig. 1. The pyrometer consists of six photomultiplier tubes which can measure the light intensity emitted from the shock heated sample at six different wavelengths ranging from 254-830 nm. Including the ultraviolet portion of the spectrum allow measurements up to 30,000 K, which is important for studying many of the materials of current high interest such as N_2[4] and He.[5] The output of the photomultiplier tubes is then recorded on fast oscilliscopes and film.

For shock temperature measurements on most planetary fluids, cryogenic specimen holders are required. Our holder uses the cryogenic design used previously for Hugoniot experiments.[6] For NH_3 experiments a variation using a cold N_2-gas flow system[7] was employed to avoid handling liquid NH_3. A sapphire window was placed at the rear of the sample cavity to permit light emitted from the shock front to exit the holder and be sent to the pyrometer by a turning mirror. Light is observed from a specimen area of 2 mm^2 on axis.

An oscillogram from the 600 nm channel of the pyrometer is shown in Fig. 2. The signal is flat as the shock traverses the sample, and the double shock off the shappire window is clearly seen as a rise in light intensity. The magnitude of the voltage signals are related to actual intensities through an in-situ calibration against a known source of spectral radiance. The data for the two sets of measurements are shown

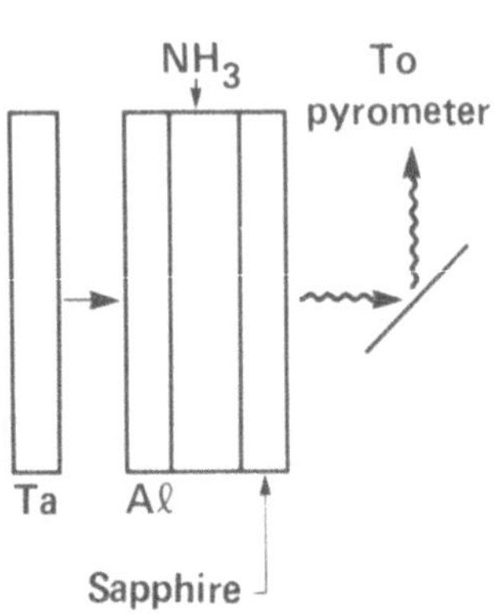

Fig. 1 Schematic diagram of the experimental configuration. Light from the shocked NH_3 escapes through the transparent shapphire window.

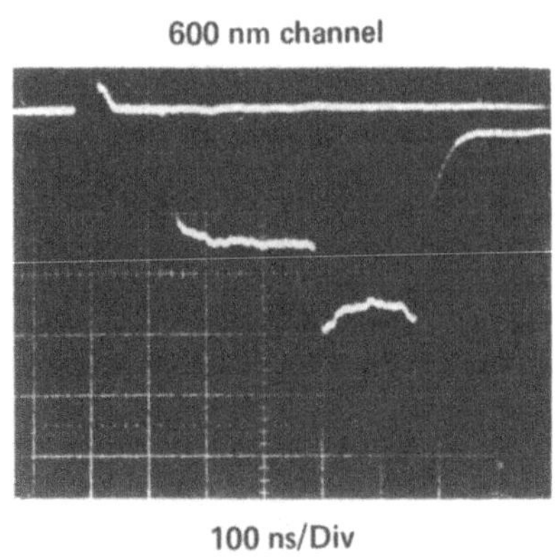

Fig. 2 Oscilligram from the 600 nm channel of the pyrometer. Record from shot NH_3 #2(T = 4380 K).

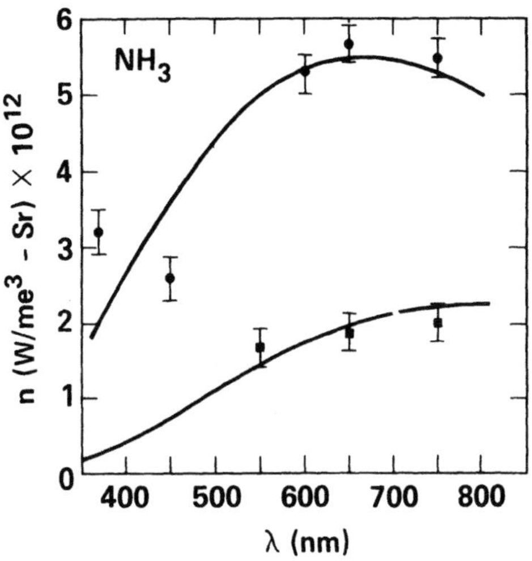

Fig. 3.

in Fig. 3. The temperature was determined by least square fitting the data to a Planck greybody spectrum,

$$n(\lambda) = \frac{\varepsilon C_1}{\lambda^5}\left(e^{C_2/\lambda T} - 1\right)^{-1} , \qquad (1)$$

where $n(\lambda)$ is the spectral radiance, λ is the wavelength, $C_1 = 1.19 \times 10^{-16}$ w - me^2/Sr, $C_2 = 1.44 \times 10^{-2}$ me-K and ε is a wavelength independent emissivity. The least squares fits to the data are shown in Fig. 3. The results, plotted as temperature vs pressure are shown in Fig. 4 and tabulated in Table 1, along with other relavent parameters. The pressure was obtained by measuring U_I and the method of shock impedence matching. The C and S parameters used in the impedence matching are tabulated in Table 2.

Table 1. Shock Temperatures and Emissivities for NH_3

Shot	T_0	ρ_0 (g/cm^3)	U_I (km/s)	P (GPa)	V/V_0	T (K)	ε
NH_3 #3	230 K	0.6933	5.60	47.6	0.414	3550 ± 200	0.95
NH_3 #2	230 K	0.6933	6.56	61.3	0.389	4380 ± 200	1.00

1 GPa = 10 kbar
V_0 = 24.6 cm^3/mole

Table 2. C and S Parameters used to obtain the pressure in shocked NH_3 using the method of shock impedence matching

	NH_3[a]	Al[b]	Ta[b]
C (km/s)	3.211	5.386	3.293
S	1.201	1.339	1.307

[a] Ref. 7. Valid for $4.0 < U_p < 7.6$ km/sec
[b] Ref. 11.

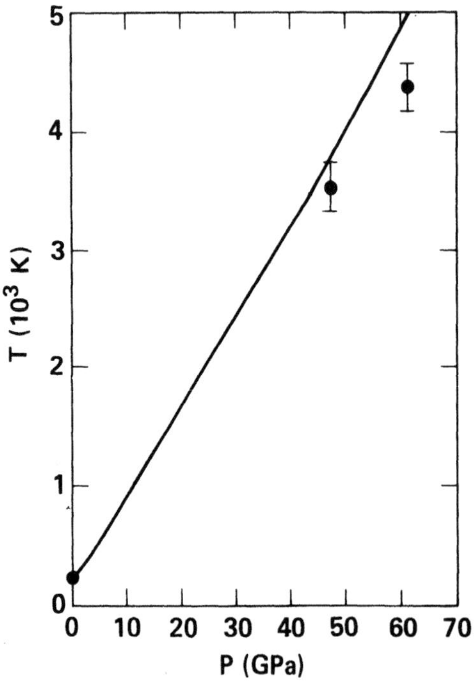

Fig. 4 NH_3 shock temperature plotted as a function of pressure. Solid line is a calculation discussed in the text.

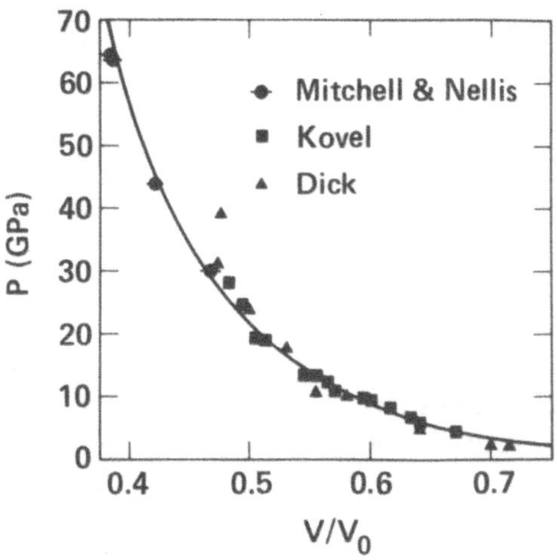

Fig. 5 Best fit to the Hugoniot data is shown for NH_3. The pair potential derived from the fit was used to calculate the temperatures in Fig. 4. A good fit to the data was obtained over the entire pressure region.

DISCUSSION

The solid line shown in Fig. 4 is the result of a calculation using fluid perturbation theory and is based on an intermolecular potential derived from a fit to the Hugoniot data. The form used was an exponential-sixth power potential,

$$\phi(r) = \frac{\varepsilon}{k} \{[\frac{6}{\alpha-6}] \exp [\alpha(1-r/r^*)] - [\frac{\alpha}{\alpha-6}] [\frac{r^*}{r}] \}^6 \quad (2)$$

The parameters chosen to best fit to the Hugoniot data were $\alpha = 12.5$, $\varepsilon/k = 150$ K, and $r^* = 3.85$ Å. The Hugoniot data[7,9-10] and the resulting fit are shown in Fig 5. The calculations include contributions from rotation and vibration.

Comparison of the calculated temperatures with the experimental results in Fig. 4 shows that the model used is in reasonably good agreement with the data. The small deviation of 10% at the highest pressure maybe an indication of an additional absorption mechanism at these high pressures. Previous electrical conductivity measurements[9] on ammonia have show that it becomes conductive in the range 7-28 GPa. The mechanism proposed to explain the high conductivity was molecular dissociation-ionization. The relatively small difference between the measured temperatures and those calculated using this non-dissociative model suggests that dissociation is not sufficient to affect the equation of state in a dramatic way. Including dissociation-ionization in the model would have the effect of lowering the calculated temperatures in the direction of better agreement with the experimental data.

ACKNOWLEDGEMENTS

This work was performed under the auspices of the U.S. Department of Energy by the Lawrence Livermore National Laboratory under contract number W-7405-ENG-48, with partial support from the U.S. National Aeronautics and Space Administration under contract No. 83-033.

We acknowledge P. C. McCandless for fabricating the cryogenic specimen holders, C. D. Wozynski, K. Peterson and R. F. Schuldheisz for firing and maintaining the two-stage gun, and I. Miller for fabricating the projectiles.

REFERENCES

1. W. L. Slattery, Icarus 32, 58, (1977).
2. W. B. Hubbard, and J. J. MacFarlane, J. Geophys. Res. 85, 225 (1980).
3. D. J. Stevenson, Annu. Rev. Earth Planet Sci. 10, 257, (1982).
4. W. J. Nellis, N. C. Holmes, A. C. Mitchell, and M. van Thiel, Phys. Rev. Lett. 53, 1661, (1984b).
5. W. J. Nellis, N. C. Holmes, A. C. Mitchell, R. J. Trainor, G. K. Governo, M. Ross, and D. A. Young, Phys. Rev. Lett. 53, 1248, (1984a).
6. W. J. Nellis and A. C. Mitchell, J. Chem. Phys. 73, 6137 (1980).
7. A. C. Mitchell and W. J. Nellis, J. Chem. Phys. 76, 6273 (1982).
8. M. Ross, J. Chem. Phys. 71, 1567 (1979).
9. M. I. Kovel, Ph.D. thesis, Lawrence Livermore National Laboratory report UCRL-51367, pp. 90.
10. R. D. Dick, J. Chem. Phys. 74, 4053 (1984).
11. A. C. Mitchell and W. J. Nellis, J. Appl. Phys. 52, 3363 (1981).

ELECTRICAL CONDUCTIVITY MEASUREMENTS IN SHOCK

COMPRESSED LIQUID NITROGEN

D. C. Hamilton, A. C. Mitchell and W. J. Nellis

Lawrence Livermore National Laboratory
Physics Department
Livermore, California 94550

ABSTRACT

The electrical conductivity of shock compressed liquid nitrogen was measured in the pressure range 20-50 GPa using a two-stage light-gas gun. The conductivities covered a range $4x10^{-2}$-$1x10^{2}$ ohm^{-1} cm^{-1}. The data are discussed in terms of a liquid semiconductor model below the onset of the dissociative phase transition at 30 GPa.

INTRODUCTION

Recent dynamic experiments indicate that liquid nitrogen begins to undergo a phase transition to the monatomic state at shock pressures of 30 GPa.[1,2] This is evidenced by an increase in the shock wave compressibility and is attributed to the combined effects of shock compression and heating. Below the transition region nitrogen is believed to remain diatomic and follow a law of corresponding states.[3]

Nitrogen is a major constituent in explosives, comprising about 30 molar percent of the detonation products.[4] Nitrogen, in its molecular form or as ammonia, is also represented to a significant extent in the interiors and atmospheres of the outer planets.[5] Issues for both planetary conditions and detonation phenomena are the importances of dissociation, phase separation and semiconduction. Electrical conductivity measurements can provide information on the partitioning of energy between internal degees of freedom like electronic excitation and thermal energy and enable an estimate of the degree of ionization and mechanisms of charge transport. For these reasons we performed the first reported electrical conductivity measurements of liquid nitrogen over a pressure range of 20-50 GPa.

EXPERIMENT

The experiments were performed using a two-stage light-gas gun. The diagnostic system[6] and sample preparation[2] are described elsewhere. The liquid samples were condensed from reagent grade nitrogen gas with measured impurity concentrations of oxygen and argon not exceeding 0.30 and 0.015

mole percent, respectively. Final pressures of the shocked samples were inferred from impedance matching using the known equations of state of aluminum[7], tantalum[7] and nitrogen[2].

The electrical conductivity was determined by methods outlined in references 8-11. A constant-voltage circuit was used for the measurements of conductivity below 0.1 ohm^{-1} cm^{-1} and constant-current circuit for higher conductivity values. In the former case a shock wave traversing perpendicular to two parallel electrodes situated in the sample acted as a switch and resistor of an RC circuit. The magnitude of the signal compared to signals produced by several standard resistors gave the resistance of the final state of the sample. In the constant-current circuit a capacitor was discharged through current-limiting resistors and a shunt resistor (across the electrodes) just prior to the shock arrival at the electrodes.[11] The shunt was chosen to match the expected resistance of the sample for maximum sensitivity. For both types of circuits and all experiments the resistances of the cells were calibrated using liquids of known conductivities.

RESULTS AND DISCUSSION

In Figure 1 the electrical conductivity of liquid nitrogen versus shock pressure is shown to cover nearly four orders of magnitude over the shock pressure range 20-50 GPa. Preliminary analysis indicates the shocked fluid behaves as a liquid semiconductor and follows the characteristic form

$$\sigma \sim c\ e^{-E/KT}$$

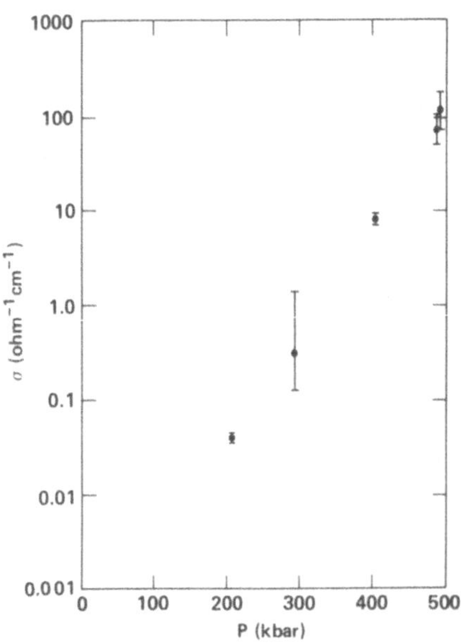

Fig. 1. Electrical conductivity of shock compressed liquid nitrogen versus pressure.

conductivity in the molecular region. Calculated shock temperatures[3] were used for this analysis. The slope of the logarithm of the electron conductivity in the molecular phase below 300 kbar versus the reciprocal of the calculated temperature gives an activation energy of about 2.7 eV ± 1.2 eV which corresponds to a semiconducting effective band gap of 5.4 eV. This suggests a density-dependent narrowing of the band gap from its value at 77 ^{o}K of around 9 eV, which is based on optical absorption measurements.[12] Proceeding with a semiconductor description we calculate the carrier concentration[13] $n = 2.4 \times 10^{18}$ cm^{-3} at 20 GPa, where we have used Ross's calculated temperature of 4800 ^{o}K and a band gap of 5.4 eV. This is about one carrier per 10^4 nitrogen molecules. The effective mobility, assuming a form $\mu_{eff} = \sigma/ne$, where σ is the electrical conductivity (4×10^{-2}) and e the electronic charge, is 0.1 cm^2/volt-sec, which is more than 10 times its value measured at 77 ^{o}K.[14] The electronic collision time $\tau = m\sigma/ne^2$, is 6×10^{-17} s, where m is the free-electron mass. Using an electron thermal velocity v_T of 4.7×10^7 cm/s, the mean free path, $\lambda = \tau v_T$, is 0.3 Å. This estimate is less than a molecular diameter and is characteristic of a percolation or Brownian-type charge transport process. In this model electrons are described as being excited into extended states near a mobility edge.[15]

Extrapolating the calculated temperatures to higher shock pressures, which probably overestimates the temperatures because of the neglect of molecular dissociation, the conductivity increases dramatically from its constant-slope value on a 'logσ vs 1/T' plot indicating a large increase in carrier mobility and/or carrier density. Further interpretation of the data is awaiting shock temperature measurements.

ACKNOWLEDGEMENTS

The author would like to thank A. K. McMahan for valuable comments, and Craig Wozynski, Kirk Pederson, Robert Schuldheisz, and Neil Holmes for their support at the two-stage gas gun.

This work was performed under the auspices of the U. S. Department of Energy at Lawrence Livermore National Laboratory under Contract No. W-7405-ENG-48.

REFERENCES

1. W. J. Nellis, N. C. Holmes, A. C. Mitchell and M. van Thiel, Phys. Rev. Lett. 53, 1661 (1984).
2. W. J. Nellis and A. C. Mitchell, J. Chem. Phys. 73, 6137 (1980).
3. M. Ross and F. H. Ree, J. Chem. Phys. 73, 6146 (1980).
4. M. Finger, E. Lee, F. H. Helm, B. Hayes, H. Hornig, R. McGuire, M. Kahara and M. Guidry, Proc. Sixth Symp. on Detonation, ACR-221, Office of Naval Research-Dept. of the Navy (1976).
5. D. J. Stevenson, Mat. Res. Soc. Symp. 22, 357 (1984).
6. A. C. Mitchell and W. J. Nellis, Rev. Sci. Instrum. 52, 347 (1981).
7. A. C. Mitchell and W. J. Nellis, J. Appl. Phys. 52, 3363 (1981).
8. A. C. Mitchell and R. N. Keeler, Rev. Sci. Instrum. 39, 513 (1968).
9. A. A. Brish, M. S. Tarasov and V. A. Tsukerman, Sov. Phys. JETP 11, 15 (1960).
10. V. V. Yakushev, Soviet Combustion, Explosion and Shock Waves 14, 131 (1978).
11. A. C. Mitchell and W. J. Nellis, J. Appl. Phys. 76, 6273 (1982).
12. P. Gurtler and E. E. Koch, Chem. Phys. 49, 305 (1980).
13. C. Kittel, Introduction to Solid State Physics (John Wiley & Sons, Inc., New York, 1971).

14. B. Halpern and R. Gomer, J. Chem. Phys., 51, 1031 (1969).
15. M. Silver and P. Smejtek, in Conduction in Low Mobility Materials, ed. by N. Klein, D. I. Tannhauser and M. Pollack (Taylor and Francis, Ltd., London, 1971), p. 435.

HUGONIOT AND ELECTRIC RESISTIVITY MEASUREMENTS ON AMORPHOUS SE

K. Kani
Faculty of Education, Okayama University, Okayama 700, Japan

T. Yamada
Niihama National College of Technology, Niihama 792, Japan

M.Abe
Shin-Nihon Steel Corporation, Hikari 743, Japan

ABSTRACT

Hugoniot and electric resistivity measurements on amorphous Se have been made in the pressure range up to 24.5 GPa by projectile impact experiments. Impact velocity was measured by magnet flyer method. A new pin contactor method which was developed by an application of anormalous thermoelectric effect of a thermocouple was used to measure transit time of shock wave in the sample. The electric resistivity of a shocked sample was evaluated by measuring the electric current flowing through the sample from a constant voltage power source. The U_s-U_p Hugoniot obtained showed a linear relation up to 24.5 GPa. The electric resistivity of the sample decreased from 10^3 Ω.cm to a few Ω.cm between 5 GPa and 7 GPa. The mechanism of the electric resistivity change was discussed.

INTRODUCTION

Amorphous selenium (Se) is a black solid and has high electric resistivity at room temperature and pressure. Many investigations have been performed on amorphous Se only under static pressures. Bridgman[1], Weir[2] and Soga et al.[3] clarified the compression characteristics. Balchan and Drickarmer[4], Witting[5], Fuhs et al.[6], Moodenbaugh et al.[7], Aoki et al.[8], Gupta and Ruoff[9], and Bundy and Dunn[10] investigated electric conduction behaviors.

On the other hand, compression or electric resistivity data are quite limited[11] under shock compression. The present study was motivated by this fact. In this paper, the shock wave velocity-particle velocity Hugoniot and electric resistivity of amorphous Se under shock compression are clarified.

EXPERIMENT

Amorphous Se disc specimens used for shock compression experiments were prepared by melting and rapid solidification of 99.9 % purity Se

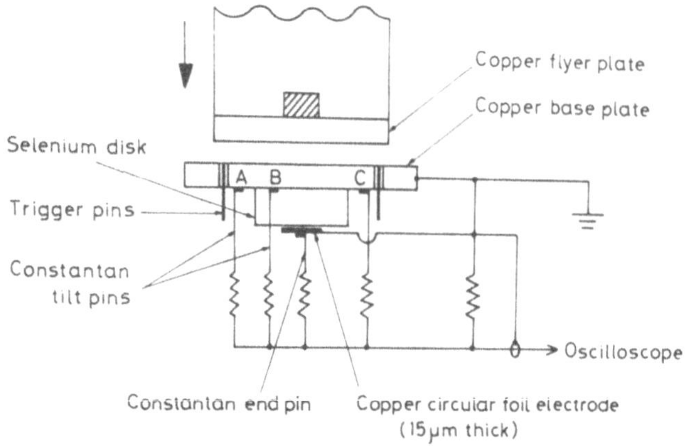

Fig.1 Schematic illustration of sample assembly and measuring circuit used for Hugoniot measurement.

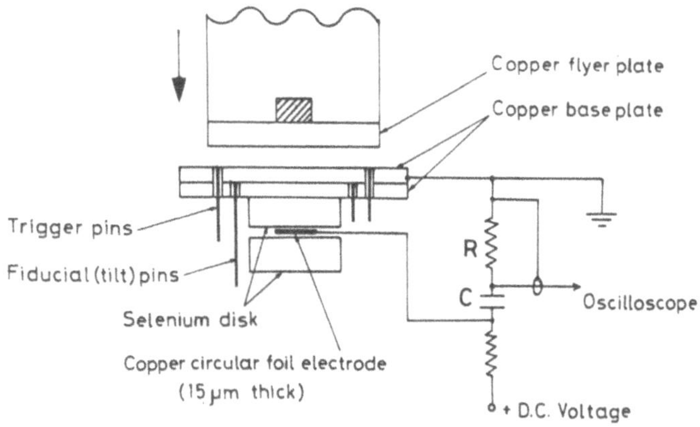

Fig.2 Schematic illustration of sample assembly and measuring circuit used for electric resistivity measurement.

powder. The diameter of all specimens is 15 mm, while their thickness was chosen according to experimental purposes: 3 ~ 4 mm for Hugoniot measurement and 1 ~ 2 mm for electric resistivity measurement. The porosity of the specimen was less than 1.9 % . Crystal structures and physical properties of Se specimens prepared by the same method had been already clarified by Soga et al.[3].

Figs.1 and 2 show schematic illustration of sample assemblies and measuring circuits employed for Hugoniot and electric resistivity measurements respectively. These assemblies are impacted by a 2 mm thick copper flyer plate glued on the front surface of a projectile which is accelerated by a propellant gun with 30 mm diam. smooth bore. Using the sample assemblies shown in Fig.1, the transit time of shock wave propagating in a Se sample was measured by newly developed electric pins. The details of the pin will be reported elsewhere[12], therefore only a brief description is given here. It is well known that, as an anomalous thermoelectric effect of thermocouples, great electromotive force (EMF) is generated when a junction of the thermocouple is subjected to a shock compression.

In Fig.1, at points A,B and C, each end of constantan strips, 0.5 mm wide and 10 μm thick, is put into contact with the rear surface of a copper base(driver) plate. That is, each contact point forms the junction of a copper-constantan thermocouple. Therefore, the arrival times of a shock wave at contact points(A,B,C) are known by measuring great EMFs generated in the thermocouples. These junctions function as tilt pins. The central region of the rear surface of a Se specimen is covered with a thin copper foil (15 μm thick). Then the same constantan strip used as the tilt pin was put into contact with the copper foil in order to detect the arrival time of a shock wave travelled through the specimen (end pin). From the signals generated by tilt pins and the end pin, shock wave velocity is determined. Then the particle velocity is evaluated from the shock wave velocity and a flyer impact velocity by the impedance-match method. The flyer impact velocity is measured by the magnet flyer method[13].

In the sample assembly shown in Fig.2, the Se specimen disc is sandwiched betweeen a copper driver plate and a circular copper-foil backing electrode. Then the electrode, 15 μm thick and 7 mm in diameter, is backed up by a thick Se disc having the same diameter as that of the sample for a shock wave to propagate through the backing electrode. Three tilt pins(ordinary type) are set to detect the inclination of shock wave front, and then to know the precise arrival time of a shock wave front at the backing electrode. In experiments, a constant voltage is supplied through the read-out resistance (50 Ω) between the driver plate and the backing electrode. The read-out voltage V_x observed on the oscilloscope is related to the specimen electric resistance as follows;

$$R_s = R_0 \, (V_0 \,/\, V_x - 1) \qquad (1)$$

where V_0 is the voltage of the constant voltage power source used, and R_s and R_0 are specimen electric resistance and read-out resistance respectively. Then, the relationship between R_s and specimen electric resistivity ρ is given by

$$R_s = \int_0^h [\rho/\pi\{r_{Cu} + (r_{Se} - r_{Cu})x/h\}^2]\, dx \qquad (2)$$

here, h is specimen thickness after shock compression, X is distance measured from a backing electrode, and r_{Cu} and r_{Se} are the radius of backing electrode and that of Se specimen respectively.

RESULTS AND DISCUSSION

Hugoniot Measurement

Fig.3 shows an example of tilt and end pin signals obtained from the copper-constantan thermocouple pins used. In the figure, numbers 1,2 and 3 indicate the shock wave arrival time at each tilt pin, and E corresponds to that at an end pin. Based on these data and the procedure mentioned in the previous section, shock wave and particle velocities in Se specimens were obtained and are shown in Fig.4. The relationship between shock wave and particle velocity obtained by the method of least squares is expressed as the following equation:

$$U_s = 1.63 + 1.38\, U_p \qquad (3)$$

here, U_s and U_p are shock wave velocity and particle velocity respectively, and both are measured in Km/s. The bulk sound speed calculated from the longitudinal and transverse wave velocity measured by Soga et al.[3] is 1.49 Km/s. This value is about 10 % lower than U_s corresponding to zero particle velocity.

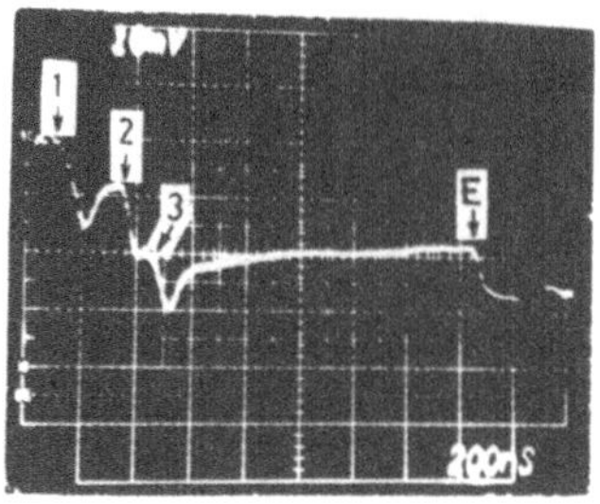

Fig.3 An example of tilt pins and end pin signal obtained using sample assembly shown in Fig.1. Vertical:10 mV/div., Time sweep:200 ns/div.

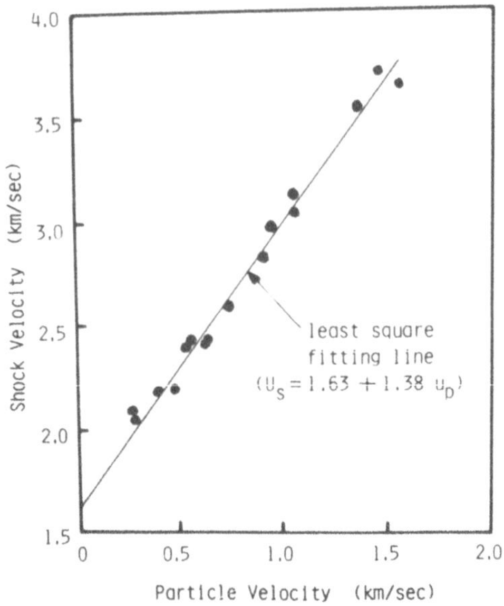

Fig.4 Shock wave velocity-particle velocity Hugoniot obtained for amorphous Se.

From the U_s - U_p Hugoniot shown above, pressure(P)-volume(V) Hugoniot is derived using the fundamental equations for a shock wave, and the room-temperature isotherm is calculated by the same method as that used by McQueen et al.[14] from the P-V Hugoniot using the Grüneisen equation of state and material constants such as specific heat at constant volume and the Grüneisen coefficient obtained by Soga et al.[3]. Fig.5 shows the comparison between the isotherm and Bridgman's data. The fairly good agreement between them suggests that shock wave parameters used for determination of the Hugoniot were precisely measured by the present authors.

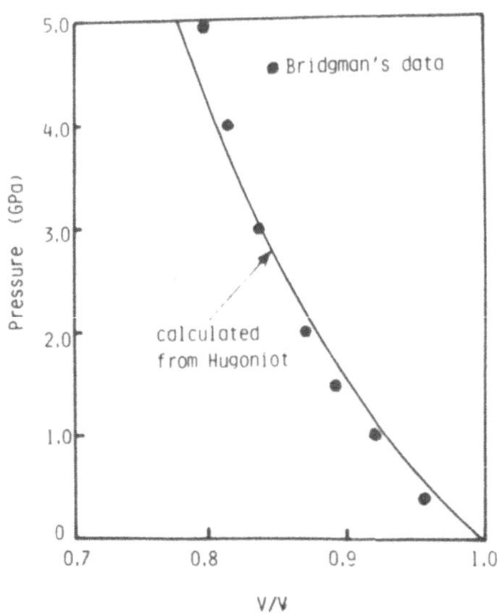

Fig.5 Comparison between isotherm derived from P-V Hugoniot and Bridgman data.

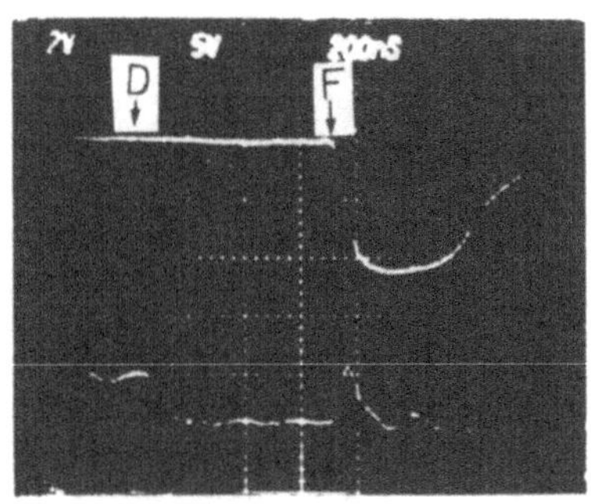

Fig.6 An example of current signal(upper trace) and tilt pin signal(lower trace) using sample assembly shown in Fig.2. Pressure:6.3 GPa, Time sweep:200 ns/div.

Fig.6 shows an example of a current signal(upper trace) and a tilt pins signal(lower trace). In the figure, D and F indicate the arrival times of a shock wave front at the front surface and the rear surface of a sample respectively. These times are evaluated from the tilt pins signal and shock wave velocity data. It is apparent from the figure that the time F agrees well with the time when conduction current begins to flow. Sample resistivity is evaluated using eq.(2) from the sample resistance data which were calculated by eq.(1).

Fig.7 shows the relationship between sample resistivity and shock pressure. It is obvious from the figure that the resistivity of amorphous Se decreases from about 10^3 Ω.cm to a few Ω.cm with the variation of the pressure from about 5 GPa to 7 GPa.

Based on the resistivity data of the amorphous Se, brief discussion on its conduction mechanism under shock compression is given below. Fig.8 shows the relationship between the reciprocal of absolute sample temperature(T) and the resistivity. The sample temperature under shock compression is evaluated by the same method used in the derivation of the isotherm mentioned above. The resistivity data indicate that the resistivity decreases with increase in temperature as is well known in intrinsic semiconductors. Fuhs et al.[6] concluded from static compression experiments that the electric resistivity of amorphous Se shows the same behavior as that of an intrinsic semiconductor. In addition, Aoki et al.[8] obtained the pressure dependence of energy gap E_g of amorphous Se by static compression experiments. From their E_g data, the following equation is derived:

$$E_g \text{ (eV)} = 0.63 - 0.059 \text{ P (GPa)} \qquad (4)$$

The coefficient, 0.059, shows fairly good agreement with those (0.04,

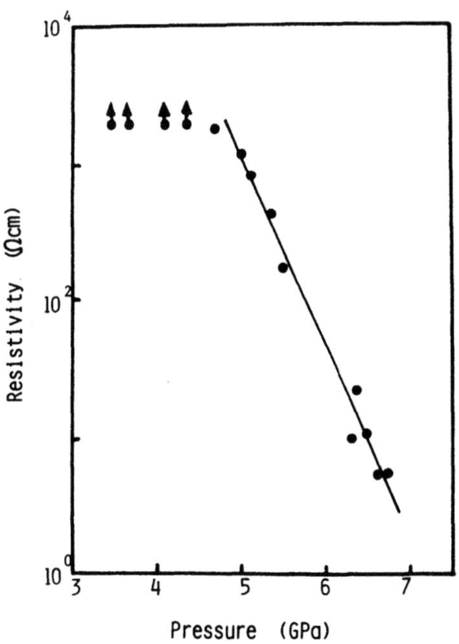

Fig.7 Pressure dependence of resistivity of amorphous Se.

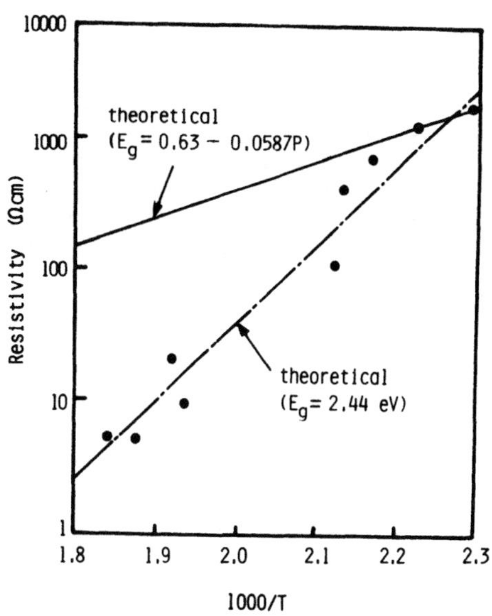

Fig.8 Relation between reciprocal of absolute sample temperature and resistivity.

0.08, 0.09) obtained by Bundy and Dunn[10].

Here, the following assumption that the resistivity of amorphous Se varies according to the next equation is made:

$$\rho = \rho_0 \exp\{(0.63 - 0.059\ P)/\ 2kT\} \qquad (5)$$

where k is Boltzmann constant and ρ_0 is determined in the present discussion as the greatest value of the resistivity in Fig.8. The calculated results are shown in the figure by a solid line. From the figure, it is obvious that the present estimation for pressure dependence of energy gap E_g by eq.(4) is deficient, and the resistivity data suggest that E_g is nearly constant and independent of pressure. For the further discussion, the following assumption that energy gap of amorphous Se is independent of both pressure and temperature is made. From this assumption, an energy gap of 2.44 eV (dashed line) is evaluated from the resistivity data. This value of E_g is almost equal to that of molten Se[15]. This fact suggests that the conduction mechanism of amorphous Se may be same as that of molten Se. This suggestion is quite probable because the two materials have an almost identical atomic arrengement.

SUMMARY

The followings are conclusions obtained in the present experimental study on amorphous Se.

1) Shock wave velocity(U_s)-particle velocity(U_p) Hugoniot is expressed by a linear equation up to 24.5 GPa as follows;

$$U_s = 1.63 + 1.38\ U_p$$

2) Electric resistivity decreased from about 10^3 Ω.cm to a few Ω.cm with pressure increase from 5 GPa to 7 GPa.

3) From the discussion on the conduction mechanism, it is suggested that amorphous Se behaves in the same way as molten Se in electric conduction.

REFERENCES

1. P.W.Bridgman, Proc. Am. Acad. Arts. Sci. 74, 41 (1940).
2. C.E.Weir, J. Res. NBS 52, 247 (1954).
3. N.Soga, M.Kunugi and R.Ota, J.Phys. Chem. Solids 34, 2143 (1973).
4. A.S.Balchan and H.G.Drickarmer, J. Chem. Phys. 34,1948 (1961).
5. J.Witting, J.Chem. Phys. 58, 2220 (1973).
6. W,Fuhs, D.Schlotter and J.Stuke, Phys. Status Solidi 57, 587 (1973).
7. A.R.Moodenbaugh, C.T.Wu and R.Viswanathan, Solid State Commun. 13, 1413 (1973).
8. K.Aoki, O.Shimomura and S.Minomura, Proc. 4th Int. Conf. on High Pressure 314 (1974).
9. M.C.Gupta and A.L.Ruoff, J. appl. Phys. 49, 5880 (1978).
10. F.P.Bundu and K.J.Dunn, J. Chem. Phys. 71, 1550 (1979).
11. T.T.Cole and J.W.Lyle, Rev. Sci. Instrum. 42, 1258 (1971).
12. Rev. Sci. Instrum. to be submitted.
13. K.Kondo, A.Sawaoka and S.Saito, Rev. Sci. Instrum. 48, 74 (1977).
14. R.G.McQueen et al., in High-Velocity Impact Phenomena, edited by R.Kinslow (Academic Press, New york and London 1970) 294.
15. R.W.Haisty and H.Krebs, in Physics of Se and Te, edited by W.C. Cooper (Pergamon 1969) 345.

FERROELECTRIC PROPERTIES AND SHOCK RESPONSE OF A POLED PVF_2 POLYMER AND OF VF_2/C_2F_3H COPOLYMERS

Francois Bauer

Institut Franco-allemand de recherches de Saint-Louis (ISL)
12 rue de l'Industrie, 68301 Saint-Louis, France

I - INTRODUCTION

When dielectrics are permanently polarized, this is the case for the electrets, ferroelectrics, piezoelectric polymers, and so forth, they turn out to become essential active components such as piezoelectric transducers,voltage generators, electrooptical modulators, energy sources used for initiating igniters containing primary or secondary explosives, probes used for measuring shock waves, and so forth.

Ferroelectrics and piezoelectrics have found widespread use. In the majority of cases these applications are based on the mechano-electrical or electromechanical energy conversion. In 1969 KAWAI [1] found out that the semi-crystalline PVF_2 or PVDF (polyvinylidene fluoride) polymer become strongly ferroelectric after having been subjected to the effects of both mechanical stretching and application of an electric field. Polyvinylidene fluoride also called PVF_2 or PVDF is a semi-crystalline polymer whose monomer is $-CH_2-CF_2-$.

As reported from KEPLER [2] we know that polyvinylidene fluoride is a crystalline polymer which exists in three stable polymorphs. In phase I , that is in the polar form, the polymer can be poled when a strong electric field is applied. PVF_2 material is of particular interest. In effect it combines the characteristics of a plastic material with those of piezoelectric elements. PVF_2 in the form of a mechanically stretched film exhibits a remanent polarization (P_r = 2 to 10 $\mu C/cm^2$) the stability of which is permanent at room temperature. The dielectric constant measured at low frequencies is high (ϵ_r = 12) at room temperature. This is an increasing function of the temperature, which diverges in the vicinity of T_f = 170°C.

Vinylidene fluoride and trifluorethylene copolymers P(VDF-TrFE) or VF_2/C_2F_3H are also known to be of ferroelectric nature. These copolymers with more than 50 mole percent vinylidene fluoride can be made piezoelectric. They exhibit a ferroelectric to paraelectric transition at a temperature level which increases with increasing vinylidene fluoride content [3, 4, 5].

It is also well-known [6] that these copolymers can be formed into plates or films (1 μm to 500 μm or more) either by cast from solution or by extrusion or compression of the melts. They cristallize directly into the polar phase or β phase [6].

Previously conducted studies have shown that adequately prepared polyvinylidene fluoride films can serve as the active material in a piezoelectric shock wave stress gauge [7,8]. These first results have led to extensive investigations of the shock response of PVF_2 and VF_2/C_2F_3H copolymers.

The studies of the behaviour of shock loaded PVF_2 and of some VF_2/C_2F_3H copolymers will be presented. We also shall analyze the contribution of the polarization and of its spatial distribution which are predominant in the piezoelectric shock response of these organic materials. The calibration curves of PVF_2 stress gauges will be shown. Comparison between experimental and computed shock data will be presented. Results achieved in the case of impact loading and explosive loading (in small dimension) will be shown.

II - FERROELECTRIC POLARIZATION OF ORGANIC POLYMERS AND COPOLYMERS

In the preparation of piezoelectric polymers and copolymers it is in general necessary to apply a high poling electric field to an essentially insulating material. During this operation three basic processes can occur. The dipoles tend to rotate and to orient themselves, ions migrate under the internal electric field and space charges can be injected at the interfaces. The relative importance of these processes depends only on a number of parameters such as poling temperature and the art of the electric field applied. The latter can be generated either with adjacent electrodes or with charges deposited on the material itself in a "Corona discharge" or by electron or ion bombardement. Following is a summary of the type of poling process applied to PVF_2 and copolymers VF_2/C_2F_3H.

II.1 Polarization of PVF_2 [8],[9]

At room temperature, an increasing sine high-voltage signal is applied to the specimen with sputtered gold electrodes at a very low frequency and the current traversing the PVF_2 sample is measured. Since D is equal to $\varepsilon E + P$, the intensity i passing through the polar dielectric is given by $i = i_D + i_R = \varepsilon \cdot dE/dt + dP/dt + E/R$ (where i_D = displacement current, i_R = current into the resistance R, D = electric induction, P = polarization of the material, R = internal resistance of the material). From the intensity i measured, the resistive component E/R as well as the capacitive component $(\varepsilon \cdot dE/dt)$ are subtracted. We obtain a cycle for i such that $i = dP/dt$.

By integrating the current we obtain the curve which indicates the total polarization of the material as a function of the electrical field applied. This polarization appears as a hysteresis loop. Typical experimental results for a biaxially oriented sample are shown in Fig. 1. The hysteresis loop was traversed many times at room temperature in order to drain the ions and local space charges. PVF_2 showing this hysteresis loop is "ferroelectric". The remanent polarization attains 10 $\mu C/cm^2$ on the samples of 23 μm made available by Rhône Poulenc. At room temperature the coercive field equals 1 MV/cm. The piezoelectric coefficient d_{33} attains 22 pC/N [7]. Our results seem to be in better agreement with BROADHURST'S model [10]. In the latter case P_r may attain 11 $\mu C/cm^2$ for a polymer containing 50% of crystallites.

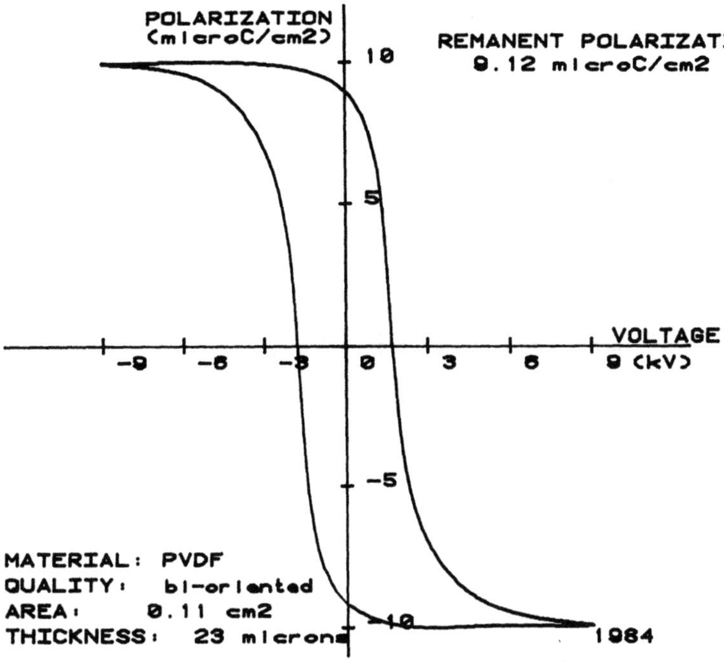

Fig. 1: Hysteresis curve for biaxially stretched PVF_2

The polarization of the material is stable between -40°C and +80°C. It is verified that the coefficient d_{33} (measured with a Berlincourt piezometer) is proportional to the remanent polarization (Fig. 2). Reproducible d_{33} values of PVF_2 samples can be obtained with this poling process. Maximum values of d_{33} are obtained for electric field just below the value of dielectric breakdown.

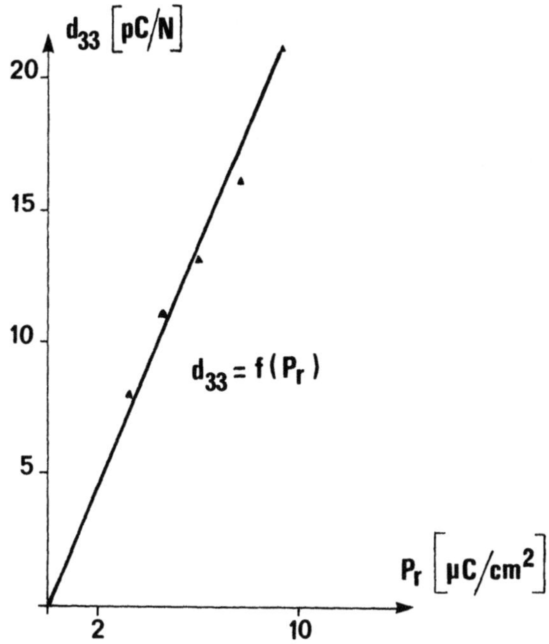

Fig. 2: d_{33} versus remanent polarizatioon P_r

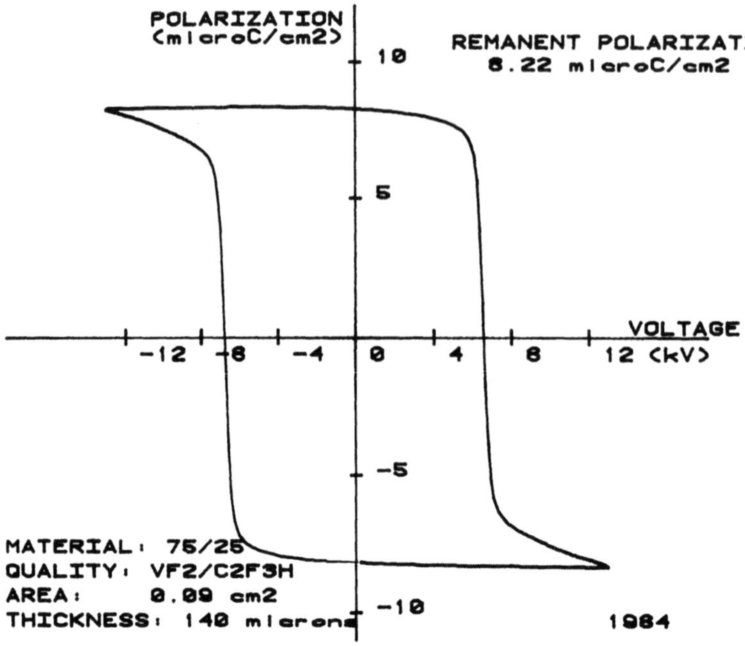

Fig. 3: Hysteresis curve for a VF_2/C_2F_3H copolymer

II.2 Ferroelectric Polarization of VF_2/C_2F_3H Copolymers

Measurements on copolymers VF_2/C_2F_3H pressed at ISL (resin available from ATOCHEM in France) was achieved thanks to the kind assistance given by Mr. Blaise. Copolymers with 60 to 80 mole % VF_2 were studied and polarized. The method used is described in the preceding section. These copolymers with such VF_2 content reveal ferroelectricity. They give square shaped P-E hysteresis curves (Fig.3) with large remanent polarization(6 to 9 $\mu C/cm^2$, table I) and lower coercive fields than those observed in PVF_2. For VF_2/C_2F_3H with 60 mole % VF_2, d_{33} values (measured with a d_{33} Berlincourt piezometer) attain 35 pC/N.

Table I

Copolymer	Mole VDF %	Remanent polarization
VF_2/C_2F_3H	60/40	P_r = 7 $\mu C/cm^2$
"	70/30	P_r = 8.8 $\mu C/cm^2$
"	75/25	P_r = 9 $\mu C/cm^2$
"	80/20	P_r = 8.5 $\mu C/cm^2$

II.3 Homogeneity of the Polarization Induced into PVF_2 and VF_2/C_2F_3H

Using the pressure wave propagation (pulse pressure of few nano-seconds duration) (PWP) C. ALQUIE and J. LEWINER [11] have performed some tests on our poled PVF_2 materials in order to examine the polarization distribution obtained by the "cycling hysteresis method". The delivered current i(t) versus time [11] is a function of a spatial distribution of the charge density after the complete penetration of the pulse pressure in the sample. In first approximation if $i(t) \simeq 0$ the ferroelectric polarization of the material is homogeneous. Results obtained show that

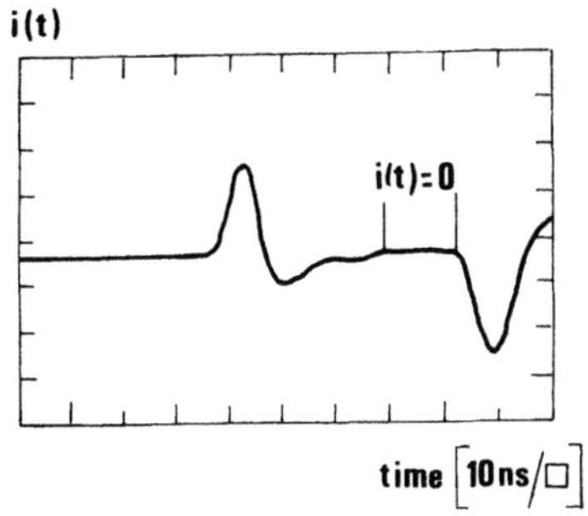

Fig. 4A: Propagation in one direction: PVF_2 110 µm thick

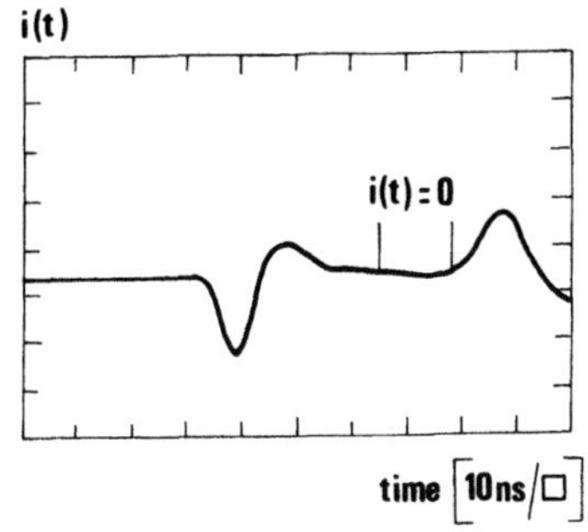

Fig. 4B: Propagation in reverse direction PVF_2 110 µm thick

our PVF_2 poled in such a manner has a good spatial polarization homogeneity as can be seen on Fig. 4 which was obtained with a 110 microns thick polyvynilidene fluoride foil. Figure 4A shows the signal obtained in short-circuit conditions for a propagation of the pressure pulse in one direction. Fig. 4B shows the corresponding signal for the reverse direction of propagation.

Currents i(t) versus time for an ISL pressed and poled VF_2/C_2F_3H (75 mole percent) with 240 µm in thickness are plotted in Fig. 5A and 5B. The polarization of the copolymer poled in the same way as PVF_2 is also uniform over the whole thickness of the sample. In all cases both the attenuation and dispersion of the Fourier components of the pressure pulse are low enough to be neglected.

III. BEHAVIOR OF PVF_2 AND VF_2/C_2F_3H UNDER ACTION OF A SHOCK WAVE: SHOCK RESPONSE

The studies of the behavior of shock loaded PVF_2 have been the subject of extensive investigations conducted at ISL [8].

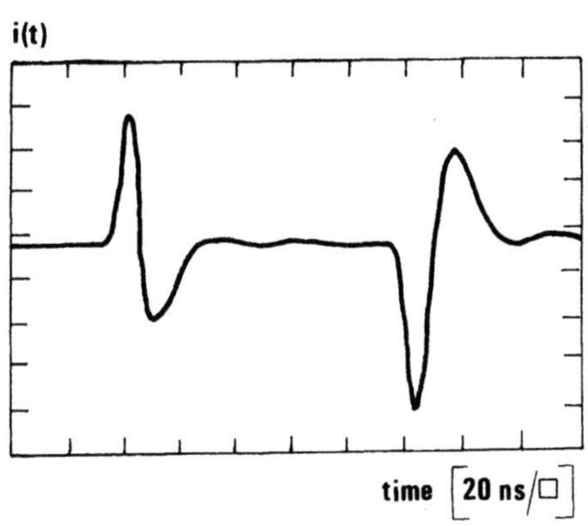

Fig. 5A: Propagation in one direction: VF_2/C_2F_3H (75/25)

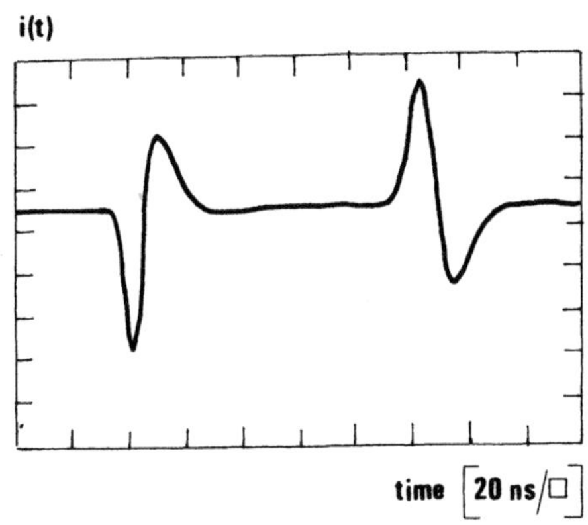

Fig. 5B: Propagation in reverse direction: VF_2/C_2F_3H (75/25)

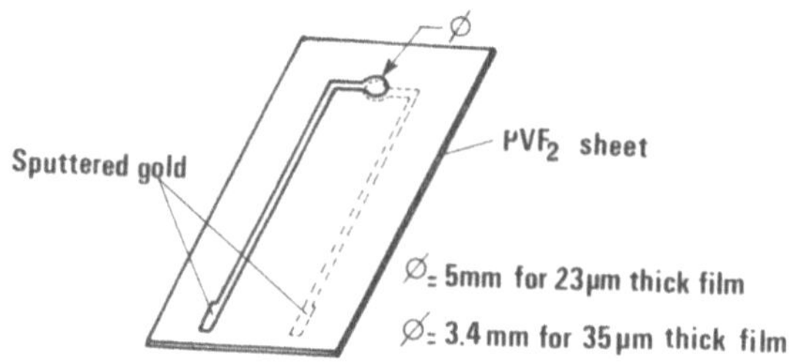

Fig. 6: ISL PVF_2 shock gauge

The results achieved in investigating shock-loaded PVF_2 polymers allowed the original piezoelectric stress gauges to be developed [8]. The original gauge configuration included thin films of biaxially stretched PVF_2 (thickness 23 µm) and of transversally stretched PVF_2 (thickness 35 µm) (Fig. 6). The poling technique applied to these active areas was the ISL process previously described in section II.1. The new gauge design has been successfully tested in very strong electrical environments [8].

If we assume the delivered charge to be proportional to the applied stress, the curve Q = f(t) (electric charge versus time) should be expected to exhibit the shock wave profile.

$$Q = \int_0^t i(t)dt, \text{ with } t = \text{experimental recording time.}$$

III.1 Experimental Technique and Results

Powder gun plate impact experiments [13] were used to determine the PVF_2 shock gauge response. The dynamic compression loading behavior of PVF_2 gauges was investigated using step compressional loading input waves (Fig. 7).

The gauge was embedded in PMMA in duraluminium or in copper targets using "Araldite" (Fig. 7). The recoverable signal i(t) was measured with a two-channel Tektronix 7612 transient digitizer. The delivered charge Q(t) was obtained by integrating current versus time i(t).

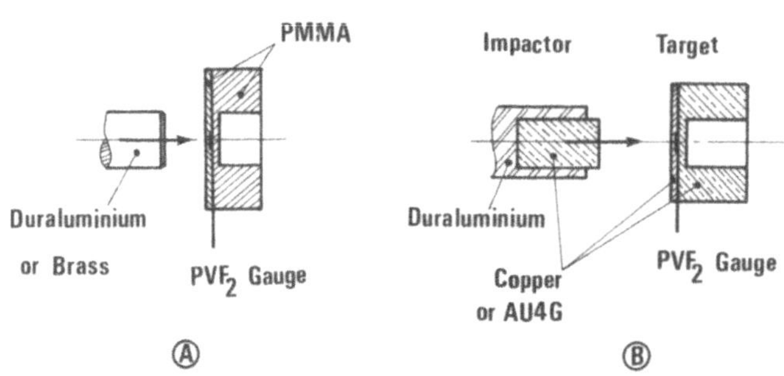

Fig. 7: Experimental arrangement

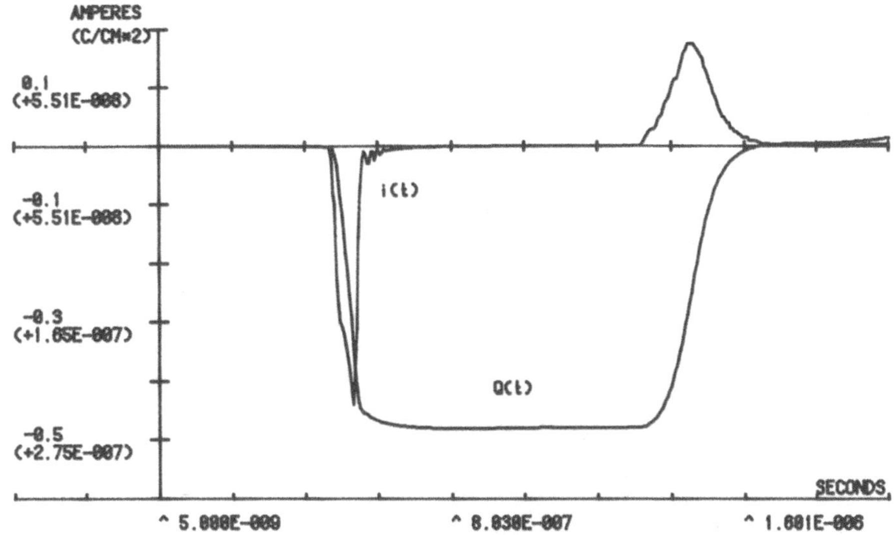

Fig. 8: Step shock compression, PVF_2 35 µm, target PMMA (2.44 kbar)

Figures 8 and 9 give examples of records obtained at 2.44 kbar and 120 kbar. We found out that the delivered electrical charge of PVF_2 describes the shock profile. We have used the published Hugoniot curves [12] for the determination of shock pressure induced in PMMA, duraluminium or copper. PVF_2 is piezoelectric in the 0 - 120 kbar range with no apparent hysteresis upon unloading. The ISL poling process achieved by running hysteresis loops is seen to eliminate the dielectric relaxation of the PVF_2 occuring under shock loading.

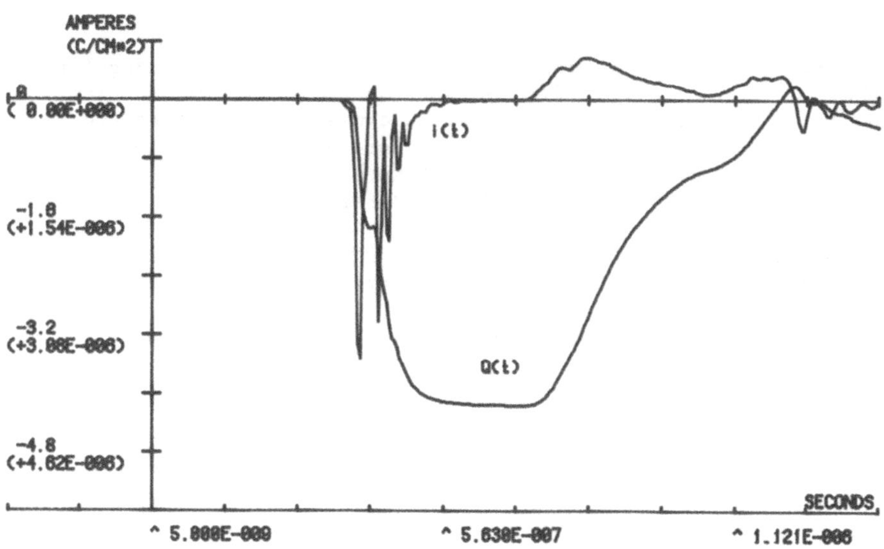

Fig. 9: Step shock compression, PVF_2 35 µm, target copper (120 kbar)

Figure 10 shows the history of the setting in equilibrium of the PVF_2 gauge (current and charge versus time) at the time of successive reflections of the shock wave at both the interface gauge and copper target.

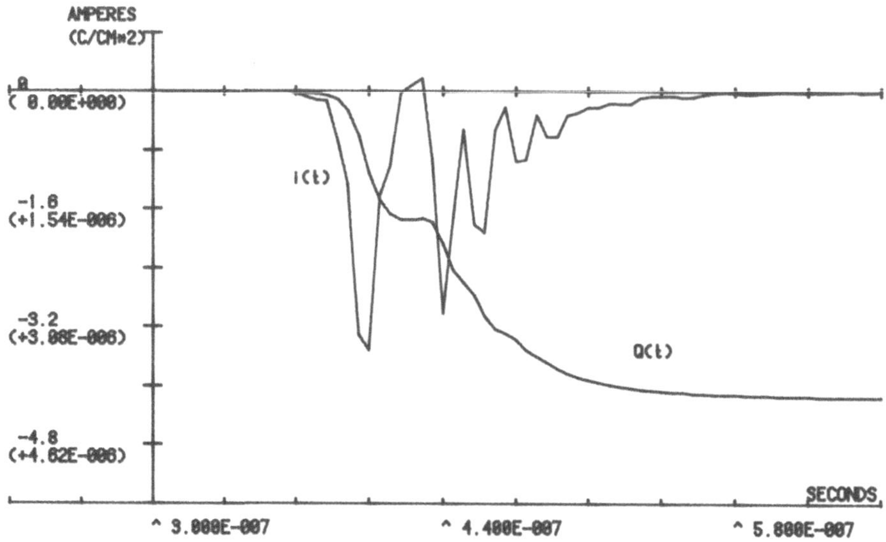

Fig. 10: History of the setting in equilibrium of the PVF_2 gauge (copper target)

These "step shock compression" trials permit plotting for each quality of PVF_2 poled at ISL the "electric charge versus pressure". Figures 11 and 12 give the calibration curves of the gauges. We observe also on VF_2/C_2F_3H with 75 mole content that the piezoelectric activity is reversible until 100 kbar. Figure 13 shows an example of results obtained with a 100 µm thick film pressed in our laboratory.

The calibration curves of the PVF_2 (Figures 11 and 12) gauges were smoothened in a linear interpolation achieved with the aid of a monomial. The law governing the electric charge Q released per unit surface as a function of the unidimensional axial stress can be expressed as follows:

$$Q = a\,\sigma^b$$

where Q is the electric charge given in $\mu C/cm^2$, a and b are constants depending on the film quality and on the piezoelectric coefficients obtained after poling (a = 0.248 and b = 0.637 for the PVF_2 specimen of 23 µm in thickness; a = 0.248 and b = 0.583 for the 35 µm thick PVF_2 specimen). The application of this law to the experimental data allows the shock pressure to be directly plotted as a function of time (Fig. 14 shows the case of shock induced compression already presented in Fig. 9).

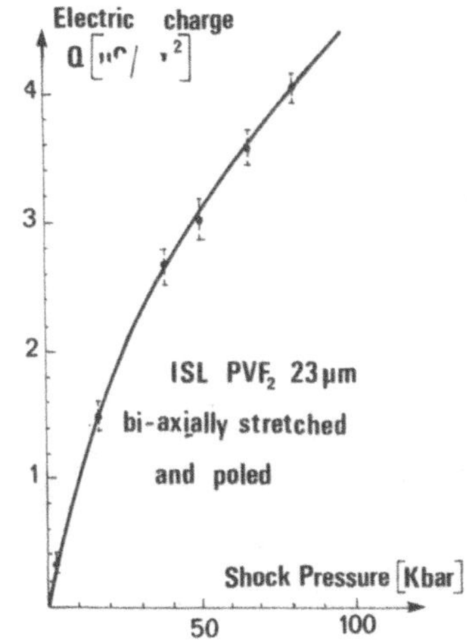

Fig. 11: Calibration curve: electric charge versus shock pressure for the ISL PVF_2 23 µm thick gauge

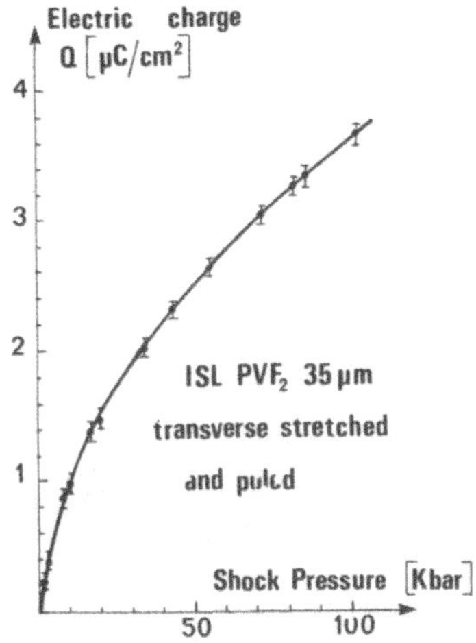

Fig. 12: Calibration curve: electric charge versus shock pressure for the ISL PVF 35 µm thick gauge

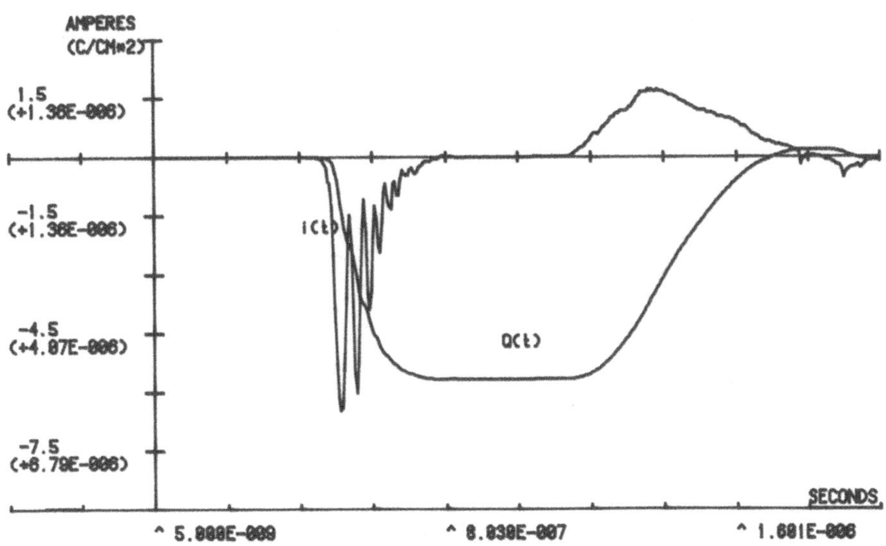

Fig. 13: Step shock compression: VF_2/C_2F_3H copolymer gauge (thickness 100 µm) in a copper target (78 kbar)

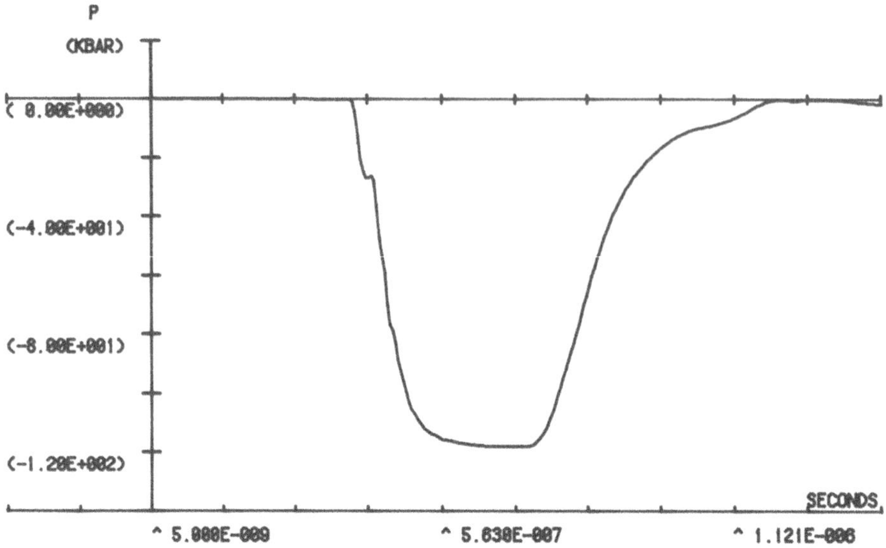

Fig. 14: Shock pressure versus time: PVF_2 gauge 35 μm in thickness

III.2 Shock Response of PVF_2: Comparison Between Experimental and Computed Data (co-worker Henri Moulard of ISL)

For a critical analysis of the shock induced response of the PVF_2, polymeric gauges were tested in the configurations shown in Figures 15A and 15B. The shock wave is generated by the impact of a 15 mm diameter projectile launched from a powder gun. Both the projectile and target are

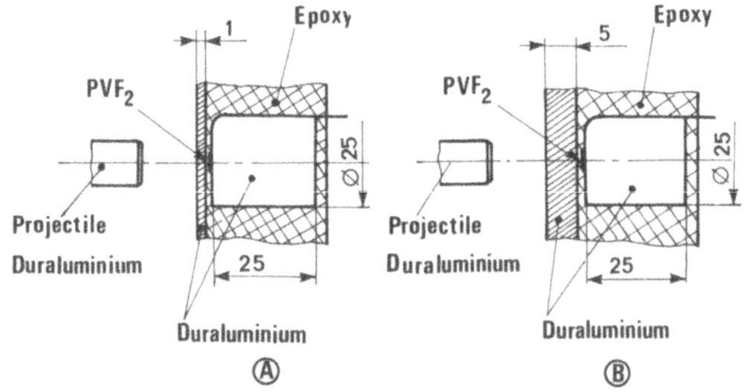

Fig. 15: Experimental arrangement. Study of PVF_2 shock response

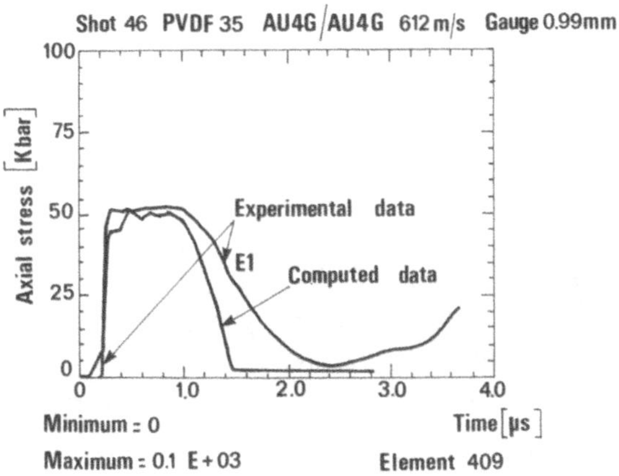

Fig. 16: Axial stress versus time: computed and experimental data

made of duraluminium whose shock parameters are well defined. In the first experiment (shot N° 46, Fig. 15A), the PVF_2 gauge of 35 µm in thickness was placed at a depth of 0.99 mm inside the target. In the second experiment (shot N° 48, Fig. 15B) the gauge was placed at a depth of 5.01 mm. The effect generated is therefore a one dimensional one followed more or less rapidly by a bidimensional flow. The pressures measured with the PVF_2 gauges are plotted in Figures 16 and 17.

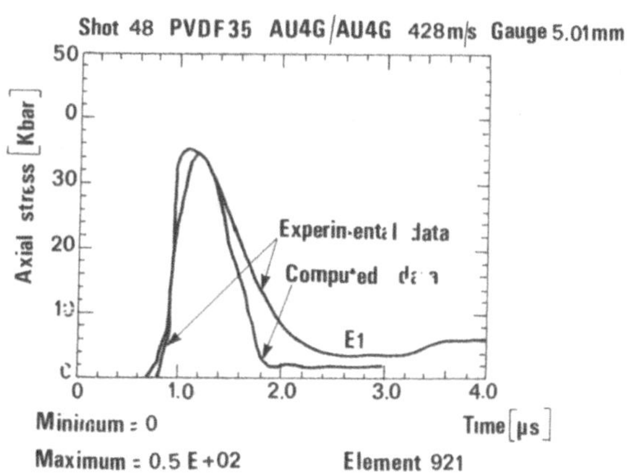

Fig. 17: Axial stress versus time: computed and experimental data

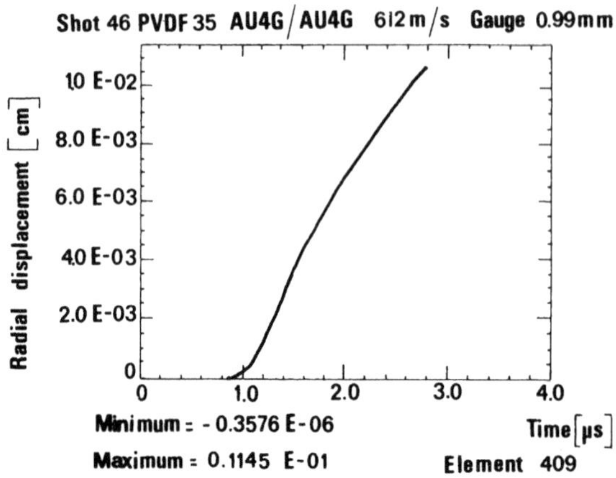

Fig. 18: Computed radial displacement versus time

With the aid of the Dyna 2 D code [14], Henri Moulard of ISL has evaluated the axial stress versus time for each gauge as well as the axial displacement which corresponds to the measurement of the bidimensional effect or border effect. The computations were made using the Dyna standard viscosity with a fine network structure.

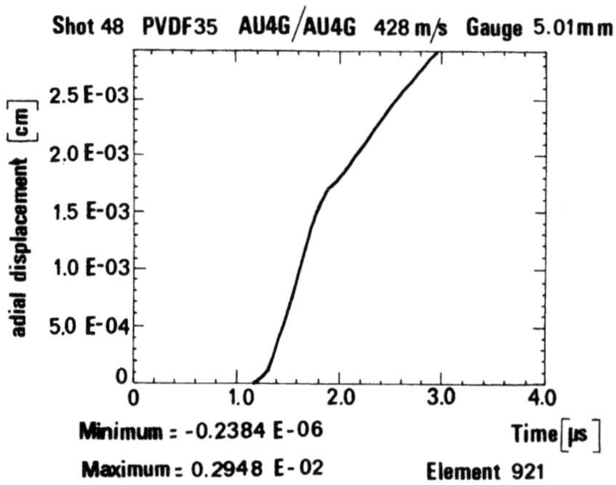

Fig. 19: Computed radial displacement versus time

The PVF_2 was found to exhibit an excellent resolution. There was a very good agreement between experimentally and theoretically determined pressures in the initial phase of the phenomenon, Figures 16 and 17. As soon as the lateral rarefaction waves arrive, some discrepancy appears between the experimental and theoretical stresses.This phenomenon is due to the conjugate action of the radial displacement (Figures 18 and 19) and of the axial stressing.

Like all ferroelectric materials PVF_2 is sensitive to bidimensional effects. The shock response of PVF_2 is quite remarkable if one takes into account the adiabatic increase in temperature up to more than 1100°K at a pressure level exceeding 100 kbar.

IV. SHOCK PRESSURE MEASUREMENTS IN THE VICINITY OF SMALL-SIZE DETONATORS [14]

From the number of particular uses of PVF_2, we have selected a typical example, that is measurements of the shock pressure in the vicinity of small size detonators. The following parameters were taken into consideration: nature of the compressed high explosive, diameter of the load: 3.6 mm, mass of the explosive: 120 mg, damming: steel of 2.5 mm in thickness [15].

The ISL probes used are PVF_2 gauges (thickness 35 µm) of 2 mm in diameter. The gauges are embedded in PMMA, duraluminium or copper. The measurements performed are analogous to those described in section III.1. One experimental record of the shock pressure induced in duraluminium is shown in Figure 20. The results achieved show the pressures of 85 kbar and 190 kbar, respectively, obtained in duraluminum and copper. By means of the shock data it is possible to evaluate the shock pressure induced in the steel cup of the detonator. This pressure ranges from 120 kbar to 130 kbar. The peak pressure is assumed to be unaffected by the rarefaction waves.

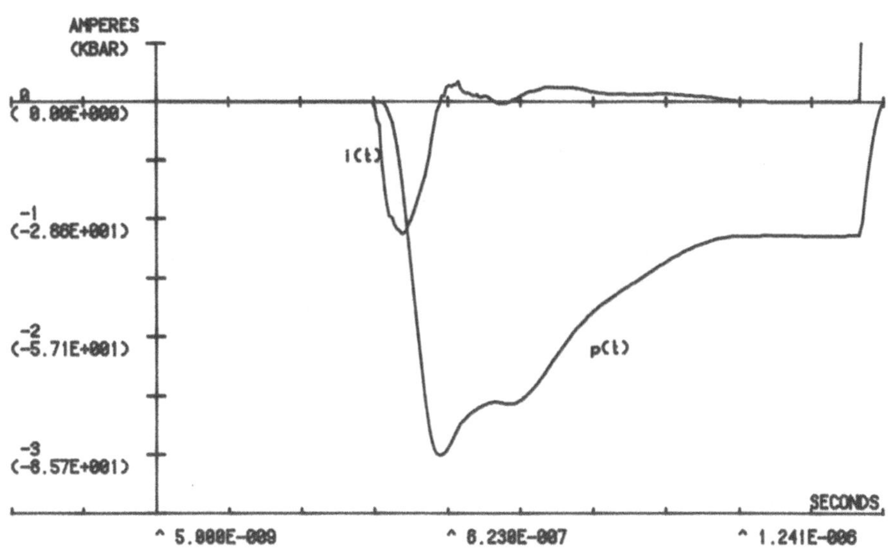

Fig. 20: Experimental record of the axial stress measured in the vicinity of a small size detonator

V. CONCLUSION

The piezoelectric PVF_2 adequately poled in a homogeneous way is a strong candidate for shock pressure transducers. This homopolymer as well as the copolymers belonging to the same family open the way to novel physical investigations of shock waves [16].

The results reported here and achieved in the framework of fundamental research work confirm the confidence placed in these organic elements provided that the laws existing in the field of ferroelectricity are taken into account. The existing technologies must be adapted to these novel transducers. This is especially true for the difficult area of shock wave physics and that of reactant media and/or explosive.

ACKNOWLEDGMENT

The author wishes to thank the APS Program Commitee, in particular R.A. Graham and J.R. Asay (Sandia Laboratories), Y.M. Gupta, chairman of the APS 1985, K. Harnischmacher and J. Riha for experimentations and especially Henri Moulard for helpful advice and discussions.

References

[1] H. Kawaï, Jpn. J. Appl. Phys. 8, 975 (1969)
[2] R.G. Kepler, Org. Coatings and Plastics Chemistry 38, 278 (1978)
[3] T. Furukawa, M. Date and E. Fukada, Jpn. J. Appl. Phys. 19, L 109 (1980)
[4] T. Yamada, T. Ueda, T. Kitayama, 26th Spring Meeting of Japan Society of Applied Physics and Related Societies, March 1979, p.126
[5] G.T. Davis, M.G. Broadhurst, A.J. Lovinger, T. Furukawa Ferroelectrics, 57, 73, 1984
[6] K. Kimura, H. Ohigashi, Appl. Phys. Lett 43, 9, 1983
[7] F. Bauer, Ferroelectrics, Vol. 49, 231, 1983
[8] F. Bauer, Shock Waves in Condensed Matter, Elsevier Science Publishers B.V. 1984
[9] F. Bauer, Patent 8221025
[10] M.G. Broadhurst, G.T. Davis, J.E. McKinney, R.E. Collins, J. Appl. Phys. 49, 10, 1978
[11] C. Alquié, J.Lewiner, Rev. Phys. Appliquée, 20, 131, 1985
[12] LASL Shock Hugoniot Data, S.P.Marsh, University of California Press 1980
[13] F. Bauer, ISL Report 131/76 unpublished I DE 77003 Thesis, Lyon 1977
[14] J.O. Hallquist "User's Manual for DYNA 2 D", UCID 18756
[15] R. Refouvelet, J. Duguet, Note REP 103-84, ATS Tarbes, unpublished
[16] L.M. Lee, D.J. Williams, R.A. Graham, F. Bauer, to be published in "Shock Waves in Condensed Matter", 1985.

STUDIES OF THE BAUER PIEZOELECTRIC POLYMER GAUGE (PVF_2) UNDER IMPACT LOADING*

L.M. Lee,[†] W. David Williams,[‡] R.A. Graham[‡] and F. Bauer[§]

[†]Ktech Corporation [‡]Sandia National Laboratories, Albuquerque [§]Institut Franco-Allemand de Recherches de Saint-Louis France (ISL)

ABSTRACT

A cooperative program has been initiated among our laboratories to provide a highly reproducible PVF_2 polymer film for widespread use as a gauge under shock-compression loading. The material source and electrical treatment are based on prior work of Bauer, but new, independent studies of gauge response are being carried out. First measurements indicate excellent agreement between the independent studies.

INTRODUCTION

It has been known for many years that appropriately treated polyvinylidene fluoride (PVF_2) exhibits piezoelectric properties suitable for a wide variety of applications. Because a sensor which functions under shock-compression loading is destroyed in use, these applications require a high degree of reproducibility in material constants. Various studies of PVF_2 film by Bauer [1-4] have shown that conventionally available piezoelectric films do not exhibit sufficient reproducibility for shock applications. However, selected PVF_2 film processed with the Bauer electrical poling method achieves the desired level of reproducibility.

In order to provide a stress gauge for widespread use in shock measurements, a cooperative program has been initiated between ISL and Sandia National Laboratories with technical support from Ktech Corporation. A large quantity of film has been acquired from which gauges will be fabricated by the Bauer method. This material will serve as the standard material for shock applications. We are carrying out an independent study of the standard gauge material to refine some of the prior work.

In the present paper we report our first measurements on shock-wave response which duplicate earlier work of Bauer, the first "impact surface" measurements with the gauge and first measurements with a new twin-cable electrical circuit.

---

*This work supported at Sandia National Laboratories by the U.S. Department of Energy under Contract No. DE-AC04-76DP00789.

EXPERIMENTAL

The gauges used in the present work were produced by Bauer in his conventional design. The Beta-phase film is produced by Rhône Poulenc as either transversely stretched (35 μm thick) or biaxially stretched (23 μm thick). The film is poled according to the Bauer method [1-3] after electroding with vapor-deposited gold. The sensing elements were 3.4 mm in diameter with 2 mm wide, 60 mm long gold-plated electrical leads. Short lead wires were soldered to the plating with a special low-melting-point InGa solder.

In the first series of experiments, the gauges were carefully embedded in Plexiglas (Rohm and Haas, UVIIA). The gauges were centered in a precisely machined slot in the Plexiglas (PMMA) back piece to which a PMMA buffer plate was attached. Hysol 2038 epoxy resin hardened with Hysol 3490 was used to bond all surfaces and the bonding process was carried out under pressure to achieve bonds of minimal thicknesses.

Both PMMA and 6061-T6 aluminum impactors were used to achieve a range of stresses. Impact experiments were carried out on the Sandia high velocity compressed gas gun [5]. Standard techniques to achieve precision in impact conditions were followed [6]. Thicknesses of all matérials, details of experimental conditions and experimental results are shown in Table 1.

Precision, nominally 0.3 Ω current-viewing resistors (CVR) were attached to the target cup after target assembly. A very low-loss coaxial cable was then connected between the CVR and transient digitizers which were set on two different gains. A Tektronix 7912 digitizer recorded the high amplitude signals while a 7612A digitizer recorded the low amplitude signals. This is the method used previously by Bauer. In this configuration one digitizer records signals in the range of volts while the other records signals in the range of tens of millivolts.

Table 1. Piezoelectric Polymer (PVF_2) Gauge Impact Experiments

Exp. No.	Experimental Configuration[a] Impactor → Target	Impact Velocity (km/s)	Equil.[b] Stress (GPa)	Shock Jump[c] Stress (GPa)	Elect. Charge (mC/m^2) Equil.	Shock
2009	CF / PMMA → PMMA /PVF2-T/PMMA (6.35)(1.48) (1.39)(0.035)(11.17)	0.613	1.20	0.77	10.7	8.8
2010	CF / Al → PMMA /PVF2-T/PMMA (6.35)(1.48) (1.47)(0.035)(11.22)	0.609	1.78	1.15	13.0	11.3
2011	CF / Al → PMMA /PVF2-T/PMMA (6.35)(1.51) (1.46)(0.035)(11.22)	0.788	2.23	2.04	16.1	14.1
2030[d]	Cu / PIL → PFA /PVF2-B/ Cu (6.32)(1.72)(0.023)(0.023)(25.33)	0.767	----	----	----	----

[a] Material code (dimensions = mm): CF = carbon foam; PMMA = plexiglas; Al = 6061-T6 aluminum; Cu = OFHC copper; PIL = Barker pillow impactor; PFA = PFA Teflon; -T = transversely stretched; -B = biaxially stretched.

[b] Equilibrium stress is taken to be the impact stress determined by the instantaneous response data on PMMA by Schuler and coworkers [8,9] and 6061-T6 data by Lundergan and Herrmann [10] and Christman, et al. [11].

[c] Shock stress is determined by the ratio of shock to peak stress by Schuler and coworkers. Values are 1/1.55 for experiments 2009 and 2010 and 1/1.0 for 2011.

[d] Impact surface measurement.

In addition to the embedded gauge measurements, an impact surface measurement was carried out, as shown in Table 1 (Exp. 2030). A gauge was placed on the impact surface of a thick copper target disk and covered with a 25 μm film of PFA Teflon. A "pillow" impactor designed by Barker [7] to achieve quasi-isentropic impact loading was then impacted on the target. The pillow was vapor-plated with aluminum to avoid problems with shock-induced polarization.

In the impact surface experiments two very low-loss coaxial cables of identical length are attached to the gauge leads. One cable is connected to a precision microwave divider (100:1) at one digitizer and the second cable is connected to a similar divider with a different division ratio (10:1) at the second digitizer. The two cables are then electrically terminated with 50 Ω at the input terminal of the digitizers. In this configuration signal levels are all in the volt range. This configuration avoids destruction of the CVRs used in the first series of experiments.

RESULTS

Details of the various experiments and results are shown in Table 1. Raw data from Experiment 2009 is shown in Figure 1, which shows both the loading and release signals. The nature of the gauge as a stress-difference (or acceleration) gauge is apparent from the response to loading, which is rapid and of large amplitude, and the response to unloading, which is slow and of low amplitude. It is this feature which poses a demanding instrumentation challenge.

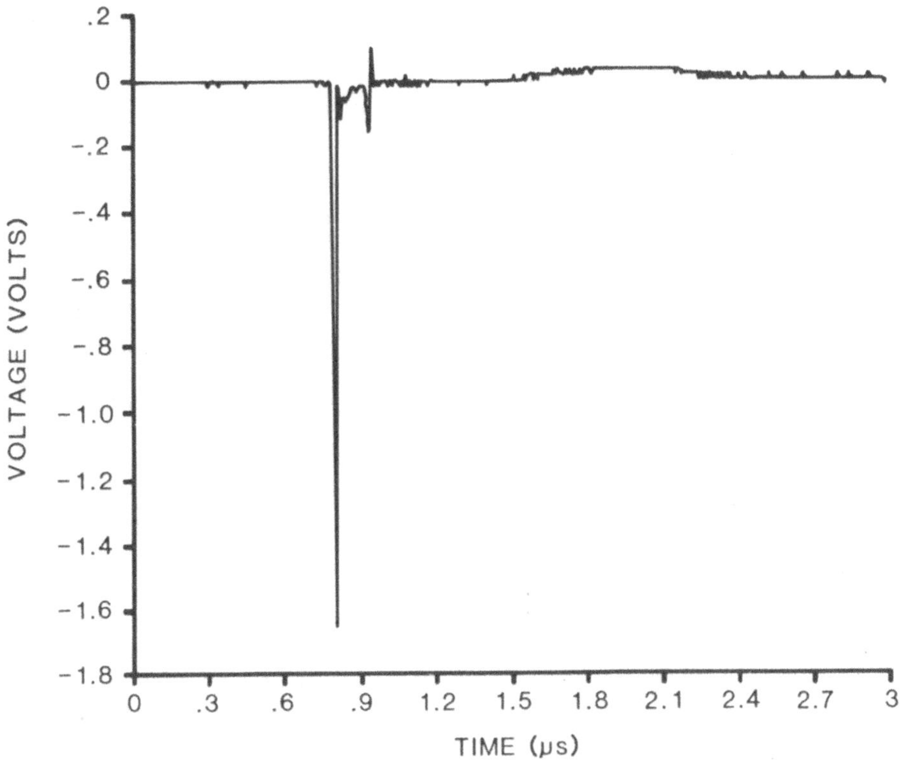

Figure 1. Raw transient digitizer data for Experiment 2009. The sharp initial spike is the shock response and is followed by a slow increase in stress, a signal reverberation in the cable and finally a release of stress.

Figure 2. Current pulse data are digitally smoothed and integrated to provide high resolution charge at stress levels of 0.77 to 2.2 GPa.

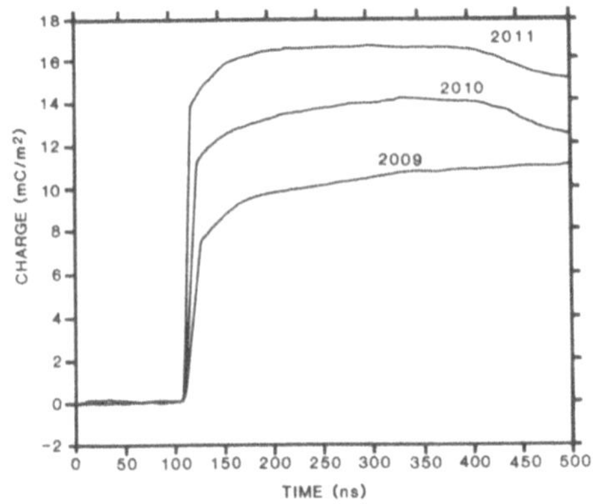

One significant feature of the raw data is evidence for a 120 ns period reverberation in the coaxial cable resulting from the electrical response of input capacitance of the digitizers for a very short rise time achieved. These pulses are readily identified and removed in data analysis.

The current pulses are digitally integrated to provide electrical charge as a function of time. Figure 2 shows the integrated charge versus time relation for the smoothed data. In every case what is observed is the arrival of a shock (within 10-20 ns resolution) followed by a smooth relaxation to an equilibrium stress state. This behavior has been carefully

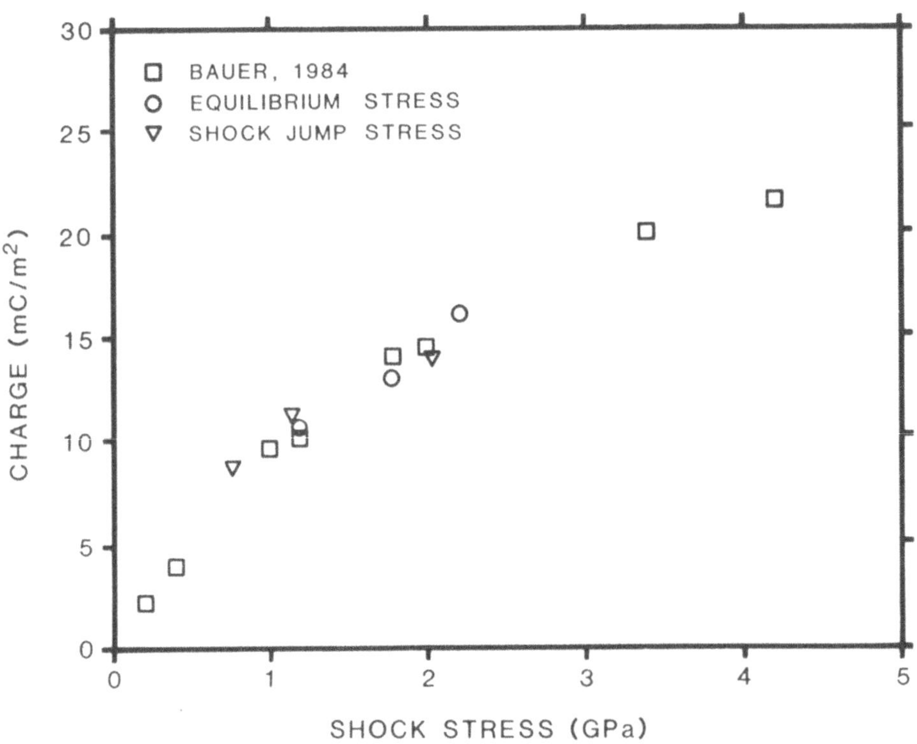

Figure 3. The piezoelectric charge determined at various stress levels is found to be in excellent agreement with prior work.

Figure 4. An impact surface gauge response for a "pillow" impactor is found to follow the general shape expected from prior work.

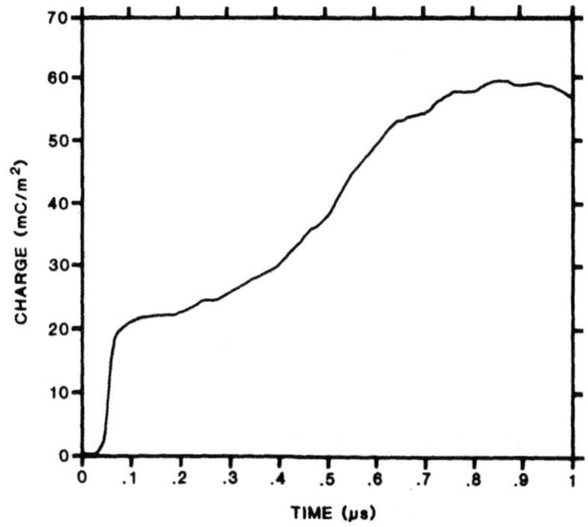

studied by Schuler and coworkers [8,9]. In determining the charge associated with a given level of stress in the present paper, both equilibrium stress values and shock jump stress are used, as indicated in Table 1.

The results of the charge determinations are plotted in Figure 3, which also shows the prior work of Bauer [4]. Although the gauges are nominally the same between the prior and present work, the experiments are entirely independent. Given the independent nature of the studies, the agreement is extraordinary. Differences, if any, are thought to arise from the stress determination based on various materials used in the experiments rather than on gauge response.

The result of the impact surface measurement of pillow response is shown in Figure 4. Here the initial shock due to the polymer face of the pillow is shown to be followed by a slowly rising stress pulse as anticipated. There are uncertainties in instrumentation and degree of nonlinearity in gauge response which delay our detailed analysis of the record. However, the qualitative behavior of the gauge is consistent with the VISAR records obtained by Barker [7].

CONCLUSIONS

The important first steps in standardizing material for the Bauer piezoelectric polymer gauge have been taken. Independent experiments verify the excellent reproducibility of gauges prepared by the Bauer methods. The gauge appears to offer a major new measurement capability not available by other means. More study is needed to specify a full range of response characteristics.

ACKNOWLEDGEMENTS

The authors are pleased to acknowledge the excellent technical assistance of Bob King, Mark Wing, Mark Anderson and Dave Wackerbarth.

REFERENCES

1. F. Bauer, Behavior of Ferroelectric Ceramics and PVF_2 Polymers Under Shock Loading, in Shock Waves in Condensed Matter-1981, edited by W.J. Nellis, L. Seaman and R.A. Graham, American Institute of Physics (1982), p. 251.

2. F. Bauer, Piezoelectric and Electric Properties of PVF_2 Polymers Under Shock Wave Action: Application to Shock Transducers, in Shock Waves in Condensed Matter-1983, edited by J.R. Asay, R.A. Graham and G.K. Straub, North Holland (1984), p. 225.
3. F. Bauer, PVF_2 Polymers: Ferroelectric Polarization and Piezoelectric Properties Under Dynamic Pressure and Shock Wave Action, Ferroelectrics 49:231 (1983).
4. F. Bauer, in press, 1984.
5. R.E. Setchell, Sandia 25-Meter Compressed Helium/Air Gun, in Nellis, et al., editors (1982), loc. cit., p. 657.
6. G.E. Ingram and R.A. Graham, in Proceedings, Fifth Symposium (International) on Detonation, 1970, Office of Naval Research Report ACR-184, p. 369.
7. L.M. Barker, High Pressure Quasi-Isentropic Impact Experiments, in Asay, et al., editors (1982), loc. cit., p. 217.
8. K.W. Schuler and J.W. Nunziato, Recent Results in Nonlinear Viscoelastic Wave Propagation, Int. J. Structures 9:1237 (1973).
9. K.W. Schuler and J.W. Nunziato, The Dynamic Mechanical Behavior of Polymethyl Methacrylate, Rheologica Acta 13:265 (1974).
10. C.D. Lundergan and W. Herrmann, Equation of State of 6061-T6 Aluminum at Low Pressures, J. Appl. Phys. 34:2046 (1963).
11. D.R. Christman, W.M. Isbell, S.G. Babcock, A.R. McMillan and S.J. Green, Measurements of Dynamic Properties of Materials, Vol. III, 6061-T6 Aluminum, Materials and Structures Laboratory, General Motors Technical Center, MSL-70-23, Vol. III, 1971.

POLYVINYLIDENE FLUORIDE TRANSDUCER FOR DYNAMIC PRESSURE MEASUREMENTS

E.R. Lemar, J.W. Forbes, and D.G. Tasker
Research Department, Naval Surface Weapons Center

A.J. Bur
Center for Materials Science, Polymers Division
National Bureau of Standards

INTRODUCTION

The purpose of this paper is to present some results on the charge output of an electrically poled Polyvinylidene Fluoride (PVDF) gage. This PVDF gage package is of interest because of its large charge output (approx. 4×10^{-11} C/N), its low shock impedance, and its nominal thickness of 0.05 mm. It is a good candidate for gage use in materials of low shock impedance because of it creating a minimum disturbance in the measured wave. The main result of this study was the measurement of the gage output (charge) as a function of stress. A few changes to improve gage accuracy are discussed.

The piezoelectric and pyroelectric properties of PVDF have been the subject of many reports during the past decade.[1,2,3] PVDF can be processed into thin sheets which become piezoelectrically active after the application of a large electric field. Upon removal of the electric field, the material remains permanently polarized and displays piezoelectric and pyroelectric activity[4]. The change from the unpoled to the poled state has been described as a ferroelectric transition[5].

The gage, as shown in Figure 1, is made from two 12 μm thick films of biaxially oriented PVDF. The shaded regions are the active areas on which electrodes consisting of 100 nm of aluminum followed by 100 nm indium have been deposited. The diameter of the active area is 0.5 cm. The material becomes piezoelectrically active by poling the electroded regions with an electric field of 2 MV/cm at room temperature. After poling, the two active regions are epoxied together with their polarization vectors opposing each other. In this bilaminate configuration, the exterior electrodes are at ground potential and provide shielding for the inner electrodes which carry the signal potential.

EXPERIMENTAL PROCEDURES

The gage package of nominal thickness 0.05 mm was epoxied in between two 1.27 cm thick by 5.08 cm diameter polymethylmethacrylate (PMMA) discs. The PMMA samples containing the PVDF gages were impacted by 6061 T6 aluminum projectiles from a 8.9 cm bore light gas gun[6].

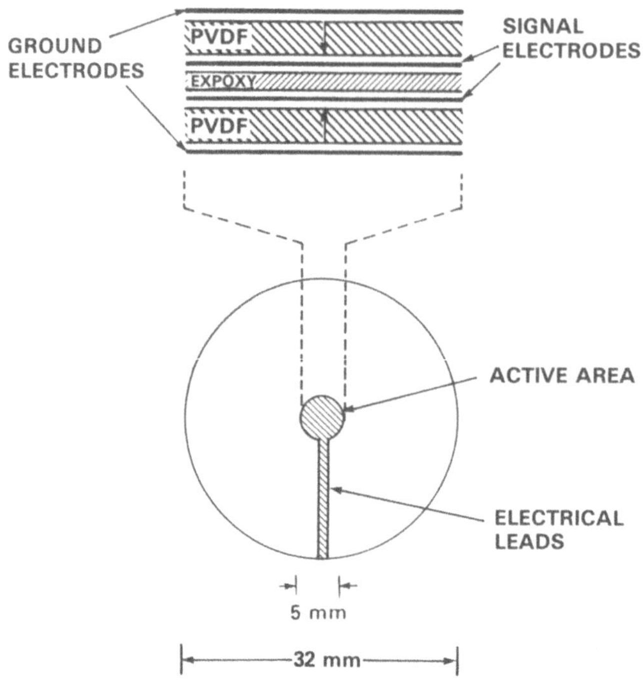

Figure 1. Schematic of PVDF Gage

These experiments were designed to only look at the PVDF gage response under shock compression. The charge generated by the gage was collected by a capacitor (120 nF) and observed by an oscilloscope. The recording circuit is depicted in Figure 2. This circuit caused some ringing at the initial step in voltage due to cable reflections and stray circuit inductances.

A typical voltage record is shown in Figure 3. The voltage record has a number of oscillations at the beginning and then increases as a function of time. The voltage goes towards the baseline after a few microseconds. The initial jump in voltage is obtained by using a straight line extrapolated from the top of the signal back to zero time where the shock entered the gage.

Table 1 gives a summary of the experiments. Listed there are the stress induced in the PMMA, the projectile velocity, the piezoelectric constant d_h for hydrostatic pressures, the voltage at the initial part of the signal, and the initial charge. The d_h is measured to within ±7%

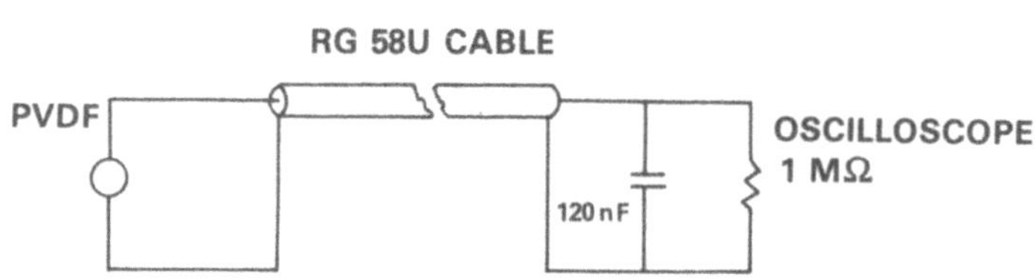

Figure 2. Charge Recording Circuit

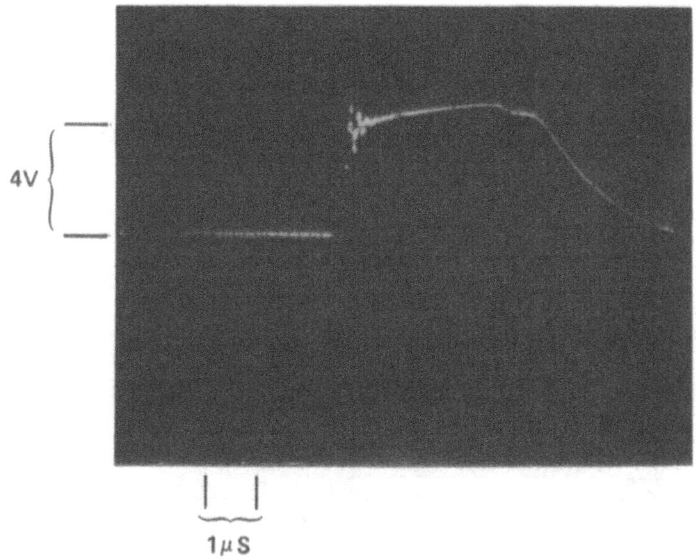

Figure 3. Typical PVDF Gage Record

at low pressure (40 bar) using a drop test which is described elsewhere[7]. The stress in PMMA is determined within ±5% from the conservation relations and the known Hugoniots of PMMA and the projectile.

The charge output from four gages, normalized to a value d_h = 40 pC/N, is plotted versus pressure in Figure 4. A nonlinear least squares fit of the data is shown by the full line which has been forced to intersect with the origin. The dashed straight line is a linear extrapolation from the low pressure measurements of d_h and emphasizes the nonlinear behavior of the gage at high pressure. Nonlinearity, which has been observed previously in static measurements[5], is

Table 1 -- PVDF Gage Results

STRESS (kbar)	PROJECTILE VELOCITY[a] (mm/μs)	d_h (pC/N)	INITIAL PEAK VOLTAGE (V)	CHARGE (μC)
4.5	0.155	44.2	2.19	0.26
7.5	0.254	----[b]	2.92	0.35
11.8	0.391	37.6	3.20	0.38
11.8	0.392	----[b]	3.76	0.45
14.6	0.480	42.0	5.64	0.67
16.7	0.480	40.0	5.77	0.69

[a] Projectiles were 6061 T-6 aluminum except for experiment at 16.74 kbar which used a copper impactor.
[b] d_h was not measured for these gages.

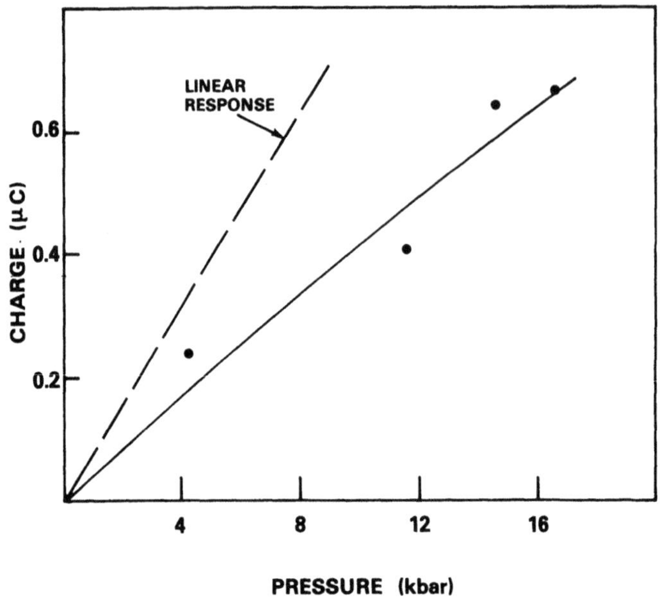

Figure 4. PVDF Gage Charge Output as a Function of Stress

attributed to a decrease in the compressibility of PVDF as pressure increases.

A single experiment was performed to observe the gage response to a one-dimentional shear wave. The PVDF gages was placed between a PMMA disc and a Y-cut quartz disc. The Y-cut quartz was impacted by an aluminum projectile at a velocity of 0.2 mm/μs. The record was extra noisy due to the recording circuit picking up signals from the Y-cut quartz electrical response to shock loading. However, the signal clearly shows a reduction in voltage of about 10% when the shear wave arrives at the gage. The shear wave stress amplitude was not measured for this experiment.

Improvement in the gage recording circuit is possible by taking special care to reduce all stray inductances and by properly terminating the cable for high frequency signal components. The reduction of stray inductance can simply be done by keeping all circuit element leads as short as possible and connecting these elements via a double clad copper circuit board. An improved circuit for good time response is given in Figure 5. A single experiment was conducted using this circuit. The record had only a few very fast oscillations at the beginning and then a smooth trace following these oscillations. The oscillations were likely due to shock waves bringing the gage to stress equilibrium.

DISCUSSION AND CONCLUSIONS

It has been suggested[8], that the charge output gives a direct image of the stress pulse. However, a voltage ramp was noted following the initial voltage jump for this gage package after it was subjected to a flat topped shock. The reasons for this voltage ramp is not understood. Speculation is that the ramp is due to either an electro-mechanical effect or a thermal effect at the PMMA-PVDF interface.

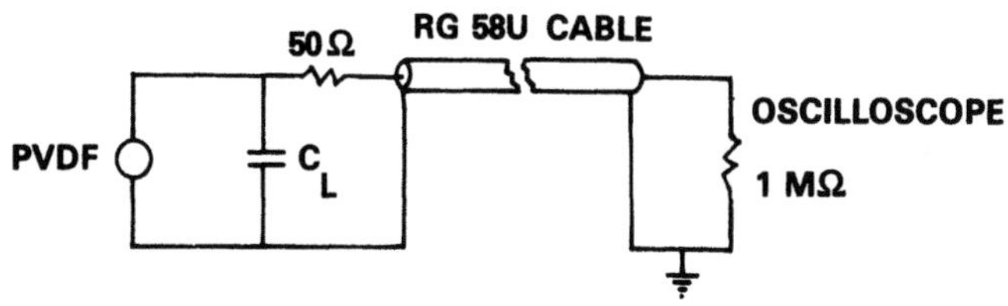

Figure 5. Improved Charge Recording Circuit

The PVDF gage package used in this study showed that the initial charge output is related to the stress. The scatter of the initial peak charge as a function of stress are likely due to differences in gage construction and non-uniform poling of the PVDF. Visual examination of the gage packages show that the ground electrode is not perfectly placed over the positive electrode. The variations in d_h in Table 1 indicate that variations in effective gage area and/or permanent polarization are present in these gages.

A slight reduction in the scatter of the voltage-stress data can be obtained by normalizing each initial voltage to a constant d_h. However, it is correct and desirable to use gages with the same piezoelectric coefficients.

The single experiment using Y-cut quartz to send a shear wave into the PVDF gage demonstrated that shear components in shock waves will affect the output of the gage. This means that placement of the gage in two-dimensional shock experiments has to be done with care and some knowledge of the shock waves orientations.

In conclusion, with some modifications, the gage package studied here shows promise for use in shock studies. More work is needed on the precise construction and poling of the gages to insure that each gage has the same piezoelectric coefficients.

ACKNOWLEDGEMENTS

A. Brown, R. Baker, K. Harrison, and J. Watt assisted in the gas gun experiments. L. C. Chhabildas of Sandia National Laboratory kindly provided the Y-cut quartz sample. The study was supported by M. Frankel of DNA.

REFERENCES

1. M. G. Broadhurst and G. T. Davis in Topics in Applied Physics, Vol. 33:Electrets, editor: G. M. Sessler, Springer-Verlag, Berlin, 1979.

2. G. M. Sessler, J. Acoust. Soc Am 70, 1596 (1981).

3. A. J. Lovinger, in Developments in Crystalline Polymers-1 Editor: D. C. Bassel, Applied Science Publ., London 1981, p. 195.

4. M. G. Broadhurst, G. T. Davis, J. E. McKinney and R. E. Collins, J. Appl. Phys. 49, 4992 (1978).

5. R. G. Kepler and R. A. Anderson, J. Appl. Phys. 49, 1232 (1978).

6. E. R. Lemar, J. W. Forbes, J. O. Erkman, and J. W. Watt, in Shock Waves in Condensed Matter-1981, Editors W. J. Nellis, L. Seaman, R. A. Graham, AIP Proceedings 78, 1982, p. 663.

7. A. J. Bur and S. C. Roth J. Appl. Phys. 57, 113 (1985).

8. F. Bauer, In Ferroelectrics, vol 49, pp 231 (1983).

PIEZORESISTANCE RESPONSE OF MANGANIN FOILS: EXPERIMENTS AND ANALYSIS [a]

Satish C. Gupta[b] and Y. M. Gupta

Shock Dynamics Laboratory
Department of Physics
Washington State University
Pullman, WA 99164-2814

INTRODUCTION

The objective of the present work was to understand the response of the manganin foil gauges,[1] oriented parallel and perpendicular to the wave front, in shock wave experiments. We carried out experiments in which the gauges were subjected to well characterized stress waves up to 100 kbar to obtain the gauge response on impact loading and to determine the hysteresis effect on unloading. We then modeled the experimental results using the theoretical model developed earlier.[2,3] A brief summary of the experimental and theoretical results is presented in this paper.

EXPERIMENTS

The experimental technique was similar to that described in our earlier paper.[4] The gauges were cut into four terminal configurations from approximately 33 micron thick manganin sheets obtained from the Hamilton Technology, Lancaster, Pennsylvania[5] These foils have been characterized in quasi-static experiments.[6] The gauges were about 1 mm wide and had about 8 mm long active element. Their initial resistance was about 110 mΩ.

The gauges were embedded into grooves, of the same size, that had been machined in the polycrystalline aluminum oxide matrix, Vistal.[7] Vistal has been examined under impact conditions[8] and it remains elastic up to 90 kbar. This feature conforms to the requirements of the existing theoretical model.[2,3] Each experiment had two gauges, one with major surface parallel to the shock front and the other with its major surface perpendicular to the shock front. In a few of the experiments, the gauges were emplaced in the polymethylmethacrylate (PMMA) matrix.

Eighteen experiments were performed using 63.5 mm gas gun at our laboratory. The thickness of the flyer was chosen to obtain complete unloading of the stress in the target. Although most of the experiments were done with symmetric impact condition, some experiments at lower stresses used fused silica flyers. The projectile velocity was measured using standard technique.

The resistance change profile was measured by sending a constant current through the outer leads of the gauge using a pulsed power supply and recording the voltage difference between the two inner leads using fast oscilloscopes.

RESULTS AND ANALYSIS

The peak resistance change and the residual resistance change for the gauges in both the orientation were measured in each experiment. The peak longitudinal stress was determined from the knowledge of the projectile velocity and the Hugoniot of Vistal[8] and fused silica.[9]

In Fig. 1, we plot the peak resistance change versus the peak longitudinal stress in the matrix for both orientations. The resistance change for both the orientations increase smoothly and, as expected, the difference between the resistance changes of the longitudinal gauge and the lateral gauge increase with increasing peak longitudinal stress. The two experiments performed

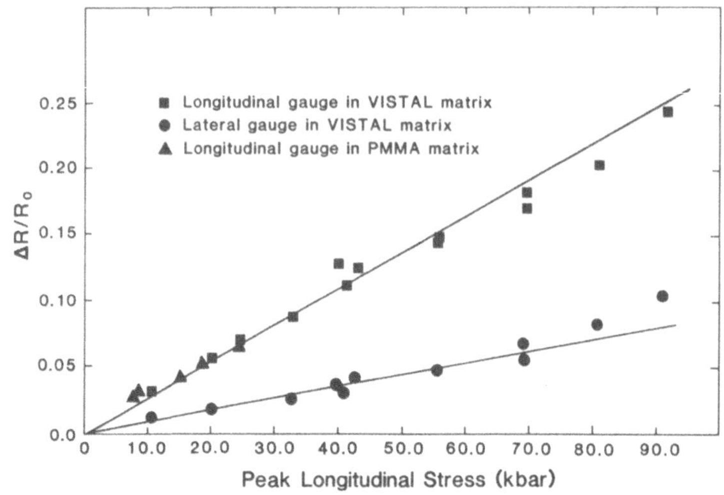

Fig. 1. The peak fractional resistance change for the two orientations as a function of peak longitudinal matrix stress. The solid curves show the calculated values.

with different matrices (Vistal and PMMA) shocked to the same longitudinal matrix stress (24.5 kbar) show that the longitudinal gauge response is very close in the two cases.

Figure 2 shows the residual resistance change versus the peak longitudinal stress. For both orientations the residual resistance change increases smoothly up to 50 kbar and becomes almost constant above this stress. Our analysis shows that even if the hysteresis effects are ignored in the plate impact experiments, the stress measured using manganin gauges may be over-estimated by at most 5 kbar.

The experimental results were analyzed using the theoretical model[2,3] and the electromechanical constants of manganin foils determined from quasi-static measurements.[6] The residual resistance change data for the longitudinal gauge was used for determining the electrical strain hardening coefficient, η , as a function of the second invariant of the plastic strain tensor, γ_p . The η values decreased with γ_p and fall on a parabolic curve with a maximum of 0.875 at $\gamma = 0.001$ and has asymptotic value of 0.075. These values of η were used for further resistance change calculations.

The calculated peak resistance change as a function of the peak longitudinal stress is shown as a dark line in Fig. 1. The agreement between the experimental and the calculated values is found to be close.

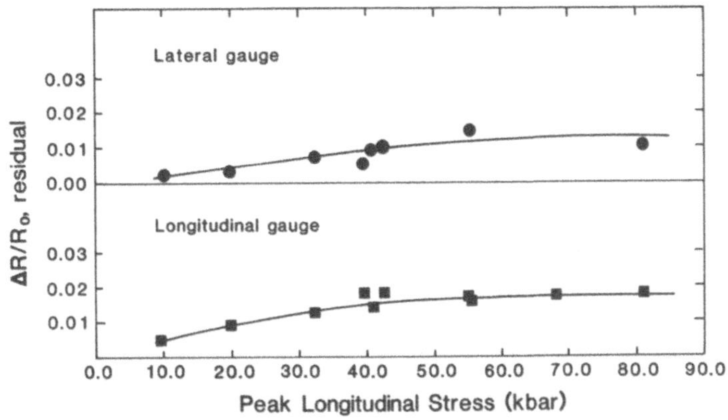

Fig. 2. The residual fractional resistance change for the two orientations.

REFERENCES

a. A brief summary of the results is presented here. A detailed account of this work will be published in the Journal of Applied Physics.

b. Present address: Neutron Physics Division, Bhabha Atomic Research Center, Bombay 400 085, India.

1. The term "gauge" here refers to the piezoresistance foil only and the two are used synonymously.
2. Y. M. Gupta, J. Appl. Phys. 54:6256 (1983).
3. Satish C. Gupta and Y. M. Gupta (unpublished).
4. Satish C. Gupta and Y. M. Gupta, J. Appl. Phys. 57:2464 (1985).
5. These foils were provided to us by SRI International (courtesy of Mr. D.D. Keough).
6. D. Y. Chen, Y. M. Gupta and M. H. Miles, J. Appl. Phys. 55: 3984 (1984).
7. Vistal is the trademark of Coors for polycrystalline Al_2O_3 and is a replacement for Lucalox manufactured by General Electric.
8. Y. M. Gupta, J. Geophys. Res. 88:4304 (1983).
9. L. M. Barker and R. E. Hollenbach, J. Appl. Physics 41:4208 (1970).

PIEZORESISTANCE RESPONSE OF DIFFERENT BATCHES OF YTTERBIUM FOILS

N.S. Brar and Y.M. Gupta

Shock Dynamics Laboratory, Department of Physics
Washington State University, Pullman, WA 99164-2814

INTRODUCTION

The piezoresistance response of ytterbium and manganin has been the subject of several recent papers by Gupta and coworkers.[1-7] Although these materials have also been studied by others,[8-10] the work in Ref. 1-7 has several novel features: (i) an electro-mechanical model to include gauge plasticity, (ii) gauge-matrix interaction analysis treating the gauge as an elastic-plastic inclusion, (iii) development of a quasi-static method to measure the electro-mechanical constants for the gauge material, (iv) dynamic experiments, designed to conform to the requirements of the theoretical analysis, (v) correlation between quasi-static and dynamic experiments, and (vi) modeling the results, including gauge hysteresis, observed in dynamic experiments.

The present work on ytterbium, part of an on going effort to understand piezoresistance response, had the following objectives: (1) to confirm the correlation between quasi-static and dynamic experiments reported earlier[6] by examining a different batch of ytterbium foils, (2) to correlate the response of different batches, and (3) to ascertain the effects, if any, of the loading rate on gauge response by performing ramp wave experiments. The first two objectives are expected to be helpful in resolving difficulties associated with variations of physical properties between different batches. The third objective is intended as a check on the theoretical model which at present does not include rate effects.

EXPERIMENTAL METHOD AND RESULTS

A. MATERIALS

Ten ytterbium foils, with nominal dimensions of 3" x 3" x 0.002" were obtained from Research Chemical Corporation, Phoenix, Arizona. To facilitate comparison with previous work,[4-6] the present batch of foils (labeled batch II) were intended to have comparable purity and preparation to the earlier foils (labeled batch I). Resistivity measurements on batch II yielded values of 44-50 $\mu\Omega$-cm in contrast to the 38-43 $\mu\Omega$-cm values reported from batch I.[4] The somewhat higher resistivity values for batch II suggest that these foils had a higher concentration of lattice defects.

PMMA (Rohm and Haas Type II UVA plexiglass) and fused silica (Dynasil 1000) were used as matrix materials in the dynamic experiments. Both of these materials have been characterized under dynamic loading conditions.[11]

B. QUASISTATIC EXPERIMENTS

To determine the electromechanical material constants (Young's modulus, E; yield stress, Y; strain hardening coefficient, M ; piezoresistive coefficients, α, β, and η) the method developed by Chen, *et al.*[4] was followed. Tensile experiments were performed on one sample from each of ten foils. The measured constants are summarized in Table I along with the

Table I - Electromechnical Constants of Ytterbium Foils

Property	Batch II[a]	Batch I[b]
Bulk Modulus[c]	148 kb	148 kb
Young's Modulus	163 kb	120 kb
Yield Stress in Simple Tension	0.78 kb[d]	0.48 kb
Strain Hardening Coefficient, M[e]	7.12 kb[d]	4.4 kb
Piezoresistive Coefficient, π_{11}[e]	$-4.196 \times 10^{-2} (kb)^{-1}$	$-4.2 \times 10^{-2} (kb)^{-1}$
Piezoresistive Coefficient, π_{12}[e]	$-0.787 \times 10^{-2} (kb)^{-1}$	$-0.788 \times 10^{-2} (kb)^{-1}$
Electrical Strain Hardening Coefficient η[e]	1.08	1.13

a. Mean of 9 measurements.
b. Chen, *et. al.* (Ref. 4); mean of 8 measurements.
c. From Grady and Ginsberg,[9] and LASL Hugoniot data.
d. Mean of 5 measurements.
e. Defined in Reference 4.

values for batch I.[4] The mechanical constants for batch II are consistently higher than those for batch I. This result reflects the difference in the tensile stress-strain curves for the two batches shown in Fig. 1. The higher tensile strength supports the inference of higher lattice defect concentration indicated earlier. The mean values of the piezoresistive coefficients, however, are nearly equal. This agreement is somewhat misleading in view of the variation of these coefficients from foil to foil within a batch and the experimental errors associated with these measurements. The value of π_{11} for batch II ranges from $-3.901 \times 10^{-2}\,(kbar)^{-1}$ to $-4.659 \times 10^{-2}\,(kbar)^{-1}$; a variation of about ten percent from the mean value. π_{12} values range from $-0.555 \times 10^{-2}\,(kbar)^{-1}$ to $-0.935 \times 10^{-2}\,(kbar)^{-1}$. The cause for the larger variation in π_{12} lies in the method of determination of these coefficients. [4] π_{11} is determined directly from our tensile measurements; π_{12} is determined by combining the π_{11} values with the resistance change measurements under hydrostatic compression.[4] We have used the hydrostatic data of DeCarli in obtaining the π_{12} values.[12]

C. DYNAMIC EXPERIMENTS

Gauge emplacement in both the PMMA and fused silica matrix was in conformity with the theoretical analysis.[3] Target assembly in the PMMA matrix was nearly identical to the earlier work;[6] there were no lateral gauges in the present work. Ytterbium foil gauges in the longitudinal orientation were embedded into a groove on the surface of a 12.7 mm thick PMMA disk. The groove dimensions were chosen to be nearly equal to that of the gauge. A very thin layer of hysol epoxy was used to emplace the gauge in the groove. The PMMA disc containing the gauge was bonded with a very thin glue bond to a 1 mm thick PMMA cover plate. The reasons for choosing this experimental design are given in Ref. 6.

In the ramp wave experiments, designed to examine the effect of loading rate on gauge response, fused silica targets containing a surface gauge (G1) and a embedded gauge (G2) were assembled as shown in Fig. 2. Both the gauges were emplaced into grooves, following the same technique as in PMMA. Gauge G1 was intended to be step loaded, and gauge G2 was intended to be loaded by a larger rise time wave due to the propagation in fused silica.[11]

We conducted six experiments, three in PMMA matrix and three in fused silica matrix using the 6.35 cm gas gun at our laboratory. Four terminal resistance measurements were made following the technique described earlier.[6] The results of all six experiments are summarized in Table II. The third experiment at 18.5 kbar peak stress in the PMMA matrix was repeated to check the reproducibility of the resistance change measurements. The results from the two experiments agreed to within 1 percent.

In all six experiments, except the 18.5 kbar shot in PMMA, a complete unloading of the stress wave was observed. For the 18.5 kbar shot in PMMA, the flyer was polycrystalline Al_2O_3, and therefore, complete unloading could not be observed. A typical profile of

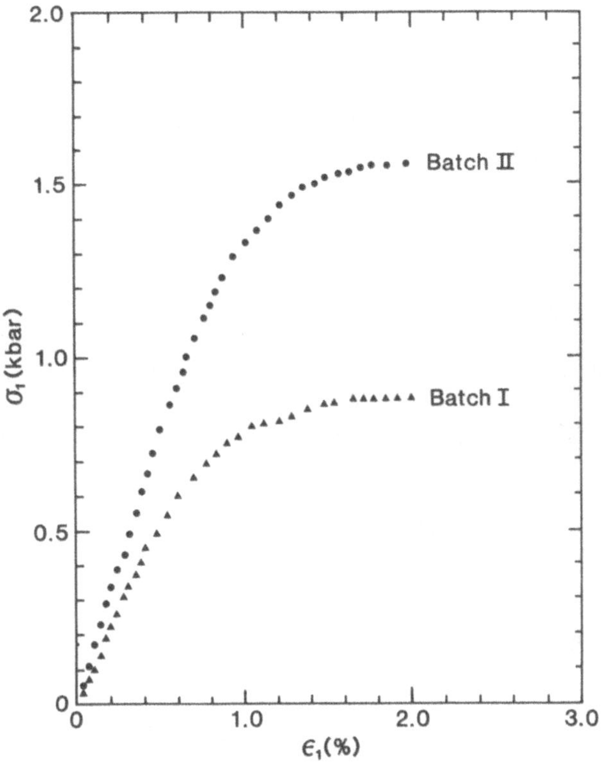

FIG. 1. Typical tensile stress-strain curves for the two batches of ytterbium foils. Measurements for batch I are from Ref. 4; batch II results are from the present work.

resistance change versus time from G2 for a peak stress of 6.5 kbar in fused silica matrix is shown in Fig. 3. The almost flat peak of the profile between 1.8 and 3.1 μsec shows the resistance change of 0.35 at peak stress. The residual resistance change value for this experiment is the mean value (0.04) in the time interval 3.1-4.0 μSec in Fig. 3.

The peak resistance change values, $\Delta R/R_o|_{peak}$, for both the matrix materials are plotted in Fig. 4a and b. The predicted resistance changes, based on the inclusion model,[3] are shown by drawing curves through the calculated points and will be discussed in the next section.

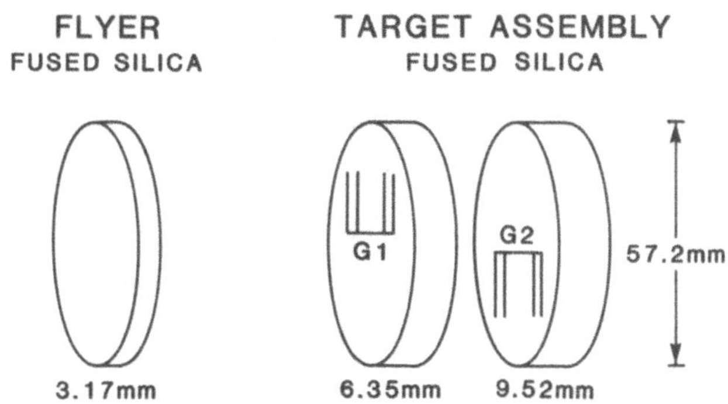

FIG. 2. Schematic view of the flyer and the target assembly for experiments in fused silica matrix. G1 refers to the surface gauge and G2 refers to the embedded gauge.

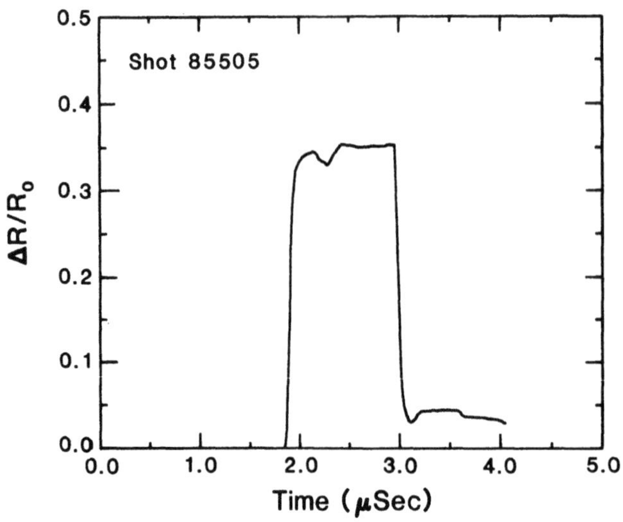

FIG. 3. Typical $\Delta R/R_o$ versus time profile for an embedded gauge (G2) in fused silica matrix. The slight dip in the value of $\Delta R/R_o$ between 2.1 and 2.4 μsec time is common to all records in the three experiments; the cause of this dip is not known.

Table II- Summary of Dynamic Experiments

(a) PMMA Matrix[a]						
Exper. No.	Impactor Velocity (mm/μs)	Peak Long. Stress (kbar)	Embedded Gauge $\Delta R/R_o\|_{peak}$	Embedded Gauge $\Delta R/R_o\|_{res.}$		
84563	0.202	3.6	0.181	0.046		
84550	0.617	11.6	1.05	0.148		
84564	0.520	18.5	1.91 ± 0.01	...		
(b) Fused Silica Matrix[b]						
			Embedded Gauge peak	Embedded Gauge res.	Surface Gauge peak	Surface Gauge res.
85505	0.100	6.4	0.35	0.041	0.29	0.056
84569	0.138	8.7	0.61	0.075	0.53	0.075
85504	0.245	15.0	1.28	0.095	1.36	0.087

a. All experiments used impactor material PMMA, with the exception of 84564.
b. All experiments used impactor material Fused Silica.

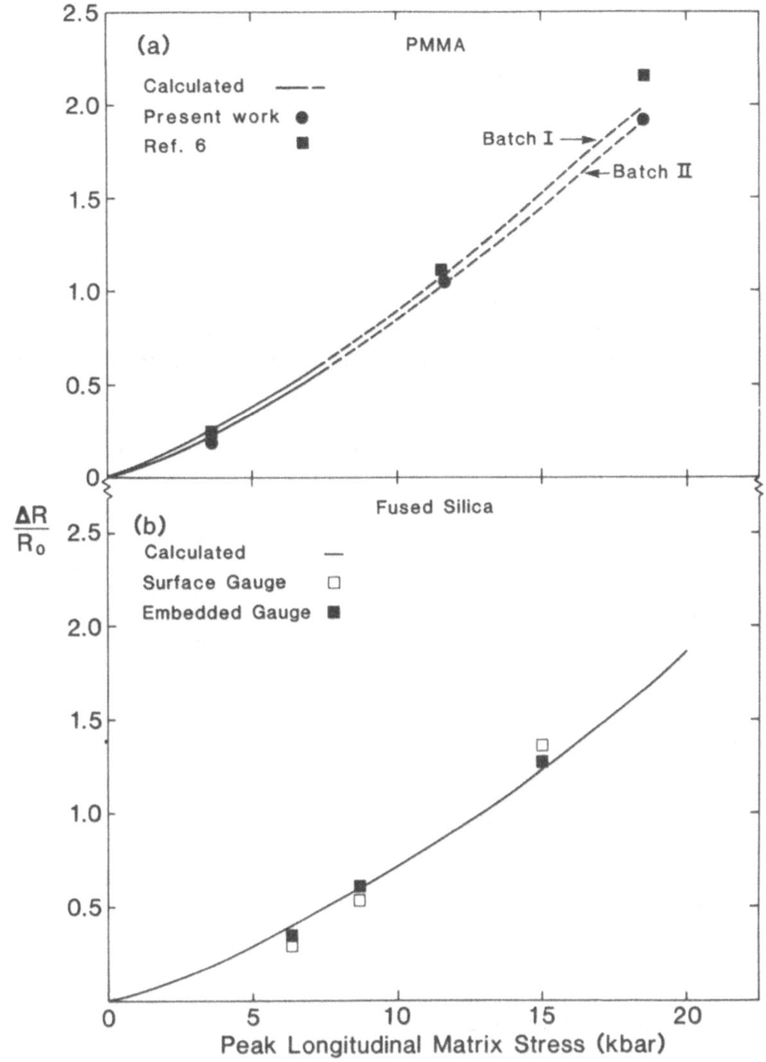

FIG. 4. Calculated and measured values of $\Delta R/R_o|_{peak}$ as a function of peak longitudinal matrix stress in (a) PMMA and (b) fused silica. See text for an explanation of the broken line for calculations in PMMA.

DISCUSSION OF THE RESULTS

The quasistatic measurements, summarized in Table I, indicate the differences in the mechanical properties of the two batches. For batch II the range of values for E from 148-180 kbar are consistently higher than the mean value of 120 kbar for batch I; the yield stress value of 0.78 kbar for batch II is higher than the mean value of 0.48 kbar for batch I. Despite the differences in the mechanical properties of the two batches, the piezoresistive coefficients are nearly equal. These findings indicate that the dynamic response of the two foil batches should differ by a small amount. Inclusion calculations [3,6] are needed to determine the effect of the mechanical properties on the gauge response.

The resistance change measurements from the dynamic experiments are compared for the two batches in Fig. 4(a). We have shown three data points from Ref. 6 that correspond to the stresses in our experiments. In all three cases, the batch I foils showed higher resistance change. To understand the role of material constants, we have calculated the resistance change for the two batches. The calculated values are shown as a solid line up to 7.6 kbar and then as a broken line beyond this stress. This representation is chosen to emphasize the fact that the calculations, based on the assumption of an elastic matrix, are strictly valid to 7.6 kbar.

The calculated curve for batch I is higher than that for batch II and reflects the effect of the mechanical constants on the gauge response. Except for the 18.5 kbar datum from batch I, the calculations are in reasonable agreement with the measurements. The agreement is better for batch II foils. Although the matrix is not elastic beyond 7.6 kbar, the agreement between theory and experiment is good in the higher stress region. This agreement is most likely due to the following factors: (i) continuity of the normal stress in the matrix and the thin foil gauge, (ii) the transverse stresses in the gauge are controlled primarily by the normal stress in the foil and gauge plasticity and (iii) at higher stresses, the stress contributions dominate the resistance change in ytterbium; strain contributions are minimal. Because of these factors, the gauge response is relatively insensitive to the transverse stresses in the matrix.

The agreement between the theory and the experiment for both batches confirms the overall validity of our approach and suggests that the quasi-static experiments can be used to characterize the dynamic response of the foils.

The results of the ramp wave experiments show a variation of approximately 6-19 percent between the surface gauge and the embedded gauge. Because there are no systematic trends for the three experiments, firm conclusions cannot be drawn from these data. Part of the scatter reflects the lower bound in measurement accuracy. This results in a larger percentage variation at the lower stresses. It is intersting to note that the calculated curve agrees quite well with the embedded gauge measurements. This result suggests that the surface gauge may be sensitive to impact conditions and may not conform to the requirements of the theoretical model. In future experiments, we will examine the response of ramp wave loading by embedding gauges at two different sample depths. For the present, we infer no strain rate effects on the gauge response from ramp wave experiments.

CONCLUSIONS

(i) The electromechanical constants determined in the quasi-static experiments are valuable in characterizing and correlating different batches of ytterbium foils.
(ii) The measured dynamic response of a gauge, from a particular batch of ytterbium foils, relates quite well to the response predicted on the basis of the theoretical model [3] using the material constants measured for that batch.
(iii) Preliminary investigation on ramp wave loading does not show any loading rate effects on the response of ytterbium gauges. Further experiments are needed to obtain firm conclusions on loading rate effects.

ACKNOWLEDGEMENTS

The excellent technical support provided by J. Thompson is acknowledged. This work was supported by AFOSR under grant 82-0132.

REFERENCES

1. Y.M. Gupta, D.D. Keough, D. Henley and D.F. Walter, Appl. Phys. Lett. 37:395 (1980).
2. Y.M. Gupta, J. Appl. Phys. 54:6094 (1983).
3. Y.M. Gupta, J. Appl. Phys. 54:6256 (1983).
4. D.Y. Chen, Y.M. Gupta, and M.H. Miles, J. Appl. Phys. 55:3984 (1984).
5. S.C. Gupta and Y.M. Gupta, J. Appl. Phys. 57:2464 (1985).
6. S.C. Gupta and Y.M. Gupta, submitted to J. Appl. Phys.
7. S.C. Gupta and Y.M. Gupta, Proceedings of the Fourth Topical Conference on Shock Waves in Condensed Matter, Spokane, Washington, (July 22-25, 1985); see this volume.
8. E. Barsis, E. Williams, and C. Skoog, J. Appl. Phys. 41:5155 (1970).
9. D. E. Grady and M.J. Ginsberg, J. Appl. Phys. 48:2179 (1977).
10. Z. Rosenberg, D. Yaziv, and Y. Partom, J. Appl. Phys. 51:3702 (1980).
11. L.M. Barker and R.E. Hollenbach, J. Appl. Phys. 41:4208 (1970)
12. P.S. DeCarli, Private communication (1984). Resistance measurements by Aidun and DeCarli on different batches, comparable to ours, yield nearly identical values. These results lead us to infer that the hydrostatic measurements are not sensitive to small variations between batches.

PIEZORESISTANT RESPONSE OF VAPOR-DEPOSITED YTTERBIUM

B. S. Holmes and J. B. Aidun

SRI International
333 Ravenswood Avenue
Menlo Park, CA 94025

BACKGROUND

Vapor-deposited ytterbium (Yb) stress gages have been used in several stress wave studies and may be especially effective when the gage thickness must be kept to a minimum.[1,2] One application for these gages is the measurement of the stress waves produced in solid targets by pulsed lasers.[2,3] In this application, the gages are usually located just beneath the surface of the target. The laser produces a surface pressure pulse with a very rapid rise time and peak pressures from tens of bars to several kilobars.

It is common practice to calibrate piezoresistance gages in dynamic uniaxial strain experiments. However, this method is costly. In addition, the laser stress measurement application presents special difficulties because the strain state in the gage is not uniaxial during the entire loading unloading cycle.[3] This effect and any plastic deformation of the gage makes interpretation of the gage record uncertain at late times in the experiment.

These problems of two-dimensional gage response and yielding of the gage have motivated the current study. Gupta and coworkers suggested that Yb foil gages can be calibrated from simple hydrostatic and tensile test data.[4,5] We have attempted to extend this idea to vapor-deposited Yb gages to (1) show that the constitutive and piezoresistant behavior of the thin vapor-deposited Yb can be described by isotropic bulk properties and (2) measure the piezoresistant coefficients suggested by Gupta[4,5] for vapor-deposited Yb and suggest a general method for gage calibration.

EXPERIMENTAL TECHNIQUE

Calibration experiments were performed to measure the resistance change for vapor-deposited Yb elements subjected to two independent loading paths. These data are sufficient to determine the piezoresistant response of Yb in the elastic region if the response is isotropic and the elasic moduli of the Yb are known. We also deformed the Yb elements plastically by further increasing the load until strong nonlinear response was observed.

The calibration experiments consisted of a compressive loading path and a tensile loading path. For the compression experiment, the Yb element was deposited onto fused silica, Figure 1(a). This specimen was compressed hydrostatically in a hydraulic pressure vessel. For the tension experiment, the Yb element was deposited onto a polyimide film that was bonded to an aluminum tensile specimen, Figure 1(b). The polyimide film insulated the element from the aluminum. In both experiments, the deformation of the Yb element is partially constrained by the substrate. Consequently, the Yb elements in the two experiments are not subjected to either hydrostatic compression or simple tension.

Figure 2 plots the measured resistance change versus pressure for the compression experiment. The same gage and substrate package was cycled three times. Note that there was a permanent change in gage resistance with each cycle. Also, the range of the truly linear response is limited to low pressures in each cycle.

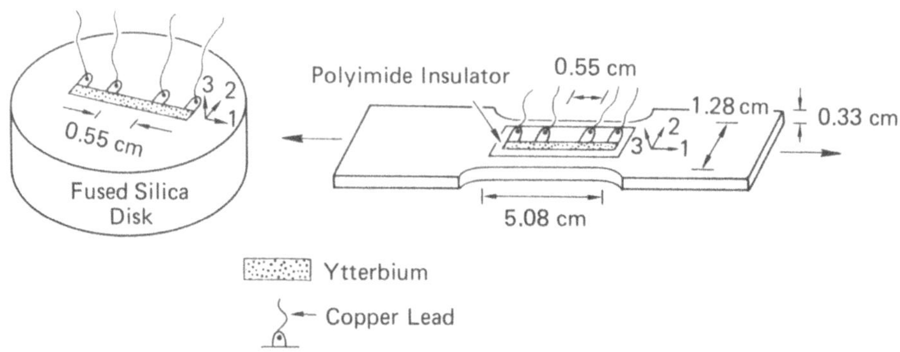

(a) Compression Test Specimen (b) Tensile Test Specimen

Fig. 1. Test specimens with vapor-deposited ytterbium elements

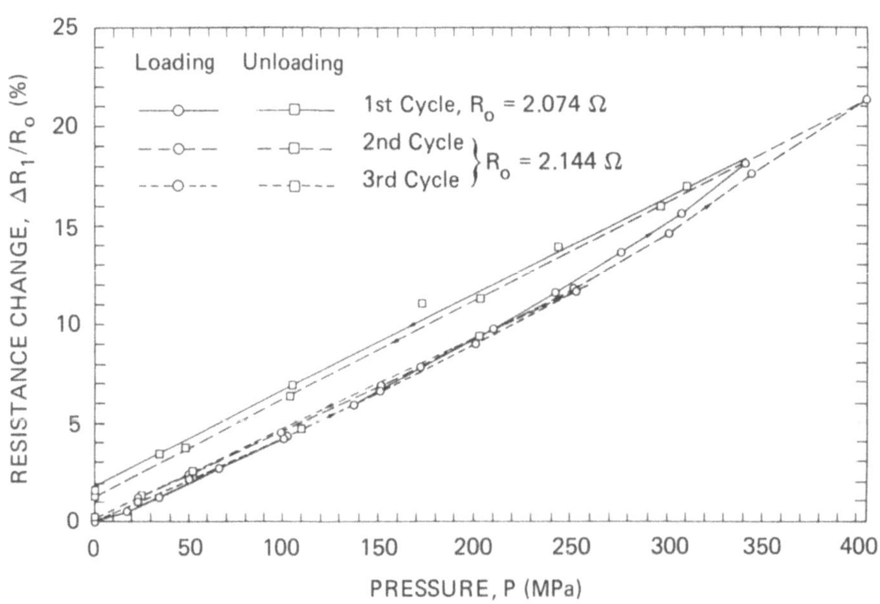

Fig. 2. Compression of ytterbium deposited onto fused silica.

In the tensile experiment, the specimen was loaded and unloaded (cycled) 12 times. The strain in the aluminum was measured by a resistance strain gage and an MTS extensometer (1-inch gage length). These two strain measures were consistent throughout the experiment. The measured resistance change in the tensile experiment is shown in Figure 3 as a function of axial strain of the tensile specimen (ε_1^A). The load-strain relationship for the aluminum tensile specimen is also shown in Figure 3.

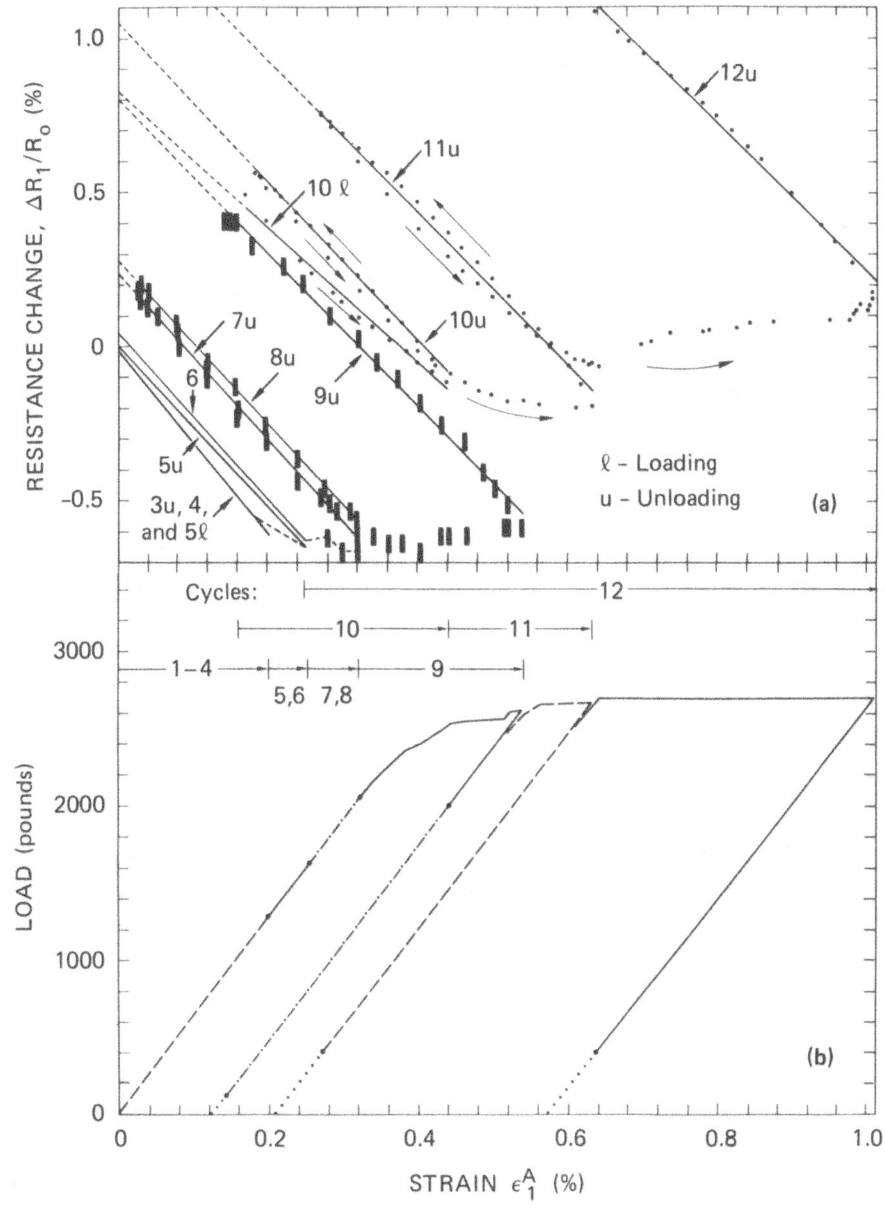

Fig. 3. Tensile test of vapor-deposited ytterbium on aluminum. (Symbol size represents estimated uncertainty in measurement.)

At small strains, the data show a linear response. Over the first eight cycles, the aluminum tensile specimen did not yield, but the Yb resistance changed slightly. A simple explanation for this behavior based on the model for piezoresistance[4] is that the Yb yielded. In cycles 9, 11, and 12 the aluminum yielded. In cycle 10, the Yb probably yielded slightly.

INTERPRETATION

For the purposes of analysis, we assign a rectangular coordinate system to the Yb elements as shown in Figure 1. The gage resistance is measured in the 1 direction. Analysis of the experiments depends on several assumptions. First, we assume that isotropic bulk properties can be assigned to the thin Yb layer. Second, we assume that the Yb is firmly bonded to the substrate and that the strains in the 1 and 2 directions are the same as the corresponding strains in the substrate. We further assume that the elastic moduli of the vapor-deposited Yb are the same as those measured in Yb foil. The material properties used in the analysis are listed below:

Ytterbium				
Young's modulus,	E	-	12.00	GPa
Shear modulus,	G	-	4.40	GPa
	λ	-	11.87	GPa
Fused Silica				
Bulk modulus,	K^s	-	36.6	GPa
Aluminum				
Poisson's ratio,	ν^A	-	0.33	
Young's modulus,	E^A	-	66.0	GPa

Elastic Response

Based on the assumption of similar strains in the Yb element and substrate, the strains in Yb in the 1 and 2 directions (ε_1 and ε_2) and the stress in the 3 direction (σ_3) are known.

$$\varepsilon_1 = \varepsilon_2 = -P/3K^s \qquad [1]$$

$$\sigma_3 = -P \qquad [2]$$

where P is the hydrostatic pressure. For our coordinate system, all the shear components are zero and the 1, 2 and 3 coordinate directions (indicated by subscripts) are principal directions in the materials. For elastic response, we use the constitutive relation.

$$\sigma_1 = 2G\varepsilon_1 + \lambda(\varepsilon_1 + \varepsilon_2 + \varepsilon_3) \qquad [3]$$

Combining [1] through [3], we obtain

$$\sigma_1 = \sigma_2 = (P/3K^s)\left[-2(G + \lambda) + \lambda\left(\frac{2\lambda - 3K^s}{\lambda + 2G}\right)\right] \qquad [4]$$

In the tensile experiment, we again assume that the Yb strains follow those in the substrate. Then,

$$\varepsilon_1 = \varepsilon_1^A \ ; \qquad \varepsilon_2 = \varepsilon_2^A = -\nu^A\varepsilon_1^A \ ; \qquad \sigma_3 = 0 \qquad [5]$$

From [3] and [5], we obtain

$$\varepsilon_3 = -\varepsilon_1 \frac{\lambda(1 - \nu^A)}{\lambda + 2G} \quad [6]$$

and

$$\sigma_1 = 2G\varepsilon_1\left[\frac{\lambda(1 - \nu^A)}{\lambda + 2G} + 1\right] \quad ; \qquad \sigma_2 = 2G\varepsilon_1\left[\frac{\lambda(1 - \nu^A)}{\lambda + 2G} - \nu^A\right] \quad [7]$$

so that all the stress and strain components in the two experiments are determined.

Following the model for piezoresistant response (4) the incremental resistance change ΔR_1 when the current and electric field are along the 1 direction is:

$$\frac{\Delta R_1}{R_o} = \alpha(\Delta\sigma_1 + \Delta\sigma_2 + \Delta\sigma_3) + 2\beta\Delta\sigma_1 + \Delta\varepsilon_1 - \Delta\varepsilon_2 - \Delta\varepsilon_3 + \eta\,\Delta\sqrt{I'^p_2} \quad [8]$$

where R_o is the reference resistance, and α, β and η are material properties. $\sqrt{I'^p_2}$ is the second invariant of the plastic deviator strain tensor and is used as a scalar measure of plastic deformation. Combining [1] through [8], we can write two equations for the linear elastic response of the Yb elements in each of the experiments in which the unknowns are the coefficients α and β.

Compression

$$S_c = \frac{\Delta R_1/R_o}{\Delta P} = \left\{(\alpha + \beta)\left[2\lambda\left(\frac{2\lambda - 3K^s}{\lambda + 2G}\right) - 4(G + \lambda)\right] - 3\alpha K^s - \frac{2\lambda - 3K^s}{\lambda + 2G}\right\}/3K^s \quad [9]$$

Tension

$$S_T = \frac{\Delta R_1/R_o}{\Delta\varepsilon_1^A} = 2\alpha G(1 - \nu^A)\left[1 + \frac{2\lambda}{\lambda + 2G}\right] + 4\beta G\left[1 + \frac{\lambda(1 - \nu^A)}{\lambda + 2G}\right] + 2\left[\frac{\lambda + (1 + \nu^A)G}{\lambda + 2G}\right] \quad [10]$$

Using values for S_c and S_T determined from the data, we solved [9] and [10] for α and β. The results are given below with the corresponding values for Yb foil.[4]

		α (GPa^{-1})	β (GPa^{-1})	S_c(GPa^{-1})	S_T
Vapor-deposited	A	-0.07	-0.15	0.43	-2.72
	B	-0.11	-0.12	0.49	-2.61
Foil[5]		-0.079	-0.171		

The first pair of coefficients (A) results from estimates of S_c and S_T of the early portion of the intial loading cycle for the respective experiments. The second pair (B) results from an S_c value from a least-squares fit to the unloading data and an S_T value from a least-squares fit to the data of cycles 2 - 8.

Yield Point

As in the elastic analysis, we attempted to explain the behavior of the Yb in terms of bulk properties. As suggested by Gupta,[4] we assumed that permanent resistance change and non-linear response in the Yb occurs with yielding. We further assume that the yielding of Yb can be described by a Mises yield condition.

$$(\sigma_1 - \sigma_2)^2 + (\sigma_2 - \sigma_3)^2 + (\sigma_3 - \sigma_1)^2 = \sigma_o^2 \qquad [11]$$

where σ_o is the yield strength in simple tension. For the tensile experiment, it appears that the Yb yields (shows non-linear response) first at $\varepsilon_1^A = 0.002$ or less. Substituting [5] and [6] into [3] and using the resulting expression for σ_3 with [7] in [11], we obtain

$$\sigma_o^2 = 2(G\varepsilon_1^A)^2 \{(1 + \nu^A)^2 + \frac{\lambda^2(5 - 8\nu^{A2} + 5(\nu^A)^2) + 4G(2\lambda + G)(1 +(\nu^A)^2) - 8G\lambda\nu^A}{(\lambda + 2G)^2}\} \qquad [12]$$

If we assume that yielding occurs in the tensile test at $\varepsilon_1^A = 0.002$, then $\sigma_o = 24$ MPa.

For the compression test, combining [11] with [2] and [4] gives

$$\sigma_o = P\ \{1 - [2(\lambda + G) - \frac{\lambda(2\lambda - 3K^s)}{\lambda + 2G}]/3K^s\} \qquad [13]$$

Substituting in the value of 24 MPa for σ_o, we obtain a linear elastic limit of 94.5 MPa for Yb in the compression experiment.

CONCLUSIONS

Piezoresistance coefficients have been determined for vapor-deposited Yb in the elastic range. The coefficients that govern response were found to be similar to those measured for Yb foil used in stress gages but evaluation of the coefficients depended on the interpretation of the average slope in the compression experiment. This suggests that the vapor-deposited material can be assigned bulk properties and implies that general elastic response can be predicted from two independent calibration experiments. However, yielding of the Yb at low stress levels will limit the applicability of such a calibration to stress levels of 100 MPa or less. We have not attempted to describe the post yield behavior seen in the experiments.

REFERENCES

1. D. D. Keough, Development of a High-Sensitivity Piezoresistive Shock Transducer for the Low Kilobar Range, Final Report., SRI International, Menlo Park, CA. (March 25, 1970).
2. B. S. Holmes, Measuring Pressure from Short Laser Pulses, in: "Shock Waves in Condensed Matter-1983," by J. R. Asay et al., eds., North Holland, Amsterdam (1983).
3. B. S. Holmes, Pressure Measurements in Laser-Supported Plasmas with Piezoresistance Gages, Experimental Mechanics 37:32 (March 1985).
4. Y. M. Gupta, Stress Measurements Using Piezoresistance Gauges: Modeling the Gauge as an Elastic Plastic Inclusion, J. Appl. Phys. 54:11 (1984).
5. D. Y. Chen, Y. M. Gupta, and M. H. Miles, Quasistatic Experiments to Determine Material Constants for the Piezoresistance Foils used in Shock Wave Experiments, J. Appl. Physics 55:11 (1984).

ON THE LATERAL STRESS MEASUREMENT IN SHOCK LOADED TARGETS WITH TRANSVERSE PIEZORESISTANCE GAUGES

Z. Rosenberg and Y. Partom

A.D.A., P.O. Box 2250
Haifa, Israel

ABSTRACT

We present an analytical approach for the calibration of transverse piezoresistance gauges in shock loaded specimens. Our approach is based on a few simple assumptions concerning the stress state of the gauge element. These assumptions lead to a rather simple calibration procedure which we applied to transverse manganin gauges. We show how our experimental data for PMMA targets can be accounted for both under loading and complete unloading conditions. In order to further demonstrate the validity of our approach we present transverse stress measurements in glass targets.

INTRODUCTION

One of the most important issues in the study of dynamic behavior of solids under shock loading is the direct determination of the lateral stresses which prevail in the shocked solid. The only direct technique which has been attempted uses thin piezoresistance stress gauges which are inserted laterally in the shocked target as shown schematically in Fig. 1. Several workers attempted to measure this stress with various foil gauges (manganin, ytterbium and carbon[1-5]). We have shown that the response of transverse manganin gauges can be very different from that of the more usual (parallel) gauge by embedding both gauges in polymethylmethacrylate (PMMA) targets[6]. In a more recent work Gupta[7] showed that this is also the case of ytterbium gauges. This motivated him to perform a detailed calculation of the interaction between gauge and matrix using an Eshelby inclusion approach[7]. One of his main conclusions is that "analytical developments are needed to obtain the calibration for a lateral stress measurement". The purpose of the present paper is to outline such an analytical calibration scheme and to apply it to manganin gauges. We shall demonstrate the validity of this calibration by comparing experimental and predicted results, for shock loaded PMMA and glass.

EXPERIMENTAL TECHNIQUES

The planar impact experiments were conducted on our two 64 mm laboratory guns described in Refs. 8 and 9. The gauges we use are

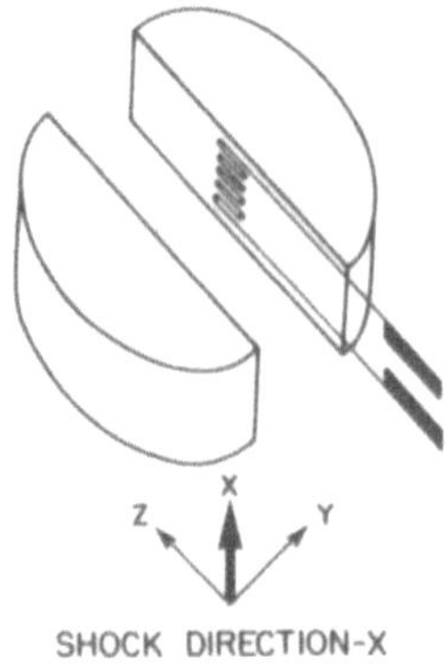

Fig. 1. Lateral stress gauge arrangement. (Gauge length is along z axis.)

manufactured by Micro Measurements (type LM-SS-125CH-048). These are 5 μm thick, grid-like foil gauges which are attached to a 40 μm thick epoxy sheet. The calibration curve of these gauges is given also in Ref. 8 in terms of their relative resistance change versus shock stress in the 0 to 200 kbar range.

The major advantage of using PMMA as an embedding material in these calibration studies is the fact that owing to the epoxy backing of the gauge and the epoxy bonding which is used in target assembly, one can consider the gauge to be embedded in a homogeneous material. This is so because epoxies and PMMA have close dynamic properties[10]. In order to enhance our results we made a few measurements with strips of manganin which are 5 μm thick, 0.5 mm wide, and 13 mm long. These strips of manganin are attached to a polymide backing and they were cut from the current leads of a different gauge, also manufactured by Micro Measurements. The idea was to check whether our results were sensitive to the geometrical shape of the gauge (grid-like versus strips) or if they depend upon the backing material of the gauge.

RESULTS AND DISCUSSION

Experiments in PMMA

Fig. 2 shows the resistance change ratio (lateral gauge to parallel gauge) versus the longitudinal shock stress in the 1 to 30 kbar stress range. Also shown in Fig. 2 some data points with the manganin strips and as is clearly seen they fall on the same curve of the grid-like gauges. Another point from the work of Keough[11] is also shown in Fig. 2 and we can see that it is close to our curve.

It is obvious from Fig. 2 that transverse manganin gauges do not respond to the lateral stresses in the same manner as parallel gauges respond to the longitudinal stress in the target. One cannot imagine that the lateral stresses in PMMA, shock loaded to 1 to 3 kbar, would be 3 to 2 times higher than the longitudinal stress. On the contrary, evidence by other techniques reveal that the lateral stresses in PMMA are substantially lower than the longitudinal in the 0 to 10 kbar range[12].

Calibration of Transverse Gauges

We first assume that the stress along the thickness of the gauge is equal to the lateral stress in the target, σ_y. This assumption is actually the basic one behind the lateral stress gauge technique. We also assume

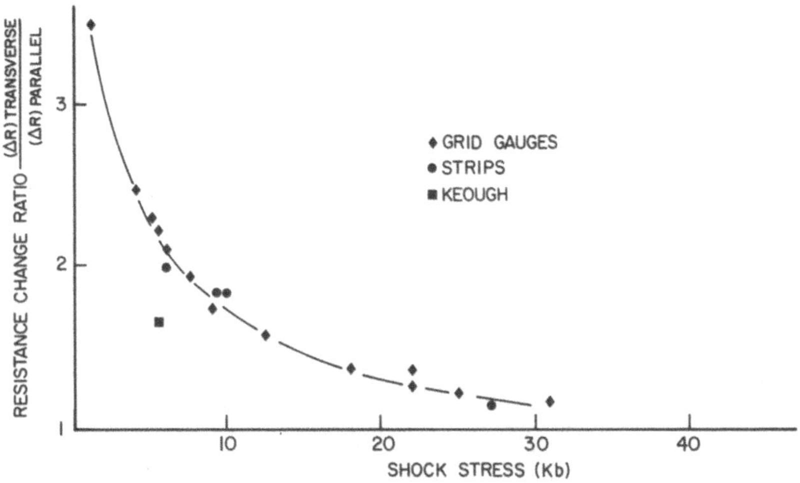

Fig. 2. Experimental data for the resistance change ratio of lateral versus parallel gauges.

that the strain along the shock direction which the gauge experiences ε_x is the same as the uniaxial strain in the target. This assumption is justified for foil gauges which have large widths. In a related work[13] we have shown that a thin wire or foil which is placed on a longitudinal configuration in the target, so that its length coincides with the shock direction, undergoes the same longitudinal strain as its surrounding. One can gain some confidence in these assumptions by observing the stress and strain states which were calculated by Gupta[7] using his elasto-plastic inclusion model. One finds that his transverse ytterbium gauges follow these two assumptions. We have to remember that the elastic constants of manganin and PMMA are very different and the assumption of equal ε_x means that the stress component in the gauge along its width, σ_x, will be very large if we assume elastic behavior.

We thus conclude that the elastic behavior of the transverse gauge is confined to a very small stress range, probably less than about 0.5 kbar in the PMMA, and that the values that σ_x can take are determined by both σ_y and the yield criterion of the gauge. In the plastic range the difference between the two stresses equals the yield strength. Thus:

$$\sigma_x = \sigma_y + Y \tag{1}$$

where Y is the yield strength of the gauge material.

There remains the question of the magnitude of the third stress component, σ_z, along the gauge length. Since this dimension is the longest of all three we may assume, as is customarily done, that the strain along this direction is zero and that the gauge's length does not change during the experiment. As for the stress along this direction we make the assumption that it is equal to the stress component in the y direction (foil thickness). This assumption can have some justification if we consider again Gupta's calculations[7] which show that the two lateral stresses in the transverse gauge are nearly equal. Thus, the stress and strain tensors which describe the transverse gauge are:

$$\sigma_{ij} = \begin{pmatrix} \sigma_y + Y & & 0 \\ & \sigma_y & \\ 0 & & \sigma_y \end{pmatrix} \qquad \varepsilon_{ij} = \begin{pmatrix} \varepsilon_x & & 0 \\ & \varepsilon_v - \varepsilon_x & \\ 0 & & 0 \end{pmatrix} \tag{2}$$

where ε_v is the volumetric strain of the gauge corresponding to the hydrostatic pressure which it experiences. This pressure is given (from Eq. (2)) by:

$$P = \frac{\sigma_x + \sigma_y + \sigma_z}{3} = \sigma_y + \frac{1}{3} Y \tag{3}$$

We next use the general expressions which we derived for the response of manganin gauges in Ref. 14. In the plastic range of the gauge its response is given by:

$$\frac{\Delta R}{R_o} = 3.45\ \varepsilon_v + 9.89\ \varepsilon_v^2 + 5.17 \cdot 10^{-4} \tag{4}$$

where ε_v is the volumetric strain which is determined by the hydrostatic compression curve of manganin as given by Barsis et al.[15]:

$$P = 1160\ \varepsilon_v + 4120\ \varepsilon_v^2 \tag{5}$$

where P is in kilobars.

In order to determine the lateral stresses - σ_y - from the measured resistance changes of the transverse gauges we start by solving Eq. (4) for ε_v and have:

$$\varepsilon_v = -0.1744 + \sqrt{0.03042 + \Delta/9.89} \tag{6}$$

where $\Delta \equiv \Delta R/R_o$ is the measured resistance change of the transverse gauge.

After having the volumetric strain we insert it in Eq. (5) and get the hydrostatic pressure, P, from which by Eq. (3) we get the desired value of σ_y. The value of Y for manganin changes with pressure, as we showed in Ref. 16, but for the low stress region we can take Y = 7.5 kbar as we found in Ref. 8.

If we carry out the above analysis for the 1 kbar experiment in PMMA we get a value of P = 3 kbar for the transverse gauge. Using the value of 1/3 y = 2.5 kbar in Eq. (3) we obtain a value of σ_y = 0.5 kbar for this experiment. Similar results are obtained for the other experiments which account for the shape of Fig. 2.

Release Experiments

In order to further enhance our understanding of the transverse gauge, we performed a few experiments in which the gauge was located near the free surface of the PMMA target. This way we have both the shock level (the upgoing steps) and the release which brings the target to a state of $\sigma_x = 0$. Fig. 3 shows a typical gauge record from such an experiment.

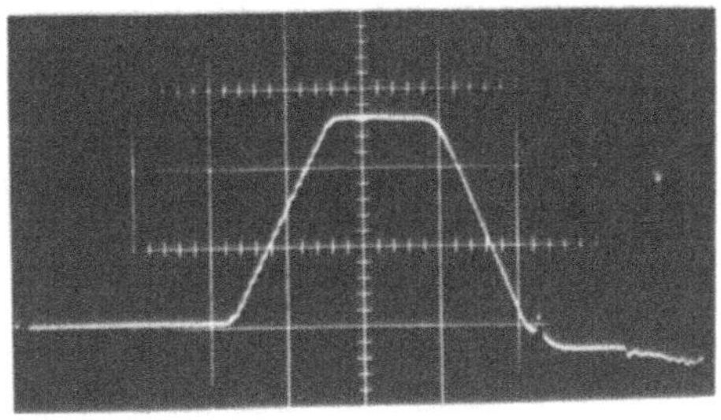

Fig. 3. A typical transverse gauge record in full loading-unloading experiment.

The most striking thing about this record is the fact that the release part of the record reaches negative values of the resistance change, which occurred in every one of these experiments (at σ_x = 5, 6, 7.5, and 9 kbar). What is even more important is that in all four experiments we obtained similar values for the negative resistance change and these values corresponded to a negative pressure of P = -2.5 kbar. Here we assumed that the hydrostatic compression curve is symmetric and can be used for tensions also. This value of negative P is in very good agreement with a stress state in which two stresses are equal to zero and one is equal to the negative value of the yield strength of manganin (-Y = -7.5 kbar). We thus assume that upon release both σ_y and σ_z go to zero and σ_x in the gauge reverses sign and reaches -Y. It is interesting to note that Gupta[7] obtained similar negative resistance changes upon stress release in his calculations for ytterbium gauges embedded in PMMA targets.

Example - Measurements in Glass

Fig. 4 shows typical oscilloscope records of parallel and transverse gauges in a soda lime glass. This material has a very high Hugoniot elastic limit (over 60 kbar) and we performed several experiments in its elastic range (for the one shown in Fig. 4 σ_x = 28 kbar). We know that in the elastic range of a shocked solid the relation between σ_y and σ_x is $\sigma_y = (\nu/1-\nu)\sigma_x$ where ν is Poisson's ratio. For soda lime glass $\nu = 0.25$ (see Ref. 16) and for the elastic range we expect $\sigma_y = 1/3\sigma_x$. The measured transverse gauge resulted in a value of σ_y = 9.2 kbar which is in excellent agreement with the expected 9.3 kbar value. Similar agreement was also obtained for the other shots. Other experiments with transverse gauges, embedded in ceramic targets, are reported in Ref. 17. The agreement between measured and expected values for σ_y, in the elastic range of the ceramics, was also very good.

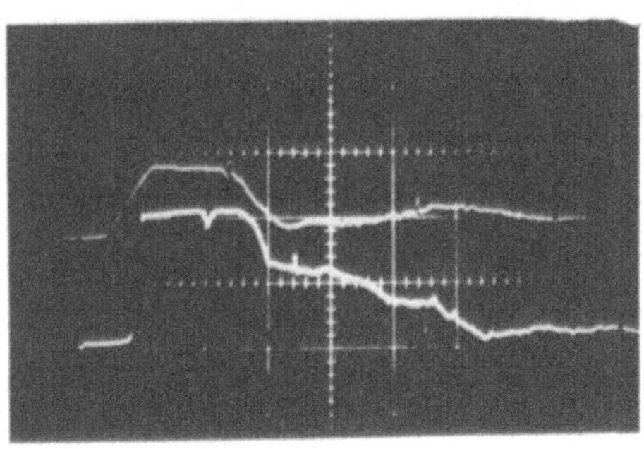

Fig. 4. Gauge records for transverse (upper beam) and parallel (lower beam) gauges in glass, (1 μs/div, 0.5 V/div).

CONCLUSIONS

We have developed an analytical approach for the calibration of transverse piezoresistance gauges in shock wave experiments. We showed how this approach can account for our experimental data of transverse manganin gauges in PMMA. We have also demonstrated the validity of the technique by comparing the expected transverse stresses in shock loaded glass with the measured values.

ACKNOWLEDGMENTS

The authors wish to thank Mr. R. Kreif and Mr. D. Mazar for their help in performing the experiments.

REFERENCES

1. D. Bernstein, C. Godfrey, A. Klein, and W. Shimin, "Behavior of Dense Media Under High Dynamic Pressures," (Gordon and Bresch, New York, 1968), p. 461.
2. P.F. Chartagnac, J. Appl. Phys. 53, 948 (1982).
3. I. Kanel, A.M. Moladets, and A.N. Dremin, Combust. Explos. and Shock Waves 13, 772 (1978).
4. P.F. Chartagnac, "Shock Waves in Condensed Matter," (APS Meeting, Menlo Park, 1981), eds. W.J. Nellis, L. Seaman, and R.A. Graham.
5. Y.M. Gupta, D.D. Keough, D. Henley, and D.F. Walter, Appl. Phys. Lett. 37, 395 (1980).
6. Z. Rosenberg, Y. Partom, and D. Yaziv, J. Appl. Phys. 52, 755, (1981).
7. Y.M. Gupta, J. Appl. Phys. 54, 6256 (1983).
8. Z. Rosenberg, D. Yaziv, and Y. Partom, J. Appl. Phys. 51, 3702 (1980).
9. Y. Porat and M. Gvishi, J. Phys. E. 13, 504 (1980).
10. L.M. Lee, J. Appl. Phys. 54, (1973).
11. D.D. Keough in the 1st Symp. on Piezoresistive Gauges, Arcachon, France, Sept. 29-30, 1981.
12. Y.M. Gupta, J. Appl. Phys. 51, 5352 (1980).
13. Z. Rosenberg, D. Yaziv, and Y. Partom, J. Appl. Phys. 51, 4790 (1980).
14. Z. Rosenberg and Y. Partom, J. Appl. Phys. 58, 1814 (1985).
15. E. Barsis, E. Williams, and C. Skoog, J. Appl. Phys. 41, 5155, (1970).
16. M.V. Swain and J.T. Hagan, J. Phys. D. 9, 2207 (1976).
17. D. Yaziv, Ph.D. Thesis, University of Dayton, July 1985.

GENERATION OF ULTRA-HIGH PRESSURE SHOCKS IN 0.26 μm WAVELENGTH LASER EXPERIMENTS

F. Cottet,[1] M. Hallouin,[1] J.P. Romain,[1]
R. Fabbro,[2] B. Faral[2] and H. Pépin[3]

[1]ENSMA, rue Guillaume 7, 86034 Poitiers, France
[2]Ecole Polytechnique, 99128 Palaiseau, France
[3]INRS-Energie, Varennes, Québec JOL2PO, Canada

The application of high-power pulsed lasers to generate ultra-high pressure shock waves has opened possibilities for studying the properties of materials under extreme pressure and temperature conditions. The traditional laboratory experiments currently permit one to obtain pressures well below 1 TPa; this pressure range has been largely extended by using laser-driven shock waves, such as early experiments have shown [1,2]. Recently, some short-wavelength laser experiments, realized at 0.35 μm [3] and at 0.26 μm [4,5] laser wavelengths, have shown the possibility of easily reaching pressures comparable to those measured in nuclear-explosion driven shock waves [6]. In this paper, we report experiments based on two techniques to increase the laser-induced shock pressure: impedance-match and flyer-plate techniques.

These two techniques, extensively used in conventional laboratory shock experiments, have been applied to laser-shock experiments. Using the advantage that the laser-shock pressure is independent of target material [4,5], the impedance-match method permits one to increase the shock amplitude by transmission of the laser-induced shock from a low impedance material, such as aluminum, into a high impedance one, such as copper or gold. Experiments by Holmes et al [7] on Al-Cu targets irradiated at 1.06 μm wavelength and 5.10^{13}-4.10^{14}W/cm^2 incident intensity have confirmed this effect in a pressure range below 1 TPa, giving a transmitted pressure in copper about twice the incident pressure in aluminum. The second type of experiment uses laser-driven ablation to accelerate a first foil, which subsequently impacts a second foil, generating a high-pressure shock wave. The potential advantage of this technique is that the target is not subjected to preheat due to the laser-matter interaction process. Experiments by Rosen et al. [8] have demonstrated the amplification power of this colliding foil technique: a pressure of 2 TPa was measured in a carbon impact foil, while the laser-induced pressure in the first foil was only 0.5 TPa.

We have made experiments using the same techniques with the GRECO ILM Nd laser quadrupled in frequency. The output energy on target at

---

Work reported performed by GRECO ILM, Ecole Polytechnique, FRANCE.

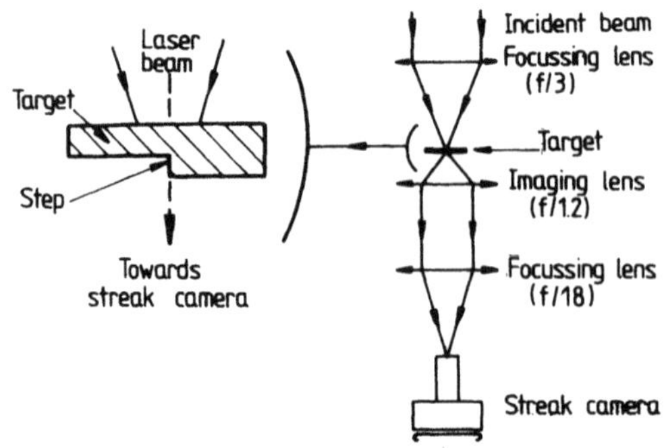

Fig. 1. Experimental set-up and scheme of stepped target. The rear side of the foil is imaged on the streak-camera slit, which is centered along a diameter of the focal spot.

0.26 µm wavelength is about 15 J in 0.5 ns full width at half maximum pulse; since 50% of this energy was inside 50 µm spot diameter, the incident intensity was about 10^{15}W/cm^2. In order to measure shock wave velocities obtained over short distances, we use targets with a step on the rear side, formed by vapor deposition. Transit times across the step are measured using a streak camera to record the arrival time of the shock wave as it reaches each free surface of the step (Fig. 1). Shock velocities are determined by dividing the step height by the transit time across the step. Time and spatial resolution of the system are about 5 ps and 5 µm respectively.

Examples of streak records obtained with the impedance-match technique (a) and the accelerated foil technique (b) are shown in Fig. 2. The experiments, corresponding to the case (a), were realized with Al-Au targets: 28 µm thick aluminum base and 6 µm high gold step. One can clearly see the delay of the luminosities corresponding to the shock emergence at the bottom and top of the gold step; this delay leads to a shock wave velocity of 28.5 mm/µs ± 5.7 mm/µs. Using Sesame equation of state data, this velocity corresponds to a pressure of 10.5^{+5}_{-4} TPa. This result can be compared with the theoretical value that we should have obtained. From previous experiments, performed in the same irradiation conditions, we know that the peak ablation pressure is equal to 5 TPa ± 0.5 TPa. So, in a pressure-particle velocity plane, we can graphically determine the transmitted shock pressure from Al and Au principal Hugoniots and the reflected Hugoniot at the Al-Au interface (Fig. 3): 10.6 TPa. The comparison shows that the experimental value is very consistent with the theoretically predicted value of shock pressure.

The typical targets used in the experiments based on the accelerated foil technique are shown in Fig. 2b. The initial spacing between the aluminum accelerated foil of 9 µm thickness and the aluminum impact foil is about 50 µm. This initial separation distance was adjusted for satisfying as well as possible two

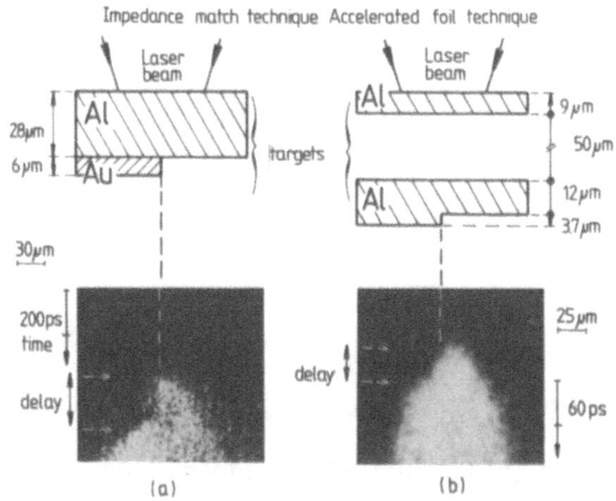

Fig. 2. Design of targets used and corresponding streak records for impedance-match experiment (a) and accelerated foil experiment (b). Streak-camera photos show optical signals correlated with shock breakout at two surfaces on stepped target. Incident laser intensity is about 10^{15} W/cm^2 in 0.5 ns FWHM at 0.26 μm.

conditions: the collision must occur at the end of the laser acceleration process, and the decompression of the accelerated foil must not be too important yet. Experiments were realized with stepped aluminum target foils of 6 μm and 12 μm thickness. The height of the step at the back face of the foil was precisely measured: 3.7 μm. Delay measurements give shock wave velocities of 135 ± 40 mm/μs and 82 ± 16 mm/μs for 6 μm and 12 μm base thicknesses of the target respectively. Assuming that the initial state of material is unperturbed in front of the shock wave, we can have a rough estimate of the induced pressure, using Sesame equation of state data for aluminum: 440 ± 240 Mbar and 160 ± 60 Mbar, respectively. The higher measured pressure, corresponding to a 6 μm thick target foil, may be overestimated considering the importance of preheating effects at this small thickness.

The main goal of these experiments was to show that UV high power lasers, associated with conventional shock techniques, offer the potential to obtain ultra-high pressure shock waves in the laboratory. Experiments, based on impedance-match technique, yield 10 TPa shock pressure in gold, and experiments using the accelerated foil method yield 15 TPa in aluminum for an incident laser-shock pressure of 5 TPa. These experiments demonstrate the possibility of significantly increasing the shock pressure range accessible to laser shock experiments at fixed irradiation conditions.

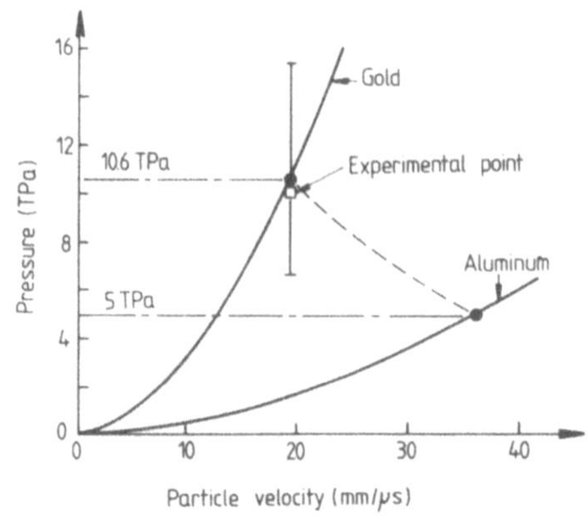

Fig. 3. Aluminum and gold principal Hugoniots (solid curves) and reflected Al Hugoniot at Al-Au interface (dashed curve). For an incident pressure of 5 TPa in Al, the impedance match technique, applied on Al-Au targets, leads to a transmitted shock pressure of 10.6 TPa in Au. The corresponding experiment gave a shock pressure of 10.5^{+5}_{-4} TPa (open square).

References

[1] L.R. Veeser and J.C. Solem, Phys. Rev. Lett, 40:1391 (1978).
[2] R.J. Trainor, J.W. Shaner, J.M. Auerbach, N.C. Holmes, Phys. Rev. Lett. 42:1154 (1979).
[3] B. Yaakobi, T. Boehly, P. Bourke, Y, Conturie, R.S. Craxton, J. Deletrez, J.M. Forsyth, R.D. Frankel, L.M. Goldman, R.C. McGrory, M.C. Richardson, W. Seka, D. Shvarts and J.M. Soures, Opt. Comm. 39:176 (1981).
[4] F. Cottet, R. Fabbro, J.P. Romain and B. Faral, Phys. Rev. Lett. 52:1884 (1984).
[5] F. Cottet, Hallouin, J.P. Romain, R. Fabbro and B. Faral, in Proceedings of XXIInd EHPRG meeting, J. Phys. C8, 45:305 (1985).
[6] C.E. Ragan III, Phys. Rev. A 21:458 (1980).
[7] N.C. Holmes, R.J. Trainor, R.A. Anderson, L.R. Veeser and G.A. Reeves, in Shock Waves in Condensed Matter, edited by W.J. Nellis, L. Seaman, and R.D. Graham (New York, AIP) 78:160 (1982).
[8] M.D. Rosen, D.W. Phillion, R.H. Price, E.M. Campbell, S.P. Obenshain, R.R. Whitloch, E.A. McLean and B.H. Ripin, in Shock Waves in Condensed Matter, edited by J.R. Asay, R.A. Graham and G.K. Straub (New York, North-Holland) 323 (1984).
[9] J. Abdallah, R.C. Albers, B.I. Bennett, F. Dowell, B.L. Holian, W.F. Huebner, J.D. Johnson, D.A. Liberman, S.P. Lyon, G.K. Straub and K.S. Trainor, Los Alamos National Laboratory Report No. 834 (1983).

LASER-DRIVEN SHOCK WAVES IN PLEXIGLASS

R.F. Schmalz[1], R. Fedosejevs[2], R. Sigel[1], Yung-Lu Teng[3]

[1]Max-Planck-Institut für Quantenoptik, 8046 Garching
Germany
[2]Department of Electrical Engineering, University
of Alberta, Edmonton, Canada T6G2G7
[3]Shanghai Institute of Laser Technology
319 Yueyang Road, Shanghai, China

EXPERIMENTAL RESULTS

When solid material is irradiated by an intense laser beam, a hot and dense plasma is formed on the irradiated surface. This plasma exerts a considerable pressure upon the neighbouring material, ranging up to ≅ 100 Mbar for present-day lasers. This pressure generates a shock wave which can be observed in trasparent material. The basic scheme of the irradiating laser beam, the solid plexiglass target and the dye laser beam used for shadowgraphy is shown in Fig. 1.

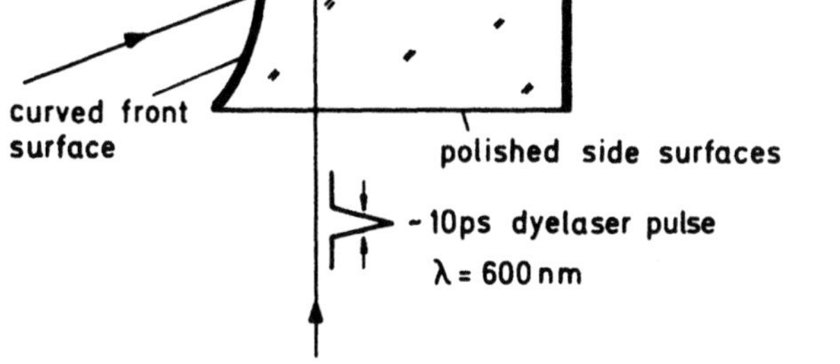

Fig. 1. Experimental arrangement

The target was irradiated by the Asterix III iodine laser, described in detail elsewhere[1]. Under typical operating conditions it delivered a 100 J pulse 280 (±60) ps (FWHM) at a wavelength of 1.3 μm. The beam was focused with an f/2 (f = 40 cm) aspheric lens. In the plane of smallest cross-section of the focused beam 50% of the incident energy was contained in a circle 85 μm in diameter. Two cases were investigated:

i) "out of focus" experiment with a 450 μm diameter of the irradiated area and a laser intensity of 1.4×10^{14} W cm^{-2},

ii) "in focus" experiment with an 85 μm diameter of the irradiated area and a laser intensity of 3×10^{15} W cm^{-2}. The target consisted of a stick of plexiglass (polymethyl-methacrylate $(C_3O_2H_8)_n$) of 6 x 6 mm cross section. The front surface was made slightly concave (radius 10 mm) in order to prevent diffraction at the edges of the target from disturbing the imaging of the vacuum-plexiglass interface underneath the irradiated area. Note that internal total reflection of the dye laser light from the curved surface causes the dark vertical band seen in the photographs. The plasma is hidden behind this dark band and, by reflection from the curved surface, its self-luminosity is partly suppressed.

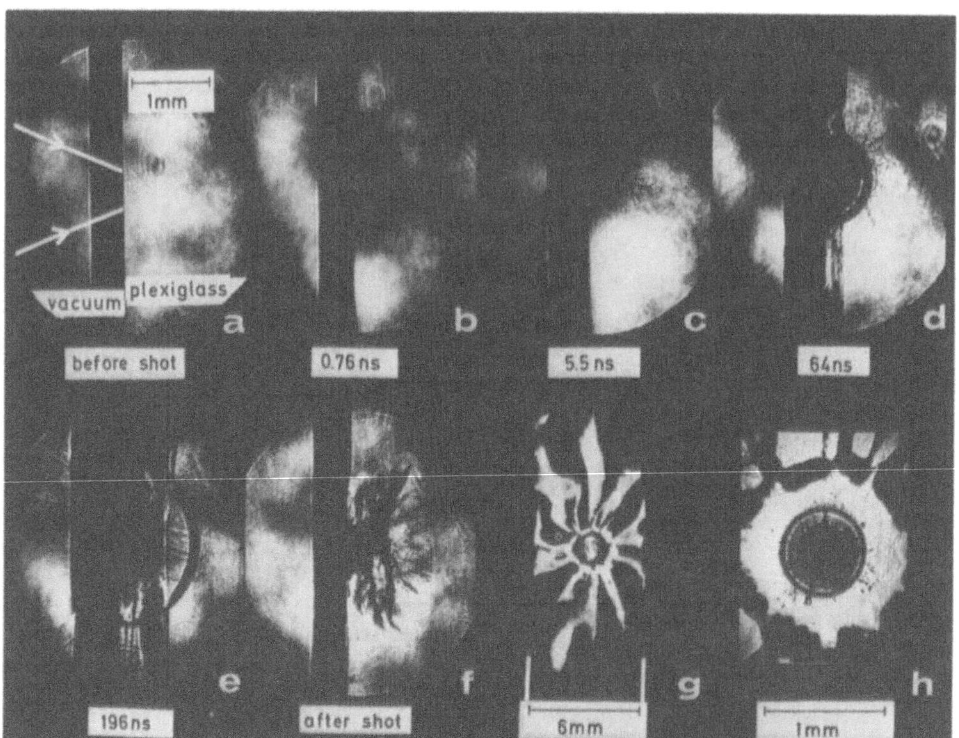

Fig. 2. Sequence of shadowgrams showing shock wave propagation in plexiglass at a laser irradiance of 3×10^{15} W cm^{-2}

Figure 2 shows a series of shadowgrams taken with the target in the "in focus" position at $\Phi \cong 3 \times 10^{15}$ W cm^{-2}. Figure 2b shows the plexiglass interior 0.76 ns after the maximum of the iodine laser pulse. A dark hemispherical cloud with diffuse boundaries has formed which is opaque to the dye laser light. It extends about 500 µm into the plexiglass. After 5.5 ns (Fig. 2c) tree-like structures are now visible which extend from the interior of the dark cloud into the clear plexiglass. These structures very much resemble the pattern of discharge channels found in corona discharges from a positive anode point. We conclude that we observe here basically an electrical breakdwon phenomenon [2,3]. After 64 ns (Fig. 2d) the shock wave released from the irradiated area has caught up with the dimensions of the opaque cloud and becomes visible. After 196 ns (Fig. 2e) the shock wave has proceeded further and separated from the opaque zone. Figure 2f shows the permanent damage to the target. When the target is taken out of the vacuum chamber and inspected (Fig. 2g, note the difference in magnification), surface damage with a star-like pattern is found around the irradiated area. The bright areas in Fig. 2g have become rough and scatter the illuminating light; dark areas are still polished and do not scatter. The damage traces bend around the stick and continue on the side surfaces. Figure 2h shows a magnified detail of Fig. 2g, namely the crater left in the target surrounded by surface damage.

After the shock front has traversed the opaque breakdown zone its motion can be measured by framing and streak photography and compared with calculated shock wave trajectories.

COMPARISON WITH SIMULATIONS

The shock wave generated in the "out of focus" experiment with the relatively large spot diameter remains approximately planar for a time of the order of 10^{-8} s. 1-D computer simulations [4] of the shock wave trajectory agree quantitatively with the experiments and suggest a peak pressure of 6 Mbar during laser irradiation.

The "in focus" experiments at 3×10^{15} Wcm^{-2} on a 85 µm diameter focal spot were simulated with the two-dimensional Eulerian MAGT code, which uses quasi-Lagrangian rezoning[5]. A study about the influence of various parameters on the shock wave trajectory was made. The following parameters were considered:

i) equation of state: The shock wave trajectory was computed for the SESAME[6,7] and ideal gas equation of state; the result is shown in Fig.3. For the ideal gas the initial temperature was chosen such that the sound velocity in the undisturbed material matches the sound velocity in plexiglass. As in the 1-D case the more realistic SESAME equation of state gives a somewhat higher velocity but the difference is slight. Note that the experimental point in Fig. 3 was obtained from Fig. 2d.

ii) diameter of the area of energy deposition: For the diameter of the area of energy deposition it is normally assumed that it corresponds to the optical spot size on the target. However, especially at small spot sizes and high intensities, lateral transport of energy could enlarge the area of energy deposition. The calculations performed with the nominal optical spot size, including flux-limited electron heat conduction, did not show a significant enlargement. Nevertheless, because the actual physical processes at high intensity may be more complex, the diameter of the area of energy deposition was arbitrarily enlarged 4 times in some runs, keeping the total absorbed energy constant. As is seen in Fig. 3, for the larger area of energy deposition the shock lags behind initially (because of the reduced laser intensity and pressure) but catches up with the nominal case later (because the planarity of the shock is preserved for a longer time). No difference is left at the time of the experiment (64 ns). Obviously, the initial conditions become "forgotten" with time (impulsive load, see Ref.8) and cannot be recovered in a unique manner from a measurement of the shock position at late time alone. There is, however, a difference in the shape of the schock wave for the two cases. As was noted above, for the nominal case the computed shock wave is nearly perfectly hemispherical, whereas it is noticeably wider (along the target surface) than deep in the case with the enlarged area of energy deposition.

iii) effective absorbed energy: The absorbed energy is measured rather accurately in the experiment, but a fraction of it could be transported away (for example by fast electrons and ions or discharge phenomena to much larger distances than characteristic for the processes under investigation. A planar wave released along the target surface shows that long-range energy transport exists in prin-

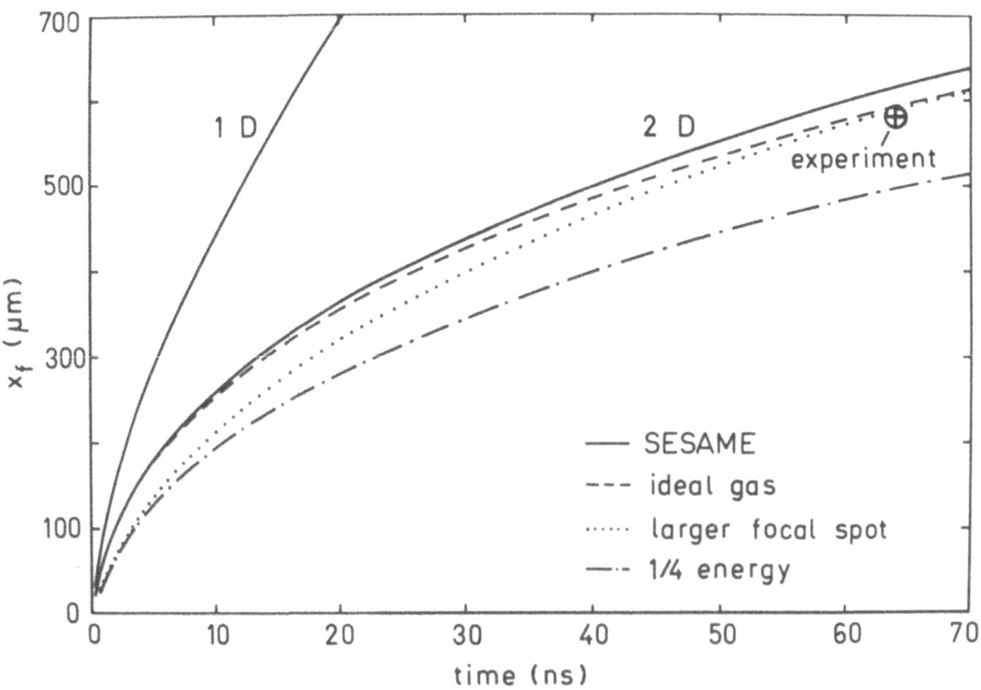

Fig. 3. Shock front position versus time from 2-D simulations

ciple; also recent shock wave studies[9] show some evidence that these losses may be energetically significant. In order to obtain bounds for this effect we have reduced the absorbed energy to 1/4 of its nominal value in some calculations. The calculated shock wave trajectory (see Fig. 3) differs too much from the experimental point at 64 ns, i.e. such a large reduction of the effective absorbed energy is not consistent with the experiment (a reduction by a factor 2 at most cannot be excluded with the present experimental accuracy). We cannot confirm therefore (at least for our experimental conditions) the much more drastic reductions in the effective energy found in Ref. 9.

The input data for the simulations were the same as those used previously for the simulation of thin foil acceleration experiments. Good agreement with the experiment in both cases suggests that a quantitative prediction of the laser-induced hydrodynamic motion is possible with such hydrocodes. The codes incorporate a flux limiter (f = 0.02) for electron heat conduction.

For the nominal "in focus" conditions the code calculations give a peak pressure of 50 Mbar. If we allow simultaneously for as much as a factor of 2 for each of the two effects - enlargement in spot size and reduction in effective energy - the peak pressure drops to 12 Mbar.

Acknowledgement

This work was supported in part by the Commission of the European Communities in the framework of the Association Euratom/IPP.

REFERENCES

1. G. Brederlow, E. Fill, and K.J. Witte, The High-Power Iodine Laser, Springer Verlag (1983)
2. R. Fedosejevs, Yung-lu Teng, R. Sigel, K. Eidmann, and R. Petsch, J. Appl. Phys. 52, 4186 (1981)
3. F. Amiranoff, R. Fedosejevs, R.F. Schmalz, R. Sigel, Yung-lu Teng, to appear in Phys. Rev. (1985)
4. R.F. Schmalz, P. Mulser, and G. Spindler, Max-Planck-Institut für Quantenoptik, Report MPQ 75 (1983)
5. G.J. Pert, J. Comp. Phys. 49, 1 (1983)
6. B.I. Bennet, J.D. Johnson, G.I. Kerley, and G.T. Rood, LANL report LA-7130 (1978)
7. R.F. Schmalz, and J. Meyer-ter-Vehn, Phys. Fluids 28, 932 (1985)
8. Ya. B. Zel'dovich, Yu. P. Raizer, Physics of Shock Waves and High-Temperature Hydrodynamic Phenomena, Academic Press, New York, (1967)
9. F. Cottet, J.P. Romain, R. Fabbro, and B. Faral, J. Appl. Phys. 55, 4125 (1984)

VELOCITY MEASUREMENTS OF LASER-DRIVEN FLYERS BACKED BY HIGH IMPEDANCE WINDOWS

S. A. Sheffield, J. W. Rogers, Jr., and J. N. Castañeda

Sandia National Laboratories
Albuquerque, New Mexico

INTRODUCTION

Because of possible applications to inertial confinement fusion, the response of free-standing foils to laser intensity levels of 10^3-10^6 GW/cm^2 has been well characterized as discussed in Ref. 1 and references therein. Our interest is in characterizing the response of free-standing and backed foils at much lower intensities, i.e., in the region of 1-20 GW/cm^2, and using these flyers to initiate insensitive secondary high explosives.

Sheffield and Fisk[2] previously reported velocity measurements on laser-driven free-standing and water-backed foils at intensity levels near 1 GW/cm^2 where velocities as high as 1.5 km/s were obtained under the proper conditions. In the present work, the output of a 3.5-J Nd:glass laser with a 16 ns pulse was used to drive 1.5-mm-diameter aluminum (Al) flyers. The velocities of both free-standing flyers and flyers backed on the driven side by windows of water, fused silica, or sapphire were measured using ORVIS (Optically Recording Velocity Interferometer System). Intensities were on the order of 10 GW/cm^2 and velocities as high as 5 km/s were measured for 13-μm-thick, Al flyers. Velocity and acceleration were observed to vary only slightly with the shock impedance of the window. Preliminary measurements indicate that PETN (pentaerythritol tetranitrate) at a density of 1.4 grams/cm^3, and HNS-FP (hexanitrostilbene fine particle) at a density of 1.6 grams/cm^3, can be promptly detonated by 66-μm-thick, Al flyers impacting at a velocity of near 2.5 km/s.

EXPERIMENTAL

The experimental setup has been previously described[2] and will be only briefly mentioned here along with changes from the earlier work. The apparatus used to obtain the data, shown schematically in Fig. 1, consisted of a driver laser, a flyer assembly, and ORVIS. The driver laser is a Q-switched, Nd:glass oscillator (Lasermetrics, Inc., Englewood, NJ) with a wavelength of 1.06 μm which was operated at a nominal output energy of 3.5 J and a Gaussian pulsewidth (FWHM) of 16 ns. For most experiments, the beam was focused with a 5-cm-diameter, 50-mm-focal-length lens through a window onto which the foil was attached with silicone grease. Some experiments were done using free-standing foils with no backing window. The windows were made of water, optical quality fused silica, or z-cut sapphire and were 1.3 to 3 cm thick. In this

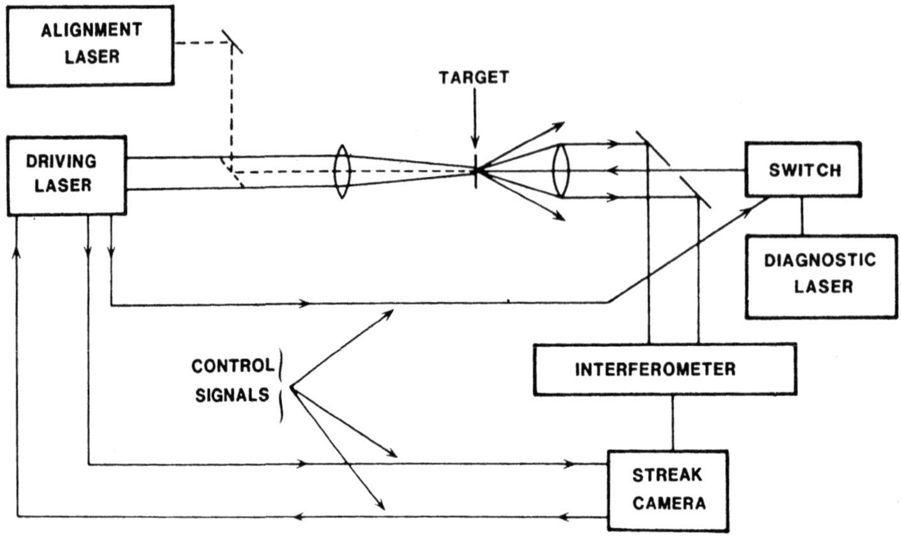

Figure 1. Schematic of apparatus used for laser-driven flyer experiments.

configuration, the spotsize on the foil was about 1.5 mm in diameter. After considering energy losses due to surface reflection of the lens and window, the delivered laser fluence was about 175 J/cm^2 and the intensity was 11 GW/cm^2.

The diagnostic system, an ORVIS velocity interferometer has been described elsewhere[3]. The heart of the ORVIS system is a 514.5-nm-wavelength argon-ion laser (Spectra Physics Model 165, Mountain View, CA), operated in single-mode at an energy of 600 to 800 mW, whose beam is passed through an acousto-optic modulator and imaged onto the back side of the flyer using a 18-cm-focal-length lens. The Doppler-shifted signal was returned through a velocity interferometer and the time-dependent fringe pattern was recorded photographically (Polaroid Type 47 film) using an Imacon Model 675 streak camera (Hadland Photonics, LTD, Herts, UK). The fringe constant for a 5-cm-thick, fused-silica etalon and a 1.2-cm-thick beam splitter was 0.66 km/s/fringe. The photographic records were digitized and analyzed on an HP 9845/9874 Computer/Digitizer System (Hewlett-Packard, Inc., Cupertino, CA).

RESULTS AND DISCUSSION

A typical velocity record for a 25-μm-thick, Al foil backed by sapphire is shown in Fig. 2a. Five fringes are clearly visible and their vertical deflection as a function of time (horizonal axis) is a direct measure of the flyer velocity[3]. One fringe deflection represents a velocity change of 0.66 km/s. The calibrated velocity-time profile for this record is shown in Fig. 2b. This flyer was accelerated to 90 percent of its final velocity in less than 50 ns and reached a peak velocity near 2 km/s.

Figure 3 shows the peak velocities attained as a function of initial foil thickness for free-standing and fused-silica-backed Al flyers. For the data points shown as circles and squares, we estimate that 3.1 J of laser energy reached the foils. The data shown as triangles were taken

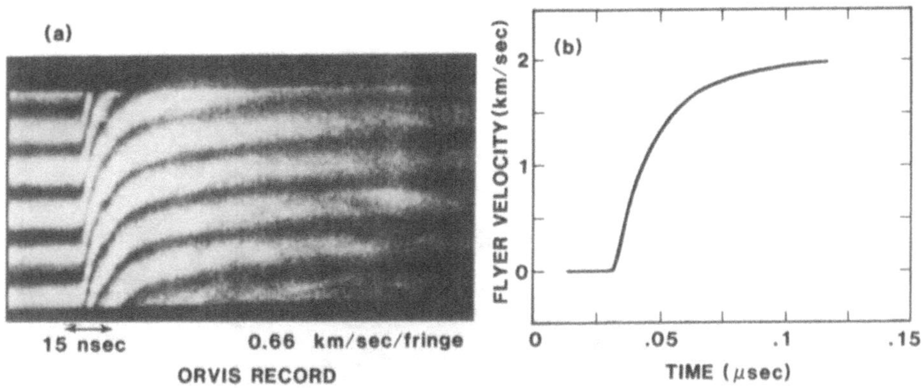

Figure 2. (a) A typical ORVIS record for a 25-μm-thick, Al foil backed by a sapphire window. One fringe deflection represents a velocity change of 0.66 km/s. (b) The corresponding velocity-time profile for this record.

from the earlier work of Fisk and Sheffield[2] where the laser energy reaching the foils was 0.6 J. (Their spotsize and pulse duration were comparable to ours). These data have been normalized to an energy of 3.1 J and the fit with the current results indicates that the final velocity for a given foil thickness is linearly dependent on laser fluence. It is apparent from the figure that the tamped foils were accelerated to peak velocities about 4.3 times higher than the free-standing foils. For both the free-standing and backed flyers, the peak velocity appears to be inversely proportional to the square root of the foil thickness. This relationship is expected from a model where total energy is conserved and the foil kinetic energy is equated to the laser energy multiplied by a coupling efficiency which is assumed to be constant.

The 25 and 66-μm-thick, fused-silica-backed flyers were still being accelerated when the streak camera stopped recording. The small arrows above these data points in Fig. 3 indicate that these are not peak velocities. The 5-μm-thick, free-standing (3.1-J pulse only) and 13-μm-thick, silica-backed flyers were ablated before reaching a constant velocity as will be discussed.

Water, fused silica, and sapphire were investigated as backing (tamping) materials for the flyers. In all cases, the tamping materials were sufficiently thick (one centimeter or greater) to keep shock waves from the free surface from interacting with the plasma formed at the tamper/foil interface during the timescale of the experiment. Flyer behavior was very sensitive to the position of this interface compared to the focal point of the driving laser making absolute comparisons between different tamper materials difficult. Since the shock impedance of water and sapphire are greatly different, a large difference in velocity might be expected due to interactions between the tamper and the plasma. Though there were some small differences in peak flyer velocities (less than 20 percent) and acceleration characteristics depending upon which backing material was used, no definitive conclusions could be drawn.

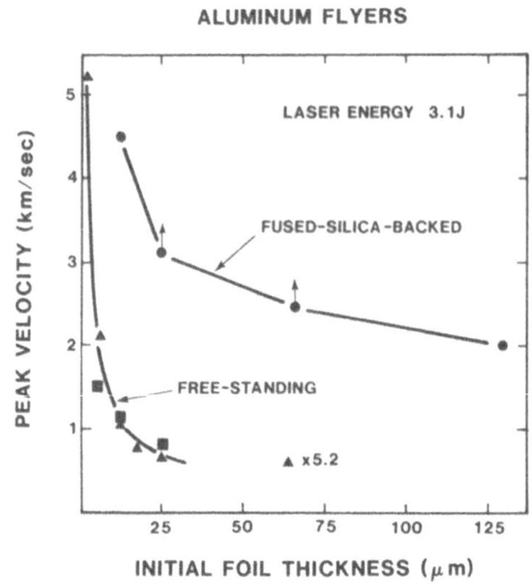

Figure 3. Peak velocities as a function of initial foil thickness for free-standing (squares) and fused-silica-backed (circles) Al flyers. The results from Ref. 2 have been normalized to 3.1 J and are included (triangles).

The amount of ablation during acceleration could be determined for some of the foils. For the 13-μm-thick, tamped flyers (3.1 J deposited laser energy), the recorded fringe pattern became very diffuse about 40 ns after the start of acceleration suggesting that the flyer was either breaking up or completely ablated. Some of the 66-μm-thick, tamped flyers were actually recovered after acceleration. These flyers were about 53 μm thick and were fairly uniform in thickness. This indicates that about 13 μm of Al is ablated during acceleration which is consistent with the observations for the 13-μm foils but further work is needed to determine if the amount of ablation is independent of initial foil thickness. For free-standing foils the amount of ablation was much less. From the fringe patterns of the thinner free-standing foils we estimate that about half as much Al was ablated as for the backed foils.

As mentioned earlier, our interest in accelerating flyers with lasers is to determine the feasibility of using the flyers to promptly initiate secondary high explosives. From empirical shock initiation experiments with electrically-driven flyers, the critical energy necessary to initiate PETN[4] (density = 1.4 g/cm^3) and HNS-FP[5] (density = 1.6 g/cm^3) can be calculated. For a given flyer diameter, this can be expressed in terms of flyer thickness and velocity as depicted in the non-detonation/detonation curves in Fig. 4. In preliminary experiments, a pellet of PETN or HNS-FP at the density shown above, was offset by 250 μm from a 66-μm-thick, Al foil attached to a fused-silica window. The foil was accelerated into the explosive pellet with 3.1 J of deposited laser energy and the final flyer velocity and thickness at impact are shown by the triangle in Fig. 4. It occurs in the detonation regime for both materials, and, as predicted, both materials were promptly initiated as measured by ORVIS.

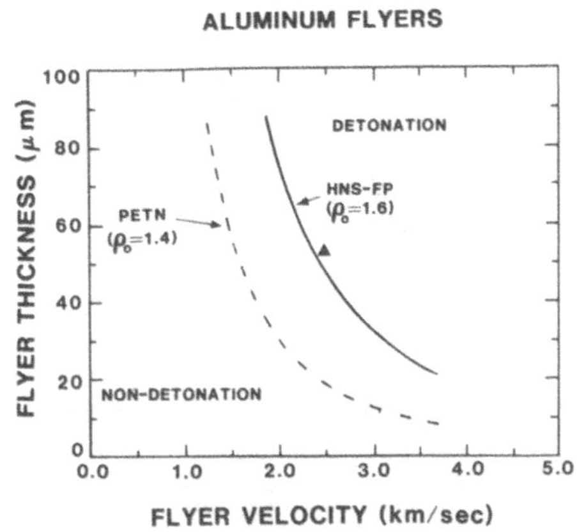

Figure 4. Initiation criteria for PETN (density = 1.4 g/cm^3) and HNS-FP (density = 1.6 g/cm^3) impacted by Al flyers. These curves are based on empirically obtained critical energies determined from Refs. 4 and 5. Prompt detonation will occur under conditions represented by the areas to the right of the curves. The triangle shows the conditions under which explosive pellets at these densities were detonated using a fused-silica-backed, Al flyer driven by a 175 J/cm^2 (11 GW/cm^2) deposited laser pulse.

The efficiency of coupling laser energy into flyer kinetic energy can be determined for these conditions. As previously mentioned, the 66-μm-thick, fused-silica-backed, Al flyers were recovered after acceleration in shots where explosives were not involved. From the recovered flyers the mass at impact was measured and the velocity at impact is known from the ORVIS records. The kinetic energy at impact, determined in this way, averaged 0.8 J. Since 3.1 J of laser energy reached the foil this gives an efficiency of about 25 percent. We estimate the coupling efficiency for the free-standing foils to be less than one percent indicating the backing materials increase the efficiency by about 25 times. If the data in Figs. 3 and 4 is compared, it is apparent that free-standing foils of sufficient thickness cannot be accelerated to velocities high enough to detonate these explosives unless a higher laser fluence is used. Future work is planned to determine shock initiation thresholds as a function of flyer thickness, diameter, and laser fluence.

SUMMARY

We have measured the velocity and acceleration behavior for both free-standing and tamped laser-driven Al flyers. For a given laser energy, the tamped foils are accelerated to about four times the velocities attained by free-standing foils. A 66-μm-thick, tamped Al foil can be accelerated to a velocity greater than 2.5 km/s and about 20 percent of the Al is ablated during the acceleration process. Under these conditions, the efficiency of coupling laser energy into flyer kinetic energy is about 25 percent. Upon impact, this flyer can detonate high-density, insensitive, secondary explosives. To accelerate free-standing foils for this purpose, the laser fluence would need to be greatly increased.

ACKNOWLEDGEMENT

This work was performed at Sandia National Laboratories, Albuquerque, New Mexico and was supported by the U.S. Department of Energy under contract DE-AC04-76DP00789.

REFERENCES

1. B. H. Ripin, R. Decoste, S. P. Obenschain, S. E. Bodner, E. A. McLean, F. C. Young, R. R. Whitlock, C. M. Armstrong, J. Grun, J. A. Stamper, S. H. Gold, D. J. Nagel, R. H. Lehmberg, and J. M. McMahon, Laser-plasma Interaction and Ablative Acceleration of Thin Foils at 10^{12} - 10^{15} W/cm^2, Phys. Fluids 23:1012 (1980).
2. S. A. Sheffield and G. A. Fisk, Particle Velocity Measurements in Laser Irradiated Foils Using ORVIS, Shock Waves in Condensed Matter - 1983, Proceedings of the American Physical Society Topical Conference held in Santa Fe, New Mexico, July 18-21, 1983, p. 244.
3. D. D. Bloomquist and S. A. Sheffield, Optically Recording Interferometer for Velocity Measurements with Subnanosecond Resolution, J. Appl. Phys. 54:1717 (1983).
4. A. C. Schwarz, Sandia National Laboratories, Albuquerque, NM, unpublished data.
5. A. C. Schwarz, Shock Initiation Sensitivity of Hexanitrostilbene (HNS), Proceedings of the Seventh Symposium (International) on Detonation, Annapolis, MD, June 16-19, 1981, p. 1024.

RECENT DEVELOPMENTS IN MICROSHELL®-TIPPED OPTICAL FIBERS AS HIGH-PRESSURE SHOCK DETECTORS

F. J. Mayer, R. L. Maynard, D. L. Musinski, N. W. Schmerberg, and M. R. Wixom

KMS Fusion
P.O. Box 1567
Ann Arbor, Michigan 48106-1567

and

R. F. Benjamin

Dynamic Testing Division, Los Alamos National Laboratory
P.O. Box 1663, Group M-6, Mail Stop J970
Los Alamos, New Mexico 87545

INTRODUCTION

This paper describes recent results from our ongoing development of Microshell®-tipped optical fibers as high pressure shock wave sensors. The sensors, which have been previously described,[1] are commonly referred to as "optical pins" by analogy to the standard electrical "pins" which have been used in the shock wave community for many years. The sensor consists of a gas-filled (usually argon) Microshell® container glued to an optical fiber. When the Microshell® container is impacted, the contained gas is shock-heated and produces a bright light pulse that is transmitted via the optical cable to a photodetector, usually a streak camera, a photodiode or photomultiplier. This device is actually a miniaturized embodiment of the "flashgap" shock detection scheme described by Walsh and Christian.[2] Arrays of optical pins have been used in a variety of shock-sensing applications including ballistic impacts, detonation velocity measurements in explosives, and magneto-explosive generators.[3]

The optical pin has a number of advantages when compared to electrical pins for sensing shock arrival. First, in pulse-power environments, optical pins are immune to the electromagnetic pulse that overwhelms electrical pin signals. Second, optical pins are safe to use directly in explosives since they are completely passive. Unlike electrical pins, optical pins require no electrical bias or current carrying cables. Additionally, optical pins have a significant economic advantage because optical data are less costly to transmit and record than are electrical data. The cost of a streak camera capable of simultaneously recording 100 channels of optical data with subnanosecond resolution is only a few hundred dollars per channel, significantly less than the cost of 100 channels of electrical data recorded on oscilloscopes or fast digitizers.

CONSTRUCTION

Figure 1 is a schematic representation of a typical device. The spherical Microshell® gas container is epoxied or glued directly over the core of a freshly cleaved optical fiber. The duration, rise-time and intensity of the emitted optical pulse are influenced by the diameter of the shell. Decreasing the diameter of the shell decreases the optical pulse rise time and improves the spatial resolution. We find that 50-200 μm diameter shells work well in our experiments. Further, we optimize the coupling to the fiber by matching the core diameter of the fiber to the size of the Microshell® vessel or by over-filling the core acceptance angle.

The Microshell® gas containers are made from either glass or polymer materials. The choice of containers for these devices is set by the choice of gas, due to the diffusive gas-filling technique. Neon and argon are available in both glass and polymer shells while krypton is limited to polymer shells. To protect the shell from damage and to provide support during the experiment we usually add a protective sleeve and embed the shell in blackened epoxy filler. The epoxy serves two functions. First, it provides mechanical strength to the Microshell® gas containers. Second, being opaque, it prevents light "leakage" down the cable both before the shock arrival and after the shock has passed. We have found that opaque epoxy improves the optical signal to noise ratio upon shock passage.

RISE-TIME MEASUREMENTS

The optical pulse rise time has been measured in a number of different experimental configurations. One measurement is shown schematically in Fig. 2, where the array of optical pins in a cylindrical PETN powdered explosive have been imbedded perpendicular to the detonation direction in order to measure the detonation speed. In this experiment, the detonation

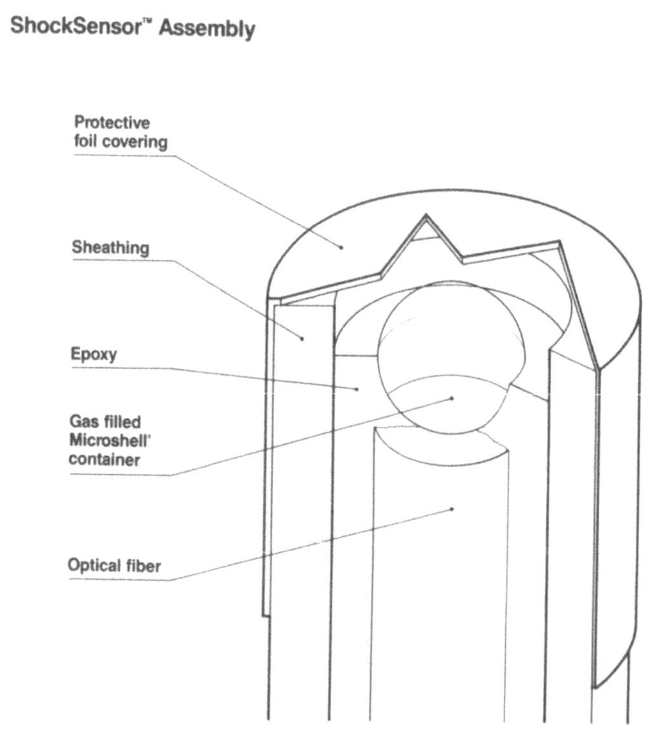

Fig. 1. A cut-away schematic of the optical pin construction.

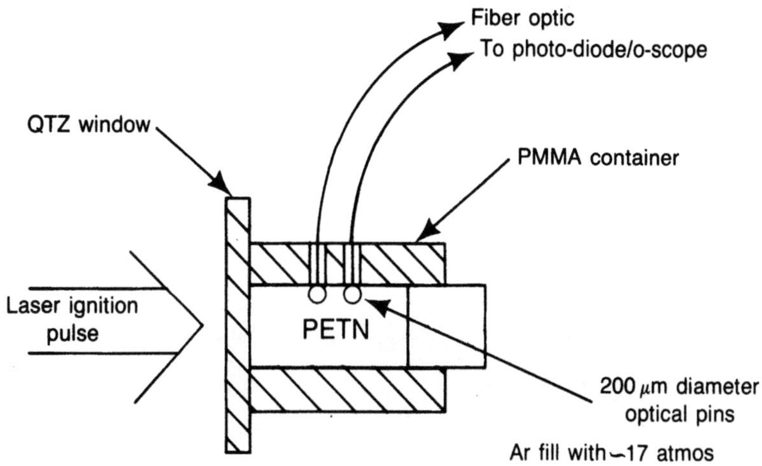

Fig. 2. Schematic of the laser-ignited PETN detonation velocity measurement showing optical pin configuration.

is launched at one end with a laser pulse. As the detonation wave advances, the optical pins are sequentially impacted causing the two optical pulses detected by a EG&G FND100 photodiode and a Tektronix 7904 oscilloscope. The photodiode/oscilloscope system has a rise time < 1 ns. Figure 3 shows the oscilloscope trace from the detonation measurement and optical pulses have a rise time of about 15 ns and an interval between them of 140 ns which, with the 1 mm pin spacing, gives a detonation velocity of about 7 km/sec. The detonation transit time across the 200 μm diameter Microshell® container is 28 ns, roughly twice the observed optical pulse rise time. Thus, the rise time of the optical pulse is considerably less than the transit time of the pressure impulse through the gas container.

A previous measurement with a much slower "BB" projectile (0.2 km/sec),[1] we had measured an optical pulse rise time of 80 ns with an associated transit time of 1 μm. In this low velocity experiment optical pulse rise time was only 8% of the transit time. We believe that these results support the physical model of the shock wave in the contained gas "reverberates" between the moving piston (the Microshell® container wall first impacted) and the down stream wall (last impacted). This interpretation is similar to the results of the experiments reported by Chaudhri, Almgren, and Person.[4]

Finally, we note that the optical signal level is quite small for low velocity impacts (requiring optical gain, a photomultiplier or image intensifier) but increases rapidly with increasing velocity as is expected for shock-heated gas radiating as a blackbody.

SPECTRAL OUTPUT

We have made a time-resolved measurement of the emission from an impacted optical pin by streaking (Hadland Photonics IMACON 790 streak camera) the output from a spectrograph (Hadland Photonics) in the region 400-700 nm. The projectile was accelerated to 1 mm/μs and allowed to impact the optical pin which was mounted in a wooden block several centimeters from the gun barrel. No emission or absorption lines were present in the spectra which supports the shock-heated gas being a blackbody radiation source. Unfortunately, the streaked spectrograph was not calibrated so we were not able to estimate an apparent temperature of the shock-heated gas. Calibrated experiments of this type are presently being pursued.

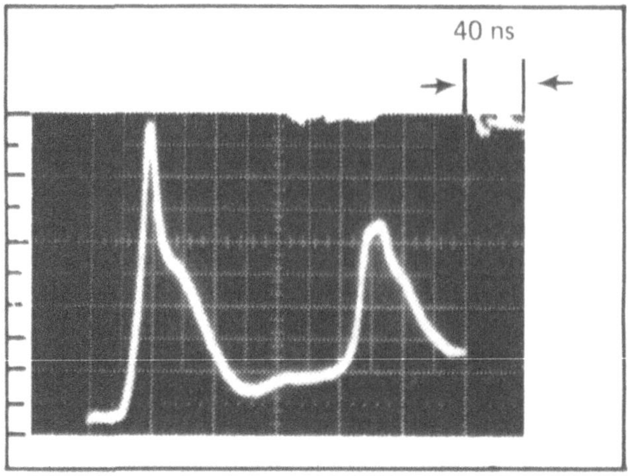

Fig. 3. Oscilloscope trace of optical pin light output from experiment of Fig. 2. The two optical fibers have been multiplexed onto one photodiode.

NONDESTRUCTIVE TESTING

Two methods have been developed for in situ verification of optical pin integrity. If the pins are to be used in an environment of low atomic number ("low Z") materials, e.g., explosives or plastics, the quartz optical fiber can be directly imaged in a contact radiograph. In higher Z environments the epoxy is doped with lead oxide or the Microshell® container is coated with a thin high Z film to indirectly image the pin. In addition to verifying the integrity of the shell and fiber tip, radiography is useful for determining pin positioning.

The second nondestructive test involves observation of an optical signal reflected from the sensor end of the fiber. Doping the epoxy or the Microshell® container itself with a fluorescent or phosphorescent dye allows one to detect respectively the frequency shifted or time shifted reflected probe signal. A most dramatic example of the fluorescence test is observed when an argon ion laser is used as the probe in conjunction with rhodamine dye as the dopant. Using only a beam splitter and optical spike filter, rhodamine fluorescence is easily detected at the receiving end of the fiber as verification of pin integrity. Fluorescent dyes are also being investigated as optical amplifiers of the shock signals whereby the dye red shifts the blackbody emission to a more sensitive region of the photodetector spectral response.

Time shifted signals can be observed if a phosphorescent dopant is excited by a pulsed optical probe given that the probe fall time is significantly faster than the phosphor decay time. If a suitable phosphor is used, i.e., one that absorbs at typical semiconductor emission wavelengths, the time shift test can be performed with commercially available optical fiber testing instrumentation (Tektronix Time Domain Reflectometer). The phosphor test is better adapted to field application since it does not require a monochromatic source or dispersing element.

USER APPLICATIONS

Optical pins are now commercially available[5] and are marketed under the trade name ShockSensor™. A variety of shock sensing applications are being explored by ShockSensor™ customers. These include applications as witness plates for detonation, measuring shock velocities within the bulk of high explosives, tracing detonation-induced or impact-induced shock profiles in structures, and measuring the arrival time of projectiles or flier plates.

REFERENCES

1. R. F. Benjamin, F. J. Mayer, and R. L. Maynard, Microshell®-tipped optical fibers as sensors of high-pressure pulses in adverse environments, in: "Fiber Optics in Adverse Environments," Roger A. Greenwell, ed., Proc. SPIE 506:116 (1984).

2. J. M. Walsh and R. H. Christian, "Phys. Rev." 97:1544 (1955).

3. R. S. Caird, R. F. Benjamin, R. G. McQueen, and D. J. Erickson, Application of optical-fiber pins to explosive, pulse-power generators, in: "Fiber Optics in Adverse Environments," Roger A. Greenwell, ed., Proc. SPIE 506:122 (1984).

4. M. M. Chaudhri, L. A. Almgren, and A. Persson, High speed photography of the interaction of shocks with voids in condensed media, in: "15th International Congress on High Speed Photography and Photonics," Lincoln L. Endelman, ed., Proc. SPIE 348:388 (1982).

5. This product is not endorsed by the Los Alamos National Laboratory, the University of California, or the U.S. Department of Energy.

APPLICATION OF PVF2 ACTIVE GAGE AS AN INITIATION DIAGNOSTIC

Michel Bouchu and Jean-Yves Guillamot

C.E.A.-C.E.V.

Boîte Postale No. 7, 77181 Courtry FRANCE

Summary

We present an initiation diagnostic using an active gage made of a biaxially stretched and polarized polyvinylidene fluoride film (23 µm thick).

The material has in this case piezo and ferroelectric properties and a remanent polarization $P_r \simeq 7$ to 8 $\mu C/cm^2$.

When loaded by one or several shocks, the gage polarization varies leading to a charge liberation, and we record a signal V = f(t) in a resistive circuit.

Theory shows that the charge variation is sensitive to the pressure variations undergone by the gage. The signal V = f(t) integration will allow us to obtain Q = g(t) and to establish an analogy with P = h(t).

Such a gage can be utilized to verify the operation of pyrotechnic systems. An example of a detonator running-order check is presented and discussed. The calibration Q = f(P) having not yet been achieved, the qualitative analysis of the Q = g(t) curve nevertheless lets us see the good running order of the different stages of a detonator and allows us to establish a diagnostic of how it operates.

Introduction

Most of the hitches observed during experiments on explosives implicate the initiation devices. An a posteriori analysis of the defects does not generally lead to a definite conclusion because the different suspected steps have been destroyed by the explosion effects.

In the case of exploding wire detonators, despite their high reliability, it is not sufficient to check the arrival on the wire of the high voltage pulse. That is why we have developed a diagnostic allowing dynamic control of all the detonator components.

Principle

A detonator is a complex device that is generally the assembly of an ignition system with several explosives. These, during their reaction, generate shock waves that are characterized by the pressures they create in the media they go through. The time evolution of pressure is representative of the operation of the detonator components.

The proposed diagnostic works with a gage using the piezoelectric and ferroelectric properties of polyvinylidene fluoride (PVF2) when polarized.

The behavior of this material when shock loaded has been studied by several authors [1-3].

A one-dimensional plane wave, the propagation axis of which is colinear to the polarization axis of PVF2 (axis 3, normal to the material surface) induces under the assumption of a uniaxial stress σ (t) in the polymer, a variation of its remanent polarization P $(0,0,P_r)$ as follows:

$$P_3(t) = P_r - P(t) = \sum_j d_{3j} \, \sigma(t) \tag{1}$$

where $P_3(t)$ stands for the variation of the remanent polarization on axis 3, d_{3j} are the piezoelectric tensor components and $\sigma(t)$ is the applied stress.

If S is the active surface of polymer, the amount of delivered charges is:

$$Q(t) = SP_3(t) \tag{2}$$

and the amplitude of the resulting signal in a resistive circuit R is given by:

$$V(t) = R \frac{dQ(t)}{dt} \tag{3}$$

Starting from relations (1) and (2), one establishes a bi-univocal relation between the stress applied to the gage and the amount of delivered charges; the expression (3) then becomes:

$$V(t) = K \frac{d\sigma(t)}{dt} \tag{3}$$

where K is a constant under the assumption that the piezoelectric tensor components are invariant with the stress.

The variations of V(t) thus give the chronometry of the phenomena that occur in the shock wave profile. The charge Q(t) is calculated by numerically integrating the signal V(t):

$$Q(t) = \frac{1}{R} \int V(t)dt = K'\sigma(t)$$

The Q(t) curve is thus representative for the pressure variation with time. The knowledge of K' goes through the experimental determination of the relation $Q = g(\sigma)$ i.e., the calibration of the gages. Having not yet realized this step, we cannot obtain quantitative measurements, but it is possible to determine the relative variations of pressure and then to characterize some phenomena, such as the operation of an exploding wire detonator.

Description of the Gage

The gage is made of a PVF2 polymer film (23 µm thick) bi-axially stretched and polarized. The electrodes in copper are deposited on each side by evaporation under vacuum. The geometry of the gage itself is the one determined by the parts of the electrodes which are facing each other. This geometry, a scheme of which is given in Fig. 1, is fitted to one of the detonators we give as an example.

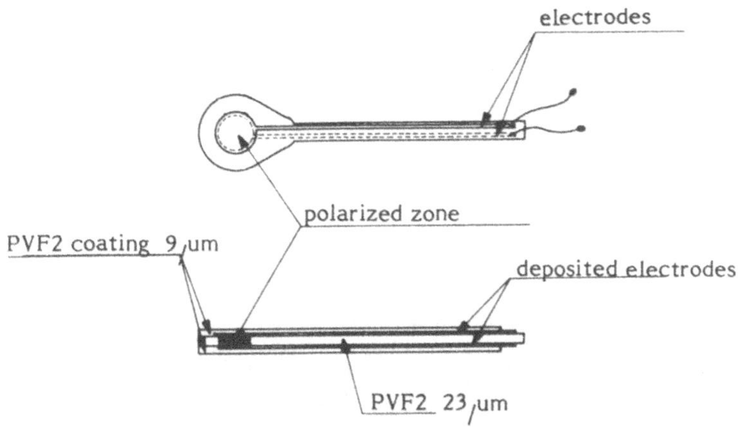

Figure 1. PVF2 gage geometry.

The gage is polarized by application, between the electrodes, of a high voltage alternating field of 2.6 10^8 v/m at very low frequency 0.05 Hz [4]. The remanent polarization thus obtained is $P_r \simeq 7$ to 8 $\mu C/cm^2$.

The polarized area is indicated in Fig. 1, the gage insulation being accomplished by two 9 µm thick non-polarized PVF2 polymer films.

Example of Utilization

Fig. 2 represents the set-up used for testing our initiation diagnostic.

The detonator has a radial output, and is constituted of an exploding wire ignition system and of three different explosives, a, b, and c. The whole thing is embedded in a metallic case, the detonator top part being cast in an epoxy resin.

The gage is held by the device that keeps the detonator in its bedding. It is charged by a 50Ω resistance and the signal record is made on a digitizer oscilloscope or on a transient recorder.

We have tested the detonator both in normal configuration and with a defect simulation.

Detonator in a normal configuration

Fig. 3 shows the record of the signal delivered by the gage (a) and the integral curve of this signal (b). On these curves are pointed the different steps of the detonator operation. From A to B, one registers the shock created by explosive (a) reaction, some tests realized with an "inert" detonator did not show the shock induced by the wire explosion.

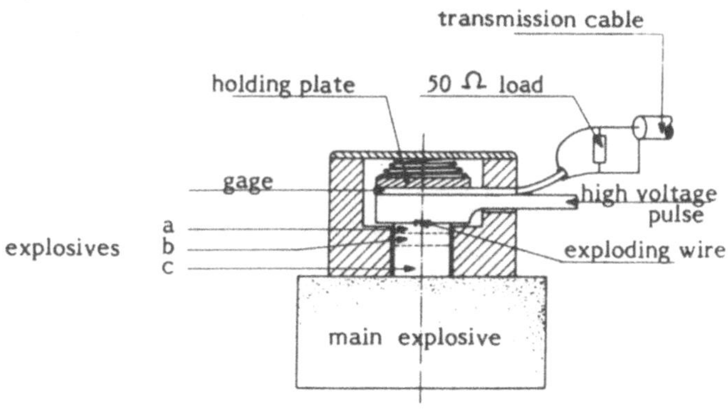

Figure 2. Experimental set-up.

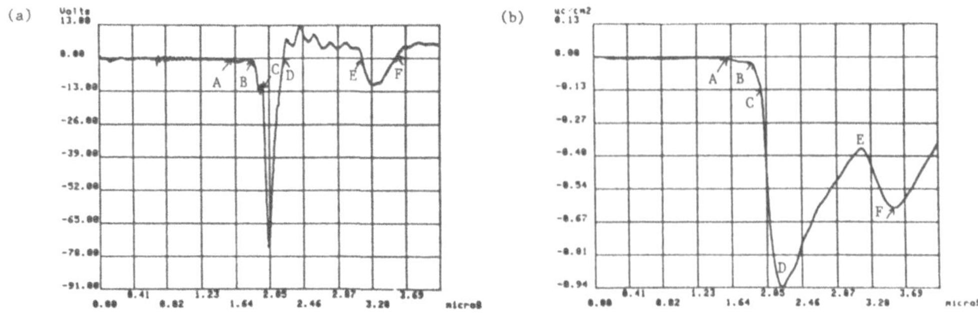

Figure 3. Typical recording (a) signal recurded on a Ω load
(b) intergrated signal.

From B to C explosive (b) reaction is observed, then from C to D the one of explosive (c), and from E to F the shock reflected by the detonator case bottom part.

This interpretation comes from the quantitative analysis of the chronometry of curve 3 (a) characteristic points. Moreover, the dynamic range is important; one easily distinguishes low amplitude phenomena that characterize the operation of detonator first stages.

Calibration Q = f(P) of the polymer gage has not yet been done, so we have realized a test where the polymer gage has been replaced by a 50Ω carbon gage (C 300-50 EK by DYNASEN), pressure calibrated. The obtained signal is shown in Fig. 4(a).

Comparing signals in Fig. 4, one can see: that characteristic points representative of the behavior of different stages of the detonator are in perfect chronometric similitude, and the pressure profile obtained with the carbon gage is different from the integral curge of the PVF2 gage signal by a quicker decay after explosive (c) stage has run. The origin of this gap can be found both in a wrong behavior of carbon gage in rarefaction mode [5] and in a geometrical difference. It can also happen as a hysteresis phenomenon in the polymer gage response. The calibration work done on this latter gage should remove the question.

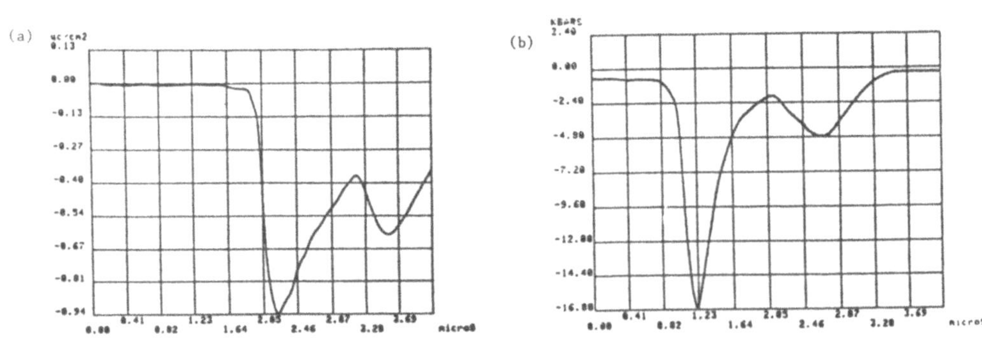

Figure 4. Comparison PVF2gage - Carbon gage.

Detonator in a wrong configuration

Two possible defects have been simulated on the detonator running mode, that is:

a) explosives (b) and (c) do not react,

b) explosive (c) does not react.

To do this, detonators have been realized with the following configurations:

A) exploding wire,
explosive (a) normal,
explosives (b) and (c) replaced by equivalent inert compositions.

B) exploding wire,
explosives (a) and (b) normal,
explosive (c) replaced by an equivalent inert composition.

Fig. 5 shows the results obtained with configurations A and B compared with those obtained with a normal detonator.

Comparison between configurations A and B on one side and the normal one on the other side shows very different dynamic ranges. These are due to the nature and to the role of the explosives constituting the detonator; they demonstrate as well the ability of this diagnostic to control not only the detonator global operation but also each one of its components.

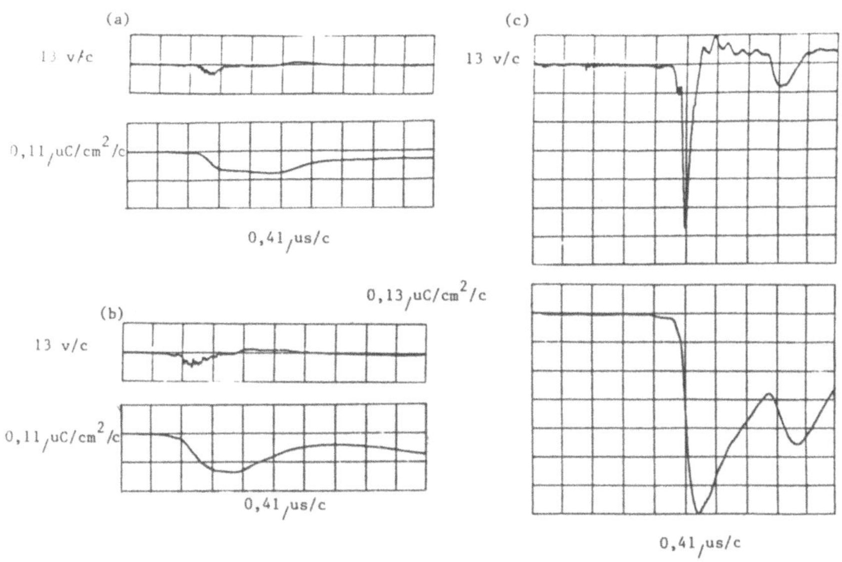

Figure 5. Defects simulation: (a) configuration A; (b) configuration B; (c) normal configuration

Conclusion

We have developed an initiation diagnostic from an active gage that utilizes ferroelectric properties of polarized PVF2.

Beyond the important dynamic range of the signal delivered by the gage, its mechanical flexibility, small thickness and geometrical adaptation possibilities are so that it can be embedded in complex initiation devices without disturbing their operation.

Moreover, as the gage is active, one need not have an external voltage source, so it is very simple in many occasions to use it.

The chosen example of application, the detonator initiation diagnostic, shows the aptitude of polymer gages for finely analyzing detonation phenomena.

References

[1] F. Bauer, Le polyflourure de vinylidène PVF2. Propriétés ferroélectriques et piézoélectriques. Comportement sous compression dynamique et sous l'action d'une onde de choc. Journées de détonique 1984, Centre d'Etudes de Gramat, France.

[2] D. Tougne, Thése, Institut National Polytechnique, Toulouse 1983.

[3] M. Perez, D. Tougne, Réponse et compression par choc du PVF2 dans la gamme 10-250 kb. Journées de détonique 1984, Centre d'Etudes de Gramat, France.

[4] F. Bauer, American Institute of Physics Proceedings of the Topical Conference on Shock Waves in Condensed Matter, Menlo Park, CA, June 1981.

[5] M. Perez, P. Chartagnac, Comportement dynamique en compression et en détente des jauges piézorésistives du carbone. Review of Scientific Instruments S1(7), 7 Juillet 1980.

ANALYSIS OF ELECTRICAL NOISE FROM SHOCK LOADING

A STEEL FLATPACK SOIL STRESS GAGE

Robert E. Tokheim

SRI International
333 Ravenswood Avenue
Menlo Park, CA 94025

INTRODUCTION

Extremely high electrical noise in the output of a carbon steel flatpack soil stress gage interferes with the desired piezoresistive shock wave signal at stresses above and, to a lesser extent, below about 10 GPa. Examples of this noise are indicated in stress histories of 17 GPa impact experiments performed by P. S. DeCarli[1] and shown in Fig. 1. Fig. 1(a) shows electrical noise on the stress signal obtained from a manganin gage experiment with a very short carbon steel flatpack, and a repeated experiment with the gage replaced by a constantan shorting element. The gage is clearly not the source of the noise. Fig. 1(b) shows results of repeated experiments with a very short stainless steel flatpack. No noise is present. The noise must therefore be due to the carbon steel flatpack. The purpose of this paper is to give a quantitative explanation for this noise, especially in the field flatpack configuration.

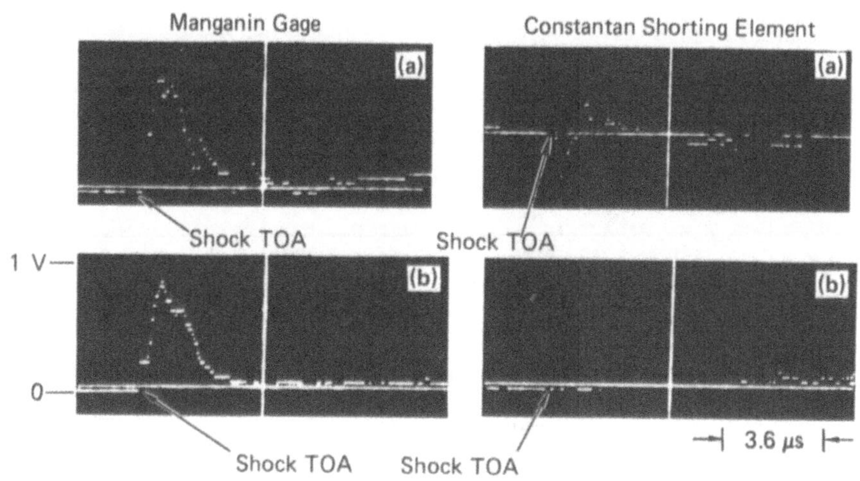

Fig. 1. Electrical noise for 17 GPa shock wave in very short length carbon (a) steel flatpack and (b) stainless steel flatpack in short pulse duration laboratory experiments.

DESCRIPTION OF FLATPACK

The approximate field configuration of the flatpack is shown in Fig. 2. The length ℓ_1, which contains the gage element, is placed normal to the direction of the shock wave. It is assumed that the duration t_d of the shock wave pulse is greater than the engulfment time of the flatpack, so that we only consider shock loading. The electronic circuit and the shielded transmission line of the flatpack are shown in Fig. 3. The capacitor C_0 at voltage V_0 discharges a large current that amplifies the circuit voltage of the piezoresistance gage arising from the shock wave (i.e., $\Delta V = I_0 \Delta R_g$).

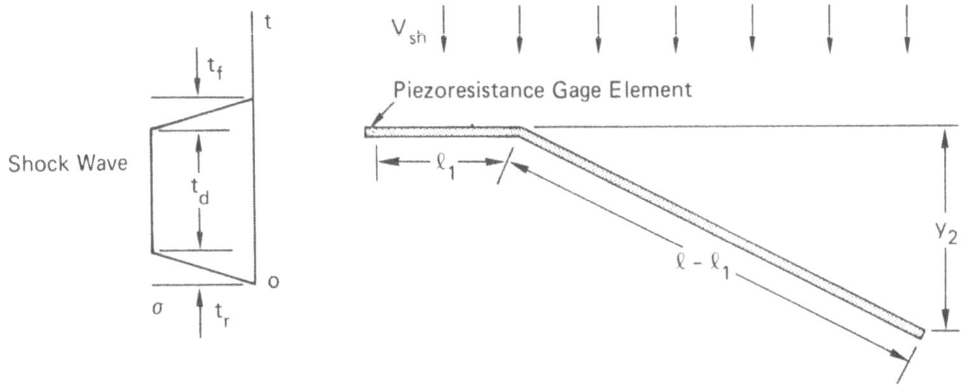

Fig. 2. Approximate field configuration of flatpack

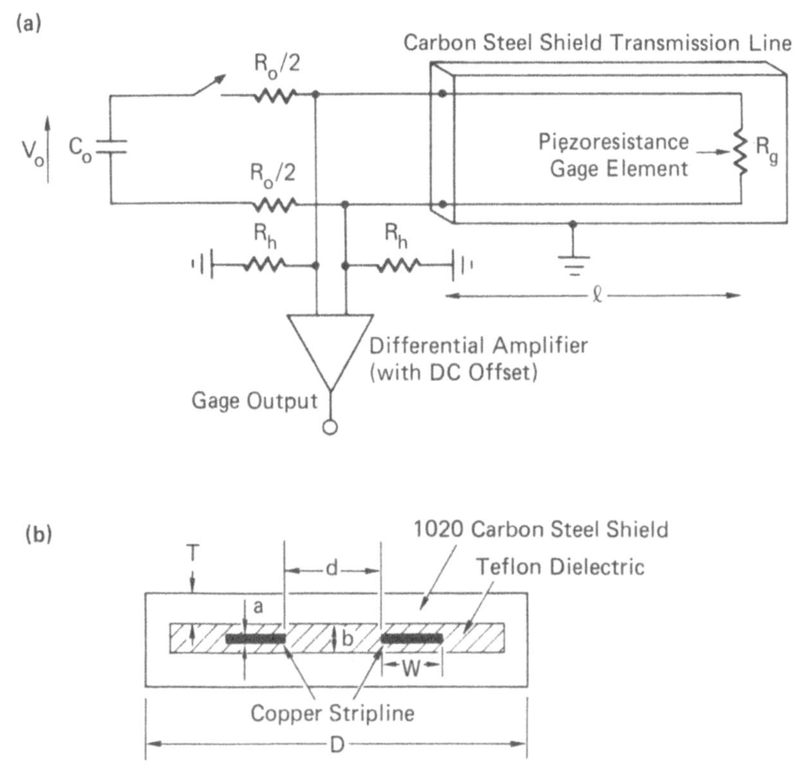

Fig. 3. Flatpack circuit considered in analysis:
(a) electronic circuit and (b) shielded transmission line

We first consider the equivalent circuit for a field flatpack as the capacitor initially discharges, and then consider the superimposed effect of shock wave loading. Typical flatpack circuit parameters are C_o = 1 mF, R_o = 50Ω, R_h = 1 MΩ, V_o = 50 V, R_g = 0.3Ω, $t_r = t_f \sim 1$ μs, t_d = 1 ms, V_{sh} = 3 mm/μs, ℓ = 5 m, ℓ_1 = 1 m, y_2 = 1 m, a = 0.051 mm, w = 3 mm, b = 0.45 mm, d = 30 mm, T = 3 mm, and D = 60 mm. In our analysis below we assume $t_r = t_f = 0$.

EQUIVALENT CIRCUIT DURING CAPACITOR DISCHARGING

The equivalent circuit during initial discharging is given by Fig. 4(a). The total resistance R_T comprises the charging resistance R_o, the gage resistance R_g, and the internal direct-current resistance R_{int} and the eddy-current resistance R_{ec} of the stripline. The total inductance L_T comprises the inductance in the iron (carbon steel) shield, the inductance in the dielectric region L_{diel}, and the internal inductance L_{int} of the stripline center conductors. Initially, the iron inductance is the eddy-current surface inductance L_{ec} of the iron shield, but later, after magnetic diffusion, it becomes the in-depth inductance of the iron. The time for the discharging current to overcome the transient effect of L_{ec} is estimated to be much less than 1 μs. The time required for magnetic diffusion (assuming a constant field at the dielectric-shield interface) is much larger and is estimated to be of the order of 10 to 50 ms for the flatpack. It is assumed for this analysis that there is sufficient time before arrival of the shock wave for magnetic diffusion to take place, and that the discharging current reaches a quasi-steady state given approximately by $I_o = V_o/R_o$.

The maximum inductance of the iron after magnetic diffusion is given by the flux per unit current in the strip (assuming no interaction between center strips because d >> b) as

$$L_{\substack{iron\\max}} = \frac{1}{I_o} \int \mu H_z \, dA = \frac{(\mu H_z)_{avg} \, \ell T}{I_o}$$

Assuming a center strip with rounded corners and w >> b >> a, we obtained an estimate for the magnetic field in the iron as

$$H_z = \frac{I}{2w + 2\pi(x - b/2)}$$

by using the method of images described by Smythe[2] for a current in a nonpermeable medium near the plane face of a semiinfinite permeable medium. Using a relative permeability of the order of 1000, based on ingot iron (similar to flatpack carbon steel) properties,[3] we estimated the maximum inductance to be about 2 mH (L_1 below). This is assumed to be the circuit inductance before the arrival of the shock wave.

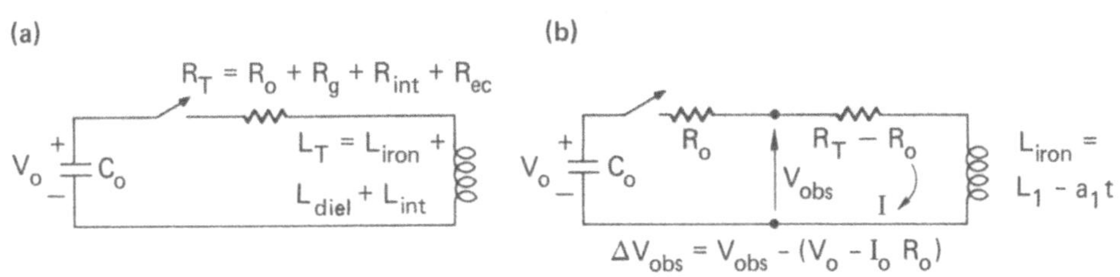

Fig. 4. Flatpack electrical equivalent circuit during (a) capacitor discharging and (b) shock loading

EQUIVALENT CIRCUIT DURING SHOCK LOADING

Rose et al.[4] investigated the effects of shock waves on the residual magnetic properties of Armco iron up to 25 GPa. At 17.5 GPa they found the permeability was reduced by a factor of 2 to 4 with respect to unshocked values. For carbon steel we assume a factor of 2 reduction during shock loading, giving an iron inductance of 1 mH (L_2 below).

Referring to the flatpack configuration of Fig. 2 we estimate the inductance of the flatpack. We assume that initially, length ℓ_1 of the gage is under simultaneous shock loading, and the remaining length ℓ_2 is in the unshocked state, giving a linearly decreasing inductance with time as

$$L = \left[L_1 - (L_1 - L_2)\frac{V_{sh}t}{2T}\right]\frac{\ell_1}{\ell} + L_1\frac{\ell - \ell_1}{\ell}$$

where L_1 is the initial unshocked inductance of the transmission line of length ℓ, and L_2 is the inductance of entire transmission line of length ℓ under shock compression.

Assuming that the engulfment time of the configuration $t_{eng} = (y_2 + 2T)/V_{sh}$ is less than the duration of the shock wave pulse t_d, then the inductance for $t > 2T/V_{sh}$ is given by

$$L = \left[L_1 - (L_1 - L_2)\frac{V_{sh}}{y_2}\left(t - \frac{2T}{V_{sh}}\right)\right]\frac{\ell - \ell_1}{\ell} + \frac{L_2\ell_1}{\ell}$$

where we have neglected fine-structure decreases in inductance in the second leg of the flatpack. At the end of the shock wave pulse we expect similar expressions for inductance recovery (in general to somewhat greater values).

From the above equations, we obtain the rate of change of inductance with time for our parameter values as $\left(\frac{dL}{dt}\right)_1 \equiv a_1 = -100$ and $\left(\frac{dL}{dt}\right)_2 \equiv a_2 = -2.4$. Thus, the initial rate of change of inductance with time is about 42 times the later rate. In this paper we assume a_2 is negligible. This assumption appears to be justifiable for our field parameter set.

Based on the above shock response of the shield inductance, we obtain the differential equation for current as

$$(L_1 - a_1 t)\frac{d^2 I}{dt^2} + (R_T - 2a_1)\frac{dI}{dt} + \frac{1}{C_o}I = 0$$

This equation shows that the effect of the time-dependent inductance is to introduce a negative resistance in the circuit that allows a transient current to exist greater than I_o. Because the capacitance C_o is so large, it is permissible to replace the capacitor with the constant voltage source V_o, thereby eliminating the last term. Using Kamke[5] we find the solution for the current I, and then the differential gage output voltage ΔV_{obs} indicated in Fig. 4(b), including the gage response. For $t < t_1 = 2T/V_{sh}$

$$\Delta V_{obs} = - \frac{I_o R_o}{\frac{R_T}{a_1} - 1} \left[1 - \left(1 - \frac{a_1 t}{L_1} \right)^{\frac{R_T}{a_1} - 1} \right] + I_o \Delta R_g$$

and for $t \gg t_1$

$$\Delta V_{obs} = - \frac{I_o R_o}{\frac{R_T}{a_1} - 1} \left[1 - \left(1 - \frac{a_1 t_1}{L_1} \right)^{\frac{R_T}{a_1} - 1} \right] e^{(t-t_1)/\tau_2} + I_o \Delta R_g$$

where $\tau_2 = (L_1 - a_1 t_1)/R_T$. An indication of flatpack output voltage and its effect on the expected shock wave response is sketched in Fig. 5. For our typical field flatpack, we find $t_1 = 2\mu s$, $\Delta V_{obs}(t_1) = -5.4V$ (noise), and $t_2 = 36\mu s$. These results are roughly consistent with SRI's observations of field flatpack electrical noise in amplitude (many volts) and duration (many tens of μs). Much shorter effective durations are obtained for the laboratory experiment in Fig. 1, consistent with the much lower inductances and much shorter relative ℓ_2. (Exact details of the laboratory configuration were not available.)

CONCLUSIONS FOR CARBON STEEL FLATPACK

Preliminary flatpack circuit model predictions of noise are consistent with general observations. A shorter controlled pre-shock charging time should reduce electrical noise by reducing magnetic diffusion. The initial and spikiest noise can be reduced by decreasing the flatpack length normal to the incident shock wave. It should be possible to eliminate this noise by a noise cancellation technique shown in Fig. 6.

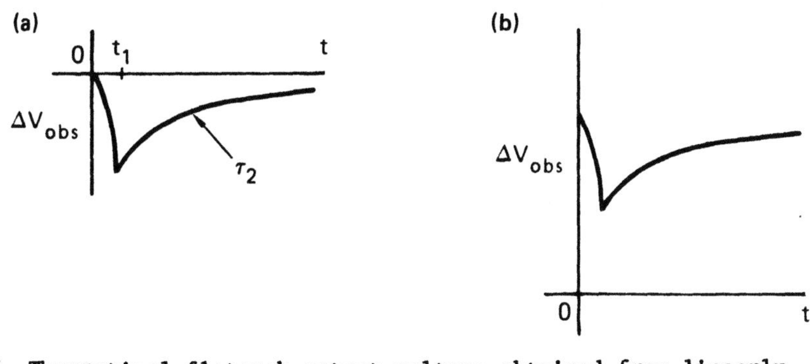

Fig. 5. Theoretical flatpack output voltage obtained from linearly decreasing inductance with (a) shorted piezoresistance element and (b) normal piezoresistance element

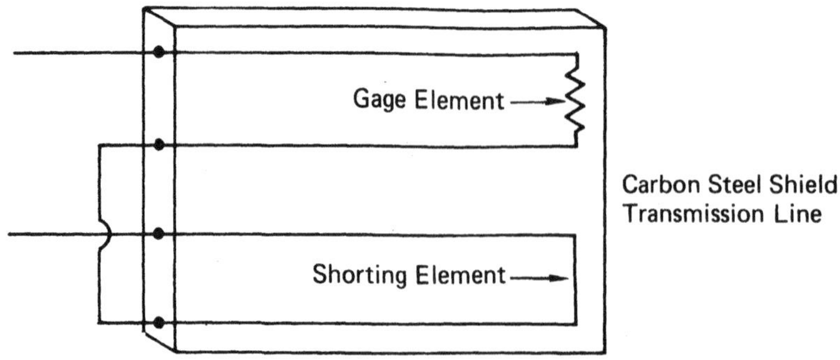

Fig. 6. Noise cancellation circuit

REFERENCES

1. P. S. DeCarli, private communication, April 1984.
2. W. R. Smythe, Static and Dynamic Electricity, Third Edition (McGraw-Hill Book Company, New York, 1968) p. 306.
3. Allegheny Ludlum Steel Corporation, Electrical Materials Handbook (Pittsburgh, Pennsylvania, 1961).
4. M. F. Rose, M. P. Villere, and T. L. Berger, "Effect of Shock Waves on the Residual Magnetic Properties of Armco Iron," Phil. Mag. 19, 39-45 (1969).
5. E. Kamke, Differentialgleichungen Losungsmethoden and Losungen (Chelsea Publishing Company, New York, 1948) p. 440.

DESCRIPTION OF A NEW 63-MM DIAMETER GAS GUN FACILITY

S. A. Sheffield and D. W. Dugan

Sandia National Laboratories
Albuquerque, N.M. 87185

INTRODUCTION

This paper describes a new light gas gun that is just being completed at Sandia National Laboratories, primarily for use in the study of shock initiation and shock properties of explosive materials. This gun was patterned after the 100-mm diameter gas gun at Washington State University[1]. Several unique features have been designed into the facility to make it easier to control and use the gun and to improve shot turnaround time. These will be discussed in some detail in the sections that follow. Initial gun performance information will also be presented.

GENERAL DESCRIPTION

A pictorial layout of the gun is shown in Fig. 1. It consists of a breech, barrel, target chamber, and catch tank along with the necessary supporting equipment. The barrel section has an inside diameter of 63.4 mm and a length of 9 m; it is made up of two 4-m long steel sections and a 1-m long titanium muzzle section. Titanium is used to minimize magnetic flux distortion and problems associated with magnetizing the barrel in electromagnetic gauging experiments. Since the barrel and breech are not allowed to recoil during a shot, the barrel is fixed with adjustable clamps that hold it from three directions at each of six separate clamping points. This makes it possible to align the barrel section using an auto-collimating telescope[2] and an alignment projectile which is pulled through the barrel as adjustments are made. The barrel is straight to within 0.2 mm over the 9-m length as a result of this procedure.

The breech has an inside diameter of 25 cm and a length of approximately 70 cm, giving a volume of about 34 l. It is capable of being pressurized to 42 MPa. The breech is designed to be quick-acting and have a quick-change capability from wrap-around to dual-diaphragm mode; these features will be described in the next section. Helium is used in wrap-around mode and dual-diaphragm mode for projectile velocities above 0.6 km/s. With this combination of breech and barrel a maximum velocity of about 1.6 km/s can be obtained.

Figure 1. Light gas gun layout.

The target chamber is larger than found in other guns to allow room to use a variety of types of instrumentation for measurements at the target, including velocity interferometry, in-situ electromagnetic gauging, Raman spectroscopy, TRISP spectroscopy, as well as high-speed electronic streak and framing camera photography. It is 1.03 m in diameter and 0.88 m long, with an elongated side port on each side measuring 15 cm by 57 cm and six 15-cm round ports at positions around the periphery and in the front of the chamber. Up to 100 g of explosive can be incorporated into an experiment because of the 1.6-cm thick steel walls. Isolation from the catch tank is provided by a metal door with a 32-cm diameter mylar diaphragm. Target alignment is accomplished by a remotely operated target manipulating system and projectile velocity is measured by a noncontact, bar-chart reading system; both the alignment system and the velocity measuring system will be described in the next section.

Catch tank design is similar to other guns with a steel bar "rag bag" filled with nylon rags used to stop the projectile-target debris and absorb the momentum by impacting on shock absorbers at the inside rear of the catch tank. The back end of the "rag bag" is a 25-cm thick steel plate to stop the high velocity fragments from impacting on the back of the tank. The tank has the capability of being evacuated.

The gun pressurization and firing sequence is controlled by a Hewlett Packard 86 computer interfaced to the valves, vacuum pumps, safety interlocks, etc. using a general purpose parallel I/O interface and an Opto-22 Pamux control board. The computer leads the operator through the various functions needed to setup and fire the gun; it continuously monitors the appropriate interlocks and opens and closes the appropriate valves to fire the gun when the fire button is depressed. The pressurization sequence is done manually but comparators control the pressure in the breech and between the diaphragms, automatically maintaining the preset pressures.

UNIQUE DESIGN FEATURES

Several unique features have been incorporated into this facility that make it easier to control and operate and to provide for noncontact measurement of the projectile velocity. They are: a quick-acting, quick-change breech, a remote target manipulating system, and a laser-based, bar-chart velocity measurement system. A description of these units follows in this section.

Quick-Acting, Quick-Change Breech

A quick-acting breech action was obtained by using a design similar to that used in large military guns. An interrupted buttress-thread nut screws into the breech, which has mating buttress threads. The breech can be opened or shut with a single 60 degree turn of the nut. This design eliminates the tedious work of tightening and removing eight or ten large bolts before and after each dual-diaphragm shot. In addition the breech was designed to accommodate two inserts (fabricated from aluminum and each weighing about 4 kg), one used in wrap-around shots and one used for dual-diaphragm shots, as shown in Fig. 2. These inserts go between the breech nut and the breech and can be easily removed by opening the breech and pulling them out. Changing from wrap-around to dual-diaphragm operation is simply a matter of changing the insert. Changing diaphragms is simply a matter of disassembly and reassembly of the insert, which takes only a few minutes. Since the inserts mate with the projectile end of the barrel, the driver gas is stored in the breech behind them and does not undergo any large changes in direction as it exits the breech; this results in higher projectile velocities for a given gas charge than might otherwise be obtained.

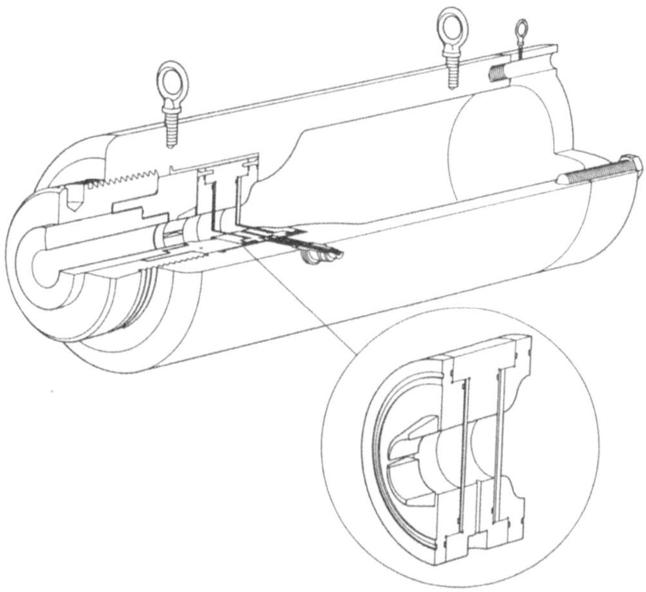

Fig. 2 Cutaway of the breech with dual-diaphragm unit shown in place as well as separately.

Remote Target Manipulating System

This system consists of a target mounting platform attached to three differential screws that can be remotely driven to accomplish target-projectile alignment. The differential screws are on the end of rods that protrude out the front of the target chamber and are each driven by a separate stepping-motor, gear-reducer, and chain-drive mechanism that is remotely controlled from the breech end of the gun. Fig. 3 shows the alignment system on the front of the target chamber. Alignment is accomplished by inserting an auto-collimating telescope into the breech end of the barrel and putting a reflecting target in the target mounting platform. Each of the differential screws is remotely adjusted, while looking into the telescope, until the alignment pattern projected from the telescope and reflected by the target is aligned over the cross-hair pattern in the telescope eyepiece. With this procedure the target can be aligned to give a tilt at impact of 0.03 milliradians or less. In addition to the tilt adjustment, the physical position of the whole target assembly can be moved 20 cm along the axis of the barrel. This allows the

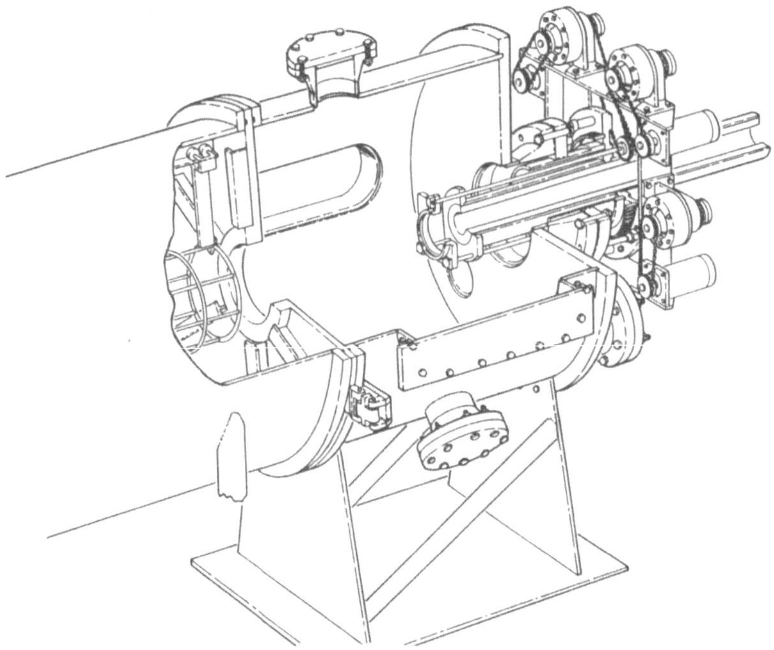

Fig. 3 Cutaway of the catch tank/target chamber assembly showing the target and target manipulating system on the front and inside the chamber.

target to be moved to accommodate various optical inputs to the target chamber and minimizes the optical setup inside the chamber. It also allows positioning of the target at the optimal point inside the magnet for in-situ electromagnetic particle velocity measurements.

Projectile Velocity Measuring System

A noncontact method of measuring projectile velocity has been designed and installed in the gun. It is composed of a laser diode that illuminates a bar chart with a fiber optic coupling and a second fiber optic that collects the reflected signal and sends it to a photodiode. Electronic circuitry connected to the photodiode discriminates the level of light returned and produces an output electrical signal when the reflective part of the bar chart is being read. The sending and receiving units are modified fiber optic communication modules purchased from Advanced Fiberoptics Corp.[3] This system is depicted in Fig. 4. The 0.1 mm diameter sending and receiving fiber optics are terminated in a special connector which is installed in a barrel extension fixture at the end of the barrel, about 0.7 mm out from the inside diameter of the barrel. The bar chart, which is mounted on the projectile, consists of alternating reflecting and black stripes. As the projectile moves past the fiberoptic connector, the sender fiber optic illuminates the chart with 850-nm light and the receiver fiber optic collects the reflected light, sends it back to the receiver unit where the appropriate electrical output, depending on the light level, is produced. This electrical signal is then sent to a recording device. With the bar spacing known and the time for each bar to pass the fiber-optic probe measured, the projectile velocity can be calculated for each bar. This technique provides up to 30 separate measurements that can be averaged to provide an accurate determination of the projectile velocity; it is also hoped that eventually information concerning projectile acceleration just prior to impact can be obtained.

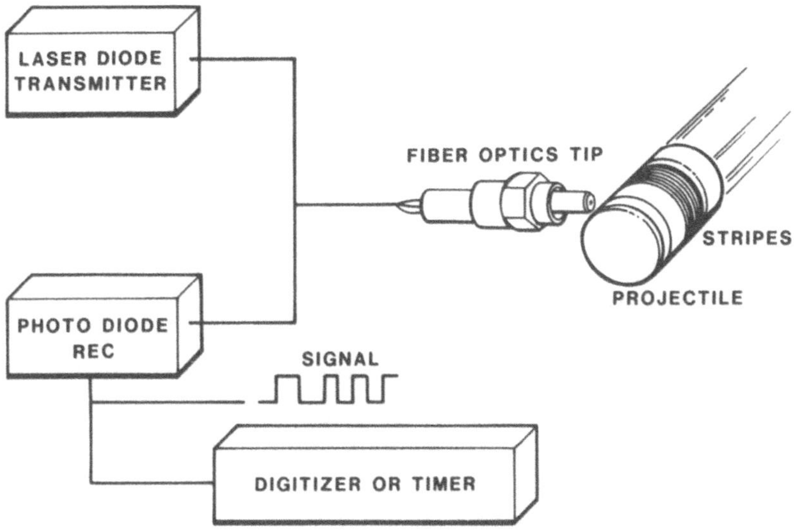

Fig. 4 Schematic of the laser-based, fiber-optic, bar-chart projectile velocity measuring system. Up to 30 separate measurements can be made on each experiment, providing information to determine acceleration of the projectile.

Optical ports have been put in the titanium muzzle section about 0.8 meter from the barrel end. The same system used for projectile velocity measurement is used to sense the projectile coming by. From this a trigger signal can be obtained to start laser flashlamps several hundred microseconds before impact in laser spectroscopy gun experiments.

FIRING CHARACTERISTICS

Although the gun has only been fired a few times, the preliminary velocity curves for both helium and nitrogen are shown in Fig. 5. Nitrogen can be used as the driver gas for experiments up to about 0.6 km/s and helium above this. It is hoped that velocities up to 1.6 km/s can be obtained with light-weight, nylon-syntactic foam projectiles.[4] Eventually we hope to heat the driver gas several hundred degrees in order to obtain higher velocities.

ACKNOWLEGEMENTS

Support for this project from Dr. James E. Kennedy is gratefully acknowledged. We appreciate the help of Walter Haskell, Larry Cowley and Roy Dickey in helping design and prepare drawings for the gun. The original design, which has been extensively modified, was done by Utah Research. Several discussions with Paul Bellamy of Washington State University were very helpful. This work was sponsored by the U.S. Department of Energy under Contract DE-AC04-76-DP00789.

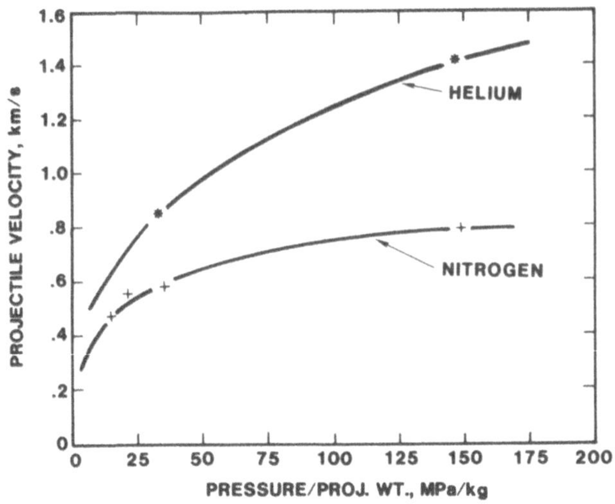

Fig. 5 Preliminary projectile velocity curves for nitrogen and helium driver gas at ambient temperature are shown.

REFERENCES

1. G. R. Fowles, G. E. Duvall, J. Asay, P. Bellamy, F. Feistmann, D. Grady, T. Michaels, and R. Mitchell, Gas Gun for Impact Studies, Rev. Sci. Instr. 41:984 (1970).
2. The auto-collimating telescope model 71-2022 was purchased from Keuffel and Esser Co., Morristown, NJ and modified by putting a sleeve on it that produces a slip fit with the barrel.
3. Advanced Fiberoptics Corp., 637 S. Hayden Rd., Tempe, AZ 85281. The sender module was modified so the laser diode was continuously on and the receiver module was modified to discriminate between the light level returned by the reflective and black stripes.
4. C. F. Peterson and J. T. Rosenberg, "Dynamic Properties of Rocks Required for Prediction Calculations," Stanford Research Institute Report for Defense Nuclear Agency DNA-3579F (1974).

APPLICATION OF SHOCK WAVE DATA TO EARTH AND PLANETARY SCIENCE

Thomas J. Ahrens

Seismological Laboratory 252-21
California Institute of Technology
Pasadena, CA 91125

ABSTRACT

Shock wave data for metallic iron, when compared with seismic pressure-density relations for the earth's molten outer core, imply that this region of the earth, extending in depth from 2900 to 5200 km or, from pressures of 136 to 330 GPa, is less dense by ~1 g/cm^3 than pure iron. Shock wave data for geochemically abundant iron sulfides and oxides suggest that the outer core is alloyed with some 11 to 13 weight % S or 7-8 wt % O. In contrast, Hugoniot sound speed data for iron are consistent with a model in which the inner core of the earth is pure iron with the temperature at the inner core/outer core boundary in the range 5200 to 6600K. Hugoniot data for silicates such as $(Mg, Fe)_2SiO_4$ and $(Mg, Fe)SiO_3$ are consistent with a lower mantle composition at depths greater than 670 km having a composition equivalent to $(Mg_{0.86},Fe_{0.14})SiO_3$. Shock wave and other equation of state data, combined with solar elemental abundance and Jeffrey's number data, have been used to construct compositional models for the major planets. These models imply that Jupiter is about 95% (by mass) hydrogen and helium, the remainder being mainly iron and silicates in its core. Saturn, in contrast, has a larger iron-silicate core making up about 20% of its mass.

Shock wave data are of critical importance in models of impact cratering and accretionary interactions in the solar system. For the infall of silicate impactors onto a body the size of the Moon, the ejecta mass lost to space (normalized to the meteoroid mass) varies from 0.01 to 12 as the impactor speed varies from 5 to 45 km/sec. The impact velocity at which the amount of ejecta escaping the planet just equals the mass of the impactor increases from 20 to 45 km/sec as the escape velocity increases from 2.4 (Moon) to 5 km/sec (Mars). The amount of rock melted and vaporized by impact may be calculated using shock wave and other data. For silicate projectiles impacting a silicate planet at 15 and 45 km/sec, a mass 10 and 100 times the mass of the projectile, respectively, is melted. However, only the 45 km/sec produces significant (10^0 times the projectile mass) silicate vapor.

I. Introduction

Shock wave data for (1) low temperature condensable gases H_2 and He, (2) H_2O, CH_4, NH_3, CO, CO_2, and N_2 ices, and (3) silicates (e.g. $MgSiO_3$), metals, oxides and sulfides (Fe, FeS, and FeO) have many applications in geophysics and planetary science. These include:

[1] The description of physical processes involved in the formation of planetesimals and their accretion to form planets. Significant issues in this area which can or have been

addressed in the case of the terrestrial planets include (1) the formation of planetary metallic cores during accretion (Ringwood, 1979; Stevenson, 1981) and (2) the production of a shock-melted "magma ocean" and concurrent impact volatilization or retention of volatiles during accretion (Abe and Matsui, 1985). Also of interest are the chemical reactions between meteorite components (e.g. H_2O and Fe: Lange and Ahrens, 1984). The formation of primitive atmospheres containing a large fraction of planetary H_2O and CO_2 budgets is also addressable using shock wave and other thermodynamic data (e.g., Lange and Ahrens, 1981, 1985). In the case of accretion of the icy satellites of Jupiter and Saturn, a strong case can be made for impact volatilization of ices giving rise to the increase density of these objects with increasing planetary size (Ahrens and O'Keefe, 1985; Stevenson, 1985).

[2] Models of the velocity, pressure, density and internal energy fields of material accelerated by meteorite impact on planetary surfaces (Bjork, 1961; O'Keefe and Ahrens, 1977a, b, 1982; Ahrens and O'Keefe, 1977). Such cratering calculations, combined with recently developed scaling relations which take into account variations in material strength and planetary gravity (Holsapple and Schmitt, 1982; Housen et al., 1983; Holsapple and Housen, 1985), can be used to obtain bounds as to the depth to which a given crater diameter can exhume to the surface samples of the planetary interior. Such calculations are of great value in obtaining planetary compositions, and evaluating mineralogic and isotopic analyses of samples brought to the earth by spacecraft (French and Maran, 1981), delivered to the earth by impact on other planets (e.g. Melosh, 1984), or studied in-situ on the planetary surfaces by landers, or by flyby or orbiting spacecraft.

Additional issues pertinent to the study of both accretion of planets and impact cratering on planets include determining (1) the amount of impact ejecta melted and vaporized, (2) the distribution of this ejecta on the planetary surface, and (3) the amount of ejecta which escapes the planet versus the type of material on the planet, type of impactor, planetary gravity, and impactor velocity.

[3] The pressure-volume-energy-temperature equation of state obtained for condensed gases and ices provide constraints on the internal structure of the major planets whereas the similar data for iron, iron oxides, and sulfides constrain the composition of the earth and other terrestrial planetary cores.

New data on the Hugoniot sound speed of iron (Brown and McQueen, 1980) suggests that iron melts along its Hugoniot at about 240 GPa. Since the outer core appears to be a liquid iron-alloy, iron Hugoniot temperatures at and above this pressure provide an independent upper bound to the temperature profile in the outer core above this pressure. Dynamic compression data for the major minerals in the mantle place constraints on the range of possible mantle compositions. Shock temperatures constrain the energy of the major phase transitions which occur in all silicates and many oxides upon compression into the range of pressures of planetary mantles (Ahrens, 1982b).

Also, recent work (Rigden et al., 1984) on the shock compression of molten silicates at upper mantle pressures provides a new constraint on the depths in the mantle from which molten lava can reach the surface as a result of its buoyancy relative to the surrounding solid.

II. Composition of Planetary Interiors

The assemblage of elements in the "cosmic abundance" (Table 1), produced by nucleosynthetic processes occuring prior to the formation of the protosolar nebulae, are believed to be the basic materials from which the planets are formed. The protosolar nebulae produced the planets, via condensation of dust, which aggregated into planetesimals which, in turn, accreted to form the planets. A major achievement of nucleosynthetic theory is the correct prediction of the relative abundance of the elements in the sun. The heavier, less volatile elements with atomic number >7 are found in

Table 1. Cosmic Abundances (after Stevenson, 1984)

Element	Abundance[a] (as a number fraction)	Likely Forms
H	0.92	H_2, H (Met)
He	8 x 10^{-2}	Dissolved in H_2, H
O	7 x 10^{-4}	H_2O, CO; dissolved
C	4 x 10^{-4}	CH_4, CO, elemental?; dissolved in H
Ne	1.5 x 10^{-4}	Dissolved in H_2, H
N	1 x 10^{-4}	NH_3, N_2; dissolved
Mg	4.6 x 10^{-5}	Silicates, oxide
Si	3.7 x 10^{-5}	SiH_4, silicates; dissolved in H
Fe	3.3 x 10^{-5}	Fe-cores (with S, O, H, ...)

[a]From Anders and Ebihara (1982)

essentially solar abundance in the most primitive type of meteorites (Fig. 1). Using the above "cosmic abundance" arguments, it is reasonable to expect that the major constituent of the planets can naturally be divided into three groups (Stevenson, 1984):

[1] Gases such as H_2 and He which are non-condensable, even at >30K at the temperatures of the solar nebulae's outer regions (~50 AU) or at the orbits of the outermost planets (Neptune and Pluto).

[2] Ices (H_2O, CH_4, NH_3, CO, CO_2, and N_2).

[3] The refractory planetary materials formed planetesimals in the hotter region around the sun. These are the rocks and metals of Figure 2 and make up the terrestrial planets.

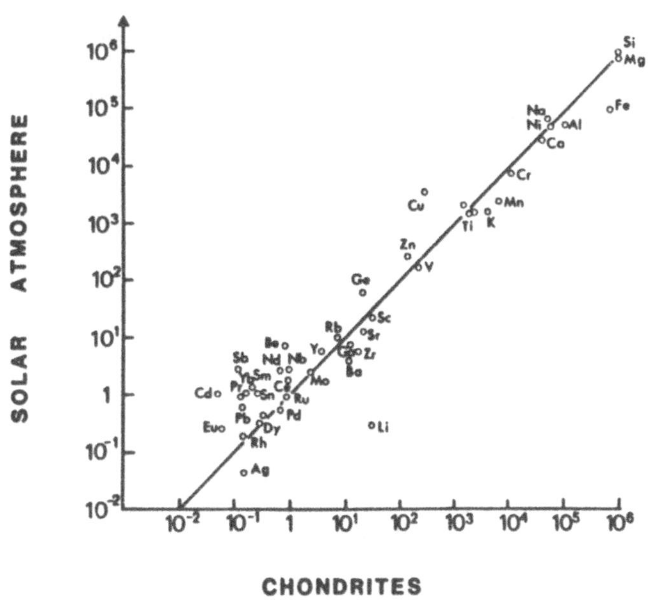

Fig. 1. Abundance of the elements in solar corona versus their abundance in chondrites (from Allegre, 1982).

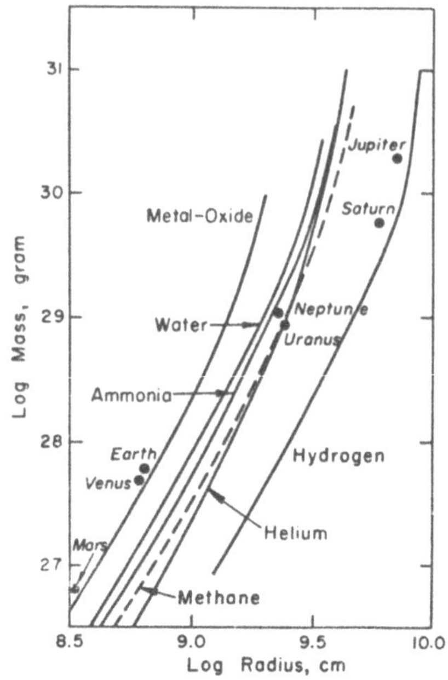

Fig. 2. Planetary radius, norms, planetary mass for hypothetical isentropic self-gravitating pure material planets. Mass and radii of solar system planets are indicated.

Using the known or inferred isentropic equation of states for the various substances shown in Figure 2, the total mass versus radius for a series of self gravitating monomineralic planets are plotted. The degree to which these describe in simplistic terms the various planetary masses and radii provide a simple guide to planetary composition.

In the case of Jupiter and Saturn (Figure 2), they are largely H-He mixtures and the major feature of their mantles is the H molecular to metallic transition. According to the models of Stevenson (1984), Jupiter (Figure 3a) and Saturn (Figure 3b) are largely H and He. Jupiter is ~95% cosmic mixture by mass, with a 5% rock-iron component

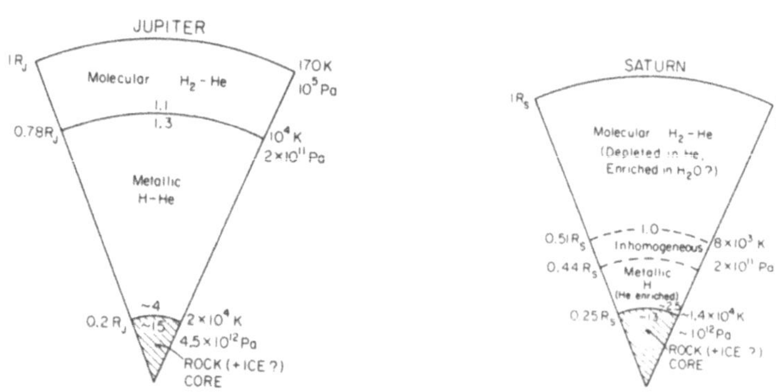

Fig. 3. Planetary interior models of Jupiter (a) and Saturn (b) after Stevenson (1984). Internal pressures and temperatures indicated on right side of sector.

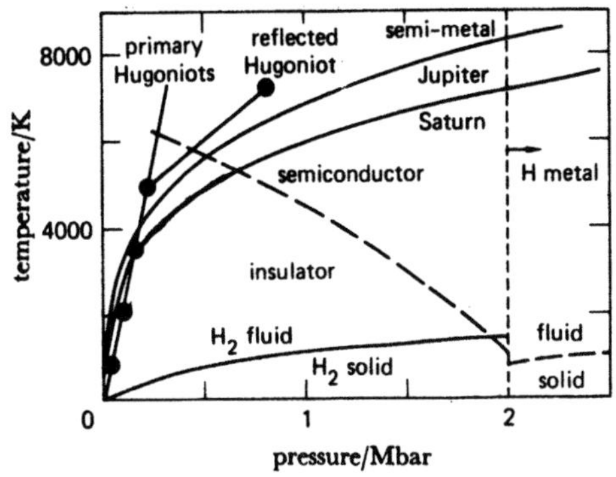

Fig. 4. Shock wave data for hydrogen. Also shown are Jupiter and Saturn isentropes and a calculated hydrogen phase diagram. Single shock data up to 0.21 Mbar and reflected points near 0.80 Mbar are indicated. Dashed lines indicate the more speculative regions of the phase diagram (after Ross et al., 1981).

core. For Saturn, a higher fraction , ~20%, of the planet is rock-metal and this comprises the planetary core. Saturn also appears to have less than a cosmic abundance of He.

Both the principal and reflected Hugoniot data for H (Fig. 4), constrain the models of the major planets. These provide pressure-density data in the pressure-temperature range which approximates the range of conditions of isentropic compression of H starting from Jovian or Saturnian gas at near-surface conditions. Although the pressure of the molecular H to metallic H transition appears in the models to be well established to be between 3 and 5 Mbars, according to the data of Grigoriev et al. (1973) (Figure 5) the pressure, in this study was actually calculated. Experimental isentropic compression of H and the study of its metallization under these conditions has yet to be conducted. Shock wave data for H_2O in both the ice and water form (Bakanova et al., 1976; Mitchell and Nellis, 1982) have been conducted. To a lesser extent, data on other molecular materials (e.g. NH_3) provide constraints on the internal structure of Uranus and Neptune which are expected to have large, mostly water-ice mantles and silicate-iron cores.

III. The Earth's Mantle

The region between 2900 and 1-10 (beneath the ocean floor) or 30-80 (beneath continents) kms depth in the earth is known as the mantle. As with all the terrestrial mantles and crusts, it is believed to be composed of silicates. Shock wave data (Simakov et al., 1961; Trunin et al., 1964; McQueen an Marsh, 1966; McQueen et al., 1967; Graham and Ahrens, 1973; Watt and Ahrens, 1983) for olivines, $(Mg,Fe)_2SiO_4$, pyroxenes, $(Mg,Fe,Ca)SiO_3$, and garnets, $(Mg,Fe,Ca)_3Al_2Si_3O_{12}$, have been used to constrain the composition of the earth's upper mantle, and in the case of olivines and pyroxenes, the composition of other terrestrial planetary (Mars, Venus, and Mercury) mantles (Zharkov and Trubitsyn, 1978). A major result of studying both the mantle minerals and minerals of the crust is that some 50% or more of the compression resulting from the application of shock pressures of 200 GPa or more is associated with phase transitions to denser phases, rather than self-compression as shown in Table 2.

Table 2. Shock Induced Phase Changes in Mantle Components

Low Pressure Phase	Density (g/cm^3)	High Pressure Phase(s)	Density (g/cm^3)	References
Mg_2SiO_4 (olivine)	3.22	MgO (periclase) + $MgSiO_3$ (perovskite)	3.94	Watt and Ahrens, 1983; Syono et al., 1981
$MgSiO_3$ (enstatite)	3.20	$MgSiO_3$ (perovskite)	4.10	Watt and Ahrens, 1985; Jakubith and Seidel, 1980
CaO (calcia)	3.35	CaO (B_2) structure	3.76	Jeanloz et al., 1979
FeO (wustite)	5.86	unknown	6.05	Jeanloz and Ahrens, 1980
SiO_2 (quartz)	2.65	SiO_2 (rutile structures)	4.29	McQueen et al., 1963

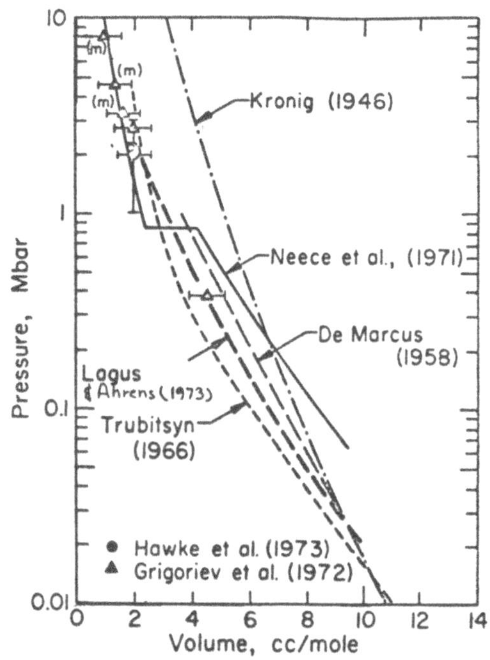

Fig. 5. Proposed 0° K isotherms of solid molecular hydrogen and 0° K isotherm of metallic hydrogen proposed by Neece *et al.* (1969). Isentropic compression data centered at 4° K due to Grigoriev *et al.* (1972), and an isentropic compression datum due to Hawke *et al.* (1972) in liquid molecular hydrogen at an unspecified initial temperature. Grigoriev *et al.* data conjectured by them to be in metallic phase labelled with (m) (after Lagus, 1973).

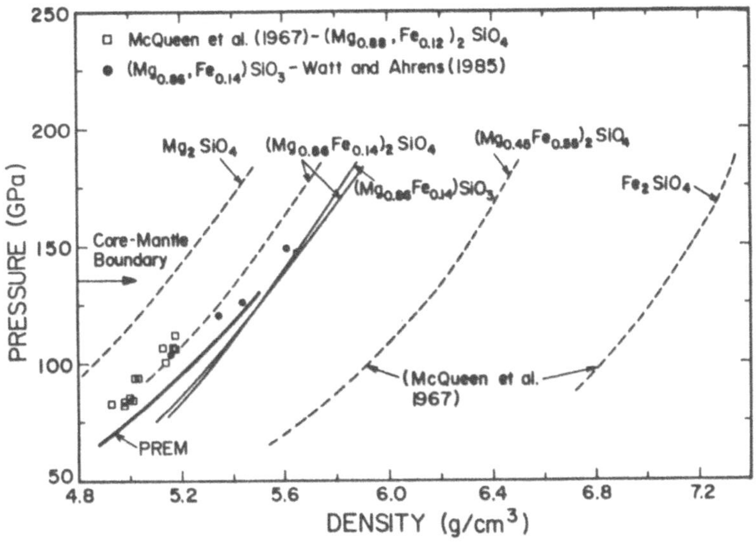

Fig. 6. Shock wave data for mantle minerals. Pressure of core-mantle boundary and of pressure density model PREM (Dziewonski and Anderson, 1981) indicated. Mantle density-pressure profile is close to $(Mg_{0.86}, Fe_{0.14})_2SiO_4$ or $(Mg_{0.86}, Fe_{0.14})SiO_3$ Hugoniots.

Recent developments in both plate tectonics (e.g. Turcotte and Schubert, 1982; Hoffman and McKenzie, 1985) and concepts of the geochemical evolution of the earth suggest that although solid state convection is taking place in the mantle and is removing both radiogenic and possibly primordial (accretionary) heat by advection from the earth's interior, mixing may not be taking place between the convective systems of the upper mantle (above depths of 670 km) and the lower mantle (extending to a depth of 2900 km). Because the upper and lower mantles of the earth may have evolved to different compositions (D. Anderson, 1982) it is useful in addition to examining the shock-induced high pressure properties of the minerals of the upper mantle to also study the component oxides. Olivines and pyroxenes when shocked to pressures of a 100 GPa (McQueen et al., 1967) demonstrated a pressure-density behavior at high pressure which indicates a 25-30% greater zero pressure-density than the starting materials. Alt'shuler and Sharipdzhanov (1971) demonstrated that the shock states above 100 GPa could be approximated using an appropriate mass-weighted sum of the Hugoniots of the equivalent oxides ---- MgO, FeO, SiO_2, and CaO. Separate studies of FeO and SiO_2 demonstrate that these oxides undergo shock-induced phase changes (Table 2). Although shown in Figure 6, the lower mantle demonstrates a density-pressure relation which can be approximated by shocked $(Mg_{0.86}, Fe_{0.14})SiO_3$ (Fig. 6), the existing pressure-density data are insufficient to differentiate between models which have the $MgSiO_3$ versus Mg_2SiO_4 stoichiometry. The present results would be more definitive if sound speed (either longitudinal or bulk) measurements were available along the Hugoniot of these materials. These types of measurements have been both difficult to carry out as well as interpret because of the effects in the shock state of both finite strength and phase changes. Recently, Chhabildas and Miller (1985) have measured the sound speed of a high pressure phase of SiO_2 (stishovite). Similar measurements on other oxides are required.

In relation to the origin of magma in the earth's mantle, Rigden et al. (1984) have measured the density-pressure relations of the molten silicate $[CaAl_2Si_3O_8]_{0.36}$,

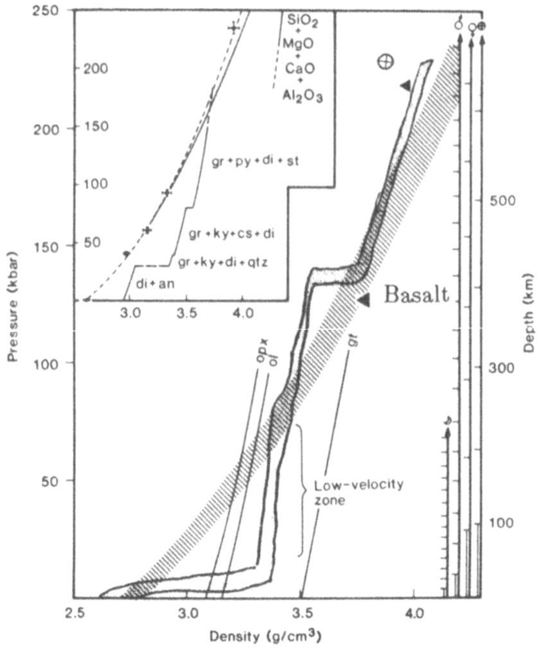

Fig. 7. Density of basic silicate melt compared with that of Earth's mantle and likely liquidus minerals olivine (ol), orthopyroxene (opx), and garnet (gt). The hachured band represents the range of melt densities expected in the compression of basaltic compositions from an average mid-ocean ridge basalt ($\rho_0 = 2.7$ g/cm^3) to komatiite ($\rho_0 = 2.8$ g/cm^3). Adiabatic compression curves are shown for olivine, pyroxene and garnet in equilibrium with the melt. The stippled band represents the range of density-depth relations in Earth derived from two recent seismological whole-Earth models (PEM and PREM) and a density model based on free oscillation data. Depth scales are shown at the right for the major terrestrial planets and Earth's moon; the tick marks on each scale are at 100-km intervals. The inset shows dynamic compression data for $An_{0.36}Di_{0.64}$ at an initial temperature of 1400° C ($\rho_0 = 2.615 \pm 0.025$). The dashed line passing through the data points is a Hugoniot fit to the experimental points. The solid line adjacent to the Hugoniot is calculated from the Birch-Murnaghan equation with $K_s^0 = 226$ kbar and $K_s' = 4.15$. This curve deviates from the Hugoniot only above ~100 kbar where shock heating becomes important. Adiabatic compression curves of the expected solid of this bulk composition are also shown for an initial temperature of 1400° C. Above 80 kbar the phase assemblage is uncertain but is probably well approximated by the lefthand curve to ~150 kbar. The rightmost curve is a maximum possible density estimate based on an assemblage of high-pressure-phase mixed oxides. Abbreviations: an, anorthite; di, diopside; qtz, quartz; cs, coesite; ky, kyanite; gr, grossular; py, pyrope; st, stishovite; ☾, moon; ⊕, Earth; ♂, Mars; ♀, Venus (after Rigden *et al.* 1984).

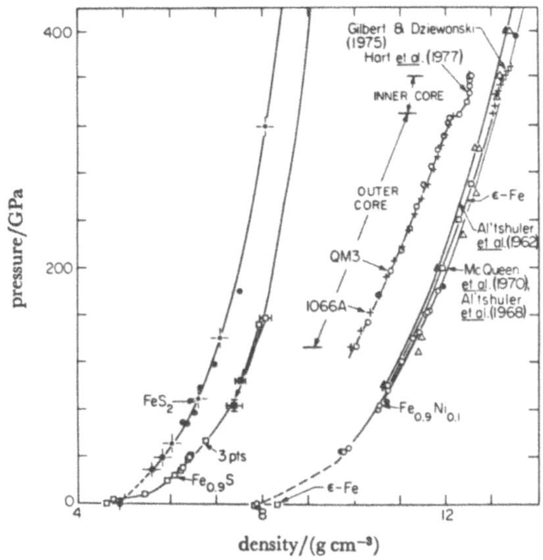

Fig. 8. Shock pressure plotted against density relations for candidate materials of the Earth's core and seismological data. Shock-wave data for FeS_2 are from Simakov *et al.* (1974) and Ahrens & Creaven, unpublished. Data for $Fe_{0.9}S$ are from Ahrens (1979) (after Ahrens, 1982a).

$[MgCaSi_2O_6]_{0.64}$ initially at 1400° C (Fig. 7). This composition is close to that of natural basalts without iron, and the determination of density at high pressure provides a measure of buoyancy of molten silicate with respect to the solid phases in equilibrium with the melt in the upper mantle of the earth. This study demonstrated that the basalt achieves a density comparable to the upper mantle of the earth and the minerals olivine and pyroxene at a depth of 120 to 240 km. This result implies that at these depths, molten basalt becomes gravitationally stable and will not migrate to the surface. The stability of a partially molten mantle in this depth range could give rise to a world-wide low seismic velocity region. At shallower depths, the basalt migrates upward through the mantle and into the crust whereas at depths >240 km, these results imply that if molten basalt is generated, it would sink down.

IV. The Earth's Core

Shock wave data have played a critical role in interpreting the nature of the earth's core which extends from a depth of 2890 km (136 GPa) to the earth's center (6371 km, 364 GPa).. The core is subdivided into two regions: the outer core, which is liquid and extends from a depth of 2890 to 5150 km (329 GPa), and the inner solid core, which extends from 5150 km to the earth's center (364 GPa).

Shock wave data for iron and iron-nickel demonstrate (Fig. 8) that at pressures of 136 GPa (at the outer part of the outer core) the density of iron or iron-nickel is 11 g/cm^3, whereas the density of the liquid core is $\sim$ 10 g/cm^3. This density deficit of the outer core is believed to result from a lighter element, like sulfur, in the amount of 12 $\pm$ 1 wt. % or, oxygen, 7-8 wt. % being dissolved in the iron (Figure 9). The density deficit cannot be attributed to much higher temperatures in the core than, for example, along the Hugoniot of iron. The sound speeds along the Hugoniot measured by Brown and McQueen (Fig. 10) have been found consistent with the solid inner core being composed of nearly pure iron (O. Anderson, 1982). Moreover, the drop in sound speed observed at

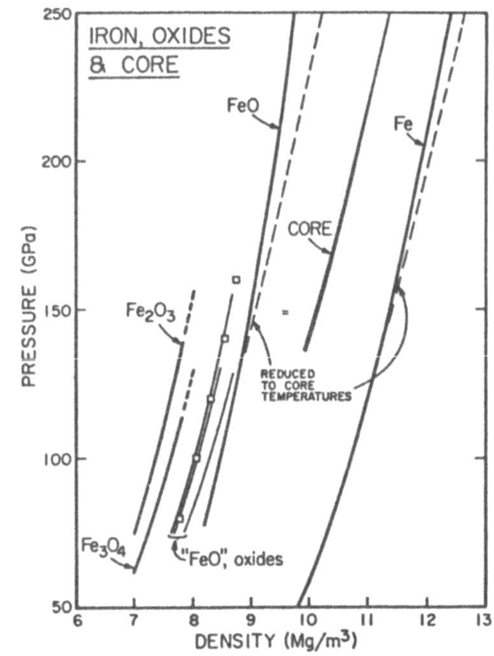

Fig. 9. Shock-wave data for FeO (after Jeanloz and Ahrens, 1980), Fe (McQueen *et al.*, 1970), Fe_2O_3 and Fe_3O_4 (McQueen and Marsh, 1966) (Heavy curves) compared with the seismologically determined compression curve of the core (Dziewonski, *et al.*, 1975; Anderson and Hart, 1976). The Fe and FeO data, corrected to core temperatures, are shown as thin dashed curves. Mixtures of oxides and Fe corresponding to FeO are given as thin solid curves: 1/3 (Fe and Fe_3O_4) and ($Fe_3O_4 - Fe_2O_3$), with increasing density; open squares are from Al'tshuler and Sharipdzhanov (1971). Hugoniots for only the high-pressure phases are shown, and the wüstite data are corrected to the stoichiometric endmember (after Jeanloz and Ahrens, 1980).

the presumed γ-iron to iron-liquid transition at 240 GPa along the solid iron Hugoniot can be used when taken thermochemical data to infer the temperature range at the inner/outer core boundary. This is found to be 5200 - 6600 K.

V. Impacts on Planetary Surfaces

Equations-of-state of earth materials are critically important for describing the energetics of flow of materials which result from impacts of planetary surfaces. Figure 11 illustrates the Hugoniot and critical release adiabats for silicate gabbroic anorthite (Ahrens and O'Keefe, 1977). These are used in finite difference calculations which result in particle velocity flow fields such as shown in Fig. 12. An impact of a silicate projectile on a solid silicate planet produces a transient cavity (Fig. 12a) which is lined with impact melt. The amount of melt is some $\sim$ 10 times the mass of the projectile. However virtually no vapor is produced. In contrast, for silicate impactor at 45 km/sec (Fig. 12b), the amount of shock melted material is increased to about 10^2 times the mass of the impactor, and about 10^0 times the mass of the impactor is vaporized. The vector particle velocity fields shown in Fig. 12 are given in terms of a normalized time which is

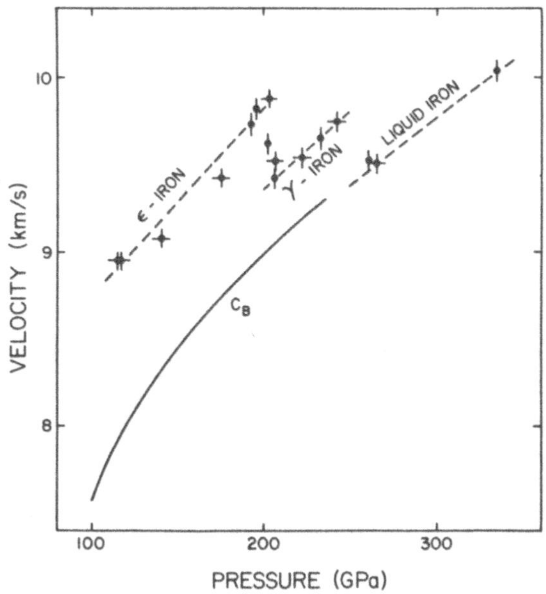

Fig. 10. Elastic wave velocities as a function of pressure along the Hugoniot of iron. The solid curve is the calculated bulk sound velocity (from Brown and McQueen, 1980).

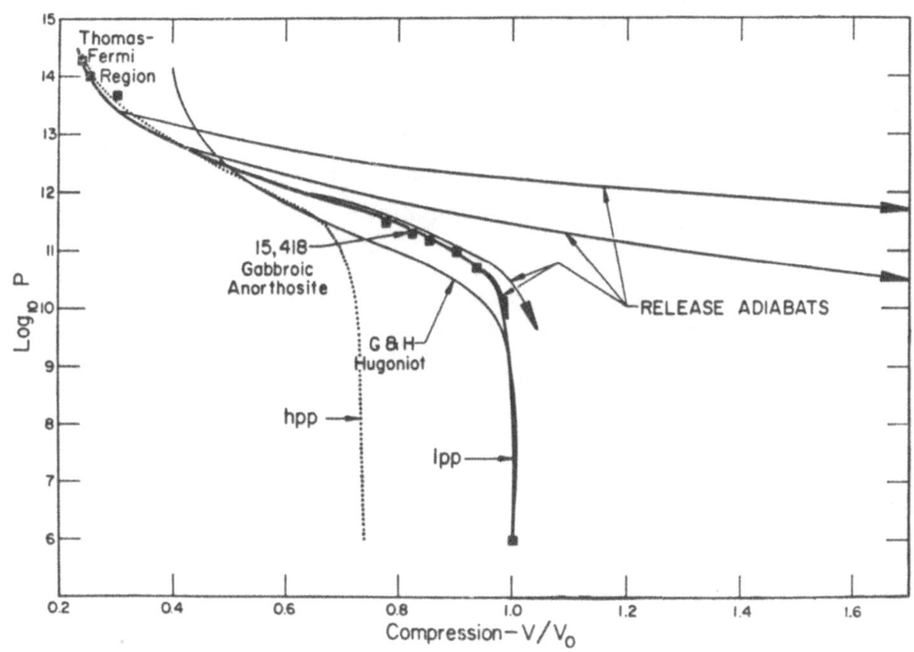

Fig. 11. Equation of state of gabbroic anorthosite in low- (lpp) and high-pressure (hpp) regimes. Log_{10} (pressure, c.g.s.) vs. compression for 2.86 g/cm^3 basalt used in Gault and Heitowit (1963) (G & H) formulation is shown also for comparison. In the G & H treatment release adiabats are assumed to coincide with Hugoniot curve (after O'Keefe and Ahrens, 1978).

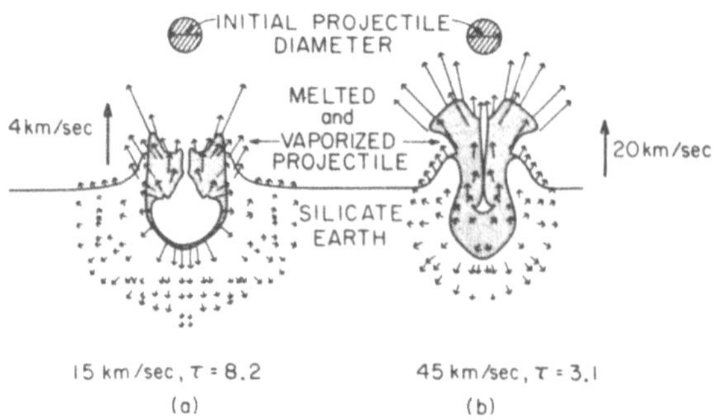

Fig. 12. Particle velocity flow field from a silicate projectile impacting a strengthless silicate planetary surface at 15 and 45 km/s. Flow fields at (a) $\tau = 8$; (b) $\tau = 3.1$, where τ is normalized time (after O'Keefe and Ahrens, 1982).

measured in the units of the time required for the projectile to travel across its own diameter.

Calculations for differing material and velocity projectiles specify the mass of ejecta, M_e, traveling at various velocities. When normalized to meteoroid mass, M_m, plots such as shown in Fig. 13 can be used to evaluate the net loss or gain of material as a function of planetary escape velocity (and hence, planet mass and radius) during accretion of the planet. We note in Fig. 13 that for a given impact velocity and planetary escape velocity, the horizontal line through $M_e/M_m = 1.0$ divides impact conditions which result in planetary erosion, $M_e/M_m > 1$, versus planetary accretion, $M_e/M_m < 1$. Thus, for anorthite impacting anorthite, at 5 km/sec for planetary escape velocity greater than 0.83 km/sec, accretion occurs. In contrast, erosion occurs when the escape velocity is less than 0.83 km/sec. At an impact velocity of 45 km/sec, net accretion occurs for planetary escape velocity of greater than 5.3 km/sec (similar to Mars) whereas for an escape velocity less than 5.3 km/sec, net erosion occurs.

VI. Impact Effects upon Planetary Accretion

In addition to the mechanical cratering aspects of planetary accretion, impact induced devolatilization and reaction between species is expected. If the planetesimals which are accreted into the earth were anything like the primitive meteorites (Ringwood, 1979; Stacey, 1977), planet forming planetesimals should have contained anhydrous silicates, metallic iron-nickel, iron sulfide as well as organic compounds. Although devolatilization reactions such as

$$Mg_3Si_2O_5(OH)_4 \text{ serpentine} \rightarrow 2\,H_2O + Mg_2SiO_4 + MgSiO_3 \quad (1)$$

have long thought to have brought to Earth, Venus, and Mars their water budgets, it has only been recently that experiments such as depicted in Fig. 14 have been conducted. The reactions of impact released water will produce Fe_2SiO_4 (and use up all the available water) via the following reaction:

$$MgSiO_3 + Fe + H_2O \rightarrow \frac{1}{2}\,Mg_2SiO_4 + \frac{1}{2}\,Fe_2SiO_4 + H_2\uparrow \quad (2)$$

Since it is usually assumed that the H_2 from the above reaction is lost to space from the accreting planet via Jean loss and hydrodynamic escape, reaction (2) is irreversible. In the case of the Earth, both the water abundance and the Fe_2SiO_4 content of the mantle

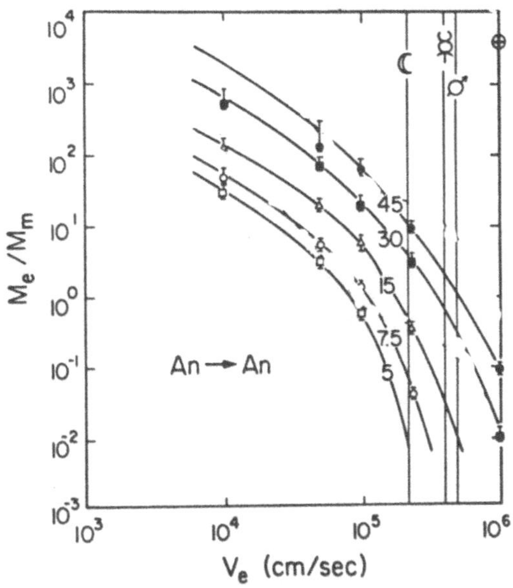

Fig. 13. Ratio of mass ejected, M_e, to meteorite mass, M_m, plotted against escape velocity, V_e. An → An indicates anorthosite impacting anorthosite (after O'Keefe and Ahrens, 1977b).

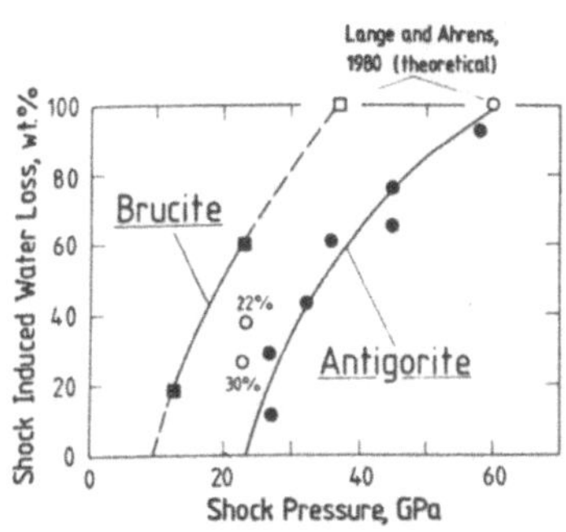

Fig. 14. Shock-induced water loss (in wt.% of total amount of initially present water) in brucite and antigorite-serpentine as a function of shock pressure in shock recovery experiments. Open circles indicate results for antigorite samples with 22% and 30% porosity (after Lange and Ahrens, 1984).

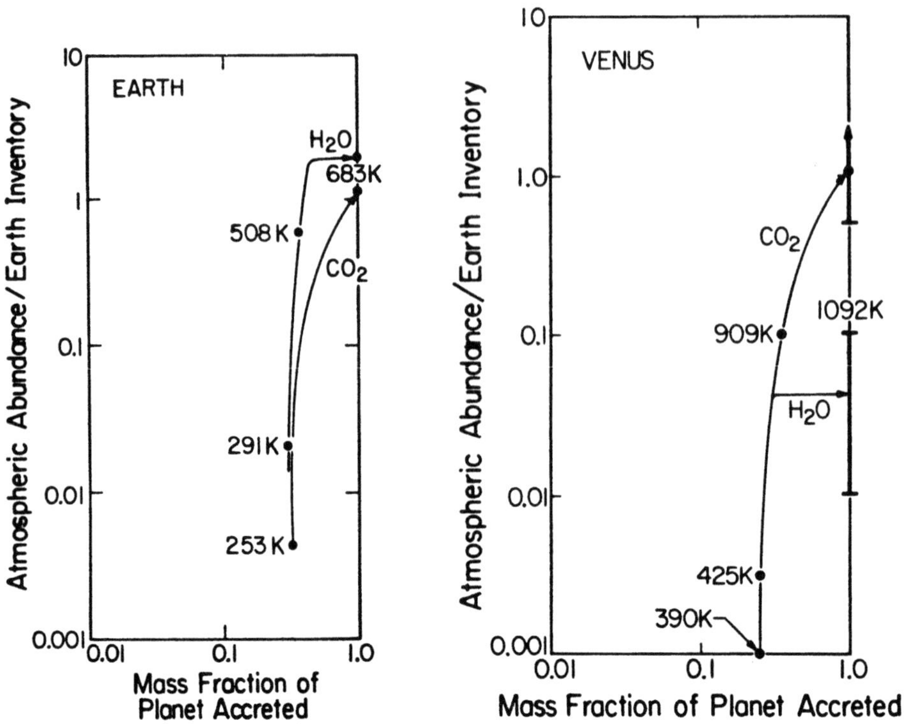

Fig. 15. Calculated primordial atmosphere accretion models for the Earth (a) and Venus (b). Formation of CO_2 and H_2O primordial atmosphere commences when impact shock pressure in accreting planetesimals achieves 20 GPa. Surface temperatures controlled by atmosphere in radiative equilibrium are indicated. Total volatile content of planetesimals 0.4 wt.%, of which 93% is assumed to be H_2O. Atmospheric losses assumed are from impact cratering and hydrodynamic escape. Earth inventory of CO_2 and H_2O assumed is $4\text{x}10^{20}$ and $1.4\text{x}10^{21}$ kg respectively (after Ahrens and O'Keefe, 1985b).

are bounded by gross earth properties (see Fig. 6), hence the degree to which the Earth accreted both water-bearing minerals and metallic iron at the same time must be limited (Lange and Ahrens, 1984). The chain of reactions following the impact release of water

$$2H_2O \rightarrow 2H_2 + O_2 \tag{3}$$

and oxidation of carbon from meteoritic graphite or hydrocarbons

$$C + O_2 \rightarrow CO_2 \tag{4}$$

$$CH_4 + 2O_2 \rightarrow CO_2 + H_2O \tag{5}$$

yield CO_2 bearing atmospheres for Venus, Earth and Mars.

Results of shock devolatilization experiments can be used in accretion calculations (Fig. 15a, b) by taking into account an initial effective radiative temperature of Venus (390K), Earth (275K), and Mars (218K). The differing starting temperatures result from successively greater distance of the terrestrial planets from the Sun (Rasool and DeBergh, 1970). As a result of shock devolatilization upon impacts of ~ 3 km/sec, both CO_2 and H_2O form from material having volatile contents of 0.4% by weight. Due to the higher radiation temperature, H_2 is preferentially lost from Venus versus the Earth as the atmosphere accumulates as a result of cratering of the atmosphere and hydrodynamic escape. In the case of Venus, the post accretional higher temperature can be modelled by assuming the same amount, 0.4×10^{21} kg of CO_2 is in the atmosphere (90 bar) as is buried in the earth's crust.

Acknowledgments. Contribution #4260, Division of Geological and Planetary Sciences. Supported under NASA grants NGL05-002-105 and NSG 7129 and NSF grants EAR8219435, 8313746, 8407784. I appreciate B. Svendsen's critical and helpful comments on this paper.

REFERENCES

Abe, Y., and Matsui, Y., 1985, The formation of an impact-generated H_2O atmosphere and its implications for the early thermal history of the earth, Proc. 15th Lunar & Planet. Sci. Conf., Part 2, J. Geophys. Res., 90:C545.

Ahrens, T. J., and O'Keefe, J. D., 1985, Shock vaporization and the accretion of the icy satellites of Jupiter and Saturn, in: "Ices in the Solar System," P. Klinger, A. Dollfus, and R. Smoluchowski, eds., Reidel,

Ahrens, T. J., O'Keefe, J. D., and Lange, M. A., 1985, Accretion of the earth's water budget and atmospheric cratering, Abstract, Lunar and Planetary Science XVI:9.

Ahrens, T. J., 1982a, Constraints on core composition from shock wave data, Phil. Trans. Roy. Soc. London, 306:37.

Ahrens, T. J., Lyzenga, G. A., and Mitchell, A. C., 1982b, Temperatures induced by shock waves in minerals, in: "High Pressure Research in Geophysics," S. Akimoto and M. H. Manghnani, eds., Center for Academic Publications, Japan.

Ahrens, T. J., 1979, Equations of state of iron sulfide and constraints on the sulfur content of the earth, J. Geophys. Res., 84:985.

Ahrens, T. J., and O'Keefe, J. D., 1977, Equations of state and impact-induced shock-wave attenuation on the moon, in: "Impact and Explosion Cratering," D. J. Roddy, R. O. Pepin, and R. B. Merrill, eds., Pergamon, New York.

Allegre, C. J., 1982, Cosmochemistry and primitive evolution of planets, in: "Formation of Planetary Systems," A. Brahic, ed., Centre National D'Etudes Spatiales, Paris.

Al'tshuler, L. V., Bakanova, A. A., and Trunin, R. F., 1962, Shock adiabats and zero isotherms of seven metals at high pressures, Soviet Physics, JETP, 15:65.

Al'tshuler, L. V., Simakov, G. V., and Trunin, R. F., 1968, On the chemical composition of the earth's core, Izv. Earth Phys., 1:3.

Al'tshuler, L. V., and Sharipdzhanov, I. I., 1971, Additive equations of state of silicates at high pressures, Izv. Earth Physics, 3:11.

Anders, E., and Ebihara, M., 1982, Solar system abundances of the elements, Geochim. Cosmochim. Acta, 46:2363.

Anderson, D. L., 1982, Isotopic evolution of the mantle: A model, Earth Planet. Sci. Lett., 57:13.

Anderson, D. L., and Hart, R. S., 1976, An earth model based on free oscillations and body waves, J. Geophys. Res., 81:1461.

Anderson, O. L., 1982, The earth's core and the phase diagram of iron, Phil. Trans. R. Soc. Lond., A306:21.

Bakanova, A. A., Zubarev, V. N., Sutulov, Yu. N., and Trunin, R. F., 1976, Thermodynamic properties of water at high pressures and temperatures, Sov. Phys. JETP, 41:544.

Bjork, R. L., 1961, Analysis of the formation of Meteor Crater, Arizona: A preliminary report, J. Geophys. Res., 77:3379.

Brown, J. M., and McQueen, R. G., 1980, Melting of iron under core conditions, Geophys. Res. Lett., 7:533.

Chhabildas, L. C., and Miller, J. M., 1985, Release-adiabat measurements in crystalline quartz, Sandia Report SAND 1092, Sandia National Laboratory, Albuquerque, N.M.

Dziewonski, A. M., and Anderson, D. L., 1981, Preliminary reference earth model, Phys. Earth Planet. Interiors, 25:297.

Dziewonski, A. M., Hales, A. L., and Lapwood, E. R., 1975, Parametrically simple earth models consistent with geophysical data, Phys. Earth Planet. Interiors, 10:12.

French, B. M., and Maran, S. P., 1981, "A Meeting with the Universe," NASA EP-177.

Gault, D. E., and Heitowit, E. D., 1963, The partition of energy for hypervelocity impact craters formed in rock, in: "Proceedings of the 6th Hypervelocity Impact Symposium," Cleveland, Ohio.

Gilbert, F., and Dziewonski, A. M., 1975, An application of normal mode theory to the retrieval of strueval of structural parameters and source mechanisms from seismic spectra, Phil. Trans. Roy. Soc. Lond., A278:187.

Graham, E. K., and Ahrens, T. J., 1973, Shock wave compression of iron-silicate garnet, J. Geophys. Res., 78:375.

Grigor-ev, F. V., Kormer, S. B., Mikhailova, O. L., Tolochko, A. P., and Urlin, V. D., 1972, Experimental determination of the compressibility of hydrogen at densities of 0.5-2 g/cm^3, metallization of hydrogen, JETP Letters, 16:201.

Hart, R. S., Anderson, D. L., and Kanamori, H., 1977, The effect of attenuation on gross earth models, J. Geophys. Res., 82:1647.

Hawke, R. S., Duene, D. E., Huebel, J. G., Keeler, R. H., and Klapper, H., 1972, Isentropic compression of fused quartz and liquid hydrogen to several Mbar, Phys. Earth Planet. Interiors, 6:44.

Hoffman, N. R. A., and McKenzie, D. P., 1985, The destruction of geochemical heterogeneities by differential fluid motions during mantle convection, Geophys. J. R. Astron. Soc., 82:163.

Holsapple, K. A., and Schmidt, R. M., 1982, On scaling of crater dimensions, 2. Impact processes, J. Geophys. Res., 87:1849.

Holsapple, K. A., and Housen, K. R., 1985, Cratering estimates for ice, submitted to Proc. Lunar & Planet. Sci. Conf..

Housen, K. R., Schmidt, R. M., and Holsapple, K. A., 1983, Crater ejecta scaling laws: Fundamental forms based on dimensional analysis, J. Geophys. Res., 88:2485.

Jakubith, M., and Seidel, P., 1982, Shock-loading experiments on eclogite, Geophys. Res. Lett., 9:408.

Jeanloz, R., and Ahrens, T. J., 1980, Equations of state of FeO and CaO, Geophys. J. R. Astron. Soc., 62:505.

Jeanloz, R., Ahrens, T. J., Mao, H. K., Bell, P. M., 1979, B1/B2 transition in CaO from shock-wave and diamond cell experiments, Science, 266:829.

Lagus, P. L., 1974, The equations of state of hydrogen and argon; applications to the Jovian interior, Ph.D. thesis, California Institute of Technology, Pasadena, California, 131 pp.

Lagus, P. L., and Ahrens, T. J., 1973, Shock wave measurements on solid hydrogen and argon, J. Chem. Physics, 59:3517.

Lange, M. A., and Ahrens, T. J., 1985, Shock-induced CO_2 loss from $CaCO_3$; implications for early planetary atmosphere, Earth Planet. Sci. Lett., submitted.

Lange, M. A., and Ahrens, T. J., 1984, FeO and H_2O and the homogeneous accretion of the earth, Earth Planet. Sci. Lett., 71:111.

Lange, M. A., and Ahrens, T. J., 1982, Impact-induced dehydration of serpentine and the evolution of planetary atmospheres, Proc. Lunar Planet. Sci. Conf. 13th, J. Geophys. Res., 87:A451.

Lange, M. A., and Ahrens, T. J., 1980, The evolution of an impact-generated atmosphere (abstract), in, "Lunar and Planetary Science XI (Lunar and Planetary Institute, Houston, TX)."

McQueen, R. G., Marsh, S. P., Taylor, J. W., Fritz, J. N., and Carter, W. J., 1970, The equation of state of solids from shock wave studies, in: "High Velocity Impact Phenomena," R. Kinslow, Academic, New York.

McQueen, R. G., Marsh, S. P., and Fritz, J. N., 1967, Hugoniot equations of state of twelve rocks, J. Geophys. Res., 72:4999.

McQueen, R. G., and Marsh, G. P., 1966, unpublished data, in: "Handbook of Physical Constants, Geol. Soc. Am. Memoir 97," G. P. Clark, ed., Geol. Soc. Am., New York.

McQueen, R. G., and Marsh, S. P., 1966, Shock wave compression of iron-nickel alloys and the earth's core, J. Geophys. Res., 71:1751.

McQueen, R. G., Fritz, J. N., and Marsh, S. P., 1963, On the equation of state of stishovite, J. Geophys. Res., 68:2319.

Melosh, H. J., 1984, Impact ejection, spallation and the origin of meteorites, Icarus, 59:234.

Mitchell, A. C., and Nellis, W. J., 1982, Equation of state and electrical conductivity of water and ammonia shocked to the 100 GPa (1 Mbar) pressure range, J. Chem. Phys., 76:6273.

O'Keefe, J. D., and Ahrens, T. J., 1982, The interaction of the Cretaceous-Tertiary extinction bolide with the atmosphere, ocean, and solid earth, Proc. Conf. on Large Body Impacts and Terrestrial Evolution, Sp. Paper #190 Geol. Soc. Am., 103.

O'Keefe, J. D., and Ahrens, T. J., 1978, Impact flows and crater scaling on the Moon, Phys. Earth Planet. Inter., 16:351.

O'Keefe, J. D., and Ahrens, T. J., 1977a, Impact-induced energy partitioning, melting and vaporization on terrestrial planets, Proc. Lunar Sci. Conf. 8th, 3357.

O'Keefe, J. D., and Ahrens, T. J., 1977b, Meteorite impact ejecta: Dependence of mass and energy lost on planetary escape velocity, Science, 198:1249.

Rasool, S. I., and deBergh, C., 1970, The runaway greenhouse and the accumulation of CO_2 in the Venus atmosphere, Nature, 226:1037.

Rigden, S. M., Ahrens, T. J., and Stolper, E. M., 1984, Densities of liquid silicates at high pressures, Science, 226:1071.

Ringwood, A. E., 1979, in: "Origin of the Earth and the Moon," Springer-Verlag.

Ross, M., Graboske, H. C., Jr., and Nellis, W. J., 1981, Equation of state experiments and theory relevant to planetary modeling, Phil. Trans. R. Soc. Lond., A303:303.

Simakov, G. V., Pavlovskii, M. N., Kalashnikov, W. G., and Trunin, R. F., 1974, Shock compressibility of twelve minerals, Izv. Phys. Solid Earth, 488.

Stacey, F. D., 1977, in: "Physics of the Earth," J. Wiley & Sons, 2nd ed.

Stevenson, D. J., 1985, Cosmochemistry and structure of the giant planets and their satellites, Icarus, 62:4.

Stevenson, D. J., 1984, Condensed matter physics of planets: Puzzles, progress and predictions, Mat. Res. Soc. Symp. Proc., 22:357.

Stevenson, D. J., 1981, Models of the Earth's Core, Science, 214:611.

Syono, Y., Goto, T., Takei, H., Tokonami, M., and Nobugai, K., 1981, Dissociation reaction in forsterite under shock compression, Science, 214:177.

Trunin, R. F., Gon'shakova, V. I., Simakov, G. V., and Galdin, N. E., 1965, A study of rocks under the high pressures and temperatures created by shock compression, Izv. Earth Physics, 9:1.

Turcotte, D. L., and Schuber, G., 1982, in: "Geodynamics," J. Wiley, ed.

Watt, J. P., and Ahrens, T. J., 1985, Shock-wave equation of state of enstatite, J. Geophys. Res., submitted.
Watt, J. P., and Ahrens, T. J., 1983, Shock compression of single-crystal forsterite, J. Geophys. Res., 88:9500.
Zharkov, V. N., and Trubitsyn, V. P., 1978, in: "Physics of Planetary Interiors," Pachert, Tucson, AZ.

HIGH-PRESSURE RELEASE-WAVE MEASUREMENTS AND PHASE TRANSFORMATION IN $CACO_3$*

D. E. Grady

Thermomechanical & Physical Division
Sandia National Laboratories
Albuquerque, NM

ABSTRACT

Release-wave measurements on theoretical-density single crystal calcium carbonate from Hugoniot states of 11.5-, 17.9- and 26.4-GPa shock pressure have been performed using time-resolved laser velocity interferometry diagnostics. Samples were subjected to planar impact with metal discs accelerated to high velocity using a powder gun facility. Motion was measured at a diffusely reflecting interface between the sample and lithium fluoride window material. Evidence for a high-pressure phase transformation between 12-14 GPa was provided from the 17.0- and 26.4-GPa experiments in terms of a well-defined rarefaction shock wave. The new phase may be a further metastable phase of calcite or identical to the high pressure state of aragonite.

INTRODUCTION

Calcium carbonate is a natural mineral occurring in considerable abundance in the earth's upper crust. It exists in two crystalline structures at ambient temperature and pressure; aragonite (ρ_o = 2930 kg/m^3) and the more common calcite I (ρ_o = 2710 kg/m^3). The latter is the stable phase below a pressure of about 350 MPa. Static volumetric compression studies on calcite I have identified two additional phases, calcite II and calcite III, with transformation pressures of approximately 1.44 and 1.77 GPa, respectively[1]. Jamieson[2] subsequently showed that the calcite II and III forms of calcium carbonate are metastable with respect to aragonite. The calcite II and III transformations have been observed under shock-wave compression and profoundly influence the stress-wave propagation properties of this material[3].

In addition to these low pressure phases of calcium carbonate, the possible occurrence of a high-pressure phase transformation in calcium carbonate was first reported by Adadurov, et al.[4]. From a slight break in shock velocity-particle velocity data, a transformation at about 15 GPa was suggested. Further data of Ahrens and Gregson[5] also indicate a high-pressure transformation, and particle velocity profile data of Murri, et al.[6] provide additional support for this transition.

*This work performed by Sandia National Laboratories supported by the U. S. Department of Energy under contract #DEAC04-76-DP00789.

In the present work we have conducted shock loading and release experiments on crystalline calcite and measured the time-resolved shock and release profiles with velocity interferometry techniques[7]. These data further support the existence of a high-pressure phase change through the occurrence of a distinctive rarefaction shock wave on release encompassing the pressure range between about 15 and 8 GPa. Although phase transformations in a number of materials have been identified under shock compression[8], the observation of a high-pressure rarefaction shock wave is unusual. Consequently, the present data is strong supporting evidence for the classical concepts of shock instabilities as put forth by Bethe[9] and others.

EXPERIMENTAL METHOD

Samples of optical grade Icelandic spar calcite crystals (ρ_o = 2710 kg/m^3) were cut on planes normal to the crystal z-axis and, consequently, the direction of shock propagation was along the z-axis. The experimental configuration is shown in Figure 1. A lithium fluoride window with a 0.2 μm diffusely-reflecting vapor-deposited silver film is epoxy bonded to the back surface of the calcite sample. The shock impedance of lithium fluoride is approximately 10% higher than calcite. The time-resolved shock-induced motion at the center of this interface is measured with laser velocity interferometry (VISAR) techniques[7].

Metal discs (aluminum or stainless steel) are mounted on a projectile body, and a void cavity immediately behind the disc provides for an unloading wave following the initial impact-induced shock loading. Projectiles are accelerated in an 88-mm diameter powder gun capable of a maximum velocity near 2.2 km/s. Electrical shorting pins are used to measure impact velocity and planarity, and to establish timing for oscilloscope and digitizer recording of the VISAR signal.

In two cases experiments at the same impact velocity, but with different calcite sample thicknesses, were performed. This allowed incremental determination of the wave velocity, and calculation of the continuous stress-volume response through integration of the appropriate Riemann invariant[10]. These experiments also confirmed centered-wave behavior which was assumed at the impact amplitude where only one profile was measured.

Aluminum (6061-T6) impactors were used on the experiments at shock pressures of 11.5 and 17.9 GPa, which provided for nearly complete unloading from the Hugoniot state. Stainless steel was used for the highest shock pressure experiments (26.4 GPa) and only partial unloading was achieved.

EXPERIMENTAL RESULTS

The impact experiments on crystalline calcite are summarized in Table 1. Hugoniot states measured in these experiments are shown in Figure 2 and compared with other Hugoniot data. The number in parenthesis indicates the number of tests performed at each pressure level.

The first series of experiments performed were CA1, CA2, and CA3. The wave profiles measured in these tests are shown in Figure 3. These profiles provided the most dramatic observation of the phase transformation induced-rarefaction shock wave on unloading. A Hugoniot pressure of 17.9 GPa was achieved through two loading waves. The 1.5 GPa precursor is due to the calcite I-II-III transformation and yielding. Unloading occurs through regions of isentropic release separated by a nearly discontinuous drop of about 7 GPa through a rarefaction shock.

To achieve greater peak stress levels, stainless steel impactors were used at near maximum velocity for experiments CA-4 and CA-5. At this impact amplitude the precursor

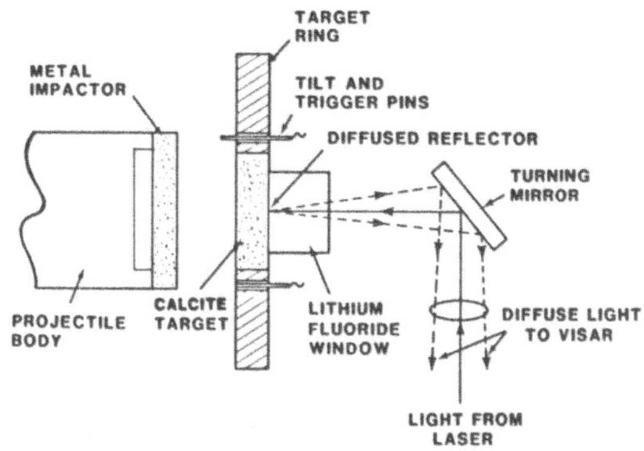

Figure 1: Impact configuration for crystalline calcite shock and release wave experiments.

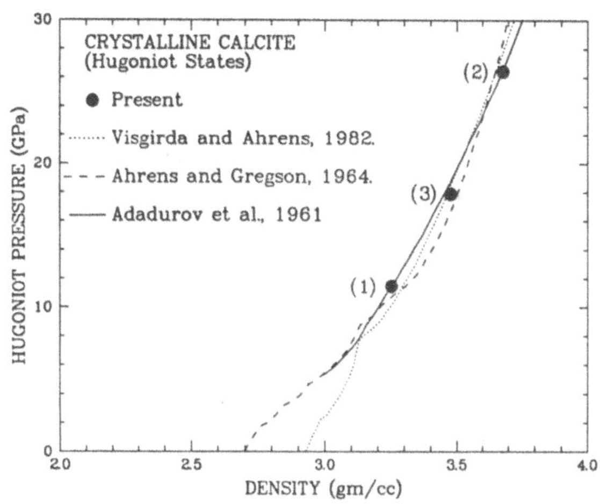

Figure 2: Hugoniot states for shock-compressed calcite.

Table 1
Crystalline Calcite Data

Exp. No.	Sample Thickness (mm)	Impactor Thickness (mm)	Impactor Material	Impact Velocity (km/s)	Window Material	First Arrival Velocity (km/s)	Maximum Pressure (GPa)	Maximum Strain
CA-1	5.054	2.645	aluminum	2.170	LiF	5.75*	17.9	0.223
CA-2	3.020	2.634	aluminum	2.152	LiF	5.75*	17.9	0.223
CA-3	8.035	2.582	aluminum	2.171	LiF	5.75*	17.9	0.223
CA-4	6.101	2.116	stainless steel	2.172	LiF	6.05*	26.4	0.266
CA-5	2.071	2.106	stainless steel	2.154	LiF	6.05*	26.4	0.266
CA-8	5.079	2.601	aluminum	1.521	LiF	5.75*	11.5	0.167

*Calculated from previous data on Z-cut calcite [12].

was overdriven, and loading to the Hugoniot pressure of 26.4 GPa occurred through a single shock wave as seen in Figure 4. Since the impedance of the stainless steel is larger than that of calcite, the unloading profiles observed in Figure 4 occurred through two wave reflections in the steel impact plate. A less pronounced rarefaction shock is again observed just prior to the plateau preceeding unloading due to the second reflection.

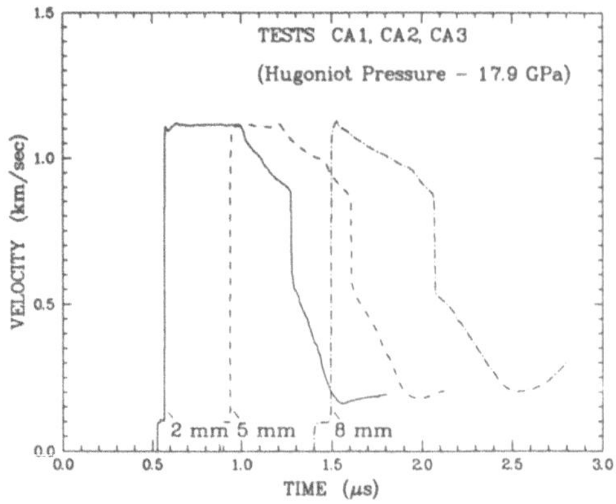

Figure 3: Wave profiles for 17.9 GPa shock compression experiments on crystalline calcite.

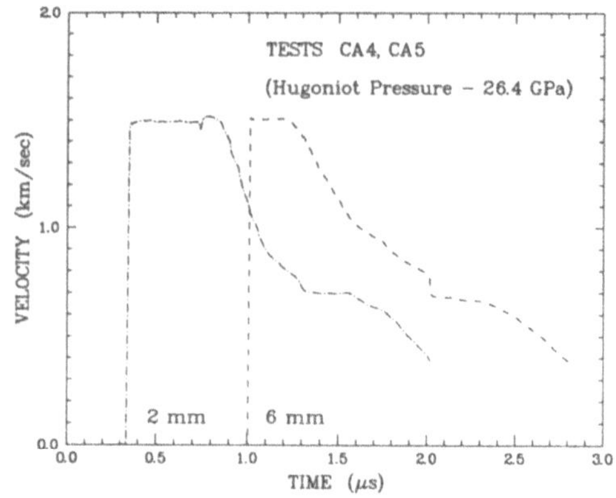

Figure 4: Wave profile for 26.4 GPa shock compression experiment on crystalline calcite.

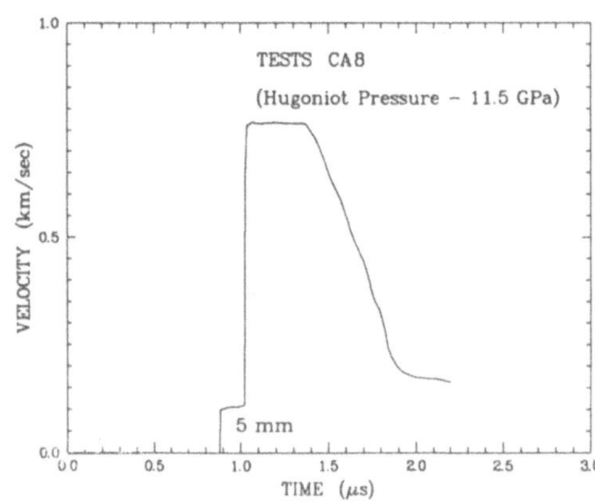

Figure 5: Wave profiles for 11.5 GPa shock compression experiments on crystalline calcite.

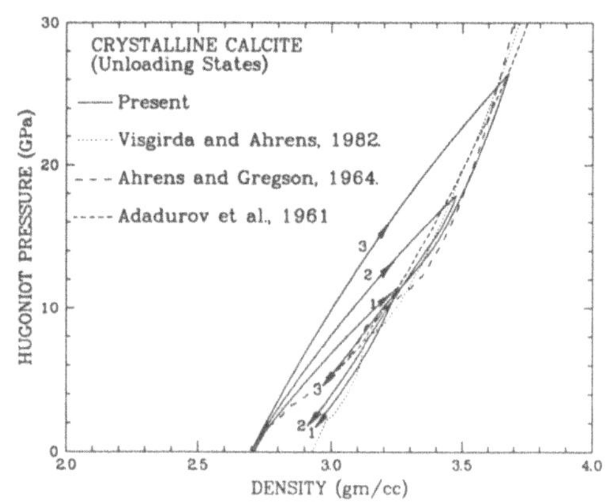

Figure 6: Compression and unloading paths for shock compressed crystalline calcite.

The final experiment (CA-8) achieved a Hugoniot pressure of 11.5 GPa. The profile in Figure 5 shows no evidence of the rarefaction shock wave, suggesting that the transformation pressure exceeds 11.5 GPa.

The loading and release wave profiles in Figures 3, 4, and 5 were used to provide continuous stress-density behavior. This was performed by determining incremental wave velocities from the data and integrating the appropriate Riemann integral. Within this process, correction for the impedance difference between the lithium fluoride window and sample material is also performed[10]. The stress-strain behavior for each amplitude are shown in Figure 6 and compared with Hugoniot data for calcium carbonate.

DISCUSSION

The Hugoniot data in Figure 6 are for calcite[4,5] and for aragonite[11]. Unloading paths determined with the present data from Hugoniot states between approximately 10-30 GPa appear to lie close to the Hugoniot. This might be construed to imply a relatively small component of thermal pressure in calcium carbonate, however, it is more likely that complexities in the phase transformation process in this range are responsible for the behavior. An estimate of the thermal pressure can be inferred from the unloading data at low pressure. At a density of 3000 kg/m^3 the pressure difference of approximately 2.0 GPa corresponds to a Grüneisen parameter of about 1.5, which is comparable with values obtained by other means. Although the rarefaction shock wave during release from pressures greater than about 15 GPa provides strong evidence for a high-pressure first-order phase change in calcium carbonate, it is important to note that no indication of this transformation is observed in the structure of the deformational shock wave. This feature is probably a consequence of the low pressure calcite I-II-III transitions which have a collective transformation volume strain of approximately 5% and also have the effect of initiating shock compression from a slightly distended state. As a consequence the deformational shock path (Rayleigh line) would miss the transformation inflection point on loading in the present experiments.

Details of the shock-induced phase change are difficult to infer from the rarefaction shock because the measured stress-volume response over this segment of anomalous compressibility follows a Rayleigh line, rather than an isentropic equation-of-state path. Our present best efforts place the transformation pressure between 12 and 14 GPa with a transformation volume between about 3 and 5% . The dramatic drop in pressure through the rarefaction shock in Figure 3 occurs in less than 10 *ns*, attesting to the rapid rate at which this transformation proceeds. There is actually a complicating segment of steady-wave just proceeding the nearly discontinuous drop (see Figure 3) of the rarefaction shock. Although not yet understood, this feature may indicate a preliminary rate-sensitive onset of the transition proceeding the final rapid transformation portion. The high-pressure phase may be equivalent to the high-pressure state of aragonite[11] or a further metastable phase of calcite.

REFERENCES

1. P. W. Bridgman, The High Pressure Behavior of Miscellaneous Minerals, Amer. J. Sci., 237, 7-18, 1939.
2. J. C. Jamieson, Phase Equilibrium in the System Calcite-Argonite, J. Chem. Phys., 21, 1385-1390, 1953.
3. D. E. Grady, R. E. Hollenbach, and K. W. Schuler, Compression Wave Studies in Calcite Rock, J. Geophys. Res., 83, 2839-2849, 1978.
4. G. A. Adadurov, D. B. Balshar, and A. N. Dremin, A Study of the Volumetric Compressibility of Marble at High Pressure, Izv. Acad. Sci. VSSR Phys. Sol. Earth, Engl. Transl., 5, 463-466, 1961.
5. T. J. Ahrens, and V. G. Gregson, Shock Compression of Crustal Rocks: Data for Quartz, Calcite, and Plagioelase Rocks, J. Geophys. Res., 68, 529-537, 1963.
6. W. J. Murri, D. E. Grady, and K. D. Mahrer, Equation of State of Rocks, Final Report., Stanford Research Inst., Menlo Park, CA, 1975.
7. L. M. Barker, and R. E. Hollenbach, Laser Interferometer for Measuring High Velocities of any Reflecting Surface.
8. G. E. Duvall, and R. A. Graham, Phase Transitions Under Shock Wave Loading, Rev. Mod. Phys., 49, 523-579, 1977.
9. H. A. Bethe, A Theory of Shock Waves for an Arbitrary Equation of State, Office of Scientific Research and Development, Report No. 545, Serial No. 237, 1942.
10. D. E. Grady, and E. G. Young, Evaluation of Constitutive Properties from Velocity Interferometer Data, Sandia National Laboratories Rept., SAND75-0650, August 1976.
11. J. Visgirda, and T. J. Ahrens, Shock Compression of Aragonite and Implications for the Equation of State of Carbonates, J. Geophys. Res., 87, 4747-4758, 1982.
12. D. E. Grady, Stress-Wave Experiments on Selected Crustal Rocks and Minerals, Sandia National Laboratories, Rept., SAND83-1898, September 1983.

ANALYSIS OF SHOCK WAVE STRUCTURE IN

SINGLE-CRYSTAL OLIVINE USING VISAR

Michael D. Furnish[1], Dennis E. Grady[2] and J. Michael Brown[3]

[1]Dept. of Geological Sciences, Cornell Univ., Ithaca, NY 14853, [2]Sandia National Laboratories, Albuquerque, NM 87185, [3]Geophysics Program, Univ. of Washington, Seattle, WA 98195

INTRODUCTION

The transition of olivine (of approximate composition $(Mg_{0.9}Fe_{0.1})_2SiO_4$) to the high-pressure γ-spinel phase, with β-spinel as a likely intermediate, is generally acknowledged to be the cause of the 400 km discontinuity in the mantle (Fig. 1). There is less general agreement concerning the mechanism by which this transition occurs. It may depend on such conditions as shear stress and temperature (Green, 1984). Two principal models have been proposed: (1) incoherent nucleation and growth of the high-pressure phase; and (2) a "martensitic" or "restacking" process in which dislocations sweep the basal (100) planes of olivine, restacking the oxygen ions from a distorted hcp arrangement to the ccp arrangement of γ-spinel, and the cations recorded either by a coherent scheme of diffusively. Experimental evidence supporting each model has been published (see, eg., Green (1984) and Furnish and Bassett (1983) for reviews), with results from longer-term experiments tending to favor incoherent nucleation and growth, and from shorter-term experiments (up to a few hours transition time) favoring a restacking mechanism.

If a restacking mechanism is operative, a phase transition in a given crystal will be strongly promoted by appropriately oriented shear stress (Poirier, 1982); superplastic deformation and shear-propagated deep-seated earthquakes are possible geophysical consequences (Parmentier, 1981, Sammis and Dein, 1974).

Considerable shock-wave research has been performed on olivine of various compositions, to aid in the interpretation of meteorite textures as well as to gain information about the equation of state of these compositions. Recent data suggest that the olivine structure is stable under shock to 45-50 GPa, then undergoes a gradual transition to some high-pressure form, finally reaching this form at a pressure of about 100 GPa. While γ-spinel has been reported in studies of several of these meteorites (eg. Binns, et al, 1969; Putnis and Price, 1979, Price, et al, 1979), only one laboratory shock wave experiment has yielded evidence for this transition. Watt and Ahrens [1983] report observing 3-wave structures in streak camera records of the surface motion of two synthetic forsterite samples shocked along the a-axis to 16 and 18.3 GPa, where the middle shock was at a pressure of 9 GPa and possibly might correspond to this transition. Although this pressure is below the transition pressure observed in static experiments, and the temperature is quite low (Fig. 1), shear processes associated with shock loading may be such as to promote the transition.

The present experiments were designed to confirm the existence of a three-wave structure in single-crystal olivine shocked to approximately 18 GPa and to look for orientation dependencies of the shock structure in the hopes of gaining new information about the transition of olivine to spinel. If a transition sensitive to (100)[013] shearing is effective, we might predict a lower pressure for the second shock, which reflects the transition, when it promotes such shearing.

TECHNIQUES AND MATERIALS

Single crystals of San Carlos peridot (compositions $Fo_{90.27}$-$Fo_{90.84}$) were oriented by back-scattered Laue reflection and cut to yield specimens 1.0-1.3 cm across and 2.88-3.86 mm thick. Two specimens were cut to have normals along the a-axis, and two to have normals approximately bisecting (100) and (013). This second orientation was chosen to maximize (100)[013] shear, and hence to promote a restacking transition mechanism. All specimens were clear and nearly free of macroscopic fractures and inclusions. Densities were measured as 3.334(5) gm/cm^3 for the a-axis slices and 3.334(10) for the oblique slices. Each crystal was mounted into a target assembly as shown in Fig. 2, with an aluminized surface of a LiF crystal held against its downstream surface by a thin film of epoxy. To minimize any effect of side rarefaction waves, the epoxy binding the sample into the holder was saturated with copper powder.

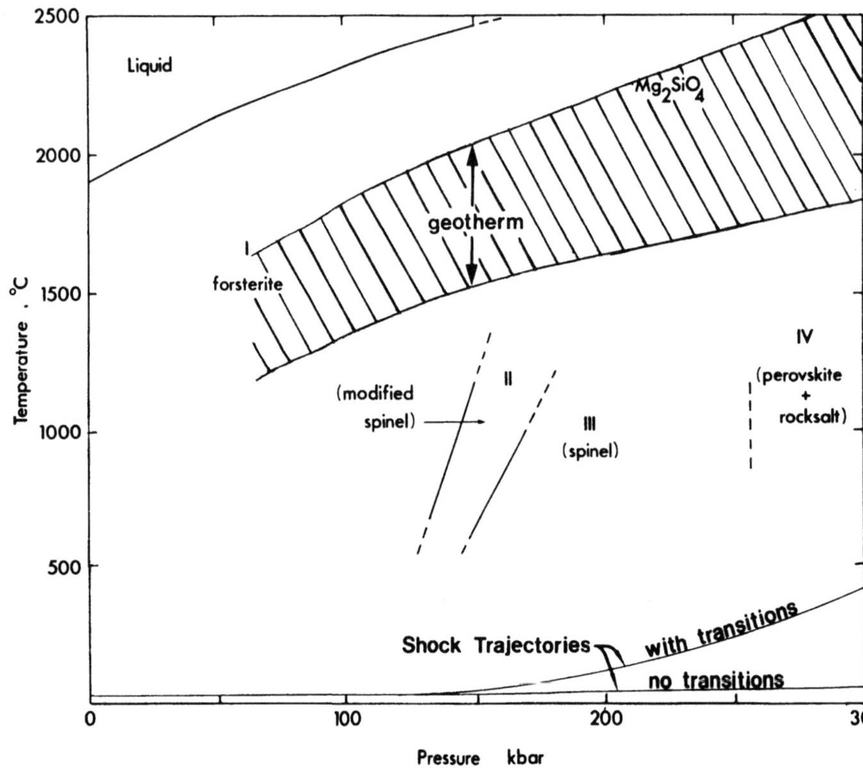

Fig. 1. Pressure vs. temperature phase diagram for forsterite, Mg_2SiO_4. Base diagram adapted from Liu and Bassett (in press) with permission; geotherm range from data in Bolt (1982). Shock trajectories are sketched to illustrate the low temperatures characteristic of shocked forsterite in this pressure range.

The motion of the aluminized surface was monitored by means of a VISAR (Velocity Interferometry System for Any Reflector). Details of this system are given by Barker and Hollenbach (1972). Two important attributes of the VISAR method of monitoring velocity are that the reflecting surface whose velocity is monitored need not be a free surface (it is not in this experiment) and the release trajectory may be observed; these do not apply to such techniques as streak camera recording of the image of a tilt mirror held against the free surface of a sample.

The single-stage powder gun at Sandia National Laboratories' Star Facility was used to launch phenolic projectiles capped with aluminum at nominal velocities of 1.9 km/sec. A flier plate (6061T6 aluminum, 2.54 cm diameter and 1.2 mm thick) and plexiglass backer were cemented into the front of each projectile, the flier plate thickness being chosen to cause a release wave to arrive at the downstream specimen surface approximately 300 nsec after the initial compressional arrival, thus allowing the mapping of the release curves for the specimens.

RESULTS

The immediate results from the four shots are summarized in Fig. 2 as a history of reflector velocity vs. time. LiF has a slightly lower shock impedance than does peridot (an 18.5 GPa wave propagating from peridot into LiF will cause a 3.8 GPa rarefaction wave to be reflected), and the particle velocities in the peridot are approximately 85% of the velocities shown.

It should be noted that the transit time of the initial wave is not accurately measured by this experiment, and must be calculated on the basis of independent sound velocity measurements. In the present experiment, Brillouin measurements were made for this purpose in the laboratory of D.J. Weidner on fragments from the preparation of samples. Measured velocities were 9.74 km/sec (a-axis) and 8.92 km/sec (oblique), and are reflected in Figure 2.

These data were reduced to yield pressure vs. density curves (Fig. 3) by techniques discussed by Grady (1977), and Grady and Young (1976). Shocking along the a-axis yielded an elastic wave of approximate amplitude 6.5 GPa (0.28 km/sec) preceeding a plastic wave of amplitude 18.5 GPa (0.90

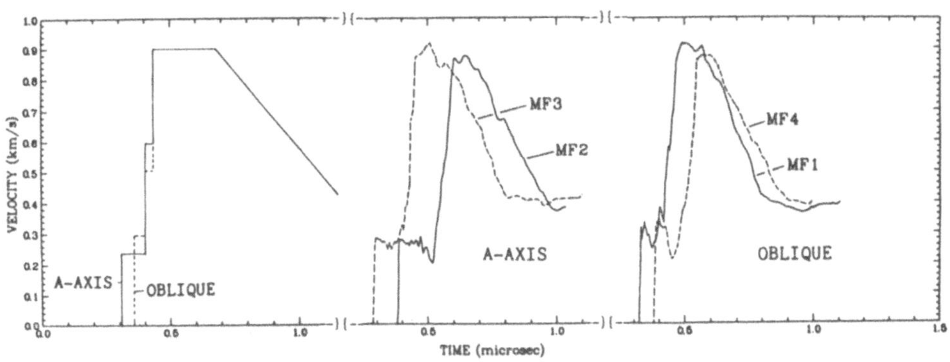

Fig. 2. (Left) Hypothetical U_p trajectory for single-crystal olivine shocked to 18.5 GPa, assuming a middle shock due to an olivine to β- or γ-spinel transition. (Middle) Observed history of the motion of the aluminized surface of the LiF held in contact with peridot shocked along the a-axis (Right) Same for shocking approximately along the bisector of the (100) and (013) directions, so as to promote (100)[013] shear. Relative horizontal magnifications of the curves are due to differing sample thicknesses.

km/sec) by 100-150 nsec. The value of 6.5 GPa agrees well with the HEL of 6.5(5) reported by Syono, et al (1981). No other shocks were observed. Following a short residence at peak pressure, the release wave arrived, manifested as a slow (400 nsec) reduction in reflector velocity to 0.4 km/sec. Shocking obliquely yielded an elastic wave of amplitude 7.8 GPa (0.33 km/sec), followed after 100 nsec by a plastic wave of amplitude 18.5 GPa (0.90 km/sec), with no other shocks observed. The release wave appeared approximately as for the a-axis experiments.

Considerable plastic relaxation appears to have occurred after passage of the elastic precursor for the oblique shock experiments, but in the a-axis experiments this effect was not nearly so pronounced. This "relaxation trough" has been observed for other materials, for example, MgO (Grady, 1977), but of greatest interest here is the orientation dependence.

Small, high frequency excursions occurring during this time may correspond to the collapse of crystalline imperfections, or may indicate heterogeneous shear, and reflect the high resolution of the VISAR technique. The use of synthetic crystals might affect the appearance of such features.

In each case, the elastic wave exhibited a very fast rise (< 10 nsec), suggesting that the epoxy bond was thin enough to have little effect on the observed wave profiles, especially in view of the compression of the bond by the elastic wave. The plastic waves, however, had much longer (70 nsec) rise times, corresponding to a shock thickness of order 70 μm; thus, a longer run distance would probably be necessary to produce a steady-state wave.

If the middle of the rise is taken as the "arrival" time for the plastic waves, their velocities are 6.476 and 6.500 km/sec (oblique shocks; shots MF1 and MF4) and 6.556 and 6.508 km/sec (a-axis shocks; shots MF2 and MF3), or 6.510 (0.0335) km/sec, that is, they are essentially the same. Also, in pressure vs. density space (Fig. 4), all of the curves are identical to within experimental precision above the HEL. Hence we discern no orientation dependence in either the form or velocity of the plastic wave.

The release curves for the four shots lie nearly together (Fig. 3), lying at densities just above those for the α-phase isentrope. On this basis, it seems unlikely that a phase transition occurred in this experiment.

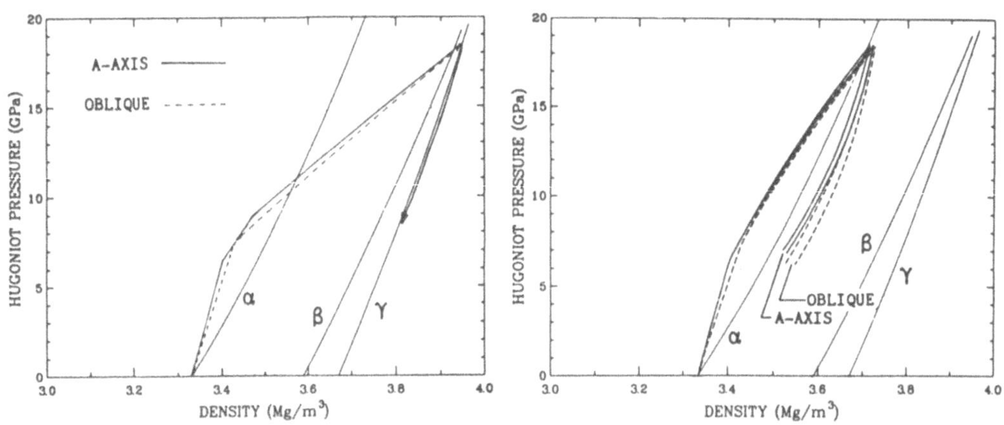

Fig. 3. (Left) Hypothetical trajectories in pressure-density space, assuming a phase transition. (Right) Trajectories calculated from observed velocity histories. The isentropes shown are for $(Mg_{.91}Fe_{.09})_2SiO_4$, and were calculated using isentropic bulk modulus values from Jeanloz and Thompson (1983).

DISCUSSION AND CONCLUSIONS

Under shocking to approximately 18.5 GPa, peridot (Fo_{90}) does not undergo a phase transition. This applies for a shock direction chosen to promote (100)[013] shearing as well as for shocking along the a-axis. The middle wave reported by Watt and Ahrens (1983) may be due to interaction between an elastic wave reflected from the free surface used and the main plastic loading wave. The lack of transition is consistent with evidence in the literature that neither pure forsterite nor rocks of dunite composition (Fo_{92}) undergo a phase transition when shocked to less than 45 GPa. Lyzenga and Ahrens (1980) have estimated the latent heat of the observed phase transition (?) which occurs in shocked forsterite between 45 and 100 GPa as 1.5 MJ/kg. Inasmuch as this energy must come from the shock wave, such a transition will not occur until the energy of the shock wave, $P(V_o-V)/2$, surpasses this quantity, which occurs at about 45 GPa for forsteritic compositions. The corresponding pressure for an olivine-spinel transition of 12.7 GPa (taking a latent heat of 0.5 MJ/kg as suggested by Davies and Gaffney (1973)) is appreciably less than the maximum pressure of 18.5 GPa used in the present experiment, so the absence of a transition in this experiment is probably due to kinetic factors.

Hence, in this material, the shearing introduced by shocking to 18.5 GPa has not effected transition. Furthermore, we find the Hugoniot state to be independent of crystal orientation.

ACKNOWLEDGEMENTS

We are grateful to D.J. Weidner of SUNY (Stony Brook) for providing his time and facilities for the Brillouin velocity measurements. Special thanks are also due W.A. Bassett of Cornell University for encouraging one of us (MDF) to continue pursuing this project, and for providing funding for incidentals and related experiments. This work was supported by the U.S. Dept. of Energy.

REFERENCES

Barker, L.M. and R.E. Hollenbach, J. Appl. Phys., 43, 4669-4675, 1972.
Binns, R.A., R.J. Davis and S.J.B. Reed, Nature, 221, 943-944, 1969.
Bolt, B.A., Inside the Earth, W.A. Freeman and Co., 1982
Davies, G.F. and E.S. Gaffney, Geophys. J.R. astr. Soc., 33, 165-183, 1973.
Furnish, M.D. and W.A. Bassett, J. Geophys. Res., 88, 10,333-10,341,1983.
Grady, D.E., pp. 389-438 in M.H. Manghnani and S-I. Akimoto (eds), High Pressure Research: Applications in Geophysics, Academic Press, 1977.
Grady, D.E. and E.G. Young, Evaluation of constitutive properties from velocity interferometer data, Sandia Nat. Labs report SAND 75-1650, 1976.
Green, H.W., Geophys. Res. Lett., 11, 817-820,1984.
Jeanloz, R. and A.B. Thompson, Rev. Geophys. Space Phys., 21, 51-74, 1983.
Liu, L-G and W.A. Bassett, Elements, Oxides and Silicates: High Pressure Phases. Oxford Univ. Press, in press.
Lyzenga, G.A. and T.J. Ahrens, Geophys. Res. Lett., 7, 141-144, 1980.
Parmentier, E.M., Geophys. Res. Lett., 8,143-146, 1981.
Poirier, J.P., pp. 361-371 in S. Akimoto and M.H. Manghnani (eds.), High-pressure research in geophysics, Adv. Earth Planet. Sci. 12, 1982.
Price, G.D., A. Putnis and S.O. Agrell, Contrib. Mineral. Petrol., 71, 211-218, 1979.
Putnis, A. and G.D. Price, Nature, 280, 217-218, 1979.
Sammis, C.G. and J.L. Dein, J. Geophys. Res., 79, 2961-2965, 1974.
Watt, J.P. and T.J. Ahrens, J. Geophys. Res., 88, 9500-9512, 1983.

SHOCK LOADING AND RELEASE BEHAVIOR OF X-CUT QUARTZ*

L. C. Chhabildas

Thermomechanical and Physical Division
Sandia National Laboratories
Albuquerque, NM

ABSTRACT

Results of release-adiabat measurements are reported for shock-loaded x-cut quartz over the stress range of 15 GPa to 110 GPa. Release paths determined from a shock stress of 15 GPa indicate a return to the original density, while measurements of release adiabats from shock stresses over 50 GPa, suggest release to be occurring primarily from and in the high density phase. Significant hysteretic effects are observed upon release and a retransformation to a low density phase is observed at lower stresses. Measurements of wave speeds in shocked quartz at stresses over 50 GPa indicate a decrease in wave speeds over the stress regime 71 to 108 GPa.

INTRODUCTION

There is considerable interest in the shock loading behavior of quartz since silica or silicate minerals are presumed to be a major constituent of the earth's crust and mantle. At room temperature and pressure, SiO_2 exists primarily in the crystalline (2.65 gm/cm^3) or fused quartz (2.2 gm/cm^3) phase. At high pressures and temperatures, the stable phase is presumed to be stishovite (4.3 gm/cm^3), although under proper thermodynamic conditions a coesite phase (2.9 gm/cm^3) can also exist. Both of these phases have been synthesized using static high pressure and high temperature techniques. Thus, shock loading techniques could provide a valuable tool for addressing the kinetics of the quartz-to-stishovite transition, while the release adiabat measurement techniques could address the reverse transformation kinetics. It is the purpose of this paper to report shock loading and release-adiabat measurements in x-cut quartz over the stress range of 15.0 GPa to 110.0 GPa.

EXPERIMENTAL TECHNIQUE

A thin plate reverberation technique[1,2] was used to measure the release adiabats from shock stress levels of 15 and 27 GPa. In this technique, a thick X-cut quartz sample impacts a thin tantalum witness plate. The multiple wave reflections that occur in the witness plate

*This work performed by Sandia National Laboratories supported by the U. S. Department of Energy under contract #DEAC04-76-DP00789.

are monitored at the free surface with a velocity interferometer, VISAR[3], from which the release states of quartz can be deduced[1,2]. These experiments were performed on a 89-mm powder gun facility.

In a further set of experiments, tantalum impactors (0.8 mm thick) were accelerated to an appropriate velocity on a two-stage light gas gun to obtain stress levels of 50, 71, and 108 GPa in quartz samples that had lithium-fluoride windows bonded to them. The lexan sabot backing to the tantalum impactor allows stress release states to be introduced into the sample after shock loading. Interface velocity measurements were made using a VISAR, for two different sample thicknesses at each stress level. Lagrangian wave analysis techniques were used to estimate stress-particle velocity and stress-volume release paths. These experiments and the corresponding wave profiles measured are described in reference[1].

RESULTS AND DISCUSSION

Release States

The stress vs specific volume release paths for x-cut quartz obtained in this study using the thin plate reverberation technique from stress levels of 15.5 GPa and 27.0 GPa are shown in Figure 1. The experimental Hugoniot for quartz is represented in the figure by circles. Also shown in Figure 1 are the room temperature isotherms for the quartz phase[4] and the stishovite phase[5,6]. The experimental isotherms determined up to approximately 15.0 GPa have been fitted to a Birch-Murnaghan equation of state and then extrapolated to higher stresses as indicated in the figure. A comparison of the quartz and stishovite isotherms with the Hugoniot for quartz suggests an onset of transformation to the high-density phase at a Hugoniot stress of approximately 15 GPa and completion by 40 GPa. The shock and release experiments at 15.5 GPa seem to confirm the lower bound. The release states from 15.5 GPa indicate complete recovery to the original density on unloading. This suggests that only a minute fraction of stishovite may be present, if any, and reverse transformation kinetics are therefore insignificant from this stress level. This is also in agreement with previous shock recovery experiments[7,8], where predominantly crystalline quartz has been recovered after being shocked to approximately 15 GPa.

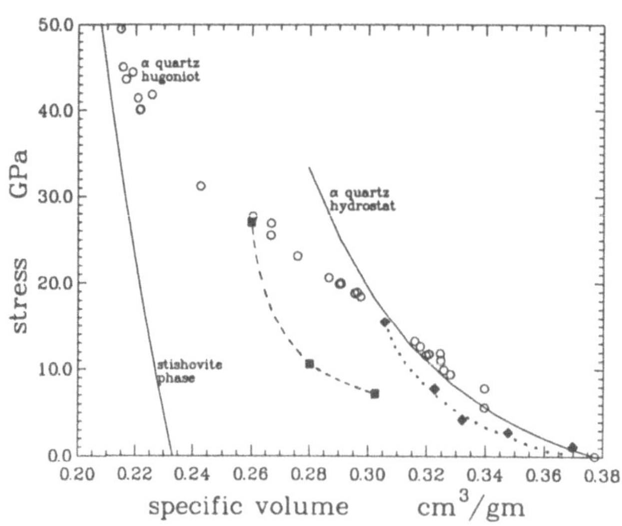

Figure 1: Release states of quartz from shock stresses of 15.5 and 27.0 GPa, respectively. The circles are Hugoniot states from previous studies.

The measured release states for x-cut quartz from 15.5 GPa lie below the Hugoniot and the experimental isotherm; this could be due to the strength of the material in the shocked state. Analysis of the data in the stress-volume plane gives an estimate of approximately 2.5 GPa for the shear stress at a shock stress of 15.5 GPa.

The shock state at 27.0 GPa is apparently in a mixed phase state. Also, by comparing the shocked state with quartz and stishovite isotherms (ignoring temperature changes and material strength), one can estimate that at least 40% of the material has transformed to stishovite. The steep release path could be due to release occurring along a frozen mixed phase and/or due to large shear strength in the shocked state. It is conceivable that the kinetics of phase transformation are intertwined with elastic-plastic effects, and cannot therefore be addressed independently. The release path also suggests reverse transformation of stishovite to a low density state at lower stresses. However, since an entire release path is not determined, the kinetics of the reverse phase transformation at these low stresses cannot be addressed definitively; the two bounding possibilities being that it transforms to a specific volume of ~0.34 gm/cm^3 or to the fused silica specific volume of 0.45 gm/cm^3. The former would indicate a mixture of quartz and stishovite phases upon recovery, whereas the latter suggests transformation to fused quartz. In recovery experiments[7], where crystalline quartz was shocked over stresses of 20.0 GPa, a fraction of fused quartz (densified) has been recovered.

The relief paths in the stress-specific volume plane for x-cut quartz shocked to stresses of 50.0 GPa are shown in Figure 2. It is presumed that upon shock loading to stresses over 50 GPa quartz is fully transformed to the high-density stishovite phase. As indicated in Figure 2, the release paths are steep suggesting release to occur from and in the high-density phase. The stress relief path of shocked quartz from 50.0 GPa indicates release in the stishovite phase to ~10.0 GPa and at that stress, a retransformation toward a low density phase. Likewise, the 71 GPa experiment seems to suggest release in the high-density phase to ~18 GPa before retransformation to the low-density phase. In the 110 GPa experiment, retransformation to the low-density phase seems to initiate at ~30 GPa. Since fused silica has been obtained in shock recovery experiments, the final density of released stishovite in Figure 2 from all shots, should be that of liquid silica. This would strongly suggest therefore that shocked quartz, *i.e.*, stishovite, melts upon unloading.

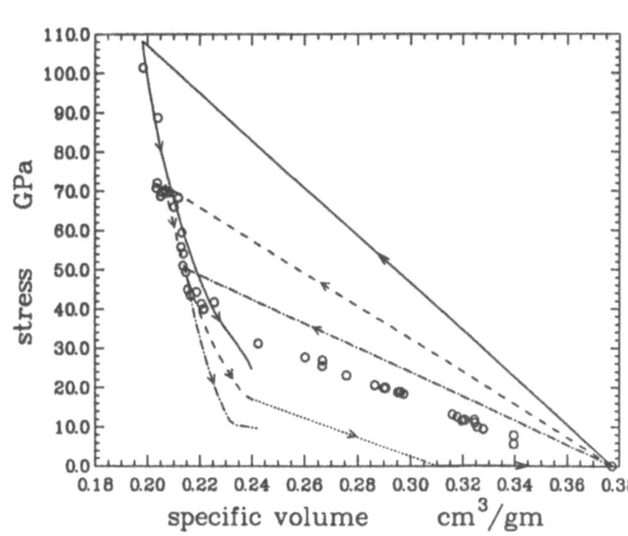

Figure 2: Release states of quartz from shock states of 49 GPa, 71 GPa, and 108 GPa.

The sound speed measurements in shocked quartz obtained from these experiments are plotted in Figure 3. The sound speeds were determined by measuring the round trip transit time of a small amplitude reshock wave which is generated in the shocked quartz sandwiched between the tantalum impactor and the lithium-fluoride window. Also shown in Figure 3 are the results of previous wave speed measurements in quartz[9–11]. As indicated, the wave speed increases with increasing stress up to 71 GPa; the 108 GPa experiment indicates a decrease in sound speed suggesting a shock-induced phase change over the stress regime 71 to 108 GPa. Melting is suspected at these high stresses due to extreme temperatures induced in the material quartz upon shock compression[12,13]. This is also consistent with Hugoniot temperature measurements where melting is thought to occur at stresses over 84 GPa[12,13]. Additional sound speed measurements over the stress regime 71 to 108 GPa would define the shock induced melt boundary more accurately.

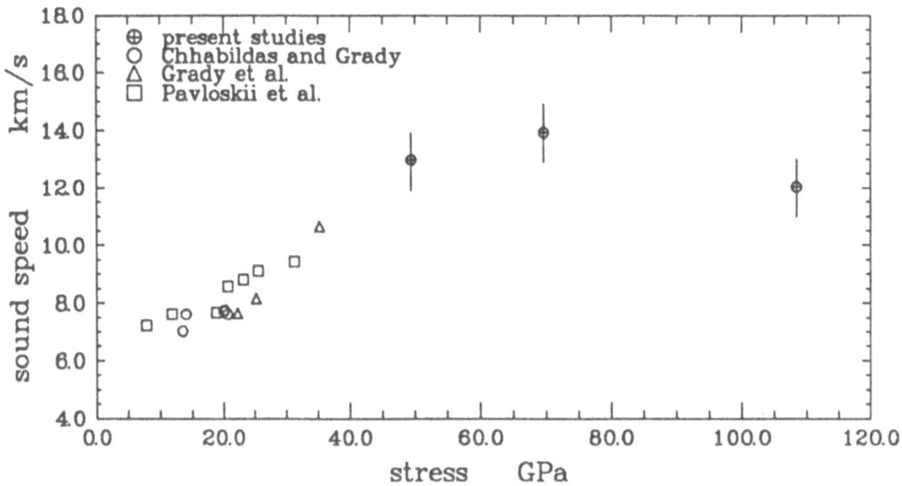

Figure 3: Results of wave speed measurements as a function of stress. A decrease in wave speed is observed over the stress regime 71 to 108 GPa, suggesting a shock-induced phase transformation. Melting is suspected over this stress range.

SUMMARY

Results of release-adiabat measurements have been reported for x-cut quartz shocked initially over the stress range of 15 GPa to 110 GPa. Hugoniot measurements of quartz suggest the transformation of quartz to the high-density phase commences at approximately 15 GPa, with completion at stresses over 40 GPa. Release paths determined from a shock stress of 15 GPa suggest a return to the original quartz density. Measurements of release adiabats from shock stresses over 50 GPa suggest release to be occurring primarily from and in the high density phase. Significant hysteretic effects are observed upon release and a retransformation to a low density phase is observed at lower stresses. The results of these experiments when combined with previous recovery studies suggest that the final state upon release from stresses above 50 GPa might be molten silica liquid, which reverts to fused silica at late times. Measurements of wave speeds in shocked quartz at stresses over 50 GPa indicate melting to occur over the stress regime 71 to 108 GPa. Additional sound speed measurements over this stress regime would define the phase/melt boundary more accurately.

REFERENCES

1. L. C. Chhabildas and J. M. Miller, "Release-Adiabat Measurements in Crystalline Quartz," Sandia National Laboratories Report, SAND85-1092 (1985), unpublished.
2. L. C. Chhabildas and D. E. Grady, "Dynamic Material Response of Quartz at High Strain Rates" in Mat. Res. Soc. Symp. Proc. Vol. 22 (Edited by C. Homan, R. K. MacCrone, E. Whalley, Elsevier Science Publishing, North Holland) p.147 (1984).
3. L. M. Barker and R. E. Hollenbach, "Laser Interferometer for Measuring High Velocities of any Reflecting Surface," J. Appl. Phys., 43, 4669 (1972).
4. D. B. McWhan, "Linear Compression of Quartz to 150 Kbar," J. Appl. Phys. 38, 347 (1967).
5. L. G. Liu, W. A. Bassett and T. Takahashi, "Effect of Pressure on the Lattice Parameters of Stishovite," J. Geophys. Res. 77, 1160 (1972).
6. H. Mizutani, Y. Hamano, and S. I. Akimoto, "Elastic-Wave Velocities of Polycrystalline Stishovite," J. Geophys. Res. 77, 3744 (1972).
7. A. V. Anan'in, O. N. Breusov, A. N. Dremin, S. V. Pershin, and V. F. Tatsii, "The Effect of Shock Waves on Silicon Dioxide 1. Quartz," Combustion, Explosion and Shock Waves, 10, 372 (1974).
8. J. Wackerle, "Shock-Wave Compression of Quartz," J. Appl. Phys., 33, 922 (1962).
9. M. N. Pavloskii, "Measurements of the Velocity of Sound in Shock Compressed Quartzite, Dolomite, Anhydrite, Sodium Chloride, Paraffin, Plexiglass, Polyethylene, and Fluoroplast-4," J. Appl. Mech. and Tech. Phys., Polyethylene, 5, 136 (1976).
10. D. E. Grady, W. J. Murri, and P. S. Decarli, "Hugoniot Sound Velocities and Phase Transformations In Two Silicates," J. Geophys. Res., 80, 4857 (1975).
11. L. C. Chhabildas and D. E. Grady, "Shock Loading Behavior of Fused Quartz, in Shock Waves in Condensed Matter," (Edited by J. R. Asay, R. A. Graham, and G. K. Straub, Elsevier Science Publishers), 176 (1983).
12. G. A. Lyzenga, T. J. Ahrens, and A. C. Mitchell, "Shock Temperatures of SiO_2 and their Geophysical Implications," J. Geophys. Res., 88, 2431 (1983).
13. R. G. McQueen, J. N. Fritz and J. W. Hopson, EOS 60, 952 (1979).

THERMAL HISTORY OF SHOCK-COMPRESSED SOLIDS

Bob Svendsen and Thomas J. Ahrens

Seismological Laboratory
California Institute of Technology
Pasadena, CA 91125

ABSTRACT

Taking the view that observed, shock-induced optical radiation from transparent solids is dependent on defect-controlled (*i.e.*, heterogeneous) properties and processes, we examine an isotropic, heterogeneous, viscous, thermoelastic model of the uniaxially shock-compressed state to determine conditions under which this radiation might be dominantly thermal or nonthermal. Assuming shock compression establishes some distribution of material properties in the initially-compressed state, perturbations in the stress field relax heterogeneously to the hugoniot state by elastic deformation for timescales short and viscous deformation for timescales long relative to the local Maxwell time. Regions of locally high temperatures producing thermal radiation then develop only where the local viscosity is low and Maxwell time short. Alternatively, regions of low elastic moduli and long Maxwell time experience sustained elastic deformation which may result in microfracture and triboluminescence. Consequently, a shock-induced defect structure dominated by long Maxwell times may not develop thermal localization. This may be the case for MgO, which produces no resolvable ($\gtrsim$ 2000K) thermal radiation below 60 GPa. Conversely, fused SiO_2 develops locally high temperatures ($\sim$ 3000K) above 10 GPa (below the stishovite transition), implying that it may develop low viscosity, short Maxwell time regions that persist in the hugoniot state.

INTRODUCTION

Traditionally, solids of geophysical interest (which Grady[1] has termed "brittle": alkali halides, carbonates, oxides, silicates, *etc.*) and others recovered from shock-compression-release experiments (*e.g.*, see Stöffler[2]) display localized, inelastic deformation (*e.g.*, planar) features representing various manifestations of mechanical localization. As well, shock-induced radiation from these materials (*e.g.*, Kormer[3], Kondo and Ahrens[4], Schmitt and Ahrens[5], Brannon *et al.*[6]) clearly demonstrates the heterogeneous, nonequilibrium nature of shock-induced deformation. These and other observations suggest that, in the simplest terms, the shock wave energy is partitioned principally between elastic deformation of the lattice and inelastic deformation of the defect structure in these materials in the low pressure range ($\lesssim$75 GPa for the above materials). In this context, if a significant fraction of the shock wave energy is partitioned into inelastic (dissipative) deformation of the defect structure, mechanical dissipation transforms this into thermal energy, resulting in localized, "anomalously high" temperatures (thermal localization). However, if most of the shock energy is partitioned into elastic (nondissipative) deformation of the lattice, inelastic deformation, dissipation, localized heating and thermal radiation are minimal. In order to gain some basic physical insight into these issues and for analytic tractability, we examine these

MODEL FORMULATION

We take the initial state of compression in the material as the reference configuration of our model and develop a Lagrangian description of processes which relax the initially-compressed material toward the near-equilibrium (hugoniot) state, assuming that any deformation produced by these processes is infinitesimal. We assume shocked brittle solids deform thermoelastically on short timescales (*i.e.*, shock wave risetime $\sim 10^{-9}$–10^{-8} sec), and inelastically (*i.e.*, viscously or plastically) on longer timescales ($\sim 10^{-8}$–10^{-7} sec), both less than that of experiments ($\sim 10^{-7}$–10^{-6} sec). We also assume that the material properties are isotropic and at most functions of position in the deformed material in order to simulate the effects of a defect structure on the relaxation of a heterogeneous, shock-induced stress field in the simplest fashion. These assumptions are embodied in the following 4 relations:

$$\dot{e}_{ij}^{(el)} = \frac{1}{2\mu}\dot{\sigma}_{ij}' - \frac{1}{3K}\delta_{ij}\dot{p} + \alpha\delta_{ij}\dot{T} \tag{1}$$

$$\dot{e}_{ij}^{(in)} = \frac{1}{2\eta_\mu}\sigma_{ij}' - \frac{1}{3\eta_K}\delta_{ij}p \tag{2}$$

$$\dot{e}_{ij} = \dot{e}_{ij}^{(el)} + \dot{e}_{ij}^{(in)} \tag{3}$$

$$\rho c_\sigma\dot{T} - [kT_{,i}]_{,i} = \sigma_{ij}\dot{e}_{ij}^{(in)} + 3\alpha T\dot{p} + \rho r \tag{4}$$

where the superposed dot stands for $\partial/\partial\tau$, τ is time relative to the initial state, and spatial derivatives are with respect to the reference coordinates x_i of the initial state. The general balance of linear momentum, *i.e.*

$$\rho\dot{v}_i = \sigma_{ij,j} + \rho f_i \tag{5}$$

closes the system of relations (1)-(4). In relations (1)-(5), v_i and f_i are the Cartesian velocity and body force vector components, $\dot{e}_{ij}^{(in)}$, $\dot{e}_{ij}^{(el)}$, σ_{ij}, σ_{ij}' and p are the inelastic and elastic deformation rate, Cauchy stress and deviatoric stress tensor components, and mechanical pressure $p \equiv -\sigma_{kk}/3$, while T and r are the absolute temperature and energy source density, respectively. Additionally, ρ, K, μ, η_K, η_μ, c_σ, k and α are the density, isentropic bulk and

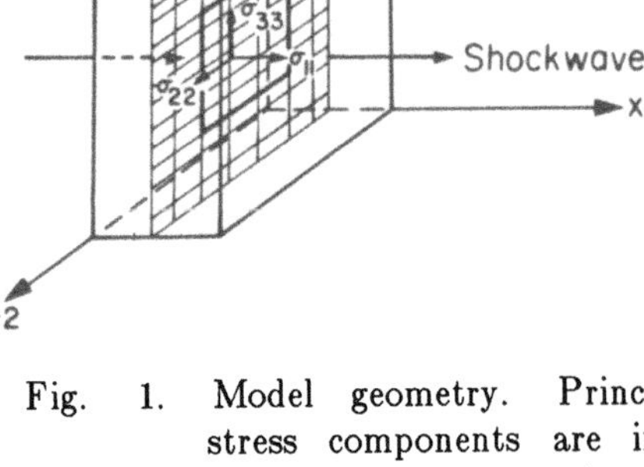

Fig. 1. Model geometry. Principal stress components are indicated on an x_2–x_3 plane in sample interior.

shear elastic moduli, bulk and shear viscosities, specific heat at constant stress, thermal plus radiative conductivity and coefficient of thermal expansion, respectively, of the initial and hugoniot states. In this case, the magnitude and distribution of material properties in the initial state influence the evolution of the σ_{ij} and T fields into the hugoniot state.

Shock compression experiments are generally designed to produce a state of uniaxial plane deformation in target materials. Assuming this and also that the initial state evolves toward a hugoniot state with the same uniform pressure and inelastic bulk deformation rate simplifies the system (1)-(5) to three relations, given by

$$\partial_\tau \dot{e}_\mu = \frac{4}{3}\,\sigma_{,11} \tag{6}$$

$$\dot{e}_\mu = \frac{1}{2}\,[\,\pi\,\partial_\tau\sigma + \mathrm{Re}\,\sigma\,] \tag{7}$$

$$\partial_\tau T - Pe^{-1}\,T_{,ii} = \frac{8}{3}Ec\sigma\dot{e}_\mu^{(in)} - 3Ecp\dot{e}_K^{(in)} + r \tag{8}$$

which we have rendered nondimensional (ND). In these relations, $\dot{e}_\mu$, $\dot{e}_K$ and σ are the effective shear and bulk deformation rate, and shear stress, respectively. Note that the second term on the right-hand side of (8) is constant by our assumption of constant pressure and inelastic deformation rate. The important parameters of our model are (1) the Reynolds number, $\mathrm{Re}\equiv\sigma_R\tau_R/\eta_\mu$, where σ_R is the reference shear stress magnitude and τ_R is the reference timescale, (2) a ND timescale $\omega\equiv\tau_R/\tau_M$, where $\tau_M\equiv\eta_\mu/\mu$ is the Maxwell time, (3) a ND stress magnitude $\pi\equiv\sigma_R/\mu$, (4) the Eckert number, $\mathrm{Ec}\equiv\sigma_R/\rho_R c_\sigma(T_m-T_R)$, where T_m is the melting temperature and T_R the reference temperature, and (5) the Peclet number, $\mathrm{Pe}\equiv\kappa_R\tau_R/l_R^2$, where $\kappa_R\equiv k_R/\rho_R c_\sigma$ is the thermal plus radiative diffusivity and l_R is the reference lengthscale. Then the ND inelastic shear deformation rate is $\dot{e}^{(in)}=\mathrm{Re}\sigma$, and the elastic shear deformation rate is $\dot{e}^{(el)}=\pi\dot{\sigma}$. We note that σ is proportional to $e^{-\omega}$, while the ND dissipation rate, Γ, is proportional to $\mathrm{Ec}\sigma\dot{e}^{(in)}=\mathrm{EcRe}\sigma^2$. Finally, T is scaled by T_R

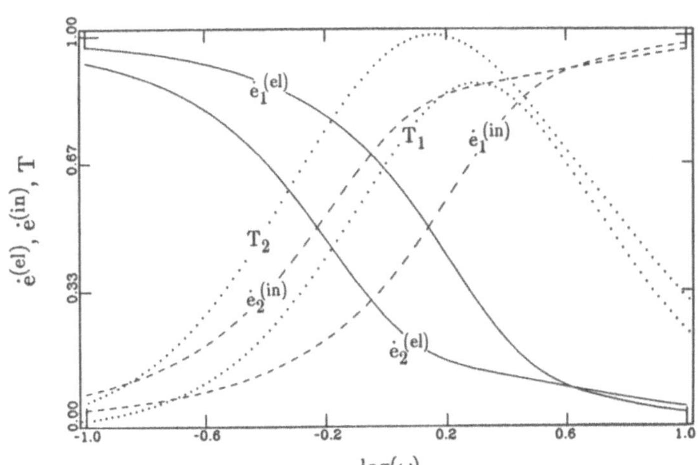

Fig. 2. Nondimensional fields at 2 positions along the direction of shock propagation as a function of log(ω). Both deformation rates are scaled by $\dot{e}$, and T is scaled by T_m. Note that the magnitude of $\dot{e}^{(el)}$ is higher and $\dot{e}^{(in)}$ (and so T) is lower for the position (denoted by subscript 1) nearer the shock front. In this case, Re (=1) is constant, so that ω varies directly with μ. See text for details.

and T_m. Characteristic values of these parameters for brittle materials in the low pressure regime are: $\omega \sim 0.1$–10, $[Re,\pi,Ec] \sim 1$–10 and $Pe \sim 10^5$–10^7. For the following calculations, we'll set Ec=10 and $Pe=10^5$, thereby maximizing the possibility for dissipation and conduction to affect the temperature field. However, since conduction is scaled by Pe^{-1}, it should not significantly relax the temperature field on the timescales of interest (10^{-7}–10^{-8} sec).

We idealize the shock front, a nonmaterial boundary of our model, by assuming that it is a propagating, impulsive plane (x_2–x_3 plane: Fig. 1) source of shear stress σ which creates some defect (material properties) structure and a perturbed shear stress field in the initial state. Using this, and assuming that σ and T relax to zero and T_R, respectively, in the hugoniot state and at the bondaries of our model (Fig. 1), we may solve (6)-(8) analytically for σ and T as a function of x_i and τ.

RESULTS AND DISCUSSION

Fig. 2 displays $\dot{e}^{(el)}$, $\dot{e}^{(in)}$ and T as a function of ω for two positions along the direction of shock propagation (x_1) at constant Re. Since Re is constant, Γ varies only with $\sigma(\omega)$ and $\dot{e}^{(in)}(\omega)$. Also, ω varies with μ and π varies with ω^{-1}. Consequently, for $\tau_R \ll \tau_M$, $\dot{e}^{(el)}$ dominates $\dot{e}^{(in)}$, and T deviations above T_R (thermostatic T) are low. As $\omega(\mu)$ increases, however, $\dot{e}^{(el)}$ decreases and $\dot{e}^{(in)}$ eventually dominates, resulting in $T \rightarrow T_m$ for $\tau_R \gtrsim \tau_M$. Further increase of ω, however, leads to decreasing T's because Γ is proportional to σ^2 and $\dot{e}^{(in)}$ is proportional to σ, which decays rapidly with increasing ω.

Note that T's farther away from the shock front are higher. This occurs primarily because we assume constant radiative and thermal conductivities equal to their respective STP values; consequently, the corresponding diffusivities decrease with pressure and these processes are then even more inefficient in the shock-compressed state of brittle materials than at STP.

Fig. 3 is the same as Fig. 2 except that π is constant so that ω varies with η_μ^{-1} and Re varies with ω. Then Γ varies directly and indirectly with ω through $\dot{e}^{(in)}$ (*i.e.*, Re) and σ, respectively. For small ω, we see again that $\dot{e}^{(el)}$ dominates. Since Re scales with ω, $\dot{e}^{(in)}$ and T increase with ω as $\dot{e}^{(el)}$ declines. Finally, as ω achieves large values, the decay of σ dominates the increase of Re and T's again fall.

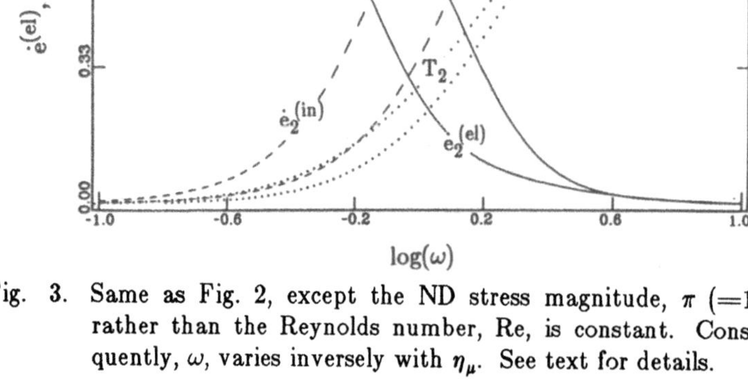

Fig. 3. Same as Fig. 2, except the ND stress magnitude, π (=1), rather than the Reynolds number, Re, is constant. Consequently, ω, varies inversely with η_μ. See text for details.

In general, then, large $\dot{e}^{(el)}$ and low T's are expected for $\omega \lesssim 1$ and large $\dot{e}^{(in)}$ and high T's for $1 \lesssim \omega < 10$. On this basis, we speculate that brittle, heterogeneous materials with average $\omega << 1$ deform locally by microfracture and radiate nonthermally, while those with average $\omega \sim 1$ deform by inelastic processes and radiate thermally. Note that the reference timescale is governed by the length of time we may "observe" the evolution of these processes and their effects behind the shock front, less than or equal to the experimental timescale. Radiation is the observed manifestation of these processes, and so, in this case, the reference timescale is proportional to the optical depth of the shocked material divided by the shock wave velocity. Consequently, transparent and opaque shock states will have relatively long and short τ_R, respectively.

An example of a material with average $\omega < 1$ may be MgO, which may retain some strength (Grady[1]) in the hugoniot state and shows no evidence of thermal radiation up to $\sim$ 60 GPa (*e.g.*, Schmitt and Ahrens[5]). In contrast, materials as diverse as fused SiO_2 and LiF (which does not undergo any known phase change in the low pressure regime) may have average $\omega \gtrsim 1$, developing localized, resolvably high (relative to thermostatic) temperatures ($\sim$ 3000K and 4000K, respectively) at pressures near 10 and 40 GPa, respectively (Schmitt *et. al.*[7]). In addition, when we add the potential of a kinetically-controlled phase change to the constitutive assumptions stated above, only regions of high dissipation rate nucleate and grow new phase(s). On this basis, we speculate that only those materials localizing significant thermal energy during shock compression have the potential to transform to high pressure phases on the timescales of shock experiments in the low pressure regime.

Acknowledgements

Supported under NSF grant EAR-8313746 and NASA grant NGL-05-002-105. Contribution #4240, Division of Geological and Planetary Sciences.

REFERENCES

1. D. E. Grady, Shock compression of brittle solids, J. Geophy. Res. 85:913 (1980).
2. D. Stöffler, Deformation and transformation of rock-forming minerals by natural and experimental shock processes, Fortschr. Miner. 49:50 (1972).
3. S. B. Kormer, Optical study of the characteristics of shock-compressed condensed dielectrics, Ups. Fiz. Nauk 94:641 (1968).
4. K. E. Kondo and T. J. Ahrens, Heterogeneous shock-induced thermal radiation in minerals, Phys. Chem. Minerals 9:173 (1983).
5. D. R. Schmitt and T. J. Ahrens, Temperatures of shock-induced shear instabilities and their relationship to fusion curves, Geophys. Res. Lett. 10:1077 (1983).
6. P. J. Brannon, C. H. Conrad, R. W. Morris, E. D. Jones and J. R. Asay, Studies of the spectral and spatial characteristics of shock-induced luminescence from x-cut quartz, J. Appl. Phys. 54:6374 (1983).
7. D. R. Schmitt, B. Svendsen and T. J. Ahrens, Shock-induced radiation from minerals, this volume.

SEISMIC VERIFICATION OF UNDERGROUND EXPLOSIONS*

Lewis A. Glenn

Lawrence Livermore National Laboratory
Physics Department
Livermore, CA 94550

INTRODUCTION

Bans or limitations on nuclear weapons tests are one of the many types of arms control measures that have been attempted over the years. The first nuclear test agreement, the test moratorium, was made in 1958 and lasted until the Soviet Union unilaterally resumed testing in the atmosphere in 1961. It was followed by the Limited Test Ban Treaty of 1963, which prohibited nuclear tests in the atmosphere, in outer space, and underwater. In 1974 the Threshold Test Ban Treaty (TTBT) was signed. This treaty limited underground tests after March 1976 to a maximum yield of 250 kilotons (kt). The TTBT was followed by a treaty limiting peaceful nuclear explosions and, although neither of these was ever ratified by the U.S. Senate, both the United States and the Soviet Union claim to be abiding by the 150-kt yield limit. A comprehensive test ban treaty (CTBT), prohibiting all testing of nuclear weapons, has also been discussed and there is widespread public interest in this idea. However, a verifiable CTBT is a contradiction in terms. No monitoring technology can offer absolute assurance that very-low-yield illicit explosions have not occurred.

VERIFICATION PROCESS

The principal tools for monitoring compliance with a CTBT are seismic networks and surveillance satellites. On-site inspections might also be required to resolve ambiguous events. Satellites are actually of limited use since it is possible to carry out low-yield explosions in buried cavities without resulting in any visible ground-surface motion. Moreover, test activities, including the required excavations, can be associated with mining or other large-scale industrial undertakngs. The critical element of the monitoring system is thus the network of seismic stations, and in particular the in-country stations.

Internal stations provide much more useful data than do stations situated outside the borders of testing nations. The external stations are generally so far from potential evasion sites that they record the motion of only a few waves from the larger events (those with seismic magnitudes, m_b, greater than about 4 on the Richter scale) and high

*Work performed under auspices of USDOE by LLNL under #W-7405-Eng-48.

frequencies have been significantly reduced by attenuation and scattering throughout the earth. The larger amplitudes recorded by the internal stations, for the same event, result in an increase in detection capability. The multiple waves and higher frequencies improve the ability to distinguish between explosions and earthquakes. These two functions, detection and identification, are the main elements in the verification process.

The detection capability of a seismic network is specified by the seismic magnitude of the smallest source that can be detected within its assigned territory, with a specified degree of confidence. Theoretical estimates of detection capabilities as low as $m_b = 2.7$ have been made for the Soviet Union, at the 90% confidence level, assuming a very extensive network of internal stations[1]; even lower estimates result if lower confidence limits are acceptable.[2]

Given that waves from an event have been detected, the source must be identified as a nuclear explosion, a chemical explosion, or an earthquake. Many discriminants have been proposed. First, the located depth of the source is useful since current drilling limits are less than about 12 km. Therefore, explosions are not likely to be responsible for signals originating at greater depths.

Second, the areas in which large magnitude earthquakes occur are fairly well known. A strong seismic signal originating at a shallow depth in say Kansas is much more likely to be the result of an explosion than an earthquake; the same would not necessarily be true for California. However, areas that are seismically inactive at high magnitude levels may not be so at the low levels demanded for CTB verification. For example, a recent study of parts of Louisiana, Oklahoma, and Texas which had shown only one earthquake of magnitude 4.5 or greater in 8 years, revealed that almost one earthquake per day occurred in the magnitude range 2.0 to 3.9.[3] Similar results can be expected in many granitic regions so that the areal discriminant is probably of limited use for small events.

For those large events that are not eliminated by depth or location, one of the most useful discriminants is based on the ratio of surface-wave to body-wave magnitudes ($M_s : m_b$). If an explosion and an earthquake have the same body-wave magnitude, the surface-wave magnitude for the earthquake is generally larger.[4,5] It has yet to be proven that $M_s : m_b$ is useful at low magnitudes, especially when explosions are set off in long tunnels or odd-shaped cavities (see below).

A number of other promising regional discriminants have been suggested, based on differences in the spectral content and radiation pattern of the recorded signal.[6]

EVASION OPPORTUNITIES

Detection of an explosion can be avoided by insuring that the signal generated is below the background noise level, by emitting "normal" signals at times of high noise, or both. If the noise only partially masks the signal, detection may occur but identification may be impaired. Reducing the signal strength at the receiving station may be accomplished either by reducing the coupling of the explosive energy into the earth at the source or by choosing (through site selection) a path from the source to the seismic station through a region that absorbs significant seismic energy.

To understand the significance of decoupling it is necessary to interpret the detection capability of a network in terms of explosive yield, as illustrated in Figure 1. The seismic body-wave magnitude is plotted against the yield in kilotons. The data are representative of U.S. experience at the Nevada Test Site (NTS).[7] Various investigators have presented evidence that a given explosion in the Soviet Union registers several tenths of a magnitude unit higher than the same explosion at the NTS[8,9] (this in fact was a principal factor in the recent controversy over alleged violations of the Threshold Test Ban Treaty).

The upper band in Figure 1 shows the seismic magnitude that would be generated by explosions that are tamped, or well-coupled to hard rock. The middle line drawn in this band is based on the "granite" representation in ref. 7; the lower (ref. 10) and higher (ref. 1) lines derive from independent analyses made for hard rock. The widening spread at low yield is due to the facts that there are very few hard-rock data below 1 kt, that these are characterized by considerable variability, and that there are problems in obtaining a consistent definition of magnitude as well. Under these conditions, if the lower limit of detectability corresponds to $m_b \approx 2.7$, the maximum explosive yield would be approximately 0.1 kt.

The middle band in Figure 1 shows the effect of detonating the explosion in the deep, dry alluvial deposits at the NTS. For the same body-magnitude, the explosive yield would be about 10 times greater in alluvium than in hard rock. The reason is that the alluvium is porous and a large fraction of the source-energy is dissipated by pore collapse.

The lower band in Figure 1 corresponds to explosions in large cavities and represents by far the most important potential for evasion.

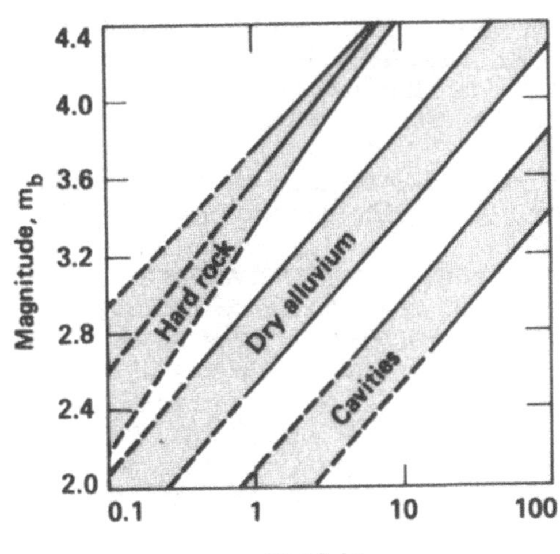

Fig. 1. Seismic body-wave amplitude vs. explosive yield for different explosion environments. The relationships are based on U.S. data from the Nevada Test Site.

CAVITY DECOUPLING

The idea of using a large cavity to muffle the seismic signal from an underground explosion was first proposed 25 years ago by Albert Latter and Hans Bethe at an early conference on the discontinuance of nuclear tests.[11] They calculated that the strength of the distant seismic signal in an elastic medium is determined primarily by the volume of the cavity created by the explosion. If the explosion is set off in a large cavity, then it is the differential increase that determines the signal amplitude. Moreover, if the cavity is just large enough so that its walls suffer only elastic deformation, the distant signal, for a given yield, is minimized and it does not pay to make the cavity any bigger. Inelastic behavior causes increased coupling because the medium is unable to support shear and can thereby flow and thus undergo large displacements.

Experimental confirmation of the cavity-decoupling theory was first obtained in a series of experiments collectively called COWBOY, that was conducted with chemical explosives in a Louisiana salt mine in 1960.[12] Salt was chosen as the test medium because of its homogeneity in situ and because of the ease in which a cavity can be constructed and maintained. The ratio of the seismic amplitude in the tamped explosion to that obtained when the same yield is generated in a excavated cavity is called the decoupling factor; a decoupling factor of 100 was measured in COWBOY.

Confirmation of cavity decoupling in a nuclear experiment was made in the STERLING event (0.38-kt yield)[13], which was fired in the Tatum salt dome (near Hattiesburg, Mississippi) on December 3, 1966, more than two years after the SALMON event (5.3-kt yield)[14] created the cavity. A maximum decoupling factor 70 ± 20 was determined for STERLING, indicating that full decoupling may not have been achieved. It was known prior to STERLING that the SALMON cavity (17.4-m radius) was too small to fully decouple the explosion; this would have required roughly 60% more volume. Moreover, although creating the cavity with an earlier explosion was cheaper than conventional mining techniques, the salt structure around the cavity was thereby weakened (the theory predicts the decoupling factor to be proportional to the rigidity modulus of the surrounding medium).

The significance of these results can be seen when they are combined with the lower limit of seismic detectability. As illustrated in Figure 1, an m_b measurement of 2.7 now corresponds to a 5-kt explosion (instead of the tamped, hard-rock value of 0.1 kt), even if the decoupling factor is only 50. Decoupling factors several times higher are predicted if the cavity can be excavated in a stiffer medium, such as granite.[15]

Notwithstanding the experimental evidence, recent studies have questioned the utility of cavity decoupling.[16,17] Figure 2 illustrates the problem. The spectral variation of the seismic amplitude is plotted, for a fixed yield, for both tamped and cavity-decoupled explosions. In either case it is observed that only modest changes occur as the angular frequency is increased to the neighborhood of the eigenfrequency, $\omega \approx c/R$, where c is the wave speed and R the elastic radius, after which severe damping occurs. For the cavity-decoupled explosion, this "corner" frequency is based on the cavity radius, R_D. For the tamped explosion, the radius at which the shock wave attenuates to an elastic wave, R_T, is much larger. A heuristic explanation for this derives from the fact that the radial momentum increase in a strong explosion is proportional to the square-root of the mass engulfed by the blast wave, which is evidently

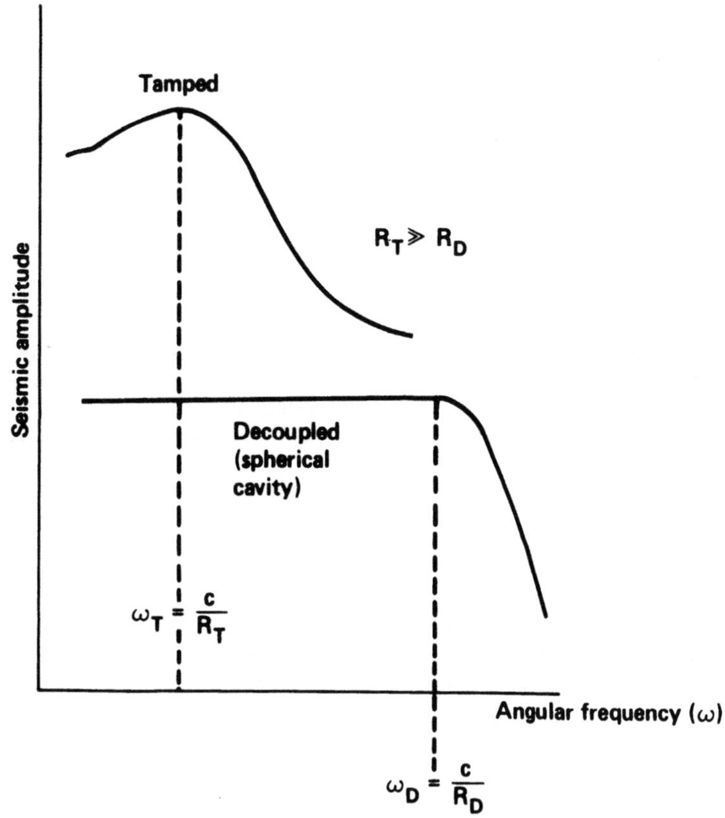

Fig. 2. Spectral variation of seismic amplitude for tamped and cavity-decoupled explosions. Both abscissa and ordinate scales are logarithmic. The ratio of the tamped to the decouple amplitude is called the decoupling factor and theoretically at frequencies above $_T$.

many times larger when the cavity is initially filled with rock than with air. Thus, even though the amplitude of the tamped explosion is 50 or more times higher for $\omega < \omega_T = c/R_T$, it may be much less in the region $\omega_T < \omega < \omega_D = c/R_D$. The decoupling factor can therefore be expected to decrease substantially in this frequency interval. The effect is of no signficance for teleseismic monitoring, since frequencies greater than ω_T are effectively attenuated by the earth. There is some evidence, however, that efficient high-frequency transmission can occur at regional distances in certain geographical regions of importance. Also, earth noise decreases with increasing frequency, aiding in the signal analysis.

One remedy would be to increase the size of the decoupled cavity. However, to fully decouple a 5-kt explosion requires a salt cavity with a radius of 49 m[18], so that it is infeasible, or at least impractical, to excavate a sphere of the required size. An alternate method of increasing the characteristic size of the cavity is to connect long tunnels to the explosion chamber. These could easily extend beyond distances corresponding to R_T. Recent calculations have shown that, by exploiting constructive interference effects, elongated cavities may serve as effective low-pass filters for seismic radiation in directions roughly parallel to the long axis.[19,20] For cylindrical tunnels, for example, the corner frequence varies inversely with length or, relative to a sphere with the same volume, with the 2/3-power of the aspect ratio.

Construction of such facilities would be easier and cheaper than spherical cavities of the same volume. Hard-rock excavations of the required size would nevertheless be enormous and costly undertakings, difficult to conceal. Because such formations are often located in regions of very active seismicity, however, cavities excavated there would seriously complicate the already difficult problem of discriminating between low-yield explosions and earthquakes.

The excavation of large underground cavities may not be hard to disguise from satellite observation if the activity is associated with mining or other large-scale industrial undertakings. In salt, the least expensive method is solution mining. A 1970 study indicated that a 50-m radius cavity could be solution-mined in the Tatum salt dome at a cost under $10M.[21] By 1977 the cost was put at $20M, but it was determined that a single cavity might be useful for up to 25 tests, performed over a period of two years.[22] Reuse of a cavity would probably be an important consideration since realistic present-day costs are undoubtedly still higher. With salt, however, there is always the SALMON option, i.e., to create a primary cavity with a tamped explosion, and then to enlarge it in steps by exploding in it larger and larger devices. It is possible that this technique may already have been employed by the Soviets in their Peaceful Nuclear Explosion (PNE) program, in which numerous nuclear explosions have been conducted for the announced purpose of storing the natural gas condensates from the Orenburg and Astrakhan fields.[23]

REFERENCES

1. W. J. Hannon, Jr., Science, 227, 251-257 (1985).
2 L. R. Sykes and J. F. Evernden, Scientific American, 247, 47-55 (1982).
3. D. Racine and P. Kiouda, "Seismicity of the Salt Areas of Texas, Louisiana, Oklahoma, and Kansas," Teledyne Geotech Rept. AL-79-3 (1979).
4. O. Dahlman and H. Israelson, Monitoring Underground Nuclear Explosions (Elsevier Scientific Publishing, New York, 1977).
5. C. E. Husebye and S. Mykkelveit, "Identification of Seismic Sources - Earthquake or Underground Explosion, NATO Advanced Study Institute Series," 74, Series C (D. Reidel Publishing Co., Boston, 1981).
6. P. Pomeroy, W. Best, and T. McEvilly, T. Bull. Seism. Soc. Amer., 72 (6B), S89-S130 (1982).
7. B. A. Bolt, Nuclear Explosions and Earthquakes (W. H. Freeman and Co., San Francisco, 1976, p. 167).
8. J. F. Evernden, Bull. Seism. Soc. Amer., 66, 549-592 (1976).
9. P. Marshall, D. Springer and H. Rodean, Geophys. J., 57, 609-638 (1979).
10. J. F. Evernden, J. Geophys. Res., 75, 1028-1032 (1970).
11. A. L. Latter and H. Bethe, in Verbatim Record of the 7th Meeting of Technical Working Group 2, Conference of the Discontinuance of Nuclear Weapons Tests, held in the Palais des Nations, Geneva, Switzerland, December 2, 1959 GEN/DNT/TWG.2/PV.7, January 15 (1960).

12. R. F. Herbst, G. C. Werth, and D. L. Springer, J. Geophys. Res., 66 959-978 (1961).
13. D. Springer, M. Denny, J. Healy, and W. J. Mickey, J. Geophys. Res., 73, 5995-6011 (1968).
14. D. Rawson, R. Taylor, and D. Springer, Naturwissenschaften, 54, 525-531 (1967).
15. A. L. Latter, R. E. LeLevier, E. A. Martinelli, and W. G. J. MacMillan, J. Geophys. Res., 66, 943-946 (1961).
16. R. R. Blandford, Bull. Seism. Soc. Amer., 72, S60-S87 (1982).
17. D. R. Larson, "Spherical Wave Propagation in Elastic Media and its Application to Energy Coupling for Tamped and Decoupled Explosions," Lawrence Livermore National Laboratory Rept. UCRL-52655 (1979).
18. L. A. Glenn, "Cavity Decoupling and the Evasion of a Comprehensive Test Ban," Lawrence Livermore National Laboratory Rept. UCRL-90321 (1984).
19. L. A. Glenn, A. J. C. Ladd, B. Moran, and K. A. Wilson, Geophys. J. R. astr. Soc., 81, 231-242 (1985).
20. L. A. Glenn, B. Moran, A. J. C. Ladd, K. A. Wilson and J. A. Rial, submitted for publication in Geophys. J. R. astr. Soc., Lawrence Livermore National Laboratory Rept. UCRL-92337 (1985).
21. Project Payette Final Summary Report, prepared by Fenix & Scisson, Inc., Tulsa, Oklahoma, March (1970).
22. L.A. Grove, H. W. Hubbard, and E. A. Martinelli, "Feasibility of a Decoupled Test Facility in a Salt Dome," R&D Associates, P.O. Box 9695, Marina del Rey, CA 90291, Rept. RDA-TR-104900-013, April (1977).
23. I. Y. Borg, Energy and Technology Review, Lawrence Livermore National Laboratory, 30-37, May (1983).

COMPARATIVE RESPONSE OF ALLUVIUM TO HOPKINSON BAR AND GAS GUN LOADING

E. S. Gaffney and J. A. Brown

Earth and Space Sciences Division
Los Alamos National Laboratory
Los Alamos, NM 87545

INTRODUCTION

Low amplitude shock waves in highly dispersive media such as dry alluvium quickly degenerate into ramp waves. One result of this behavior is that standard explosive techniques used for dynamic testing of such materials do not produce data under conditions relevant to most real applications such as confinement of buried explosions or assessment of the effects of explosions on buried structures. We have developed techniques for testing soils dynamically at strain rates ranging from about 10^2 to about 10^4 sec^{-1} in uniaxial strain using a Hopkinson bar. This admits direct comparison with data from gas gun tests where strain rates are in the range of 10^4 to 10^5 sec^{-1}. This, in turn, permits the separation of inertial effects from direct strain-rate effects. In order to assist in evaluation of the results we have also developed a one dimensional microphysical model of soil.

EXPERIMENTAL METHODS

A Hopkinson bar is shown schematically in Figure 1. It consists essentially of two long, round bars. A cylindrical specimen is placed between the two bars and a compressive stress pulse is generated at the far end of one of the bars. This pulse travels the length of the bar where it is incident on the sample (hence the term incident bar). At the interface between the incident bar and the specimen, part of the energy is reflected back into the incident bar as a reflected pulse and part is transmitted into the specimen. This latter continues through the specimen to the other surface where a portion of it continues into the other bar (called the transmitter bar) as a transmitted pulse. The stress pulses in the bars are recorded with strain gauges mounted sufficiently distant from either end to prevent overlap of the wavetrains. A momentum trap at the end of the transmitter bar prevents multiple reflections of the transmitted stress wave. The bars remain elastic during the passage of the stress wave so the strain gauges accurately reproduce the stress histories in the bars.

Our Hopkinson bar consists of incident and transmitter bars supported by a reaction frame. The bars are 60.3 mm diameter, 1.22 m long, maraging steel with a yield stress of about 2 GPa. Teflon bushings in the cross members of the reaction frame support the bars at about half meter intervals. Stress waves in the bars are monitored by strain gauge pairs

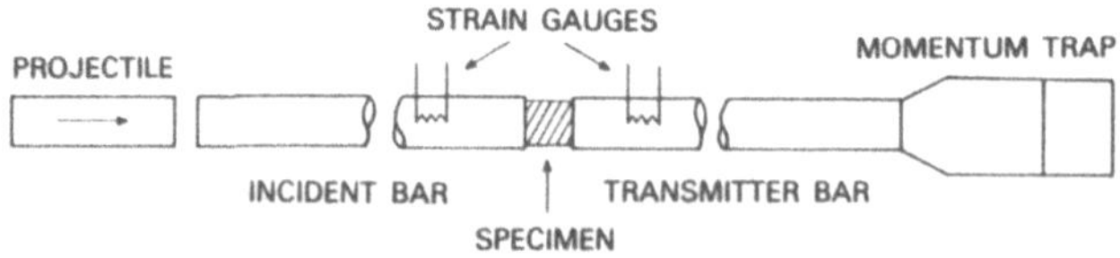

Figure 1. Schematic of Hopkinson bar apparatus.

mounted 0.61 m from the specimen ends. The pairs are used in half bridge configuration to null bending strains. The incident stress pulse is generated by impact of a projectile from a small gas gun. Projectile velocity is measured just prior to impact by three pairs of diode lasers and photodetectors mounted in the muzzle of the gun.

The soil studied was alluvium from near Yuma, AZ. The water content was 3.5 percent of dry weight, and the initial bulk density was 1770 kg/m^3. It was necessary to provide radial support merely to keep the specimen in position for the test. We have used a thick-walled cylinder of bronze or steel with an axial hole just slightly larger than the diameter of the bar, as shown in Figure 2. This very stiff support has the advantages of being easy to implement and providing a strain path that can be duplicated at higher and lower strain rates - uniaxial strain. This strain path also facilitates data analysis, as we shall show below. Tests with no samples have shown that the sleeve carries substantially less than 0.1 percent of the stress pulse during loading. Radial strains in this configuration are less than 0.01 percent while there is still substantial gas-filled porosity remaining in the soil.

The gas gun used in these tests has a 200 mm bore and is capable of firing a 6.5 kg projectile at velocities up to 350 m/sec. The gas gun target is illustrated in Figure 3. The sample was compacted in three lifts 183 mm in diameter and 12.7 mm thick with two embedded piezoresistive carbon stress gauges. The sample was encapsulated in plexiglas. Pins to measure the projectile velocity were secured in the plexiglas rings surrounding the radial surface of the target. The plexiglas also contained embedded carbon gauges to measure the stress delivered to the sample and transmitted through it. The gauges in the plexiglas were separated from the sample by 1.3 to 1.6 mm. In-material gauge packages were made by encapsulating the carbon gauge (with 13 μm of Kapton on each side of the gauge element) in 0.38 mm of electronic grade mica sheet on each side, providing heavy leads (#20 copper) through the plexiglas and wedges of plexiglas (as shown in Figure 2) to spread the shear caused by the transition from soil to plexiglas.

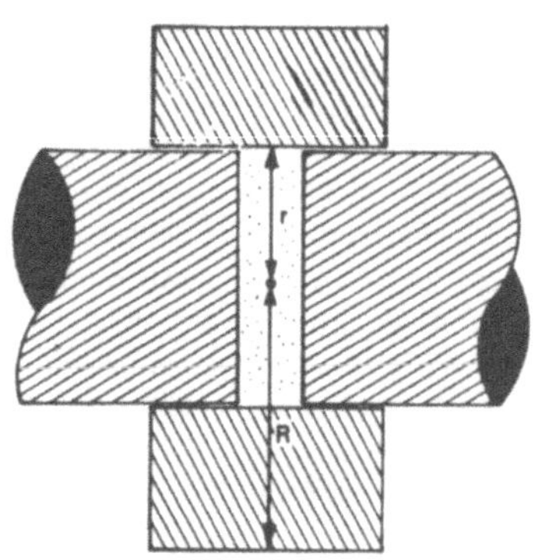

Figure 2. Hopkinson bar sample holder for soil.

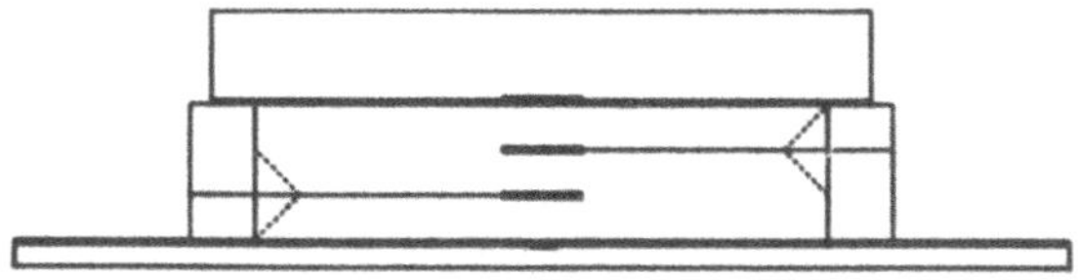

Figure 3. Gas gun target.

Both types of data were recorded by multichannel CAMAC-based waveform digitizers (IEEE-583 Std.), with 8-bit resolution. Sampling rates were 2 MHz for the Hopkinson bar and 4 MHz for the gun. In both cases the analog bandwidth of the recording system was about 1 MHz. The recorded data were automatically read by a small microcomputer and stored on flexible disk for later processing on a larger machine.

ANALYTICAL METHODS

Raw data from both the gas gun experiments and the Hopkinson bar tests were in the form of stress histories - at Lagrangian locations for the former and at the interfaces between the bars and the samples for the latter. We have converted both types of data to stress-strain paths using the Lagrangian analysis algorithms of Seaman [1]. Since the gas gun data is indeed Lagrangian, their treatment is straightforward and will not be discussed further. However, the treatment of the Hopkinson bar data will be described in some detail in the remainder of this section.

The strain pulses are shifted in time to coincide with the transit time to or from the center of the sample. We have used the method of Follansbee and Frantz [2] to correct for rod wave dispersion. This procedure also removes the high frequencies in the data which are noise.

The velocity measurement at the transmitted interface can not be considered to be a Lagrangian measurement in an unperturbed flow field. However, we have been able to convert the data to a Lagrangian, free-field equivalent using a technique which relies on observation of the change in wave speed between the first wave transmitted through the soil and the first wave reflected off the rear interface. It is usually possible to determine the velocity of these two waves from the data; we have sometimes detected the second reflection arriving at the transmitter bar.

Consider the experiment in pressure-particle velocity space as illustrated in Figure 4. The stress wave in the bar loads the steel from the origin to the peak of the steel load-unload curves. When the wave encounters the interface, the steel unloads to the state designated u_f, P_f which is also the state in the soil. However, when that wave (or its attenuated counterpart) arrives at the rear of the sample, it encounters steel that is not moving, and a reflected state is measured which lies at the intersection of the reflected loading curve for the soil and the initial loading curve for the steel at u_m, P_m. If the speed of the reflected wave is n times that of the incident wave, we find that the ratios are

$$(u_f/u_m) = (\rho_b D_b/\{n+1\}\rho_s D_s) + n/(n+1)$$

and

$$(P_f/P_m) = n(\rho_s D_s/\{n+1\}\rho_b D_b) + 1/(n+1).$$

The technique just outlined will work well so long as the loading curve of the soil is fairly smooth and all waves are compressive. If the loading curve is not smooth, the reflection coefficients just derived will not be nearly constant, and a simple multiplicative factor will not relate the free-field and measured wave forms. This might occur if the soil has a

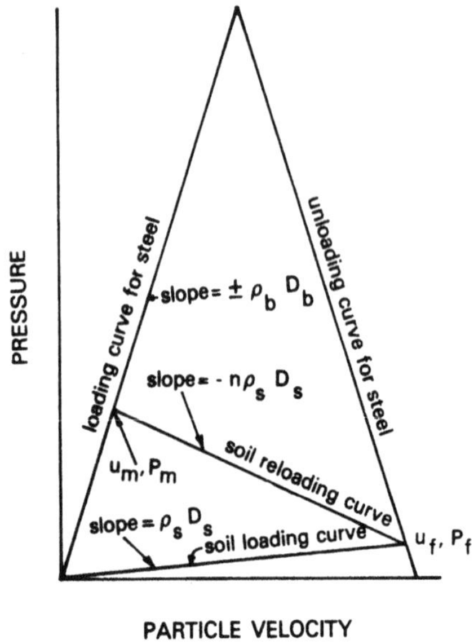

Figure 4. Hopkinson bar test of soil in pressure-particle velocity space.

very sharply defined yield point or when all the gas-filled porosity has been eliminated. In the former case, the reflection coefficient will decrease drastically as the yield point is exceeded, whereas in the latter case it will increase sharply. If the incident wave has a rarefaction phase, this will likely have a much higher speed than the wave which would occur upon its reflection from the transmitter bar. This would result in a very complicated geometric relation in the pressure-particle velocity plane. Although a scheme comparable to the one just described could probably be developed, this has not been accomplished to date. Consequently we treat only the initial loading portion of the experiments here.

To illustrate this method, consider the data from the four experiments shown in Figure 5. Two of the experiments used a sample 25 mm thick (22 and 24) and the other two used 13 mm samples. Experiments 21 and 22 used the same impactor and 23 and 24 used a second impactor. We have analysed both experiments with the same impactor as a single set, scaling the amplitude of the data from the 25 mm experiment up or down by the amount necessary to bring the incident pulses into coincidence. Incident wave amplitudes for each pair of experiments are within 2 percent of the same value. Data from experiments 21, 23 and 24 show that reflections off the transmitter bar travel about 1.6 times faster (Lagrangian wave speed) than those incident on that bar; a second reflection seen on experiment 24 shows a similar increase. Since the density is changing by only about 10 percent at the most, we have used n = 1.6 in our analysis. With the initial soil density of 1770 kg/m^3 and an incident wave speed of 220 m/s and the corresponding values for the bar (8100 kg/m^3 and 4860 m/s) we find that u_f/u_m = 40. and P_f/P_m = 0.39. With this scale factor for the velocities at the transmitted interface, we can use Seaman's technique to calculate the free-field stress histories and find satisfactory agreement with the calculated stress ratio.

RESULTS AND DISCUSSION

The stress-strain paths observed in the gas gun experiment and the four Hopkinson bar experiments are shown in Figure 5 along with a quasistatic uniax. The strain rates in the Hopkinson bar tests reach 5000 s^{-1} and the strain accelerations reach about 2.5×10^8 s^{-2}. The Hopkinson bar data are indistinguishable from the static loading. The gas gun data,

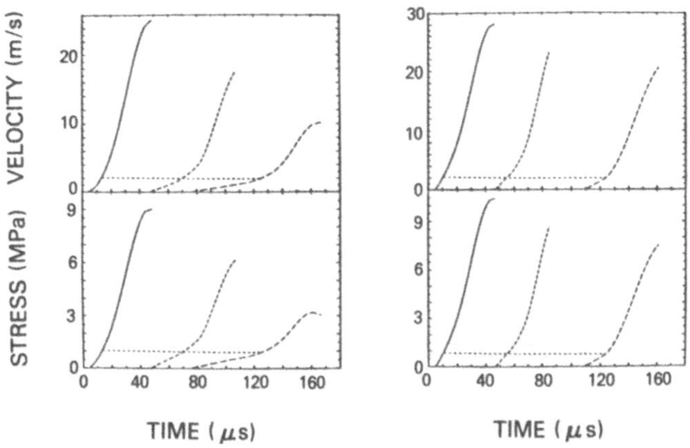

Figure 5. Velocity (upper) and stress (lower) histories measured at sample interfaces. (a) Tests 21 and 22. (b) Tests 23 and 24.

which represent strain rates up to 10^4 and 3×10^4 s^{-1} and accelerations up to 6×10^9 s^{-2}, lie substantially above the other curves. Because of the similarity of strain rates at the lower end of the gas gun regime and the upper end of the Hopkinson bar regime, the distinct behavior of the soil in the gas gun tests is probably due to inertial effects rather than to direct strain-rate effects.

The loading behavior of this soil can be reproduced by a simple one-dimensional soil model which treats the soil as a collection of cells, each composed of three elements. A typical cell consists of two rigid half cubes of a sand grain ℓ on a side separated by void and a pillar of a Bingham material of dimension $\ell\times\ell\times\ell\phi_c/(1-\phi_s)$, where ϕ_c and ϕ_s are the volume fractions of clay and sand, respectively. By considering the strength, "viscous" flow, and horizontal and vertical inertia of the deforming clay pillar, along with the inertia of the sand, the stress-strain relation for this cell can be written as

$$\sigma = \frac{\pm2\phi_c\sigma_o}{(1-\phi_s-\varepsilon)} + \frac{3\mu\phi_c^3\phi_s^2\dot{\varepsilon}}{4(1-\phi_s-\varepsilon)^6} + \frac{\phi_c\rho_c\ell^2}{2(1-\phi_s-\varepsilon)^4}\left(\frac{(\phi_c\phi_s\dot{\varepsilon})^2}{(1-\phi_s-\varepsilon)} + \frac{\ddot{\varepsilon}}{2}\right) + \ell^2\left(\frac{\rho_s}{\phi_s} + \frac{\rho_c\phi_c}{\phi_s^2}\right)\ddot{\varepsilon},$$

where ρ_s and ρ_c are the densities of the sand and clay, respectively, and σ_o is the strength and μ the "viscosity" of the Bingham material.

At first glance, this appears a rather fearsome formulation. However, of seven material properties in the expression (σ_o, μ, ϕ_c, ϕ_s, ρ_c, ρ_s and ℓ), only the first two cannot be determined by standard soil analysis methods. The value of ρ_s will be the grain density of the sand fraction of the soil. The value of ρ_c can be calculated from the grain density of the silt and clay fraction plus the water content. Values for ℓ, ϕ_c and ϕ_s will have distributions in a typical soil, but their mean values can be determined readily from the standard grain-size analysis. For the soil studied, the mean values of ϕ_s and ϕ_c are 0.57 and 0.15, respectively, and the mean grain size of the sand fraction was assumed to be 0.5 mm. The sand was predominantly quartz and feldspar, so we have taken ρ_s to be 2650 kg/m^3. We estimate ρ_c to be about 2400 kg/m^3.

The low strain portion of the static behavior is most sensitive to the distribution of ϕ_c and to σ_o. We have assumed a beta-distribution for ϕ_c

$$P(\phi_c) = B(1.5,2.25)\left(\frac{\phi_c}{1-\phi_s}\right)^{0.5}\left(1-\frac{\phi_c}{1-\phi_s}\right)^{2.25},$$

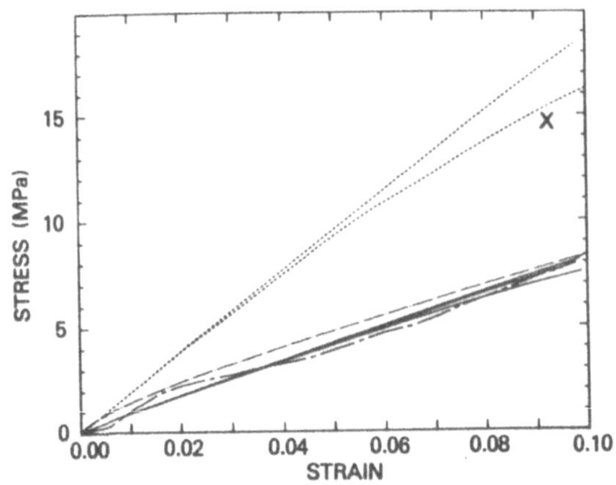

Figure 6. Stress-strain paths in dry alluvium: ——— static, ——— Hopkinson bar, ... gas gun, —·— calculated static, X calculated gas gun.

where B is the beta function. With a yield stress of 10 MPa we get a static stress-strain curve as shown in Figure 6. With $\mu = 10^3$ Pa•s, neither the Hopkinson bar nor the gas gun response would be appreciably above the static curve in the absence of inertial effects. Although we have not yet calculated the complete loading path using the real strain histories, the inertial effects would bring the stress-strain path close to the measured gas gun response with the major inertial contribution being from the sand fraction. For example, the stress-strain point indicated by the symbol in Figure 6 would be predicted for the nominal strain, strain rate and strain acceleration of the gas gun tests.

In simulating the behavior of this soil we have merely selected σ_o and μ to fit the data. There is an extensive literature on the rheology of clay-water mixtures. For most such mixtures, the strength and "viscosity" used here would be high, but these properties are very sensitive to the solids:water ratio at high solid concentrations [4,5]. We have been unable to locate any data for mixtures as dry as the one considered, and extrapolation from four times our water contents can give values on the order of those used here. However, we have little confidence in the extrapolations.

This soil model is clearly very simplified and requires extensive further development before it could be considered predictive. Nevertheless, it does permit some evaluation of the relative importance of inertial effects and direct strain-rate effects in soil. The model supports our conclusion based solely on the data that the difference between the response of dry alluvium to testing in the gas gun and testing at lower rates is primarily attributable to inertial effects.

REFERENCES

[1] Seaman, L., Lagrangian analysis for multiple stress or velocity gauges in attenuating waves. J. Appl. Phys. 45, 4303-4310 (1974).
[2] Follansbee, P. S., and C. Frantz, Wave propagation in the split Hopkinson pressure bar. J. Eng. Mater. Tech. 105, 61-66 (1983).
[3] Gaffney, E. S., Preliminary constitutive model for alluvium. Los Alamos National Laboratory memo ESS-5:85-OU-410, June 1985.
[4] Norton, F. H., A. L. Johnson and W. G. Lawrence, Fundamental study of clay: VI, Flow properties of kaolinite-water suspensions. J. Amer. Ceram. Soc. 27, 149-164 (1944).
[5] Nguyen, Q. D., and D. V. Boger, Yield stress measurements for concentrated suspensions. J. Rheol. 27 321-349 (1983).

DETERMINING A SOIL MODEL FROM WAVE PROPAGATION DATA*

L. Seaman

SRI International
333 Ravenswood Avenue
Menlo Park, California 94025

INTRODUCTION

A procedure is being developed for determining a computational soil model from stress measurements obtained during a high-explosive test. Gages at several depths in the soil measure the stress normal and parallel to the direction of propagation. Effects of the finite gage size and its frequency and damping are removed from the stress records. A Lagrangian analysis then provides the stress-strain and pressure-volume paths taken. With these paths, a model for the soil is constructed.

The Waterways Experiment Station has conducted small in-situ material property tests with explosive loadings and with the geometry shown in Fig. 1.[1] Small diaphragm stress gages have been placed in the soil to measure the vertical or horizontal component of the stress at several locations. Our purpose is to process the gage records to obtain the stress-strain paths followed by the soil during the experiment. These paths should guide development of a soil model.

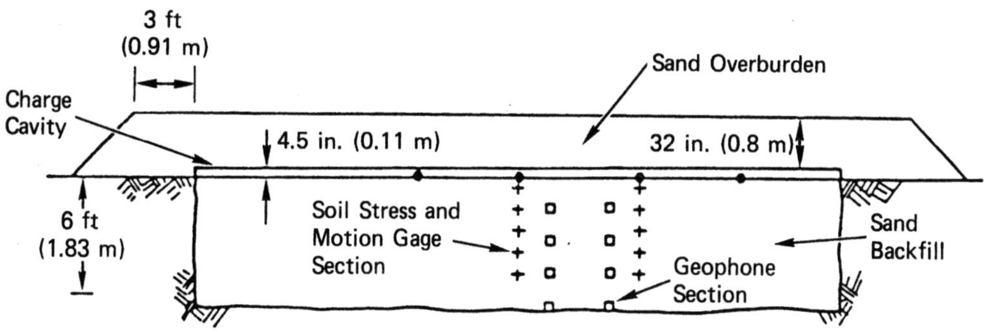

Fig. 1. Cross section of the DISKO-1 small-scale, high-explosive simulation test conducted by WES.

Source: Reference 1.

*Supported by U.S. Army Waterways Experiment Station

The determination of the stress-strain paths presents a series of challenges. The stress gages have a natural period of 25 to 35 μs, yet the rise time of the stress pulses is 5 to 10 μs. Hence, spurious oscillations occur on the records and the wave front is obscured (a sample record is shown in Figure 2). The stress calibration of the gages appears to be different during loading and unloading. In addition, the finite time for the wave to traverse the face of the horizontal stress gages causes a smearing of their records. From the records we can obtain only the stress acting on the gage face, not the free-field stress (stress that would have been acting at the gage position in the absence of the gage): we are not accounting for the inclusion effect. There is some scatter in the records from the gages at each depth; this scatter will strongly affect the strain histories derived from the records. All these challenges except the inclusion effect are addressed in the current analysis.

SOLUTION STEPS

The steps we are taking to determine the stress-strain paths are shown in Fig. 3. The first step is the calibration of the records, using the calibration curve shown in Fig. 4. The loading and unloading paths were taken from static measurements using a small sample chamber at WES.[2] The reloading paths are estimates. It is not known whether a similar calibration is appropriate for dynamic loading. However, strain histories based on calibrated stress histories have a much more reasonable character than those based on uncalibrated stresses; hence we conclude that the calibration is at least qualitatively correct.

The next step is the processing of the stress records to eliminate the effects of the gage frequency and damping. Our purpose here is to construct a stress history such that when this history is applied to a one-degree-of-freedom oscillator with the characteristics of the stress gage, the original record is obtained. In the procedure we begin with a simple smoothing which eliminates most of the ringing. Then we compute the oscillator response to this history and compare the response to the actual record. Usually, the

76.5 cm depth in the DISKO-1 event.

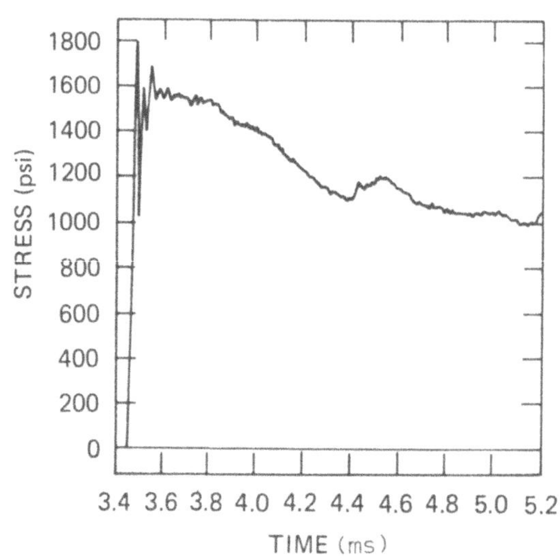

Figure 2. Scanning electron micrographs taken from an outer region of specimen 1G856, showing (a) substantial reactions, (b) a detailed feature of the reacted zone, and (c) a variety of phases.

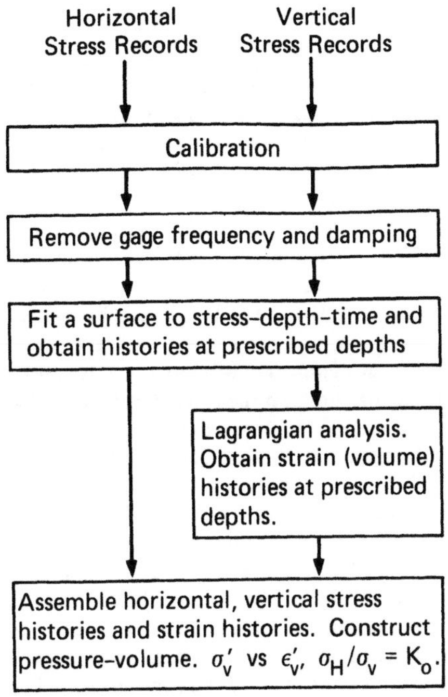

Fig. 3. Steps for constructing stress-strain paths from horizontal and vertical stress records.

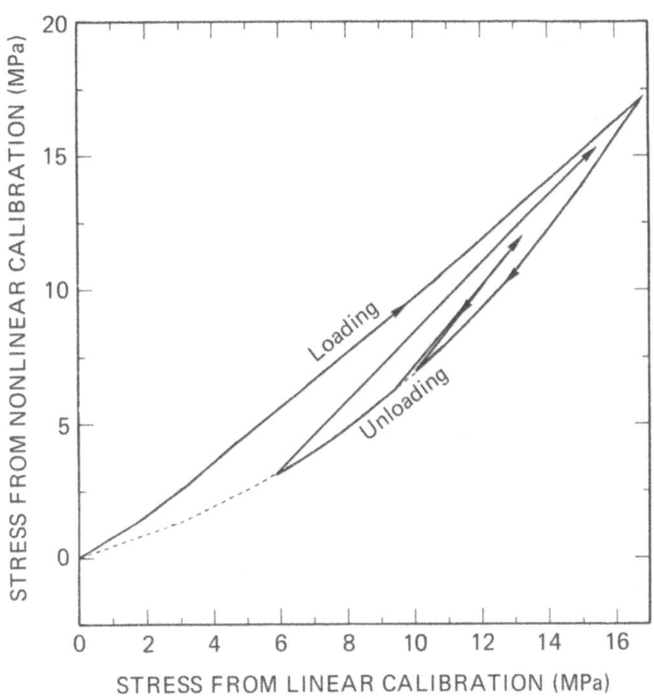

Fig. 4. Calibration curve for loading and unloading of low range diaphragm stress gage embedded in sand.

(Curve provided by WES.)

record and the response are very similar except near the wave front. Here the constructed history is modified, and new oscillator responses are computed until the comparison is judged satisfactory. For the horizontal stress gages, it is also necessary to remove the effects of the time for the wave to sweep across the gage face. The removals of both the frequency and sweeping effects are basically unstable because we are trying to recover information that has been partially lost. We have forced stability on the method by constructing a history with only four points per period, instead of a continuous function. The constructed histories appear to be stable and repeatable. Thus, we seem to have a method for smoothing the stress records, while also improving the fidelity of the wave fronts.

The development of a surface in stress-depth-time space requires a fit of the family of vertical or horizontal histories. Such a surface is shown in Fig. 5. The least-squares fitting method used was described by Seaman and Keough at the previous Shock Waves Conference.[3] From the fitted surfaces, vertical and horizontal stress histories at selected depths in the soil are computed. By reconstructing records at chosen depths the analyst can make a direct comparison of horizontal and vertical stresses at a single depth, although the gages were not placed at the same depths in the experiment. During the fitting process, we have the option of forcing the peaks and valleys of the surfaces to coincide. Such forcing implies a rate-independent representation for the soil and thus limits the fidelity of the data. However, for scattered data, such forcing may be necessary to permit computation of physically reasonable strain histories.

With the assembled vertical records we perform a Lagrangian analysis following the method in Reference 3. Strain (specific volume) histories are obtained at the selected depths, in addition to particle velocities, energies, and other quantities.

The final step is collection of the vertical and horizontal stress histories and the strain histories at each selected depth and computation of relationships between the stresses and strains. The two stresses provide the complete tensor, so we can compute the deviatoric stress in the direction of propagation, σ'_v, and the mean stress (or pressure), P:

$$\sigma'_v = \sigma_v - P$$

$$P = \frac{\sigma_v + 2\sigma_h}{3}$$

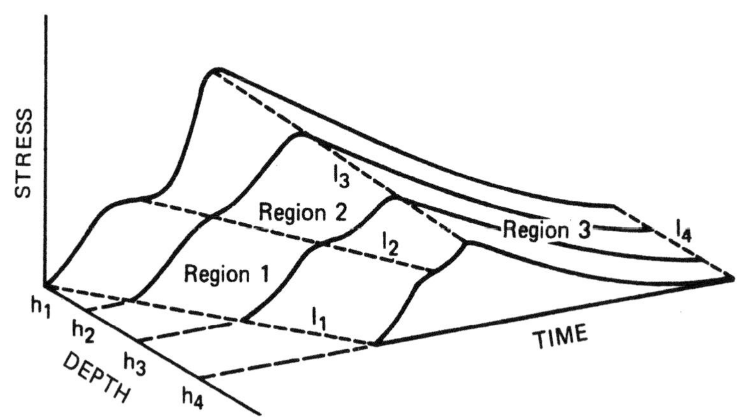

Fig. 5. A stress-depth-time surface from stress histories.

where σ_v and σ_h are the vertical and horizontal stress components. The strain is defined from the volume change:

$$\varepsilon_v = \frac{V}{V_o} - 1$$

where V is the current specific volume, and V_o is the initial value. The deviatoric strain in the direction of propagation is $\varepsilon_v' = 2\varepsilon_v/3$. With these stress and strain quantities we can now produce the following relationships:

1. Histories of K_o, the ratio of horizontal to vertical stress
2. Stress difference, $\sigma_v - \sigma_h$ versus mean stress, P,
3. Deviator stress, σ_v', versus deviatoric strain, ε_v, and
4. Mean stress, P, versus specific volume, V.

These functions are sufficient to characterize a rate-independent model for the case of planar, one-dimensional loadings and unloadings. Because we obtain these functions at several depths and thus at a range of strain rates, we also have some estimate of the strain-rate effect in the material.

RESULTS

Sample results from the DISKO-1 experiment are shown in Figures 6 and 7. Strange features in the paths may be associated with irregularities in the data or with material rate effects. Note that hysteresis is more important in the deviatoric part of the stress-strain relations than in the pressure part. This result emphasizes the importance of determining the deviatoric stress in soils in preparation for developing a material model.

This exploration of an approach for determining a soil model from dynamic experimental data appears promising.

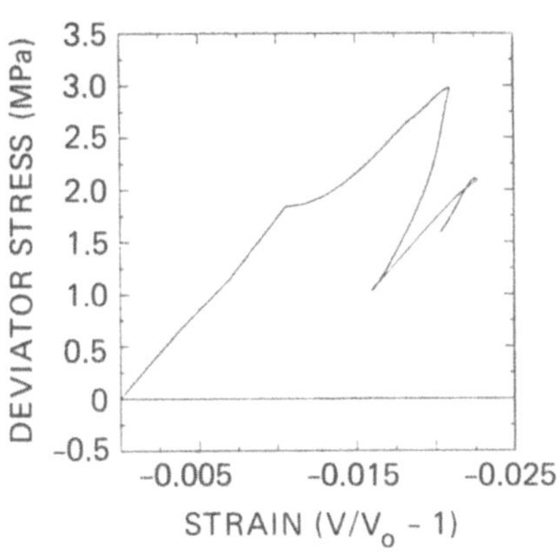

Fig. 6. Path of deviator stress versus strain obtained from DISKO-1.

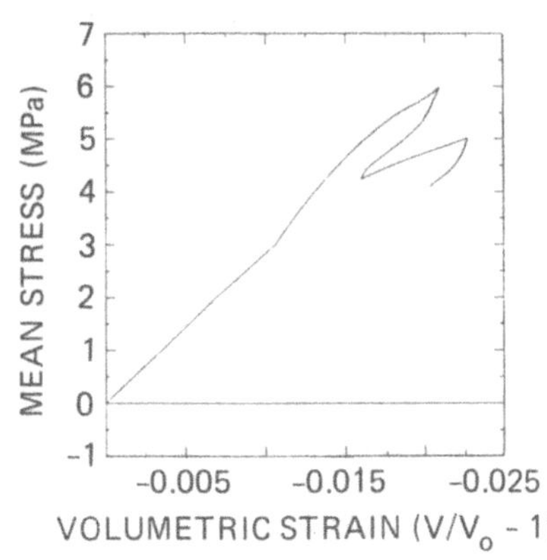

Fig. 7. Pressure-strain path obtained from DISKO-1.

REFERENCES

1. John E. Windham, "Dynamic In Situ Property Test Program: Results of DISKO-1," by U.S. Army Engineer Waterways Experiment Station, for Defense Nuclear Agency, Washington, DC 20305 (May 1984).
2. John E. Windham, private communication (January 1985).
3. Lynn Seaman and Douglas D. Keough, "Lagrangian Analysis with an Emphasis on Unloading Processes," 3rd Topical Conference on Shock Waves in Condensed Matter, Eds. J. R. Asay, R. A. Graham, and G. K. Straub, North-Holland Physics Publishing, Amsterdam, The Netherlands (1983), p. 117.

MEASUREMENTS AND CALCULATIONS OF SHOCK PROPAGATION IN DRY DESERT ALLUVIUM

Robert C. Bass

Sandia National Laboratories
Albuquerque, New Mexico 87185

INTRODUCTION

Data obtained on shock propagation from underground explosive detonations are often "contaminated" at larger distances by surface reflections and spall signals. This is a result of the usual placement of instrumentation vertically above the detonation point in a single satellite hole; the use of multiple satellite holes that would permit measurement near the detonation horizon being prohibitively expensive. A recent underground nuclear detonation in alluvium utilized a large diameter device emplacement hole and the entire instrumentation array was installed in the backfill material. No significant spall or surface reflections were noted in the data obtained in this large cased hole and the shock structure was clearly detailed to locations close to the surface.

Several smaller conventional high explosive detonations located closer to the surface have also been studied in detail. In these instances several satellite instrumentation holes were available and measurements were made at or near the detonation horizon minimizing surface effects. Data from two explosive detonations and the data from the nuclear detonation are compared with computer code calculations made to describe underground explosions in alluvium. Supportive data from a plane wave surface detonation are also discussed.

MEASUREMENT TECHNIQUE

Similar measurement techniques were used in the four experiments. In the first high explosive test considered, SCOOTER(1), a charge of 0.5 kilotons of TNT was detonated at a depth of 38.1 meters. Instrumentation consisting of accelerometers and velocity gauges installed in five satellite holes was all located at the charge elevation. The second explosive test, VUGS II, involved the detonation of 175 kilograms of TNT at a depth of 10.7 meters. Instrumentation consisted of lithium niobate stress gauges located in three satellite holes. All gauges were located near the charge elevation.

The nuclear event reported was instrumented with ytterbium stress gauges, accelerometers, and a SLIFER(2) cable system to continuously record shock arrival over a large distance. The fourth experiment reported was a planar surface detonation of several centimeters of nitromethane over alluvium. This experiment, BACKBURNER, was a portion of a continuing gauge development study and instrumentation included stress gauges, accelerometers, and several low crush strength SLIFER cables.

CALCULATIONS

Underground explosive detonations have been the subject of extensive code calculations in several laboratories. The knowledge of equation of state parameters for a variety of geologic materials has led to reliable results in the high and medium pressure regions. Material modeling coupled with experimental data gathered in wet tuff and several hard rocks has led to a good description of propagation in the lower stress region in these materials. The modeling of the low stress behavior of porous materials such as dry tuff and alluvium has not yet reached the state required for reliable calculations. The calculation considered here is one made with the CHART-D hydrocode(3) and reported in Bass and Larsen(4). The input to the CHART-D equation of state routine, ANEOS, was based on data for alluvium given in Bass, Hawk, and Chabai(5). No attempt was made to describe the low pressure behavior of the material; the calculation was totally hydrodynamic. The results have been cube root scaled to appropriate values for the comparisons presented.

EXPERIMENTAL RESULTS

A data record typical of those obtained on the four detonations is shown in Figure 1, a plot of the velocity derived from an accelerometer located at a scaled range of 125 meters above the nuclear detonation. We see an elastic precursor arriving at 65 milliseconds followed by a larger plastic wave at 290 milliseconds. Similar results were obtained at other locations on this event and on the two underground high explosive detonations. The arrival times of the plastic and elastic waves seen in these data are plotted versus range for the three events in Figures 2, 3, and 4. The scaled results of our hydrodynamic calculation are included in each figure. In each case the plot of the calculated arrival curve is terminated when the calculated velocity matched that observed for the terminal portion of the data. This velocity was 343 meters per second in each case. The pressure values shown (right ordinate) are those obtained from the calculation and they agree well with the limited pressure values obtained in the experiments.

Similar propagation of the terminal portion of the plastic wave was observed in the BACKBURNER data. A constant velocity of 343 meters per second was determined from SLIFER and stress gauge arrival time data beyond the position where the data indicated a pressure of less than 0.4 kilobars.

OBSERVATIONS, COMMENTS, AND CONCLUSIONS

We note good agreement between the data and the calculation of hydrodynamic shock propagation in dry desert alluvium. The use of the hydrodynamic model appears valid as long as the velocity of the plastic wave is greater than the speed of sound in air. Beyond this point, at pressure of less than 0.4 kilobars, a constant propagation velocity is observed. This seems to indicate that the air contained in this weak porous material is the factor controlling shock propagation. The porosity of the dry alluvium in the four explosions studied was approximately 40%. We have not yet developed a satisfactory physical model for the effects noted.

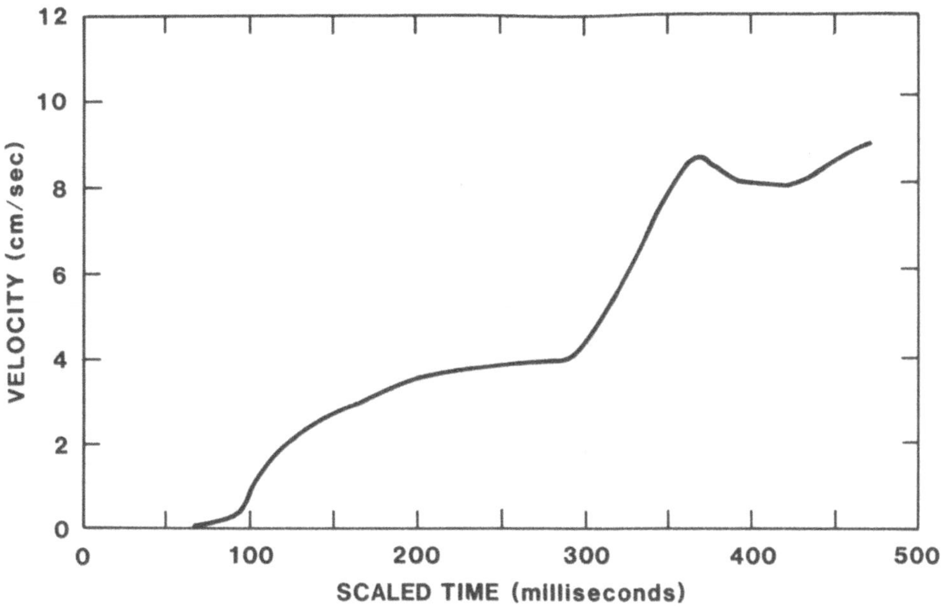

Figure 1. Velocity Profile Derived from an Acceleration Measurement at a Scaled Range of 125 Meters on the Nuclear Event.

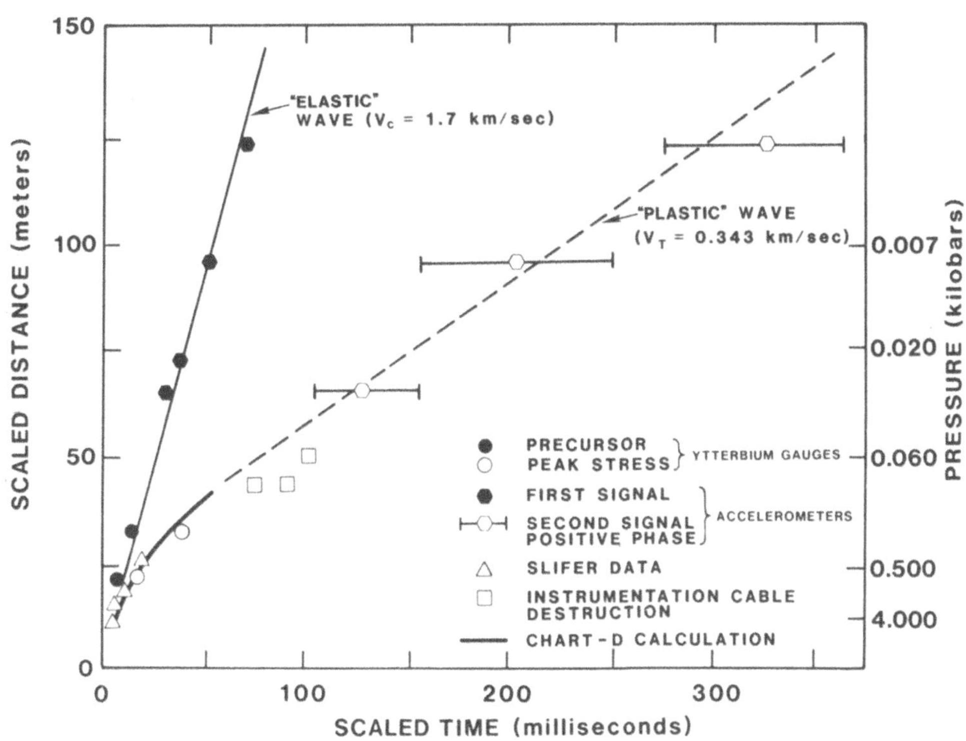

Figure 2. Elastic and Plastic Wave Arrival Time Data for the Nuclear Event.

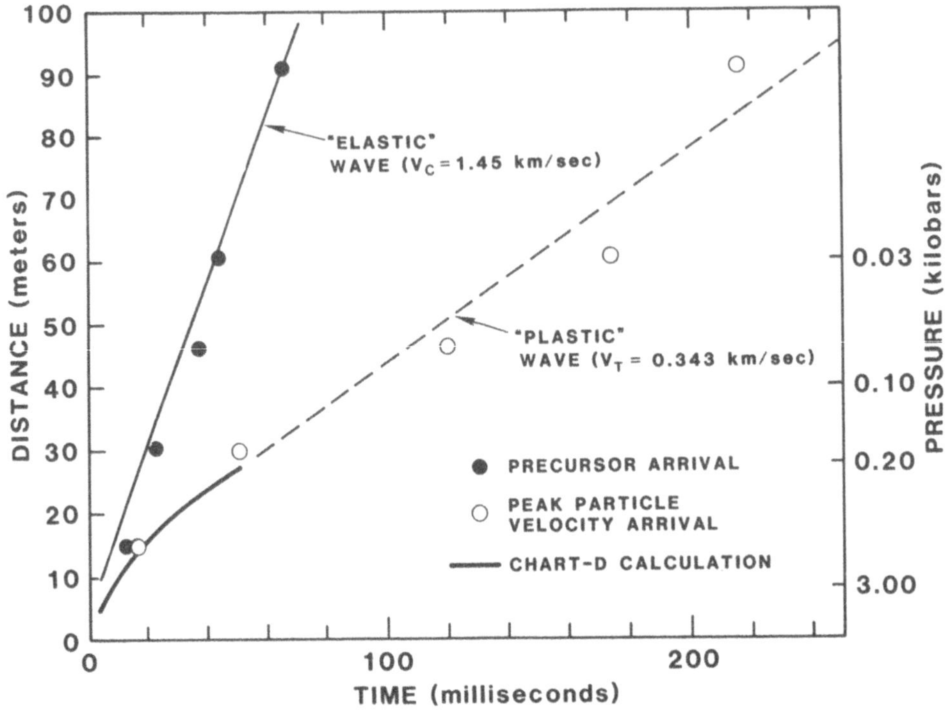

Figure 3. Elastic and Plastic Wave Arrival Time Data for the SCOOTER Event.

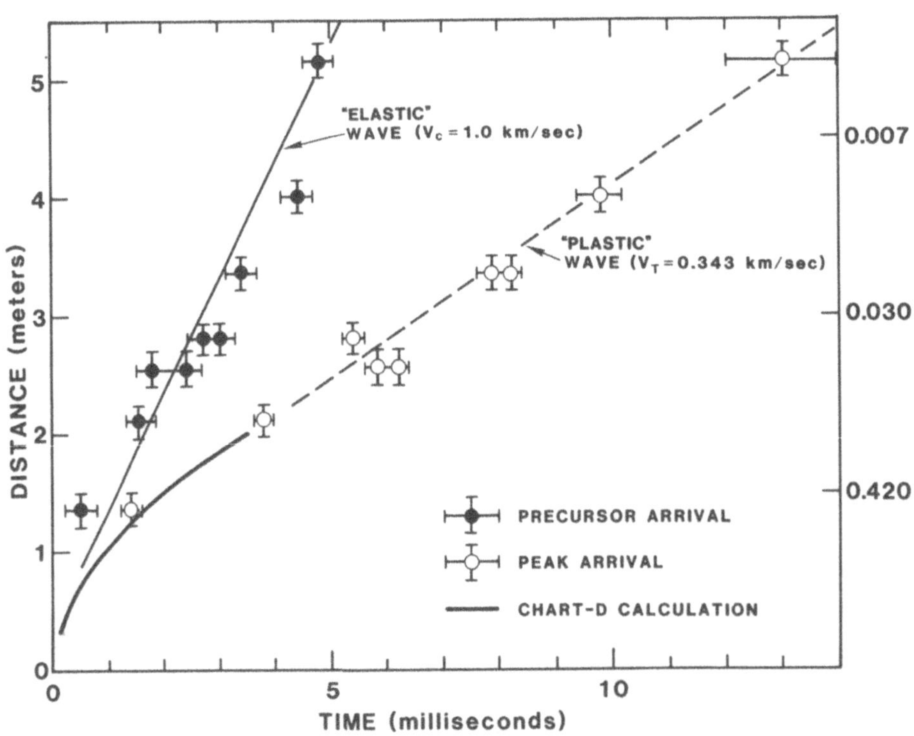

Figure 4. Elastic and Plastic Wave Arrival Time Data for the VUGS II Event.

ACKNOWLEDGMENT

This work was sponsored by the Department of Energy.

REFERENCES

1. W. R. Perret, A. J. Chabai, J. W. Reed, and L. J. Vortman, "Project Scooter," SC-4602(RR), Sandia Laboratory, Albuquerque, NM, October 1963.
2. M. Heusinkveld and F. Holzer, "A Method of Continuous Shock Front Position Measurement," Review of Scientific Instruments, 35-9.
3. S. L. Thompson and H. S. Lauson, "Improvements in the CHART-D Radiation-Hydrodynamic Code II: A Revised Program," SC-RR-0713-71, Sandia Laboratories, Albuquerque, NM, February 1972.
4. R. C. Bass and G. E. Larsen, "Shock Propagation in Several Materials of Interest in Hydrodynamic Yield Determinations," SAND77-0402, Sandia National Laboratories, March 1977.
5. R. C. Bass, H. L. Hawk, and A. J. Chabai, "Hugoniot Data for Some Geologic Materials," SC-4903-RR, Sandia Laboratory, Albuquerque, NM, June 1963.

GROUND MOTION MEASUREMENTS FOR THE QUESO NUCLEAR EVENT

H. David Glenn
Lawrence Livermore National Laboratory
Livermore, CA 94550

T. F. Stubbs and J. A. Kalinowski
EG&G Inc., San Ramon, CA 94583

QUESO was a low yield nuclear event detonated at a depth of 216.4 m in hole U10bf at the Nevada Test Site on August 11, 1982. Shock wave histories containing the peak residual stress values were obtained for the first time in free field measurements. Radial and tangential stress measurements indicated a strong residual stress field was induced above the cavity by 200 ms after the nuclear detonation, following rebound of that region after passage of the ground shock. Free field particle velocity profiles corresponded closely with measurements made at comparable ranges in the emplacement hole.

INTRODUCTION

For over a decade numerical simulations of underground nuclear explosions have predicted, following passage of the shock wave, the region around the cavity rebounds locking in a strong tangential stress field.[1-4] Measurements have shown the presence of a residual tangential stress field around the cavity for high-explosive laboratory,[5] and field[6] experiments. This paper describes the first successful free-field measurements of the radial or tangential residual stress fields for a nuclear test.

QUESO offered the opportunity to make free field measurements in a 0.305 m diameter satellite hole (UE10bf), 15.24 m west of the emplacement hole (U10bf).[7] Normally satellite holes are not present on an underground nuclear test and subsurface shock measurements are limited to the emplacement hole. Since the emplacement hole is very difficult to accurately represent in numerical simulations, there arises the question "How representative are these measurements to those for the free-field at the comparable radial distance from the explosion?" This report addresses this question.

GEOLOGIC SITING, STEMMING PLAN AND GROUND MOTION DIAGNOSTICS

The device canister was located in a weak low density (ρ_o=1.62 Mg/m^3) layer of tuffaceous alluvium. The region above the working point (WP), 207m to the surface, was a strong high density (ρ_o=2.13Mg/m^3) layer of mixed alluvium. The mixed alluvium layer contained all the measurements described in this report. Figure 1 shows the QUESO stemming plan. Fines layers and four plastic plugs inhibit the flow of cavity gases up the emplacement hole. The three lowest plugs were keyed to the geology via hole enlargements.

Piezoresistant (ytterbium) stress gages were located in the emplacement hole (Fig. 1) and the satellite hole. Table 1 summarizes the types of stress gages, their locations and orientations. For positioning and orientation requirements, stress gages in the emplacement hole were supported by brackets bolted to the emplacement pipe. The two stress gages in the satellite hole were supported by brackets bolted to the fiberglas pipe used to backfill the hole with medium matching grout. A cable pendant in the emplacement hole supported a motion canister located in the middle of each of the four plastic plugs. Each motion canister contained a vertical velocity gage, a vertical acceleration gage, and a thermistor. The thermistor provided a temperature reading for correction of the velocity gage record. A similar motion canister was emplaced at three different radial distances from the WP in the satellite hole corresponding to the locations of a motion canister in the top three plastic plugs of the emplacement hole. A surface motion canister (Station 62) was bolted to the wall of the surface casing at a depth of 0.6 m. This velocity measurement will be compared with that from a surface motion canister (Station 61) buried at a depth of 1.22 m and a horizontal distance of 15.24 m from surface ground zero.

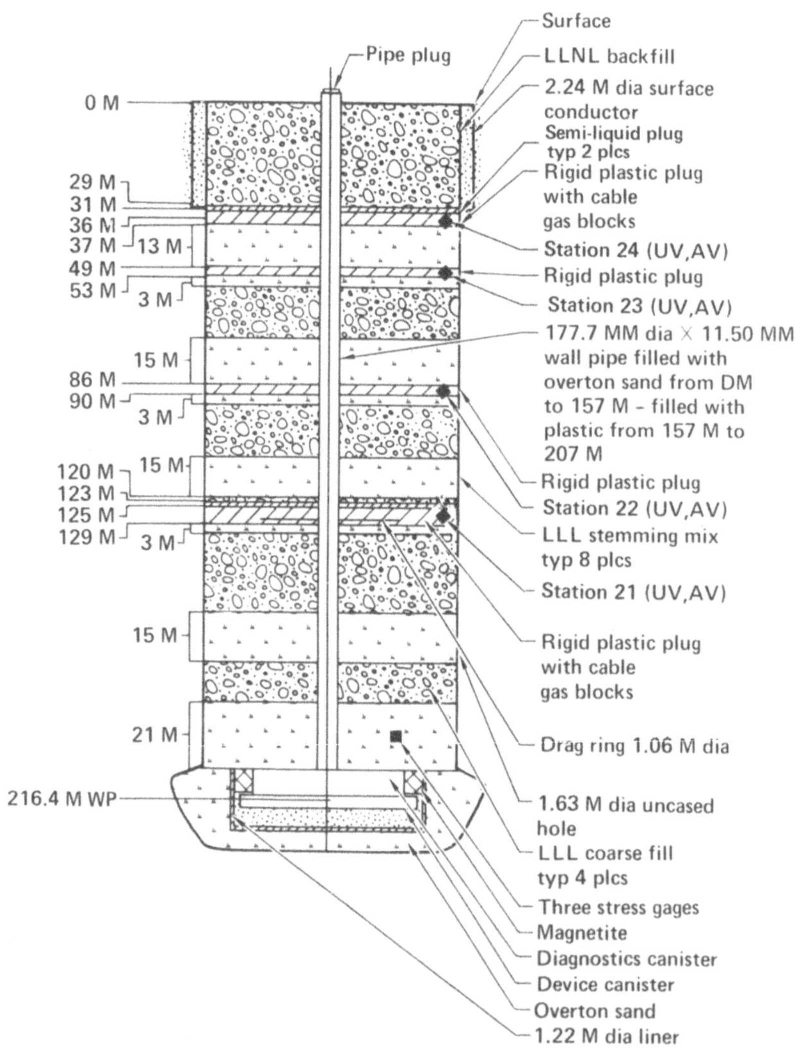

Fig. 1. The QUESO (as built) stemming plan.

Table 1. Piezoresistant (ytterbium) stress gages installed in the emplacement hole (U10bf) and the satellite hole (UE10bf).

Hole I.D.	Gage Number	Gage Design[a]	Radial Distance to WP (m)	Orientation of Measurement
U10bf	11	Composite	22.0	Radial
	12	Composite	22.0	Tangential
	13	Flat-pack	22.0	Tangential
UE10bf	52	Flat-pack	22.0	Radial
	53	Flat-pack	22.0	Tangential

[a]The Composite gage was constructed by Dynasen, Inc., Goleta, CA, and the Flat-pack gage was constructed by SRI international, Menlo Park, CA.

GROUND MOTION RESULTS

Stress Profiles in the Free-Field and Emplacement Hole Configuration

Figure 2 presents the radial stress profile for the flat pack[8] (gage 52) located in the satellite hole (UE10bf). First shock time-of-arrival (TOA) occurs at 13 ms and peaks at 65 MPa by 18 ms. By 35 ms the pulse has decayed to half maximum and then to 2.5 MPa at 125 ms. Then the medium rebounds to a peak radial stress of 6 MPa at 192 ms. Afterward the medium relaxes to a steady state in the 2-4 MPa range until the record is terminated at 653 ms. Figure 3 gives the tangential stress profile for the flat pack (gage 53) located in the satellite hole. First shock TOA occurs at 14 ms and peaks at 64 MPa by 18 ms. The tangential stress decays to half maximum at 30 ms and then to 4.5 MPa at 140 ms. Then the medium rebounds to a peak of 9.0 MPa at 192 ms. Shortly after 200 ms the reading becomes erratic due to what we believe are problems with the junction between the cables and gage. Orientation of the flat pack and cables are important in reducing differential motions induced by the shock wave at the junction. This problem may be corrected in the future with a 90 degree twist in the steel strips of the flat pack. Then the cables would exit the flat pack in a plane parallel with the radial stress gage.

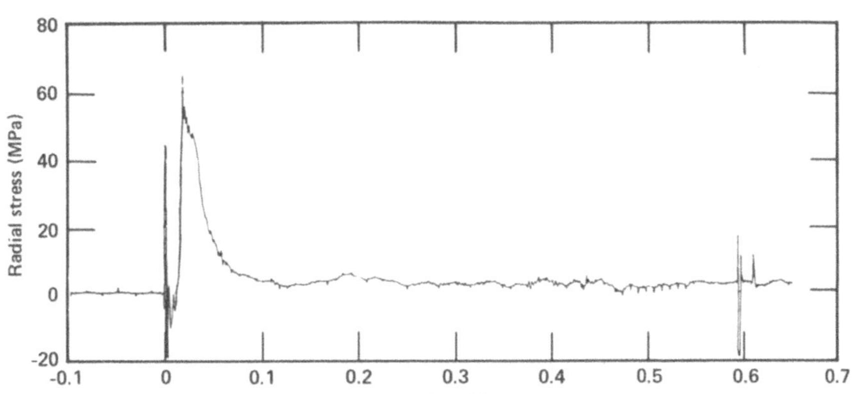

Fig. 2. Radial stress profile in the satellite hole at a range of of 22 m from the WP.

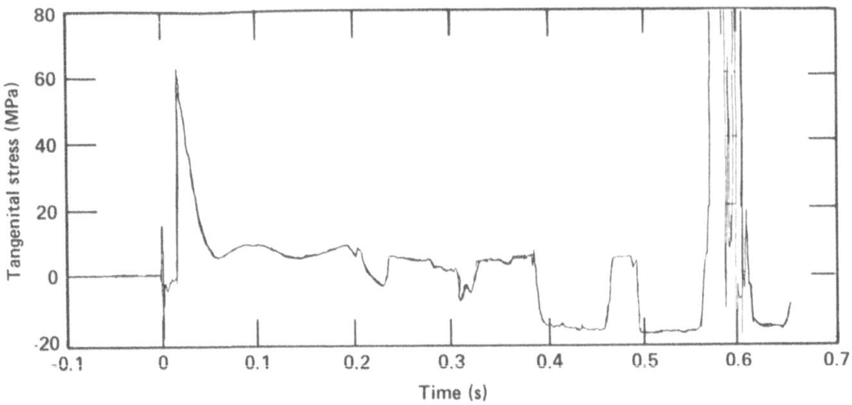

Fig. 3. Tangential stress profile in the satelite hole at a range of 22 m from the WP.

Comparing radial and tangential stress measurements in the satellite hole (Figs. 2 and 3), reveals some important features. The peak radial stress slightly exceeded the peak tangential stress at 18 ms. Calculational results normally predict[9] an even higher peak value for the radial stress than for the tangential stress in the shock front. The relatively high peak reading for the tangential stress may be a gage interaction effect and/or inclusion effect.[10] Rebound appears complete by 192 ms with peak tangential and radial stresses of 9.0 and 6.0 MPa, respectively. These values are consistent with the model proposed[2,4] to characterize the residual stress field following a nuclear explosion.

Of the three gages (Table 1) in the emplacement hole only the composite tangential gage[11] (number 12) survived the ground shock. Figure 4 illustrates the tangential composite gage records for the ytterbium before and after correction for longitudinal strain. First shock TOA occurs at 20 ms and the peak of 19 MPa at 25 ms. The pulse decays relatively slowly and reaches half maximum at about 110 ms. The pulse reaches a minimum of 8.5 MPa at about 140 ms during a transient noise disturbance. Following rebound, the tangential stress peaks at 9.5 MPa at 193 ms and then relaxes to a steady state in the 8.3-8.7 MPa range until the record terminates at 653 ms. A composite gage on the VIDE event, located in the emplacement hole, measured about 16 MPa for the residual stress field.[11] The geologic medium in the vicinity of the gage was tuff, which normally has a greater shear strength than alluvium[12] and may explain the higher residual stress field for VIDE.

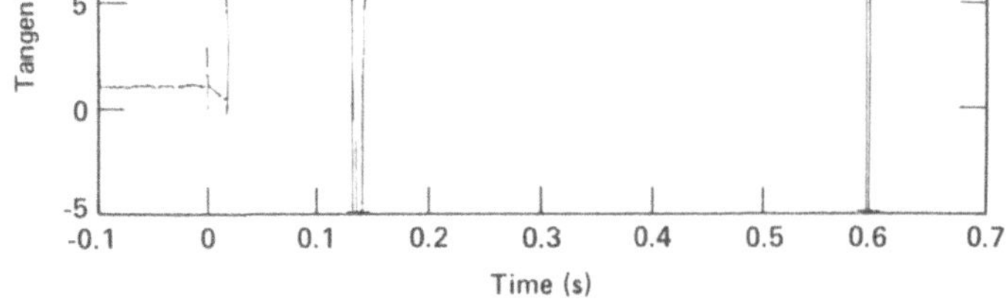

Fig. 4. Tangential stress profile (lower curve) at a range of 22 m above the WP. The upper curve shows the signal following correction for longitudinal strain.

The tangential stress measurement in the emplacement hole indicated a later TOA for the shock wave and factor of three lower peak stress value than was recorded in the satellite hole. The shock wave arriving at the gages was coupled in from the medium. The gage was located at 0.41 m, or half the radius from the center of the emplacement hole. The LLL stemming mix has much higher air-filled porosity than the mixed alluvium medium. Consequently as the shock propagates from the medium into the hole it is delayed and attenuated by compaction of the stemming. The delay and attenuation effects are felt to be the explanation for the later TOA and lower peak stress reading for the tangential gage (number 12) with respect to that observed in the satellite hole.

Comparison of Vertical Velocity Profiles

Figure 5 presents particle velocity profiles from the velocity gages in the four plastic plugs and the three velocity gages in the satellite hole. The particle velocity profiles of the top three plugs in the emplacement hole track closely the corresponding three free-field measurements in the satellite hole. This agreement can be better understood if we examine the time frame of the profiles. The 15-20 ms rise times at Station 22 for the first two velocity peaks correspond to the most rapidly changing segment of any of

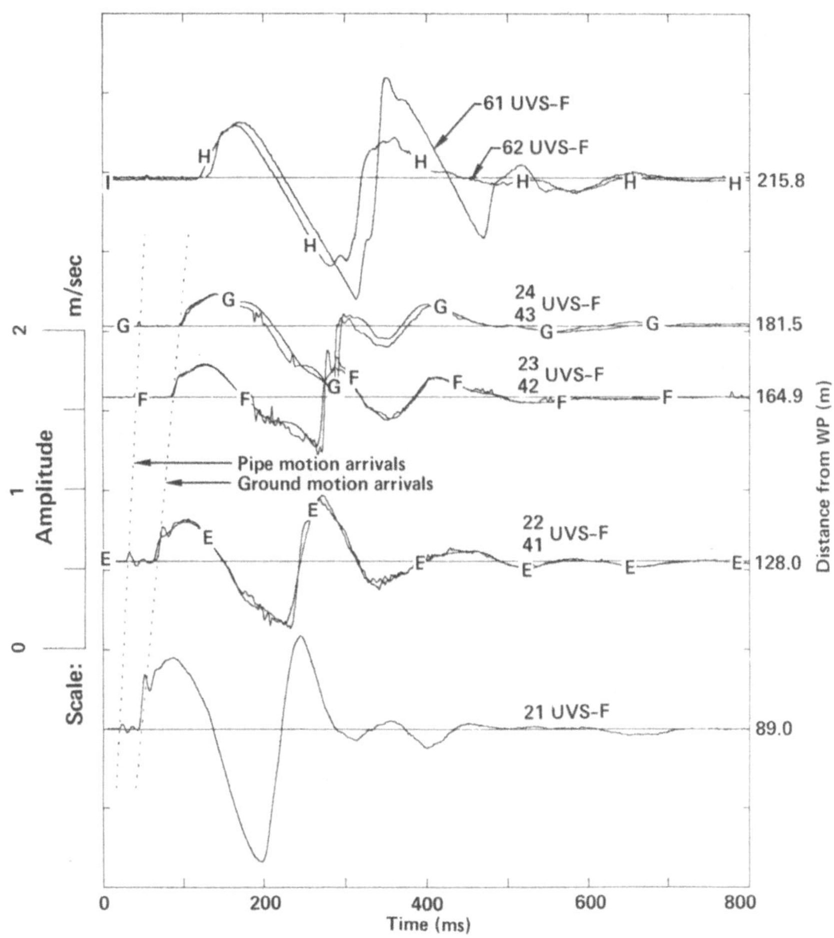

Fig. 5. Particle velocity profiles in the emplacement hole (Stations 21-24 and 62), satellite hole (Stations 41-43) and at the surface (Station 61). There exists no velocity measurements in the satellite hole corresponding to Station 21 in the emplacement hole.

the profiles. Even for these short times the shock has propagated 30-40 m, a distance an order of magnitude greater than the radius (0.82 m) of the emplacement hole. Consequently, there was adequate time for the shock front to transit the 0.82 m radius many times to bring the gage into equilibrium with the surrounding medium. The comparison of the particle velocity profile for the motion canister attached to the surface casing (Station 62) with the surface motion canister (Station 61) is an example of how the emplacement procedure can affect the measurement. The 34.9 m long surface casing tends to move somewhat like a rigid body and thus provides a restraining or damping force. This damping effect is particularly evident on the timing and magnitudes of the initial fall back and subsequent oscillatory behavior for the gage attached to the surface casing.

ACKNOWLEDGMENTS

Work performed under the auspices of the U.S. DOE by LLNL under contract number W-7405-ENG-48. The authors are indebted to W. Webb for field electronic support; R.S. Salazar and C.A. Cordill for mechanical engineering support, D.E. Burton and J.B. Bryan for their critical reviews, R.M. Glenn and L.D. Grabowski for preparation of this paper.

REFERENCES

1. J.S. Kahn, R.W. Terhune, H.D. Glenn, and B.K. Crowley, "Increasing DNA's Test Site Real Estate," Lawrence Livermore National Laboratory, Internal Document UCON 74-32, Memo to Carl Keller, Defense Nuclear Agency Headquarters, Albuquerque, NM, (Apr. 1975). Readers outside the lab who wish further information on LLNL internal documents should address inquiries to Technical Information Dept., LLNL, Livermore, CA 94550.
2. R.W. Terhune, H.D. Glenn, D.E. Burton, and J.T. Rambo, "Containment Analysis for the Simultaneous Detonation of Two Nuclear Explosives," Lawrence Livermore National Laboratory Rept. UCRL-52268 (Apr. 1977).
3. N. Rimer, "The Influence of Clay and Water in Rocks on Containment of Underground Nuclear Explosions," Systems, Science, and Software Rept. SSS-R-77-3339 (June 1977).
4. R.W. Terhune, H.D. Glenn, D.E. Burton, H.L. McKague, and J.T. Rambo, "Numerical Simulation of the BANEBERRY Event". Nuclear Technology 46, (November 1979).
5. A.L. Florence and R.W. Gates, "Explosive Cavity Residual Stress Measurements", Stanford Research Institute Rept. TR-76-3702-3 (1976).
6. C.W. Smith, "PUFF TOO: A Residual Stress Experiment", Sandia National Laboratory Rept. SAND 79-1674 (1980).
7. H.D. Glenn, T.F. Stubbs, J.A. Kalinowski, and E.C. Woodward, "Ground Motion Data and Containment Analysis for the QUESO Nuclear Event," LLNL Rept. UCRL-53457, (Oct. 1983), (Title U, Report CFRD).
8. H.D. Glenn, "Diagnostics Techniques Improvement Program", Defense Nuclear Agency Report, Systems, Science & Software Rept DNA2978T (1972).
9. R.W. Terhune, LLNL, private communication (1982).
10. A.L. Florence, D.D. Keough, and R. Mak, "Calculational Evaluation of the Inclusion Effects on Stress Gage Measurements in Rock and Soil", Proc. Interaction of Non-Nuclear Munitions in Structures Symp., 10-13 May 1983, USAF Academy, Colorado Springs, CO.
11. C.W. Olsen, "Stress Measurement Development at LLNL", Proc. Monterey Containment Symp., Vol. II, 26-28 August 1981, Naval Postgraduate School, Monterey, CA, (Title U, Report CFRD).
12. R.W. Terhune and H.D. Glenn, "Estimate of Residual Shear Strength Around Explosive Cavities in Earth Media", Proc. 19th U.S. Symp. Rock Mech., Stateline, NV, (1978).

EXPERIMENTAL TECHNIQUES FOR MEASURING MASS EJECTION

FROM SHOCK-LOADED METALLIC SAMPLE

Philippe Elias, Pierre Chapron

Commissariat à l'Energie Atomique
Centre d'Etudes de Vaujours
77181 Courtry, France

1) INTRODUCTION

The evidence and the understanding of the predominant physical mechanisms with control the matter ejection driven by intense shcok waves reflecion on metallic sample free surface, require a quite large experimental approach [1,2,3]. This invloves diagnostics carefully fitted to achieve such experimental studies. In order to quantify localized or uniformly distributed ejected mass, different measurement techniques such as optical shadowgraphy, X-Ray radiography and thin foil Doppler laser interferometry (D.L.I.) have been developed and carried out in the laboratory [4,5].

2) OPTICAL SHADOWGRAPHY

Shadowgraphy consists of optical recording with Barr and Stroud CP5 ultra-high speed framing camera associated with an electronic flash which lights up the back-ground of the object and gives thereby shadow pictures (figure 2). A 4 000 turns per second rotation speed of the camera mirror allows 20 µs framing time with 170 ns between two pictures and 60 ns time-exposure per picture.

For distributed ejected mass, shadowgraphy is able to discriminate qualitatively on the same experiment two different free surface preparations, for instance finely-roughly machined (figure 1).

Time history of the evolution of localized ejecta contours (jetting volume and maximum velocity) is also recorded through this method.

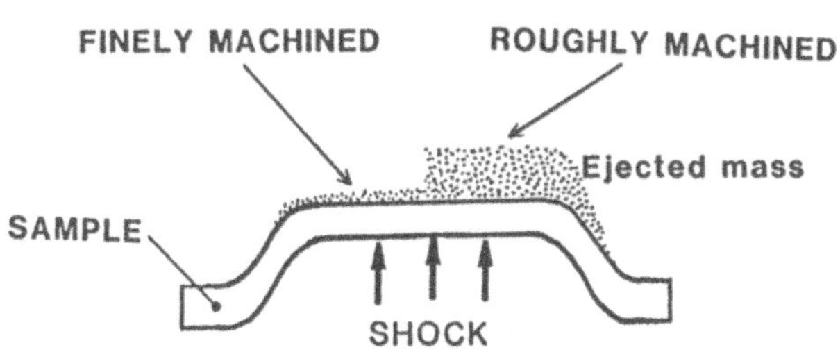

Fig.1 : Shadowgraphy. Discrimination of two different free surface preparations.

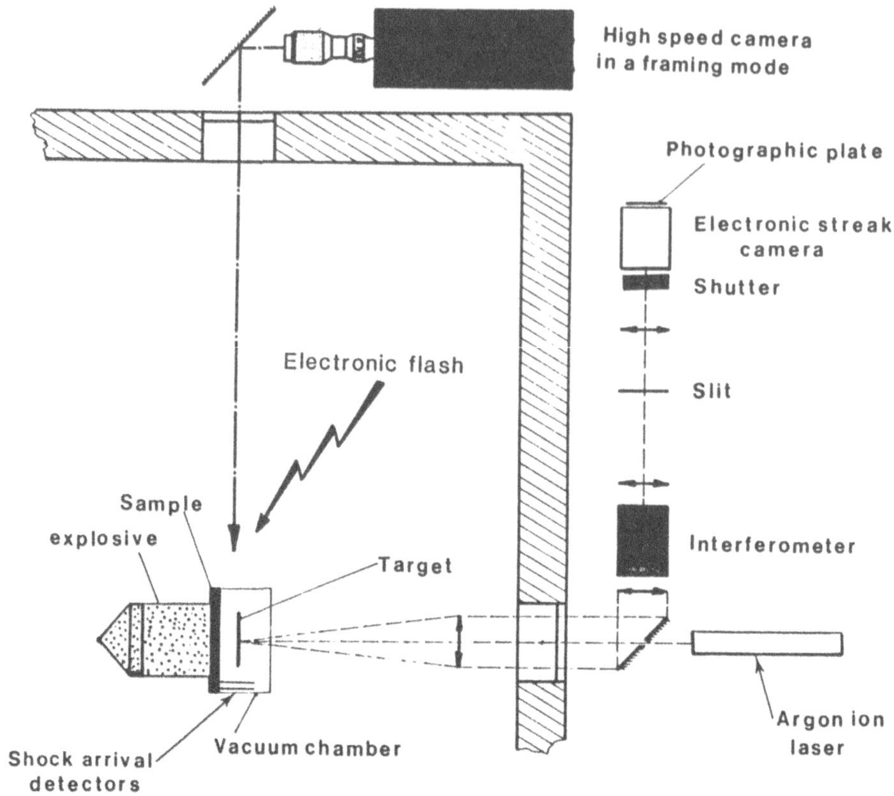

Fig.2 : Experimental device for optical shadowgraphy and D.L.I. thin foil.

3) X-RAY FLASH RADIOGRAPHY

This technique uses a 400 KV X-Ray pulse generator Scandiflash 600 providing a 15 mR exposure during 20 ns (FWHM) with a 1,2mm focalspot size in diameter. One pulse X-Ray picture is taken parallely to the shocked free surface a few microseconds after the shock arrival. With these experimental arrangements, an assessment of geometrical softfocus gives 200 µm, the actually measured softfocus on the image (with diffusion effects taken into account) is 800 µs in diameter, and the penetrated maximum ejected mass per surface area is about 10 g/cm^2 in copper (figure 3).

For distributed ejected mass measurement, assuming an uniform spatial distribution in the direction parallel to the free surface and using suitable penetrameters which give the relation between film optical density and thickness of penetrated matter, ejected material density distribution can be simply derived from an optical density record perpendicularly to the surface (figure 4).

For localized ejection studies, the same method with wire penetrameters is applied to determine the density distribution inside material jets and the total ejected mass (figure 5).

Images processing is made with a JOYCE and LOEBL microdensitometer and a more precise analysis of these distributions is developed in the laboratory using numerical image computing and tomography.

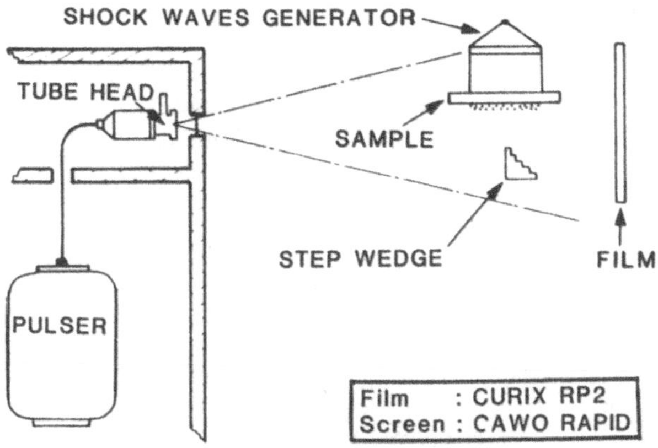

Fig.3 : Experimental device for radiographic investigation.

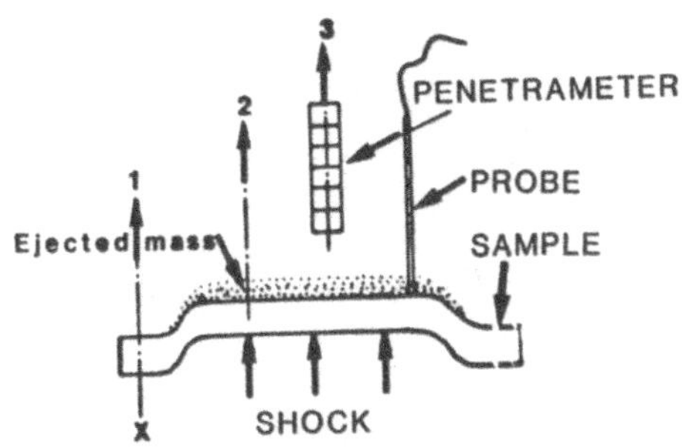

Fig.4 : Distributed ejected mass.

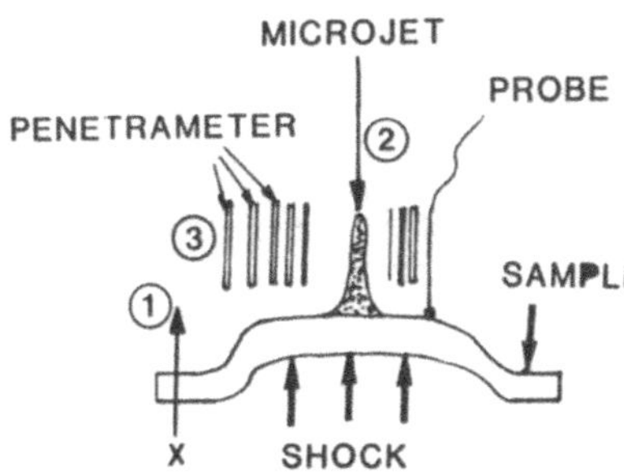

Fig.5 : Localized ejected mass.

4) THE THIN FOIL D.L.I. TECHNIQUE

This experimental technique is based upon the principle of momentum-transfer between matter ejection and a thin foil ahead of the free surface whose distance is initially specified. A D.L.I. diagnostic, associating a Fabry-Perot interferometer and a streak electronic camera TSN 503, records the thin foil velocity. The main necessary assumptions to deduce the ejected mass curve Me (amount of specific mass travelling faster than Ve) versus Ve (particule velocity) from the thin foil velocity time history are : (i) instantaneous ejection during shock reflection, (ii) uniform distribution of ejected mass parallel to the free surface, (iii) completely inelastic collision between ejecta and the thin foil.

$$Ve = \frac{d_o + \int_o^t V_f \, dt}{t} \qquad \text{particule velocity}$$

$$Me\,(Ve) = \frac{M_f \int_o^{V_f} t dV_f}{d_o - \int_o^{V_f} t dV_f} \qquad \text{ejected mass}$$

M_f : foil mass $\qquad$ V_f : foil velocity

d_o : initial distance between the foil and the sample surface.

In the experimental arrangement, the sample free surface and the thin foil are placed in a vacuum chamber (evacuated from 10^{-2} to 10^{-3} Torr) closed by a window, to suppress atmospheric pressure influence on the ejected particules.

According to the amount of expected ejected mass, the foil thickness (in aluminum, steel or nickel) range from 0,3 µm to 400 µm, and the measurable ejected mass from 1 µg/cm² to 100 mg/cm² (figure 2, 6, 7 and 8)

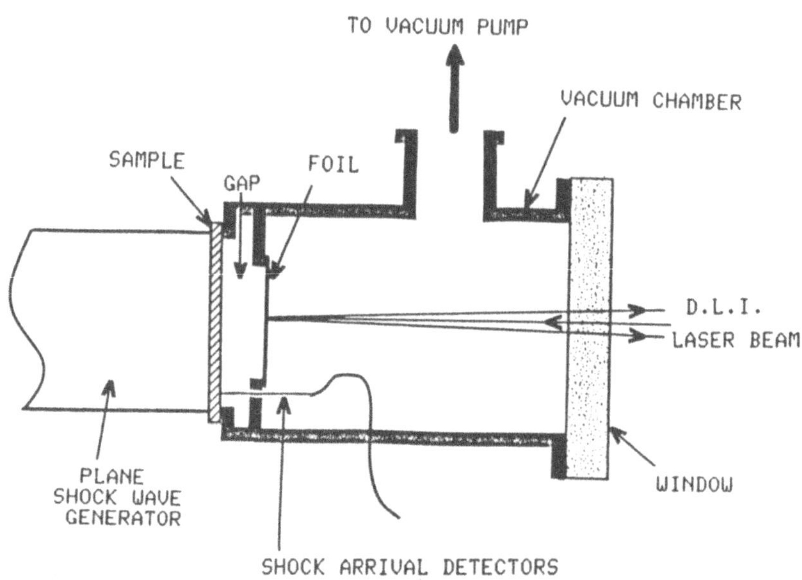

Fig.6 : Mass measurement experimental device.

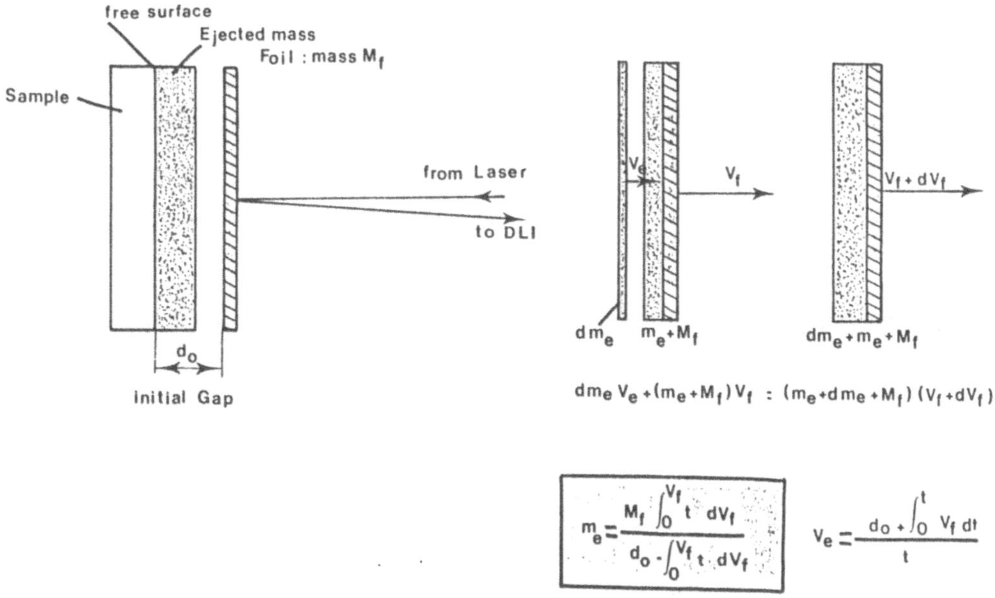

Fig.7 : Principle of mass measurement.

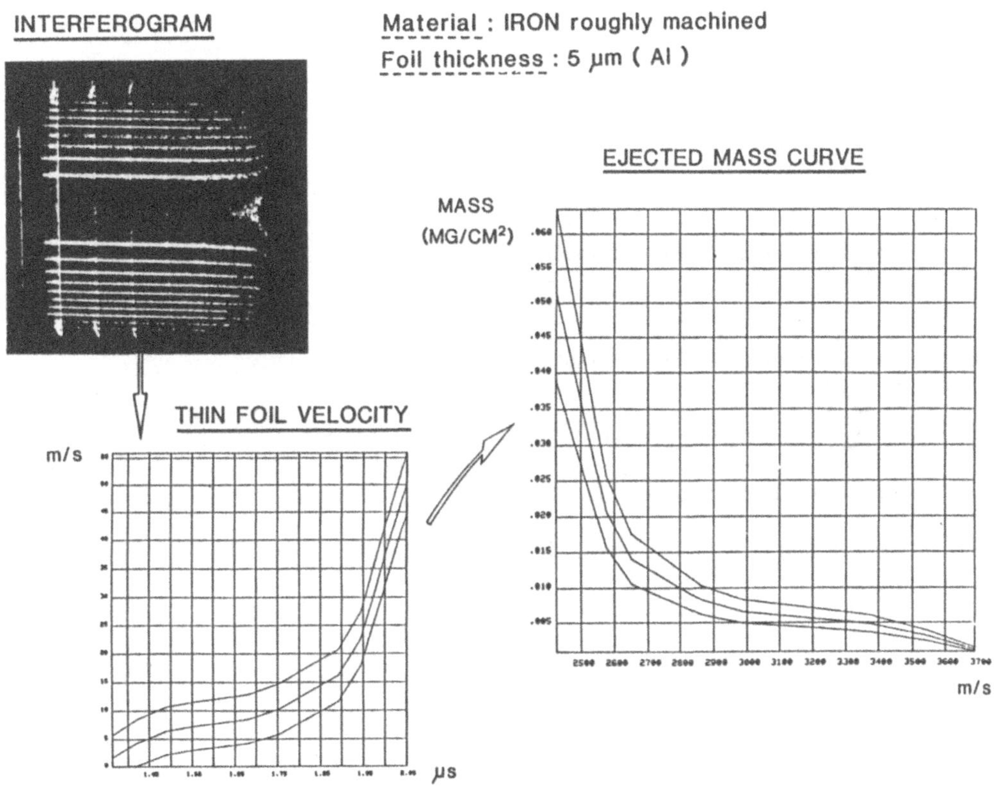

Fig.8 : D.L.I. technique for mass measurement. Example.

5) SHOCK GENERATORS

In the range of 50 GPa in the studied metallic sample, most of the experiments related in this paper use explosive plane generators. However, a powder gun 60 mm in diameter and a large 2 stage light gas gun 80 mm in diameter are used for studying shock pressure and shock melting influences on matter ejection. The shocks generated by the powder gun range from 10 GPa to 40 GPa in tin (projectile velocities from 1000 m/s to 1800 m/s with 200 g weight), while the 2 stage light gas gun reaches 50 to 250 GPa in tantalum (projectile velocities range from 1800 m/s to 4500 m/s with 900 g weight).

6) CONCLUSION

To explore this large experimental program and understand the different phenomena related to matter ejection, we have developed (associated with a micro-contact chronometry) three complementary diagnostic : optical shadowgraphy which gives a qualitative idea of distributed ejection and permits a velocity measurement of localized ejection, X-Ray radiography which records spatial density distribution inside localized or distributed ejecta, and the ejected mass can be quantitatively measured (only for distributed ejecta) through the D.L.I. thin foil method.

Acknowledgements

The authors wish to thank C. BIZEUIL, M. JOLY, B. LAURENT for diagnostics developpement, and J. FOUCART, J. MATHIAS for experimental technical support.

References

[1] P.ANDRIOT - P.CHAPRON - F.OLIVE
Shock waves in condensed matter
MENLO PARK 81.

[2] P.ANDRIOT - P.CHAPRON - V.LAMBERT - F.OLIVE
Shock waves in condensed matter,
SANTA FE 83

[3] R.CHERET - P.CHAPRON - P.ELIAS - J.MARTINEAU
Shock waves in condensed matter
SPOKANE 85

[4] C.BIZEUIL - P.CHAPRON - P.ELIAS - B. LAURENT
16ème Congrès International de photographie rapide et de photonique
STRASBOURG, Août 84.

[5] M.DURAND - P.LAHARRAGUE - P.LALLE - A.LE BIHAN - J.MORVAN - H.PUJOLS
Rev. Sci. Instrum. 48 (1977) 275.

MASS EJECTION FROM THE FREE SURFACE OF SHOCK-LOADED

METALLIC SAMPLES

Roger Cheret, Pierre Chapron, Philippe Elias
and Jacques Martineau

Commissariat à l'Energie Atomique
Centre d'Etudes de Vaujours - 77181 Country - France

Introduction

When a shock wave emerges from a solid metallic sample, it often happens that the expanding free surface is actually moving behind a cloud of so-called ejected mass. Many features of these ejection phenomena have been described during the past few years[1,2,3] by varying

. the bulk nature of the sample,
. the state and geometrical aspect of the surface,
. the level and profile of the incident shock.

Some thorough insight has been achieved by combining various diagnostic techniques[3]. From this research program, this paper has selected those results which deal with the ejection from pure metallic samples carefully prepared in order to ensure bulk-state homogeneïty and geometrically well-defined free surface.

The function M_j

When a metallic thin foil, initially ahead of the free surface of the sample is impacted by the ejected grains, it begins to move forward and accelerates through inelastic momentum transfer. A Doppler-laser interferometry record of the movement of the foil enables -as already shown in details[2,3] - to measure the total ejected mass M_e (per unit area of emetting surface) whose velocity exceeds a given value v_e. From this curve M_e (v_e), one may extract many valuable informations, among which is the total ejected mass M_j whose velocity exceeds the classical free velocity v_s.

In order to get to an easier comparison of the values of M_j for samples spread in a wide range of initial density, we generally consider the ratio M_j/M_{ref}, where M_{ref} is defined as the average lack of mass correlated to the volume of the "defects" (per unit area of emitting surface) computed from the actual measured rugosity.

The first series of results is related to samples on the finely-machined free surface of which parallel grooves are machined 130° in angle, 10 µm in depth, 40 µm in spacement. Eight different metals were tested, whose standard melting temperature $T°_f$ ranges from 232C (tin) to 2997C (tantalum). Table 1 and Figure 1 summarize results which are related to the same explosive plane wave generator : the mentioned

Table 1. Values of M_j/M_{ref} for eight materials with same surface preparation shocked by same explosive lens

	T^o_f C	P GTa	M_j/M_{ref}
Sn..........	232	48	1,0 ± 0,2
Zn	420	51	4,1 ± 1,5*
Al	660	40	0,69 ± 0,15
Cu	1 084	56	0,57 ± 0,15
Fe	1 535	55	0,005 ± 0,002
Zr	1 852	49	0,005 ± 0,002
Nb	2 497	56	0,003 ± 0,001
Ta	2 997	63	0,003 ± 0,001

* average of two shots

shock induced pressure p results from the usual impedance mismatch procedure.

In spite of rather close shock impedance, the eight materials exhibit scaled values of M_j spread over three orders of magnitude. This statement invites to look for a deeper view of the role of the so-called incident pressure, and to investigate the possible role of shock induced melting.

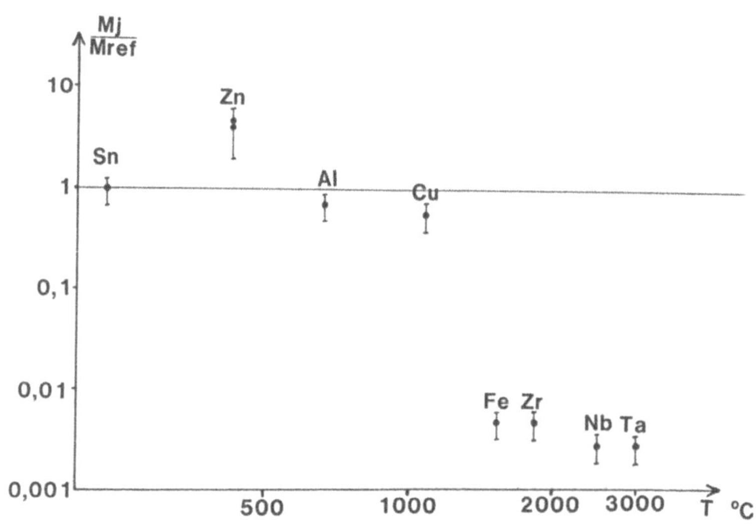

Fig. 1 . Values of $M_j/M_{ref.}$ for eight materials, with same surface preparation, shocked by same explosive lens, as a function of the standard melting temperature T^o_f C.

The control variable : p.

For the tantalum and copper samples, with the same surface preparation, the shock-induced pressure p was used as a control variable by completing impact experiments with a two-stage-light gas gun. Thus, pressures ranging from 62 to 209 GPa in tantalum and reaching 80 GPa in copper, were achieved. The DLI records show a smooth overall increase of M_j and of the maximum particle velocity, when p increases. These records lead to a simple linear relationship between M_j/M_{ref} and p (cf.Fig.2) for tantalum.

For tin samples, with again the same surface preparation, some other experiments were completed with a powder gun, involving lower pressures. A drastic decrease of M_j appears (cf. Fig. 2) when pressure drops from 48 GPa to 30, whose order of magnitude (a factor of 10) is very similar to that evidenced for a finely machined surface when decreasing the pressure from 50 to 20 GPa.

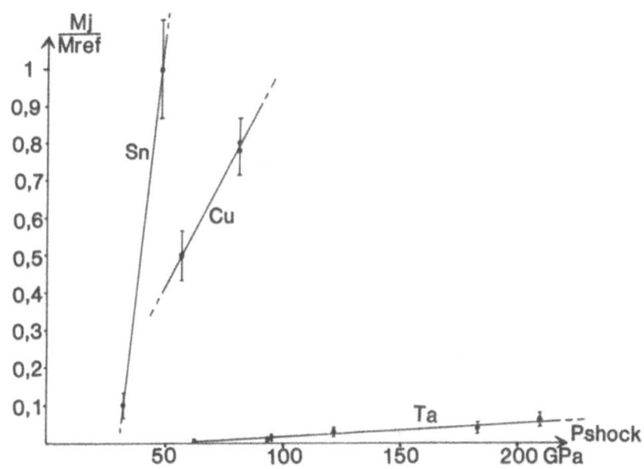

Fig. 2 Values of M_j/M_{ref} for grooved finely-machined samples as a function of the so-called shock induced pressure p.

Frame of overall interpretation

The above experimental results, complemented by X-Ray measurements[3,4,5], together with previous ones, allow to put forward the following assumptions :

(i) the *initiation* of jets when the incident shock interacts with the slopes of the *main* geometrical defects play the dominant role in the ejection phenomenon,

(ii) the scattered dependance of M_j/M_{ref} on the shock-induced compression is due to the fact that the *feeding* of these jets is highly dependant on the spatial occurrence of melting.

The second assumption may be explicited as follows. When the incident compression does not involve any melting or involves only heterogeneous melting correlated with the defects distribution (for instance p $>$ 209 GPa in Ta), then the slow increase of M_j with pressure reflects the independant extension of the domains concerned with the feeding of the jet issued by each defect. When the incident compression does involve homogeneous melting (for instance p $\gtrsim$ 48 GPa in Sn), then the rapid increase of M_j with pressure reflects the coalescence of the above mentioned domains.

In order to test these assumptions, a scaled compressible fluid model has been set up[6] for the initiation and feeding of a micro-jet. This model involves two parameters which are relevant to the local mechanism of jet formation.Fig. 3a and 3b show the theoretical fitting to copper samples experiments in two quite differents cases : isolated conical hole (aperture 2α ranging from 30° to 160°) and "grooved finely machined" (p = 56 and 80 GPa). These examples and other ones support the validity of the basic assumptions (i) and (ii).

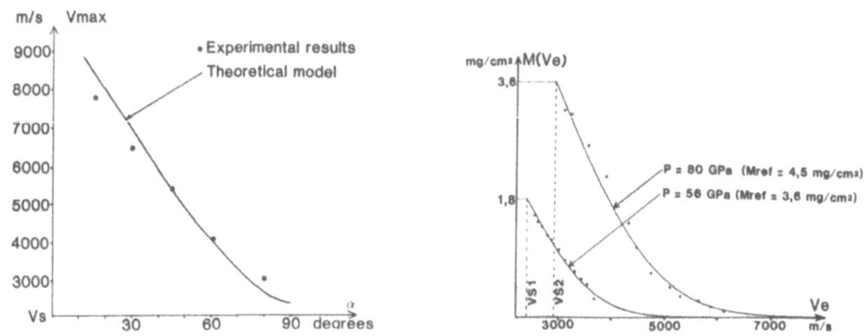

Fig. 3. Copper samples (a) grooved finely machined (b) unique conical hole, 2α in aperture. Comparaison of fitted theoretical model and of experimental data obtained through (a) DLI thin foil velocimetry (b) optical shadowgraphy and X-Ray picture.

This frame of interpretation may be easily extended to alloys where surface state heterogeneïty exists together with surface geometry heterogeneïty. For Sn Pb polished samples (pure Sn grains in an eutectic matrix of higher density) undergoing a shock leading to homogeneous fusion of the phases, the increase of M_j when % Pb increases[2] is due to heterogeneous compression field which enhances the ejection of the most compressed one, i.e. the eutectic. For Cu Pb polished samples (pure Cu grains in an eutectic matrix of lower melting curve) undergoing a shock leading to melting of the only matrix, the increase of Mj when % Pb increases[2] is due to heterogeneous melting field which enhances the ejection from the melting one, i.e. the eutectic.

REFERENCES

1. J. R. ASAY, L.P. MIX, M.C. PERRY, Appl. Phys.Lett, 29, 5 (1976).

2. P. ANDRIOT, P. CHAPRON, F. OLIVE, Ejection of material from shocked surfaces of tin, tantalum and lead-alloys, in "Shock waves in condensed matter (1981), A.I.P., New York (1982).

3. P. ANDRIOT, P. CHAPRON, V. LAMBERT, F. OLIVE, Influence of melting on shocked free surface behavior using Doppler-laser-interferometry and X-Ray densitometry, in "Shock waves in condensed matter 1983", Elsevier Science publishers, (1984).

4. P. ELIAS, P. CHAPRON, Experimental techniques for measuring mass ejection from shock-loaded metallic samples, in "Shock waves in condensed matter 1985", Plenum, (1986).

5. R. COUCH, L. SHAW, R. BARLETT, L. STEINMENTZ, W. BEHRENDT, C. FIRPO. Dymat, Paris (1985), à paraître.

6. P. ELIAS, P. CHAPRON, J. MARTINEAU -unpublished results.

HETEROGENEOUS RESPONSE OF ALUMINUM AND LITHIUM NIOBATE TO SHOCK WAVES

Frank C. Perry and Donald D. Noack

Sandia National Laboratories
P. O. Box 5800, Organization 1263
Albuquerque, NM 87185

ABSTRACT

Heterogeneous shock effects were studied in aluminum (6061-T6) and single crystal lithium niobate (Z-cut) in the range 0.7-4.5 GPa by means of holographic interferometry. Heterogeneous yielding in lithium niobate was observed directly for the first time. In addition, nucleation or activation of defects were observed in both aluminum and lithium niobate. The single crystal data indicate that a defect-dominated fracture process constitutes a reasonable model for shock compression. The polycrystalline data suggest that a similar pattern will prevail at a sufficiently high pressure.

INTRODUCTION

Heterogeneous shock effects in materials have been observed for many years. Both recovery experiments as well as anomalies in wave profile and Hugoniot data have led investigators[1] to speculate on microscopic models of shock compression that include aspects of heterogeneity.

In an effort to understand the behavior of real materials under shock loading, we have concentrated on a diagnostic technique which is resolved in both time and space: holographic interferometry. It was recently shown[2] by this technique that the variation in orientation between grains in a polycrystalline metal leads to heterogeneous deformations in response to particle beam-induced shock waves. These deformations contain dominant wavelengths equal to or greater than the grain size corresponding to individual or aggregate grain motion. Copper demonstrated heterogeneous behavior for both compressive and reverse plastic loading for pressures $\leqslant$ 2.0 GPa. Aluminum (6061-T6) deformed heterogeneously only under reverse plastic straining[3] and deformed homogeneously under compressive straining for pressures less than the Hugoniot elastic limit (HEL = 0.6 GPa).

The primary purpose of the present study is to extend the measurements for aluminum to higher pressures (above the HEL). However, since all the polycrystalline data have shown the strong influence of grain boundaries, it was important also to investigate the response of a single crystal. We chose Z-cut lithium niobate because (1) it was readily available and easy to handle, (2) has a low HEL (2.4 GPa) so that the particle velocities might not be too high for our holographic system, and (3) has been shown to exhibit a strong reduction in shear strength[4] for shock amplitudes just above the HEL.

We will present data for aluminum in the range 0.7–3.3 GPa and for lithium niobate in the range 1.0–4.5 GPa. All of the results will be discussed in the context of the probable physical mechanisms which dominate the shock process.

EXPERIMENTAL TECHNIQUE

The holographic technique has been adequately described in Refs. 2 and 3. The double-pulse system,[5] employing a ruby laser with a 3 ns pulse-width, produces a series of four exposures of a target rear surface with a time separation of 26 ns during the shock event. The results presented here were obtained using the standard double pulse holographic interferometry where the first pulse occurred prior to the experiment and the second pulse occurred during the shock event. The interference pattern was then proportional to the surface deflection, each fringe corresponding to a change in position of one-half the ruby wavelength or about 0.35 μm.

A single-stage compressed light gas gun[6] was used in the planar plate impact experiments. Both the gun and the projectile-impactor construction are described in Ref. 6. The interferometer system requires two triggers—one for firing the laser flash lamps (approximately 1.4 msec prior to impact) and one for firing the Pockels cell. In order to facilitate the flash lamp trigger a CW laser was directed perpendicular to the direction of projectile motion, and the cutting of the laser beam by the impactor generated a photo-diode signal to the flash lamp circuit. Pockels cell triggering was accomplished by contact of the impactor with a standard pin assembly positioned in the target mounting structure.

All the experiments were conducted at a projectile velocity of 0.3 km/sec. The pressure range was varied by using impactors of PMMA, aluminum (6061-T6), and single crystal saphire (Z-cut). The Pockels cell trigger described above was used to sequence in time the holographic images of the target back surface after passage of the shock wave.

RESULTS

A holographic interferogram of the back surface of an aluminum target after being shocked by a composite impactor is shown in Fig. 1(a). The center disk of the impactor was made of PMMA (diam = 11.4 mm) and generated a 0.7 GPa shock. The outer aluminum ring (outer diam. = 5.1 cm) generated a 2.2 GPa shock. The interferometer was aligned 5.7 mm off the impactor axis of symmetry so that the effects at 0.7 and 2.2 GPa could be viewed simultaneously. The timing of the interferogram corresponds to that required for interaction of the plastic wave with the free surface. One notes that there is a qualitative difference between the effects at 0.7 and 2.2 GPa. At 0.7 GPa, just above the HEL of 0.6 GPa, one observes a mottled pattern—perhaps indicative of slipped regions corresponding to activation or nucleation of individual defects or defect clusters. At the higher pressure of 2.2 GPa, well above the HEL, one notes the pattern of grain aggregates and single grains that were described in detail in Ref. 2. The pattern evolves slowly before the unloading wave and is similar to that in Fig. 1(a). In Fig. 1(b) we show for comparison the effect of shocking lithium niobate to a pressure of 1.0 GPa. Note the strictly homogeneous deformation indicative of shock tilt. Using a single component impactor at a pressure of 3.3 GPa, an interferogram of the rising-plastic wave in aluminum shows a mottled pattern, along with possible incipient fracture

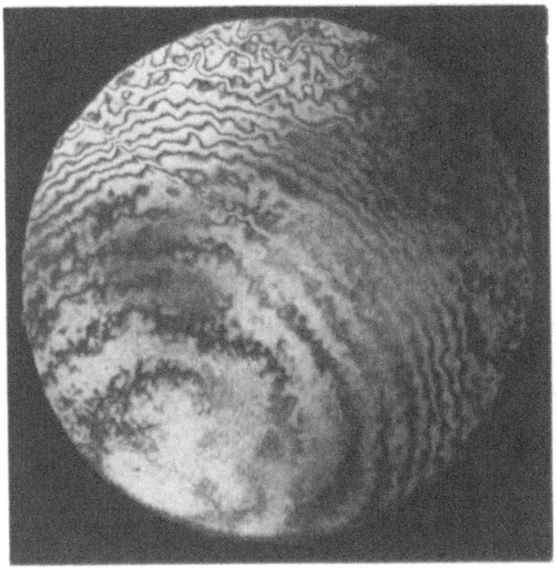

1(a)

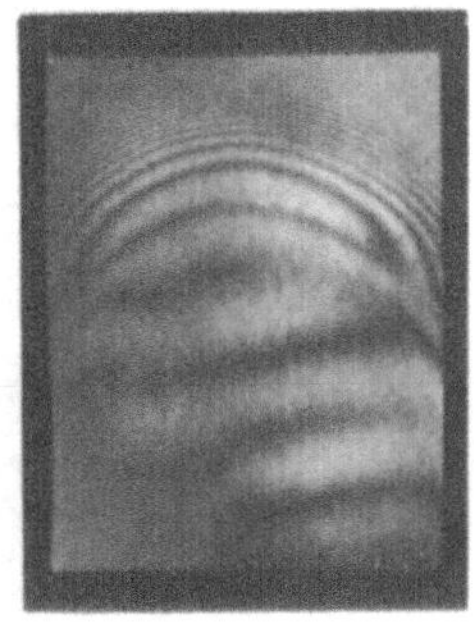

1(b)

Fig. 1(a). Interferogram of aluminum back surface after being shocked by PMMA and aluminum impactor. The lower left corner corresponds to the PMMA impactor (0.7 GPa); the upper right corner corresponds to the aluminum impactor (2.2 GPa); the center region with more closely-spaced fringes corresponds to a pressure gradient (non uniaxial strain) arising from the PMMA-aluminum boundary. The circle diameter is 11.8 mm.

Fig. 1(b). Interferogram of lithium niobate back surface after being shocked by PMMA and aluminum impactor. The linear fringes indicate homogeneous deformation with shock tilt at a pressure of 1.0 GPa. The closely spaced circular fringes correspond to the PMMA-aluminum boundary.

lines in Fig. 2. One notes in Fig. 2 that the homogeneous shock tilt evident in Fig. 1(b) is radically altered. Also, the individual grain outlines observed at 2.2 GPa are not evident in the interferogram of Fig. 2.

With these polycrystalline results as a background, we next examined single crystal lithium niobate. An interferogram of the back surface of a lithium niobate target after being shocked to 2.7 GPa (HEL = 2.4 GPa) by an aluminum impactor is shown in Fig. 3. One notes here the complex pattern of cavities and ridges (dominant wavelength $\sim$ 1 mm) indicative of slipped regions of the crystal. This pattern is maintained, except for small ripples, for at least 0.080 μs. In another experiment at 2.7 GPa, but somewhat earlier in time, the lithium niobate demonstrated a mottled pattern similar to that for aluminum at 0.7 GPa (see Fig. 1(a) - lower left corner). At a pressure of 4.5 GPa (well above the HEL of 2.4 GPa) nearly homogeneous deformation, except for a small ripple, is the experimental result as shown in Fig. 4. Therefore, it appears that only in a relatively small pressure region above the HEL are strongly heterogeneous deformations observed.

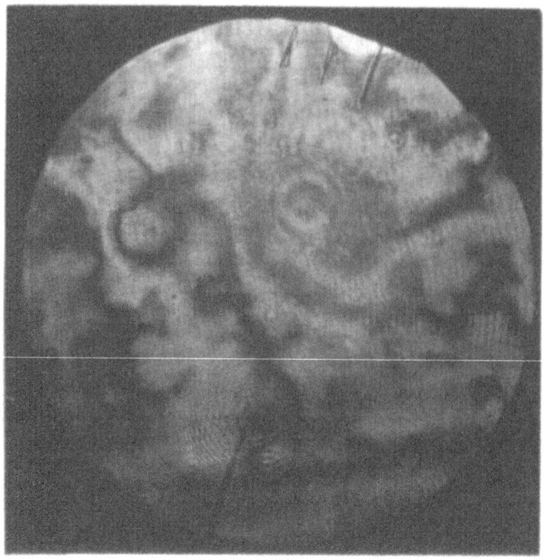

Fig. 2. Interferogram of aluminum back surface after being shocked by single crystal saphire impactor. The pressure is 3.3 GPa. Note the long wavelength perturbations—suggesting the onset of fracture. The circle diameter is 11.8 mm.

Fig. 3. Interferogram of lithium niobate back surface after being shocked by aluminum impactor. The pressure is 2.7 GPa. The circle diameter is 11.8 mm.

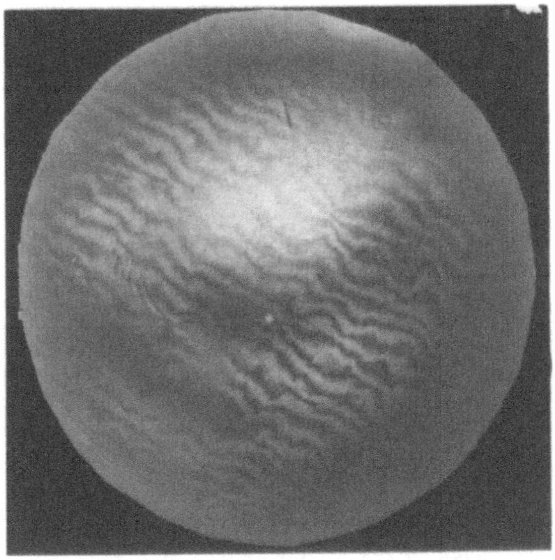

Fig. 4. Interferogram of lithium niobate back surface after being shocked by saphire impactor. The pressure is 4.5 GPa. Note the nearly homogeneous behavior except for a small ripple on the linear fringes. The linear fringes are due to shock tilt. The circle diameter is 11.8 mm.

CONCLUSIONS

In spite of the large differences in properties between the two materials under investigation a pattern of physical processes seems to be operative. In both cases homogeneous deformation occurs for shock pressures below the HEL. For aluminum at a pressure only slightly above the HEL one observes a mottled pattern suggesting the activation or nucleation of defects or clusters of defects. At a pressure of 2.2 GPa, aluminum demonstrates the characteristic heterogeneous grain structure with no visible structure within the grains. A pattern containing longer wavelengths occurs in the rising portion of plastic waves of significantly higher peak pressure (3.3 GPa). For lithium niobate at 2.7 GPa the defect dominated (mottled) pattern occurs early in time, and the stationary ridges and cavities occur after the wave is more fully developed. At a pressure of only 4.5 GPa nearly homogeneous deformation is evident in the inelastic wave response.

The picture for lithium niobate suggests a defect-dominated fracture process in the pressure range 2.7–4.5 GPa. This view is supported by the recent results of Brannon et al[7,8] where heterogeneous luminescence patterns in Z-cut lithium niobate and x-cut quartz were observed at pressures exceeding the HEL. In addition, band-like emission spectra (rather than black body) were observed suggesting a molecular fracture process.

The data for aluminum also indicates shock processes which are dominated by defects. At a pressure of only 2.2 GPa the grain boundaries are the location of heterogeneous response. However, at 3.3 GPa possible fracture lines and other effects are observed. This response is likely related to the well-known, non-ideal plastic behavior of aluminum at these levels. At sufficiently high pressures, where non-ideal effects become less important, the shock response may also be dominated by a fracture process.

ACKNOWLEDGEMENTS

The authors thank the members of the STAR facility for their assistance in these experiments. We especially thank C. Konrad for the use of his photo-diode trigger system and J. R. Asay for comments and suggestions regarding many phases of this work. This work was supported by the U. S. Department of Energy under Contract Number DEAC04-76-DP00789.

REFERENCES

1. L. Davison and R. A. Graham, Shock Compression of Solids, Physics Reports 55:255 (1979).
2. F. C. Perry, Dynamic Heterogeneous Response of Aluminum and Copper to Stress Waves, in "Shock Waves in Condensed Matter-1983," Chap. VI:10, 255, Elsevier Science Publishers B.V., Amsterdam, 1984.
3. F. C. Perry and L. P. Mix, Electron Beam Driven Instability in a Solid, Appl. Phys. Lett. 27:194 (1975).
4. P. L. Stanton and R. A. Graham, Shock-Wave Compression of Lithium Niobate from 2.4 to 44 GPa, J. Appl. Phys. 50:6892 (1979).
5. L. P. Mix and R. W. Kessler, Four Exposure Holographic System, Sandia Labs Report SAND77-0209, March 1977.
6. J. R. Asay, The Sandia National Laboratories Shock Thermodynamics Applied Research (STAR) Facility, SAND81-1901, August 1981.
7. P. J. Brannon, R. W. Morris, C. H. Konrad and J. R. Asay, Shock-Induced Luminescence from X-Cut Quartz and Z-Cut Lithium Niobate, in "Shock Waves in Condensed Matter-1983," Chap. VII:3, 303, Elsevier Science Publishers B.V., Amsterdam, 1984.
8. P. J. Brannon, R. W. Morris, C. H. Konrad and J. R. Asay, Shock-Induced Luminescence from Z-Cut Lithium Niobate, J. Appl. Phys. 57:1676 (1985).

ASYMMETRIC SUPERSONIC COLLISIONS

Wang Jihai

Beijing Institute of Technology
Department of Engineering Mechanics
P.O. Box 327, Beijing, China

INTRODUCTION

There are two regimes in high velocity oblique impact of metal surfaces: collision with jet formation and collision with attached shock waves. The calculation of pressure behind the attached shock waves and direction of interface in jetless impact and establishment of criteria separating these two regimes are actual problems in practical applications, for instance, in explosive welding and in collapsing shaped charges. Many authors have investigated the criteria of jetless impact of symmetric collision [1-3]. In this paper, the asymmetric oblique collision of various metals is treated by means of the shock polar method, which was applied to the interaction of detonation wave with metals [4].

SHOCK POLAR METHOD [4]

We assume that metals at high pressure are considered as inviscid compressible fluid. The relations connecting the parameters behind and ahead of the shock front will be given. Let P_o, ρ_o, $\vec{q}_o$ and P_1, ρ_1, $\vec{q}_1$ denote pressure, density and velocity of fluid ahead and behind the oblique shock front OS respectively (Fig. 1), and N_i, L_i (i = 1, o) denote the corresponding normal and tangential components of velocities to the shock front.

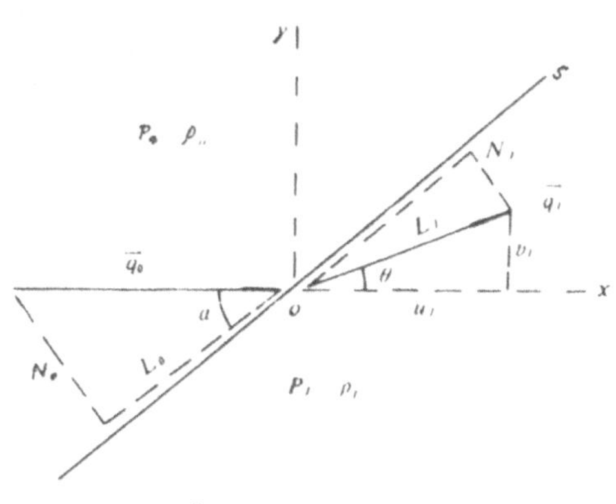

Fig. 1.

The conservation of mass and momentum are

$$\rho_o N_o = \rho_1 N_1$$

$$P_o + \rho_o N_o^2 = P_1 + \rho_1 N_1^2 \tag{1}$$

The continuity of tangential component of velocity across the shock front is

$$L_o = L_1 \tag{2}$$

From equation (1), the expression

$$N_o - N_1 = \sqrt{(P_1 - P_o)[(1/\rho_o) - (1/\rho)]} \tag{3}$$

is obtained. Putting the x axis with $\vec{q}_o$ and y perpendicular to it, we have

$$\vec{q}_o = q_o \vec{i}$$

$$\vec{q}_1 = u_1 \vec{i} + v_1 \vec{j}$$

where $\vec{i}$ and $\vec{j}$ denote the unit vectors of the x and y axes. Using (2), the momentum equation is rewritten as

$$P_1 - P_o = \rho_o N_o (N_o - N_1) =$$

$$\rho_o \{N_o (N_o - N_1) + L_o (L_o - L_1)\}$$

$$= \rho_o \vec{q}_o \cdot (\vec{q}_o - \vec{q}_1) = \rho_o q_o (q_o - u_1)$$

thus

$$u_1 = q_o - \frac{P_1 - P_o}{\rho_o q_o} \tag{4}$$

The square of difference of velocities is

$$(\vec{q}_o - \vec{q}_1)^2 = (N_o - N_1)^2 = (q_o - u_1)^2 + v_1^2$$

From (3) and (4) we have

$$v_1^2 = (N_o - N_1)^2 - \left(\frac{P_1 - P_o}{\rho_o q_o}\right)^2 = (P_1 - P_o)\left(\frac{1}{\rho_o} - \frac{1}{\rho_1}\right) - \left(\frac{P_1 - P_o}{\rho_o q_o}\right)^2 \tag{5}$$

Therefore the angle between the $\vec{q}_1$ and $\vec{q}_o$ is θ; we have

$$\tan\theta = \frac{v_1}{u_1} = \pm \frac{\sqrt{(P_1 - P_o)\left(\frac{1}{\rho_o} - \frac{1}{\rho_1}\right) - \left(\frac{P_1 - P_o}{\rho_o q_o}\right)^2}}{q_o - \frac{P_1 - P_o}{\rho_o q_o}} \tag{6}$$

If the form of internal energy function is known, the relation between ρ_1 and P_1 may be determined by Hugoniot equation; the equation (6) represents the relation between P_1 and θ. The curve indicating this relation in (P, θ) plane is called shock polar. For condensed matter the Hugoniot equation may be given approximately by [5]

$$D = C_o + \lambda U$$

Where D is the velocity of the shock wave, U is $N_o - N_1$, and C_o and λ are constant. For $P_o = 0$, the shock polar relation may be written as

$$P = \rho_o U(C_o + \lambda U)$$

$$\tan\theta = \pm \frac{U\sqrt{q_o^2 - (C_o + \lambda U)^2}}{q_o^2 - U(C_o + \lambda U)} \tag{7}$$

where the U may be considered as a parameter. The condition of existence of the shock wave is

$$q_o > C_o \tag{8}$$

OBLIQUE COLLISION OF MOVING SURFACE WITH TARGET

We consider a collision of a moving metal surface with a target. The materials of the flyer and the target may be different. The angle between these surfaces at the moment of impact is denoted by a_o. We suppose that the velocity of moving surface $\vec{u}_o$ is perpendicular to this surface and is large enough with a sufficiently small angle a_o, so that the shock waves attached to the collision point are produced. In a coordinate system moving with the collision point, the velocities of the flyer and target surface before collision are

$$q_o = \frac{u_o}{\sin a_o}$$

$$q_1 = \frac{u_o}{\tan a_o} = q_o \cos a_o \tag{9}$$

Let the x axis parallel the $\vec{q}_o$; in (P, θ) plane we have two initial points (0,0) and $(0, a_o)$. If both velocities are supersonic, the related shock polar will be established for every media. The points of intersection of these shock polars are the solutions of our problem. In Fig. 3, the point c with lower pressure, corresponding realized case, is

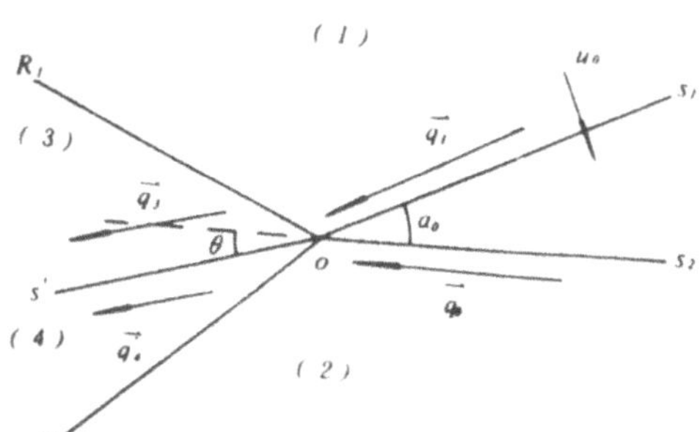

Fig. 2.

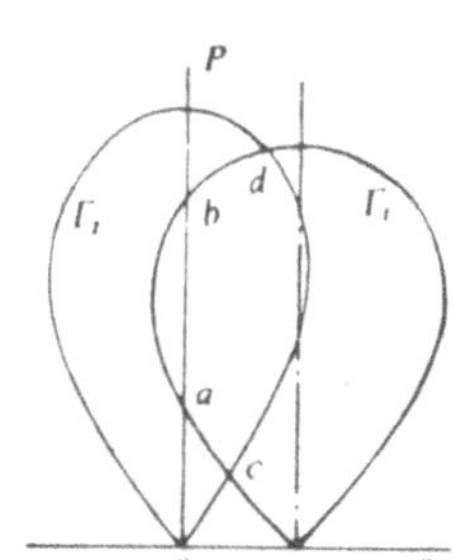

Fig. 3.

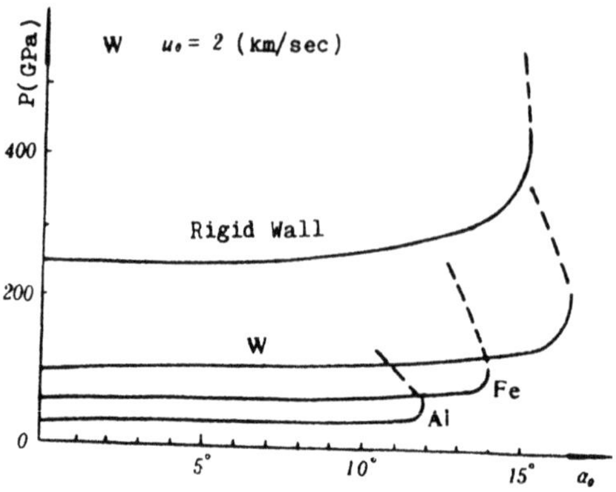

Fig. 4.

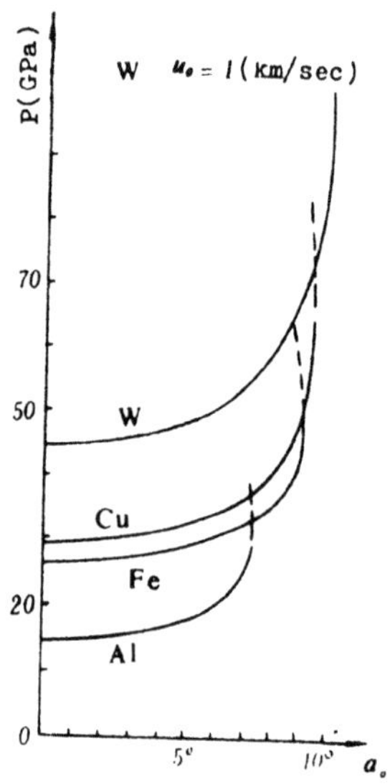

Fig. 5.

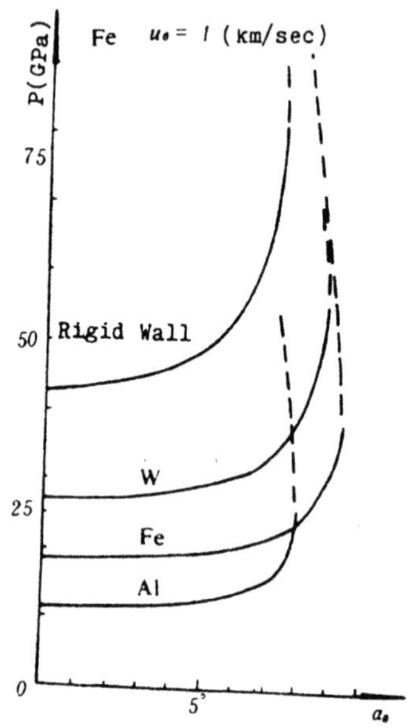

Fig. 6.

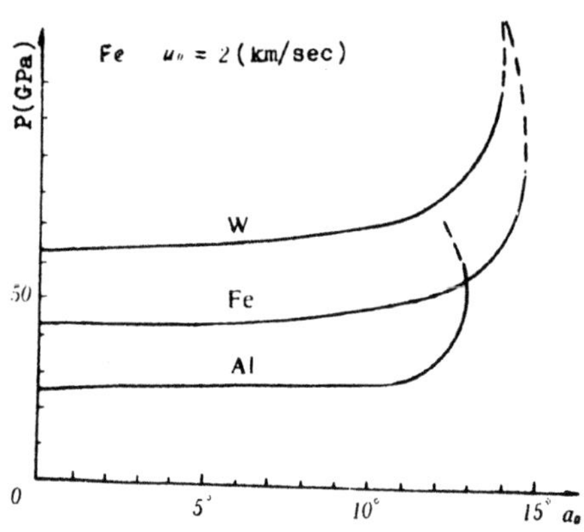

Fig. 7.

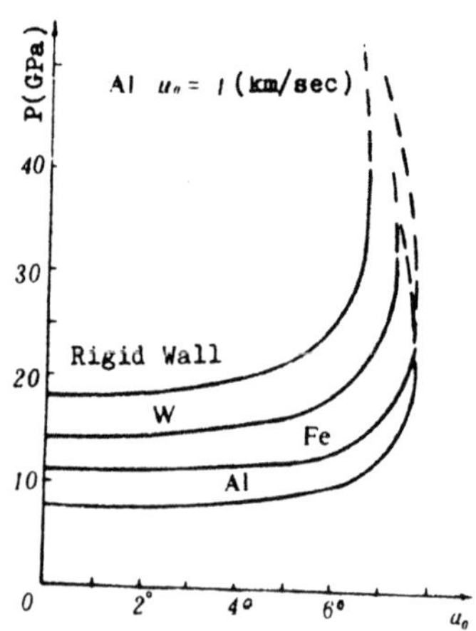

Fig. 8.

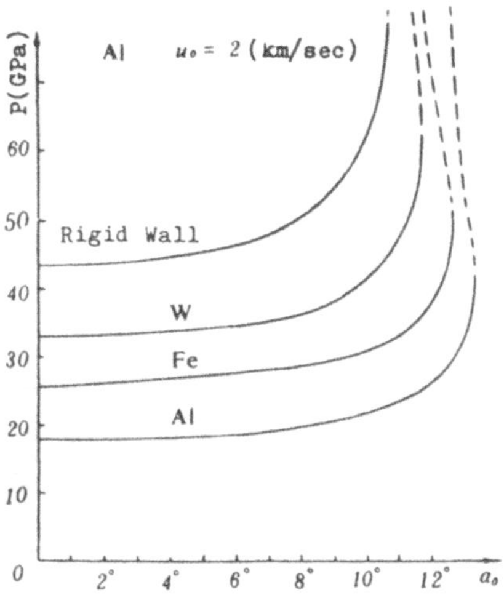

Fig. 9.

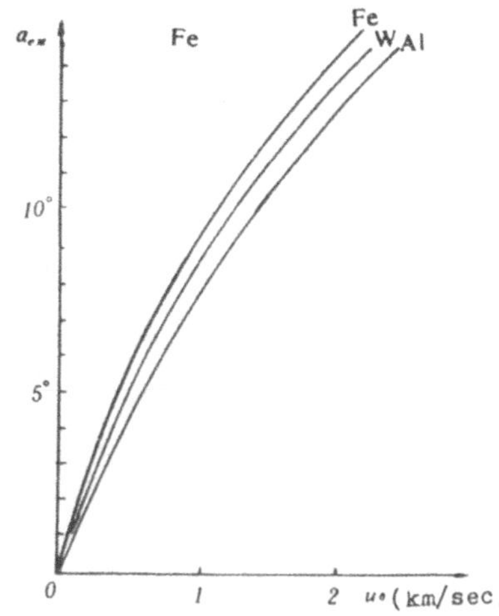

Fig. 10.

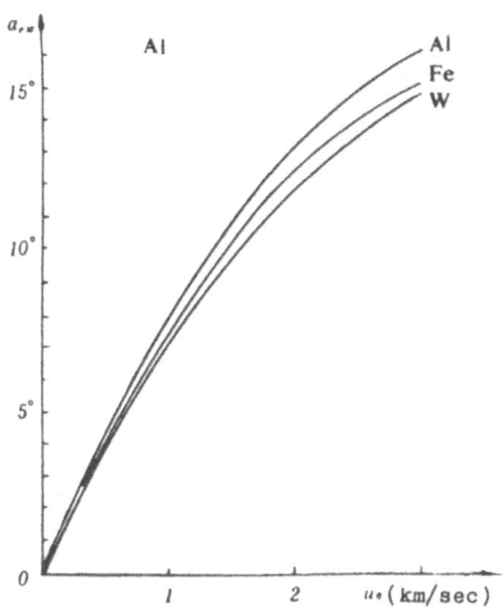

Fig. 11.

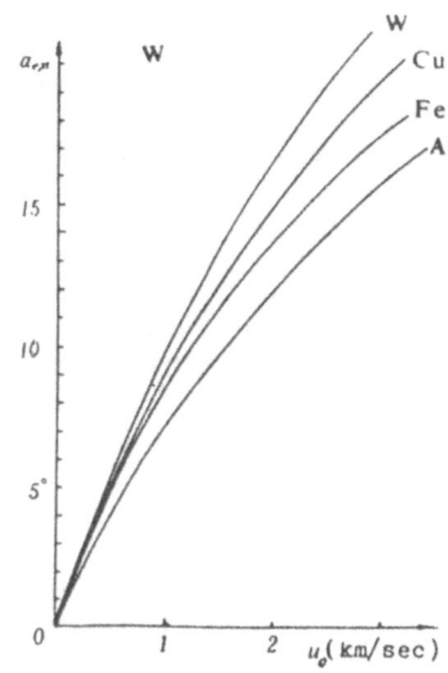

Fig. 12.

called weak solution. Another point, d, with higher pressure is called strong solution, which is not commonly realized in practice. Thus for given u_0 and colliding materials, the pressure behind formed shock waves is a function of collision angle a_0. The curves describing the above function are calculated and given in Figs. 4-9 for given u_0 and different flyer materials. The material of the target is denoted at every curve. The points of intersection of polar Γ_1 with the axis in Fig. 3 are a and b, which correspond to the collision with a rigid wall (symmetric impact). The dotted lines are the strong branch of solution and the continuous lines represent the weak solution. When a_0 increases, both branches will be approached. When a_0 is larger than a critical value, the above shock polars do not intersect and a gel is produced near the collision point. The critical value of the collision angle depends on the velocity of impact and properties of colliding materials. The curves of critical angle a_{exp} are given in Figures 10-12.

It is found that when the flyer and target consist of the same material, the critical angle takes the largest value, because in this case the two above shock polars are similar to each other and have a good coupling.

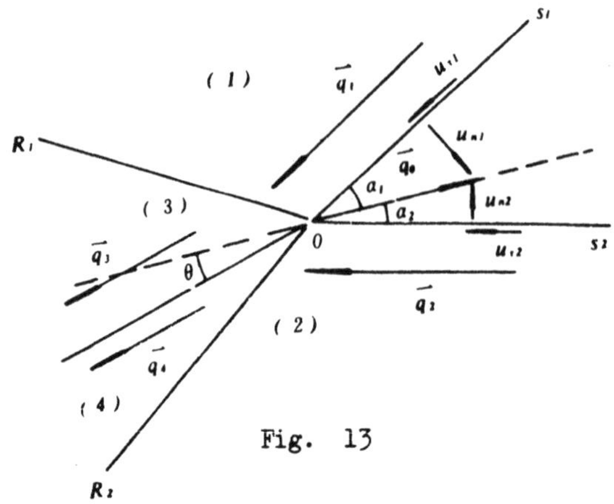

Fig. 13

GENERAL CASE

The above treatment can be extended to a more general case in which both surfaces are moving with different velocities. Related normal and tangential velocities for corresponding surfaces are denoted by u_{n1}, $u_{\Gamma 1}$, u_{n2}, $u_{\Gamma 2}$(see Fig. 13). The angle between these surfaces is a_o. The velocity of the collision point is

$$q_o = \frac{u_{n^1}}{\sin a_1} = \frac{u_{n^2}}{\sin a_2}$$

where a_1 and a_2 are determined by

$$\cot a_1 = \frac{1}{\sin a_o} \left(\frac{u_{n^2}}{u_{n^1}} + \cos a_o \right)$$

$$a_2 = a_o - a_1$$

In a system moving with collision point, related velocities are

$$q_1 = u_{n^1} \cot a_1 + u_{\Gamma 1}$$

$$q_2 = u_{n^2} \cot a_2 + u_{\Gamma 2}$$

If both velocities are supersonic, the shock polar are built by formula (7). Calculations will be analogous to above.

REFERENCES

1. J.M. Walsh, R.G. Shreffler, and F.J. Willig, Limiting Condition for Jet Formation in High Velocity Collisions. J. Appl. Phys. 24:349 (1953).
2. F.H. Harlow and W.E. Pracht, Formation and Penetration of High-Speed Collapse Jets, Phys. Fluids 9:1951 (1966).
3. P.C. Chou, J. Carleone, and R.R. Karpp, Criteria for Jet Formation from Impinging Shells and Plates, J. Appl. Phys. 47: 2975 (1976).
4. Wang Jihai, Refraction of Detonation Wave at Material Interface (in Chinese), Explosion and Shock Waves 2:1 (1981).
5. Appendix, in "Shock Waves and High-Strain-Rate Phenomena in Metals," M.A. Meyers and L.E. Murr, eds., Plenum Press, New York, (1981).

AN INVESTIGATION OF THE SHOCK STRUCTURES AND

CONDITIONS FOR JETTING DURING LIQUID IMPACT

J.P. Dear and J.E. Field

Cavendish Laboratory
Madingley Road
Cambridge, U.K.

INTRODUCTION

As is well known, the collision between a liquid mass and a solid surface can generate high transient pressures and cause significant damage, particularly if the impact velocity is high. The high pressures have been attributed to compressible behaviour in the liquid. The fact that numerous situations exist in which such collisions occur has made the study of liquid impact crucial for the further development of related technologies. However, despite considerable activity in the subject there exists a large number of unexplained and poorly understood phenomena associated with liquid impact[1]. The reasons for this are connected with the fact that the high pressures and much of the damage formation takes place during a very short time period following initial contact, and as we will show, involve a complex dependence on the geometry of the liquid and solid surface in the contact area.

Considerable progress has been made in overcoming these obstacles by making extensive use of high-speed photographic techniques and rapid response crystal gauges. One result of this work is that when a curved liquid drop strikes a solid, the highest pressures do not occur at the centre of impact where the normals to both the liquid and solid surfaces are parallel but on a nearly circular boundary at a distance which depends on the drop radius and the velocity of impact[1-4]. Another result is that after an interval comparable to the time it takes the drop to spread out to the maximum pressure radius, a "jetting" motion develops. In certain impact speed ranges this jet has been observed to have a speed up to ten times the impact speed[5-8].

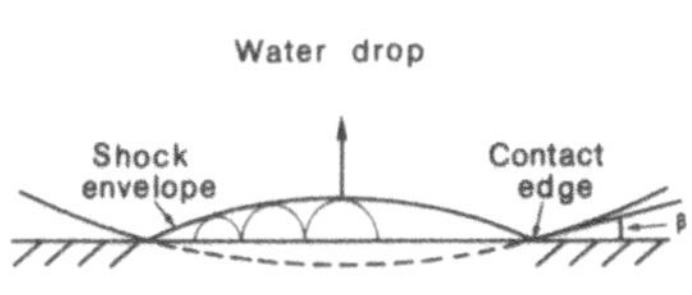

Fig. 1. Impact of a drop with a target. For a drop β constantly changes.

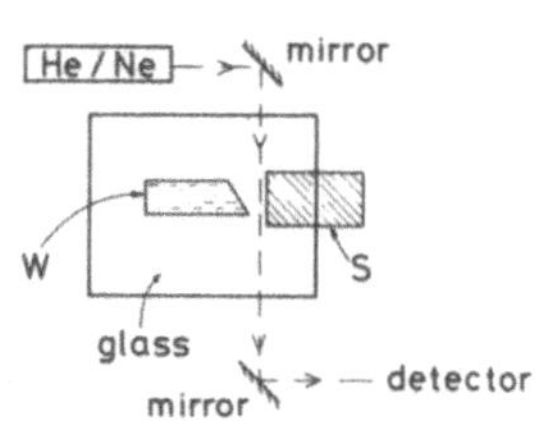

Fig. 2. Impact geometry, W, is the liquid/gelatine wedge & S the impacting striker

The study of these events has been considerably complicated by the fact that for the drop sizes normally used in experiments, which are typically up to a few millimetres in diameter, both the "jetting" and high pressure thresholds are reached when the intercept angle between drop edge and solid surface is, at most, a few degrees. Thus it becomes difficult to interpret the photographic evidence, and to observe the detailed behaviour of the liquid in the critical zone.

EXPERIMENTAL

A disadvantage of impact with a drop, both from the viewpoints of analysis and experimental interpretation, is that there is a constantly changing angle between the drop boundary and the solid (angle β in figure 1). This can be overcome by using wedges of liquid which have a constant, prechosen β.

In the present experiments, two-dimensional wedges and other geometries were prepared by first mixing 12% by weight of gelatine with water, allowing it to set into a thin sheet and then cutting into the required shape. We have shown that the flow properties are not significantly different to that of pure water once impact velocities exceed a few metres per second.

The water/gelatine sheets were made by casting into a 200 x 200 x 3 mm vertical mould to reduce evaporation. Both liquid and mould were at 330 K. Each of the large mould faces had been lightly greased and covered with a thin plastic film. After slow cooling, to reduce shrinkage, the mould was disassembled and the sheets placed horizontally. The layers, with plastic sheets attached, could be kept for several days. After some practice it was possible to cut out wedges with flat and smooth edges.

The liquid/gelatine layer is placed between spaced glass blocks, and a striker, fired from a rectangular bore gas gun, projected between them. Velocities of up to ca. 300 m s^{-1} have been achieved. The striker triggers an Imacon framing image converter camera by intersecting a laser beam just before impact. A schematic diagram of the apparatus is shown in figure 2. Schlieren optics are used to make the stress waves visible. This basic situation can be readily adapted for a wide variety of other impact and shock collapse studies.

RIGID SLIDER IMPACTS

Figure 3 shows the impact geometries for a rigid slider striking a wedge of liquid. Two situations can be distinguished depending upon the contact angle, β, of the wedge. In case (a), β is less than a particular critical angle β_c, the contact point moves faster than the shock wave and

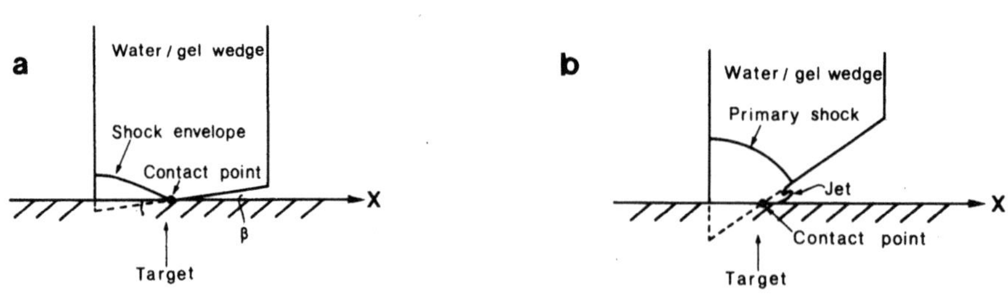

Fig. 3. Impact geometries for a target striking a liquid wedge. α is the angle of inclination of the shock and β the contact angle.

the shock envelope is attached to the contact point and elongated in the x-direction. For this case, there is no jetting. In case (b), with β greater than β_c, the primary shock wave can move ahead of the more slowly moving contact edge and deform the liquid surface to produce a jet. The value of β_c which determines when jetting starts is dependent upon the shock speed in the liquid and the impact velocity of the rigid target. If the impact velocity of the rigid target is U_i, then the impact Mach number M_i is defined as $M_i=U_i/C_o$ where C_o is the acoustic wave velocity. Lesser[1] has shown that for low impact velocities, the value of β_c which determines when jetting occurs is given by

$$\beta_c = \tan^{-1}\left[M_i/\sqrt{(1+4.762M_i^{2/3})}\right] \qquad (1)$$

For $M_i<0.12$, this gives results which are within 10% of the full numerical solution and also agree with Heymann[2]. For an impact velocity of 150 m s^{-1}, (1) gives β_c for a rigid target as 4°. For a full discussion, see references[9,10,11].

Figures 4, 5, 6 and 7 are all at an impact velocity of 150 m s^{-1}, with wedge angles below, equal and above the critical jetting angle. In figure 4, with a β-value of 3° which is less than β_c, a strong shock envelope develops which expands much faster in the x-direction than vertically. No jet can be detected in the air wedge though a high-speed jet, labelled J, of velocity ca. 1300 m s^{-1} forms when the shock wave reflects at the right-hand edge. This is to be expected and is discussed later for the case of the "double-angled" wedge. Figure 5 is for β slightly greater (β ca. 4°) but still less than the critical value. The angle of inclination of the shock wave has steepened but it is still attached to the contact edge and no jet can be detected in the air wedge. Jet formation at the right-hand edge does occur and the jet has a velocity ca. 1800 m s^{-1}. Figure 6 is for a contact angle, β, approximately equal to β_c. Infact, for this sequence β is made to vary so that initially, the contact angle is slightly less than β_c but then it increases to pass through the critical situation and produces jetting. The initial contact angle is ca. 4° (increasing to ca. 6°) and the contact edge is clearly supersonic in frames 1 to 3. Jetting occurs between frames 3 and 4. The angle α (see figure 3) increases and should approach 90° when the shock is about to overtake the contact point and this takes place in frame 4. Careful examination of frame 4 suggests jetting has started. The jet can be seen just visible above the slider in frame 5 but is more clearly seen in frame 6. Its velocity is about 2500 m s^{-1}. In figure 7 the contact angle is ca. 12° and is considerably greater than the critical jetting angle. The primary shock wave is now part of a circle, it moves ahead of the contact point and a jet can be clearly detected in the air wedge from frame 4 onwards. The jet velocity in this case was ca. 1000 m s^{-1}. Measurements of the angle of inclination of the shock, α, are given in table 1.

These results confirm the theoretical predictions and give an experimental value for β_c at M_i = 0.1 of 5±1° in good agreement with the theoretical value of 3.8°. It should be remembered that the experimental value will be slightly greater than the theoretical one since the critical condition occurs when the shock wave just detaches from the contact edge. The contact angle will need to increase slightly before the detachment of the shock or jetting are clearly in evidence.

PMMA SLIDER IMPACTS

Wedge impact studies have also been carried out with plastic slider mainly polymethylmethacrylate (PMMA), to study the effect which slider

Table 1. Experimental and theoretical values (Lesser[9]) for the angle of inclination of the shock, α, for different β.

Figure	$\beta/^{o}$	$\alpha_{experimental}/^{o}$	$\alpha_{theoretical}/^{o}$
4	3	39	38
5	4	61	60
6	4 - 6	51 - 83	60 - jetting

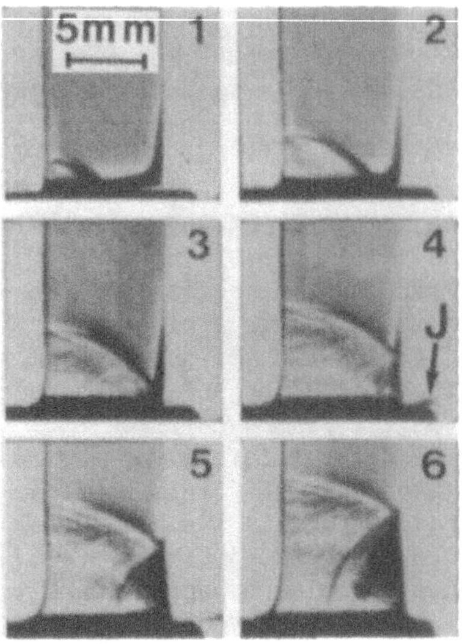

Fig. 4. Impact of wedge with β ca.3°. U_i = 150 m s^{-1}. Interframe time = 0.96μs.

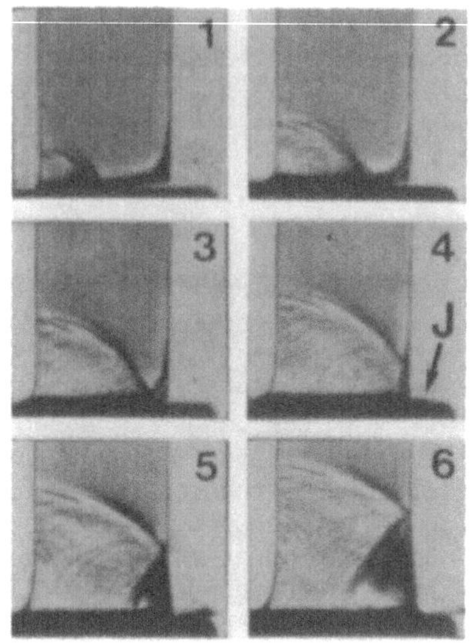

Fig. 5. Impact of wedge with β ca.4°. U_i = 150 m s^{-1}. Interframe time = 0.96μs.

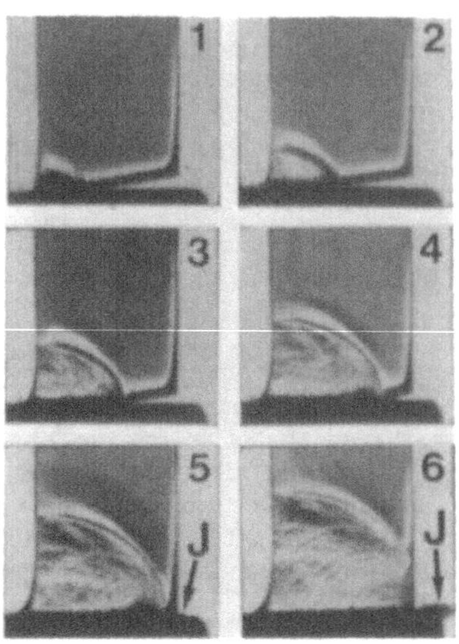

Fig. 6. Impact of wedge with β ca.β_c. U_i = 150 m s^{-1}. Interframe time = 0.96μs.

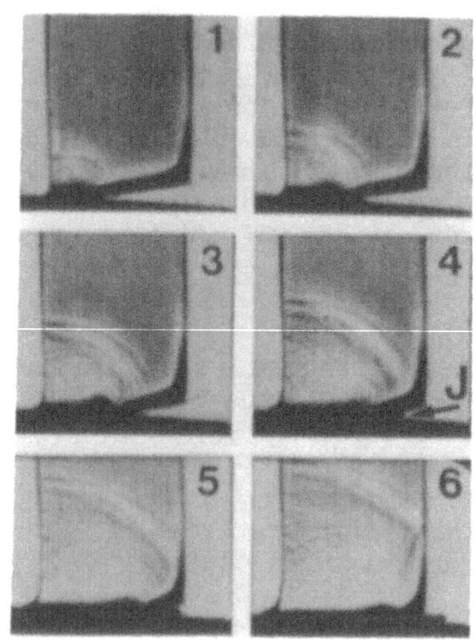

Fig. 7. Impact of wedge with β ca.12°. U_i = 150 m s^{-1}. Interframe time = 0.96μs.

slider compliance has on the critical angle for jetting. Lesser[9] has suggested that this critical angle will increase if the impacting solid is non-rigid (i.e. compliant). For a PMMA slider impacting at $M_i = 0.2$, β_c is predicted(see Lesser[11] for treatment) to be 8.3° (over two degrees greater than the value of 6° for a rigid target). PMMA sliders were fired at a velocity of 300 m s^{-1} at liquid/gel wedges with different contact angles to determine the critical angle for jetting. Figure 8 is an example near the critical angle with the front surface of the slider again slightly curved so that initially β is less than β_c, but by frame 6, the critical condition has been achieved giving a critical angle of 10±1°. This confirms Lesser's prediction that target compliance increases β_c. However, the experimental value of β_c appears to be slightly greater than that predicted. A more detailed analysis of the compliant target would have to take into account surface motion of the target material ahead of the contact edge caused by the compressive shock wave in the solid.

DOUBLED—ANGLED WEDGE

It is possible for theoretical predictions to be made of the velocity of the jets produced during liquid/solid impact. One configuration which has been studied in some detail by Finnström[12] is the double-angled wedge as depicted in figure 10. Impact first occurs at O and two wedge angles β_0 and β are involved, both of which can be varied. However, the angle β_0 is kept below the critical angle for jetting so that the contact point moves supersonically along OA. The angle β takes values from the critical angle up to 90°. Under these conditions Finnström has determined theoretically the jet velocity with very good agreement with our experimental results (see[11] for a more detailed discussion). Figure 9 gives an example with $\beta = 35°$ and β_0 just below the critical angle. A jet starts to form in frame 5 when the shock wave overtakes the contact edge and moves up the free surface. Its velocity averaged over frames 6, 7 and 8 is ca. 1700 m s^{-1} which agrees well with the theoretical pre- of 1820 m s^{-1} for the jet velocity for this particular arrangement. The jet velocities for figures 4,5 and 6 where β_0 is variable and $\beta=90°$ are also in good agreement with theoretical predictions. In general, theory and experiment show high jetting velocities are to be expected if the first contact angle, β_0, is just less than the critical angle for jetting.

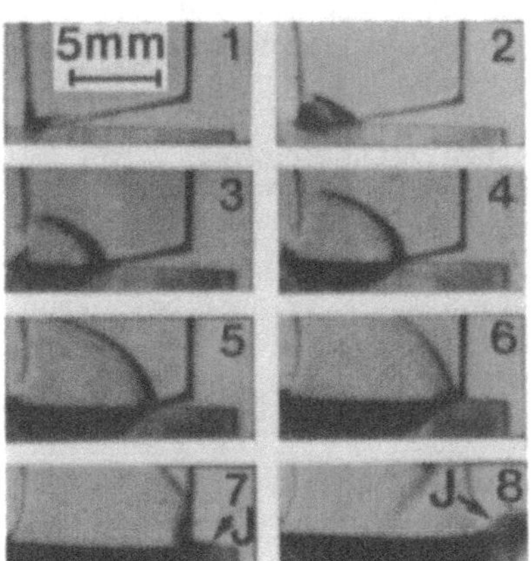

Fig. 8. PMMA slider impact of wedge with β ca. β_c.

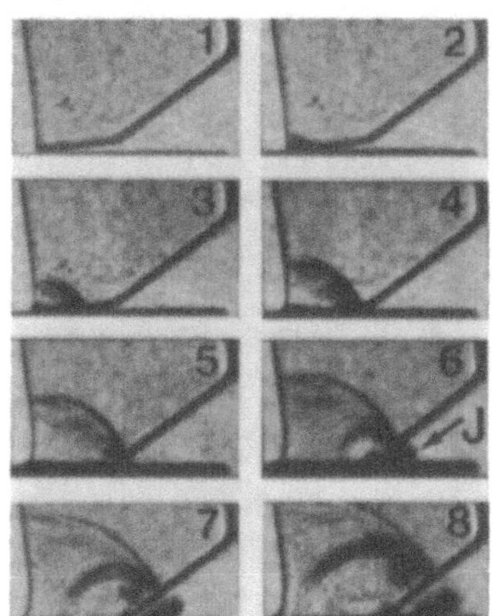

Fig. 9. Impact with double-angled wedge. U_i = 150 m s^{-1}.

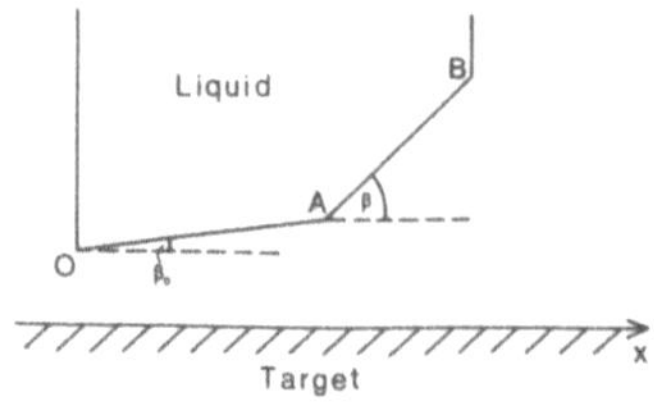

Fig. 10. Impact with double-angled wedge with $\beta_o < \beta_c$ and β variable from β_c to 90.

CONCLUSIONS

The work presented in this paper is part of a much wider study of liquid/solid impact for which the experimental equipment was developed. Using the two-dimensional gel technique, it has been shown that liquid wedge geometries can be constructed which provide detailed information on shock structures, high "edge" pressures and "jetting" effects. The techniques developed are also very convenient for examining the collapse of single cavities and the interactions of arrays of cavities. Further work is being performed on the above and related topics.

ACKNOWLEDGEMENTS

We thank Professor M.B. Lesser for his theoretical contributions and the Ministry of Defence (Procurement Executive) for support.

REFERENCES

1. M. B. Lesser and J. E. Field, Ann. Rev. Fluid. Mech. 15:97 (1983).
2. F. J. Heymann, J. Appl. Phys. 40:511 (1969).
3. J. H. Brunton and M. C. Rochester, Erosion of solid surfaces by the impact of liquid drops, in: "Erosion", C.M. Preece, ed., Academic Press, New York (1979), 185.
4. W. F. Adler, (ed.), "Erosion: prevention and useful applications", Philadelphia, Penn. ASTM (1979).
5. D. C. Jenkins and J. D. Booker, in: "Aerodynamic Capture of Particles", E. G. Richardson, ed., Pergamon, Oxford, 97.
6. A. A. Fyall, "Proc. 2nd. Int. Conf. on Rain Erosion and Associated Phenomena", Meersburg, West Germany, A. A. Fyall and R. B. King, ed., RAE Farnborough, U.K. (1967), 563.
7. J. H. Brunton and J-J. Camus, "Proc. 3rd Int. Conf. on Rain Erosion and Associated Phenomena", Meersburgh, West Germany, A. A. Fyall and R.B. King, ed., RAE Farnborough, U.K. (1970).
8. J-J. Camus, Ph.D. Thesis, University of Cambridge (1971).
9. M. B. Lesser, Proc. Roy. Soc. Lond. A377:289 (1981).
10. J. E. Field, J. P. Dear, P. N. H. Davies and M. Finnström, "Proc. 6th Int. Conf. on Erosion by Liquid and Solid Impact", J. E. Field and N.S. Corney, ed., Paper 19, Cambridge (1983).
11. J. E. Field, M. B. Lesser and J. P. Dear, Proc. Roy. Soc. Lond., in press.
12. M. Finnström, Licentiate Thesis, Luleå, Sweden (1984).

COMPUTER SIMULATION OF THE EFFECT OF FREE SURFACE REFLECTION ON SHOCK WAVE PROPAGATION IN WATER*

M. Kamegai, L. S. Klein, and C. E. Rosenkilde

Lawrence Livermore National Laboratory
Livermore, CA 94550

A computer simulation is used to study the behavior of a fluid which has been set into motion by a point explosion below a bounding free surface. Of particular interest is the interaction between the incident shock wave and the reflected rarefaction signal which, under certain conditions, can overtake the shock front. Onset criteria for the envelope which separates the region of irregular from that of regular rarefaction, the nonspherical shape of the shock front, and the pulse profile for several depths of burst covering both strong and weak shock interactions with the surface will be discussed.

The shock wave, which propagates away from a shallow explosion, remains spherically symmetric until it either interacts with the surface or with a reflected signal from the surface. When certain conditions are met, the surface rarefaction signal can overtake the directly propagating shock front, reduce its overpressure and retard its propagation, so that the wave front appears to be "pulled back" from its otherwise spherical shape (see Fig. 1). The phenomena is referred to as anomalous[1] surface rarefaction, and is analogous to Mach stem formation in the case of shock reflection from a rigid wall. Regular oblique reflection from a free surface is described by a Prandtl-Meyer rarefaction wave.

The conditions[2] for the onset of anomalous rarefaction can be derived from Fig. 2. They have the form:

$$S \tan\theta^* = [c^2 - (S - u)^2]^{1/2} \quad \text{and} \quad \sin\theta^* = D/R^*.$$

These conditions first obtain when the phase velocity, c_{ph}, of the surface rarefaction signal <u>along the shock front</u> becomes equal to the phase velocity, S_{ph}, of the surface along the shock front. Thereafter, the conditions are satisfied along a locus of points <u>below</u> the surface. Regular rarefaction obtains in the region below and to the <u>left of</u> this locus.

We made computer simulations of strong subsurface explosions with a total energy output of about 10^{15} Joules in water. The initial detonation phase was calculated with a Lagrangian hydrocode, which subsequently was linked to an

*This work was performed under the auspices of the U.S. Department of Energy by Lawrence Livermore National Laboratory under contract No. W-7405-Eng-48.

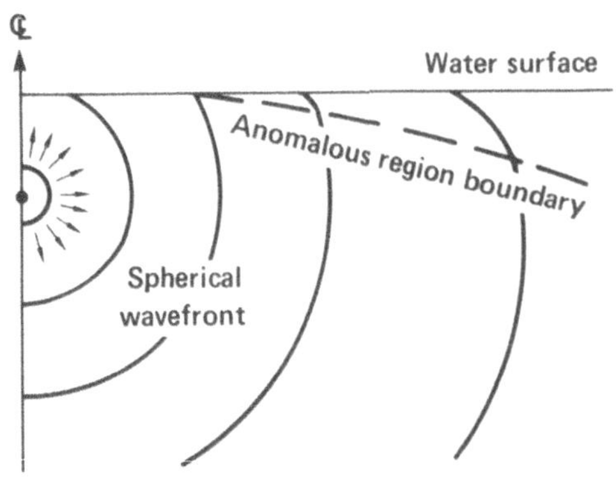

Fig. 1. Anomalous surface rarefaction.

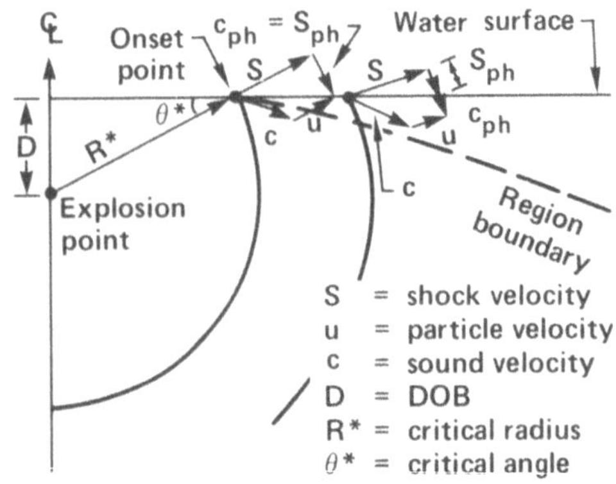

Fig. 2. Onset conditions for the anomalous region.

arbitrary Lagrangian-Eulerian code (ALE) to compute the wave propagation until the peak overpressure decayed to less than 1 Kbar. We computed three problems with the same energy source placed at three different depths of burst (DOB): 3 m, 21 m, and 66.5 m. Fig. 3 shows, as an example, the mesh and material

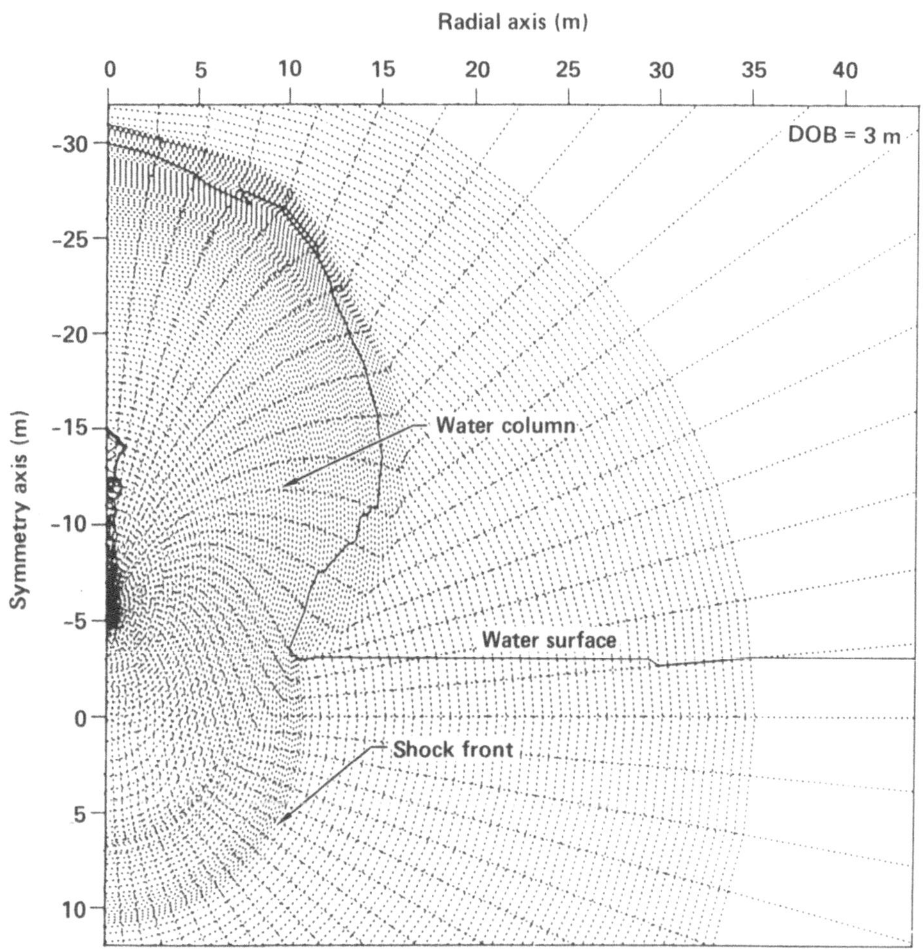

Fig. 3. Mesh at 252 μsec from computer simulation of an underwater explosion.

boundaries at 252 μsec for the 3 m DOB case.. A spherical shock front and a water column are clearly discernible. The debris in the central region has risen above the original water surface.

An elementary theoretical model, which used a Grüneisen equation of state for water, was employed to predict the onset point and to calculate the envelope, which separates the region of anomalous rarefaction from that of regular rarefaction. The theoretical results are compared with the computational results for the strong-shock case (DOB = 3 m) in Fig. 4 and for the weak-shock case (DOB = 66.5 m) in Fig. 5.

For the strong-shock case (Fig. 4), the anomalous rarefaction enters at a horizontal range of about 10 m, where the shock overpressure is 2.5 Mbar. The theory predicts the onset point reasonably well even at these high overpressures (and short distances), but the computational and theoretical predictions of the region boundary deviate greatly at large distances. Also shown in the same figure are the radial wave profiles at three points. Point A is in the region of regular rarefaction, where the rarefaction wave is unable to catch and perturb the shock front. Point B is situated directly on the boundary, where the rarefaction signal is just able to catch up to and slightly attenuate the shock front. Point C is well within the anomalous region, where the rarefaction signal has significantly reduced the shock overpressure.

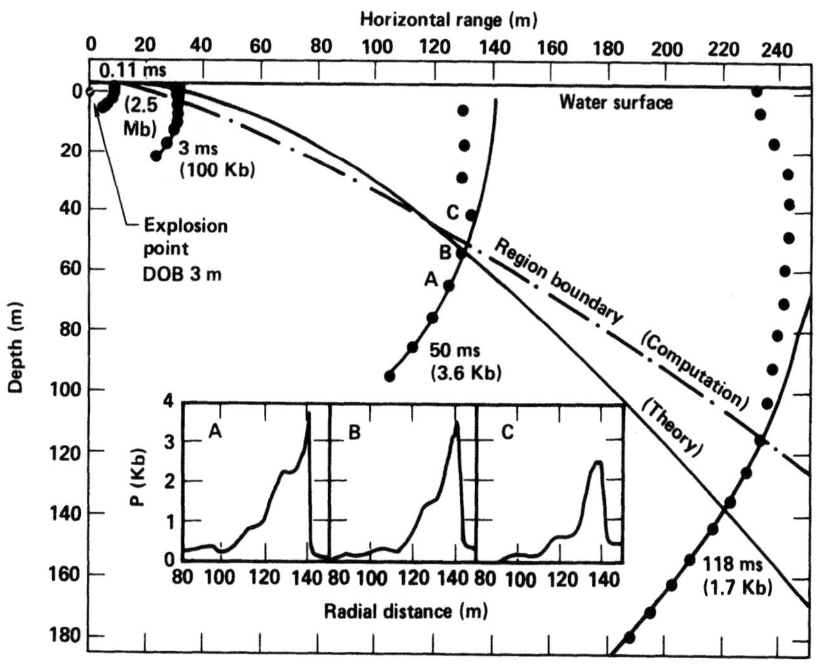

Fig. 4. Wave profiles and anomalous region boundary in a strong-shock regime.

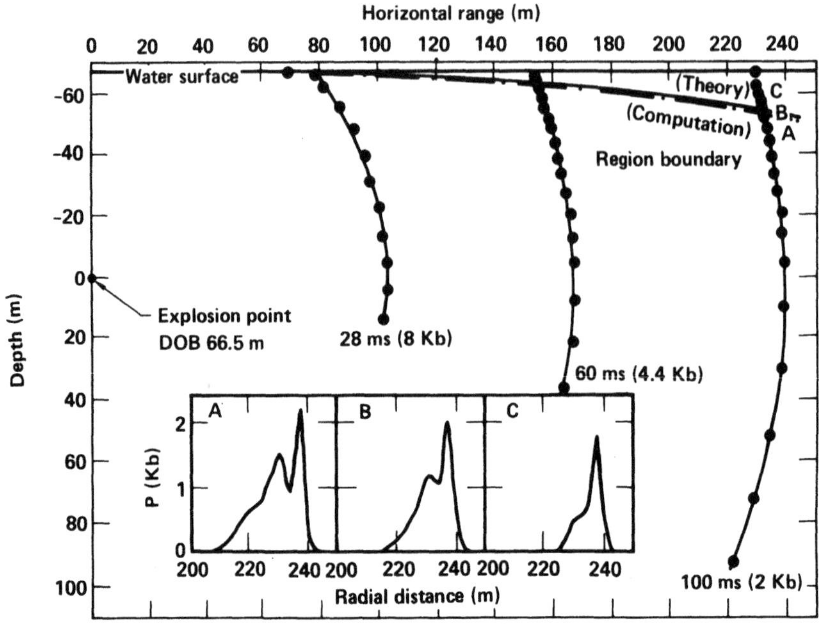

Fig. 5. Wave profiles and anomalous region boundary in a weak-shock regime.

For the weak-shock case (Fig. 5), the anomalous rarefaction starts at a horizontal range of about 80 m, where the shock overpressure is 8 Kbar. The theoretical predictions of the onset point and the region boundary agree very well with the computational predictions. Also shown in the same figure are the radial wave profiles at three points: point A in the region of regular rarefaction, point B on the region boundary, and point C within the region of anomalous rarefaction. The same descriptions apply to the profiles at these points, as were associated with the corresponding ones in Fig. 4.

We were able to compute the anomalous surface rarefaction from strong subsurface explosions. The surface interaction is a complex phenomena. There is an elementary theoretical model for the anomalous surface rarefaction, which does predict the region boundary in the weak-shock regime, but fails in the strong-shock regime.

REFERENCES

1. H. G. Snay and A. R. Kriebel, "Surface Reflection of Underwater Shock Waves," Naval Ordinance Laboratory Report, NOLTR 70-31, 13 October, 1970.

2. A. Falade and M. Holt, "Interaction of a Blast Wave With a Free Surface," Phys. Fluids 21, 1702-1708 (1978).

THE CALCULATION OF DIMENSIONS OF THE STEADY MACH WAVE

A.I. Martynov and S.S. Batsanov

State Standard Comm. USSR,

Leninsky Prospect 9, Moscow II7049, USSR

The Mach interaction of shock waves is employed in physics and chemistry of dynamic pressures as an effective means to obtain high parameters. The theory of that phenomenon, however, is far from being completed. In particular, the dimensions of the Mach wave have not been evaluated so far.

There are two basic types of the Mach wave formed as a result of the Mach interaction of shock waves. In the first case its size grows in proportion with the distance travelled whereas in the second the Mach wave retains its size and velocity constant along its path. For the Mach wave of the first type the approximate solution has been found and its dimensions evaluated either in a short wave approximation[1] or using the cumulative jet theory[2]. The theoretical studies of the steady Mach wave are still limited with search for possible calculation methods[3,4].

Here we propose the method to calculate the size of a steady Mach wave as well as pressure drop behind its front when such a wave is formed in a cylinder of condensed matter as a result of collapse of a cone shock wave. The latter is generated by an explosive surrounding the cylinder.

Let us consider the flow pattern in the reference frame fixed to the Mach wave. In Fig. 1 the flow parameters behind the incident wave are denoted by subscript 1 , behind the reflected wave by 2 , and behind the Mach wave by 3 . The initial values of flow parameters are subindexed by o and those at the tangent discontinuity boundary are marked with a tilde.

When the curvature of the Mach wave front is slight, one can use a quasi-one-dimensional approximation allowing the flow velocity q_3 to be related to its cross-section area by the Hugoniot equation as

$$(M_3^2 - 1) = \frac{dq_3}{q_3} = \frac{dS_3}{S_3} \quad , \tag{1}$$

where M_3 is the Mach flow number and S_3 the cross-section area of the flow.

Expressing the fractional change of cross-section area of the flow in terms of its radius r and the angle of the flow boundary with respect to the symmetry axis Θ_3, we obtain the equation

$$r = \frac{2\tilde{q}_3 \Theta_3}{(1 - M_3^2)(dq_3/dx)} , \qquad (2)$$

where x is the coordinate coinciding with the symmetry axis of the flow.

Applying Eq. (2) to the parameters at the cross-section through the Mach wave front, we get the relation from which the size of the Mach wave can be found. Indeed, in this case Eq. (2) contains the parameters $M_{3,o}$, $q_{3,o}$, $\Theta_{3,o}$ whose values are easy to calculate provided the intensity of a cone wave entering the cylinder, incidence angle of its front relative to the symmetry axis, and dynamic properties of the substance of which the cylinder is made are all known. Some complications may arise while calculating the angle $\Theta_{3,o}$ by the shock polar method in the case when the shock polar curves of the incident and reflected waves do not cross. This difficulty can be evaded though, if the angle $\Theta_{3,o}$ is calculated by the method reported in[4].

The calculation of velocity gradient involves examination of the whole flow pattern behind the Mach wave since this gradient emerges due to pressure equalization behind the Mach wave front and in the substance around the cylinder. Differentiating Eq. (2) and taking into account that $dr/dx = -\theta_3$, we discover that the magnitude of velocity gradient is determined by the shape of flow boundaries behind the Mach wave. In its turn, the shape of flow boundaries is defined by the behavior of the flow in the space between the Mach wave and

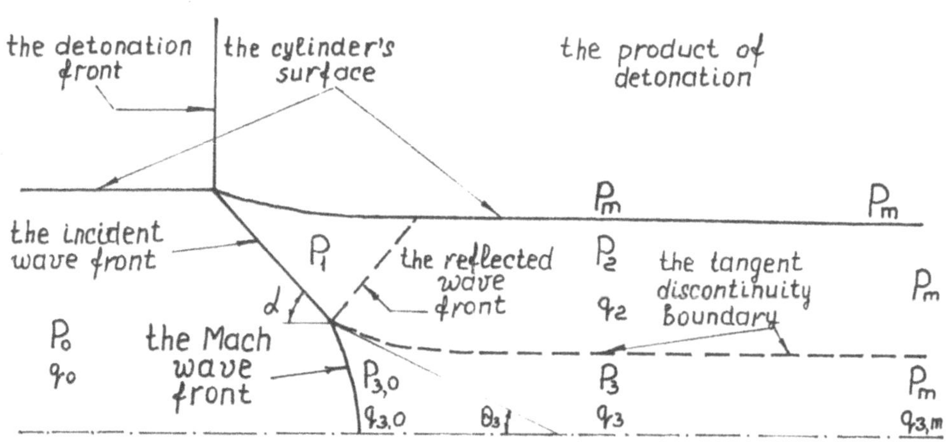

Fig. 1

the cylinder's surface (designated by subindex 2 in Fig. 1). To calculate the parameters of that flow, we used the equation

$$\frac{dq_2\theta_2}{dx} = -\frac{dq_2}{dy} \quad , \tag{3}$$

which is valid in the absence of vortices; y is the radial coordinate.

A few simplifying assumptions are appropriate to make now. The function $q_2 = q_2(y)$ is supposed to be linear, the pressure exerted on the cylinder's surface is taken to be constant and equal to $P_m = (P_1 + P_d)/2$, where P_d is the detonation pressure. The parameters of flow 2 and flow 3 are interrelated by the equalities holding at the tangent discontinuity boundary : $P_2 = P_3$ and $\theta_2 = \theta_3$.

Thus, differentiating Eq. (2) and taking into account Eq. (3) and the abovementioned simplifying assumptions, we obtain the differential equation describing the flow behind the Mach wave front

$$\frac{(1-M_3^2)d\dot{q}_3^2}{2 \quad dq} - \frac{(1-M_3^2)}{2q_3}\dot{q}_3^2 - 2M_3\frac{dM_3}{dq_3}\dot{q}_3^2 - \frac{2\tilde{q}_{3,0}(q_3-q_{3,m})\sin(\alpha-\theta_{1,0})}{\tilde{q}_{2,0}\,r\left(R_0\sqrt{\frac{\sin(\alpha-\theta_{1,0})}{\sin\alpha}}-r\right)\sin\alpha} \tag{4}$$

where $\dot{q}_3 = dq_3/dx$, $r = r_0\sqrt{\rho_{3,0}q_{3,0}/\rho_3 q_3}$, r_0 the radius of the Mach wave, $R = R_0\sqrt{\frac{\sin(\alpha-\theta_{1,0})}{\sin\alpha}}$, R_0 the radius of the cylinder, and $q_{3,m}$ the equilibrium velocity of the flow in which the pressure equals P_m . The velocity $q_{3,m}$ is calculated from the Bernoulli's theorem and equation of isentropy for the cylinder's substance according to the given parameters $P_{3,0}$ and P_m .

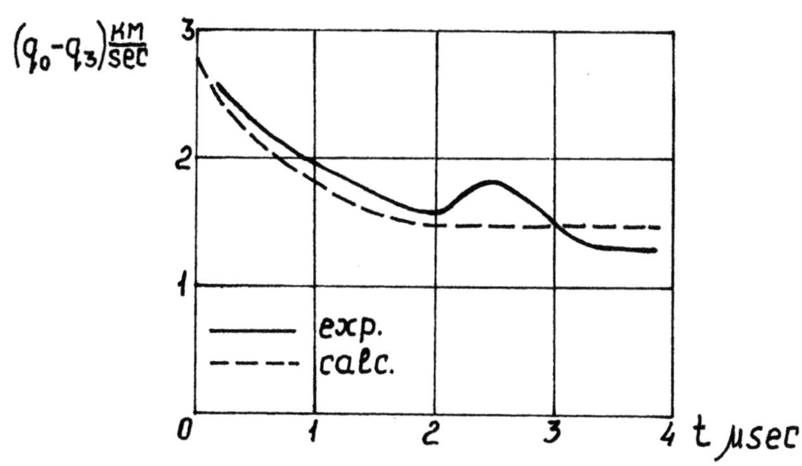

Fig. 2

As the value $M_3 = I$ is a singular point of this equation, its solution depends on whether the flow passes the critical point. When $M_3 \approx M_{3,0} < I$, the coefficient in Eq.(4) can be regarded as constant and the general solution can be found. After having obtained the arbitrary constant appearing in the solution from the condition $q_3 = 0$ at $\dot{q}_3 = q_{3,0}$ and put the found solution in Eq.(2) we come to the following expression:

$$r_0/R_0 = \frac{2\tilde{\theta}_{3,0}^2 \tilde{q}_{3,0} \tilde{q}_{2,0} \sqrt{\sin\alpha \cdot \sin(\alpha - \theta_{I,0})}}{2\tilde{\theta}_{3,0}^2 \tilde{q}_{3,0} \tilde{q}_{2,0} \sin\alpha + (I - M_{3,0}^2)(q_{3,m} - q_{3,0})^2 \sin(\alpha - \theta_{I,0})}$$

Table I lists the calculated values of relative dimensions of the Mach wave in plexiglas, aluminum, copper and nitromethane cylinders surrounded by explosives generating taken from[4-7].

Table I. Calculated and Experimental Values of Relative Dimensions of the Mach Wave

Cylinder's material	P_d, kbar	q_0, km/s	$\frac{r_0}{R_0}$ calc.	$\frac{r_0}{R_0}$ exp.	Ref.
Plexiglas	260	7.8	0.60	0.62	5
Plexiglas	I45	6.75	0.64	0.6	5
Plexiglas	55	4.9	0.57	0.57	5
Aluminum	250	7.65	0.58	0.6	4
Copper	290	7.9	0.I0	0.I5	6
Nitromethane	250	7.65	0.64	0.625	7

Fig.2 shows the calculated curve of mass velocity $(q_0 - q_3)$ behind the Mach wave in a plexiglas cylinder and corresponding experimental curve by Adadurov et al[8].

Although the available experimental data are insuffient for checking reliability of the obtained equation, the satisfactory agreement between the experimental and calculated dimensions of the Mach wave testifies to the correctness of the developed method.

REFERENCES

I. S.A.Khristianovich and O.S.Ryzhov, Prikl.Mekhan. i Matem. 22: 586 (I958)
2. A.I.Martynov, I.N.Temnitsky, and S.S.Batsanov, in: "Fizika Impulsnykh Davlenii", VNIIFTRI, Moscow, I979. p.20
3. A.I.Martynov, E.V.Lazareva, and I.N.Temnitsky, in: "Tezisy Dokladov III Vsesoyuznogo Simpoziuma po Impulsnym Davleniyam", VNIIFTRI, Moscow, I979, p.IOI
4. G.S.Doronin, V.P.Stupnikov, V.V.Roman'kov, V.Ya.Belen'ky, R.I.Zaslavsky, and S.S.Batsanov, Zh.Tekhn.Fiz. 43: I059 (I973)
5. G.A.Adadurov, A.N.Dremin, and G.I.Kanel, Zh.Prikl.Mekhan. i Tekhn.Fiz. No 2: I26 (I969)
6. G.R.Fowles and W.M.Isbell, J.Appl.Phys. 36: I377 (I965)
7. M.F.Gogulya, I.M.Voskoboinikov, and I.G.Demchenko. Fiz. Goreniya i Vzryva I7: I53 (I98I)
8. G.A.Adadurov, A.N.Dremin, G.I.Kanel, and S.V.Pershin, Fiz.Goreniya i Vzryva 3: 28I (I967)

SPHERICAL SHOCKS IN SOLIDS

M. P. Ranga Rao

Department of Mathematics
Indian Institute of Technology
Powai, Bombay 400 076, India

INTRODUCTION

The blast wave propagation in solid media is of interest in the study of impact phenomena[1,2], underground nuclear explosions[3,4] etc. After the instantaneous energy release a sufficiently strong shock is generated in the medium whose material strength may be neglected compared with the pressure associated with the shock processes. In this state (hydrodynamic regime) the shock propagation phenomenon is nonself-similar because of the equation of state of the shocked medium. This phenomenon is studied by many investigators[1-7] using quasi-similar method and Porzel-Zaker method etc.

In this paper the study of strong spherical shock wave propagating in a solid medium is carried out. We consider the shock wave due to point explosion and also the impact generated shocks. In the hydrodynamic regime the Hugoniot characterising the shocked state of the medium is assumed to be in general form so that it includes the solids like metals, porous media and rocks. The shock propagation is studied by using quasi-steady method of Rae[2]. An additional assumption in this method is proposed by which one can get the results in a close analytical form. The dependence of the shock pressure, the shock decay coefficient and shock Mach number on shock radius is investigated. The results corresponding to c,s medium are derived as particular cases of our solutions. A comparative study of these results, which could be obtained quickly without much numerical computation, with those obtained by other investigators shows a remarkable agreement.

BASIC EQUATIONS

The basic conservation equations of mass, momentum and energy behind the shock governing the unsteady spherical motion due to either point explosion or impact of the shocked state of solid media in the hydrodynamic regime are given by

$$\frac{\partial \rho}{\partial t} + \rho \frac{\partial u}{\partial r} + u \frac{\partial \rho}{\partial r} + \frac{2\rho u}{r} = 0, \tag{1}$$

$$\frac{\partial u}{\partial t} + u\frac{\partial u}{\partial r} + \frac{1}{\rho}\frac{\partial p}{\partial r} = 0, \tag{2}$$

$$\frac{\partial e}{\partial t} + u\frac{\partial e}{\partial r} - \frac{p}{\rho^2}\left(\frac{\partial \rho}{\partial t} + u\frac{\partial \rho}{\partial r}\right) = 0, \tag{3}$$

where p, ρ and u are the pressure, density and velocity of the solid particle, respectively. In the above equations, t is the time, r is the Eulerian space coordinate and e is the specific internal energy. Neglecting counter-pressure, the Rankine-Hugoniot conditions at the shock moving into a medium at rest are given by

$$\rho_0\dot{R} = \rho_1(\dot{R} - u_1), \qquad p_1 = \rho_0 u_1 \dot{R}, \tag{4,5}$$

$$e_1 - e_0 = \frac{1}{2} p_1 \left(\frac{1}{\rho_0} - \frac{1}{\rho_1}\right), \tag{6}$$

where the suffixes 0 and 1 indicate the quantity just ahead and behind the shock respectively, R is the shock radius and $\dot{R} = (dR/dt)$ is the shock velocity. The Hugoniot of the medium is assumed in a general form

$$p_1(\eta) = \rho_0 c^2 F(\eta), \qquad \eta = (\rho_1/\rho_0) - 1 \tag{7}$$

where $F(\eta))$ is a known function of η. In this equation, c is a characteristic constant with dimension of velocity. For example, in the case of a porous medium in which the equation of state is assumed to be Mie-Gruneisen equation for non-porous c, s medium ($\dot{R} = c + su_1$), the function $F(\eta)$ takes the form (see Rae[2])

$$F(\eta) = \frac{\eta(\eta + 1)}{[s - (s - 1)(\eta + 1)]\,[s - m(s - 1)(\eta + 1)]}, \qquad m = \rho_0/\rho_p, \tag{8}$$

where ρ_p is the undisturbed density of the porous medium. The case $m = 1$ corresponds to non-porous c, s medium. Expressions similar to (8) can be derived for solids like rocks[4] and metals[6]. The shock conditions (4) and (5) with the help of (7) can be written as

$$\dot{R} = c\,[(\eta + 1) F(\eta)/\eta]^{1/2}, \tag{9a}$$

$$u_1 = c\,[\eta F(\eta)/(\eta + 1)]^{1/2}. \tag{9b}$$

APPROXIMATE METHOD FOR SHOCK PROPAGATION

In order to investigate the shock profile we need the energy balance equation given by

$$E = \sigma \int_0^R \left[e - e_0 + \frac{u^2}{2}\right] \rho r^2 dr, \tag{10}$$

where $\sigma = 4\pi$ for a point explosion problem, $\sigma = 2\pi$ for impact generated shock problem and E is the total energy assumed to be constant. One can write equation (10) in the form

$$\left(\frac{R}{R_0}\right)^{-3} = M^2 I(M), \qquad R_0 = (E/\sigma\rho_0 c^2)^{1/3}, \tag{11}$$

where

$$I(M) = \int_0^1 [g(x,M) + \frac{1}{2} f^2(x,M)]\, h(x,M)\, x^2 dx \tag{12}$$

$$f(x,M) = u/\dot{R}, \qquad h(x,M) = \rho/\rho_0, \qquad g(x,M) = (e - e_0)/\dot{R}^2, \tag{13}$$

$$x = \frac{r}{R}, \qquad M = \frac{\dot{R}}{c}, \qquad \gamma = \frac{\eta + 2}{\eta}. \tag{14}$$

Now we use the basic assumption of the quasi-steady method (see Rae[2]), namely,

$$I(M) = J[\gamma(M)], \tag{15}$$

where J is given by (A6) of the Appendix at the end of this paper. The relation (15) is taken to be valid for variable γ so chosen as to give correct values of the pressure, density and particle velocity at the shock. The relation between γ and M using (9) and (14) is given by

$$\gamma = \frac{2M^2 - F(\eta)}{F(\eta)}. \tag{16}$$

By differentiating (11) and using (15) and (16) we get an expression for shock decay coefficient γ as

$$-\frac{R\ddot{R}}{\dot{R}^2} \equiv \lambda = \frac{3}{2 + \phi(\eta)\dfrac{d\ln J}{d\ln\gamma}}, \tag{17}$$

where

$$\phi(\eta) = \frac{4F(\eta)(\eta + 1)}{(\eta + 2)\left[F(\eta) - \eta(\eta + 1)\dfrac{dF}{d\eta}\right]}. \tag{18}$$

The shock trajectory can be found by

$$\frac{ct}{R_0} = \int_0^{R/R_0} \frac{1}{M}\, d(R/R_0). \tag{19}$$

For a given value of M, the corresponding value of γ is calculated using (9) and (16). Then for that value of γ the integral J is calculated either from the table prepared by Rae[2] or from (A6). For $\gamma > 10$, the integral J is approximated by the relation (see Rae[2])

$$J = 0.366\,\gamma^{-1.7}. \tag{20}$$

One can fit a similar relation for $1.2 \le \gamma \le 10$ from the table given by Rae[2]. However, we assume that (20) is almost valid for all values of γ. Using (17) and (20) we get an expression for λ in the form

$$\lambda = 3/(2 - 1.7\phi). \tag{21}$$

Using (11), (16) and (20), we get a relation connecting M with R/R_0 in the form

$$(R/R_0)^{-3} = 0.366\,M^2\left[\frac{2M^2 - F(\eta)}{F(\eta)}\right]^{-1.7} \tag{22}$$

In particular, for c, s medium the equations (16), (21) and (22) reduce to

$$\gamma = [(2s - 1)M + 1]/(M - 1) \tag{23}$$

$$\lambda = \frac{3(M - 1)[(2s - 1)M + 1]}{2(M - 1)[(2s - 1)M + 1] + 3.4sM} \tag{24}$$

$$(R/R_0)^{-3} = 0.366\,M^2\left[\frac{M - 1}{(2s - 1)M + 1}\right]^{1.7} \tag{25}$$

These relatiosns are in simple explicit analytical form which can be used for knowing quickly the dependence of γ, λ and R/R_0 on M for a specified value of s. Also one can easily find the shock trajectory from (19) and (25). For s = 1.2 a comparative study of our analytical solution (24) for λ versus 1/M with the results obtained by Bach and Lee[5] (see Fig. 1). The variation of 1/M with R/R_0 using (25) is compared with the results obtained by Rae[2] in Fig. 2. One can note a remarkable agreement between these results. Here it can also be noted that the modification we proposed is that in quasi-steady method the energy integral J depends upon γ in an explicit form (20) for all values of γ. This assumption enabled us to get the results in a closed analytical form.

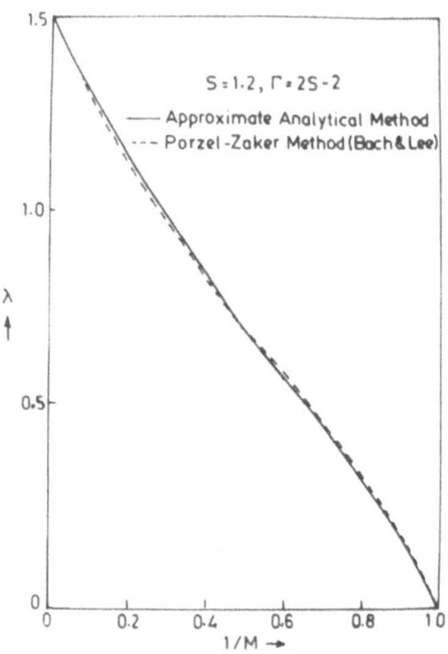

Fig. 1. Variation of shock decay coeeficient with shock mach number.

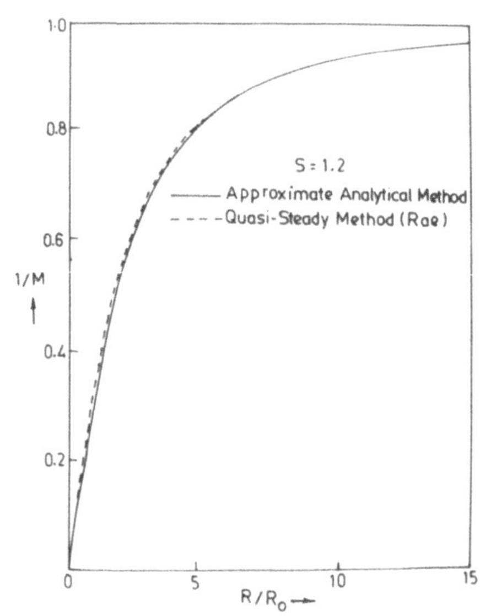

Fig. 2. Shock speed-shock radius relation.

APPENDIX

From the basic equations (1) to (3) in the case of strong spherical shock propagating in a perfect gas ($e = p/(\gamma - 1)\rho$) with the transformations

$$u = \dot{R}V(x), \qquad \rho = \rho_0 H(x), \qquad p = \rho_0 \dot{R}^2 G(x), \qquad x = r/R, \tag{A1}$$

we obtain

$$\frac{dG}{dx} = \frac{(3G - 2\gamma GV/x)(V - x)H - 1.5\gamma VGH}{H(V - x)^2 - \gamma G} \tag{A2}$$

$$\frac{dV}{dx} = \frac{1.5V - \frac{dG}{dx}/H}{V - x} \tag{A3}$$

$$\frac{dH}{dx} = - \frac{2HV + xH\frac{dV}{dx}}{x(V - x)}. \tag{A4}$$

The shock conditions are given by

$$V(1) = G(1) = \frac{2}{\gamma + 1}, \qquad H(1) = \frac{\gamma + 1}{\gamma - 1}. \tag{A5}$$

The expression for $J(\gamma)$ is given by

$$J(\gamma) = \int_0^1 \left(\frac{G}{\gamma - 1} + \frac{HV^2}{2} \right) x^2 \, dx. \tag{A6}$$

For a given value of γ the functions G, V and H are obtained by integrating numerically the differential equations (A1), (A2) and (A3) starting from the shock conditions (A5). After knowing these functions, the relation (A6) is used to obtain $J(\gamma)$ for the assumed value of γ.

REFERENCES

1. R. Kinslow, "High-velocity impact phenomena", Academic Press, New York (1970).
2. W. J. Rae, "Non-similar solution for impact generated shock propagation in solids", NASA CR-54251 CAL Report AI-1821-A-2 (1965).
3. T. A. Zaker, "Point source explosions in a solid", Armour Research Foundation, Illinois Inst. of Tech. Rep. ARF 4132-6 (1959).
4. S. C. Gupta, S. K. Sikka and M. P. Ranga Rao, "A semi-analytical nonself-similar technique for shock waves from underground peaceful nuclear explosions", Proc. Indian Acad. Sci., 89:3:347 (1980).
5. G. G. Bach and J. H. Lee, "Shock propagation in solid media", AIAA, 67-141 (1967).
6. M. H. Rice, R. G. McQueen and J. M. Walsh, Compression of solids by strong shock waves, in: "Solid state physics", F. Seitz and D. Turnbull, eds., Academic Press, New York, 6:1 (1958).
7. H. L. Yadav, Some studies on blast wave propagation, Ph.D. Thesis (Unpublished), I.I.T., Bombay (1978).

THE ATTENUATION OF SHOCK WAVES IN PU FOAM AND ITS APPLICATION

M. P. Lee, G. M. Wang, P. H. Sung, W. L. Chang, Y. L. Lee, and K. Lin

P. O. Box 1-4
CSIST, Lung-Tan
Taiwan, China

INTRODUCTION

In the explosive field, it is imperative that there be a safe method for shipping small quantities of explosives. These explosives may be uncharacterized materials sent for laboratory evaluation or standard explosives for test comparisons. For this purpose, Hildner and Urizar (1981) have designed a shipping container for small samples of high explosives. They used vermiculite as a filler to absorb explosive energy. PU (polyurethane) foams are widely used as impact energy absorbers for long time. The attenuation of shock wave in condensed media and porous solids has been studied enormously (Meyers et al., 1983). In contrast, the dissipation of detonation energy in plastic foams has not attracted much attention yet. In the propagation of shock wave in porous materials, the dissipation of detonation energy may be influenced by the matrix strength and porosity (Dunin et al., 1979). In the present paper, the attenuation capability of different polymers are assessed. From the preliminary results we found the PU is a suitable material to reduce the shock energy. In order to investigate the porosity or density effect on the shock pressure attenuation, we have prepared PU foams with different densities to test their effectiveness. And it shows that the PU foam with density of 0.15 g/cm^3 is a satisfactory material for the explosion-proof purpose.

EXPERIMENTAL

Apparatus

Shock wave attenuation was assessed by the apparatus shown in Fig. 1. A Technoproducts PT-3 piezoelectric force transducer was screwed into a piston assembly, till the sensor element was slightly attached to a pressure piston which transmits the pressure as a force against the transducer element. The piston was lubricated with a thin layer of silicone grease and could move in and out the piston chamber freely. The assembly was then inserted into a large hollow housing block which was welded to a large base plate. Attached to the piston, a testing sample of polymer rod was fastened to the base plate with three metal plates, one on the top and the other two on each side, by long bolts. At the opposite end of the polymer rod, a tetryl pellet and a No. 8 blasting cap were

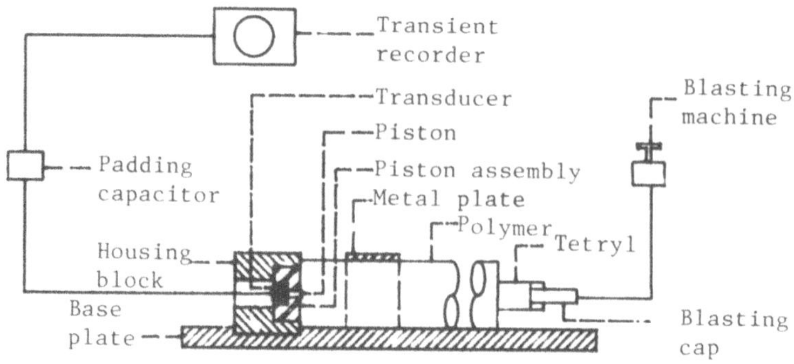

Fig. 1 Testing assembly for shock attenuation measurement.

taped and the blasting cap was initiated by a blasting machine.

The transducer output is fed to a padding capacitor of 0.1 μfd in parallel with the transducer, and then the charge on the capacitor is recorded by Eisenman MTR-80 transient recorder with sampling rate of one point per microsecond. The transducers were calibrated by means of Techno-products TC-2 calibrator.

Sample Preparation

PU foams with densities from 0.05 to 1.05 g/cm^3 were prepared by conventional foaming method. The basic materials are listed in Table 1.

Table 1. The materials and their sources used for PU rod and foams.

Designation	Description	Source
PPO	Polypropylene Oxide ($\overline{Mn}$:1000 g/mole)	Chiun-Long
MDI	4,4'-Diphenylmethane Diisocyanate	Mitsui-Nisso
L-3600	Silicone Surfactant	Union Carbide
DABCO	Triethylene Diamine	Fluka
T-12	Dibutyltin Dilaurate	M & T
Freon 11	CCl_3F	Asahi

All materials were used as received without further purification. They were premixed into two components. Component A includes PPO, L-3600, DABCO and T-12, and component B includes MDI and Freon 11. Components A and B were mixed at 2000 rpm for one minute, then poured into a 60 cm long mold preheated at 50^{0}C. Other materials including PN (nylon-6), PP (polypropylene), ABS (acrylonitrile-butadiene-styrene) and polyester were obtained from local manufacturers.

Stress-Strain Properties

The measurements of tensile properties were carried at room temperature by means of Instron Model 1137 Tester at a constant crosshead speed of 5 cm/min using a 50 Kg load cell. Specimens with dumbbell shape were used for this purpose.

Dynamic Properties

Dynamic properties were determined by applying a sinusoidal force to a rectangular specimens of dimensions 6.50×1.25×3.00 cm^3 by Rheometric

Mechanical Spectroscopy Model 605. The measurements were taken at temperature range of -120°C to 230°C with heating rate of 5°C/min and frequency of 10 rad/sec.

RESULTS AND DISCUSSION

The variation of decayed pressure with distance studied for five polymers, ABS, PN, PP, PU and polyester with densities about 1 g/cm^3 are plotted in Fig. 2. It is observed that the peak pressures decrease as the length of test sample increases. Furthermore, the PU series could reduce the decayed pressure to the lowest level.

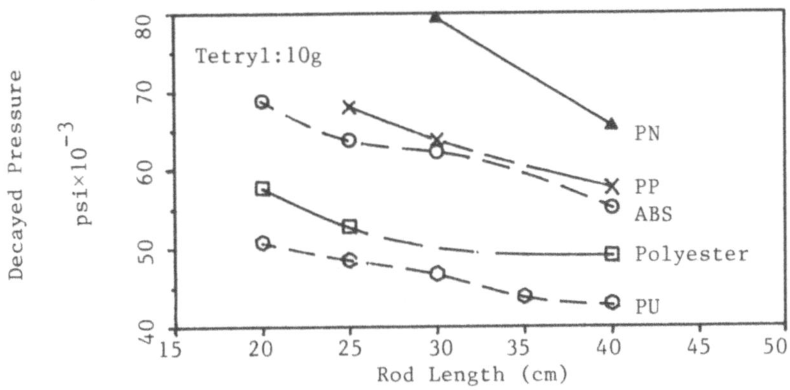

Fig. 2 Decayed pressure vs polymer rod length.

Shock wave propagation in porous media has a number of specific features associated with the effect of porosity. The pressures in porous PU foams as a function of its density are presented in Fig. 3 with minimum around 0.15 g/cm^3.

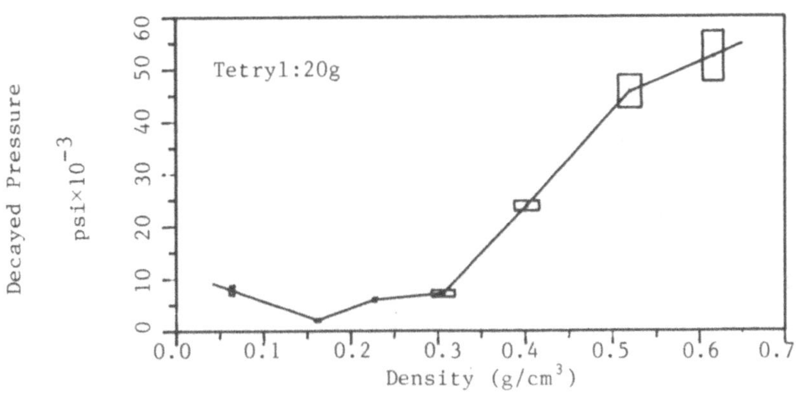

Fig. 3 Effect of PU foam density on shock attenuation for rod length of 20 cm.

In order to investigate whether there is any correlation between shock absorption capability and mechanical properties in PU foams, the tensile strength, Young's modulus, elongation, storage and loss modulus at different densities were measured. These results are presented in Fig. 4 to 8. Generally, the relationship between the mechanical properties of a foam product and its densities is expressed by an empirical equation (Hilyard, 1982):

$$E = k \cdot \rho^n$$

where k and n are empirical constants and E is the mechanical properties. Figs. 4 and 5 show the tensile strength and Young's modulus obey the above equation. However, the density dependence of elongation exhibits a distinct maximum at density 0.15 g/cm^3. The higher percentage elongation means the PU sample can sustain higher deformation without rupture. Dunin et al. (1979) indicated that the porosity changes significantly only when the entire mass of the solid substance enters into the ductile state. In this study it shows the actual nature of pressure attenuation is largely

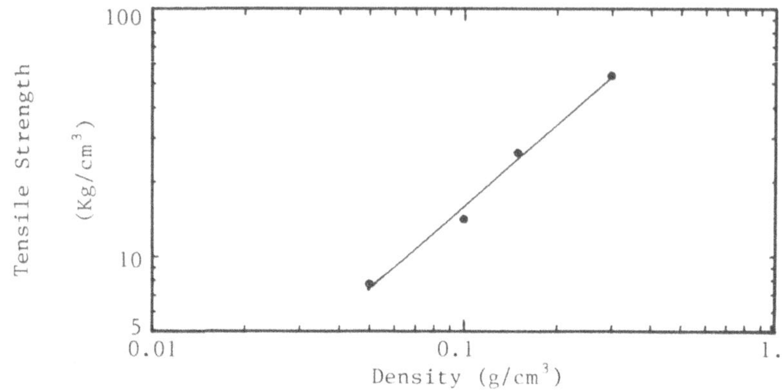

Fig. 4 Tensile strength vs PU foam density.

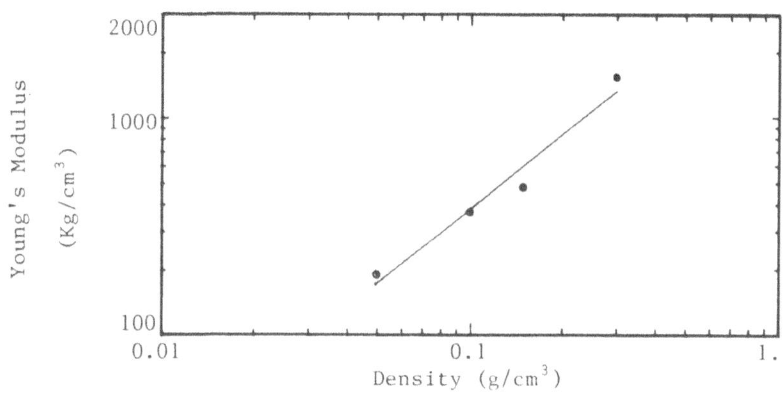

Fig. 5 Young's modulus vs PU foam density.

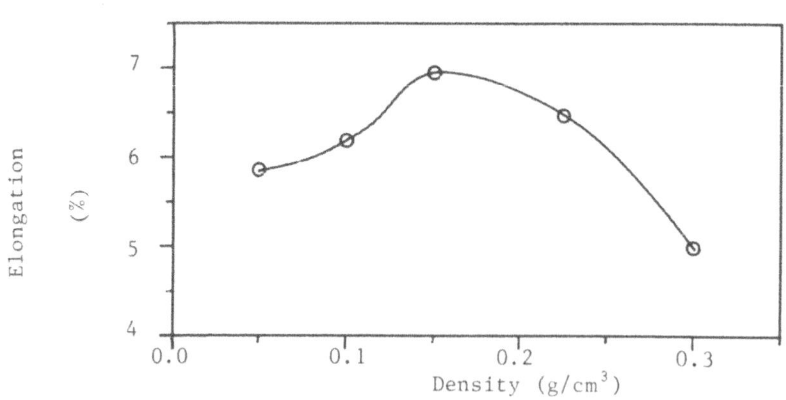

Fig. 6 Elongation vs density of PU foam.

determined by the elongation property of the porous solid.

The energy response of plastic foams can be divided into two parts, storage energy and loss energy. The storage energy is defined as the energy stored and recovered after deformation. The loss energy is the energy dissipated to heat after such deformation. In other words, sample with higher loss modulus should have better shock absorbing capability. The storage and loss modulus are generally varied with the change of foam density. But Figs. 7 and 8 show the PU foams at density 0.15 g/cm^3 possess unusual behaviors. Its storage modulus and loss modulus are higher than that of density 0.225 g/cm^3 PU foams.

A shipping container based on Hildner and Urizar's design constructed with PU foam liner of density 0.15 g/cm^3. This container survived the explosion of 45 g TNT (1.6 g/cm^3) without moving and cracking. The container without the PU foam liner was bursted into halves and thrown to a distance about twenty meters as shown in Fig. 9.

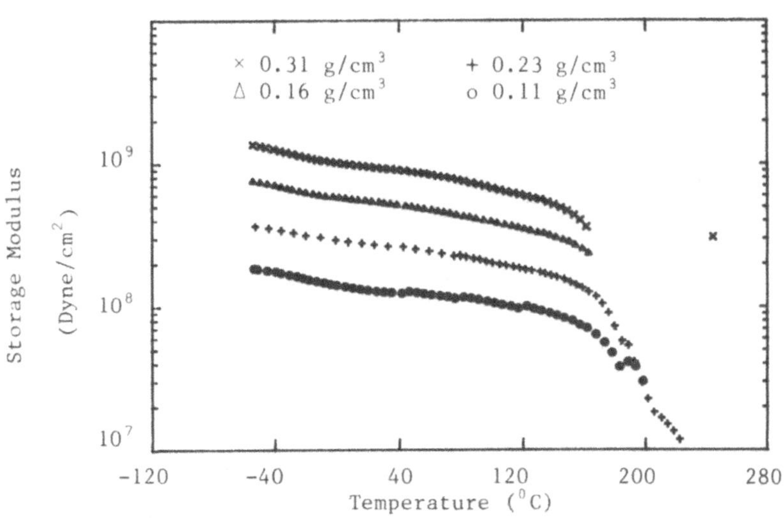

Fig. 7 Storage modulus vs density of PU foam.

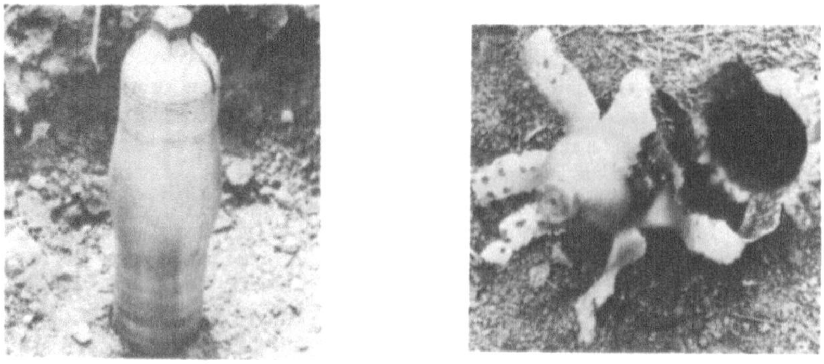

(a) (b)

Fig. 9 Survived test of shipping container, (a) with PU foam liner (b) without PU foam liner.

REFERENCES

Dunin, S. Z., and Surkov, V. V., 1979, structure of a shock wave front in a porous solid, J. Appl. Mech. & Tech. Phys. (USA), 20:612.

Hildner, R. A., and Urizar, M. J., 1981, shipping container for small samples of high explosives, LA-9107-MS, DE82-010042, Los Alamos National Laboratory, New Mexico.

Hilyard, N. C., 1982, "Mechanics of Cellular Plastics", Macmillan Publishing Co., Inc., New York.

Meyers, M. A., Hsu, K. C., and Couch-Rodino, K., 1983, the attenuation of shock waves in nickel : second report, Mater. Sci. & Eng., 59:235, and references cited in.

CHEMICAL SYNTHESIS UNDER HIGH PRESSURE SHOCK LOADING[+]

R. A. Graham*, B. Morosin*, Y. Horie**, E. L. Venturini*,
M. Boslough*, M. J. Carr*, and D. L. Williamson***

*Sandia National Laboratories, Albuquerque, NM
**North Carolina State University, Raleigh, NC
***Colorado School of Mines, Golden, CO

ABSTRACT

Many reports of chemical synthesis under high pressure shock loading have been published over the past thirty years, but little progress has been made in developing an understanding of the mechanisms responsible for the observed effects. Indeed, there remains significant skepticism as to the credibility of many experiments. Certain of the skepticism is based on a "benign shock compression" paradigm in which chemical changes are not expected to occur in a perfect lattice description of solids in thermodynamic equilibrium on the microsecond duration of the shock-compression event. This paper discusses the current status of studies in shock-induced solid state chemistry. It is observed that it is relatively easy to initiate and control chemical reactions in shock-loaded powder materials. In solids it is well known that chemical changes are controlled by defects at all levels from atomic and microstructural to continuum. A major program to study shock-induced modification of inorganic powders provides explicit evidence for unusually large plastic deformation and resulting defect structures. Shock-modified powders also show major enhancements in solid state reactivity in post-shock experiments. Currently, two systems are found to provide a basis for studying controlled chemical synthesis. Intermetallic alloys have been found to be produced in a controlled manner from mechanical mixtures of aluminum with either nickel or titanium. Optical measurements show that a strong chemical reaction is induced on a 100 nanosecond time scale in such aluminum-nickel mixtures under shock compression. The nonferrous metal oxide-ferric oxide system is another promising chemical system in which mixtures of zinc oxide and ferric oxide readily form a zinc-deficient zinc ferrite. Shock-induced chemical synthesis appears to provide a significant opportunity for the synthesis of new materials. Study of chemical effects associated with shock loading provides a new probe of shock-compression processes and current descriptions of the conditions under which they can occur suggest that chemical effects may be ubiquitous in fully dense materials encountered in shock compression studies.

INTRODUCTION

Shock-induced solid state chemistry has recently become a very active area of research. Observation of chemical changes in shocked solids raised particularly profound questions concerning the state of shock-compressed matter. The mobility of ions in the solid state is so restricted that processes in thermodynamic equilibrium in a perfect lattice will not permit significant changes in chemical state. Based on this perfect lattice paradigm, "benign shock compression," it has proven easy to dismiss observations of reactions in solids as due to uncontrolled post-shock effects of unknown but extreme shock states. It has also proven easy to dismiss experimental observations because the vast majority of the observations are under such uncontrolled, unknown conditions that they cannot bear the weight of critical scientific examination. However, since the inception of modern solid state chemistry in the early twentieth century, the central role of defects for solid state reactions has been recognized [1,2]. Accordingly, a description of shock-compressed matter in a "catastrophic shock compression" paradigm which incorporates the influences of defects at all levels of description, is fundamental for a description of shock-induced solid state chemistry. The most casual examination of the literature in solid state reactivity, solid state chemistry or mechanochemistry and consideration of shock-compressed matter in the defect state reveals that the question is not whether solid state chemistry can occur under shock compression, but how much can occur. The question is not qualitative but quantitative. To what extent are chemical effects operative in particular materials under particular shock-compression circumstances? Are yields and thermodynamic consequences significant?

The question of the behavior of low density powder compacts as a higher temperature analog of processes in solid density materials is also a fundamental issue. Much of the chemical synthesis has been carried out in powder compacts and it is well known that significantly higher mean bulk temperatures are encountered than in fully dense solids. Is chemical reaction a direct consequence of higher temperature consistent with a benign shock view, or are there defect processes inherent in powders not encountered in solids that lead to greatly enhanced solid state reactivity?

These issues were recently considered in the Duvall committee on Shock Compression Chemistry in Materials Synthesis and Processing sponsored by the National Materials Advisory Board of the National Research Council of the National Academy of Sciences [3]. A key finding of the report was "as a result of its deliberations, the committee finds that there are substantial grounds for believing that novel physical and chemical processes are involved in the compression and heating of materials by shock waves and that some of these processes may ultimately serve as the basis for useful technologies."

In the present paper an overall assessment of recent work on shock-induced solid state chemistry in metallic and inorganic substances is presented.

BACKGROUND

The shock-induced solid state chemistry program in which the present authors are involved was initiated in about 1980 after a recognition that chemistry in shocked solids, like electronic effects, is controlled to a large degree by defect-centered rather than perfect-lattice centered models. A major program in the Soviet Union, as summarized by Gol'danskii and coworkers [4,5], raised intriguing questions about the shock process but has provided few answers. After our first experiments it became apparent that it was very easy to cause chemical changes under shock compression. It was clear, however, that complex and unique processes were involved and that a long range, systematic, quantitative study was required. Two factors were thought to be critical: 1) methods for quantitative, reproducible shock-modification and preservation of samples for post-shock analysis over a wide range of shock conditions and, 2) study of shock-induced defects and microstructural features and their relationship to shock-enhanced solid state reactivity.

Recovery Fixtures

The study of samples which have been subjected to controlled shock loading and preserved for post-shock analysis provides the major foundation upon which shock-induced chemistry rests. Although other probes, particularly optical methods, afford an opportunity to directly probe the shock-compressed state, the limited information content is not sufficient to define a chemical process in most solids. The shock-recovery experiment provides quite explicit and detailed information but the sample is produced as the result of a complex process of loading and unloading. As a result, although the material held in the hand is the essence of all materials science, it is not easy to identify particular material features with particular features of the shock process. Nevertheless, the recovered samples contain vital information not available from other sources and many studies of materials-centered processes (for example rapid quenching) suffer similar limitations. The recovery experiments mimic in many ways the early days of chemistry in which substances were mixed and the resulting reaction products studied. In the hands of skilled scientists of different disciplines, a great deal can be learned and what is learned will not come from other experiments.

Careful study of shock compression conditions in sample recovery fixtures has shown that quantitative specification requires two-dimensional numerical simulation [6-9]. Much of the earlier work on chemical synthesis has been carried out under conditions which cannot be independently confirmed and it is now apparent that statements about pressure and temperature achieved based on one-dimensional analyses are substantially in error. The fixtures used to acquire the data reported by our group provide all the features needed for quantitative, reproducible scientific studies. It should be recognized, however, that the loading conditions are considerably more complex than that encountered in uniaxial strain in a solid density sample. Although shock conditions are stated with significantly greater errors than in wave profile measurements, they are sufficiently accurate to draw stated conclusions and are subject to refinement if required.

Enhanced Reactivity

It is possible to develop quantitative descriptions of the extent that shock modification increases the solid state reactivity by studies in the post-shock condition. Examples studied include: solid-solid reactions, solid-liquid reactions, solid-gas reactions, structural transformations and powder-consolidation processes. In Fig. 1, examples of four of the processes are shown. Figure 1a shows differential thermal analysis (DTA) traces recording the reaction of mixed powders of zirconia and lead oxide [10]. The upper trace is for unmodified zirconia, the middle trace is for a sample shock-modified at a peak pressure of about

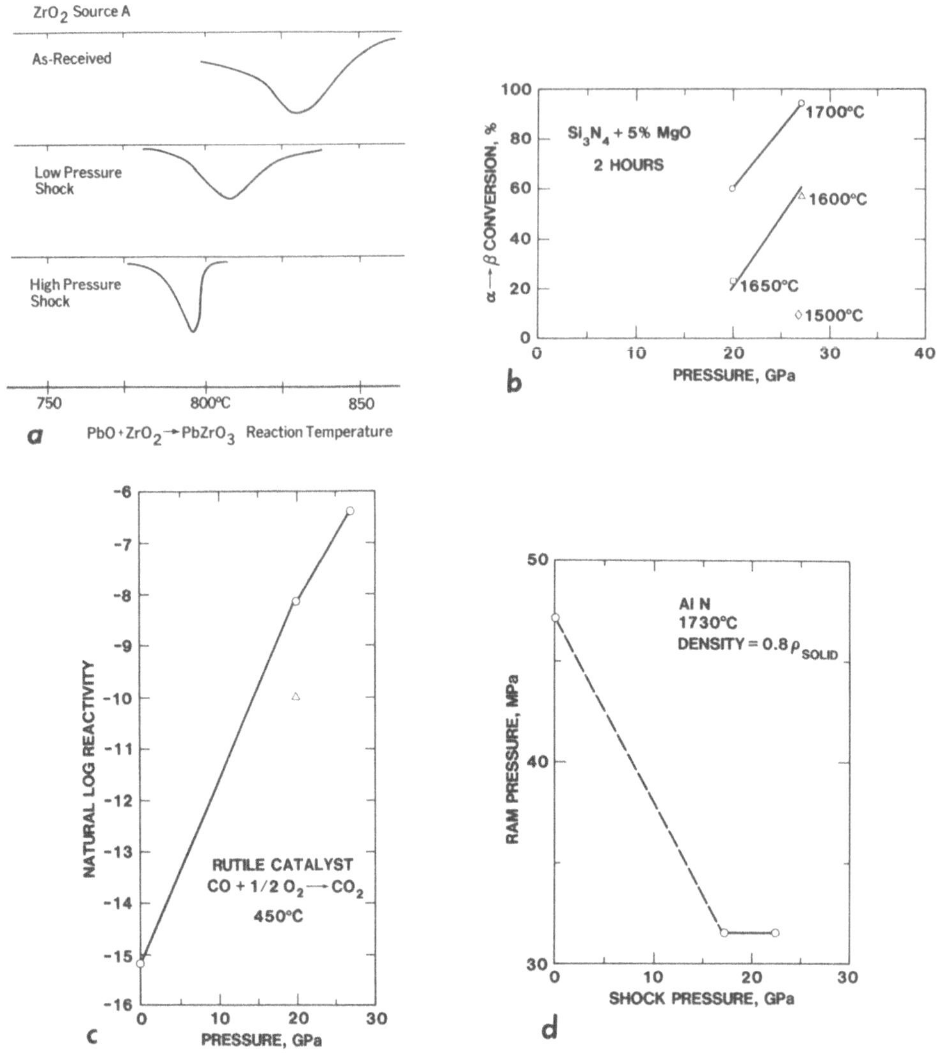

Figure 1. Examples of shock-enhanced solid state reactivity of powders preserved for post-shock study are shown. a) is a differential thermal analysis trace of the reaction of lead oxide and zirconia. (Exothermic temperature is downward.) b) shows the structural conversion of silicon nitride upon heating observed for modification at different shock pressures. The conversion is due a dissolution-precipitation process. c) shows the catalytic activity of a shock-modified rutile catalyst in a flow reactor used for the oxidation of carbon monoxide. d) shows the ram pressure required to achieve a density of 80% of solid density in a hot pressing experiment on aluminun nitride.

22 GPa and the bottom trace is for a powder modified at a pressure of about 27 GPa. The DTA trace for the most strongly modified zirconia has a similar appearance to that of a zirconia sample with a very large specific surface. These measurements show a strongly shock pressure dependent increase in solid state reactivity such that the shocked zirconia, with a modest specific surface, reacts much like a more conventional powder with a large specific surface.

In Figure 1b, the conversion of shock-modified α-phase silicon nitride to a β-phase upon heating is shown [12]. This conversion can only result from a dissolution-precipitation process and the shock modification can only affect the dissolution. The experiments indicate a greatly enhanced dissolution rate for the shock-modified powder. The extent of the effect is a strong function of the shock-modification conditions.

In Figure 1c, the catalytic activity of shock-modified rutile is compared to standard material for the oxidation of carbon monoxide in a flow reactor [11]. The enhanced catalytic activity is found to be substantial and controlled by the shock conditions. The enhanced catalytic activity is a result of shock-induced defects which greatly affect surface states.

Finally, in Figure 1d the consolidation rate under hot pressing is shown for shock-modified aluminum nitride at 17 and 22 GPa shock pressure [13]. The consolidation rate is found to be substantially higher in the shocked material. Control experiments have demonstrated that a recrystallization process of the highly defective shock-modified powder under the stress of hot pressing is responsible for the higher rates of consolidation. The resulting microstructure of the hot-pressed, shock-modified aluminum nitride is unique [13].

The four explicit, quantitative examples of greatly enhanced solid state reactivity serve to demonstrate that shock-induced defects and microstructural features are readily introduced by shock loading and that they can have a dramatic effect on chemical changes in the solid state. In none of the examples, or others which have been studied, are there changes in particle specific surface large enough to affect reactivity.

Shock Modification

Our early synthesis work led us to believe that modeling of shock-induced chemistry required substantially improved understanding of shock modification of inorganic materials. Accordingly, a significant effort in shock modification of inorganic powders involving numerous materials has been carried out. Table 1 lists materials which have been studied to a significant degree. Other materials have been studied less extensively or are currently being studied. A range of material probes has been employed to characterize the shock-modified samples. Studies of particle morphology have been carried out with scanning electron microscopy (SEM), particle distributions from computer processing of SEM photographs (Quantimet), specific surface measurements and optical microscopy. Atomic level defects and microstructural features were studied with x-ray diffraction line broadening, electron spin resonance (ESR) and transmission electron microscopy (TEM). For magnetic materials, static and dynamic magnetization and Mössbauer measurements were carried out. Differential thermal analysis of energy release on heating and cooling was employed in a number of cases and consolidation behavior upon sintering was sometimes studied. Although observations with the unaided human eye are not quantitative, it proved obvious from visual inspection of many materials that very substantial changes are induced by shock compression. All materials were not studied with all probes, but typical analysis included a number of the different material characterizations. The present shock-modification effort is probably the most extensive ever undertaken. In most cases, shock conditions of pressure and temperature were varied over a wide range in a quantitative, reproducible manner.

Table 1. Shock-Modified Materials Studied

Minerals	**Borides**
Pyrite	TiB_2
Oxides	**Antiferromagnets**
TiO_2 rutile	Fe_2O_3
TiO_2 anatase	**Ferrites**
ZrO_2	Fe_3O_4
Al_2O_3	$CuFe_2O_4$
Nitrides	$NiFe_2O_4$
AlN	$CoFe_2O_4$
Si_3N_4	**Crystals**
Carbides	Quartz
TiC	Lithium niobate
	Rutile

Mixed Powders
- PbO + ZrO_2
- ZnO + Fe_2O_3
- 30 % Al-Ti
- 30 % Al-Ni
- 6 Al-Mn

Complete results of the shock modification program are too extensive for inclusion in the present paper and many have been published [14-20]. Several overall features are of particular note. It is well known that greatly enhanced solid state reactivity can be achieved by particle size reduction and increase in specific surface. Careful, consistent specific surface measurement show that shock-modified materials exhibit only modest changes in specific surface. It can be confidently stated that shock-enhanced solid state reactivity is not the result of increases in specific surface.

X-ray diffraction provides one of the most direct and universally available tools for the study of residual strain and crystallite size reduction caused by shock compression. The method also provides a direct indication of structural or chemical composition changes. Morosin and coworkers have carried out numerous detailed analyses of a large number of shock-modified materials [14,20-24]. The studies show that care must be given to the analysis as shock-modified materials differ from those derived from other processes. The overall material response is illustrated in Figure 2. Although most of these materials are brittle and rarely undergo plastic deformation in conventional environments, it is observed that plastic deformation is readily produced under shock loading. The shock pressure required to initiate residual strain is low, of the order of 1 GPa or less. Only in the case of titanium carbide and titanium diboride is a significant critical pressure required to initiate deformation.

The material-dependent and pressure-dependent behavior that is most prominent is the pressure to achieve a saturation level of residual

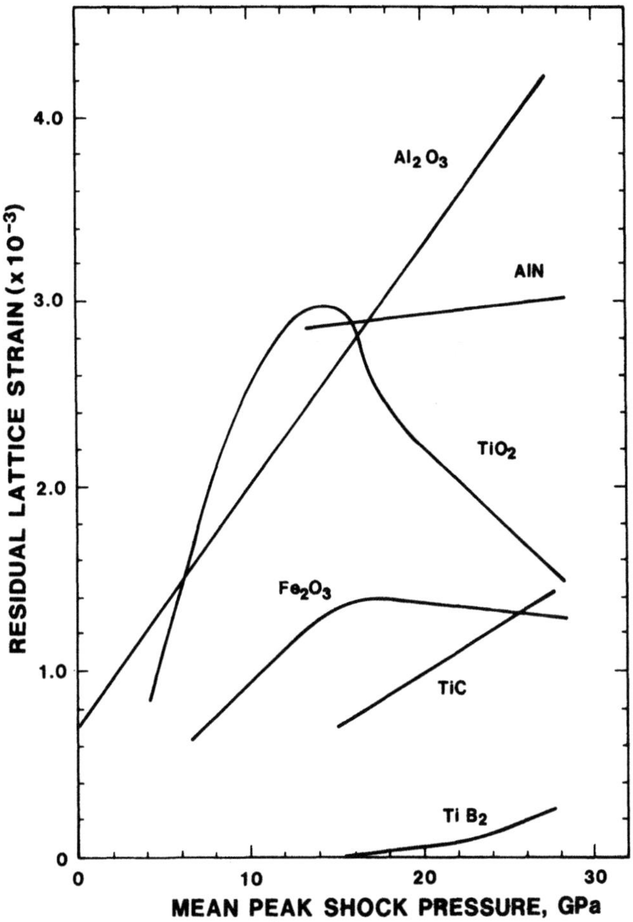

Figure 2. Residual strain observed in shock-modified inorganic powders shows that they are readily deformed plastically under shock compression. Only titanium carbide and titanium dibodride show a significant pressure to initiate plastic strain.

strain. This saturation strain value is about 5×10^{-3}, a value characteristic of heavily worked metals. TEM observations confirm the extensive plastic deformation and typically show such a high defect density that individual features of dislocations cannot be resolved. At higher pressures, rutile is observed to show less residual strain than at lower pressures. This behavior is thought to be the result of post-shock annealing.

The plastic deformation observed in these brittle refractory materials is extraordinary and is a shock-modification feature not found in other processes.

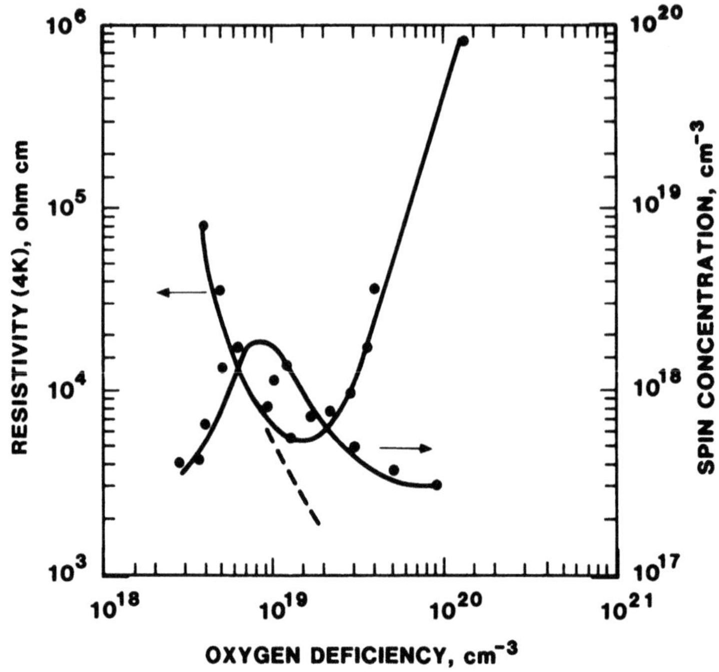

Figure 3. Prior work by Hasiguti [27] in relating oxygen deficiency to resistivity and paramagnetic spin concentration shows a critical behavior at a deficiency of about 10^{19}. Shock-modified rutile achieves larger spin concentrations (about 10^{19}) and lower resistivities (about 10^{-4} ohm-cm) at microwave frequencies.

Shock-induced point defect concentrations provide a particularly sensitive probe of defect processes. It has long been recognized that shocked metals contain unusually large concentration of vacancies due to the high speed motion of dislocations [25]. Venturini and coworkers (15,26) have carried out an unusually broad and thorough study of shock-induced paramagnetic defects with electron spin resonance (ESR). In general what is seen is very large defect concentrations indicative of large plastic deformation and defect states produced in a highly nonequilibrium manner. Defect concentrations are highly dependent on the particular material and on the particular pressures and temperatures achieved. In rutile, reductions in defect concentrations at higher shock temperatures indicate that there is a significant post-shock annealing in some cases and that during the shock process concentrations are significantly higher.

In the case of rutile, where there is substantial prior work [27] on defects induced by other processes, the unusual nature of shock defects is readily apparent. Based on prior work, it is known that as oxygen vacancies are introduced that low temperature resistivity is reduced and paramagnetic defect concentrations of Ti^{3+} interstitials are increased. However, as shown in Figure 3, at a critical oxygen deficiency the behaviors reverse. The minimum in resistivity and the maximum in paramagnetic spin concentration is unlike that observed in other narrow gap semiconductors such as germanium and silicon. Prior studies have indicated that the critical behavior is due to the formation of higher-order point defect complexes which eventually lead to a "crystallographic shear" configuration involving a planar array of oxygen vacancies.

Shock-modified rutile shows a substantially different behavior. Paramagnetic spin concentrations of Ti^{3+} interstitials are $10^{19} cm^{-3}$, far in excess of those achieved in prior work. This higher concentration defect state leads to a metallic-like conductivity at microwave frequencies. TEM studies show no evidence for crystallographic shear. It is apparent from ESR and TEM studies of rutile that the intense, rapid nature of shock deformation has led to a defect state not previously achieved.

As a final example of the role of shock deformation in inducing defects, studies on α-hematite are particularly revealing. Well characterized powders were shock modified over a wide range of shock pressures and temperatures and subjected to detailed static and dynamic magnetization study. Magnetization measurements provide a particularly sensitive probe of shock-induced defects as α-hematite is an antiferromagnet with zero net magnetization in the perfect lattice. The antiferromagnetism arises from iron magnetic moments which are ferromagnetically coupled within a (111) sheet and directed perpendicular to this sheet. In successive atomic layers the moments are antiparallel. Because of this arrangement, any disorder affecting parallel and antiparallel domains can have a major effect on magnetization. The nature of the disorder on the atomic scale cannot necessarily be determined from net magnetization, but Mossbauer effect measurements are in progress which reveal interesting and unusual features [28].

Magnetization observed for shock-modified hematite as a function of peak shock temperature (which involves shock pressure) is shown in Figure 4. The effect of shock modification is seen to be very large; a

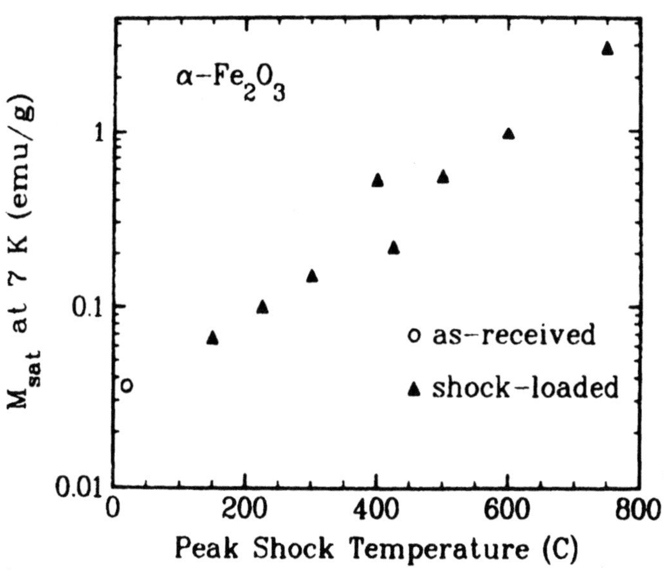

Figure 4. The low temperature magnetization of hematite is strongly altered by shock temperatures (and pressures). At the two highest temperatures, the measurements indicate that a 1% and 3% conversion to magnetite is responsible for the magnetizations shown.

two order of magnitude increase in magnetization is observed. Shock deformation has proven to be extraordinarily effective in introducing defects in hematite. At the two highest temperature points the magnetization results from formation of about 1% and 3% magnetite.

SHOCK-INDUCED CHEMICAL SYNTHESIS

The shock-induced modification studies leave us with explicit physical and chemical characteristics of materials from which we can infer whether conditions in various parts of the shock process can lead to chemical reaction. If chemical reactions are to occur on the submicrosecond or microsecond duration of the shock process when the material is at pressure and accelerating to high speed, the shock-compression must induce a state of greatly enhanced solid state reactivity. In the pressure release phase, events may occur which are not possible at pressure and the rapid decrease in material velocity provides accelerations whose overall directions are opposite but less severe than those in the loading process. Finally, if the sample is well contained in a fixture in the immediate post-shock state, the material will cool through thermal conduction and may be at a modest residual pressure in the fixture.

The shock-modification study shows that, in most cases, significant increases in reactivity due to increases in particle specific surface area are not to be expected. On the other hand, it appears quite likely that enhanced solid state reactivity induced by defects will be maximized. This optimization of defect reactivity is not encountered in other processes and is a unique feature of the shock process. Unfortunately, because this defect state cannot be fully achieved otherwise, a limit to the enhanced reactivity cannot be assessed from other independent studies at present. Greatly enhanced diffusion and significantly increased ease in nucleation will result from the highly defective state. Local stress and temperature gradients can be extreme and lead to enhanced diffusion as will chemical concentration gradients. The very large particle deformation and increase in defect density would be expected to produce surface conditions in which oxides and other impurities have been removed. Such atomically clean surfaces are known to be highly reactive. The overall effect of the shock modification would be expected to achieve very large enhancements in solid state reactivity, but the relative contribution of different aspects of the modification and quantitative specification of the processes must await further experimentation and analysis.

Two systems have been chosen for detailed chemical synthesis study: a nonferrous metal oxide (zinc oxide) plus ferric oxide (hematite) and intermetallic reactions between aluminum and nickel and aluminum plus titanium. The specific materials chosen have interesting properties in themselves and represent systems in which many other candidate reactions can be studied. Both synthesis systems have been carefully studied under other conditions. Both systems give large yields of reaction products under shock loading which can be readily controlled by varying the shock conditions. Both systems lend themselves to straightforward but revealing analysis of the reaction products. The intermetallic reactions are strongly exothermic while the ferrite reaction involves essentially no significant energy release.

Zinc Ferrite Synthesis

Ferrites as a class have been extensively studied and the relationships between magnetic properties and atomic level configurations are well understood [29]. Stoichiometric zinc ferrite is not usually used in device applications but it is studied in conjunction with other compositions. Because the stoichiometric composition is paramagnetic and the starting zinc oxide is nonmagnetic, the synthesis products can be explicitly studied with magnetization. Prior work [30,31] has shown a zinc ferrite could be synthesized under unknown shock conditions from starting mixed powders of zinc oxide and hematite. Typical conventional synthesis is carried out as a solid state reaction between mixed powders heated in a controlled oxygen environment. It is known that the formation of oxygen vacancies in hematite is critical to the promotion of diffusion of zinc into the hematite.

Four material probes were used to study the shock-synthesized products. Of these, magnetization provided the most sensitive measurements, X-ray diffraction the most explicit structural data, Mossbauer the most explicit atomic level data and, although TEM was used on only a limited number of samples, it provides the most explicit chemical composition and microstructural data. All four probes reveal a somewhat different material description. The differences emphasize the need for a wide variety of materials characterization before detailed models of unique materials can be confidently developed.

Mechanically-mixed powders in a stoichiometric ratio to form $ZnFe_2O_4$ were subjected to a wide range of shock conditions of pressure and temperature and detailed by Morosin, et al [32]. In this work, structural properties of the synthesized products and unreacted components are shown based on X-ray diffraction. Magnetization measurements on certain of the samples are reported by Venturini, et al [33], Mossbauer measurements by Williamson, et al [34] and TEM by Carr and Graham [35]. The results of determinations of yield of products by the various techniques are summarized in Figure 5. In the figure, bar graphs are used beside the points in pressure and temperature at which yields were determined by the different techniques. X-ray diffraction yields are based on the relative amount of material in the spinel structure. Yield based on magnetization is based on prior magnetization studies. Mossbauer measurement show two spinel phases, one ferrimagnetic and one paramagnetic. Both products are shown. In this figure only results based on Composition B explosive loading are shown as the yields obtained from baratol explosive were consistently lower. (This observation is thought to provide explicit evidence that the amplitude of the initial low pressure wave has a significant effect on the chemical reaction because the early loading history is strongly affected by the type of explosive.)

From Figure 5 it is notable that magnetization provides the most sensitive detection of reaction products. Even at the lowest pressure and temperature a detectable magnetization occurred. This magnetization is over and above that due to hematite modification. Detectable yield at such modest loading conditions, which is also observed in the intermetallic synthesis, provides explicit evidence for the extraordinary efficiency of mechanically-forced reaction due to shock compression.

As indicated in Figure 5, X-ray diffraction showed that the yield of spinel-structure zinc ferrite was a strong function of shock pressure and temperature. Yields were very large, up to 85% in the most extreme pressure and temperature case. Line broadening due to strain, reduced crystallite size and possible inhomogeneities was significant in all

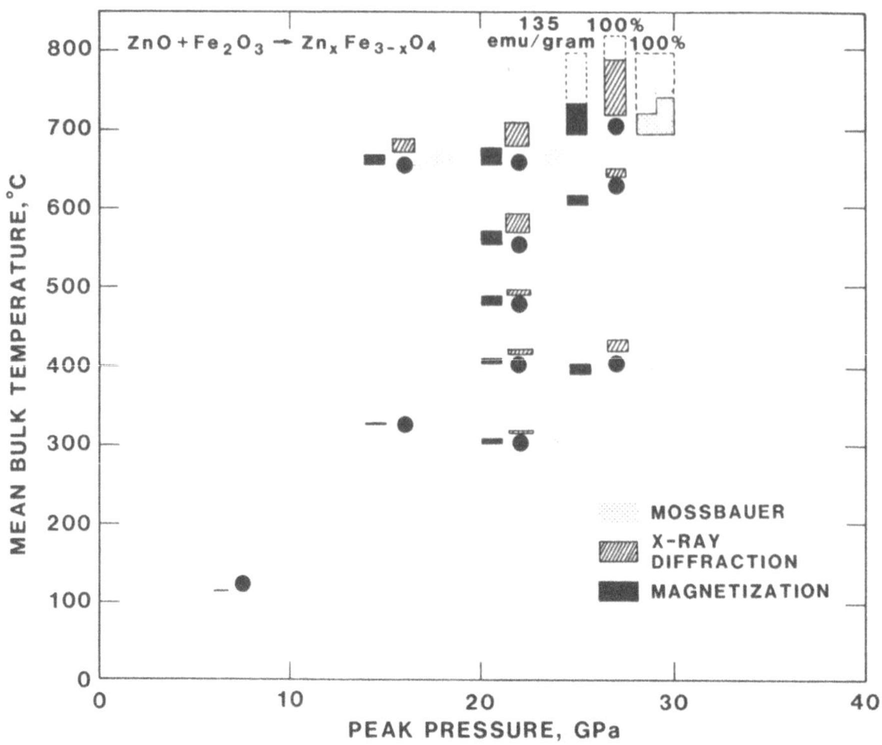

Figure 5. The yield of shock-synthesized zinc ferrite is shown to be strong function of shock pressure and shock temperature. Bar graphs beside the points at which experiments were conducted show the yields determined by the three techniques. Magnetiztion shows detectable yields at the lowest pressure. Mossbauer includes two different spinel-structure components.

samples except at the most extreme shock condition at which a dramatic sharpening of diffraction lines occurred. Although there is a hint of a threshold for reaction of 15 GPa pressure and a temperature of about 300 C, the detection limit is about 5% and, as indicated below, magnetization showed ferrite yield at the most modest shock condition. The yields of spinel are in good agreement with the Mossbauer data. X-ray diffraction indicated that the starting powders were not fully consumed in the reaction. More hematite was consumed relative to zinc oxide, suggesting that the spinel is zinc deficient. The sharp lines observed in the most extreme loaded sample permitted a determination of a zinc compositiom of 0.92.

Mossbauer spectroscopy showed three distinct iron resonances in the shocked material. One of these is due to iron in the shocked but unreacted hematite and its characteristics match those of hematite shocked in independent experiments to the same shock conditions. The Mossbauer data are explicit in demonstrating that changes in magnetic properties are due to the reaction products, not some defect-induced structural modification of hematite.

The second magnetic phase is identified as a zinc-deficient ferrite $Zn_xFe_{3-x}O_4$ with x = 0.2. The data indicate that the composition is independent of shock condition; it is the yield which is strongly affected.

The third phase is paramagnetic in character and present indications are that it is superparamagnetic due to a relaxation phenomenon for domain sizes in the range of about 500 nm.

Based on the combination of data shown in Figure 5, it is apparent that magnetization measurements provide the most sensitive detection of reaction products. Even at the lowest pressure and temperature, a detectable magnetization was observed.

Based on the magnetizations observed, the ferrite is composed of a zinc-deficient ferrite with zinc composition between 0.4 and 0.6. As is observed with Mossbauer, yield is strongly dependent on shock condition but composition remains the same.

The TEM study was more limited in scope than the others and included only the starting powder mixture, a shocked but only modestly reacted sample and the sample subjected to the most extreme conditions. As indicated earlier, the most extreme conditions appear to have produced a ferrite with significantly different characteristics. The starting powder was found to be well blended down to the individual grain level. Grain sizes were about 200 nm. Both powders showed low defect density. No ferrite was found in the modestly shocked sample. Only hematite grains could be easily located; the softer zinc oxide flowed between the harder hematite grains. X-ray energy dispersion spectroscopy scanning (EDS) across "grain boundaries" showed heavy concentrations of zinc between hematite grains. Some hematite grains were found to contain up to 5% zinc.

At the most extreme shock condition ferrite grains about 500 nm in size were readily apparent and were found to be defect free. They appear in clusters up to 10,000 nm across. The appearance of the grains demonstrates they have gone through a recrystallization process. There was no evidence for compositional gradients within the grains. The material between the clusters was observed to be a well mixed conglomerate of the starting materials. As this sample is the only sample showing recrystallization and the yield is much greater than other samples, it is tempting to relate the recrystallization process to higher yields. The composition is found to have a zinc value between 0.89 and 0.93 which agrees with that determined from X-ray diffraction lattice parameters.

The various material probes provide what is, on the surface, a major disagreement concerning the chemical composition of the synthesized ferrite. EDS provides a direct measurement of the composition but the technique has only been used on the sample subjected to the most extreme shock conditions. At this extreme condition both X-ray diffraction and TEM provide zinc concentrations in agreement with each other and close to a stoichiometric composition. Neither of these determinations agree with the magnetization or Mossbauer data which, depending on interpretation, can give zinc values between 0.2 and 0.6. It should be recognized however that concentrations with the latter probes do not provide a direct measure of the concentration. Values are derived indirectly from published reports in which the relationships between magnetization or Mossbauer parameters were established on ferrites synthesized by conventional methods. Resolution of the differences must involve consideration of the possibility that the shock-synthesized material is substantially different from other compositionally similar material.

Intermetallic Synthesis

Intermetallic alloys or compounds are a particularly interesting class of materials which have been widely studied because of important technological applications. The aluminides have been identified as compounds with highly desirable characteristics of high temperature strength and resistance to corrosion. It has proven difficult to prepare suitable aluminides with good properties as impurities at grain boundaries have acted to greatly reduce the ductility. Our group has carried out the first synthesis of nickel and titanium aluminides under shock compression [36] as well as a detailed study of the relationship of reaction products on shock compression conditions. The experiments and analysis are extensive and only an overall summary can be presented. Other papers in this proceedings by Horie and coworkers [37-40] contain more detail.

Although most synthesis in the binary system is carried out from the melt or by rapid quenching or sustained high temperature synthesis, intense mechanical working by cold rolling or high energy ball milling are known to produce amorphous alloys. Thus it is to be expected that intense mechanical deformation of the shock process will have a significant effect on reaction.

Three powder mixtures have been studied. They were a mechanically-blended mixture of aluminum and nickel in an approximate stoichiometric ratio to form Ni_3Al and an equivalent mixture of aluminum and titanium. A less extensively studied composite of spherical particles of aluminum coated with nickel was also examined. For the composite particles the constituents were about 80 weight percent nickel and 20 weight percent aluminum. Starting and shock-modified materials were characterized by x-ray diffraction, TEM, optical microscopy and electron-probe microanalysis.

The mechanically-blended aluminum-nickel powders provided the most definite results. Strong reactions were observed with yields of up to 80 to 90%. Many reaction products were observed and their strong dependence on shock conditions allow identification of local reactions as precursors to the principal reaction product. At the higher shock pressures and temperatures a characteristic pattern of reaction products were observed. In the outer region of the sample, where numerical simulation shows a higher temperature, a large area of almost fully reacted Ni_3Al is observed. In the center part of the same sample where the temperature is lower and the pressure is higher, local reaction regions of NiAl are observed. In a thin transition region next to the Ni_3Al there is a thin region of Ni_2Al_3 which surrounds a eutectic system of $NiAl_3$ and Al. No amorphous materials are detected although there are numerous regions of extremely well mixed substituents in unreacted regions.

As indicated in Figure 6, the yield of reaction products is strongly dependent on shock pressure and temperature. The NiAl product is formed at very modest pressures and temperatures. As in the case of the zinc ferrite synthesis, this observation is thought to be clear evidence for initiation of reaction due to mechanical action. As pressures and temperatures are increased past a critical value of about 500 C and 15 GPa, the Ni_3Al product is observed over an area encompassing the entire region which was above the critical temperature. Due to the large extent of this area and the strongly exothermic nature of the reaction,temperatures of thousands of degrees are expected and the final product would have resulted by pressure quenching and cooling by thermal conduction from the melt. There is evidence from the sample shocked to the highest pressure state that excess pressure (above about 20 GPa) inhibits the initiation of strong reaction.

All reaction products identified have been seen in prior synthesis; what is different is the absence of amorphous material and the presence of Al_3Ni and Al_3Ni_2 which has not been seen in the extensive work on fast quenching of ion-implanted layers with pulsed-electron-beam heating [39]. The shock-synthesized Ni_3Al appears to have microstructural features not

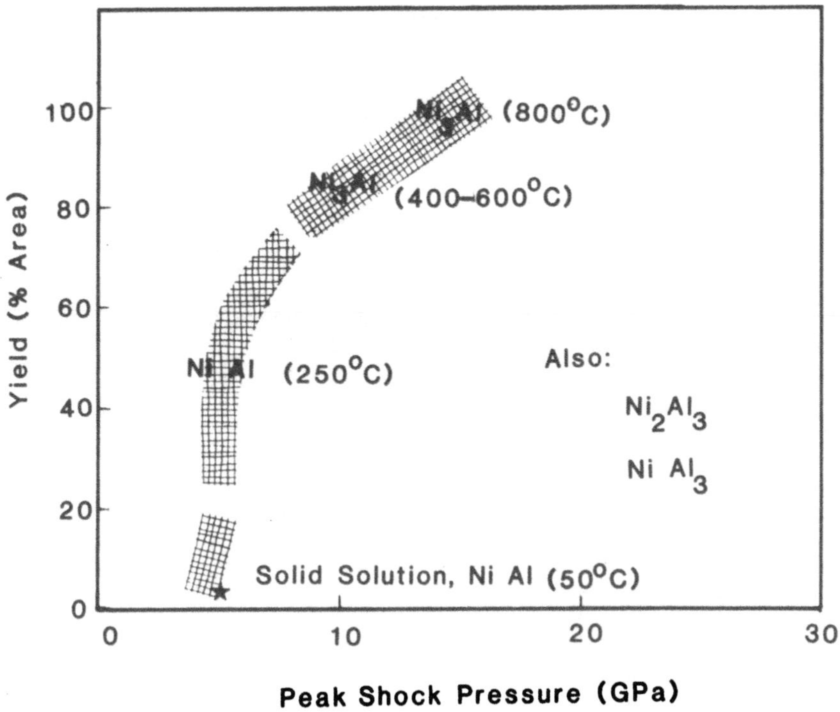

Figure 6. The yield of the indicated reaction products is found to be significantly dependent on shock pressure and temperature. Specific locations of yield determinations are shown with the compound designated. Regions between are shaded by inference. Not shown is the reduction of yield at pressures above 20 GPa.

previously observed. The microhardness is in the range of a heavily cold worked composition even though TEM observations show few dislocations. The hardness is apparently due to an unusually small crystallite size (about 2 nm.) Such material would be expected to have ideal properties for a wide variety of applications.

The smooth sequence of shock-modified features and reaction products leads to a description of the reactions in which early in the shock process the individual grains are heavily deformed leading to a highly defective state and clean surfaces. There is then an extremely efficient mixing process which involves fluid-like flow of aluminum around and into open areas of nickel. Nickel is also mixed into aluminum rich areas. The nickel-aluminum reaction appears locally but is not sustained over large areas. These locally-reacted regions can then serve as nucleation sources when sufficiently high temperatures and larger mixing intensity are achieved. The products in the transition region next to the strongly reacted regions likely result from the high temperature achieved due to a strong exothermic energy release.

In the composite powder, there appears to be a different initiation threshold for strong reactions. A major change in particle morphology is seen resulting from flattening of the composite particles at relatively low pressures. Local regions of Ni-Al are found within nickel-aluminum grains at their interfaces and 100 nm size regions of Ni_3Al are found inside the NiAl.

The titanium-aluminum powder mixtures react far less strongly. Yields are limited and products more difficult to identify. As in the Ni-Al system, the largest region of reaction was found in the outer part of the sample where the temperature is higher than in the center. The product in this region is hard to identify but apparently corresponds to $TiAl_3$ which is difficult to distinguish from the ordered superstructures Ti_9Al_{23} and Ti_8Al_{24}.

The chemical synthesis of intermetallic alloys or compounds from mixed powders of aluminum-nickel and aluminum-titanium has provided an extraordinary rich storehouse of reaction products and shock-activated but unreacted products on which to build models of shock-induced solid state chemistry. There is no evidence that shock-induced melting of the aluminum is necessary to initiate the reaction.

OPTICAL OBSERVATIONS

Recent observations of Boslough and coworkers [41,42] appear to provide the most significant result achieved to date in shock-induced solid state chemistry. The study used optical pyrometry and high speed optical streak observations and provided explicit time-resolved and spacially-resolved data on temperatures associated with shock-induced reaction between aluminum and nickel powder mixtures described above. The observations are particularly significant in that they clearly demonstrate that chemical reaction in shock-compressed solids can be initiated within about 100 nsec after critical shock conditions are achieved. The observations also tie together the numerical simulation of shock processes in recovery fixtures and the recovery experiments which show strong reaction initiating in the same part of the sample in which large yields of intermetallic compounds were observed. Once it is recognized that solid state chemical reactions are possible on such a short time scale, consideration of shock conditions in solid density materials establishes that chemical reactions are likely to be ubiquitous in a wide range of conventional materials. Optical measurements promise to provide explicit data on heat of reaction in shock-induced chemical reactions in solids.

SUMMARY

Although considerable progress has been made in identifying the processes important to shock-induced solid state chemistry, it is clear that there are few quantitative data on which to assess their relative importance. The work summarized in the present paper provides a major source of material response data on shock activation, shock modification and shock synthesis.

The available data indicate that shock-induced solid state chemistry must be based on a catastrophic shock description involving the defect state, nonequilibrium thermodynamics and changes in material configuration occurring at high speed. Chemical reaction under shock compression involves generation of large numbers of defects which can act to greatly enhance solid state reactivity. Plastic deformation of particles has a major effect on configuration and morphology which will lead to a surface cleaning essential for significant solid reaction. The intense material velocities will lead to very efficient mixing. It is clear that there are processes encountered in low density powder compacts that do not occur at solid density.

Optical observations of chemical reactions in powders compressed to solid density while under shock compression remove a conceptual barrier that has prevented serious consideration of chemical reaction in solids on the submicrosecond time scale. With the recognition of the possibility of fast reaction in shock-compressed solids, it is well to inquire as to whether similar effects are possible in solid density samples. What is required for reaction is the presence of different chemical species, a localized increase in temperature and a material mixing. Different chemical species are present in most materials as precipitates or impurities concentrated at grain boundaries. Localized deformation at grain boundaries, at precipitate interfaces and in deformation bands in the bulk will lead to high local temperatures on a transient time scale. The high speed motion of defects and localized deformation at different velocities can then lead to sufficient mixing to initiate chemical reaction and cause it to grow to critical size [40]. Indeed it appears that evidence for such shock-induced solid density chemistry is already published but previously unidentified.

Hill and Rack [43] have reported a TEM study of a Nitronic stainless steel in which an unidentified phase was observed at the intersections of deformation bands. This phase appears to be a sigma phase characteristic of chemical reaction which normally requires hours at elevated temperature but the rate of reaction is known to be greatly accelerated by cold working.

In his interesting historical account of the development of shock compression science at Los Alamos [44], John Taylor closes the article relating a conversation with John Walsh concerning elastic-plastic behavior in which it was agreed "We're in the metallurgical mud." It appears now that we are also in the chemical ooze. If we are to extract ourselves from this ooze perhaps we need to recognize the situation "of the hypnotic power of mental habitudes" posed by the eminent Swedish solid state chemist J. Arvid Hedvall when he related the resistance he encountered in developing solid state chemistry in the early twentieth century [45].

+Work supported at Sandia National Laboratories under U. S. Dept. of Energy contract # DE-AC04-76DP00789.

REFERENCES

1. H. Schmalzried, Solid State Reactions, Verlag Chemie, Basel (1981).
2. A. R. West, Solid State Chemistry and its Applications, Wiley, NY (1984).
3. G. E. Duvall, Chairman, "Shock Compression Chemistry in Materials Synthesis and Processing ", National Materials Advisory Board, National Research Council, National Academy of Sciences (1984).
4. G. A. Adadurov, V. I. Gol'danskii and P. A. Yampolskii, Mendeleev Chemistry Journal 18:92 (1973).
5. G. A. Adadurov and V. I. Gol'danskii, Russian Chemical Reviews, 50:948 (1981).
6. L. Davison, D. M. Webb and R. A. Graham, in Shock Waves in Condensed Matter-1981 (Menlo Park), edited by W. J. Nellis, L. Seaman and R. A. Graham, American Institute of Physics (1982) p. 67.
7. R. A. Graham and D. M. Webb, in Shock Waves in Condensed Matter-1983, edited by J. R. Asay, R. A. Graham and G. K. Straub, North Holland, NY (1984) p. 211.
8. R. A. Graham and D. M. Webb, this proceedings.
9. F. R. Norwood, R. A. Graham and A. Sawaoka, this proceedings.
10. D. L. Hankey, R. A. Graham, W. F. Hammetter and B. Morosin, J. Mat. Sci Lett. 1:445 (1982).

11. J. Golden, F. Williams, B. Morosin, E. L. Venturini and R. A. Graham, in Nellis, et al, eds (1982) loc cit, p. 72.
12. E. K. Beauchamp, R. E. Loehman, R. A. Graham, B. Morosin and E. L. Venturini, in Emergent Process Methods for High-Technology Ceramics, edited by R. F. Davis, H. Palmour III and R. L. Porter, Plenum, NY (1984) p. 735.
13. E. K. Beauchamp, R. A. Graham and M. J. Carr in Defect Properties and Processing of High Technology Nonmetallic Materials, edited by J. H. Crawford Jr., Y. Chen and W. A. Sibley, North Holland (1984) p. 281.
14. B. Morosin and R. A. Graham, Mat. Science and Engr. 66:73 (1984).
15. E. L. Venturini and R. A. Graham in Crawford, et al, eds (1984) loc cit, p. 383.
16. Y. K. Lee, F. L. Williams, R. A. Graham and B. Morosin in Asay, et al, eds (1984) loc cit, p. 399.
17. Y. K. Lee, F. L. Williams, R. A. Graham and B. Morosin, J. Mat Sci., in press.
18. W. J. Hammetter, J. R. Hellmann, R. A. Graham and B. Morosin, in Asay, et al, eds (1984) loc cit, p. 391.
19. J. R. Hellmann, K. Kuroda, A. H. Heuer and R. A. Graham, in Asay, et al, eds (1984) loc cit, p. 387.
20. B. Morosin and R. A. Graham, in Crawford, et al, eds (1984) loc cit p. 335.
21. B. Morosin and R. A. Graham, in Asay, et al, eds (1984) loc cit, p. 355.
22. B. Morosin, E. J. Graeber and R. A. Graham in Advances in X-ray Analysis, Vol. 27, edited by J. B. Cohen, J. C. Russ, D. E. Leyden, C. S. Barrett and P. K. Predecki, Plenum (1984) p. 369.
23. B. Morosin, E. J. Graeber and R. A. Graham in Cohen, et al, eds (1984) loc cit, p. 379.
24. B. Morosin and R. A. Graham, Proceedings, 1985 Explomet, Portland, OR, July 29-31, 1985, to be published.
25. M. A. Mogilevsky, Physics Reports 97:358 (1983).
26. E. L. Venturini, B. Morosin and R. A. Graham, in Nellis, et al, eds (1982) loc cit, p. 77.
27. R. R. Hasiguti in Ann. Rev. Mat. Sci., Vol. 2, edited by R. A. Huggins, R. H. Bube and R. W. Roberts, Annual Reviews, Palo Alto (1972) p. 69.
28. D. L. Williamson, private communication.
29. see the compendium in Landolt-Bornstein, Numerical Data and Functional Relationships in Science and Technology, New Series, Group III, Vol. 4, Part b, Editor-in-Chief, K. H. Hellwege, Springer-Verlag, NY (1970).
30. Y. Kimura, Japan. J. Appl. Phys. 2:312 (1963).
31. G. A. Adadurov, G. V. Novikov, N. S. Ovanesyan, V. A. Trukhtanov and V. M. Shekhtman, Soviet Physics-Solid State 11:1601 (1970).
32. B. Morosin, E. L. Venturini and R. A. Graham, this proceedings.
33. E. L. Venturini, B. Morosin and R. A. Graham, this proceedings.
34. D. L. Williamson, B. Morosin, E. L. Venturini and R. A. Graham, this proceedings.
35. M. J. Carr and R. A. Graham, this proceedings.
36. Y. Horie, R. A. Graham and I. K. Simonsen, Materials Lett., in press.
37. I. K. Simonsen, Y. Horie and R. A. Graham, this proceedings.
38. Y. Horie, D. E. P. Hoy, I. Simonsen, R. A. Graham and B. Morosin, this proceedings.

39. S. A. Mayers, C.C. Koch, Y. Horie and R. A. Graham, this conference.
40. H. R. Pak, Y. Horie and R. A. Graham, this conference.
41. M. B. Boslough and R. A. Graham, Chem. Phys. Lett., submitted.
42. M. B. Boslough, R. A. Graham and D. M. Webb, this proceedings.
43. C. R. Hill and H. J. Rack, Mat. Sci. and Engr. 51:321 (1981).
44. J. Taylor in Asay, et al, eds (1984) loc cit, p. 3.
45. J. Arvid Hedvall, Solid State Chemistry: Whence, Where and Whither, Elsevier (1966).

DYNAMIC CONSOLIDATION OF ALUMINUM POWDERS

J. E. Flinn, G. E. Korth, R. N. Wright and R. C. Green

Idaho National Engineering Laboratory
EG&G Idaho Inc.
Idaho Falls, ID

INTRODUCTION

Dynamic consolidation of metallic powders, with an emphasis on rapidly solidified particulate, has received considerable attention over the past five years.[1-12] Some potential advantages of dynamic consolidation for rapidly solidified powders are relatively low bulk temperatures, nearly random texture, and near neighbor particle retention during densification, as well as hardening and possibly beneficial phase transformations. Dynamic consolidation, as a relatively low bulk temperature process, can permit retention of some of the metastable features obtained from rapid solidification processing (RSP) of the powder.

Densification, particle interaction, and bonding of the particles during dynamic loading are difficult to ascertain experimentally, particularly in situ, or to model. Extensive two- and three-dimensional computer models are required, which eventually must contain appropriate phenomenological correlations for parametric influences associated with dynamic consolidation. While modeling efforts are receiving attention,[7-9,13-15] considerably more effort needs to be devoted to modeling, particularly to the development of dynamic models that can be utilized for near-net-shape consolidation.

One of the near-term objectives for dynamic consolidation of RSP powders at the Idaho National Engineering Laboratory (INEL) is to produce fully dense, bonded, and crack-free monoliths of suitable size for structure/property correlations. This has been realized for RSP Type 304 SS powders[10] and the dynamic consolidation techniques are being applied to other RSP alloy systems. The dynamic consolidation behavior of powders of commercially pure aluminum and a dispersion strengthening aluminum alloy is reported in this paper.

EXPERIMENTAL METHODS

Commercially pure aluminum and Al + 1.4 Co powders were obtained through the courtesy of G. J. Hildeman, ALCOA Technical Center. Both powder materials were produced by air atomization and had been screened through <325 mesh prior to their receipt. No further screening was performed, and the particle size distribution for the two powder materials was not determined. Scanning electron micrographs of "grab" samples are shown in Figure 1. The shape of both powder materials is irregular and it appears that the aluminum powder has a significantly smaller average particle size than the Al + 1.4 Co powder. The densities of the two powder materials were determined to be:

	Al	Al + 1.4 Co
Pour (g/cm^3)	1.206	1.264
Tap (g/cm^3)	1.504	1.556

For the aluminum powder, the percent packing densities are 45 and 56%, for pour and tap, respectively, of wrought commercial aluminum (2.71 g/cm^3). Packing percents for the Al + 1.4 Co powder are not certain since a wrought counterpart was not available as a baseline.

X-ray diffraction (XRD) analysis was performed on both of the powder materials. Indications of an Al_9Co_2 phase was observed in the Al + 1.4 Co alloy powder. XRD peak intensities of the powders are compared with those of the dynamic consolidated monoliths.

The dynamic consolidation of the aluminum powders was performed with explosives using a cover plate arrangement. Vibrogel^R 3, a commercial geophysical prospecting dynamite, was used for the consolidation tests. The experimental configuration for the shots, shown in Figure 2, involves four rectangular powder cavities symmetrically offset from the center of the target. The primary cover plate, die target, lateral spall plates, and pedestal were 6061-T6 aluminum. The powder containing die/punch plate was 1100 (H14) aluminum. A 2.5 cm thick Type 304 SS plate was placed between the 6061 Al cover plate and the explosive.

The powders were poured into the cavities and settled by tapping the side of the die/target. Two shots were performed, using 3.8 (shot 1) and 6.3 kg (shot 2) of explosive. Multipoint detonation was used for the two shots, with two monoliths from each powder material obtained per shot. The rectangular monoliths were ∿ 3.8 cm x 5.0 cm x 0.36 cm thick.

The examinations performed on the dynamic consolidated aluminum powder monoliths were visual, including optical (light) microscopy. Measurements included immersion density, XRD, and microhardness.

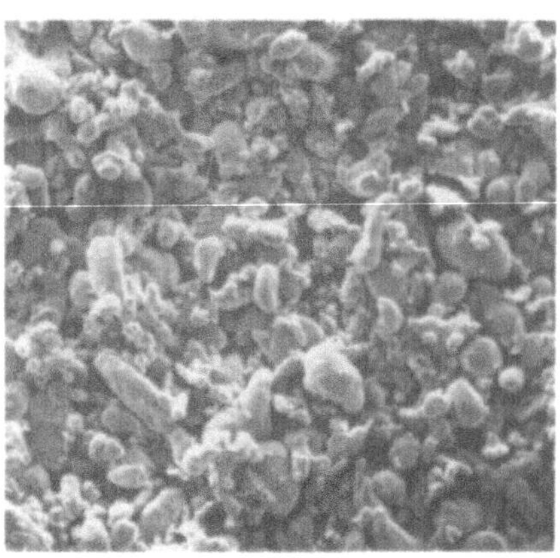

a. Aluminum 50 μm

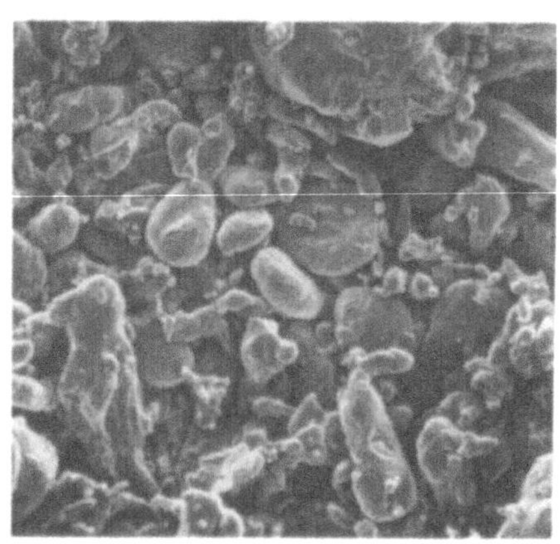

b. Al + 1.4 Co 50 μm

Figure 1. SEM photomicrographs of air atomized aluminum powders.

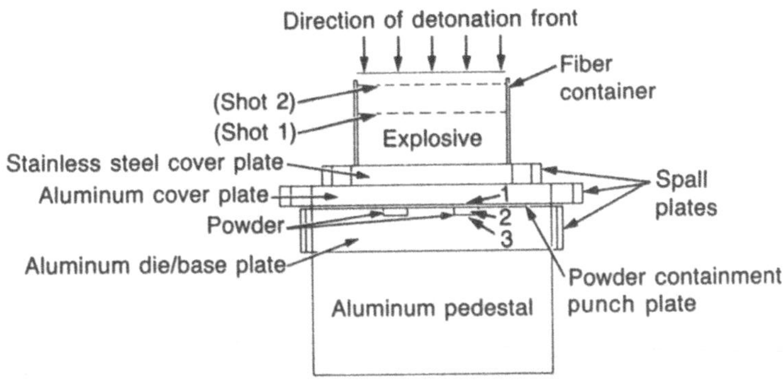

Figure 2. Dynamic consolidation configuration for aluminum powders.

RESULTS AND DISCUSSION

The explosive test configuration (Figure 2) was evaluated for pressure histories at discrete locations using the CSQ II computer code.[15,16] The pressure calculations for the 3.8 kg explosive shot are shown in Figure 3 for tracer particle locations just above, below, and in the center of the powder region. The time scale is referenced to initiation of explosive detonation. The peak pressure calculated for the center of the powder is ∿12.5 GPa, and occurs about 23 μs after detonation of the explosive. The model predicts near equivalent pressure time behavior after 25 μs for the three tracer particles. The peak pressure for the 6.3 kg explosive shot is the same except for a larger Taylor wave contribution following the peak pressure.

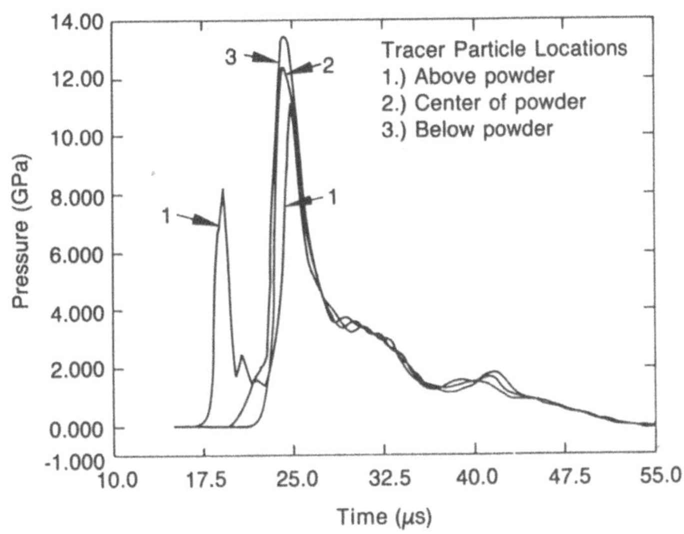

Figure 3. Calculated pressure histories at several tracer particle locations for aluminum powder using 3.8 kg of Vibrogel^R explosive.

The two explosive shots produced eight monoliths, four of each powder type, that appeared to be consolidated. The aluminum powder monoliths from shot 1 (3.8 kg of explosive) showed evidence of some regions with delaminated layers through its thickness, whereas the Al + 1.4 Co powder appeared to be intact. No apparent layering was observed for the aluminum powder monoliths from the second shot, i.e., 6.3 kg of explosive. There appeared to be some evidence of melting on the bottom surfaces of the monoliths. In addition, a rather large gas pocket region was observed on the bottom surfaces for most of the monoliths.

Several 1 cm x 1 cm specimens were sectioned from the monoliths for examinations and measurements. Typical optical (light) microscopy photographs are shown in Figure 4. Some particles boundaries are evident in the as-polished aluminum and Al + 1.4 Co consolidated powders (Figs. 4a & d). Some particle extrusions are observed after etching (Figs. 4b & d), and it appears that the particles for the Al + 1.4 Co alloy are more compressed in the primary direction of the dynamic stress wave. Figure 4c shows an area of delamination for the aluminum powder compact, and Figure 4f a highly porous region near the bottom of the Al + 1.4 Co compact that has apparently melted. This porous pattern is attributed to entrapped gas during consolidation.

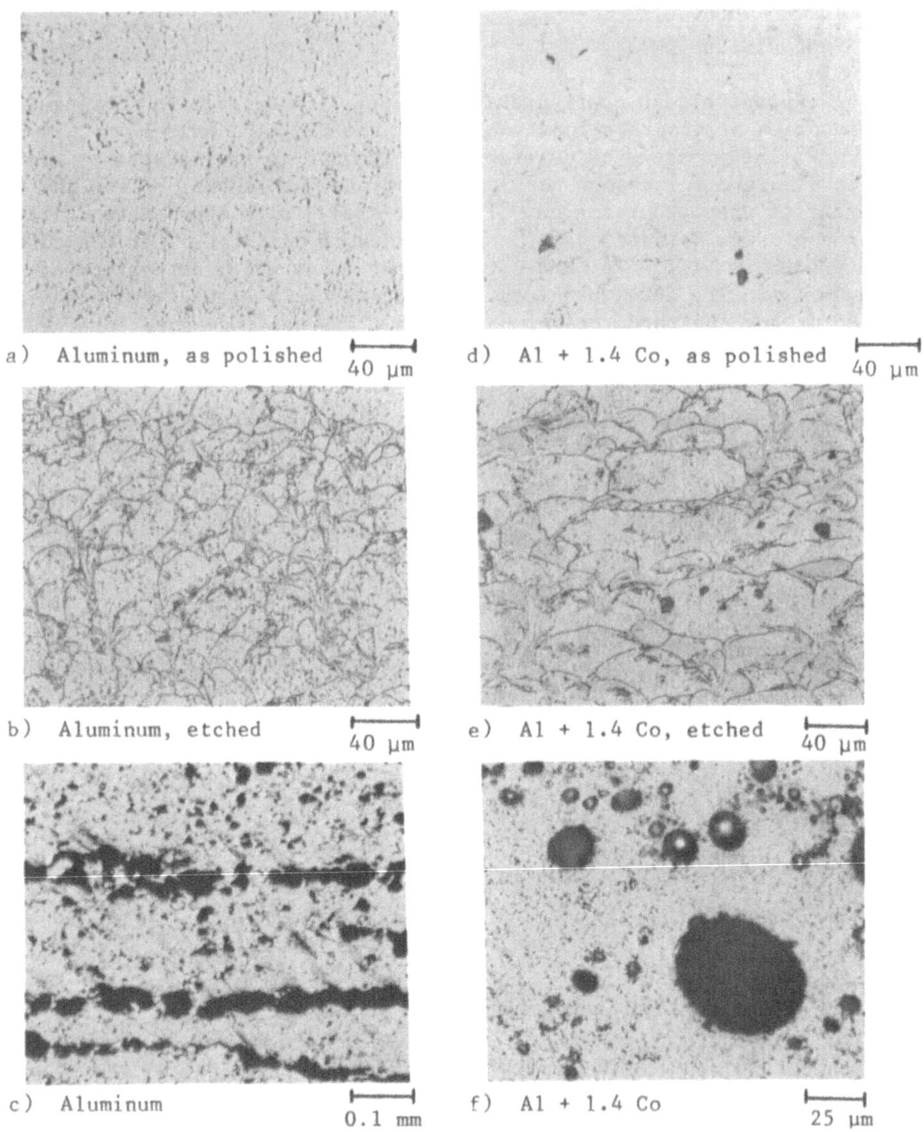

Fig. 4. Explosive consolidated aluminum (a-c) and Al + 1.4 Co (d-f) powders. c) delaminated region, f) gas porosity/melting.

The results obtained for microhardness and immersion density are shown in Table 1. From the microhardness measurements, it appears that some shock hardening of the powder compacts has occurred when compared with the wrought materials. However, the XRD results on the compacted powder specimens show only a small degree of line broadening when compared to the precompacted powders. The fixtures were fairly warm after the shots, and could have achieved high enough temperatures to promote some microstructural recovery. Transmission microscopy on these specimens is planned and, hopefully, will resolve the apparent discrepancy between the hardness and X-ray results. The densities observed for the specimens appear to be less than theoretical, e.g., the consolidated commercially pure aluminum powder is 98.7 percent of theoretical. Optical microscopy showed no evidence of internal porosity, void regions, or cracks for the specimens used for the density measurements. Interparticle bonding will be further evaluated by pulling specimens in tension to failure and examining the fracture surfaces for evidence of ductility.

Even though fairly good monoliths were obtained from the two explosive shots, it is speculated that the two aluminum powders used for the study were not optimum for dynamic consolidation studies, particularly in terms of powder size, shape, and possibly surface films. The influence of these intrinsic parameters on dynamic consolidation of powders has not received much attention, and requires detailed study. We have observed in dynamic consolidation of RSP Type 304 SS powders that spherical powders with relatively clean surfaces produce good monoliths which are fully dense and metallurgically bonded.[10,17] Also it appears that larger particles (∿100 μm) aid in interparticle bonding. Our observations on the RSP Type 304 SS powders show high levels of particle extrusion into the interstice regions. The microlevel model calculation by Williamson and Berry[14] shows that very high localized temperatures can result from high

Table 1. Microhardness and Immersion Density

		Microhardness (DPH, 25 g Load)		
Shot No.	Material	Sample	Wrought-handbook.[a]	Density (g/cm^3)
1	Al powder	46	44 (H18)	2.674
2	Al powder	40	38 (H16)	2.676
1	Al + 1.4 Co	53[b] 41[c]	-- --	2.711
2	Al + 1.4 Co.	65[b] 55[d]	-- --	2.699
2	1100 Al (H14) Punch plate	46	32(H14)	--
2	6061-T6 Al Target	110	95 (T6)	2.714

a. Metals Handbook, Desk Edition, eds. H. E. Boyer and T. L. Gall, American Society for Metals, Metals Park, OH. (1985).

b. Bulk of the sample.

c. Melt/porous zone.

d. Bottom zone of the sample.

particle deformation rates. This model appears to be consistent with the results on the RSP Type 304 SS powders. The rather fine, irregularly-shaped aluminum powders did not exhibit the degree of the particle extrusion observed for the stainless steel powders. Also, it is suspected that the aluminum powders may possess a fairly thick film of aluminum hydroxide which may inhibit effective particle bonding.

ACKNOWLEDGMENTS

The authors greatly appreciate the assistance of M. D. Harper in the explosive efforts; K. Messick, D. V. Miley, and V. Smith for examinations and measurements; M. M. Taylor for reviewing the manuscript; and B. Moreno for typing the manuscript. This work was supported by the U.S. Department of Energy under DOE Contract Number DE-AC07-76ID01570.

REFERENCES

1. D. Raybould, The properties of stainless steel compacted dynamically to produce cold interparticle welding, J. of Mat. Sci. 16:589 (1981).
2. D. G. Morris, Dynamic compaction of metallic glasses and microcrystalline materials, in: "Rapid Solidification Processing: Principles and Technologies-II," R. Mehrabian, B. H. Kear, and M. Cohen, eds., Claitor's Publ. Div., Baton Rouge, LA (1980).
3. D. G. Morris, Rapid-solidification phenomena, Met. Sci. 16:457 (1982).
4. D. G. Morris, Bonding processes during the dynamic compaction of metallic powders, Mat. Sci. and Eng. 57:187 (1983).
5. M. A. Meyers, B. B. Gupta, and L. E. Murr, Shock-wave consolidation of rapidly solidified superalloy powders, J. of Met., 32:21 (1981).
6. W. H. Gourdin, Local microstructural modifications in dynamically consolidated metal powders, Met Trans A 15A:1653 (1984).
7. W. H. Gourdin, Energy deposition and microstructural modification in dynamically consolidated metal powders, J. Appl. Phys. 55:172 (1984).
8. P. Kasiraj, T. Vreeland Jr., R. B. Schwarz, and T. J. Ahrens, Shock consolidation of a rapidly solidified steel powder, Acta Metall. 32:123 (1984).
9. R. B. Schwarz, P. Kasiraj, T. Vreeland Jr., and T. J. Ahrens, A theory for the shock-wave consolidation of powders, Acta Metall. 32:12 (1984).
10. G. E. Korth, J. E. Flinn, and R. C. Green, Dynamic consolidation of rapidly solidified Type 304 SS powders, to be presented and published in the conference proceedings for EXPLOMET 85, July 28-August 1, 1985, Portland, OR.
11. Dynamic compaction of metal and ceramic powders, NMAB-394, National Academy Press, Washington, D.C. (1983).
12. J. E. Flinn, "Rapid Solidification Technology for Reduced Consumption of Strategic Materials," Noyes Publications, Park Ridge, NJ (1985).
13. R. A. Berry and R. L. Williamson, A multiphase model for the shock induced collapse of interstices in metal powders, proc. of this conference.
14. R. L. Williamson and R. A. Berry, Microlevel numerical modeling of the shock wave induced consolidation of metal powders, ibid.
15. R. A. Berry and R. L. Williamson, Numerical modeling of the experimental dynamic consolidation of rapidly solidified stainless steel powders, see Reference 10.
16. S. L. Thompson, CSQ-II, an Eulerian finite difference program for two-dimensional materials response; part 1: material sections, SAND 77-1339, (1979).
17. J. E. Flinn, Y. Gupta, and M. Williams, unpublished results on gas gun consolidation tests.

PROPERTIES OF NIOBIUM RECOVERED FROM MEGABAR DYNAMIC PRESSURES

W. J. Nellis, W. C. Moss, H. B. Radousky, A. C. Mitchell,
L. T. Summers and E. N. Dalder

Lawrence Livermore National Laboratory
Livermore, California 94550

M. B. Maple and M. McElfresh

University of California, San Diego
La Jolla, California 92093

ABSTRACT

Nb was processed and recovered intact from peak dynamic pressures in the range 0.6-1.2 Mbar. Pressure and effective-plastic-strain histories were calculated using a finite-element hydrodynamic computer code. Recovered specimens were characterized by X-ray diffraction, metallography, Vicker's hardness, and superconducting transition temperature T_c. The maximum change observed in the T_c of 9.18 K for the unshocked specimen was a decrease of 0.035 K in the specimen shocked to a maximum pressure of 0.6 Mbar. Shock processing of V_3Si at 1 Mbar was recently reported by Stewart et al. to depress T_c 1.8 K from the initial value of 16.4 K. These results indicate that shock-induced defects and disorder have a relatively weak effect on the T_c of equilibrium phases and suggest that shock-synthesized metastable superconductors might have T_c's close to intrinsic values for the ordered material.

I. INTRODUCTION

High dynamic pressures are known to produce a wide variety of interesting phenomena in condensed matter. For example, metastable stoichiometric A15 Nb_3Si with a high superconducting transition temperature T_c=18 K has been synthesized only by processing the Ti_3P phase at Mbar shock pressures.[1] Other specimens of Nb_3Si with $T_c \approx 18$ K have also been produced by high dynamic pressure.[2-4] By comparison, the maximum T_c obtained in metastable A15-phase thin films is about 9 K at a maximum achievable Si concentration of about 20%.[5] The higher T_c in the shock-synthesized Nb_3Si is thought to be caused by the higher Si content and smaller lattice parameter. Since the dynamic pressure process has not yet been studied systematically, it is possible that still higher T_c's and/or upper critical fields H_{c2} can be achieved. The T_c of stoichiometric A15 Nb_3Si has been predicted to be in the range 25-38 K, based on systematics for other Nb compounds.[6-9]

The work we report here was undertaken to develop the experimental and computational techniques to subject materials to shock pressures up to 1.5 Mbar and to recover them for investigations of material structure and physical properties. Our goal is to utilize fast application and release of high densities and temperatures to synthesize new metastable phases. Although most shock processing and synthesis have been done using high explosives,[1-3] two-stage light-gas guns have been utilized,[4,10] as well. We are using a two-stage gun because pressures and loading times are tuneable by selection of impactor material, velocity, and thickness.

We chose to study Nb to investigate the effects of shock-induced defects on the T_c of an elemental Type II superconductor. The effects of shock-induced defects in Nb are expected to be greatest in specimens processed at Mbar pressures, because Nb has been predicted to have maximum shock-induced hardness at about 0.9 Mbar.[11] Cold rolling Nb at low strain rates ($\sim 10^{-1}$/s) causes an increase in T_c of 0.1 K or less.[12,13] Strong shock waves produce much higher strain rates of at least 10^8/s or larger[14] and might produce different defect structures and have a different effect on T_c. Of course, the higher strain rates are accompanied by high shock and residual temperatures which can anneal defects.

II. GENERATION OF HIGH DYNAMIC PRESSURE

Strong shock waves were generated by using a two-stage light-gas gun[15] to accelerate planar, 3 mm-thick steel impactors to velocities in the range 2.5-4 km/s and impacting specimens in a Pb/steel target fixture illustrated in Fig. 1. This fixture was chosen to be sufficiently strong and massive to withstand Mbar shock pressures and to contain the specimen, which flows plastically at late-time. A design which transfers momentum to materials other than the specimen and minimizes plastic flow is clearly very desirable also. In three experiments impact pressures were 0.6, 1.0, and 1.2 Mbar in the Nb specimens and were obtained by measuring impactor velocity U_I and shock-impedance-matching using the known Hugoniot equations of state for steel and Nb.[16]

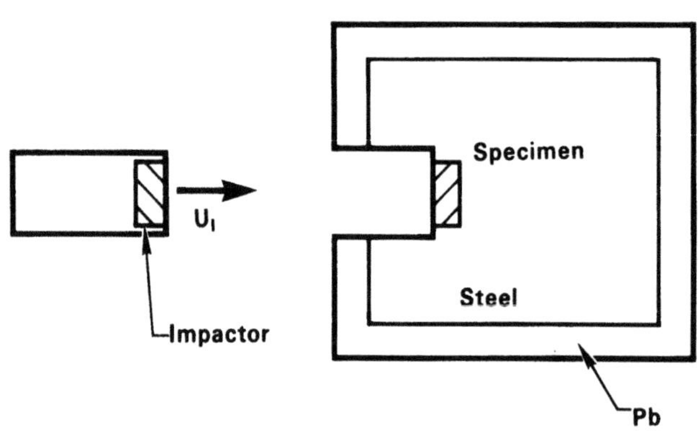

Fig. 1. Illustration of specimen recovery fixture for processing of materials at Mbar dynamic pressures. Strong shock wave is generated on impact of specimen by impactor at velocity U_I.

In order to characterize the loading histories of the specimens, two-dimensional hydrodynamic calculations were performed using the DYNA2D finite element hydrodynamic computer code.[17] The Nb specimen, Pb/steel recovery fixture, steel impactor and polycarbonate sabot were included. Calculations were performed for each of the three measured impact velocities. Mie-Grüneisen equations of state and pressure-and temperature-dependent yield strengths were used.[18] The calculations matched the observed shape change of the specimen, but only if the yield strength was allowed to decrease with temperature. The pressure dependence of the yield strength has a weak effect on the calculation, presumably because most of the flow occurs at late time when the pressure is reduced substantially from the strong impact-generated shock. The calculation shows an impact-driven initial stress pulse of ~0.8 μs duration, followed by late-time ringing that includes a strong tensile release wave near the axis. Figure 2 is a plot of the effective plastic strain versus time for a zone initially near the half-height and half- radius of the cylindrical Nb specimen. Figure 2 shows that only about 20% of the plastic strain is generated by the impact shock and that it is actually dominated by late-time stress ringing.

III. EXPERIMENTAL RESULTS

All three Nb specimens were recovered intact from maximum shock pressures of 0.6, 1.0, and 1.2 Mbar, respectively. Initial Nb sample shapes were cylinders 13 mm in diameter and 9 mm thick. Final dimensions were about 20 mm in diameter and 4 mm thick, indicating substantial plastic flow. The Nb was nominally of 99.9% purity. Slow-scan-rate X-ray diffraction was performed on the as-received and 1 Mbar-shocked

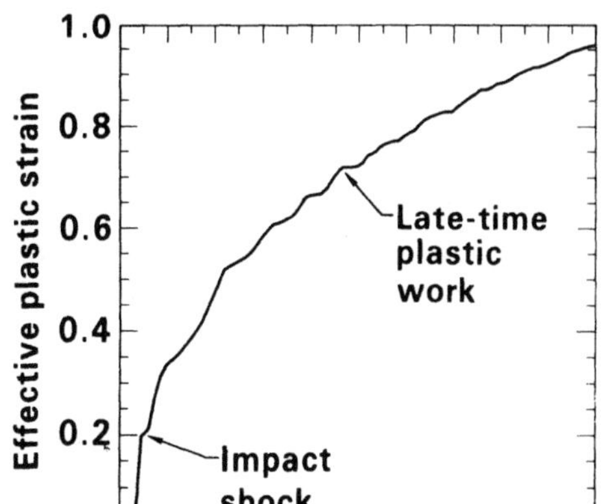

Fig. 2. Calculated effective-plastic-strain history for a zone initially near half-height and half-radius of cylindrical Nb specimen shocked to 1 Mbar.

specimens. Comparison showed a one-to-one matching of the bcc Nb diffraction peaks before and after shocking, with no indication of a shock-induced phase introduction.

Photomicrographs were obtained for all specimens and are shown in Fig. 3. The as-received sample has a rather isotropic grain structure in the sense that the long and short dimensions of the grains are comparable. The samples shocked to 1.0 and 1.2 Mbar have pronounced wavy bands of grains, which are long, thin, and oriented preferentially with their long dimensions perpendicular to the shock-wave front. Similar features are observed for Nb cold-rolled to more than 40% deformation.[13]

Vicker's hardness for the as-received and the three shocked specimens are shown in Fig. 4. Pie-sections were cut from each specimen and measurements were made along the radius at the half height of each. The as-received specimen was harder near the outer radius than on axis and permitted investigation of material response as a function of initial hardness. In the center where the Nb was less hard initially, the 0.6 Mbar-shocked specimen is substantially harder than the as-received. Hardness then decreases with increasing P_{max}, as for 304 stainless steel and Ni above 0.6 Mbar.[19] Since shock temperature increases with P_{max}, these results suggest that increased annealing occurs with increasing shock and residual temperature. At the hardest, outer edge of the specimens, hardness is only weakly dependent on P_{max} and the shock processing decreases hardness somewhat. Least-squares parabolic fits through the data indicate broad, weak maxima with radial position, with the largest effect in the 0.6 Mbar-shocked specimen. The prediction of maximum hardness at 0.9 Mbar[11] was based on the assumption of

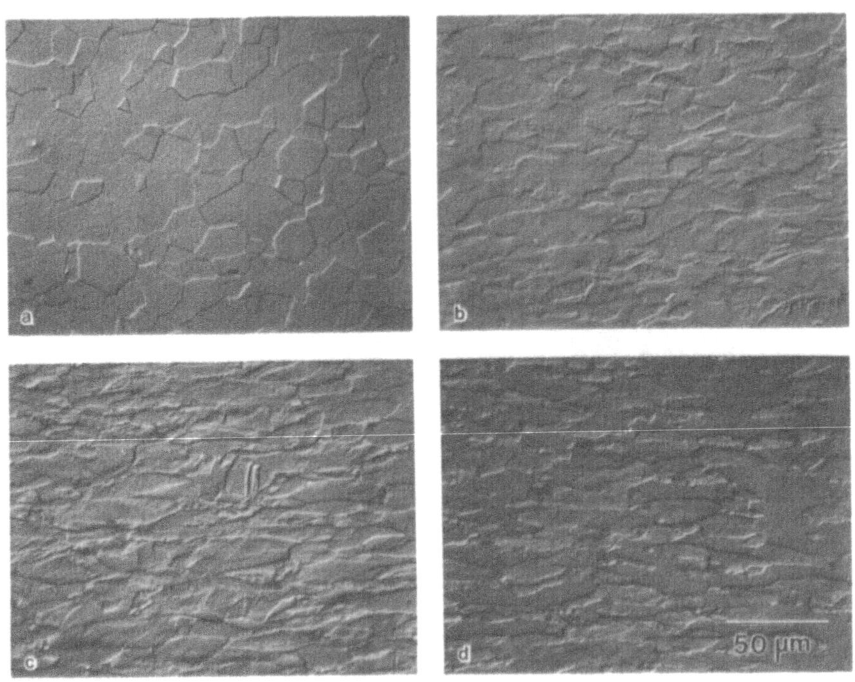

Fig. 3. Photomicrographs near half-radius for as-received (a) and specimens shocked to 0.6, 1.0, and 1.2 Mbar (b–d). For shocked specimens orientation is such that impact shock traveled in direction from top to bottom.

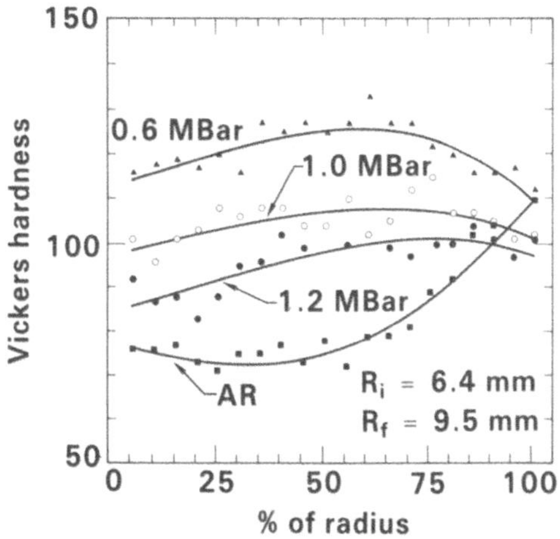

Fig. 4. Vicker's hardness, along half-height of cylindrical specimens, as function of radius for four specimens of Fig. 3. Radii of shocked specimens are about 1.5 times that of unshocked specimens.

negligible plastic deformation, a condition that clearly was not the case here.

The superconducting transition temperatures of the Nb samples were determined by means of ac magnetic susceptibility χ_{ac} measurements at 20 Hz in a He^4 cryostat. The superconducting transition temperature T_c is defined at the 50% value of the change $\Delta\chi_{ac}$ in χ_{ac} between the normal and superconducting states. The widths of T_c were taken as the difference between the temperatures for the 10% and 90% value of $\Delta\chi_{ac}$. Sections about 1mm thick were cut from the center to the outer radius of the disks and these were then further cut at various radii to obtain the final samples. Up to six samples were measured in a single run; 33 specimens were measured in all from the as-received and three shocked cylinders. For the as-received specimens T_c = 9.18 ± 0.02 K; widths of the transitions were 0.01-0.03 K. The maximum a decrease in T_c occurred for the 0.6 Mbar-shocked specimen. This decrease is less than 0.035 K, occurs at 50% of the outer radius and correlates roughly with the weak maximum observed in the hardness data for this specimen at the same radial position.

IV. SUMMARY

In conclusion, the T_c of Nb plastically deformed by maximum shock pressures of 0.6, 1.0, and 1.2 Mbar was found to decrease at most 0.035 K from the value of 9.18 K for the unshocked specimen. Shock processing V_3Si at ~1 Mbar depresses the transition temperature 1.8 K from the initial value of 16.4 K.[20] These results indicate that shock-induced defects and disorder have a relatively weak effect on the superconducting transition of equilibrium phases and suggest that shock-synthesized metastable superconductors might have T_c's close to intrinsic values for the ordered material.

The largest decrease in T_c occurs at a radial position near 50% of the outer radius of the Nb disk shocked to 0.6 Mbar. At this same location the hardness has a maximum. Both these observations are consistent with maximum shock-induced defects at this radial position.

Hydrodynamic computer calculations can now be used to design loading profiles to investigate the response of material structure and properties to different stress histories.

ACKNOWLEDGEMENTS

Work performed under the auspices of the U.S. Department of Energy by the Lawrence Livermore National Laboratory under contract number W-7405-ENG-48 and supported by the U.S. DoE Office of Basic Energy Science, Division of Materials Science. Research at UCSD was supported by the U.S. DoE under contract number DE-AT03-76ER70227 and by the LLNL Branch of the University of California Institute of Geophysics and Planetary Physics.

REFERENCES

1. B. Olinger and L. R. Newkirk, Solid State Comm. 37, 613 (1981).
2. V. M. Pan, V. P. Alekseevskii, A. G. Popov, Yu. I. Beletskii, L. M. Yupko, and V. V. Yarosh, JETP Lett. 21, 228 (1975).
3. D. Dew-Hughes and V. D. Linse, J. Appl. Phys. 50, 3500 (1979).
4. S. Ohshima, T. Wakiyama, T. Goto, and Y. Syono, Japan. J. Appl. Phys. 22, 264 (1983).
5. R. H. Hammond and S. Hazra, in Proceedings of Thirteenth International Conference on Low Temperature Physics, edited by K. D. Timmerhaus, W. J. O'Sullivan, and E. F. Hammel (Plenum, New York, 1974), Vol. 3, p. 465; R. D. Feldman, R. H. Hammond, and T. H. Geballe, Appl. Phys. Lett. 35, 818 (1979).
6. L. Gold, Phys. Stat. Sol. 4, 261 (1964).
7. D. Dew-Hughes and V. G. Rivlin, Nature 250, 723 (1974).
8. S. Geller, Appl. Phys. 7, 321 (1975).
9. J. Noolandi and L. R. Testardi, Phys. Rev. B 15, 5462 (1977).
10. A. Sawaoko, Japanese Scientific American, October, 1984 (in Japanese).
11. A. A. Deribas, V. F. Nesterenko, and T. S. Teslenko, Soviet Combustion, Explosion, and Shock Waves 18, 670 (1982).
12. R. B. Zubeck, T. W. Barbee, Jr., T. H. Geballe, and F. Chilton, J. Appl. Phys. 50, 6423 (1979).
13. S. J. Thompson and P. E. J. Flewitt, J. Less-Common Met. 40, 269 (1975).
14. L. C. Chhabildas and J. R. Asay, J. Appl. Phys. 50, 2749 (1979).
15. A. C. Mitchell and W. J. Nellis, Rev. Sci. Instrum. 52, 347 (1981).
16. R. G. McQueen, S. P. Marsh, J. W. Taylor, J. N. Fritz, and W. J. Carter, in High-Velocity Impact Phenomena, edited by R. Kinslow (Academic, New York, 1970), p. 293.
17. J. O. Hallquist, User's Manual for DYNA2D, Lawrence Livermore National Laboratory, unpublished report, 1984.
18. D. J. Steinberg, S. G. Cochran, and M. W. Guinan, J. Appl. Phys. 51, 1498 (1980).
19. L. E. Murr, in Shock Waves and High-Strain-Rate Phenomena in Metals, ed. M. A. Meyers and L. E. Murr (Plenum, New York, 1981), p. 668.
20. G. R. Stewart, B. Olinger, and L. R. Newkirk, Phys. Rev. B 31, 2704 (1985).

DYNAMIC COMPACTION OF A MONOSIZED SPHERICAL TUNGSTEN POWDER

W. H. Gourdin and S. L. Weinland

University of California
Lawrence Livermore National Laboratory
Livermore, CA

ABSTRACT

We present a limited number of measurements of the Hugoniot of a 45-74-μm spherical tungsten powder in the stress range 0 to 7 GPa. From these data we infer that compaction to the solid density occurs at about 4 GPa. Examination of recovered specimens shows that interparticle bonding does not occur at stresses as high as 6.2 GPa. This is consistent with calculations based on the energy flux at powder particle surfaces during compaction, which suggest that interparticle melting will not occur at stresses less than about 9.0 GPa.

INTRODUCTION

Dynamic compaction has received much attention in recent years as a possible means of densifying rapidly solidified powders without substantially altering their initial microstructures.[1-5] The method is also of potential interest in the processing of refractory metal powders, however, because their high melting points make conventional processing difficult and expensive. Boade[6] studied the dynamic response of a sintered porous tungsten with a density of 12.64 x 10^3 kg/m^3 over a stress range of 1.2 GPa to over 100 GPa. Emphasis was placed on the wave propagation and equation-of-state aspects of the compaction process, which were analogous to those observed in sintered copper specimens.[7,8] Similarly, early Russian work[9] focused on equation-of-state studies at high stresses (>10 GPa) and low initial densities (<50% of solid). Later studies extended this work[10] but also explored some aspects of plastic flow, densification, and energy deposition at lower stress levels.[11]

In this paper we report the results of a limited number of Hugoniot experiments on unsintered, carefully sized spherical tungsten powder. Our objectives were to define the dynamic densification behavior as a function of stress and to seek evidence for interparticle melting or other local thermal modification of the microstructure in the final compact.[12]

EXPERIMENTAL PROCEDURE

All of the experiments in this study were performed on a 6.35-cm-diam smooth-bore helium-driven gun.[13] The specimen assembly is shown in Fig. 1.[13] Piezoresistive carbon film sensors[14] were affixed to the centers of a 0.3-cm-thick cover plate and a 1.27-cm-thick back plate, both of 6061-T6 aluminum, with a thin (<0.0013 cm) layer of epoxy. Leads from

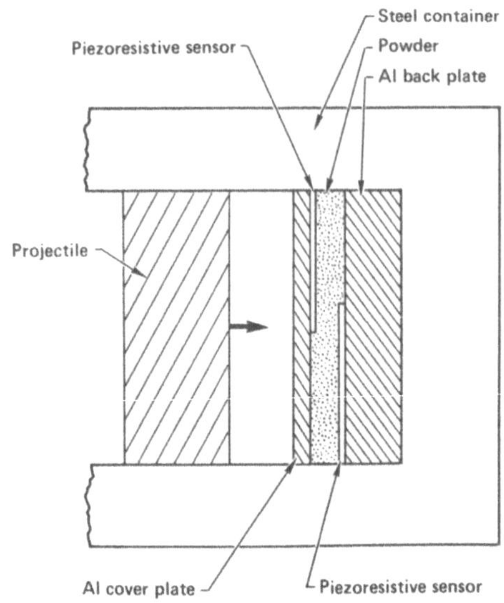

Fig. 1. Schematic of the experimental assembly.

the sensors were brought out the rear of the assembly to bridge circuits used to measure the change in resistance of the sensing element during powder consolidation. Although designed to give a direct reading of the stress, the calibration of these sensors is somewhat uncertain.[13] Hence the primary data of this study were derived from impedance matching relative to the aluminum 6061-T6 cover and back plates. The constancy of wave speeds in porous tungsten and copper,[6-8] the apparent steadiness of wave profiles in copper,[8] as well as the success of the approach in describing a variety of metal powders[13] indicate that this approach is justified. In the configuration used here a shock is produced in the cover plate by a projectile impact, which then releases into the powder along the isentrope rather than the Hugoniot. At the low stresses of concern here, however, the two are very close and the use of the Hugoniot to describe the release path is an excellent approximation. The Hugoniot of aluminum 6061-T6 is complicated by the presence of the Hugoniot elastic limit at about 0.41 GPa that produces an elastic precursor in the cover plate of this amplitude.[15,16] This precursor impinges on the powder specimen about 80 to 90 ns sooner than the slower plastic wave at the lowest projectile speeds we used. Thus the initial loading of the powder specimen is accomplished through a complex series of interactions between reflected and transmitted waves. In practical terms, however, the stress wave introduced into the porous specimen by the precursor in the cover is of small amplitude (0.1-0.2 GPa) and has a correspondingly small propagation speed. In contrast to Boade's observations,[6] we find no evidence for precursor propagation in the powder, and so conclude that this initial wave will be overtaken by the main compaction wave. We conservatively estimate the speed of the initial wave to be about 0.33 km/s, half of the wave speed we observed at the lowest stress, 0.65 km/s. The distance two waves travel in the powder before the compaction wave overtakes the initial wave is given by

$$d = \Delta t/(1/v_1 - 1/v_2) \qquad (1)$$

where Δt is the difference in arrival times at the cover-powder interface, and v_1 and v_2 are the wave speeds of the initial and compaction wave in the powder, respectively. Taking Δt = 90 ns, v_1 = 0.33 km/s, and v_2 = 0.65 km/s, we find that d is approximately 6 x 10^{-5} m (60 μm). This is of the same order as the powder particle size, which defines the minimum width of the compaction shock in the powder.[12,17] Hence we conclude that the details of the initial loading of the powder produced by the aluminum cover plate are lost in the slow rise time of the compaction shock and are therefore not important for our experiments. We note also that d is significantly less than the thickness of the carbon sensors (1.27×10^{-4} m), a principal source of systematic error.

The Hugoniot we have used for aluminum 6061-T6 is described by the shock speed - material speed relationship

$$v_{Al} = 5.35 + 1.34U_p \quad \text{(km/s)} \quad . \tag{2}$$

This was derived from data obtained at stresses in excess of 7.0 GPa,[18] and when extrapolated to low stresses does not describe the details of the elastic response. However, Wallace[15] has noted that a relationship similar to Eq. (2) provides an adequate description of the plastic wave speeds at low stresses as well. A comparison of the proper aluminum Hugoniot with that given by Eq. (2) shows that for the ranges of stress and particle speed of interest here, the two are very close. In practical terms, then, we feel (2) is adequate for our experiments.

The dominant source of systematic error is the relatively large thickness of the carbon sensors and the long rise times associated with them (100-200 ns in most experiments), affecting both the transit time and the powder thickness measurements. We chose the powder thickness to be the distance between the midpoints of the sensors and the transit time to be that between the trigger points. While this approach is somewhat arbitrary, it has nevertheless been successfully applied to more ductile metals that compact to near-solid density.[13] Comparisons of our data with that of Boade[6] and the approach of our data to the proper curve in the limit of complete densification indicate that the systematic errors introduced are not large. Random errors were calculated in the usual manner,[19] and are shown as error bars in Fig. 3.

The material used in these experiments is a commercially available plasma sprayed powder.[20] Coarse sizing was accomplished by seiving, and final sizing and elimination of irregular particles was carried out on a vibrating "riffler" table. The characteristics of the resulting 45-74-μm powder are summarized in Table 1 and Fig. 2. The measured solid density is 2% less than the nominal density of 19.3×10^3 kg/m³. We attribute this to a small amount of included porosity, and we refer our results to the measured value.

RESULTS

The Hugoniot data are given in Table 2 and Fig. 3. The Hugoniot derived under the assumption of complete compaction is also shown in Fig. 3 and it

Table 1. Tungsten Powder Characteristics

Characteristic	Value
Nominal composition	99.5 wt% W, 0.15 wt% O
Solid density[a]	18.93×10^3 kg/m³
Solid specific volume	5.2826×10^{-5} m³/kg
Packing density	11.45×10^3 kg/m³
Relative density	60%
Particle size range	45-74 μm
Specific surface[b]	13.8 m²/kg
Area equivalent diameter	23 μm

[a] Air comparison pyncnometer.
[b] BET method.

Table 2. Hugoniot Data, Tungsten Powder 11.45 x 10^3 kg/m^3

Expt.	Projectile speed (km/s)	Shock speed (km/s)	Stress (GPa)	Material speed (km/s)	V (10^{-3} m^3/kg)
160	0.166	0.65	0.82	0.11	0.07258
159	0.424	0.76	2.33	0.27	0.05605
158	0.657	0.98	4.26	0.38	0.05321
157	0.789	1.11	5.55	0.44	0.05299
156	0.821	1.15	5.90	0.45	0.05332
161	0.854	1.20	6.24	0.46	0.05329

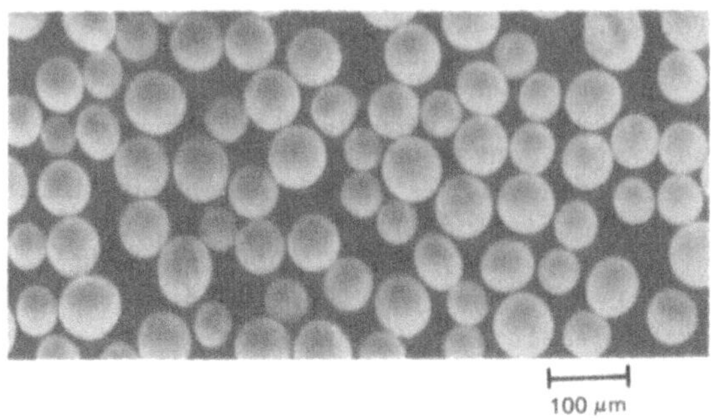

Fig. 2. Scanning electron micrograph of the tungsten powder.

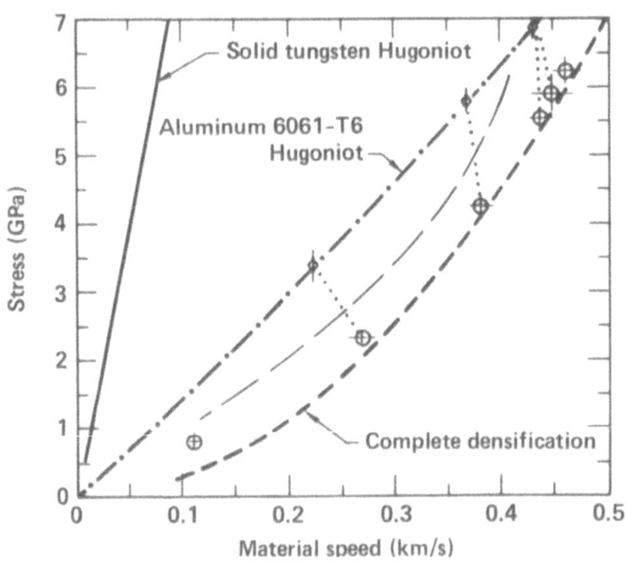

Fig. 3. Hugoniot data for unsintered tungsten powder, 11.45 x 10^3 kg/m^3 (circles). Error bars errors of measurement. Dotted lines connect the primary shock data to the reshock states (diamonds) that lie on the Hugonoit of 6061-T6. The porous Hugonoit derived under the assumption of complete densification is shown. The Hugonoit of solid tungsten and the data of Boade[6] for sintered porous tungsten, 12.65 x 10^3 kg/m^3 (shown approximately by the lighter dashed line), are given for reference.

is apparent from this, as well as the specific volume data in Table 2, that the powder reaches the solid density for compaction stresses in excess of about 4.0 GPa, in agreement with the observations of Boade.[6] Reshock data for several experiments were obtained directly from the back plate sensor readings, and are shown in Fig. 3. Because of the high stresses produced upon reshock, however, these readings cannot reliably be corrected for errors in the sensor calibration,[13] and are therefore not expected to be accurate. As a result, they are not included in Table 2. Nevertheless, these data clearly demonstrate that the reshock Hugoniot becomes increasingly steep as the compact density behind the primary shock increases.

The cross section of a recovered specimen compacted at approximately 6.2 GPa (projectile speed 0.856 km/s) is shown in Fig. 4. Although badly fractured, it is apparent that the powder was compacted to solid density. The specimen had virtually no cohesive strength, however, and only the small amount shown cohered well enough to survive handling. We found no evidence for any thermal modification of the microstructure as a result of dynamic consolidation.

DISCUSSION

The microstructure of the recovered specimen (Fig. 4) is characteristic of dynamically compacted spherical powders of ductile metals.[21] There is little mechanical interlocking in such compacts, so the lack of substantial compact strength indicates the absence of the interparticle bonding found in dynamically formed specimens of less refractory materials.[1-3,12] From the data in Table 2, we calculate that the energy deposited behind the primary compaction wave in the compact of Fig. 4 is approximately 1.06 x 10^5 J/kg. With the assumption that all of this energy appears at powder particle interfaces, we estimate the maximum surface temperature to be 2150°C,[12] well below the melting point of 3410°C and consistent with the observed lack of specimen cohesion. Extrapolating the Hugoniot in Fig. 3, we find that surface melting is not expected until the stress exceeds about 9.0 GPa. This corresponds to a projectile speed of over 1.1 km/s, beyond the capabilities of our apparatus.

SUMMARY AND CONCLUSION

We have reported a limited number of Hugoniot measurements for a carefully sized, 45-74-μm spherical tungsten powder in the stress range 0 to 7 GPa. From this work we conclude the following:

(1) Tungsten behaves as a ductile metal and is compacted to the solid density for compaction stresses in excess of about 4.0 GPa.
(2) Local thermal modification and interparticle bonding (melting) do not occur within the stress range required for complete densification.
(3) Calculations indicate that melting at particle interfaces will not occur for stresses less than about 9.0 GPa.

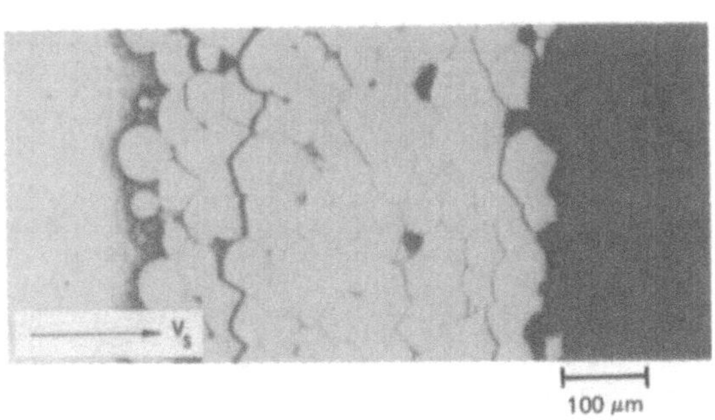

Fig. 4. Cross section of a recovered specimen compacted at a projectile speed of approximately 0.856 km/s. The arrow indicates the propagation direction of the primary compaction wave.

ACKNOWLEDGMENT

Special thanks to Cynthia Talaber for preparing the manuscript. Work performed under the auspices of the U.S. Dept. of Energy by the Lawrence Livermore National Laboratory under Contract W-7405-Eng-48.

REFERENCES

1. W.H. Gourdin and J.E. Smugeresky, in "Proceedings of the Third Conference on Rapid Solidification Processing: Principles and Technologies," R. Mehrabian, ed., National Bureau of Standards (1983), pp. 565-571.
2. W.H. Gourdin, Metall. Trans. A, 15A: 1653-1664 (1984).
3. P. Kasiraj, T. Vreeland, Jr., R.B. Schwarz, and T.J. Ahrens, Acta Metall. 32: 1235-1241 (1984).
4. D.G. Morris, J. Mater. Sci. 17: 1789-1794 (1982).
5. D.G. Morris, Metal Science 14: 215-220 (1980).
6. R.R. Boade, J. Appl. Phys. 40: 3781-3785 (1969).
7. R.R. Boade, Ibid. 39: 5693-5702 (1968).
8. R.R. Boade, Ibid. 41: 4542-4551 (1970).
9. K.K. Krupnikov, M.I. Brazhnik, and V.P. Krupnikova, Soviet Phys. JETP 15: 470-476 (1962).
10. A.A. Bakanova, I.P. Dudoladov, and Y.N. Sutulov, Zhurnal Prikladnoi Mek. i Tek. Fiz. 2: 117-122 (1974).
11. O.V. Roman, V.F. Nesterenko, and I.M. Pikus, Fizika Goreniya i Vzryva 15: 102-107 (1979).
12. W.H. Gourdin, J. Appl. Phys. 55: 172-181 (1984).
13. W.H. Gourdin and S.L. Weinland, in "Shock Waves in Condensed Matter--1983," J. R. Asay, G. K. Straub, and R. A. Graham, eds., North Holland, New York (1984), pp. 99-102.
14. Dynasen, Inc., Goleta, CA.
15. D.G. Wallace, Phys. Rev. B 22: 1487-1494 (1980).
16. J.N. Johnson and L.M. Barker, J. Appl. Phys. 40: 4321-4334 (1969).
17. L. Davison, Ibid. 42: 5503-5512 (1971).
18. S.P. Marsh, ed., "LASL Shock Hugoniot Data," University of California Press, Berkeley (1980).
19. W.H. Gourdin and S.L. Weinland, submitted to J. Am. Ceram. Soc.; also University of California, Lawrence Livermore National Laboratory, report UCRL-91836 Rev 1 (1985).
20. Sylvania Chemical/Metals, GTE Products Corporation, Hawes Street, Towanda, PA 18848, type SD-351.
21. W.H. Gourdin, Materials Science and Engineering 67: 179-184 (1984).

SHOCK CONSOLIDATION OF IN-100

NICKEL-BASE SUPERALLOY POWDER

S.L. Wang, M.A. Meyers and R.A. Graham*

Center for Explosives Technology Research, New Mexico Institute of Minning and Technology, Socorro, NM 87801
*Sandia National Laboratories, Albuquerue, NM 87185

INTRODUCTION

Shock consolidation is a technique of some potential in producing bulk parts from rapidly-solidified powders[1-3]. Although successful reports of shock consolidation of several rapidly-solidified powders have been made, the shock-compression conditions needed are typically not well known. The objectives of the experiments described herein were to establish the stress, stress duration, and temperature conditions required for consolidation of a nickel-base superalloy. In order to accomplish this, the fixtures developed by Graham, Webb, and Davison[4,5] at Sandia National Laboratories were used. These fixtures were developed to produce quantitative, reproduceable experiments with well-characterized stress and temperature histories, determined by extensive studies with a two-dimensional code.

The alloy used in this investigation was a nickel-base superalloy powder prepared in powder form by the rotating-electrode process. The powders were spherical and had a mean diameter of 74 μm. The nominal composition of this IN-100 powder is (in wt. pct.): Ni - 60; Cu - 10; Co ∸ 15: Mo - 3; Al - 5.5; Ti - 4.7; C - 0.18; B - 0.014; Zr - 0.06. These powders exhibit a high strength and have been proven to be difficult to consolidate. The tensile yield stress of wrought IN-100 (bulk) exceeds 1 GPa.

The shock consolidation fixtures used are listed in Table 1. The stresses and stress durations vary throughout the capsules and the range is shown. The average stress and stress duraction can be considered as figures of merit, for comparison purposes between the various fixtures. It can be seen that both the pressure and stress duration were varied in the present experiments. The powders were kept under an argon atmosphere prior, during, and immediately following the experiments.

RESULTS AND DISCUSSION

The degree of consolidation was not uniform throughout the capsules. It is generally accepted for metals that consolidation requires interparticle melting, while densification can be achieved in the solid state. These differences are illustrated in Figure 1.

Table 1. Shock Consolidation Fixtures Used

Fixture	Explosive	Pressure, GPa Range	Pressure, GPa Average	Duration μs
Baby Bear	Baratol	13-26	20	2.5
Baby Bear	Comp B	22-32	27	2.6
Momma Bear	Baratol	5-10	7.5	4.5
Momma Bear A	Baratol	14-20	16	4.5
Momma Bear A	Comp B	19-26	22	4.7
Bertha A	Comp B	14-26	17	9.8

Figure 1 shows the microstructure of the powder consolidated at an average stress of 22 GPa and duration of 4.7 μs (Momma Bear A Comp. B). Figure 1(a) shows the densified material. Notice that the particles are deformed, filling most of the voids. This material exhibits, however, very low strength. The white regions in Figure 1(b) are thought to be produced by interparticle melting which is an important consolidation parameter. Meyers et al.[2] showed, for Mar M 200, that the white regions were microcrystalline. It is generally thought that melting occurred, although crystallization cannot be discarded.

In Figure 1(c) hardness indentations made in the center of the particles (where a microdendritic structure may be observed) and in the interparticle melting regions can be seen. The hardness in the interior of the particles was found to be higher than in the white regions. The small indentations gave a hardness of HVN 646, while the larger indentation gave a hardness of HVN 451. Figure 1(d) is a scanning electron micrograph showing a fracture in a region where compaction occurred, but no "interparticle melting" was evident. The particles are not welded together, but simply densified. Comparing Figures 1(a) and 1(b), one can see that 1(b) shows more apparent interparticle melting.

Another interesting aspect of the consolidated material is that very often a void is observed in the center of the interparticle melt region. These voids are thought to be formed by contraction during solidification of the molten material, and are potential weakening elements, acting as nucleation sites for cracking. Figure 2 shows such a void, in the center of a white region. Another possibility for the formation of these voids is the entrapped gases. The voids between the particles were filled initially with argon.

Figure 3 shows photographs of cross-sections cut through the various fixtures. The black indicates material which had enough integrity not to flow out of capsules when they were opened and can be called "compacted" material. The three systems were drawn in different scales, so that they appear to have identical dimensions; this was done to help the comparison. The different series have different dimensions; the volumes of the capsules are 1, 5, and 60 cm^3 for the Baby Bear, Momma Bear, and Bertha systems, respectively. Momma Bear Baratol did not yield any compaction. It is evident from Figure 3 that there is a preference for compaction along the edges, although the pressure is somewhat lower at the edges. The explanation for this is given in Figure 4, where the temperature is plotted as a function of pressure at both the center and the edge of the sample. At the edge it is known from numerical simulations and experiments that the temperature is higher[6]. Thus, one can conclude that the shock temperature is an important parameter.

Figure 5 shows that the amount of interparticle (apparent) melting and therefore, the quality of the consolidate, depends on the temperature. Interparticle (apparent) melting was measured by the point counting tech-

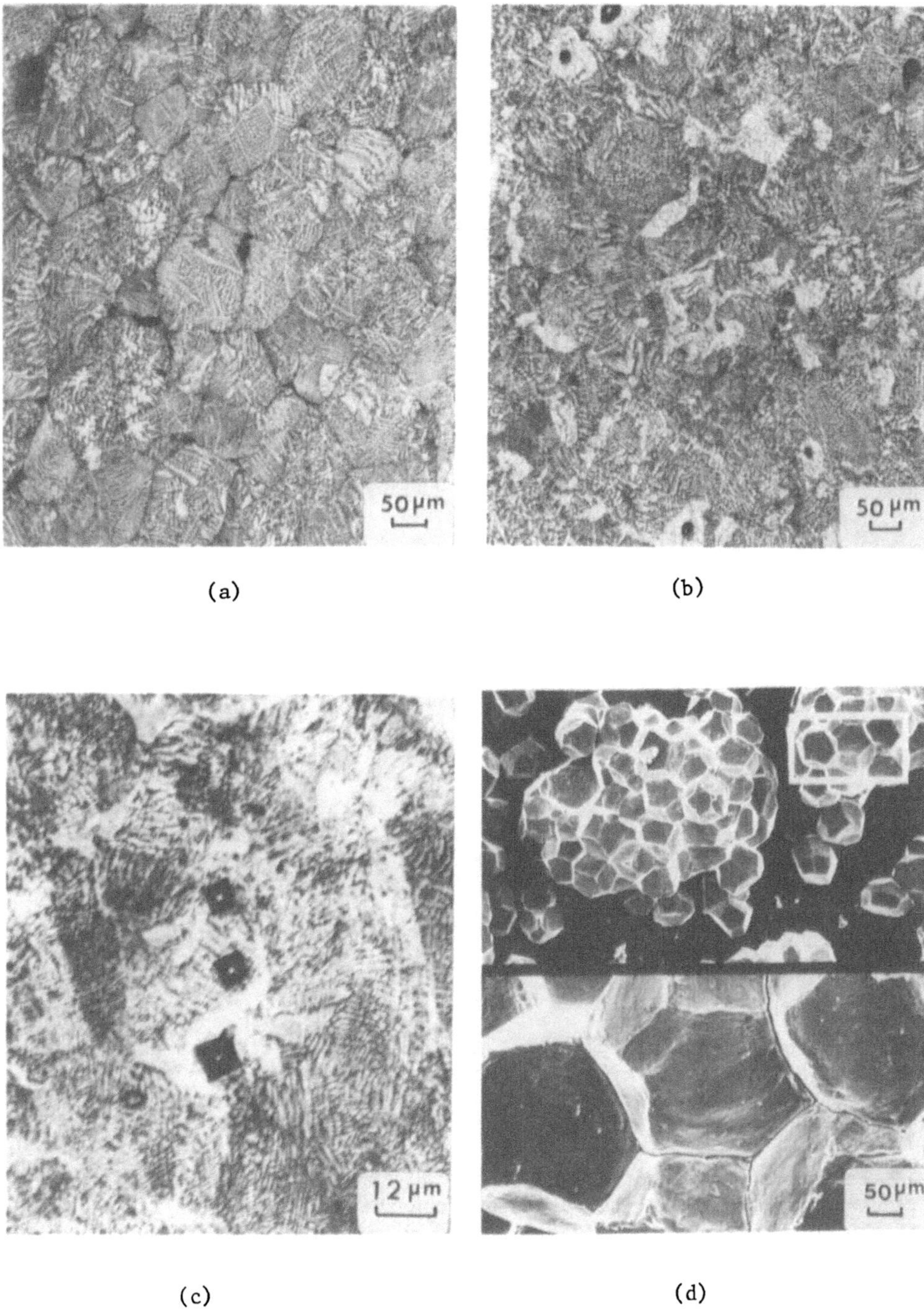

Fig. 1. Micrographs of compacted samples from Momma Bear A, Comp. B; (a) Optical micrograph of center; (b) Optical micrograph of edge; (c) Microindentations in particle center and apparent melt regions; (d) Scanning electron micrographs of a fracture.

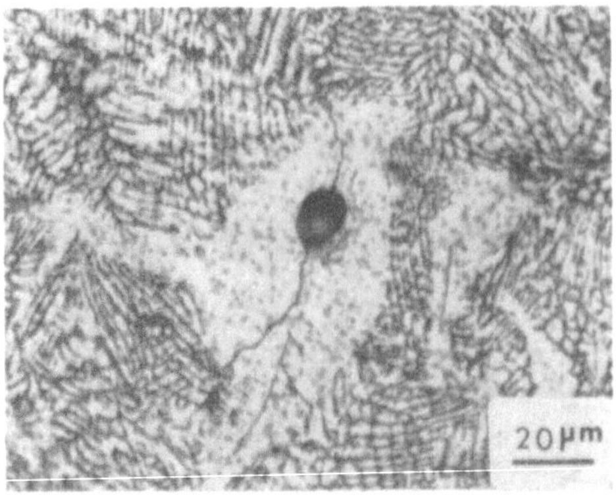

Fig. 2. White interparticle region with void at center.

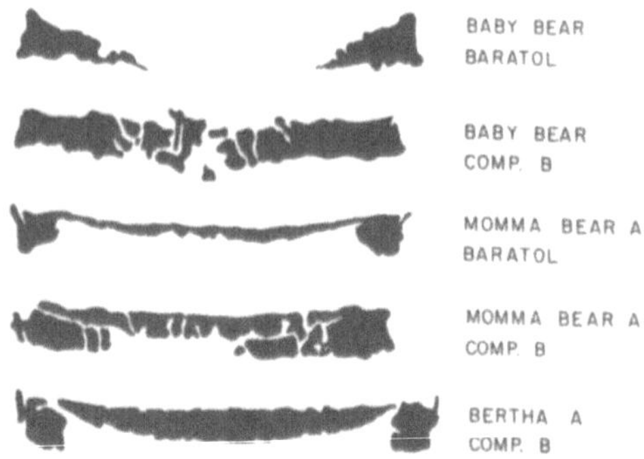

Fig. 3. Compacted sections from different fixtures, no consolidation in Momma Bear Baratol.

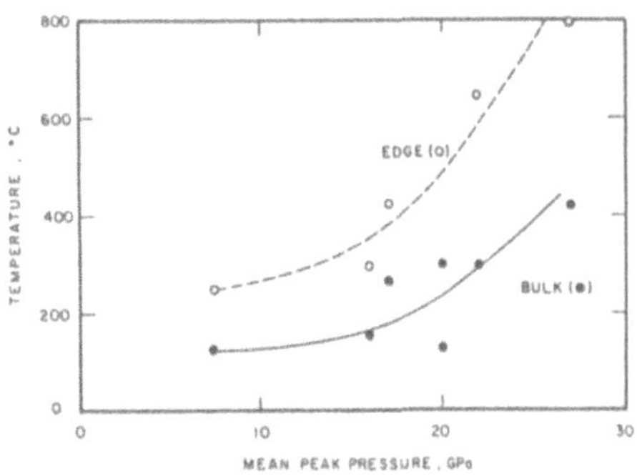

Fig. 4. Calculated maximum temp. as a function of peak pressure for various systems at edge and in center[6].

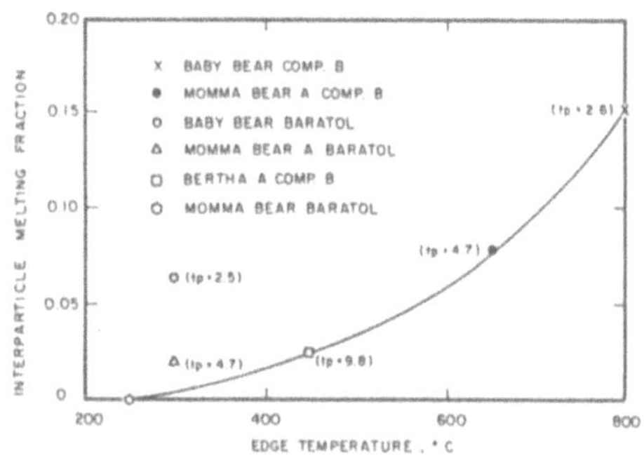

Fig. 5. Interparticle melting fracture as a function of edge temperature for the different systems. In the figure t_p is the stress pulse duration.

nique applied to photomicrographs. As the temperature is increased, the fraction of interparticle melting increases. Figure 6 shows that the pressure at the edge is also important; hence, it is shown that both pressure and temperature are important.

Table 2 shows the fractions compacted for the various systems. The fractions compacted are the ratios between the black areas in Figure 3 and the total rectangular areas. It can be seen that the pressure is an important parameter. There is also an increase in fraction compacted with pulse duration, at a constant (average) pressure. For 16-20 GPa, the fractions compacted are 0.27, 0.48, and 0.64, for average durations of 2.5, 4.5, and 9.8 µs, respectively. The interparticle melting fraction, however, did not show any significant dependence on pulse duration, at a constant pressure.

From the above observations, one can infer that compaction takes place primarily at the shock front and that an increase in pulse duration, from 2.5 to 9.8 µs, does not result in a systematic increase in the fraction consolidated.

Table 2. Fraction Compacted for the Various Fixtures

Fixture	Average Pressure GPa	Average Pulse Duration μs	Fraction Compacted
Baby Bear Baratol	20	2.5	0.27
Baby Bear Comp B	27	2.6	0.68
Momma Bear Baratol	7.5	4.5	0
Momma Bear A Baratol	16	4.5	0.48
Momma Bear A Baratol	22	4.7	0.78
Bertha A Comp B	17	9.8	0.64

The effect of temperature can be qualitatively described by Figure 7. In a shock front, the energy irreversibly dissipated is ΔE_1, (Fig. 7(a)). If the pressure reaches its maximum after the material has been densified by a lower stress wave, the temperature increase is much lower. In a shock front, the jump conditions establish the irreversible work, most of which manifests itself as a temperature increase; a small fraction is stored as defects in the material. A gradual increase in pressure, on the other hand, allows densification at lower pressures, with less energy release. If the high-pressure pulse travels in a material that is already densified, the residual temperature increase is much less significant.

SUMMARY AND CONCLUSIONS

Shock consolidation of IN-100 powders was studied in Sandia Bear and Bertha fixtures over a pressure from 7.5 to 27 GPa. In addition, the influence of pulse duration from 4.5 to 10 μs was investigated at 16 GPa. From the metallographic observations it was established that:

a) the system subjected to a 7.5 GPa average pressure did not exhibit any consolidation. From Figure 6, it can be seen that a local pressure of approximately 20 GPa is required to produce a 10 pct interparticle melting fraction. Microindentation hardness tests showed that interparticle melting is required to assure bonding between the particles.

b) while the fraction compacted increases with increasing pulse duration, at a constant pressure, the apparent melting fraction does not show a systematic variation with pulse duration.

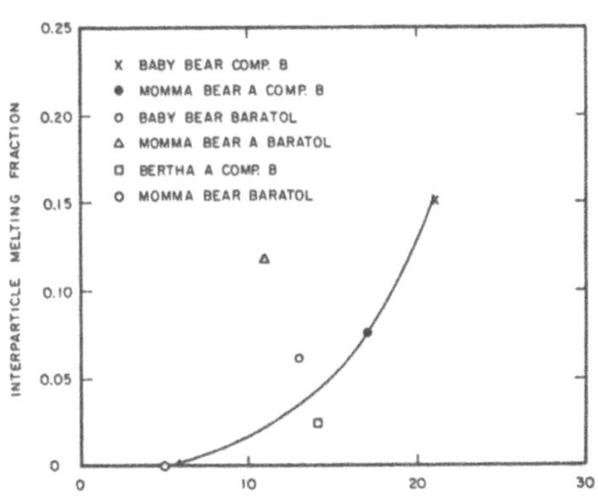

Fig. 6. Interparticle melting fraction as a function of pressure at edge.

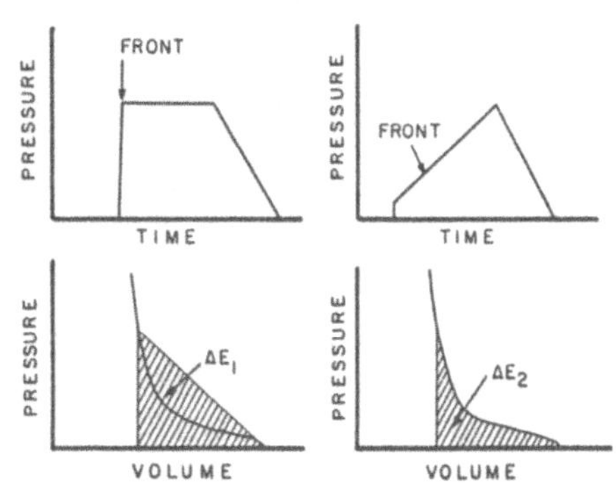

Fig. 7. Irreversible energies (heat) ΔE_1, and ΔE_2, for different stress-wave configurations; (a) rapidly rising stress; (b) slowly rising stress.

c) the shock temperature is an important parameter. It significantly enhances interparticle melting. At identical peak pressures and pulse durations, the regions undergoing the highest temperatures exhibit the greatest amount of (apparent) interparticle melting.

d) in order to specify the requirements for a good shock consolidation, the two most important parameters that have to be specified are peak pressure and peak temperature.

ACKNOWLEDGEMENTS

We wish to thank Mr. L.H. Yu for help in metallography and microhardness testing, and the Nuclear Metals Inc. for supplying IN-100 powder.

REFERENCES

1. D. Raybould, "The Cold Welding of Powders by Dynamic Compaction," Intl. Journal of Powder Met. & Powder Tech., 16: 9 (1980).
2. M.A. Meyers, B.B. Gupta, and L.E. Murr, "Shock-Wave Consolidation of Rapidly Solidified Superalloy Powders," J. of Metals 33: 21 (1981).
3. D.G. Morris, "Rapid Solidification Phenomena," Metal Sci., 16: 457 (1982).
4. L. Davison, D.M. Webb, and R.A. Graham, "Analysis of Capsules for Recovery of Shock-Compressed Matter," in Shock Waves in Condensed Matter, W.J. Nellis, L. Seaman, and R.A. Graham (eds.) AIP, 1982, p. 67-71.
5. R.A. Graham and D.M. Webb, "Fixtures for Controlled Explosive Loading and Preservation of Powder Samples," in Shock Waves in Condensed Matter, eds. J.R. Asay, R.A. Graham, and J.K. Straub, North Holland, 1984, pp 211.
6. R.A. Graham and D.M. Webb, "Shock-Induced Temperature Distributions in Powder Compact Recovery Fixtures," The Fourth APS Topical Conference on Shock Waves in Condensed Matter, Spokane, Washington, July 22-25, 1985.

MACH STEM CHARACTERIZATION IN MBAR DESIGNS USING RSR POWDER

K. P. Staudhammer and K. A. Johnson

Los Alamos National Laboratory
Materials Science and Technology Division
P.O. Box 1663, Los Alamos, NM, 87545

INTRODUCTION

The experimental studies used in this investigation are based in part on the mach stem lens formation work by C. E. Morris, R. G. McQueen and S. P. Marsh[1]. One of the fortuitous aspects of the cylindrical explosion design is its suitability as a screening tool with its very high success rate of recovery. The designs allow the experimenter in one shot to observe a wide variety of pressure and strain variables. The authors[2] have shown the effectiveness of this technique using solid specimens of 304 stainless steel. The effect on powder studies here is more complex, and it will be shown that the mach stem melt zone size and location is a function of several additional variables. The unique microstructure of the Rapid Solidification Rate (RSR) powder was chosen to help characterize and deliniate the effects. It should be emphasized that the powder was selected to help in the modeling of the formation of the mach stem zone and that this technique is not particularly used or recommended to compact RSR powder. In addition, the conditions explored here are in excess of those normally used to compact metallic powders.

EXPERIMENTAL

The material employed in this study was a nickel based super alloy powder produced by Pratt and Whitney and designated as RSR-834. This powder possessed an average particle size of <100μm. The as received powder is shown in Figure 1 and is characterized by a fine dendritic microstructure internally with many particles containing a thin outer amorphous layer (marked A in Figure 1). These powders were produced by centrifugal atomization and forced convection cooling ($\sim 10^5$ °C/s). A more detailed description of these powders is given by reference 4. Chemical analysis of this powder indicated 10 Mo, 6.8 Al, 6 W, 3 Cr, 1.5 Ta, 0.1 Y, 0.2 Hf, balance Ni (all wt/o). Its melting temperature is 1375°C. Prior to shock loading the powder was vibra packed into a 0.95cm diameter by 6.7cm long tube and welded in vacuum. No additives were employed. The initial density of the powder was 49% of the theoretical density of RSR-834.

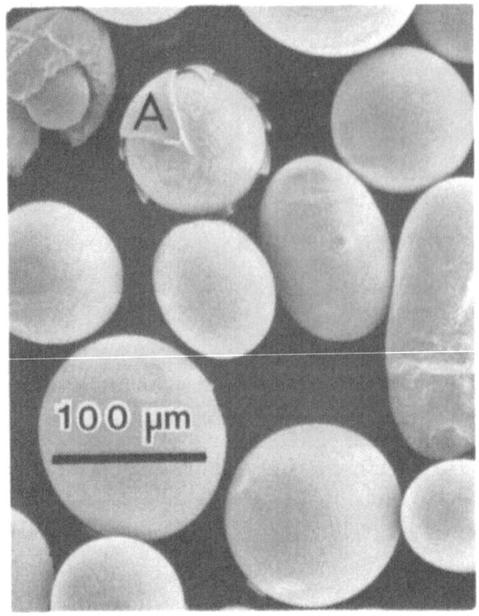

Figure 1. Scanning electron micrograph of as received RSR-834 powder.

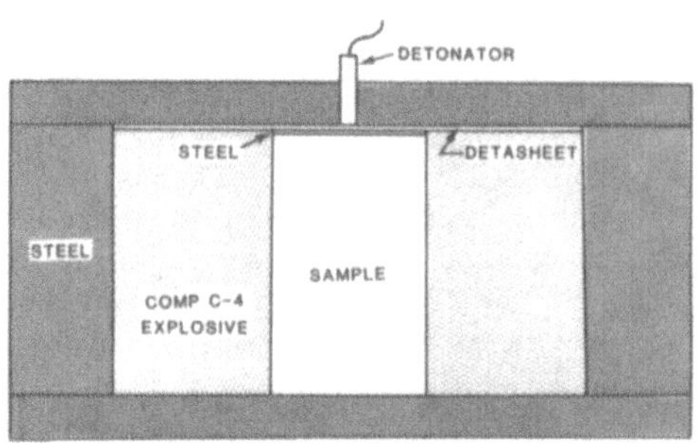

Figure 2. Shock loading assembly schematic.

The shock loading assembly is shown in figure 2. This design and its modified versions are discussed in several papers[1,2,5,6]. The shock assembly consists of a tube welded in vacuum containing the sample of interest (capsule dimensions given above) inserted into the central axis of a 0.95cm diameter right cylinder 6.7cm long. The test sample was surrounded by composition C-4 explosive (having a density of 1.59 g/cc and a detonation velocity of 8.04 mm/µs) and then encased in a steel recovery assembly. Initiation of the explosive occurs around the outer edge of the top of the sample recovery cylinder. The detonation wave moves down along the outer edge of the cylinder and out into

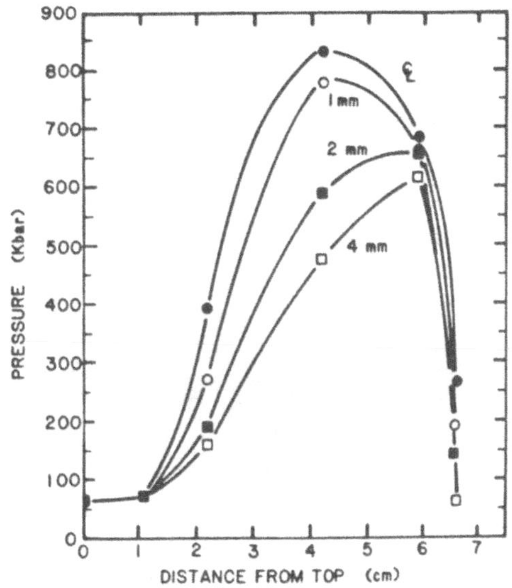

Figure 3. Pressure vs distance along central axis for 49% theoretical density powder, calculated values.

the explosive. The sweeping shock wave generated by the explosive converges at the center of the cylinder which creates a high pressure region called the Mach stem. The terminal shock velocity achieved is the detonation velocity of the explosive and this determines the maximum pressure which can be attained. During the time of formation of the Mach stem, the powder in the central tube experiences increasing shock pressures with increasing distance down the tube length. At the top the pressure is approximately 60 Kbar increasing to 830 Kbar at a position 64% of the way down the tube, then decreasing slightly with further axial travel. The RSR-834 powder thus experienced shock pressures from 60 to 830 Kbar. Hydrocode computer calculations were employed to determine shock pressures experienced by the RSR-834 powder. These calculations employed a two-dimensional Eularian code which has the ability to incorporate multi-materials, material strengths, equation-of-state, and programmed burn rate for the high explosives. The calculated pressure profiles both axially and radially are shown in Figure 3.

RESULTS

The calculated pressures for this RSR powder compaction shown in Figure 3 take into account the initial density of the powder. The Mach stem region (along the central axis) and the radial distances up to 4mm are shown. Radial distances beyond 4mm are significantly below peak pressure and are not shown. They are radially beyond the powder specimen holder and of little relevance here, and are discussed in reference 2. For comparison the optical micrographs of the Mach stem region are shown in Figure 4. As is obvious, the Mach stem zone exceeds the melt temperature of the RSR-834 powder. However, the radial melt zone far exceeds the high pressure regime. Figure 4a is a cross section

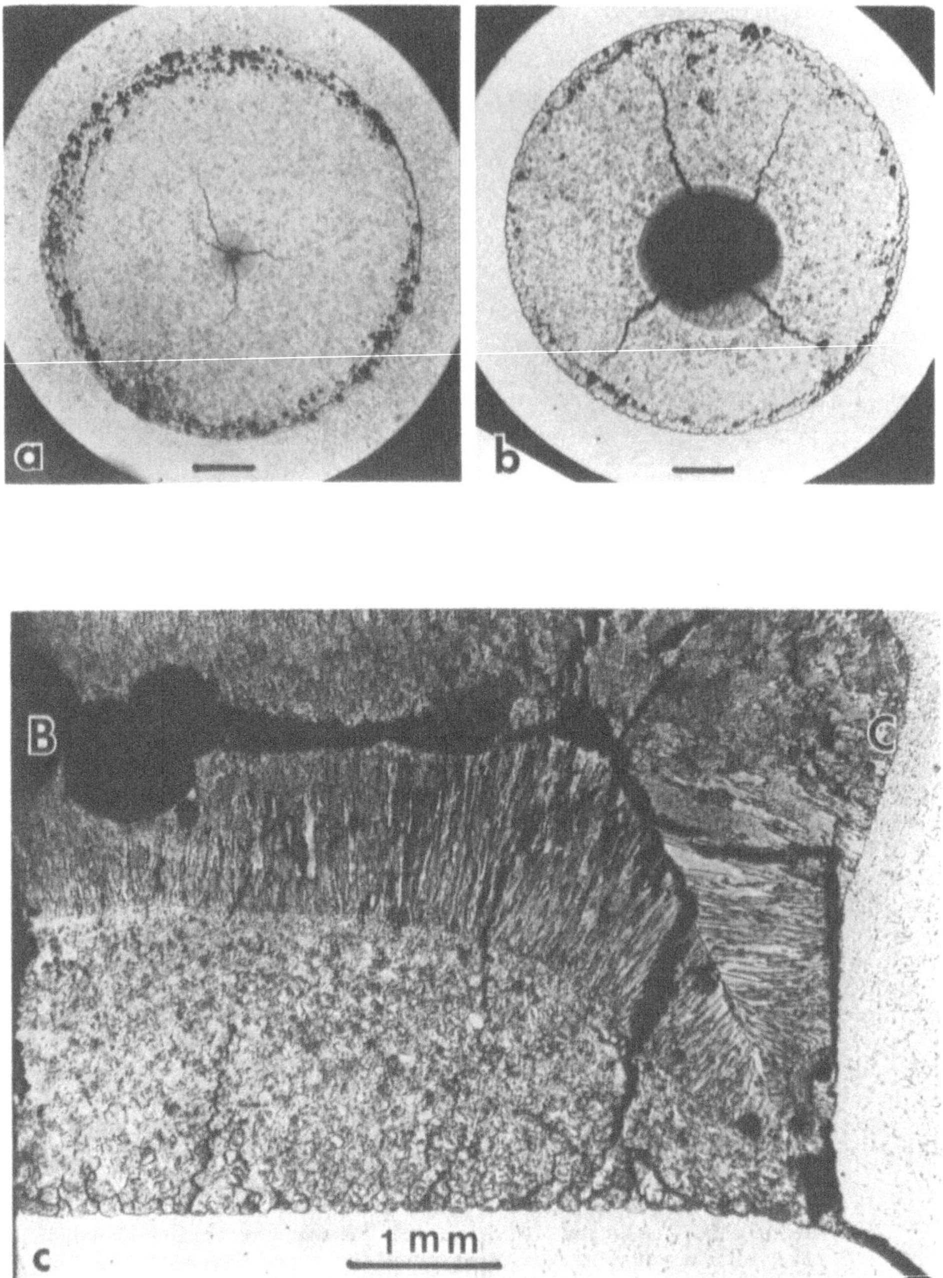

Figure 4. Optical microscopy of compacted powder, pressures along central axis a) 60 Kbar position, b) 830 Kbar position, c) bottom end pressure at B, 600 Kbar to C ∿100 Kbar.

normal to the specimen holder of the 60 Kbar pressure regime (central axis) while Figure 4b is a cross section at the 830 Kbar pressure regime, same orientation. Figure 4c is a cross section parallel to the

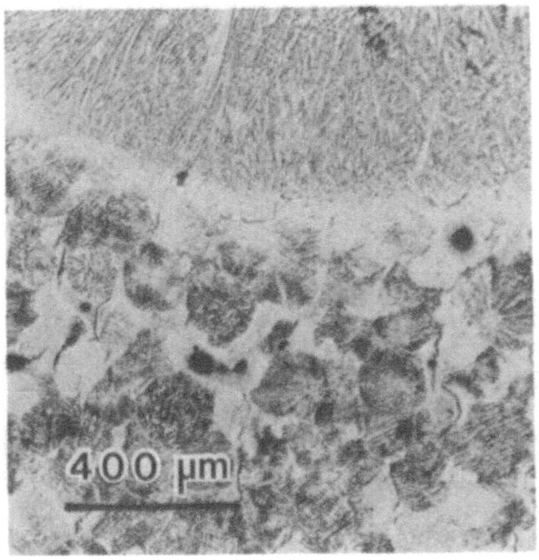

Figure 5. Melt zone interface in compacted RSR-834 powder.

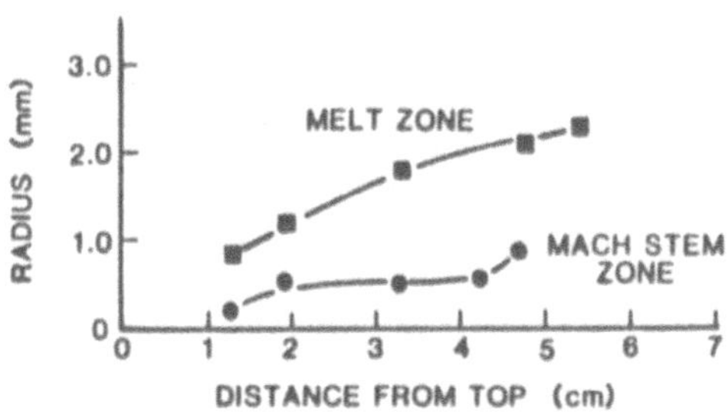

Figure 6. Melt zone radius vs distance for compacted RSR-834 powder.

central axis just below Figure 4b. The pressure along the central axis ranges from 600 Kbar at point B to 100 Kbar at point C. The pressure along the outer diameter is approximately 100 Kbars from top to bottom of Figure 4c. Evident in Figure 4c is the "classical" solidification pattern evident in as cast ingots and weld zones. In this regime the entropic ΔT plus adiabatic ΔT had obviously exceeded the melt temperature of the alloy (1375°C). The dendritic structure spacing of the as received powder spacing was 2.6µm and the dendrite spacing in the mach stem region was 1.5µm. The spacing of shock melted resolidified powder had a smaller dendritic spacing and the dendrites were significantly shorter. This dendritic structure is extremely fine and is a consequence of the cooling rate, thus indicating that the cooling rate of the shock melted portion was equal to or faster than the rate for the as fabricated powder. In essence, the melt zone has retained the dendritic morphology while the macro structure has been altered to a columnar growth indicative of the heat flow. EDX analysis across dendrites in the melt zone were identical to those in the compacted outer non-melt zone. This indicates that no chemical segregation of constituents occurred at the micron level.

Further analysis of the compaction/melt phenomena in this design revealed that compaction occurs first then melting. This observation is illustrated in Figure 5. Here, the melt zone interface is in a smooth radial curve about the central axis and many particles at the melt zone perimeter are only partially melted. This implies prior compaction. Had they not been compacted first then entropic heat would have melted a larger, more irregular zone with greater melt zones in and around each particle. Figure 6 shows the mach stem and melt zone size from top to bottom of the specimen holder taken from areas at different distances along the holder. The mach stem zone (discounting end effects, past 4.3cm from the top) does not exceed 0.8mm radius along the central axis and is consistent with the mach stem region calculated in the hydrocode. The melt zone, however, is not accounted for in the hydrocode calculations and does represent an area for further modeling work. The increase in the melt zone radius with increased travel of the shock wave is not surprising in light of the increase in pressure and a change from primarily surface deformation of particles at low pressures, to more massive non-homogeneous particle deformation at the higher pressures. This, consequently, is indicative of an increasing entropic temperature as well as adiabatic heating resulting in a large melt zone.

SUMMARY

Suitable selection of powders can be used as a modeling device for complicated experimental designs. The powder melt zone is clearly defined and the RSR-834 powder is reasonably well behaved. This experiment was with only one composition, size and distribution. However, it is believed that other morphologies, composition, and size distributions could result in a more complete modeling of the shock compaction process.

ACKNOWLEDGEMENTS

The authors wish to thank S. P. Marsh for the hydrocode calculations.

REFERENCES

1. C. E. Morris, R. G. McQueen, and S. P. Marsh, Shock Waves In Condensed Matter - 1983, Proceedings of the American Physical Society Topical Conference, Santa Fe, New Mexico, 1983, North Holland, pp. 207-10, 1984.
2. K. A. Johnson, L. E. Murr, and K. P. Staudhammer, Acta Metall., Vol. 33, No. 4, pp. 677-84, 1985.
3. "Rapid Solidification Processing Principles and Technologies, III" ed., Mehrabian, NBS, 1982.
4. E. H. Aigeltinger, A. F. Giamei and D. M. Dimiduk, "Rapid Solidification Processing Principles and Technologies, III", ed. Mehrabian, NBS, pp. 186, 1982.
5. K. P. Staudhammer, K. A. Johnson, and B. W. Olinger, Shock Waves in Condensed Matter - 1983, Proceedings of the APS Topical Conference, Santa Fe, New Mexico, 1983, North Holland, pp. 419-22, 1984.
6. J. J. Petrovic, B. W. Olinger, and R. B. Roof, Shock Waves in Condensed Matter - 1983, Proceedings of the APS Topical Conference, Santa Fe, New Mexico, 1983, North Holland, pp. 463-66, 1984.

SHOCK SYNTHESIS OF NICKEL ALUMINIDES

I. K. Simonsen, Y. Horie and R. A. Graham*

North Carolina State University
*Sandia National Laboratories

INTRODUCTION

Recently we reported the synthesis of nickel aluminides in a mixture of elemental aluminum and nickel powders under high pressure shock wave loading [1]. The following are selected features of the reactions observed by post-shock analysis of those specimens: 1. The reaction product and yield were sensitively affected by the shock-loading conditions within the specimens. 2. Broadly speaking, there were three regions of product formation: a region of no reaction or sparse local reactions, a region of substantial multiple reactions containing all the products, and a region of Ni_3Al in bulk containing unreacted Ni grains. 3. This bulk Ni_3Al was found in a region of high shock-induced mean bulk temperature (above about 500°C), but relatively low pressure (less than 15 GPa) predicted by careful numerical simulation. 4. The microhardness of Ni_3Al (440 DPH at 30 pd load) is close to that produced by rapid solidification method, and suggests a unique character to the shock process and its potential for producing materials with non-equilibrium microstructures and new compounds.

In this paper we present the results of further investigation of the observed synthesis with particular emphasis on metallography.

EXPERIMENTS

Since our first report [1], three additional recovery experiments have been conducted to cover a wider range of loading conditions. The shock loading and recovery was carried out in the Sandia Bear recovery fixtures [2]. The extent of our studies is shown in Fig. 1.

The starting elemental powders were CERAC A-1189, -325 mesh aluminum and CERAC N-1095, -200+325 mesh nickel. The mechanically blended powders were pressed in place in each recovery fixture to a density of 4.25 Mg/m^3, about 60% of solid density. After loading radial sections were cut from the shocked specimen compacts and examined with x-ray diffraction, a Reichert metallograph, and electron microprobe analysis.

RESULTS AND DISCUSSION

Compound formation was observed in all the specimens. Intermediate phases were identified by quantitative EMPA measurements and, where enough compound material was available (Ni_3Al, NiAl, and $NiAl_3$), by x-ray diffraction. In multiphase regions the identification of nickel aluminides was based upon the comparable hardness and colors of those in large quantities which had been analyzed by EMPA and x-ray diffraction.

The distribution of products was reasonably symmetric with respect to the center line of the cylindrical specimens in spite of somewhat poor mixing of powders. However, the pattern of product formation was very complex resulting from the strong interaction of two-dimensional wave propagation and the reaction kinetics. The general features summarized above were also observed in the new specimens. In what follows we shall discuss the results by examining a series of metallographs corresponding to specific shock wave loading conditions.

One unique feature, common to all the specimens, is the strong reactions in the high temperature region near the disk periphery [3]. We now know from the numerical simulation that this region is also a region of relatively low pressures (10 ~ 15 GPa). This pressure effect will be discussed again in relation with the formation of bulk Ni_3Al.

There are other distinct macroscopic features of compound formation in the Ni-Al system. 1. Significant reaction zones shown in Fig. 2 occurred under the weakest loading conditions: calculated temperature in the region is about 250°C and peak pressure 5.0 GPa. This is certainly an indication of greatly enhanced solid-state reactivity under shock wave loading. In this figure Region A is basically unreacted, but contains some localized reactions. In contrast, the reacted region B contains all the known intermediate phases of the Ni-Al system. 2. Figures 2 through 5 show some evidence of temperature effect on the progression of the observed chemical reactions. Particularly, the calculated temperature of above 400°C appears to be necessary for the final formation of bulk Ni_3Al. The bulk Ni_3Al is always found in the high temperature regions. 3. In experiment 8G856 shown in Fig. 5, the calculated temperature reaches about 800°C. But there still exists a zone containing a significant amount of unreacted Ni. Our current interpretation based upon the numerical simulation is that the high pressure (above 15 GPa) strongly inhibits the reactions. More direct evidence of this negative pressure effect is based upon optical observations [4].

Optical micrographs in Fig. 6 through 8 were selected to illustrate the spatial progression of the chemical reactions from Region A to Region C. Basic features are found to be common to all the specimens.

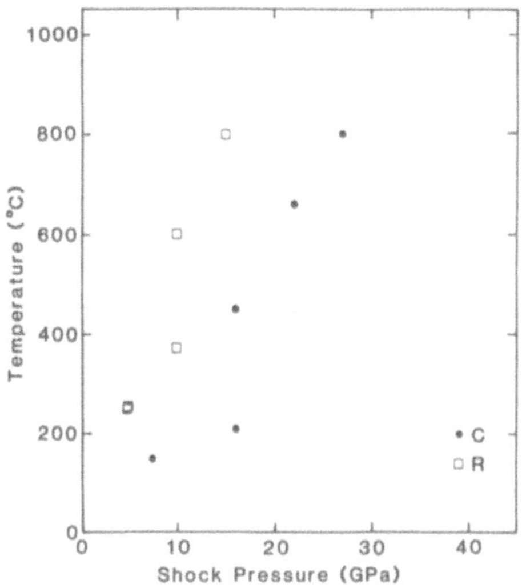

Fig. 1 Calculated peak loading conditions in Ni-Al powder mixtures. C represents the conditions in central regions and R reacted regions.

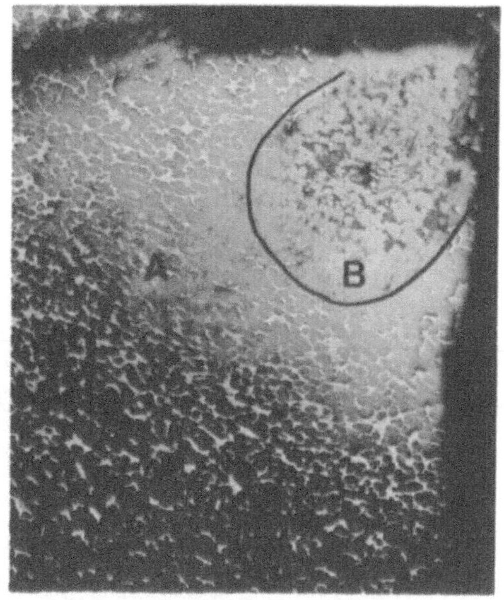

Fig. 2 Experiment 93G846. The weakest loading conditions to date: Mean pressure = 5.0 GPa and peak temperature = 250°C. Region A is mostly unreacted. Region B is a multiple reaction zone containing all the products.

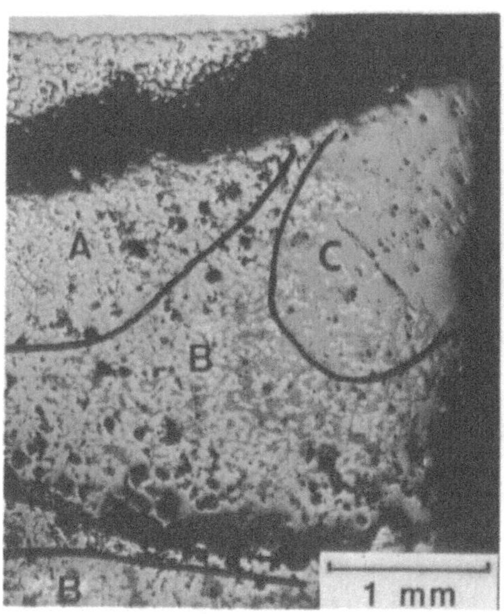

Fig. 3 Experiment 96G846. Mean pressure = 10 GPa and peak temperature = 375°C. Bulk Ni_3Al is found in Region C next to the periphery of the disk specimen.

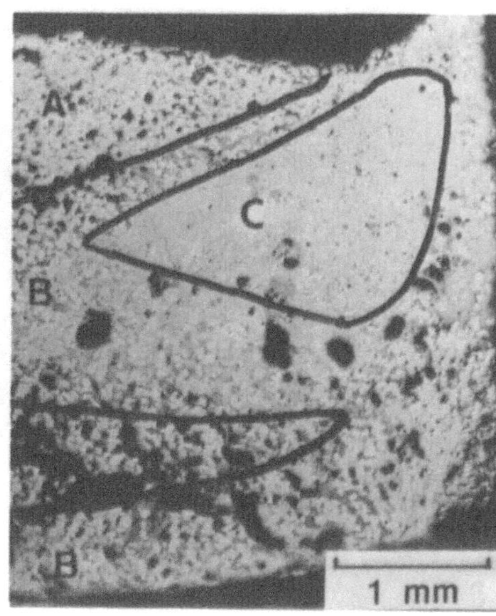

Fig. 4 Experiment 72G846. Mean pressure = 10 GPa and peak temperature 600°C. Bulk Ni_3Al extends further toward the center of the specimen than that in 96G.

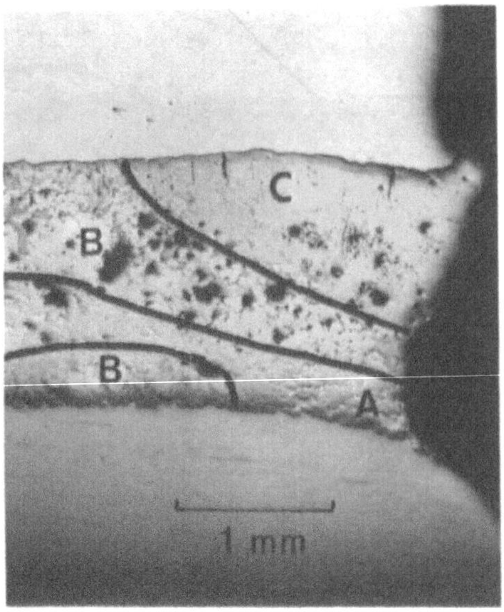

Fig. 5 Experiment 8G856. Highest loading conditions to date: Mean pressure = 15 GPa and peak temperature = 800. High pressures (above 20 GPa) develop from bottom r.h. corner to upper l.h. corner.

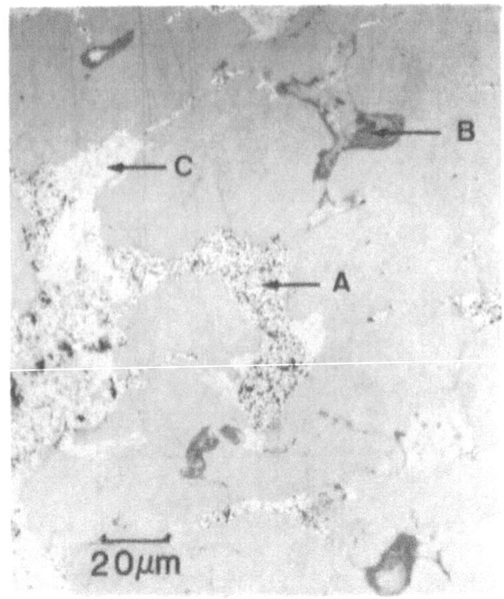

Fig. 6 Experiment 96G846. Localized beginning reactions in Region A of Fig. 2. A = Al, B = NiAl: The color darkens as the content of aluminum increases. C = solid-solution.

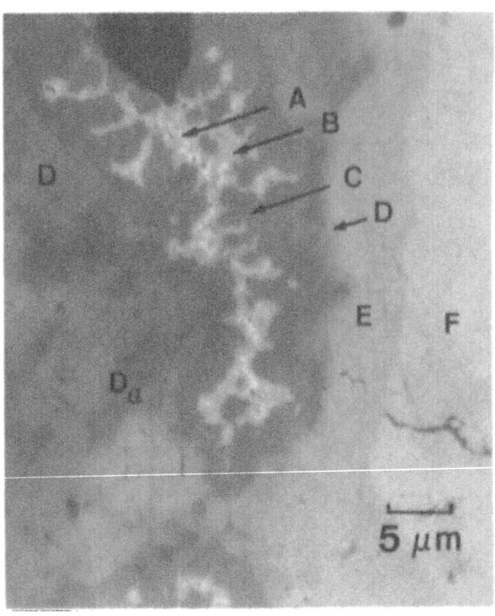

Fig. 7 Experiment 96G846. Multiple reactions seen in Region B of Fig. 3. A = Al, B = $NiAl_3$, C = Ni_2Al_3, D = nickel rich NiAl, E = Ni_3Al, F = Ni, D_a = aluminum rich NiAl.

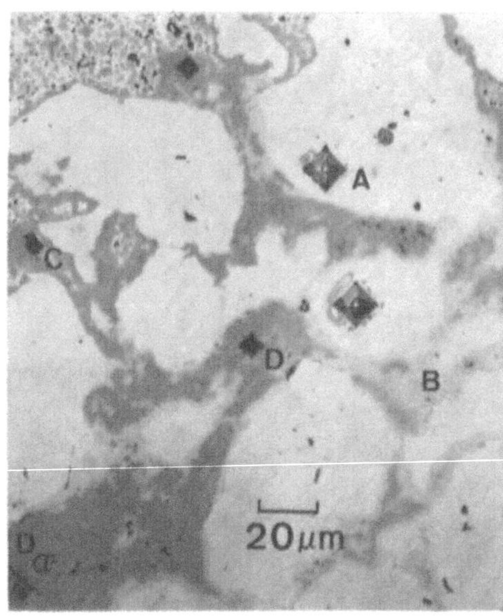

Fig. 8 Progression of compound formation in 96G846. A = Ni, B = Ni_3Al, C = Ni_2Al_3, D = NiAl, D_a = aluminum rich NiAl. See Figs. 6 and 7 for details.

Figure 6 shows the morphology of Region A containing localized reactions: the formation of solid solutions and NiAl. The identification of NiAl is based upon the comparable hardness and colors that are predetermined from those observed in larger areas which were corroborated by EMPA and X-ray analysis. No indication of $NiAl_3$ and Ni_2Al_3 was found in regions of widely separated localized reactions. Interestingly, a similar observation has been made with electron beam pulse melted Al implanted with Ni [5]. Both processes involve localized, highly non-equilibrium states. The shocked state was created by a combination of mechanisms such as inhomogeneous deposition of deformation energy and enhanced diffusivity due to mechanical mixing and activation. We have, however, no information about the existence in the shocked specimens of the amorphous phase observed in the Ni-implanted Al when the concentration of Ni was above 6 at %. Solid solutions in aluminum grains are easily recognized by hardness and the lack of grainy appearance after polishing. We speculate that they are supersaturated solid solutions.

Figure 7 shows a common morphology seen in Region B. The transition from Region A to Region B is relatively sharp and can be marked within the accuracy of a few tenth of a millimeter.

There are several unique features of Region B that earmark the progression of the observed reactions: 1. The occurrence of $NiAl_3$ in Al grains. This occurs almost uniformly over the entire aluminum grains. This could occur as a result of a fast dissolution of nickel into liquefied aluminum from grain boundaries and/or of trapped fine nickel particles in aluminum. 2. Formation of Ni_2Al_3. This phase appears always as a thin layer either on the original boundary of nickel particles which might be transformed to NiAl or inside the precipitated $NiAl_3$. The amount of Ni_2Al_3 is always limited. 3. Formation of NiAl and Ni_3Al. There is a larger yield of NiAl which is produced by either localized reactions or the dissolution of nickel into aluminum. Ni_3Al appears between NiAl and Ni. In some areas whole Ni_3Al grains may be seen.

Figure 8 illustrates the overall progression of the reactions from the stage where $NiAl_3$ is seen in Al grains to the stage where Ni grains are mostly transformed into Ni_3Al or NiAl. This progression is in the order of increasing temperatures and is consistent with the intensive study by Janssen and Riech [6] on the nickel aluminum diffusion couple. In their study products at 600°C were Ni_2Al_3 and NiAl; above 900°C the products were Ni_3Al, NiAl, and Ni_2Al_3. In another study [7] it is shown that within the temperature range of 870-1000°C the NiAl grew by consuming Ni_2Al_3.

Although the static data are not directly comparable with that observed in the shock-loaded specimens, the observed sequence of product phases provides a clue to the kinetics of compound formation in the Ni-Al system under shock-wave loading. One simplified scenario among many possibilities is the following: 1. Initially, supersaturated solid-solutions and NiAl are formed through non-equilibrium processes under shock compression. Such processes include inhomogeneous plastic deformation, scuffing of oxides exposing fresh metallic surfaces, formation of extensive defects, dissolution of fine Ni-particles in Al, etc. 2. Secondly, if the imposed temperature or the heating due to initial exothermic reactions are large enough, then there is a rapid formation of NiAl and Ni_3Al from nickel-rich melt or liquid-like reaction products which requires almost total involvement of nickel. It is reported [5] that the nucleation of Ni_3Al from a non-equilibrium state could occur within the time scale of 1 μsec. In every stage, pressure will be a significant factor not only in the diffusion of Ni and Al, but also in the formation of NiAl and Ni_3Al. Even a moderate level of pressure (a few gigapascal) is said to affect the diffusion rate by a factor of 100 [8]. 3. The transition zone between the unreacted but shock-modified powder mixture and the strongly reacted (hot) zone contains the aluminum-rich products Al_3Ni_2 and Al_3Ni. Such products are the result of less complete reaction of nickel. 4. Finally, those products resulting from strong, exothermic reactions will be in a liquid state at some stage of the process and final composition and microstructure will be determined by pressure unloading, reduction in particle velocity and the cooling process.

CONCLUSIONS

Localized formation of solid solutions and NiAl were observed in a specimen subjected to the weakest loading conditions (peak pressure of 5.0 GPa and peak temperature of 250°C). This is an indication of greatly enhanced solid state reactivity encountered under shock wave loading. The formation of bulk Ni_3Al appears to require a temperature of about 400°C, but the reactions are inhibited by high pressures (above 15 GPa).

REFERENCES

1. Y. Horie, R. A. Graham and I. K. Simonsen, Materials Letters, in press.
2. R. A. Graham and D. M. Webb, in Shock Waves in Condensed Matter-1983, J. R. Asay, R. A. Graham and G. K. Straub, Eds., North Holland, New York, 1985.
3. R. A. Graham and D. M. Webb, this proceedings.
4. M. B. Boslough, R. A. Graham and D. M. Webb, this proceedings.
5. S. T. Picraux and D. M. Follstaedt in Surface Modification and Alloying, Ed. J. M. Poate, G. Foti and D. C. Jacobson, Plenum, New York, 1981.
6. M. M. P. Janssen and G. D. Rieck, Trans. TMS-AIME 239, 1372 (1967).
7. A. J. Hickl and R. W. Heckel, Met. Trans. A 6, 431 (1975).
8. D. Lazarus and N. H. Nachtrieb, in Solids Under Pressure, Ed. W. Paul and D. Warschauer, McGraw Hill, New York, 1963.

The work performed at Sandia National Laboratories was supported by the U. S. Dept. of Energy under contract # DE-AC04-76DP00789.

SHOCK-WAVE SYNTHESIS OF TITANIUM ALUMINIDES

Y. Horie, D. E. P. Hoy, I. Simonsen;
R. A. Graham,* and B. Morosin*

North Carolina State University
*Sandia National Laboratories

INTRODUCTION

This paper presents the post-shock analysis of compound formation in an elemental titanium-aluminum powder mixture subject to a wide range of explosive shock-wave loading conditions. Insofar as we are aware, this is the first attempt to synthesize titanium aluminides through use of shock loading. This investigation is a part of a continuing effort to gain improved insight into shock-induced chemical reactions in solids [1].

Titanium aluminides like nickel aluminides have special properties which "intrigue the imagination and challenge the ingenuity of the metallurgist" [2]. Therefore, a variety of processing techniques have been used for producing titanium alloys. There is even a process which may be considered in a certain sense as a static counterpart of shock-wave loading [3]. This process involves blending of elemental powders, mechanical pressing to form compacts, vacuum sintering at temperatures in excess of 1200°C, and finishing steps which include hot-isostatic pressing (HIP), final machining, etc.

Shock-wave processing contains steps which are to some degree similar to the above described elemental powder process, but the material states achieved in shock synthesis at very high pressures and high rates of loading are not accessible by any other process.

EXPERIMENT

Experimental arrangements are identical to those used for the synthesis of nickel aluminides [4], and will not be reported here. The powders were CERAC T-1146, -200 mesh titanium, and CERAC A-1189, -325 mesh aluminum of the nominal purities supplied by the manufacturer 99.5% (packed under argon) and 99.99% (thin oxide coating), respectively. The mechanically blended powder mixtures containing 30 vol. pct. Al and 70 vol. pct. Ti were cold pressed in place in the recovery fixtures to a density of about 60% theoretical (mole fraction x metal densities). A protective oxide coating on the powders is expected from exposure to laboratory atomosphere.

Six recovery experiments were conducted to cover a wide range of loading conditions shown in Table 1. These conditions indicate only relative peak magnitudes within the powder specimens. Actual local loading conditions are very difficult to determine. They depend not only upon continuum wave propagation, but also upon its wave interactions with constituent particles involving plastic flow, comminution, mixing, diffusion, etc. [5].

Table 1. Bulk loading conditions at selected locations within disk specimens

	Conditions in reacted region[a]		Condition in center region	
Exp.	p_r(GPa)[b]	T_r(°C)[c]	p_c(GPa)[d]	T_c(°C)[e]
9G856	15	800	27	800
3G856	10	600	22	660
2G856	10	375	16	450
73G846	5	250	16	210
6G856	5	250	7.5	150

[a]Reacted regions are found in an outer portion of a cylindrical specimen where numerical simulation predicts high temperatures but relatively low peak pressures. A simultaneous application of high temperatures and high pressures seems to retard reactions.

[b]Peak shock pressure in reacted region.

[c]Peak mean bulk temperature in reacted region.

[d]Mean shock pressure in center region.

[e]Peak mean bulk temperature in center region.

RESULTS AND DISCUSSION

A reaction with significant yields was observed in experiments 2G856, 3G856, and 9G856 for which the peak bulk temperature in reacted regions was above 450°C. In others, reactions were mostly local. The location of the strong reactions within the powder compacts is similar to that found in the Ni-Al powder experiments [4,6] and is a region of high temperatures and low pressures (see Table 1 and Ref. [5]).

There are many explanations for the lack of strong reactions in low temperature experiments compared to observations on Al-Ni. The following is a partial listing of possible explanations. (1) Deformation of Al or Ti particle was not severe as that found in a Al-Ni elemental powder mixture because of the shape and relative hardness of Ti. The Ti powder was crushed from sponge and has hardness between Ni and Al [7]. In the Ni-Al system plastic deformation is mostly taken up by Al particles [4]. But as shown in Fig. 1, the deformation in the Ti-Al system appears to be

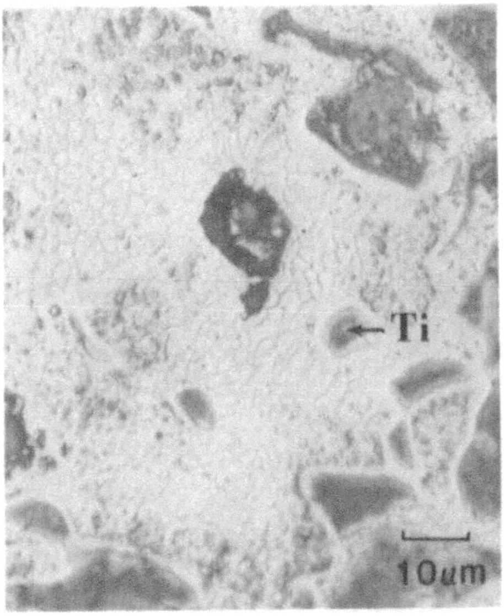

Fig. 1 Microphotograph of 9G856 showing the particulate mixing, deformation and some chemical reactions.

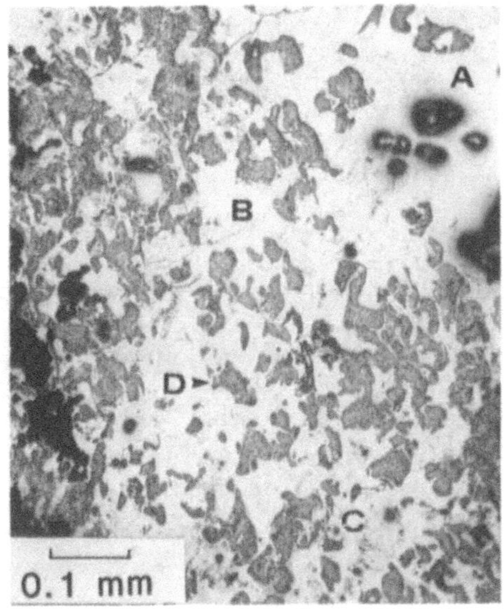

Fig. 2 Microphotograph of 9G856 showing a strongly reacted region. EMPA shows that A=Ti_3Al, B=Ti_2Al, C=TiAl, and D=Ti(etched).

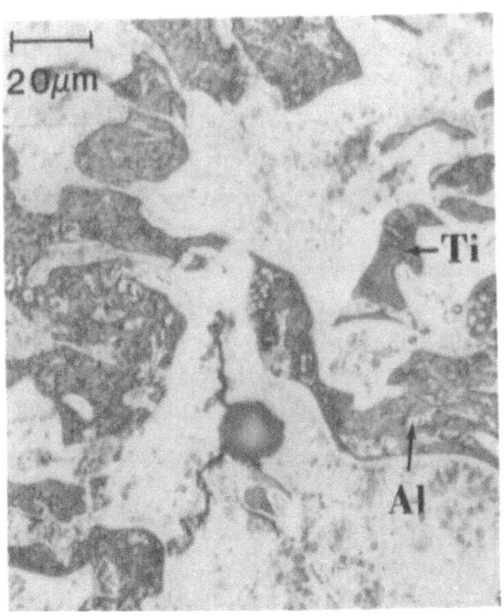

Fig. 3 Microphotograph of 9G856 two microstructural features: a cellular freezing and a liquation

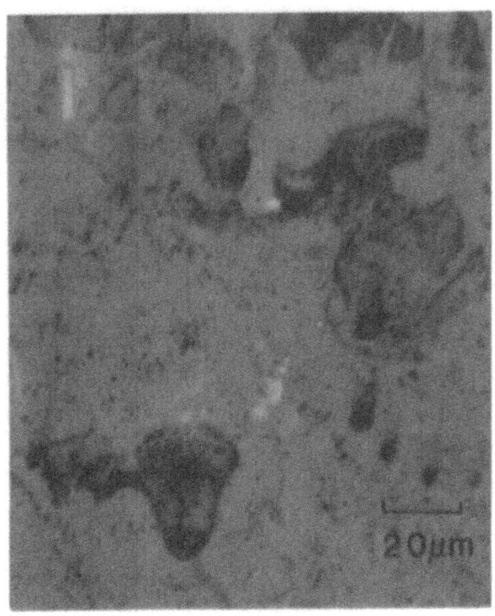

Fig. 4 The appearance of inclusions in the grain boundaries of a new phase. This photograph is located in a strongly reacted region.

partitioned in some manner between the two powders. Post-shock analysis of hardness shown in Table 2 may be used as a qualitative guide for relative deformation. Also observed in Fig. 1 is the prevalent embedding of Al in Ti powders. This phenomenon is not seen in the Ni-Al system. (2) The melting temperature of Ti is 200°C higher than that of Ni [7]. (3) Reactivity of the Ti-Al system is lower than that of the Ni-Al system as seen through electron negativity values [8]. (4) There is lack of miscibility. (5) Oxide layers are thicker than those on the Ni-Al particles.

The optical micrograph in Fig. 2 shows a strongly reacted outer edge of a radial section of 9G856. This section was etched by a reagent that attacks α-Ti [9], but not titanium aluminides. Residual Al was mostly washed away. Hence in the figure 'compounds' were found in smooth unetched white areas. Their compositions based upon electron microprobe analysis, EMPA, showed discrete elemental distributions which can be interpreted as Ti_3Al, Ti_2Al, and TiAl. Their locations can be approximately identified in terms of topographical Ti distribution. For instance, Ti_3Al is found in Region A where Ti particles are mostly disappeared. In contrast the composition in Region C is approximately that of TiAl.

X-ray diffraction analysis indicated, besides the initial elements, only $TiAl_3$: it is difficult to distinguish the $TiAl_3$ pattern from the ordered superstructures, Ti_9Al_{23} and Ti_8Al_{24} in such combinations. Apparently, such a discrepancy between X-ray diffraction, which determines crystal structure type and EMPA, which determines elemental composition, is not uncommon [10]. Transmission electron microscopy may allow more microscopic resolution of material in order to clarify these results.

The following static data may be of some relevance to not only the above described discrepancy but also induced chemical reactions. The interaction of Ti and Al in sandwich diffusion couples prepared by applying a pressure of 1.5 GPa at 420°C for 30 min. showed no diffusion layer [11]. When annealed for 2 to 150 hours at temperatures from 580 to 640°C, a single diffusion layer of composition $TiAl_3$ was observed in all cases. At temperatures of about 585°C the layer consisted of a mixture of two superstructures of $TiAl_3$, Ti_8Al_{24} and Ti_9Al_{23}. In contrast the hot dipping of a strip of Ti at 850°C into molten Al for 10 sec formed no diffusion layer. In the study of a sputtered Al film on a Ti substrate [12], it is found (1) that $TiAl_3$ occurs very rapidly on heating to 635°C while Ti_3Al is formed after heating to 900°C, but (2) that further annealing at 900°C causes the Ti_3Al to decompose with substantial amounts of Al dissolving in α-Ti.

Upon comparison of the static and shock wave data, we may suggest (1) that the alloy formation under shock loading takes place at temperatures that are much lower than the static values and at enhanced rates, and (2) that the residual compound is Ti_3Al (within the resolution of X-ray diffraction) and the Ti-rich compositions found with EMPA are caused by the limit of its spatial resolution (~5μm).

Microhardness data are given in Table II. The hardness of the starting powders is consistent with handbook values. Also, the hardness of shock-loaded elemental powders in an unreacted region are comparable to those for cold-worked specimens. Since there was an uncertainty about the exact composition of the new compounds, no differentiation was made in hardness measurements. However, all the readings of 'new compounds' fell in a narrow range shown in the table. This magnitude is approximately two times those reported for Ti-rich components in the Ti-Al system [13]. The hardness of residual Ti in reacted regions is close to that of δ'(Ti,Al).

We have no information on the sources for such a high hardness, but this is evidence that shock synthesis has potentials for producing special microstructures for improved properties. In the case of a Ni-Al mixture, the hardness of bulk product Ni_3Al is thought to be caused by unusually fine grain sizes (50 ~ 100Å) [14].

Figures 3 and 4 show metallographic features that have not been seen in the Ni-Al system [4]. Figure 3 shows two features of peritectic alloys: (1) a cellular structure, a form of interfacial instability said to be caused by the inhomogeneous distribution of solute during freezing [15]; (2) a liquation also known as 'burning', which is the formation of a continuous film of liquid along the grain boundaries of the α-Ti. In the case of the Ti alloys, which are peritectic in fusion, the liquation can occur at temperatures below that of the peritectic [15]. Since $TiAl_3$ has the lowest melting temperature, the material in the ring around the grain boundaries is probably $TiAl_3$.

Figure 4 shows inclusions or particles distributed around the grain boundaries of a product compound. They could be precipitated particles or remnant of oxide layers from starting powder. Whatever the composition of these particles, small cracks seen in Fig. 3 appear to be propagating along the grain boundaries. No further information is available on the nature of the particles in the grain boundaries.

Table 2. Hardness data (DPH) of various phases in the Ti-Al system.

Ti			Al		Alloy
Unshocked powders	Particles in unreacted region	Particles in reacted region	Unshocked powders	Shocked particles in unreacted region	Varying particles
133±26	153±26	202±17	35±6	53±11	594±71

Hardness measurements were made on a Reichert metallograph with various loads from 10-100g. Influence of loads on hardness was within statistical errors.

CONCLUSION

The initiation of strong chemical reactions in a mixture of Ti and Al powders subject to high pressure shock-wave loading was observed when the bulk peak temperature of the region calculated from numerical simulation was above 450°C. A high temperature with a high pressure (~above 15 GPa) does not promote reactions with significant yield. A reaction product (or products) has hardness twice the values for Ti-rich equilibrium compounds in the Ti-Al system. There was lack of agreement between x-ray analysis and EPMA regarding the composition of the products. X-ray diffraction showed the pattern for $TiAl_3$ or its two ordered superstructures while EPMA favored Ti-rich comounds-Ti_3Al, Ti_2Al and TiAl.

ACKNOWLEDGMENTS

The work at Sandia was partially supported by the U. S. Department of Energy under Contract number DE-AC04-76DP00789.

REFERENCES

1. R. A. Graham, Am. Phys. Soc. 1985 Topical Conference on Shock Waves in Condensed Matter, Spokane, WA, July 22-25, 1985.
2. E. M. Grala in: "Mechanical Properties of Intermetallic Compounds," J. H. Westbrook, Ed., John Wiley, New York 1959.
3. C. A. Kelto, B. A. Kosmal, D. Eylon and F. H. Froes in: "Powder Metallurgy of Titanium Alloys," F. H. Froes and J. E. Smugeresky, ed., AIME, New York, 1981.
4. Y. Horie, R. A. Graham, and I. Simonsen, in press, Materials Letters.
5. R. A. Graham and D. M. Webb in: "Shock Waves in Condensed Matter - 1983," J. R. Asay, R. A. Graham and G. K. Straub, ed., North Holland, New York, 1985.
6. H. R. Pak, Y. Horie and R. A. Graham, Am. Phys. Soc., loc cit.
7. E. Rabinowicz in: "Friction and Wear of Materials," John Wiley, New York, 1965.
8. L. Pauling in: "College Chemistry," 3rd ed., Freeman, San Francisco, 1964.
9. G. Petcow in: "Metallographic Etching," ASM, Metals Park, Ohio, 1971.
10. Private communication with M. J. Carr of Sandia National Laboratories.
11. F. J. J. van Loo and G. D. Reick, Acta Met. 21:61 (1973).
12. V. B. Rao and C. R. Houska, Met. Trans. 14A:61 (1983).
13. E. Ence and H. Margolin, Trans. Met. Soc., AIME 221:151 (1961).
14. S. Myers, et al., Am. Phys. Soc., loc cit.
15. G. A. Chadwick in: "Metallography of Phase Transformation," Butterworths, London, 1972.

TEM OF NICKEL ALUMINIDES PRODUCED BY SHOCK COMPACTION

S. A. Myers[1], C. C. Koch[1], Y. Horie[1] and R. A. Graham[2]

[1]North Carolina State University, Raleigh, N. C.

[2]Sandia National Laboratory, Albuquerque, N. M.

INTRODUCTION

Nickel aluminides were produced by high pressure shock loading. X-ray diffraction and electron microprobe microanalysis show large regions of the intermediate phase Ni_3Al resulting form chemical synthesis of the elemental nickel and aluminum powders. Microhardness tests of this Ni_3Al show mechanical properties similar to Ni_3Al alloys prepared by rapid solidification techniques. To understand the structure/property relationships TEM was used to analyze the Ni_3Al regions in two samples fabricated by different explosives, mean pressures and bulk temperatures. Increasing pressure and resulting peak temperatures gives rise to a much finer grain size of less than 5 nm with a corresponding increase in hardness.[2] No antiphase boundaries were seen as is common in rapidly solididied material[3] and very little defect damage was seen as is common in shock loaded samples.

EXPERIMENTAL

Shock synthesis tests were performed at Sandia National Laboratories using recovery fixtures. The composition of the blended powder was 30 volume percent aluminum with the balance nickel powder. The details have been reported previously.[1,2] Thin sections were cut from the shocked butttons, and rough polished on 240-600 grit SiC to a thickness of .004". Final polishing with 0.05 m Al_2O_3 abrasive was used to remove any residual scratches. A GATAN ion mill was used to thin the sample to a thickness suitable for examination in the TEM. Only samples from the outer periphery of the button, shown to be Ni_3Al by electron microprobe and x-ray diffraction, were examnined on a Hitachi 800 Scanning Transmission Electron Microscope.

RESULTS AND DISCUSSION

Table 1 lists the processing parameters used for sample 71 and sample 72. Both samples were synthesized in the same fixture and the duration of the shock loading was similar. The explosives used were not the same, therefore, the samples experienced different pressures and temperatures.

Table 1 Shock Conditions in Powder Compacts

Experiment	Fixture	Peak Pressure Range (GPa)	Duration (μsec)	Peak Temp. (°C)
71G846	Momma Bear A	14-16	4.5	450
72G846	Momma Bear A	19-22	4.7	660

The peak pressure of 16 Gpa and the peak temperature of 450 C attained by sample 71 is considerably less than that of sample 72. The effect of these parameters may be seen in the unetched optical micrograph in figure 1. The regions marked Ni_3Al on the micrographs were identified by electron microprobe and x-ray diffraction.[1] The size of the Ni_3Al reacted zone in sample 72 is much larger than that of sample 71. In sample 72, the reacted zone is plainly visible after polishing and can be measured in terms of centimeters. In sample 71, the reacted zone is on the order of millimeters and is very difficult to see without an optical microscope.

Transmission electron microscopy was used to examine the microstructure of the Ni_3Al reacted area. Figure 2 is a micrograph from a representative Ni_3Al region in sample 71. The average grain size for the Ni_3Al grains observed in this sample was 1.4μm. The accompanying diffraction pattern is from the 100 pole. The arrows indicate the superlattice spots identifying a fcc ordered structure ($L1_2$). Figure 3 is a higher magnification picture of the same Ni_3Al grain showing few dislocations and very little defect damage, which is not usual in shock loaded samples.

Figure 4 from sample 72 shows a markedly different microstructure than sample 71. The representative area shown was taken at a magnification of 300,000X and yields an average grain size of 5 nm. The diffraction pattern in figure 5 shows a ring pattern typical of a sample with small grains. The "d" spacings confirm a $L1_2$ Ni_3Al sample with a slightly expanded lattice parameter.

Loading pressure and temperature conditions greatly changed the apparent reacted yield and microstructure of the reacted zone in these two shock loaded samples. The volume percent of Ni_3Al produced is much larger in sample 72 which had higher peak temperatures and pressures during processing than sample 71. It has been suggested that 400 C is necessary for the formation of bulk Ni_3Al[1]. The 660 C experienced by sample 72 is much greater than the threshold temperature and correspondingly, there is a much larger reacted area of Ni_3Al. In sample 71, the peak temperaure is much lower, so the reacted zone is predictably much smaller. The grain size in these samples can not be expained by the temperatures these samples experienced. Sample 72, with a high er peak temperature, should have had larger grains than sample 71 according to standard grain growth mechanisms. Therefore the higher pressure in sample 72 must have greatly influenced grain growth, or lack of it, in these shock loaded samples.

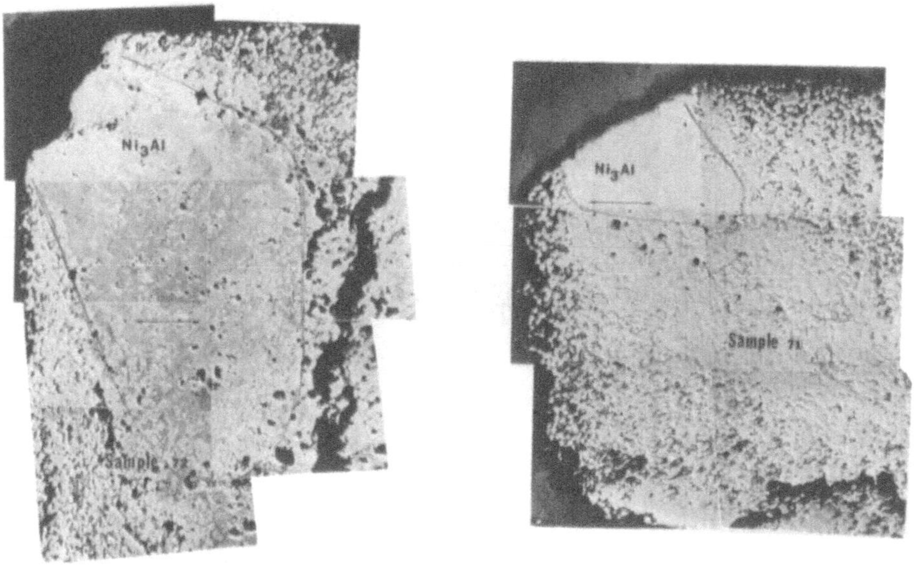

Figure 1: Optical micrograph of unetched shock loaded buttons-sample 72 and sample 71. The length of the bar is 0.5 mm.

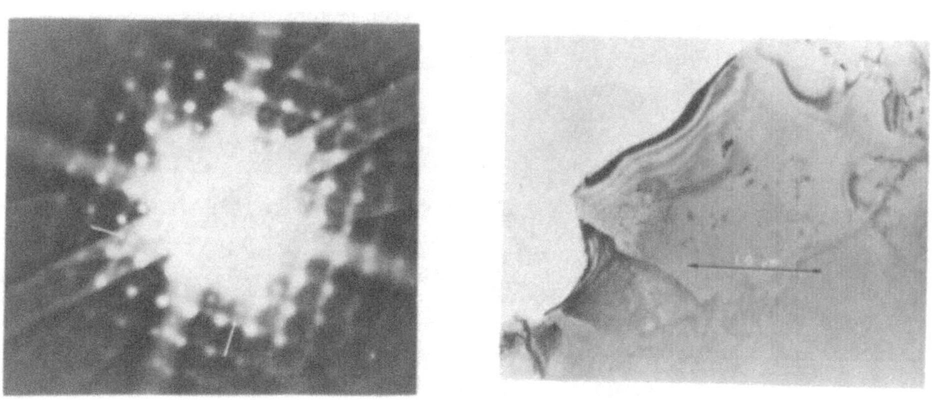

Figure 2: TEM micrograph from sample 71 showing large Ni_3Al grain size and corresponding diffraction pattern. The length of the bar is 1μm.

Figure 3: Higher magnification of the Ni_3Al grain.

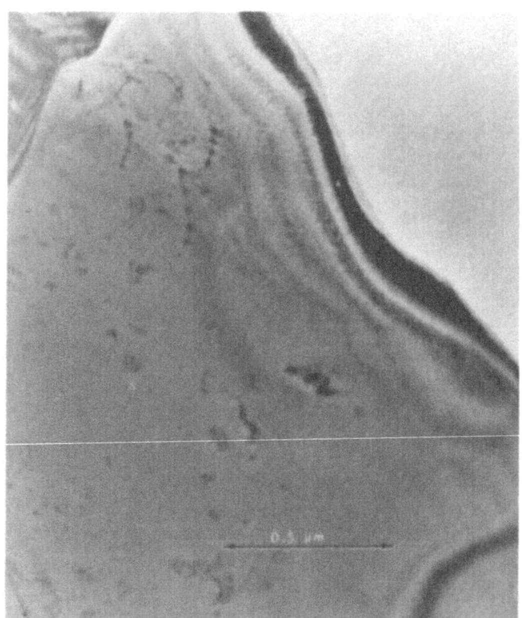

Figure 4: TEM microgragh from sample 72 of "average" Ni_3Al grains.

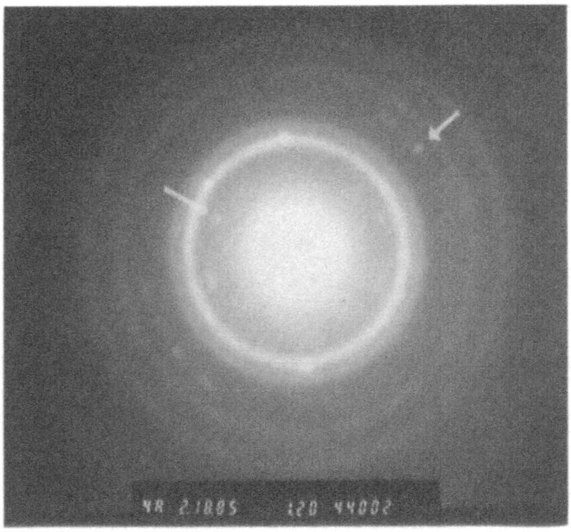

Figure 5: Diffraction pattern from Ni_3Al grains shown in figure 4. The ring pattern is indicative of very small grains.

CONCLUSIONS

1. Sample 72, which was synthesized with a peak pressure of 19-22 GPa and a peak temperature of 660 C, had an average grain size of less than 5 nm and a hardness similar to rapidly solidified materials, which is attributed to the extremely small grain size.

2. Sample 71, which was synthesized with a peak pressure of 14 GPa and a peak temperature of 450 C, had an average grain size of $1.4 \mu m$.

3. The differences in grain size are believed to be a result of the different pressures each sample was subjected to during the shock compaction. By increasing the pressure, the degree of grain growth is dramatically decreased.

REFERENCES

1. I.K. Simonsen, Y. Horie and R.A Graham, "Shock Synthesis of Nickel Aluminides", this proceedings.

2. Y. Horie, R.A. Graham and I.K. Simonsen, "Synthesis of Nickel Aluminides Under High Pressure Shock Loading", to be published Materials Letters.

3. J.A. Horton and C.T. Liu, "Anisotropic Antiphase Boundaries in Rapidly Solidified Ni_3Al", to be published Acta Met.

SYNTHESIS OF NICKEL-ALUMINUM ALLOYS BY SHOCK COMPRESSION OF COMPOSITE PARTICLES

Han-ryong Pak*, Y. Horie** and R.A. Graham***

*Center for Explosives Technology Research, New Mexico Institute of Mining and Technology, Socorro, NM 87801
**North Carolina State University, Raleigh, NC 37650
***Sandia National Laboratories, Albuquerque, NM 87185

ABSTRACT

We report electron microscope observations of compound formation in Ni/Al composite powders by shock loading. Reactions with substantial yield are seen in outer areas of a recovered specimen, while unreacted or locally reacted regions are seen in the center and middle areas of the specimens. Ni_3Al is formed in a gamma solid-solution of Ni. The microstructure is very complex.

INTRODUCTION

Enhancement of chemical reactivity has been known to occur in organic and inorganic materials subjected to shock loading[1,2]. Such enhancement is widely believed to be caused by a large number of defects produced in crystalline lattices. Recent studies[3,4] have shown shock synthesis in mechanically blended elemental powders of Ti and Al, and Ni and Al. In the present study Ni/Al composite powders are used to study chemical synthesis in post-shocked specimens by electron microscopy and x-ray spectroscopy. We report identification of shock-synthesized Ni-Al alloys (having exothermic reaction) and their microstructures.

EXPERIMENTAL

Starting material was an 80wt.%Ni-20wt.%Al composite powder (with impurities of 0.25%) supplied by Sheriff Gorden, Canada. The powder consists of spherical core of aluminum and an outer layer of nickel deposited on aluminum by hydrogen reduction in a metal salt solution. The average size of the powder was 53-88 μm.

Shock loading conditions in the experimental fixtures are listed in Table 1 (5). After loading, radial specimens were cut from the consolidated specimens and examined with a Hitachi HHS-2R scanning electron microscope equipped with a Tracor Northern TN-5500 energy dispersive X-ray analyzer. For transmission electron microscopy a Hitachi HU-200F microscope operated at 200 keV was used. Thin plates (0.5 mm thick) were cut from the central and middle regions of the specimen and reduced to a thickness of 0.25 mm by polishing with emery papers. Thin foils were then made with a Fishione jet electropolisher. The electropolishing conditions are: a solution of 75 vol pct C_2H_5OH and 25 vol pct $HClO_4$; the temperature ranging between -40 and -60°C.

Table 1. Bulk loading conditions at selected locations within disk specimens (Momma Bear A fixtures used)

Conditions in:		reacted region[a]		middle region		axial region	
Specimen	Explosive	P(GPa)[b]	T(°C)[c]	P(GPa)[d]	T(°C)	P(GPa)[d]	T(°C)[e]
1G856[f]	Comp. B	19	660	22	--	48	300
7G856[f]	Baratol	11	210	16	--	30	150

a. Reacted regions are found in an outer region of specimen 1G856.
b. Peak shock pressure in a reacted region.
c. Peak mean bulk temperature in a reacted region.
d. Mean peak pressure in middle and axial region.
e. Peak mean bulk temperature in an axial region.
f. Initial density: 4.14 Mg/m^3 (67% of solid density), 1G356; 4.06 Mg/m^3 (66%), 7G856.

RESULTS AND DISCUSSIONS

Surface observations were made of radial section from specimens, 1G856 and 7G856. Regions of sparse local reactions were observed in broad areas of both specimens. The exception is the central region of specimen 7G856 where no reactions took place. Generally speaking, reactions occurred more strongly in specimen 1G856 than in specimen 7G856. No regions with substantial yield were observed in specimen 7G856. In this paper, we report results which are obtained from specimen 1G856.

SEM observations and EDS analyses

Figure 1(a) is a micrograph taken from the center region of the as-mechanically-polished specimen, showing regions of no reactions and those of local reactions. The latter is indicated by arrows. The elements and phases could be qualitatively identified by differences in emission of electrons from unlike elements if backscattered electron mode or similar mode obtainable in the operation at low accelerating voltages is used[6]. In this study the latter (~5kV) was used. Taking the phase diagram of the Ni-Al system into account, it is considered that bright regions are Ni or a gamma solid solution of Ni with low Al content; that dark regions are Al; and that gray regions are a mixture of compounds or a mixture of compounds and a gamma solid solution with high Al content. These gray regions indicate

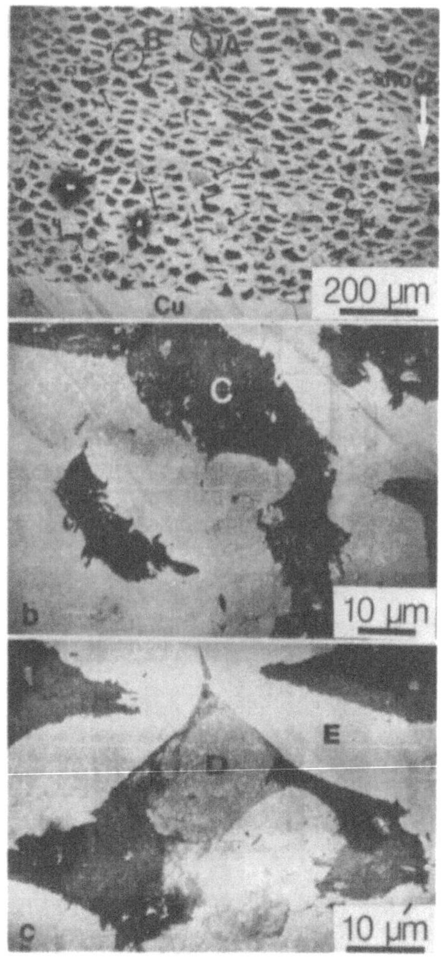

Figure 1. Scanning electron micrographs taken from a center region of disk specimen 1G856, showing (a) local reactions; (b) no-reaction but ragged interface; (c) some local reactions.

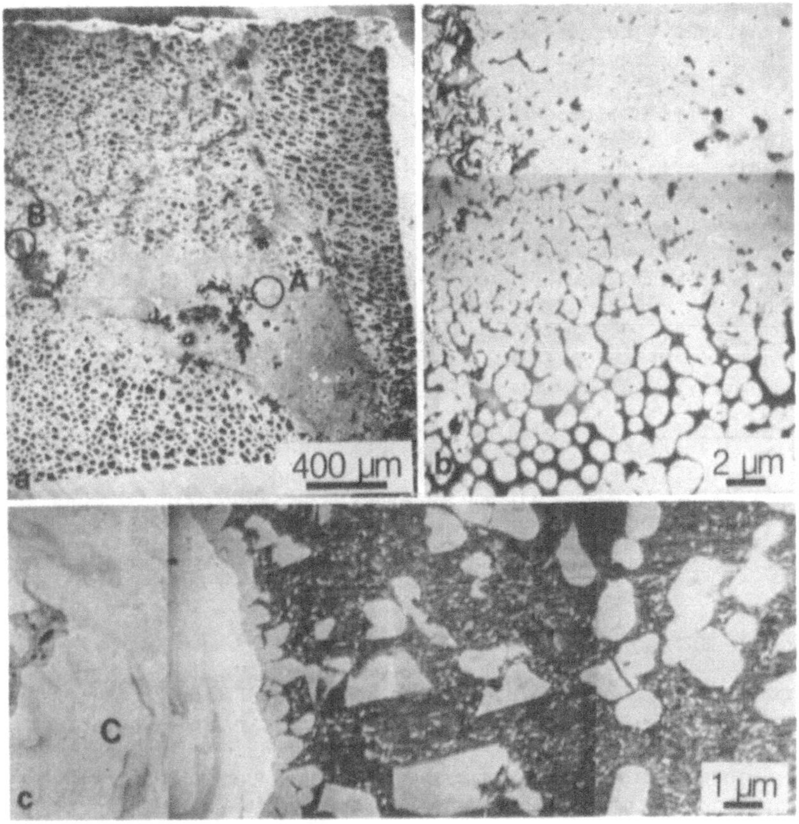

Figure 2. Scanning electron micrographs taken from an outer region of specimen 1G856, showing (a) substantial reactions, (b) a detailed feature of the reacted zone, and (c) a variety of phases.

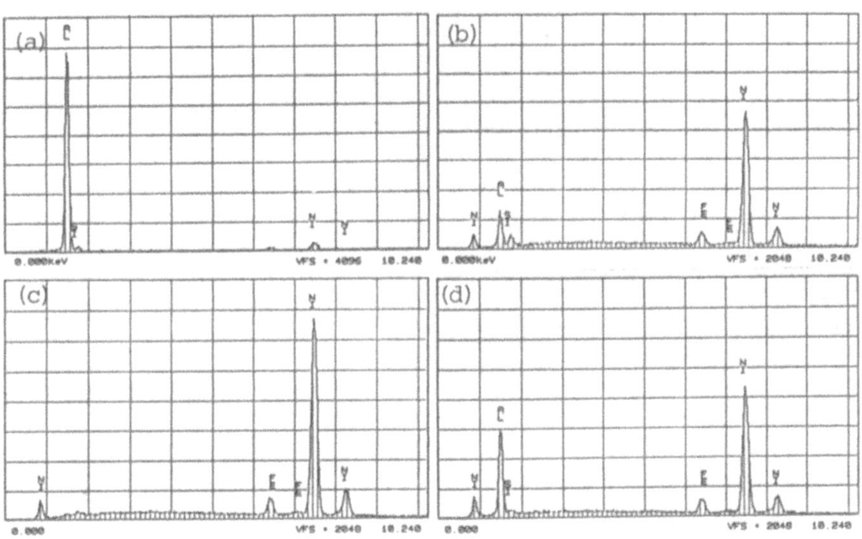

Figure 3. EDS spectrums. (a) from region C in Figure 1(b); (b) from region D in Figure 1(c); (c) from region E in Figure 1(c); (d) from region C of a reacted area in Figure 2(c).

reactions of nickel with aluminum. Figures 1(b) and (c) are high magnification micrographs taken from Regions A and B. Similar observations were obtained in the middle region of the same specimen.

Figure 2(a) is a micrograph taken from the outer region of the specimen, showing a zone of substantial reactions. A high magnification micrograph of Region A in the reacted zone (Figure 2(b) reveals a typical structure of powders sintered by static means, suggesting that the phases seen in Region A are the equilibrium phases, Ni_3Al ($L1_2$-ordered fcc) and NiAl(B_2 -ordered bcc). The predicted temperature of this region is about 660°C (Table 1). Since this temperature is not high enough for a substantial formation of such compounds, a strong shock enhanced reactivity is anticipated. Figure 2(c) is a microstructure of Region B in Figure 2(a), showing a variety of phases. No transmission electron microscope observation was made for this zone. Hence, it is not known if $NiAl_3$ exists in Region B.

Figures 3(a-d) are results of elemental analysis using spot analysis made with the EDS system. Figure 3(a) corresponds to Region C (dark) of Figure 1(b); Figure 3(b) to Region D (gray) of Figure 1(c); Figure 3(c) to Region E (light) of Figure 1(c); and Figure 3(d) to Region C to Figure 2(c). The results are consistent with the color inspection based on electron emission.[6]

TEM Observations

Figure 4 is a bright field electron micrograph of an unreacted zone in a central region of specimen 1G856, showing a typical microstructure of recrystallized aluminum. No other phases were found in such recrystallized regions. Figure 5 is another microstructure of a central region of specimen 1G856, showing local reactions. In Figure 5(a) a clear interfacial separation is seen between the aluminum and reacted areas. Figure 5(b) shows the reacted region in Figure 5(a), revealing a complex internal structure. Inspection of diffraction patterns taken from the reacted region showed that no phases but Ni (γ-solid solution) and Ni_3Al could consistently explain the diffraction pattern. The diffraction pattern taken from the whole area of Figure 5(a) is shown in Figure 5(e), showing that all the reflection spots can be explained only for Al, Ni (or a γ-solid solution) and Ni_3Al. Since no ordered reflections were observed, the latter two phases were not distinguishable from the diffraction pattern in Figure 5(e). However, the microstructure in Figure 5(b) suggests that there are two phases in the reacted region.

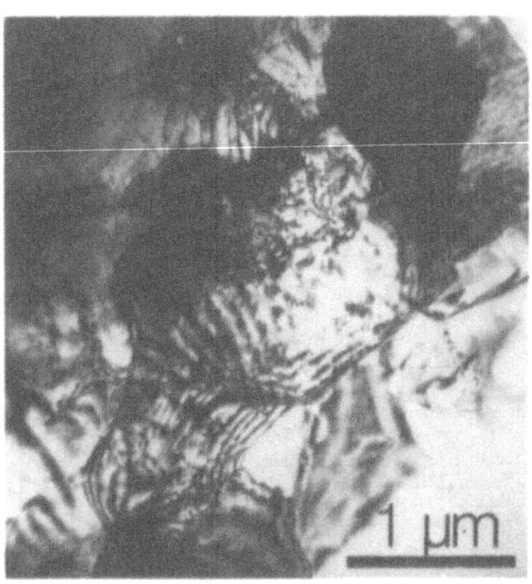

Figure 4. Electron micrograph of recrystallized Al, taken from a central region of specimen 1G856. No other phase was in the recrystallized Al.

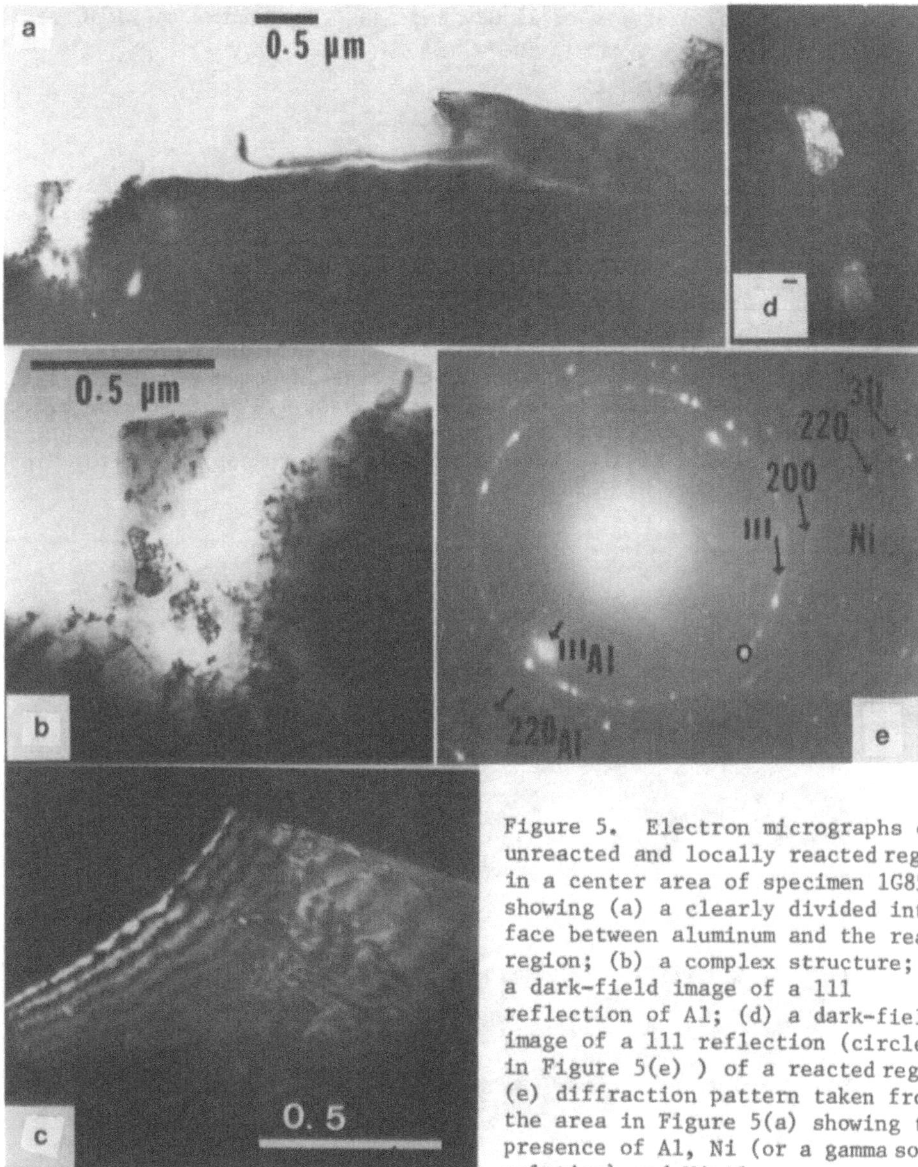

Figure 5. Electron micrographs of unreacted and locally reacted regions in a center area of specimen 1G856, showing (a) a clearly divided interface between aluminum and the reacted region; (b) a complex structure; (c) a dark-field image of a 111 reflection of Al; (d) a dark-field image of a 111 reflection (circled in Figure 5(e)) of a reacted region; (e) diffraction pattern taken from the area in Figure 5(a) showing the presence of Al, Ni (or a gamma solid solution) and Ni_3Al.

Figures 6(a) and (b) show a variety of microstructures of locally reacted regions, often seen in middle regions of specimen 1G856. The diffraction pattern (Figure 6(c)) corresponding to the area of Figure 6(b) clearly show the presence of Ni_3Al.

SUMMARY

1. A gamma solid-solution of nickel (or Ni) and Ni_3Al were found in the center and middle regions of specimen 1G856. The axial region was subjected to the loading conditions (mean peak pressure of 48 GPa and peak temperature of 300°C).
2. No evidence was obtained for the formation of compounds in recrystallized aluminum.
3. Shock-synthesized Ni_3Al had a complex microstructure.

4. Localized reactions were observed in specimen 7G856 subjected to the conditions (peak pressure of 11 GPa and peak temperature of 210°C).

REFERENCES

1. Dodson, B.W. and Graham, R.A., in Shock Waves in Condensed Matter-1981 (Menlo Park) edited by Nellis, W.J., Seamen, L. and Graham, R.A., American Institute of Physics, NY (1982).
2. Morosin, B. and Graham, R.A., in Nellis et al editors, loc. cit., p. 4.
3. Horie, Y., Graham, R.A. and Simonsen, I.K., Materials Letters, in press.
4. Hoy, D.E.P., Horie, Y. and Graham, R.A., in this proceedings.
5. Graham, R.A. and Webb, D.M., in Shock Waves in Condensed Matter-1983 (Santa Fe, NM), edited by Asay, J.R., Graham, R.A. and Straub, G.K., North-Holland, NY (1984), p. 211.
6. Goldstein, J.I., et al., Scanning Electron Microscopy and X-ray Microanalysis, Plenum, NY (1981), p. 162.

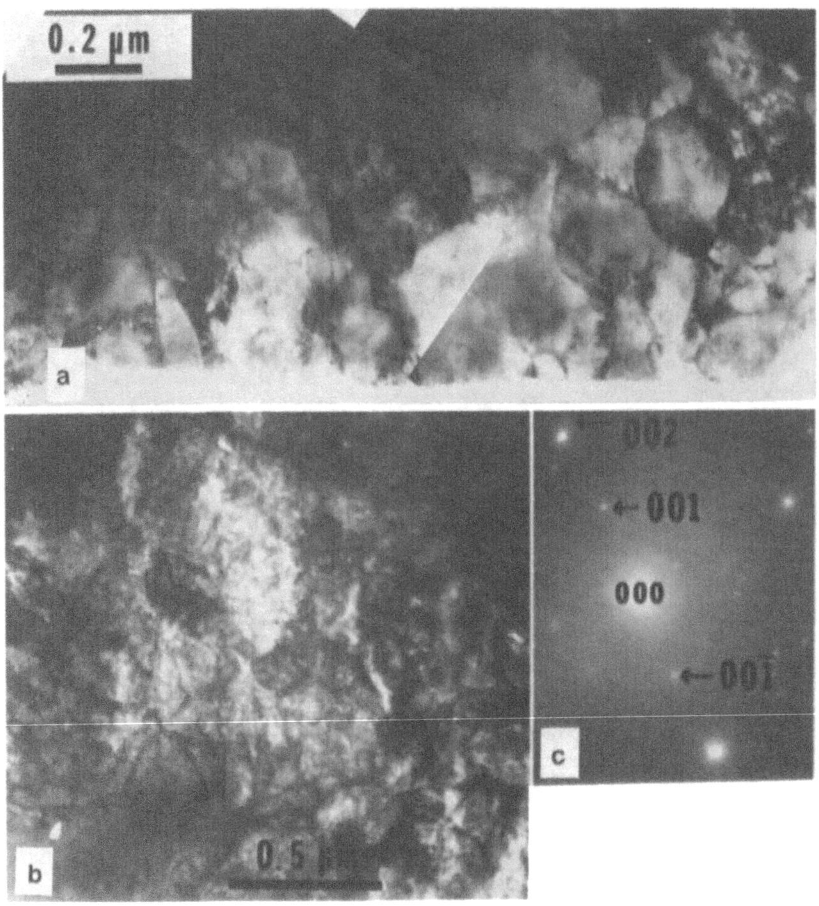

Figure 6. Electron micrographs of a thin foil taken from a middle region of specimen 1G856. (a) Growth of precipitates in a solid-solution of Ni; (b) a dislocated region consisting of a solid-solution of Ni and Ni_3Al; (c) a diffraction pattern taken from the area of Figure 6(b) showing 001 reflection of Ni_3Al.

OPTICAL MEASUREMENTS OF SHOCK-INDUCED CHEMICAL REACTIONS IN MIXED ALUMINUM-NICKEL POWDER*

M. B. Boslough, R. A. Graham and D. M. Webb
Sandia National Laboratories
Albuquerque, NM 87185

ABSTRACT

Optical pyrometer and high-speed photography experiments have been carried out to determine the temperature and temporal-spatial evolution of a chemical reaction zone in shock-loaded, nickel-aluminum powder mixtures. The temperature generated is about 5000 K, and the reaction occurs on a 100 nanosecond or faster time scale. Experimental results are compared to numerical calculations, which suggest that there are threshold shock pressures and temperatures of about 15 GPa and 500°C which are required to initiate the strong exothermic reaction. Further, excessive pressure appears to inhibit reaction. This observation is in agreement with previous experiments in which nickel-aluminum alloys were recovered from shock-loaded mixtures.

INTRODUCTION

Solid state reaction under shock loading has been recognized as important to an understanding of the shock deformation process and to synthesis of new materials. Previous experiments have shown that when aluminum-nickel powder mixtures are shock-loaded to sufficiently high pressure and preserved for post-shock analysis the recovered materials are in the form of chemical reaction products [1-3]. The well-reacted product shows both similarities and differences to those derived from other rapidly quenched processes; however, neither the time scale nor the energetics of the reaction is obtainable from post-shock study. In a previous publication [4] we reported optical observations of intense light emission characteristic of a strong exothermic reaction within a few hundred nanoseconds after critical shock conditions were achieved. In the present article we report on further analysis of the optical data and a more detailed comparison of the onset of reaction with pressure and temperature conditions in the powder.

*This work performed at Sandia National Laboratories supported by the U.S. Dept. of Energy under contract # DE-AC04-76DP00789.

EXPERIMENTAL

The experiment made use of the precisely controlled impact loading of an aluminum-nickel mixture (Figure 1). A projectile carrying a 50 mm diameter copper plate at 1.0 km/sec impacted a copper target fixture which contained a mechanical mixture of 30 vol. pct. aluminum and 70 vol. pct. nickel [1], pressed to 60% of solid density (4.25 Mg/m^3). A sapphire anvil allowed light to escape from the powder-sapphire interface, and an edge mask ensured that spurious light was not observed. Light emitted from the sample was reflected to one of two optical instruments: 1) an optical pyrometer which measured spectral radiance at four wavelengths in the visible and near infrared, 2) an image converter streak camera which generated a time-resolved image of a narrow strip of the sample. The pyrometer measured absolute light intensity (spectral radiance) radiated from the entire 25.4mm aperture of the target fixture. The four wavelengths were 0.5, 0.9, 1.4, and 1.8 microns, with bandwidths of about 0.2 microns. The Imacon 500 streak camera was used with an image intensifier system. The aperture of the target was imaged onto a slit, which collected light from a strip about 1 mm wide and 25.4 mm long. This image was swept across the film at a rate of about 1 μs/cm, to produce a streak containing both spatial and temporal information. These instruments are described in more detail by Boslough and Graham [4].

RESULTS

Time-resolved spectral radiance data were fit with a greybody spectral distribution (Planck function with a wavelength independent emissivity). A two parameter least squares fit was carried out to determine time-resolved temperatures and effective emissivities (Figure 2). Because, as determined by streak photography, the sample does not radiate uniformly, the brightest regions of the sample dominate the spectral radiance measurements. To first order, the brightness distribution can be approximated as bimodal; i.e., a uniformly bright area and a uniformly dark (non-radiating) background. The effective emissivity is therefore the true emissivity of the radiating material multiplied by its fractional area. The temperature of the material reaches about 5000 K, compared to the 1200 K peak temperature calculated for the shocked but unreacted material. The large amount of energy required to reach the observed temperature must come from an exothermic reaction. The reaction occurs within a few hundred nanoseconds after appropriate critical values of shock pressure and temperature are achieved.

Further evidence for the reaction comes from high speed streak photography. The streak photograph in Figure 3 shows a radially converging reaction zone, with a schematic representation of the radiating surface. It is clear from the streak that the large increase in effective emissivity (see Figure 2) during the first microsecond of light emission is mainly due to the increasing amount of radiating area. It is interesting to note that, despite an order of magnitude increase in effective emissivity, the temperature remains nearly constant during this time interval. Another major characteristic of the streak record is the discontinuous onset of light on the axis of the sample. This appears to occur before the radial waves converge at that location.

The credibility of the chemical reaction interpretation is further enhanced by comparing the streak results to numerical calculations. Because the powder layer is relatively thick and there is a large shock velocity mismatch between it and the surrounding materials, two dimensional effects dominate the shock loading history. To better understand

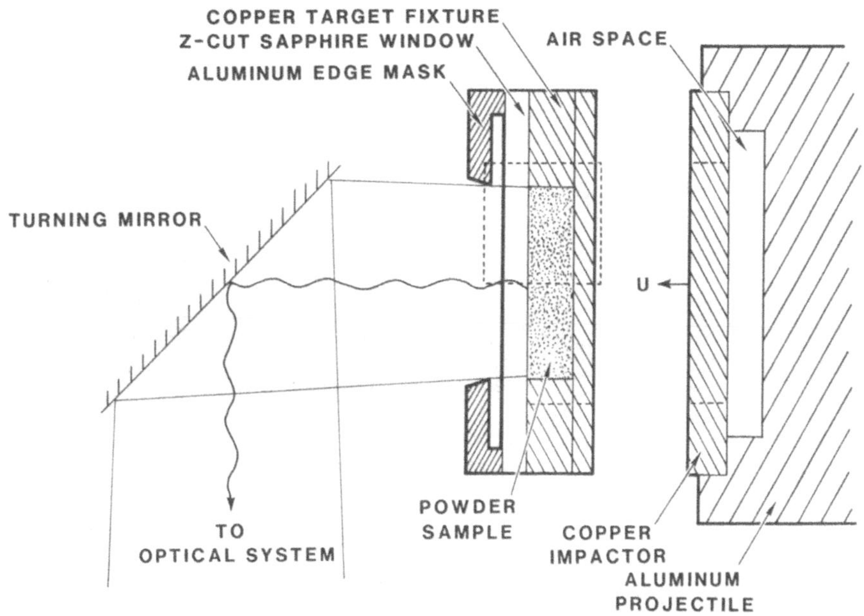

Figure 1 Experimental configuration. Light radiated from powder-sapphire interface is reflected to one of two optical systems.

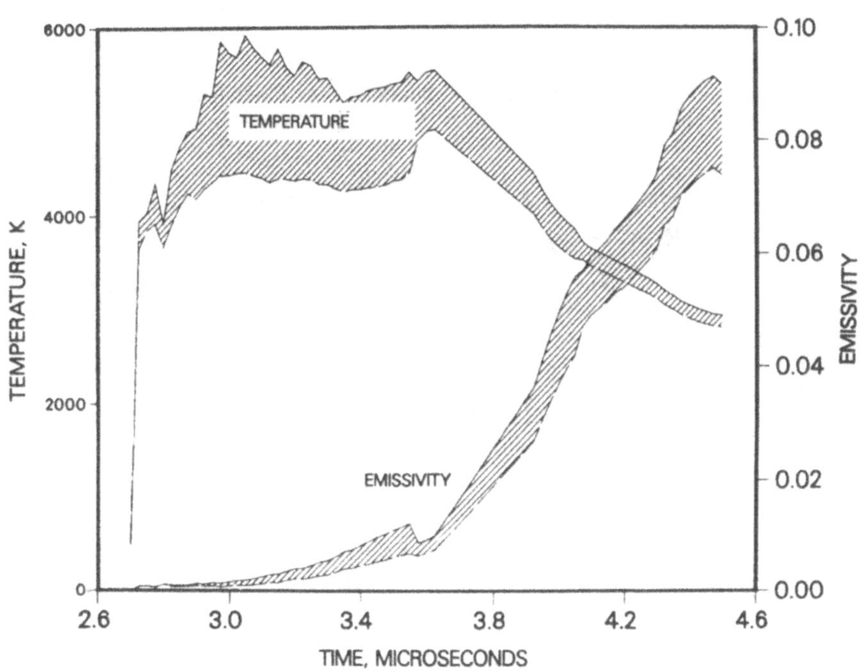

Figure 2 Time-resolved greybody temperature and effective emissivity of aluminum-nickel reaction zone. The ranges given are indicative of the experimental uncertainty.

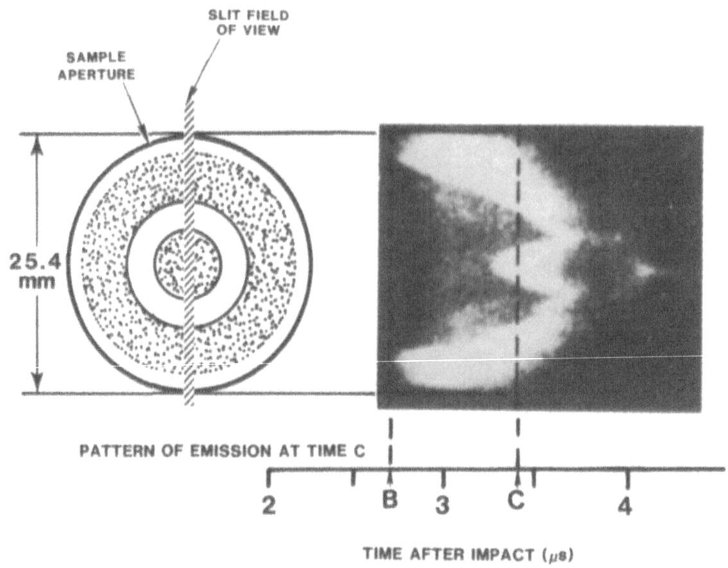

Figure 3 Streak record shows a converging reaction zone. The relative aperture size is shown at left, with a superimposed slit field of view (hatched). B is the time at which light emission begins. The approximate pattern of emission at time C is shown at left.

these effects, it was necessary to carry out two dimensional numerical calculations. Several results of such calculations are given by Graham and Webb [5] and by Boslough and Graham [4]. The results of the calculation are plotted as pressure contours as a function of time and radial distance, for the powder-sapphire (observed) interface, in Figure 4. Superposed on these contours is an outline of the bright area from the streak in Figure 3. It appears from this figure that there is a correspondence between the arrival of a given pressure state and the onset of reaction. Lack of an absolute time scale on the streak experiment prevents our ability to correctly position the light emission zone along the time axis with certainty, but the relative scales are correct. As plotted, the curve representing the onset of light emission lies within 5 GPa of the 15 GPa contour for nearly its entire length. In the temperature plane, the onset in light emission occurs after the material has reached at least approximately 500°C everywhere except at the extreme edge of the sample. The fact that the onset of reaction is delayed by several hundred nanoseconds after critical values are achieved for radii greater than 3 mm suggests that the reaction is inhibited by the high pressure, and only occurs upon release.

In the inner 3 mm of the sample the reaction begins sooner than expected for the arrival of a radial loading wave, apparently because the pressure begins to drop as soon as critical values are reached. This is illustrated qualitatively by the similarity in trend in the curve of light emission onset and the curve of maximum pressure in the time-radial distance plane (Figure 4). The curve which defines the trailing edge of the streak, where the light extinguishes, is subparallel to calculated pressure release contours, which is consistent with adiabatic cooling, as would be expected. Until accurate absolute timing for the experimental and calculated results is achieved, precise pressure and temperature thresholds cannot be obtained from these plots because of the steep

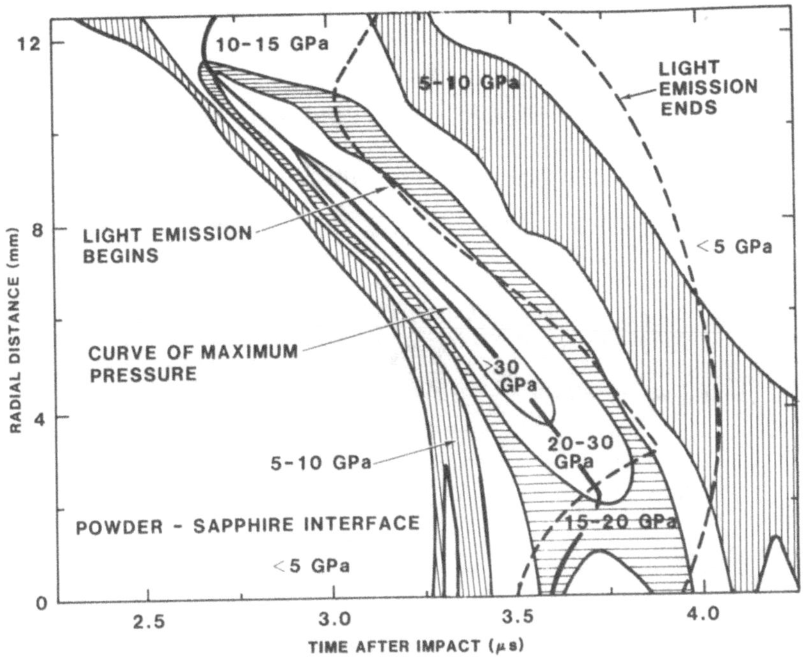

Figure 4 Calculated pressure contours at the powder-sapphire interface. Dashed curve outlines the zone of light emission obtained by the streak camera, which observes this time-radial distance plane directly. Heavy curve follows a ridge of maximum pressure.

gradients in the calculation; however, it is clear from the present work that such thresholds exist and are measurable.

The numerical simulations used in this study are based on a "rutile-like" description of the powder compact response which has been found to provide a good overall description for sample recovery fixtures [5]. Nevertheless, in the present case where events are followed closely in time the powder response may not be as accurate as desired. To supplement the calculations additional simulations have been carried out with a "snow-plow" type model for distended aluminum and for distended nickel. Although the calculated peak pressures and temperatures are significantly higher in the latter case than for the rutile description, the timing of critical observed events is not changed and the overall conclusions are not altered by this radically different sample description.

CONCLUSIONS

Optical pyrometric measurements of a shock-compressed aluminum-nickel powder mixture show that a chemical reaction occurs within a few hundred nanoseconds after appropriate shock conditions of temperature and pressure (about 15 GPa and about 500°C) are achieved. Optical streak records show that the intense light originates at the hot outer edge of the specimen, a location where chemical reaction products have been observed in sample recovery experiments. The speed at which this reaction moves radially inward is about the same at that of loading and release from high pressure and indicates that high pressure inhibits the reaction. The isolated initiation of light emission near the axis of the sample provides supporting evidence for this interpretation.

We wish to thank Y. Horie, W. F. Hammetter and R. E. Setchell for helpful discussions and M.U. Anderson, K. Elsner and D. E. Wackerbarth for technical assistance.

REFERENCES

1. Y. Horie, R. A. Graham and I. K. Simonsen, in press, Materials Letters.
2. I. K. Simonsen, Y. Horie and R. A. Graham, this proceedings.
3. Y. Horie, D. E. P. Hoy, I. Simonsen, R. A. Graham and B. Morosin, this proceedings.
4. M.B. Boslough and R.A. Graham, submitted, Chem. Phys. Lett.
5. R. A. Graham and D. M. Webb, in: "Shock Waves in Condensed Matter - 1983", J. R. Asay, R. A. Graham and G. K. Straub, Eds., North Holland, New York (1984) p. 211.

SHOCK WAVES PROFILES IN POWDER COMPACTS OF α Al_2O_3

Takashi Taniguchi, Ken-ichi Kondo and Akira Sawaoka

Research Laboratory of Engineering Materials, Tokyo

Institute of Technology, Midori, Yokohama, 227 Japan

1. Introduction

In general, the shock state of porous material is estimated under the thermodynamical equilibrium condition, as based on the Mie-Grüneisen equation of state.[1] However, since the mechanical and/or thermal equilibrium states are not actually achieved in the porous material under shock loading, it is necessary to know the precise constitutive relation for a porous system in order to determine the shock state. Although a dynamic compression process for porous ductile metal has been actively studied,[2,3] there are few studies on the shock compression path for brittle material of ceramic powder compact, which is accompanied with fracture and/or fragmentation of the particles. In order to characterize the actual shock-compression behavior and the heterogeneous shock state for ceramic powders, it is important to measure directly the shock wave profiles at multiple Lagrangian positions of the material.

In this paper we report the experimental results of shock wave profile measurements on two kinds of unsintered α-Al_2O_3 powder compacts, with grain sizes of 2 μm and 100-350 μm, using multiple particle velocity gauges and a pair of stress gauges up to a stress level of 45GPa.

2. Experimental

The particle velocity measurement is based on the principle of a moving conductor in a magnetic field. Details of the experiment are given elsewhere.[4] A sample assembly for the experiment is shown in Fig. 1. The samples used are two disk-shaped and unsintered compacts of α-Al_2O_3 powder with different grain size (2 μm and 100-350 μm for samples (A) and (B), respectively). The porosity is 42 ± 4% for a coarse compact and 45 ± 2% for a fine compact. The coarse particles were enclosed by a polycarbonate guardring as shown in Fig. 1. The coarse powder of α-Al_2O_3 was prepared by crushing and milling of sapphire crystals and is, therefore, highly irregular in shape.

The sample assembly was impacted by a sapphire flyer plate accelerated by a two-stage-light gas gun. The flyer plate and a driver plate were of sapphire crystal whose compression curve was

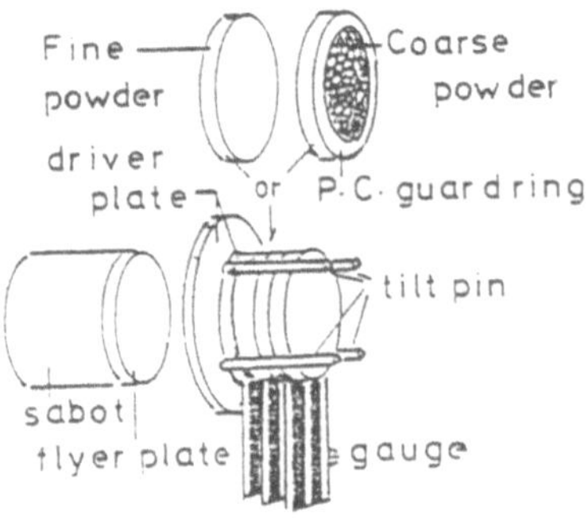

Fig 1. Sample Assembly.

established.[5] Thus we can determine the time-resolved Hugoniot state of α-Al_2O_3 compact at the boundary between the sample and the driver plate by using the impedance match method described elsewhere.[6] The stress histories were also measured by using manganin stress gauges. The experimental arrangement is similar to the particle velocity measurement as mentioned above and is also given elsewhere in detail.[7]

3. Results and Discussion

Typical profiles of particle velocity (Up) versus time are shown in Fig. 2. The profiles (A) and (B) are for samples of fine and coarse particle α-Al_2O_3, respectively. As shown in Fig. 2, the particle velocity profiles in both samples, though differing in particle size, resemble each other at the shock front and do not show an elastic precursor wave. This lack of an elastic precursor wave is probably attributed to poor initial bonding in the case of unsintered compacts.

At both profiles, (A) and (B), significant attenuation of the shock wave front is observed: a decrease in the maximum value of the particle velocity at the shock front, and the inclining of the shock front. According to an analysis of these experiments in the x-t diagram, this attenuation is not a hydrodynamic attenuation due to rarefaction wave. Furthermore, the particle velocity behind the shock front in both the profiles (A) and (B) are decreasing with time, as shown in Fig. 2. This phenomenon is interpreted as a relaxation of the shock state. Such typical attenuation and relaxation of the particle velocity profiles in the porous material are generally related to the mechanism of pore-collapsing, which requires additional energy and results in energy absorption.

In order to determine the shock state for the porous sample using the measured attenuating particle velocity profiles, the concept of the path-line system was employed in this paper. The concept of this system is introduced by Grady[8] and is described in detail by Seaman.[8]

The calculated shock compression path for the porous α-Al_2O_3 in shots #321 and #317 are shown in Fig. 3. The Hugoniot P-V relation for the sapphire crystal is also shown. Moreover, the calculated thermodynamical equilibrium state of a shock-compressed porous sample with an initial porosity of 45% is also shown in Fig. 3, calculated with the Mie-Grüneisen equation of state. Since the observed attenuation of shock wave results from the spatial- and time-dependent heterogeneities of the shock state, the compression path of the sample depends on the position of the Up gauges. The compression paths, denoted by 1-2, 1-3, and 2-3 in Fig. 3, represent the average shock

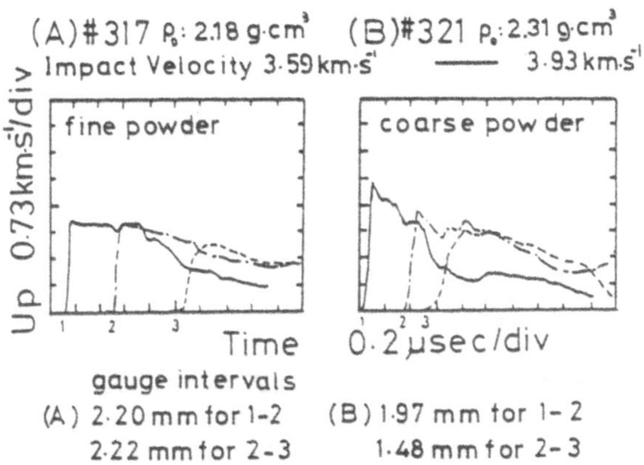

Fig. 2. Upgauge profiles.

condition in the sample space between the first and second UP gauges, the first and the third Up gauges, and the second and third Up gauges, respectively. It is noticeable that the compression paths go down to the denser side through the first to the third shock-compression state (i.e., 1-2, → 2-3) and are convex in shape. The calculated specific internal energy for the corresponding compression path is also shown in Fig. 3.

It can be seen that the porous material is well densified with the increase in propagation distance of the shock wave, as shown in Fig. 3. This is consistent with the fact that a deceleration rate of the shock wave velocity is much larger than that of the particle velocity in the attenuating shock wave. Figure 4 shows the deceleration rate of the shock wave velocity and particle velocity as a function of impact velocity. In this figure, the shock wave velocity (Us) is determined by the time difference between the shock front of the maximum particle velocity at each pair of particle velocity gauges. In the Rankine-Hugoniot jump condition, the volume-compression of the material under shock state is given simply by Up/Us. Although the spatial and time-derivatives of particle velocity must be taken into account in the attenuation of shock wave, the calculated compression paths are

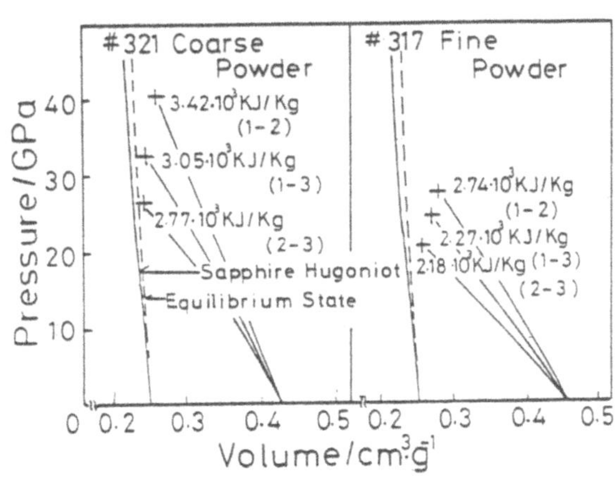

Fig. 3. P-V.

essentially characterized by the large deceleration rate of the shock wave velocity, which results from the inclining of the shock wave front as shown in Fig. 2.

Therefore, it seems that the inclined shock wave front forms an effective structure to shrink the powder compacts with the increase of the propagation distance of shock wave. As shown in Fig. 3, the specific internal energy is decreasing through the first compression path to the third one. Such difference of the specific internal energy among the compression paths is probably due to formation of the shock wave front with a long rise time structure. A possible explanation for the formation of such a shock wave front structure is as follows: since the particle bonding is weak in the unsintered sample compacts, the flow direction of the constituent particles accelerated by the shock wave should spread out to collapse the voids, accompanied by rearrangement and/or fragmentation of the particles. If the structure of the shock wave front depends on the flow direction of the constituent particles, the formation of the long-rise-time shock wave front is probably due to microscopic spreading of flow direction of the constituent particles as three dimensional flow. Consequently, the mechanical equilibrium state for shock-compressed powder compact is easily achieved by such a shock wave front structure, which includes three dimensional stress components. This is consistent with the fact that the third downward compression path (2-3) seems to approach the thermodynamical equilibrium state, as shown in Fig. 3.

On the other hand, another Hugoniot P-V-E relation is also obtained by using the impedance match method. The shock state determined from the impedance match method is compared with that determined directly from the multiple gauges, as shown in Fig. 5. It is clearly seen that the data calculated from the sapphire pressure for a corresponding particle velocity represents higher pressure and lower density than those for the multiple Up gauge analysis. At the driver-sample interface, the porous sample seems to behave as a rigid material instantaneously.

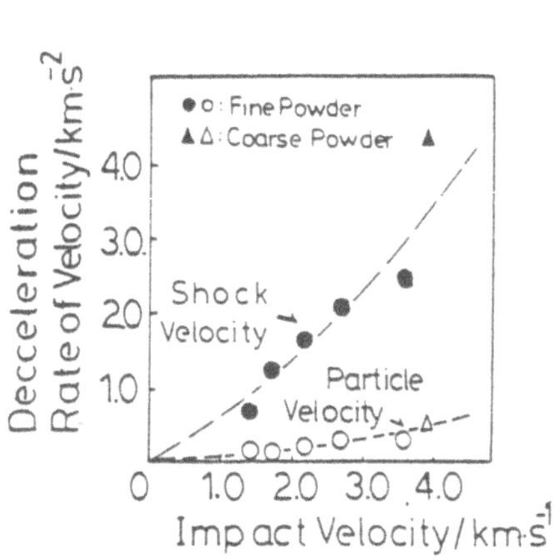

Fig. 4. Decceleration Rate of US and UP vs Impact Velocity

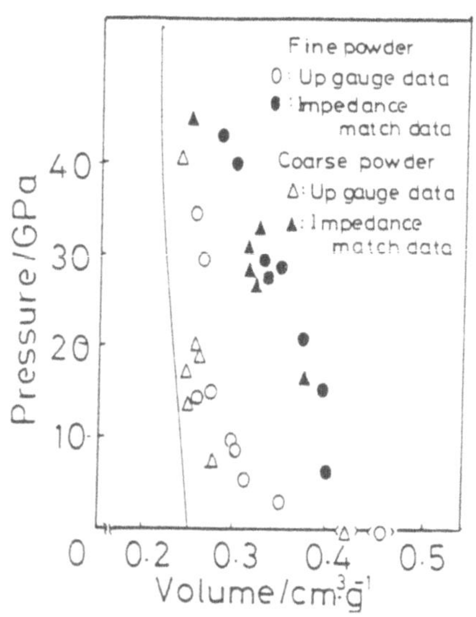

Fig. 5. P-V.

Typical stress wave profiles for the α-Al_2O_3 compact of fine particle are shown in Fig. 6. The profiles A-1 and A-2 correspond to the first and the second manganin stress gauges, respectively. It is noticeable that an initial current spike is seen in the first stress gauge, which was embedded between the sapphire driver plate and the porous sample (indicated by arrow (A) in Fig. 6). We recognize that the observed instantaneous response of the porous sample determined from the particle velocity profile should correspond to the observed initial spike in the stress profile. The shock state of the porous sample determined from the stress wave profiles is compared with that determined from the particle velocity profiles in Fig. 7. As shown in Fig. 7., the porous sample instantaneously behaves as a rigid material at the driver-sample interface, and successively the shock state of porous sample passes into an denser structure so as to crush and condense the porous structure. The possible explanation for the instantaneous response at the driver-sample interface is that, since the granules of α-Al_2O_3 are directly touched by the driver plate, the effect of the porous structure is not immediately realized at the interface.

On the other hand, according to particle velocity histories behind the shock front, as shown in Fig. 2, the difference between particle velocities among the gauge positions becomes smaller with time and they seem to approach the same velocity. Therefore, we conclude that the mechanical process toward equilibrium would follow the initial heterogeneous shock-compressed state so as to reach a more consolidated state, because the relaxation of the particle velocity could be caused by the change from uniaxial flow to three-dimensional flow. As compared in Fig. 2 and Fig. 6, the change of pressure in the shock-compressed sample is accompanied by a change of particle velocity.

A significant difference in shock-compression pathway between the samples of fine particle and coarse particle was not observed because of the limited analyses. However, the mechanical and thermal relaxation process and stress release paths should differ between fine powder and coarse powder, because the final density of the shock

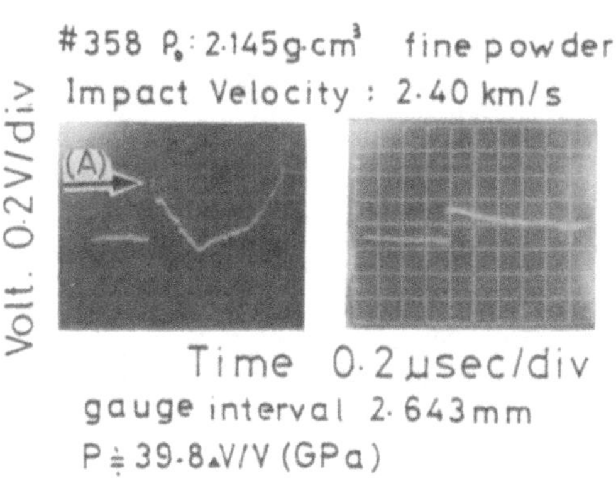

Fig. 6. Stress gauge profiles.

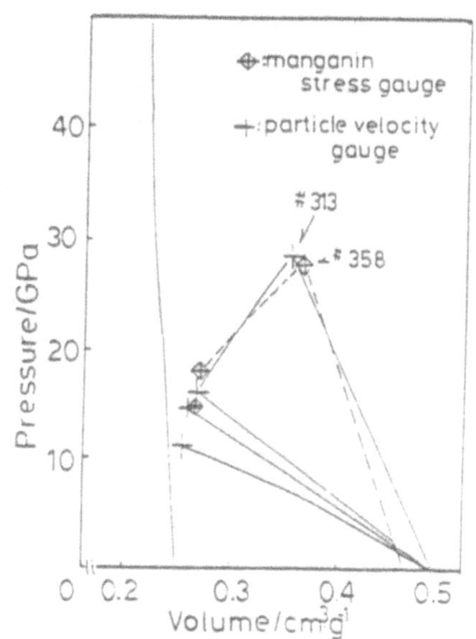

Fig. 7. P-V

recovered sample depends on the initial particle size.[10] It can be seen that duration time for the relaxation process is a rather important parameter to achieve the equilibrium shock state in porous material.

In order to discuss further these relaxation and release processes of shock-compressed porous materials, it is necessary to establish a technique for recording the clear stress profiles and particle velocity profiles behind the shock front in which several kinds of disturbances exist (i.e., extreme local high temperature at the collided particle interface, hydrodynamic attenuation due to rarefaction wave, and macroscopic yielding of material structure, etc.). Such data would enable us to derive the constitutive model for these processes.

4. Conclusions

(1) The difference of the shock wave profile between the fine powder compact of α-Al_2O_3 and that of coarse powder is not observed. Neither sample shows an elastic precursor wave. This is probably due to poor initial bonding of the particles in unsintered compacts.

(2) The compression path for the porous compact depends on the distance of shock wave passing. As the shock wave propagates in the porous sample, the shock wave front seems to change with time so as to form the suitable shock front structure to effectively shrink the powder compacts.

(3) The porous compact instantaneously behaves as a rigid material at the interface of the sapphire drive plate.

Acknowledgement

The authors thank Prof. K. Kani, Okayama University, for his useful advice in the experiments.

References

[1] R.K. Linde and D.N. Schmidt, Shock Propagation in Nonreactive Porous Solids, J. Appl. Phys. 37: 3259 (1966).

[2] W. Hermann, Constitutive Equation for the Dynamic Compaction of Ductile Porous Materials, J. Appl. Phys. 43: 2494 (1969).

[3] M.M. Carrol and A.C. Holt, Suggested Modification of the P-α Model for Porous Materials, J. Appl. Phys. 43: 1629 (1972).

[4] K. Kondo, T. Yasumoto, H. Sugiura, and A. Sawaoka, Multiple Shock Reverberations in a Layer Structure Observed by Particle-Velocity and Pressure Gauges, J. Appl. Phys. 52: 772 (1981).

[5] S.P. Marsh, LASL Shock Hugoniot Data (Univ. Calif. Press, Berkeley, 1980).

[6] A. Sawaoka and K. Kondo, Proceedings of the 21st High Pressure Symposium, Tokyo, 25 (1980).

[7] Y. Yasumoto, K. Kondo, and A. Sawaoka, A Convenient Technique for Shock-Pressure Measurement Using a Manganin Gauge, Jpn. J. Appl. Phys. 19: 1805 (1980).

[8] D.E. Grady, Experimental Analysis of Spherical Wave Propagation, J. Geophys. Res. 78: 1299 (1973).

[9] L. Seaman, Lagrangian Analysis for Multiple Stress or Velocity Gauges in Attenuating Waves, J. Appl. Phys. 45: 4303 (1974).

[10] R.A. Prummer and G. Ziegler, Structure and Annealing Behavior of Explosively Compacted Alumina Powders, Powder Meta. Int. 9: 11 (1977).

DYNAMIC COMPACTION OF SiC POWDER

Tamotsu Akashi, Victor Lotrich, and Akira Sawaoka

Center for Explosives Technology Research
New Mexico Institute of Mining and Technology
Socorro, New Mexico

and

Edwin K. Beauchamp

Sandia National Laboratories
Albuquerque, New Mexico

INTRODUCTION

Dynamic compaction techniques have considerable potential in the consolidation and fabrication of hard-to-densify powder materials such as non-oxide ceramic materials. Consolidation of the powder materials by this technique involves the densification of the powder at the shock wave front followed by interparticle bonding which occurs during the shock-compression process. In the dynamic compaction of a porous material, most of the energy of shock compression has been assumed to be deposited at the surfaces of the particles (1, 2) due to the heterogeneous deformation of material during shock loading (3). This preferential energy deposition, in some cases, allows the particle surfaces to melt and results in interparticle bonding.

Recently, a series of shock-compaction experiments on fine SiC powder have been performed by Kondo et al. (4). Sintered compacts with densities of 97% of theoretical and micro-Vicker's hardness values of 2700 kg/mm^2 were produced. They suggested that the disruptive effects, indicated by microstrain increase and crystallite size reduction, must remain in the compacted sample in order to obtain a well-bonded compact.

The purpose of this work is to study the compaction mechanism of coarse SiC powder in the dynamic compaction process, which may be different from that of the fine powder.

EXPERIMENTAL PROCEDURES

SiC powder provided by Shinano Denki Seiren Co. Ltd. was used as the starting material. The powder was a mixture of

α-phase and a few percent of β-phase. The powder consisted of angular particles with an average particle size of 10 μm. The powder was pressed into stainless steel capsules at 350 MPa to form discs 5 mm thick and 12 mm in diameter with densities of 60% of theoretical. Dynamic compaction experiments were performed using a mouse trap type plane wave generator and momentum trap recovery system. An iron flyer plate, 4.3 mm thick, was impinged upon the capsules at velocities of 1.1, 1.6, and 2.1 km/sec. The shock pressures induced in the powder compacts were estimated to be 4, 8, and 13 GPa, respectively, using a one-dimensional impedance matching method.

The relative densities of the compacts were measured by a water displacement method. The micro-Vicker's hardness values were measured using a 500 gr load with a loading time of 20 seconds. The values of microstrain and crystallite size for compacted samples were obtained based on X-ray line-broadening analysis. The microstructures of the compacts were examined using scanning electron microscopy.

RESULTS AND DISCUSSION

Samples were recovered as whole discs containing a few cracks after shock treatments. The relative densities and micro-Vicker's hardness values of the compacts produced at 4, 8, and 13 GPa as a function of shock pressure are shown in Figure 1. The relative density of the compacts increased from 96.2% to 98.6% of theoretical with shock pressure increase from 4 to 8 GPa, and reached a constant, 98.6%, above the shock pressures of 8 GPa. On the other hand, microhardness continued to increase with increasing shock pressure throughout the pressure range employed in this

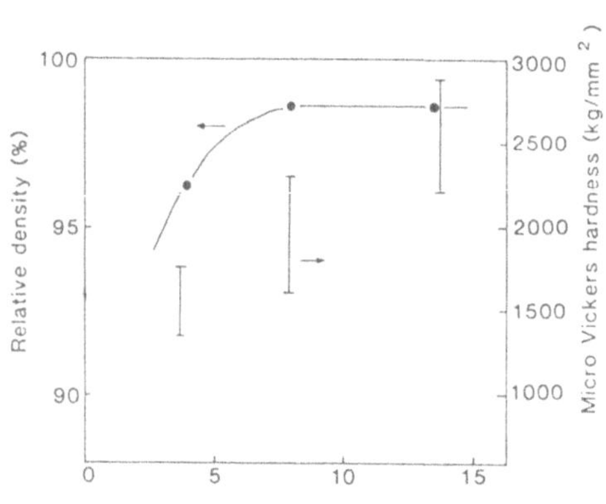

Fig.1 Relative density and micro-vicker's hardness of the SiC compacts.

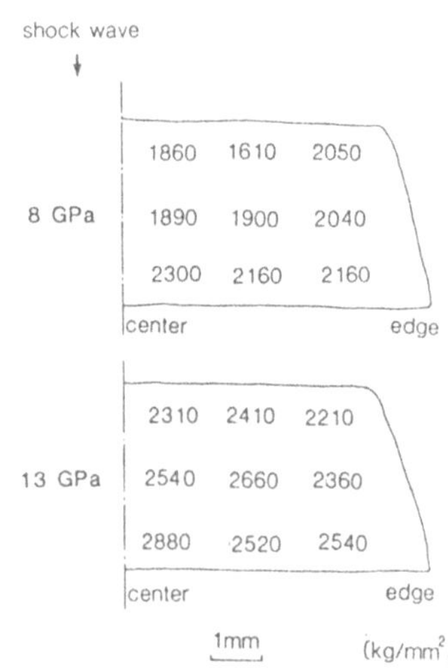

Fig.2 Distribution of micro-vicker's hardness values within the cross section of the compacts.

experiment. The results show that a shock pressure of 8 GPa was high enough to densify coarse SiC powder to near solid density but the shock temperature reached at 8 GPa was not high enough to produce strong interparticle bonding. The distribution of the microhardness within the cross section of the compacts obtained at 8 and 13 GPa is illustrated in Figure 2. Each point shown in this figure represents the average of three measurements. Microhardness values ranged from 1610 to 2300 kg/mm^2 and 2210 to 2880 kg/mm^2 for the compacts obtained at 8 and 13 GPa, respectively. The fairly wide distribution in microhardness values is not random as seem in Figure 2. In both compacts, microhardness values in the bottom regions of the samples, where the shock wave left the powder compacts, appeared considerably higher than those in the upper regions. This is apparently due to differences in the pressure and temperature history within the powder compacts during the dynamic-compaction process; pressure and temperature in the bottom regions might be higher than in the upper regions. In dynamic compaction of fine SiC powder with an average particle size of 0.28 μm, sintered compacts with 97% theoretical density and average micro-Vicker's hardness of 2700 kg/mm^2 have been produced at 22 GPa using a similar experimental arrangement as used in these experiments(4). They obtained microhardness values of 2700 kg/mm^2, which approximates the average micro-hardness value in the bottom region of the compact obtained at 13 GPa in the present work.

The microstructure of a fracture surface of the SiC green compact with a porosity of 40% is shown in Figure 3

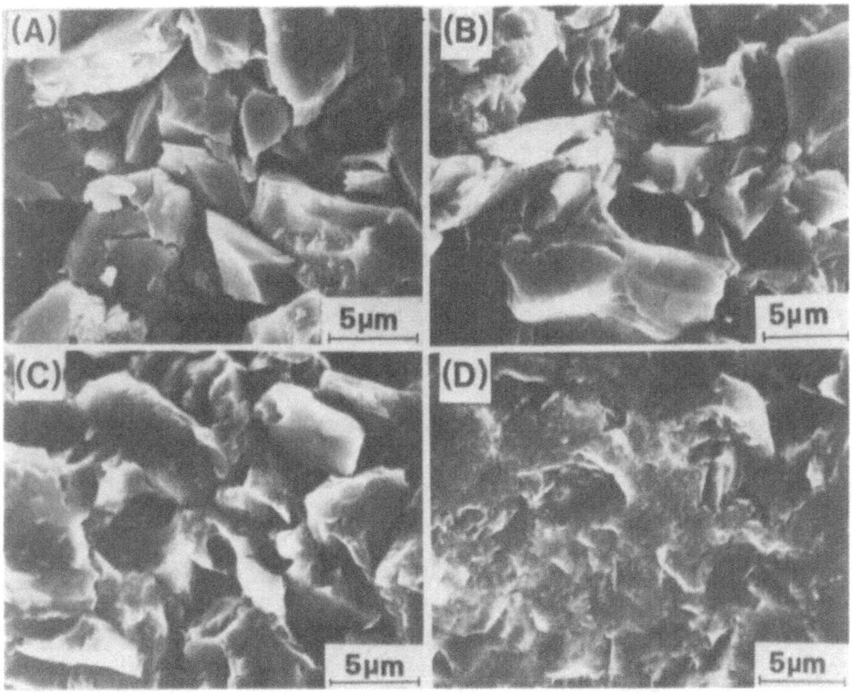

Fig. 3 SEM photographs of the fracture surfaces of the SiC compacts. (A); SiC green compact: (B), (C) and (D); dynamically compacted samples at 4, 8 and 13 GPa.

Table 1 CRYSTALLITE SIZE AND LATTICE STRAIN OF THE SiC COMPACTS

Shock pressure (GPa)	Crystallite size (nm)	Lattice Strain
4	1210	6.5×10^{-3}
8	1010	7.5×10^{-3}
13	1310	5.7×10^{-3}

(A), which reveals that under static compaction, the coarse SiC powder densified by fracture at contact points and rearrangement of the large particles. Figures 3 (B), (C), and (D) show the microstructures near the center of the bottom regions of the compacts obtained at 4, 8 and 13 GPa, respectively, corresponding to the regions where microhardness values were maximum in each compact. Recrystallization behavior is seen to differ significantly in these compacts. In the compacts produced at 4 and 8 GPa, recrystallization occurred only in the regions between large particles. The predominance of intergranular fracture in these compacts indicates that interparticle bonding was not strong. On the other hand, the fracture surface of the compact obtained at 13 GPa exhibited mostly transgranular fracture and recrystal-lization extended to the interiors of the large particles.

Table 1 shows coherent scattering dimension and lattice strain in the compacts after shock treatments at 4, 8, and 13 GPa determined using the Hall equation referring to the starting SiC powder (5). Table 1 shows that in the sample compacted at 13 GPa, lattice strain decreased and coherent scattering dimension increased as compared to the compacts obtained at 4 and 8 GPa. This result contrasts with a general tendency of increasing lattice strain and decreasing crystallite size, with increasing shock pressure (6, 7). This may be due to the higher residual temperature produced at 13 GPa than at 4 and 8 GPa causing the compact at 13 GPa to be more annealed. This behavior is consistent with the recrystallization in these compacts. However, this result differed from that of dynamically compacted fine SiC powder (4), where annealing effects were not recognized in the well bonded compacts. Recrystallization and grain growth were seen to have occurred in the compacts which had low microhardness. This seems to be attributed to the rapid

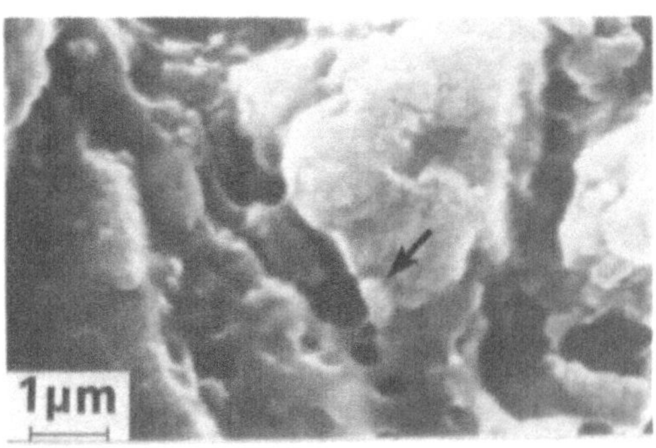

Fig.4 Microstructure of the crack region in the compact obtained at 13 GPa.

grain growth of the fine particles at high temperature. Sufficiently high residual temperatures are considered to act effectively in annealing the compacts, resulting in the elimination of cracks in the resulting compacts.

Slight particle size reduction and large lattice strain in the compacted samples suggest that coarse SiC powder densified mainly by plastic deformation accompanied by pulvarization and scraping of the corners of the angular particles during the shock-compression process. Grady has suggested that deformation of a brittle material under shock compression is a heterogeneous process due to localized yielding in shear zones, and localized deformation can result in a non-uniform distribution of thermal energy after passage of the shock wave (3). Thus, it is considered from the results that a large part of the heat in the SiC powder compact is generated as a result of heterogeneous deformation of the particles, especially the contact surfaces of the particles. Figure 4 shows the microstructure of a crack region in the compact obtained at 13 GPa. Particle morphology, in particular, rounding of the edges, suggests local flow and implies that the crack opened while the compact was still at high temperature. Additional evidence of a fluid layer at the grain boundaries was found in stringers which bridged microcracks between grains. With 1.0 wt. pct Oxygen in the starting material, it was assumed that this was SiO_2 glass. However, emmersion in HF acid did not remove this material. Possibly, the molten material was $Si_x C_y O_z$ or even SiC. This is evidence that very high temperatures have been generated at particle surfaces during the shock-compression process. Although the shock duration is very short, $\simeq 1$ μsec., thin surface layers of the SiC particles in the compact produced at 13 GPa were melted and/or sublimated due to the heterogeneous heating in the powder compact. After release of the shock pressure, heat was transmitted to the interiors of the large particles causing the heavily deformed particles to recrystallize. On the other hand, recrystallization in the compacts produced at 4 and 8 GPa was observed only in the regions between large particles as shown in Figure 3 (B) and (C). This implies that the amount of material melted in the compacts obtained at 4 and 8 GPa was considerably less than in the compact obtained at 13 GPa. Strong interparticle bonding was produced by solidification of the molten interparticle layers. The strength of interparticle bonding is represented by the microhardness value of the compact and may be affected by the residual strain in the resulting compact. It is therefore believed that the difference in the microhardness values of the samples compacted at 8 and 13 GPa is due to the differences in the amount of material melted between particles during shock compression and the annealing effects due to the high residual temperature.

The major difficulties in the conventional sintering of SiC powder without densification aids are the formation and growth of interparticle necks and particle coarsening before densification (8). In dynamic compaction, powder compacts of coarse SiC powder can be densified to near solid density at the shock wave front and consolidated to a strongly bonded compact by melting and subsequent solidification of material during and after shock-compression process.

REFERENCES

1. D.G. Morris, "Melting and Solidification During Dynamic Compaction of Tool Steel," Metal Science, 5:116 (1981).
2. D. Raybould,"The Properties of Stainless Steel Compacted Dynamically to Produce Cole Interparticle Welding," J. Mat. Sci., 16:589 (1981).
3. D.E. Grady, "Shock Deformation of Brittle Solids," J. Geophs. Res., 85:913 (1980).
4. K. Kondo, S. Soga, A. Sawaoka and M. Araki, "Shock Compaction of Silicon Carbide Powder," J. Mat. Sci., 20:1033 (1985).
5. H.P. Klug and L.E. Alexander, <u>X-Ray Diffraction Procedures</u>, John Wiley and Sons, New York, 1975, p. 491.
6. R. Prummer and G. Zielger, "Structure and Annealing Behavior of Explosively Compacted Alumina powders," Powder Metall. Int., 9:11 (1977).
7. B. Morosin and R.A. Graham, "X-Ray Diffraction Line-Broadening Studies on Shock-modified Rutile and Alumina," Mat. Sci. Engr., 66:73 (1984).
8. C. Greskovich and J.H. Rosolowski, "Sintering of Covalent Solid," J. Am. Ceram. Soc., 59:336 (1976).

SHOCK-RECOVERY EXPERIMENTS ON PZT 95/5*

L. C. Chhabildas, M. J. Carr, S. C. Kunz, and B. Morosin

Sandia National Laboratories
Albuquerque, NM

ABSTRACT

Recovery experiments on porous PZT 95/5 ferroelectric have been performed on a single-stage compressed gas gun over the stress range of 0.3 to 4.6 GPa. Density measurements on the recovered specimens suggest permanent pore compaction. X-ray diffraction analysis indicates that the shock-induced FE-to-AFE phase transformation is reversible, with a reduction in the residual lattice strain, as compared with the unshocked material. SEM and TEM microscopy show evidence of ferroelectric domains with fewer numbers at the higher stress. The microstructure of the 3.0 GPa sample is significantly affected, as evidenced by dislocation slip bands and oval-shaped voids.

INTRODUCTION

PZT 95/5 is a ferroelectric ceramic material which has a stress-induced ferroelectric-to-antiferroelectric (FE-to-AFE) phase transformation. Recently, shock and pressure-shear loading experiments performed on the porous ceramic (9% porous)[1] over the stress range of 0.8 GPa to 4.6 GPa, have indicated complex material behavior during shock compression. Under a hydrostatic pressure environment, the material shows a complex pressure-temperature phase field with considerable hysteresis[2] (Figure 1). It is expected that the behavior under shock compression will be more complex, and stress, strain and pore dynamics will play important intermingled roles. Transformation of the ceramic from the FE phase to the AFE phase commences near 0.5 GPa with a mixture of phases indicated over the stress range of 0.9 GPa to 2.6 GPa. Above 2.6 GPa, the kinetics of pore compaction dominate the dynamic response. Upon release, a remnant strain is measured which may be due to irreversible phase transformation and/or permanent densification. Pressure-shear experiments suggest a constant shear stress being sustained over the mixed-phase region of 0.9 to 2.6 GPa.

These hypotheses were pursued with experiments on shock-recovered material. The samples were embedded in an OFHC copper capsule, chosen because its shock impedance is similar to the PZT sample. Impact of copper flyer plates accelerated, on the single-stage compressed gas gun, yielded stress levels of 0.3, 0.8, 1.7, 3.0, and 4.6 GPa, respectively, in the PZT sample. The dimensions of the sample and the copper assembly were chosen to prevent spallation, either in the capsule, or in the PZT sample. Numerical simulations of

*This work performed by Sandia National Laboratories supported by the U. S. Department of Energy under contract #DEAC04-76-DP00789.

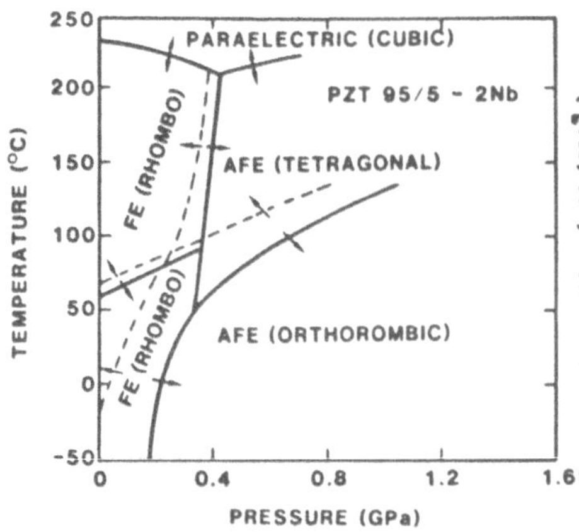

Figure 1: Phase Diagram for 95/5 PZT Ceramic.

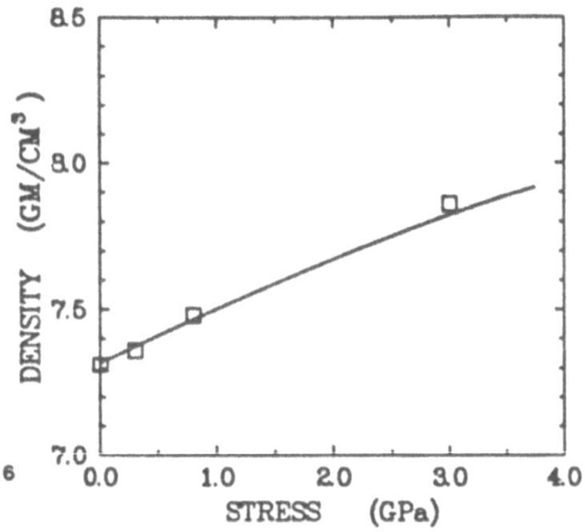

Figure 2: Measured density of shock recovered specimens.

the recovery fixture, using the two dimensional code CSQ,[3] suggest similar loading histories at the center of the sample for early shock loading times, based on either one-dimensional or two-dimensional loading assumptions. Density measurements, x-ray diffraction techniques, scanning and transmission electron microscopy techniques were used to determine the nature of the recovered phase and the deformation features induced in the ceramic due to shock loading. The results are summarized in this paper.

RESULTS AND DISCUSSION

Density Measurements and X-ray Analysis

Density measurements on the recovered specimens are shown in Figure 2, which indicate permanent shock-induced compaction. The material investigated has ~9% porosity, arising from incomplete volume filling during sintering and from 0.8 weight percent of 100 μm lucite spheres added to the ceramic powder to control the final density. (Since the recovered samples were cracked after shock loading, the density measurements obtained by immersion techniques are of limited accuracy.) The increase in density is thought to result primarily from the shock-induced compaction process. X-ray diffraction analysis was also performed to determine if part of the densification resulted from irreversible phase transformation. The density of the 4.6 GPa sample was not measured since the recovered sample was pulverized. The recovered 3.0 GPa sample suggests the presence of ~3% porosity when compared to its theoretical density of 8.028 gm/cm^3 calculated on the basis of its chemical composition $Pb_{.9897}Nb_{.0206}(Zr_{.96}Ti_{.04})_{.9794}O_3$ and the x-ray lattice constants determined in this investigation.

X-ray diffraction generally allows detection of different structural phases of material. When the differences between the crystal structures are small, x-ray lines of a higher symmetry phase are split for the phases of lower symmetry. Should the splitting be sufficiently large (greater than the line width), a precise determination can be made. One may then predict how the cubic line would split under tetragonal, rhombohedral, and orthorhombic symmetry, both with respect to line position and relative intensity. Using several selected lines and avoiding cubic degenerate lines (*e.g.* 300 and 221), an unambiguous choice can be made. In the case of perovskites, and in particular for this composition of PZT 95/5, some of this splitting is small. It should be noted that the actual cell dimensions may be more complex than the simple cell deduced because of weak lines. Furthermore, in shock-loaded materials, the x-ray line profiles may be broadened by two effects: reduction of crystallite domain size and retained residual lattice strain. Generally, these two effects are not very severe in inorganic materials at modest pressures of 5 GPa[4].

X-ray diffraction studies employed Cu Kα and Cr Kα radiation, the latter to take advantage of obtaining non-degenerate cubic lines at large scattering angles. A standard 114.5 nm Norelco powder camera was used in order to obtain data at large angles. No line broadening studies for size and strain were carried out since these lines were expected to be split from structure considerations; however, an estimate can be made from the present data using an approximate method developed in previous studies. The x-ray specimens were generally taken from edges of the fractured specimen. In samples purposely broken, no differences were seen in the x-ray diffraction characteristics as to their origin, *i.e.*, either near the center of the disk or the edge. The most striking observation was that the unshocked PZT showed broader x-ray line profiles at higher scattering angles than the shocked material. This is consistent with a marked residual lattice strain (near 10^{-3}) in the as-received unshocked material. Upon recovery from shock loading the lines are much sharper and allowed the determination of the lattice parameter for this material.

The lattice parameters determined using the 3 GPa recovered sample and Cr Kα radiation are a = 4.138(2) $A°$, and α = 89.83(4)°. This rhombohedral distortion corresponds to a 10-minute angular departure from cubic symmetry, a value sufficiently small that distinct separation of certain x-ray lines, which in the cubic form appear as single lines, is not achieved. Nevertheless, the appropriate lines which are split or broadened allow verification of the lattice symmetry of the material.

Optical Microscopy

Disc-shaped test specimens of unshocked and shock-recovered material were encapsulated under vacuum in a room temperature cured epoxy resin. The mounted specimens were cut diametrically, through the disc thickness and were polished through a 1-μm diamond slurry on a vibratory polisher. A Leitz metallograph was used to examine the specimens for pore compaction and to compare the microstructures after recovery from shock loading to various stresses.

Except for the unshocked material, all of the specimens had cracks across the width and thickness which typically, but not necessarily, interconnected the pores. The degree of cracking increased markedly with shock level. The pores in the unshocked, 0.3 GPa and 0.8 GPa shocked specimens were uniformly circular and approximately 100 μm in diameter. In the material shocked to 1.6 GPa and higher, many of the pores were flattened to an oval shape in the plane parallel to the shocked surface. The number of oval-shaped pores increased with increasing shock level such that the majority of the pores in the 4.6 GPa shocked material were deformed. This compaction by permanent pore deformation substantiates the measured increase in density with shock level.

Scanning Electron Microscopy

The unshocked and shocked materials were examined using a Hitachi S-500 scanning electron microscope (SEM) at 25 KeV in two conditions *i.e.*, as-polished, and as-etched. In the first, a thin conductive carbon coating was evaporated onto the freshly polished surface, and a backscattered electron imaging mode was used to view the topography of the polished surface. In the second, an etchant[5] (0.5% to 1% by volume of a $HF - HNO_3$ mixture in water) was used to reveal the domain structure. The etched specimens were also carbon coated for SEM examinations. The as-polished surfaces of both unshocked and shocked materials revealed no topographical feature such as grain boundaries or domain walls in

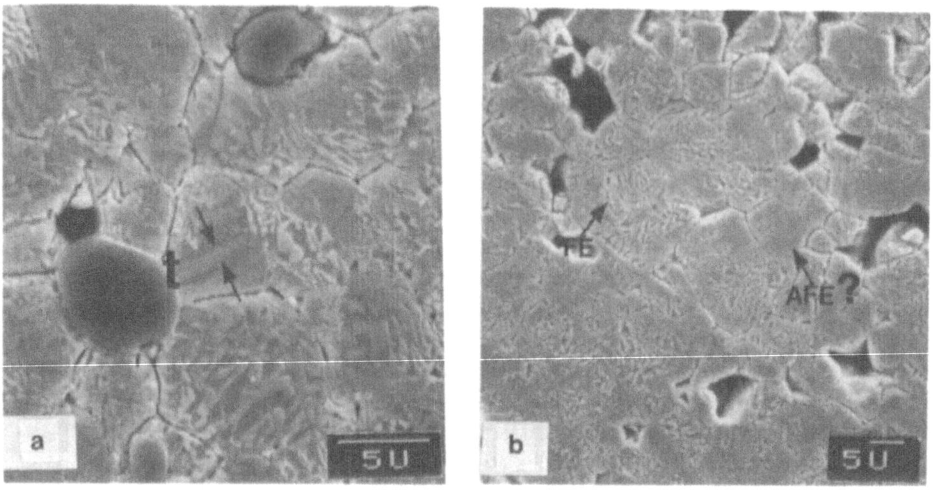

Figure 3: Scanning electron micrographs of the 3.0 GPa shock-recovered PZT 95/5.

either a secondary electron or a topographical backscattered imaging mode. This is consistent with other attempts at microstructural examination of unetched barium titanate and lead zirconate titanate[6].

In contrast, etching of the unshocked, as well as the shocked specimens, clearly delineates 5 - 10 μm diameter grains and reveals FE domain walls within the grains. Figure 3(a) shows typical etched microstructure. The porosity observed at the crystallite triple junctions is inherent to the material as a result of the sintering process. A classical "herringbone" domain pattern is clearly visible in which the narrow light and dark stripes are alternating pairs of 90 degree domains[6]. The thickness, t, of the domains (Figure 3(a)) is of the order of 1.0 μm which agrees with that reported for the finely textured domains seen in other niobium-modified PZT. The width of the domains varies from tenths of microns to tens of microns, although, the domains typically do not extend across the entire grain diameter.

A comparison of SEM micrographs of unshocked and shocked specimens indicates the absence of a clearly defined domain structure in some grains of the material shocked to high stress levels. The 3-GPa sample in Figure 3(b) exhibited the most "blank" grains, whereas virtually all grains in the unshocked and the low level shocked material showed evidence of FE domains. The absence of visible FE domains in some grains subjected to high shock levels suggests the presence instead of an AFE phase. This suggests, in turn, that the degree of reversibility of the shock-induced FE to AFE phase transformation depends on the shock level.

Transmission Electron Microscopy

The microstructure of unshocked starting material and recovered material shock loaded to 0.3 GPa, 0.8 GPa and 3.0 GPa was examined. Specimens were prepared by ion milling in a cold stage to avoid heating the specimens above the AFE transition. All specimens were examined at 200 Kev with a JEOL 200CX.

The structure of the unshocked material consisted of well-annealed grains exhibiting very low dislocation density (10^6 per cm^2), as shown in Figure 4. Substructural features of the ferroelectric domains, typically <100 nm wide, were present in all grains. These fine structures are not resolved in SEM micrographs of etched surfaces. An intergranular phase filled some smaller grain boundary triple points. This phase was observed to be crystalline, but was not identified. The 0.3 and 0.8 GPa material also indicated features similar to the unshocked material, indicating that loading to that point was elastic.

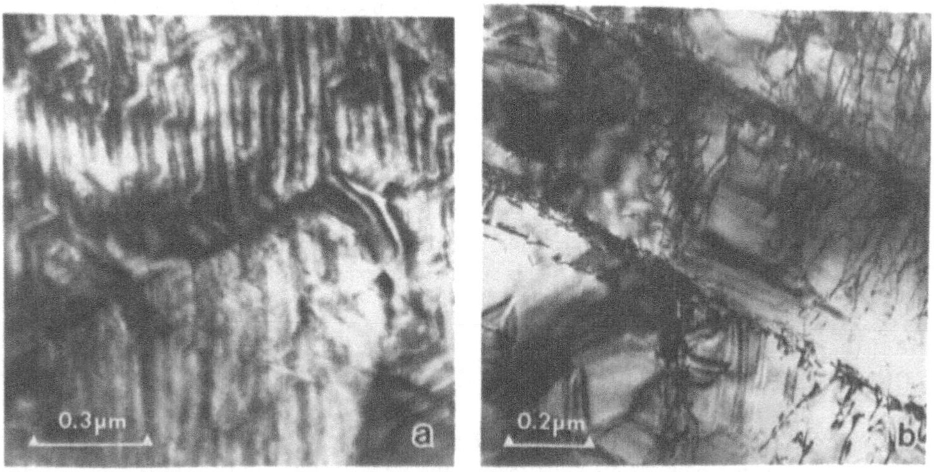

Figure 4: Transmission electron micrographs of unshocked and 3.0 GPa shock-recovered PZT 95/5.

The microstructure of the 3.0-GPa material was substantially affected by the shock loading process; the deformation was inhomogeneous. Optical micrographs showed a network of lighter bands crisscrossing the thickness of the specimen. Deformation was found to be concentrated in these bands. Grains outside these bands showed the same low dislocation density as the starting material. The bands were found to consist of fractured and deformed grains. Deformed material exhibited dislocations, often lying along well defined slip planes (Figure 4). Some slip bands terminated at fractures suggesting that slip may precede cleavage or fracture in some grains. Dislocation density varied from grain to grain; typical dislocation densities in deformed grains were about 10^9 per cm^2. In all grains, deformed and undeformed, no ferroelectric domains like those in the unshocked material were found. Initially, it appeared that there were no ferroelectric domains at all, but more detailed work showed what appeared to be widely spaced domain boundaries pinned to substructural features. The domain size in these materials ranged from 500 to 1000 nm, which agrees with that observed in SEM. The microstructural changes which occur in the 3.0-GPa shock-loaded material are consistent with plastic flow by dislocation motion and fracture.

SUMMARY

Density measurement, x-ray diffraction, scanning and transmission electron microscopy have been used to characterize the microstructure and the phase morphology of shock-recovered 95/5 PZT ceramic. Density increases suggest pore-compaction as a densification mechanism, which is also substantiated by observations of permanently deformed oval-shaped pores.

X-ray diffraction analysis suggests that the shock-induced FE-to-AFE phase is reversible, and also that the rhombohedral crystal structure of the FE phase is only a slight distortion from the cubic structure.

Ferroelectric domains observed by electron microscopy on shock-recovered specimens complement the results of the x-ray analysis which suggest that the shock-induced FE-to-AFE phase is reversible. The degree of reversibility, however, appears to depend on stress, since a decreasing number of FE domains are observed in material recovered from high stresses.

Inhomogeneous deformation of the material at higher stresses is indicated by a higher density of dislocations concentrated in a network of criss-crossing slip bands.

Shock-recovery experiments on PZT 95/5 complement the instrumented (time-resolved wave profile) studies[1] performed on the material. The results deduced from the analysis of the recovered material in general agrees with the observations inferred from the instrumented studies. Considering that the time scales of these two types of experiments are vastly different, this apparent agreement is indeed gratifying. It lends credence to the assumption that the microstructure of shock-recovered specimens reflects complex deformation mechanisms that occur at the shock front. This seems to be particularly true for shock-recovered 95/5 PZT.

ACKNOWLEDGMENTS

We would like to acknowledge R. H. Dungan for providing the PZT samples, and D. M. Webb for performing numerical simulations of the recovery fixture. Special thanks are also due to J. M. Miller, M. E. McAllaster and L. J. Storz for their assistance during the course of this study.

REFERENCES

1. L. C. Chhabildas, Sandia National Laboratories Report, SAND84-1729, (1984).
2. I. J. Fritz and J. D. Keck, J. Phys. Chem. Solids, 39, 1163 (1978).
3. S. L. Thompson, Sandia National Laboratories Report, SAND77-1339 (1979).
4. B. Morosin and R. A. Graham, in "Shock Waves in Condensed Matter," ed. by J. R. Asay, R. A. Graham, and G. K. Straub (Elsevier Science Publishers), 355 (1984).
5. D. Kontoleon and J. Tomilson, Metal Progress, 69, 98 (1955).
6. B. Jaffe, W. R. Cook, and J. Jaffe in "Piezoelectric Ceramics," Academic Press, NY (1971).

CATALYTIC ACTIVITY OF SHOCK MODIFIED ZNO FOR CO OXIDATION AND METHANOL SYNTHESIS

F.L.Williams and Y.K.Lee

Chemical and Nuclear Engineering
University of New Mexico
Albuquerque, New Mexico 87131

and

B.Morosin and R.A.Graham

Sandia National Laboratories
Albuquerque, New Mexico 87175

INTRODUCTION

Alteration of the catalytic activity of inorganic powders by dynamic shock compression has been reported in the literature for over twenty years Recently Golden et.al.(1,2) and Lee (3) have extended the early pioneering work of Horiguchi and Nomura(4) and Batsanov et.al.(5,6,7). Previous work reported by Batsanov et.al.(5) for the effect of shock compression on ZnO catalysis revealed an enhanced activity for CO oxidation that decreased during the rate measurements. The purpose of this paper is to report steady state rate measurements of CO oxidation and methanol synthesis over shock compressed ZnO powders under reaction conditions believed to be transport limitation free.

EXPERIMENTAL

Pure powders of ZnO (99.9%) were obtained from Cerac. The powders were pressed to an initial packing density of 3.01 Mg/m^3 (55% of theoretical) in Sandia "bear" series recovery fixtures under ambient atmosphere(8). Three ZnO samples were shock loaded at mean peak shock pressures of 7.5, 16 and 22 GPa. Another sample was shock loaded at a 16 GPa shock pressure using a different explosive resulting in significantly different shock temperature.

The peak pressure and temperature experienced by the ZnO powders during shock compression are presented in Table 1. The pressures and temperatures are a result of calculations based on a two-dimensional numerical simulation of the shock compression conditions of pressure(8) and temperature(9) within the powder compacts.

Table 1. Schedule of Mean Peak Shock Pressure and Temperature for the shock modified ZnO

Sample	Fixture	Explosive	Peak Shock Pressure (GPa)	Peak Shock Temperature (K)
49G836	Momma bear	Baratol	7.5	500
48G836	Momma Bear	Comp B	16.0	575
50G836	Momma Bear	Baratol	16.0	700
51G836	Momma Bear	Comp B	22.0	675

Even though contamination was not totally avoided during the shock compression experiment, special precaution was taken to minimize it. A gold foil was placed on the planar interfaces of the powder compacts before shock compression to help prevent contamination with debris which could be formed during shocking or when the sample capsule was opened after the shock modification experiment. Any gold contamination would be catalytic inactive for CO oxidation or methanol synthesis.

Post-shock analysis the samples by emission spectroscopy revealed contamination of the shocked samples with gold and low levels of copper. The concentrations did not follow any trend with respect to the catalytic activity changes and therefore are not considered to influence these results. Specific surface areas of the powders were measured using a standard six point BET analysis of nitrogen adsorption isotherms. Recovered powders had an x-ray diffraction pattern characteristic of a line broadened, shock modified material with no extra lines.

Two reactors were used for the rate measurements. The CO oxidation reaction was carried out in a pyrex reactor of the same design as used by Golden et al(1). Improvements in controlling reactant flows were accomplished by the addition of electronic flow controllers. The methanol synthesis reaction was carried out in a high-pressure stainless steel reactor of the Berty design. Reactants were compressed and mixed in a large stainless tank from which a stoichiometric mixture of hydrogen and carbon monoxide could be drawn into the reactor. Reaction products were continuously drawn from the reactor and fed to a gas chromatograph after passing through pressure reducing valves.

The ZnO powders were pretreated in the reactor before being subjected to reaction conditions. Prior to measurement of CO oxidation rates fresh ZnO powder was heated to 723 K in a mixture of 2 percent excess oxygen in carbon monoxide at atmospheric pressure for 18 hours. Prior to measurement of methanol synthesis rates previously unused ZnO powder was heated to 673 K in ultrahigh purity hydrogen at atmospheric pressure for 12 hours. The temperature was adjusted to that desired and stoichiometric reactant mixtures were introduced to the reactors. The oxidation reaction was followed for over 2 hours until steady state rates were observed and then the temperature dependence of the reaction was determined. The synthesis reaction pressure was increased to 130 atmospheres over 2 hours and then the reaction was followed for more than 12 hours until steady state rates were observed.

The reactions were followed by conventional gas chromatography. While the pyrex reactor did not contribute to the oxidation of CO, the stainless reactor did convert some of the reactant mixture to methane. The methanol synthesis rates reported here have been corrected for the reactor induced conversion. Both experiments and calculations were carried out of determine that the rate measurements were free of heat and mass transfer alteration of the observations.

RESULTS and DISCUSSION

The rate of CO oxidation is enhanced by shock compression of ZnO. The rate of CO oxidation for the four shocked powders and the one unshocked powder is shown in Figure 1 as a function of temperature. This Arrhenius type plot shows that the activation energy for all the powders is similar and that the degree of rate enhancement depends upon shock loading conditions. A summary of the rate data in Table 2 reveals that the shock-modified ZnO catalysts have a lower activation energy than the unshocked ZnO in agreement with previously reported trends and values (5,6).

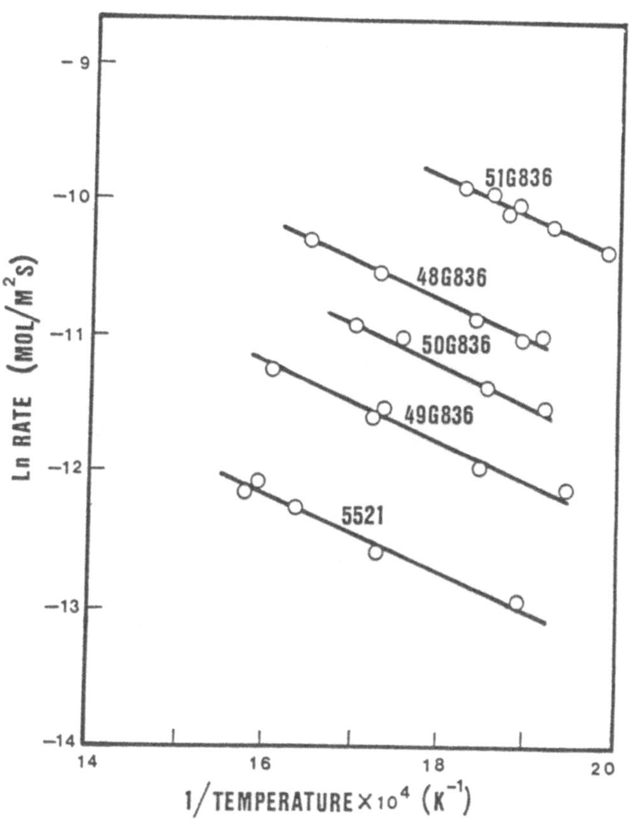

Fig. 1 Arrhenius plots for CO oxidation rates over ZnO powders. Stoichiometric CO/O_2 feed, 1 atm.

Table 2. Activation Energy and Rate of CO Oxidation Reaction over ZnO

Sample	Activation Energy (kcal/mol)	Reaction Rate at 550 K ($mol/m^2 sec$)
5521*	12.8	2.9×10^{-6}
48G836	10.8	8.9×10^{-6}
50G836	9.6	1.2×10^{-5}
49G836	11.8	2.1×10^{-5}
51G836	9.0	5.0×10^{-5}

* as received powder

The effect of peak shock pressure can be seen in Figure 2. The rate of reaction is increased by a factor of 17 at the highest peak shock pressure with what might appear to be a linear dependence for the intermediate pressures. However, the four shocked powders each experienced a different peak shock temperature. At 16 GPa we find that a higher shock temperature leads to a lower oxidation rate. Therefore, the actual shock pressure dependence is likely to be non-linear with dramatic enhancement at the higher shock pressures if the shock temperatures could be kept low.

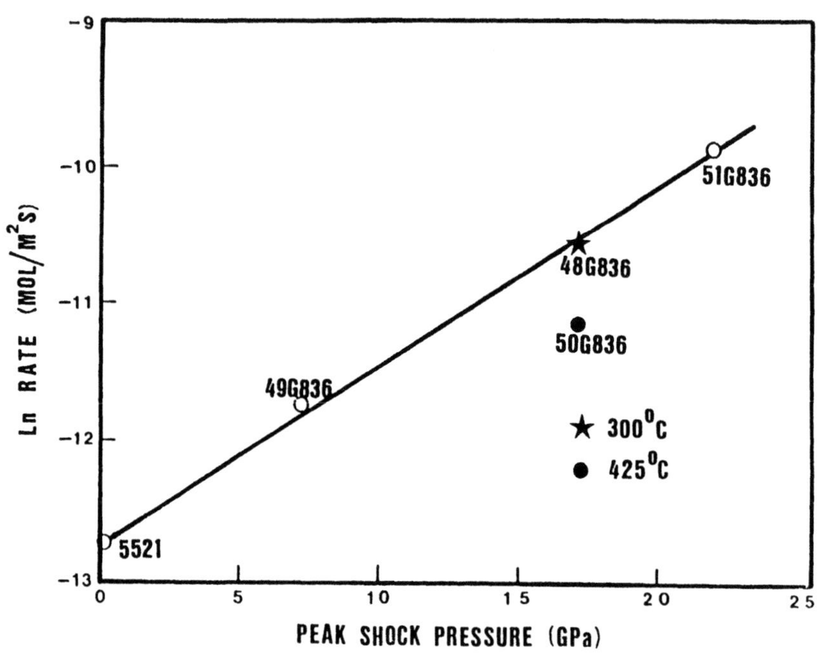

Fig. 2 Apparent shock pressure dependence of CO oxidation rates over ZnO powders. Stoichiometric CO/O_2 feed, 555 K, 1 atm.

The rate of methanol synthesis is also enhanced by shock compression of ZnO. The rate of reaction for each of the major species formed is shown as a function of peak shock pressure in Figure 3. The most immediate observation from Figure 3 is that all the rates of reaction first decrease at the lowest shock pressure and then increase as the peak shock pressure increases. For example, the rate of methanol formation decreased by a factor of 1.4 at the lowest shock pressure (7.5 GPa) compared to the unshocked ZnO. At 16 GPa shock pressure, the rate of methanol formation increased by a factor of 3.8 and continued to increase to a factor of almost 18 at the highest shock pressure of 22 GPa.

Noting that the rate dependencies parallel each other, one can observe that no substantial selectivity change occurred at the different peak shock pressures. In fact the selectivity profile is the same for the shocked and unshocked ZnO powders. Even Arrhenius plots for each product species reveal that the activation energies are the same for any one species produced over any of the five powders. The rate dependence upon peak shock pressure is more dramatic than indicated in Figure 3 because, just as the case for CO oxidation, higher shock temperatures give lower degrees of catalytic rate enhancement.

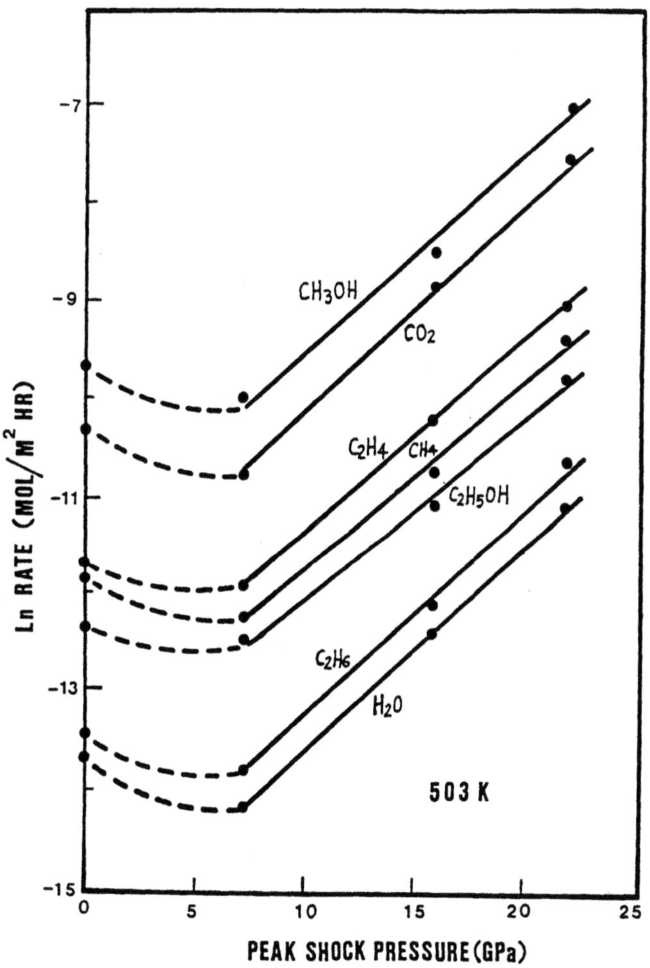

Fig. 3 Product species reaction rates versus mean peak shock pressure for methanol synthesis over ZnO. Stoichiometric CO/H_2 feed, 503 K, 130 atm.

Post shock, residual stain was considered as an explanation for the observed results. However, there is at best an inverse relationship between residual strain and reaction rate. That relationship does not hold when the magnitude of strain differences is compared to the magnitude of rate change.

Based on the information to date, our best explanation of the enhanced catalytic activity of ZnO powders is that shock compression modifies the ability of the particle surface to form active centers under reaction conditions. The shock compression probably does not itself put long lived active centers at the surface or major changes in activation energies or selectivity would be observed. However, if local strain or defects were induced in the surface, it would be possible for atomic rearrangement of the surface to result in additional sites like those on unshocked ZnO. In that way, the catalytic chemistry of all the powders would only be different in the number of sites available for catalysis. The minima in Figure 3 may then result from a surface distortion in the unshocked powder that is released by a relative mild shock compression.

REFERENCES

1. J.L. Golden, F.L. Williams, B. Morosin, E.L.Venturini and R.A. Graham, in: "Shock Waves in Condensed Matter - 1981 (Menlo Park)," p.72, American Institute of Physics, Conference Proceedings No.78, New York (1983).
2. J.L. Golden, M.S. Thesis, University of New Mexico (1982).
3. Y.K. Lee, Ph.D. Dissertation, University of New Mexico (1985).
4. Y. Horiguchi and Y. Nomura, Carbon 2:436 (1965).
5. S.S. Bastsanov, G.K. Boreskov, G.V. Gridasova, N.P. Keier, L.M. Kefeli, V.M. Kudinov, V.I. Mali and I.S. Sazonova, Kinet. and Catal. 8:1140 (1967).
6. G. Boreskov, I. Sazonova, N. Keyer, V. Kudinov, G. Gridasova, V. Maly and L. Kefely, in: "Behavior of Dense Media under High Dymanic Pressures," p.389, Gordon and Breach, New York (1968).
7. S.S. Batsanov, V.P. Bokarev, I.A. Kostenshuk, Yu.S. Mardashov, and I.N. Temnitskii, React. Kinet. Catal. Lett., 20:43 (1982).
8. R.A. Graham and D.M. Webb, in: "Shock Waves in Condensed Matter -1983," p.211, North Holland, Amsterdam (1984).
9. R.A. Graham and D.M. Webb, this proceedings.

X-RAY DIFFRACTION STUDIES OF SHOCK-SYNTHESIZED ZINC FERRITE†

B. Morosin, E. L. Venturini, and R. A. Graham

Sandia National Laboratories
Albuquerque, New Mexico 87185

ABSTRACT

Mechanically-blended stoichiometric powder mixtures of zinc oxide and ferric oxide have been subjected to high pressure shock wave loading between 7.5 and 27 GPa. By varying the initial packing densities between 2.15 and 3.38 Mg/m^3 in recovery fixtures, peak shock temperatures between 125 and 1,100°C were achieved. X-ray diffraction shows formation of a spinel structure type. The amount depends on shock parameters; unequal consumption of initial oxides suggests a defective spinel $Zn_{1-x}Fe_xFe_2O_4$ rather than stoichiometric $ZnFe_2O_4$ is formed. A comparison is made between materials prepared by shock loading and by furnace firing.

INTRODUCTION

Greatly enhanced solid state reactivity is observed while a solid is under shock compression. This reactivity is believed to be the result of the large concentrations of defects which have been introduced by the shock process [1,2] along with the mechanical mixing and surface cleaning. The highly nonequilibrium nature of this process creates a possibility to synthesize solids with chemical and physical characteristics different from those prepared by other processes. Prior studies on the use of shock compression for material synthesis have been reported [3,4]. In particular, the synthesis of a zinc ferrite from mechanical blends of zinc oxide and ferric oxide has been previously studied [5,6]. The material prepared by shock-loading differs substantially from that prepared by more conventional ceramic processing.

BACKGROUND

The solid state reaction of ZnO and Fe_2O_3 powders at elevated temperatures indicates a complex process [6]. Between 400-500°C in an

†This work performed at Sandia National Laboratories supported by the U.S. Dept. of Energy under contract # DE-AC04-76DP00789 and in part by ONR under N00014-84-F-0151.

inert gas atmosphere, oxygen is evolved. The product formed contains excess ZnO and the mixed spinel $Zn_{0.85}Fe_{2.15}O_4$. Stoichiometric zinc ferrite can be obtained by using excess zinc oxide or an oxygen atmosphere.

$ZnFe_2O_4$ crystallizes in the normal spinel lattice in which the tetrahedral or A sites are occupied by divalent zinc ions while the octahedral or B sites contain trivalent iron ions [7]. In such an arrangement, stoichiometric $ZnFe_2O_4$ is paramagnetic; however, placing the magnetic ions partially in the A sites results in a ferrimagnetic material. Magnetic measurements have shown that under shock compression, a mixed site and/or zinc deficient spinel is obtained [6].

In the $ZnFe_2O_4$ - Fe_3O_4 system, Miyata has reported lattice constants as a function of composition [8]. The values of lattice constants increase essentially linearly. The difference in values for the two end members of this alloy system is ~ 5.5×10^{-3} Å, using Miyata's values or the slightly different ones compiled in Crystal Data [9]. The lattice constant of our standard $ZnFe_2O_4$ spinel powder prepared in large bulk sample under oxygen atmosphere of 8.441 Å is consistent with these values.

EXPERIMENTAL DETAIL

The ZnO and Fe_2O_3 powders used in these experiments were commercial grade (99.9% purity, CERAC, Inc., Milwaukee, WI). A stoichiometric mechanical mixture for all experiments was prepared by ball milling the powders together for 24 hours. Scanning electron microscopy showed that the dispersal was at grain level with typical particle size of 0.3 μm. The stoichiometric batch powder was pressed into compacts in place within the copper capsules. Various initial packing densities were used. Each sample was then subjected to controlled, quantifiable explosive loading. The Sandia "Bear" fixtures together with various explosives were used as indicated in Table I. The shock-loading conditions for the experiments are also included in Table I. Quantification of pressure and temperature conditions within the compact is achieved with an extensive program of numerical simulation described elsewhere [10,11].

The copper capsules are designed to seal the compacts in place in the capsule so that after explosive loading there is confidence that the loading conditions did not cause any extraneous, uncharacterized effects. This gas-tight seal is maintained throughout the shock-wave loading and unloading. Hence there is no path for oxygen evolution during the shock synthesis in contrast to heating a mixture of oxides in an inert atmosphere where rapid evolution was observed above 400 C. Following the shock experiment the copper capsule slowly cools to room temperature before the gas-tight seal is cut open and the product removed. The recovered powder is lightly ground to assure complete dispersion of any reacted material within an x-ray sample.

In order to compare the properties of the shock-synthesized material with that formed by a short time furnace firing, the identical mechanically blended, stoichiometric mixture was placed in a furnace in air for two hours at a particular temperature and then quenched to room temperature. This material served as a comparison with that obtained in the microsecond time frame at peak shock temperature under shock compression.

Powder x-ray diffraction studies employed Fe K_α radiation. In order to quantify the compositional determination, standard powders were prepared by mixing the unreacted stoichiometric powder with that of our standard $ZnFe_2O_4$ spinel powder. Shock-modified powders typically show broadened x-ray diffraction profiles due to residual lattice strain and reduced coherent crystal domain size so that the detailed diffraction profiles of the x-ray pattern of our standards are not identical to the resultant shock-modified powders. The two-hour furnace reacted samples also show broadened profiles at high scattering angle, but for a different reason, that of compositional inhomogeneity.

RESULTS AND DISCUSSION

X-ray diffraction on the shock-loaded material shows formation of a spinel-structure type in amounts as indicated in Table I. In addition, the unreacted, recovered Fe_2O_3 and ZnO powder shows broadened line profiles resulting from reduction in crystallite size and/or residual lattice strain. As the data in Table I show, both shock-loading pressure and temperature appear to influence formation of the spinel material. Shock pulse signature may also be important. On samples with larger amounts of ferrite formed, a greater amount of Fe_2O_3 is consumed relative to that of ZnO; this suggests that a defective, iron rich spinel is being synthesized of the form $Zn_{1-x}Fe_xFe_2O_4$.

Our 14G846 sample showed a radial variation in color upon opening of our recovery fixture. This is consistent with radial changes in pressure and temperature history for the "Baby Bear" fixture used in this experiment [10]. The amount of ferrite formed in various capsule locations is quite large, ranging from 33% in the bulk to 85% on the edge. The edge sample showed relatively sharp, high angle profiles, consistent with lattice constants of 8.437Å. Employing the end member difference 5.5×10^{-3}Å noted above, this suggests our ferrite for this sample is consistent with a composition of $x = 0.08$. The difference in the edge material is due to the higher temperature experienced in that region [11].

We have compared our shock synthesized material with that prepared by our furnace treatment. The initial stoichiometric blend as well as the furnace material subjected to temperatures as high as 600°C but which remain unreacted show identically sharp diffraction profiles. The very small amount of broadening due to milling is just detectible. By 625°C, reaction is occurring with about 30% of the material in the form of a ferrite with large compositional range; by 650°C, 60% has reacted with a narrower range of compositions which correspond to lattice constant differences near 2×10^{-3}Å; by 690°C, 80% is spinel with 2×10^{-4}Å lattice constant spread. Even at 800°C, about 5% of the original oxides remain; however, the ferrite profiles are as sharp as the initial unreacted powders. The lattice constant for this latter material is 8.440Å. Magnetic measurements show that the material is not stoichiometric even though the cell parameter is within 0.002Å of our standard ferrite material.

Table I
Schedule of Experiments -- ZnO+ Fe_2O_3 Mixtures

Expr.(a)	Fixture(b)	Explosive	Density Mg/m^3	Peak Pressure GPa (c)	Mean Temp.(c) C°	Spinel(d) (%)
-	-	-				
44G836	(MB)	Baratol	3.3	7.5	125	ND
43G836	(MB)	Comp B	3.3	16	320	ND
30G846	(MB)	Comp B	2.15	16	650	15
42G836	(MBA)	Baratol	3.3	16	150	ND
28G846	(MBA)	Baratol	2.75	16	400	ND
32G846	(MBA)	Baratol	2.33	16	500	<4
21G846	(MBA)	Baratol	2.15	16	550	<4
12G846	(BB)	Baratol	3.3	20	175	<4
29G846	(BB)	Baratol	2.75	20	350	<5
31G846	(BB)	Baratol	2.33	20	540	<4
41G836	(MBA)	Comp B	3.3	22	300	<6
24G846	(MBA)	Comp B	3.03	22	400	8
33G846	(MBA)	Comp B	2.75	22	475	8
25G846	(MBA)	Comp B	2.48	22	550	25
8G846	(MBA)	Comp B	2.15	22	655	27
36G846	(BB)	Comp B	3.38	27	400	15
9G846	(BB)	Comp B	2.33	27	625	8
14G846	(BB)	Comp B	2.15	27	725	75
14G846 ctr	(BB)	Comp B	2.15	27	725	33
14G846 edg	(BB)	Comp B	2.15	27	1100	85

(a) all samples are from the bulk except two indicated taken at center (ctr) and edge (edg).
(b) MB = Momma Bear, MBA = Momma Bear A, BB = Baby Bear.
(c) The peak shock pressure and temperature produced in the powder compact are based on realistic two-dimensional simulations (refs. 10 and 11).
(d) ND = none detected.

CONCLUSIONS

X-ray diffraction studies show the formation of a spinel structure type product when mechanically-blended stoichiometric powder mixtures of zinc oxide and ferric oxide are subjected to high pressure shock loading. The amount of spinel formed is a function of the shock compression parameters. The spinel formed is a defective, iron rich spinel of the form $Zn_{1-x}Fe_xFe_2O_4$. The lattice constant on the sample with the highest peak shock pressure and shock temperature is consistent with a composition of x = 0.08.

REFERENCES

1. E. L. Venturini and R. A. Graham in Defect Properties of Processing of High Technology Nonmetallic Materials, ed. J. H. Crawford, Jr., Y. Chen and W. A. Sibley (North-Holland, NY, 1984) p. 383.

2. B. Morosin and R. A. Graham in Shock Waves in Condensed Matter - 1983, ed. by J. R. Asay, R. A. Graham and G. K. Straub (North-Holland, NY, 1984) p. 355.

3. B. Morosin and R. A. Graham in Shock Waves in Condensed Matter - 1981, ed. by W. J. Nellis, L. Seaman, and R. A. Graham (AIP, NY, 1982) p. 4.

4. S. S. Batsanov, ibid, p. 1.

5. Y. Kimura, Japan, J. Appl. Phys. 2, 312 (1963).

6. E. L. Venturini, B. Morosin and R. A. Graham, Magnetic and Structural Properties of Shock Synthesized Zinc Ferrite, J. Appl. Phys. 57, 3814 (1985); references therein.

7. J. M. Hastings and L. M. Corliss, Rev. Mod. Phys. 25, 114 (1953).

8. N. Miyata, J. Phys. Soc. Japan 16, 1291 (1961).

9. J. D. H. Donnay and H. M. Ondik, Crystal Data, Determinative tables, 3 Ed., Vol II: Inorganic compounds, U.S. Dept. of Comm., National Bureau of Standards and Joint Committee on Powder Diffraction Standards (1973).

10. R. A. Graham and D. M. Webb in Shock Waves in Condensed Matter, 1983, ed. J. R. Asay, et al. loc cit. p. 211.

11. R. A. Graham and D. M. Webb, this proceedings.

ANALYTICAL ELECTRON MICROSCOPY STUDY OF

SHOCK SYNTHESIZED ZINC FERRITE

R. A. Graham and M. J. Carr

Sandia National Laboratories*
Albuquerque, New Mexico 87185

ABSTRACT

Mixtures of zinc oxide and hematite were shock loaded to various conditions, some of which resulted in the formation of zinc ferrite. Selected samples of reacted and unreacted mixtures were examined for the presence of zinc ferrite. Shock synthesized ferrite, having the spinel crystal structure, was found to occur as recrystallized regions within a heavily deformed matrix of well mixed reactants. The synthesized ferrite was not stoichiometric, but rather was found to be $Zn_{1-x}Fe_{2+x}O_4$, where x=0.09+/-0.02.

INTRODUCTION

Chemical synthesis has been observed in a number of inorganic systems,[1] but detailed microstructural and microchemical characterizations of the reactants and products have rarely been reported. The mechanisms by which chemical synthesis may occur during a shock pulse are not well understood, although shock compression of ceramic powders has been shown to produce large numbers of point, line, and surface defects while simultaneously violently mixing the reactants on a microscopic level at high pressure and elevated temperature. These conditions have been cited as contibuting to enhanced chemical reactivity. Zinc ferrite has been synthesized by shock compression, and the physical and chemical properties of the products have been characterized by several different analytical techniques[2-4] in an effort to study the changes which result from the shock loading.

The purpose of this paper is to present the results of an analytical electron microscopy (AEM) study where microstructural, crystallographic, and microchemical information were obtained from selected samples of reacted and unreacted mixtures of zinc oxide (ZnO) and hematite (Fe_2O_3).

* This work performed at Sandia National Laboratories supported by the U. S. Department of Energy under Contract Number DE-AC04-76DP00789.

MATERIALS AND METHODS

The starting powder was a well blended 1:1 mixture of zinc oxide and hematite, the correct stoichiometry for $ZnFe_2O_4$, lightly ground to break up coarse agglomerates and passed through a 100 mesh sieve. The powders were shock loaded in the Sandia Bear fixtures described by Graham and Webb.[5] Samples of the starting blend were pressed into recovery capsules, taking special care to control the density of the compact, which has a strong effect on the peak temperature experienced during the shock loading. Although a wide range of shock compression conditions were examined, only two shocked samples and the unshocked starting material were characterized by AEM. The two shocked samples were selected to represent conditions where ferrite synthesis did and did not occur, based on x-ray diffraction measurements.[2] The reacted powder was shocked to 27 GPa and 725 degrees centigrade (exp. # 14G846), while the unreacted powder was shocked to 22 GPa and 300 degrees (exp. # 41G846). The pulse duration in both shocks was about 5 microseconds.

These two shock loaded and the initial blended, unreacted powder samples studied by AEM were prepared by centrifuge compaction of epoxy-powder mixtures, followed by ion micromilling, a method[6] shown to reliably produce a true representation of powder microstructures. Electron transparent areas were first examined in a JEOL 200CX transmission electron microscope operated at 200 KeV. The higher voltage produces better images of thicker regions without noticeable beam damage. The presence of zinc oxide, hematite, and/or ferrite was determined by electron diffraction analysis. Computed[7] diffraction patterns were based on the crystallographic data summarized in Table I. The same areas were then examined at 100 KeV with a JEOL 100C analytical electron microscope, equipped for scanning transmission electron microscopy (STEM) and x-ray energy dispersive spectroscopy (EDS). Quantitative compositional data for zinc and iron were obtained by processing the spectra obtained with the standard Sandia thin film analysis algorithm.[8] The k-factor ($k_{Zn/Fe}$=.735) used to quantify the data was obtained by analyzing a spinel oxide standard of known composition ($ZnFe_2O_4$+MnO, 95:5 wt%) in the same instrument prior to analyzing the shocked powders. The precision of the compositional measurement was limited by the counting statistics to +/- 3.7% relative.

TABLE I: Crystallographic parameters for the phases in this study

ZINC OXIDE[9]: hexagonal, a=.3250 nm, c=.5207 nm
(ZnO) space group $P6_3mc$, 2 atoms/unit cell

HEMATITE[9]: hexagonal, a=.5035 nm c=1.375 nm
(Fe_2O_3) space group $R\bar{3}c$, 30 atoms/unit cell

FERRITE[9,10]: cubic (spinel), a=.8440 nm
($ZnFe_2O_4$) space group Fd3m, 56 atoms/unit cell

RESULTS

The starting powder, shown in Figure 1a, was found to be a mixture of well annealed zinc oxide and hematite (i.e., low dislocation density). The grain size was approximately 200 nm. The original particles were only loosely bonded and broke up easily into individual grains when the mixture was blended. Blending was complete, down to the individual grain level. Both zinc oxide and hematite were identified by electron diffraction.

The shocked but unreacted powder mixture was found to contain no ferrite. The microstructure of the unreacted powder, shown in Figure 1b, consists of heavily dislocated, macroscopically deformed but unbonded grains. Electron diffraction polycrystalline ring patterns showed the presence of both compounds in all areas examined. In experimental polycrystalline electron diffraction ring patterns from zinc oxide and hematite, many overlaps occur so that only the {1010} ZnO ring and the {0112} Fe_2O_3 rings are useful for analysis. The {111} ring from ferrite is well separated from other rings, and was taken as diagnostic of ferrite, although none was found in this material. Structurally, only hematite grains could be easily located; the softer zinc oxide was comminuted, smeared, or flowed between the harder hematite grains. EDS analyses across 'grain boundaries' also showed zinc concentrated mainly between hematite grains, although most hematite grains measured appeared to contain at least some zinc. Some hematite was found to contain as much as 5 wt% zinc.

The shocked and reacted powder was found to contain ferrite as well as residual zinc oxide and hematite. Ferrite grains were distinct from the unreacted material, exhibiting a strain and defect free recrystallized microstructure (Figure 1c). The ferrite grain size was about 500 nm. Good single crystal diffraction patterns were obtained which allowed a firm identification to be made. The recrystallized grains were found in isolated clusters of many grains, extending typically 10,000 nm across. EDS data showed that the Zn/Fe stoichiometry of the ferrite to be $Zn_{1-x}Fe_{2+x}O_4$, where $x=0.09+/-0.02$, in good agreement with the composition inferred from lattice parameter data[2]. Although there was some scatter in the compositional measurements, no evidence was found for compositional gradients within ferrite grains.

DISCUSSION

The samples studied in this investigation were selected from a larger study,[2] and represent two extreme conditions. The unreacted sample experienced the highest peak pressure and temperature condition of any sample that did not show significant conversion to ferrite. (As discussed elsewhere,[2,3,4] most 'unreacted' samples contained small fractions of ferrite detectable by sensitive magnetic or nuclear techniques.) The reacted sample contained the highest fraction of ferrite of any shock loaded sample. The unshocked sample provided a baseline for comparison. These samples were found to have distinct and unique structural characteristics; it is possible that in other samples produced in the full study, intermediate structures and/or mixtures may exist.

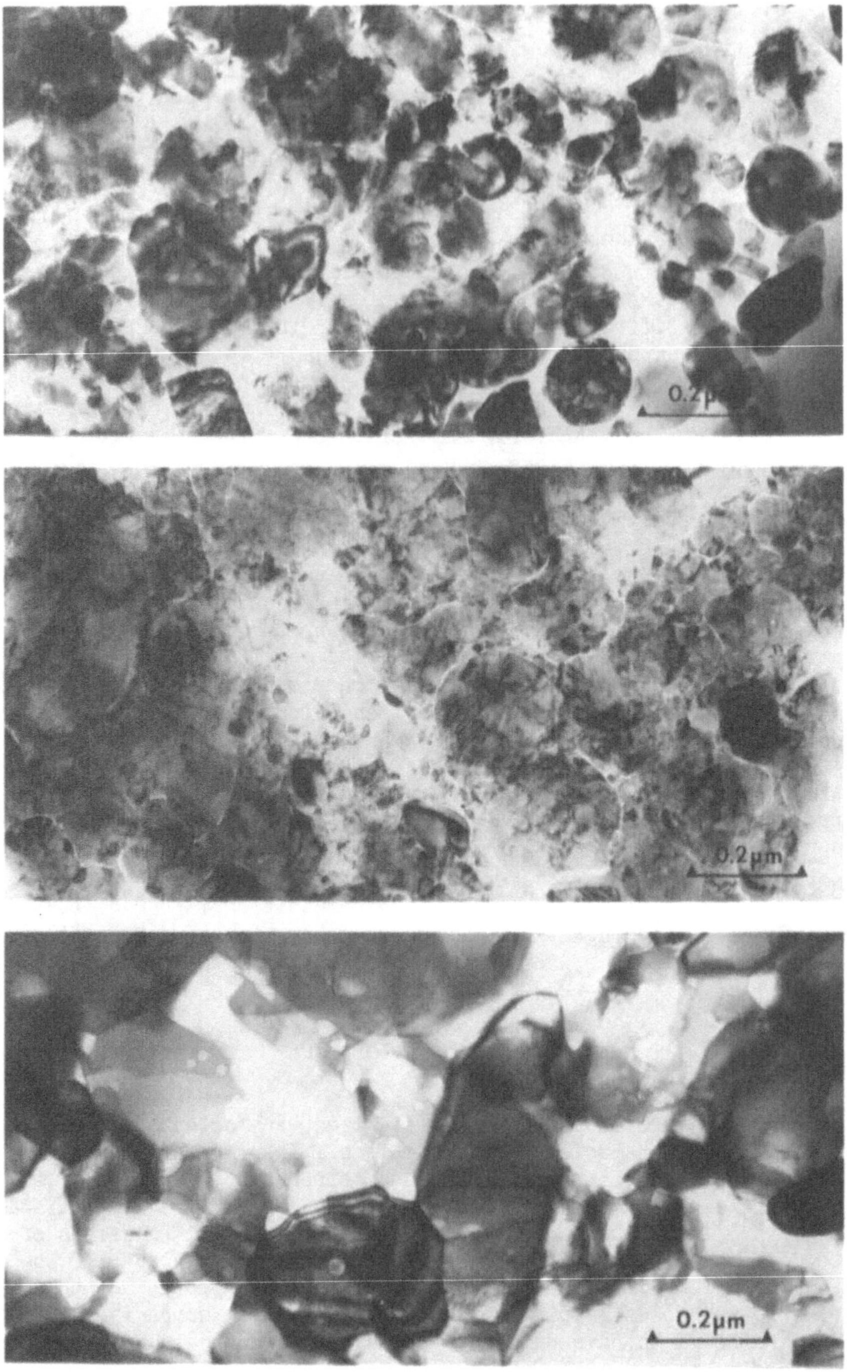

a

b

c

Fig. 1. TEM brightfield micrographs showing the microstructures of: a. Unshocked starting mixture of powders. b. Shock loaded, but unreacted powders. Note the macroscopic deformation of the individual powders. c. Shock-synthesized zinc ferrite. Note defect-free, recrystallized structure.

The formation of ferrite was found to be associated with a recrystallized microstructure. Such microstructures generally result when stored strain energy in a lattice, due to defects, is relatively high and sufficient thermal energy is available to allow strain free grains to nucleate and grow by diffusional processes. The strain energy of the shocked powders is high, a result of the high density of defects. The unreacted mixture reached a calculated bulk mean temperature of 300 degrees centigrade, while the reacted powder reached 725 degrees centigrade, in each case for only a few microseconds during the shock pulse followed by a rapid quench lasting at most a few seconds. Both temperatures are rather low fractions of the melting points of these materials and would not be expected to cause recrystallization under normal conditions. However conditions achieved during shock loading are far from normal. Very high defect densities have generally been shown to lead to recrystallization at lower temperatures,[11] and recrystallization has been observed in shock loaded ceramic materials, usually at higher shock pressures.[12] There is evidence[13] that this reaction does not involve significant exothermic energy release so localized heating at the reaction front does not seem plausible in this case. Defect enhanced diffusion has been shown to substantially lower the activation energy for diffusion,[14] and the effect may be large enough to play a role in this reaction.

Ferrite synthesis is presumed to occur during the shock pulse and possibly for brief interval afterward while the temperature remains elevated. Enhanced diffusion at the reaction boundary is required to account for the redistribution of elements from the reactant mixture to the uniform product phase.

CONCLUSIONS

At the highest peak pressure and temperature studied, the synthesis of zinc ferrite by shock compression was found to be recrystallized. The composition of the shock synthesized ferrite was found to be $Zn_{1-x}Fe_{2+x}O_4$, with x=0.09+/-0.02. The microstructure of the shock loaded but unreacted powder was found to consist of heavily deformed grains of hematite, well mixed with zinc oxide. The defective microstructure clearly resulted from the shock pulse, since the starting powder mixture was well blended but relatively low in observable defects.

REFERENCES

1. B. Morosin and R. A. Graham, "Shock-Induced Inorganic Chemistry," in Shock Waves in Condensed Matter-1981, ed. by W. J. Nellis, L. Seaman, and R. A. Graham, North-Holland (pub.), 1982, p. 4.
2. B. Morosin, E. L. Venturini, and R. A. Graham, "X-Ray Diffraction Studies of Shock-Synthesized Zinc Ferrite," this proceedings.
3. D. L. Williamson, B. Morosin, E. L. Venturini, and R. A. Graham, "Mossbauer Studies of Shock-Synthesized Zinc Ferrite," this proceedings.
4. E. L. Venturini, B. Morosin, D. L. Williamson, and R. A. Graham, "Magnetic Studies of Shock-Synthesized Zinc Ferrite," this proceedings.
5. R. A. Graham and D. M. Webb, "Fixtures for Controlled Explosive Loading and Preservation of Powder Samples," in Shock Waves in Condensed Matter-1983, ed. by J. R. Asay, R. A. Graham, and G. K. Straub, North-Holland (pub.), 1984, p. 211.
6. M. J. Carr, "A Method for Preparing Powdered Specimens for Transmission Electron Microscopy," J. Electron Microscopy Technique, in press.
7. M. J. Carr and W. F. Chambers, "A Review of Crystallographic Calculational Methods Used in the RAD group of Computer Programs for Analytical Electron Microscopy," J. of Microscopy, **134**, p. 55.
8. A. D. Romig, Jr., "X-Ray Microanalysis in the Analytical Electron Microscope," Sandia National Laboratories Report SAND82-2938, Albuquerque, New Mexico, 1983.
9. W. B. Pearson, "A Handbook of Lattice Spacings and Structures of Metals and Alloys," Vol. 2, Pergamon Press, Oxford, 1967, p. 403.
10. Landolt-Bornstein, Numerical Data and Functional Relationships in Science and Technology: Volume 4--Magnetic and Other Properties of Oxides and Related Compounds, ed. by K. H. Hellwege and A. M. Hellwege, Springer-Verlag, Berlin, 1970, p. 315.
11. R. E. Reed-Hill, in Physical Metallurgy Principles, Van Nostrand Rheinhold Company (pub.), New York, 1964, p. 192.
12. M. J. Carr, "Deformation Microstructures in Shock Modified Ceramics," in High Pressure Explosive Processing of Ceramics, ed. by R. A. Graham and A. Sawaoka, in preparation.
13. J. Beretka and A. J. Marriage, "The Initial Formation Temperature of Ferromagnetic Spinels," Aust. J. Appl. Sci., **15**, 1964, p. 321.
14. A. D. Romig, Jr. and J. I. Goldstein, "The Diffusivity of Ni in Fe-Ni and Fe-Ni-P Martensites," Met. Trans. A, **12A**, 1981, p. 243.

MOSSBAUER STUDY OF SHOCK-SYNTHESIZED ZINC FERRITE

D. L. Williamson*, B. Morosin**, E. L. Venturini**, and R. A. Graham**

*Physics Department, Colorado School of Mines, Golden, CO.
**Sandia National Laboratories, Albuquerque, NM

ABSTRACT

Mössbauer spectroscopy has been used to characterize ZnO-Fe_2O_3 powder mixtures subjected to shock loading at 22 GPa with associated peak shock temperatures ranging from 300°C to 655°C. The ^{57}Fe spectra for the shock-synthesized zinc ferrite yield similar values for the isomer shift, quadrupole splitting and magnetic hyperfine field suggesting that the same composition is produced over a wide range of shock conditions. A comparison is made to furnace-reacted material prepared from the same powder mixture.

INTRODUCTION

Recent reports [1,2] describe the use of carefully controlled shock-wave loading to synthesize ferrimagnetic zinc ferrites from equimolar powder mixtures of ZnO and α-Fe_2O_3 (hematite). Companion papers in this proceedings describe further characterization of the shock-modified material by x-ray diffraction [3], magnetic measurements [4], and electron microscopy [5]. Here we report results of ^{57}Fe Mössbauer spectroscopy analysis of one group of samples from the extensive set of shock experiments [3]. Previous Mössbauer studies of $Zn_{1-x}Fe_{2+x}O_4$ for $x = 0$ to $x = 1$ show the ability of this local Fe environment probe to provide information on magnetic structure [6] and relaxation [7], octahedral (A) vs tetrahedral (B) Fe-site occupation [8], charge transfer processes [8] and thermal formation mechanisms [9]. The Mössbauer spectral parameters established by these studies provide a basis for comparison of the shock-synthesized spinel parameters obtained here and allow an estimate of the stoichiometry factor x. The results are also compared to the x-ray and magnetization measurements of powders from the same shock experiments and to Mössbauer data from furnace-fired samples prepared from the same starting mixture of ZnO-Fe_2O_3.

EXPERIMENTAL

The shock experiment details are described elsewhere [1,10]. The furnace-reacted samples were prepared by compressing the equimolar ZnO-Fe_2O_3 mixed powder into 0.5 g disks and heating in air at various temperatures for 2 h. Absorbers for Mössbauer transmission measurements were prepared by dispersing 30 mg of the shocked powder or furnace-reacted powder in an inert binder and compressing into a disk 1.3 cm in diameter. Spectra were acquired in the conventional constant acceleration mode with a 25 mCi ^{57}Co:Rh source. Each spectrum was computer fitted with a superposition of Lorentzian resonance lines to extract the following spectral parameters: δ = isomer shift relative to metallic Fe; Δ = quadrupole splitting (computed here for a six-line magnetic pattern as $(V_2-V_1 + V_5-V_6)/2$ where $V_1(V_6)$ is the lowest (highest) velocity line position; H = magnetic hyperfine field; Γ = full linewidth at half maximum resonance; F = fractional resonance area due to each phase (corrected for thickness saturation effects [11]). The six lines of a magnetic phase were restricted by the computer to have the same linewidth and pair-wise equal intensities.

RESULTS AND DISCUSSION

Figure 1 compares Mössbauer spectra from the starting ZnO-Fe_2O_3 mixture, a powder shocked at 22 GPa/655°C and a powder furnace-fired at 650°C. The subspectra in Fig. 1 show the presence of three distinct Fe resonances in the shocked material, two magnetic phases, M1 and M2, and a central paramagnetic doublet, P. Comparison with the starting powder shows M1 to be from hematite. The furnace fired sample does not show any evidence of the M2 phase. The spectral parameters of M1, M2 and P are listed in Table 1 for the 22 GPa-shocked series and three furnace-reacted samples.

Comparison of the P-phase parameters of the furnace-reacted material with those of stoichiometric, paramagnetic $ZnFe_2O_4$ [12] (δ = 0.358(5)mm/s, Δ = 0.336(5)mm/s) suggests that near-stoichiometric zinc ferrite is formed with a starting temperature near 600°C. The slightly larger Δ for the 650°C sample may be an indication of a deviation from stoichiometry and would be consistent with the x-ray analysis [3] which shows a reduced lattice constant in this ferrite compared to $ZnFe_2O_4$. A previous determination of the initial formation temperature of zinc-ferrite of 585°C [13] agrees with the detection of only 4% of the P-phase at 600°C and is consistent with the limits of detection in the x-ray determination of spinel in this same sample [3].

A separate study of pure hematite shocked to pressures up to 27 GPa [14] demonstrated that no new phases were formed except a very small fraction (~1%) of Fe_3O_4 (magnetite) at the maximum 27 GPa/600°C conditions. The only significant changes in the room temperature spectral parameters were a decrease in Δ and an increase in Γ, both of which are seen here (compare M1 in shocked and starting material, Table 1). Thus, the M2 and P phases do not result from some shock-induced modification of hematite and can be fully ascribed to a shock-synthesized ferrite. Examination of Table 1, the x-ray [3] and magnetic measurements [4] show that this reaction is already initiated at peak shock temperatures of only 300°C (Exp. 41G836) compared to about 600°C for the simple thermal reaction.

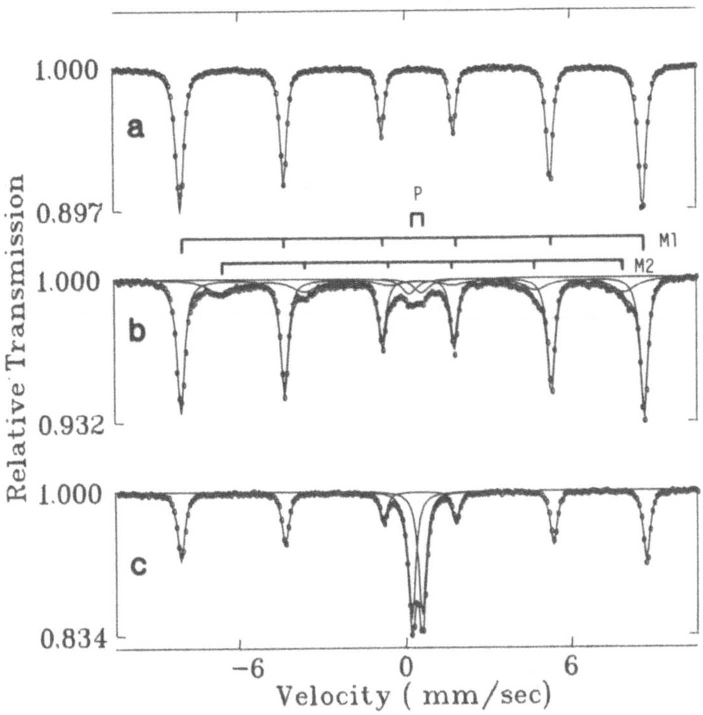

Fig. 1 Mössbauer spectra of : (a) starting mixture of ZnO and α-Fe_2O_3; (b) Sample 8G846, shocked at 22 GPa/655°C; (c) Furnace-fired sample, heated at 650°C in air for 2 h. Solid line passing through data points is least-square fit of sum of Lorentzian lines shown as subspectra.

The M2 parameters δ, Δ, H and Γ in Table 1 are the same within experimental error for all shocked samples. This suggests that the same phase is being generated by the 22 GPa peak shock pressure and that the effect of the increasing peak temperature is simply to increase the fraction of M2. A comparison of the M2 parameters with literature values from thermally prepared $Zn_{1-x}Fe_{2+x}O_4$ [8] shows close agreement for the octahedral (B)-site with x = 0.8. Based on our experimental uncertainties and those in the literature [8] we estimate an uncertainty in x of ± 0.1. For example, our M2 values clearly do not agree with x=0.6 where H = 41.2(1.0) T and Δ = -0.14(5) mm/s [8]. The rather broad linewidth of $\Gamma \simeq 0.9$ mm/s is evidence for a range in H of about ± 10 kOe, which would correspond to a range in x of approximately ± 0.1 relative to x = 0.8 [8]. The tetrahedral (A)-site resonance is not readily observable in the shocked material because of strong overlap with the M1 (hematite) resonance at room temperature. Studies at other temperatures may help to resolve this component. The experimental and theoretical A-site populations [8] for x = 0.8 are 25% and 27.8% respectively.

On the other hand, one should exercise caution in extracting an x-value for our shock-modified materials based on thermally prepared samples. The exact nature of the shock-synthesized products is unknown and their relation to samples synthesized by more conventional chemical reactions is only recently being evaluated.

Table 1. Room Temperature Mössbauer Spectral Parameters

Conditions (Exp. #)	Phase	δ (mm/s)	Δ (mm/s)	H (T)	Γ (mm/s)	F (%)
Starting Mixture						
	M1	0.368(2)[a]	-0.216(3)	51.63(3)	0.305(6)	100
Shocked Samples						
22 GPa/300°C (41G836)	M1	0.365(3)	-0.207(4)	51.58(3)	0.323(6)	93(1)
	M2	0.6(1)	+0.1(1)	45.2(6)	0.7(2)	6(1)
	P	0.3(1)	b	0	0.7(2)	1.0(5)
22 GPa/400°C (24G846)	M1	0.366(3)	-0.202(4)	51.51(3)	0.336(6)	90(1)
	M2	0.7(1)	+0.1(1)	45.4(5)	0.8(2)	9(1)
	P	0.3(1)	b	0	0.7(3)	1.0(5)
22 GPa/475°C (33G846)	M1	0.366(3)	-0.206(4)	51.56(3)	0.323(6)	84(1)
	M2	0.63(6)	+0.10(6)	45.3(3)	0.7(1)	14(1)
	P	0.3(1)	b	0	0.7(2)	2(1)
22 GPa/550°C (25G846)	M1	0.367(3)	-0.204(4)	51.52(3)	0.322(6)	76(1)
	M2	0.64(8)	+0.11(6)	45.1(3)	0.9(2)	18(1)
	P	0.37(8)	0.4(1)	0	0.5(2)	6(1)
22 GPa/655°C (08G846)	M1	0.367(3)	-0.207(4)	51.51(3)	0.319(6)	70(1)
	M2	0.60(5)	+0.09(5)	44.9(3)	1.0(1)	23(1)
	P	0.37(9)	0.4(1)	0	0.6(2)	7(1)
Furnace Reacted Samples						
490°C/2 h	M1	0.363(3)	-0.227(4)	51.46(3)	0.302(4)	100
600°C/2 h	M1	0.363(3)	-0.226(3)	51.49(3)	0.298(4)	96(1)
	P	0.36(3)	0.38(4)	0	0.29(5)	4(1)
650°C/2 h	M1	0.364(3)	-0.217(4)	51.57(3)	0.291(6)	53(1)
	P	0.353(3)	0.372(5)	0	0.314(8)	47(1)

(a) Probable error in last significant figure, based on statistics and deconvolution uncertainties, is given in parentheses after each parameter.
(b) Fitted with single broad line so Δ is undetermined.

It is tempting to attribute the doublet P resonance in the shocked samples to paramagnetic $Zn_{1-x}Fe_{2+x}O_4$ which occurs for $x \leq 0.2$ [7,8]; however, this would imply the unlikely co-formation of two distinct phases with $x \leq 0.2$ and $x \simeq 0.8$. Another possibility is that P corresponds to Fe which is exhibiting superparamagnetic or electronic relaxation [7,15]. Cooling of sample 8G846 and another sample with a much higher P fraction (14G846) [3] shows a gradual decrease in F(P). This behavior is consistent with relaxation phenomena as observed by the Mössbauer technique [16] so that the P phase material very likely is the identical defective zinc ferrite composed of smaller crystallites.

An additional test that the same type of zinc ferrite is produced for this series of samples is the correlation between the fractional ferrite resonance, F(P) + F(M2), and the low temperature saturation magnetization, M_s[4]. Since M_s varies with x, a linear correlation would be good evidence for constant x for the series. Such a correlation is indeed observed in Fig. 2. If we take x = 0.8, as estimated above, and

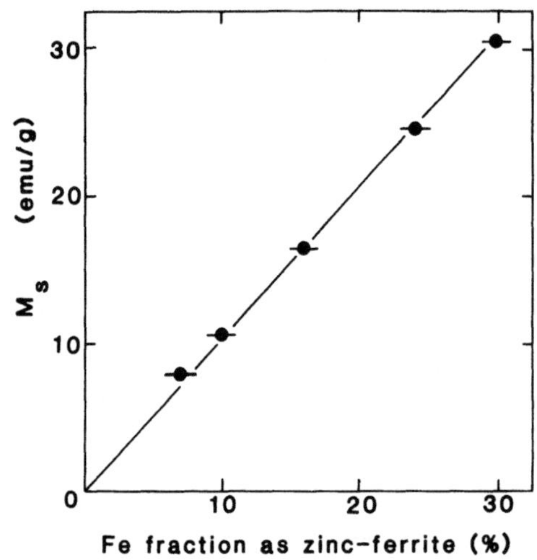

Fig. 2 Correlation of low temperature (7 K) saturation magnetization, M_s, and the fraction of Fe as the zinc ferrite phase as determined from the Mössbauer spectra (F(P) + F(M2) from Table 1).

correct for a 25% A-site resonance [8] which is assumed unresolved from M2, then we obtain a $Zn_{0.2}Fe_{2.8}O_4$ weight fraction of 26% (volume fraction: 27%) for sample 8G846 and 21% (23%) for sample 25G846 which are in good agreement with the x-ray values [3] of 27% and 25%, respectively. Finally, if we use the slope of Fig. 2 and again correct for the A-site resonance, we obtain a value of 117 emu/g for M_s of the zinc ferrite. Comparing this value with $M_s(x)$ of thermally-reacted $Zn_{1-x}Fe_{2+x}O_4$ [6] yields $x \simeq 0.85$ or $x \leq 0.4$. The former choice provides consistency of the magnetization data and Mössbauer analysis. Again this is based on thermally prepared materials as "standards" for data evaluation which in themselves report wide variances in M_s data [17].

CONCLUSIONS

The zinc ferrite produced by solid-state reaction at shock pressures of 22 GPa and a range of peak shock temperatures from 300°C to 655°C has Mössbauer hyperfine parameters that match those of $Zn_{1-x}Fe_{2+x}O_4$ for x = 0.8 ± 0.1. The quantitative fractions of this phase are in good agreement with x-ray analysis and show a linear correlation with low temperature saturation magnetization data. The apparent paramagnetic Mössbauer doublet in the shocked material is attributed to a superparamagnetic or electronic relaxation effect associated with the presence of small crystallites of the same shock-produced phase rather than to the formation of paramagnetic $ZnFe_2O_4$. The latter material (with x close to 0) is produced by thermal reaction above about 600°C, in sharp contrast to the composition and initial reaction temperature (< 300°C) of the shock-synthesized zinc ferrite.

REFERENCES

1. E. L. Venturini, B. Morosin and R. A. Graham, J. Appl. Phys. 57:3814 (1985).
2. E. L. Venturini, B. Morosin and R. A. Graham, Matl. Lett. (to be published).
3. B. Morosin, E. L. Venturini and R. A. Graham, this proceedings.
4. E. L. Venturini, B. Morosin and R. A. Graham, this proceedings.
5. M. J. Carr and R. A. Graham, this proceedings.
6. P. A. Dickof, P. J. Schurer and A. H. Morrish, Phys. Rev. B22:115 (1980).
7. C. M. Srivastava, S. N. Shringi and R. G. Srivastava, Phys. Rev. B14:2041 (1976).
8. D. C. Dodson, J. W. Linnett and M. M. Rahman, J. Phys. Chem. Solids 31:2727 (1970).
9. G. Chol, Y. Gros, J. C. Peday-Peyroula and J. P. Barbe, Matl. Res. Bull. 2:753 (1967).
10. R. A. Graham and D. M. Webb in: "Shock Waves in Condensed Matter," J. R. Asay, R. A. Graham and B. K. Straub, eds., North Holland, New York (1983).
11. D. L. Williamson, T. W. Guettinger and D. W. Dickerhoof in: "Mössbauer Spectroscopy and Its Chemical Applications," J. G. Stevens and G. K. Shenoy, eds., American Chemical Society, Washington, D.C. (1981).
12. M. Robbins, G. K. Wertheim, R. C. Sherwood and D. N. E. Buchanan, J. Phys. Chem. Solids 32:171 (1971).
13. J. Beretka and A. J. Marriage, Aust. J. Appl. Sci. 15:321 (1964).
14. D. L. Williamson, E. L. Venturini, R. A. Graham and B. Morosin, in preparation.
15. J. K. Srivastava, K. Muraleedharan, and R. Vijayaraghaven, Phys. Letters 104A:482 (1984).
16. S. Morup, J. A. Dumesic and H. Topsoe in: "Applications of Mössbauer Spectroscopy, Vol. II", R. L. Cohen, ed., Academic Press, New York (1980).
17. R. Parker, C. J. Rigden and C. J. Tensley Trans. Faraday Soc. (London) 65:219 (1969).

MAGNETIC PROPERTIES OF SHOCK-SYNTHESIZED AND FURNACE-REACTED ZINC FERRITE*

E. L. Venturini, B. Morosin and R. A. Graham

Sandia National Laboratories*
Albuquerque, NM 87185

ABSTRACT

The static magnetic properties of spinel structure zinc ferrite are compared for products from shock synthesis and conventional isothermal reaction using a stoichiometric mixture of zinc oxide and ferric oxide powders. The high-pressure shock-wave loading produces a ferrimagnetic material with a substantial saturation magnetization, while the furnace reaction product is paramagnetic at room temperature. Since stoichiometric zinc ferrite is paramagnetic, these results indicate that a zinc-deficient ferrite is formed by shock processing in contrast to a nearly stoichiometric compound obtained by heating in air. X-ray diffraction data confirm that less zinc oxide is consumed than ferric oxide during shock loading.

INTRODUCTION

The dynamic, highly nonequilibrium nature of high-pressure shock-wave loading affords an opportunity to synthesize solids with chemical and physical properties different from those prepared by more conventional processing techniques.[1,2] Prior work[3,4] has demonstrated the synthesis of zinc ferrite by shock compression of a mechanical mixture of zinc oxide and ferric oxide. Numerous investigations[5-8] of the solid state reaction of ZnO and α-Fe_2O_3 (hematite) powders at elevated temperatures indicate a complex process. Below 1100 C the rate of formation of the spinel phase is determined by zinc diffusion.[7,8] As neither of the starting materials is strongly magnetic while the synthesized ferrite may be ferrimagnetic, magnetization data provide an effective probe of the synthesized material. Further, magnetic properties are very sensitive to atomic level defects, site disorder and deviations from stoichiometry.

$ZnFe_2O_4$ is paramagnetic at temperatures above an antiferromagnetic transition near 10 K.[9] It crystallizes in the normal spinel structure

*This work performed at Sandia National Laboratories supported by the U. S. Department of Energy under Contract No. DE-AC04-76DP00789.

where the tetrahedral or A sites are occupied by divalent zinc ions while the octahedral or B sites contain trivalent iron ions.[10] The static magnetic properties of nonstoichiometric ferrimagnetic zinc ferrite $Zn_{1-x}Fe_{2+x}O_4$ where the A sites contain both Fe and Zn have been studied extensively.[5,8,11,12] The low temperature saturation magnetization increases from 25 emu/gram at x = 0.10 to 125 emu/gram at x = 0.30, and remains above 100 emu/gram for x between 0.3 and 1.0.[8]

In the present paper we compare the static magnetic properties of a mixture of ZnO and α-Fe_2O_3 powders subjected to a peak shock pressure of 22 GPa at various shock temperatures with the properties of the same starting mixture reacted in air at elevated temperature in a furnace. Both processing techniques yield a powder containing substantial amounts of spinel-structure zinc ferrite. During shock synthesis, however, the few microsecond duration of the event and the highly nonequilibrium conditions prevailing allow the formation of a new compound only through defect controlled diffusion and mechanically driven mass motion. Hence the zinc ferrite produced by high-pressure shock-wave loading may differ substantially from material prepared by more conventional furnace processing. The dramatic differences are revealed by comparing the static magnetic properties of the respective products.

EXPERIMENTAL DETAILS

The ZnO and α-Fe_2O_3 powders used in these experiments were commercial grade (99.9% purity, CERAC, Inc., Milwaukee, WI). A stoichiometric batch for all experiments was prepared by ball milling the powders together for 24 hours. Scanning electron microscopy showed the dispersal at grain level with typical particle size of 0.3 microns.[13] A portion of the stoichiometric batch powder was pressed into compacts in place within the copper capsules to various initial packing densities, and subjected to controlled, quantifiable explosive loading. The conditions for the five shock experiments discussed in this paper are included in Table I. By using initial powder packing densities of 2.15 to 3.30 Mg/m^3 the peak shock temperature was varied from 655 to 300 C at the same peak shock pressure of 22 GPa.[14] Quantification of pressure and temperature conditions within the compact is achieved with an extensive program of numerical simulation described elsewhere.[14] An important feature of these capsules is the gas-tight seal of the powder compact which is maintained throughout the shock-wave loading and unloading. A separate portion of the batch powder was pressed into 1/2 gram disks for heating in air in a furnace. The reaction temperatures of the four disks used in this work are shown in Table I.

Static magnetization measurements for samples consisting of 20 to 100 mg of loose powder were made on a SQUID susceptometer (S.H.E. Corporation, San Diego, CA) for temperatures between 7 and 400 K and magnetic field strengths from 5 to 40 KOe. There was negligible field hysteresis observed above 5 kOe for any powder in this study, and no change with time in the magnetization up to 400 K, i.e., no annealing of the shock-synthesized material. In general the magnetization of a material can be separated into a ferromagnetic (or ferrimagnetic) term which saturates (reaches a constant value) at sufficiently large field strengths and a paramagnetic term which increases linearly with applied field. The simplest empirical expression[4] containing both terms is:

$$M = M_s/[1 + 4\pi A/H] + \chi \cdot H \qquad (1)$$

where M is the measured sample magnetization for a given applied magnetic field H, M_s is the saturation magnetization for the ferrimagnetic term, A is a magnetic "hardness" which determines the ferrimagnetic approach to saturation, and χ is the paramagnetic susceptibility. Eq. (1) was used to analyze all magnetization data in this study.

RESULTS AND DISCUSSION

Fig. 1 compares the static magnetization versus applied field at 7 and 300 K for two powders, one recovered from a shock synthesis experiment with a peak shock pressure of 22 GPa and a peak temperature of 655 C (experiment 08G846 indicated by open circles), and the other prepared by reacting the oxide mixture in a furnace for two hours at 690 C (shown as solid triangles). At both measurement temperatures the samples are quantitatively different: the shock-synthesized material has a magnetization nearly independent of magnetic field, while the furnace-reacted sample shows a dominant linear increase with applied field. This demonstrates a major difference in materials prepared by shock-loading and furnace reaction using the same starting mixture: the magnetic response of the shocked powder is predominantly ferrimagnetic, while that of the furnace material is paramagnetic. The solid curves in Fig. 1 are calculated using the values for M_s and χ given in Table I. A separate study[15] of the magnetization of shock-loaded hematite (α-Fe_2O_3) powder shows that the increases in M_s and χ in the shocked oxide mixture are not due to changes in the unreacted hematite.

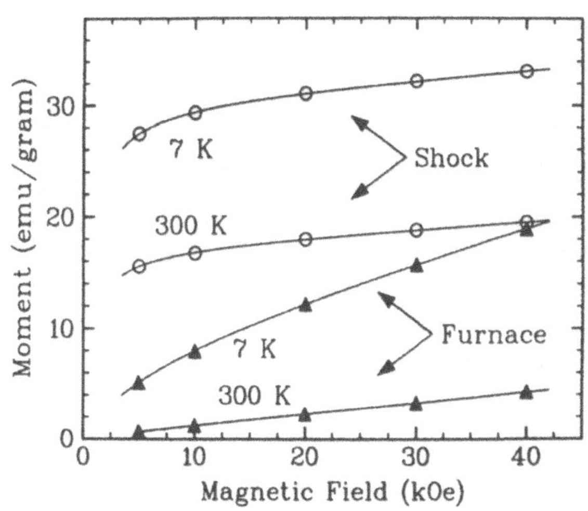

Fig. 1. Magnetization versus applied magnetic field at 7 K and 300 K for powder recovered from shock-compression experiment 08G846 (open circles) and powder prepared by furnace reaction (solid triangles). The solid lines are calculated from Eq. (1) using parameters given in Table I.

Table I. Comparison of Shock-Loaded and Furnace-Reacted Mixtures

Experiment	Summary of Conditions	M_s (emu/gram) at 7 K	M_s (emu/gram) at 300 K	χ (10^{-5} cm^3/gram) at 7 K	χ (10^{-5} cm^3/gram) at 300 K	% spinel (X-ray)
		starting mixture				
--	--	0.015	0.17	1.1	1.3	--
		shock-synthesized				
41G836	22 GPa/300 C	8.2	6.4	2.0	2.1	≈6
24G846	22 GPa/400 C	10.7	8.3	2.5	2.1	8
33G846	22 GPa/475 C	16.5	12.6	2.7	2.3	8
25G846	22 GPa/550 C	24.6	15.8	5.3	4.1	25
08G846	22 GPa/655 C	30.5	17.2	7.7	6.6	27
		furnace-reacted				
---	490 C/2 hrs	0.034	0.21	1.2	1.3	--
---	600 C/2 hrs	0.47	0.28	2.9	1.7	--
---	690 C/2 hrs	8.8	0.34	29.	10.	80
---	800 C/2 hrs	16.0	0.49	33.	13.	95

The values for M_s and χ at 7 and 300 K for the shock-loaded and furnace-reacted samples are given in Table I. Also shown are values for the starting stoichiometric mixture of zinc and ferric oxide. The decrease in M_s by an order of magnitude between 300 and 7 K for the starting powder is due to the Morin transition[16] in α-Fe_2O_3 near 250 K. This decrease is also evident in the furnace-reacted material held at 490 C for 2 hours where little reaction of the oxides is detected.

The M_s values for the shock-synthesized powders are substantial at 300 K, increasing from 6 to 17 emu/gram as the peak shock temperature rises from 300 to 655 C for a constant peak shock pressure of 22 GPa. Furthermore, M_s for a given shocked material increases by only 1/3 to 2/3 between 300 and 7 K, suggesting that the magnetization of these samples is dominated by a ferrimagnetic component with a magnetic ordering temperature well above 300 K. Additional evidence for this conclusion is the small increase in χ for all five shocked powders between 300 and 7 K.

Stoichiometric $ZnFe_2O_4$ is paramagnetic[9,17] above 10 K, so the ferrimagnetic component in the shock-loaded powder must be substoichiometric $Zn_{1-x}Fe_{2+x}O_4$[8,11,12] or stochiometric $ZnFe_2O_4$ with an extremely large amount of site disorder[18] (i.e., substantial Fe ions on A sites and Zn ions on B sites). For the shock-synthesized samples with larger amounts of ferrite formed, x-ray diffraction data indicate that a greater amount of α-Fe_2O_3 is consumed relative to ZnO, suggesting that the ferrimagnetism is primarily due to the formation of a defective, iron-rich zinc ferrite. In addition stoichiometric zinc ferrite with 4% site disorder[18] has a five-fold increase in magnetization between 300 in 7 K in contrast to the 1/3 to 2/3 rise observed in Table I.

Table I also shows the % spinel phase present in the recovered powder from x-ray diffraction data, reaching 27% in the highest shock temperature experiment (08G846). Note that M_s and at both 7 K and 300 K and the % spinel phase increase by the same factor of 3 to 4 between shock experiments 41G836 and 08G846 with peak temperatures of 300 and 655 C, respectively. This strongly suggests that the average stoichiometry of the spinel product is the same for these shocked powders despite the wide peak temperature variation. The powder from shock experiment 08G846 with 27% spinel phase has the highest M_s of 30.5 emu/g at 7 K. Since this M_s arises almost entirely from the spinel component, it implies an effective magnetization near 113 emu/g (30.5/0.27) for this constituent. Magnetization versus temperature curves in a 7 kOe field[11] and magnetization versus field curves curves at 4.2 K[12] have been reported for $Zn_{1-x}Fe_{2+x}O_4$ samples with x from 0.2 to 1. The data for $x = 0.4$ and $x = 0.6$ have values for both M_s and χ comparable to those in Table I if the latter are adjusted using the % spinel phase. This analysis suggests that the spinel component in the shock-loaded powder is a nonstoichiometric zinc ferrite with x near 0.5. X-ray determination of deviations in composition from the stoichiometric $x = 0$ is difficult for samples with relatively small spinel concentrations.

In contrast to the shock-synthesized powder, none of the furnace-reacted material shows a significant M_s value at 300 K despite the high conversion to spinel phase at the highest annealing temperatures. The furnace-reacted mixture shows no detectable spinel phase using x-ray diffraction following 2 hour anneals at temperatures up to 600 C. This is in good agreement with data[19] showing that the reaction begins above 580 C. The powder held in air at 800 C for 2 hours is 95% spinel phase, and has a 32-fold rise in M_s between 300 and 7 K and a 250% rise in χ. This behavior is consistent with data[18] for a stoichiometric zinc ferrite formed by a similar furnace treatment, and is attributed to site disorder with 4% of the Zn on B sites. Such site disorder forms ferrimagnetic Fe clusters within the paramagnetic zinc ferrite lattice, and results in a small value for M_s at 300 K and large increases in M_s and χ as the temperature decreases.[18] Hence the furnace-reacted powder is nearly stoichiometric and weakly ferrimagnetic zinc ferrite with nominal site disorder, while the shock-synthesized powder is substoichiometric and strongly ferrimagnetic.

A final point concerns the reaction mechanisms in the two preparation techniques described above. An important distinction besides the obvious major differences in magnetic properties is the reaction temperatures. The two-hour furnace anneals produce little change in the starting mixture for temperatures up to 600 C, while the powder shock-loaded to a peak temperature of only 300 C at a peak pressure of 22 GPa contains ≈6% spinel phase, resulting in a large ferrimagnetic response. This clearly demonstrates an enhanced solid state reactivity due to shock loading.[1,2]

Shock compression conditions of forced relative mass motion, local stress and temperature gradients and rapid quenching from high pressure, high temperature states produce highly nonequilibrium conditions in the powders. Comparable mechanisms are not present in conventional material synthesis techniques such as furnace reaction, and one expects the physical properties of a product from these distinct preparations to differ considerably. The static magnetization of zinc ferrite powder offers striking evidence for the validity of this argument. New opportunities in materials technology appear to be possible based on the unique nature of the shock process. Such synthesis studies also provide a revealing probe of the shock process.

REFERENCES

1. S. S. Batsanov, Perspectives on inorganic chemistry, in: "Shock Waves in Condensed Matter - 1981 (Menlo Park)", W. J. Nellis, L. Seaman, and R. A. Graham, ed., American Institute of Physics, New York (1982), pp. 1-3; S. S. Batsanov, Inorganic synthesis under shock-wave compression, ibid., pp. 14-26.
2. B. Morosin and R. A. Graham, Shock-induced inorganic chemistry, loc. cit., pp. 4-13.
3. Y. Kimura, Formation of zinc ferrite by explosive compression, Japan. J. Appl. Phys. 2, 312 (1963).
4. E. L. Venturini, B. Morosin and R. A. Graham, Magnetic and structural properties of shock-synthesized zinc ferrite, J. Appl. Phys. 57 3814-3816 (1985) and references therein.
5. E. W. Gorter, Saturation magnetization and crystal chemistry of ferrimagnetic oxides, Philips Res. Repts. 9, 295-443 (1954).
6. J. F. Duncan and D. J. Stewart, Kinetics and mechanism of formation of zinc ferrite, Trans. Faraday Soc. 63, 1031-1041 (1967).
7. J. Beretka and M. J. Ridge, Formation of zinc ferrite at low temperatures, Nature 216, 473-474 (1967).
8. R. Parker, C. J. Rigden, and C. J. Tinsley, Solid-state reaction between zinc oxide and ferric oxide, Trans. Faraday Soc. 65, 219-224 (1969).
9. F. K. Lotgering, The influence of Fe^{3+} ions at tetrahedral sites on the magnetic properties of $ZnFe_2O_4$, J. Phys. Chem. Solids 27, 139-145 (1966).
10. J. M. Hastings and L. M. Corliss, Neutron diffraction studies of zinc ferrite and nickel ferrite, Revs. Mod. Phys. 25, 114-121 (1953).
11. C. M. Srivastava, S. N. Shringi, R. G. Srivastava, and N. G. Nanadikar, Magnetic ordering and domain-wall relaxation in zinc-ferrous ferrites, Phys. Rev. B 14, 2032-2040 (1976).
12. P. A. Dickof, P. J. Schurer, and A. H. Morrish, Magnetic structure of zinc-substituted magnetite at T = 4.2 K, Phys. Rev. B 22, 115-127 (1980).
13. M. J. Carr, private communication.
14. R. A. Graham and D. M. Webb, Fixtures for controlled explosive loading and preservation of powder samples, in: "Shock Waves in Condensed Matter - 1983 (Santa Fe)", J. R. Asay, R. A. Graham and G. K. Straub, ed., North Holland, New York (1984), pp. 211-214.
15. D. L. Williamson, E. L. Venturini, R. A. Graham, and B. Morosin, On the Morin transition of shock-modified hematite, to be published.
16. F. J. Morin, Magnetic susceptibility of α-Fe_2O_3 and α-Fe_2O_3 with added titanium, Phys. Rev. 78 819-820 (1950).
17. S. Ligenza, A study of the $^6S_{5/2}$-term splitting of an Fe^{3+} ion in zinc ferrite by neutron spectroscopy, Phys. Stat. Solidi (b) 75, 315-326 (1976).
18. J. M. Hastings and L. M. Corliss, An antiferromagnetic transition in zinc ferrite, Phys. Rev. 102, 1460-1463 (1956).
19. J. Beretka and A. J. Marriage, The initial formation temperature of ferrimagnetic spinels, Australian J. Appl. Sci. 15, 321-327 (1964).

THE EFFECT OF SHOCK COMPRESSION ON GRAPHITE-LIKE BORON NITRIDE

T. Akashi and A. Sawaoka

The Center for Explosives Technology Research
New Mexico Institute of Mining and Technology
Socorro, New Mexico, 87801

and

R.A. Graham

Sandia National Laboratories
Albuquerque, New Mexico, 87115

ABSTRACT

Well crystallized graphite-like boron nitride (g-BN) powder particles were shock treated at a nominal peak pressure of 27 GPa in the Sandia Baby Bear fixture. The microstructure of the recovered BN compacts was found to be sensitive to the pressure and temperature histories within the fixture. G-BN powder densified mainly by particle fracture in the central region of the fixture and by plastic deformation in the outer region of the fixture.

INTRODUCTION

One of the effects of shock compression on graphite-like boron nitride (g-BN) is the initiation of the phase transformation of g-BN to the high dense form of BN with wurtzite type structure (w-BN). The transformation of g-BN to w-BN is induced by shock compression above 12 GPa (1). Most of the work done on the transformation of g-BN to the high dense form BN by shock compression has been done on systems of mixtures of g-BN and metal powder, in order to produce higher shock pressure and quickly cool the BN powders after shock treatment (2,3).

In the present work, well crystallized g-BN powder, without of metal powder, was subjected to controlled shock loading. Microstructures of recovered BN compacts were examined to gain insight into the deformation of BN powder during shock compression.

EXPERIMENTAL PROCEDURES

G-BN powder (Grade GP) provided by Electro Chemical Industry Co., Ltd. (Tokyo) was used as the starting material.

Fig.1 SEM photographs of the g-BN starting powder (a) and green compact (b).

Scanning electron micrographs of the starting powder and the g-BN green compact with a density of about 60% are shown in Fig. 1. The powder was pressed into copper recovery fixtures to densities of 40% and 60% of theoretical and subjected to controlled shock loading. Details of the shock treatment fixture and numerical analysis of the shock loading have been reported elsewhere (4,5). The fixture used in this experiment is designated "Baby Bear B" in reference 5. The explosive used was composition B. Estimation of the shock pressures and temperatures were not carried out on g-BN powder but similar computer calculation on rutile (TiO_2) powder showed a peak pressure of 22 to 32 GPa for the outer and center regions within the powder compact and a mean bulk temperature during shock compression of about 800 K and 1000 K for the initial densities of 60% and 40%, respectively.

Characterization of the recovered BN compacts was performed by x-ray diffraction and scanning electron microscopy. To investigate the influence that the orientation of the shock wave front had on the microstructure of the resulting BN compacts, the orientation of the shock wave front during post-shock examination was retained.

RESULTS AND DISCUSSION

The conversion ratios of g-BN to w-BN in the recovered samples were estimated from the relative intensities of the x-ray diffraction lines (002) for g-BN and (100) for w-BN. The g-BN to w-BN ratio differed throughout the position within the compacts from about 10% at the outer region to about 25% at the center region. However, those ratios were almost the same in the compacts with initial densities of 40% and 60%. This may be attributed to the characteristic powder-compaction process in the shock treatment fixture used, predicted by the two-dimensional simulation. The powder is densified to near solid density during the passage of the initial planar shock wave and subsequently shock-compressed by the passage of the radial shock wave. Peak pressures induced by the radial shock wave are estimated to be about 22 and 32 GPa for the outer and center region, respectively, while shock pressures caused by the initial planar shock wave is only about 3 GPa. Therefore, the transformation of g-BN to w-BN took place during the passage of the radial shock wave into the already compacted g-BN powder. The residual temperatures in both compacts with initial densities of 40 and 60% were too low to re-transform from w-BN to g-BN, except locally. Thus, the conversion ratio is seen to be independent of the initial density of the g-BN powder compacts in this work.

Microstructures of the compacted BN samples differed significantly at different positions within the compacts. Fig. 2(a) and (b) show typical SEM photographs of fracture surfaces in the center and outer regions of the compacts with initial densities of 40 and 60%. The lamellar structure consisting of plate-like BN particles, seen in the g-BN green compacts, was retained after shock treatment in the center regions of both compacts even though these particles were subjected to fracture and deformation as shown in Fig. 2. Fig. 2 (A) reveals that plate-like g-BN particles were broken into small domains of relatively uniform size and shock-compressed. Although some of the BN particles appeared to have been plastically deformed, especially in the compact with initial densities of 40%, the g-BN powder in the center region seemed to be densified mainly by particle fracture perpendicular to the flat surfaces of the plate-like particles rather than by plastic deformation due to the low shock temperature. The large amount of particle deformation seen in the compact with an initial density of 40 % compared to the amount of particle deformation seen in the compact with an initial density of 60 % was probably caused by the high porosity of this compact.

On the other hand, in the outer region of both compacts, macro- and micro-scopic flow was observed. Fig. 3, a low magnification view of the outer region of the compact with an

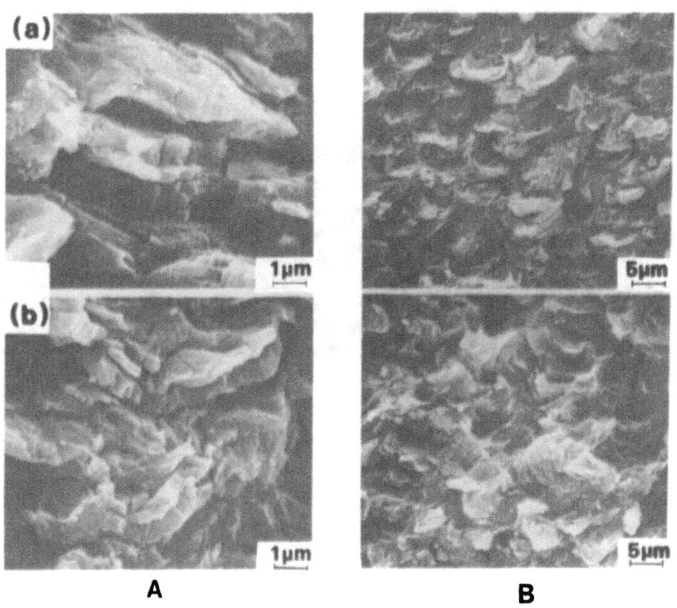

Fig.2 SEM photographs of the fracture surfaces in the center (A) and outer region (B) for the compacts with an initial density of 40 (a) and 60 % (b).

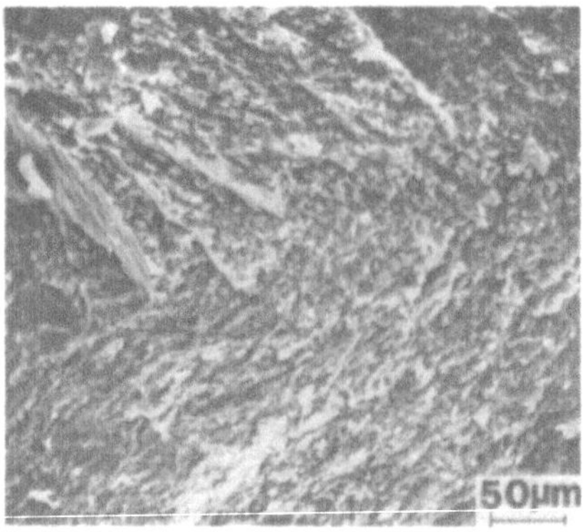

Fig.3 Material flow observed in outer region of the compact with an initial density of 40 %.

initial density of 40 %, shows two oblique mass flows; one is from the left bottom side to the center, the other is from the left top side to the center in the figure. Fig. 2 (B) is a high magnification view of the flow regions in the compacts with initial densities of 40 and 60 %. The figure corresponds to the flow from the left bottom side to the center in Fig. 3. It is difficult to distinguish between individual plate-like particles. The microstructure was not uniform as compared to that seen in the center regions of the compacts. It is probable from these results that the g-BN powder in the outer region was shock-compressed by at least two shock waves from different directions almost instantaneously resulting in the different material flows mentioned above, rather than simply by a radial shock wave. Furthermore, in the outer region localized recrystallization, judging from relatively uniform size and fine grains, took place. This suggests that shock and residual temperatures in this region were higher than in the center region, which may be attributed to the temperature rise resulting from heavy deformation in the outer region. The existence of the high temperatures seen in the outer regions of the powder compacts during shock compression in this recovery fixture have been predicted by a two-dimensional computer simulation on this fixture (5). Recently, M.B.

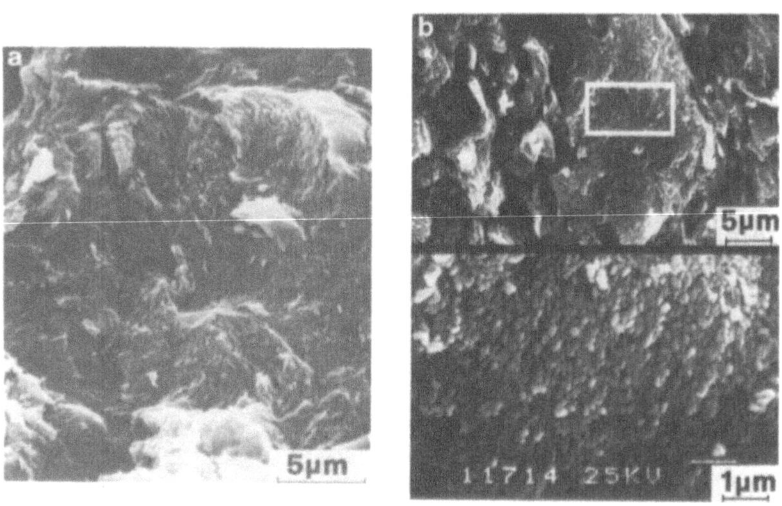

Fig.4 Recrystallization observed in the outer region of the compact with an initial density of 40%.

Boslough and R.A. Graham have reported that an exothermic reaction in Ni-Al powder was ignited in the outer region of the compact during shock compression using this fixture (6). They suggested the existence of multi-dimensional shock waves in the outer region. Fig. 4(a) clearly shows microscopic mass flow which was observed in the outer region of the compact with an initial densitiy of 40%. This reveals that the g-BN particles deformed plastically rather than by particle fracture and that recrystallization occured in this heavily deformed region. Fig. 4(B) also shows such a recrystallized region in the compact with an initial density of 40 %. The shape of the recrystallized grains was uniform and spherical, being about 0.3 μm in diameter. Intense material flow and recrystallization in the outer region implies that the multi-dimensional shock wave configuration caused by the difference in the shock impedances of copper and g-BN powder and the reflection of the shock wave at the boundary of the copper and the powder compact induced localized heavy deformation and high temperatures.

The microstructure of shock-compressed BN powder was found to be sensitive to the pressure and temperature histories within the powder compact. In the shock treatment fixture employed in this experiment, g-BN powder densified mainly by particle fracture in the center region and by plastic deformation in the outer region. The different densification processes are attributed to differences in the shock history at different positions in the powder compact.

REFERENCES

1. G.A. Adadurov, Z.G. Aliev, L.O. Atovmyan, R.V. Bavina, Yu.G. Borod'ko, O.N. Breysiv, A.N. Dremin, A.Kh. Muranevich and S.V. Pershin, "Formation of a Wurtzite-like Modification of Boron Nitride in Shock Compression," Soviet Phys.-Dokl., 12:173 (1967).
2. A. Sawaoka, T. Sōma and S. Saito, "Structure Determination of Boron Nitride transformed by Shock Compression," Japan J. Appl. Phys., 13:891 (1974).
3. T. Sōma, A. Sawaoka and S. Saito, "Characterization of Wurtzite Type Boron Nitride Synthesized by Shock Compression," Mat. Res. Bull., 9:755 (1974).
4. R.A. Graham and D.M. Webb, "Fixture for Controlled Explosive Loading and Preservation of powder Samples," pp. 211 in Shock Waves in Condensed Matter-1983, ed. by J.R. Asay, R.A. Graham and G.K. Staub, North Holland, New York (1984).
5. L. Davison, D.M. Webb and R.A. Graham," Analysis of Capsules for Recovery of Shock-Compressed Matter," pp. 67 in Shock Waves in Condensed Matter-1981, ed. by W.J. Nellis, L. Seaman and R.A. Graham, American Institute of Physics, New York (1982).
6. M.B. Boslough and R.A. Graham, "Submicrosecond Shock-induced Chemical Reactions in Solids: First Real-time Observations," submitted to Chem. Phys. Lett.

EFFECT OF EXPLOSION ON THE CATALYTIC PROPERTIES OF STOCHIOMETRIC INORGANIC SUBSTANCES

V.P. Bokarev, I.N. Temnitsky, Yu. M. Bondarev,
Yu. S. Mardashev, S.S. Batsanov

State Standard Comm. USSR, Leninsky Prospect 9
Moscow 117049, USSR

Earlier we have found that shock treatment of non-stochiometric oxide semiconductors affects their electronic properties and causes a sharp increase of catalytic activity of oxides with electronic conductivity [1].

Shock compression of natural fluorite, a stochiometric compound, not only increases its catalytic activity, but also affects the selectivity of the catalyzed reaction [2]. Nevertheless, we observed no changes of chemical or phase composition of fluorite, but only an increase of density of dislocations and other lattice defects.

In the present work, we have studied the effect of shock compression on the catalytic properties of stochiometric crystalline compounds, namely CaF_2, LiF, and also MGF_2 and MgO on which bis(acetylacetonate) copper, $Cu(Acac)_2$ was placed.

The shock treatment was carried out usually in plane steel recovery systems. To obtain different density of defects in the catalysts we used various schemes of dynamic loading or heating of the samples before compression to different temperatures.

The catalytic activity of CaF_2 and LiF was studied in model reactions of isopropanol transformations. The activity of $Cu(Acac)_2$ placed on the surface of shocked MgF_2 and MgO was studied in model reaction of heptene-1 transformations. Catalytic studies were carried out on an impulse micro-catalytic device with chromatographic analysis of reaction products [3] using pure hydrogen as a carrier gas.

The phase composition and real crystal structure of the starting and shocked catalysts were studied by the X-ray method. The dimensions of coherent X-ray scattering areas D (Å) and residual strains $E_{1/2D}$ were calculated by harmonic analysis of X-ray line profiles. The specific surface of the catalysts was determined from low-temperature adsorption of nitrogen.

As we have shown [2], in the presence of fluorite, isopropanol undergoes dehydration and dehydrogenization. The ability of fluorite to catalyze two different reactions indicates that two types of active centers are present on its surface. Changing of selectivity of fluorite after compression reflects changing of relative amounts of these centers. We found that under certain conditions the shocked samples of fluorite can exhibit a 100% selectivity in dehydrogenization reaction, as well as a 100% selectivity in dehydration. The samples that are more selective in dehydrogenization preserve this property at temperatures up to 800°C. This means that the active centers of this reaction are located where dislocations or grain boundaries intersect with the surface. This supposition is confirmed by the fact that both activity and selectivity of such samples decrease after heating to temperatures above 1200°C, when, according to X-ray data, fluorite undergoes an intense recrystallization.

Our experiment revealed some dependence between the magnitude of plastic deformation of fluorite and shift of selectivity towards dehydrogenation. The selectivity is approximately proportional to the number of dislocations generated by plastic deformation. These data were confirmed by static compression of fluorite using Bridgman anvils [4]. A shift of selectivity towards dehydrogenization takes place at the pressure of 3.5 GPa, due to an increase in plastic deformation and hence the density of dislocations. The shift of selectivity towards the reaction of dehydration is more scarce, and such samples are less thermally stable. Thus we may suppose that active centers of this reaction are connected with some kind of point defects. Up to 700°C such catalysts are stable, therefore active centers are probably the vacancy complexes.

In Fig. 1 is shown the dependence of total catalytic activity on temperature for the starting natural fluorite and shocked samples.

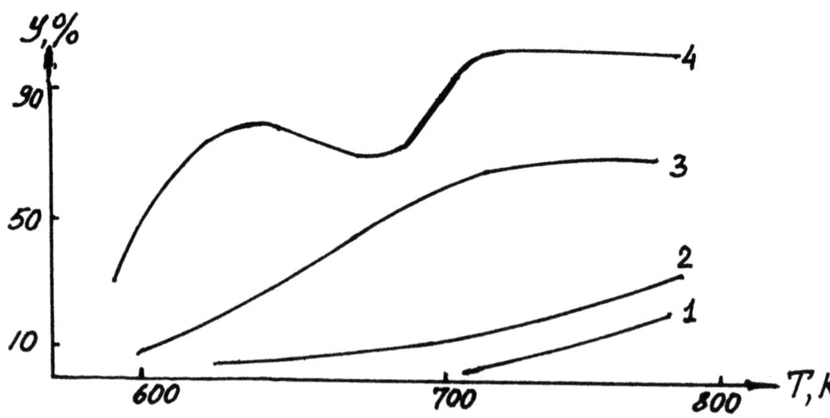

Fig. 1: Catalytic activity of shocked CaF_2 in iso-propanol conversion. 1, unshocked fluorite; 2, absolutely selective in dehydration; 3, absolutely selective in dehydrogenation; 4, highly active fluorite.

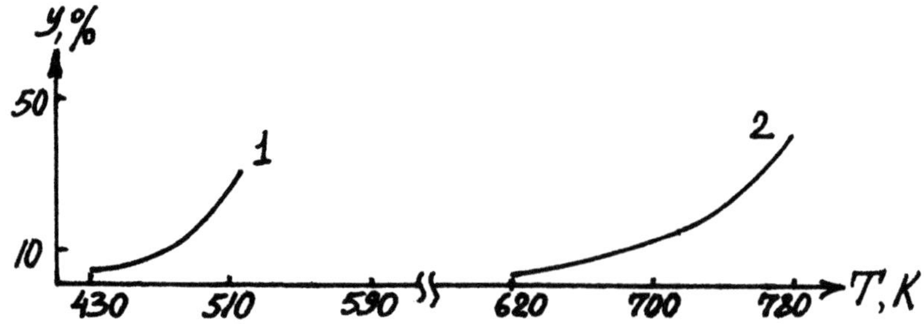

Fig. 2: Shocked LiF in the reaction of i-propanol dehydrogenation. 1, shocked at 300 K; 2, shocked at 770 K.

It can be seen from the figure that the temperature at which the conversion of isopropanol begins is lower by 200°C for the shocked samples. Thus, the activation energy of the reaction is reduced, due to an effect of high dynamic pressure on the energetic properties of active centers.

In Fig. 2 is shown the temperature dependence of transformation i-propanol for shocked LiF by 3.5 GPa at 300 and 770 K.

Catalytic activity of $Cu(Acac)_2$ placed on MgO and MgF_2 shocked at 3.5 GPa have been studied also.

In Fig. 3 is shown the dependence of catalytic activity of $Cu(Acac)_2$ in heptene-1 isomerization on dislocation density for MgO- and MgF_2-supports.

Thus the variation of dislocation density of the support crystal leads not only to a change in the activity but also affects the selectivity of catalysts.

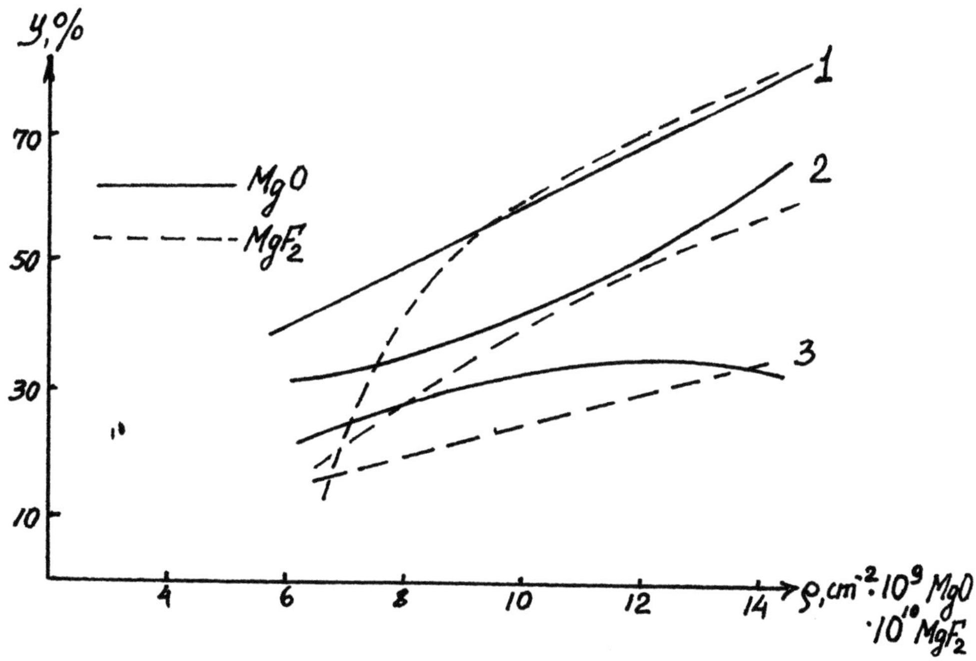

Fig. 3: $Cu(Acac)_2$ on shocked MgO and MgF_2 in heptene-1 isomerization. 1, selectivity of 0.1% $Cu(Acac)_2$; 2, selectivity of 0.01% $Cu(Acac)_2$; 3, conversion on 0.1% $Cu(Acac)_2$.

References

1. S. Batsanov, G.K. Boreskov, G.V. Gridasova, N.P. Keier, L.M. Kefeli, V.M. Kudinov, V.I. Mali, I.S. Sazonova, Kinet. Catal. 8:1140 (1967).
2. S.S. Batsanov, V.P. Bokarev, I.S. Kostenchuk, Y.S. Mardashev, I.N. Temnitsky, React. Kinet. Catal. Lett. 20:43 (1982).
3. Yu. C. Mardashev, O.G. Grebenshikova, Russian J. Phys. Chem. 48:461 (1974).
4. S.S. Batsanov, V.P. Bokarev, O.M. Bokareva, Yu. S. Mardashev, V.P. Semin, Russian J. Phys. Chem. 56:1265 (1982)

SHOCK-INDUCED TEMPERATURE DISTRIBUTIONS IN POWDER COMPACT RECOVERY FIXTURES*

R. A. Graham and D.M. Webb

Sandia National Laboratories
Albuquerque, NM

ABSTRACT

Sandia "Bear" and "Bertha" powder compact recovery fixture systems for explosive loading have been widely used to preserve powder compact samples for post-shock analysis. Shock conditions within the compacts are determined from an extensive program of numerical simulation which reveals that, even though the initial loading is planar, the sample conditions are controlled by a radial-mode compression. The wave-trapping phenomenon which occurs in the powder samples leads to relatively uniform shock-induced temperatures except for a very hot area in the outer edge region of some fixtures.

INTRODUCTION

In prior work [1,2] results of an extensive numerical simulation program to study the response of explosively-loaded sample recovery fixtures were reported. The fixtures are the Sandia "Bear" and "Bertha" systems which have been used extensively to study many materials in a shock-induced solid state chemistry program [3-10]. Ten combinations of fixture geometry and loading arrangement provide the capability to routinely perform recovery experiments in a quantitative, reproducible manner over a wide range of pressure and temperature. The prior work has summarized the principal results of the numerical simulations of pressures achieved in typical samples. In the present paper, principal results of numerical simulation of shock-induced temperature will be summarized.

EXPLOSIVE RECOVERY FIXTURES

The general configuration used for the recovery fixture system is shown in Figure 1. A typical arrangement includes an explosive loading system, a sample capsule placed inside a larger steel block and a steel wave-shaping disk between the capsule and the explosive. The two basic loading methods incorporate either a 56 mm diameter (Bear system) or a 102 mm diameter (Bertha system) plane-wave explosive generator designated as P-22 or P-40, respectively. Two different initial pressures are generated by using the plane-wave generators to initiate either baratol

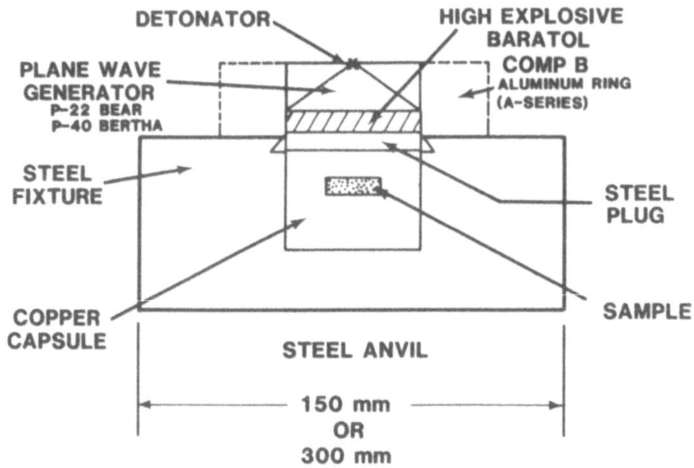

Figure 1. The Sandia Bear and Bertha explosive recovery systems as shown provide the capability to carry out quantitative, reproducible shock compression experiments in powder compacts at pressures between 7.5 and 27 GPa and shock-induced temperatures between 50 and 1100°C.

or Composition B explosives which are 12.5 or 25.4 mm thick for Bear or Bertha, respectively, and of the same diameter as the plane wave generators.

The plane-wave generators are standard designs used for many years at Department of Energy laboratories. The explosive materials themselves and the plane-wave generators are subjected to rigorous manufacturing control and their performance has been subjected to unusually comprehensive study [11]. In these studies, experimental data on explosive performance are incorporated into computer codes for extensive numerical simulation studies of shock processes and the results of the simulations are carefully checked against precise experimental results. Because of this very substantial background study, considerable confidence can be placed in the accuracy, reproducibility and realism of numerical simulation of explosive performance.

All other critical components of the fixture systems are constructed of materials of known shock-compression properties so that their shock-wave response can be realistically modeled in numerical simulation. The detonation wave from the explosive impinges on a mild steel plate and the shock-induced 13 GPa phase transition acts to shape the sharply decaying detonation wave to a shock wave of more constant amplitude. The copper capsule holding the powder sample not only is easy to machine and flows well under shock to provide seals, but its very low Hugoniot elastic limit prevents any multiple wave effects. Thus, all aspects of the recovery system can be accurately described in numerical simulation with computer codes. The principal unknown is the shock response of the powder compacts. Here there is little data to guide us but one fortunate aspect of the wave-trapping phenomenon is to minimize the importance of the shock response of the porous sample. As long as the density of the powder compact is in the range of 35 to 65% of solid density, various numerical simulations have shown that the peak shock pressures and temperatures in the samples are relatively insensitive to details of the shock compression properties of the material.

Not only are materials carefully selected and modeled, but precise assembly procedures are followed to minimize undesirable variations in local interfaces. All surfaces in the path of the incident shock-loading are lapped to optical flatness to eliminate "gaps". The powder samples are pressed in place within the capsules to eliminate radial and planar gaps at interfaces between the powder and capsule. By pressing the powder in place, it is also a routine procedure to prepare powder compact samples with densities ranging from about 40 to 65% of solid density. The powder compact density controls the shock-induced temperature rise.

TEMPERATURE DISTRIBUTIONS

In a porous solid the sample "crush-up" behavior to solid density controls the increase in mean bulk temperature (MBT). Such crush-up has been modeled by various modifications to the "P-α" model originally developed by Herrmann [12]. As the increase in MBT is controlled by the pressure-volume compression work, lower density samples will experience higher MBT than higher density samples.

One-dimensional numerical simulations have been used to provide insight into the crush-up behavior characteristics of inorganic powders. Based on shock-compression data on rutile as documented in prior work [2], the response to one-dimensional explosive loading in the Baby Bear configuration has been studied. What is observed is an initial low pressure wave followed by a "ring-up" between copper interfaces. The first reflection causes a major increase in pressure due to the large compression of the incident wave. The corresponding temperature (MBT) shows a quite different behavior. The major increase in temperature occurs for the low pressure initial wave as the major volume change occurs under those conditions.

A full two-dimensional simulation of temperature is required to predict MBT in the recovery fixture systems. These have been carried out for all the Sandia fixtures and a typical result showing the development of the temperature distribution during shock compression is shown in Figure 2. In the figure, temperature contours for an incident wave moving from top to bottom are shown in a Momma Bear configuration loaded with Composition B explosive. In Figure 2a the incident wave is shown approaching the back side of the sample and a hot region has developed in the peripheral area . This hot region is due to the interaction of waves moving into the sample from the wave in the surrounding copper. In Figure 2b the incident wave has reached the planar copper interface and reflects, causing an expansion of the hot region. In 2c and 2d there are subsequent small changes in temperature distribution but, as the principal part of the crush-up was completed earlier in time, the later time changes are not substantial.

The hot region in the peripheral area is a major feature to be considered in shock recovery experiments. If the material property or chemical reaction is sensitive to temperature differences in the ranges encountered, great care must be used in segregating the material from the fixture for subsequent analysis. If the material is carefully segregated and the numerical simulation is sufficiently detailed, the hot region can give an experimental point in a temperature range different from the bulk of the sample.

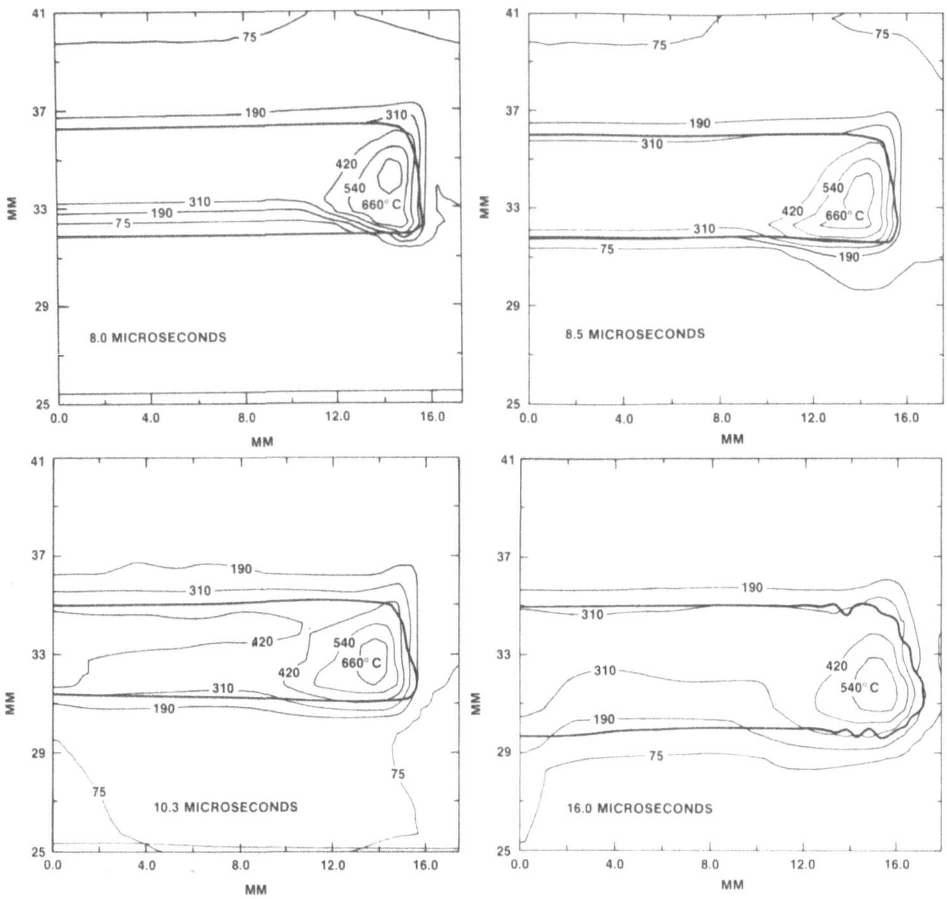

Figure 2. Two-dimensional numerical simulation of temperatures for a rutile powder compact at 55% density are displayed as contours at various times in the vicinity of the compact. The hot, outer region is apparent in this simulation of a Momma Bear fixture with Composition B explosive loading. The heavier solid line delineates the compact boundary.

A summary of temperatures encountered in the various Sandia fixtures is shown in Table 1. In most of the two-dimensional simulations, the base calculation was performed for powder compacts at 55% solid density. Temperatures at other densities are estimated by the differences observed from one-dimensional analysis. In a number of cases, the temperatures were calculated at various densities in two-dimensional analyses and found to check well with temperature differences estimated from one-dimensional analysis. Various estimates of temperature based on different material models lead to a conclusion that, in most cases, the temperatures are known to an absolute accuracy of + or - 20%. For the same material in a given fixture studied at various loading conditions, the relative temperature accuracy is thought to be significantly improved.

As can be observed by inspection of the temperature data of Table 1, not all fixtures develop hot outer edge regions. Those which do not have sample diameters large compared to the outside diameter of the loading system. In these cases fast arrival of unloading prevents the development of strong edge interations.

TABLE 1 - SHOCK-INDUCED MEAN BULK TEMPERATURES*

Baratol Explosive	Powder Compact Density**		
Fixtures	62%	55%	45%
Bertha	50-50	75-75	275-275
Papa Bear	80-50	150-75	350-275
Big Bertha A	150-150	200-200	450-450
Mama Bear	125-150	225-250	425-450
Bertha A	100-100	200-200	425-425
Momma Bear A	150-210	225-310	500-675
Baby Bear	100-150	250-310	450-525
Composition B Explosive			
Papa Bear	275-200	375-300	475-400
Big Bertha A	325-450	425-550	525-650
Bertha A	200-325	300-425	400-525
Momma Bear	300-450	425-550	525-650
Momma Bear A	300-550	400-650	550-750
Baby Bear	425-800	525-950	625-1050

*Temperatures are in °C. **Densities are in % of solid density. Temperature values are: first value = center region; second value = outer region

In situations in which more detail on temperature is required, a detailed examination of the full two-dimensional response is available in the various simulations.

There is strong experimental evidence for the existence of a hot region in the outer part of certain recovery fixtures. From our earliest synthesis work (e.g. lead oxide plus zirconia leading to lead zirconate and other products [13]) major differences in chemical composition or defect structure have often been observed. In all cases the differences seem to be consistent with the existence of the high temperature region. The specific published instances are: 1) conversion of zirconia from monoclinic to tetragonal [14], 2) shock-induced changes in powder specific surface [15,16], 3) reaction of zinc oxide with hematite [17], 4) reaction of aluminum with nickel [18] and reaction of aluminum with titanium [19].

Perhaps the most quantitative assessment comes from our extensive studies of paramagnetic defects in high purity rutile [20]. These defects have been found to be highly sensitive to shock-induced temperature. Noting a distinctively different color in the outer region of shock-modified rutile powder compact, which has been found to correlate with paramagnetic defect concentration, a special experiment was performed at a lower compact density which would achieve the same MBT in the bulk of the sample as was calculated in the outer region. The measured defect density between the two regions was identical within experimental error. The most compelling of the observations is that of Boslough and coworkers in which optical observations show strong chemical reaction initiating in the outer region [21].

Finally it should be observed that it is a routine observation to find a thin region of bonding between powder samples and the copper plugs in the outer region but not in other locations. Thus, in agreement with calculations, there is strong and compelling evidence for a hot region in the outer part of the powder samples

CONCLUSIONS

Numerical simulations show that the wave-trapping phenomenon which occurs when low density powder compacts are subjected to planar loading in metal capsules is found to lead to a region of much higher temperature in the outer region of the sample than in the bulk in some fixtures. There is strong experimental evidence for its existence and detailed quantitative studies of shock modification or shock synthesis must give consideration to inhomogeneous temperature distributions within typical powder samples.

*This work is performed at Sandia National Laboratories supported by the U. S. Department of Energy under contract #DE-AC04-76DP00789.

REFERENCES

1. L. Davison, D. M. Webb and R. A. Graham in: "Shock Waves in Condensed Matter - 1981," W. J. Nellis, L. Seaman and R. A. Graham, eds., American Institute of Physics, New York (1982).
2. R. A. Graham and D. M. Webb in: "Shock Waves in Condensed Matter - 1983," J. R. Asay, R. A. Graham and G. K. Straub, eds., North Holland (1984).
3. B. Morosin and R. A. Graham, Materials Science and Engineering, 66:73 (1984).
4. E. K. Beauchamp, R. A. Graham and M. J. Carr in: "Defect Properties and Processing of High-Technology Nonmetallic Materials," J. H. Crawford, Jr., Y. Chen and W. A. Sibley, eds., North Holland (1984).
5. E. L. Venturini and R. A. Graham in Crawford, et al., eds., loc cit. (1984).
6. M. J. Carr, R. A. Graham, B. Morosin and E. L. Venturini in Crawford, et al. eds., loc cit. (1984).
7. J. Golden, F. Williams, B. Morosin, E. L. Venturini and R. A. Graham in Nellis, et al., eds., loc cit.(1982).
8. Y. K. Lee, F. L. Williams, R. A. Graham and B. Morosin in Asay, et al., eds., loc cit.(1984).
9. Y. Horie, R. A. Graham and I. K. Simonsen, Materials Letters, in press.
10. E. L. Venturini, B. Morosin and R. A. Graham, J. Appl. Phys. 57:3814 (1985).
11. T. R. Gibbs and A. Popolato, eds., LASL Explosive Property Data, University of California Press (1980).
12. W. Hermann, J. Appl. Phys. 40:2490 (1969).
13. J. D. Keck, D. L. Hankey, R. A. Graham and B. Morosin in Nellis et al., eds., loc cit.(1982).
14. B. Morosin, R. A. Graham and J. R. Hellmann in Asay, et al., eds., loc cit.(1984).
15. Y. K. Lee, F. L. Williams, R. A. Graham and B. Morosin, J. Mat. Science, in press.
16. Y. K. Lee, F. L. Williams, R. A. Graham and B. Morosin, in Asay et al., eds. (1984) p. 399.
17. B. Morosin, R. A. Graham, E. L. Venturini and D. L. Williamson, this proceedings.
18. I. K. Simonsen, Y. Horie and R. A. Graham, this proceedings.
19. Y. Horie, D. E. P. Hoy, E. Simonsen, R. A. Graham and B. Morosin, this proceedings.
20. E. L. Venturini and R. A. Graham in Crawford et al., eds., loc cit.(1984).
21. M. B. Boslough, R. A. Graham and D. M. Webb, this proceedings.

NUMERICAL SIMULATION OF A SAMPLE RECOVERY FIXTURE FOR HIGH VELOCITY IMPACT*

F. R. Norwood and R. A. Graham
Sandia National Laboratories
Albuquerque, NM

A. Sawaoka
Tokyo Institute of Technology, Japan and
Center for Explosives Technology Research,
Socorro, New Mexico

ABSTRACT

Numerical simulations of shock compression conditions in the Tokyo Institute of Technology sample recovery fixture for high velocity impact have been carried out with two-dimensional (CSQ) and three-dimensional (HULL) computer codes. For the impact of stainless steel on the stainless steel capsule at a velocity of 2.5 km/sec, it is found that two-dimensional effects within the low density powder samples dominate the loading process and that interactions between the four samples are negligible. Because of the much higher pressure encountered, the two-dimensional effects are qualitatively different than those found for the Sandia "Bear" explosive fixtures.

INTRODUCTION

Studies of shock-induced materials synthesis and materials modification are based on samples which have been subjected to high pressure shock loading and preserved for post-shock analysis.[1,2] Although interpretations of characteristics of "recovered" samples are limited by the range of complex processes encountered, many properties of the material and the process can only be determined by recovery experiments. Nevertheless, if the processes are to be controlled and compared to other processes, quantitative specification of shock conditions is essential. In prior work[3,4] in which realistic, quantitative, two-dimensional simulation was carried out on recovery fixtures subjected to planar explosive loading, it was discovered that a wave-trapping phenomenon dominates the loading of a porous powder and

---

*This work performed at Sandia National Laboratories supported by the U.S. Department of Energy under contract number DE-AC04-76DP00789.

converts the initially planar loading to a radial-mode loading in the sample. The two-dimensional radial effects then dominate the shock pressures and temperatures such that one-dimensional analysis is incorrect and typically leads to substantial errors.[3,4]

The fixture used for planar impact loading at the Tokyo Institute of Technology is one of the most widely used; yet, determination of shock conditions has been limited to one-dimensional analysis (see e.g. 5,6). In the present paper, we report two- and three-dimensional numerical simulation of the fixture under loading conditions used in the past for many experiments.

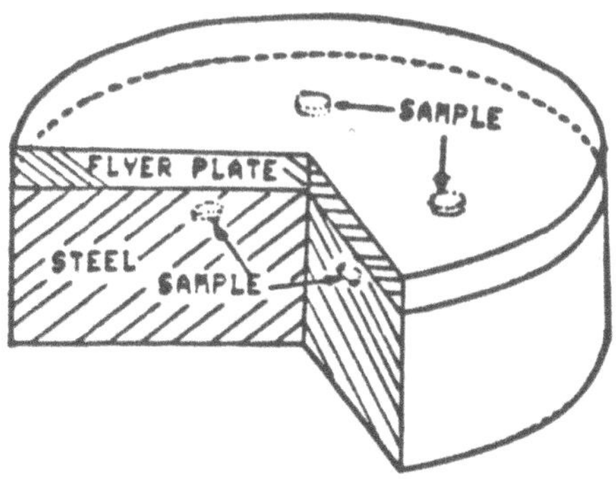

Figure 1: The Tokyo Institute of Technology recovery fixture for high velocity impact has four sample cavities in a configuration requiring three-dimensional numerical simulation. The plate impacts at 2.5 km/sec.

The fixture configuration is shown in Figure 1. There are four sample cavities located 90 degrees from each other at a radius of 22 mm from the center of the fixture. Each cavity is located 3 mm from the impact interface, with a thickness of 5 mm and a radius of 6 mm. For the calculation, the samples consisted of rutile powder at a packing density of 2.6 Mg/m^3 and all other parts, including the flyer plate were of stainless steel. The impacting plate was 3.2 mm thick and had a radius of 60 mm. The initial velocity of the flyer plate was taken as 2.5 km/sec. As indicated by Graham and Webb, [4] rutile is probably the most extensively studied porous inorganic powder and the numerical results may be compared with existing experimental data. In the present case, the shock compression data of Trunin and coworkers[7] directly overlap the pressures encountered in the recovery fixture.

RESULTS

CSQ RESULTS. In References 3 and 4, the wavecode CSQ was found to be quite useful for the plane-wave investigation of shock propagation effects in recovery fixtures. For the present work, in addition to the state of stress, the temperature was also determined by CSQ. Also, by using the code in its axisymmetric mode, preview information was obtained inexpensively for the three-dimensional results obtained with HULL. For the axisymmetric CSQ calculation, only one sample, placed along the axis of the fixture at a depth of 3 mm from the impact interface, was used to model the process.

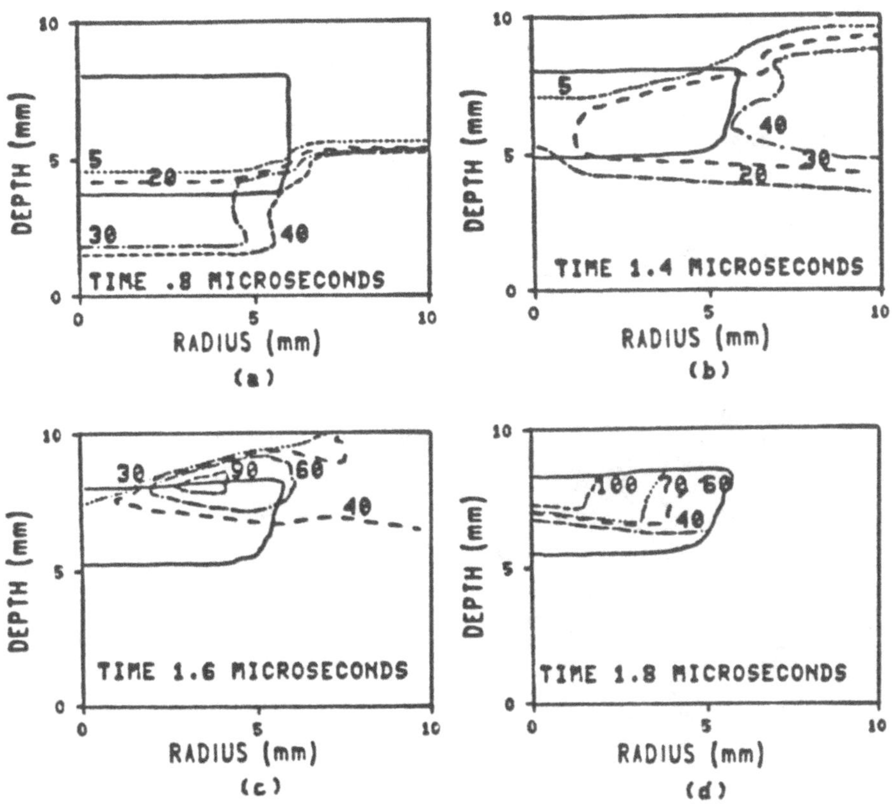

Figure 2: Contours of pressure in GPa are shown at various times within and in the vicinity of the sample cavity for loading moving from bottom to top. The solid line shows the sample boundary. Because of the high pressure, the wave speed in the powder at the radial interface with the steel is about that of steel. As the wave reaches the back outer region, a pressure approaching 100 GPa is produced and propagates radially inward along the back side. After focusing at the axis, the high pressure propagates toward the front of the fixture.

Figure 2a shows an early time condition in which a plane state prevails for radial values away from the edge of the sample (r=6 mm). The wavefront is slowed within the sample. One can see from this figure that the wavefront at the edge of the sample travels almost as fast as the wave front in the steel due to the pressure of over 40 GPa. This behavior is significantly different from that observed at lower pressure where the wave speed in the sample was so slow that the wave in the metal capsule reached the back of the sample long before the direct wave through the sample.

Figure 2c shows pressure contours at 1.6 microseconds. At this time, high pressures are found at the rear of the sample, with much lower stresses in the interior of the sample. The pressure concentration propagates toward the axis of the fixture along the back surface of the sample. After focusing along the axis this pulse then propagates toward the front of the sample. The result of this high pressure focusing is shown in Figure 2d.

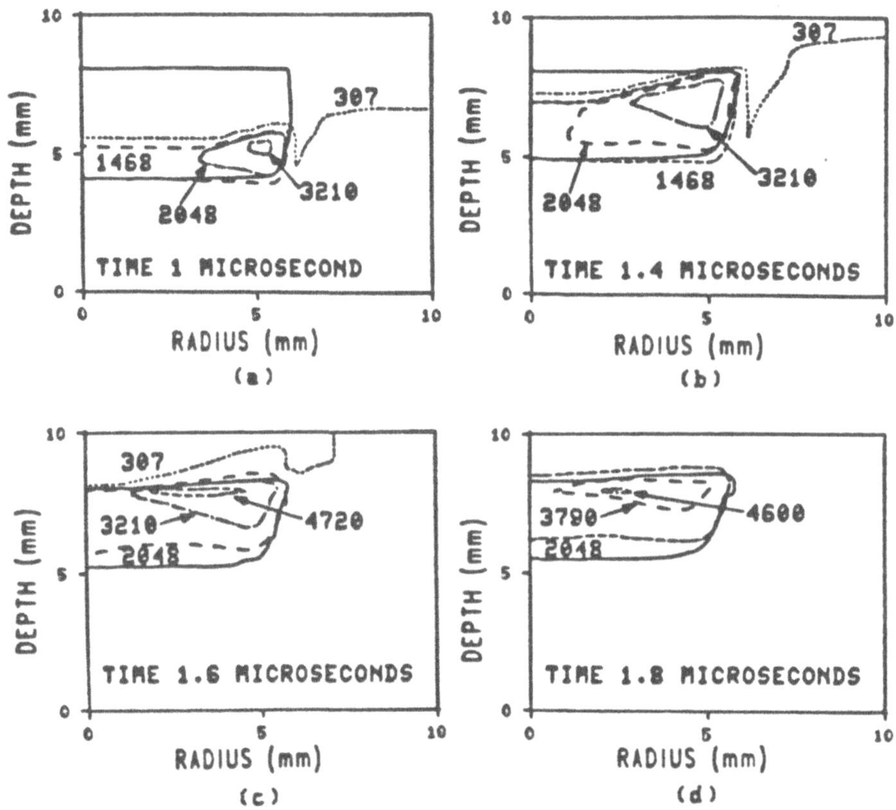

Figure 3: Temperature distributions in °C are shown as the wave propagates through the sample. As at lower pressure, a hot region develops in the outer part of the sample. A hot region also is produced along the back of the sample.

Temperature distributions are shown in Figure 3. In 3a, high temperature regions form at the corner r=6 mm, depth= 3 mm. As shown in Figure 3b, these regions expand into the center and rear of the sample. However, still higher temperatures are reached at the rear of the sample. This is evident in Figures 3c and 3d.

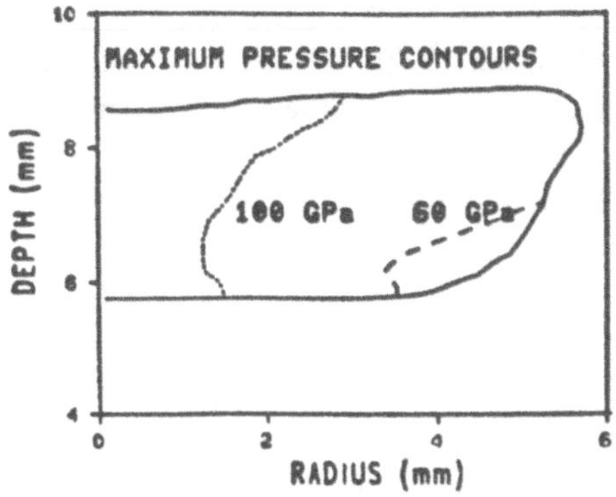

Figure 4: The maximum pressures produced by the loading are shown to be significantly different throughout the sample and substantially greater than predicted by one-dimensional analysis.

Figure 4 shows contours of the maximum pressure achieved within the sample during the entire loading process. This figure shows three regions which differ by the range of the maximum pressure felt. These regions are separated by the contour lines at which maximum pressures of 100 GPa and 60 GPa are achieved. Post-shock examination of samples from this fixture under conditions simulated in the present analysis show deformation patterns corresponding to the regions indicated in Figure 4.

HULL RESULTS. The calculations with HULL modelled the configuration shown in Figure 1, a fully three-dimensional problem, in contrast to the axisymmetric geometry employed with CSQ. The numerical results agreed with the CSQ results and showed that, for the previously indicated dimensions, the sample to sample interaction and also the interaction with the outside boundary were second order effects. Thus, the conditions modelled in CSQ provide a good simulation of the three-dimensional geometry as the shock conditions are controlled by the wave-trapping in the sample.

SUMMARY AND CONCLUSIONS

The present work clearly shows that, in both the axisymmetric and the three-dimensional geometry, the radial effects are the dominant feature. These two-dimensional radial effects dominate the shock pressures and temperatures such that one-dimensional analysis is incorrect and would lead to substantial errors.

REFERENCES

1. G. A. Adadurov and V. I. Gol'danskii, Russian Chemical Reviews, 50:948 (1981).
2. B. Morosin and R. A. Graham, in "Shock Waves in Condensed Matter–1981," W. J. Nellis, L. Seaman and R. A. Graham, eds., American Institute of Physics (1982), p. 4.
3. L. Davison, D. M. Webb and R. A. Graham, in Nellis *et al.*, eds., loc. cit. p. 67, (1982).
4. R. A. Graham and D. M. Webb, in "Shock Waves in Condensed Matter–1983," J. R. Asay, R. A Graham and G. K. Straub, eds., p. 211, North Holland (1984).
5. T. Akashi and A. Sawaoka, Materials Letters 3:11 (1984).
6. A. Sawaoka, K. Kondo, N. Hashimoto and S. Saito, in "Proc. International Symposium of Factors in Densification and Sintering of Oxide and Non-oxide Ceramics," p. 339, Japan (1978).
7. R. F. Trunin, G. V. Simakov and M. A. Podurets, Isvestiya, Earth Physics, 12:789 (1974).

RATE-DETERMINING STEPS FOR IGNITION OF SHOCKED NITROMETHANE

Richard D. Bardo

Naval Surface Weapons Center
White Oak Laboratory
Silver Spring, Maryland 20903-5000

ABSTRACT

Rates of intermolecular-to-intramolecular energy transfer and chemical reaction were determined by quantum mechanical calculations on systems of strongly interacting molecules at high pressure. The energy transfer rate was estimated by utilizing a theory of radiationless transitions which describes the coupling of a sparse set of internal mode energy levels with a quasicontinuum of intermolecular mode energy levels. An important quantity in this theory is the time t_{pv} required for activation of the internal modes. Calculated values of t_{pv} <1 μsec are identified with the most efficient transfer of energy when the molecules interact and react bimolecularly in the head-to-tail configuration. These times are consistent with those found in shock experiments, and those calculated by other techniques. A comparison of t_{pv} with chemical reaction half-lives indicates that energy transfer and reaction alternate as rate-determining steps over large ranges of P and T.

I. INTRODUCTION

A complete theoretical analysis of the ignition of shocked liquids and solids requires a knowledge of the rate-determining steps corresponding to vibrational energy transfer processes and chemical reactions. Unfortunately, past studies of these steps have been hindered by the use of unrealistic reaction kinetics, and by the complexity of the methods used for calculating rates of energy exchange. In this paper, a simpler quantum mechanical method is introduced and utilized for the calculation of rates of intermolecular-intramolecular vibrational energy transfer in shocked, condensed nitromethane. These rates are compared with those for different chemical reactions. The rate-determining steps are then identified over large pressure and temperature ranges.

Shock-induced vibrational activation processes have been studied much more extensively for gases than for liquids and solids. For the low-lying vibrational states of diatomic and triatomic molecules, Zel'dovich and Raizer[1] have reviewed in detail the theory of relaxation of vibrational energy toward complete thermodynamic equilibrium before and during molecular dissociation. They have discussed vibrational relaxation corresponding to vibrational-translational (V-T), vibrational-vibrational (V-V), and vibrational-rotational (V-R) energy exchange processes. For diatomics, these authors have discussed the results of shock tube experiments which indicate that relaxation does not interfere with reaction, since equilibrium in the vibrational degrees of freedom is established within a period of time which is short compared to that for dissociation. In general, theory and experiment have shown that dissociation and relaxation rates for the low-lying vibrational levels of gases in shock tubes follow the order, dissociation < V-T < V-R < V-V.

For the high levels of vibrational excitation characteristic of dissociation, Tardy and Rabinovitch[2] have reviewed the collisional activation and deactivation of polyatomic molecules as well as the applicability of RRKM (Rice-Ramsperger-Kassel-Marcus) theory to the study of unimolecular reactions. These authors have indicated that RRKM behavior, which requires that intramolecular and intermolecular V-V transfer must be substantially faster than decomposition, may be violated in shock tubes at very high temperatures. Here, the time scale for randomization of vibrational energy could be significantly larger than 10^{-12} sec. for collision-activated systems. In this case, the order given above for relaxation rates would not be obeyed.

When the dissociation rate is faster than the rate of excitation of the high-lying vibrational levels, the kinetics of both types of processes must be considered together. In order to address this situation for polyatomic molecules, Eyring[3] introduced "starvation kinetics" in which he assumed that the rate-determining step would be the flow of energy from an energy-deficient reservoir of vibrational degrees of freedom into a hidden bond to be broken. Here, he considered the breaking bond to be in equilibrium with the reservoir, which itself is not in equilibrium with the translational degrees of freedom of the gas in the shock tube. An earlier theory of gas kinetics was developed by Pritchard[4] who also showed that anomalously-low Arrhenius activation energies for gases in shock tubes could be ascribed to nonequilibrated vibrational degrees of freedom. More recently, Tarver[5] has introduced nonequilibrium models for the one-dimensional, steady-state detonation wave in gases, liquids, and solids. In these models, he has considered the detonation wave to consist of four zones. In one of the zones for condensed systems, he applied "starvation kinetics" to describe the slow, equilibrated excitation of endothermic bond breaking reactions.

Other methods, which are different from the ones mentioned above, have been utilized to analyze the activation and deactivation of intramolecular and intermolecular vibrational degrees of freedom. In the case of shock initiation, Pastine, et al.[6] suggested that the shock pulse duration required for the growth to detonation may be identified with the time required to achieve a critical internal temperature of the molecules comprising an explosive. Subsequently, Zerilli and Toton[7] (ZT) and Coffey and Toton[8] (CT) developed quantum mechanical methods to calculate vibrational excitation rates arising from the coupling of "cold" molecular and "hot" lattice modes. In the ZT model, the study of the excitation of the low-lying states of shocked crystalline nitromethane was based on a mathematical framework which is similar to the one employed by Lin[9] to describe vibron-phonon relaxation in condensed materials. For nitromethane, ZT concluded that the 0-1 transition rate was slower than the higher excitations 1-2, 2-3,... for the lowest-frequency internal mode. In the CT model, multiphonon excitation arising from compressive waves was studied. This model was applied to the excitation of the entire manifold of internal vibrational levels leading to molecular dissociation. In Section II of the present paper, another quantum mechanical theory of energy transfer is presented which incorporates some of the features from the ZT and CT models.

At the high shock pressures which may be generated in liquids and crystals, unimolecular bond scission reactions are often so slow that other, faster reactions are dominant.[10] The rates of these reactions and the intermolecular-to-intramolecular energy transfer processes giving rise to them must then be compared in order to permit identification of the rate-determining step. In Section III, comparisons will be made between reaction half-lives of a new initial reaction for nitromethane and the times obtained from the theory described in Section II.

II. THEORY OF VIBRON-PHONON ENERGY TRANSFER

A. Relaxation Processes and the Master Equation Approach

Quantum relaxation processes in atoms, molecules, or condensed materials are generally treated by considering the interaction between two or more zero-order subsystems which comprise the complete physical system. For isolated atoms and molecules, the relaxation or radiationless processes are autoionizations, unimolecular chemical dissociation reactions as treated by RRKM theory, molecular predissociation, and intramolecular radiationless transitions which involve a change in the electronic and vibronic states of a large molecule.[11] All of these processes are interpreted in terms of energy transfer between two subsystems or sets of levels with the fastest flow occurring into the denser set. For molecular liquids and solids, these subsystems may be taken to be the manifold of intramolecular vibrational levels, and a denser set of intermolecular levels which, in the case of crystals, corresponds to lattice modes. For strongly-interacting molecules in shocked, condensed systems such as the nitric oxides, nitroaliphatics, and nitroaromatics, it has been suggested that intramolecular radiationless transitions can also occur as a consequence of large red shifts of the low-lying excited electronic states.[12]

For the study of vibrational relaxation in condensed systems, Lin[13] has employed the generalized master equation in the form

$$-\frac{d\rho_{mm}}{dt} = \sum_n k_{mn}(t)[\rho_{nn}-\rho_{mm}] + \sum_n \sum_{k=1}^{\infty}\left(\frac{d^k\rho_{nn}}{dt^k} - \frac{d^k\rho_{mm}}{dt^k}\right)\int_0^t dt_1 \frac{(-t_1)^k}{k!} K_{mmnn} \tag{1a}$$

where ρ_{mm} etc. are the diagonal elements of the density matrix, $K_{mmnn}(t)$ is a memory kernel, and $k_{mn}(t)$ is a rate constant defined by

$$k_{mn}(t) = \int_0^t dt_1\, K_{mmnn}(t_1) \ . \tag{1b}$$

By using perturbation theory, he has expanded Eq. (1a) to include contributions through fourth order. With this approach, he has shown that (1) the first order contribution vanishes, (2) the second order approximation may be identified with the conventional Pauli master equation with the rate constants expressed in the "golden rule" form, (3) the third order approximation may still be identified with Pauli type master equations with the inclusion of perturbations of the initial and final molecular oscillator states, and (4) the fourth and higher-order approximations contain the so-called memory effects.

All of these orders of approximation address the totality of energy exchange events for strong coupling of the molecular oscillators to the lattice modes. These events include both sequential and direct processes. Sequential processes involve at least two molecular or vibronic modes which are coupled to the lattice or phonon modes. In the case of vibron deactivation, intramolecular and vibron-phonon coupling results in conversion of high to lower frequency vibron modes followed by vibron-phonon induced relaxation of the lower frequency vibron modes and excitation of phonons. Direct processes, on the other hand, involve only one vibron mode which is coupled to the phonon modes. This situation occurs for vibrational relaxation of diatomics and for the lowest frequency modes of larger polyatomics. For activation of the vibron modes through vibron-phonon coupling, the reverse of the above two-stage process occurs in which phonon modes are deactivated followed by vibronic activation. This is the situation described by ZT[7] and CT[8] who used methods which may be shown to be equivalent to the various types of Pauli master equations corresponding to the first three orders of approximation to the generalized master equation.

B. Relaxation Processes and the Density of States Approach

Bardo and Wolfsberg[14] have shown that the nonrelativistic Schrödinger wave equation for a set of coordinates fixed in a translating and rotating molecular system can be written in the form

$$(T+V-\varepsilon)\Psi_s^J + T_+\Psi_{s+1}^J + T_-\Psi_{s-1}^J + T_{++}\Psi_{s+2}^J + T_{--}\Psi_{s-2}^J = 0\,,\quad s=-J,-J+1,\ldots J. \tag{2a}$$

In this set of (2J + 1) coupled equations, J and s are the quantum numbers of the total angular momentum and the projection of the total angular momentum onto the z-axis of a rotating coordinate system, respectively. T, T_+, T_-, T_{++}, and T_{--} are kinetic energy operators, the forms of which depend on the choice of coordinate system rotating with the molecular system. T_+, T_-, T_{++}, T_{--} arise from the Coriolis-like interactions between the rotational, electronic, and vibrational motions. V is the usual expression for the potential energy which includes the electrostatic potential energy terms among electrons and the N nuclei. ε is the total energy (excluding the energy of the center-of-mass motion) of the rotational and internal motions of the system.

When J=s=0 in Eq. (2a), the electronic state of the molecular system is nondegenerate, and all molecular rotations within the system may be assumed to be quenched so that Eq. (2a) reduces to

$$(T+V-\varepsilon)\Psi = 0 \quad . \tag{2b}$$

This approximation, which will be used in the sequel, is an excellent one even at high temperatures in condensed systems which have been subjected to high pressures.

Equation (2b) is applicable to an entire assemblage of molecules in condensed systems. For these systems, each wavefunction Ψ may be expanded in terms of a complete set of orthonormal electronic wavefunctions Φ_K,

$$\Psi = \sum_K \Phi_K F_K \quad , \tag{3}$$

where F_K is a function of the 3N-6 coordinates of the nuclei, and Φ_K is a function of these nuclear coordinates and the coordinates of the electrons. Here, the set of wavefunctions $\{\Phi_K\}$ determines the ground and excited state electronic behavior for the complete system. In molecular solids, for example, $\{\Phi_K\}$ defines the band structure. The set of wavefunctions $\{F_K\}$ in Eq. (3) determines the vibrational behavior for the complete system and, for solids, includes all vibron and phonon contributions. Thus, when Eq. (3) is a solution of Eq. (2b), the electronic-vibrational description of the condensed system is complete.

The solution of Eq. (2b) is greatly simplified if $\{\Phi_K\}$ and $\{F_K\}$ are chosen in such a way that the number of terms in Eq. (3) is as small as possible. In the so-called adiabatic solution, of which the Born-Oppenheimer (BO) approximation and the adiabatic correction are special cases,[14] Eq. (3) includes a single term

$$\Psi_{Kr} = \Phi_K F_{Kr} \quad , \tag{4}$$

where F_{Kr} identifies the r-th vibrational state in the K-th electronic state Φ_K. These approximations are excellent ones when Φ_K varies slowly with nuclear displacement. In the so-called nonadiabatic solution, the variation of Φ_K is large and at least two terms are needed in Eq. (3) to describe vibronic coupling in intramolecular radiationless transitions, and in the other electronic relaxation processes mentioned previously in Section II. A. In order to simplify the subsequent analysis, the pressure-induced electronic nonadiabatic processes described in Ref. 12 will be ignored so that Eq. (4) applies. Also, it will be assumed that the shocked molecular system exists in its ground state only. In the sequel, the subscript K in Eq. (4) will be dropped.

The large ratio R of the electronic "frequency" to vibrational frequency, $100<R<1000$, allows the separation of electronic and vibrational motions to be made in the adiabatic approximation, Eq. (4). Analogously, the large ratio R' of vibron frequency to phonon frequency, $R'\approx 10$, permits the adiabatic separation of the intramolecular and intermolecular vibrational motions to be made in the lowest order of approximation.[6,7,9] In this approximation, the wavefunction F_r in Eq. (4) becomes the simple product

$$F_{mr} = \phi_m(q,Q)\,\chi_{mr}(Q) \, , \tag{5}$$

where $\phi_m(q,Q)$ and $\chi_{mr}(Q)$ are vibron and phonon wavefunctions, respectively. The sets of coordinates for the intramolecular and intermolecular configurations of the nuclei are given by q and Q, respectively. Here, the functions $\phi_m(q,Q)$ are taken to be solutions of the Schrödinger equation for the vibron subsystem,

$$[T(q) + U(q,Q)]\,\phi_m(q,Q) = E_m(Q)\,\phi_m(q,Q) \, , \tag{6}$$

where T(q), U(q,Q), and Em(Q) are, respectively, the kinetic energy operator for the molecular oscillators, the total potential energy for electronic and nuclear motions, and the total vibron energy at the nuclear configuration Q. The potential energy U(q,Q) includes the electronic ground state energy which may be obtained by solving the standard BO electronic problem for the entire system of molecules.[14] The conditions under which the adiabatic approximation for vibron-phonon modes is valid have also been examined by Lin[9] who used a similar analogy to electronic-nuclear separability. For coupled vibron and phonon modes, ZT[7] analyzed the breakdown of the adiabatic approximation in shocked nitromethane by means of a nonadiabatic operator formalism.

The nonadiabatic solution for coupled vibron-phonon oscillators is now obtained by expanding the wavefunction F_r in Eq. (4) in terms of the complete set of eigenfunctions $\phi_m(q,Q)$ of Eq. (6),

$$F_r = \sum_m \phi_m(q,Q)\,\chi_{mr}(Q) \quad . \tag{7}$$

The phonon wavefunctions $\chi_{mr}(Q)$ are solutions of the set of coupled differential equations,

$$\begin{aligned}&[T(Q) + E_m(Q) + \langle\phi_m|T(Q)|\phi_m\rangle - \varepsilon]\,\chi_{mr}(Q)\\ &= -\sum_{ns\neq mr}\left[\langle\phi_m|T(Q)|\phi_n\rangle - \hbar^2\sum_\alpha M_\alpha^{-1}\langle\phi_m|\partial/\partial Q_\alpha|\phi_n\rangle\,\partial/\partial Q_\alpha\right]\chi_{ns}(Q)\,.\end{aligned} \tag{8}$$

In Eq. (8), the bra-ket notation indicates integration over the vibronic coordinates q, and T(Q) denotes the kinetic energy operator for the phonons,

$$T(Q) = -\frac{\hbar^2}{2}\sum_\alpha M_\alpha^{-1}\,\partial^2/\partial Q_\alpha^2 \quad . \tag{9}$$

Equation (8) has the same form as that for vibron-vibron coupling between different electronic states.[11]

The magnitude of the right hand side (RHS) of Eq. (8) determines the validity of the adiabatic approximation. If the coupling matrix elements on the RHS of Eq. (8) are small, the adiabatic solution is a good one. On the other hand, if the coupling is large, that approximation breaks down. In either case, a perturbation approach to the solution of Eq. (8) will require the following matrix elements between the adiabatic wavefunctions of Eq. (5),

$$\langle F_{mr}|H|F_{ns}\rangle = \int\chi_{mr}^*\,T_{mn}\,\chi_{ns}\,dQ - \hbar^2\sum_\alpha M_\alpha^{-1}\int\chi_{mr}^*\,D_{mn}\frac{\partial\chi_{ns}}{\partial Q_\alpha}\,dQ\,, \tag{10a}$$

where

$$T_{mn} = \langle\phi_m|T(Q)|\phi_n\rangle \tag{10b}$$

and

$$D_{mn} = \langle\phi_m|\partial/\partial Q_\alpha|\phi_n\rangle = \frac{\langle\phi_m|\partial U/\partial Q_\alpha|\phi_n\rangle}{E_n(Q) - E_m(Q)} \quad . \tag{10c}$$

[Equation (10c) is obtained from the differentiation of the Schrödinger equations (6) corresponding to the orthonormal solutions ϕ_m and ϕ_n. It may also be shown from Eq. (9) that T_{mn}, Eq. (10b), reduces to a sum of terms containing the matrix elements D_{mn}, Eq. (10c).]

According to Eq. (10c), the adiabatic approximation applies when the energy difference between the vibron states is large relative to the phonon matrix elements connecting these states. If the energy difference in Eq. (10c) is small, the adiabatic approximation completely fails and radiationless transitions occur.

The models of ZT[7] and Lin[9] require that the adiabatic approximation is valid in the zeroth order, and that the nonadiabatic corrections give energy transfer rates within the context of the golden rule[7,9] and the conventional Pauli master equation.[9] With the latter method, Lin[9] considered vibrational activation and deactivation to consist of single-step or direct processes for one vibron mode, and sequential processes involving two vibron modes. In the case of nitromethane, ZT[7] considered direct processes in which the activation of NO_2 rocking parallel to the NO_2 plane was chosen to be the fast step. Another approach, which includes the activation and deactivation steps described by ZT[7] and Lin[9], but which also includes memory or feedback effects corresponding to the fourth and higher orders of approximation of the master equation, will now be considered.

Shocking a molecular system mixes the stationary vibrational states F_r of Eq. (7) so that

$$F_r(t) = \sum_s a_{rs}(t)\, F_s = \sum_m \sum_s a_{rs}(t)\, \phi_m \chi_{ms} \, , \tag{11}$$

where the evolution of the system among the basis states $\phi_m \chi_{ms}$ is contained in the time-dependent coefficients $a_{rs}(t)$. The time derivatives of the probabilities $|a_{rs}|^2$ give the rates of activation and deactivation. This formulation of vibrational relaxation which involves the explicit time dependence of the wavefunction in Eq. (11) constitutes what is known as the Schrödinger picture of quantum dynamics. For the purposes of the present formulation of relaxation, an equivalent, but more convenient approach is presented which is based on the Heisenberg picture. Part of the formalism to be used here has already been developed and utilized by Jortner and coworkers[15] in connection with the problem of vibrational relaxation of a guest molecule in a host matrix consisting of either atoms or polyatomic molecules.

Nitzan, et al.[15] have studied the vibrational relaxation of a molecule in a dense medium (or phonon bath) by employing a model which is based on the assumptions that (1) the molecular vibrations are harmonic, (2) the molecular vibron modes are not coupled to each other by the vibron-phonon interaction, (3) the vibron-phonon interaction is linear in the intramolecular displacements, and (4) dissipative processes involving the phonon bath are fast on the time scale of vibrational relaxation so that the bath is in thermal equilibrium, which is an assumption also made by ZT[7] and Lin.[9] With these assumptions, Nitzan, et al.[15] have shown that the vibrational relaxation rate of a vibron with frequency ω is

$$W(\omega) = 2\pi \sum_{(\nu)} |G(\nu)|^2\, n(\nu)\, \rho(\nu) \, , \tag{12}$$

where $G(\nu)$ is a vibron-phonon coupling matrix element, $n(\nu)$ depends on the temperature-dependent phonon occupation number $\langle n_\nu \rangle_T$ so that

$$n(\nu) = \sum_{\nu'} \langle n_1 \rangle_T \langle n_2 \rangle_T \cdots \langle n_{\nu'-1} \rangle_T \left(\langle n_{\nu'+1} \rangle + 1 \right) \cdots \left(\langle n_N \rangle_T + 1 \right) \, , \tag{13}$$

and $\rho(\nu)$ is the compound many-phonon density of states

$$\rho(\nu) = \int d\epsilon_1 \int d\epsilon_2 \cdots \int d\epsilon_{N-1}\, \rho_1(\omega-\epsilon_1)\rho_2(\epsilon_1-\epsilon_2)\cdots\rho_N(\epsilon_{N-1}), \quad (14)$$

which gives the convolution of single phonon densities of state $\{\rho_\nu(\epsilon)\}$. The collection of phonon states satisfies the energy conservation law for a polyatomic molecular solid,

$$\omega = \sum_\alpha \omega_\alpha + \sum_\beta \omega_\beta \ , \quad (15)$$

where ω_α and ω_β correspond to collections of phonon modes (α) and other vibron modes (β), respectively, where $\omega_\beta < \omega$. Each of the terms in Eq. (12) corresponds to a given order of multi-phonon processes. In the sequel, the assumption of harmonic molecular oscillators will be relaxed as a consequence of Eq. (15), which depends on the condition that $\omega_\beta < \omega$.

The $|G(\nu)|^2$ terms rapidly decrease with increasing order of the multiphonon processes since $G(\nu)$ is identified with Eq. (10a). Consequently, the dominant contribution to the rate in Eq. (12) is the term corresponding to the set of phonon states $(\bar{\nu}) = 1,2,\ldots,N$ where N is the smallest number of phonons $N \equiv N(\bar{\nu})$ which can result in a vibrational relaxation process constrained by energy conservation, Eq. (15). Equation (12) may then be written as

$$W(\omega) \simeq 2\pi |G(\bar{\nu})|^2\, n(\bar{\nu})\, \rho(\bar{\nu}) \ , \quad (16)$$

which is the usual form of the golden rule. For the polyatomic molecular solid, Eq. (16) is satisfied when the minimum number of phonons $N_\alpha(\bar{\nu})$ is

$$N_\alpha(\bar{\nu}) = \frac{\omega - N_\beta \langle\omega_\beta\rangle}{\omega_D} \ , \quad (17)$$

where ω_D is of the order of the Debye frequency, $\langle\omega_\beta\rangle$ is the average of vibron frequencies other than ω, and N_β, the most probable number of vibrons, lies in the range

$$\frac{\omega}{\langle\omega_\beta\rangle} > N_\beta > \frac{\omega}{\langle\omega_\beta\rangle} - 1 \ . \quad (18)$$

The total rate W_T for all of the intramolecular modes is obtained from the sum

$$W_T = \sum_\omega W(\omega) \simeq \bar{W}(\bar{\omega}) = 2\pi |\bar{G}(\bar{\nu})|^2\, \bar{n}(\bar{\nu})\, \bar{\rho}(\bar{\nu}) \quad (19)$$

where $W(\omega)$ corresponds to Eq. (16), and $\bar{W}(\bar{\omega})$ may be taken to be the rate associated with the average of the largest terms, which correspond to the most efficient low frequency modes. In this regard, the assumption that the vibron-phonon interaction is linear in the intramolecular displacements has been shown by Lin[9] to be a good one for multiphonon relaxation processes which involve only three or fewer phonons. This will be the case for relaxation of low frequency vibron modes where

$N_\alpha(\bar{\nu}) < 3$, if $\omega/\omega_D \simeq 3$ is substituted into Eq. (17). In nitromethane under pressures of 5-18 kbar, for example, $\bar{\omega} \simeq \omega = 485\ cm^{-1}$ for the low frequency mode with NO_2 rocking parallel to the NO_2 plane. At the shock pressure of 80 kbar, say, the frequency ω_D may be estimated from a power-law relationship[7] to be $\omega_D \simeq 193\ cm^{-1}$ so that $\omega/\omega_D \simeq 3$, where it has been assumed that ω is essentially unchanged from its lower pressure value of $485\ cm^{-1}$. Here, then, the above assumption concerning linearity in the vibron-phonon interaction should provide a reasonably good approximation.

Equation (19) may be expressed in terms of the total relaxation time t_{vp},

$$t_{vp} = \frac{\hbar}{2\pi \bar{v}^2 \bar{\rho}} \quad , \qquad (20)$$

where $\bar{W}(\bar{\omega}) = t_{vp}^{-1}$, $\bar{\mathbf{v}}^2 = \hbar^2 |\bar{G}(\bar{\nu})|^2$ and $\bar{\rho} \equiv \bar{n}(\bar{\nu})\bar{\rho}(\bar{\nu})/\hbar$ is a "temperature-dependent" density of states. In Eq. (20), $\pi \bar{\mathbf{v}}^2 \bar{\rho}$ is the vibrational relaxation half-width γ. Equation (20) may be rewritten as

$$t_{vp} = \frac{\hbar \bar{\rho}}{2\bar{n}} \quad , \qquad (21)$$

where $\bar{n}$ is the total number of phonon states $\bar{n} = \bar{\rho}\gamma$. In Eq. (21), $\hbar\bar{\rho}$ may be identified with the time t_{pv} required for feedback from the dense set of phonon levels into the vibron states,

$$t_{pv} = \hbar \bar{\rho} \quad . \qquad (22)$$

This identification is made in analogy to the recurrence time $\hbar\rho$ established by Bixon and Jortner[11] for the nonradiative decay of a vibron for an excited electronic state into a quasicontinuum of vibonic states of density ρ corresponding to a lower electronic state.

The above discussion pertaining to Eqs. (21) and (22) deals with the sequence of events which involves the deactivation of a "hot" molecule followed by feedback from the phonon bath. For systems which initially contain "cold" molecules, the events are reversed when shocking preferentially excites the phonon states followed by activation of the vibron states. In the latter sequence, Eqs. (21) and (22) are still valid where now t_{pv} is the time for activation and t_{vp} is identified with feedback. Here, it is important to note that the shock pressure must be maintained at sufficiently high levels to avoid substantial deactivation of the vibrons since, in general, $t_{vp} \ll t_{pv}$ for low levels of phonon excitations. For the shocked system at high temperatures where $\bar{\rho}$ is quite large, it is reasonable to expect that $\bar{\rho}\mathbf{v} > 1$ so that $t_{vp} < t_{pv}$. In this context, it will be recalled that, for any two vibron levels i and j where $E_j > E_i$, the transition probabilities $P(i \to j)$ and $P(j \to i)$ obey $P(i \to j) < P(j \to i)$ for interacting molecules. Under these conditions, t_{pv} will provide a convenient and reliable measure of the rate of energy transfer and activation of vibrons following the excitation of the phonon bath. The only quantity needed for the determination of t_{pv} is $\bar{\rho}$, the calculation of which will be considered below. It will be shown in another paper that the introduction of Eq. (22) into the present analysis is equivalent to the inclusion of fourth and higher orders of approximation in the perturbation expansion of the generalized master equation.[16]

The well-known classical and semiclassical approximations for calculations of densities of states are not sufficiently accurate, since anharmonicities are not treated correctly. A much better approximation due to Haarhoff[17] is employed here, where

$$\bar{\rho} = \left(\frac{2}{\pi n}\right)^{1/2} \frac{(1-1/12n)\lambda}{h\langle\nu\rangle(1+\eta)} \left[\left(1+\frac{\eta}{2}\right)\left(1+\frac{2}{\eta}\right)^{\eta/2}\right]^{n} \left[1-\frac{1}{(1+\eta)^2}\right]^{\beta_0} , \quad (23a)$$

where n is the number of vibrational degrees of freedom, $\langle\nu\rangle$ is the average frequency, ν_i are the n frequencies, and λ, β_0 and η are defined by the equations

$$\lambda^{-1} = \prod_i \left(\nu_i/\langle\nu\rangle\right) , \quad (23b)$$

$$\beta_0 = [(n-1)(n-2)\alpha_2 - n^2]/6n , \quad (23c)$$

$$\alpha_2 = \langle\nu^2\rangle/\langle\nu\rangle^2 , \quad (23d)$$

$$\eta = E/E_0 . \quad (23e)$$

In Eq. (23e), E is the internal energy interval and E_0 is the zero-point energy. Equation (23a) incorporates very important anharmonic vibrations which are excited at high shock temperatures.

III. CALCULATIONS OF RATES OF ENERGY TRANSFER AND REACTION

A. Energy Transfer

The known lattice-mode frequencies required for use in Eq. (23a) are limited to Raman spectra which have been obtained for polycrystalline nitromethane at ambient pressure.[18] As a rough approximation, it may be assumed that these frequencies and the "Debye" frequency ω_D given earlier are increased by the same factor at the shock pressure P_s so that Eqs. (23b) and (23d) remain unchanged. Then, Eq. (23a) will depend primarily on the variation of Eq. (23e). For example, at Ps = 80 kbar, the power-law pressure relationship[7] for ω_D gives $W_D(80) = 193\ cm^{-1}$ so that $W_D(80)/W_D(1atm)=1.8$. As a result, the six lattice-mode frequencies given in Ref. 18 are increased to 90, 129, 151, 178, 200, and 269 cm^{-1}. At P_s = 80 kbar and T_s = 850 K along the shock Hugoniot of nonreacting nitromethane, the initial state of which is at the melting point of 244.6 K, Hardesty and Lysne[19] have determined the internal energy E' to be about 17 kcal/mol or 5947 cm^{-1}. If it is assumed that all of this energy is initially available in the phonon bath, E' may be equated with E in Eq. (23b)-(23e), one then obtains $\bar{\rho} = 2\times10^{19}$/erg and, from Eq. (22), $t_{pv} = 2.6\times10^{-8}$ sec. After vibrational relaxation, the energy of 17 kcal/mole will be distributed among the vibron and phonon modes.

At P_0 = 1 atm and T_0 = 244.6 K, E' = 3.5 cm^{-1} from Ref. 19 so that $\bar{\rho} = 2\times10^{13}$/erg and $t_{pv} = 2.5\times10^{-14}$ sec. At P_0 = 1 atm and the higher temperature of T = 850 K, E' = 3249 cm^{-1}, which is estimated from the harmonic approximation,

$$E' \simeq \sum_i \omega_i \left(e^{\omega_i hc/kT} - 1\right)^{-1} . \quad (24)$$

Here, $\bar{\rho} \simeq 10^{19}$, which is close to the value obtained previously at P_s = 80 kbar and T_s = 850 K. These results indicate that $\bar{\rho}$ depends much more strongly on temperature than on pressure when T is sufficiently large.

A graph of log t_{pv} vs. P_s is given in Fig. 1 for more data of Ref. (19) corresponding to the range of shock pressures 1 atm < P_s < 80 kbar. It should be mentioned here that the $\bar{\rho}$ values calculated above and the values of t_{pv} plotted in Fig. 1 are underestimations of the actual density of states, since the frequencies corresponding to the IR-active lattice modes are unknown and not included in Eqs. (23a)-(23e). It is expected that the actual values for $\bar{\rho}$ and t_{pv} should be at least an order of magnitude larger.

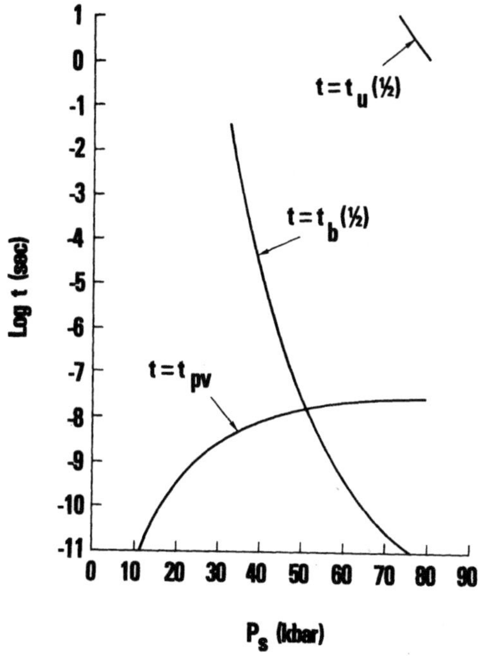

Figure 1. Comparison of energy transfer times t_{pv} and reaction half-lives $t_b(1/2)$ and t_u (1/2) for biomolecular and unimolecular processes.

B. Chemical Reactions

The energy transfer times t_{pv} plotted in Fig. 1 will now be compared with reaction half-lives. In Ref. 20, an initial, highly-exothermic (-48 kcal/mol) reaction step for shocked nitromethane is presented which is bimolecular and which is, therefore, strongly accelerated by pressure. This step, which is the fastest possible bimolecular reaction for nitromethane, is represented by the head-to-tail reaction

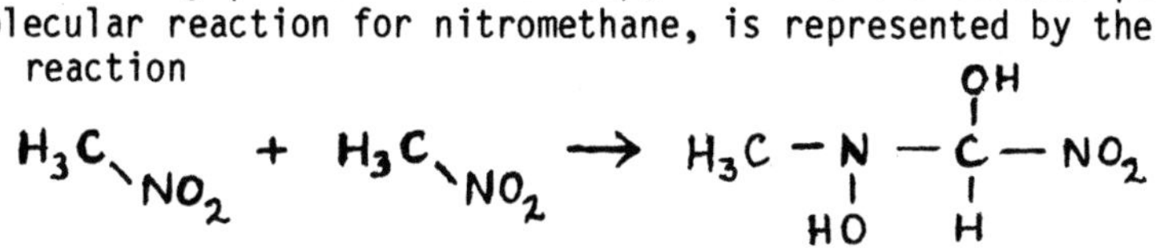

(25)

where the product is a dimer. At P_s = 80 kbar and T_s = 850 K, for example, the entropy, volume, and energy of activation for this step are given in Ref. 20 as S_a = -10 eu/mol, V_a = -12 cm^3/mol, and E_a = 22.5 kcal/mol respectively. In order to obtain the reaction half-life t(1/2), these parameters are substituted into the following expression for the rate coefficient k,

$$-\ln t(1/2) = \ln k = \ln \nu + \frac{S_a}{R} - \frac{PV_a}{RT} - \frac{E_a}{RT} , \qquad (26)$$

where $\nu = k_B T/h$. From Eq. (26), $t_b(1/2) = 5 \times 10^{-12}$ sec for the reaction in Eq. (25). A comparison of this value with $t_{pv} = 2.6 \times 10^{-8}$ sec calculated previously shows that energy transfer from the phonon bath into the vibron states is the slow or rate-determining step at $P_s = 80$ kbar and $T_s = 850$ K. These results are displayed in Fig. 1. Also plotted in Fig. 1 are log $t_b(1/2)$ for $P_s < 80$ kbar and $T_s < 850$ K. The ranges of activation parameters which are utilized for these pressures and temperatures are $-32 < S_a < -10$ eu/mol, $-30 < V_a < -12$ cm^3/mol, and $32.5 > E_a > 22.5$ kcal/mol. These ranges were calculated from the estimated pressure-dependent intermolecular distances and the equations given in Ref. 20.

There are two important results which are discerned from Fig. 1. First, the graphs of log $t_b(1/2)$ and log t_{pv} intersect at about $P_s = 50$ kbar. This is about the pressure at which there is a change in rate-determining step. For $P_s < 50$ kbar, the reaction of Eq. (25) is the slow step. For $P_s > 50$ kbar, energy transfer dominates in achieving ignition of nitromethane. The interesting possibility exists here that the ability of nitromethane to achieve a stable high-velocity detonation may be determined by this change of rate-determining steps in the neighborhood of 50 kbar. Second, at sufficiently high pressure, $P_s > 80$ kbar, t_{pv} approaches a constant value of $t_{pv}^{max} = 3 \times 10^{-8}$ sec. At these pressures, the density of phonon states also becomes constant at $\rho_{max} = 3 \times 10^{19}$/erg.

The graph of log t_{pv} in Fig. 1 will be displaced to longer times with improved calculations of $\bar{\rho}$. The largest contribution to an increase of ρ would come from the inclusion of the IR-active lattice-mode frequencies along with the Raman frequencies used previously. These additional frequencies could increase $\bar{\rho}$ by at least a factor of 10. It is possible then, that t_{pv}^{max} would approach more closely a value of 10^{-6} sec which corresponds to the measured induction time t(ind) = 2×10^{-6} sec for ignition and growth to detonation at $P_s = 80$ kbar.[21] If t_{pv}^{max} were to be equated to t(ind), $\rho(\text{ind}) = 1.9 \times 10^{21}$/erg. In any case, the curves for log t_{pv} and log $t_b(1/2)$ will intersect within a relatively narrow range about $P_s = 50$ kbar, indicating the existence of a critical pressure for a change of rate-determining step.

Also plotted in Fig. 1 are the values for log $t_u(1/2)$ for the bond-scission reaction

$$H_3C-NO_2 \longrightarrow H_3C + NO_2 \qquad (27)$$

for which $3 > S_a > 2$ eu/mol, $3 > V_a > 1$ cm^3/mol, and $54 > E_a > 50$ kcal/mol in the ranges 1 atm $< P_s \leq 80$ kbar and $244.6 < T_s \leq 850$ K. These ranges of the activation parameters were estimated on the basis of the effects of compression on the "activated complex". It is seen from Fig. 1 that the bimolecular reaction, Eq. (25), is the overwhelmingly dominant reaction at all high pressures of interest for the ignition of nitromethane. In fact, the half-life for C-N bond scission, or for any unimolecular bond breaking process in CH_3NO_2, is much larger than the times associated with shock initiation which are often less than a μsec. These results show that, if one were to arbitrarily select C-N bond scission as the initial reaction, he would wrongly conclude that it is the dominant step for ignition at high shock pressure.

IV. CONCLUSIONS

A quantum mechanical theory has been introduced for the description of energy exchange between the lattice modes or phonon states and the molecular modes or vibron states. This theory, which is simpler and more general than the ones previously employed to describe energy transfer in condensed systems, permits the relatively straightforward calculation of the time t_{pv} for activation of the vibron states. The time t_{pv}, which is directly proportional to the density of states of the phonon bath, is a measure of the rate of energy transfer from the bath into the intramolecular modes. The same density of states is contained in the rate expression or golden rule for vibron-to-phonon energy dissipation. Activation of the vibrons is shown to proceed most efficiently for the low-frequency molecular modes. This feature of the theory will be important for all energetic materials containing NO_2 groups which, being located at the periphery of each molecule, give the lowest frequencies.

The half-life $t_b(1/2)$ of a new pressure-accelerated, head-to-tail, exothermic bimolecular reaction was compared with t_{pv}. Above a critical pressure of about 50 kbar, this reaction, which involves the fast reaction of the NO_2 and CH_3 groups of neighboring molecules, becomes faster than the activation of the internal levels, a result of possible significance for the interpretation of the transition of shocked, ignited nitromethane to steady detonation. While this result may suggest that the present description is equivalent to the one for "starvation kinetics" proposed by Eyring,[3] a crucial difference is that the present theory requires no ad hoc assumption concerning a hidden bond, since unimolecular bond breaking is disallowed at high shock pressure.

ACKNOWLEDGEMENTS

This work was supported by the NSWC Independent Research Program. Special thanks are extended to Drs. S. Coffey, P. Miller, and F. Zerilli for helpful discussions.

REFERENCES

1. Y. B. Zel'dovich and Y. P. Raizer, "Physics of Shock Waves and High Temperature Hydrodynamic Phenomena," Vol. I, Academic Press, New York, 1966, Chapter VI.

2. D. C. Tardy and B. S. Rabinovitch, Intermolecular Vibrational Energy Transfer in Thermal Unimolecular Systems, Chem. Rev. 77:369 (1977).

3. H. Eyring, Starvation Kinetics, Science 199:740 (1978).

4. An elementary discussion of H. O. Pritchard's theory is presented in the article "Theory Explains Gas Kinetics Problem," Chem. Eng. News, 23 Apr 1962.

5. C. M. Tarver, Chemical Energy Release in One-Dimensional Detonation Waves in Gaseous Explosives, Combust. Flame 46:111 (1982); C. M. Tarver, Chemical Energy Release in Self-Sustaining Detonation Waves in Condensed Explosive, Combust. Flame 46:157 (1982).

6. D. J. Pastine, D. J. Edwards, H. D. Jones, C. T. Richmond, and K. Kim, Some New Concepts Relating to the Initiation and Failure of Detonable Explosives, in "High Pressure Science and Technology," Vol. 2, Plenum, New York, 1979.

7. F. J. Zerilli and E. T. Toton, Shock-induced Molecular Excitation in Solids, Phys. Rev. B 29:5891 (1984).

8. C. S. Coffey and E. T. Toton, A Microscopic Theory of Compressive Wave-induced Reactions in Solid Explosives, J. Chem. Phys. 76:949 (1982).

9. S. H. Lin, Theory of Vibrational Relaxation and Infrared Absorption in Condensed Media, J. Chem. Phys. 65:1053 (1976).

10. R. D. Bardo, T. N. Hall, and M. J. Kamlet, Energies and Volumes of Activation for Condensed Detonating Explosives, J. Chem. Phys. 77:5858 (1982).

11. M. Bixon and J. Jortner, Intramolecular Radiationless Transitions, J. Chem. Phys. 48:715 (1968).

12. R. D. Bardo, The Roles of Excited Electronic States in the Bond-Scission Proces in Detonation Reactions, in "Proceedings of the Seventh Symposium (Int.) on Detonation," Office of Naval Research, Washington, DC, 1981, pp. 93-103.

13. S. H. Lin, On the Master Equation Approach of Vibrational Relaxation in Condensed Media, J. Chem. Phys. 61:3810 (1974).

14. R. D. Bardo and M. Wolfsberg, The Wave Equation of a Nonlinear Triatomic Molecule and the Adiabatic Correction to the Born-Oppenheimer Approximation, J. Chem. Phys. 67:593 (1977).

15. A. Nitzan and J. Jortner, Vibrational Relaxation of a Molecule in a Dense Medium, Mol. Phys. 25:713 (1973); A Nitzan, S. Mukamel, and J. Jortner, Some Features of Vibrational Relaxation of a Diatomic Molecule in a Dense Medium, J. Chem. Phys. 60:3929 (1974).

16. R. D. Bardo, paper to be published in J. Chem. Phys.

17. P. C. Haarhoff, The Density of Vibrational Energy Levels of Polyatomic Molecules, Mol. Phys. 7:101 (1963).

18. I. Kanesaka, H. Naka, and K. Kawai, The Methyl Torsion in Toluene (h_8,d_8) and Nitromethane (h_3,d_3) in the Polycrystalline State, J. Chem. Phys. 70:5773 (1979).

19. D. R. Hardesty and P. C. Lysne, "Shock Initiation and Detonation Properties of Homogeneous Explosives," Sandia Report SLA 74-0165, May 1974.

20. R. D. Bardo, Calculated Reaction Pathways for Nitromethane and Their Role in the Shock Initiation Process, in "Proceedings of the Eighth Symposium (Int.) on Detonation," Office of Naval Research, Washington, DC, to be published.

21. D. R. Hardesty, An Investigation of Shock Initiation of Liquid Nitromethane, Combust. Flame, 27:229 (1976).

MOLECULAR ORBITAL CALCULATION OF INDICES OF IMPACT AND SHOCK INDUCED REACTIVITY IN TRINITROAROMATIC MOLECULES

F.J. Owens and P. Politzer*

Energetic Materials Laboratory
ARDC, Dover, New Jersey 07801
University of New Orleans*

INTRODUCTION

The phenomena of impact and shock induced explosion has been mostly modeled by continuum ,thermodynamic macroscopic approaches emphasizing the influence of physical properties on the process. Unfortunately these approaches while satisfying engineering design give little insight into the fundamental cause of explosion, the chemical reactions In the case of impact or shock induced explosion mechanical failure generates heat which causes the molecule to react. Both the innate reactivity of the molecules and the overall mechanical properties of the explosive will relate to the internal molecular structure of the constituent molecules of the materials.

The purpose of this work is to examine the role of molecular structure in determining explosive behavior ,particularly shock and impact sensitivity using parameters obtained by molecular orbital calculations of a series of trinitroaromatic molecules. It has previously been shown in trinitroaromatics that empirically determined parameters such as the electron donating ability of the non nitro substituents on the ring do correlate to impact and shock sensitivity as well as explosive yield (1) This work examines further these correlations using both approximate and ab initio molecular orbital calculations Parameters indexing reactivity are derived from the molecular orbital methods and their correlation to sensitivity is examined. One index of sensitivity used here is impact height which causes explosion. Since experimental configuration and other physical parameters influence sensitivity measurements only data from one laboratory is used and it is only used for relative ordering of reactivity.(2)

APPROXIMATE MOLECULAR ORBITAL RESULTS

The approximate molecular orbital calculations employ the Closed Shell Intermediate Neglect of Differential Overlap (CINDO) approximation (3)Table I lists the molecules for which the calculations were made. Also shown are the calculated charge on the nitrogen and carbon of the C-NO_2 bond and the electrostatic potential between the carbon and nitrogen.The calculations were performed for molecular geometries obtained from the crystal state.It is seen that the charge on the ring increases as the total donating effect of the non nitro substituents increases. Interestingly the variation of the charge on the nitrogens is small but not the charge on the carbon of the $C\text{-}NO_2$ bond. The charge on the carbon correlates well to impact sensitivity. Figure 1 is a plot of impact sensitivity versus the electrostatic potential between the carbon and the nitrogen of the $C\text{-}NO_2$ bond which is an index of bond strength. This supports previous empirical correlations between $C\text{-}NO_2$ bond strength and sensitivity. (1) Previous experimental work has shown that shock and impact sever the $C\text{-}NO_2$ bond in trinitroaromatics. (4).

Table I CINDO calculated charges on carbon and nitrogen of weakest $C\text{-}NO_2$ bond, bond length, electrostatic between C and N of bond and impact sensitivity of a series of trinitroaromatic molecules

molecule	Q_n	Q_c	R (A)	Q_nQ_c/R	impact sensitivity (cm)
TATB	0.490	-0.183	1.422	-0.063	341
DATB	o.480	-0.152	1.51	-0.048	320
MATB	0.460	-0.0569	1.474	-0.018	170
TNB	0.450	·0.040	1.50	0.012	100
TNT	0.621	-0.058	1.496	-0.024	160
TENA	0.601	0.114	1.487	0.046	41
TETRYL	0.597	0.002	1.497	0.001	32

TATB 2,4,6 triaminotrinitrobenzene, DATB 2,4,6 diaminotrinitrobenzene, MATB 2,4,6 trinitroaniline, TNB 2,4,6 trinitrobenzene, TNT 2,4,6 trinitro toluene, TENA 1,2,3,5 tetranitroaniline, TETRYL 2,4,6 trinitrophenyl-methylnitramine

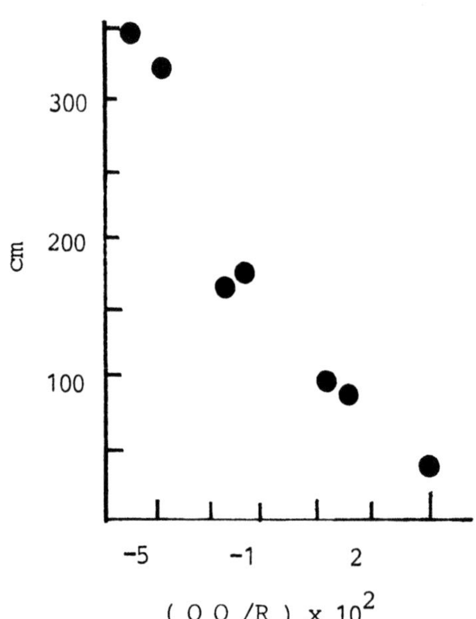

Figure 1 Plot of impact sensitivity versus potential between the C and N of $C\text{-}NO_2$ bond using charges obtained from CINDO calculations

To further substantiate the above correlations ab initio molecular orbital calculations were carried out on a series of trinitro aromatic molecules using a GAUSSIAN 80 program at the STO-3G level. The calculations were performed for molecular geometries determined from the crystal state. Table II lists the molecules for which the SCF calculations were performed ,as well as the calculated charge on the nitrogen and carbon of the $C\text{-}NO_2$ bond and the electrostatic potential at the mid point of the $C\text{-}NO_2$ bond.

TABLE II GAUSSIAN 80 SCF ab initio calculated charges on the C and N of the $C\text{-}NO_2$ bond and the electrostatic potential at the mid point of the the weakest bond.

molecule	Q_N	Q_C	V_{mid} (au)
TENA	0.135	0.203	0.240
TNP	0.076	0.194	0.193
TNB	0.092	0.172	0.187
TNT	0.092	0.177	0.192
TNA	0.055	0.176	0.166
DATB	0.010	0.144	0.114
TATB	-0.020	0.158	0.103

TNP 2,4,6 trinitrophenol

The ab initio results confirm the previous conclusions . The total charge on the carbon of the weakest $C\text{-}NO_2$ bond correlates to shock and impact sensitivity. From the calculated charge distributions the electrostatic potential at the mid point of the $C\text{-}NO_2$ bond was calculated. Figure 2 shows a plot of the impact sensitivity versus the potential at the mid point of the bond showing a correlation. The correlation coefficient is 0.94.

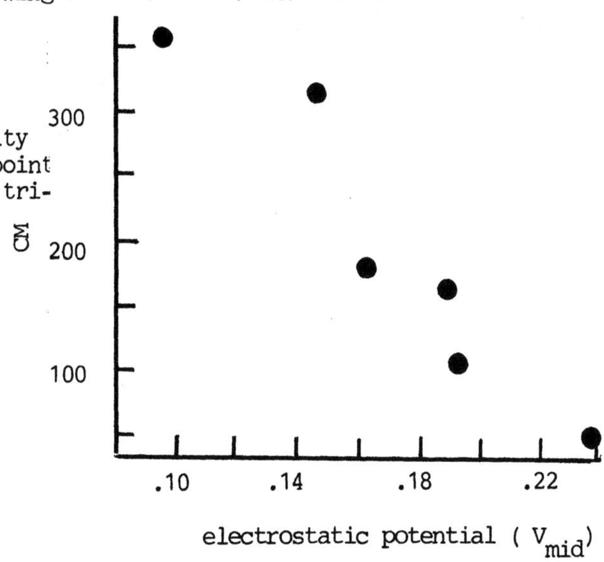

Figure 2 Impact sensitivity versus potential at mid point of weakest $C\text{-}NO_2$ bond in trinitroaromatics listed in Table II

The above discussion has assumed that the innate reactivity of the nitroaromatic molecules solely determines explosive sensitivity. However, it is known that explosive sensitivity is also influenced by physical properties such as mechanical strength which in turn is determined by molecular properties. Therefore it is necessary to determine to what extent the molecular properties that correlate to sensitivity also relate to mechanical properties. Unfortunately mechanical properties data is not available for all the molecules for which calculations were made. Thus under the assumption of a Lindeman like relationship for these materials the melting temperature will be taken as an index of mechanical strength. Figure 3 is a plot of the melting temperature of a series of nitroaromatics versus the ab initio calculated charge on the nitrogen of the $C\text{-}NO_2$ bond showing a clear correlation.

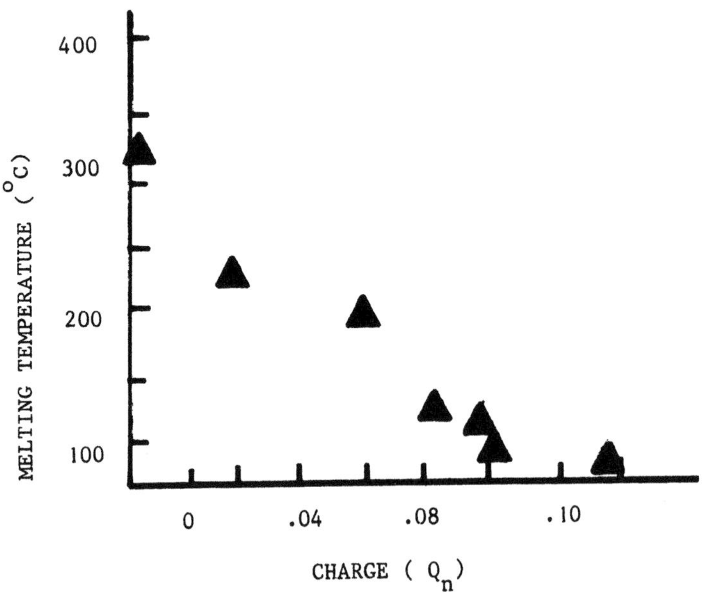

Figure 3 Melting temperature of nitroaromatics versus charge on nitrogen of $C\text{-}NO_2$ obtained from ab initio molecular orbital calculations.

This means that the larger the donating effect of the non nitro substituents the higher the melting temperature and the stronger the material. A similar relationship is also observed for crystal density and thus detonation velocity. Thus not only does increasing donating effect on the ring make the molecules more insensitve but it makes them have a higher energy output.

CONCLUSION

The results of this work show that the susceptibility to reaction in an homologous series of nitroaromatic molecules decreases as the total donating ability of the substituent groups on the molecular ring increases. In particular it is shown that the donor groups increase the charge on the carbon of the $C\text{-}NO_2$ bond which in turn causes the bond to strengthen. The fact that the strength of the C-N bond correlates to sensitivity is somewhat surprising as explosion is the result of a long chain of reactions. This correlation suggests that rupture of the C-N bond may be a rate controlling step in the nitro-

The same molecular parameters which influence sensitivity also influence physical properties. An increase in the total donating ability of the non nitro substituents correlates to an increase in the melting temperature, the crystal density and thus the detonation velocity. The reason for this latter effect is because in the crystals of the trinitroaromatics the molecules pack such that the NO_2 groups point toward the donor groups of the adjacent molecule and the H atoms of the donor groups form H bonds with the oxygens of the NO_2. Thus the intramolecular potential of the lattice is strengthened.

REFERENCES

1. F.J. Owens J. Mol. Structure 121, 213 (1985)

2. M.J. Kamlet, Proceedings Sixth International Symposium on Detonation San Deigo CA, 1976

3. J.A. Pople,D.L. Beveridge and P.A. Dobosh, J. Chem. Phys. 47, 2026 (1967)

4. J. Sharma and F.J. Owens Chem. Phys. Letters 61, 280 (1979)

INVESTIGATION OF SHEAR INDUCED REACTION IN COMPOSITION B3

M. Cowperthwaite and Y. M. Gupta*

SRI International, Poulter Laboratory

Menlo Park, California 94025

INTRODUCTION

Shear-induced reaction (SIR) in explosives was postulated in the 1940s by Bridgman[1] and the results of explosive sensitivity tests, such as the drop-weight[2,3] and fragment impact tests, provide some evidence for its occurrence. Such tests cannot, however, be used to establish the occurrence of SIR because they are not well instrumented and the explosives are subjected to different combinations of deformations that are not analyzed quantitatively. These limitations prompted Gupta[4] to propose the use of the one-dimensional compression and shear wave technique[5,6] to study SIR in explosives. In a subsequent study, DeCarli and Gupta[7] examined the response of a composite explosive (with 70-wt.% RDX), using stress gages, under combined compression and shear loading. The experimental results were ambiguous and the onset of SIR was not well established. In the present work, we have chosen to examine SIR in Compositon B3 using particle velocity measurements to avoid some of the ambiguities encountered in the previous work.

EXPERIMENTAL APPROACH

In the one-dimensional compression and shear technique, the gas gun is used to perform parallel-inclined impact experiments with multiple Lagrange particle velocity gages embedded in the target.[6] The shear loading can be varied by changing the impact angle. In a series of experiments, the magnetic field is orientated to allow the gages to record either the longitudinal particle velocity u_1 or the shear particle velocity u_2. These measured Lagrange particle velocity histories can then be used in a Lagrangian Analysis for Compression and Shear Waves[8] to integrate the one-dimensional equations of motion and generate compressive stress-strain relationships, shear stress-strain relationships, and the corresponding internal energy.

---

*Permanent Address:
Department of Physics
Washington State University
Pullman, Washington 99164-2814

In using the one-dimensional compression and shear technique to investigate SIR, longitudinal particle velocity histories are measured in normal and parallel-inclined impact experiments with the projectile velocities chosen to produce the same initial compressional loading conditions in the explosive. Projectile velocities must also be chosen so that the compression wave by itself does not initiate significant reaction in the explosive in the ti
loading conditions differ only by a shear component, significant differ ences between the sets of particle velocity histories recorded in the normal and parallel-inclined impact experiments can be taken as evidence for the onset and occurrence of SIR.

EXPERIMENTAL METHODS

Composition B3 (60/40 by weight RDX/TNT) was chosen as the test explosive because it is a material with known shock properties that is readily available and can be cast around gage assemblies to provide appropriate samples for the gas gun experiments. The gage assemblies used in the present work were designed to make four cylindrical discs of explosive, 40 mm in diameter and 10 mm high, each containing three gages located 0.5 mm, 2.5 mm, and 4.5 mm from the impact surface with the gage leads coming out of the back of the samples. Each gage was constructed by soldering the active gage element, a copper wire 10 mm long and 0.002-inches in diameter, onto the gage leads made from solid copper posts (0.008-0.01) inches in diameter. Because it was necessary to validate the casting procedure before performing the gas gun experiments, a set of Composition B samples were cast and sectioned to verify (1) that the explosive was uniform and contained no cavities and (2) that the gages were firmly embedded in the explosive and not significantly disturbed by the casting process. The explosive discs were backed by 6-8 mm of epoxy before the experiments were performed, and fused silica discs were used as flyer plates.

Two sets of experiments with different loading conditions were performed to investigate the occurrence of SIR in Composition B3. The first set of experiments, labelled Shot 26, Shot 27, and Shot 28, were designed to produce an initial compressional stress in Composition B3 in the 20 kbar region. The second set of experiments labelled, Shot 33 and Shot 34, and 35 were designed to produce an initial compressional stress in the 10 kbar region.

The magnetic field was orientated to measure the longitudinal particle velocities produced by normal impact in Shots 26 and 33, to measure the longitudinal particle velocities produced by parallel-inclined impact (15° angle) in Shots 28 and 34, and to measure the shear particle velocities produced by parallel-inclined impact (15° angle) in Shots 27 and 35.

RESULTS AND CONCLUSIONS

Figure 1a shows the longitudinal particle velocity histories pro- duced in Shot 26 by the normal impact of a 1/4-inch-thick fused silica flyer traveling at 0.558 mm/μs, and Figure 1b shows those produced in Shot 28 by the parallel-inclined impact (15° angle) of a 3/8-inch-thick fused silica flyer traveling at 0.575 mm/μs. Figure 2a shows the longitudinal particle velocity histories produced in Shot 33 by the normal impact of a 3/8-inch-thick flyer traveling at 0.295 mm/μs, and Figure 2b shows those produced in Shot 34 by the parallel-inclined impact (15° angle) of a 3/8-inch-thick flyer traveling at 0.292 mm/μs. Because of lead effects data beyond 3.5μs are questionable.

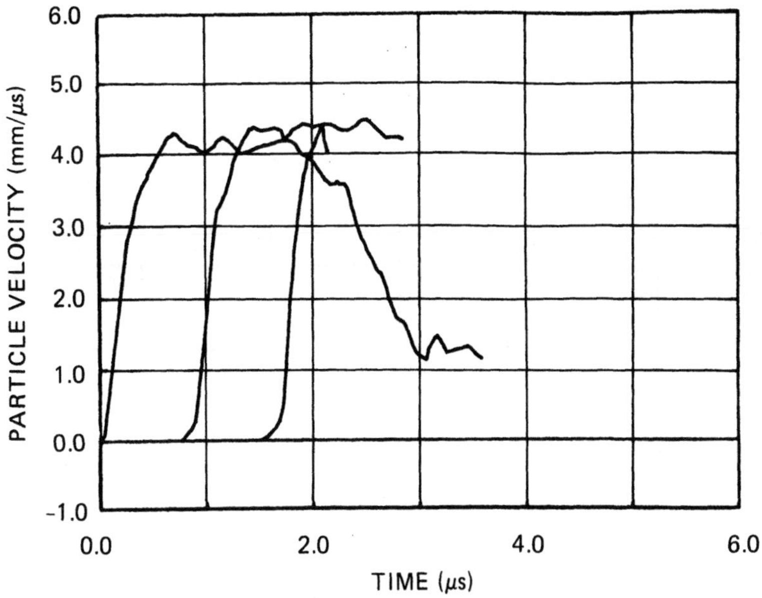

(a) Shot 26, Normal Impact

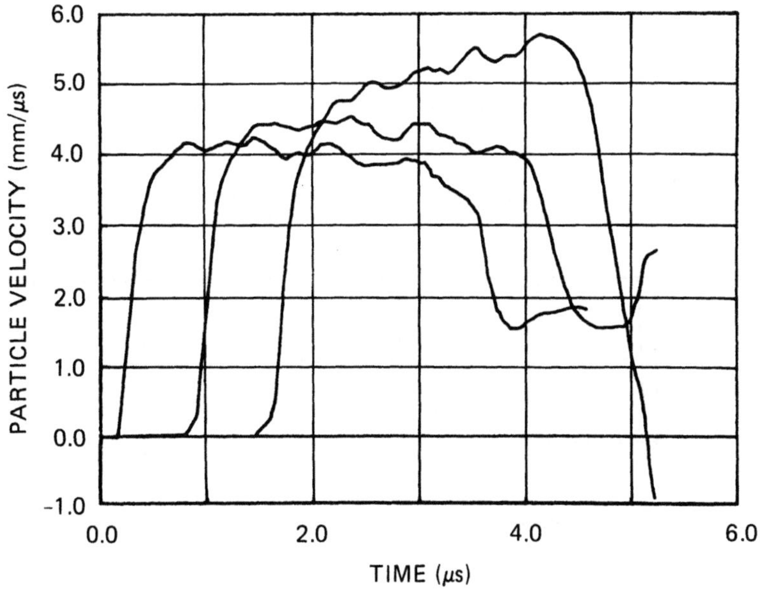

(b) Shot 28, Parallel-Inclined Impact (15° Angle)

Figure 1. Composite Plot of the Longitudinal Particle Velocity Histories Produced by an initial Compressional Loading in the 20-kbar Region.

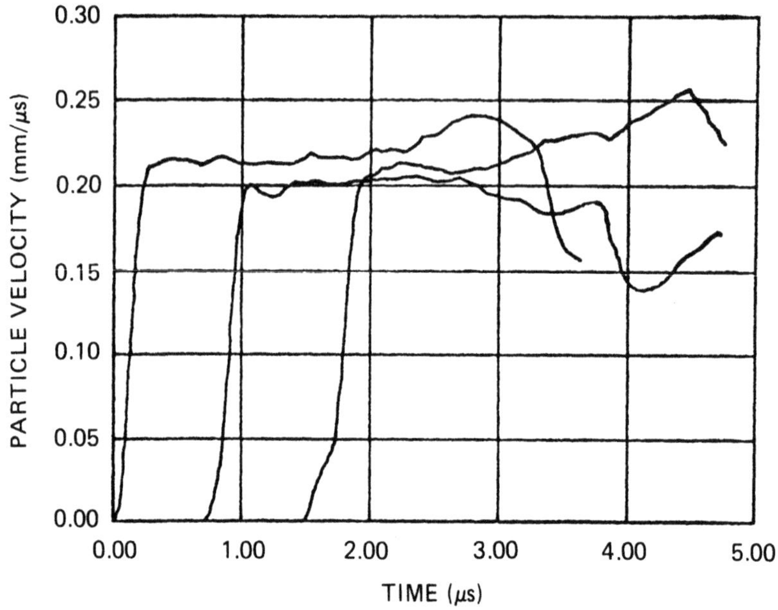

(a) Shot 33, Normal Impact

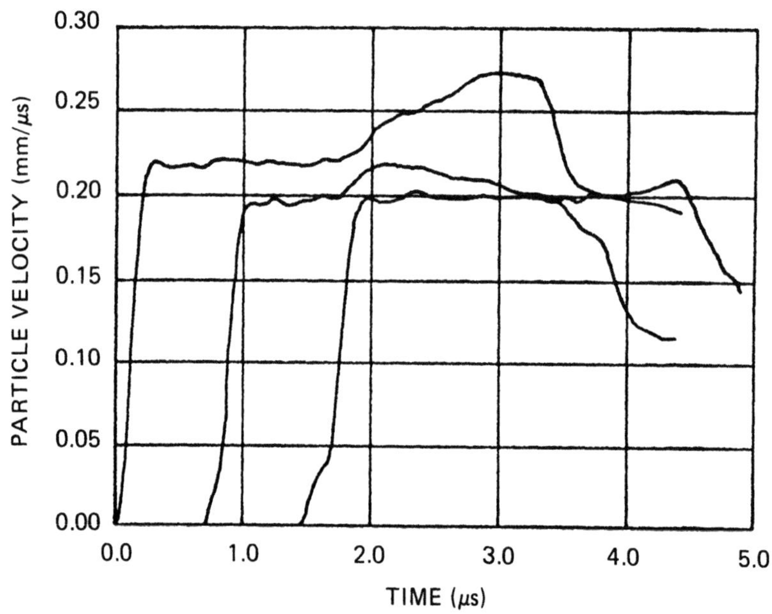

(b) Shot 34, Parallel-Inclined Impact (15° Angle)

Figure 2. Composite Plot of the Longitudinal Particle Velocity Histories Produced by an Initial Compressional Loading in the 10-kbar Region.

The records from Shots 27 and 35 (designed to measure shear wave profiles) are not presented because no significant shear particle velocity was recorded beyond the first gage position in both experiments. This result suggests that the shear deformation is located near the impact surface.

The sets of longitudinal particle velocity histories produced by similar compressional loading conditions in normal and parallel-inclined impact experiments were compared by superposing one set of histories on the other. Superposition of Figures 1a and 1b showed (1) that these sets of histories cannot be compared at all three gage positions because the compression wave broke the third gage in the normal impact experiment, and (2) that the histories recorded by the first and second gages in both types of impact experiment are essentially the same when the flyer plate thicknesses are taken into account. Comparison of the histories in Figure 1b reveals significant features of the developing flow. The peak particle velocity at the wave front increases, and until the flow behind the wave front is influenced by the rarefaction from the rear of the flyer plate, the stress gradient is zero at the first and second gage positions but negative at the third gage position. These flow features suggest that the compressional loading in the 20 kbar region initiates a reactive flow which develops as a growing "square" wave in the stress-Lagrange distance plane until the stress gradient becomes negative in the neighborhood of the third gage position.

We conclude that these 20 kbar experiments are inconclusive about SIR because the particle velocity histories are essentially the same at the first two gage positions and indicate that the compression waves initiate significant reaction in the explosive in the time frame of the experiment.

Superposition of Figure 2a and 2b shows that the flows recorded in Shots 33 and 34 initially develop in the same way but then become significantly different. Properties of the particle velocity histories recorded in both shots will be discussed below in some detail to specify the similarities and differences between the flows, and associate them with SIR.

Qualitatively, the particle velocity histories recorded at the first gage plane are similar and exhibit, behind the wave front, a region of constant particle velocity followed by a region where the particle velocity rises to a peak and then falls. Quantitatively, these histories are essentially identical for 1.76 μs and then become different. After 1.76 μs, the particle velocity recorded in Shot 34 rises to a peak value of 0.275 mm/μs at a time of 3.0 μs, but the particle velocity recorded in Shot 33 remains essential constant for another 0.5 μs before rising to a peak value of 0.239 mm/μs at a time of 2.78 μs.

Qualitatively, the particle velocity histories recorded at the second gage plane are different in the sense that the history from Shot 34 exhibits a well-defined particle velocity peak, but the history from Shot 33 does not. Quantitatively, however, these histories are essentially identical for 1.06 μs. After 1.06 μs, the particle velocity recorded in Shot 34 rises to a peak value of 0.219 mm/μs at a time of 2.12 μs but the particle velocity recorded in Shot 33 remains essentially constant for most of the duration.

The particle velocity histories recorded at the third gage plane are both qualitatively and quantitatively different. The particle velocity in Shot 34 rises to a value of 0.2 mm/μs at the wave front and remains constant for 2 μs, but the particle velocity recorded in Shot 33 rises to a value of 0.207 mm/μs at the wave front, continues to rise and attains a value of 0.213 mm/ μs at 2.25 μs, then falls to a value of 0.207 mm/μs at 2.68 μs, and finally rises to a value of 0.25 mm/μs at 4.33 μs.

Further consideration of these sets of particle velocity histories leads to the conclusion that both flows develop as the same decaying step shock in the stress-Lagrange distance plane until the particle velocity starts to increase in the parallel-inclined impact experiment. Thereafter, the stress gradient becomes negative at the first and second gage planes in Shot 34 and at the first gage plane in Shot 33, and the stress increases at these gage positions as the flows develop differently.

The similarities and differences exhibited by the particle velocity histories recorded at the first two gage planes in the parallel-inclined and normal impact experiments are taken in this paper as evidence of the onset of SIR. Additional support for the onset of SIR is obtained by associating the acceleration and corresponding negative stress gradient observed in such gas gun experiments with the chemical energy release rate. Because both flows develop initially as decaying step shocks in the stress-Lagrange distance plane, an increase in the particle velocity is accompanied by an increase in stress when the stress gradient becomes negative. This being the case, we associate the negative stress gradient with an increasing energy release rate and recall first, that the particle velocity at the first gage position starts to accelerate first in Shot 34, and second, that the particle velocity accelerates for a longer period of time and attains a significantly higher peak value in Shot 34 than in Shot 33.

Based on the above discussion of the particle velocity data we conclude that the increase in the energy release rate at the first gage position is larger in the parallel-inclined impact experiment than in the normal impact experiment, and attribute the difference in the energy release rates to the occurrence of SIR. More gas gun experiments and quantitative analyses are planned to support this conclusion.

ACKNOWLEDGEMENTS

D. R. Henley and D. F. Walker are thanked for their participation in the experiments. The samples were cast by H. E. Hanna. B. Y. Lew is thanked for her assistance in the data reduction. This work was supported by U. S. Army Research Office under Contract DAA929-84-K-0018.

REFERENCES

1. In 1940, P. W. Bridgman studied the response of explosives under superposed shear and pressure, and reported them in J. Chem. Phys. 15, 311 (1947). In 1960, in the introduction given in his Collected Experimental Papers, Vol. I (Harvard University Press, 1964) he noted the difficulty of detecting a true chemical instability under the combination of compression and shear. He recommended the need for further experimental efforts to confirm his conjectures in the 1940's.

2. C. S. Coffey, M. J. Frankel, T. P. Liddiard, S. J. Jacobs, "Experimental Investigation of Hot Spots Produced by High Rate Deformation and Shocks," Seventh Symposium (International) on Detonation, NSWC MP82-334 (1982) pp 970-975.

3. R. Frey, U.S. Ballistic Research Laboratory, Aberdeen Proving Ground, Aberdeen, MD 21005, Private Communication.

4. Y. M. Gupta, Presentation at the DOD Workshop on "New Opportunities for Research in Energetic Materials," Wrightsville Beach, North Carolina (November 1977).

5. Y. M. Gupta, Appl. Phys. Letters 29, 694 (1976).

6. Y. M. Gupta, D. D. Keough, D. F. Walter, K. C. Dao, D. Henley, and A. Urweider, Rev. Sci. Instr., 51, 183 (1980).

7. P. S. DeCarli and Y. M. Gupta, "High Strain Rate Mechanical Deformation of Composite Explosives," Final Technical Report to NSWC, White Oak, MD, under Contract N60921-78-G0110, SRI International, Menlo Park CA 94025 (August 1979).

8. Y. M. Gupta, "Development of a Method for Determining Dynamic Shear Properties," Final Technical Report to DNA Headquarters, Washington DC, under Contract 001-76-C-D384, SRI International, Menlo Park CA 94025 (May 1978); also see Ploymer Engineering and Science 24, 851 (1984).

9. E. R. Lemar, J. W. Forbes, J. W. Watt, and W. L. Elban, J. Appl Phys 58, 3404 (1985).

DYNAMIC AND STATIC COMPRESSIBILITY OF POROUS GRANULAR HEXANITROSTILBENE†

R.E. Setchell and P.A. Taylor

Sandia National Laboratories
Albuquerque, New Mexico 87185

INTRODUCTION

Hexanitrostilbene (HNS) is a binderless, granular explosive prepared over wide ranges of initial grain size and final bulk density. During shock initiation of HNS, ignition initially occurs at discrete hot spots formed during the dynamic compaction of the porous material. In order to develop a predictive understanding of the shock initiation mechanisms occurring in this explosive, it is necessary to formulate constitutive relations modeling the mechanical response of HNS materials under compressive loading. In an early study of HNS properties, Sheffield et al. [1] established shock Hugoniot properties and formulated an equation of state for bulk densities of 1.0 - 1.7 g/cc (crystal density is 1.74 g/cc). Subsequently, Hayes and Mitchell [2] and Hayes [3] developed a model for the shock compaction and chemical decomposition of HNS. In these studies, the non-reactive material response was treated as depending only on the initial bulk density (related to initial porosity) of the HNS sample. Our recent studies [4,5] have established that the shock sensitivity of HNS samples pressed to a common density depends critically on the initial grain size distribution of the material. In the present study, we have undertaken both experimental and analytical studies of the compressibility behavior of HNS under non-reactive conditions. In particular, we have examined the compressibility of HNS materials having a range of initial grain size distributions.

EXPERIMENTAL RESULTS

In order to examine dynamic compressibility properties, a series of planar impact experiments were performed in which the evolution of ramp waves through different HNS materials was examined. The ramp waves were generated in 19.1 mm thick samples of fused silica through the impact of aluminum projectiles at different velocities. Waves having a rise time of approximately 0.60 μs and a peak axial stress of 15 or 24 kbars were introduced into HNS samples from 2 to 4 mm thick. The stress histories

†This work performed at Sandia National Laboratories supported by the U.S. Department of Energy under contract #DE-AC04-76DP00789.

for the initial waves produced in HNS samples are shown in Figure 1. The resultant waveforms, after various propagation distances, were observed at an interface with a fused silica window using velocity interferometry (VISAR) instrumentation. A complete description of this experimental approach has been reported previously [6]. Three types of HNS materials were used in these experiments denoted here as HNS-II, -I, and -FP. HNS-II samples were pressed from a powder having typical grain dimensions of 10 - 20 µm and an average specific surface area of 0.44 m^2/g. HNS-I samples were pressed from a powder having typical grain dimensions of 2 - 10 µm and a specific surface area of 2.1 m^2/g. HNS-FP samples were pressed from a powder having typical grain dimensions of 1 - 2 µm and a specific surface area of 8.2 m^2/g. All HNS samples were pressed to a common bulk density of 1.60 ± .01 g/cc (8% net porosity).

Figure 2 shows the waveforms recorded in HNS-I and HNS-FP samples for the 24 kbar ramp wave case. In HNS-I materials, the initial ramp wave has steepened to form a shock within a propagation distance of 2 mm. Waveforms measured after distances of 3 and 4 mm appeared to be the same as the 2 mm waveform. The initial ramp wave also steepens rapidly in HNS-FP material, but an elastic precursor wave is apparent in waveforms measured after a propagation distance of 2 and 3 mm. The peak amplitude of this precursor wave in HNS-FP samples is approximately 3 kbars. Figure 3 shows the waveforms recorded in all three HNS materials after a propagation distance of 2 mm in the 15 kbar ramp wave case. Surprisingly, the HNS-FP experiment did not show evidence of elastic behavior up to 3 kbar as had the samples in the 24 kbar ramp wave case. The explanation for this inconsistency has not been established. However, the degree of steepening of the waveforms does correlate inversely with the initial grain size of the HNS material. The waveform recorded for the HNS-FP case has nearly reached that of a shock, whereas the waveform in the HNS-II case still has a significant rise time. These waveforms are clear evidence that the dynamic compressibility of HNS materials having a common porosity can have discernable differences if the microstructural features of that porosity differ between the materials. Thus, although microstructural features have been ignored in

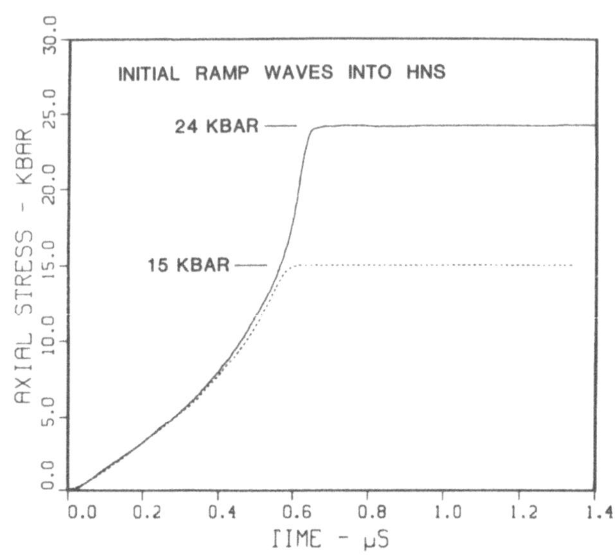

Figure 1. Initial ramp waves introduced into HNS samples during dynamic compressibility experiments. These waveforms result from impedance-matching calculations on measured waveforms produced in 19.1 mm thick discs of fused silica.

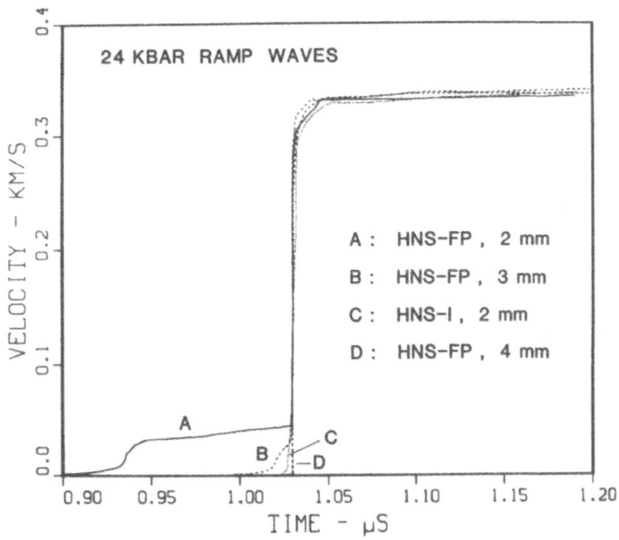

Figure 2. Waveforms recorded in HNS-I and HNS-FP samples for the 24 kbar ramp wave case. The velocity histories shown are the waves transmitted into the fused silica window at the HNS/window interface, and have been shifted in time to facilitate comparisons.

fairly successful models of the dynamic compressibility of other porous materials (see e.g., [7]), an accurate description of HNS behavior will have to account for these features.

In order to examine the influence of microstructural features on the compressibility of HNS materials in greater detail, we have also examined the static compression of these materials under the conditions of uniaxial strain loading. Starting with loose powders under vacuum, a hydraulic press was used to observe the variation of volume occupied by a given mass of powder with an applied axial stress up to 20 kbars. The results of these static tests are shown in Figure 4 as curves A, B, and C. These compression curves, associated with the three different initial

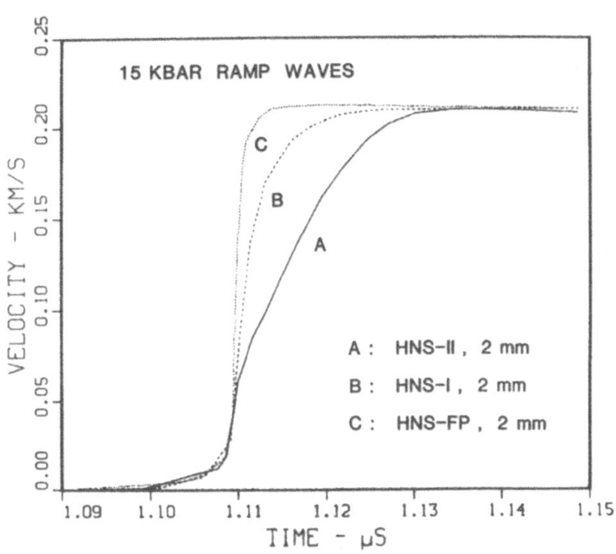

Figure 3. Waveforms recorded in three HNS materials for the 15 kbar ramp wave case.

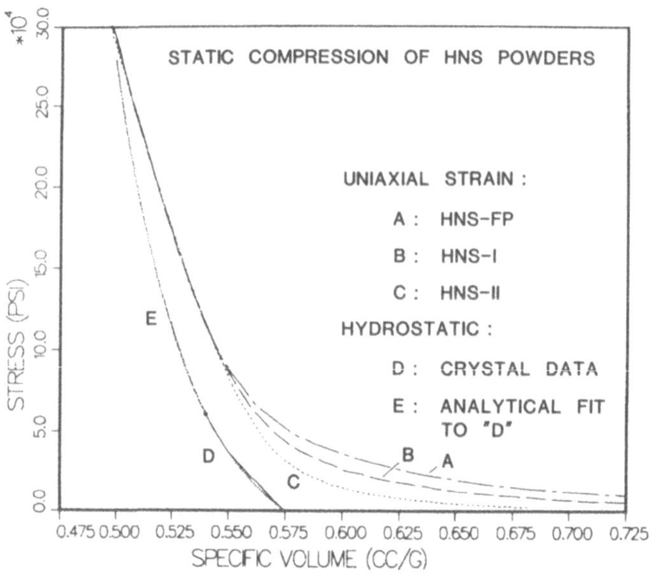

Figure 4. Static uniaxial (curves A-C) and hydrostatic (curves D and E) compression of HNS powders under vacuum. The hydrostatic curves represent compressibility of the HNS crystal material in absence of porosity.

grain sizes, are well separated at low stresses and do not converge until an axial stress of roughly 6 kbars is reached. Beyond this level, the static uniaxial compressive response of HNS displays no dependence on initial grain size. Also shown on Figure 4 is an approximate curve (curve D) for the hydrostatic compressibility of crystalline HNS up to a pressure of 4 kbars. This curve was obtained by performing a mercury penetration test on samples of loose HNS-II powder. The applicablity of mercury penetration methods in determining crystal bulk compressibility lies in the fact that above a pressure of approximately 1000 psi, the mercury is assumed to have filled all of the available void space within the loose powder sample. As pressure is increased, an apparent increase in mercury penetration may then be taken to be a result of the hydrostatic compression of the HNS grains. Curve D in Figure 4 represents the contribution of crystalline compressibility to the compaction of the powders for pressures up to 4 kbars.

DISCUSSION

The results of the present experiments indicate that the compressibility of HNS materials depends not only on the initial porosity, but also on the microstructural features associated with that porosity. HNS samples pressed to some initial density from a fine-grained powder will have a larger number of smaller voids and a correspondingly higher internal specific surface area than samples pressed to the same bulk density from a coarser-grained powder. The experiments at 24 kbars indicated observable elastic strength in the finest-grained material, but this behavior was not consistent. As seen from the static compression curves A-C in Figure 4, the initial grain size of the powder influences the compression of the pressed material up to an axial stress of approximately 6 kbars. Beyond the stress of 6 kbars, the three material types exhibit identical compressive stress-specific volume responses indicating an absence of the grain size effects apparent in their responses at higher volumes. The convergence of the curves above 6 kbars may be explained in view of data collected on particle size distributions for the three HNS powder types before pressing and after particle dispersion of the pressed pellets. Due to the pressing process, the in-situ particle size distributions of both HNS-II and -I are shifted towards smaller particle sizes. Specifically, the in-situ particle size distribution for HNS-II shows an increase in particles displaying sizes in the range of 2 - 9 μm with a corresponding reduction of particle sizes in the range of 10 - 20 μm. Similarly, HNS-I displays an increase in particle sizes in the range of 0.7 - 3.0 μm with a reduction of sizes in the range of 2 - 10 μm. The pressing process, however, appears to have little effect on the fine-grained HNS-FP. Thus, during the pressing process, the larger-grained HNS-II and -I materials experience grain fracture and breakup thereby modifying their in-situ particle size distributions. An explanation of this phenomena may lie with the fact that the loose powder particles of HNS-II and -I possess aspect ratios of 3 and 2 respectively whereas a value of 1 is displayed by the HNS-FP particles. Apparently, the materials composed of particles possessing the higher aspect ratios tend to fracture and break-up more during the pressing process than the HNS-FP material which is composed of roughly spherical particles. Thus, the merging of the uniaxial compression curves A-C in Figure 4 suggests that the process of grain fracture occurs in the uniaxial tests up to an applied stress of 6 kbars. Above this value, the particle size distributions of the three HNS types are sufficiently similar that their responses become identical. Although the pressing process does reduce the in-situ particle sizes of the coarser-grained materials to the point where the static responses converge, further analysis based on mercury intrusion data indicate that the three HNS types retain distinct pore size distributions. In particular, the pore size distributions directly correlate with the initial grain size of the material. This point is crucial in understanding the shock sensitivity of HNS where it was found [8] that the difference in average pore sizes displayed by the three material types is responsible for the reversal in sensitivity of HNS as observed previously [4]. The coalesce of the uniaxial compression curves, however, does not necessarily imply that the porosity has been completely eliminated. Using the free energy function proposed for homogeneous HNS by Sheffield et al. [1], the following isothermal hydrostatic compressibility relation applies:

$$P = \frac{K}{N}[(V_0/V)^N - 1],$$

where V is the specific volume (V_0 =0.575 cc/g), K is the isothermal bulk modulus at room temperature, N is an empirically determined exponent, and P is the hydrostatic pressure. The modulus K was determined by sound speed measurements and a conversion from the isentropic modulus. It assumes the value of 41.7 kbars. The exponent N was chosen so as to give a best fit of this equation to the HNS hydrostatic compressibility data curve D shown in Figure 4. The resulting value of N is 14.4. The compressibility curve predicted by this equation is plotted as curve E in Figure 4. An estimate of the stress state of a pressed sample in the case of uniaxial strain gives the intrinsic hydrostatic pressure as being approximately equal to the applied uniaxial stress. In view of this, a comparison of curve E with the uniaxial compression curves A-C in Figure 4 shows that for a particular axial stress, a small amount of porosity persists up to large values of stress in the uniaxial experiments.

Both the static and dynamic data bases reported herein, illustrating grain size and porosity effects, are currently being used to determine the constitutive response of pressed samples of HNS. The resulting contitutive equations will provide a means by which the influence of porosity and/or granularity on the initiation characteristics of pressed explosives may be taken into account by numerical simulation wavecodes.

ACKNOWLEDGEMENTS

The authors would like to thank Ms. M. Lewis, Mr. M. Anderson, and Mr. D. Wackerbarth for their assistance with the dynamic and static experiments and to Dr. R. K. Wilson for an expedient review of the manuscript. The material analyses and mercury penetration tests were performed by Mr. A. Duncan, Pantex Plant, Amarillo, TX.

REFERENCES

1. S. A. Sheffield, D. E. Mitchell, and D. B. Hayes, The Equation of State and Chemical Kinetics for Hexanitrostilbene (HNS) Explosive, in: Proceedings of the Sixth Symposium (International) on Detonation, Office of Naval Research, ACR-221, Arlington, VA, 748 (1976).
2. D. B. Hayes and D. E. Mitchell, A Constitutive Equation for the Shock Response of Porous Hexanitrostilbene (HNS) Explosive, in: "Symposium (International) on High Dynamic Pressures," Commissariat a l'Energie Atomique, Paris, 451 (1978).
3. D. B. Hayes, Shock-Induced Hot-Spot Formation and Subsequent Decomposition in Granular, Porous Hexanitrostilbene Explosive, in: "Shock Waves, Explosives, and Detonations," J. R. Bowen, N. Manson, A. K. Oppenheim, and R. I. Soloukin, eds., 87:445 (AIAA, New York, 1983).
4. R. E. Setchell and P. A. Taylor, The Effects of Grain Size on Shock Initiation Mechanisms in Hexanitrostilbene (HNS) Explosive, in: "Dynamics of Shock Waves, Explosions, and Detonations," J. R. Bowen, N. Manson, A. K. Oppenheim, and R. I. Soloukin, eds., 94:350, (AIAA, New York, 1984).
5. R. E. Setchell, Experimental Studies of Chemical Reactivity During Shock Initiation of Hexanitrostilbene, to appear in: Proceedings of the Eighth Symposium (International) on Detonation, Albuquerque, NM (July, 1985).
6. R. E. Setchell, Ramp-Wave Initiation of Granular Explosives, Combust. Flame 43:225 (1981).
7. W. Herrmann, Constitutive Equation for the Dynamic Compaction of Ductile Porous Materials, J. Appl. Phys. 40:2490 (1969).
8. P. A. Taylor, The Effects of Material Microstructure on the Shock Sensitivity of Porous Granular Explosives, to appear in: Proceedings of the Eighth Symposium (International) on Detonation, Albuquerque, NM (July, 1985).

RAMAN SCATTERING TEMPERATURE MEASUREMENT BEHIND A SHOCK WAVE

A STUDY OF SOLID EXPLOSIVES

A. Delpuech, A. Mentil, and B. Pouligny

C.E.A., C.E.S.T.A.
B.P. n°, Le Barp, 33830 Belin Beliet, France

I - INTRODUCTION

We have previously shown (1) the possibility to determine experimentally, from Raman scattering, the temperature reached by a material compressed by a shock wave. In order to obtain sustained shock profiles we realized an experimental arrangement fit to use in field conditions with important explosives structures.

The study of liquids explosives with this experimental set-up was illustrated in the 1983 APS Topical Conference on Shock Waves in Condensed Matter (2). The results obtained with nitromethane, and shock pressures up to 8,5 GPa, were in agreement with those deduced from several state equations models. The best agreement was obtained with the values computed with an equation in which the specific heat C_v is considered increasing with temperature.

We present here, the extension of this method to the study of solid explosives. The first explosive chosen is RDX, studied in the shape of single crystal with shock pressure up to 14 GPa. In order to verify the conclusions obtained in the case of nitromethane, the experimental results are compared with values of temperature given by different state equations.

II - PRINCIPLES OF THE METHOD

When a monochromatic radiation of frequency ν_0 passes through a transparent medium, the spectrum of scattered light includes in addition to the ν_0 frequency (Rayleigh Scattering), some other lines due to the inelastic collisions between photons and particles. These lines, of frequencies $\nu_0 \pm \nu_i$, form the Raman spectrum. The lines $\nu_0 - \nu_i$ are called "Stokes", the lines $\nu_0 + \nu_i$ are termed "anti-Stokes".

For a Raman frequency ν_i, the intensity of the Stokes line is given by :

$$I_S = C_i \; I_0 \; (\nu_0 - \nu_i)^4 \; (1 - e^{-h\nu_i/kT})^{-1} \qquad [1]$$

For the anti-Stokes line we have :

$$I_{AS} = C_i \; I_0 \; (\nu_0 + \nu_i)^4 \; (e^{+h\nu_i/kT} - 1)^{-1} \qquad [2]$$

Where I_0 is the incident intensity on the observed area and C_i is a parameter including Raman cross section for the ν_i frequency, medium density, optics aperture, etc ...

I_0 and C_i are factors which vary or may vary during shock compression. Therefore we only investigated the possibility of temperature estimation from the Stokes, anti-Stokes intensities ratio.

Indeed from (1) and (2), we obtain :

$$R = \frac{I_{AS}}{I_S} = \left|\frac{\nu_0 + \nu_i}{\nu_0 - \nu_i}\right|^4 e^{-h\nu_i/kT} \qquad [3]$$

Accordingly temperature determination will require simultaneous measurement of I_S and I_{AS} at a given value of the frequency ν_i, previously measured in another way.

Such a measurement implies :

1 - to check the existence of the Raman spectrum of the explosive under shock,
2 - to determine the possible modifications in Raman frequencies ν_i that the shock may induce,
3 - to measure Stokes and anti-Stokes lines intensities, using the frequencies obtained under shock as guides,
4 - to check the absence of stray light in signals obtained under shock,
5 - to establish the corrections to apply to the signals, to take into account the differences between Stokes and anti-Stokes receptors.

III - EXPERIMENTAL SET-UP

The used set-up meets a double purpose. At first we have to evaluate the possible changes of the Raman spectrum induced by the shock, then to measure simultaneously the intensities of the Stokes and anti-Stokes lines for the chosen frequency.

Except the chronometrical device, the whole experimental arrangement is represented in figure 1.

1 - Optical device

The excitation source is a two-stage YAG laser, frequency doubled and modulated by a POCKELS Cell. The flash emitted at 532 nm corresponds to 100 mJ energy with a duration of 10 nanoseconds.

The laser beam is perperdicular to the shock front.

Scattering light at 90° from the incident beam is collected on each side of the sample by wide aperture lenses. The scattered intensity, in the form of parallel beams, is sent on one side in the "bichromator", on the other side in the optical spectrum analyser.

Preliminary optical alignement are made with a continuous ionized argon laser (λ = 514,5 nm, 1 W).

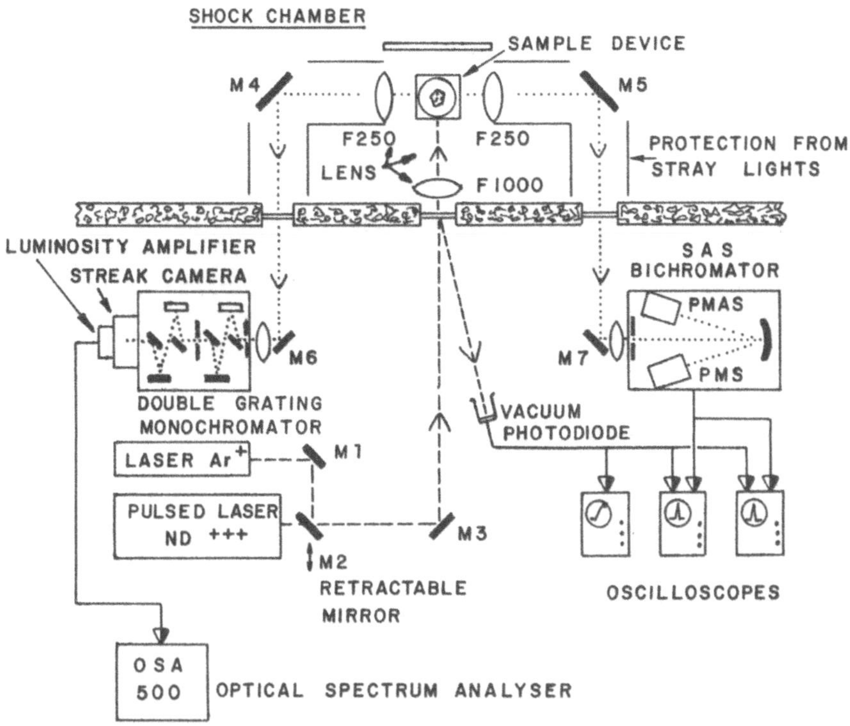

Figure 1 : Experimental arrangement for Raman spectroscopy.

2 - Morphological analysis of the spectrum

One of the scattered light beams is focused on the entrance slit of a double monochromator. The exit slit is suppressed and the spectrum is focused on the photocathode of an electronic streak camera. This camera is only used as a shutter against stray lights. Coupled with the phosphor at the rear of the camera is a multichannel light amplifier. The spectrum is formed again on the target of a vidicon SIT tube.

Then the image is processed by a spectral analyser (O.S.A. 500 B & M Spektronik). The spectrum can be restored in the classic form I = f (λ) by plotting on a pen recorder.

The ranging used during these experiments limits the spectral area to 450 cm^{-1}.

3 - Measurement of Stokes and anti-Stokes lines intensities

The principle scheme of the S.A.S. bichromator is given in figure 2. It's mainly constitued of a concave holographic grating which enables to get nearly symmetrical spectra on the order + 1 and - 1. Two photomultipliers are settled so as to be centred on the frequency of the Raman line to measure, Stokes for one, anti-Stokes for the other.

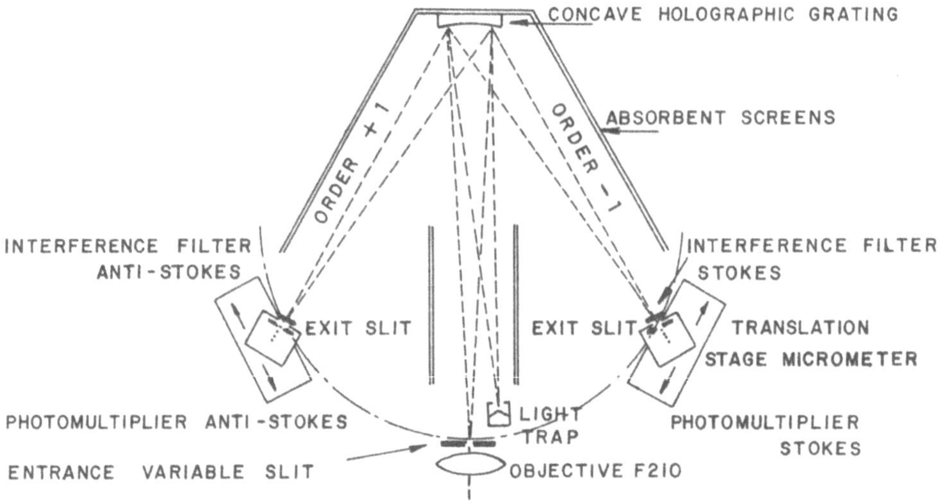

Figure 2 : Bichromator used for temperature measurement.

In order to take into account the broadening and the shifting of the lines under shock, which are more or less well determined, wide slits (0,8 mm) are placed in front of the photomultipliers. This limits the spectral selectivity to 1,6 nm but presents the advantage of allowing to include the possible position uncertainties in wave length.

Each photomultiplier and its slit can be moved giving possibility for change of the Raman frequency.

To improve the contrast of our signal we have been led to place in front of each photomultiplier an interferential filter adapted to the wavelenght to be measured.

The signals provided by the P.M. are either displayed on oscilloscopes or analyzed and memorized in a two channels transient analyser.

IV - SHOCK WAVE GENERATION

Two types of experiments have been realized :

1 - Shock wave is created by a small amount of explosive, up to 5 g. The shock is generated by a detonator with a cylinder of plastic bonded HMX booster.

The explosive is put in a metallic tube in order to mask light emitted during detonation. The tube is obtured by a copper plate through which shock wave is transmitted to the single crystal.

2 - In order to obtain sustained shock profiles, the shock is generated by the detonation of a cartridge of plastic bonded HMX (Ø = 100 mm, h = 95 mm) initiated by a plane wave generator. The value of generated pressure depends of the nature of

the metal gap inserted between the cartridge and the crystal. In this case the pressure decreases very slowly with time.

For all the set-ups used the generated pressure and its evolution with time have been deduced from measurement of free surface velocities, made by laser Doppler interferometry (3).

V - RESULTS

The choice of the Raman frequency depends on the intensity of the band but it's more importance to obtain a good matching with the temperature to be measured. In the case of temperature between 300 and 1000 K, the precision will be better when using a Raman frequency line in the range from 500 to 1000 cm^{-1}.

In the case of RDX the Raman line chosen is the vibration frequency located at 884 cm^{-1}.

The shock pressures were 9,5 and 13,5 GPa.

1°) - Spectrum morphological evolution under shock

In the case of nitromethane, the modifications under shock of the Raman line at 917 cm^{-1} were of two types : broadenning and shifting (1)(2).

Figure 3 shows the shape of the Stokes lines of RDX located at 884 cm^{-1} at rest and the same line under shock. No morphological evolution under shock is observed.

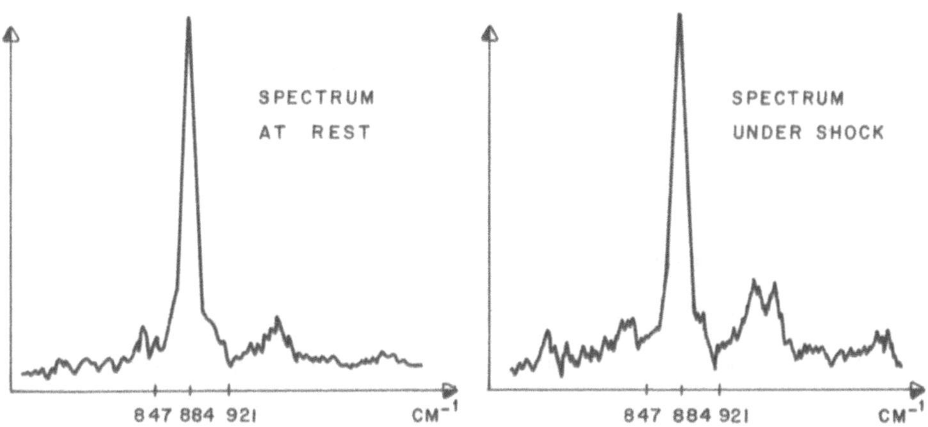

Figure 3 : Morphological evolution under shock of the Stokes line at 884 cm^{-1}

2°) - Temperature determination

The temperature values obtained in the case of a shock generated by a detonator with a cylinder of plastic bonded HMX booster are :

. for a pressure of 9,5 GPa, T = 405 K,
. for a pressure of 13,5 GPa, T = 485 K.

Four samples have been fired for each pressure. The reproducibility of the measurements for a given type of device is of the order of ± 5 %.

3°) - Discussion of the results

We want to compare the results obtained with evaluations made by using a state equation in which $(\frac{\partial P}{\partial T})_V$ = Ct and C_v = Ct or C_v = f (T) :

Table I

		9,5	13,5
	Induced pressure GPa	9,5	13,5
Temperature (K)	Experimental value	405	485
	State equation with C_v^* = 0,261 (4)	510	664
	State equation with C_v^* = 0,229 (5)	540	715
	State equation with Cp^* = 0,236 + $6{,}9.10^{-4}$ T (6)	415	490
	State equation with Cp^* = 0,232 + $7{,}2.10^{-4}$ T (6)	414	487

(*)$Cal.g^{-1}{}^{\circ}C^{-1}$

This table shows that agreement is quite satisfactory between the experimental values and those computed when the specific heat is considered increasing with temperature.

VI - CONCLUSION

The first conclusion of this work is that the experimental set up that we have realized allows to determine, from Raman scattering, the temperature reached by a solid explosive compressed by a shock wave.

The reproducibility of the measurement for a given type of device is of the order of ± 5 %.

The results obtained with single crystal of RDX, and with shock pressures up to 14 GPa, confirm the conclusions stated in the case of nitromethane. The best agreement between experimental temperatures and values deduced from several state equations models, is obtained when the specific heat is considered increasing with temperature.

VII - REFERENCES

(1) - F. BOISARD, C. TOMBINI, A. MENIL
7th Symposium on Detonation - ANNAPOLIS - p. 1010 (1981)

(2) - A. DELPUECH, A. MENIL
1983 APS Topical Conference on Shock Waves in Condensed Matter SANTA FE - p. 309 (1983)

(3) - DURAND et Coll.
Rev. Sci. Inst. 48 (3), (1973)

(4) - G. KRIEN, H. LICHT, J. ZIERATH
Therm. Chem. Acta 6, 465, (1973)

(5) - M. REY LAFON, E. BONJOUR
Mol. Cryst., 24, 191 (1974)

(6) - B.M. DOBRATZ
"L.L.N.L. Explosive Handbook" U.C.R.L. 52 997 (1981).

EQUILIBRIUM SHOCK AND RELEASE PROPERTIES OF POST DETONATION MIXTURES OF PBX-9404 LX-14, RDX, AND TNT

M. van Thiel, F. H. Ree, and J. Sayer

University of California
Lawrence Livermore National Laboratory
Livermore, California, 94550

ABSTRACT

The statistical mechanical model, CHEQ, has been used to calculate equilibrium properties of the detonation products of PBX-9404, LX-14, RDX, and TNT. The results are compared with experiment. The theoretical-experimental comparison indicates good agreement for the HMX mixtures and RDX, except for CJ pressures. Such differences may be related to rate effects. The behavior of TNT products is qualitatively different because of the lower density range and larger carbon content. The shock data indicates some characteristics of carbon as formed in these hot mixtures. These characteristics are discussed.

INTRODUCTION

Recent work[1] on HMX rich explosives has shown peculiarities in the empirical equations of state (EOS). To fit both shock and release wave data one needs to introduce a nearly discontinuous change in slope at the Chapman-Jouguet point for 9404. This indicates that even for these high performance formulations rate processes may be present. This discontinuous slope has since been confirmed with additional overdriven shock wave data.[2] The near or total absence of this pattern in the data for PETN[3] suggests that the discontinuity is related to the amount of carbon present. Recent chemical equilibrium calculations[4] show that, aside from some disagreement in the CJ pressure, good agreement can be obtained with experiments on HMX rich explosives. The chemical equilibrium study was done with CHEQ,[5] a statistical mechanical theory of multiphase mixtures. The purpose of this paper is to extend the study to a high-carbon content low-energy explosive to study the effect of carbon on the agreement between equilibrium theory and experiment.

THEORY

The CHEQ code combines an accurate statistical mechanical theory of mixtures with an advanced method for calculating the equilibrium composition of these mixtures. We use the Ross theory, which is an improved version[6] of the Mansoori-Canfield-Rasiah-Stell theory. In this theory the Helmholz free energy (A) of a fluid is found by minimizing the right hand side of

$$A < AHS + F(\eta)NkT + (\rho N/2) \int \phi(r) gHS(r,\eta) dr \qquad (1)$$

with respect to the hard-sphere packing fraction, η. The significance of the hard sphere free energy, AHS, the hard-sphere radial distribution function, gHS, and $F(\eta)$, a small correction factor is discussed in Ref. 5.

Early high pressure Hugoniot[7] work showed that the exp-6 potential

$$\phi(r) = \epsilon\{(6/(\alpha-6))\exp[\alpha(1-r/r^*)] - (\alpha/(\alpha-6))(r^*/r)6\}, \qquad (2)$$

is the best form for high pressure studies. Subsequent Hugoniot studies on a number molecules by Nellis and others has confirmed this.[8] Because α is constant for many molecules before ionic processes come into play, a corresponding-states principle can be used to derive the potentials. For hydrogen, a relatively small value of α destroys the correspondence. Also, highly polar molecules (water and ammonia) require a dipole term, λ/T,[8] resulting in a potential.

$$\phi(r,T) = (1+\lambda/T)\phi(r) \quad . \qquad (3)$$

This improves the theoretical calculations of these Hugoniots at temperatures below λ.

The present theory uses the improved van der Waals one fluid (vdWlf) model to describe the average potential for the mixture. The interaction between unlike pairs necessary for this purpose is obtained from the combination rules[9]:

$$r^*_{ij}=k_{ij}(r^*_{ii}+r^*_{jj})^{1/2}; \; \epsilon_{ij}=l_{ij}(\epsilon_{ii}\epsilon_{jj})^{1/2};$$

$$\alpha_{ij}=m_{ij}(\alpha_{ii}\alpha_{jj})^{1/2} \quad . \qquad (4)$$

The factors, k_{ij}, l_{ij}, m_{ij}, provide for possible mix defects. We use $k_{ij}=l_{ij}=m_{ij}=1$ for all species' interactions except that $k_{H_2O,N_2} = 1.03$ and $k_{H_2O,CO_2} = 0.965$. This small adjustment has a negligible effect on the EOS but shifts a fluid-fluid phase line into a region where the phase separation affects the CJ properties of high density explosives. HMX explosives have CJ volumes that correspond to those of the phase separated mixture.

To produce the effective potential of the improved vdWlf model needed for the Helmholz free energy minimization, we use the mole fraction(x)-weighted potential parameters

$$(r^*)3 = \sum_{ij} x_i x_j (r^*_{ij})^3, \qquad \epsilon = \sum_{ij} x_i x_j \, \epsilon_{ij}(r^*_{ij})^3/(r^*)^3,$$

$$\text{and} \quad \alpha = \sum_{ij} x_i x_j \, \alpha_{ij} \, \epsilon_{ij} \, (r^*_{ij})^3/\epsilon(r^*)^3. \qquad (5)$$

These provide us with the composition-dependent potential with which each molecule interacts with its neighbors. One of us tested the accuracy of mixture rules with Monte Carlo calculations of a hydrogen-helium mixture.[10] The statistical mechanical results closely follow the Monte Carlo simulations.

While the HMX mixtures discussed here contain only small amounts of solid carbon, TNT contains a considerable amount. The EOS of such solids are handeled separately. In this work we use an EOS of carbon based on experimental data. Two different solid state equations for both graphite and diamond have been examined in this work. One is a Murnahan form

$$V(P,T) = V0(1+B0'P/B0)(-1/B0') \quad , \tag{6}$$

where V_0 varies linearly with temperature. The other is a modified Grüneisen form in which V_0 is a constant and P is replaced by P-aT where a is the room temperature value of $(dP/dT)_v$. This was done to test the thermal effects in the EOS of the carbon forms. Both forms give very similar results. The chemical potentials $\mu_i = dG/dn_i$ (G = Gibbs free energy) for the chemical species are now used in the chemical equilibrium calculations.[8]

The equilibrium composition is found by minimizing the Gibbs free energy with respect to the chemical composition of the mixture. This part of the model uses the method of univariant descent[11] to maximize the speed of the convergence. The computational procedure uses numerous iterations before an equilibrium point is established. The data is computed as a series of isotherms in (P,T) space. These may be interpolated to yield Hugoniot points or isentropes for any particular entropy in the T, ρ, P, E, S grid of the CHEQ data base. The Hugoniot calculations interpolate on the zero solution of the Hugoniot function:

$$FH = E - E_0 - P(V-V_0)^{1/2} \quad . \tag{7}$$

The minimum in the shock speed vs. density curve defines the value of the Chapman-Jouguet (CJ) point.

RESULTS

It must be noted here that RDX and HMX have nearly the same EOS, since the compositions are identical and the heats of formation nearly the same. The same can be said for LX-14 and PBX-9404. HMX is not discussed here, for the sake of brevity.

TABLE 1. Chapman Jouguet data

Source		ρ_0 g/cc	D km/s	T °K	P GPa	Carbon Phase
TNT	Exp.	1.63	6.93		21.0	
"	"	1.64	6.95		19.0	
"	CHEQ	1.63	6.73	3850	17.5	diamond
"	"	1.63	7.24	3900	19.5	graph.
"	WCA4	1.64	6.908	3595	18.3	diamond
"	"	1.64	7.11	3618	20.1	graph.
"	Exp.	1.061	5.25		11.0	
"	CHEQ	1.061	5.18	3760	9.15	diamond
"	WCA4	1.061	5.312	3333	7.7	diamond
RDX	Exp.	1.80	8.754		34.7	
"	CHEQ	1.80	8.71	3690	32.0	diamond
"	"	1.77	8.61	3717	31.7	diamond
"	WCA4	1.80	8.754	4074	34.7	diamond
9404	Exp	1.84	8.80		37.0	
"	CHEQ	1.84	8.81	4109	33.2	diamond
LX-14	Exp.	1.81	8.76			
"	"	1.835	8.79		34-37	
"	CHEQ	1.835	8.74	4012	32.2	diamond

A number of CJ points were calculated for the explosives shown in Table 1. Some recent results of the WCA4 model[12] are included for comparison. The experimental results are those of Dobratz[13] and Mader,[14] except the pressure of LX-14, which is a preliminary data point determined by Quirk and Kury.[15] This work is still in progress. The CJ pressures as calculated by CHEQ tend to be low. The percentage difference is less for RDX and the HMX mixtures than it is for TNT. The pressures computed by WCA4 are generally higher, except for low density TNT.

The detonation velocities agree more closely with the experimental values than the CJ pressures. Except for TNT, where the discrepancy is surprisingly large. A test of the effect of the carbon phase on the CJ detonation velocity showed that allowing only the graphite phase over-corrects the results. The WCA4 code shows qualitatively the same result but with a smaller difference in detonation velocity, probably due to the use of a denser amorphous graphite.

The effect also shows up in comparison to the overdriven data near the CJ point. Figure 1 shows the computed CHEQ Hugoniots for the two HMX mixtures and TNT. The HMX calculations show reasonable agreement with experiment[3,16] though the CJ pressures are low. LX-14 is an explosive with slightly more HMX than PBX-9404 but with a less energetic binder and slightly higher heat of formation. It should therefore behave similar to 9404. See Fig. 1. No shock data for LX-14 is available yet. At higher pressures the CHEQ results are slightly stiffer, due to physics that is not now included in the fluid model. The situation for TNT is quite different. Here we note again that the usual calculation with the diamond form of carbon yields a low pressure and detonation velocity, but at higher pressures the data[16] cross the experimental curve and show a relationship to calculation that is similar to that of the HMX explosives. Also plotted are the empirical JWL equations. It was noted above that the JWL form requires a correction[2,3] to fit overdriven experimental data. This correction introduces a sharp increase in the slope at the CJ point of 9404, but is not needed in the TNT equation untill a pressure of about 40 GPa is reached on the overdriven Hugoniot. This is consistent with the deviations from the standard JWL form plotted in Fig. 1.

The importance of the carbon EOS in TNT is underscored by the principal (or CJ) isentropes in Fig. 2. Here we compare the JWL fits (dashed lines) to the CHEQ results (solid lines). Again the results for 9404 and LX-14 are similar. The fluctuations in curvature due to the nitrogen phase separation near 0.2 Mbar (20 Gpa) and the graphite-diamond transition near .1 Mbar (10 GPa), are not reproducible by the JWL form. The effect of the carbon transition is much larger in the case of TNT. As a result, the CHEQ data do not agree as well with the relatively smooth JWL fit as do the HMX mixture calculations.

If the pressure of the CHEQ calculations is low, in the region between the diamond and fluid-fluid phase transitions, it should also be low in comparison to porous Hugoniot data. This is confirmed by the results shown in Fig. 3. Here the calculated detonation velocities of porous RDX and TNT are plotted against initial density as are the corresponding experimental data.[17,18,19] The region of (P,ρ_0) space shown for RDX is the region of largest deviation. At lower density and near normal density the differences are smaller. The temperatures on the RDX curve are somewhat higher than the temperature of the normal density CJ point, while those along the TNT curve are similar (see Table 1). Again we note that the presence of graphite in the post detonation mixture yields a higher detonation velocity than measured and therefore a higher pressure, while diamond yields a lower value. In Fig. 3 the upper TNT curve (dotted lines) was obtained by raising the heat of formation of diamond by 2.44

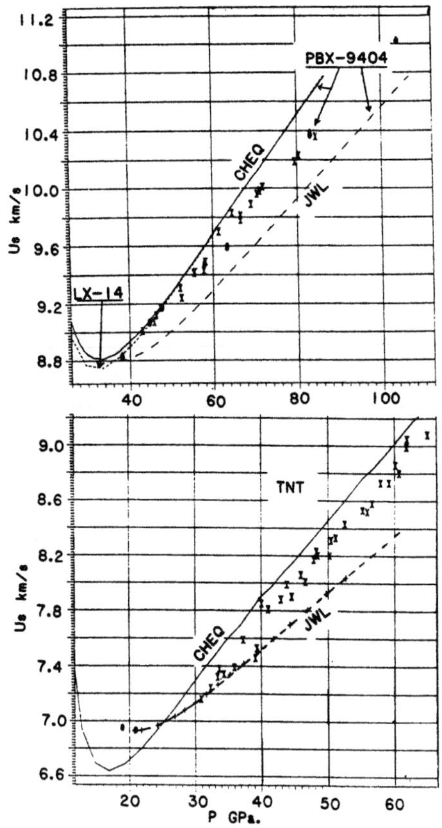

Fig. 1. Hugoniots for PBX-9404, LX-14 (upper figure) and TNT (lower). Curves are as indicated. Experimental data: X Kineke, ⊕ Green.

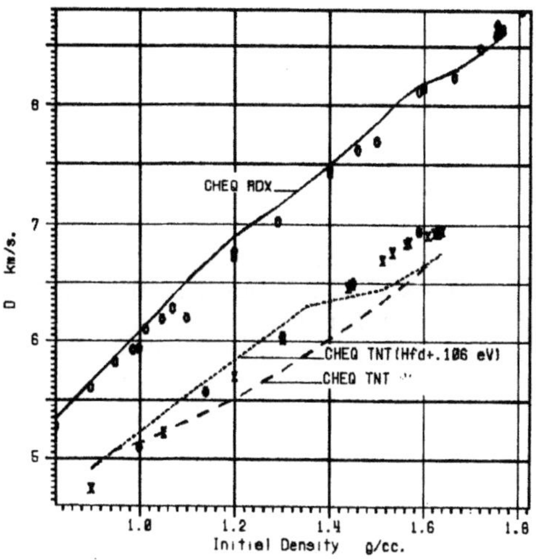

Fig. 3. Detonation velocity vs. initial density. O RDX data; ⊕ Dremin; X Urizar. ....CHEQ with higher heat of formation of diamond.

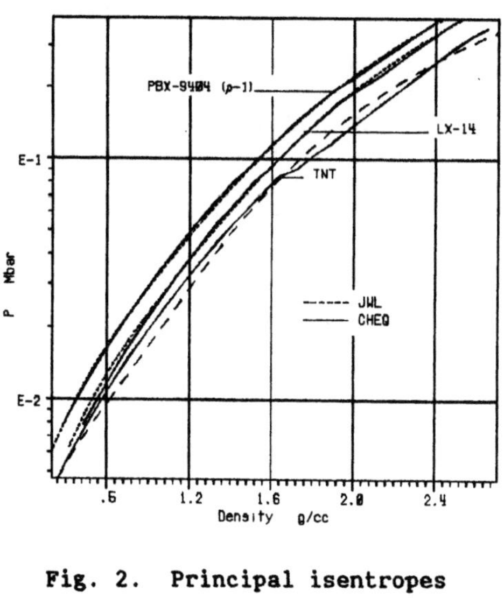

Fig. 2. Principal isentropes

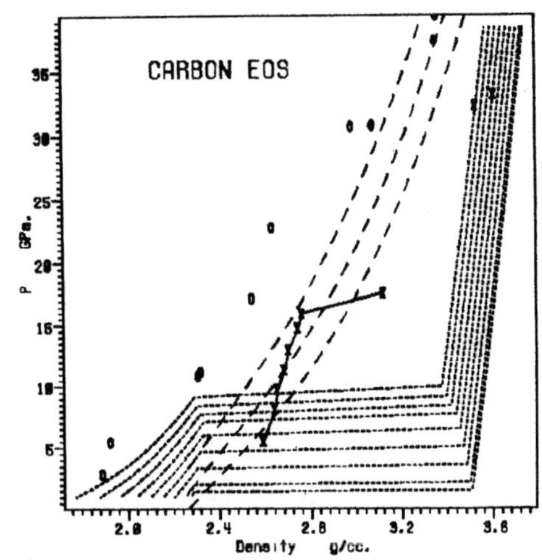

Fig. 4. Carbon EOS:.........CHEQ (T=.025, .05, .1, .15, .2, .25, .3, .35, .4 eV);- - - - Cowan-Mader (T=.025, .28, .57 eV); —X— from TNT analysis; X 9404 and LX-14.

kcal/mol to simulate the amorphous nature of diamond formed during the detonation. This exercise shows the differences in heats of formation or chemical potentials needed to produce a significant shift in the phase line. We may now use this figure to determine the density of carbon that is needed to make the calculation fit the data.

The result of this calculation is shown in the carbon (P,V) space in Fig. 4. The result is in reasonable agreement with the Cowan form to a pressure of 16 GPa, after which the curve bends towards the diamond phase. This form of amorphous carbon seems therefore to be slightly denser than normal graphite. The vitrious data of Gust[20] on the other hand was taken on a sample with a lower density (1.5 g/cc). X-ray examination of vitreous carbon, however, shows graphitic particles of about 50 by 15 Å with 5% smaller interlayer spacing, supporting a higher density for small carbon particles. The apparent shift in the graphite phase transition to about 20 GPa decreases the slope near the CJ point and masks the slope increase noticed in PBX-9404.
The temperatures in these porous shocked samples is high. Table 1 shows temperatures close to the 4000 K at the diamond-graphite-liquid-phase triple point of pure carbon. At high pressure the melting temperature may be lowered in the detonation products considered here. The behavior observed above may therefore be due to liquid carbon.

ACKNOWLEDGMENTS

Work performed under the auspices of the U.S. Department of Energy by the Lawrence Livermore National Laboratory under contract number W-7405-ENG-48.

REFERENCES

1) M. van Thiel, E. L. Lee, and S. Cochran, J. Appl. Phys. 54, 6760 (1983).
2) L. Green, E. L. Lee, A. C. Mitchell, F. H. Ree, R. Tipton, and M. van Thiel, "High Pressure Science and Technology," eds. C. Homan, R. K. MacCrone and E. Whalley, Proceedings of the 9th AIRAPT Conference, Albany, NY, July 1983 (North Holland Publ., NY, 1984) V. III, p. 293.
3) E. L. Lee, M. van Thiel, L. G. Green and A. Mitchell, "Shock Waves in Condensed Matter-1983," eds. J. R. Assay, R. A. Graham, and G. K. Straub, Proceedings of the APS Topical Conference, July 1983, (North Holland Publ. Co., NY, 1984) 617; see also "Supercompression of LX-07, LX-17, PBX-9404, RX-26AB, and the EOS of Detonation Products," by L. G. Green, E. L. Lee, A. Mitchell and C. Tarver, in Proceedings of the Eighth Symposium International on Detonation, Albuquerque, NM, July 1985.
4) F. H. Ree and M. van Thiel, ibid.
5) F. H. Ree, J. Chem. Phys. 81, 1251 (1984).
6) M. Ross, J. Chem. Phys. 71, 1567 (1979).
7) M. van Thiel and B. J. Alder, J. Chem. Phys. 44, 1056 (1966); M Ross, Phys. Rev. A 8, 1466 (1973).
8) F. H. Ree, J. Chem. Phys. 81, 1251 (1984). (For a list of references to experimental data, see p. 1255.)
9) T. M. Reed and K. E. Gubbins, Applied Statistical Mechanics (McGraw Hill, New York, 197) p. 32.
10) F. H. Ree, J. Chem. Phys. 78, 409 (1983).
11) H. B. Levine, J510-82-008/5041 (Jaycor, Del Mar, 1982). A report on a program developed under contract with LLNL.
12) R. Chirat and J. Baute, in Proceedings of the Eighth Symposium Int. on

Detonation, Albuquerque, 1985 (to be published, by Dept. of the Navy); J. Baute and R. Chirat, ibid.

13) B. M. Dobratz, LLNL Explosive Handbook, Lawrence Livermore National Laboratory, Livermore, CA, UCRL-52997 (1981).

14) C. L. Mader, Numerical Modeling of Detonations (U. of Cal. Press., Berkeley, 1979).

15) W. J. Quirk, J. W. Kury and R. D. Breithaupt, Lawrence Livermore National Laboratory, Livermore, CA, private communication (1985).

16) J. H. Kineke and C. E. West, in Proceedings of the Fifth Symp. (Int.) on Detonation, Pasadena, CA, 1970 (ACR-184, ONR, #0851-0057 U.S. Gov. Printing Office, Wash. DC, 1970) p. 533f. J. H. Kineke Jr., "Shocked States of Zeven Overdriven Explosives," Meeting report, April 1971, BRL Proving Grounds, Aberdeen Md.

17) D. J. Steinberg, in Proceedings of the Eighth Symposium Int. on Detonation, Albuquerque, 1985 (to be published, by Dept. of the Navy)

18) M. J. Urizar, E. James, Jr., and L. C. Smith, Phys. of Fluids 4, 262 (1960)

19) A. N. Dremin and P. F. Pokhil, Proceedings of the Academy of Science USSR, Phys. Chem. Section 128, 839 (1959)

20) W. H. Gust, Phys. Rev. B 10, 4744 (1980).

IMPACT EXPERIMENTS ON THIN LAYERS OF POLYMERS AND INTERMEDIATE EXPLOSIVES

G.M. Swallowe and J.E. Field
R.A.R.D.E., Fort Halstead, Sevenoaks, TN14 7BP, Kent, UK

C.D. Hutchinson
Cavendish Laboratory, Madingly Road, Cambridge, CB3 OHE, UK

Recent papers from the Cavendish Laboratory have described impact experiments on small masses of explosives and polymers. This work has been extended to examine the effect of dynamic compression at pressures in the region of 0.5 - 1.0 GPa on PTFE and Kel-F 800 and to study the effects that these materials have upon the ignition behaviour of PETN by impact. Similar experiments have also been performed on tetryl, 3-aminotetryl and 3,5-diaminotetryl. It is shown that both polymers have a sensitising effect on PETN and this is explained in terms of their thermomechanical properties. Under impact, ignition frequently occurs with the aminotetryls but does not lead to propagation. Rapid flow precedes the ignition of 3,5-diaminotetryl but not with 3-aminotetryl. Ignition was not found with tetryl. These results are found to be consistent with previous work describing the behaviour of other energetic materials and are explained in terms of the mechanical stength properties and mode of failure of the explosives.

INTRODUCTION

The ignition mechanisms of thin layers of explosives undergoing mechanical deformation, either as pure materials or as mixtures with certain polymers have been examined by several workers[1-5]. It is thought that initiation of explosives by mechanical shock is thermal in origin, the impact energy being concentrated into localised zones or 'hot spots'. The main mechanisms whereby this energy-localisation occurs are adiabatic compression of trapped gas spaces, localised adiabatic shear, friction between the impacting surfaces, the explosive crystals and additives, and viscous heating of material rapidly extruded between the impacting surfaces[3]. High-speed photographic studies showed that various phenomena occurred during the impact of small samples of pure explosives including plastic flow, jetting, shear banding and melting, with ignition occurring at a few 'hot spots' following severe sample deformation[2]. The introduction of polymeric grits with softening points well below the expected ignition temperatures of pure explosives was found to cause sensitisation in certain cases [eg polycarbonate] but not in others [eg polypropene][4]. Mechanically strong materials which failed catastophically under impact were found to sensitise and this was attributed to their ability to localise energy on failure either at crack tips or in shear bands. Strain gauge measurements of both pure explosives and explosive/polymer mixtures produced evidence for sharp pressure drops during impact when the explosives underwent rapid flow or when the polymer failed. If ignition occurred, it did so after this pressure drop.

In several plastic bonded explosives of current interest, PTFE and Kel-F 800 [a copolymer of chlorotrifluoroethene and vinylidene fluoride] are used as binders. Although the sizes of polymer particles in the moulding powders are considerably smaller than those used in previous studies[4], the S.E.M. showed that the particle size of PTFE in particular may be many tens of microns. It was considered worthwhile to investigate the effect of impact upon these two polymers and to determine what effect they have on the sensitiveness of explosives in the drop-weight test.

Studies into the development of heat-resistant intermediate (booster) explosives have led to the synthesis of 3-aminotetryl (3AT) and 3,5-diaminotetryl (3,5DAT)[6]. Their explosiveness, as measured from small-scale safety certificate tests, was found to be greater than that of tetryl; they also showed greater impact sensitiveness compared with this latter material. It was therefore of interest to examine these explosives in the drop-weight tests, not only to extend the scope of compounds thus far examined, but also to see if differences in their behaviour compared with tetryl could be detected.

EXPERIMENTAL

The two drop-weight machines, in which strain rates of ca $10^3 s^{-1}$ are attained, have been described previously[2]. An AWRE C4 continuous access rotating mirror camera was used to record the impact of the 5.5kg mass falling from 1m on to the test sample. Two toughened glass anvils allowed the sample to be viewed by transmitted light (Fig 1). In the instrumented drop-weight machine,(Fig 2), two semiconductor strain gauges in series were attached to the lower of the three steel rollers. The output of these gauges was analysed according to the method described in Refs. 1 & 2. For the explosive samples, a simple circuit was made between roller R2 and R3 which monitored the electrical resistance of the sample which changes significantly if ignition occurs. This apparatus provided useful evidence on the relation between mechanical failure of the sample and the instant of ignition.

Impacts on pure polymers were carried out on both irregular and disc-shaped samples up to 5mm in diameter and 2mm thick. 10 - 15mg samples of PETN containing a small pieces of the polymers were pressed at 650MPa into 3mm diameter pellets for the photographic investigation and for a quantitative determination of the effects of the two polymers on PETN, H_{50} heights were determined for the pure explosive and the explosive + polymer in the

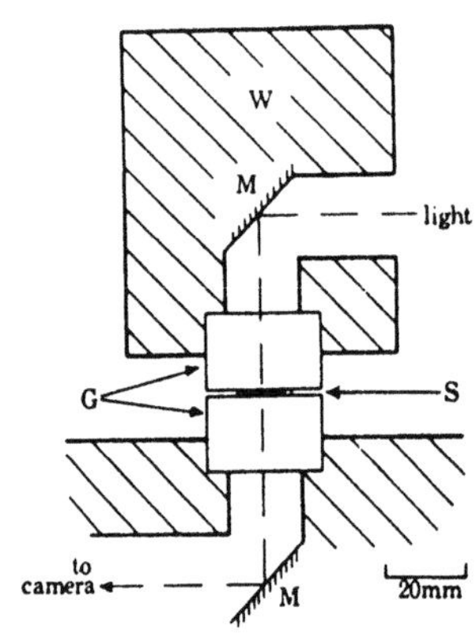

Fig. 1. Experimental arrangement at instant: W, drop weight; G, glass blocks; M, minor; S, sample. The upper glass block is attached to the drop weight.

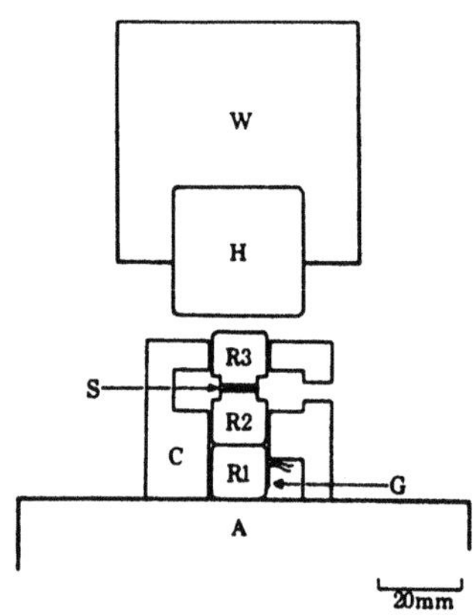

Fig. 2. Experimental arrangement for obtaining pressuretime curves: W, drop-wight; H, R1. R2, and R3, hard steel rollers; C, cylindrical guiding sleeve; S, sample; G, strain gauge; A, cast steel anvil.

form of a loosely compacted pellet containing 10mg PETN. These samples were placed on the top of roller R2 and a single, small piece of polymer placed on top of the PETN. 25 drops on each sample type were performed. For the tests on the tetryls, 20 - 25mg samples were pressed at 750MPa to form 5mm diameter pellets with densities of 1.50 - 1.65Mg m^{-3}. In extending the examination of the deformation of polymers in the glass anvil drop-weight machine, the technique of placing the sample on heat-sensitive film (3M Type 570) and then impacting, was employed[7,8]. By combining this technique with high-speed photography, estimates of the temperatures attained in the polymers during impact were made.

RESULTS AND DISCUSSION

Drop-weight Experiments on Polymers

Photographic records of the behaviour of PTFE and Kel-F 800 under the impact of a 5.5kg mass from 1m showed that the polymers underwent plastic deformation at velocities of 20 - 40m s^{-1}. Neither failed as catastrophically as polysulphone or polycarbonate[4], but the remains of the impacted PTFE showed that extensive fracture had occurred. Kel-F 800 showed evidence for bulk plastic flow associated with extensive softening together with a minor amount of fracture. Both polymers caused extensive darkening of the heat-sensitive film. With PTFE, the darkest areas were located in the regions of the fracture whereas with Kel-F 800, the film was uniformly darkened. Analysis of the photographic and heat-sensitive film records according to the method outlined in Ref. 8 gave temperatures of 720 +/- 50K for PTFE and 730 +/- 50K for Kel-F 800. The temperature attained by PTFE was slightly lower than that found for polymers which underwent massive deformation by cracking and shearing; the high value for Kel-F 800 was, however, not unusual for a material that undergoes bulk plastic flow[8].

Extensive drop-weight tests on PETN[2] showed that the explosive first underwent plastic flow followed by the material becoming translucent and finally transparent at which stage the velocity of flow between the anvils attained 300m s^{-1}. Shear bands were observed in the explosive before the sample melted and if ignition of the material occurred, it invariably took place at a limited number of locations after the high flow velocity regime. When polymers were mixed with PETN, localisation of the impact energy into 'hot spots' occurred when the polymer underwent catastrophic failure and led to ignition sites developing before the PETN underwent rapid flow[4].

Fig 3 shows an impact sequence of PETN + Kel-F 800. The PETN/polymer pellet was impacted in frame 2 and deformed plastically; shear bands in the PETN were apparent (labelled A in frame 5). By this stage the PETN had become partially translucent but no rapid flow was observed. Despite this, an ignition site (labelled B in frame 6) developed at the junction of PETN and Kel-F 800 which propagated throughout the explosive, consuming it. This behaviour was repeatable and observed with both polymers. Confirmation of the sensitising effect that PETN and Kel-F 800 imparted to PETN was obtained from the 2.4kg drop-weight apparatus (Table 1).

Table 1. 50% Heights for PETN and PETN + Polymer Samples.

SAMPLE	H_{50} / cm
PETN	17.2 +/- 3.6
PETN + Kel-F 800	10.8 +/- 0.6
PETN + PTFE	9.5 +/- 1.4

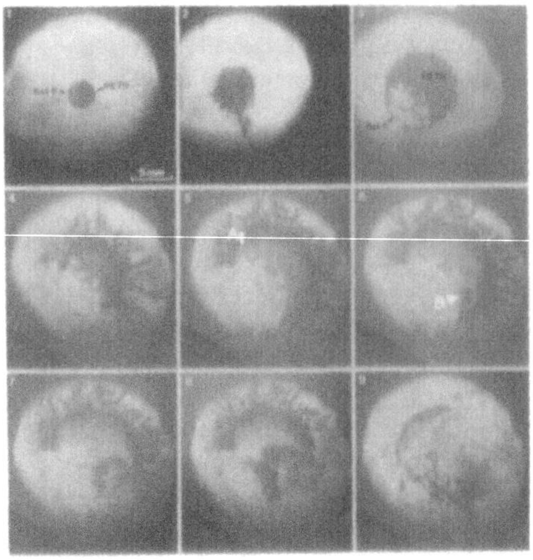

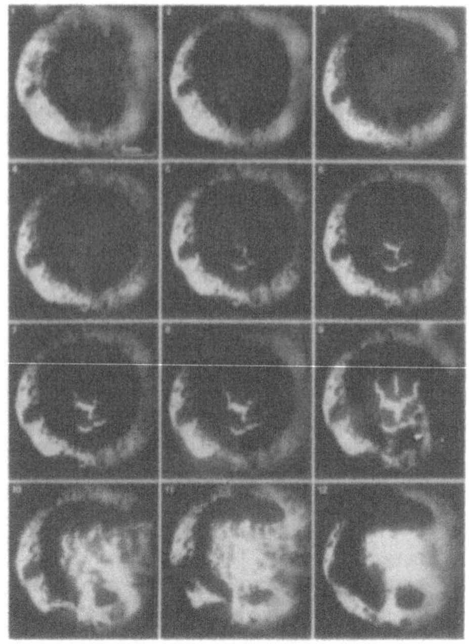

Fig 3. Impact on PETN + Kel-F 800. See text for details. Frame time (μ s) (1) 0; (2) 168; (3) 295; (4) 395; (5) 456; (6) 462; (7) 469; (8) 476; (9) 496.

Fig 4. Impact on 3-aminotetryl. See text for details. Frame time (μ s) (1) 0; (2) 27; (3) 40; (4) 54; (5) 74; (6) 81; (7) 87; (8) 94; (9) 101; (10) 114; (11) 134; (12) 315.

Stress-strain curves for the two polymers at initial strain rates of $1.1 \times 10^3 s^{-1}$ are shown in Fig 5. Both polymers exhibited a small region of elastic behaviour followed by the major plastic deformation regime. Kel-F 800 showed considerable strain softening and was capable of greater strains than those achieved by PTFE. Although there was a difference in the failure mode of the two polymers, neither failed as catastrophically as polysulphone or polycarbonate[4]. The ability of PTFE and Kel-F 800 to sensitise explosives to impact was found to agree with the theory proposed by Swallowe & Field[4]. PTFE was seen to undergo extensive fracturing upon failure, thereby concentrating the stored impact energy along discrete zones which were found to attain high temperatures. With Kel-F 800, the impact energy was not localised but the whole of the material attained elevated temperatures by means of bulk plastic flow to high strains, together with viscous heating before failure occurred. Because of the low latent heat value (3.9kJ kg^{-1}), little energy was absorbed when the Kel-F 800 softened

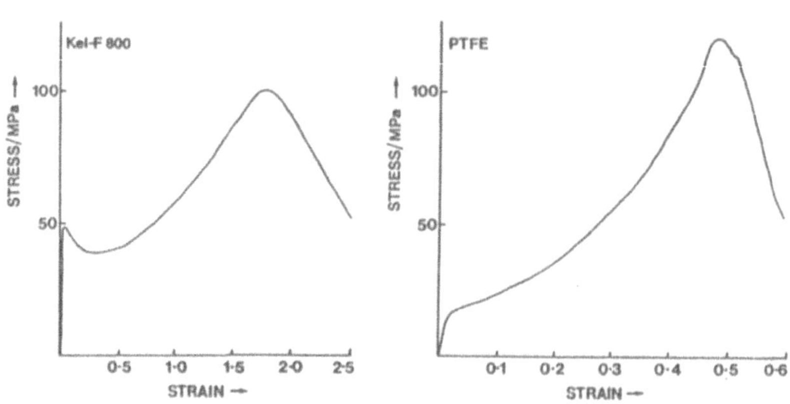

Fig 5. True stress/true strain curves for samples of Kel-F 800 and PTFE.

and therefore only minor assistance was afforded in relieving the energy build up in the sample. Consequently both polymers were found to exert a sensitising influence on PETN, but by different mechanisms.

Drop-weight Experiments on the Tetryl Family of Explosives

Photographic sequences obtained during the impact of tetryl showed that although material flow velocities as high as 170m s^{-1} occurred, ignition did not take place. Examination of the anvils after impact showed evidence of melting of the sample in contact with the glass indicating that temepratures greater than ca 400K had occurred. Since ignition was not found with tetryl, we conclude that there was no mechanism acting whereby the impact energy could be localised in this size sample in this test.

Fig 4 illustrates the C4 sequence of the impact process on 3AT. Impact had commenced in frame 1 and jetting and circumferential material ejection were evident in frames 3 - 5. In frame 5, the compacted explosive began to break up with the formation of three cracks. These cracks continued to widen and extend as the material flowed at a velocity of ca 80m s^{-1} . In frame 9 (arrowed), ignition was seen to occur along a fissure which developed between a crack and the circumference of the explosive layer. With the cracks having generated a considerable amount of new surface area, the combustion front rapidly propagated towards the centre of the explosive consuming the fractured material. The black crescent-shaped mass of 3AT seen in frame 12 was recovered after the test. With 3,5DAT a similar pattern of reaction was observed. There was, however, very little jetting and although material flow at a velocity of 240m s^{-1} was observed, it did not lead to ignition. Subsequent fracture led to an ignition site developing, the propagation of which caused limited consumption of the explosive; the bulk of the explosive was recovered unconsumed. The behaviour of the aminotetryls was similar to that of ammonium perchlorate[9] in which jetting and fracturing preceded ignition which led to very limited propagation and consumption of the material.

The instrumented drop-weight test was used to produce force/time curves for the three tetryls. These were obtained from samples 0.75mm thick at initial strains of $5.3 \times 10^3 s^{-1}$. The upper traces in Fig. 6 are the strain gauge records and the lower traces record the instant of ignition by monitoring the change in electrical resistance of the sample. The upper records for the three impacts showed that the stress within the sample rose to a critical value whereupon the sample failed and the pressure dropped before rising again due to the elastic collision of the anvils. The relative magnitudes of the forces attained before failure increased from tetryl through 3AT to 3,5DAT, but this may not be significant if the rate of increase in area of the samples differed considerably. The instant of ignition has been shown to be associated with this pressure drop[1,9,10] and this was clearly illustrated by the lower oscilloscope traces. For tetryl, which was found to

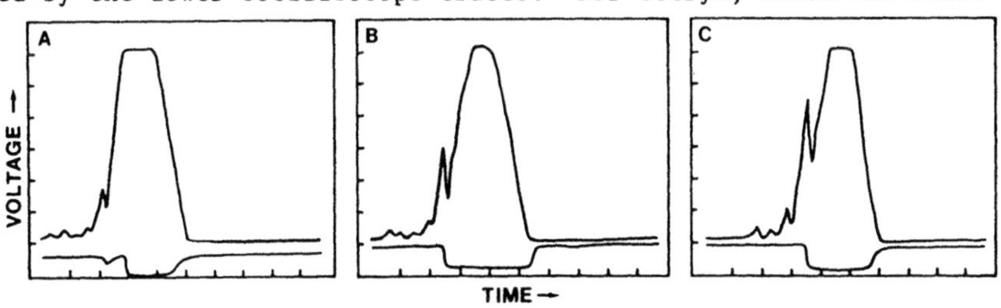

Fig 6. Force/time curves for the impact of (A) tetryl,(B) 3-aminotetryl and (C) 3,5-diaminotetryl. Mass of hammer 2.4kg; drop height 0.82m; full vertical deflection 1V; horizontal scale 100μs per division. The lower trace records the instant of ignition by monitoring the change in electrical resistance in the sample. Gauge calibration factor 81.8N mV^{-1}.

undergo limited ignition and no propagation, the pressure rise before failure and the change in resistance upon ignition were small. This was in accordance with the conclusion that this material, having a low melting point (402K) showed only minor deformation upon impact[5] and consequently was difficult to ignite in this test. On the other hand, the aminotetryls have been observed to flow under the drop-weight and then after the flow stage, when presumably they have work-hardened to some extent, fracture occurred leading to ignition. The oscilloscope traces clearly showed larger pressure drops associated with failure, (Fig 6), and the large resistance changes at this instant indicated ignition.

CONCLUSIONS

High-speed photographic and pressure monitoring of the drop-weight test have allowed us to extend previous work on the effect of polymers on the impact sensitiveness of explosives and to examine two new polymers and two new explosives. We have found that both PTFE and Kel-F 800 cause sensitisation of PETN and by using heat-sensitive film find that temperatures in the order of 700 - 750K may be generated during the deformation of these polymers. In PTFE these high temperatures are localised along fracture planes whereas with Kel-F 800 bulk heating of the polymer occurs. In conjunction with the thermomechanical properties of the polymers we are able to explain these sensitising effects. In PBX-explosives, the binder thicknesses would most likely be less than the 'hot-spot' dimensions required for ignition. However, this paper has shown that particles of these polymers can effect an explosives impact sensitiveness.

Three tetryls have been examined and it is found that the mode of failure of 3AT and 3,5DAT is by fracture. Although rapid flow may occur, ignition does not accompany this phase. Ingress of air occurring during the fracture stage causes ignition to be restricted to areas where adiabatic compression of the gas could occur. However, frictional effects and triboluminescent discharge may also contribute to the formation of 'hot spots' leading to ignition. In neither aminotetryl does extensive propagation occur. These results confirm earlier findings and go some way towards explaining the variation in sensitiveness and explosiveness of the three tetryls.

ACKNOWLEDGEMENTS

C.D. Hutchinson thanks the Director, RARDE, for study leave to work at Cambridge. This work was supported by grants from AWRE, Aldermaston, the U.S. Navy, Grant No. N60921-81-C-0045 and Los Alamos National Laboratory.

REFERENCES

1. Afanas'ev G T & Bobolev V K, Initiation of Solid Explosives by Impact. Jerusalem: Israel Program for Scientific Translations, 1971.
2. Heavens S N & Field J E, Proc. R. Soc., Lond. A338, 77, 1974.
3. Field J E, Swallowe G M & Heavens S N, Proc. R. Soc., Lond. A382,231,1982
4. Swallowe G M & Field J E, Proc. R. Soc., Lond. A379, 389, 1982.
5. Mohan V K & Field J E, Comb. & Flame, 56, 269, 1984.
6. Hutchinson C D, Mohan V K & Millar R W, Prop. Expl. Pyro.,9,(5),161,1984.
7. Coffey C S & Jacobs S J, J. Appl. Phys., 52, (11), 6991, 1981.
8. Swallowe G M, Field J E & Walley S M, Proc. 3rd. Conference on Mechanical Properties at High Rates of Strain, Inst. Phys. Conf. Series 70,p443,1984
9. Heavens S N, Ph.D. Thesis, University of Cambridge.
10. Ljungberg S, Nobel Hefte, 24, 40, 1958.

SUBTHRESHOLD GENERATION OF FREE RADICALS IN SHOCK-LOADED ORGANIC SOLIDS

Brian W. Dodson, E.L. Venturini, and J.W. Rogers, Jr.

Sandia National Laboratories
Albuquerque, NM 87185

INTRODUCTION

A wide variety of chemical reactions have been observed in organic compounds subjected to shock loading [1]. A shock wave in condensed media produces a unique and violent environment which is characterized by high pressure, elevated temperature, strong shear forces, and large rates of change of these parameters. The resulting conditions on the molecular size scale are not well understood at present, but seem to produce both static and dynamic damage to both the lattice structure and the molecular structure which serves, in some cases, to greatly accelerate reaction rates relative to equivalent, but static, conditions of temperature and pressure [1]. For instance, there are examples of shock-induced polymerizations occurring perhaps ten orders of magnitude faster than is seen in conventional processes, and also reactions which only occur conventionally via the action of a catalyst which occur with large yield under shock loading. Such unusual reactive processes provide a probe into the molecular-scale environment of a shocked material, since chemistry almost always takes place between neighboring molecules.

The mechanisms through which a shock wave can induce chemical reactivity are currently unknown in any detail. However, there does exist some evidence [2,3] and considerable speculation [4,5] that generation of free radicals in the passage of the shock wave may be an important step in many shock-induced reaction mechanisms. We have been studying the generation of free radicals in organic solids, both non-energetic and explosive, at pressures below the threshold for gross reactivity, in an attempt to sort out the importance of this step in shock-induced reactivity in organic solids.

EXPERIMENTAL PROCEDURES

The technique utilized here is to shock-load an organic sample, recover the sample, and measure the electron-spin resonance (ESR) spectrum to determine the post-shock free radical density. In some cases, other analytical techniques were also used. Results from such techniques will be identified when referred to.

The recovery technique is described in detail elsewhere [3], and thus will only be outlined here. A sample, in powder form, is pressed into a copper recovery fixture using 0.1-0.3 GPa static pressure. The nonenergetic samples were roughly 250 mg in weight, but the explosive samples were limited to 60 mg to reduce handling difficulties. The shock wave is introduced by a gas-gun propelled impactor, which allows close control over the impact conditions. The sample chamber is protected by a system of spall rings and momentum traps, and is then recovered in a balsa wood catcher.

A few words are necessary regarding the stress history which the sample experiences using the experimental geometry described in [3]. Characterization of the stress history in recovery capsules is difficult owing to the dominance of two-dimensional effects, which results from the interaction of the planar impact wave with two different materials, the copper and the organic sample, which have very different wave speeds. Webb and Graham [6] have performed wavecode calculations of a similar recovery capsule with a polymethylmethacrylate sample, which reveals the qualitative pattern of loading. They find that the initially planar shock wave is transformed into a converging radial wave upon interaction with the sample material. In a simulation with a 400 m/s impact velocity (Cu-Cu), giving an initial stress of 7.6 GPa, the sample experiences an average peak stress of 6.5 ± 1.2 GPa. Thus, the average peak stress seen by the sample is slightly less than that of the initial impact-generated shock. However, since the capsule design, sample materials, and impact conditions differ considerably from those used in this simulation, only the qualitative picture of planar impact resulting in radial-mode loading can be trusted. We will therefore refer to a given experiment by the initial impact stress in the copper to avoid potentially misleading statements. For the Webb-Graham example above, the pressure would be described as 7.6 GPa(Cu).

EXPERIMENTAL RESULTS

Nonenergetic Organics

In the course of this work we have studied shock-generated free radicals in a number of organic compounds. The sample materials all show a reaction threshold of 10-20 GPa(Cu). Thus, the experimental pressures (1-7 GPa(Cu)) represent significantly subthreshold conditions. To confirm this, post-shock chemical studies were carried out, revealing that no gross chemical activity had taken place, within the sensitivity of the analytical techniques (IR spectroscopy and HPLC), in any sample. However, we did find that free radicals are produced in significant quantities at these subthreshold stresses. In Table 1 are displayed the stress values and radical descriptions for the recovery experiments to date. A broad summary of these data shows that shock pressures over a considerable range (10-60% of threshold) will produce radical densities of 1-10 ppm. These radicals have very similar ESR spectra, and are consistent with the formation of peroxy radicals resulting from reaction of the shock-induced radical with atmospheric oxygen [7]. The isolation of the radical electron from the organic molecule virtually eliminates molecule-specific information from the ESR spectra.

Nitroaromatic Explosives

We have recently begun adaptation of the above technique to the study of a series of nitroaromatic explosives. The sample explosives chosen are trinitrobenzene (TNB), trinitrotoluene (TNT), diaminotrinitrobenzene (DATB), and triaminotrinitrobenzene (TATB), all being

Table 1. Subthreshold Radical Generation in Nonenergetic Organic Compounds

Compound	Stress	Radical Density	g-value	Resonance FWHM
Acrylamide	4.8	1	2.0024	2
	6.7	1	2.0024	2
1,6-Diphenoxy-	1.2	1.5	2.0025	3.2
2,4-Hexadiyne	2.7	2	2.0025	3.2
Anthracene	2.5	1	2.0024	1.5
9-Cyanoanthracene	1.8	7	2.0034(10)	7.7
	4.0	10	2.0025	2.8
		1	2.0025	0.7

Stress in GPa(Cu); Radical density in ppm, FWHM in Oe.

substitutional derivatives of TNB. Of course, TNT and TATB are both explosives of considerable practical interest.

Recovery of shocked explosive samples in the closed capsules described above presents a difficult design problem, since one expects a significant degree of reaction and gas evolution even under shock conditions less extreme than those leading to detonation. A system of locking rings was developed which allows sample retention of small (60 mg) explosive samples to peak stresses at least in the 1-2 GPa(Cu) region. Insufficent trials have yet been completed to determine the limits of application of the current capsule.

Reasonably complete analysis has only been completed on one set of samples (TNB, TNT, and TATB only) subjected to 0.6 GPa(Cu) shock loading. Only the TATB sample showed any sign of radical generation. ESR measurements revealed the production of 0.1 ppm of a radical having an isotropic g-factor of 2.0027 and a FWHM of 0.1 Oe, which is some 10-30 times narrower than the shock-induced radicals in the nonreactive organics. This difference is not understood, but may result from the graphite-like crystal structure of TATB. However, all other indications suggest that the observed radical is again a peroxy radical. Due to the somewhat unusual nature of the observed resonance, we are currently repeating this recovery experiment.

All of the samples were examined using x-ray photoelectron spectroscopy (XPS) by J. Sharma at NSWC. Although the observed chemical changes are near the sensitivity threshold of this technique, reactivity was observed to have occured in two of the samples. A 4% reduction in the NO_2 line was seen in the TATB sample, which suggests some fragmentation of the TATB molecule. In the TNT sample, a new line was seen in the XPS spectrum which was identified with the production of a small amount of anthranile, which is produced from TNT by abstraction of a water molecule. We are currently processing samples resulting from higher shock pressures.

DISCUSSION

There are several interesting points concerning the generation of free radicals in the nonenergetic organics, most of which have been discussed in detail in past publications [2,3]. The most significant one for the current paper is the comparison of the anthracene and 9-cyano-anthracene (9-CA, a simple substitutional derivative) samples. At 2.5 GPa(Cu) we generate 1 ppm of residual radicals in anthracene. However, when 9-CA is subjected to 1.8 GPa(Cu), the radical density is 7 ppm, with a different g-value and a 5 times larger linewidth. At 4.0 GPa(Cu), the shape of the resonance is best explained as a pair of radicals, which may correspond to peroxy radicals on the cyano group and on the polyconjugated ring structure. Since the narrower 4.0 GPa(Cu) resonance seems to be identified with the radical observed in the anthracene structure, we conclude that the substituent group is much more sensitive to radical formation than is the anthracene structure itself. The addition of the electron-withdrawing cyano group makes the resulting 9-CA system more sensitive to shock-induced damage, which reinforces earlier claims [8] that electron accepting substituents on aromatic compounds (specifically benzene) increases gross chemical reactivity under higher (10-20 GPa(Cu)) stresses.

An important question which is still unanswered is that of the relation between the shock-induced generation and retention of free radicals at subthreshold pressures and the sensitivity of the sample towards undergoing gross chemical reactions. That is, if sample A shows more residual radicals at a given pressure than does a chemically related sample B, may one reasonably expect that sample A will have a smaller gross chemical reaction threshold than sample B, or that at a given large pressure, that sample A will show a higher product yield than sample B? Assuming that the production of radicals is a significant step in the shock-generation of gross chemical reactions, it seems likely that there is some correlation between these facts, but there are too many other variables, such as crystal structure and other aspects of the microstructure of the material, to make a strong quantitative statement. A notable example of such a variable is the crystal property known as topochemistry, which is the influence of crystal structure on solid-state chemical reactivity. The reaction threshold for acrylamide polymerization is 10-12 GPa(Cu), despite the fact that radicals are produced at much lower pressures, and this reaction takes place via a radical mechanism [9]. In a perfect single crystal, the polymerization reaction is not allowed by the crystal structure, or is topochemically forbidden, and thus is defect-mediated and thermally activated in nature. However, in the case of trioxane [10], polymerization is topochemically allowed, resulting in large yields of high-molecular-weight polymer beginning at about 3 GPa(Cu). Thus topochemical factors, which may have no influence on generation and retention of radicals. can determine the reaction threshold pressure.

The new information on generation of radicals and minor chemical reactivity in nitroaromatic explosives, although incomplete at this time, is suggestive on this topic. By conventional standards of initiation sensitivity of hetrogeneous explosives, TNB would be the most sensitive, followed by TNT, and then by TATB, which is much less sensitive to initiation [4]. This order of sensitivity is not reflected in the 0.6 GPa(Cu) recovery experiments. Here we see residual radicals only in the TATB sample, which also shows some decomposition in a reduction of the NO_2 group. The TNT sample shows formation of small quantities of a new chemical compound, whereas the TNB sample was essentially inert under these loading conditions. It will be interesting to see if this inverted pattern continues at higher pressures.

We have demonstrated that substantial densities of active molecular species are formed by subthreshold shock loading of organic compounds. The generation of such active centers may be of great importance for the shock-induced reactivity observed in large classes of organics. In most cases, however, it would seem that topochemical and other factors inhibit a reaction from taking place merely in the presence of such defects. In such cases, we suggest that the generation of active species, which can take place at rather low stresses, is necessary but not sufficent to initiate a large-scale chemical reaction. An additional factor which would play a major role in overcoming such topochemical and other structural restraints is a shear-driven molecular transport mechanism which would become important in organic compounds in the 10-20 GPa(Cu) region. The details of such a mechanism are currently unknown, but evidence of its importance does exist [11]. In any case, shock-induced chemical reactions in organics seem to be multi-step processes, starting with generation of active centers and proceeding via currently unidentified mechanisms, among which may be shear-driven molecular transport, to a significant product yield.

ACKNOWLEDGEMENTS

We wish to acknowledge the contribution of J. Sharma of NSWC for providing the XPS analyses of the explosive samples. This work was supported by the U.S. Department of Energy under contract number DE-AC04-76DP00789.

REFERENCES

1. B.W. Dodson and R.A. Graham in "Shock Waves in Condensed Matter-1981", W.J. Nellis, L. Seaman, and R.A. Graham, Eds., American Institute of Physics, New York, 1982.
2. B.W. Dodson and C. Arnold, J. Phys. Chem. 87, 3023, 1983.
3. B.W. Dodson et al, in "Shock Waves in Condensed Matter-1983", J.R. Asay, R.A. Graham, and G.K. Straub, Ed., Elsevier, New York, 1984.
4. F.E. Walker and R.J. Wasley, Prop. and Exp. 1, 73, 1976.
5. B.W. Dodson, J. Phys. Chem. (to appear).
6. D.M. Webb and R.A. Graham, unpublished.
7. W. Gordy, "Theory and Aplications of Electron Spin Resonance", Wiley, New York, 1980.
8. L.V. Babare et al, in 2nd All-Union Symposium on Combustion and Explosion, Chernogolovka, 1971. Translation in SNL report RS3140/79/43, 1979.
9. I.N. Ignatovish et al, High Ener. Chem. 4, 394, 1970.
10. L.V. Babare et al, Comb., Exp., and Shock Waves. 5, 364, 1969.
11. R.A. Graham, in "Shock Waves and High-Strain-Rate Phenomena in Metals", ed. M.A. Meyers and L.E. Murr, Plenum, New York, 1981.

POP PLOT AND ARRHENIUS PARAMETERS FOR <110> PENTAERYTHRITOL TETRANITRATE SINGLE CRYSTALS

J. J. Dick

Los Alamos National Laboratory
M-9, MS P952
Los Alamos, New Mexico 87545

ABSTRACT

Plane shock initiation sensitivity measurements for <110> single crystal pentaerythritol tetranitrate were extended to 20 GPa. The data of run distance versus input stress do not fall on the usual straight line on the Pop plot. From these data Arrhenius parameters were calculated. Computed activation energy was 6.4 kcal/mole, and the frequency factor was $5X10^6$/s. These are much lower than values obtained from isothermal heating experiments. These results can be used to support the idea that the uniaxial strain due to the shock leads to heterogeneous heating of the homogeneous material.

Explosives have traditionally been separated into two types: Homogeneous and heterogeneous. Homogeneous explosives include liquids and single crystals. Heterogeneous explosives include powder compacts, mixtures and slurries; hot spots are important in these materials. Initiation of detonation and detonation failure are believed to proceed by different mechanisms in homogeneous and heterogeneous explosives. Initiation of detonation in homogeneous explosives is believed to proceed by thermal explosion at the rear boundary, and detonation failure proceeds by transverse failure waves.

Data on plane shock initiation of <110> single crystal pentaerythritol tetranitrate (PETN) are presented here. The plane shock wave was generated by a 203-mm diameter explosive plane-wave lens, a pad of booster explosive, and a polymethylmethacrylate attenuator plate. Shock trajectories in the wedge samples were recorded with a rotating mirror camera. Hugoniot points and time and distance to detonation were obtained from the records. Shock sensitivity has been measured to 20 GPa. The data represent by far the largest range of input stresses used on a homogeneous explosive. The data in the Pop plot (Fig. 1) are not fitted by the usual straight line, unlike heterogeneous explosives. Since no other homogeneous explosive has been studied over such a large range of input stresses, it is hard to say if it is unique among homogeneous explosives.

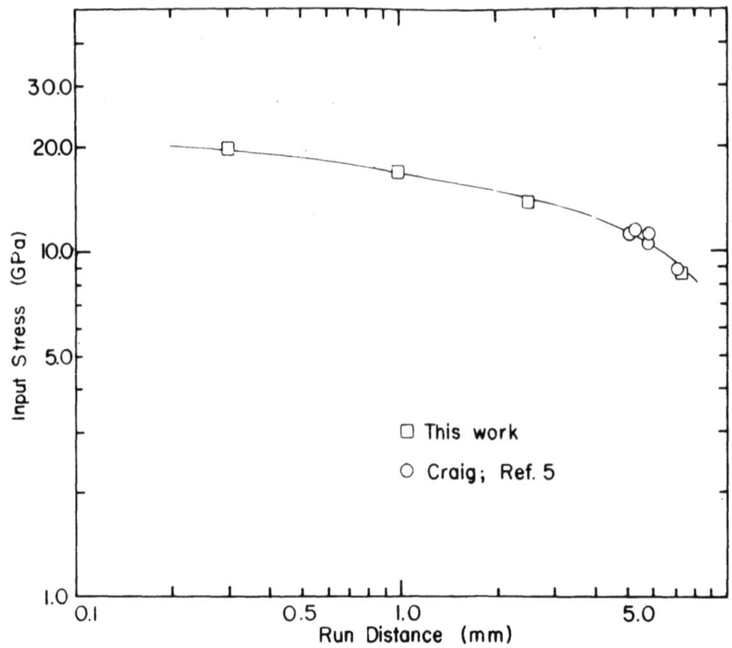

Figure 1. Pop plot for <110> PETN single crystals.

Hardesty did an extensive study of liquid nitromethane.[1] He found Arrhenius parameters obtained from induction time measurements for the thermal explosion in shock experiments to be much smaller than the gas phase constants (Table 1). It seemed worthwhile to extract Arrhenius parameters from the PETN wedge data in the usual way,[2,3,4] using

$$t_i = \nu^{-1} \frac{T_o^2}{T_F T_A} \exp(T_A/T_o) \quad ,$$

where ν is the frequency factor, T_F is the temperature rise through the reaction zone, T_o is shock temperature, and T_A is activation temperature. Induction time t_i was extracted from the wedge data as

$$t_i = \left(1 - \frac{U-u}{D}\right) t^* \quad ,$$

where t^* is time to detonation at the shock front, D is detonation velocity at the precompressed density, and U and u are shock and particle velocities respectively. The energy parameter calculated is the Gibbs free energy of activation. One is assuming that the same activation energy is operative at all densities and temperatures sampled. An Arrhenius plot is shown in Fig. 2. A nonlinear least squares fitting program was used to obtain parameters from the thermal explosion equation. An activation energy of 6.4 kcal/mole and a frequency factor of $5X10^6/s$ were found. In comparison, the results from isothermal heating experiments on liquid PETN are 47 kcal/mole and $6X10^{19}/s$.[6] This is a much larger difference than that found for nitromethane (Table 1). The discrepancy will be discussed in later paragraphs.

Table 1. Arrhenius Parameters

Material	Shock Activation Energy (kcal/mole)	Shock Frequency Factor (s^{-1})	Isothermal Activation Energy (kcal/mole)	Isothermal Frequency Factor (s^{-1})
PETN[(a)]	6.4	$5X10^6$	47	$6X10^{19}$
Nitromethane[(b)]	23.	$3X10^9$	59	$4X10^{15}$

(a) Shock data is for <110> crystals; isothermal data is for liquid phase (Ref. 6).

(b) Shock data from Ref. 1; Isothermal data is for gas phase (Ref. 7).

One notes that the straight line fit to PETN data in Fig. 2 yields nonrandom residuals, perhaps implying that the decomposition is not unimolecular. It could also mean that there are different regimes of behavior at low and high stresses. It should be noted that Lee and coworkers showed that static high pressure suppresses the thermal decomposition of PETN.[8] It is also true that the assumption of the thermal explosion model that the ratio of the activation energy to the shock thermal energy is much greater than one, is not well satisfied for the PETN value computed here.

Calculated shock temperatures for PETN are shown in Fig. 3. The Hugoniot of single crystal PETN was taken as $U = 2.74 + 1.81u$, density as $1.778\ g/cm^3$, heat capacity as 1000 J/kg-K, and the Gruneisen constant as 1.15. Note the position of the melt temperature relative to the shock temperatures. The melt phase line is not known, but in drop-hammer impact

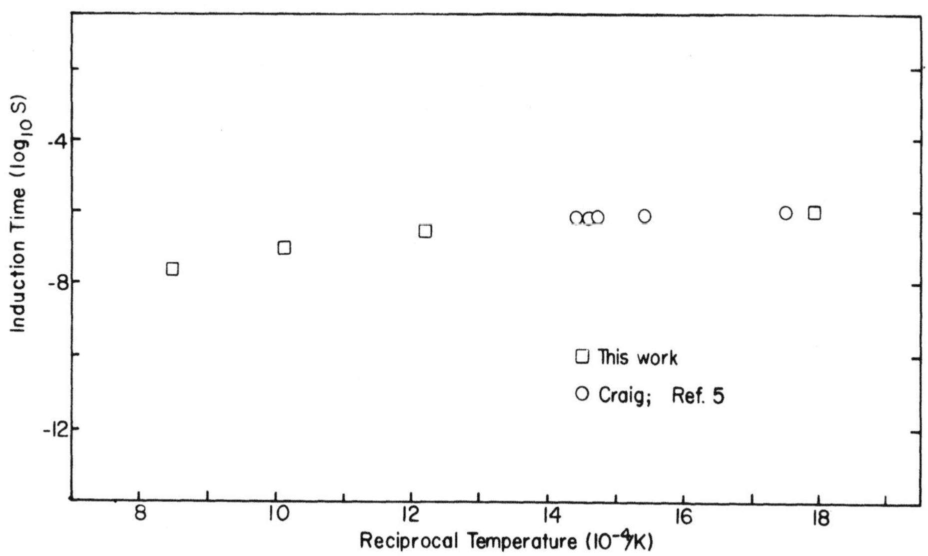

Figure 2. Arrhenius plot for <110> PETN single crystals.

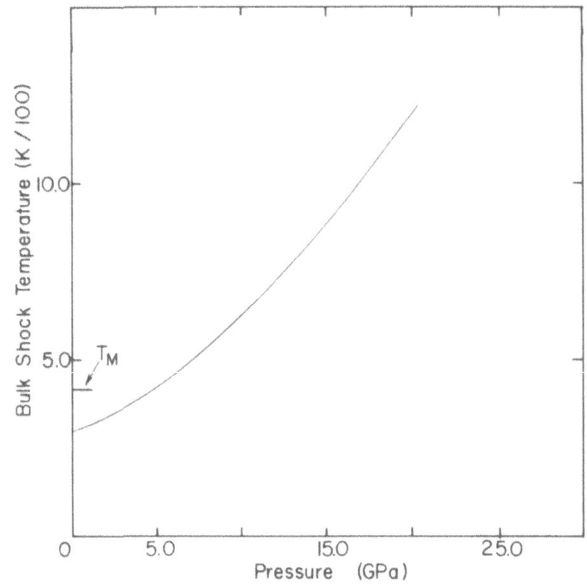

Figure 3. Bulk shock temperature for PETN.

tests at Cambridge it is found that PETN melts before ignition occurs.[9] This could also be the case in uniaxial shock compression.

Another aspect of PETN must be addressed. It has been shown that shock sensitivity depends on crystal orientation.[10] This suggests that plasticity plays a role in initiation of detonation. Adiabatic shear bands have been observed in PETN during high strain rate deformation.[11] This means that the temperature distribution is probably not homogeneous in shocked PETN. The initial decomposition may be taking place in hot shear bands. If ignition occurs in shear bands we can say that the Arrhenius parameters we calculated using the bulk temperature are low for that reason and in serious error; the initial decomposition is taking place at temperatures much higher than the bulk temperature. One may summarize by observing that the explosive PETN is a homogeneous, anisotropic material subject to plastic instability in uniaxial strain.

These results affect our picture of what we call homogeneous explosives. Even in a liquid there will be shear flow in uniaxial shock compression. Whether instability of the viscous flow occurs and leads to inhomogeneous heating is not known, but it is something to consider.

REFERENCES

1. D. R. Hardesty, Combust. Flame 27, 229 (1976).

2. A. W. Campbell, W. C. Davis, and J. R. Travis, Phys. Fluids 4, 498 (1961).

3. T. L. Jackson and A. K. Kapila, SIAM J. Appl. Math. 45, 130 (1985).

4. D. R. Kassoy and Justin Poland, SIAM J. Appl. Math. 39, 412 (1980).

5. B. G. Craig (unpublished); compiled in LASL Explosive Property Data, edited by T. R. Gibbs and A. Popolato (University of California, Berkeley, 1980), p. 318.

6. A. J. B. Robertson, J. Soc. Chem. Ind. (London) 67, 221 (1948).

7. S. W. Benson and H. E. O'Neal, Kinetic Data on Gas Phase Unimolecular Reaction, NSRDS-NBS21, (U. S. Government Printing Office, Wash., D. C., 1970).

8. E. L. Lee, R. H. Sanborn, and H. D. Stromberg, Fifth Symposium (International) on Detonation (Office of Naval Research, Wash. D. C. , 1970), p. 331.

9. S. N. Heavens and J. E. Field, Proc. R. Soc. Lond. A338, 77 (1974).

10. J. J. Dick, Appl. Phys. Lett. 44, 859 (1984).

11. J. E. Field, G. M. Swallowe, and S. N. Heavens, Proc. R. Soc. Lond. A382, 231 (1982).

CHEMICAL REACTION OF ENERGETIC MATERIALS DURING HIGH ACCELERATION

M.Y.D. Lanzerotti and J. Pinto

Energetic Materials Division, Large Caliber Weapon Systems Laboratory, U.S. Army Armament Research and Development Center, Dover, New Jersey 07801-5001

The fracture phenomena of energetic materials during high acceleration have been studied in an ultracentrifuge. The recovered samples have been studied by x-ray photoelectron spectroscopy (XPS) and electron paramagnetic resonance (EPR). The results of these measurements indicate that chemical reaction has occurred in TNT, TNT-1% HNS, and M30 propellant as a result of fracture and impact during high acceleration.

INTRODUCTION

Knowledge of the fracture phenomena of energetic materials subjected to high acceleration is important in such fields as ordnance, industry, and space applications where energetic materials are often subjected to high, fluctuating, and/or sustained accelerations. In particular, if energetic materials react during high acceleration, understanding this phenomena may be important for solving the problems of abnormal propellant burning and undesirable ignition or initiation of explosives.

Experiments using an ultracentrifuge have been initiated in order to investigate the behavior of energetic materials during high acceleration. Experimental results on the fracture behavior of explosives during high acceleration have previously been published[1,2] as have results on the deformation and failure of M30 propellant[3]. Chemical reactions have been discovered to occur in TNT, TNT-1% HNS, and M30 propellant as a result of fracture and impact during high acceleration in the ultracentrifuge[2]. This discovery is discussed further herein with additional diagnostic measurements of the reacted materials. The reacted samples have been studied with electron paramagnetic resonance (EPR) and x-ray photoelectron spectroscopy (XPS) to verify that reaction has occurred and to investigate properties of the reaction products.

EXPERIMENT

A Beckman preparative ultracentrifuge (Model L8-80) with a swinging bucket rotor (Model SW 60 Ti) is used to rotate the sample under study up to 60,000 rpm. The distance of the specimen from the axis of rotation can be chosen as a variable between 6 and 12 cm.

Samples are prepared in two ways. Cylindrical polycrystalline plugs of explosive are prepared by pouring about one-half gram of molten material into 9 mm i.d. polycarbonate tubes and allowing it to crystallize. The open-ended sample tube is then joined to a short closed-end polycarbonate centrifuge tube as shown in Fig. 1. The as-cast surface

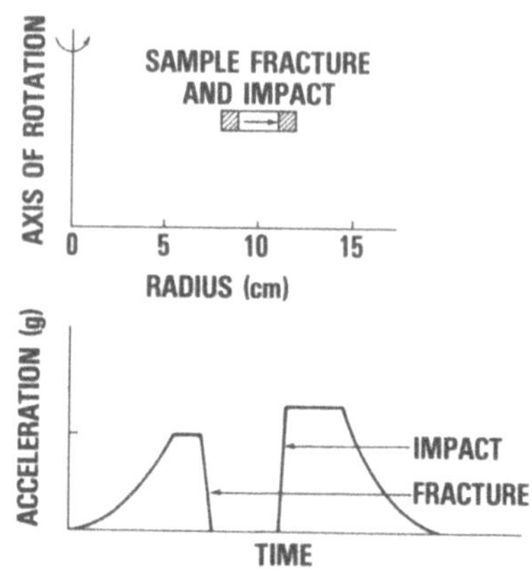

Fig. 1. Fracture and Impact of Melt-cast TNT During High Acceleration

of the explosive faces away from the axis of rotation. In a second experimental method, small pieces (0.05 g) of propellant are glued to individual aluminum disks. An individual disk with the sample facing away from the axis of rotation is joined to a short, closed-end polycarbonate tube, which supports the disk at a fixed distance from the axis of rotation.

The sample experiences a time rate of change of the acceleration up to a maximum acceleration. The sample then remains at this maximum acceleration for a time such that for each run there is a combined total elapsed time of 5 minutes. The sample then decelerates smoothly to zero acceleration. The initial maximum acceleration is less than the fracture acceleration for the material. The maximum acceleration for the sample is then increased systematically in each successive 5-minute run until the sample fractures. In these experiments essentially all the sample mass fractures and transfers to the closed-end tube during one run as shown in Fig. 1. When the sample fractures its acceleration decreases to zero in a very short time (the travel time of the speed of sound through the sample); when the sample impacts the end of the tube its acceleration increases to the maximum acceleration in a very short time (the travel time of the stress wave through the sample)[4,5].

RESULTS

Chemical reaction has been discovered to occur in TNT and TNT-1% HNS as a result of the fracture and the impact of the material at the end of the tube. Extensive black (to the eye) discoloration is observed in the recovered material. The pressure produced by the reaction is sufficient to cause a 2-cm long crack perpendicular to the axis of rotation in the polycarbonate collector tube.

The range of conditions for which chemical reaction has been observed has been estimated from the following equations. The acceleration at fracture or impact is the radial acceleration

$$a = \omega^2 r,$$

where ω is the angular velocity and r is the radius of the sample with respect to the axis of rotation. The velocity at fracture is the tangential velocity

$$v = \omega r.$$

The maximum velocity at impact and the minimum time between fracture and impact are calculated by solving Newton's law of motion using Coriolis' theorem[6]. The minimum time between fracture and impact, t_{min}, is given by the expression

$$t_{min} = \frac{1}{\omega} \ln \left(\frac{r_f}{r_i} \pm \left(\frac{r_f^2}{r_i^2} - 1\right)^{1/2}\right),$$

where r_i and r_f are the initial and final radius of the sample, respectively. The maximum velocity at impact, v_{max}, is given by the expression

$$v_{max} = \frac{r_i}{2} \omega \left(e^{\omega t_{min}} - e^{-\omega t_{min}}\right).$$

The estimated conditions at fracture are: acceleration 20,000-59,000 g; velocity 140-230 m/s; time 2.0 μs; load rate 1.0-2.9 x 10^{10} g/s. The estimated conditions at the moment of impact at the end of the polycarbonate tube are: acceleration 26,000-77,000 g; velocity 110-190 m/s; time 2.3 μs; load rate 1.2-3.4 x 10^{10} g/s. The estimated time between fracture and impact is 300-510 μs. These experiments have been performed at 0,25 and 50°C; chemical reaction has occurred at all three temperatures.

The XPS N 1s spectra of melt-cast TNT before[7-9] and after acceleration and impact is displayed in Fig. 2. A new line centered at about 401 eV is observed in the spectrum of the reacted TNT. This line has also been reported from experiments when TNT is subjected to heat[8-10], uv photolysis[7-9], and x-rays[10]. There is also evidence of the 401 eV line in the XPS N 1s spectrum of reacted TNT-1% HNS from the acceleration experiment (not shown here).

The EPR spectrum of the TNT after acceleration and impact is displayed in Fig. 3. This is the EPR spectrum of the reacted TNT. A similar spectrum has also been observed when TNT is uv[11] and thermally[11,12] decomposed. The radical associated with the EPR spectrum has not as yet been identified.

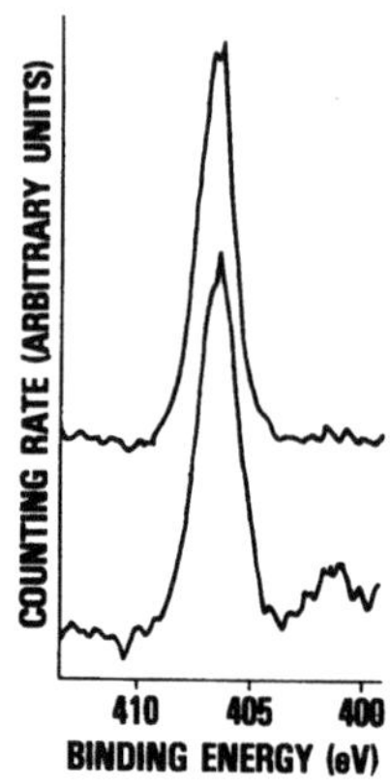

Fig. 2 XPS N 1s Spectra of Melt-cast TNT Before (Upper Spectrum) and After (Lower Spectrum) Acceleration and Impact

Chemical reaction has also been discovered to occur in M30 propellant as a result of the fracture and the impact of the material at the end of the tube during high acceleration. Yellow (to the eye) discoloration is observed in the recovered material. The estimated conditions at fracture are: acceleration 130,000-240,000 g; velocity 340-460 m/s; time 1.0 μs; load rate 1.3-2.4 x 10^{11} g/s. The estimated conditions at the moment of impact at the end of the polycarbonate tube are: acceleration 170,000-310,000 g; velocity 290-390 m/s; time 1.0 μs; load rate 1.7-3.1 x 10^{11} g/s. The estimated time between fracture and impact is 150-200 μs. These experiments have been performed at 25°C.

The XPS N 1s spectra of M30 propellant before[13] and after acceleration and impact is displayed in Fig. 4. The intensities of the ONO_2 and NO_2 lines decrease relative to the intensity of the amine/imine line in the reacted M30 propellant. Nitrocellulose and nitroglycerin contain

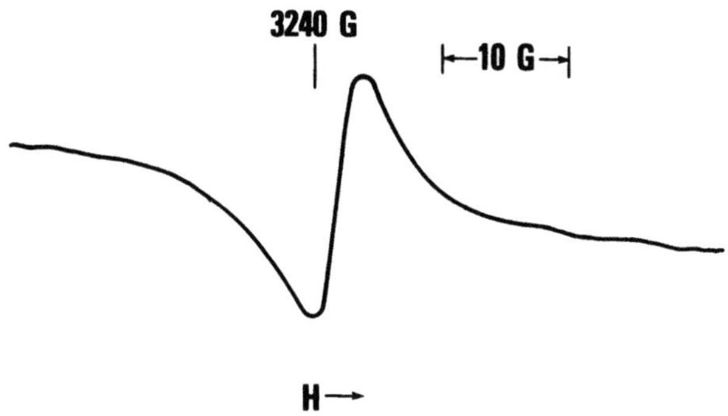

Fig. 3 EPR Spectrum of Melt-cast TNT After Acceleration and Impact

ONO_2 groups and nitroguanidine contains NO_2 groups. Thus, reaction may have occurred in these three components of M30 propellant[14].

DISCUSSION

The fracture phenomena of TNT, TNT-1% HNS, and M30 propellant during high acceleration have been studied in an ultracentrifuge. The recovered samples have been studied by XPS and EPR. The results of these measurements indicate that chemical reaction has occurred in these materials as a result of fracture and impact.

There are several mechanisms discussed in the literature which can be considered for producing the reaction. One mechanism is shock initiation. For example, the pressure at impact can be estimated by matching the hugoniots of polycarbonate and TNT[5]. From this, the temperature rise due to the pressure at impact, 0.3 GPa, is seen to be insufficient to ignite TNT[15]. Therefore, it is unlikely that the reaction occurs during the time of impact.

Another mechanism is rapid shear. According to Frey[16], the impact pressure and a shear velocity produce a maximum temperature in the absence of reaction. If the shear velocity is the same order of magnitude as the impact velocity (110-190 m/s) a maximum temperature of the

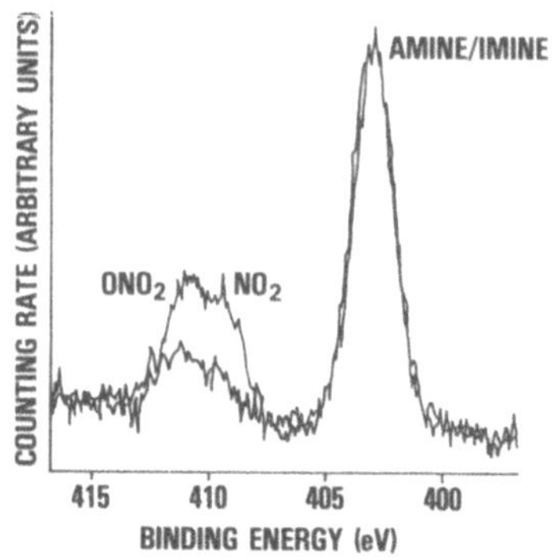

Fig. 4 XPS N 1s Spectra of M30 Propellant Before (Upper Spectrum) and After (Lower Spectrum) Acceleration and Impact

order of 250-350°C would be produced in the absence of reaction. For a temperature range of 250-350°C, the e-folding time of the reaction which produces the free radical (shown in the EPR spectrum) would range between 21 minutes and 12 seconds, respectively[12]. If the impact velocity is amplified by a factor of three[17], the maximum temperature would be 700°C and the e-folding time of the reaction would be about a millisecond. Even one millisecond is too long by almost two orders of magnitude for Frey's mechanism to cause the reaction during the time of impact.

Another mechanism is frictional heating. A normal force is produced by the Coriolis acceleration as the sample slides along the collector tube after fracture to impact. The resulting pressure of the sample against the polycarbonate tube is < 0.01 GPa. According to Frey[16], this pressure and a shear velocity of the order of the impact velocity would produce a negligible temperature rise. If the time between fracture and impact is assumed to be the e-folding time of the reaction, a TNT temperature of about 800°C would be expected[12].

In some other types of experiments (e.g., impact, compaction) Sandusky, Coffey, and Liddiard[18] have reported reaction thresholds in energetic materials at lower pressures than obtained here (< 0.1 GPa versus the ~ 0.3 GPa here) and at rates of bulk deformation (~ 80-200 m/s) which are similar to the impact velocity of the present experiment. The reaction-producing mechanisms involved in the Sandusky et al. process may be similar to that for the present experiment and need to be further studied. In the propelled impingement test, chemical reaction of TNT has been reported at an impingement velocity[19] which is within the range of impact velocities found to cause chemical reaction here. It is also interesting to note that a reaction in propellants has been reported at impact velocities similar to those of this experiment[19, 20].

ACKNOWLEDGMENTS

The authors are indebted to F.J. Owens for the EPR spectrum of reacted TNT, to W.L. Garrett for the XPS N 1s spectra of M30 propellant before and after acceleration, and to J. Sharma, B. Fishburn, and D. Wiegand for helpful discussions.

REFERENCES

1. M.Y.D. Lanzerotti and J. Sharma, "Brittle Behavior of Explosives During High Acceleration," Applied Physics Letters, 39, 455 (1981).

2. M.Y.D. Lanzerotti, "Fracture Phenomena of Energetic Materials During High Acceleration," Symposium on Thermomechanical Properties of Energetic Materials and Their Effects on Munitions Survivability, TTCP W-1/W-1 Joint Meeting, Naval Weapons Center, China Lake, CA, 18-20 September 1984.

3. M.Y.D. Lanzerotti, "Mechanical Behavior of Gun Propellant During High Acceleration," 21st JANNAF Combustion Meeting, Johns Hopkins University, Applied Physics Laboratory, Laurel, MD, 1-4 October 1984, Volume I, pp 275-280.

4. A.S. Argon, "Brittle Fracture," in Mechanical Behavior of Materials, ed. F.A. McClintock and A.S. Argon, Addison-Wesley Publishing Company Inc., Reading, MA, 489, 1966.

5. S.P. Marsh, Editor, LASL Shock Hugoniot Data, University of California Press, Berkeley, CA, pp 483, 648, 1980.

6. K.R. Symon, Mechanics, Addison Wesley Publishing Company Inc., Reading, MA, 238, 1953.

7. J. Sharma, T. Gora, S. Bulusu, and D.A. Wiegand, "X-ray Photoelectron Spectroscopy of Explosive Solids," Proceedings U.S. Army Science Conference, West Point, NY, 1974.

8. F.J. Owens and J. Sharma, "X-ray Photoelectron Spectroscopy and Paramagnetic Resonance Evidence for Shock-Induced Intramolecular Bond Breaking in Some Energetic Solids," Journal of Applied Physics, 51, 1494 (1980).

9. J. Sharma, "X-ray Photoelectron Spectroscopic (XPS) Detection and Identification of Explosives Residues," Proceedings of the International Symposium on Analysis and Detection of Explosives, FBI Academy, Quantico, VA, 29-31 March 1983, p. 181.

10. J. Pinto (unpublished).

11. F.J. Owens, "Paramagnetic Resonance Studies of uv and Thermally Induced Radicals in Trinitrotoluene and Dinitrotoluene," Mol. Cryst. Liq. Cryst., 101, 235 (1983).

12. R. Guidry and L.P. Davis, "Thermochemical Decomposition of Explosives. I. TNT Kinetic Parameters Determined from ESR Investigations," Thermochimica Acta, 32, 1 (1979).

13. P.P. Ostrowski, J.F. Grant, J. Sharma, W.L. Garrett, D.S. Downs, and S. Krasner, "Laser Ignition of Double and Triple Based Gun Propellants," 17th JANNAF Combustion Meeting, NASA Langley Research Center, Hampton, VA, 22-26 September 1980, Volume II, p. 175.

14. B.T. Fedoroff and O.E. Sheffield, "Cannon Propellant," in: Encyclopedia of Explosives and Related Items, PATR 2700, Volume 2, Picatinny Arsenal, Dover, NJ, USA, C34, 1962.

15. C.L. Mader, Numerical Modeling of Detonations, University of California Press, Berkeley, CA, 153, 1979.

16. R.B. Frey, The Initiation of Explosive Charges by Rapid Shear," Proceedings Seventh Symposium (International) on Detonation, Annapolis, 36, 1981.

17. M. Rosenblatt, L.A. DeAngelo, G.E. Eggum, and K.N. Kregenhagen, "Numerical Investigation of Water Drop Erosion Mechanisms in Infrared - Transparent Materials," California Research and Technology, Inc., Technical Report AFML-TR-76-193, 38, December 1976.

18. H.W. Sandusky, C.S. Coffey, and T.P. Liddiard, "Rate of Deformation as a Measure of Reaction Threshold in Energetic Materials," in Mechanical Properties of Materials at High Rates of Strain, 1984, ed. J. Harding, Adam Hilger, London, 373, 1984.

19. T.W. Ewing and W.A. Cabbage, "A Compilation of Hazards Test Data for Propellants and Related Materials," Final Engineering Report on Production Engineering Project PE-489, Radford Army Ammunition Plant, Hercules, Inc., Radford, VA, pp 69, 73, September 1976.

20. R.C. Jensen, E.J. Blommer, and B. Brown, "An Instrumented Shotgun Facility to Study Impact Initiated Explosive Reactions," Proceedings Seventh Symposium (International) on Detonation, Annapolis, 299, 1981.

A METHOD OF PARAMETERISING HYDROCODE KINETIC MODELS FOR NON-IDEAL EXPLOSIVES BASED ON THE VARIATION OF DETONATION VELOCITY WITH CHARGE DIAMETER

G.A. Leiper and I.J. Kirby

Detonation Research Group
Imperial Chemical Industries PLC
Stevenston, Ayrshire, UK

ABSTRACT

The experimental detonation velocity charge diameter relationship is analysed using a two-dimensional divergent flow model to yield energy release rates at the Chapman-Jouguet (CJ) plane in the explosive. These rates are fitted to a first order pressure dependent Arrhenius rate law and extrapolated to zero reaction, giving the absolute reaction rate. The dependence of the absolute rate on detonation velocity and hence shock pressure gives the hydrocode model calibration. The Forest Fire rate model, calibrated thus, is used to predict the shock-to-detonation (SDT) behaviour of an air sensitised commercial explosive as a function of initial voidage and shock pressure. A comparison with experimental data is given.

INTRODUCTION

The need to estimate the response of explosives to shock impulse has lead to the development of numeric models, which when calibrated, enable the user to predict the initiation behaviour of an explosive under a variety of initial conditions.[1,2] Such models offer explosives manufacturers a means of assessing both the hazard and performance potential of their compositions without protracted experimental programmes. In the case of 'military' type explosives calibration data can be readily obtained. Commercial blasting agents cannot be so easily investigated: the non-ideality of these compositions gives rise to a large detonation velocity decrement close to the critical point, and to critical diameters which make conventional plane wave SDT experiments difficult.

It has been shown that the variation of detonation velocity with charge diameter is a function of both the hydrodynamics and chemical kinetics of the explosive[3]. It should therefore be possible to use this variation to derive reaction rate data for hydrocode rate models.

ANALYSIS OF DETONATION VELOCITY - DIAMETER DATA

Wood Kirkwood theory of slightly divergent axisymmetric flow[4] has been used with a generalised chemical reaction rate model and an empirical relationship between shock front curvature and charge diameter, to model the diameter effect in condensed explosives[3]. This approach can be formulated as an inverse method: namely given both the diameter effect and shock front curvature relationship, the chemical reaction rate and extent of reaction at the CJ plane can be calculated. In deriving this method, the work of Chan has been followed closely, with extension to a more general particle trajectory between shock and CJ planes.

From the first CJ condition, the equation of state and the quasi-one dimensional flow equations as defined in (4), a set of equations describing the CJ parameters can be formulated:-

$$\left(\frac{u}{D}\right)_{CJ} = \frac{\gamma A}{\gamma+1}\left(1 - \frac{\rho_0}{\rho_s}(A-1)\right) \quad 1.1$$

$$\left(\frac{\rho_0}{\rho}\right)_{CJ} = \frac{\gamma A^2}{\gamma+1}\left(1 - \frac{\rho_0}{\rho_s}(A-1)\right) \quad 1.2$$

$$\left(\frac{P}{\rho_0 D^2}\right)_{CJ} = \frac{1}{\gamma+1}\left(1 - \frac{\rho_0}{\rho_s}(A-1)\right) \quad 1.3$$

$$\lambda_{CJ} = \left(\frac{D}{D_\infty}\right)^2\left(\left(\frac{u}{D}\right)^2(\gamma+1)^2 - (\gamma^2-1)\right) \quad 1.4$$

where u is particle velocity; D detonation velocity; A streamtube area; ρ_0 initial density; ρ_s shock density; p pressure; and γ the adiabatic exponent in the polytropic equation of state. The explosive is further defined by the hugoniot, the ideal detonation velocity and the reaction thermicity.

$$D = H_1 + H_2 u + H_3 u^2 + \ldots\ldots \quad 1.5$$

$$D_\infty = 2(\gamma^2-1)q \quad 1.6$$

$$\sigma = (\gamma-1)qc^2 \quad 1.7$$

where H_i are coefficients of the shock hugoniot, q the heat released by chemical reaction, and c is the local sonic velocity. Equations 1.1-1.7 provide an explicit relationship between the CJ reaction co-ordinate, λ_{CJ} and the CJ streamtube area, A_{CJ}, for any given detonation velocity. The second CJ condition is added to solve for the two unknowns.

$$\sigma \dot{\lambda}_{CJ} = 2\omega_r \quad 1.8$$

$$\omega_r = \frac{u}{2A} \cdot \frac{dA}{dx} \quad 1.9$$

where ω_r is the flow divergence. The final equation to close the solution is the variation of streamtube area with distance behind the shock, A(x). The approach of (4) is extended by writing the A(x) function to allow a general shock - CJ particle trajectory. In a shock fixed frame, a particle originally travelling at a velocity D and angle θ to the shock normal will be deflected through the shock to a lower velocity u at angle ϕ such that

$$D\sin\theta = u\sin\phi \quad 2.1$$

For a spherically symmetric shock of radius R, the streamtube area will be increased on going through the shock such that

$$A = \left(1 + \int_0^x \frac{\sin(\theta - \phi)}{R\sin\theta}\,dx\right)^2 \qquad 2.2$$

Equations 2.1 and 2.2 are combined to give a general function which is soluble if the variation of the isobar radius of curvature with distance is known. Hydrocode modelling and flash x-ray experiments[5,6] have suggested that the isobar curvature through the CJ zone is approximately constant at the shock front radius of curvature i.e. $R(x) = R_s$, giving

$$A = \left(1 + \frac{1}{R}\int_0^x \frac{D-u}{u}\,dx\right)^2 \qquad 2.3$$

These equations now describe the locus of $\lambda_{CJ}(t_{CJ})$ for any set of detonation velocity data. The solution strategy starts at the critical point, finding the detonation parameters at which the streamtube area is a maximum, and subsequently incrementing diameter and iterating on CJ streamtube area and CJ time to provide a self consistent solution to the flow equations. Typical results are shown below for an air-sensitised emulsion explosive at various initial densities.

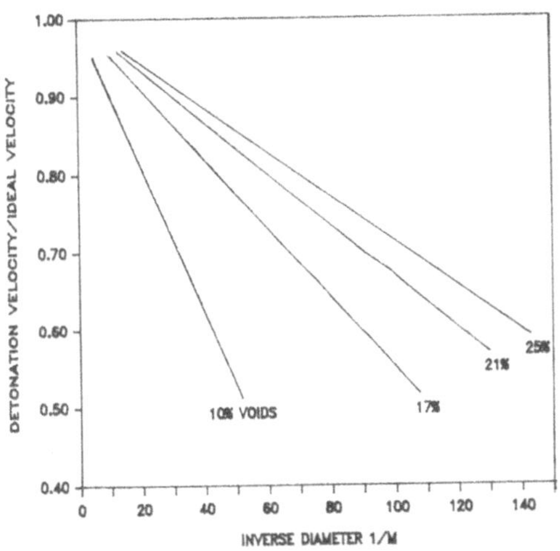

Fig. 1. V.o.D. - Diameter

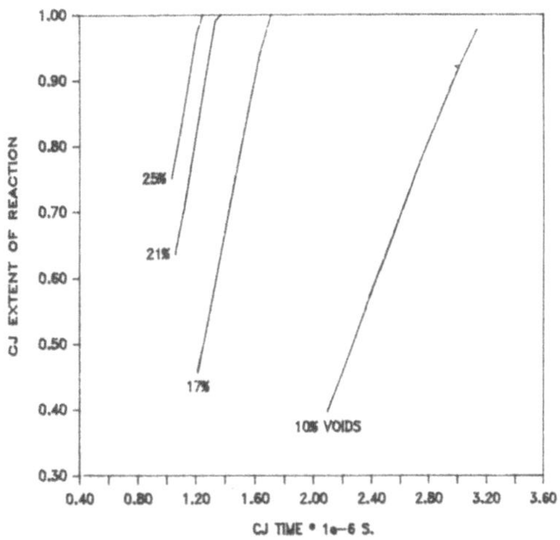

Fig. 2. CJ Reaction vs CJ Time

CALIBRATION TO HYDROCODE RATE MODEL

The variation of $(\lambda, \dot{\lambda}, t)_{CJ}$ with detonation velocity provides the fundamental data for hydrocode calibration. In the case of the well known Forest Fire rate Model, a further step is required to reduce the basic data to a suitable form, namely the dependence of absolute ($\lambda=0$) rate on shock pressure. This is accomplished by fitting the CJ triplet data and pressure profile in the CJ region to a suitable kinetic law and evaluating the function at the shock front. A first order pressure dependent rate law, used previously to predict the behaviour of several condensed explosives3, is defined as

$$\dot{\lambda} = Z(1-\lambda)\left(\frac{p}{p_r}\right)^2 e^{-\frac{p'}{p}} \qquad 3.1$$

where Z is a preexponential constant, p_r a reference pressure; and p' the activation constant for the reaction. The pressure in the reaction zone is approximated by a linear profile to permit analytic solution of 3.1

$$p = p_s - (p_s - p_{CJ})\left(\frac{t}{t_{CJ}}\right) \qquad 3.2$$

The reference pressure is taken to be the initial shock pressure, p_s. Equations 3.1 and 3.2 combine and integrate to give

$$\ln(1-\lambda_{CJ}) = \frac{Z\,t_{CJ}}{p_{CJ}-p_s}\left[\frac{e^{-p'v}}{3v}\left(\frac{1}{v^2}-\frac{p'}{2v}+\frac{p'^2}{2}\right)+\frac{p'^3}{6}\left(\ln(v)+\sum_{n=1}^{\infty}\frac{(-p'v)^n}{n.n!}\right)\right]_{v=1/p_{CJ}}^{v=1/p_s} \qquad 3.3$$

Equation 3.1 and 3.3 form a simultaneous set in two unknowns, and Z, which can be solved using the $(\lambda,\dot{\lambda},t)_{CJ}$ data and pressure profiles from the velocity diameter analysis. Plots of the absolute rate versus shock pressure are shown below for the four initial voidages used.

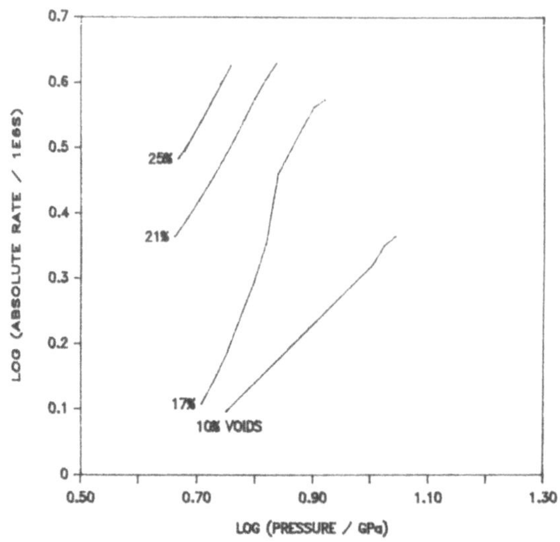

Fig. 3. Forest Fire Pilot

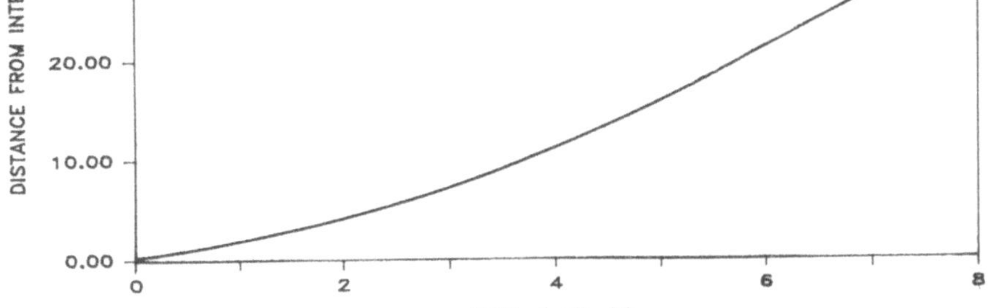

Fig. 4. Experimental Streak Record

VERIFICATION OF FOREST FIRE CALIBRATION

A 55 mm explosive plane wave lens was used with PMMA attenuators to create shocks of various intensities in 25 mm diameter cartridges of test explosive. Streak photography was used to examine the SDT process, recording the displacement time history of the shock. A typical streak record, as shown below after smoothing to remove small variations in shock velocity with local composition, shows the shock accelerating smoothly to stable detonation. No clear break point, as would be seen with military compositions, is evident. Records were obtained over a range of pressures from 1.3 GPa to 4.0 GPa. At 1.1 GPa no evidence of initiation was found.

The experiments were simulated using Forest Fire rates and the 1D hydrocode SIN[7] to model a PMMA plate inpacting a slab of emulsion. The initial plate velocity and thickness were chosen such that the particle velocity time profile at the interface was close to that measured experimentally using an electromagnetic gauge[8]. A HOM equation of state was used for both materials. Emulsion porosity was modelled by modifying the shock hugoniot. Ideal detonation states were calculated using the BKWR[9] equation of state.

In general there was good agreement between SIN predictions and experiment, justifying the calibration procedure used. Agreement was best at shock pressures over twice the critical pressure. In this regime initiation was governed by the strength of the incident planar shock, edge generated rarefaction failed to perturb the system and a one dimensional approximation as modelled with SIN was valid. At lower shock pressures edge rarefactions were seen to be a dominant factor on the experimental streak records, and reaction was extinguished by them, before SDT could occur. In this two dimensional regime SIN was not expected to be an adequate representation and did not predict the experimentally observed initiation failure at 1.1 GPa.

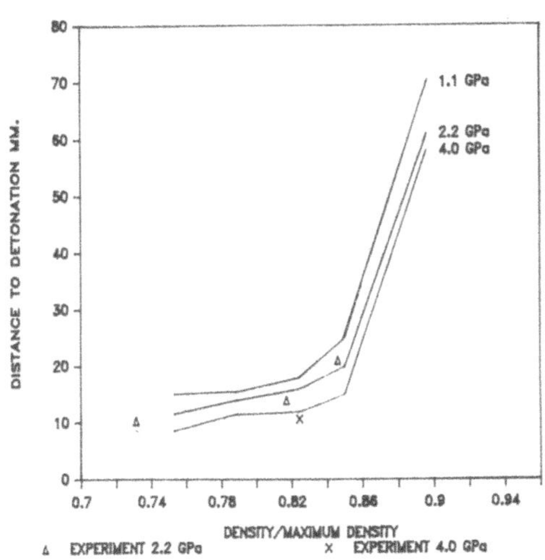

Fig. 5. Distance to SDT

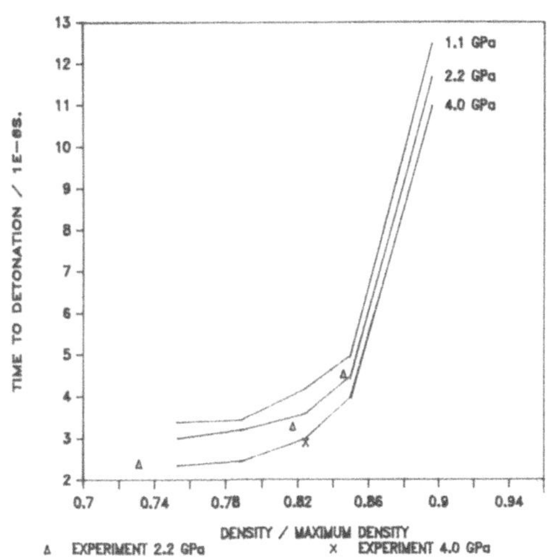

Fig. 6. Time to SDT

CONCLUSION

A method has been developed for deriving hydrocode kinetic model parameters for non-ideal explosives. The advantage over standard techniques is one of experimental simplicity in obtaining the relevant data. In cases where standard SDT data cannot be obtained, this technique can be used to give calibrations which enable hydrocode users to model explosive performance with some confidence.

REFERENCES

1. Forest and Mader, Los Alamos Report, LA7423.
2. Lee and Tarver, Phys. Fluids 23 (12), 1980, p2362-72.
3. Chan, 7th Symp. on Detn. 1981, p589-610.
4. Wood and Kirkwood, J. Chem. Phys. 22 (11), 1954, p1920-4.
5. Yoshida and Tanaka, J. Expl. Soc. Jap, 1982.
6. Freud and Geiger, Acta Astro 1, 405-16, 1974.
7. Mader, Los Alamos Report LA7213.
8. Zaitsev et al, Dokl. Acad. Nauk. SSSR, 132, p1339-40.
9. Finger et al, 6th Symp. on Detn. 1981, p710-22.

HIGH CURRENT ELECTRICAL CONDUCTION OF PRESSED CONDENSED DETONATING EXPLOSIVES

Douglas G. Tasker and Richard J. Lee

Naval Surface Weapons Center
White Oak Laboratory
Silver Spring, MD 20903-5000

INTRODUCTION

The work reported here is part of a continuing theoretical and experimental study of the mechanisms and effects of the conduction of large electrical currents in detonating high explosives.

The reaction zone of a detonating high explosive is an electrical conductor whereas the unreacted explosive and the product gases are insulators.[1-5] The conductivity is of great interest, by measurement and theoretical analysis valuable insights into the mechanisms of detonation can be obtained.

Conductivity data are sought at current densities approaching 10^{10} A/m^2. This paper reports on recent conductivity measurements at high current densities, the experimental methods necessary to obtain the data and the extraordinary experimental problems that pertain. Two particular problems are highlighted: (1) the effect of an air gap on the resistance measurements and (2) the problems of making accurate high voltage measurements.

ELECTRICAL RESISTANCE MEASUREMENTS

Experiment

The resistance of the conduction zone was measured in a coaxial configuration similar to that of earlier experiments.[1] In such high current experiments unwanted capacitive and inductive effects can be significant. To minimize inductive effects a new stripline configuration is employed for electrical connection to the explosive. A significant contribution to resistance due to shocked air, between the explosive and the outer cylindrical electrode, has been observed. This has been minimized by slightly tapering the explosive charge and electrode to eliminate air gaps.

Current measurement. Current flow in the conduction zone was measured at the source end of the circuit using a calibrated current transformer. The equivalent circuit of the capacitor, cables and stripline were obtained prior to each shot by discharging the capacitor through a short circuit as described previously.[1] These data are used to verify the validity of the V-I data.

The frequency response of all instrumentation has been improved from earlier work.[1] Risetimes are within 50 ns for all records.

Voltage Measurement

The accurate measurement of voltage in these high current experiments is extremely difficult. A detailed discussion of these problems is warranted.

Mutual inductance effects. Rapidly changing electric currents that exist in these conductivity experiments, can produce large voltage signals comparable to the true voltage data required. A voltage probe, almost by definition, measures voltage with a negligible flow of current; consequently stray resistive or inductive effects are rarely considered. However stray inductances in both the source and measurement circuits can lead to significant mutually inductive effects.

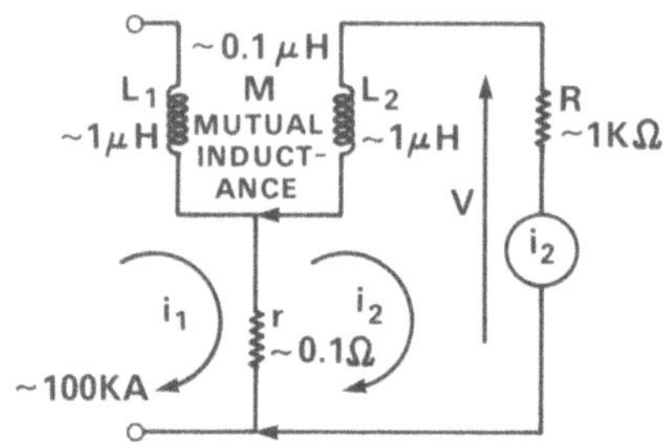

Figure 1. An idealized circuit, the resistance r is to be measured using a four probe technique. Stray inductances L_1 and L_2 exist in the high current and voltage probe circuits respectively, a mutual inductance M exists between them. From Kirchoff's laws:

$$i_2(r+R) + L_2 \frac{di_2}{dt} - M \frac{di_1}{dt} - i_1 r = 0$$

Here r 1/10Ω, R>>r and $i_1 >> i_2$. Hence the measured voltage $V \approx i_1 r + M\, di_1/dt$.

In these experiments $di_1/dt \approx 10^{11}$ A/s, $i_1 \approx 100$ kA and M could be 100 nH unless great care is taken. Thus the error from the mutual inductance $M\, di_1/dt \approx 10^4$V is comparable to the desired signal $i_1 r \approx 10^4$V. Note that these effects are independent of R and thus apply to all types of voltage probes.

The only satisfactory method of eliminating these effects is by reduction of M. The pulsed power circuits are built using stripline techniques so that $M < 0.1$ nH and mutual inductance effects are negligible.

Copper sulphate resistors. Reactance effects in the voltage probe resistor can change its impedance at high frequencies. To minimize these effects resistors of 100 Ω to 1000 Ω are used; note that these resistors are very much larger than the 1/10 Ω load to be measured. These resistors are fabricated by filling plastic tubes with saturated

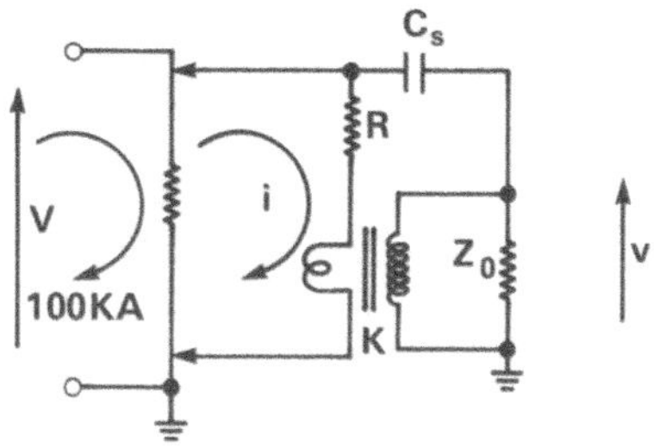

Figure 2. Voltage Probe Circuit.

solutions of copper sulphate (ρ=25 Ω cm). They have excellent high voltage stability[6] but must be calibrated immediately prior to their use to eliminate errors due to thermal drift. Current in the resistor is measured with a calibrated current transformer to eliminate ground loops.

Capacitive effects. The voltage probe circuit is shown in Figure 2. The reactance of the current transformer primary is negligible so that $V = iR$. The signal detected in the secondary is $v = Ki$. Stray capacitance C_s will exist between primary and secondary circuits, both within and without the transformer. Typical values are $K = 0.05$ A/V, $V = 1$ kV, $dV/dt = 10^9$ V/s, $R = 500\ \Omega$, $C_s = 4$pF and $Z_0 = 25\ \Omega$. So the error from the capacitive signal $(C_s dv/dt)Z_0 = 100$ mV can be comparable to the true signal $Ki = K(V/R) = 100$ mV. Clearly great care must be taken to minimize these errors by minimizing stray capacitances, reducing R and increasing K. If $C_s = 1$pF, $K = 0.5$ A/v and $R = 100\ \Omega$ then the effect is reduced to 1/2%.

Measurement circuit. The apparatus for measuring the resistance of detonating explosives is shown schematically in Figure 3. The stripline is 1m long with an inductance from input to load of 4nH. Coaxial high voltage connectors join the capacitor bank and the voltage probe to the stripline. The explosive is fitted into a brass tube of 14.0 mm internal diameter and has a 3.125 mm diameter brass central electrode. Brass plumbing compression fittings have been adapted to connect the electrodes to the stripline. Metal to metal contact resistances are less than 10 $\mu\Omega$ throughout.

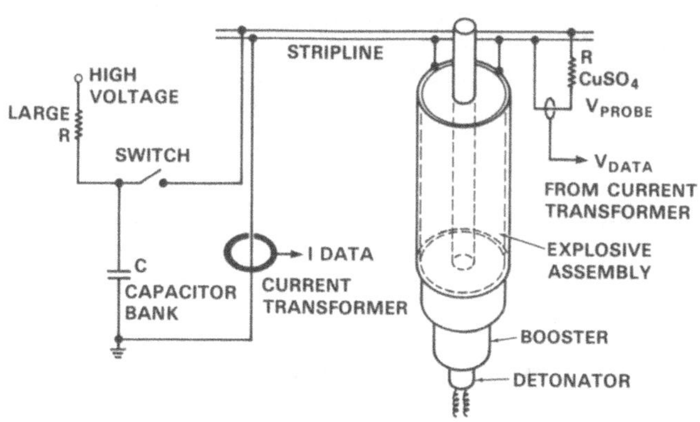

Figure 3. Electrical Circuit.

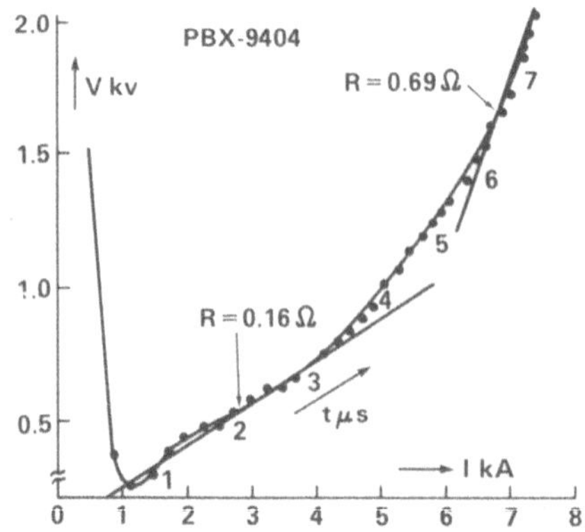

Figure 4. Effects of Air Gap.

Inductive effects in the explosive tube. The magnetic flux, ϕ, stored between the coaxial electrodes introduces a small voltage, $d\phi/dt$. As the detonation wave progresses the inductance between diminishes from 34 nH to zero. These effects cannot be eliminated but have been accurately quantified by calculation and direct measurement; all voltage records are corrected for this effect which is always less than 10% of the true data.

RESULTS

The results for one PBX-9404 charge are shown in Figure 4. Before the detonation wave enters the electrodes the explosive acts as an insulator and no current flows. When the wave enters current begins to flow; the voltage drops from V_{cap} to $V_{cap} - L\, di/dt$. Then both the current and voltage rise sinusoidally, these are plotted in the figure. From the slope of the curve the explosive resistance R and thus the product $\sigma\Delta$ are obtained where σ is the conductivity and Δ the zone width. Here $R = \dfrac{\ln r_2/r_1}{2\pi\sigma\Delta}$.

Air resistance effects. The explosive cylinder of Figure 4 had inadvertently been manufactured with a slight taper of 1 μm diameter per mm length. The largest end was closest to the detonator, the brass tube was still perfectly cylindrical so that an air gap between the tube and explosive increased from 0 at the detonator end to 75 μm along the length of the explosive. The results show that the resistance increased with length commensurate with the increase in air gap. The resistance of the explosive, obtained from other experiments, is approximately 0.14 Ω without the air gap. The slope of Figure 4 starts at 0.16 Ω and then increases to 0.69Ω. Consequently the air resistance of the 75 μm air gap (d) is 0.53 Ω, (R_{air}). If a planar conduction front is assumed and no jetting into the void occurs then the conductivity of the air σ_{air} can be estimated from $R_{air} = \dfrac{d}{2\pi r \sigma_{air}\Delta}$ where r is the radius and Δ the conduction zone width. Hence $\sigma_{air}\Delta = 2\text{x}10^{-3}$ mhos. If $\Delta = 0.1$ mm for PBX-9404 then $\sigma_{air} \sim 20$ mhos/m. Clearly several crude assumptions have been made here, nevertheless these data are consistent with air resistance effects we have observed in other experiments and in fair agreement with data extrapolated from published work.[7]

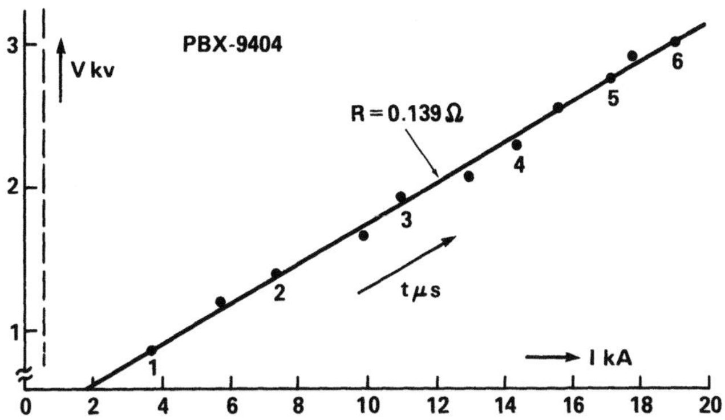

Figure 5. PBX-9404 Data without Air Effects.

Elimination of air resistance. The results show that the air gap must be reduced to a minimum. This has been achieved by careful machining of the explosive and tube so that they have matching tapers of approximately 1 μm diameter per mm length. The explosives are forced into the tubes until they stop, thus intimate contact between explosive and tube is achieved. It has not been possible to taper the center electrode; although it is tightly fitted into the explosive. Some records have shown irregularities in the voltage records that can be traced to gaps around the center electrode. Such irregularities are readily detected.

Results for tapered electrodes; PBX-9404. The results for the tapered coaxial assembly are shown in Fig. 5. As reported previously[1] the V-I plot shows excellent linearity. Here R = 0.139 Ω with a standard error of 0.003 Ω. Using $R = \frac{\ln r_2/r_1}{2\pi\sigma\Delta}$ we find the conductivity, conduction zone width product to be $\sigma\Delta$ = 1.70 mho (with an estimated standard error of 0.05 mho) which is higher than previously reported, i.e. 1.25 mho. In the earlier work no taper was used and the explosive had to be made with an air gap of approximately 10 μm to facilitate assembly. A 10 μm gap would increase the resistance by approximately 0.06 Ω which is in excellent agreement with the data, i.e. previously R = 0.206 Ω and not 0.139 Ω as above. Moreover the early work incorporated no corrections for the effects described previously due to mutual inductance, capacitance and flux changes in the coaxial electrodes. The use of the tapered electrodes has clearly eliminated the air resistance problem.

CONCLUSIONS

Improved electrical conductivity data have been obtained for PBX-9404. Great care has been taken to identify and eliminate stray electrical effects. Moreover the effects of air between surfaces have been identified, quantified and satisfactorily eliminated. The work is to be extended to study the properties of other explosives.

In keeping with previous observations the voltage and current records support the hypothesis that conduction is confined to a narrow zone behind the detonation front. Other work, not reported here, shows that the shock front and the conduction front are coincident to within 0.5 ns. Experiments are planned so that the conduction zone width Δ can be measured directly. However the voltage data obtained in these experiments suggest that conduction turns-off within 5 ns of the wave exit from the coaxial electrode assembly which corresponds to an effective conduction zone width of less than 0.44 mm.

ACKNOWLEDGEMENTS

We are indebted to H. Gillum, W. Freeman, J. Roscher, R. Baker and the late C. Sorrels for the excellence of their experimental support. J. Goforth (LANL) and M. Ginsberg (LANL) deserve credit for their support and many helpful discussions. Special thanks are due to J. W. Forbes who has made significant contributions both with his technical expertise and his enthusiasm. Our wives are to be commended for their support and forbearance. This work was supported by the Naval Sea Systems Command, task 61153N, SR024-03.

REFERENCES

1. D. L. Demske, J. Forbes, and D. G. Tasker, Shock Waves in Condensed Matter - 1983, Elsevier Science Publ., p. 551.

2. Brish, A. A., Tarasov, M. S., and Tsukerman, V. A., Soviet Physics JETP 34(10), No. 6, Jun 1960, p. 1095.

3. Hayes, B., Fourth Symposium (International) on Detonation, Proceedings, Office of Naval Research, Department of the Navy, Washington, DC, Oct 1965, p. 595.

4. Ershov, A. P., Zubkov, P. I., and Lukjantshikov, L. A., Combustion, Explosion and Shockwaves 10 (6), November-December 1974, p. 776.

5. V. A. Datsenko, et al., Sov. Phys. Tech. Phys. 23(1), Jan 1978, p. 33.

6. C. R. McClenahan, Air Force Weapons Laboratory, unpublished.

7. B. S. Punkevich, B. M. Stepanov, Fizika Goreniya: Vzryva 16 (2), p. 109, March-April 1980.

CALCULATIONS OF LIQUID PLANE-WAVE GENERATORS

Katsumi Tanaka, Shyuzo Fujiwara, Masao Kusakabe*
and Masatake Yoshida

National Chemical Laboratory for Industry,
Tsukuba,Ibaraki 305, Japan
* present address ,Midori Anzen Kogyo Co. Ltd.,
Inari, Soka, Saitama 340, Japan

INTRODUCTION

The plane-wave generator (PWG) composed of liquid explosives has been developed by Fujiwara et al.(1978) since early 1970s. Nitromethane (NM) and hydrazine hydrate solution of hydrazine nitrate (HHS) are used as shown in Fig. 1. NM encased in a right circular cone of polystyrene is commercial grade and is not purified. The measured time difference of detonation arrival at the bottom of PWG in use at NCLI was approximately 10 nanoseconds for diameters of both 60 mm (PL-060) and 100 mm (PL-100). A typical streak camera record of argon flash gap on the bottom of PL-060 is shown in Fig. 1.

Both the flatness and the pressure gradient across the PWG and behind the detonation front are important parameters for the plate acceleration and other practical uses as reported by Pimbley et al. for Baratol PWG. The properties of PL-060 are calculated by using 2DL (two dimensional Lagarangian hydrodynamic code by finite difference method using artificial viscosity) and the pressure profiles in PWG are presented.

NUMERICAL

The angle of cone for PWG is generally given by the relation, $D_1 = D_2 \cos \theta/2$, where D_1 and D_2 are detonation velocity and θ is the cone angle. This relation is true, if both the point initiation on cone apex and the sharp transition to steady state detonation are possible. Detonation in PWG, however, is unsteady state. Numerical calculation is useful to study the unsteady hydrodynamic phenomena in this work.

All the calculations are axisymmetrical. Number of cells used in the 2DL calculation are 150 and 200 to the radial and axial direction respectively.

Parameters of the equation of state (EOS) used for non reactive NM and solid polystyrene are from Mader and Marsh respectively. Tanaka's KHT results of HOM form by Mader are used for the EOS of detonation products. Chapman-Jouget (C-J) detonation properties are given in Table 1. Though calculated detonation velocities are slightly different from experimental results, these differences may be negligible for the estimation of the planarity of PWG comparing with the used mesh size in this work.

The initiation of HHS by a pressed tetryl booster is assumed to be plane. We used two reaction rate formulas for the detonation initiation and

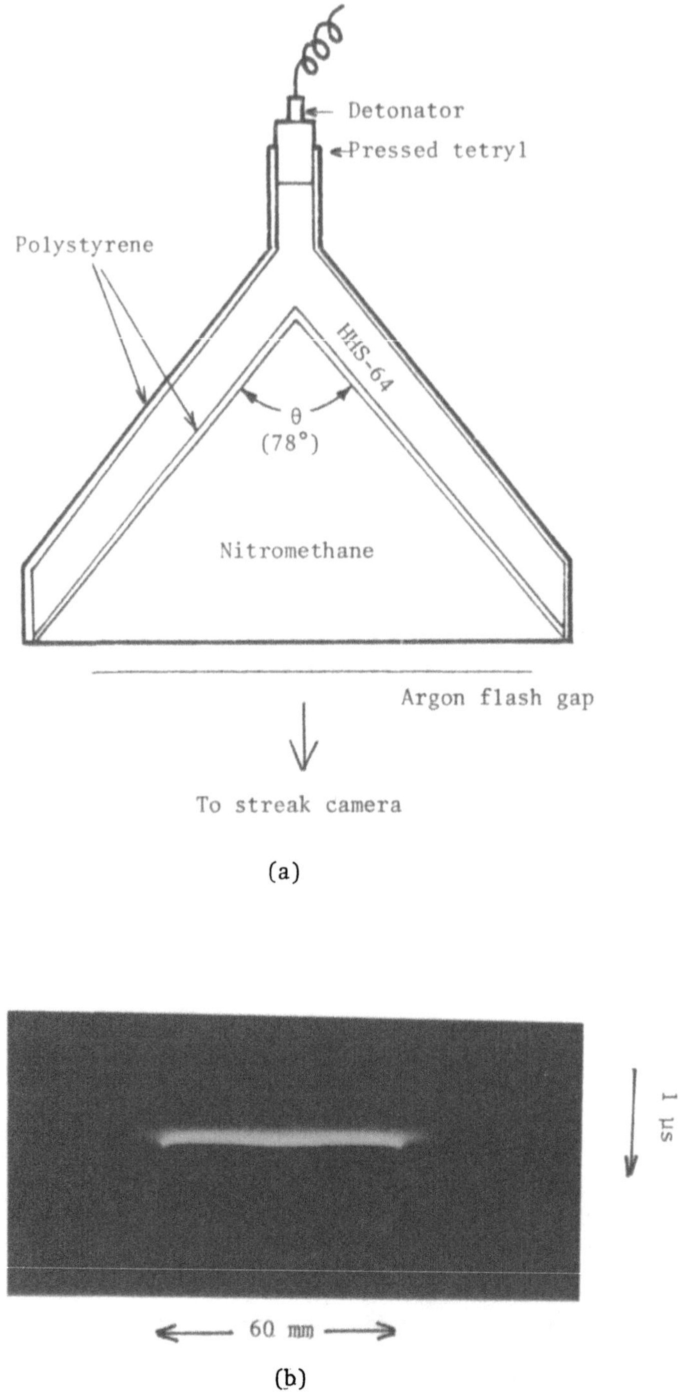

Fig. 1 (a) Schematic of the plane wave generator.
(b) A streak camera record of the argon flash gap on the bottom surface of the plane wave generator with the diameter of 65 mm.

propagation. C-J volume burn technique by Mader is used for the reaction in explosives except for the reactive NM in which the Arrhenius reaction rate law was assumed. C-J volume burn, which can save the computation time, is difficult to reproduce the exact curved detonation front by the lateral rarefaction wave.

When the detonation wave propagates 10 cells, the piston on the boundary is stopped.

Shock Wave in Non Reactive Nitromethane. The incident pressure of oblique shock wave in non-reactive NM from a detonation of HHS-64 (64 % hydrazine nitrate solution) propagating toward parallel direction to the boundary is estimated to be 6.9 GPa which is lower than the initiation pressure of NM. For a one dimensional slab geometry, the calculated incident pressure in non-reactive NM is 14 GPa from the detonation of HHS-64 if the effect of polystyrene plate between both explosives is neglected.

Both the pressure and the temperature are calculated by assuming the shocked NM to be nonreactive. The detonation wave in HHS-64 is disturbed by the lateral boundary condition assuming continuum as shown in Fig. 2. The incident pressure in NM on the apex of cone is approximately 20 GPa which gives the shock temperature of about 2,300 K sufficient to initiate NM. Implosion effects around the cylinder axis are significant. Shock wave in NM is decaying during propagation.

Initiation of Nitromethane. The initiation of NM encased in polystyrene cone was calculated by using Arrhenius reaction rate law and by assuming pressure and temperature equilibrium in reacted NM as shown in Fig. 3. Temperature in shocked NM is sufficiently high and the induction zone is too small to be resolved by the calculation here. Detonation wave near the cylinder axis is overdriven and is decaying to C-J detonation pressure during the propagation.

Pressure Configuration in Plane Wave Generator. Calculated result of liquid PWG is shown in Fig. 4. Since reaction zone is negligibly small comparing with the cell size, C-J volume burn technique was used for both NM and HHS-64. A polystyrene cone is neglected. Curvature of detonation front in NM is greater than the observed one due to the overdriven detonation at initial stage as shown in Fig. 5. The calculation for HHS-70(70 % of hydrazine nitrate) gives better planarity of detonation front, while delay of the detonation arrival of 100 nanoseconds at the center of PWG was observed by Fujiwara et al.(1977). For PL-060 lens, calculated result shows that pressure gradient in NM across the lens and behind the detonation front was high but at final stage it was not significant across the front comparing with results of P-040 Baratol lens by Pimbley et al.. However the pressure gradient near the boundary between NM and HHS is still significant.

DISCUSSION

The curvature of detonation front in HHS before the initiation of NM in PWG was considerably effective to the decay of the overdriven detonation by the implosion. Calculated results in present work overestimate the implosion effect. Other two dimensional calculations for the case of non reactive NM show that curved ,i.e., diverging detonation wave in HHS makes the pressure in shocked NM decrease. As the initiation diameter of HHS-64 becomes smaller, a decay of overdriven-detonation pressure in NM becomes faster.

The induction time of used NM is not known. The induction time of the initiation of purified NM at a room temperature is an order of several nanoseconds for the pressure of 20 GPa by the extrapolation of experimental data. It seems that the effect of the polystyrene cone on the initiation of NM is small.

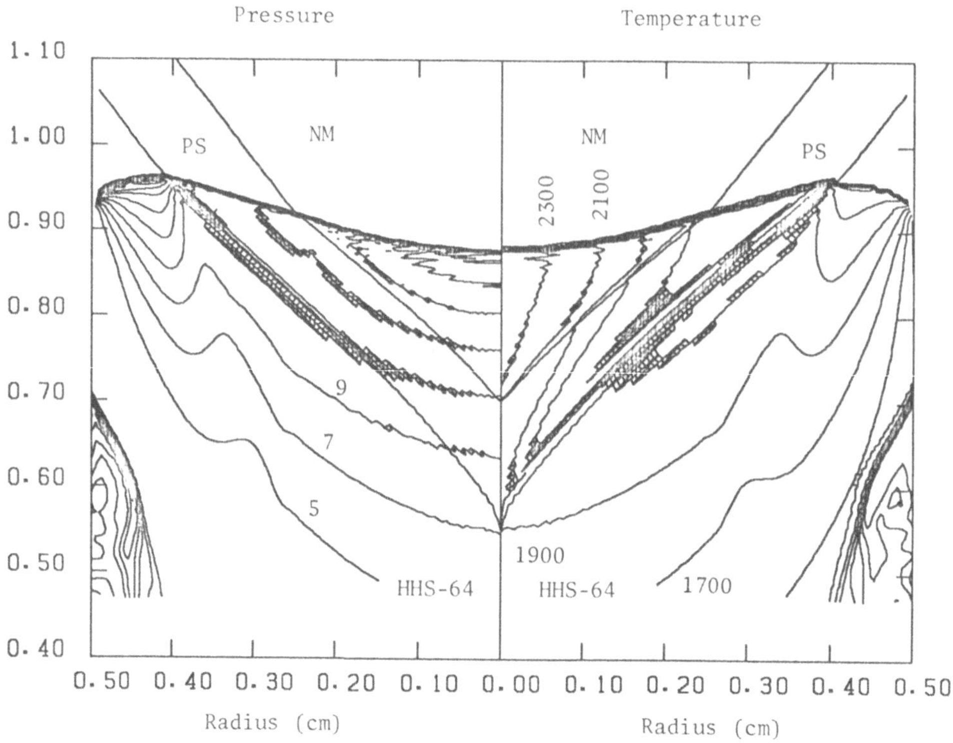

Fig. 2 Isobar and isotherm of the plane wave generators using 2-DL hydrodynamic code. Nitromethane is assumed to be nonreactive. C-J volume burn is used for the detonation of liquid explosive of HHS-64. Intervals are 2 GPa and 200 K for isobar and isotherm respectively.

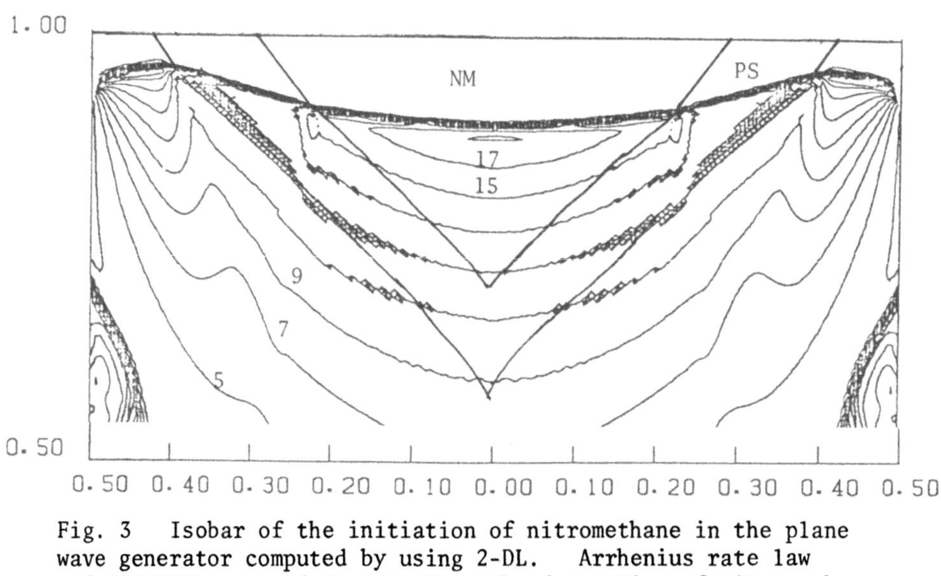

Fig. 3 Isobar of the initiation of nitromethane in the plane wave generator computed by using 2-DL. Arrhenius rate law and KHT-HOM are used to calculate the detonation of nitromethane. Interval between isobar is 2 GPa.

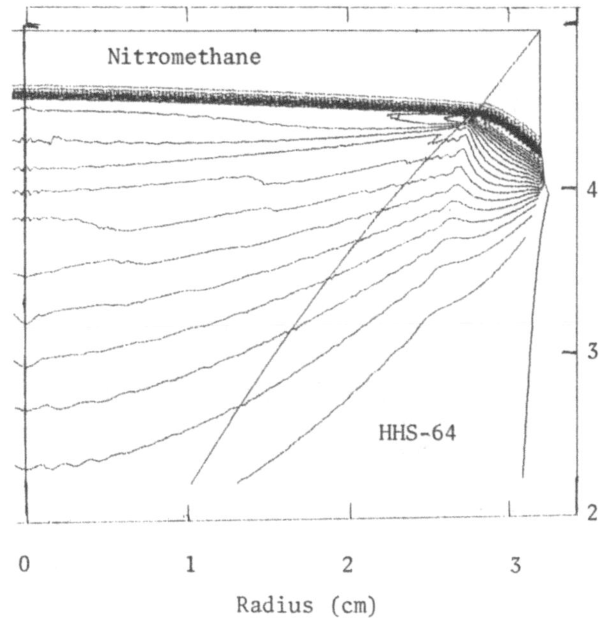

Fig.4 Isobar of the plane wave generator. Polystyrene cone is neglected. Interval is 1 GPa. C-J volume burn is used.

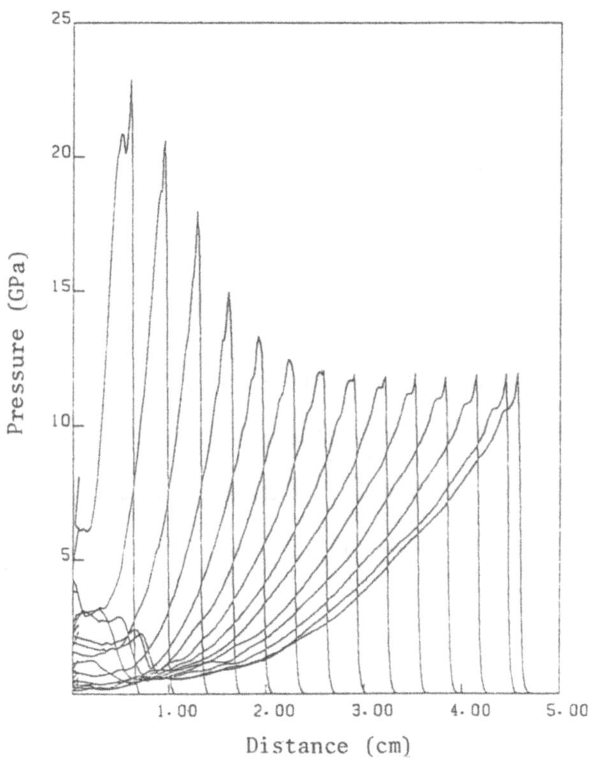

Fig. 5 Pressure profiles for the nitromethane in the plane wave generator at the axis of the cone.

Table 1. Detonation properties of liquid explosives.

		ρ_0 (g/cm^3)	D (m/s)	P_{C-J} (GPa)	T_{C-J} (K)
NM	calc.	1.128	6263	12.2	3296
	exp.(a)	1.133	6260		
HHS-64	calc.	1.314	8011	20.7	2347
	exp.(b)	1.314	8024		
HHS-70	calc.	1.356	8235	23.4	2617
	exp.(b)	1.356	8200		

NM ; Nitromethane.
HHS-64 ; Hydrazine Nitrate 63.6/Hydrazine Hydrate 36.6.
HHS-70 ; Hydrazine Nitrate 70/Hydrazine Hydrate 30.
(a) At 293 K and in 41 mm diameter polyvinyl chloride tube (Fujiwara et al.(1977)). Purity 95 % (Kodak Co. Ltd).
(b) Extrapolated to infinite diameter (Kusakabe et al.).

CONCLUSION

Calculated results show that the pressure gradient of NCLI PWG across the detonation front is relatively small except near the NM/HHS boundary. Overdriven detonation in NM was appeared for first several centimeters. The initial configuration of detonation wave in high-velocity explosive before the arrival at explosive with slow detonation velocity is possibly an important factor for the design of PWG due to the generation of overdriven detonation.

REFERENCES

Fujiwara, S. and M. Kusakabe, unpublished (1977).

Fujiwara, S. and M. Kusakabe, Japan Kokai Tokkyo Koho ,Patent No. 78-29673 (1978).

Kusakabe, M., S. Fujiwara, K. Shiino and K.Tanaka, Kogyo Kayaku, 41, 23 (1980).

Mader,C. L., 'Numerical Modeling of Detonation' Univ. Calif. Press (1979).

Marsh, S. P.,'LASL Hugoniot Data ',Univ.Calif. Press(1980).

Pimbley, G. H., C. L. Mader and A. L. Bowman,'Plane Wave Generator Calculations', LA-9119, Los Alamos National Laboratory ,(1982).

Tanaka, K.,'Detonation Properties of Condensed Explosives Computed by Kihara-Hikita-Tanaka Equation of State', NCLI Report (1983).

INDEX

GPSR Compliance

The European Union's (EU) General Product Safety Regulation (GPSR) is a set of rules that requires consumer products to be safe and our obligations to ensure this.

If you have any concerns about our products, you can contact us on ProductSafety@springernature.com

In case Publisher is established outside the EU, the EU authorized representative is:

Springer Nature Customer Service Center GmbH
Europaplatz 3
69115 Heidelberg, Germany

Batch number: 10684222

Printed by Printforce, the Netherlands